HEAT TRANSFER

Professional Version

Lindon C. Thomas

PTR PRENTICE HALL, Englewood Cliffs, New Jersey 07632

Library of Congress Cataloging-in-Publication Data

Thomas, Lindon C.
Heat transfer / Lindon C. Thomas. — Professional version.
p. cm.
Includes bibliographical references and index.
ISBN 0-13-382748-8
1. Heat—Transmission. 2. Heat—Convection. I. Title.
QC320.T493 1993
621.402'2—dc20 92-18663
CIP

Acquisitions editor: Mike Hays
Cover design: Joe DiDomenico
Cover design director: Eloise Starkweather
Copyeditor: Bill Thomas
Art production manager: Gail Cocker-Bogusz
Manufacturing buyer: Mary E. McCartney
Illustrations by Vantage Art

A Simon & Schuster Company
Englewood Cliffs, New Jersey 07632

The publisher offers discounts on this book when ordered in bulk quantities.
For more information, contact Corporate Sales Department,
PTR Prentice Hall, 113 Sylvan Avenue, Englewood Cliffs, NJ 07632.
Phone: 201-592-2863; FAX: 201-592-2249.

Printed in the United States of America

10 9 8 7 6 5 4 3 2 1

ISBN 0-13-382748-8

Prentice-Hall International (UK) Limited, *London*
Prentice-Hall of Australia Pty. Limited, *Sydney*
Prentice-Hall Canada Inc., *Toronto*
Prentice-Hall Hispanoamericana, S.A., *Mexico*
Prentice-Hall of India Private Limited, *New Delhi*
Prentice-Hall of Japan, Inc., *Tokyo*
Simon & Schuster Asia Pte. Ltd., *Singapore*
Editora Prentice-Hall do Brasil, Ltda., *Rio de Janeiro*

To Tisha, *and our family,*
Sarah *and* Arnold, Stephen, Mary, Mark, *and* Elissa

CONTENTS

PREFACE

A basic understanding of *heat transfer* has long been required of aerospace, chemical, environmental, mechanical, and nuclear engineers. The importance of heat transfer has in recent years become more widely recognized in other fields such as architectural, civil, electrical, and petroleum engineering.

This *professional version* of *Heat Transfer* was developed for the use of practicing engineers and for use as a text in introductory engineering heat-transfer courses that are intended to provide more extensive treatment of convection-heat transfer. The book should also serve as a useful reference for graduate students and others interested in energy-related problems.

The book features (1) the thorough presentation of fundamental concepts in the context of simple one-dimensional analyses; (2) the analysis of multidimensional conduction heat-transfer processes, with special consideration given to the modern numerical finite-difference method; (3) the analysis of radiation heat transfer; and (4) the coverage of convection heat transfer, with the theory of convection followed by the practical analysis approach. Although a basic knowledge of ordinary differential equations, partial derivatives, and elementary thermodynamics and fluid mechanics is generally necessary for a proper understanding of the material, these subjects are reviewed as they are needed.

The presentation throughout the text generally involves the use of brief developments followed by examples. Fundamental concepts pertaining to the mechanisms of heat transfer are introduced in Chap. 1. One-dimensional analyses are developed in Chap. 2, with emphasis given to the importance of limiting criteria for common systems, such as fins, that are inherently multidimensional. Standard approaches for solving multidimensional conduction problems are considered in Chap. 3, with attention focused on the simple but accurate numerical finite-difference method in Chap. 4. Fundamentals of thermal radiation and the practical analysis of radiation heat transfer, which involves the use of the thermal network concept, are presented in

Chap. 5. The topic of convection heat transfer is introduced in Chap. 6, with the theory of convection presented in Chap. 7. The practical solution approach, which involves the use of modern convection correlations and simple lumped formulations, is developed in Chaps. 8 through 11 for forced and natural convection, boiling and condensation, and heat exchangers, with attention given to both hydraulic and thermal aspects. Concerning the treatment of convection heat transfer, the format is such that the practical approach of Chaps. 8 through 11 can be studied before or after (or independent of) the material in Chap. 7, which deals with the theory of convection.

The unique approach to the study of convection heat transfer developed in this book, which separates the theory of convection (Chap. 7) and the practical analysis approach (Chaps. 8–11), has been motivated by the fact that students often encounter difficulties in the study of this important topic. The traditional approach involves the presentation of the theory of convection, which is fairly complex and sometimes appears somewhat bewildering, interspersed with practical aspects. The topics generally covered in the theory of convection include (1) the development of the boundary layer (partial differential) equations for laminar and turbulent flows, (2) the development of the related integral equations, (3) the introduction of the similarity solution approach for laminar forced and natural convection external boundary layer flows, and (4) the development of approximate integral solutions for laminar and turbulent external boundary layer flows, and more. These topics are important, but are not essential to the development of the practical analysis approach used in heat exchanger analysis and design and in the solution of most basic convection heat transfer problems that engineers encounter in practice. The presentation of the practical analysis approach in a self-contained unit enables beginning students to gain a basic understanding of the key elements of the evaluation and design of convection systems with or without the study of the details of the theory of convection, and provides practicing engineers with an effective format for efficient review of practical or theoretical aspects of particular interest.

In this connection, the practical hydraulic and thermal analysis approach to analyzing heat exchangers and other internal flow processes involves the use of bulk-stream characteristics. This approach has traditionally been developed in the context of a defining equation for bulk-stream (or mixing cup) temperature T_b. Unfortunately, this critical issue is frequently blurred by imprecise mathematical formulations. The approach taken in Chaps. 6 through 11 in this text features a less restrictive fundamental formulation, which involves the useful concept of bulk-stream enthalpy $\dot{H}_b$ (and bulk-stream momentum rate $\dot{M}_b$). The basic relationships among mass flow rate $d\dot{m}$, momentum rate $d\dot{M}$, enthalpy rate $d\dot{H}$, velocity, temperature, and thermophysical properties used to develop this more general perspective are introduced in Chap. 6 and Appendixes K and M. In addition to providing a basis for the development of a clearer and more general practical hydraulic and thermal analysis approach, the concept of enthalpy rate and momentum rate is used in Chap. 7 and Appendix I to develop what is believed to be a simpler less tedious approach to the formulation of the differential and integral equations for boundary layer flow. Because of its versatility and relative simplicity, the approximate integral solution approach is featured in Chap. 7 for external flows.

The *professional version* of *Heat Transfer* provides approximately 75% additional coverage of the topic of convection heat transfer. This extended coverage includes practical aspects such as (1) adaptation of the practical thermal analysis approach to variable property flow and flow over finned surfaces in Chap. 8; (2) development of comprehensive thermal/hydraulic analysis of heat exchangers which features the evaluation of $\overline{U}$ and f for shell-and-tube exchangers as well as double-pipe and compact crossflow exchangers in Chap. 11; (3) introduction of computer analysis of heat exchangers in Chap. 11; (4) development of bulk-stream relations for variable property flow in Appendix L; (5) development of practical hydraulic analysis for flow in heat exchanger cores in Appendix M; and (6) presentation of practical heat exchanger solution results and characteristics in Appendixes N and O; also covered are theoretical aspects such as (1) expanded treatment of the integral solution method for laminar thermal boundary layer flow in Chap. 7; (2) development of theoretical solutions for basic turbulent boundary layer flows in Chap. 7; (3) introduction to computer analysis of convection heat transfer in Chap. 7; (4) development of boundary layer equations and similarity transformation analysis for laminar flow in Appendix I, and (5) development of a two-parameter integral method for laminar natural convection in Appendix K.

To further aid the reader in understanding the heat transfer literature, the numerical finite-element method for solving heat transfer problems is introduced in Appendix F.

A mass transfer supplement is available, which primarily deals with issues pertaining to heat and mass transfer in chemically nonhomogeneous substances, with emphasis placed on basic low mass-flux processes. This material is organized into three chapters. Chapter 12 provides a general introduction to the topic; Chapter 13 deals with the theory of diffusion mass transfer, which involves the formulation and solution of diffusion transport equations; and Chapter 14 presents the practical analysis approach to convection heat and mass transfer, which features the use of convection coefficients. The organization is such that the practical analysis approach to convection heat and mass transfer featured in Chap. 14 can be studied before, after, or independent of the theoretical details developed in Chap 13. In addition to providing insight into the analysis of low mass-flux processes, the presentation also provides a foundation for the analysis of moderate to high mass-flux (concentration and/or injection driven) processes.

Because the changeover to metric units continues in many countries, both the international system of units (SI) and the English engineering system are used, with the SI system being used in the body of the text and in approximately 80% of the examples.

With regard to the examples, a consistent methodology is employed that involves the following format:

EXAMPLE

(statement)

Solution

Objective (concise statement of what is to be done)

Schematic (sketch of system showing all knowns and basic conditions)

Assumptions/Conditions (listing of assumptions and conditions not shown in schematic)

Properties (listing of pertinent thermophysical properties)

Analysis (presentation of mathematical formulation/solution, calculations, and comments)

It is believed that this systematic approach will aid the reader in recognizing and understanding the specific concepts presented in each example.

The *professional version* of *Heat Transfer* has been written and published with the hope that it will contribute to the effective study and practice of engineering heat transfer.

Lindon C. Thomas

ACKNOWLEDGMENTS

I wish to express my sincere appreciation to the many individuals who offered suggestions and criticism during the development of the book. Reviewers of the *professional version* of *Heat Transfer* include Benjamin T. F. Chung, University of Akron; Vincent Frisina, HTFS; Ki-Lun Lui, Koch Engineering Company; Ekkehard Marschall, University of California at Santa Barbara; Mohammad H. N. Naraghi, Manhattan College; John McCutchen, Yuba Heat Transfer; C. S. Reddy, Union College; Craig Saltiel, University of Florida Center for Advanced Study; Richard L. Shilling, Brown Fintube Company; E. M. Sparrow, University of Minnesota; and Jerry Taborek, consultant. In this connection, the author has particularly benefited from the very extensive input on practical aspects of heat exchanger analysis and design provided by Dr. Taborek.

Other reviewers who contributed to the development of this book in the context of their input on the *standard version* of *Heat Transfer* include Robert F. Boehm, University of Utah; Jerry E. Drummond, University of Akron; Ronald D. Flack, University of Virginia; Ray Knight, Auburn University; Wilbert Pulkrabak, University of Wisconsin at Plattville; Syed A. M. Said, King Fahd University of Petroleum and Minerals; and John W. Sheffield, University of Missouri at Rolla.

Work on the book was initiated while I was on the faculty of King Fahd University of Petroleum and Minerals in Dhahran, Saudi Arabia, and was completed while I was visiting the University of Bahrain in Isa Town, Bahrain. I very much appreciate the support provided by these institutions for my professional activities during the course of this work.

I also am pleased to acknowledge the important contributions to the development and production of the book provided by the editor, Mike Hays; the production editors, John Morgan and Jean Lapidus; and the copy editor, Bill Thomas. In addition, I am appreciative of the support for this project provided by Doug Humphrey and Bernard Goodwin of Prentice Hall.

Finally, I am thankful to God, who provided me with energy and good health, freedom and an education, encouraging parents, a loving and patient wife, and a growing family. I believe that what we are able to contribute professionally and otherwise is a result of what has been given to us by our creator and by those who love, encourage, teach, and support us, rather than something for which we alone should be credited.

Lindon C. Thomas

CHAPTER 1

INTRODUCTION

Heat transfer is defined as the transfer of energy across a system boundary caused solely by a temperature difference.

The study of heat transfer has long been a basic part of engineering curricula because of the significance of energy-related applications. For example, the transfer of heat in power plants from the energy source, be it fossil, nuclear, solar, or other, to the working fluid is one of the most basic processes in such systems. Similarly, the operation of refrigeration and air-conditioning units depend on the effective transfer of heat in condensers and evaporators. Other applications pertaining to environmental control, which are of particular interest currently, include the minimization of building-heat losses by means of improved insulating techniques and the use of supplemental energy sources, such as solar radiation, heat pumps, and fireplaces. Heat transfer is also very important in the operation of electrical machinery and transformers, and is often the controlling factor in the miniaturization of electronic systems.

Today the generation of electrical energy in power plants is primarily derived from fossil fuels—coal, natural gas, and oil. The cross section of a typical large coal-fired boiler used in power stations is shown in Fig. 1–1. In this system, high-pressure preheated water flowing in the vertical tubes surrounding the combustion chamber is heated by radiation and convection heat transfer, which results from the high temperatures produced by the combustion of coal. After separating the liquid and vapor in the upper steam drum, the vapor is further heated by a series of heat exchangers (known as superheaters), which operate at higher temperatures, and is then delivered to the steam turbine. The cold air supplied to the combustion chamber is first forced through a steam-heated coil in order to remove excess moisture and then through an exhaust gas/air preheater.

Although traditional fossil fuels such as coal and oil will be utilized for many years to come, a new era has been brought upon us by rising costs and environmental

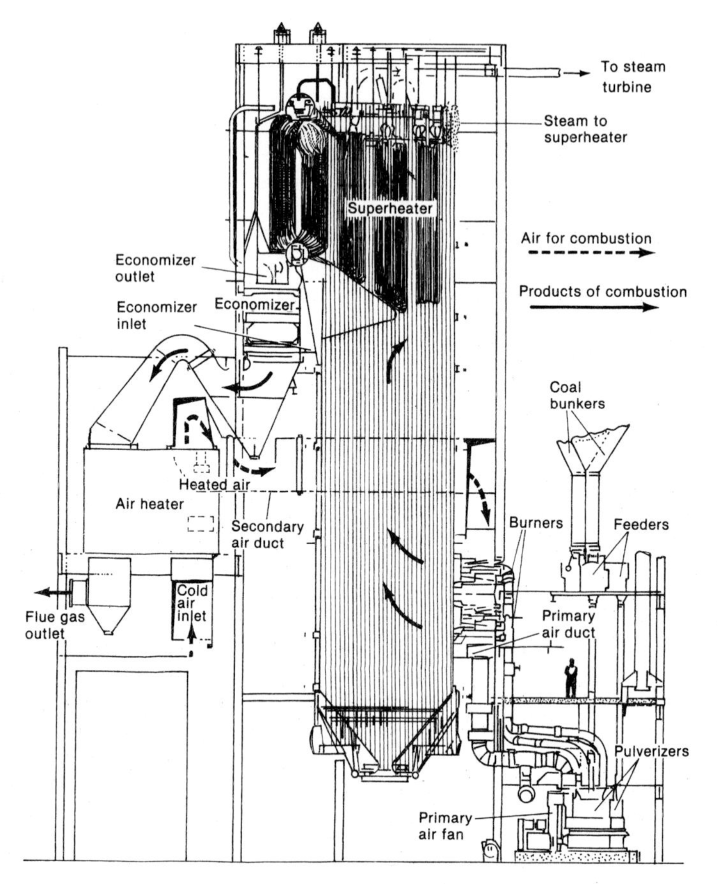

FIGURE 1–1 Coal-fired steam boiler for electric power station. (Courtesy Babcock and Wilcox.)

considerations associated with the use of these resources. As solutions are sought to our energy problems, there is no doubt that heat transfer will be a key factor. The most obvious example is the generation of power by nuclear fission, which involves (1) the transfer of heat from the reactor core to fluid circulating in a primary loop, and (2) the transfer of energy by a heat exchanger to convert secondary water to steam to power an electric generator. Although safety and environmental concerns persist, the use of nuclear power continues to increase in many countries.

Another example is the use of solar energy in heating, cooling, and energy generation. The schematic of a residential solar heating system is shown in Fig. 1–2. In this and other solar-energy applications, thermal radiation from the sun is captured in collectors, transferred to a working fluid such as air or water and stored. The operating modes in this particular system, which utilizes air as the circulating fluid and involves the automatic positioning of motorized dampers in the air-handling unit, include the following: mode A (*heating from collector*)—air circulates through the solar collector, through the air-handling unit to direct the hot air to the space, and

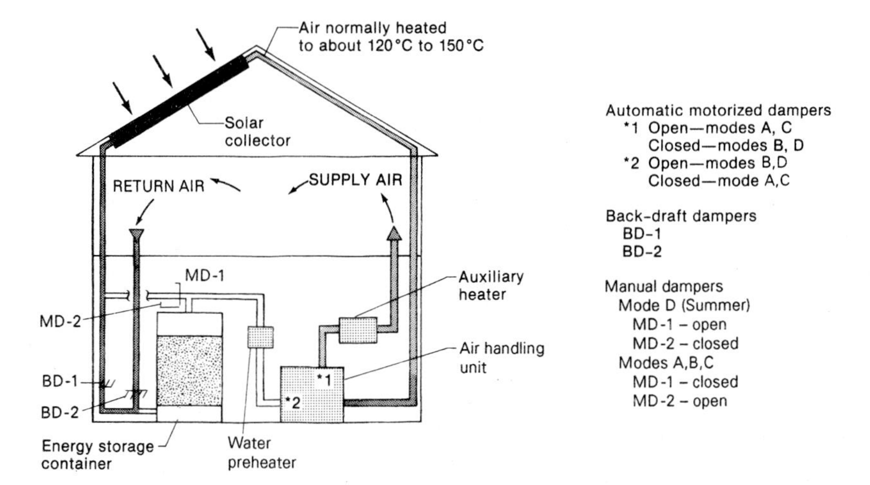

FIGURE 1–2 Schematic of solar heating system with air used as the working fluid: mode A—*heating from collector.* (Courtesy of Solar Laboratory, Colorado State University.)

back to the solar collector; mode B (*storing heat*)—air circulates through the solar collector, through the heat-storage unit, where the heat is absorbed by the pebble bed, and back to the solar collector; mode C (*heating from storage*)—at night or on cloudy days when solar energy is unavailable and when heat is needed in the space, air circulates into the bottom of the energy-storage unit, up through the pebbles where the air is heated, through the air-handling unit into space, and back to the energy-storage unit; and mode D (*summer hot water heating*)—in the summer, when space heating is not required, air is circulated through the collector where it is heated, and through the heat exchanger, where heat is transferred through the coil to the water. Of course, the effective use of solar energy requires that the system be carefully insulated.

The primary objective in the analysis of most heat-transfer problems is either to (1) determine the temperature distribution within the system and the rate of heat transfer for specified operating conditions (the function of evaluation), or (2) prescribe the necessary configuration (size and shape) in order to accomplish a given heat-transfer rate and/or temperatures (the function of thermal design). Although emphasis in our study will be placed on the evaluation function, the concepts of thermal analysis that will be developed and the expanded treatment of convection provide the basis for the actual design of systems involving heat transfer.

Because thermodynamics involves the study of heat and work for systems in equilibrium, a thermodynamic analysis can only provide us with predictions for the total quantity of heat transferred during a process in which a system goes from one equilibrium state (uniform temperature) to another. However, the length of time required for such processes to occur cannot be obtained by thermodynamics alone. On the other hand, the study of heat transfer involves a consideration of the mechanics of the transfer of thermal energy and is not restricted to equilibrium states. It is the science of heat transfer that enables us to perform the critical evaluation and design functions.

Analysis of heat-transfer processes requires the use of several *fundamental laws*, all of which are already familiar to us. These fundamental principles are of a general nature and are independent of the mechanism by which heat is transferred. In addition, *particular laws* pertaining to each of the mechanisms by which heat transfer can be accomplished must be satisfied. These fundamental and particular laws are reviewed in the following two sections, after which brief consideration is given to the very useful analogy between heat transfer and the flow of electric current. A review of basic mathematical concepts, dimensions, and units that pertain to our study of heat transfer is presented in Appendixes A and B.

1–1 FUNDAMENTAL LAWS

As in the case of thermodynamics, the *first law of thermodynamics* (conservation of energy) is the cornerstone of the science of heat transfer. This fundamental law takes the form

$$\text{rate of creation of energy} = 0$$

$$\Sigma\, \dot{E}_o - \Sigma\, \dot{E}_i + \frac{\Delta E_s}{\Delta t} = 0 \tag{1–1}$$

where $\dot{E}_i$ and $\dot{E}_o$ represent the rate of energy transfer into and out of the system, respectively, and $\Delta E_s/\Delta t$ is the rate of change in energy stored within the system.

Two other fundamental laws are required in the analysis of heat transfer in fluids that are in motion: (1) the *principle of conservation of mass* (for nonrelativistic conditions),

$$\text{rate of creation of mass} = 0$$

$$\Sigma\, \dot{m}_o - \Sigma\, \dot{m}_i + \frac{\Delta m_s}{\Delta t} = 0 \tag{1–2}$$

and (2) *Newton's second law of motion*, which is represented in terms of the x-component by

$$\text{rate of creation of momentum} = \text{sum of forces}$$

$$\Sigma\, \dot{M}_{o,x} - \Sigma\, \dot{M}_{i,x} + \frac{\Delta M_{s,x}}{\Delta t} = \Sigma\, F_x \tag{1–3}$$

where the momentum M_x represents the product of the x-component of velocity u and the mass m.

Three supplementary fundamental principles are required in the analysis of all heat-transfer processes: (1) *the second law of thermodynamics*, which provides us with the very critical conclusion that heat is transferred in the direction of decreasing temperature; (2) the principle of *dimensional continuity*, which requires that all equations be dimensionally consistent; and (3) *equations of state*, which provide information in equation, tabular, or graphical form pertaining to the thermodynamic properties at any state.

Concerning the thermodynamic properties, the symbols U and H ($H = U + PV$) will be used to designate the *internal energy* and *enthalpy* of a given mass m of a substance. Information pertaining to the *specific internal energy* e ($e = U/m$) and *specific enthalpy* i ($i = H/m$) is tabulated for common substances (e.g., steam tables). In addition, e and i can be expressed in terms of the *constant-volume specific heat* c_v and the *constant-pressure specific heat* c_P by

$$c_v = \left.\frac{\partial e}{\partial T}\right|_v \qquad c_P = \left.\frac{\partial i}{\partial T}\right|_P \tag{1–4,5}$$

For important practical applications involving ideal fluids (i.e., ideal gases and incompressible liquids), de and di are given by

$$de = c_v\,dT \qquad di = c_P\,dT \tag{1–6,7}$$

or, for systems with constant mass m,

$$dU = mc_v\,dT \qquad dH = mc_P\,dT \tag{1–8,9}$$

1–2 BASIC TRANSPORT MECHANISMS AND PARTICULAR LAWS

Conduction and thermal radiation represent the two fundamental mechanisms by which heat transfer is accomplished. These heat-transfer mechanisms occur in both solids and fluids. Transfer of heat by conduction (and sometimes by thermal radiation) from a solid surface to a moving fluid is known as *convection heat transfer*. These three modes of heat transfer and the particular laws that govern these phenomena are introduced in the following sections.

1–2–1 Conduction Heat Transfer

From the thermodynamic view, *temperature* T is a property that is an index of the kinetic energy possessed by the building-block particles of a substance (i.e., molecules, atoms, and electrons); the greater the agitation of these basic components of which matter is made, the higher the temperature. In this light, *conduction heat transfer* is the transfer of energy caused by physical interaction among molecular, atomic, and subatomic particles of a substance at different temperatures (level of kinetic energy). To expand upon this point, conduction in gases involves the collision and exchange of energy and momentum among molecules in continuous random motion. This same molecular transport mechanism occurs in liquids, but is complicated by the effects of molecular force fields, and can be augmented by the transport of free electrons in liquids that are good electrical conductors. On the other hand, conduction in solids occurs as a result of the movement of free electrons and vibrational energy in the atomic lattice structure of the material.

Fourier Law of Conduction

On the basis of experimental observation, the rate of heat transferred by conduction in the x direction through a finite area A_x for the situation in which T is a function only of x can be expressed by

$$q_x = -kA_x \frac{dT}{dx} \tag{1–10}$$

where A_x is normal to the direction of transfer x, and k is the *thermal conductivity*. This equation was first used to analyze conduction heat transfer in 1822 by a French mathematical physicist named J. Fourier [1] and has come to be called the *Fourier law of conduction*. An example of a one-dimensional molecular conduction-heat-transfer problem for which this equation applies is illustrated in Fig. 1–3, which shows a plate with surface temperatures T_1 and T_2. Because no temperature differences occur in the y and z directions, q_y and q_z are both zero. For this case in which T_1 is greater than T_2, the temperature gradient is negative. (As shown in Chap. 2, the temperature distribution is linear in this application because q_x, k, and A_x are all constants.)

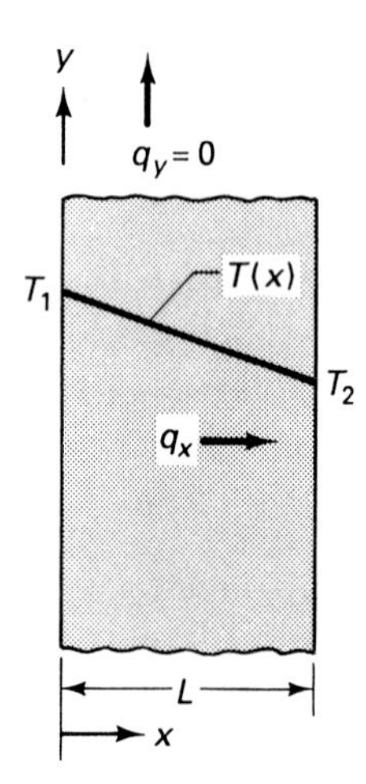

FIGURE 1–3
One-dimensional conduction heat transfer in a plate with $T_1 > T_2$; $q_y = 0$ and $q_z = 0$.

The consequence of the minus sign in Eq. (1–10) is that q_x is positive for situations such as this in which the temperature gradient is negative. This result is consistent with the second law of thermodynamics, which stipulates that heat is transferred in the direction of decreasing temperature.

For situations in which the temperature is a function of time t and one space variable, such as x, the Fourier law of conduction is written as

$$q_x = -kA_x \frac{\partial T}{\partial x} \tag{1–11}$$

where q_x also is a function of t and x. Expressions of the form of Eqs. (1–10) and (1–11) can also be written for conduction heat transfer in the y, z, or r directions, as will be illustrated in Chap. 2. For applications in which the temperature T is a function of more than one spatial dimension, the heat transfer in each direction must be accounted for. For example, for the case in which T is a function of x and y, we must write expressions for both q_x and q_y. To accomplish this task, we utilize a more general form of the Fourier law of conduction. This general Fourier law of conduction is presented in Chap. 3, which deals with multidimensional conduction heat transfer.

Thermal Conductivity The *thermal conductivity* k is a thermophysical property of the conducting medium that represents the rate of conduction heat transfer per unit area for a temperature gradient of 1 °C/m (or 1 °F/ft). The units for k are W/(m °C) [or Btu/(h ft °F)]. (Note that °C = 1 K and °F = 1 °R.) The thermal conductivities of various common substances are listed in Table 1–1 for standard atmospheric conditions. More extensive tabulations of thermal conductivities and other properties are given in Tables A–C–1 through A–C–5 of the Appendix and in references 2 through 6.

At room temperature, k ranges from values in the hundreds for good conductors of heat such as diamond and various metals to less than 0.01 W/(m °C) for some gases. Materials with values of k less than about 1 W/(m °C) are classified as insulators. As a rule of thumb, metals with good electrical conducting properties have higher thermal conductivities than do dielectric nonmetals or semiconductors. This is because the molecular interaction in good electrical conductors is enhanced by the movement of free electrons. Exceptions to this rule include dielectric crystals, such as diamond, sapphires, and quartz, and electric semiconductors, such as silicon and germanium. A second rule is that solid phases of materials generally have higher thermal conductivities than do liquid phases. An exception to this rule is bismuth, which has a higher thermal conductivity for the liquid phase than for the solid phase.

The variation of k with temperature is shown in Fig. 1–4 for several representative substances and in Figs. A–C–1 through A–C–4 for various other common materials. The thermal conductivity of many of these substances varies by a factor of 10 or more for an order-of-magnitude change in temperature. On the other hand, the variation in k with temperature for some materials over certain temperature ranges is small enough to be neglected. We also note that exceptionally high thermal conductivities occur among the solid materials that were judged to be good conductors at room temperature. For example, the thermal conductivity of aluminum reaches a maximum value of about 20,000 W/(m °C) at 10 K. This is over 100 times as large as the value that occurs at room temperature. Substances under low-temperature conditions that have such exceedingly high thermal conductivities are known as *superconductors*.

In homogeneous materials, k can generally be assumed to be independent of direction (i.e., isotropic). However, some pure materials and laminates have thermal conductivities that are dependent upon the direction of heat flow. For example, the

TABLE 1–1 Thermal conductivity of various substances at room temperature

Substance	*k* W/(m °C)	*k* Btu/(h ft °F)
Metals		
Silver	420	240
Copper	390	230
Gold	320	180
Aluminum	200	120
Silicon	150	87
Nickel	91	53
Chromium	90	52
Iron (pure)	80	46
Germanium	60	35
Carbon steel (0.5% C)	54	31
Nonmetallic Solids		
Diamond, type 2A	2300	1300
Diamond, type 1	900	520
Sapphire (Al_2O_3)	46	27
Limestone	1.5	0.87
Glass (Pyrex 7740)	1.0	0.58
Teflon (Duroid 5600)	0.40	0.23
Brick, building	0.69	0.399
Plaster	0.13	0.075
Cork	0.040	0.023
Liquids		
Mercury	8.7	5.0
Water	0.6	0.35
Freon F-12	0.08	0.046
Gases		
Hydrogen	0.18	0.10
Air	0.026	0.015
Nitrogen	0.026	0.015
Steam	0.018	0.01
Freon F-12	0.0097	0.0056

Source: From references 2 through 5.

thermal conductivity of wood is different for heat conduction across the grain than for heat transfer parallel to the grain. Other materials with such nonisotropic characteristics include crystalline substances, laminated plastics, and laminated metals. For an introduction to the topic of conduction heat transfer in nonisotropic materials, the textbook by Eckert and Drake [7] is recommended.

Heat-transfer applications involving the more familiar metallic conductors such as copper and aluminum and insulators such as rock wool and cork are known to us

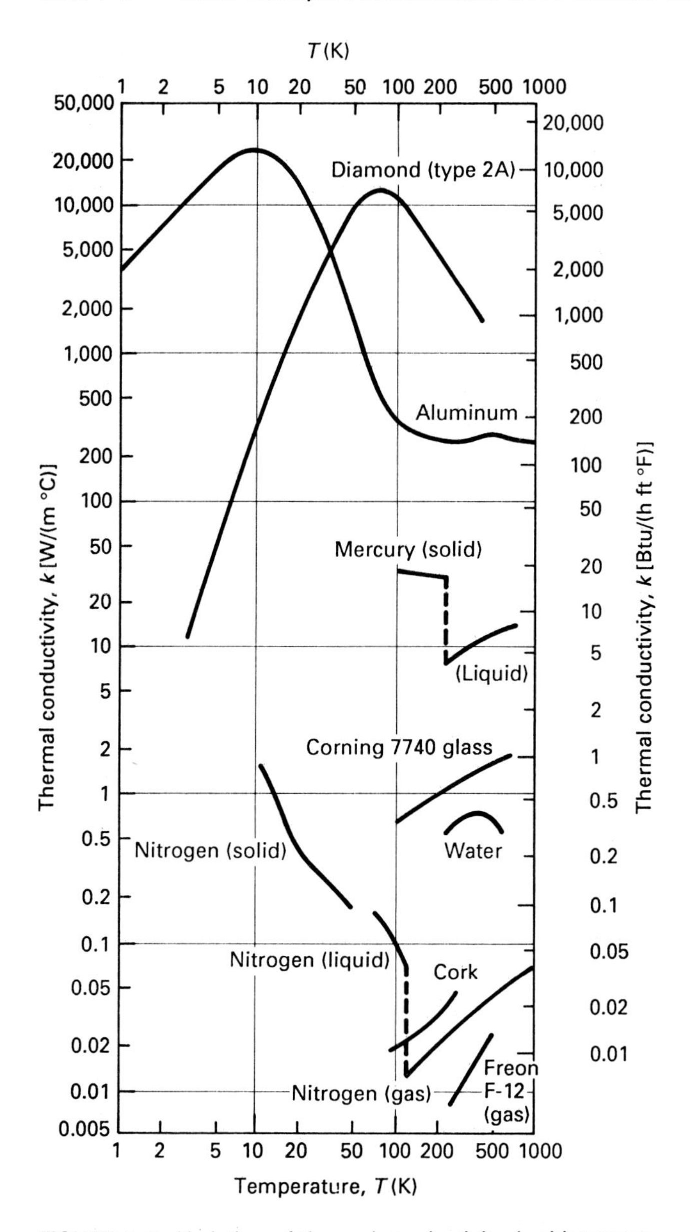

FIGURE 1–4 Variation of thermal conductivity k with temperature for representative substances. (From Touloukian et al. [2]. Used with permission.)

all. To name only a few, tubes made of copper, stainless steel, aluminum, and other metals are used in boilers, evaporators, and condensers to transmit energy from one fluid to another; metal pans are used to cook food; double glass panes are used to minimize heat loss through windows; and glass fibers, bricks, and other insulative materials are used to reduce building heat losses in the winter and heat gains in the summer.

Because of their high thermal conductivities and large electrical resistivities, silicon and diamond also find application, especially in the field of electronics. For example, silicon greases, pastes, and gaskets are often used in the construction of electronic systems in order to increase the rate of heat transfer while maintaining good electrical insulation between components. As another example, Fig. 1–5 shows a gold-plated diamond (type 2A) cube diode. Diodes such as these, ranging in size from below 0.1 mm to a few millimeters across, are used to generate high-frequency radio waves that relay telephone conversations and television broadcasts. These small diodes are characterized by very high power density operation, with operating temperatures in the range 150°C to 200°C. Because type 2A diamond has the highest thermal conductivity of all known materials in this temperature range, the diamond cube is used to remove the energy generated in the electronic semiconductor chip. Because of its effectiveness as a heat conductor, the diamond cube reduces the operating temperature of the diode, thereby increasing its lifetime and reliability.

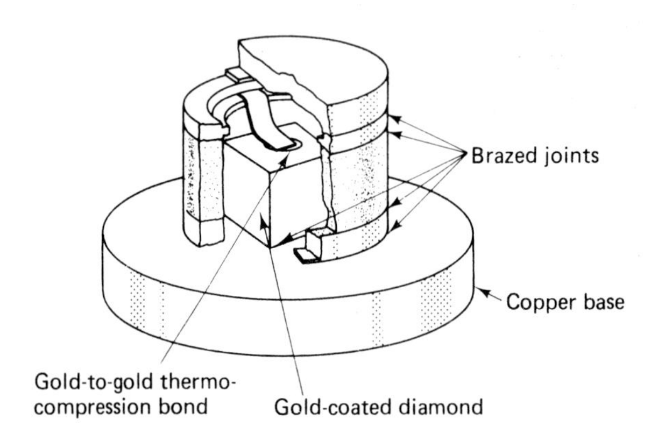

FIGURE 1–5
Microwave oscillator diode with diamond heat sink. (Courtesy of Bell Telephone Laboratories and D. Drukker & ZN. N.Y.)

Analysis of Conduction Heat Transfer

Consideration is now given to the analysis of conduction heat transfer in solids or stationary fluids. The important topic of conduction heat transfer in moving fluids will be considered separately in Sec. 1–2–3, which deals with convection.

The general theoretical analysis of conduction-heat-transfer problems involves (1) the use of (a) the fundamental first law of thermodynamics and (b) the Fourier law of conduction (particular law) in the development of a mathematical formulation that represents the energy transfer in the system; and (2) the solution of the resulting system of equations for the temperature distribution. Once the temperature distribution is known, the rate of heat transfer is obtained by use of the Fourier law of conduction. The basic concepts involved in the theoretical analysis of conduction-heat-transfer problems will be presented in Chap. 2 in the context of fairly simple one-dimensional systems. These fundamentals will then be extended to multidimensional systems in Chap. 3.

Conduction Shape Factor A simple practical approach to the analysis of basic steady-state conduction-heat-transfer problems has been developed that involves the use of an equation derived from the fundamental and particular laws. This practical equation for conduction heat transfer takes the form (for systems with uniform thermal conductivity)

$$q = kS(T_1 - T_2) \tag{1-12}$$

where q is the rate of heat transfer conducted from a surface at temperature T_1 to a surface at T_2, and S is known as the *conduction shape factor*; the unit for S is m (or ft). The conduction shape factor S is dependent upon geometry. Representative conduction shape factors are listed in Table 1–2 for several basic geometries. This practical approach to the analysis of conduction-heat-transfer problems is illustrated by several examples in this chapter. The theoretical basis for this simple method will be developed for one-dimensional systems in Chap. 2 and will be extended to more complex multidimensional systems in Chap. 3.

TABLE 1–2 Conduction shape factors S

Geometry	S
Flat plate Cross-sectional area A Thickness L	A/L
Hollow cylinder Radii r_1 and r_2 Length L	$\dfrac{2\pi L}{\ln (r_2/r_1)}$
Hollow sphere Radii r_1 and r_2	$\dfrac{4\pi r_1 r_2}{r_2 - r_1}$

EXAMPLE 1–1

Determine the rate of heat loss per unit area through a 10-cm-thick brick wall with surface temperatures of 15°C and 75°C.

Solution

Objective Determine the heat flux q''.

Schematic Plane wall.

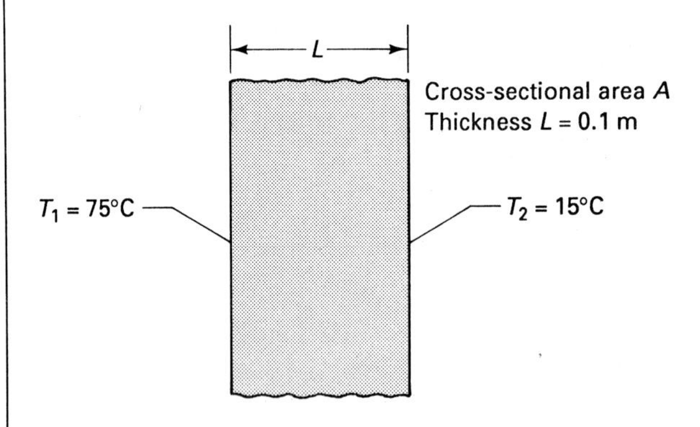

Assumptions/Conditions

steady-state
one-dimensional
uniform properties

Properties Building brick (Table 1–1): $k = 0.69$ W/(m °C).

Analysis Utilizing Eq. (1–12) with $S = A/L$, we write

$$q = kS(T_1 - T_2) = \frac{kA}{L}(T_1 - T_2)$$

or

$$q'' = \frac{q}{A} = \frac{k}{L}(T_1 - T_2) = \frac{0.69\ \text{W/(m °C)}}{0.1\ \text{m}}(75°\text{C} - 15°\text{C}) = 414\ \text{W/m}^2$$

1–2–2 Radiation Heat Transfer

The second fundamental mechanism of heat transfer involves the transfer of electromagnetic radiation emitted from a body as a result of vibrational and rotational movements of molecules, atoms, and electrons. As we have already noted, temperature is an index of the level of agitation of these microscopic particles. Because molecules and their components are continuously in motion, thermal radiation is always emitted by physical matter. Furthermore, the rate at which internal energy associated with the motion of molecules, atoms, and subatomic particles (indicated by temperature) is converted into thermal radiation increases with temperature. As discussed in Chap. 5, the manner in which thermal radiation, which encompasses ultraviolet (UV), visible light, and infrared (IR), is generated distinguishes it from other types of electromagnetic waves, such as γ rays and X rays on the one hand, and microwaves and broadcasting waves on the other.

The medium through which thermal radiation passes can be a vacuum or a gas, liquid, or solid. Objects within the path absorb, reflect, and, if they are transparent, transmit incident thermal radiation. Electromagnetic radiation that is absorbed by matter is converted into internal energy, which can be stored, transferred by conduction, and converted back into electromagnetic radiation that is given off by the material itself. The absorptivity α, reflectivity ρ, and transmissivity τ represent the fractions of incident thermal radiation that a body *absorbs*, *reflects*, and *transmits*, respectively. It follows that

$$\alpha + \rho + \tau = 1 \tag{1–13}$$

These properties are primarily dependent upon the temperature of the emitting source and the nature of the surface that receives the thermal radiation. To illustrate, most incoming thermal radiation is absorbed by a surface coated with lampblack paint (i.e., $\alpha \simeq 0.97$, $\rho \simeq 0.03$, $\tau \simeq 0$), but is reflected from the surface of a polished aluminum plate (i.e., $\alpha \simeq 0.1$, $\rho \simeq 0.9$, $\tau \simeq 0$). As another example, a thin glass plate will transmit most of the thermal radiation from the sun, but will absorb much of the thermal radiation emitted from the low-temperature interior of a building, such as a greenhouse.

Thermal radiation generally passes through gases such as air with no significant absorption taking place. Such gases with $\tau \simeq 1$ are known as *nonparticipating gases*. Gases and transparent liquids that absorb significant quantities of the thermal radiation are known as *participating fluids*. Carbon dioxide and water vapor are examples of participating gases and water is a participating liquid. Of course, many liquids, such as mercury, are opaque to thermal radiation.

Stefan-Boltzmann Law

A body continually emits radiant energy in an amount that is related to its temperature and the nature of its surface. An object that absorbs all the radiant energy reaching its surface ($\alpha = 1$) is called a *blackbody*. Such ideal absorbers emit radiant energy at a rate that is proportional to the fourth power of the absolute temperature of the surface. The *Stefan-Boltzmann law* for blackbody thermal radiation takes the form

$$E_b = \sigma T_s^4 \tag{1–14}$$

where the *total emissive power* E_b for a blackbody is the total rate of thermal radiation emitted by a perfect radiator per unit surface area, σ is the *Stefan-Boltzmann constant*

[$\sigma = 5.670 \times 10^{-8}$ W/(m^2 K^4) $= 0.1714 \times 10^{-8}$ Btu/(h ft^2 °R^4)], and T_s is the *absolute* surface temperature. The experimental basis for this famous fourth-power law was first established by the Austrian scientist J. Stefan in 1879. This discovery was followed in 1884 by a theoretical development by another Austrian, L. Boltzmann. This equation indicates that the thermal radiation energy content rapidly falls from a very substantial level for temperatures of the order of 500 K to relatively small values for common environmental temperatures of the order of 300 K.

For nonblackbody surfaces that absorb less than 100% of the incident radiant energy, the total emissive power E (i.e., the rate of thermal radiation emitted per unit surface area) is generally expressed by

$$E = \epsilon E_b = \epsilon \sigma T_s^4 \tag{1–15}$$

where the *emissivity* ϵ lies between zero and unity. For example, the emissivities of polished aluminum and lampblack paint at room temperature are of the order of 0.1 and 0.96, respectively. The emissivities of various substances are listed in Table A–C–7 of the Appendix. The relationship between the emissivity ϵ and the absorptivity α will be considered in Chap. 5.

Radiation-Heat-Transfer Rate

By definition, the rate of radiation heat transfer q_R between two bodies is equal to the *net* rate of exchange of thermal radiation. For two infinite parallel blackbody plates that are separated by a vacuum or a nonparticipating gas, all the thermal radiation emitted from one surface reaches and is absorbed by the other body. The rate of thermal radiation leaving A_s that is absorbed by A_R is $A_s E_{bs}$, and the rate of radiant energy leaving A_R that is absorbed by A_s is $A_R E_{bR}$. With A_s and A_R being equal, it follows that the rate of radiation heat transfer q_R from A_s to A_R is simply

$$q_R = A_s(E_{bs} - E_{bR}) \tag{1–16}$$

or

$$q_R = \sigma A_s(T_s^4 - T_R^4) \tag{1–17a}$$

Taking the opposite orientation, the rate of radiation heat transfer from surface A_R to surface A_s is

$$q_R = \sigma A_s(T_R^4 - T_s^4) \tag{1–17b}$$

A more general relation, which applies to other geometries, is given by

$$q_R = A_s F_{s-R}(E_{bs} - E_{bR}) = \sigma A_s F_{s-R}(T_s^4 - T_R^4) \tag{1–18}$$

where the *shape factor* (or *view factor*) F_{s-R} represents the fraction of thermal radiation leaving surface s that arrives directly on surface R. The shape factor varies from unity (for infinite parallel plates or enclosures) to zero. Further information pertaining to the development of Eq. (1–18) and to relations that enable us to determine F_{s-R} for particular geometric configurations can be found in Chap. 5.

EXAMPLE 1–2

The electrical cabinet shown in Fig. E1–2 is exposed to blackbody radiation with the surrounding walls, which are at 25°C. Determine the rate of radiation heat transfer if the cabinet temperature is 125°C.

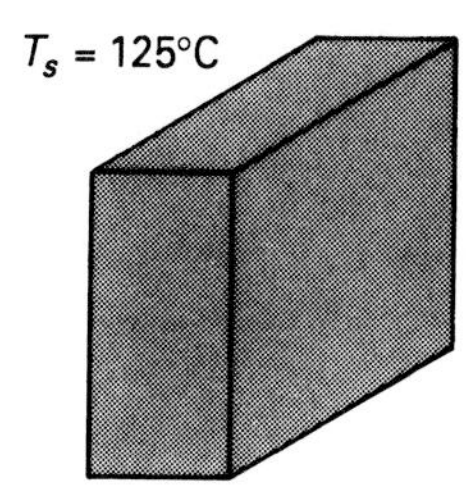

FIGURE E1–2
Cabinet mounted on a vertical wall.

Solution

Objective Determine the rate of radiation heat transfer q_R.

Assumptions/Conditions

steady-state
blackbody thermal radiation

Analysis The energy emitted per unit area from the cabinet and the walls is calculated first. Utilizing the Stefan-Boltzmann law, Eq. (1–14), we have

$$E_{bs} = \sigma T_s^4 = 5.67 \times 10^{-8} \frac{\text{W}}{\text{m}^2\,\text{K}^4} (398\ \text{K})^4 = 1420\ \text{W/m}^2$$

$$E_{bR} = \sigma T_R^4 = \sigma(298\ \text{K})^4 = 447\ \text{W/m}^2$$

The rate of blackbody radiation heat transfer from A_s to the enclosure is given by Eq. (1–18), with the radiation shape factor F_{s-R} equal to unity; that is,

$$q_R = A_s F_{s-R}(E_{bs} - E_{bR}) = 0.368\ \text{m}^2\,(1)\left(1420 \frac{\text{W}}{\text{m}^2} - 447 \frac{\text{W}}{\text{m}^2}\right) = 358\ \text{W}$$

If the space between the cabinet and the enclosure is evacuated, the total rate of heat transfer from the cabinet will be 358 W. By performing an energy balance on the cabinet for these conditions, we conclude that the power generated by the electrical system within the cabinet is 358 W. (If the cabinet is surrounded by air or another fluid, heat will also be removed by convection, which is the topic of the next section.)

1–2–3 Convection Heat Transfer

As we have already indicated, *convection* is the transfer of heat from a surface to a moving fluid. The conduction-heat-transfer mechanism always plays a primary role in convection. In addition, thermal radiation heat transfer (associated with absorbing and emitting fluids) and diffusion mass transfer (associated with chemically nonhomogeneous substances) are also sometimes involved.† In addition to the transfer of energy via the basic heat-transfer-mechanisms, convection heat transfer also involves the transfer of energy by macroscopic fluid motion. The process by which energy (or mass) is transferred by bulk-fluid motion is known as *advection*.

An introductory theoretical treatment of convection-heat-transfer processes is presented in Chap. 7. The theoretical analysis of convection requires that the fundamental laws of mass, momentum, and energy and the particular laws of viscous shear and conduction be utilized in the development of mathematical formulations for the fluid flow and energy transfer. The solution of these equations provides predictions for the velocity and temperature distributions within the fluid, after which calculations are developed for the rate of heat transfer into the fluid by the use of the Fourier law of conduction.

Newton Law of Viscous Stress

In regard to the theoretical analysis of convection-heat-transfer processes, the particular law for viscous stress in Newtonian fluids can be written for simple steady one-dimensional shear flows as

$$\tau = \frac{dF}{dA} = \mu \frac{du}{dy} \tag{1–19}$$

where the viscosity μ is a property of the fluid with units kg/(m s) [or lb_m/(ft s)], y the distance from the wall, dA the differential area normal to y, dF the differential shear force acting on the area dA, τ the shear stress, and u the axial velocity. The Newton law of viscous stress is supported by experimental observation, kinetic theory, and statistical mechanics. The viscosity is often coupled with the density and written as $\nu = \mu/\rho$, where ν is called the *kinematic viscosity*; the units for ν are m^2/s (or ft^2/s). The viscosity and kinematic viscosity of several Newtonian fluids are shown in Tables A–C–3 to A–C–5. The resistance to Newtonian fluid motion that results from shear stress is proportional to the viscosity, such that highly viscous fluids flow much less readily than do fluids with low viscosities such as water and air.

Newton Law of Cooling

The engineer is generally primarily concerned about the rate of convection heat transfer rather than the temperature distribution within the fluid. Therefore, a practical approach

† Attention will be restricted to fluids in which thermal radiation can be neglected. The topic of diffusion mass transfer is considered in the accompanying *mass transfer supplement*.

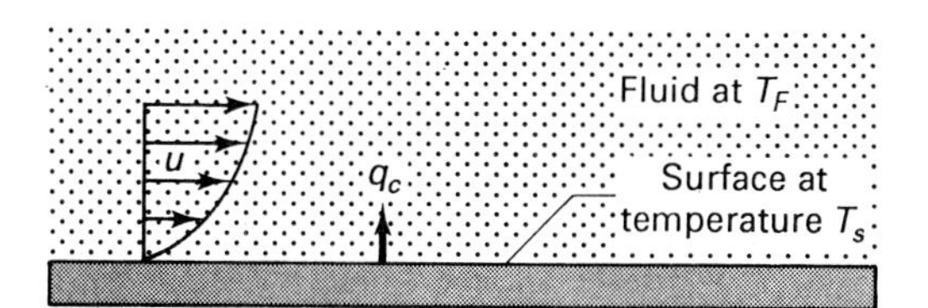

FIGURE 1–6
Convection heat transfer.

to the analysis of convection heat transfer from surfaces such as the flat plate shown in Fig. 1–6 has been developed which employs an equation of the form

$$q_c = \bar{h} A_s (T_s - T_F) \tag{1–20}$$

where q_c is the rate of heat transferred from a surface at uniform temperature T_s to a fluid with reference temperature T_F, A_s is the surface area, and $\bar{h}$ is the *mean coefficient of heat transfer*; the units for $\bar{h}$ are W/(m^2 °C) [or Btu/(h ft^2 °F)]. Equation (1–20) is often referred to as the *Newton law of cooling* in honor of the British scientist Sir Isaac Newton.

A more general form of the Newton law of cooling, which applies to cases in which T_s and/or T_F are not uniform, is given by

$$dq_c = h_x \, dA_s \, (T_s - T_F) \tag{1–21}$$

where dq_c represents the rate of heat transfer from a differential surface area dA_s and h_x is the *local coefficient of heat transfer*. The mean and local coefficients of heat transfer are related by

$$\bar{h} = \frac{1}{A_s} \int_{A_s} h_x \, dA_s \tag{1–22}$$

The Newton law of cooling in its simple or general form will be seen to be very useful in the analysis of heat-transfer processes involving convection combined with conduction and radiation in Sec. 1–2–4 and Chaps. 2 through 5, and in the evaluation and design of convection-heat-transfer systems in Chaps. 8 through 11.

Coefficient of Heat Transfer Approximate ranges of $\bar{h}$ are shown in Table 1–3 for forced and natural convection in air and water. (For forced convection, the fluid motion is caused by mechanical means such as pumps and fans. On the other hand, natural convection is caused by temperature-induced density gradients within the fluid.†) The actual value of $\bar{h}$ depends upon the hydrodynamic conditions as well as on the thermodynamic and thermophysical properties of the fluid. The details of evaluating h_x and $\bar{h}$ for standard fluid-flow systems will be considered in Chaps. 6 through 11.

† Natural convection can also occur in chemically nonhomogeneous mixtures as a result of density gradients caused by concentration gradients.

TABLE 1–3 Convection-heat-transfer coefficients—range for representative applications

System	$\bar{h}$	
	W/(m² °C)	Btu/(h ft² °F)
Natural convection		
Air	5–30	0.9–5
Water	200–600	30–100
Forced convection		
Air	10–500	2–100
Water	$100–2 \times 10^4$	$20–4 \times 10^3$
Oil	$60–2 \times 10^3$	10–400
Boiling water at 1 atm	$2 \times 10^3–5 \times 10^4$	$300–9 \times 10^3$
Condensation of steam	$5 \times 10^3–10^5$	$900–2 \times 10^4$

EXAMPLE 1–3

Determine the rate of convection heat transfer from the cabinet in Example 1–2 if it is surrounded by air at 25°C with a mean coefficient of heat transfer of 6.8 W/(m² °C).

Solution

Objective Determine the rate of convection heat transfer q_c.

Schematic Cabinet.

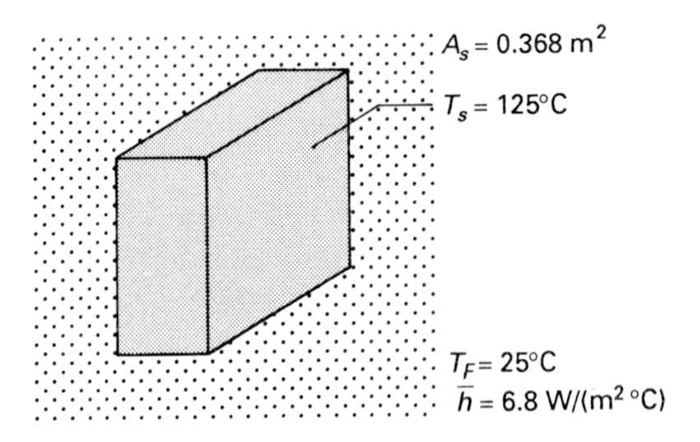

Assumptions/Conditions

steady-state

convection and blackbody thermal radiation

Analysis Utilizing the Newton law of cooling, Eq. (1–20), we have

$$q_c = \bar{h}A_s(T_s - T_F) = 6.8 \frac{\text{W}}{\text{m}^2\ ^\circ\text{C}} (0.368\ \text{m}^2)(125^\circ\text{C} - 25^\circ\text{C}) = 250\ \text{W}$$

As indicated in Example 1–4, this level of convection cooling contributes significantly to the total rate of heat transfer from the cabinet.

1–2–4 Combined Modes of Heat Transfer

Many heat-transfer processes encountered in practice involve combinations of conduction, thermal radiation, and convection. A situation in which combinations of these basic heat-transfer modes occur simultaneously is illustrated in Fig. 1–7. In this wall, which consists of two plates separated by a vacuum, heat is (1) convected from the fluid at T_{F1} to plate 1, (2) conducted through plate 1, (3) radiated from plate 1 to plate 2, (4) conducted through plate 2, (5) and convected from plate 2 to the fluid at T_{F2}. Similar composite walls have been developed for the storage of cryogenic liquids, which consist of multiple layers of highly reflective materials that are separated by evacuated spaces. These *superinsulations* provide very efficient insulative walls with apparent thermal conductivities that are as low as 2×10^{-5} W/(m °C).

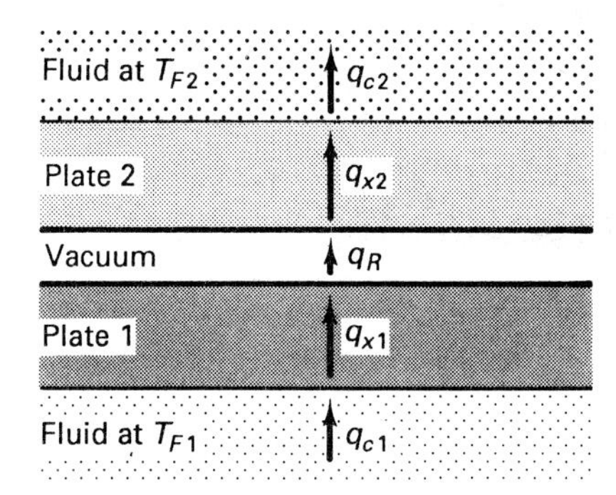

FIGURE 1–7
Heat-transfer process involving combined conduction, thermal radiation, and convection.

As illustrated by Examples 1–4 and 1–5, the analysis of basic multimode heat-transfer processes requires the use of the fundamental first law of thermodynamics and the particular laws for conduction, radiation, and convection.

EXAMPLE 1–4

An electronic cabinet made of anodized aluminum is cooled by natural convection and radiation. The surface area of the cabinet is 0.368 m^2, the temperature of the surrounding fluid and surface is 25°C, and $\bar{h}$ = 6.8 W/(m^2 °C). Estimate the rate of heat transfer from the cabinet if its surface temperature is to be maintained at 125°C.

Solution

Objective Determine the total rate of heat transfer q.

Schematic Cabinet.

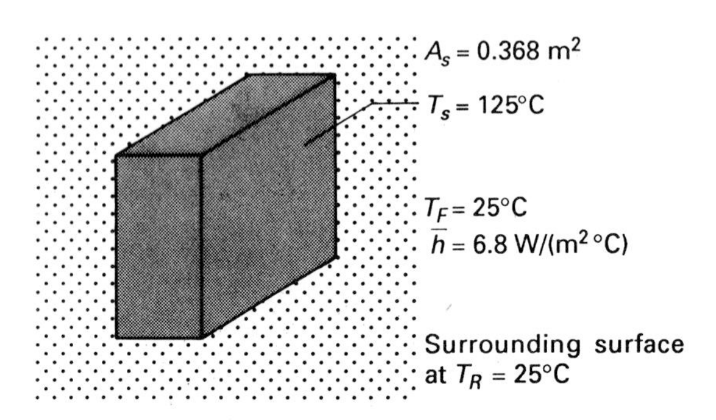

Assumptions/Conditions

steady-state
convection and blackbody thermal radiation

Analysis The total rate of heat transfer q from the cabinet is

$$q = q_c + q_R$$

where q_c is given by Eq. (1–20). Because the cabinet is made of black anodized aluminum, we will utilize the blackbody approximation given by Eq. (1–18). Referring to Examples 1–2 and 1–3, we obtain

$$q_R = A_s F_{s-R}(E_{bs} - E_{bR}) = \sigma A_s F_{s-R}(T_s^4 - T_R^4) = 358 \text{ W}$$

$$q_c = \bar{h} A_s (T_s - T_F) = 250 \text{ W}$$

such that the total rate of heat transfer q is

$$q = 608 \text{ W}$$

Note that the radiation heat transfer accounts for a very significant 59% of the total. Whereas radiation is important for systems such as this involving natural convection, radiation is often not a factor in forced-convection systems. For example, if the cabinet is cooled by forced air with $\bar{h}$ = 500 W/(m^2 °C), the resulting surface temperature will be about 28.3°C for the same rate of energy dissipation (i.e., q = 608 W), for which case only about 1.2% of the heat transfer is accomplished by radiation. (Radiation is also often insignificant in systems involving bright metal surfaces with low emissivities.)

EXAMPLE 1–5

Liquid petroleum gas (LPG), which is widely used for domestic and industrial heating, consists of propane, butane, and mixtures of the two. LPG must be kept in liquid form because of its high volatility. The liquefication is often accomplished by refrigeration so that LPG can be stored at atmospheric pressure. Butane is stored at about 0°C and propane at −40°C.

With this brief background, we consider a cryogenic fluid that is to be stored in the 2-m-diameter spherical chamber shown in Fig. E1–5. A temperature of −40°C must be maintained by the refrigerant, which flows within the 1-cm-thick shell surrounding the inner storage chamber. The outer part of the vessel consists of a 10-cm-thick insulating material with $k = 0.60$ W/(m °C). The fluid surrounding the vessel is at 35°C with $\bar{h} = 150$ W/(m^2 °C). Determine the amount of refrigeration that is required to maintain the fluid at −40°C.

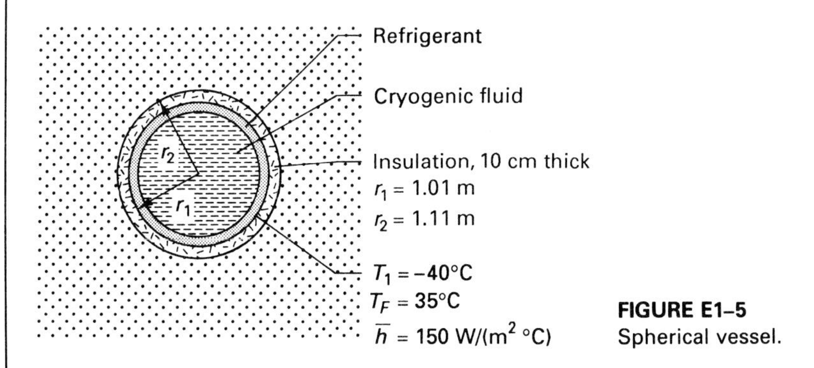

FIGURE E1–5
Spherical vessel.

Solution

Objective Determine the total rate of heat transfer that must be compensated for by refrigeration.

Assumptions/Conditions

steady-state
one-dimensional
uniform properties

Properties Insulation: $k = 0.60$ W/(m °C).

Analysis By applying the first law of thermodynamics to the outer surface of the insulating wall, we obtain

$$\sum \dot{E}_i = \sum \dot{E}_o + \cancel{\frac{\Delta E_s}{\Delta t}}$$

$$q_c = q_r \tag{a}$$

The rate of heat transfer by convection q_c is given by Eq. (1–20),

$$\begin{aligned} q_c &= \bar{h}A_s(T_F - T_2) = \bar{h}4\pi r_2^2(35°\text{C} - T_2) \\ &= 150\frac{\text{W}}{\text{m}^2\,°\text{C}}(4\pi)(1.11\text{ m})^2(35°\text{C} - T_2) = 2320\frac{\text{W}}{°\text{C}}(35°\text{C} - T_2) \end{aligned} \tag{b}$$

where the surface temperature is represented by T_2. The rate of heat transfer by radial conduction q_r is given by Eq. (1–12) with $S = 4\pi r_1 r_2/(r_2 - r_1)$; that is,

$$q_r = kS(T_2 - T_1) = \frac{4\pi r_1 r_2 k}{r_2 - r_1}(T_2 + 40°\text{C}) = 84.5\frac{\text{W}}{°\text{C}}(T_2 + 40°\text{C}) \tag{c}$$

Similarly, the application of the first law of thermodynamics to the inside surface of the insulating wall gives

$$q_r = q_{\text{ref}} \tag{d}$$

such that the rate of heat transferred through the insulation by conduction q_r must be taken out by refrigeration.

To determine q_r or q_c, we first solve for the surface temperature T_2 by utilizing Eqs. (a) through (c).

$$84.5\frac{\text{W}}{°\text{C}}(T_2 + 40°\text{C}) = 2320\frac{\text{W}}{°\text{C}}(35°\text{C} - T_2)$$

$$T_2 = \frac{(2320\text{ W/°C})(35°\text{C}) - (84.5\text{ W/°C})(40°\text{C})}{2320\text{ W/°C} + 84.5\text{ W/°C}} = 32.4°\text{C}$$

Combining this result with Eqs. (c) and (d), we obtain

$$q_{\text{ref}} = q_r = kS(T_2 - T_1) = 84.5\frac{\text{W}}{°\text{C}}(32.4°\text{C} + 40°\text{C}) = 6120\text{ W}$$

Hence, approximately 6.12 kW of refrigeration is required.

As suggested in Sec. 1–2–4, superinsulation-type arrangements involving at least one evacuated space are often used in such cryogenic applications.

EXAMPLE 1–6

Although energy transfer by microwaves is generally not classified as thermal radiation because of the manner in which electromagnetic waves of this type are generated (see Chap. 5), microwave ovens are now widely and effectively used to cook many foods. As illustrated by Fig. E1–6, microwave cooking is made possible by the fact that energy supplied by electric current and converted to microwaves reflects from metal surfaces, passes through glass, ceramic, and plastic, but is readily absorbed and converted into internal energy by food molecules. Compare the heating (i.e., cooking) principles and characteristics of a microwave oven with conventional methods of cooking.

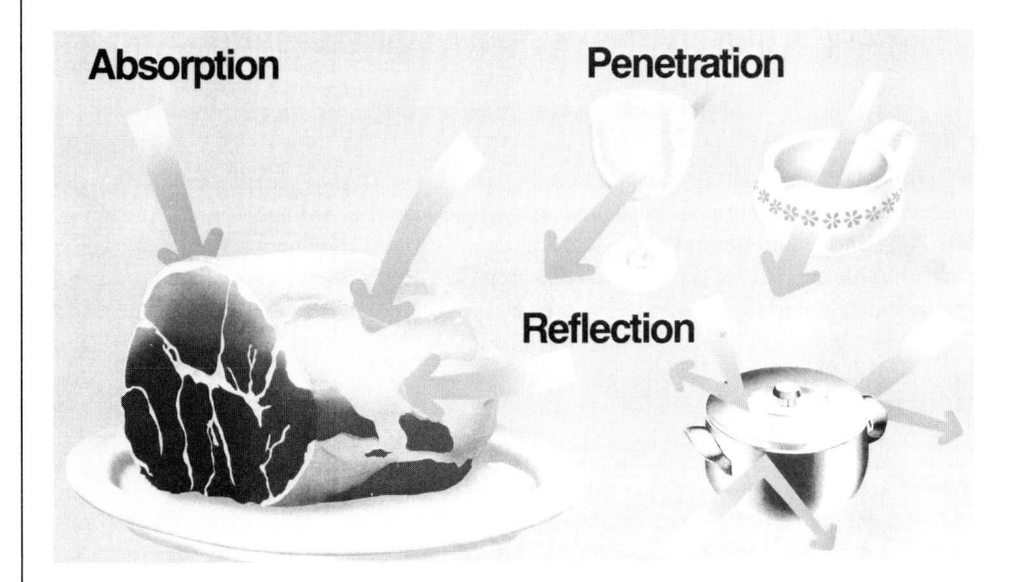

FIGURE E1–6 Characteristics of microwaves. (Courtesy of Sharp Electronics Corporation.)

Solution

Objective Develop an understanding of the cooking principles of a microwave oven.

Schematic Heating principles and characteristics of range top, conventional oven, and microwave oven.

Cooking Method	Mechanism	Characteristics
Range Top 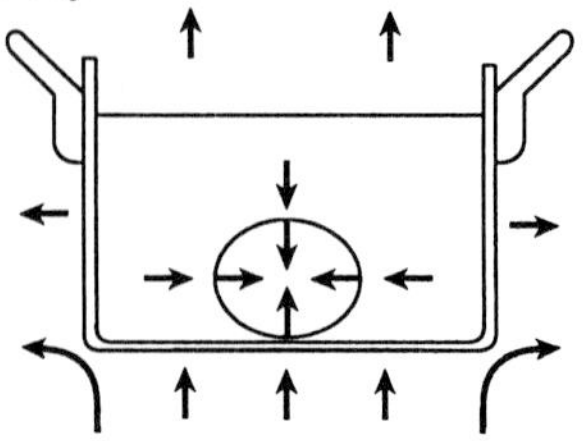	Conduction	• Large energy loss • Container and surrounding air are heated.
Conventional Oven 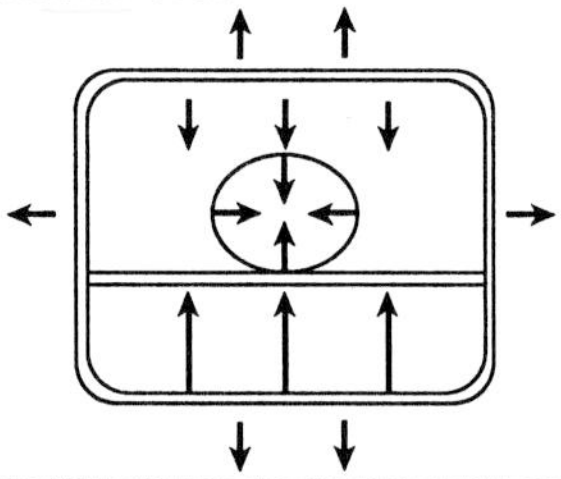	Radiation/ conduction	• Preheating required • Container and surrounding air are heated.
Microwave Oven 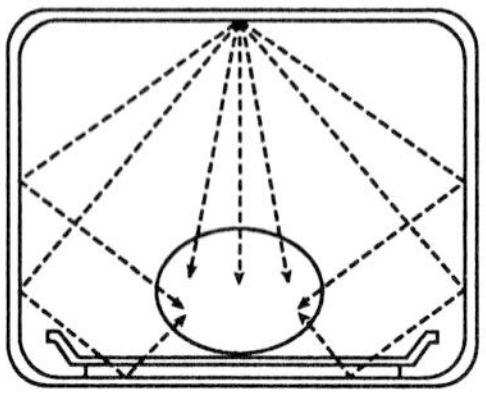	Microwaves/ conduction	• No preheating required • Food heated directly for high thermal efficiency.

Analysis Microwave ovens cook food by (1) the transfer of energy to the exposed surface of the food item via microwave radiation, (2) the absorption and conversion of microwave radiation into internal energy by the molecules of water, sugar, and fat near the surface of the item, and (3) transfer of energy into the interior of the item by conduction heat transfer. Because microwaves are reflected from metal surfaces and pass through glass and plastic cookware, the food is heated while the interior of the oven and the cooking tray stay relatively cool, with the result that no preheating is required and significantly greater cooking efficiency is achieved. On the other hand, conventional range top and oven cooking by means of conduction and thermal radiation involve large heat loss to the container and surrounding air and require oven preheating.

1–3 ANALOGY BETWEEN HEAT TRANSFER AND THE FLOW OF ELECTRIC CURRENT

The basic laws of conduction, thermal radiation, and convection heat transfer often lead to relationships for the heat transfer rate q from a surface with area A_s of the form†

$$q = \frac{\Delta T}{R} \tag{1–23}$$

where the *thermal resistance* R is a function of the thermal and geometric characteristics. This equation is similar in form to the relationship for the flow of electric current I_e through an electrical resistance R_e with voltage drop ΔE_e; that is,

$$I_e = \frac{\Delta E_e}{R_e} \tag{1–24}$$

A comparison of Eqs. (1–23) and (1–24) indicates that

q is analogous to I_e,

T is analogous to E_e,

and

R is analogous to R_e.

The thermal resistance R and the potential difference ΔT are dependent upon the heat-transfer mechanism considered. For example, Eq. (1–12) indicates that one-dimensional conduction heat transfer through a stationary medium with surface temperatures equal to T_1 and T_2 can be represented by a simple series circuit with resistance

$$R_k = \frac{1}{kS} \tag{1–25}$$

and potential difference $\Delta T = T_1 - T_2$. Similarly,

$$R_R = \frac{1}{\sigma A_s F_{s-R}(T_s + T_R)(T_s^2 + T_R^2)} \tag{1–26}$$

and $\Delta T = T_s - T_R$ for blackbody thermal radiation, and

† The corresponding relation for the rate of heat transfer dq from a *differential area* dA_s with finite temperature difference ΔT will be represented by

$$dq = \frac{\Delta T}{R'}$$

where R' is the *local thermal resistance*.

$$R_c = \frac{1}{\bar{h}A_s} \tag{1-27}$$

and $\Delta T = T_s - T_F$ for convection. As illustrated in Examples 1–7 and 1–8, heat-transfer processes can be represented by electrical circuits. Such analogous electrical circuits often facilitate the solution of rather complex heat-transfer problems. As will be seen in Chaps. 2 through 4, electrical circuits can also be set up for unsteady and multidimensional systems.

EXAMPLE 1–7

One hundred watts are generated by the flow of electric current in a 1-in.-diameter copper cable of 3-ft length. The surrounding air temperature is 75°F and $\bar{h}$ is equal to 40 Btu/(h ft² °F). Determine the surface temperature of the cable for negligible thermal-radiation losses.

Solution

Objective Determine the surface temperature T_s.

Schematic Electric conducting copper cable.

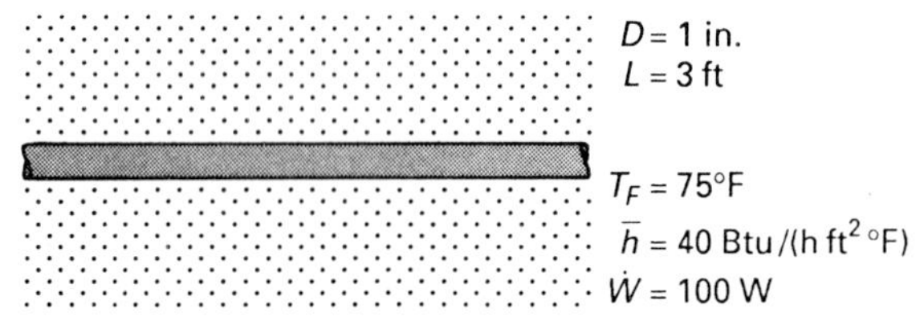

Assumptions/Conditions

steady-state
one-dimensional
negligible thermal radiation

Analysis Applying the first law of thermodynamics to the cable, we have

$$q_c = \dot{W} = 100 \text{ W} = 341 \text{ Btu/h}$$

The analogous thermal circuit is shown in Fig. E1–7 for the case in which the radiation loss from the cable is negligible. The thermal resistance to convection R_c is given by

$$R_c = \frac{1}{\bar{h}A_s} = \frac{1}{[40 \text{ Btu/(h ft}^2 \text{ °F)}][\pi(1 \text{ ft}/12)(3 \text{ ft})]} = 0.0318 \text{ h °F/Btu}$$

To obtain T_s, we write

$$q_c = \dot{W} = \frac{T_s - T_F}{R_c}$$

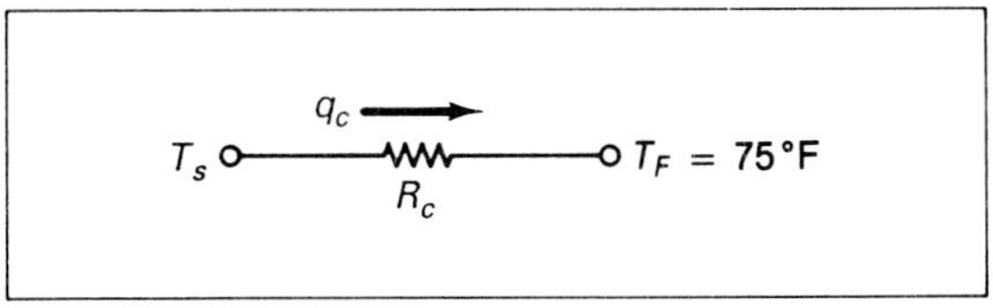

$q_c \rightarrow$

T_s o—R_c—o $T_F = 75°F$

FIGURE E1–7
Thermal circuit; $R_c = 1/(\bar{h}A_s)$.

$$T_s = \dot{W}R_c + T_F = 341\ \frac{\text{Btu}}{\text{h}}\left(0.0318\ \frac{\text{h °F}}{\text{Btu}}\right) + 75°\text{F} = 85.8°\text{F}$$

EXAMPLE 1–8

Because excessive temperature shortens the life of transistors, the heat transfer in transistors and compact integrated circuits (ICs, often called chips) that contain transistors, resistors, and capacitors is a critical design consideration. Consequently, manufacturers of transistors and ICs generally specify the maximum allowable internal junction temperature T_J and/or case temperature T_C, and the thermal resistance from the junction to case R_{JC} and from the case to ambient R_{CA}, assuming natural convection and radiative cooling. Whereas some silicon transistors and ICs can operate at temperatures as high as 200°C, the maximum temperature of components made of germanium is usually less than 100°C. In general, the life of a transistor or IC increases as its operating temperature decreases.

To illustrate, the thermal characteristics of the low-power silicon voltage regulator IC shown in Fig. E1–8a are specified by the manufacturer for an ambient temperature of 25°C as follows:

Maximum operating junction temperature: 150°C
Thermal resistance: $R_{JC} = 60$ °C/W
Thermal resistance: $R_{CA} = 90$ °C/W
Maximum power rating: 600 mW

We want to calculate the steady-state junction and case temperatures of this device at its maximum power output of 600 mW.

FIGURE E1–8a Integrated-voltage regulator (1/2 × 1 1/2 cm).

Solution

Objective Determine the temperature of the junction T_J and case T_C for steady-state conditions.

Assumptions/Conditions

steady-state
natural convection and radiation cooling

Analysis The thermal circuit for this system is shown in Fig. E1–8b. To calculate the junction temperature T_J for 600-mW output, we write

$$q = \frac{T_J - T_F}{R_{JC} + R_{CA}}$$

$$T_J = q(R_{JC} + R_{CA}) + T_F = 0.6\ \text{W}\left(60\frac{°\text{C}}{\text{W}} + 90\frac{°\text{C}}{\text{W}}\right) + 25°\text{C} = 115°\text{C}$$

This temperature is well within the safe operating range for the unit.

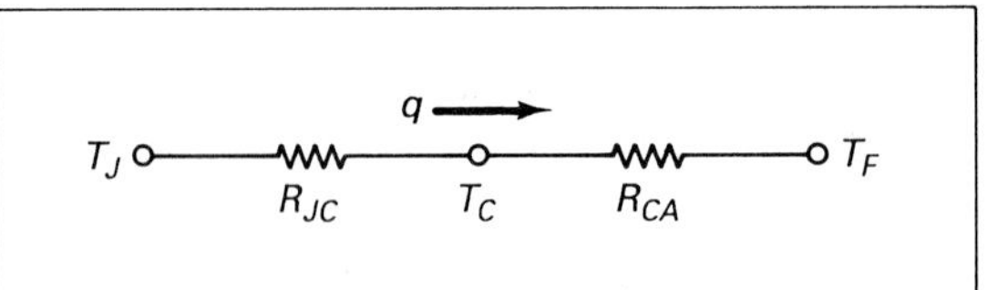

FIGURE E1–8b
Thermal circuit.

The case temperature T_C at this maximum level of power dissipation is now calculated.

$$q = \frac{T_J - T_C}{R_{JC}}$$

$$T_C = T_J - qR_{JC} = 115°\text{C} - 0.6\ \text{W}\left(60\frac{°\text{C}}{\text{W}}\right) = 79°\text{C}$$

1–4 SUMMARY

In this chapter we have reviewed the familiar fundamental laws upon which the science of heat transfer is based and we have introduced particular laws associated with conduction, thermal radiation, and convection heat transfer. As we have seen, the basic laws of conduction, thermal radiation, and convection provide the basis for a simple analogy between heat transfer and the flow of electric current and for the development of the practical analysis of basic heat-transfer problems. The principles introduced in this chapter will be used throughout the remainder of our study of heat-transfer processes involving chemically homogeneous substances. The related topic of heat transfer and diffusion mass transfer in chemically nonhomogeneous substances is treated in references 7–9 and in the accompanying *mass transfer supplement*.

CHAPTER 2

ONE-DIMENSIONAL HEAT TRANSFER

2–1 INTRODUCTION

The primary objective of this chapter is to establish the fundamental concepts involved in analyzing basic one-dimensional heat-transfer problems. These principles will first be demonstrated in the context of steady-state conduction heat transfer in a plane wall. One-dimensional analyses will then be developed for several other representative problems. This study will provide the foundation for our analysis of more complex multidimensional conduction, thermal radiation, and convection heat transfer in Chaps. 3 through 11.

2–2 CONDUCTION HEAT TRANSFER IN A PLANE WALL

We consider one-dimensional steady conduction heat transfer in the plane wall shown in Fig. 2–1(a). The classic differential approach to solving this problem is presented for several standard boundary conditions, after which an efficient but less general short method will be introduced.

2–2–1 Differential Formulation

To develop mathematical equations that represent the energy transfer within the wall, attention is focused upon the differential strip $A\ dx$ shown in Fig. 2–1(a). First, we recognize that $\Delta E_s/\Delta t$ is zero because of steady-state conditions, and conduction heat transfer in the y and z directions is zero because no temperature gradients occur in

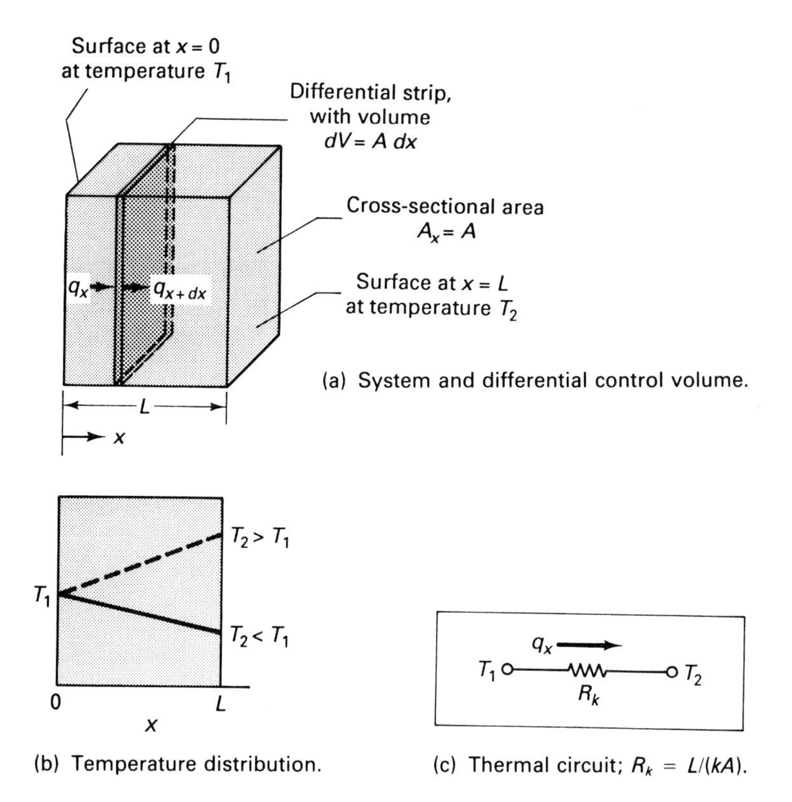

(a) System and differential control volume.

(b) Temperature distribution.

(c) Thermal circuit; $R_k = L/(kA)$.

FIGURE 2–1 One-dimensional conduction heat transfer in a flat plate.

these directions. Thus, the temperature is a function of x alone. With these points in mind, the application of the first law of thermodynamics given by Eq. (1–1) to the element $A\ dx$ gives

Step 1

$$q_x = q_{x+dx} \tag{2–1}$$

This equation indicates that q_x does not change with x.

The next step in the development of the differential formulation requires the use of a basic relationship between q_x and q_{x+dx}. This relationship is established on the basis of the definition of the derivative,

$$\frac{dq_x}{dx} = \lim_{\Delta x \to 0} \frac{q_{x+\Delta x} - q_x}{\Delta x} \tag{2–2}$$

By replacing Δx by the infinitesimal quantity dx, this equation gives rise to

$$q_{x+dx} = q_x + \frac{dq_x}{dx}\, dx \tag{2–3}$$

Utilizing this expression, we eliminate q_{x+dx} in Eq. (2–1) to obtain

Step 2

$$\frac{dq_x}{dx} = 0 \tag{2–4}$$

The quantity q_x is now expressed in terms of the temperature distribution T by introducing the Fourier law of conduction, Eq. (1–10); that is,

Step 3

$$\frac{d}{dx}\left(-kA_x\frac{dT}{dx}\right) = 0 \tag{2–5}$$

Because the cross-sectional area A_x is independent of x for this application ($A_x = A$), this equation reduces to

$$\frac{d}{dx}\left(k\frac{dT}{dx}\right) = 0 \tag{2–6}$$

For situations in which the thermal conductivity is essentially uniform, this expression takes the simple form

$$\frac{d^2T}{dx^2} = 0 \tag{2–7}$$

The differential formulation is completed by writing the boundary conditions. As the term implies, a boundary condition is a mathematical statement pertaining to the behavior of the dependent variable (T in our case) at the system boundary. It should be recalled that the number of boundary conditions is equal to the highest-order derivative in an ordinary differential equation. Hence, two boundary conditions must be specified in order to solve steady one-dimensional conduction heat-transfer problems. For the moment, the surface temperatures are simply identified by T_1 and T_2,

$$T(0) = T_1 \qquad T(L) = T_2 \tag{2–8,9}$$

The treatment of several common types of boundary conditions will be considered later in this section.

2–2–2 Solution

The solution to Eq. (2–7) is written on the basis of a simple double integration as

$$T = C_1x + C_2 \tag{2–10}$$

By applying the boundary conditions, the constants of integration C_1 and C_2 are written in terms of the surface temperatures as follows:

$$T_1 = C_2 \qquad T_2 = C_1L + C_2 \tag{2–11,12}$$

or

$$C_1 = \frac{T_2 - T_1}{L} \tag{2–13}$$

Therefore, the temperature distribution takes the form

$$T = (T_2 - T_1)\frac{x}{L} + T_1$$

or

$$\frac{T - T_1}{T_2 - T_1} = \frac{x}{L} \tag{2–14}$$

This linear relationship is shown in Fig. 2–1(b) for situations in which $T_1 > T_2$ and $T_1 < T_2$. Because Eq. (2–14) does not involve k, we conclude that the temperature distribution is totally independent of the material (steel, wood, concrete, etc.).

An expression is obtained for the rate of heat transfer in the wall by utilizing the Fourier law of conduction together with Eq. (2–14); that is,

$$q_x = -kA\left.\frac{dT}{dx}\right|_x = \frac{kA}{L}(T_1 - T_2) \tag{2–15}$$

Consistent with Eq. (2–1), this equation indicates that the rate of heat transfer in the plate is independent of x.

Equation (2–15) provides the basis for the practical expressions given by Eqs. (1–12) and (1–23). For example, the thermal resistance R_k is simply

$$R_k = \frac{L}{kA} \tag{2–16}$$

for a flat plate with uniform thermal conductivity. The thermal circuit associated with one-dimensional conduction heat transfer in a plate is shown in Fig. 2–1(c).

EXAMPLE 2–1

Determine the temperature distribution and rate of heat transfer across a copper plate with cross-sectional area 1 m^2 and thickness 5 cm and with surface temperatures of 130°C and 15°C.

Solution

Objective Determine the distribution in T and q_x for the plane wall shown in the schematic.

Schematic Temperature distribution in a plane wall.

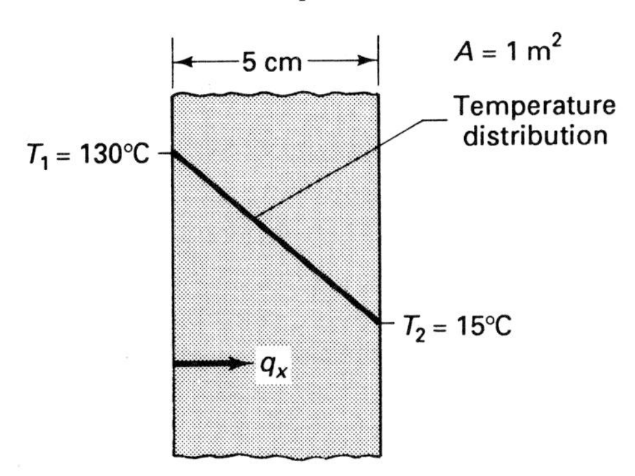

Assumptions/Conditions

steady-state
one-dimensional
uniform properties

Properties Copper at room temperature (Table A–C–1): $k = 386$ W/(m °C).

Analysis The temperature distribution is given by Eq. (2–14),

$$T = (15°\text{C} - 130°\text{C})\frac{x}{0.05\text{ m}} + 130°\text{C}$$

This linear distribution is shown in the schematic. The rate of heat transfer obtained from Eq. (2–15) is

$$q_x = 386\frac{\text{W}}{\text{m °C}}\left(\frac{1\text{ m}^2}{0.05\text{ m}}\right)(130°\text{C} - 15°\text{C}) = 888\text{ kW}$$

Notice that this heat-transfer problem can be represented by the thermal circuit shown in Fig. 2–1(c), with the thermal resistance given as $R_k = L/(kA) = 1.30 \times 10^{-4}$ °C/W.

2–2–3 Boundary Conditions

For situations in which the surface temperatures T_1 and T_2 are known, the boundary conditions given by Eqs. (2–8) and (2–9), together with the differential equation, provide a complete mathematical model of the problem. However, for the many applications encountered in practice in which one or both surface temperatures are not known, the actual heat transfer at the boundaries must be accounted for. For such cases, the solutions for T and q developed above still apply, but T_1 and T_2 must be evaluated. Examples of other common types of boundary conditions include convection, thermal radiation, specified heat flux, and combinations of these. These standard types of boundary conditions will be considered in this section, as will another important type of boundary condition that involves interfacial conduction in composites.

Problems involving these other types of boundary conditions can be solved by any of several approaches. One way is to replace Eq. (2–8) and/or Eq. (2–9) by the formal mathematical statement of the boundary condition(s) and to then solve for the constants of integration. Thus, for a flat plate with uniform thermal conductivity, the boundary conditions are utilized to evaluate C_1 and C_2 in Eq. (2–10). In another somewhat simpler approach, Eqs. (2–8) and (2–9) are retained with the unknown surface temperatures T_1 and T_2 being determined by the use of the electrical analogy concept. For example, R_k is given by Eq. (2–16) for a flat plate with uniform thermal conductivity. Both the formal and the thermal circuit approaches will be demonstrated in the sections that follow.

Standard Boundary Conditions

For the case illustrated in Fig. 2–2(a), in which the surface at $x = 0$ is exposed to fluid with temperature T_F, the heat convected from the fluid is conducted into the wall. Consequently, the boundary condition at $x = 0$ is written as

$$q_c = q_x$$

$$\bar{h}[T_F - T(0)] = -k\left.\frac{dT}{dx}\right|_0 \tag{2–17}$$

where the surface temperature $T(0) = T_1$ is unknown. The other surface at $x = L$ is maintained at a known temperature T_2, such that the second boundary condition is given by Eq. (2–9).

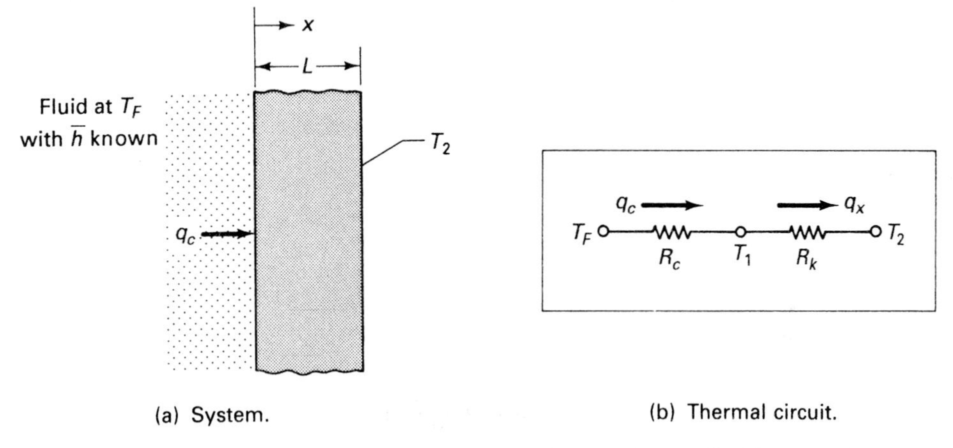

(a) System. (b) Thermal circuit.

FIGURE 2–2 One-dimensional conduction heat transfer in a plane wall with convection at one surface.

At this point in our study, $\bar{h}$ is assumed to be known. Details concerning the evaluation of $\bar{h}$ are presented in Chaps. 6 through 11, which deal specifically with the study of convection heat transfer.

Substituting the solution for T given by Eq. (2–10) into Eqs. (2–9) and (2–17), we obtain Eq. (2–12),

$$T_2 = C_1L + C_2 \tag{2–12}$$

and

$$-kC_1 = \bar{h}(T_F - C_2) \tag{2–18}$$

The solution of these two equations for the two constants of integration gives rise to an expression for T of the form

$$\frac{T - T_2}{T_F - T_2} = \frac{L - x}{L + k/\bar{h}} \tag{2–19}$$

Hence, the unknown surface temperature T_1 is given by

$$\frac{T_1 - T_2}{T_F - T_2} = \frac{L}{L + k/\bar{h}} = \frac{1}{1 + k/(\bar{h}L)} \tag{2–20}$$

With T_1 now known, the temperature distribution within the wall can also be conveniently expressed by Eq. (2–14),

$$\frac{T - T_1}{T_2 - T_1} = \frac{x}{L} \tag{2–14}$$

A more efficient approach to the solution of this convection-heat-transfer problem involves the use of the electrical analog. The series thermal circuit is shown in Fig. 2–2(b). An analysis of this elementary circuit indicates that the rate of heat transfer can be expressed as

$$q = \frac{T_F - T_2}{R_c + R_k} = \frac{T_F - T_2}{\Sigma R} \tag{2–21}$$

where $R_c = 1/(\bar{h}A)$, $R_k = L/(kA)$, $\Sigma R = R_c + R_k$ represents the sum of the resistances, and $q = q_x = q_c$. This thermal circuit can also be solved for the surface temperature T_1 by writing

$$q = q_c = q_x$$

$$\frac{T_F - T_2}{R_c + R_k} = \frac{T_F - T_1}{R_c} = \frac{T_1 - T_2}{R_k} \tag{2–22}$$

By equating q and q_c, we have

$$\frac{T_1 - T_F}{T_2 - T_F} = \frac{R_c}{R_c + R_k} = \frac{1}{1 + L\bar{h}/k} \tag{2–23}$$

Alternatively, by equating q and q_x, we obtain Eq. (2–20), and by equating q_c and q_x, we get

$$T_1 = \frac{T_F/R_c + T_2/R_k}{1/R_c + 1/R_k} = \frac{T_F\bar{h} + T_2k/L}{\bar{h} + k/L} \tag{2–24}$$

All three of these equations for T_1 are equivalent. Note that Eq. (2–24) can also be obtained directly from Eq. (2–17) by replacing q_x by $(T_1 - T_2)/R_k$.

It should also be noted that the convective resistance is negligible for large values of $\bar{h}$. For this situation, Eq. (2–23) indicates that the surface temperature T_1 is simply

$$T_1 = T_F \tag{2–25}$$

Standard boundary conditions for one-dimensional steady-state convection, blackbody thermal radiation, and uniform heat flux are summarized in Table 2–1. Representative problems involving these boundary conditions are illustrated in Examples 2–2 through 2–4.

TABLE 2-1 Standard boundary conditions for one-dimensional steady-state conduction heat transfer in a flat plate

Condition at boundary	*Boundary condition—Mathematical statement*
Convection	
at $x = 0$	$-k \left.\frac{dT}{dx}\right\rvert_0 = \bar{h}[T_F - T(0)]$
at $x = L$	$-k \left.\frac{dT}{dx}\right\rvert_L = \bar{h}[T(L) - T_F]$
Blackbody thermal radiation	
at $x = 0$	$-k \left.\frac{dT}{dx}\right\rvert_0 = \sigma F_{s-R}[T_R^4 - T(0)^4]$
at $x = L$	$-k \left.\frac{dT}{dx}\right\rvert_L = \sigma F_{s-R}[T(L)^4 - T_R^4]$
Specified heat flux q_s''†	
at $x = 0$	$-k \left.\frac{dT}{dx}\right\rvert_0 = q_s''$
at $x = L$	$-k \left.\frac{dT}{dx}\right\rvert_L = q_s''$
Insulated surface, $q_s'' = 0$	
at $x = 0$	$-k \left.\frac{dT}{dx}\right\rvert_0 = 0$
at $x = L$	$-k \left.\frac{dT}{dx}\right\rvert_L = 0$

† q_s'' positive in x direction.

EXAMPLE 2-2

Two fluids are separated by a 2-in.-thick stainless steel plate [AISI 302] with an area of 10 ft^2. The fluid temperatures and mean coefficients of heat transfer are T_{F1} = 50°F, T_{F2} = 0°F, $\bar{h}_1$ = 200 Btu/(h ft^2 °F), and $\bar{h}_2$ = 150 Btu/(h ft^2 °F). Utilize the electrical analogy approach to determine the surface temperatures and the rate of heat transfer through the plate for negligible thermal radiation at the surfaces.

Solution

Objective Determine the surface temperatures T_1 and T_2 and q for a plate with conditions shown in the schematic.

Schematic Two fluids separated by stainless steel plate.

$T_{F1} = 50°F$
$\bar{h}_1 = 200$ Btu/(h ft^2 °F)

Stainless steel plate
$L = 2$ in.
$A = 10$ ft^2

$T_{F2} = 0°F$
$\bar{h}_2 = 150$ Btu/(h ft^2 °F)

Assumptions/Conditions

steady-state
one-dimensional
uniform properties
negligible thermal radiation

Properties Stainless steel AISI 302 near room temperature (Table A–C–1):

$$k = 15.1 \frac{\text{W}}{\text{m °C}} \frac{0.578 \text{ Btu/(h ft °F)}}{\text{W/(m °C)}} = 8.73 \text{ Btu/(h ft °F)}$$

Analysis The thermal circuit for this arrangement is shown in Fig. E2–2. The thermal resistances are

$$R_{c1} = \frac{1}{\bar{h}_1 A} = 0.0005 \text{ h °F/Btu} \qquad R_{c2} = \frac{1}{\bar{h}_2 A} = 0.000667 \text{ h °F/Btu}$$

$$R_k = \frac{L}{kA} = \frac{2 \text{ in./(12 in./ft)}}{[8.73 \text{ Btu/(h ft °F)}](10 \text{ ft}^2)} = 0.00191 \text{ h °F/Btu}$$

Solving for q, we obtain

$$q = \frac{T_{F1} - T_{F2}}{R_{c1} + R_k + R_{c2}} = \frac{50°\text{F} - 0°\text{F}}{(0.0005 + 0.00191 + 0.000667) \text{ h °F/Btu}}$$

$$= 16{,}200 \text{ Btu/h} \qquad q'' = \frac{q}{A} = 1620 \text{ Btu/(h ft}^2)$$

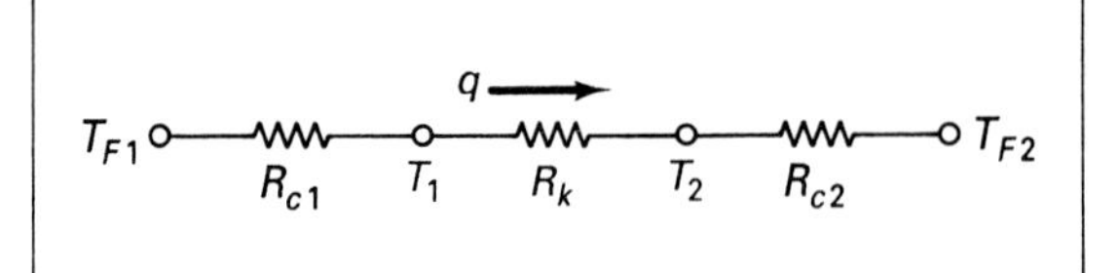

FIGURE E2–2
Thermal circuit.

The temperature distribution within the plate is given by Eq. (2–14),

$$\frac{T - T_1}{T_2 - T_1} = \frac{x}{L}$$

T_1 and T_2 are obtained as follows:

$$q = \frac{T_{F1} - T_1}{R_{c1}}$$

$$T_1 = T_{F1} - R_{c1}q = 50°\text{F} - \frac{16{,}200 \text{ Btu/h}}{[200 \text{ Btu/(h ft}^2 \text{ °F)}](10 \text{ ft}^2)} = 41.9°\text{F}$$

$$q = \frac{T_2 - T_{F2}}{R_{c2}}$$

$$T_2 = T_{F2} + R_{c2}q = 0°\text{F} + \frac{16{,}200 \text{ Btu/h}}{[150 \text{ Btu/(h ft}^2 \text{ °F)}](10 \text{ ft}^2)} = 10.8°\text{F}$$

EXAMPLE 2–3

One surface of the plate shown in Fig. E2–3a is exposed to blackbody thermal radiation with $T_R = 1000°\text{C}$. The other surface is maintained at a temperature $T_2 = 15°\text{C}$. Obtain a solution for the unknown surface temperature at the radiating surface and the heat transfer rate for uniform property conditions.

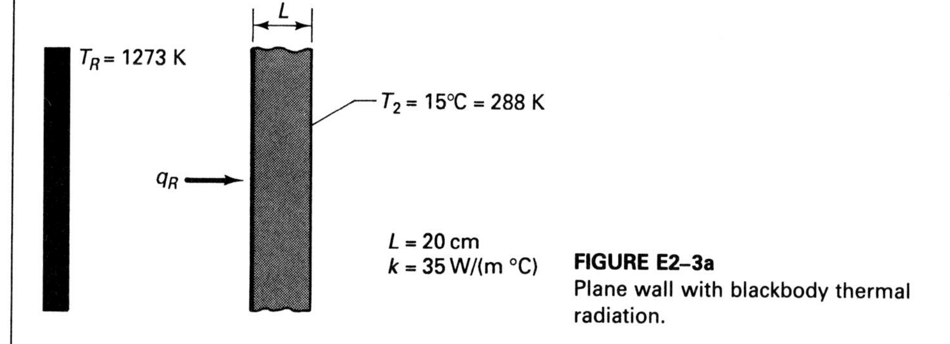

FIGURE E2–3a
Plane wall with blackbody thermal radiation.

Solution

Objective Determine the surface temperature T_1 and q_x for a plate with conditions shown in Fig. E2–3a.

Assumptions/Conditions

steady-state
one-dimensional
uniform properties
blackbody thermal radiation

Properties Plate material: $k = 35$ W/(m °C).

Analysis The boundary condition at the surface which is exposed to thermal radiation is given by

$$q_R = q_x \qquad \sigma F_{s-R} A_s (T_R^4 - T_1^4) = -kA \left.\frac{dT}{dx}\right|_0 \tag{a}$$

where $T(0)$ is denoted by T_1 and $A_s = A_x = A$. The thermal circuit for this problem is shown in Fig. E2–3b, where $R_R = [\sigma F_{s-R} A_s (T_R + T_1)(T_R^2 + T_1^2)]^{-1}$.

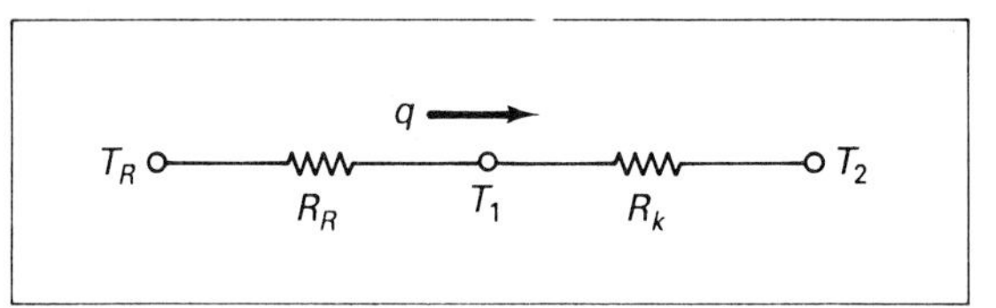

FIGURE E2–3b
Thermal circuit.

We have shown that the rate of heat transfer through the wall by conduction q_x can be written as

$$q_x = \frac{kA}{L}(T_1 - T_2) \tag{b}$$

Thus, Eq. (a) reduces to

$$\sigma F_{s-R}(T_R^4 - T_1^4) = \frac{k}{L}(T_1 - T_2) \tag{c}$$

This nonlinear equation can be solved for the unknown $T(0) = T_1$ by the successive-approximation method or the Newton–Raphson method.

To employ the successive-approximation method, Eq. (c) is rewritten in the form (to four significant figures)

$$T_{1,i+1} = \frac{T_2 + F_{s-R}(\sigma L/k)T_R^4}{1 + F_{s-R}(\sigma L/k)T_{1,i}^3} = \frac{1139\text{ K}}{1 + 3.24 \times 10^{-10}\, T_{1,i}^3/\text{K}^3} \tag{d}$$

where i is the iteration index. The right side of this equation is referred to as the *arrangement function*. This equation can be iteratively solved for a physically realistic root T_1 by starting an iterative sequence with the initial estimate $T_{1,1}$ set equal to a value between 288 K and 1273 K. With $T_{1,1}$ set equal to 1000 K, the first three values of $T_{1,i+1}$ are $T_{1,2} = 860.3$ K, $T_{1,3} = 944.2$ K, and $T_{1,4} = 894.9$ K. Continuing this iteration procedure, our calculations for $T_{1,i+1}$ converge to 913.4 K after 18 iterations.

It should be noted that Eq. (c) can be written in the alternative forms

$$T_{1,i+1} = 1139\text{ K} - \frac{3.24 \times 10^{-10}}{\text{K}^3} T_{1,i}^4 \tag{e}$$

which converges much slower than Eq. (d), and

$$T_{1,i+1} = \left(\frac{1139\text{ K} - T_{1,i}}{3.24 \times 10^{-10}/\text{K}^3}\right)^{1/4} \tag{f}$$

which diverges. Therefore, the arrangement function employed in Eq. (d) is most suitable for this problem.

The Newton–Raphson method provides a means of achieving quicker convergence. Following this approach, Eq. (c) is put into the form

$$\frac{3.24 \times 10^{-10}}{\text{K}^3} T_1^4 + T_1 - 1139\text{ K} = f(T_1) = 0$$

and $T_{1,i+1}$ is represented by the Newton–Raphson formula [1],

$$T_{1,i+1} = T_{1,i} - \frac{f(T_{1,i})}{f'(T_{1,i})} \tag{g}$$

$$= T_{1,i} - \frac{(3.24 \times 10^{-10}/\text{K}^3)T_{1,i}^4 + T_{1,i} - 1139\text{ K}}{4(3.24 \times 10^{-10}/\text{K}^3)T_{1,i}^3 + 1} \tag{h}$$

Starting with $T_{1,1} = 1000$ K, we obtain $T_{1,2} = 919.4$ K, $T_{1,3} = 913.5$ K, and $T_{1,4} = 913.4$ K, such that only three iterations are required.

To complete the solution, the rate of heat transfer per unit area is calculated by writing

$$q_x'' = \frac{q_x}{A} = \frac{k}{L}(T_1 - T_2) = \frac{35\text{ W/(m °C)}}{0.2\text{ m}}(913.4\text{ K} - 288\text{ K})$$

$$= 1.094 \times 10^5\text{ W/m}^2 \tag{i}$$

To double-check, we calculate q_R''.

$$q_R'' = \frac{q_R}{A} = \sigma F_{s-R}(T_R^4 - T_1^4)$$

$$= 5.67 \times 10^{-8}\frac{\text{W}}{\text{m}^2\text{ K}^4}[(1273\text{ K})^4 - (913.4\text{ K})^4] = 1.094 \times 10^5\text{ W/m}^2 \tag{j}$$

The agreement between q_x'' and q_R'' assures us of a proper solution.

EXAMPLE 2–4

A 1-in.-thick carbon steel plate with a 1-ft^2 cross-sectional area is exposed to a uniform heat flux of 5000 Btu/(h ft^2). (This type of boundary condition can be achieved by heating the surface with an electrical heating plate or by thermal radiation with $T_R \gg T_s$.) The other surface of the plate is maintained at a temperature of 212°F. Determine the unknown surface temperature.

Solution

Objective Determine the surface temperature T_1 for a plate with conditions shown in the schematic.

Schematic Plane wall with uniform heat flux.

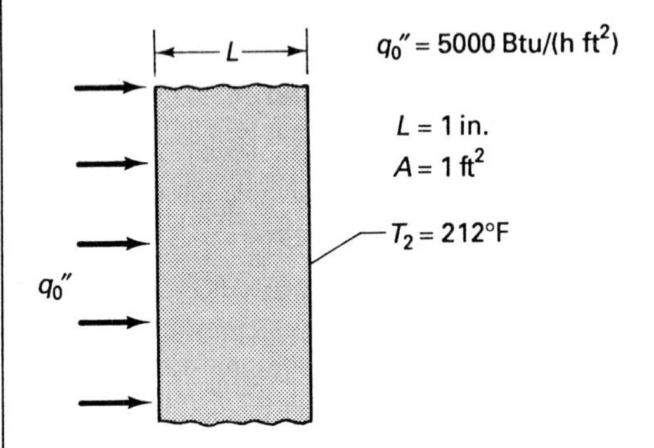

Assumptions/Conditions

steady-state
one-dimensional
uniform properties

Properties Carbon steel (1% C) (Table A–C–1):

$$k = 43 \frac{\text{W}}{\text{m °C}} \frac{0.578 \text{ Btu/(h ft °F)}}{\text{W/(m °C)}} = 24.8 \text{ Btu/(h ft °F)}$$

Analysis The thermal circuit for this problem is shown in Fig. E2–4, where $q_x = q_0$ and where $T(0) = T_1$ is unknown. Based on this simple circuit, we write

$$q_0'' = \frac{T_1 - T_2}{L/k}$$

such that T_1 is given by

$$T_1 = \frac{L}{k} q_0'' + T_2$$

Substituting into this equation, we obtain

$$T_1 = \frac{1 \text{ ft}/12}{24.8 \text{ Btu/(h ft °F)}} \frac{5000 \text{ Btu}}{\text{h ft}^2} + 212\text{°F} = 229\text{°F}$$

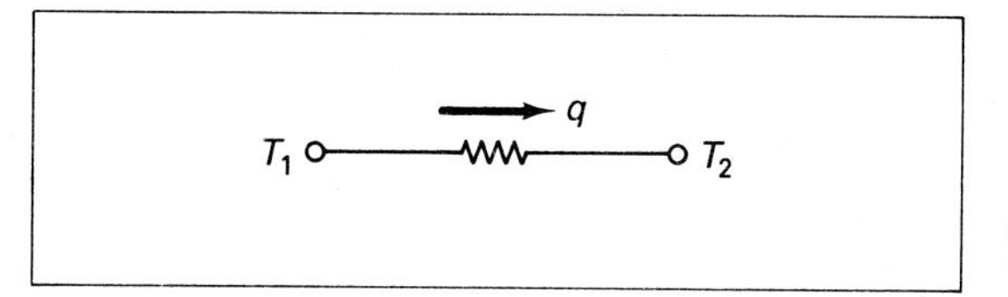

FIGURE E2-4
Thermal circuit.

Composite Walls

We now consider the interfacial condition associated with conduction heat transfer within a wall that is composed of several layers of different materials. This situation is illustrated in Figs. 2–3(a) and 2–4(a) for two types of joints. We designate the temperature distributions in these two materials by T_I and T_II. Because this composite actually involves two systems, materials I and II, we must write two boundary conditions for each material. Consequently, because the interface is a part of both materials, two interfacial boundary conditions are prescribed. Hence, for composite solids with perfect thermal contact, one interfacial boundary condition is

$$T_\text{I}(0) = T_\text{II}(0) \tag{2–26}$$

This situation in which the temperature distribution in the entire composite is continuous is illustrated in Fig. 2–3(b). The second boundary condition is written on the basis of the first law of thermodynamics, which states that the rate of energy conducted into the interface must be equal to the rate of energy conducted out; that is,

$$q_{\text{I}x} = q_{\text{II}x}$$

$$-k_\text{I} \left.\frac{dT_\text{I}}{dx}\right|_0 = -k_\text{II} \left.\frac{dT_\text{II}}{dx}\right|_0 \tag{2–27}$$

This statement is synonymous with the Kirchhoff law for electrical current flow at a junction. The thermal circuit for this composite is shown in Fig. 2–3(c).

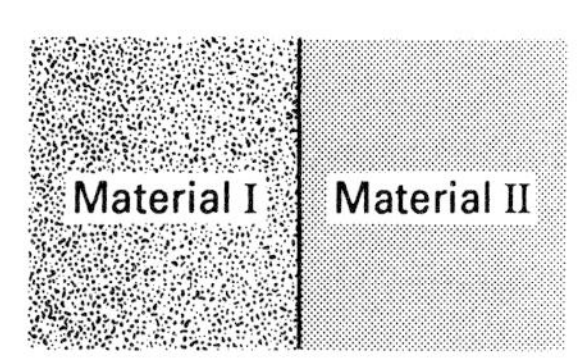

(a) System interface.

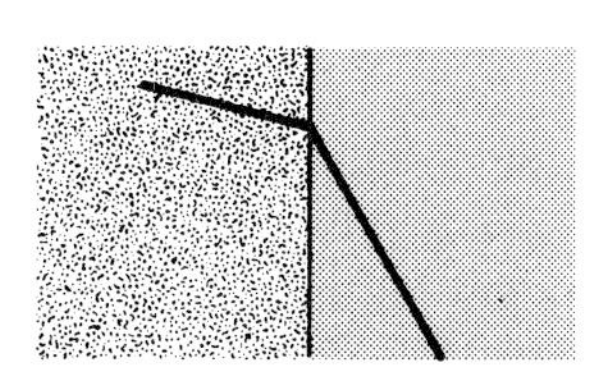

(b) Representative temperature distribution.

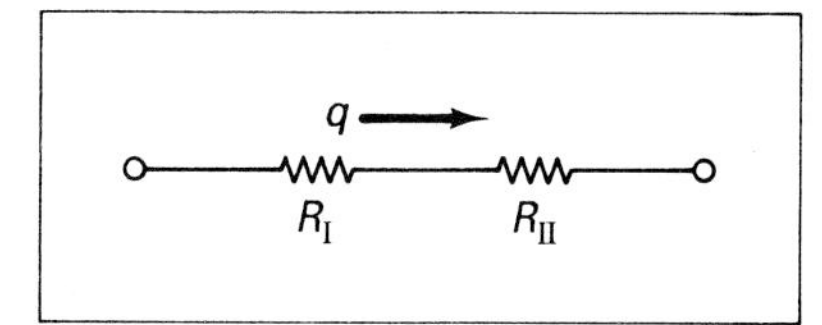

(c) Thermal circuit.

FIGURE 2-3 Composite with perfect thermal contact.

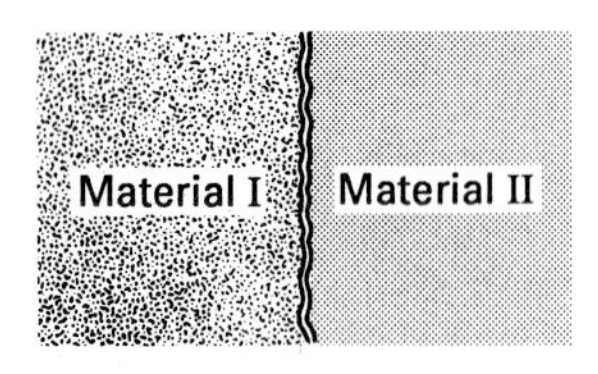

(a) System interface.

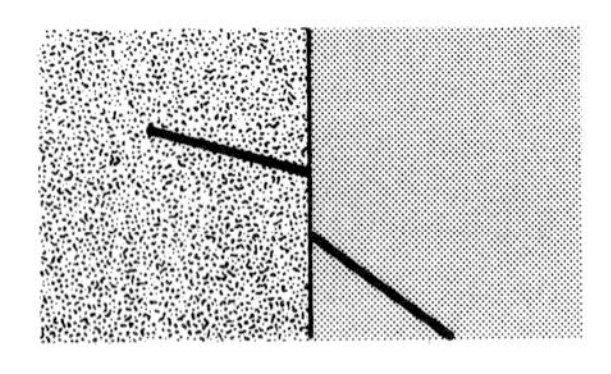

(b) Representative temperature distribution.

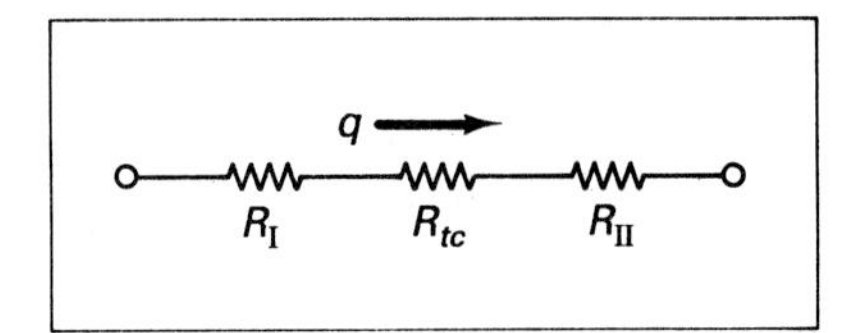

(c) Thermal circuit.

FIGURE 2–4 Composite with imperfect thermal contact.

For situations in which an imperfect mechanical joint is made because of surface roughness, a discontinuity occurs in the temperature distribution at the interface, as shown in Fig. 2–4(b). For such cases, the interfacial temperatures in materials I and II are related empirically through an equation of the form

$$q_{tc} = h_{tc}A[T_{I}(0) - T_{II}(0)] \tag{2–28}$$

where h_{tc} is called the *thermal contact coefficient*. The first law of thermodynamics still applies, such that we obtain the following two interfacial boundary conditions:

$$-k_{I}A \left.\frac{dT_{I}}{dx}\right|_{0} = h_{tc}A[T_{I}(0) - T_{II}(0)] \tag{2–29}$$

$$h_{tc}A[T_{I}(0) - T_{II}(0)] = -k_{II}A \left.\frac{dT_{II}}{dx}\right|_{0} \tag{2–30}$$

The thermal circuit for this type of problem is shown by Fig. 2–4(c); the thermal contact resistance R_{tc} is equal to $1/(h_{tc}A)$. As R_{tc} becomes small, this thermal circuit reduces to the circuit shown in Fig. 2–3(c).

The thermal contact coefficient h_{tc} is dependent upon the material, surface roughness, contact pressure, and temperature. Experimental data for h_{tc} are available in references 2 and 3 for standard materials such as aluminum, copper, and stainless steel. For example, h_{tc} ranges from 5 to 50 kW/(m^2 °C) for various aluminum surfaces at 200°C with contact pressure between 1 and 30 atm. Thermal contact coefficients are generally much smaller for stainless steel [order of 3 kW/(m^2 °C)] and much larger for copper [order of 150 kW/(m^2 °C)].

A practical means of reducing the thermal contact resistance is to insert a material of good thermal conductivity between the two surfaces. Thermal greases containing silicon have been developed for this purpose. Thin soft metal foil can also be used for certain applications.

EXAMPLE 2–5

A germanium power transistor is capable of operating at up to 5 W. The manufacturer's rating for the case-to-ambient thermal resistance of the transistor is 30 °C/W and the case temperature is not to exceed 80°C. In order to lower its operating temperature for a given power input, the transistor is to be mounted to a black anodized aluminum frame which serves as a heat sink, as shown in Fig. E2–5a. The frame provides an additional 1600-mm^2 surface area for cooling. To minimize the thermal contact resistance between the transistor and the heat sink and at the same time maintain proper electrical insulation, the surfaces are first cleaned and coated with a silicon grease (such as Dow–Corning Silicon Heat Sink Compound 340). A special electrical insulating gasket made of mica, beryllium oxide, or anodized aluminum is then inserted and the transistor is bolted tightly to the frame. As a rule of thumb, for proper thermal contact the thermal contact resistance is of the order of 0.5 °C/W.

FIGURE E2–5a
Power amplifier mounted on aluminum frame.

Determine the maximum power at which the transistor can safely be operated if the surrounding walls and air are at 25°C and $\bar{h}$ = 10 W/(m^2 °C).

Solution

Objective Determine the maximum safe operating power of the transistor.

Schematic Germanium power transistor on anodized aluminum frame.

T_C = 80°C
$T_F = T_R$ = 25°C
$\bar{h}$ = 10 W/(m^2 °C)
Exposed frame surface
A_s = 1600 mm^2
R_{CA} = 30 °C/W
R_{tc} = 0.5 °C/W

Assumptions/Conditions

steady-state
convection and blackbody thermal radiation cooling

Analysis With the transistor case temperature T_C equal to 80°C, the rate of heat transferred directly to the surroundings is

$$q_{CA} = \frac{T_C - T_F}{R_{CA}} = \frac{80°\text{C} - 25°\text{C}}{30\ °\text{C/W}} = 1.83\ \text{W}$$

In addition, heat is dissipated through the heat sink. The thermal circuit for the heat transfer through the heat sink (assuming negligible resistance to conduction) is shown in Fig. E2–5b. Assuming that blackbody conditions are approximated, R_R is given by

$$R_R = \frac{1}{\sigma A_s F_{s-R}(T_s + T_R)(T_s^2 + T_R^2)}$$

where T_s is the surface temperature of the frame. Setting T_s equal to T_C as a first approximation, we have

$$R_R = \frac{1}{\sigma(0.0016\ \text{m}^2)(1)(353\ \text{K} + 298\ \text{K})[(353\ \text{K})^2 + (298\ \text{K})^2]} = 79.3\ °\text{C/W}$$

R_c, is given by

$$R_c = \frac{1}{\bar{h}A_s} = \frac{1}{[10\ \text{W/(m}^2\ °\text{C)}](0.0016\ \text{m}^2)} = 62.5\ °\text{C/W}$$

To calculate the equivalent resistance R_{HS} for the heat transfer from the heat sink, we write

$$\frac{1}{R_{HS}} = \frac{1}{R_R} + \frac{1}{R_c} = \frac{1}{79.3\ °\text{C/W}} + \frac{1}{62.5\ °\text{C/W}}$$

$$R_{HS} = 35\ °\text{C/W}$$

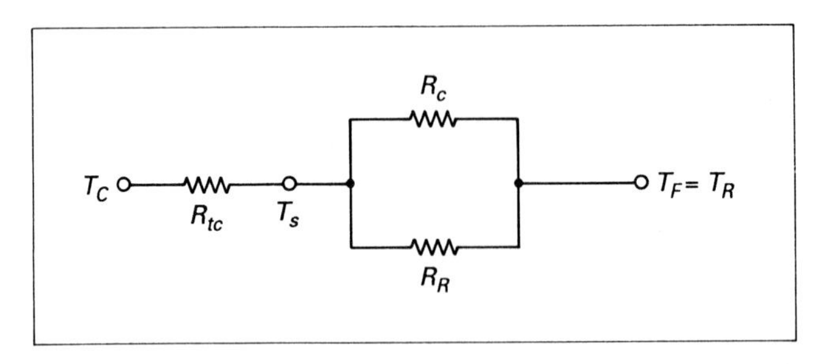

FIGURE E2–5b
Thermal circuit.

It follows that the rate of heat transferred from the power amplifier through the heat sink is

$$q_{HS} = \frac{T_C - T_F}{R_{tc} + R_{HS}} = \frac{80°\text{C} - 25°\text{C}}{0.5\ °\text{C/W} + 35\ °\text{C/W}} = 1.55\ \text{W}$$

Based on this result, the temperature of the heat sink is calculated.

$$q_{HS} = \frac{T_C - T_s}{R_{tc}}$$

$$T_s = T_C - q_{HS}R_{tc} = 80°\text{C} - 1.55\ \text{W}\left(0.5\frac{°\text{C}}{\text{W}}\right) = 79.2°\text{C}$$

Because the difference between T_s and T_C is so small, our approximation for R_R is quite adequate.

Thus, the total power that can be safely dissipated from the transistor/heat-sink unit is estimated to be

$$q = q_{CA} + q_{HS} = 1.83\ \text{W} + 1.55\ \text{W} = 3.38\ \text{W}$$

Without the heat sink, the transistor could only be operated at 1.83 W. For operation at the 5-W level, a larger more efficient heat sink would be required.

To compare the rate of heat transfer by radiation and convection from the heat sink, we write

$$q_R = \frac{T_s - T_F}{R_R} = \frac{79.2°\text{C} - 25°\text{C}}{79.3\ °\text{C/W}} = 0.683\ \text{W}$$

$$q_c = \frac{T_s - T_F}{R_c} = \frac{79.2°\text{C} - 25°\text{C}}{62.5\ °\text{C/W}} = 0.867\ \text{W}$$

Thus, radiation accounts for about 44% of the heat transfer from the heat sink.

The remainder of our study of composites will be restricted to cases in which the thermal contact resistance is assumed to be negligible. Three basic situations to be considered include series, parallel, and combined series–parallel arrangements.

Series Arrangements A composite wall that is composed of two materials in series is shown in Fig. 2–5(a). For situations in which the surface temperatures are uniform, the heat transfer in composite walls that consist of two or more materials in series is always one-dimensional. The analogous electrical circuit for this two-wall series arrangement is shown in Fig. 2–5(b). The heat-transfer rate is simply

$$q_x = \frac{T_1 - T_3}{R_{\text{I}} + R_{\text{II}}} \tag{2–31}$$

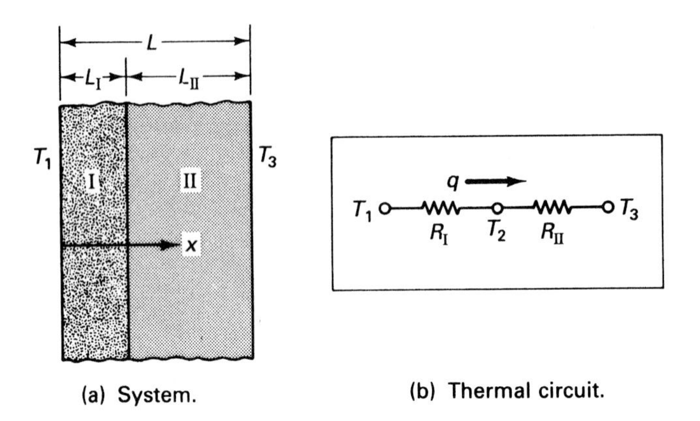

(a) System. (b) Thermal circuit.

FIGURE 2–5 One-dimensional conduction heat transfer in a composite plane wall: series arrangement.

where $R_I = L_I/(k_I A)$ and $R_{II} = L_{II}/(k_{II} A)$. The unknown interfacial temperature T_2 can be expressed as

$$T_1 - T_2 = q_x R_I = (T_1 - T_3)\frac{R_I}{R_I + R_{II}} \tag{2–32}$$

Equations can also be written for the temperature profile within each material as

$$\frac{T_{Ix} - T_1}{T_2 - T_1} = \frac{x}{L_I} \qquad \frac{T_{IIx} - T_2}{T_3 - T_2} = \frac{x - L_I}{L_{II}} \tag{2–33,34}$$

These equations both come directly from Eq. (2–14).

Parallel Arrangements A simple parallel composite of two materials is shown in Fig. 2–6(a). For this situation, the heat transfer is one-dimensional and can be rep-

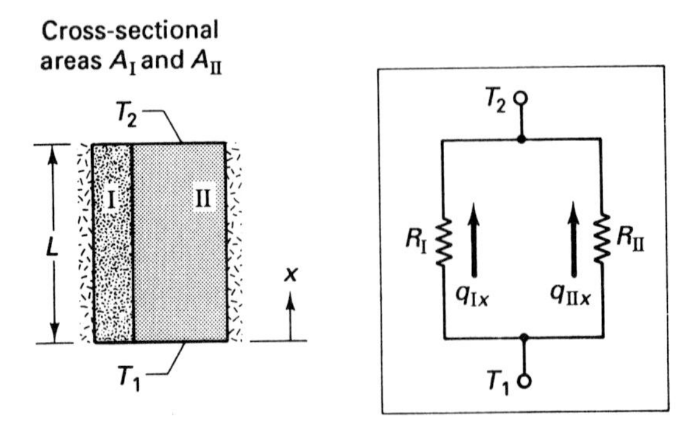

(a) System. (b) Thermal circuit.

FIGURE 2–6 One-dimensional conduction heat transfer in a composite plane wall: parallel arrangement.

resented by the electrical circuit shown in Fig. 2–6(b). The solution for the total heat-transfer rate in this parallel network can be written as

$$q_x = \frac{T_1 - T_2}{R_k} \tag{2–35}$$

where the equivalent parallel resistance is

$$\frac{1}{R_k} = \frac{1}{R_\text{I}} + \frac{1}{R_\text{II}} = \frac{k_\text{I} A_\text{I}}{L} + \frac{k_\text{II} A_\text{II}}{L} \tag{2–36}$$

This same result is achieved by recognizing that q_x is equal to the sum of the heat-transfer rates in the individual materials; that is,

$$q_x = q_{\text{I}x} + q_{\text{II}x} = (T_1 - T_2)\left(\frac{1}{R_\text{I}} + \frac{1}{R_\text{II}}\right) \tag{2–37}$$

The temperature profile in each material is given by [from Eq. (2–14)]

$$T_{\text{I}x} - T_1 = (T_2 - T_1)\frac{x}{L} \qquad T_{\text{II}x} - T_1 = (T_2 - T_1)\frac{x}{L} \tag{2–38,39}$$

The fact that these equations indicate that $T_{\text{I}x} = T_{\text{II}x}$ is consistent with the observation that this is a one-dimensional heat-transfer problem.

Combined Series–Parallel Arrangements A composite wall that provides combined series and parallel paths for the heat transfer is illustrated in Fig. 2–7(a). Such heat-transfer problems are almost always two- or three-dimensional. However, approximate solutions to these types of problems can be obtained under certain conditions

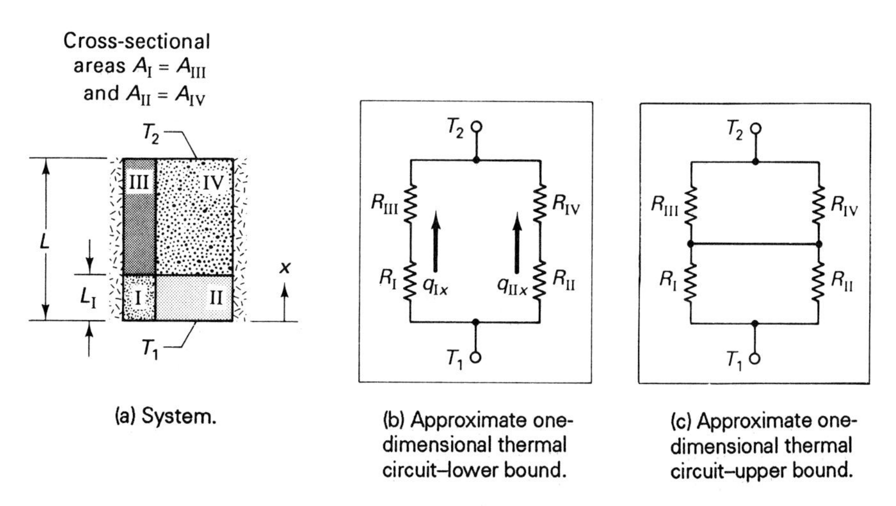

(a) System.

(b) Approximate one-dimensional thermal circuit–lower bound.

(c) Approximate one-dimensional thermal circuit–upper bound.

FIGURE 2–7 Heat transfer in a composite plane wall: series–parallel arrangement.

by assuming one-dimensional heat transfer. For approximate one-dimensional heat transfer, the process can be represented by the thermal circuit shown in Fig. 2–7(b) or Fig. 2–7(c). The solution associated with the thermal circuit shown in Fig. 2–7(b) provides a conservative estimate for q_x (i.e., *lower bound*) due to the fact that this arrangement confines the energy flow to two separate series paths, thereby prohibiting energy flow along other paths of less resistance. On the other hand, because the thermal circuit shown in Fig. 2–7(c) permits the flow of energy along ideal paths of least resistance, the solution corresponding to this circuit represents a practical *upper bound*.

With the heat-transfer process approximated by the thermal circuit shown in Fig. 2–7(b), the total rate of heat transfer q_x can be expressed by†

$$q_x = q_{\mathrm{I}x} + q_{\mathrm{II}x} \tag{2–40}$$

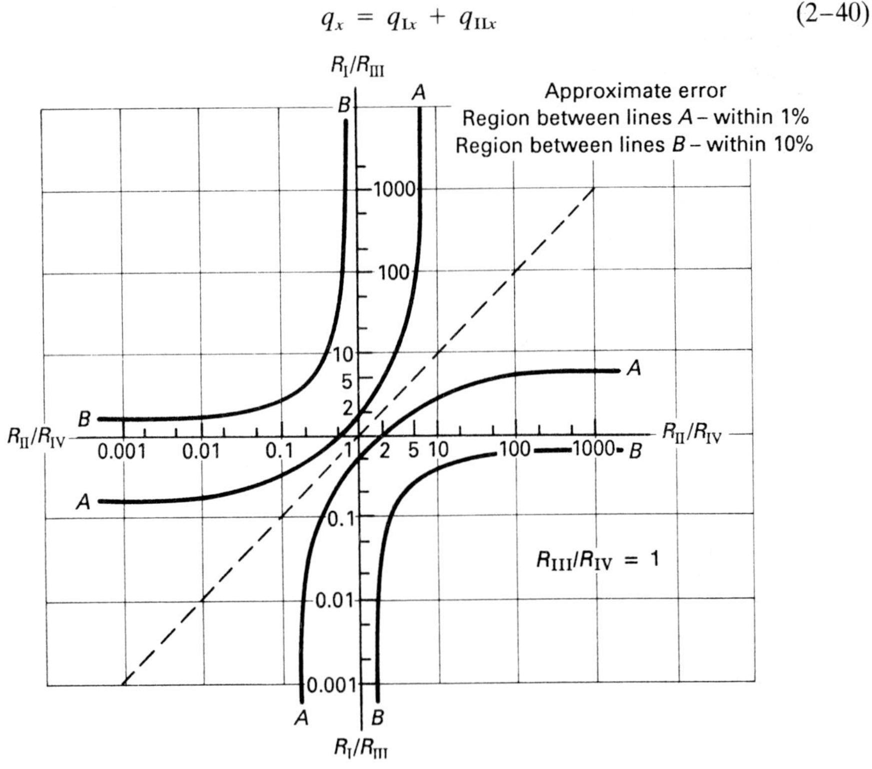

FIGURE 2–8 Criterion for applicability of one-dimensional lower bound solution given by Eqs. (2–40) and (2–41) to composite wall with four series–parallel thermal resistances.‡

† or equivalently

$$q_x = \frac{\Delta T}{R_k} = \Delta T\left(\frac{1}{R_\mathrm{I} + R_\mathrm{III}} + \frac{1}{R_\mathrm{II} + R_\mathrm{IV}}\right) \qquad R_k = \frac{1}{\dfrac{1}{R_\mathrm{I} + R_\mathrm{III}} + \dfrac{1}{R_\mathrm{II} + R_\mathrm{IV}}}$$

‡ This figure is based on a numerical finite difference solution for a problem that involves a parallel composite with convection at one surface. The numerical technique introduced in Chap. 4 can be used to develop criteria for other arrangements.

where q_{Ix} and q_{IIx} are given by

$$q_{Ix} = q_{IIIx} = \frac{T_1 - T_2}{R_I + R_{III}} \qquad q_{IIx} = q_{IVx} = \frac{T_1 - T_2}{R_{II} + R_{IV}} \tag{2–41a,b}$$

A criterion which provides a basis for judging the accuracy of this one-dimensional lower bound solution is given in Fig. 2–8 for a representative system involving four thermal resistances. This figure indicates that the heat transfer is essentially one-dimensional for situations in which $R_I/R_{III} \simeq R_{II}/R_{IV}$. In addition, this criterion is satisfied for small values of both R_I/R_{III} and R_{II}/R_{IV}, and for large values of both R_I/R_{III} and R_{II}/R_{IV}. For example, with $R_{II}/R_{IV} = 10$, the error is of the order of 1% for $R_I/R_{III} > 3$ and about 10% for $R_I/R_{III} > 0.4$. For the case in which R_I/R_{III} and R_{II}/R_{IV} are both very small, the analogous electrical circuit essentially reduces to a simple parallel circuit involving R_{III} and R_{IV} alone. Similarly, for the case in which R_I/R_{III} and R_{II}/R_{IV} are very large, the circuit reduces to a parallel network which only involves R_I and R_{II}. As indicated earlier, parallel heat-transfer systems involving only two thermal resistances are one-dimensional.

It should be noted that the alternative one-dimensional thermal circuit represented by Fig. 2–7(c) indicates an approximate upper bound solution for q_x of the form†

$$q_x = \frac{T_1 - T_2}{\dfrac{1}{1/R_I + 1/R_{II}} + \dfrac{1}{1/R_{III} + 1/R_{IV}}} \tag{2–42}$$

This relation is equivalent to Eqs. (2–40) and (2–41) for one-dimensional conditions which correspond to $R_I/R_{III} = R_{II}/R_{IV}$. For situations in which significant two-dimensional effects occur, the rate of heat transfer can be approximated by taking the average value obtained from Eqs. (2–40) and (2–41) and Eq. (2–42).

EXAMPLE 2–6

One surface of the 1-m-thick composite plate shown in Fig. E2–6a is kept at 0°C. The other surface is exposed to a fluid with $T_F = 100°C$ and $\bar{h} = 10$ W/(m² °C). Develop an approximate solution for the rate of heat transfer through this wall.

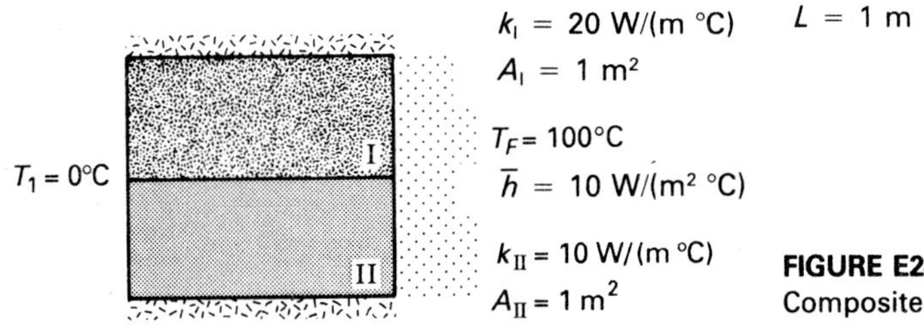

FIGURE E2–6a
Composite plane wall.

† or equivalently

$$q_x = \frac{\Delta T}{R_k} \qquad R_k = \frac{1}{1/R_I + 1/R_{II}} + \frac{1}{1/R_{III} + 1/R_{IV}}$$

Solution

Objective Estimate q for the composite wall shown in Fig. E2–6a.

Assumptions/Conditions

steady-state
negligible two-dimensional effects
uniform properties in each material
negligible thermal contact resistance

Properties

Material I: $k_I = 20$ W/(m °C).
Material II: $k_{II} = 10$ W/(m °C).

Analysis Because this composite wall/fluid system provides both series and parallel paths, this heat-transfer problem is actually two-dimensional. Assuming for the moment that the two-dimensional effects are secondary, we sketch an approximate thermal circuit in Fig. E2–6b. Calculating R_I, R_{II}, R_{III}, and R_{IV}, we have

$$R_I = \frac{L}{k_I A_I} = \frac{1 \text{ m}}{[20 \text{ W/(m °C)}](1 \text{ m}^2)} = 0.05 \text{ °C/W}$$

$$R_{II} = \frac{L}{k_{II} A_{II}} = \frac{1 \text{ m}}{[10 \text{ W/(m °C)}](1 \text{ m}^2)} = 0.1 \text{ °C/W}$$

$$R_{III} = \frac{1}{\bar{h} A_{III}} = \frac{1}{[10 \text{ W/(m}^2 \text{ °C)}](1 \text{ m}^2)} = 0.1 \text{ °C/W} \qquad R_{IV} = R_{III}$$

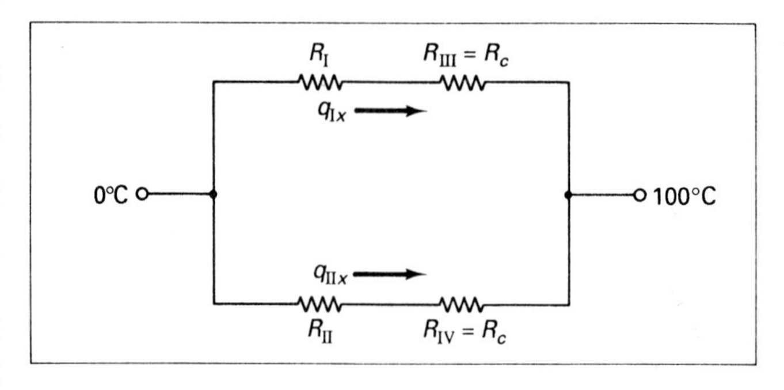

FIGURE E2–6b Thermal circuit.

and

$$\frac{R_{\text{I}}}{R_{\text{III}}} = \frac{0.05\ \text{°C/W}}{0.1\ \text{°C/W}} = 0.5 \qquad \frac{R_{\text{II}}}{R_{\text{IV}}} = \frac{0.1\ \text{°C/W}}{0.1\ \text{°C/W}} = 1$$

The rate of heat transfer indicated by this thermal circuit is

$$q_x = \frac{T_1 - T_F}{R_{\text{I}} + R_{\text{III}}} + \frac{T_1 - T_F}{R_{\text{II}} + R_{\text{IV}}} = (0°\text{C} - 100°\text{C})\left(\frac{1}{0.05 + 0.1} + \frac{1}{0.1 + 0.1}\right)\frac{\text{W}}{°\text{C}} = -1170\ \text{W} \tag{a}$$

As we have noted, the solution associated with the thermal circuit shown in Fig. E2–6b represents a lower bound. Referring to Fig. 2–8, we find that the error for a one-dimensional analysis for this system is about 1%. It follows that the actual rate of heat transfer should be about −1180 W [i.e., $q_x = 1.01(-1170\ \text{W})$].

Using the alternative one-dimensional circuit shown in Fig. 2–7(c) with $R_{\text{III}} = R_{\text{IV}} = R_c$, the approximate solution for the rate of heat transfer becomes

$$q_x = \frac{T_1 - T_2}{\dfrac{1}{1/R_{\text{I}} + 1/R_{\text{II}}} + \dfrac{1}{1/R_{\text{III}} + 1/R_{\text{IV}}}} = \frac{0°\text{C} - 100°\text{C}}{\dfrac{1}{1/0.05 + 1/0.1} + \dfrac{1}{1/0.1 + 1/0.1}} = -1200\ \text{W} \tag{b}$$

which represents an upper bound on the rate of heat transfer. In this connection, by averaging the values obtained from Eq. (a) and Eq. (b), we obtain $q_x = 1185$ W, which is very close to the numerical solution to this problem.†

Other Types of Boundary Conditions

Several other types of boundary conditions are sometimes required in the analysis of heat-transfer problems. Examples include energy dissipation caused by relative interfacial motion, and a moving interface associated with phase change. Reference 5 is suggested as an introduction to the analysis of systems with moving boundaries.

2–2–4 Differential Formulation/Solution: Summary

To recapitulate, the development of the differential formulation/solution for conduction heat transfer involves (1) the application of the first law of thermodynamics to

† The finite-difference numerical method indicates $q_x = -1184$ W.

a differential element within the system, (2) the utilization of the definition of the derivative in relating the energy entering the differential element to the energy exiting, (3) the use of the appropriate particular law(s) (in this case the Fourier law of conduction), (4) a consideration of the conditions that exist at the boundaries, and (5) the solution of the resulting system of equations for the temperature distribution and the rate of heat transfer.

2-2-5 Short Method

The full differential formulation/solution approach outlined above can be utilized to analyze any steady one-dimensional conduction-heat-transfer problem. However, a much shorter method can be developed for one-dimensional problems in which q_x is constant. This approach involves the direct integration of the one-dimensional Fourier law of conduction, Eq. (1–10). To develop this short method in the context of the flat-plate geometry, we merely separate and integrate Eq. (1–10) as follows:

$$\int \frac{q_x}{A_x}\, dx = -\int k\, dT \tag{2–43}$$

Because q_x and A_x are constant for this problem, this equation reduces to

$$\frac{q_x}{A}\int dx = -\int k\, dT \tag{2–44}$$

which can be integrated. For the case in which the thermal conductivity is uniform, Eq. (2–44) can be written as

$$\frac{q_x}{A}\int dx = -k\int dT \tag{2–45}$$

The integration of this equation from $x = 0$ to L and $T = T_1$ to T_2 gives rise to Eq. (2–15),

$$q_x = \frac{kA}{L}(T_1 - T_2) \tag{2–15}$$

On the other hand, the integration from $x = 0$ to x and $T = T_1$ to T provides a relationship for the temperature distribution of the form

$$q_x = \frac{kA}{x}(T_1 - T) \tag{2–46}$$

Replacing q_x from Eq. (2–15), we obtain Eq. (2–14),

$$\frac{T - T_1}{T_2 - T_1} = \frac{x}{L} \tag{2–14}$$

This short method can be generalized for any steady one-dimensional conduction heat-transfer system by writing

$$q_\xi \int \frac{d\xi}{A_\xi} = -\int k\,dT \tag{2–47}$$

where ξ is equal to x, y, or z in Cartesian coordinates and r in cylindrical or spherical coordinates; q_ξ is constant but A_ξ can be a function of ξ. The integration from one boundary to the other provides us with an expression for q_ξ. The integration from one boundary to an intermediate location ξ gives rise to a prediction for the temperature distribution. Because of its directness, this method will be used in the following sections whenever possible. However, the more general differential formulation/solution approach or numerical techniques will be required later when problems involving more complex situations are encountered.

2–3 CONDUCTION HEAT TRANSFER IN RADIAL SYSTEMS

Whereas the cross-sectional area A_x for conduction heat transfer in a plane wall is constant, many situations are encountered in which the area through which the heat is transferred is dependent upon the spatial coordinate. For example, for radial conduction heat transfer in the hollow cylinder shown in Fig. 2–9(a), A_r is equal to $2\pi rL$. This classic problem is analyzed in this section, after which an interesting concept known as the critical radius will be introduced.

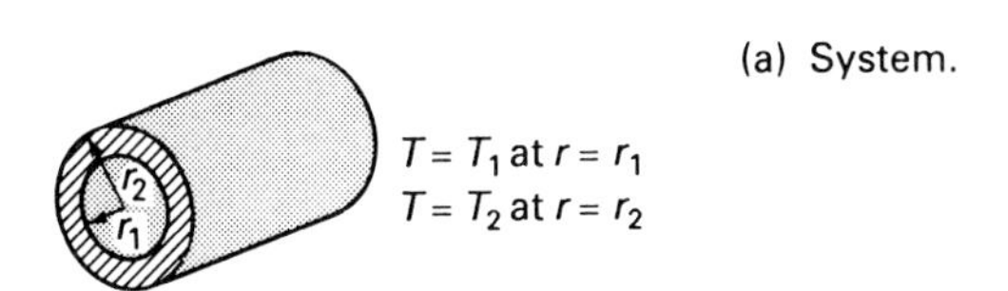

(a) System.

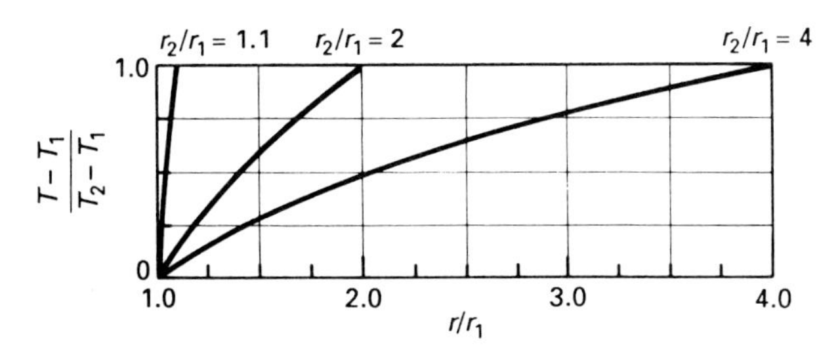

(b) Temperature distribution, Eq. (2–52).

FIGURE 2–9 One-dimensional radial heat transfer in a hollow cylinder.

2–3–1 Analysis

For one-dimensional steady-state radial conduction heat transfer in a hollow cylinder, the Fourier law of conduction takes the form

$$q_r = -kA_r \frac{dT}{dr} \tag{2–48}$$

where $A_r = 2\pi rL$. Because $q_r = q_{r+dr}$ for this problem, the short method can be used to obtain the heat transfer by writing

$$q_r \int_{r_1}^{r_2} \frac{dr}{2\pi rL} = -k \int_{T_1}^{T_2} dT \tag{2–49}$$

for uniform thermal conductivity, or

$$q_r = \frac{2\pi Lk}{\ln (r_2/r_1)} (T_1 - T_2) \tag{2–50}$$

Thus, the thermal resistance to radial conduction in a hollow cylinder is

$$R_k = \frac{\ln (r_2/r_1)}{2\pi Lk} \tag{2–51}$$

Similar to the analysis of heat transfer in a plane wall, the temperature distribution in this hollow cylinder can be obtained by integrating from r_1 to r and T_1 to T. The resulting expression for T is given by

$$\frac{T - T_1}{T_2 - T_1} = \frac{\ln (r/r_1)}{\ln (r_2/r_1)} \tag{2–52}$$

This profile is shown in Fig. 2–9(b) for three values of r_2/r_1. It is observed that the temperature distribution is nearly linear for values of r_2/r_1 of the order of unity, but decidedly nonlinear for the larger values of r_2/r_1. This nonlinearity is caused by the variation in A_r with respect to r.

For cases in which the surface temperatures are not specified, T_1 and T_2 are determined by utilizing the appropriate thermal boundary conditions. This point is illustrated in Example 2–7.

For one-dimensional radial heat transfer in a hollow sphere, A_r is equal to $4\pi r^2$. Utilizing the short method, the rate of heat transfer in a hollow sphere with surface temperatures T_1 and T_2 is easily shown to be

$$q_r = \frac{4\pi r_1 r_2 k}{r_2 - r_1} (T_1 - T_2) \tag{2–53}$$

and the temperature profile is given by

$$\frac{T - T_1}{T_2 - T_1} = \frac{r - r_1}{r_2 - r_1}\frac{r_2}{r} \tag{2–54}$$

Note that the thermal resistance for this system is

$$R_k = \frac{r_2 - r_1}{4\pi r_1 r_2 k} \tag{2–55}$$

EXAMPLE 2–7

Refrigerant flows in a 1.9-in.-O.D. copper tube with 0.281-in. wall thickness. The inside surface temperature is 5°F and the room temperature is 70°F. Determine the thickness of insulative pipe covering [k_i = 0.428 Btu/(h ft °F)] required to reduce the heat gain to the pipe by 25% for the case in which forced convection heat transfer occurs with $\bar{h}$ = 10 Btu/(h ft^2 °F). Assume that the thermal radiation effects are negligible.

Solution

Objective Determine the required insulation thickness.

Schematic Copper tube with insulation.

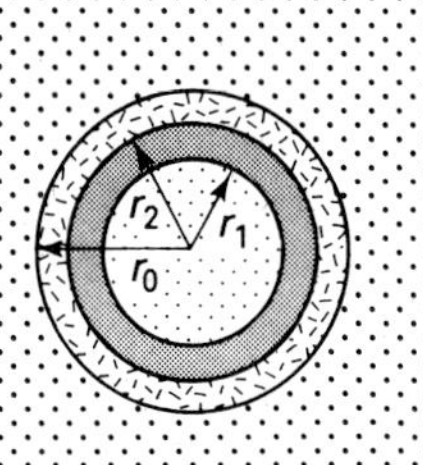

Assumptions/Conditions

steady-state
one-dimensional
uniform properties in each material
forced convection cooling
negligible thermal radiation

Properties

Copper at room temperature (Table A–C–1):

$$k = 386 \frac{\text{W}}{\text{m °C}} \frac{0.578 \text{ Btu/(h ft °F)}}{\text{W/(m °C)}} = 223 \text{ Btu/(h ft °F)}$$

Insulation: k = 0.428 Btu/(h ft °F).

Analysis Referring to the thermal circuit shown in Fig. E2–7, the resistances are

$$R_{k1} = \frac{\ln(r_2/r_1)}{2\pi Lk} = \frac{1}{L}\frac{\ln(0.95/0.669)}{2\pi}\frac{\text{h ft °F}}{223\text{ Btu}} = \frac{2.50 \times 10^{-4}}{L}\frac{\text{h ft °F}}{\text{Btu}}$$

$$R_{k2} = \frac{\ln(r_0/r_2)}{2\pi Lk_i} \qquad R_c = \frac{1}{\bar{h}A_s} = \frac{0.1}{A_s}\frac{\text{h ft}^2\text{ °F}}{\text{Btu}}$$

With no insulation, q_r is given by

$$q_r = \frac{5°\text{F} - 70°\text{F}}{\dfrac{2.50 \times 10^{-4}}{L}\dfrac{\text{h ft °F}}{\text{Btu}} + \dfrac{0.1}{2\pi L(0.95\text{ ft}/12)}\dfrac{\text{h ft}^2\text{ °F}}{\text{Btu}}} = q_c$$

$$\frac{q_r}{L} = \frac{-65°\text{F}}{(2.50 \times 10^{-4} + 0.201)\text{ h ft °F/Btu}} = -323\text{ Btu/(h ft)}$$

Note that the thermal resistance R_{k1} of the pipe wall is negligible.

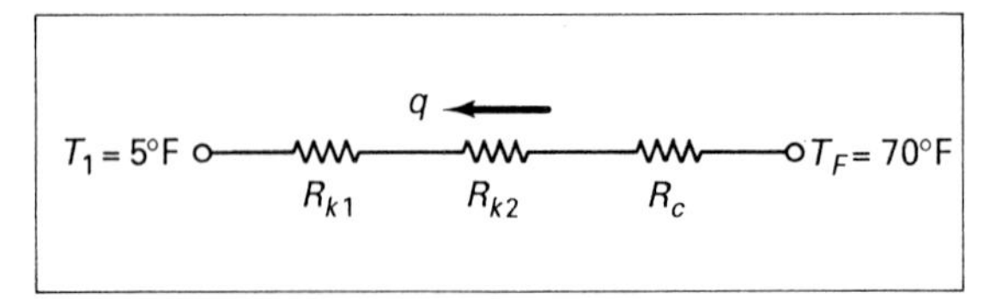

FIGURE E2–7
Thermal circuit.

A 25% reduction in the rate of heat transfer results in $q_r/L = -242$ Btu/(h ft). It follows that

$$q_r \simeq \frac{5°\text{F} - 70°\text{F}}{R_{k2} + R_c} = \frac{-65°\text{F}}{\dfrac{\ln(r_0/r_2)}{2\pi k_i L} + \dfrac{1}{\bar{h}2\pi r_0 L}}$$

$$\ln\left(\frac{r_0}{r_2}\right) + \frac{k_i}{\bar{h}r_0} = \frac{-65°\text{F}(2\pi k_i)}{q_r/L} = \frac{-65°\text{F}(2\pi)[0.428\text{ Btu/(h ft °F)}]}{-242\text{ Btu/(h ft)}} = 0.722$$

This nonlinear equation is solved for r_0 by iteration as follows:

$$\ln\left(\frac{r_0}{r_2}\right) = 0.722 - \frac{0.428}{10}\frac{\text{ft}}{r_0}$$

$$\frac{r_0}{r_2} = \exp\left(0.722 - \frac{0.0428}{r_0}\text{ ft}\right) \tag{a}$$

Starting with an assumed value for r_0/r_2 of 2, we have

$$\frac{r_0}{r_2} = \exp\left[0.722 - \frac{0.0428}{2(0.95/12)}\right] = 1.57$$

Substituting this value back into Eq. (a), we obtain

$$\frac{r_0}{r_2} = 1.46$$

Continuing this iteration sequence, we arrive at

$$\frac{r_0}{r_2} = 1.4$$

after only three more steps. Thus, the thickness of insulation required to reduce the rate of heat loss by 25% is

$$\delta = 1.4r_2 - r_2 = 0.4(0.95 \text{ in.}) = 0.38 \text{ in.}$$

EXAMPLE 2-8

Utilize the differential formulation approach to obtain expressions for the temperature distribution, rate of heat transfer, and thermal resistance for the cylindrical section shown in Fig. E2-8.

FIGURE E2-8
Cylindrical section.

Solution

Objective Develop relations for T, q, and R for this cylindrical section.

Assumptions/Conditions

steady-state
one-dimensional
uniform properties

Analysis Because no heat is transferred across the surfaces at $\theta = 0$ and θ_1 and because T_1 and T_2 are uniform, the heat transfer in this system is one-dimensional. This one-dimensional (r direction) conduction-heat-transfer problem can be analyzed by either the differential formulation approach or the short method.

The differential formulation is developed as follows:

Step 1

$$q_r = q_{r+dr}$$

Step 2

$$q_r = q_r + \frac{dq_r}{dr}dr \qquad \text{or} \qquad \frac{dq_r}{dr} = 0$$

Step 3

$$\frac{d}{dr}\left(-kA_r\frac{dT}{dr}\right) = \frac{d}{dr}\left(-k\theta_1 rL\frac{dT}{dr}\right) = 0$$

or, for uniform thermal conductivity,

$$\frac{d}{dr}\left(r\frac{dT}{dr}\right) = 0 \tag{a}$$

Finally, the boundary temperatures can be designated by $T(r_1) = T_1$ and $T(r_2) = T_2$.

Next, Eq. (a) is solved for the temperature distribution. Integrating, we have

$$r\frac{dT}{dr} = C_1$$

Continuing,

$$\int_{T_1}^{T} dT = \int_{r_1}^{r} \frac{C_1\,dr}{r} \qquad T - T_1 = C_1 \ln\left(\frac{r}{r_1}\right)$$

Utilizing the boundary condition at r_2, C_1 is given by

$$C_1 = \frac{T_2 - T_1}{\ln(r_2/r_1)}$$

and the temperature profile becomes

$$\frac{T - T_1}{T_2 - T_1} = \frac{\ln(r/r_1)}{\ln(r_2/r_1)}$$

To obtain an expression for the rate of heat transfer, we utilize the Fourier law of conduction as follows:

$$q_r = -k\theta_1 rL\frac{dT}{dr} = -k\theta_1 LC_1 = k\theta_1 L\frac{T_1 - T_2}{\ln(r_2/r_1)}$$

or

$$q_r = \frac{T_1 - T_2}{R_k}$$

where $R_k = \ln(r_2/r_1)/(\theta_1 kL)$. These results are seen to be identical to those which were obtained on the basis of the short method for the case of a hollow circular cylinder with $\theta_1 = 2\pi$.

2-3-2 Critical Radius

As shown in Example 2-7, convection and composite wall boundary conditions associated with geometries for which the area is not constant are handled in the same way that these complications were treated for the plane-wall geometry. For the case of a plane wall exposed to a fluid, an increase in the thickness of the wall results in an increase in the internal resistance $R_k = L/(kA)$ but does not change the surface resistance R_c. Hence, such an increase in the thickness of a plane wall always reduces the rate of heat transfer through the wall. Of course, a reduction in heat transfer is most easily accomplished by the use of an insulating material of low thermal conductivity. On the other hand, an increase in the wall thickness or the addition of an insulating material does not always bring about a decrease in the heat-transfer rate for geometries with nonconstant cross-sectional area.

To see this point, consider the hollow cylinder of radii r_1 and r_2 shown in Fig. 2–10(a). The inside surface is maintained at temperature T_1, and the temperature T_F of the surrounding fluid is specified. The thermal circuit for this problem is shown in Fig. 2–10(b). Now we ask, what will be the effect of increasing the outside radius r_2, with T_1, T_F, and $\bar{h}$ held constant? The rate of heat transfer from the outside surface to the fluid is

$$q_r = \frac{T_1 - T_F}{R_k + R_c} = \frac{T_1 - T_F}{\dfrac{\ln(r_2/r_1)}{2\pi Lk} + \dfrac{1}{\bar{h}2\pi L r_2}} \tag{2–56}$$

An increase in r_2 is seen to increase R_k but to decrease R_c. Therefore, the addition of material can either decrease or increase the rate of heat transfer, depending upon the change in the total resistance $R_c + R_k$ with r_2.

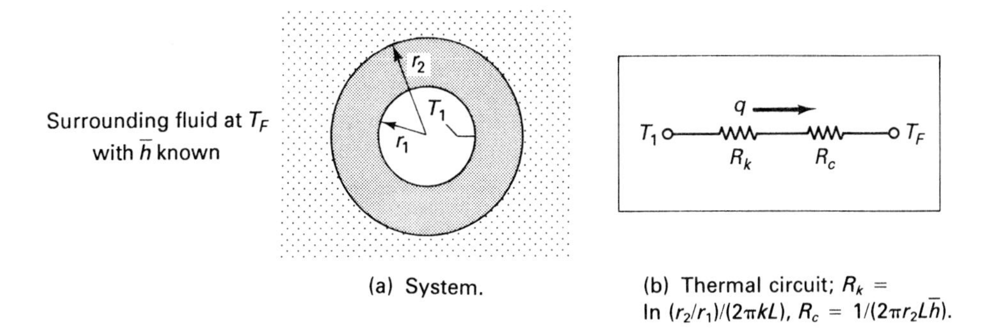

(a) System.

(b) Thermal circuit; $R_k = \ln(r_2/r_1)/(2\pi kL)$, $R_c = 1/(2\pi r_2 L\bar{h})$.

FIGURE 2–10 Convection heat transfer from a hollow circular cylinder.

To see the effect of r_2 on q_r, this equation is plotted in Fig. 2–11 for various values of $k/(\bar{h}r_1)$, with r_2/r_1 taken as the independent variable. For $k/(\bar{h}r_1)$ less than unity, the rate of heat transfer q_r continuously decreases as r_2 increases from a value of r_1. But, for $k/(\bar{h}r_1)$ greater than unity, q_r increases to a maximum and then decreases. To determine the radius r_c at which q_r is maximized for $k/(\bar{h}r_1)$ greater than unity, we set dq_r/dr_2 equal to zero,

$$\left.\frac{dq_r}{dr_2}\right|_{r_c} = -(T_1 - T_F)2\pi Lk\left.\left(\frac{1}{r_2} - \frac{k}{\bar{h}r_2^2}\right)\right|_{r_c} = 0 \tag{2–57}$$

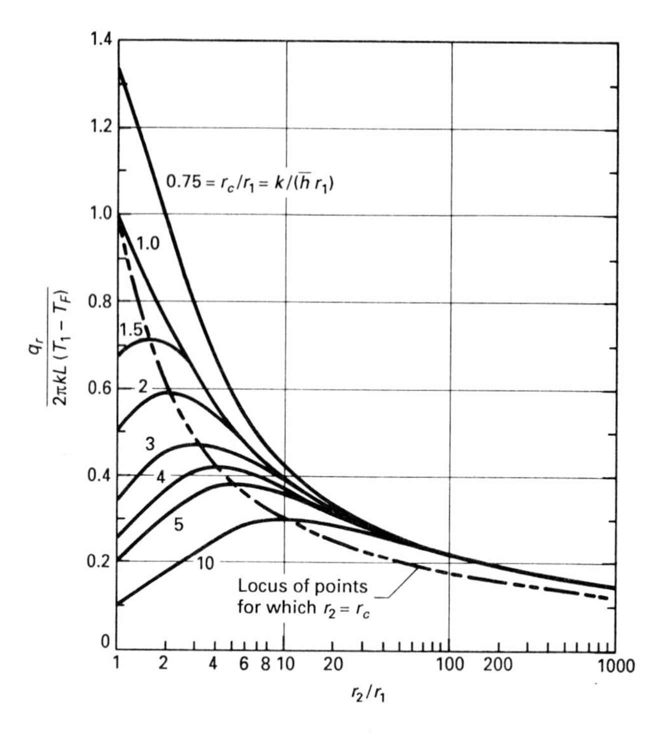

FIGURE 2–11 Effect of increase in outside radius r_2 on heat transfer from a pipe.

with the result

$$r_c = \frac{k}{\bar{h}} \tag{2–58}$$

r_c is referred to as the *critical radius*. If $r_2 > r_c$, the addition of material to the outside surface will decrease the rate of heat transfer, as in the case of Example 2–7, where $r_c = 0.514$ in. and $r_2 = 0.95$ in. This is often the case for conditions of forced convection for which $\bar{h}$ is large and r_c is small. But if $r_2 < r_c$, the addition of material will increase the heat-transfer rate until $r_2 = r_c$, after which additional increases in r_2 will decrease q_r. This situation occurs more often for natural convection than for forced convection because of the low values of $\bar{h}$, especially in gases.

In contrast, if insulation is added to the inside surface, both R_c and R_k increase. Hence, the addition of insulation to the inside surface always reduces the heat-transfer rate and the critical radius concept has no significance.

Heat transfer in a sphere is affected by the addition of insulation to the outside or inside surface in much the same way as in a cylinder. However, for a sphere, the critical radius is given by

$$r_c = \frac{2k}{\bar{h}} \tag{2–59}$$

Likewise, a critical radius would be expected to be found for thermal radiation heat transfer or combined convection and radiation from the outside surface of a cylinder or sphere.

EXAMPLE 2–9

Determine the thickness of insulative pipe covering [k_i = 0.428 Btu/(h ft °F)] required to reduce the convective heat loss from the 1.9-in.-O.D. pipe of Example 2–7 by 25% for the case in which natural convection cooling occurs with $\bar{h}$ = 3.6 Btu/(h ft^2 °F).

Solution

Objective Determine the required insulation thickness.

Schematic Copper tube with insulation.

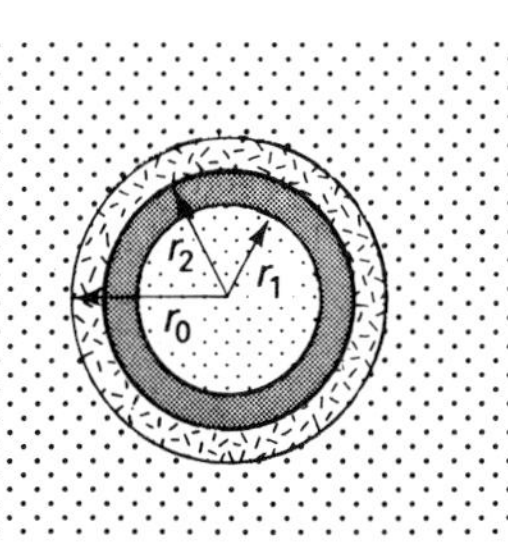

Assumptions/Conditions

steady-state
one-dimensional
uniform properties in each material
natural convection cooling
negligible thermal radiation

Properties

Copper tube (from Example 2–7): k = 223 Btu/(h ft °F).
Insulation: k = 0.428 Btu/(h ft °F).

Analysis The critical radius for this situation is

$$r_c = \frac{k_i}{\bar{h}} = \frac{0.428 \text{ Btu/(h ft °F)}}{3.6 \text{ Btu/(h ft}^2 \text{ °F)}} = 0.119 \text{ ft} = 1.43 \text{ in.}$$

Thus, we have

$$\frac{r_c}{r_2} = \frac{1.43 \text{ in.}}{0.95 \text{ in.}} = 1.5$$

Because r_c/r_2 is greater than unity, the effect of insulation for r_0 less than r_c will be to increase the rate of heat transfer q_r. In contrast, for the forced convection conditions of Example 2–7, r_c is equal to 0.514, such that the effect of insulation is to reduce q_r for all values of r_0.

Referring to Fig. 2–11, the dimensionless heat flux $q_r/[2\pi k_i L(T_1 - T_F)]$ increases from a value of 0.667 at $r_0/r_2 = 1$ to a maximum value of approximately 0.71 at r_0/r_2 equal to $r_c/r_2 = 1.5$, and then begins to fall toward zero. Note that the rate of heat transfer for $r_0/r_2 = 1$ is given by the simple Newton law of cooling,

$$q_r = 2\pi r_2 L\bar{h}(T_1 - T_F)$$

or

$$\frac{q_r}{2\pi k_i L(T_1 - T_F)} = \frac{\bar{h} r_2}{k_i} = \frac{r_2}{r_c} = \frac{1}{1.5} = 0.667$$

Our problem calls for a 25% reduction in the rate of heat transfer; that is,

$$\left.\frac{q_r}{2\pi k_i L(T_1 - T_F)}\right|_{r_0} = 0.5$$

Utilizing Fig. 2–11, we find that the dimensionless heat flux reaches this value for r_0/r_2 equal to approximately 5. Thus, the thickness of insulation is

$$\delta = r_0 - r_2 = 5(0.95 \text{ in.}) - 0.95 \text{ in.} = 3.8 \text{ in.}$$

2–4 VARIABLE THERMAL CONDUCTIVITY

As indicated in Chap. 1, the thermal conductivity of most materials is at least somewhat dependent upon temperature. The variation of thermal conductivity with temperature is shown in Fig. 2–12 for several common metals for the temperature range −100°C to 300°C. Whereas the assumption of uniform thermal conductivity is generally acceptable for problems involving small temperature differences, many situations are encountered in which the variation in k with T cannot be neglected. Referring to Fig. 2–12, we see that a linear approximation for k can be utilized over limited temperature ranges for these materials; that is,

$$k(T) = k_0(1 + \beta_T T) \tag{2–60}$$

where β_T is known as the *temperature coefficient of thermal conductivity*. The units for β_T are 1/°C (or 1/°F).

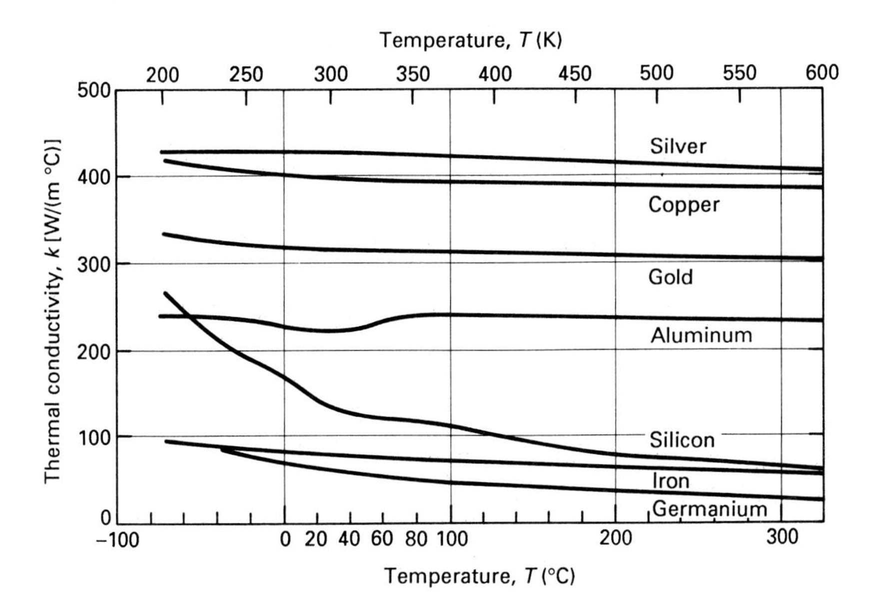

FIGURE 2–12 Variation of thermal conductivity k with temperature for several metals.

2–4–1 Analysis

Utilizing the short method, k must be retained within the integral in Eq. (2–47) for cases in which the variation of the thermal conductivity with temperature is significant. By integrating from one boundary to the other, Eq. (2–47) gives

$$q_\xi \int_{\xi_1}^{\xi_2} \frac{d\xi}{A_\xi} = -\int_{T_1}^{T_2} k\, dT \tag{2–61}$$

where $\xi_1 = 0$ and $\xi_2 = L$ for a flat plate, and $\xi_1 = r_1$ and $\xi_2 = r_2$ for a hollow cylinder or sphere.

Introducing the mean thermal conductivity $\bar{k}$,

$$\bar{k} = \frac{1}{T_2 - T_1} \int_{T_1}^{T_2} k\, dT \tag{2–62}$$

the rate of heat transfer q_ξ is given by

$$q_\xi = \frac{\bar{k}(T_1 - T_2)}{\displaystyle\int_{\xi_1}^{\xi_2} d\xi/A_\xi} \tag{2–63}$$

Based on this expression, equations can be written for the thermal resistances for flat plates, and hollow cylinders and spheres with variable thermal conductivity of the forms

$$R_k = \frac{L}{A\bar{k}} \qquad \text{flat plate} \tag{2–64}$$

$$R_k = \frac{\ln (r_2/r_1)}{2\pi L\bar{k}} \qquad \text{hollow cylinder} \tag{2–65}$$

$$R_k = \frac{r_2 - r_1}{4\pi r_1 r_2 \bar{k}} \qquad \text{hollow sphere} \tag{2–66}$$

EXAMPLE 2–10

The surfaces of a 10-cm-thick plate are maintained at 0°C and 100°C. The thermal conductivity varies with temperature according to $k = k_0(1 + \beta_T T)$ with $k = 50$ W/(m °C) at 0°C and $k = 100$ W/(m °C) at 100°C. Determine the temperature distribution and heat-transfer flux in the plate.

Solution

Objective Determine T and q'' for the plate.

Schematic Plate with variable thermal conductivity.

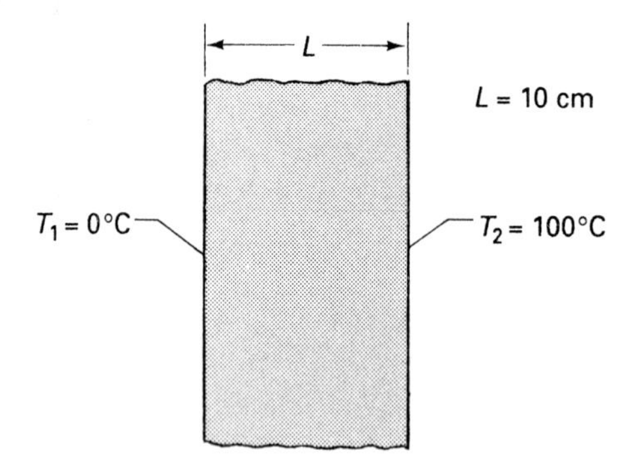

Assumptions/Conditions

steady-state
one-dimensional
variable thermal conductivity

Properties The thermal conductivity is

$$k = k_0 (1 + \beta_T T)$$

with k specified at 0°C and 100°C. To evaluate k_0 and β_T, we set k equal to 50 W/(m °C) at 0°C and 100 W/(m °C) at 100°C, with the result

$$k_0 = 50 \text{ W/(m °C)} \qquad \beta_T = 0.01/\text{°C}$$

Analysis To calculate the rate of heat transfer, we utilize Eq. (2–63), with the result

$$q = \frac{\bar{k}A}{L}(T_1 - T_2)$$

where

$$\bar{k} = \frac{1}{T_2 - T_1}\int_{T_1}^{T_2} k\,dT = \frac{k_0}{100°\text{C}}\int_{0°\text{C}}^{100°\text{C}} (1 + \beta_T T)\,dT$$

$$= \frac{k_0}{100°\text{C}}\left(T + \frac{\beta_T}{2}T^2\right)\Bigg|_{0°\text{C}}^{100°\text{C}} = \frac{k_0}{100°\text{C}}\left[100°\text{C} + \frac{10^{-2}}{2°\text{C}}(100°\text{C})^2\right] = 1.5k_0$$

Following through with the calculation, we have

$$q'' = 1.5\frac{k_0}{L}(T_1 - T_2) = \frac{1.5}{0.1\text{ m}}\left(50\frac{\text{W}}{\text{m °C}}\right)(0°\text{C} - 100°\text{C}) = -75\text{ kW/m}^2$$

To obtain the temperature distribution, the Fourier law of conduction is integrated from 0 to x and from 0°C to T as follows:

$$q = -kA\frac{dT}{dx} \qquad \frac{q}{A}\int_0^x dx = -\int_{0°\text{C}}^{T} k\,dT$$

$$\frac{q}{A}x = -k_0\left(T + \frac{\beta_T}{2}T^2\right)$$

where $q/A = q'' = -75\text{ kW/m}^2$. It follows that

$$T + \frac{\beta_T}{2}T^2 + 1.5(T_1 - T_2)\frac{x}{L} = 0$$

Solving this quadratic equation for T, we obtain

$$T = \frac{-1 \pm \sqrt{1 - 4(\beta_T/2)(1.5)(T_1 - T_2)(x/L)}}{\beta_T}$$

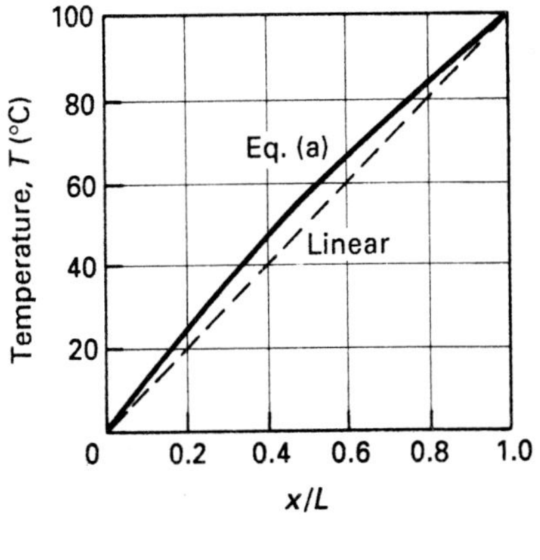

FIGURE E2–10
Temperature distribution in plane wall with variable thermal conductivity.

where the positive sign must be utilized to satisfy the boundary conditions. Substituting for β_T, T_1, and T_2, T becomes

$$T = \frac{-1 + \sqrt{1 + 3(x/L)}}{0.01} \ ^\circ\mathrm{C} \tag{a}$$

This temperature distribution is compared with the linear profile associated with uniform thermal conductivity in Fig. E2–10. Notice the distinct nonlinear behavior of T for this case in which k is a function of temperature.

2–5 INTERNAL ENERGY SOURCES

Heat-transfer systems involving internal energy sources include chemical and nuclear reactors in which energy is generated by chemical reaction or by the interaction of nuclear particles. Other important heat-transfer applications involving internal energy sources occur in electrical and electronic circuits. Examples include electrical resistance heaters made of nickel/chromium alloys, incandescent electric lamps with tungsten filaments, and semiconductor chips and transistors made of silicon or germanium. The dissipation of heat is also important in the operation of electric motors, generators, transformers, and relays.

The strength of an internal energy source is generally represented by the rate of energy generated per unit volume $\dot{q}$. For electrical and electronic systems, $\dot{q}$ is given by

$$\dot{q} = \frac{I_e^2 R_e}{V} = \frac{I_e^2 \rho_e}{A^2} \tag{2–67}$$

where the current I_e is assumed to be uniform and ρ_e $(= R_e A/L)$ is the *electrical resistivity*. The resistivity is generally a linear function of temperature; that is,

$$\rho_e = \rho_0(1 + \alpha_0 T) \tag{2–68}$$

where ρ_0 is the resistivity at a reference temperature such as 0°C, and α_0 is the *temperature coefficient of resistance* at the reference temperature. Accordingly, the power dissipation per unit volume associated with the flow of electric current can be approximated by an equation of the form

$$\dot{q} = \dot{q}_0(1 + \alpha_0 T) \tag{2–69}$$

where $\dot{q}_0 = I_e^2 \rho_0/A^2$. Data for ρ_0 and α_0 are tabulated in reference 6 for several common metals and alloys. For example, $\rho_0 = 1.8 \times 10^{-8}\ \Omega$ m and $\alpha_0 = 0.004$/°C for copper wire, and $\rho_0 = 20 \times 10^{-8}\ \Omega$ m and $\alpha_0 = 0.005$/°C for carbon steel. In comparison, the electrical resistivities of electrical insulators such as mica and electrical semiconductors such as germanium are of the order of $10^{15}\ \Omega$ m and 0.4 Ω m, respectively. In general, α_0 is quite small for metals and alloys, such that the internal energy generation can be assumed to be uniformly distributed (i.e., independent of the spatial coordinates) for moderate temperatures. However, the variation of $\dot{q}$ becomes an important factor for high-temperature operation.

2-5-1 Analysis

Application of the first law of thermodynamics to the problem of conduction heat transfer in a flat plate with internal energy generation (see Fig. 2-13) leads to

Step 1
$$\dot{q}\,dV + q_x = q_{x+dx} \tag{2-70}$$

(The energy generation can be considered as energy brought into the differential element.) Unlike the case of steady-state conduction heat transfer with no energy generation [represented by Eq. (2-1)], Eq. (2-70) indicates that q_x is not equal to q_{x+dx} for energy-generation problems. Therefore, the short method, which involves the mere integration of the Fourier law, cannot be utilized, unless the dependence of q_x on x is known. Hence, we proceed with the development of the differential formulation.

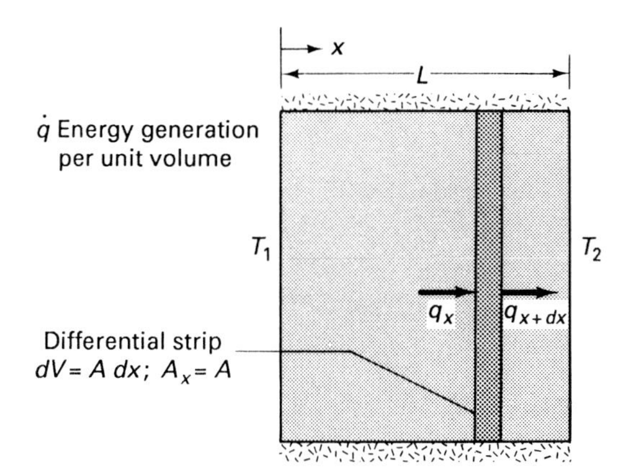

FIGURE 2-13
Conduction heat transfer in a plane wall with internal energy generation.

The substitution of Eq. (2-3) into Eq. (2-70) gives

Step 2
$$\dot{q}A\,dx = \frac{dq_x}{dx}\,dx \tag{2-71}$$

where A_x is constant ($A_x = A$) and $dV = A\,dx$. With q_x given by the Fourier law of conduction, Eq. (2-71) becomes

Step 3
$$\frac{d}{dx}\left(k\frac{dT}{dx}\right) + \dot{q} = 0 \tag{2-72}$$

For the case of uniform thermal conductivity and energy generation, this equation reduces to the form

$$\frac{d^2T}{dx^2} + \frac{\dot{q}_0}{k} = 0 \tag{2-73}$$

Notice that Eq. (2–73) is identical to Eq. (2–7) for $\dot{q}_0 = 0$.

Because this is a second-order differential equation, two boundary conditions are required; they are

$$T(0) = T_1 \qquad T(L) = T_2 \tag{2–74,75}$$

This simple differential equation can be solved by two integrations as follows:

$$\frac{dT}{dx} = -\frac{\dot{q}_0}{k}x + C_1 \qquad T = -\frac{\dot{q}_0}{k}\frac{x^2}{2} + C_1 x + C_2 \tag{2–76,77}$$

The boundary conditions require that $C_2 = T_1$ and $C_1 = (T_2 - T_1)/L + \dot{q}_0 L/(2k)$, such that Eq. (2–77) becomes

$$T - T_1 = (T_2 - T_1)\frac{x}{L} + \frac{\dot{q}_0}{k}\frac{L^2}{2}\left[\frac{x}{L} - \left(\frac{x}{L}\right)^2\right] \tag{2–78}$$

Focusing attention on the case in which symmetrical cooling takes place with $T_2 = T_1$, the temperature distribution is given by

$$T - T_1 = \frac{2\dot{q}_0\ell^2}{k}\left[\frac{x}{L} - \left(\frac{x}{L}\right)^2\right] \tag{2–79}$$

or

$$T - T_1 = \frac{1}{2}\frac{\dot{q}_0\ell^2}{k}\left[1 - \left(\frac{\xi}{L/2}\right)^2\right] \tag{2–80}$$

where $\ell = V/A_s = L/2$ and $\xi = x - L/2$. This temperature distribution is shown in Fig. 2–14. The maximum temperature within the plate for symmetrical cooling occurs at the center where $dT/dx = 0$.

Setting $x = L/2$ in Eq. (2–79) or $\xi = 0$ in Eq. (2–80), we have

$$T_{\max} = T_1 + \frac{\dot{q}_0}{k}\frac{\ell^2}{2} = T_1 + \frac{\dot{q}_0}{k}\frac{L^2}{8} \tag{2–81}$$

Because materials break down above certain temperatures, this maximum temperature represents a critical design consideration.

To obtain the rate of heat transfer for the case in which $T_2 = T_1$, we apply the Fourier law of conduction as follows:

$$q_x = -kA\left.\frac{dT}{dx}\right|_x = \dot{q}_0\frac{V}{2}\left(\frac{2x}{L} - 1\right) \tag{2–82}$$

Hence, the rates of heat transfer at the surfaces $x = 0$ and $x = L$ are

$$q_0 = -\dot{q}_0\frac{V}{2} \qquad q_L = \dot{q}_0\frac{V}{2} \tag{2–83,84}$$

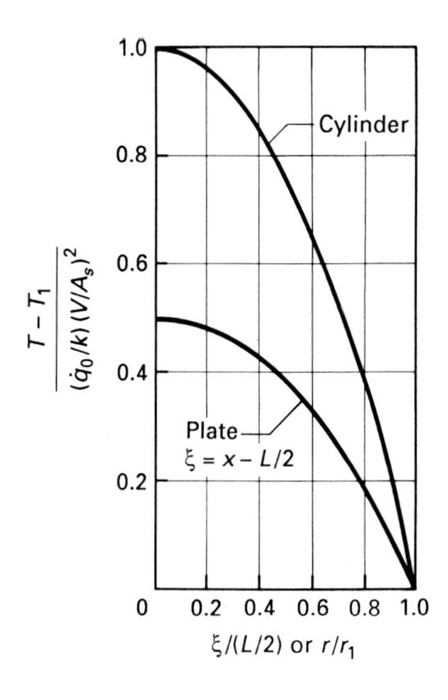

FIGURE 2–14
Temperature distributions for plate and solid circular cylinder with uniform energy generation and surface temperature T_1.

such that the total rate of energy generated within the plate $\dot{q}_0 V$ is transferred out of the two surfaces.

For the more general case in which $T_2 \neq T_1$, Eq. (2–78) is coupled with the Fourier law of conduction to obtain

$$q_x = \frac{kA}{L}(T_1 - T_2) + \frac{\dot{q}_0 V}{2}\left(\frac{2x}{L} - 1\right) \tag{2–85}$$

It follows that q_0 and q_L are

$$q_0 = \frac{kA}{L}(T_1 - T_2) - \frac{\dot{q}_0 V}{2} \tag{2–86}$$

and

$$q_L = \frac{kA}{L}(T_1 - T_2) + \frac{\dot{q}_0 V}{2} \tag{2–87}$$

Thus, we see that the total rate of heat transfer is found by simply superimposing the heat transfer through a nongenerating plate with surface temperatures T_1 and T_2 upon the heat transfer in a generating plate with both surfaces at T_1.

Solutions are also available for the temperature distribution in heat-generating cylinders and spheres. For example, the temperature distribution in a solid circular cylinder with radius r_1, surface temperature T_1, and uniform heat generation takes the form

$$T - T_1 = \frac{\dot{q}_0 \ell^2}{k}\left[1 - \left(\frac{r}{r_1}\right)^2\right] \tag{2–88}$$

where $\ell = V/A_s = r_1/2$. This expression is shown in Fig. 2–14. The maximum temperature in the cylinder occurs at the center.

$$T_{\max} = T_1 + \frac{\dot{q}_0 \ell^2}{k} = T_1 + \frac{\dot{q}_0}{k}\frac{r_1^2}{4} \tag{2–89}$$

EXAMPLE 2–11

Four amperes of current flow in a 1-mm-diameter copper wire with resistivity equal to about 30 μΩ cm. The insulation is 2 mm thick and has a thermal conductivity of 0.05 W/(m °C). Determine the surface temperature of the insulation and the maximum operating temperature of the wire if the system is cooled by blackbody thermal radiation with $T_R = -200°\text{C}$ and $F_{s-R} = 1.0$.

Solution

Objective Determine the surface and centerline temperature of the wire.

Schematic Electric conducting copper wire with insulation.

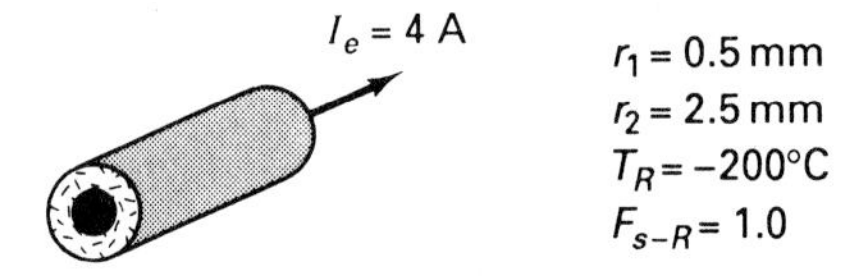

Assumptions/Conditions

steady-state
one-dimensional
uniform internal energy generation
blackbody thermal radiation cooling

Properties

Copper at room temperature (Table A–C–1): $k = 386$ W/(m °C), $\rho_e = 30$ μΩ cm.

Insulation: $k = 0.05$ W/(m °C).

Analysis The power generated within the wire is represented by

$$\dot{W} = I_e^2 R_e = I_e^2 \rho_e \frac{L}{A}$$

such that

$$\frac{\dot{W}}{L} = (4\ \text{A})^2(30 \times 10^{-6}\ \Omega\ 0.01\ \text{m})\frac{4}{\pi(10^{-3}\ \text{m})^2} = 6.11\ \text{W/m}$$

Thus, the energy generated per unit volume is

$$\dot{q}_0 = \frac{\dot{W}}{LA} = \frac{6.11\ \text{W/m}}{\pi(10^{-3}\ \text{m})^2/4} = 7.78 \times 10^6\ \text{W/m}^3$$

Because this energy must be radiated away,

$$\dot{W} = q_R = \sigma A_s F_{s-R}(T_s^4 - T_R^4)$$

or

$$T_s^4 = T_R^4 + 6.11L\,\frac{\text{W}}{\text{m}}\left(\frac{1}{\sigma A_s F_{s-R}}\right)$$

such that the surface temperature $T_s = T_2$ is found to be 288 K (=15.1°C).

To find the interfacial temperature T_1, we write

$$q_r = \frac{T_1 - T_s}{\dfrac{\ln(r_2/r_1)}{2\pi k_i L}} = q_R$$

$$T_1 = T_s + \frac{q_R}{L}\frac{\ln(r_2/r_1)}{2\pi k_i} = 15.1°\text{C} + 6.11\,\frac{\text{W}}{\text{m}}\,\frac{\ln(0.0025/0.0005)}{2\pi[0.05\ \text{W/(m °C)}]} = 46.4°\text{C}$$

Finally, utilizing Eq. (2–89), T_{max} is calculated.

$$T_{\text{max}} = T_1 + \frac{\dot{q}_0}{k}\left(\frac{r_1}{2}\right)^2 = 46.4°\text{C} + \frac{7.78 \times 10^6\ \text{W/m}^3}{386\ \text{W/(m °C)}}\left(\frac{0.5 \times 10^{-3}\ \text{m}}{2}\right)^2$$
$$= 46.4°\text{C} + 1.26 \times 10^{-3}\ °\text{C} \simeq 46.4°\text{C}$$

Thus, we see that the temperature throughout the wire is essentially uniform.

2–6 EXTENDED SURFACES

Situations often arise in which means are sought for increasing the heat convected from a surface. A consideration of the Newton law of cooling, Eq. (1–20),

$$q_c = \bar{h}A_s(T_s - T_F) \tag{2–90}$$

suggests that q_c can be increased by increasing $\bar{h}$, $T_s - T_F$, or A_s. As already indicated, $\bar{h}$ is a function of the geometry, fluid properties, and flow rate. The modulation of $\bar{h}$ through the control of these factors provides a means by which q_c can be increased or decreased, as will be discussed in Chaps. 6 through 11. With regard to the effect of $T_s - T_F$ on the rate of heat transfer, difficulties are often encountered in automobile cooling systems in very hot weather because T_F is too high. Concerning the third factor, which is the object of this section, the area of a surface that is exposed to the

(a) Longitudinal fins and spines in a fired heater. Fluid flowing in the annulus is heated by a gas- or oil-fired flame inside the fintube. (Courtesy of Brown Fintube Company, Houston, Texas.)

(b) Fins on a high-power voltage regulator. (Courtesy of RS Components Ltd.)

FIGURE 2–15 Typical applications of extended surfaces.

fluid is often "extended" by the use of fins or spines, as illustrated in Fig. 2–15. Familiar applications of such extended surface heat-transfer devices include automobile radiators, power transistors, and high-voltage electrical transformers. In addition, extended surfaces are commonly used to increase the rate of thermal radiation from surfaces.

Referring to the surface extension of the plane wall illustrated in Fig. 2–16,

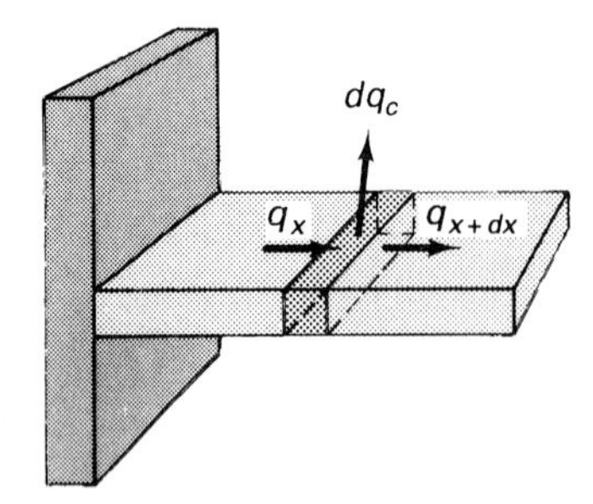

FIGURE 2–16
Fin with rectangular profile.

heat is transferred from the wall into the fin itself by conduction and from the fin surface by convection. Hence, the decrease in the surface convection resistance R_c brought about by the increase in surface area A_s is accompanied by an increase in the conduction resistance R_k. In order for the rate of heat transfer from the wall to be increased by the use of a surface extension, the decrease in R_c must be greater than the increase in R_k. As a matter of fact, the surface resistance must be the controlling factor ($R_k < R_c$, or preferably $R_k << R_c$) in practical fin applications.

In order to provide a guide for the design of extended surface systems, we introduce the *Biot number Bi*,

$$Bi = \frac{\bar{h}\ell}{k} \tag{2–91}$$

The characteristic length ℓ is equal to V/A_s, where V is the fin volume. The Biot number is generally taken as a rule-of-thumb approximation for the ratio between the conduction resistance and the surface resistance; that is,

$$Bi \simeq \frac{R_k}{R_c} \tag{2–92}$$

Because R_k must be considerably smaller than R_c for a fin to be effective, we conclude that the Biot number should be small for fin applications. In practice, the following criterion is generally maintained in the design of convective surface extensions:†

$$Bi = \frac{\bar{h}\ell}{k} \lesssim 0.1 \tag{2–93}$$

In keeping with this criterion, we conclude that fins should generally be considered for applications involving small values of $\bar{h}$ and ℓ and large values of k. Referring back to Table 1–3 we see that likely fin applications occur for natural convection, forced convection of gases, and, to a lesser extent, forced convection of liquids. However, because of the very high values of $\bar{h}$ associated with two-phase fluids, fins are generally not useful for applications involving boiling or condensation.

2–6–1 Analysis

Referring to the extended surface shown in Fig. 2–16, the heat transfer is clearly two-dimensional. However, for situations in which the Biot number is much less than unity, the temperature is essentially a function of x alone, such that an approximate one-dimensional analysis of the heat-transfer process can be developed. For values of the Biot number less than 0.1, the use of a one-dimensional analysis generally leads to an error less than about 10%, as will be demonstrated in Chap. 3. However, for values of the Biot number much greater than 0.1, the multidimensionality must be

† An alternative less restrictive criterion given by $Bi \lesssim 1.0$ is preferred by some analysts.

accounted for in the analysis. For such situations in which the Biot number is not small, the concept of using fins to increase the heat transfer must be questioned.

Assuming that the temperature distribution in the extended surface shown in Fig. 2–16 is essentially one-dimensional, the first law of thermodynamics is applied to the differential volume $A\ dx$, with the following result:

Step 1
$$q_x = q_{x+dx} + dq_c \tag{2–94}$$

Step 2
$$\frac{dq_x}{dx}\,dx + dq_c = 0 \tag{2–95}$$

where q_c represents the rate of heat convected from the fin.

Whereas q_x is given by the one-dimensional Fourier law of conduction, Eq. (1–10), dq_c requires the use of the *general Newton law of cooling*, which is given by Eq. (1–21),

$$dq_c = h_x\,dA_s\,(T_s - T_F) \tag{2–96}$$

where h_x is the *local* coefficient of heat transfer. Incidentally, the integration of this equation for situations in which T_s and T_F are uniform gives rise to the simpler form of the Newton law of cooling, Eq. (2–90), where $\bar{h}$ is defined by Eq. (1–22),

$$\bar{h} = \frac{1}{A_s}\int_{A_s} h_x\,dA_s \tag{2–97}$$

Utilizing the one-dimensional Fourier law of conduction and the general Newton law of cooling with $dA_s = p\ dx$, Eq. (2–95) takes the form

Step 3
$$\frac{d}{dx}\left(kA\,\frac{dT}{dx}\right)dx - h_x p\,dx\,(T - T_F) = 0 \tag{2–98}$$

where T_s is set equal to T in this approximate one-dimensional analysis. For our particular application, the cross-sectional area A, perimeter p, and k are uniform such that Eq. (2–98) reduces to

$$\frac{d^2T}{dx^2} - \frac{h_x p}{kA}(T - T_F) = 0 \tag{2–99}$$

For the case in which the base temperature T_0 is specified, one boundary condition is written as

Step 4
$$T = T_0 \qquad \text{at } x = 0 \tag{2–100}$$

The second boundary condition is written by recognizing that the fin loses heat from the tip by convection; that is,

$$-k\frac{dT}{dx} = h_x(T - T_F) \qquad \text{at } x = L \tag{2–101}$$

where the coefficient h_x at the tip is not necessarily equal to the coefficient along the perimeter. For the case in which the fin is very long, the temperature of the fin approaches T_F as x increases, such that Eq. (2–101) reduces to

$$\frac{dT}{dx} = T - T_F = 0 \qquad \text{as } x \to \infty \tag{2–102}$$

Of course, other boundary conditions can be written at $x = 0$ or at $x = L$, depending upon the dictates of the actual problem under consideration.

Equation (2–102) will now be used to obtain predictions for the temperature distribution and heat transfer in a very long fin. With h_x approximated by $\bar{h}$ and with T_F assumed to be uniform, Eq. (2–99) takes the form

$$\frac{d^2\psi}{dx^2} - m^2\psi = 0 \tag{2–103}$$

where $\psi = T - T_F$ and $m^2 = \bar{h}p/(kA)$. Recognizing that an exponential function satisfies Eq. (2–103), the substitution of

$$\psi = Ce^{cx} \tag{2–104}$$

into this differential equation gives

$$c^2 - m^2 = 0 \tag{2–105}$$

Hence, $c = \pm m$, and the solution is

$$\psi = C_1 e^{mx} + C_2 e^{-mx} \tag{2–106}$$

The constants C_1 and C_2 are evaluated on the basis of the boundary conditions. Based on Eq. (2–100), $\psi = T_0 - T_F$ at $x = 0$ such that

$$T_0 - T_F = C_1 + C_2 \tag{2–107}$$

As suggested above, the second boundary condition can be written in the form of Eq. (2–102) for cases in which L is very long. For this situation, $\psi = 0$ as $x \to \infty$ and Eq. (2–106) gives

$$0 = \lim_{x \to \infty} (C_1 e^{mx} + C_2 e^{-mx}) \tag{2–108}$$

This equation requires that C_1 be equal to zero, such that C_2 is equal to $T_0 - T_F$ [from Eq. (2–107)]. Therefore, the solution for this case is

$$\psi = T - T_F = (T_0 - T_F)e^{-mx} \tag{2–109}$$

To obtain the total rate of heat transfer from the fin into the fluid q_F, we perform a lumped energy balance on the entire fin. It follows that $q_F = q_b$ where the rate of heat transfer at the base q_b is obtained from the Fourier law of conduction,

$$q_b = -kA \left.\frac{dT}{dx}\right|_0 \tag{2–110}$$

With the temperature distribution given by Eq. (2–109), our prediction for q_F becomes

$$q_F = q_b = \sqrt{\bar{h}pkA}\,(T_0 - T_F) \tag{2–111}$$

Parenthetically, this same result can also be obtained by equating q_F to the total rate of heat convected from the surface,

$$q_F = \int dq_c = \int_0^\infty h_x p(T - T_F)\, dx \tag{2–112}$$

where h_x is again approximated by $\bar{h}$. Substituting for $T - T_F$ and integrating, we arrive at Eq. (2–111).

To determine the minimum fin length for which this solution applies, we merely require that T be approximately equal to T_F at x equal to L; that is,

$$\frac{T_L - T_F}{T_0 - T_F} = e^{-mL} < \epsilon \tag{2–113}$$

where ϵ is a small number. With ϵ equal to 0.01, mL must be greater than a value of about 4.6 in order for our analysis to be reasonable.

For the case in which mL is significantly less than 4.6, the heat transfer through the tip can be accounted for by the use of Eq. (2–101). The use of this more general boundary condition for "short" convecting fins gives rise to an expression for the temperature distribution of the form

$$\frac{T - T_F}{T_0 - T_F} = \frac{\cosh\,[m(L - x)] + [\bar{h}/(km)]\sinh\,[m(L - x)]}{\cosh\,(mL) + [\bar{h}/(km)]\sinh\,(mL)} \tag{2–114}$$

To obtain q_F for this type of fin, we apply the Fourier law of conduction as follows:

$$\begin{aligned} q_F = q_b &= -kA\left.\frac{dT}{dx}\right|_0 \\ &= -kA\left\{\frac{-m\sinh\,(mL) - m[\bar{h}/(km)]\cosh\,(mL)}{\cosh\,(mL) + [\bar{h}/(km)]\sinh\,(mL)}\right\}(T_0 - T_F) \\ &= \sqrt{\bar{h}pkA}\left\{\frac{\sinh\,(mL) + [\bar{h}/(km)]\cosh\,(mL)}{\cosh\,(mL) + [\bar{h}/(km)]\sinh\,(mL)}\right\}(T_0 - T_F) \end{aligned} \tag{2–115}$$

Other types of boundary conditions are sometimes encountered in fin applications. For example, for a fin with insulated tip, T and q_F are given by

$$\frac{T - T_F}{T_0 - T_F} = \frac{\cosh\,[m(L - x)]}{\cosh\,(mL)} \tag{2–116}$$

and

$$q_F = \sqrt{\bar{h}pkA}\tanh\,(mL)\,(T_0 - T_F) \tag{2–117}$$

Expressions can also be obtained for T and q_b (or q_F) for fins with specified tip temperature; that is,

$$T - T_F = (T_0 - T_F)\frac{\sinh\,[m(L - x)]}{\sinh\,(mL)} + (T_1 - T_F)\frac{\sinh\,(mx)}{\sinh\,(mL)} \tag{2–118}$$

and

$$q_b = \frac{(T_0 - T_F)\sqrt{\bar{h}pkA}}{\sinh(mL)}\left[\cosh(mL) - \frac{T_1 - T_F}{T_0 - T_F}\right] \quad (2\text{–}119)$$

where $q_F \neq q_b$ (see Table 2–2 for q_F).

2-6-2 Fin Resistance

To express the rate of heat transfer from a fin in terms of a thermal resistance, we write

$$q_F = \frac{T_0 - T_F}{R_F} \quad (2\text{–}120)$$

For example, by introducing Eq. (2–111), the thermal resistance for a very long fin with small Biot number and negligible radiation effects is

$$R_F = \frac{1}{\sqrt{\bar{h}pkA}} \quad (2\text{–}121)$$

The thermal resistances of several standard fin geometries are summarized in Table 2–2. It should be noted that the manufacturers of fin units for electronic devices and other systems generally provide a rating for the thermal resistance of the unit, which accounts for conduction, natural convection, and radiation effects. Table 2–3 gives the thermal resistance rating for several standard anodized aluminum heat-sink fin units for electronic applications.

2-6-3 Fin Efficiency

Traditionally, the rate of heat transfer from surface extensions is generally presented in the literature in terms of the *fin efficiency* η_F. Fin efficiency is defined by

$$\eta_F = \frac{q_F}{q_{\max}} = \frac{q_F}{\bar{h}A_F(T_0 - T_F)} \quad (2\text{–}122)$$

where $q_{\max}$ is the rate of heat transfer for the idealistic situation in which the Biot number is equal to zero (i.e., $R_k \simeq 0$) and the entire surface area of the fin A_F is at the base temperature T_0. Thus, q_F is expressed in terms of η_F by

$$q_F = \eta_F \bar{h} A_F (T_0 - T_F) \quad (2\text{–}123)$$

The fin efficiency is easily expressed in terms of the fin resistance by writing

$$\eta_F = \frac{T_0 - T_F}{R_F}\,\frac{1}{\bar{h}A_F(T_0 - T_F)} = \frac{1}{R_F\bar{h}A_F}$$

Thus

$$R_F = \frac{1}{\eta_F \bar{h} A_F} \quad (2\text{–}124)$$

TABLE 2–2 Thermal resistance: Fins with small Biot number†

System	*Thermal resistance R_F and/or R_b*	*Comment*
Fin with uniform cross section: Very long with $Bi < 0.1$ (T_0; T_F and $\bar{h}$ known)	$\dfrac{1}{\sqrt{\bar{h}pkA}}$	$R_F = R_b$
Fin with uniform cross section: Insulated tip with $Bi < 0.1$ (T_0; $q_c = 0$; T_F and $\bar{h}$ known)	$\dfrac{1}{\sqrt{\bar{h}pkA}\tanh(mL)}$	$R_F = R_b$
Fin with uniform cross section: Tip at T_1 with $Bi < 0.1$ (T_0; T_1; T_F and $\bar{h}$ known)	$\dfrac{\sinh(mL)}{\sqrt{\bar{h}pkA}\,[\cosh(mL) - (T_1 - T_F)/(T_0 - T_F)]}$	R_b
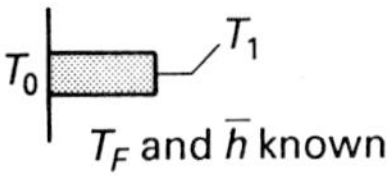	$\dfrac{\sinh(mL)}{\sqrt{\bar{h}pkA}\,[\cosh(mL) - 1][1 + (T_1 - T_F)/(T_0 - T_F)]}$	R_F
Fin with uniform cross section: Convection from tip with $Bi < 0.1$ (T_0; $q_x = q_c$; T_F and $\bar{h}$ known)	$\dfrac{1}{\sqrt{\bar{h}pkA}\,\dfrac{\sinh(mL) + [\bar{h}/(mk)]\cosh(mL)}{\cosh(mL) + [\bar{h}/(mk)]\sinh(mL)}}$	$R_F = R_b$
Blackbody fin with convection and radiation and uniform cross section: Very long with $Bi < 0.1$ (T_0; $T_F = T_R = T_\infty$; $\bar{h}$ and F_{s-R} known)	$\left(\bar{h}pkA + \dfrac{2}{5}kA\sigma pF_{s-R}\dfrac{T_0^5 - 5T_0T_\infty^4 + 4T_\infty^5}{(T_0 - T_\infty)^2}\right)^{-1/2}$	$R_F = R_b$

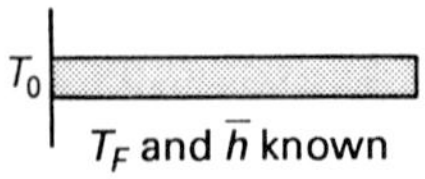

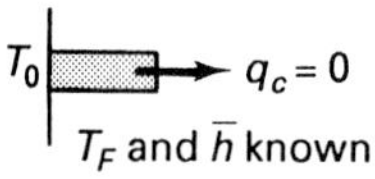

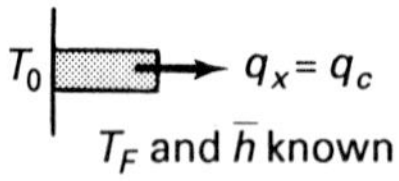

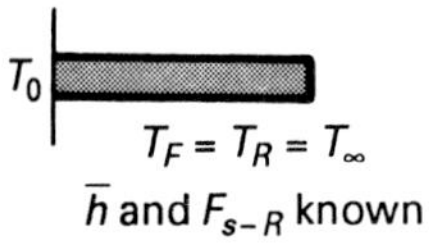

† $m^2 = \bar{h}p/(kA)$; $q_b = (T_0 - T_F)/R_b$.

TABLE 2–3 Thermal resistance for natural convection and radiation from anodized aluminum fin heat-sink units†

Description	Thermal resistance R_F
 RS 401–778 Predrilled to accept T0–3 semiconductor case. 44.5 mm × 31.7 mm × 13.7 mm	14 °C/W
 RS 401–863 Predrilled to accept variety of plastic packaged semiconductor cases. 30 mm × 25 mm × 12.5 mm	19 °C/W
 RS 401–964 Predrilled to accept variety of plastic packaged semiconductor cases. 38 mm × 27 mm × 22.5 mm	10.5 °C/W
 RS 401–497 100 mm length, overall cross section 64.5 mm × 15 mm	4 °C/W (with fins vertical)
 RS 401–403 100 mm length, overall cross section 123.8 mm × 26.7 mm	2.1 °C/W (with fins vertical)
 RS 401–807 152 mm length, overall cross section 130 mm × 32 mm	1.1 °C/W (with fins vertical)
 RS 401–958 115 mm length, overall cross section 120 mm × 120 mm	0.5 °C/W (with fins vertical)

† Courtesy of RS Components Ltd.

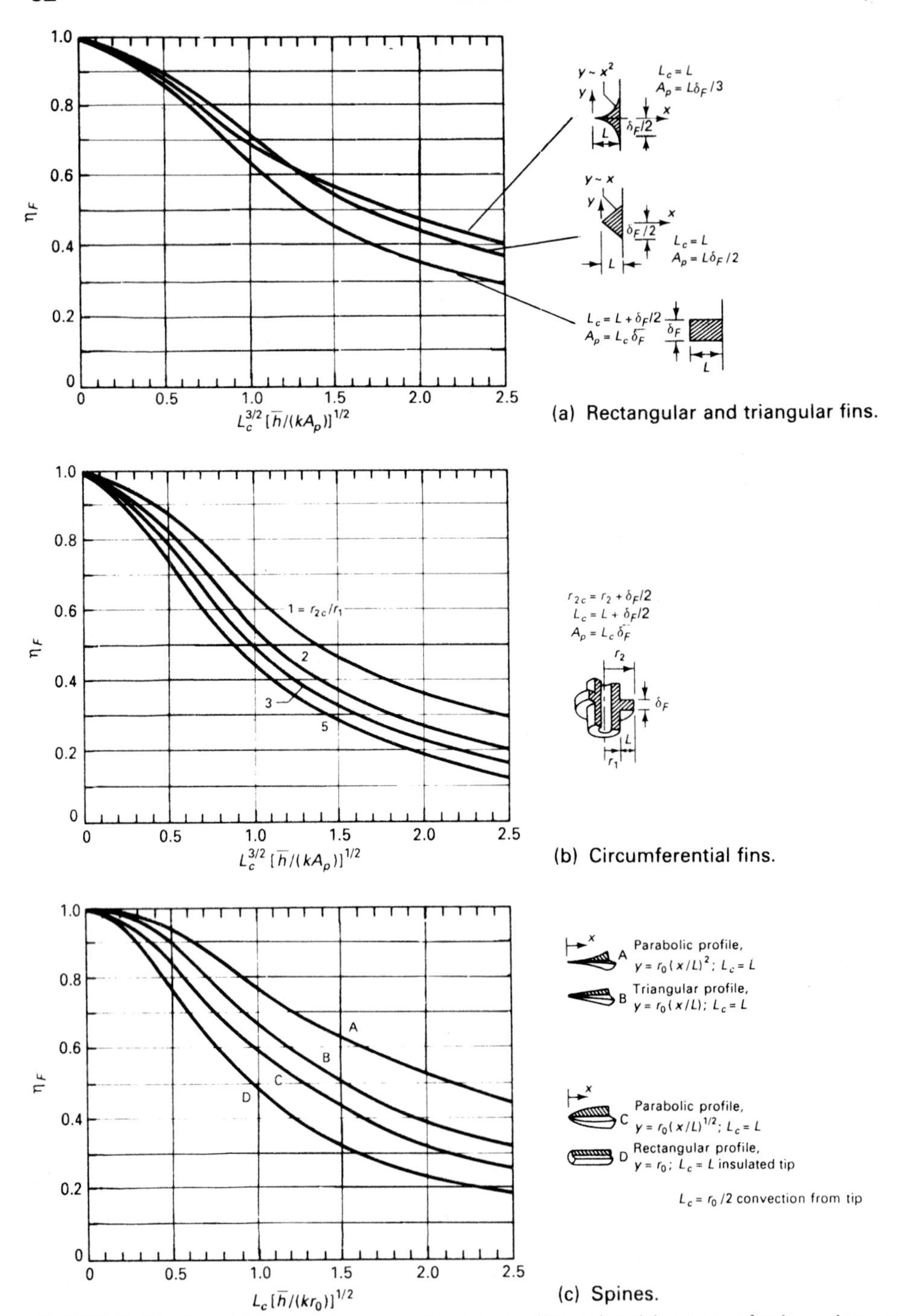

FIGURE 2-17 Fin efficiencies for approximate one-dimensional heat transfer in various extended surfaces for small Biot number (Gardner [7]).

Relations for η_F are given by Schneider [6] and Gardner [7] for a number of fin geometries. Fin efficiencies are shown in Fig. 2–17 for several standard-type fins. Notice that the parabolic and triangular profiles are more efficient than rectangular profiles. What is more, these nonuniform profiles contain less material than fins with rectangular profiles, which cuts down on weight and cost. Nevertheless, fins with uniform cross-sectional area are commonly employed in heat exchangers and other systems.

It should also be observed that η_F decreases with fin length. This happens because the temperature of a fin approaches T_F as x increases. Of course, the local rate of heat transfer dq_c convected from a fin decreases as $T - T_F$ falls. Since the Biot number Bi is proportional to $\bar{h}/k$, these figures also indicate the importance of restricting the use of fins to applications in which the Biot number is small.

EXAMPLE 2–12

A 1-cm-diameter, 3-cm-long carbon (1%) steel fin transfers heat from a wall at 200°C to a fluid at 25°C with $\bar{h}$ = 120 W/(m² °C). Determine the rate of heat transfer from the fin for the case in which the tip is insulated and thermal radiation effects are negligible.

Solution

Objective Determine q_F.

Schematic Fin with insulated tip.

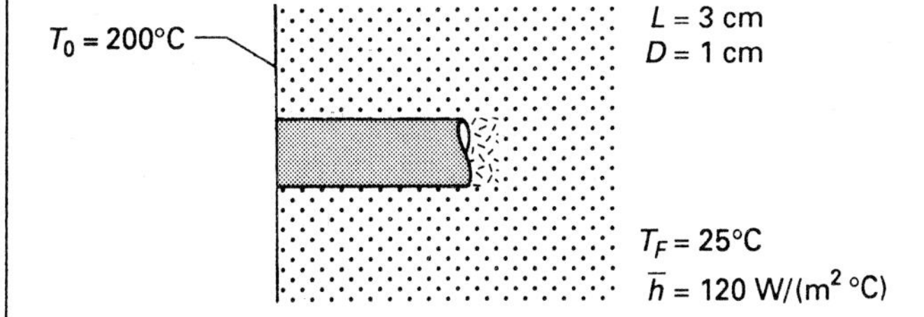

Assumptions/Conditions

steady-state
one-dimensional fin
negligible thermal radiation

Properties Carbon steel (1% C) at room temperature (Table A–C–1): k = 43 W/(m °C).

Analysis We first compute the Biot number by writing

$$Bi = \frac{\bar{h}\ell}{k} = \frac{\bar{h}}{k}\frac{D}{4} = 0.00698$$

Because the Biot number is much less than 0.1, we are justified in using an approximate one-dimensional analysis.

For this situation in which the tip is insulated, q_F is given by Eq. (2–117),

$$q_F = \sqrt{\bar{h}pkA}\tanh(mL)(T_0 - T_F)$$

where

$$m = \sqrt{\frac{\bar{h}p}{kA}} = 33.4/\text{m}$$

and

$$\sqrt{\bar{h}pkA} = 0.113\ \text{W/°C}$$

Following through with the calculation, we obtain

$$q_F = 15.1\ \text{W}$$

2–6–4 Finned Systems

Referring to Fig. 2–18, to characterize the heat transfer from surfaces with one or more fins, we distinguish between the surface area of an individual fin A_F and the area of the prime (i.e., unfinned) surface A_p adjacent to the fin. The rate of heat convected per fin from the combined fin and prime surfaces is expressed in terms of the fin efficiency η_F by

$$q_o = \bar{h}A_p(T_s - T_F) + \eta_F\bar{h}A_F(T_s - T_F) = \bar{h}(A_p + \eta_F A_F)(T_s - T_F) \qquad (2\text{–}125)$$

where the coefficient of heat transfer over the fin and prime surfaces is approximated by the local mean coefficient of heat transfer $\bar{h}$. With the combined areas of the fin and prime surfaces represented by A_o $(= A_p + A_F)$, Eq. (2–125) is rearranged to obtain

$$q_o = \eta_o\bar{h}A_o(T_s - T_F) \qquad (2\text{–}126)$$

where the *net surface efficiency* η_o† is

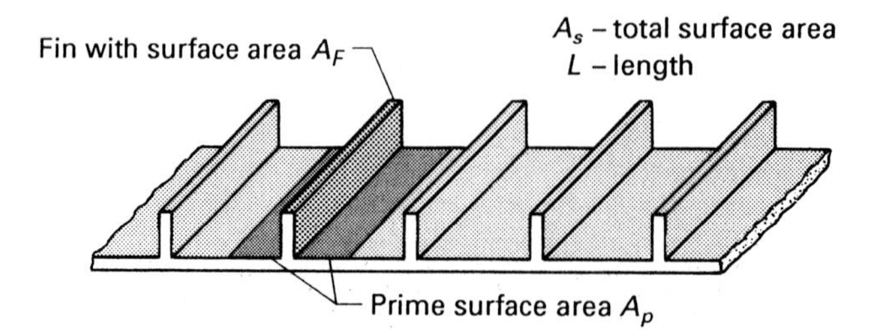

FIGURE 2–18 Representative fin system with rectangular fins.

† η_o is also commonly referred to as the *temperature effectiveness*.

$$\eta_o = \frac{A_p + A_F\eta_F}{A_o} = 1 - \frac{A_F}{A_o}(1 - \eta_F) \tag{2-127}$$

and A_F/A_o is referred to as the *finned area fraction*. It follows that the resistance to convection heat transfer for a finned surface can be represented by

$$R_o = \frac{T_s - T_F}{q_o} = \frac{1}{\eta_o \bar{h} A_o} \tag{2-128}$$

per fin.

For applications in which the local mean coefficient of heat transfer $\bar{h}$, the net surface efficiency η_o, and temperatures T_s and T_F are essentially uniform over the entire surface area A_s of a finned surface with N uniformly spaced fins, the total rate of heat transfer q_c is simply written as

$$q_c = Nq_o = N[\eta_o \bar{h} A_o(T_s - T_F)] = \eta_o \bar{h} A_s(T_s - T_F) \tag{2-129}$$

such that the thermal resistance becomes

$$R_c = \frac{T_s - T_F}{q_c} = \frac{1}{\eta_o \bar{h} A_s} \tag{2-130}$$

On the other hand, for situations in which one or more of the four parameters $\bar{h}$, η_o, T_s, and T_F change along the surface, as in the case of finned tube heat exchangers, more general relationships should be employed. To deal with this problem we represent the finned surface by an effective unfinned surface with equivalent surface area A_s and effective perimeter $p = A_s/L$, and simply approximate the local heat flux q_c'' ($\equiv dq_c/dA_s$) by the local heat flux per fin q_o''; that is,

$$q_c'' = \frac{dq_c}{dA_s} = q_o'' = \frac{q_o}{A_o} \tag{2-131}$$

or

$$dq_c = q_o''\, dA_s = \eta_o \bar{h}(T_s - T_F)\, dA_s \tag{2-132}$$

Following this practical approach, the local convective resistance R_c' for a finned surface can be approximated by

$$R_c' = \frac{T_s - T_F}{dq_c} = \frac{1}{\eta_o \bar{h}\, dA_s} \tag{2-133}$$

This relationship is useful in the analysis of finned tube banks and finned tube heat exchangers.

EXAMPLE 2–13

The power transistor of Example 2–5 is to be mounted on an RS 401–778 anodized aluminum heat-sink unit, as shown in Fig. E2–13a. Determine the maximum power that can be dissipated safely by this transistor for a case temperature of 80°C, ambient temperature of 25°C, and $\bar{h}$ = 10 W/(m^2 °C), assuming that the thermal conductivity of anodized aluminum is about 200 W/(m °C). (This unit provides two fins plus an additional 800 mm^2 of prime surface area.)

FIGURE E2–13a
Power transistor mounted on RS 401–778 heat sink. (Approximate fin dimensions: L = 12 mm, p = 64 mm, and A = 31 mm^2.)

Solution

Objective Determine the maximum operating power of the transistor.

Schematic Transistor/heat sink unit.

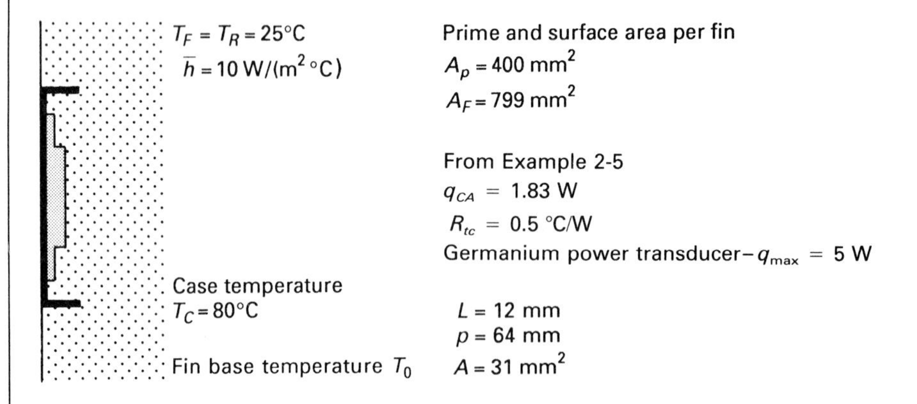

Assumptions/Conditions

steady-state
short one-dimensional fins
uniform properties
convection and blackbody thermal radiation from heat sink

Properties Anodized aluminum: k = 200 W/(m °C).

Analysis Neglecting radiation effects for the moment and focusing attention on one of the two fins, we have

$$m = \sqrt{\frac{\bar{h}p}{kA}} = \sqrt{\frac{[10\ \text{W/(m}^2\ °\text{C)}](0.064\ \text{m})}{[200\ \text{W/(m °C)}](31 \times 10^{-6}\ \text{m}^2)}} = 10.2/\text{m}$$

$$mL = \frac{10.2}{\text{m}}(0.012\ \text{m}) = 0.122$$

Because mL is much less than 4.6, we should account for the heat loss through the tip. Therefore, we employ Eq. (2–115),

$$q_F = \sqrt{\bar{h}pkA}\left\{\frac{\sinh(mL) + [\bar{h}/(km)]\cosh(mL)}{\cosh(mL) + [\bar{h}/(km)]\sinh(mL)}\right\}(T_0 - T_F)$$

where

$$\sqrt{\bar{h}pkA} = \left[10\frac{\text{W}}{\text{m}^2\ °\text{C}}(0.064\ \text{m})\left(200\frac{\text{W}}{\text{m °C}}\right)(31 \times 10^{-6}\ \text{m}^2)\right]^{1/2} = 0.063$$

and

$$\frac{\bar{h}}{km} = \frac{10\ \text{W/(m}^2\ °\text{C)}}{[200\ \text{W/(m °C)}](10.2/\text{m})} = 0.0049$$

It follows that

$$q_F = 0.00795(T_0 - T_F)\ \text{W/°C}$$

To calculate the fin efficiency, we write

$$\eta_F = \frac{q_F}{q_{\max}} = \frac{0.00795\ (T_0 - T_F)\ \text{W/°C}}{\bar{h}A_F(T_0 - T_F)}$$

Setting the surface area of each fin A_F equal to 799 mm^2 and $\bar{h} = 10$ W/(m^2 °C), the fin efficiency is found to be $\eta_F = 99.5\%$.

After mounting the power transducer, the area A_p of the primary surface of the fin unit, which is exposed to convective cooling, is approximately 400 mm^2 per fin. Therefore, the net surface efficiency η_o can be computed as follows:

$$A_o = A_p + A_F = 400\ \text{mm}^2 + 799\ \text{mm}^2 = 1199\ \text{mm}^2$$

$$\eta_o = 1 - \frac{A_F}{A_o}(1 - \eta_F) = 1 - \frac{779}{1199}(1 - 0.995) = 99.7\%$$

Substituting this result into Eq. (2–129), the total rate of heat convected from the heat-sink fin unit with surface area $A_s = 2A_o = 2398$ mm^2 is

$$q_c = \eta_o \bar{h} A_s (T_0 - T_F) = 0.997\left(10 \frac{\text{W}}{\text{m}^2\ °\text{C}}\right)(0.0024\ \text{m}^2)(T_0 - T_F)$$

$$= 0.0239(T_0 - T_F)\ \text{W/°C}$$

Thus, the thermal resistance to convection for the heat-sink unit is

$$R_{HS} = \frac{1}{0.0239\ \text{W/°C}} = 41.8\ °\text{C/W}$$

(Note that R_{HS} for this heat-sink fin unit is about 49% lower than for the simple flat-plate frame system of Example 2–5.)

Assuming that the power transistor is properly attached to the heat sink (see Example 2–5), the thermal circuit for heat transfer through the heat sink is shown in Fig. E2–13b. It follows that the power dissipated through the heat sink is

$$q_{HS} = \frac{T_C - T_F}{R_{tc} + R_{HS}} = \frac{80°\text{C} - 25°\text{C}}{0.5\ °\text{C/W} + 41.8\ °\text{C/W}} = 1.3\ \text{W}$$

We also note that this thermal circuit indicates a calculation for base fin temperature T_0 given by

$$T_0 = R_c q_{HS} + T_F = 41.8 \frac{°\text{C}}{\text{W}}(1.3\ \text{W}) + 25°\text{C} = 79.3°\text{C}$$

Combining the rates of heat transfer from the case and from the heat-sink fin unit, we obtain a total power-dissipation rate of

$$q = q_{CA} + q_{HS} = 1.83\ \text{W} + 1.3\ \text{W} = 3.13\ \text{W}$$

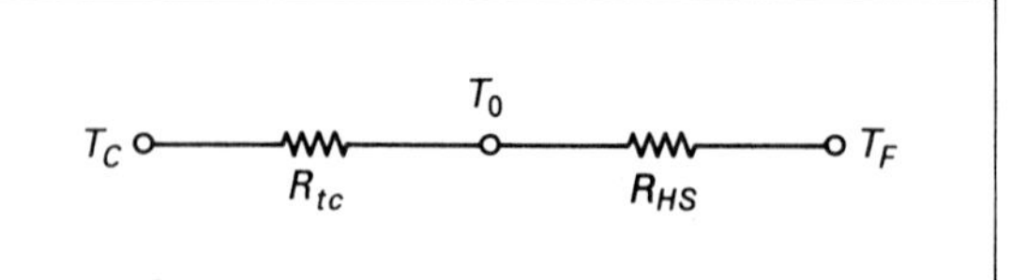

FIGURE E2–13b
Thermal circuit for case to heat-sink fin unit with negligible thermal radiation.

Because we have neglected the radiation losses, this will be a conservative estimate. The combined effects of convection and radiation heat transfer from a fin such as this can be fairly easily analyzed by means of the numerical techniques introduced in Chap. 4. However, because we are dealing with a very short fin with fin efficiency η_F approximately equal to unity, the total rate of heat transferred from the fin by convection and radiation can be closely approximated by assuming that the temperature over the entire surface of the fin is equal to the base temperature T_0. (This conclusion is reinforced by the numerical analysis of a similar convecting and

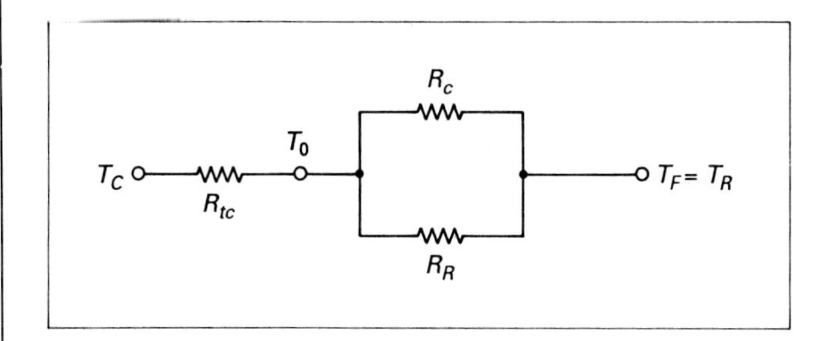

FIGURE E2–13c
Thermal circuit for case to heat-sink fin unit.

radiating fin in Example 4–7.) To implement this simplifying assumption, we sketch the thermal circuit shown in Fig. E2–13c, with R_c and R_R represented by

$$R_c = \frac{1}{\bar{h}A_s} = \frac{1}{[10\ \text{W/(m}^2\ ^\circ\text{C)}](0.0024\ \text{m}^2)} = 41.7\ ^\circ\text{C/W}$$

$$R_R = \frac{1}{\sigma A_s F_{s-R}(T_0 + T_R)(T_0^2 + T_R^2)}$$

Setting $F_{s-R} = 1$ and assuming that $T_0 \simeq T_C = 80°\text{C}$, we obtain $R_R \simeq 52.9$ °C/W. The resulting thermal resistance for the heat-sink fin unit is

$$R_{HS} = \frac{1}{1/R_c + 1/R_R} \simeq \frac{1}{1/41.7 + 1/52.9} \frac{^\circ\text{C}}{\text{W}} = 23.3\ ^\circ\text{C/W}$$

Thus,

$$q_{HS} \simeq \frac{80^\circ\text{C} - 25^\circ\text{C}}{0.5\ ^\circ\text{C/W} + 23.3\ ^\circ\text{C/W}} = 2.31\ \text{W}$$

such that the thermal limit on the total power dissipation from the transistor/heat-sink unit is estimated to be

$$q = q_{CA} + q_{HS} \simeq 1.83\ \text{W} + 2.31\ \text{W} = 4.14\ \text{W}$$

Therefore, we conclude that the power transducer can be operated safely with this heat-sink unit at up to within about 17% of its maximum power rating of 5 W.

Finally, we note that for design purposes the manufacturers of heat-sink fin units generally specify the approximate thermal resistance for combined natural convection and radiation. Referring to Table 2–3, we find that the resistance of the fin unit in this example is rated by the manufacturer at 14 °C/W. But it should be noted that this rating is based on laboratory measurements in free air (at about 25°C) at an unspecified transistor frame temperature T_C. In actuality, R_{HS} is strongly dependent upon T_C. Therefore, this value of R_{HS} should be used as a rough estimate in design work. Replacing R_c in Fig. E2–13b by this value, we obtain

$$q_{HS} \simeq \frac{80^\circ\text{C} - 25^\circ\text{C}}{14.5\ ^\circ\text{C/W}} = 3.79\ \text{W}$$

and

$$q = q_T + q_{HS} \simeq 1.83 \text{ W} + 3.79 \text{ W} = 5.62 \text{ W}$$

EXAMPLE 2–14

Determine the rate of heat transfer from the long anodized aluminum fin shown in Fig. E2–14. The base temperature T_0 is 80°C, the air and surrounding walls are 25°C, and the coefficient of heat transfer due to natural convection cooling is 10 W/(m² °C).

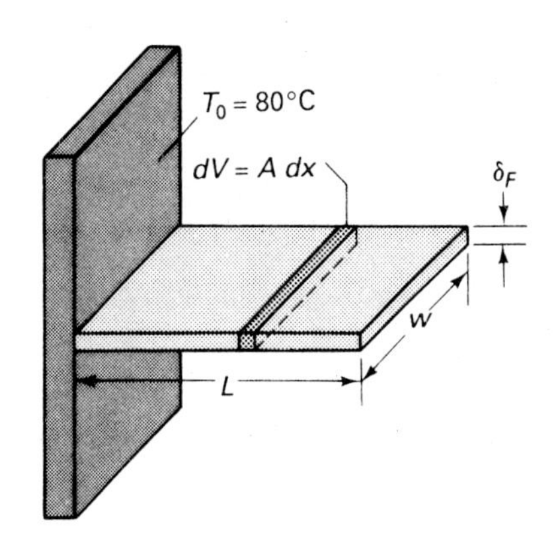

Solution

Objective Determine q_F from this convecting and radiating fin.

Assumptions/Conditions

steady-state
long one-dimensional fin
uniform properties
natural convection and blackbody thermal radiation cooling
$F_{s-R} = 1$ for fin to surroundings

Properties Anodized aluminum (Example 2–13): $k = 200$ W/(m °C).

Analysis The heat transfer in this fin can be approximated by a one-dimensional analysis if the Biot number R_k/R_s is very small. Assuming blackbody conditions, the Biot number for combined convection and radiation is approximated by

$$Bi = \frac{R_k}{R_s} = \frac{\ell}{k}\,[\bar{h} + \sigma F_{s-R}(T_s + T_R)(T_s^2 + T_R^2)]$$

where ℓ is set equal to V/A_s for the fin and $F_{s-R} = 1$. As a conservative measure, T_s and T_R are both set equal to 353 K (= 80°C). The Biot number is then found to be 4.32×10^{-5}, such that a one-dimensional analysis can be safely used.

Referring to the lumped-differential element shown in Fig. E2–14, the application of the first law of thermodynamics gives

$$q_x = q_{x+dx} + dq_c + dq_R$$

It follows that

$$0 = \frac{dq_x}{dx}\,dx + dq_c + dq_R \tag{a}$$

Because the temperature along the fin is a function of x, we utilize the general Newton law of cooling as well as a generalized form of Eq. (1–18); that is,

$$dq_R = \sigma F_{dA_s - A_R}\, dA_s\,(T_s^4 - T_R^4)$$

where $F_{dA_s-A_R}$ represents the fraction of radiant energy leaving dA_s that reaches A_R. Substituting these particular laws into Eq. (a), we obtain

$$\frac{d}{dx}\left(kA\,\frac{dT}{dx}\right) dx = h_x p\,dx\,(T - T_F) + \sigma p\,dx\,F_{dA_s-A_R}\,(T^4 - T_R^4)$$

or

$$\frac{d^2T}{dx^2} = \frac{h_x p}{kA}(T - T_F) + \frac{\sigma p F_{dA_s-A_R}}{kA}(T^4 - T_R^4) = m^2(T - T_F) + m_R^2(T^4 - T_R^4)$$

where h_x is approximated by $\bar{h}$, $F_{dA_s-A_R}$ is approximated by F_{s-R}, $m^2 = \bar{h}p/(kA)$, and $m_R^2 = \sigma p F_{s-R}/(kA)$. The boundary condition at the base is

$$T = T_0 \qquad \text{at } x = 0$$

For the case in which T_F and T_R are both equal to T_∞ and the fin is very long we also have

$$T = T_\infty \quad \text{or} \quad \frac{dT}{dx} = 0 \qquad \text{as } x \to \infty \tag{b}$$

The simplest way to solve this nonlinear system of equations for T is to use the numerical finite-difference approach introduced in Chap. 4. However, we can obtain an analytical solution for the temperature gradient by making the substitution

$$\psi = \frac{dT}{dx} \tag{c}$$

This puts our nonlinear differential equation into the form

$$\frac{d\psi}{dx} = m^2(T - T_\infty) + m_R^2(T^4 - T_\infty^4) \tag{d}$$

Based on Eq. (c), $dx = dT/\psi$. Therefore, Eq. (d) can be written as

$$\psi \frac{d\psi}{dT} = m^2(T - T_\infty) + m_R^2(T^4 - T_\infty^4)$$

We now separate the variables and integrate to obtain

$$\frac{\psi^2}{2} = m^2\left(\frac{T^2}{2} - T_\infty T\right) + m_R^2\left(\frac{T^5}{5} - T_\infty^4 T\right) + C_1$$

$$\psi = \frac{dT}{dx} = \pm\sqrt{2}\left[m^2\left(\frac{T^2}{2} - T_\infty T\right) + m_R^2\left(\frac{T^5}{5} - T_\infty^4 T\right) + C_1\right]^{1/2} \tag{e}$$

where the negative sign is retained because the gradient is known to be negative for the case in which $T_0 > T$.

The constant C_1 can be evaluated by introducing the boundary condition given by Eq. (b); that is,

$$\left.\frac{dT}{dx}\right|_\infty = 0 = \sqrt{2}\left[m^2\left(\frac{T_\infty^2}{2} - T_\infty^2\right) + m_R^2\left(\frac{T_\infty^5}{5} - T_\infty^5\right) + C_1\right]^{1/2}$$

where $T(\infty) = T_\infty$. Hence, C_1 is given by

$$C_1 = m^2\frac{T_\infty^2}{2} + \frac{4}{5}m_R^2 T_\infty^5$$

and Eq. (e) becomes

$$\frac{dT}{dx} = -\sqrt{2}\left[m^2\left(\frac{T^2}{2} - T_\infty T + \frac{T_\infty^2}{2}\right) + m_R^2\left(\frac{T^5}{5} - T_\infty^4 T + \frac{4}{5}T_\infty^5\right)\right]^{1/2}$$

An expression can now be obtained for the rate of heat transfer from the fin by writing

$$q_F = -kA\left.\frac{dT}{dx}\right|_0$$

$$= kA\sqrt{2}\left[\frac{m^2}{2}(T_0^2 - 2T_\infty T_0 + T_\infty^2) + \frac{m_R^2}{5}(T_0^5 - 5T_\infty^4 T_0 + 4T_\infty^5)\right]^{1/2}$$

$$= \left[\bar{h}pkA(T_0 - T_\infty)^2 + \frac{2}{5}kA\sigma pF_{s-R}(T_0^5 - 5T_\infty^4 T_0 + 4T_\infty^5)\right]^{1/2}$$

or

$$q_F = \sqrt{q_c^2 + q_R^2}$$

where

$$q_c = \sqrt{\bar{h}pkA}\,(T_0 - T_\infty)$$

and

$$q_R = \left[\frac{2}{5} kA\sigma pF_{s-R}(T_0^5 - 5T_\infty^4 T_0 + 4T_\infty^5)\right]^{1/2}$$

Calculating q_c and q_R, we have

$$q_c = \left[10\frac{\text{W}}{\text{m}^2\,°\text{C}}(0.064\text{ m})\left(200\frac{\text{W}}{\text{m }°\text{C}}\right)(31 \times 10^{-6}\text{ m}^2)\right]^{1/2}(353°\text{C} - 298°\text{C})$$

$$= 3.46\text{ W}$$

$$q_R = \left\{\frac{2}{5}\left(200\frac{\text{W}}{\text{m }°\text{C}}\right)(31 \times 10^{-6}\text{ m}^2)\left(5.67 \times 10^{-8}\frac{\text{W}}{\text{m}^2\text{ K}^4}\right)(0.064\text{ m})(1)\right.$$

$$\left.\times\,[(353\text{ K})^5 - 5(298\text{ K})^4(353\text{ K}) + 4(298\text{ K})^5]\right\}^{1/2} = 2.94\text{ W}$$

Thus, the total rate of heat transfer from the fin is

$$q_F = \sqrt{(3.46\text{ W})^2 + (2.94\text{ W})^2} = 4.54\text{ W}$$

The fact that q_F is about 31% greater than q_c reinforces our earlier conclusion that radiation can play a very significant role in heat transfer from fins with near blackbody surfaces.

2-7 UNSTEADY HEAT-TRANSFER SYSTEMS

We now turn our attention to unsteady heat-transfer processes such as the cooling of a billet, which is illustrated in Fig. 2-19. The heat transfer in unsteady systems like this is actually multidimensional because the temperature within the body is a function

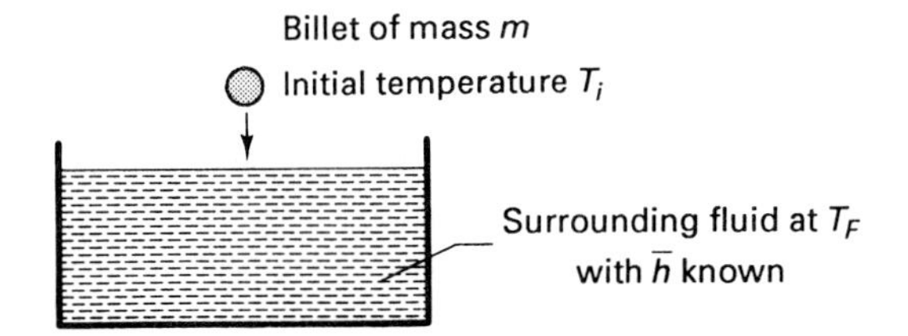

FIGURE 2-19
Convective cooling of a billet initially at temperature T_i, which is dropped into a bath at temperature T_F.

of time t and at least one space dimension. However, for problems such as this that involve convection and/or radiation, approximate lumped analyses can be utilized if the Biot number ($Bi = R_k/R_s$) is small (i.e., $Bi \lesssim 0.1$). Under these circumstances, the variation in temperature with the spatial coordinates will be very slight, such that the temperature can be taken as a function of time alone. Approximate lumped analyses are developed for representative unsteady convection heat-transfer processes in this section. Unsteady heat-transfer systems involving multidimensions and radiation will be considered in Chaps. 3 through 5.

2–7–1 Analysis

We consider the situation illustrated in Fig. 2–19 in which a body of mass m with uniform initial temperature T_i is suddenly exposed to an environmental temperature T_F. Applying the first law of thermodynamics to the lumped volume V, we obtain

$$\Sigma \dot{E}_i = \Sigma \dot{E}_o + \frac{\Delta E_s}{\Delta t} \qquad (2\text{–}134)$$

$$\bar{h}A_s(T - T_F) + \frac{dU}{dt} = 0 \qquad (2\text{–}135)$$

The internal energy U can be expressed in terms of the specific heat at constant volume for this lumped system by

$$c_v = \left.\frac{\partial e}{\partial T}\right|_v = \left.\frac{\partial}{\partial T}\left(\frac{U}{m}\right)\right|_v = \left.\frac{1}{m}\frac{\partial U}{\partial T}\right|_v \qquad (2\text{–}136)$$

Since the volume is essentially constant for heat transfer in solids, dU can be written as

$$dU = mc_v\, dT = \rho V c_v\, dT \qquad (2\text{–}137)$$

Therefore, Eq. (2–135) takes the form

$$\frac{dT}{dt} + \frac{\bar{h}A_s}{\rho V c_v}(T - T_F) = 0 \qquad (2\text{–}138)$$

The initial condition associated with this equation is

$$T = T_i \qquad \text{at } t = 0 \qquad (2\text{–}139)$$

This completes our formulation for situations in which the mass of the surrounding fluid is large and T_F is essentially independent of time. However, for systems in which the mass of the surrounding fluid is not large, the variation in T_F with time must be accounted for. This is done by applying the first law of thermodynamics to the surrounding fluid itself, as illustrated in Example 2–17.

For the case in which T_F is constant, Eq. (2–138) can be written in the form

$$\frac{d\psi}{dt} + \frac{\bar{h}A_s\psi}{\rho V c_v} = 0 \tag{2–140}$$

where $\psi = T - T_F$. With $\bar{h}$ approximated by a constant, this homogeneous first-order equation can be separated and integrated to obtain

$$\int_{\psi_i}^{\psi} \frac{d\psi}{\psi} = \frac{-\bar{h}A_s}{\rho V c_v} \int_0^t dt \tag{2–141}$$

where $\psi_i = T_i - T_F$. Continuing, we have

$$\ln\left(\frac{\psi}{\psi_i}\right) = -\frac{\bar{h}A_s t}{\rho V c_v} \tag{2–142}$$

Hence, the temperature history is represented by

$$\frac{T - T_F}{T_i - T_F} = \exp\left(\frac{-\bar{h}A_s t}{\rho V c_v}\right) \tag{2–143}$$

The rate of heat convected from the surface at any instant t can be written as

$$q_c = \bar{h}A_s(T - T_F) = \bar{h}A_s(T_i - T_F)\exp\left(\frac{-\bar{h}A_s t}{\rho V c_v}\right) \tag{2–144}$$

To obtain the total heat transferred from the surface over the time interval 0 to t, we simply integrate as follows:

$$Q_c = \int_0^t q_c\,dt = \bar{h}A_s(T_i - T_F)\int_0^t \exp\left(\frac{-\bar{h}A_s t}{\rho V c_v}\right) dt$$

$$= \rho V c_v(T_i - T_F)\left[1 - \exp\left(\frac{-\bar{h}A_s t}{\rho V c_v}\right)\right] \tag{2–145}$$

The maximum total amount of energy that can be convected to the fluid is obtained by merely allowing t to become very large; that is,

$$Q_{max} = \rho V c_v(T_i - T_F) \tag{2–146}$$

Notice that Q_{max} is equal to the relative internal energy possessed by the body at time $t = 0$. The dimensionless ratio Q_c/Q_{max} is given by

$$\frac{Q_c}{Q_{max}} = 1 - \exp\left(\frac{-\bar{h}A_s t}{\rho V c_v}\right) \tag{2–147}$$

The quantity $\rho V c_v/(\bar{h}A_s)$ appearing in Eq. (2–143) and associated equations is known as the *thermal time constant* t_c of the system. The thermal time constant provides an indication of the length of time required for a system to approach thermal equilibrium; that is, the smaller the thermal time constant, the quicker the system response. Notice that $(T - T_F)/(T_i - T_F) = 0.368$ for $t = t_c$.

EXAMPLE 2–15

An aluminum ball 0.5 cm in diameter initially at 250°C is dropped into a large tank of fluid (with $T_{\text{sat}} = 100°\text{C}$), which is maintained at 25°C. The mean coefficient of heat transfer is approximately equal to 3000 W/(m^2 °C) for boiling and 250 W/(m^2 °C) for nonboiling. Determine the temperature history of the ball and the instantaneous rate of convection heat transfer.

Solution

Objective Determine the temperature and rate of heat transfer from the ball as a function of time.

Schematic Cooling of aluminum ball in extensive fluid bath.

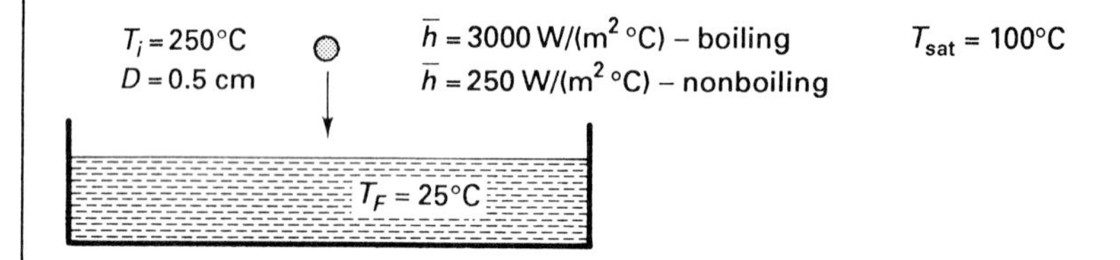

Assumptions/Conditions

unsteady-state
negligible resistance to conduction
constant bath temperature with boiling occurring for ball temperature ≥ 100°C
constant properties of aluminum ball
negligible thermal radiation

Properties Aluminum at room temperature (Table A–C–1): $\rho = 2700$ kg/m^3, $c_v = 0.896$ kJ/(kg °C), $k = 236$ W/(m °C).

Analysis Calculating the Biot number and assuming negligible radiation effects, we find that $Bi \ll 0.1$ for the entire process. Hence, the lumped differential formulation given by Eqs. (2–138) and (2–139) can be used.

The solution to Eq. (2–138) with the initial condition $T(0) = 250°\text{C}$ is given by Eq. (2–143) with $\bar{h} = 3000$ W/(m^2 °C); that is,

$$\frac{T - T_F}{T_i - T_F} = \frac{T - 25°\text{C}}{250°\text{C} - 25°\text{C}} = \exp\left(-\frac{\bar{h}A_s t}{\rho V c_v}\right) = \exp\left(-\frac{1.49t}{\text{s}}\right) \tag{a}$$

This equation applies to the period of time for which $T \geqslant 100°C$. To find the time t_1 at which boiling ceases, T is set equal to 100°C in this equation, with the result

$$t_1 = \frac{s}{1.49} \ln\left(\frac{225}{75}\right) = 0.737 \text{ s}$$

The solution to Eq. (2–138) for the condition $T(t_1) = 100°C$ and $\bar{h} = 250$ W/(m^2 °C) is given by

$$\frac{T - 25°C}{100°C - 25°C} = \exp\left(-0.124 \frac{t - t_1}{s}\right) \tag{b}$$

Equations (a) and (b) are shown in Fig. E2–15.

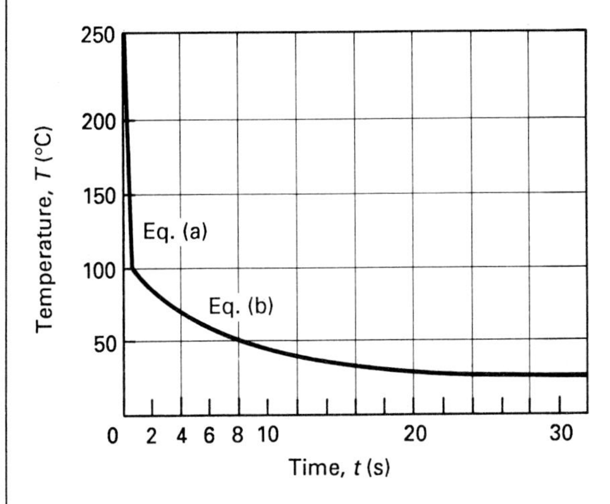

FIGURE E2–15
Temperature history.

To calculate the rate of convection from the surface, we write

$$q_c = \bar{h}A_s(T - T_F)$$

$$q_c'' = 3000 \frac{\text{W}}{\text{m}^2\ °\text{C}} (225°\text{C}) \exp\left(-\frac{1.49t}{s}\right)$$

for $t \leqslant 0.737$ s, and

$$q_c'' = 250 \frac{\text{W}}{\text{m}^2\ °\text{C}} (75°\text{C}) \exp\left(-0.124 \frac{t - 0.737 \text{ s}}{s}\right)$$

for $t \geqslant 0.737$ s.

To estimate the levels of radiation heat transfer in this problem, one can easily calculate the instantaneous rate of thermal radiation emitted by a blackbody ball with temperature given by Eqs. (a) and (b). This calculation indicates that the maximum rate of radiation from the ball is less than 1% of the rate of convection heat transfer.

EXAMPLE 2–16

Electric current of 5 A is suddenly passed through a 1-mm-diameter copper wire initially at temperature 25°C. Develop an expression for the instantaneous temperature of the wire for uniform internal energy generation ($\rho_e = 1.8 \times 10^{-8}\ \Omega$ m), constant surrounding fluid temperature T_F of 25°C, coefficient of heat transfer $\bar{h}$ of 25 W/(m² °C), and negligible radiation effects.

Solution

Objective Determine the temperature of the wire as a function of time.

Schematic Unsteady heat transfer from a wire; $I_e = 5$ A.

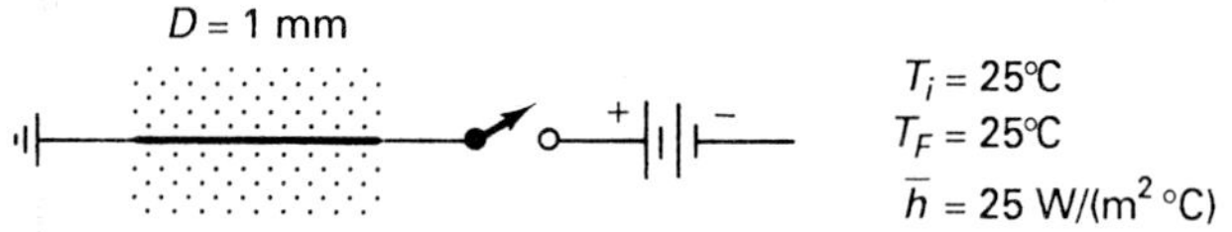

$D = 1$ mm

$T_i = 25$°C
$T_F = 25$°C
$\bar{h} = 25$ W/(m² °C)

Assumptions/Conditions

unsteady-state
constant bath temperature
uniform internal energy generation
constant properties of copper wire
negligible thermal radiation

Properties Copper at room temperature (Table A–C–1): $\rho = 8950$ kg/m³, $c_v =$ 0.383 kJ/(kg °C), $k = 386$ W/(m °C), $\rho_e = 1.8 \times 10^{-8}\ \Omega$ m.

Analysis Because the Biot number for this system is much less than 0.1, an approximate one-dimensional formulation is developed. Utilizing the first law of thermodynamics, we obtain

$$\Sigma\, \dot{E}_i = \Sigma\, \dot{E}_o + \frac{\Delta E_s}{\Delta t}$$

$$\dot{q}_0 V = \bar{h} A_s (T - T_F) + \rho V c_v \frac{dT}{dt}$$

or

$$\frac{d\psi}{dt} + \frac{\bar{h} A_s \psi}{\rho V c_v} - \frac{\dot{q}_0}{\rho c_v} = 0 \tag{a}$$

where $\psi = T - T_F$. The initial condition is

$$\psi(0) = 0$$

To obtain the homogeneous solution ψ_H, we write

$$\frac{d\psi_H}{\psi_H} = -\frac{\bar{h}A_s}{\rho V c_v}\,dt$$

and integrate, with the result

$$\ln\left(\frac{\psi_H}{C_1}\right) = -\frac{\bar{h}A_s t}{\rho V c_v} \qquad \psi_H = C_1 \exp\left(\frac{-\bar{h}A_s t}{\rho V c_v}\right)$$

To obtain the particular solution ψ_p, we assume that

$$\psi_p = C_3 + C_4 t$$

This expression is substituted into Eq. (a) to obtain

$$C_4 + \frac{\bar{h}A_s}{\rho V c_v}(C_3 + C_4 t) = \frac{\dot{q}_0}{\rho c_v}$$

Hence, we see that $C_4 = 0$ and $C_3 = \dot{q}_0 V/(\bar{h}A_s)$; that is,

$$\psi_p = \frac{\dot{q}_0 V}{\bar{h}A_s}$$

Thus, ψ takes the form

$$\psi = C_1 \exp\left(\frac{-\bar{h}A_s t}{\rho V c_v}\right) + \frac{\dot{q}_0 V}{\bar{h}A_s}$$

Utilizing the initial condition,

$$C_1 = -\frac{\dot{q}_0 V}{\bar{h}A_s}$$

and

$$\psi = T - T_F = \frac{\dot{q}_0 V}{\bar{h}A_s}\left[1 - \exp\left(\frac{-\bar{h}A_s t}{\rho V c_v}\right)\right]$$

To compute the rate of internal energy generation per unit volume $\dot{q}_0$, we write

$$\dot{q}_0 = \frac{I_e^2 \rho_e}{A^2} = \frac{(5\text{ A})^2(1.8 \times 10^{-8}\ \Omega\text{ m})}{[\pi(0.001\text{ m})^2/4]^2} = 730\text{ kW/m}^3$$

It follows that

$$\frac{\dot{q}_0 V}{\bar{h}A_s} = \frac{730\text{ kW/m}^3}{25\text{ W/(m}^2\ {}^\circ\text{C)}}\ \frac{0.001\text{ m}}{4} = 7.3^\circ\text{C}$$

$$\frac{\bar{h}A_s}{\rho V c_v} = \frac{25\text{ W/(m}^2\ {}^\circ\text{C)}}{(8950\text{ kg/m}^3)[0.383\text{ kJ/(kg }{}^\circ\text{C)}]}\ \frac{4}{0.001\text{ m}} = 0.0292/\text{s}$$

Hence, the instantaneous temperature of the wire is given by

$$T = 25°\text{C} + 7.3°\text{C}\left[1 - \exp\left(-\frac{0.0292\,t}{\text{s}}\right)\right] \tag{b}$$

This equation is shown in Fig. E2–16. Notice that the wire temperature approaches a steady-state value of 32.3°C after about 160 s.

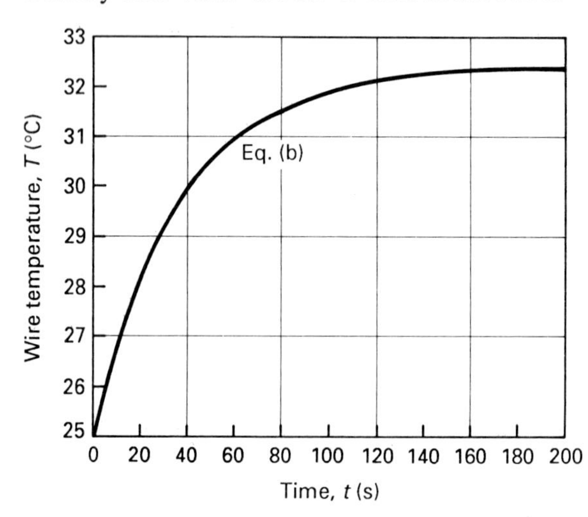

FIGURE E2–16
Temperature history.

The instantaneous convection heat-transfer flux is obtained by coupling the Newton law of cooling and Eq. (b).

$$q_c'' = \bar{h}(T_s - T_F) = 25\,\frac{\text{W}}{\text{m}^2\,°\text{C}}\,(7.3°\text{C})\left[1 - \exp\left(-\frac{0.0292\,t}{\text{s}}\right)\right]$$

$$= 183\,\frac{\text{W}}{\text{m}^2}\left[1 - \exp\left(-\frac{0.0292\,t}{\text{s}}\right)\right]$$

The heat flux increases from zero at time zero to a maximum steady-state value of 183 W/m^2.

EXAMPLE 2–17

The aluminum ball of Example 2–15 is dropped into a 0.0012-kg oil bath, which is initially at 25°C; the boiling point of the oil is 400°C and the properties are $\rho = 82.5$ kg/m^3, $c_v = 2.2$ kJ/(kg °C). The oil is well stirred such that $\bar{h} = 350$ W/(m^2 °C) and the temperature throughout the bath is uniform, but time-dependent. The container is insulated. Determine the final steady-state temperature of the ball and oil and the time required to reach steady state for the case in which radiation effects are small.

Solution

Objective Determine the temperature and time to reach steady state.

Schematic Cooling of aluminum ball in finite oil bath.

Ball (I)
$D = 0.5$ cm
Oil (II)
$m_{II} = 0.0012$ kg
$T_{IIi} = 25°C$, $T_{Ii} = 250°C$
$\bar{h} = 350$ W/(m^2 °C)
$T_{sat} = 400°C$

Assumptions/Conditions

unsteady-state
time dependent bath temperature
constant properties of the aluminum ball and oil
negligible thermal radiation

Properties

(I) Aluminum ball (Table A–C–1): $\rho = 2700$ kg/m^3, $c_v = 0.896$ kJ/(kg °C), $k = 236$ W/(m °C).

(II) Oil: $\rho = 82.5$ kg/m^3, $c_v = 2.2$ kJ/(kg °C).

Analysis We develop lumped energy balances on both the ball and the oil as follows:

$$\Sigma \dot{E}_i = \Sigma \dot{E}_o + \frac{\Delta E_s}{\Delta t}$$

Ball
$$0 = \bar{h}A_s(T_I - T_{II}) + (\rho V c_v)_I \frac{dT_I}{dt} \tag{a}$$

Oil
$$\bar{h}A_s(T_I - T_{II}) = (\rho V c_v)_{II} \frac{dT_{II}}{dt} \tag{b}$$

$$T_I(0) = T_{Ii} = 250°C \qquad T_{II}(0) = T_{IIi} = 25°C \tag{c,d}$$

These differential equations are put into the operator format

$$(D + K_I)T_I = K_I T_{II} \tag{e}$$

$$(D + K_{II})T_{II} = K_{II} T_I \tag{f}$$

where $K_I = \bar{h}A_s/(\rho V c_v)_I = 0.174/\text{s}$ and $K_{II} = \bar{h}A_s/(mc_v)_{II} = 0.0104/\text{s}$. We now eliminate T_{II} by combining Eqs. (e) and (f).

$$(D + K_I)T_I = K_I K_{II} T_I \frac{1}{D + K_{II}}$$

or

$$\frac{d}{dt}\left(\frac{dT_{\mathrm{I}}}{dt}\right) + (K_{\mathrm{I}} + K_{\mathrm{II}})\frac{dT_{\mathrm{I}}}{dt} = 0$$

Separating the variables, we obtain

$$\frac{d(dT_{\mathrm{I}}/dt)}{dT_{\mathrm{I}}/dt} = -(K_{\mathrm{I}} + K_{\mathrm{II}})\,dt$$

A first integration gives

$$\ln\left(\frac{dT_{\mathrm{I}}/dt}{C_{\mathrm{I}}}\right) = -(K_{\mathrm{I}} + K_{\mathrm{II}})t$$

or

$$\frac{dT_{\mathrm{I}}}{dt} = C_{\mathrm{I}}\exp\left[-(K_{\mathrm{I}} + K_{\mathrm{II}})t\right]$$

A second integration gives

$$T_{\mathrm{I}} = \frac{C_{\mathrm{I}}}{K_{\mathrm{I}} + K_{\mathrm{II}}}\{1 - \exp\left[-(K_{\mathrm{I}} + K_{\mathrm{II}})t\right]\} + C_2$$

Referring to Eq. (a), we see that

$$\frac{dT_{\mathrm{I}}}{dt} = -K_{\mathrm{I}}(T_{\mathrm{I}i} - T_{\mathrm{II}i}) \qquad \text{at } t = 0$$

Thus, $C_{\mathrm{I}} = -K_{\mathrm{I}}(T_{\mathrm{I}i} - T_{\mathrm{II}i})$. To evaluate C_2, we employ the initial condition given by Eq. (c); that is, $C_2 = T_{\mathrm{I}i}$. Hence, our solution for T_{I} is

$$\frac{T_{\mathrm{I}} - T_{\mathrm{I}i}}{T_{\mathrm{I}i} - T_{\mathrm{II}i}} = \frac{K_{\mathrm{I}}}{K_{\mathrm{I}} + K_{\mathrm{II}}}\{\exp\left[-(K_{\mathrm{I}} + K_{\mathrm{II}})t\right] - 1\}$$

The final steady-state temperature T_{ss} of the system is

$$T_{ss} = 250°\mathrm{C} - 225°\mathrm{C}\frac{K_{\mathrm{I}}}{K_{\mathrm{I}} + K_{\mathrm{II}}} = 37.7°\mathrm{C}$$

The time for the temperature to reach 99% of its steady-state value is obtained by writing

$$0.01 = \exp\left[-(K_{\mathrm{I}} + K_{\mathrm{II}})t_{ss}\right]$$

$$t_{ss} = \frac{\ln 0.01}{-(K_{\mathrm{I}} + K_{\mathrm{II}})} = \frac{\ln 0.01}{-0.184} = 25.0\ \mathrm{s}$$

2-7-2 Electrical Analogy

As suggested in Chap. 1, an analogy exists between unsteady one-dimensional heat transfer and the unsteady flow of electric current. For unsteady heat-transfer systems, we represent the thermal capacity C ($\equiv mc_v$) of a mass by an electrical capacitor C_e. The electric current passing through a capacitor is proportional to the time rate of change of voltage,

$$I_e = -C_e \frac{dE_e}{dt} \tag{2-148}$$

With this in mind, the thermal network for the problem under consideration is shown in Fig. 2–20. The thermal capacitor is initially charged at the potential T_i with the switch in position 1. The process is then initiated by throwing the switch to position 2, with the energy stored in the capacitor being dissipated through the resistance. With the product R_eC_e set equal to R_cC [$= \rho c_v V/(\bar{h}A_s)$] (i.e., equal time constants) and with $E_{ei} - E_{e0}$ set equal to $T_i - T_F$, the flow of electric current in this circuit is perfectly analogous to the flow of heat in the thermal system. Hence, instantaneous measurements in I_e and $E_e - E_{e0}$ correspond to the instantaneous rate of heat transfer q and temperature difference $T - T_F$, respectively.

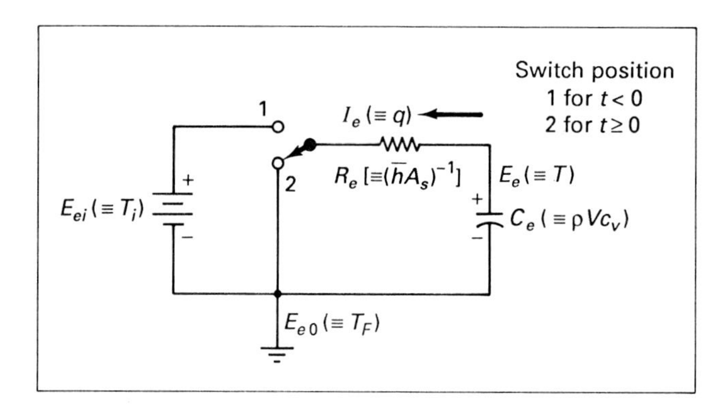

FIGURE 2–20 Analogous electrical circuit for unsteady lumped-capacitance convection heat transfer.

2-8 PRACTICAL SOLUTION RESULTS

For systems without internal energy generation, we have found that the rate of heat transfer can be expressed in terms of practical equations taking either of the two formats,

$$q = \frac{\Delta T}{R} = kS\,\Delta T \tag{2-149}$$

where R is the *thermal resistance*, S is the *conduction shape factor*, and ΔT is the *temperature difference* that characterizes the particular problem. The thermal resistances for the key steady one-dimensional nongenerating systems treated in this chapter are summarized in Table 2–4. Effects of convection, thermal radiation, and composite materials on steady-state heat transfer in these systems can be accounted for by the network approach. The thermal resistances for several fin units are given in Tables 2–2 and 2–3.

TABLE 2–4 Thermal resistance: Steady one-dimensional systems

System	*Thermal resistance* R
Plane wall: Surfaces at $x = 0$ and $x = L$ at T_1 and T_2. 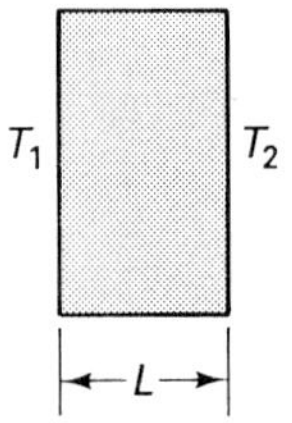	$\dfrac{L}{kA}$
Hollow Circular Cylinder: Surfaces at $r = r_1$ and $r = r_2$ at T_1 and T_2. 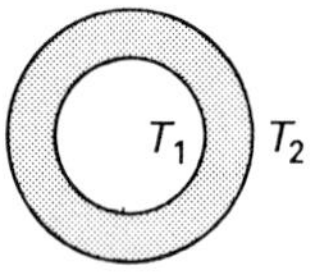	$\dfrac{\ln (r_2/r_1)}{2\pi Lk}$
Hollow Circular Cylinder Section: Surfaces at $r = r_1$ and $r = r_2$ at T_1 and T_2; surfaces at $\theta = 0$ and $\theta = \theta_1$ insulated. 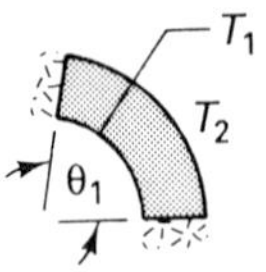	$\dfrac{\ln (r_2/r_1)}{\theta_1 Lk}$
Hollow Sphere: Surfaces at $r = r_1$ and $r = r_2$ at T_1 and T_2. 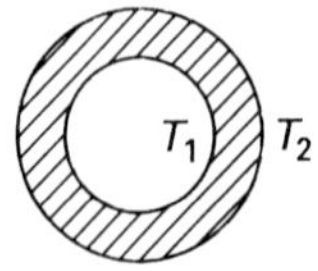	$\dfrac{r_2 - r_1}{4\pi r_1 r_2 k}$

For situations involving nonuniform conditions, the local rate of heat transfer dq is expressed in terms of a *local thermal resistance* R' by

$$dq = \frac{\Delta T}{R'} \tag{2–150}$$

The local thermal resistance associated with convection heat transfer is of the form

$$R_c' = \frac{1}{h_x \, dA_s} \tag{2–151}$$

for unfinned surfaces [which follows from the general Newton law of cooling, Eq. (2–96)], and by Eq. (2–133),

$$R_c' = \frac{1}{\eta_o \bar{h} \, dA_s} \tag{2–133}$$

for finned surfaces. These relationships prove to be very useful in the analysis of heat exchangers.

In regard to the problem of heat transfer with internal energy generation, the maximum temperature T_{max} is a very important design parameter. T_{max} is given in Table 2–5 for two standard geometries.

TABLE 2–5 T_{max} for systems with uniform internal energy generation

System	T_{max}
Flat Plate 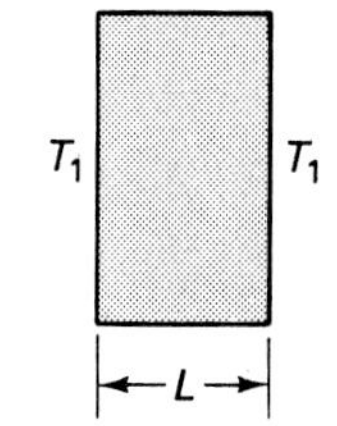	$T_1 + \frac{\dot{q}_0}{k}\frac{L^2}{8}$
Solid Circular Cylinder with radius r_1 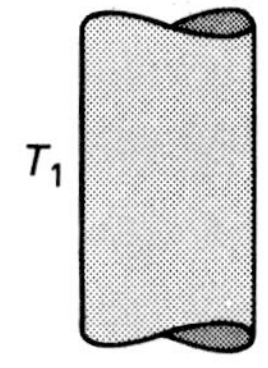	$T_1 + \frac{\dot{q}_0}{k}\frac{r_1^2}{4}$

For unsteady convection systems with small values of Biot number we found that the instantaneous rate q and total Q are given by

$$q_c = hA_s(T_s - T_F)\exp\left(\frac{-\bar{h}A_st}{\rho Vc_v}\right) \tag{2–144}$$

$$Q_c = \rho Vc_v(T_i - T_F)\left[1 - \exp\left(\frac{-\bar{h}A_st}{\rho Vc_v}\right)\right] \tag{2–145}$$

2–9 SUMMARY

In this chapter we have dealt with classic one-dimensional heat-transfer systems. These include conduction in flat plates, hollow cylinders and spheres, and composites; conduction with convection and other types of boundary conditions; conduction with variable thermal conductivity; and conduction with internal energy generation. In addition, approximate analyses have been developed for steady heat transfer in extended surfaces and unsteady heat transfer in systems with small values of Biot number. The principles set forth in the chapter can generally be adapted to the analysis of any heat-transfer problem involving one independent variable for which solutions are not already available. These principles are also applied in the development of analyses for more complex multidimensional conduction heat-transfer systems, which are considered in Chaps. 3 and 4.

CHAPTER 3

CONDUCTION HEAT TRANSFER: MULTIDIMENSIONAL

3–1 INTRODUCTION

As we have seen, one-dimensional heat transfer sometimes occurs in flat-plate, cylindrical, and spherical geometries. We have also seen that the heat transfer in multidimensional thermal systems such as series–parallel composites, extended surfaces, and unsteady convective or radiative heating and cooling of bodies can sometimes be approximated by one-dimensional analyses. But many heat-transfer systems encountered in practice are distinctly multidimensional and cannot be approximated by one-dimensional mathematical models. Therefore, we now turn our attention to the analysis of multidimensional conduction heat transfer. Our first order of business will be to present the general Fourier law of conduction for multidimensional conduction heat transfer. Differential formulations and solutions will then be developed for representative systems. In addition, practical solution results will be given to aid in the design of standard multidimensional conduction-heat-transfer systems. Throughout our study, attention will be restricted to systems involving isotropic materials.

3–2 GENERAL FOURIER LAW OF CONDUCTION

As indicated in Chap. 1, a more general form of the Fourier law must be utilized in the analysis of heat transfer when more than one spatial dimension is involved. For such multidimensional problems, the Fourier law of conduction takes the form (in Cartesian coordinates; see Fig. 3–1)

$$\boldsymbol{q}'' = q''_x\,\mathbf{i} + q''_y\,\mathbf{j} + q''_z\,\mathbf{k} = -k\left(\frac{\partial T}{\partial x}\mathbf{i} + \frac{\partial T}{\partial y}\mathbf{j} + \frac{\partial T}{\partial z}\mathbf{k}\right) \tag{3–1}$$

where the heat-flux vector $\boldsymbol{q}''$ is the resultant heat transfer per unit area, with components

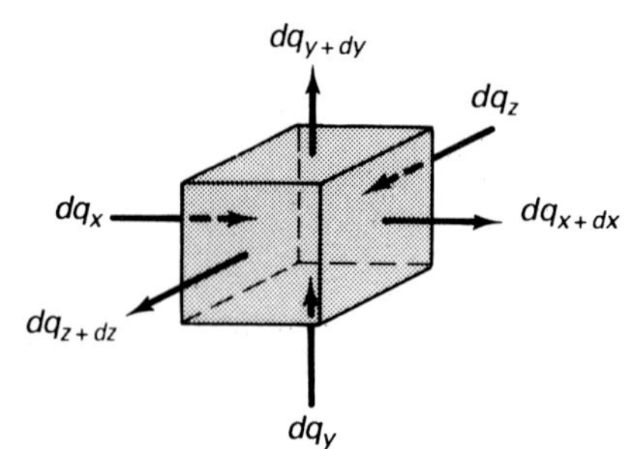

FIGURE 3–1
Differential control volume for rectangular solid; $dA_x = dy\,dz$, $dA_y = dx\,dz$, $dA_z = dx\,dy$.

represented by†

$$q''_x = \frac{dq_x}{dA_x} = -k\frac{\partial T}{\partial x} \tag{3–1a}$$

$$q''_y = \frac{dq_y}{dA_y} = -k\frac{\partial T}{\partial y} \tag{3–1b}$$

and

$$q''_z = \frac{dq_z}{dA_z} = -k\frac{\partial T}{\partial z} \tag{3–1c}$$

The terms dq_x, dq_y, and dq_z represent the rates of heat transfer through the differential areas $dA_x = dy\,dz$, $dA_y = dx\,dz$, and $dA_z = dx\,dy$, respectively. It follows that the total rate of heat transfer through a finite area, say $A_x = w\delta$, can be written as

$$q_x = \int_{A_x} -k\frac{\partial T}{\partial x}\,dA_x = \int_0^{\delta}\int_0^{w} -k\frac{\partial T}{\partial x}\,dy\,dz \tag{3–2}$$

Notice that the term $\partial T/\partial x$ must be retained within the integral(s) in Eq. (3–2) for situations in which T is a function of two or three spatial dimensions. However, for the case in which T and k are functions of t and/or x only, Eq. (3–2) reduces to the simple form of the Fourier law given by Eq. (1–11),

$$q_x = -kA_x\frac{\partial T}{\partial x} \tag{3–3}$$

3–3 MATHEMATICAL FORMULATIONS

Of the various methods that have been developed for the mathematical modeling of energy transfer in conduction-heat-transfer systems, the differential, finite-difference, and finite-element formulations are the most frequently used. The differential for-

† The general Fourier law of conduction is also commonly represented by the vector equation $\boldsymbol{q}'' = -k\,\nabla T$ and the tensor equation $q''_j = -k(\partial T/\partial x_j)$.

mulation approach, which was introduced in Chap. 2, is used in this chapter. Whereas the differential formulation approach provides the basis for establishing exact analytical solutions to many fundamental problems, the finite-difference and finite-element approaches lend themselves to the analysis of the more complex problems often encountered in practice for which exact solutions are difficult, if not impossible, to obtain. The finite-difference approach is featured in Chap. 4. The somewhat more involved finite-element method is introduced in Appendix F.

A fourth method that is commonly employed in the analysis of conduction heat transfer involves the use of integral formulations. The integral approach will be introduced in the context of the theoretical analysis of convection heat transfer in Chap. 7.

3–4 DIFFERENTIAL FORMULATION

Differential formulations are developed in this section for multidimensional conduction heat transfer in Cartesian and standard radial coordinate systems. The differential formulation for a multidimensional system is developed by following the same steps outlined in Sec. 2–2–4 for one-dimensional systems. The boundary conditions to be considered in this chapter include the basic type, which were introduced in Sec. 2–2–3 (i.e., specified temperature, convection, blackbody thermal radiation, specified heat flux, and conduction in composites).

3–4–1 Cartesian Coordinate System

Consider heat transfer in the rectangular solid shown in Fig. 3–1. The initial temperature is T_i and the internal energy generation per unit volume is given by $\dot{q}$. The conditions at the six surfaces are, of course, prescribed by the boundary conditions. Applying the first law of thermodynamics to the differential element *dx dy dz*, we obtain

Step 1
$$dq_x + dq_y + dq_z + \dot{q}\, dV = dq_{x+dx} + dq_{y+dy} + dq_{z+dz} + \frac{\partial e}{\partial t}\, dm \tag{3–4}$$

where $\Delta E_s/\Delta t = dm\, \partial e/\partial t = \rho c_v\, dV\, \partial T/\partial t$ with $dV = dx\, dy\, dz$. (Note that differential rates of heat are transferred through the differential areas dA_x, dA_y, and dA_z.) Utilizing the definition of the partial derivative, we have [see Eq. (A–6) in the Appendix]

$$dq_{x+dx} = dq_x + \frac{\partial (dq_x)}{\partial x}\, dx \tag{3–5}$$

plus similar expressions relating dq_y and dq_z to dq_{y+dy} and dq_{z+dz}. Thus, Eq. (3–4) becomes

Step 2
$$\dot{q}\, dV = \frac{\partial(dq_x)}{\partial x} dx + \frac{\partial(dq_y)}{\partial y} dy + \frac{\partial(dq_z)}{\partial z} dz + dV\, \rho c_v \frac{\partial T}{\partial t} \tag{3–6}$$

Introducing the Fourier law of conduction, we have

Step 3
$$\frac{\partial}{\partial x}\left(k\, dA_x \frac{\partial T}{\partial x}\right) dx + \frac{\partial}{\partial y}\left(k\, dA_y \frac{\partial T}{\partial y}\right) dy + \frac{\partial}{\partial z}\left(k\, dA_z \frac{\partial T}{\partial z}\right) dz + \dot{q}\, dV = dV\, \rho c_v \frac{\partial T}{\partial t} \tag{3–7}$$

Dividing through by the differential volume, this equation reduces to

$$\frac{\partial}{\partial x}\left(k \frac{\partial T}{\partial x}\right) + \frac{\partial}{\partial y}\left(k \frac{\partial T}{\partial y}\right) + \frac{\partial}{\partial z}\left(k \frac{\partial T}{\partial z}\right) + \dot{q} = \rho c_v \frac{\partial T}{\partial t} \tag{3–8}$$

for general variable property conditions, or

$$\frac{\partial^2 T}{\partial x^2} + \frac{\partial^2 T}{\partial y^2} + \frac{\partial^2 T}{\partial z^2} + \frac{\dot{q}}{k} = \frac{1}{\alpha}\frac{\partial T}{\partial t} \tag{3–9}$$

for constant thermal conductivity, where the *thermal diffusivity* α is equal to $k/(\rho c_v)$.

The formulation is completed by writing

$$T = T_i(x,y,z) \qquad \text{at } t = 0 \tag{3–10}$$

plus six conditions in x, y, and z. These boundary conditions are expressed in the same way as those that were developed for one-dimensional systems in Chap. 2, except for the fact that partial derivatives are used instead of total derivatives. This point is illustrated in Examples 3–1 through 3–3.

With $\dot{q} = 0$, Eq. (3–9) reduces to the *Fourier equation*,

$$\frac{\partial^2 T}{\partial x^2} + \frac{\partial^2 T}{\partial y^2} + \frac{\partial^2 T}{\partial z^2} = \frac{1}{\alpha}\frac{\partial T}{\partial t} \tag{3–11}$$

For conditions of steady state, this equation reduces further to the *Laplace equation*,

$$\frac{\partial^2 T}{\partial x^2} + \frac{\partial^2 T}{\partial y^2} + \frac{\partial^2 T}{\partial z^2} = 0 \tag{3–12}$$

If we have steady-state conditions with internal energy generation, Eq. (3–9) reduces to the *Poisson equation*,

$$\frac{\partial^2 T}{\partial x^2} + \frac{\partial^2 T}{\partial y^2} + \frac{\partial^2 T}{\partial z^2} + \frac{\dot{q}}{k} = 0 \tag{3–13}$$

The steady one-dimensional form of Eq. (3–9) or Eq. (3–13) takes the form

$$\frac{d^2T}{dx^2} + \frac{\dot{q}}{k} = 0 \qquad (3\text{–}14)$$

This equation was solved for various conditions in Chap. 2.

EXAMPLE 3-1

Write the differential formulation for the energy transfer in the rectangular plate shown in Fig. E3–1.

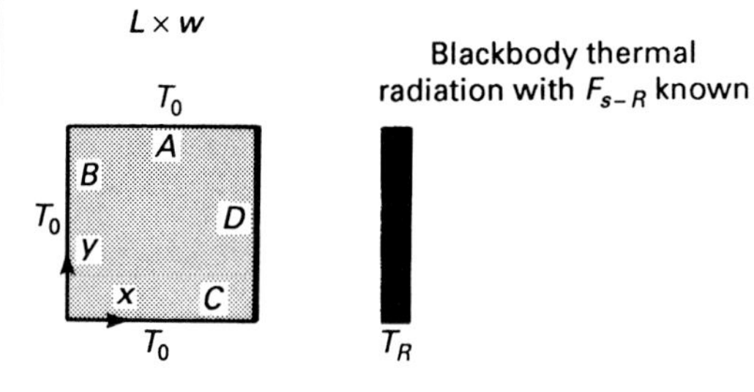

FIGURE E3-1
Conduction in rectangular plate with blackbody thermal radiation.

Solution

Objective Write the energy equation and boundary conditions for this two-dimensional plate.

Assumptions/Conditions

steady-state
two-dimensional
uniform properties
conduction with blackbody thermal radiation

Analysis Because heat is conducted from radiating surface D to surfaces A and C as well as to surface B, the temperature distribution within this plate is a function of both x and y. Thus, the two-dimensional Laplace equation applies; that is,

$$\frac{\partial^2T}{\partial x^2} + \frac{\partial^2T}{\partial y^2} = 0$$

The boundary conditions are

$$T = T_0 \quad \text{at } x = 0 \qquad -k\frac{\partial T}{\partial x} = \sigma F_{s-R}(T^4 - T_R^4) \quad \text{at } x = L$$

$$T = T_0 \quad \text{at } y = 0 \qquad T = T_0 \quad \text{at } y = w$$

Because of symmetry, either one of the y conditions can be replaced by

$$\frac{\partial T}{\partial y} = 0 \qquad \text{at } y = \frac{w}{2}$$

EXAMPLE 3–2

The flat plate shown in Fig. E3–2 is initially at a temperature T_i. It is suddenly exposed to a fluid at temperature T_F with h known. Write the differential formulation for the energy transfer, assuming uniform properties.

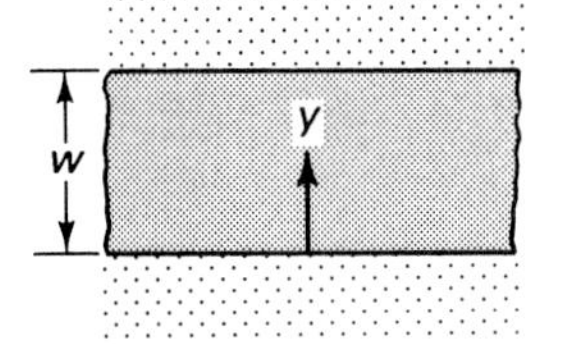

$T = T_i$ at $t = 0$

FIGURE E3–2
Unsteady conduction in plane wall; surfaces at $y = 0$ and $y = w$ suddenly exposed to a fluid at T_F with h known.

Solution

Objective Write the energy equation and boundary conditions for this plane wall.

Assumptions/Conditions

unsteady-state
one spatial dimension
uniform coefficient of heat transfer
uniform properties

Analysis The temperature distribution for this problem is a function of both distance y and time t. The energy equation for this unsteady conduction heat-transfer system takes the form

$$\alpha \frac{\partial^2 T}{\partial y^2} = \frac{\partial T}{\partial t}$$

This is the one-dimensional form of the Fourier equation. The initial and boundary conditions are

$$T = T_i \qquad \text{at } t = 0$$

$$h(T_F - T) = -k \frac{\partial T}{\partial y} \qquad \text{at } y = 0$$

$$-k \frac{\partial T}{\partial y} = h(T - T_F) \qquad \text{at } y = w$$

Note that because of symmetry, either one of the y conditions can be replaced by

$$\frac{\partial T}{\partial y} = 0 \qquad \text{at } y = \frac{w}{2}$$

Strictly speaking, the coefficient of heat transfer h is a function of time for problems such as this. However, approximate solutions are generally obtained by taking the average steady-state value of h. For the case in which h is very large, the boundary conditions reduce to

$$T = T_F \qquad \text{at } y = 0 \text{ and } y = w$$

For steady-state conditions, the energy equation reduces to

$$\frac{\partial^2 T}{\partial y^2} = 0$$

such that

$$T = T_F$$

The rate of heat transfer through the plate is zero for this case.

Another limiting condition occurs for small values of the Biot number. As indicated in Chap. 2, the temperature is nearly independent of location for values of the Biot number less than 0.1, such that the energy equation can be approximated by

$$hA_s(T - T_F) + \rho V c_v \frac{dT}{dt} = 0$$

with

$$T = T_i \qquad \text{at } t = 0$$

EXAMPLE 3–3

Write the formal differential formulation for energy transfer in the composite plate shown in Fig. E3–3.

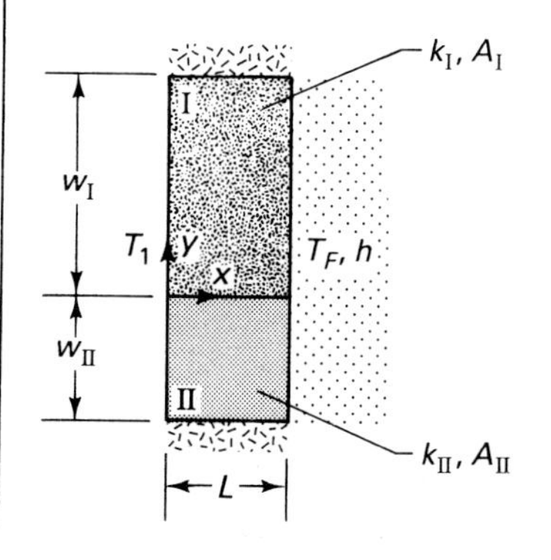

FIGURE E3–3
Composite system with convection at one surface.

Solution

Objective Write the energy equations and boundary conditions for this composite system.

Assumptions/Conditions

- steady-state
- two-dimensional
- uniform coefficient of heat transfer
- uniform properties in each material
- negligible thermal conduction resistance

Analysis Even though the surfaces at $y = w_{\mathrm{I}}$ and $y = -w_{\mathrm{II}}$ are insulated, the temperature distribution is two-dimensional because the system involves a combination of series and parallel heat-transfer paths. The energy equation is therefore written for each material as follows:

$$\frac{\partial^2 T_{\mathrm{I}}}{\partial x^2} + \frac{\partial^2 T_{\mathrm{I}}}{\partial y^2} = 0 \qquad \text{for } 0 \le y \le w_{\mathrm{I}}$$

for material I, and

$$\frac{\partial^2 T_{\mathrm{II}}}{\partial x^2} + \frac{\partial^2 T_{\mathrm{II}}}{\partial y^2} = 0 \qquad \text{for } -w_{\mathrm{II}} \le y \le 0$$

for material II. Because we have two second order partial differential equations in x and y, we require eight boundary conditions. The boundary conditions can be represented by

$$T_{\mathrm{I}} = T_{\mathrm{II}} \qquad -k_{\mathrm{I}}\frac{\partial T_{\mathrm{I}}}{\partial y} = -k_{\mathrm{II}}\frac{\partial T_{\mathrm{II}}}{\partial y} \qquad \text{at } y = 0$$

$$\frac{\partial T_{\mathrm{I}}}{\partial y} = 0 \qquad \text{at } y = w_{\mathrm{I}}$$

$$\frac{\partial T_{\mathrm{II}}}{\partial y} = 0 \qquad \text{at } y = -w_{\mathrm{II}}$$

$$T_{\mathrm{I}} = T_1 \qquad T_{\mathrm{II}} = T_1 \qquad \text{at } x = 0$$

$$-k_{\mathrm{I}}\frac{\partial T_{\mathrm{I}}}{\partial x} = h(T_{\mathrm{I}} - T_F) \qquad -k_{\mathrm{II}}\frac{\partial T_{\mathrm{II}}}{\partial x} = h(T_{\mathrm{II}} - T_F) \qquad \text{at } x = L$$

These equations represent the general mathematical formulation for a combined series–parallel composite with perfect thermal contact at the boundary. The conditions for which these equations can be approximated by the simple one-dimensional formulation are considered in Sec. 2–2–3 (see Example 2–6).

3-4-2 Radial Coordinate Systems

Attention is given next to systems that lend themselves to cylindrical or spherical coordinates.

The differential control volume for a cylindrical coordinate system is shown in Fig. 3–2. The cylindrical coordinates r and ϕ are related to x and y by

$$x = r\cos\phi \qquad y = r\sin\phi \tag{3–15a,b}$$

The general Fourier law of conduction for heat transfer in the r, ϕ, and z directions takes the form

$$\mathbf{q}'' = q_r''\,\mathbf{i_r} + q_\phi''\,\mathbf{j_\phi} + q_z''\,\mathbf{k_z} = -k\left(\frac{\partial T}{\partial r}\mathbf{i_r} + \frac{1}{r}\frac{\partial T}{\partial \phi}\mathbf{j_\phi} + \frac{\partial T}{\partial z}\mathbf{k_z}\right) \tag{3–16}$$

where $\mathbf{i_r}$, $\mathbf{j_\phi}$, and $\mathbf{k_z}$ are unit vectors, the heat-flux components are given by

$$q_r'' = \frac{dq_r}{dA_r} = -k\frac{\partial T}{\partial r} \tag{3–16a}$$

$$q_\phi'' = \frac{dq_\phi}{dA_\phi} = -k\frac{1}{r}\frac{\partial T}{\partial \phi} \tag{3–16b}$$

$$q_z'' = \frac{dq_z}{dA_z} = -k\frac{\partial T}{\partial z} \tag{3–16c}$$

and $dA_r = r\,d\phi\,dz$, $dA_\phi = dr\,dz$, and $dA_z = r\,d\phi\,dr$.

Consider, for example, the cylindrical body shown in Fig. 3–3. The initial temperature is given by T_i and the conditions at the boundaries $r = r_0$, $z = 0$, and

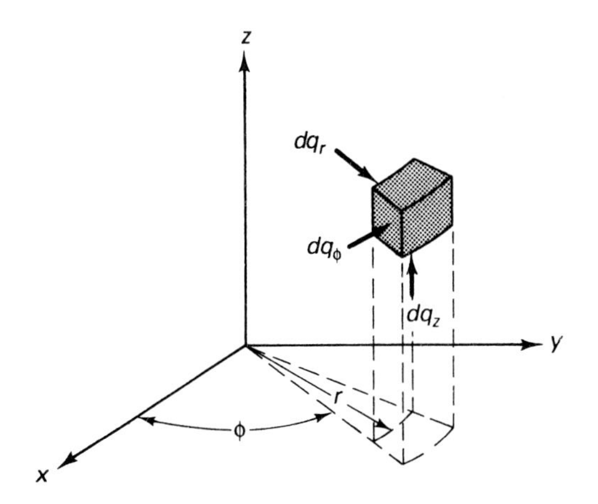

FIGURE 3–2 Heat transfer associated with a differential-control volume for a cylindrical system; $dA_r = r\,d\phi\,dz$, $dA_z = r\,d\phi\,dr$, $dA_\phi = dr\,dz$, $dV = r\,d\phi\,dr\,dz$.

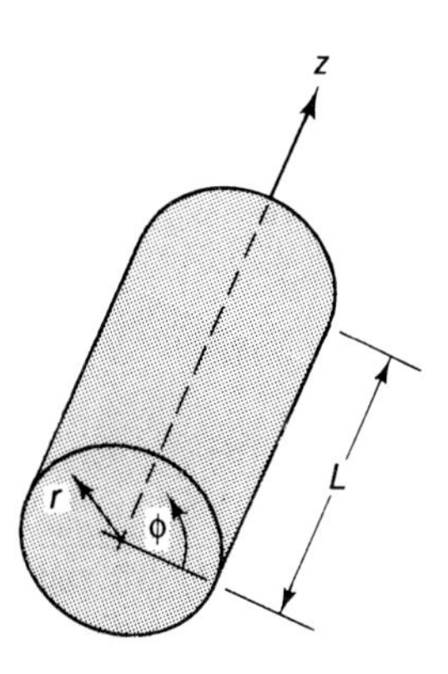

FIGURE 3–3 Cylindrical body.

$z = L$ are assumed to be known. Applying the first law of thermodynamics to a differential volume dV ($= r\, d\phi\, dr\, dz$) such as that shown in Fig. 3–2, we obtain

Step 1

$$dq_r + dq_\phi + dq_z + \dot{q}\, dV = dq_{r+dr} + dq_{\phi+d\phi} + dq_{z+dz} + \frac{\partial e}{\partial t}\, dm \tag{3–17}$$

where $\Delta E_s/\Delta t = dm\, \partial e/\partial t = \rho c_v\, dV\, \partial T/\partial t$. Utilizing the definition for the partial derivative, Eq. (3–17) takes the form

Step 2

$$\dot{q}\, dV = \frac{\partial}{\partial r}(dq_r)\, dr + \frac{\partial}{\partial \phi}(dq_\phi)\, d\phi + \frac{\partial}{\partial z}(dq_z)\, dz + \rho c_v\, dV \frac{\partial T}{\partial t} \tag{3–18}$$

Then,

Step 3

$$\frac{\partial}{\partial r}\left(kr\, d\phi\, dz \frac{\partial T}{\partial r}\right) dr + \frac{\partial}{\partial \phi}\left(k\, dr\, dz \frac{1}{r}\frac{\partial T}{\partial \phi}\right) d\phi + \frac{\partial}{\partial z}\left(kr\, d\phi\, dr \frac{\partial T}{\partial z}\right) dz + \dot{q}\, dV = \rho c_v\, dV \frac{\partial T}{\partial t} \tag{3–19}$$

In simplifying Eq. (3–19), the first term must be handled carefully when differentiating because dA_r ($= r\, d\phi\, dz$) is a function of r. Taking note of this point, Eq. (3–19) reduces to

$$\frac{1}{r}\frac{\partial}{\partial r}\left(kr\frac{\partial T}{\partial r}\right) + \frac{1}{r^2}\frac{\partial}{\partial \phi}\left(k\frac{\partial T}{\partial \phi}\right) + \frac{\partial}{\partial z}\left(k\frac{\partial T}{\partial z}\right) + \dot{q} = \rho c_v \frac{\partial T}{\partial t} \tag{3–20}$$

or, for uniform properties,

$$\frac{1}{r}\frac{\partial}{\partial r}\left(r\frac{\partial T}{\partial r}\right) + \frac{1}{r^2}\frac{\partial^2 T}{\partial \phi^2} + \frac{\partial^2 T}{\partial z^2} + \frac{\dot{q}}{k} = \frac{1}{\alpha}\frac{\partial T}{\partial t} \tag{3–21}$$

The formulation is completed by writing

$$T = T_i(r, \phi, z) \qquad \text{at } t = 0 \tag{3–22}$$

together with six conditions in r, ϕ, and z.

The steady one-dimensional form of Eq. (3–21),

$$\frac{1}{r}\frac{d}{dr}\left(r\frac{dT}{dr}\right) + \frac{\dot{q}}{k} = 0 \tag{3–23}$$

was solved in Chap. 2 for several conditions.

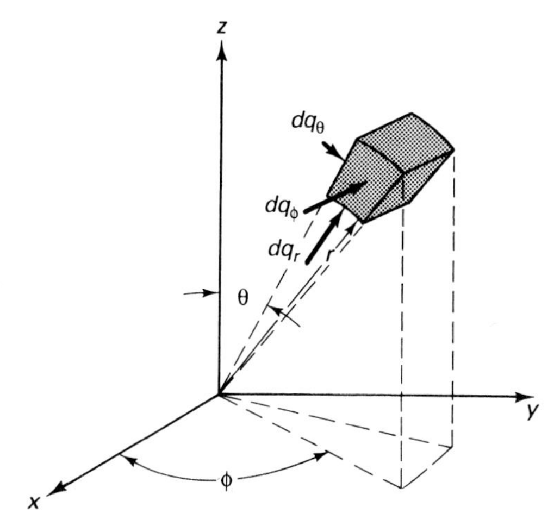

FIGURE 3-4
Heat transfer associated with a differential control volume for a spherical system; $dA_r = r^2 \sin\theta\, d\theta\, d\phi$, $dA_\phi = r\, d\theta\, dr$, $dA_\theta = r \sin\theta\, d\phi\, dr$, $dV = r^2 \sin\theta\, d\phi\, d\theta\, dr$.

Referring to Fig. 3-4, the unsteady three-dimensional energy equation for conduction heat transfer in spherical geometries is given by

$$\frac{1}{r^2}\frac{\partial}{\partial r}\left(kr^2\frac{\partial T}{\partial r}\right) + \frac{1}{r^2 \sin^2\theta}\frac{\partial}{\partial \phi}\left(k\frac{\partial T}{\partial \phi}\right) + \frac{1}{r^2 \sin\theta}\frac{\partial}{\partial \theta}\left(k \sin\theta \frac{\partial T}{\partial \theta}\right) + \dot{q} = \rho c_v \frac{\partial T}{\partial t} \tag{3-24}$$

The derivation of this equation follows the pattern established in the formulation of the energy equation for cylindrical systems.

EXAMPLE 3-4

Write the differential formulation for steady heat transfer in the cylindrical fin with circular cross section shown in Fig. E3-4.

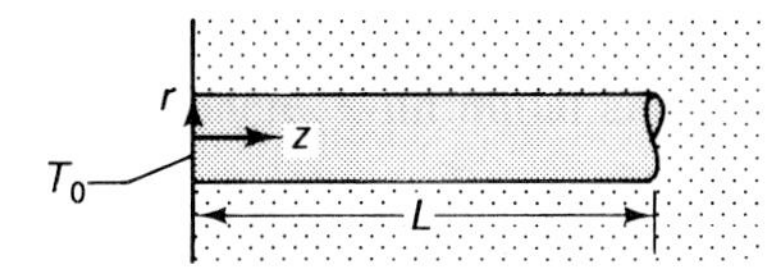

FIGURE E3-4
Circular fin.

Solution

Objective Write the energy equation and boundary conditions for this fin.

Assumptions/Conditions

steady-state
two-dimensional fin
uniform properties
uniform coefficient of heat transfer

Analysis The differential formulation for this two-dimensional problem takes the form

$$\frac{\partial^2 T}{\partial z^2} + \frac{1}{r}\frac{\partial}{\partial r}\left(r\frac{\partial T}{\partial r}\right) = 0$$

$$T = T_0 \quad \text{at } z = 0 \qquad -k\frac{\partial T}{\partial z} = \bar{h}(T - T_F) \quad \text{at } z = L$$

$$\frac{\partial T}{\partial r} = 0 \quad \text{at } r = 0 \qquad -k\frac{\partial T}{\partial r} = \bar{h}(T - T_F) \quad \text{at } r = r_0$$

As indicated in Chap. 2, for small values of the Biot number the variation of temperature with r is very slight, such that an approximate one-dimensional formulation can be utilized. The approximate one-dimensional energy equation is written as

$$\frac{d^2 T}{dz^2} - \frac{\bar{h}p}{kA}(T - T_F) = 0$$

(The axial coordinate is commonly designated by x instead of z.)

EXAMPLE 3–5

A spherical ball initially at temperature T_i is suddenly exposed to a convecting fluid at temperature T_F with h known. Develop the differential formulation.

Solution

Objective Write the energy equation and boundary conditions for this spherical ball.

Schematic Differential volume for unsteady heat transfer in a sphere.

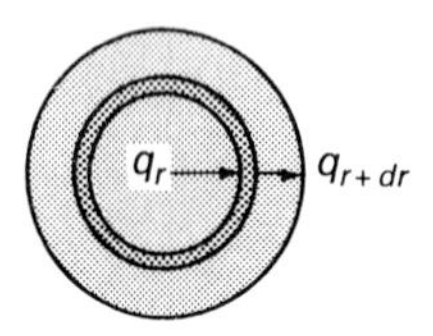

Assumptions/Conditions

unsteady-state
one spatial dimension
uniform coefficient of heat transfer
uniform properties

Analysis Referring to the schematic, the energy balance gives

Step 1 $$q_r = q_{r+dr} + \frac{\Delta E_s}{\Delta t}$$

(where $dV = 4\pi r^2\, dr$).

Step 2 $$q_r = q_r + \frac{\partial q_r}{\partial r}\, dr + \rho c_v\, dV \frac{\partial T}{\partial t}$$

Step 3 $$0 = \frac{\partial}{\partial r}\left(-kr^2 \frac{\partial T}{\partial r}\right) + r^2 \rho c_v \frac{\partial T}{\partial t}$$

For uniform properties, this equation reduces to

$$\frac{1}{r^2}\frac{\partial}{\partial r}\left(r^2 \frac{\partial T}{\partial r}\right) = \frac{1}{\alpha}\frac{\partial T}{\partial t}$$

or

$$\frac{\partial^2 T}{\partial r^2} + \frac{2}{r}\frac{\partial T}{\partial r} = \frac{1}{\alpha}\frac{\partial T}{\partial t}$$

The initial-boundary conditions are

$$T = T_i \qquad \text{at } t = 0$$

$$\frac{\partial T}{\partial r} = 0 \qquad \text{at } r = 0$$

$$-k\frac{\partial T}{\partial r} = h(T - T_F) \qquad \text{at } r = r_0$$

As indicated in Chap. 2, for small values of the Biot number, the temperature is essentially independent of r such that the energy equation can be approximated by

$$\frac{dT}{dt} + \frac{hA}{\rho V c_v}(T - T_F) = 0$$

3–5 SOLUTION TECHNIQUES: INTRODUCTION

Several techniques are available for the solution of the equations produced by our differential formulation. These solution techniques include analytical, analogical, and graphical methods, all of which are introduced in the following sections of this chapter. The differential formulation can also be solved by the numerical finite-difference approach, as will be shown in Chap. 4.

Solutions obtained by these various methods for standard two- and three-dimensional conduction-heat-transfer problems are available in formula and chart form as a convenience for design calculations. These practical results will be presented in Sec. 3–9.

3–6 ANALYTICAL APPROACHES

Among the various methods that can be utilized to develop analytical solutions for multidimensional heat-transfer problems, the *separation-of-variables* technique and the *approximate integral* approach are the simplest. The classical separation-of-variables approach is introduced in this section. Although the integral approach can be used to solve many conduction-heat-transfer problems, this popular method will be introduced in Chap. 7 in the context of convection heat transfer. Other analytical solution techniques for multidimensional conduction-heat-transfer problems are available in the literature.

3–6–1 Separation-of-Variables Method

The *separation-of-variables* technique is particularly well suited to the solution of steady and unsteady multidimensional conduction-heat-transfer problems that can be represented by linear equations. To introduce this solution concept, we consider the steady-state two-dimensional conduction-heat-transfer problem illustrated in Fig. 3–5.

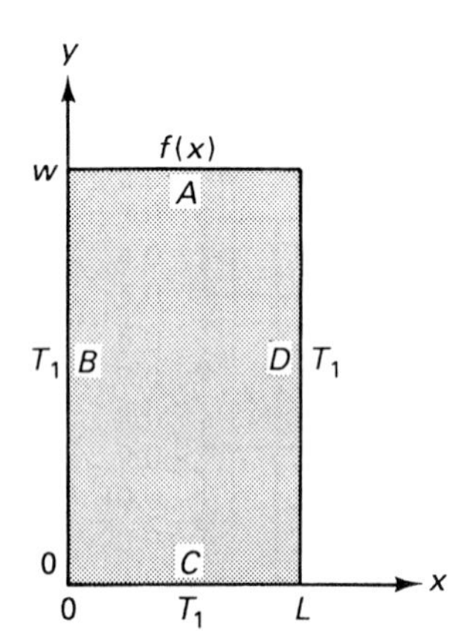

FIGURE 3–5
Steady two-dimensional conduction heat transfer in a rectangular plate.

The differential formulation for this problem for the case in which k is uniform, $\dot{q} = 0$, $\partial T/\partial z = 0$, and $\partial T/\partial t = 0$ is written as

$$\frac{\partial^2 T}{\partial x^2} + \frac{\partial^2 T}{\partial y^2} = 0 \tag{3–25}$$

$$T = T_1 \quad \text{at } x = 0 \qquad T = T_1 \quad \text{at } x = L \tag{3–26,27}$$

$$T = T_1 \quad \text{at } y = 0 \qquad T = f(x) \quad \text{at } y = w \tag{3–28,29}$$

The use of the separation-of-variables method requires that only one nonhomogeneous term appear in the linear differential formulation. Therefore, we eliminate three of the nonhomogeneous boundary conditions by utilizing the substitution $\psi = T - T_1$. This simple substitution transforms Eqs. (3–25) through (3–29) to

$$\frac{\partial^2 \psi}{\partial x^2} + \frac{\partial^2 \psi}{\partial y^2} = 0 \tag{3–30}$$

$$\psi = 0 \quad \text{at } x = 0 \qquad \psi = 0 \quad \text{at } x = L \tag{3–31,32}$$

$$\psi = 0 \quad \text{at } y = 0 \qquad \psi = F(x) \quad \text{at } y = w \tag{3–33,34}$$

where $F(x) = f(x) - T_1$. Here x is referred to as the homogeneous direction and y is the nonhomogeneous direction.

The solution to Eqs. (3–30) through (3–34) is developed in Appendix E–1 by assuming a product solution of the form

$$\psi(x,y) = X(x)\,Y(y) \tag{3–35}$$

where $X(x)$ is a function of x and $Y(y)$ is a function of y. The general solution is

$$T - T_1 = \sum_{n=1}^{\infty} c_n \frac{\sinh(n\pi y/L)}{\sinh(n\pi w/L)} \sin\frac{n\pi x}{L} \tag{3–36}$$

where

$$c_n = \frac{2}{L}\int_0^L F(x)\sin\frac{n\pi x}{L}\,dx \tag{3–37}$$

For the case in which $f(x)$ is equal to a uniform temperature T_2, we have

$$c_n = \frac{2}{L}\int_0^L (T_2 - T_1)\sin\frac{n\pi x}{L}\,dx = -\frac{2}{L}(T_2 - T_1)\frac{L}{n\pi}[\cos(n\pi) - 1]$$

$$= (T_2 - T_1)\frac{4}{n\pi} \qquad n = 1, 3, 5, \ldots \tag{3–38}$$

$$= 0 \qquad n = 2, 4, 6, \ldots$$

Hence, Eq. (3–36) reduces to

$$\frac{T - T_1}{T_2 - T_1} = \frac{2}{\pi}\sum_{n=1}^{\infty} \frac{1 - (-1)^n}{n} \frac{\sinh(n\pi y/L)}{\sinh(n\pi w/L)} \sin\frac{n\pi x}{L} \tag{3–39}$$

This series is uniformly convergent (see Kreyszig [1]) in the region $0 \le y < w$ for all values of x. Isotherms obtained on the basis of Eq. (3–39) are shown in Fig. 3–6.

For the case in which $f(x) = T_1 + T_2 \sin(\pi x/L)$ [i.e., $F(x) = T_2 \sin(\pi x/L)$], we find that

$$c_1 = T_2 \qquad c_n = 0 \qquad n = 2, 3, 4, \ldots \tag{3–40}$$

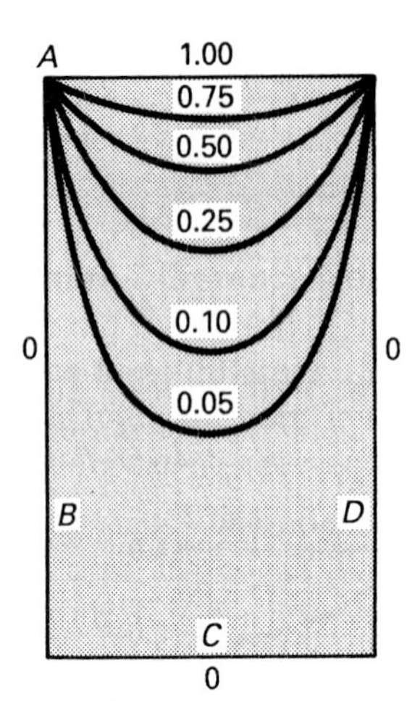

FIGURE 3–6
Dimensionless isotherms $(T - T_1)/(T_2 - T_1)$ for steady two-dimensional conduction heat transfer in a rectangular plate; $f(x) = T_2$ and $L/w = 0.6$.

such that the temperature distribution is given by

$$T - T_1 = T_2 \frac{\sinh (\pi y/L)}{\sinh (\pi w/L)} \sin \frac{\pi x}{L} \tag{3–41}$$

Once the temperature distribution for the problem of interest is known, the rate of heat transfer can be obtained by utilizing the appropriate particular law. To illustrate, we select the situation for which $f(x) = T_1 + T_2 \sin (\pi x/L)$. To obtain the local rate of heat transfer in the y direction, the general Fourier law of conduction is utilized; that is,

$$dq_y = -k\, dA_y \frac{\partial T}{\partial y} = -k\delta\, dx\, T_2 \frac{\pi}{L} \frac{\cosh (\pi y/L)}{\sinh (\pi w/L)} \sin \frac{\pi x}{L} \tag{3–42}$$

The total rate of heat transfer q_y is now obtained by integrating from x equal to 0 to L as follows:

$$\begin{aligned} q_y &= \int_0^L -k \frac{\partial T}{\partial y} \delta\, dx = -k\delta T_2 \frac{\pi}{L} \frac{\cosh (\pi y/L)}{\sinh (\pi w/L)} \left(-\frac{L}{\pi} \cos \frac{\pi x}{L} \right) \Bigg|_0^L \\ &= -2k\delta T_2 \frac{\cosh (\pi y/L)}{\sinh (\pi w/L)} \end{aligned} \tag{3–43}$$

Thus, the rate of heat transfer across surface A at $y = w$ is

$$q_A = -2k\delta T_2 \coth \frac{\pi w}{L} \tag{3–44}$$

and the rate of heat transfer across surface C at $y = 0$ is

$$q_C = -2k\delta T_2 \frac{1}{\sinh (\pi w/L)} \tag{3–45}$$

Similar expressions can be developed for the rate of heat transfer in the x direction.

As illustrated in Examples 3–6 and 3–7, the same approach is used to develop expressions for the rate of heat transfer for the more complex steady and unsteady systems in which the temperature field is expressed in terms of infinite series.

The separation-of-variables approach has been utilized in the solution of a large number of standard steady and unsteady conduction heat-transfer problems. The results of some of these analyses will be presented in Sec. 3–9.

EXAMPLE 3–6

Consider the two-dimensional problem shown in Fig. 3–5 for the case in which $f(x) = T_2 = 100°C$, $T_1 = 0°C$, $k = 100$ W/(m °C), and $\delta = 1$ cm. The solution to this problem for uniform property conditions is given by Eq. (3–39). First, demonstrate that this equation satisfies the energy equation. Then utilize this result to obtain predictions for the rates of heat transfer across surfaces A and C of the plate for the case in which $L = w$.

Solution

Objective Demonstrate the validity of Eq. (3–39) and use this equation to determine q_A and q_C.

Schematic Steady two-dimensional heat transfer in a rectangular plate.

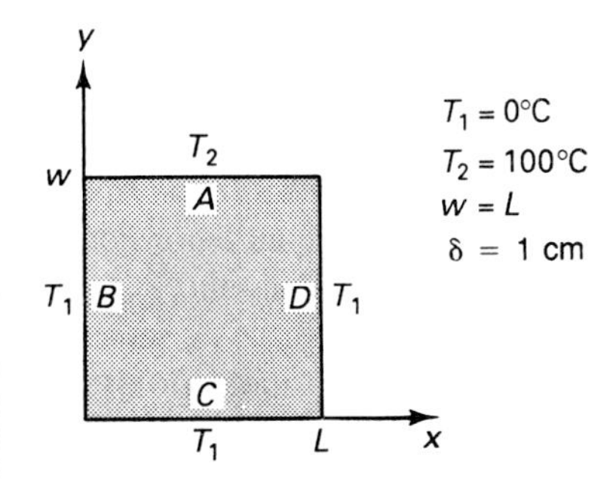

Assumptions/Conditions

steady-state
two-dimensional
uniform properties

Properties $k = 100$ W/(m °C)

Analysis As already mentioned, this infinite series converges for all values of x and y within the domain of the body. Predictions for the temperature at various locations within the body are shown in Fig. 3–6. Because Eq. (3–39) is uniformly convergent in the region $0 \le y < w$ for all values of x, this equation can be differentiated term by term in this domain.

To demonstrate that Eq. (3–39) satisfies the energy equation, Eq. (3–25), we obtain $\partial^2 T/\partial x^2$ and $\partial^2 T/\partial y^2$ as follows:

$$\frac{\partial T}{\partial x} = (T_2 - T_1)\frac{2}{\pi}\sum_{n=1}^{\infty}\frac{1-(-1)^n}{n}\frac{\sinh(n\pi y/L)}{\sinh(n\pi w/L)}\frac{n\pi}{L}\cos\frac{n\pi x}{L}$$

$$\frac{\partial^2 T}{\partial x^2} = (T_2 - T_1)\frac{2}{\pi}\sum_{n=1}^{\infty}\frac{1-(-1)^n}{n}\frac{\sinh(n\pi y/L)}{\sinh(n\pi w/L)}\left(\frac{-n^2\pi^2}{L^2}\right)\sin\frac{n\pi x}{L}$$

and

$$\frac{\partial T}{\partial y} = (T_2 - T_1)\frac{2}{\pi}\sum_{n=1}^{\infty}\frac{1-(-1)^n}{n}\frac{n\pi}{L}\frac{\cosh(n\pi y/L)}{\sinh(n\pi w/L)}\sin\frac{n\pi x}{L}$$

$$\frac{\partial^2 T}{\partial y^2} = (T_2 - T_1)\frac{2}{\pi}\sum_{n=1}^{\infty}\frac{1-(-1)^n}{n}\left(\frac{n\pi}{L}\right)^2\frac{\sinh(n\pi y/L)}{\sinh(n\pi w/L)}\sin\frac{n\pi x}{L}$$

Substituting these results for $\partial^2 T/\partial x^2$ and $\partial^2 T/\partial y^2$ into Eq. (3–25),

$$\frac{\partial^2 T}{\partial x^2} + \frac{\partial^2 T}{\partial y^2} = 0$$

we find that this equation is indeed satisfied.

To determine the local rate of heat transfer in the y direction, we apply the general Fourier law of conduction, with the result

$$\begin{aligned} dq_y &= -k\,dA_y\frac{\partial T}{\partial y} \\ &= -k(T_2 - T_1)\left\{\frac{2}{L}\sum_{n=1}^{\infty}[1-(-1)^n]\frac{\cosh(n\pi y/L)}{\sinh(n\pi w/L)}\sin\frac{n\pi x}{L}\right\}\delta\,dx \end{aligned}$$

Integrating with respect to x, the total rate of heat transfer over the region $0 \le x \le L$ becomes

$$\begin{aligned} q_y &= k\delta(T_2 - T_1)\frac{2}{L}\sum_{n=1}^{\infty}[1-(-1)^n]\frac{\cosh(n\pi y/L)}{\sinh(n\pi w/L)}\frac{L}{n\pi}[\cos(n\pi) - 1] \\ &= -k\delta(T_2 - T_1)\frac{4}{\pi}\sum_{n=1}^{\infty}\frac{1-(-1)^n}{n}\frac{\cosh(n\pi y/L)}{\sinh(n\pi w/L)} \end{aligned} \qquad \text{(a)}$$

Setting y equal to zero, the total rate of heat transfer across face C is given by

$$q_C = -k\delta(T_2 - T_1)\frac{4}{\pi}\sum_{n=1}^{\infty}\frac{1-(-1)^n}{n}\frac{1}{\sinh(n\pi)}$$

where $w = L$ for the square body. The infinite series

$$\frac{4}{\pi}\sum_{n=1}^{\infty}\frac{1-(-1)^n}{n}\frac{1}{\sinh(n\pi)}$$

converges fairly rapidly to approximately 0.221. Hence, q_C is given by

$$q_C = -0.221k\delta(T_2 - T_1) = -22.1 \text{ W}$$

To obtain the rate of heat transfer across face A, we allow y to approach w in Eq. (a).

$$q_A = -k\delta(T_2 - T_1)\frac{4}{\pi}\sum_{n=1}^{\infty}\frac{1-(-1)^n}{n}\coth(n\pi)$$

The infinite series

$$\sum_{n=1}^{\infty}\frac{1-(-1)^n}{n} = 2\left(1 + \frac{1}{3} + \frac{1}{5} + \frac{1}{7} + \dots\right)$$

is divergent [1,2]. This, together with the fact that coth $(n\pi)$ is always greater than unity, indicates that q_A is unbounded. This rather startling result stems from the fact that we have the drastic situation in which the two surfaces A and B at different temperatures T_2 and T_1 are in direct contact. If we were to attempt to set up an electrical analog for this problem, we would have burn out because of shorting at the corner sections.

EXAMPLE 3–7

The surfaces of a flat plate initially at uniform temperature T_i are suddenly brought to a temperature T_1. The differential formulation takes the form

$$\alpha\frac{\partial^2\psi}{\partial x^2} = \frac{\partial\psi}{\partial t}$$

$$\psi = T_i - T_1 \qquad \text{at } t = 0$$

$$\psi = 0 \qquad \text{at } x = 0$$

$$\psi = 0 \qquad \text{at } x = L$$

where $\psi = T - T_1$. These equations can be solved by means of the separation-of-variables solution approach [3–5], with the result†

$$\frac{T - T_1}{T_i - T_1} = \frac{2}{\pi}\sum_{n=1}^{\infty}\frac{1-(-1)^n}{n}\exp(-n^2\pi^2\alpha t/L^2)\sin\frac{n\pi x}{L}$$

which is uniformly convergent for $t > 0$ over the entire plate $0 \leq x \leq L$. Develop expressions for the instantaneous rate of heat transfer q_s from the plate and the total heat transferred Q over a period of time t.

† This system of equations can also be conveniently solved by means of the approximate integral technique and the Laplace-transform method.

Solution

Objective Develop relations for q_s and Q.

Schematic Unsteady one-dimensional heat transfer in a flat plate.

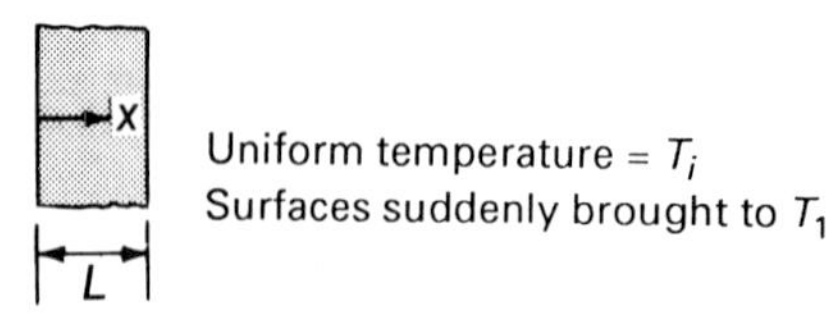

Assumptions/Conditions

unsteady-state
one spatial dimension
uniform properties

Analysis Taking q_x in the x direction as positive, the instantaneous rate of heat transfer q_0 at surface $x = 0$ is obtained by introducing the Fourier law of conduction.

$$q_0 = -kA_x \left.\frac{\partial T}{\partial x}\right|_0$$

$$= -kA(T_i - T_1)\frac{2}{\pi}\sum_{n=1}^{\infty}\frac{1-(-1)^n}{n}\exp(-n^2\pi^2\alpha t/L^2)\frac{n\pi}{L}\cos\left.\frac{n\pi x}{L}\right|_0 \qquad \text{(a)}$$

$$= -kA(T_i - T_1)\frac{2}{L}\sum_{n=1}^{\infty}[1-(-1)^n]\exp(-n^2\pi^2\alpha t/L^2)$$

Taking into consideration the symmetrical nature of the problem, we obtain

$$q_s = -q_0 + q_L = -2q_0$$

To obtain an expression for the accumulated heat transfer Q_0 over the period 0 to t, we write

$$Q_0 = \int_0^t q_0\,dt = -kA(T_i - T_1)\frac{2}{L}\sum_{n=1}^{\infty}[1-(-1)^n]\frac{-L^2}{n^2\pi^2\alpha}\exp(-n^2\pi^2\alpha t/L^2)\Big|_0^t$$

$$= -kA(T_i - T_1)\frac{2L}{\alpha\pi^2}\sum_{n=1}^{\infty}\frac{1-(-1)^n}{n^2}[1-\exp(-n^2\pi^2\alpha t/L^2)] \qquad \text{(b)}$$

The accumulated heat transfer for the entire plate Q is

$$Q = -2Q_0$$

The infinite series in Eqs. (a) and (b) are both convergent. These infinite series are generally evaluated by digital computer, except for large values of $\alpha t/L^2$ or $\alpha\tau/L^2$ for which cases hand calculations can be quickly made. Calculations for T, q_s, and Q for this problem are given in chart form in Sec. 3–9.

EXAMPLE 3–8

Solutions to heat-transfer problems involving more than one nonhomogeneous term can be developed by applying the *principle of superposition* if the equations are linear (see Appendix A). In this approach, solutions to more involved problems are obtained by summing up or superimposing solutions of sets of simpler problems. To demonstrate this important superposition concept, consider steady two-dimensional conduction heat transfer in the rectangular solid shown in Fig. 3–5 for the case in which surfaces A and D are both at T_2.

Solution

Objective Use the superposition principle to develop a solution for the temperature distribution in the rectangular plate shown in the schematic.

Schematic Steady two-dimensional heat transfer in a rectangular plate.

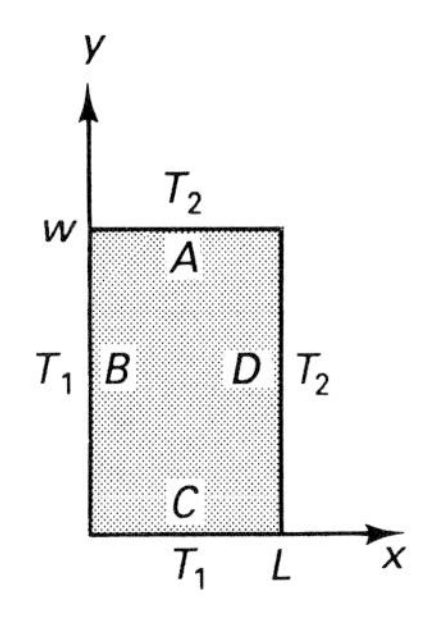

Assumptions/Conditions

steady-state
two-dimensional
uniform properties

Analysis The differential formulation for this problem can be written as

$$\frac{\partial^2 \psi}{\partial x^2} + \frac{\partial^2 \psi}{\partial y^2} = 0 \tag{a}$$

$$\psi = 0 \quad \text{at } x = 0 \qquad \psi = T_2 - T_1 \quad \text{at } x = L \tag{b,c}$$

$$\psi = 0 \quad \text{at } y = 0 \qquad \psi = T_2 - T_1 \quad \text{at } y = w \tag{d,e}$$

where the use of ψ ($\equiv T - T_1$) has created homogeneous conditions at $x = 0$ and $y = 0$. This mathematical formulation is represented by Fig. E3–8a. However, we are left with two nonhomogeneous boundary conditions. Because the use of the separation-of-variables technique requires that only one nonhomogeneous condition occur, we turn to the principle of superposition to achieve a solution.

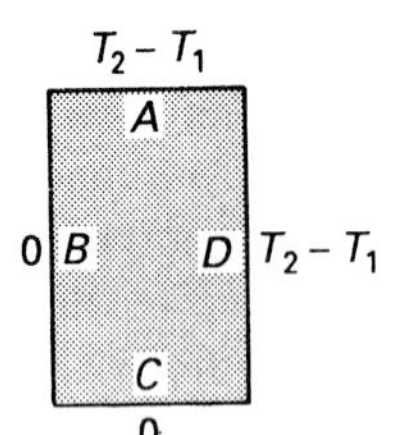

FIGURE E3–8a
Rectangular plate with two nonhomogeneous boundary conditions.

Utilizing the superposition concept, ψ is assumed to be equal to the sum of two simpler solutions; that is,

$$\psi = \psi_I + \psi_{II} \tag{f}$$

Our differential formulation can be written in terms of ψ_I and ψ_{II} by merely substituting Eq. (f) into Eqs. (a) through (e).

$$\frac{\partial^2\psi_I}{\partial x^2} + \frac{\partial^2\psi_{II}}{\partial x^2} + \frac{\partial^2\psi_I}{\partial y^2} + \frac{\partial^2\psi_{II}}{\partial y^2} = 0$$

$$\psi_I + \psi_{II} = 0 \quad \text{at } x = 0 \qquad \psi_I + \psi_{II} = T_2 - T_1 \quad \text{at } x = L$$

$$\psi_I + \psi_{II} = 0 \quad \text{at } y = 0 \qquad \psi_I + \psi_{II} = T_2 - T_1 \quad \text{at } y = w$$

Based on this system of equations, the following two simpler problems are formulated such that the differential formulation for ψ $(= \psi_I + \psi_{II})$ remains satisfied:

$$\frac{\partial^2\psi_I}{\partial x^2} + \frac{\partial^2\psi_I}{\partial y^2} = 0 \qquad \frac{\partial^2\psi_{II}}{\partial x^2} + \frac{\partial^2\psi_{II}}{\partial y^2} = 0$$

$$\psi_I = 0 \qquad \text{at } x = 0 \qquad \psi_{II} = 0$$

$$\psi_I = 0 \qquad \text{at } x = L \qquad \psi_{II} = T_2 - T_1$$

$$\psi_I = 0 \qquad \text{at } y = 0 \qquad \psi_{II} = 0$$

$$\psi_I = T_2 - T_1 \qquad \text{at } y = w \qquad \psi_{II} = 0$$

The formulations for ψ_I and ψ_{II} are represented by Figs. E3–8b and E3–8c. Note that Fig. E3–8a is equivalent to superimposing Fig. E3–8b on Fig. E3–8c. The solution for ψ_I is given by Eq. (3–39),

$$\psi_I = T_I - T_1 = (T_2 - T_1)\frac{2}{\pi}\sum_{n=1}^{\infty}\frac{1 - (-1)^n}{n}\frac{\sinh(n\pi y/L)}{\sinh(n\pi w/L)}\sin\frac{n\pi x}{L}$$

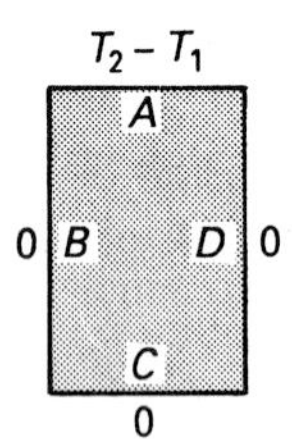

FIGURE E3–8b
Rectangular plate with one nonhomogeneous boundary condition at $y = w$.

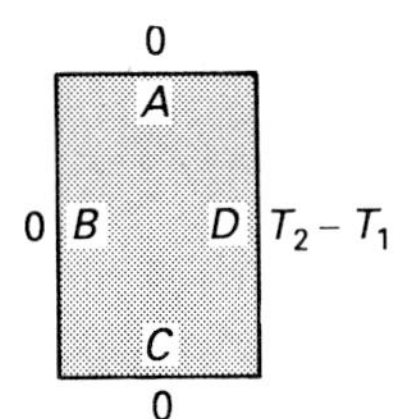

FIGURE E3–8c
Rectangular plate with one nonhomogeneous boundary condition at $x = L$.

By inspection, we see that the solution for ψ_{II} can be obtained from Eq. (3–39) by merely interchanging y and x and w and L; that is,

$$\psi_{II} = T_{II} - T_1 = (T_2 - T_1)\frac{2}{\pi}\sum_{n=1}^{\infty}\frac{1-(-1)^n}{n}\frac{\sinh(n\pi x/w)}{\sinh(n\pi L/w)}\sin\frac{n\pi y}{w}$$

Returning to Eq. (f), our full solution is

$$T - T_1 = \psi = \psi_I + \psi_{II} =$$

$$(T_2 - T_1)\frac{2}{\pi}\sum_{n=1}^{\infty}\frac{1-(-1)^n}{n}\left[\frac{\sinh(n\pi y/L)}{\sinh(n\pi w/L)}\sin\frac{n\pi x}{L} + \frac{\sinh(n\pi x/w)}{\sinh(n\pi L/w)}\sin\frac{n\pi y}{w}\right]$$

The principle of superposition can be utilized to solve a wide range of steady and unsteady heat-transfer problems, as long as the differential equation and boundary-initial conditions are linear.

3–7 ANALOGY APPROACHES

Analogies have been found to exist between heat transfer and (1) the flow of electric current, (2) fluid flow, and (3) membrane behavior, which provide useful tools for developing predictions for the temperature distribution and rate of heat transfer in multidimensional systems. Because of its prominence in heat-transfer work, the electrical analogy will be presented in this section. (An introduction to the fluid flow and membrane analogies is given by Schneider [6].)

3–7–1 Electrical Analogy

As shown in Chaps. 1 and 2, electrical circuits can be designed in which the flow of current simulates the rate of heat transfer in one-dimensional systems. The electrical/heat-transfer analogy developed to this point is summarized in Table 3–1. However, the analogy between current flow and heat transfer is more far reaching in that it extends to multidimensional electric fields and thermal systems. The differential and numerical finite-difference formulations provide the basis for this broader electrical analogy.

TABLE 3–1 Electrical analogy

Electrical	*Thermal*
E_e Voltage (V)	T Temperature (°C)
I_e Current (A)	q Rate of heat transfer (W)
R_e Resistance (Ω)	R Thermal resistance (°C/W)
C_e Capacitance (F)	C Thermal capacitance (J/°C)

To expand upon this point, we first compare the Fourier equation (in Cartesian coordinates), Eq. (3–11),

$$\frac{\partial^2 T}{\partial x^2} + \frac{\partial^2 T}{\partial y^2} + \frac{\partial^2 T}{\partial z^2} = \frac{1}{\alpha}\frac{\partial T}{\partial t} \tag{3–46}$$

with the following differential equation that governs the distribution of voltage in an electrically conducting multidimensional system:

$$\frac{\partial^2 E_e}{\partial x^2} + \frac{\partial^2 E_e}{\partial y^2} + \frac{\partial^2 E_e}{\partial z^2} = \frac{R_e C_e}{L^2}\frac{\partial E_e}{\partial t} \tag{3–47}$$

Based on the similarity between these two equations, we conclude that an electrical field can be set up within an electrical conductor which corresponds to the thermal field in the modeled heat-transfer problem, with equipotential lines and orthogonal paths of electric current flow in the voltage field being representative of isotherms and paths of heat flow, respectively. These relations provide the basis for an experimental method known as the *analog field plotter*, and a resistance and capacitance (*R/C*) *network approach*, which is introduced in Chap. 4.

Analog Field Plotter

A convenient experimental arrangement by which the electric field within steady multidimensional systems can be measured is shown in Fig. 3–7. For many two-dimensional systems, electrical conducting paper, thin strips of certain metals such as Inconel, or a shallow saline bath can be patterned after the conduction system of interest. Such experimental arrangements are known as *analog field plotters*. In regard to boundary conditions, a uniform surface temperature is modeled by maintaining a uniform voltage at the surface, and insulated surfaces correspond to surfaces that are not connected to voltage sources. Lines of constant voltage, which correspond to isotherms, are then found by utilizing a millivoltmeter. Using the resulting measured system of equipotential lines, orthogonal current flow lines can be sketched in. (The flow lines can sometimes be determined by reversing the electrical boundary conditions.) The resulting network of equipotential lines and current flow lines represent the isotherms and heat-flow paths in the analogous heat-transfer system. The usefulness

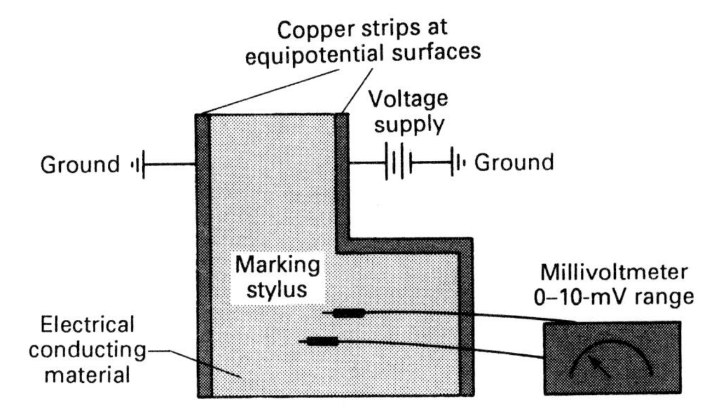

FIGURE 3–7 Representative analog field plotter arrangement.

of this information will be seen in the next section when we consider the graphical solution technique.

EXAMPLE 3–9

In Chap. 2 it was concluded that one-dimensional heat transfer occurs in composite walls consisting of only two dissimilar sections in parallel. It was also concluded that the heat transfer in composite walls consisting of combined series–parallel resistances is not one-dimensional. Demonstrate the validity of these conclusions by the use of an analog field plotter.

Solution

Objective Develop a laboratory experiment using the analog field plotter in order to show that heat transfer in composite walls is (1) one-dimensional for two sections in parallel, and (2) two-dimensional for sections arranged in series–parallel.

Assumptions/Conditions

steady-state
one- and two-dimensional systems
uniform properties

Analysis Electrical circuits that are analogous to conduction heat transfer in (a) a simple parallel composite and (b) a series–parallel composite were constructed of strips of 0.635 mm Inconel (No. 600 cold rolled), as shown in Figs. E3–9a(i) and (ii). Voltages were then applied to the ends of these two composite strips and the equipotential lines shown in the two sketches were measured. These lines are clearly one-dimensional in Fig. E3–9a(i) and two-dimensional in Fig. E3–9a(ii). Therefore, we conclude that the heat transfer in simple parallel composites consisting of two thermal resistances is one-dimensional, but that the heat transfer in combined series–parallel composites is two-dimensional.

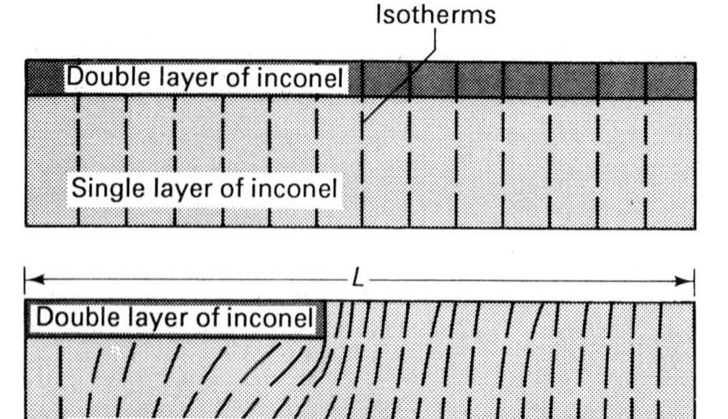

(i) Parallel arrangement.

(ii) Series-parallel arrangement.

FIGURE E3–9a
Composite wall (double layer – 5 mm wide, total wall width – 32 mm).

It follows that the heat transfer in the parallel composite can be represented by the one-dimensional thermal circuit shown in Fig. E3–9b. Representing the electrical resistivity of Inconel by ρ_e, the electrical resistance of the single layers and double layers are given by ($\delta = 0.635$ mm)

$$R_{e1} = \frac{\rho_e L}{\delta(0.027\text{ m})} \qquad R_{e2} = \frac{\rho_e L}{2\delta(0.005\text{ m})} = 2.7\, R_{e1}$$

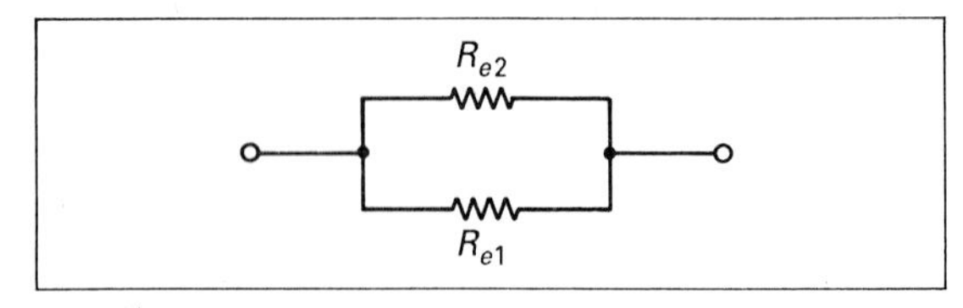

FIGURE E3–9b
Thermal circuit for parallel arrangement.

Referring to Sec. 2–2–4 of Chap. 2, because of the inherent two dimensionality of series–parallel composites, the heat transfer in the system shown in Fig. E3–9a(ii) can be approximated to within about 10% by use of a one-dimensional thermal circuit when the criterion given in Fig. 2–8 is satisfied. Otherwise, the rate of heat transfer can be estimated by taking the average of results obtained from one-dimensional lower- and upper-bound thermal circuits.

3–8 GRAPHICAL APPROACHES

Graphical solution techniques have been developed for both steady and unsteady multidimensional conduction-heat-transfer problems. The use of these approaches in analyzing two-dimensional conduction heat transfer provides further insight into these fairly complex processes. The graphical approach is presented in this section for steady two-dimensional systems.

3–8–1 Steady Two-Dimensional Systems

The key to the graphical approach to solving steady two-dimensional problems is the fact that isotherms and heat-flow lines are orthogonal (perpendicular). This point is reflected in the Fourier law of conduction itself and was touched upon in our brief

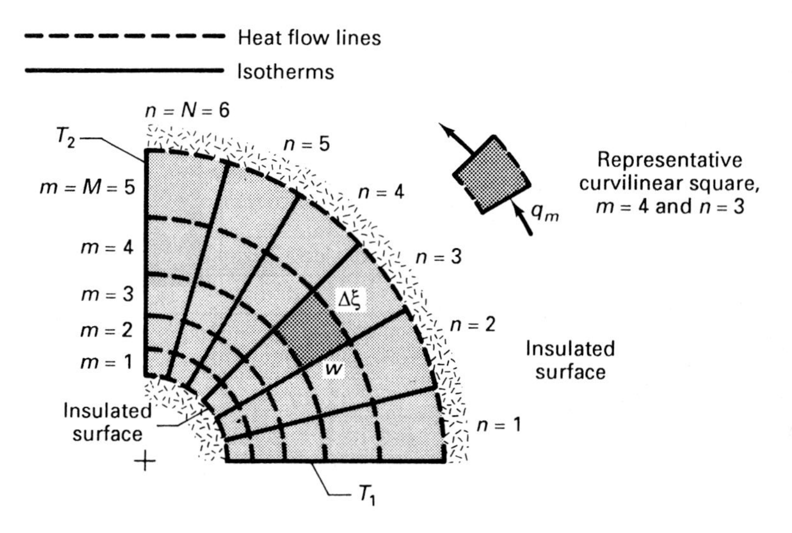

FIGURE 3–8 Cylindrical section with $r_1 = 2$ cm and $r_2 = 7.4$ cm.

study of the electrical analogy. To illustrate this point, isotherms and heat-flow lines are sketched in Fig. 3–8 for a fairly simple two-dimensional problem, such that a network of curvilinear squares is constructed.

Because heat is transferred along the M paths formed by adjacent heat-flow lines, the heat transfer in a single heat-flow lane is essentially one-dimensional (with respect to the curved coordinates, which follow the path ξ taken by individual heat-flow lanes) and is given by the following form of the Fourier law:

$$q_\xi = -kA_\xi \frac{dT}{d\xi} \tag{3–48}$$

The rate of heat transfer across each curvilinear square within the mth heat-flow path can be approximated by

$$q_m = -kL\frac{w}{\Delta\xi}\Delta T = -\frac{\Delta T}{R_n} \tag{3–49}$$

where ΔT is the mean temperature drop, $\Delta\xi$ the mean length, w the mean width, and L the mean depth associated with the nth curvilinear square. Because $\Delta\xi = w$ for full curvilinear squares, the quantity R_n is equal to $1/(kL)$.

Assuming uniform thermal conductivity k and plate length L, the rate of heat transfer in the mth heat-flow path can also be expressed in terms of the total temperature drop $T_1 - T_2$ by

$$q_m = \frac{T_1 - T_2}{\sum_{n=1}^{N_m} R_n} = \frac{kL}{N_m}(T_1 - T_2) \tag{3–50}$$

where N_m is the number of curvilinear squares in the mth heat-flow lane. The total rate of heat transfer in the system can now be obtained by summing the rates for each heat-flow path; that is,

$$q = kL(T_1 - T_2) \sum_{m=1}^{M} \frac{1}{N_m} \tag{3–51}$$

for uniform k and L. For simple problems in which the same number of full curvilinear squares N occur in each heat-flow path, Eq. (3–51) reduces to

$$q = kL \frac{M}{N}(T_1 - T_2) \tag{3–52}$$

where N and M take on integer values.

Suggestions for developing freehand plots of curvilinear networks have been developed by Bewley [7] and are summarized by Kreith and Black [8]. Although the development of such sketches is an art, one can be guided by experimental electrical analogy measurements.

EXAMPLE 3–10

Referring to the hollow quarter cylinder shown in Fig. 3–8 with $L = 1$ m, $r_1 = 2$ cm, $r_2 = 7.4$ cm, and $k = 125$ W/(m °C), determine the rate of heat transfer for (a) $T_1 = 150$°C and $T_2 = 35$°C; and (b) $T_1 = 150$°C and surface at $\theta = 0$ rad exposed to a convecting fluid with $T_F = 35$°C and $\bar{h} = 100$ W/(m^2 °C).

Solution

Objective Determine q within the cylinder section for cases (a) and (b).

Schematic Cylindrical sections.

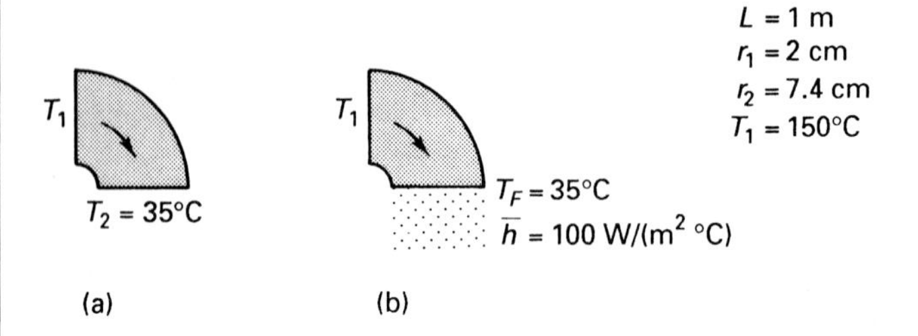

Assumptions/Conditions

steady-state
two-dimensional
uniform properties

Properties $k = 125$ W/(m °C).

Analysis Referring to Fig. 3–8, in which a network of curvilinear squares has already been developed, we see that $N = 6$ and $M = 5$.

(a) The rate of heat transfer for the case in which the surface temperatures are specified is given by Eq. (3–52),

$$q = kL\frac{M}{N}(T_1 - T_2) = 125\frac{\text{W}}{\text{m °C}}(1\text{ m})\left(\frac{5}{6}\right)(150°\text{C} - 35°\text{C}) = 12{,}000\text{ W}$$

Note that the equivalent thermal resistance for this problem is

$$R_k = \frac{N}{MkL} = \frac{1}{104}\frac{°\text{C}}{\text{W}}$$

(b) For the system with convection at one surface, we sketch the analogous electrical circuit in Fig. E3–10. Noting that the convective resistance $R_c = 1/(\bar{h}A_s)$ is equal to 1/5.4 °C/W, the rate of heat transfer through the circuit is

$$q = \frac{150°\text{C} - 35°\text{C}}{(1/104 + 1/5.4)\ °\text{C/W}} = 590\text{ W}$$

In addition, the surface temperature T_s is obtained by writing

$$T_s = T_F + R_c q = 35°\text{C} + \frac{1}{5.4}\frac{°\text{C}}{\text{W}}(590\text{ W}) = 144°\text{C}$$

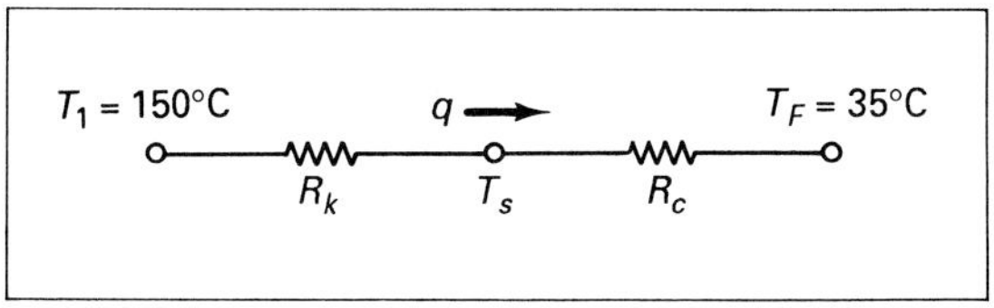

FIGURE E3–10
Thermal circuit; $R_c = 1/(\bar{h}A_s)$.

3–9 PRACTICAL SOLUTION RESULTS

The several approaches introduced in this chapter have been utilized to develop design equations for the temperature distribution and heat transfer for a number of standard conduction-heat-transfer problems that are commonly encountered in practice. Representative practical solution results are presented in this section. This useful design information is cast in the form of (1) tables for thermal resistance R, and (2) charts and analytical relations for unsteady one-dimensional processes.

3–9–1 Steady-Multidimensional Heat-Transfer Systems

The thermal resistance R for several representative steady two- and three-dimensional conduction-heat-transfer systems are given in Tables 3–2 and 3–3. A comprehensive summary of conduction shape factors S [$R = 1/(kS)$] is given by Hahne and Grigull [12].

TABLE 3–2 Thermal resistance: Steady two-dimensional systems

System	*Thermal resistance* $R = 1/(kS)$
Hollow Circular Cylinder Section: Surfaces at $\theta = 0$ and $\theta = \theta_1$ at T_1 and T_2; surfaces at $r = r_1$ and $r = r_2$ insulated. 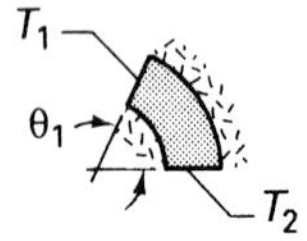	$\dfrac{\theta_1/(Lk)}{\ln(r_2/r_1)}$
Circular Cylinder Buried Horizontally in Semi-infinite Medium: $L >> r_1$. 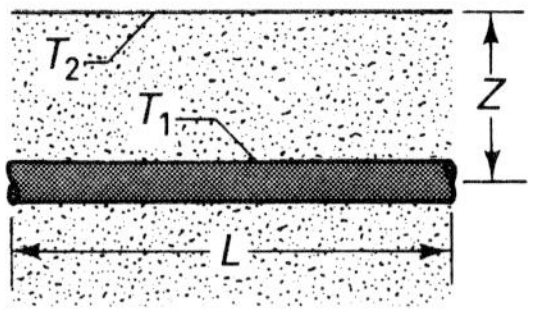	$\dfrac{\cosh^{-1}(Z/r_1)}{2\pi Lk}$
Circular Cylinder Buried Vertically in Semi-infinite Medium: $L >> r_1$. 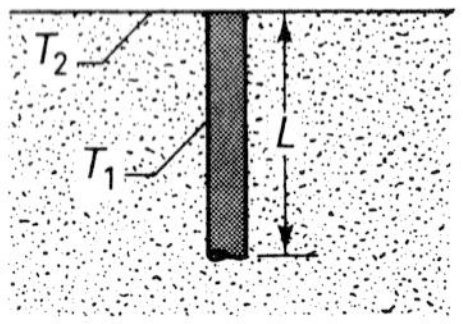	$\dfrac{\ln(2L/r_1)}{\pi Lk}$
Two Circular Cylinders Buried Horizontally in Infinite Medium: $L >> r_1, r_2$. 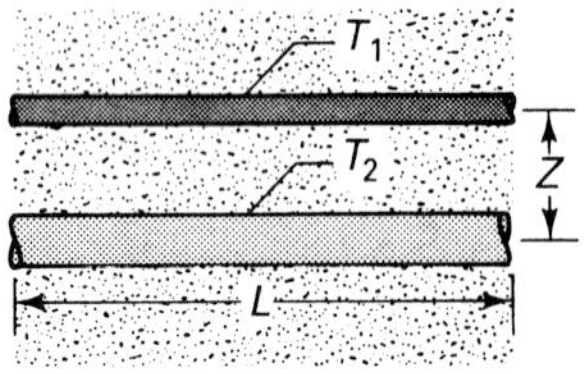	$\dfrac{\cosh^{-1}\dfrac{Z - r_1 - r_2}{2r_1r_2}}{2\pi Lk}$
Sphere Buried in Semi-infinite Medium. 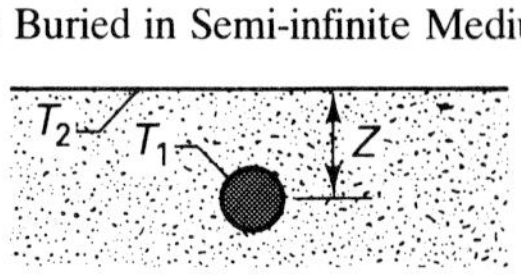	$\dfrac{1 - r_1/(2Z)}{4\pi r_1 k}$

Source: Summarized from references 9 through 11.

Note: $\cosh^{-1}(x/a) = \ln[(x + \sqrt{x^2 - a^2})/a]$ for $x/a \leqslant 1$

TABLE 3–3 Thermal resistance: Steady three-dimensional systems

System	*Thermal resistance* $R = 1/(kS)$
Circular Cylinder Buried Horizontally in Semi-infinite Medium: L short. 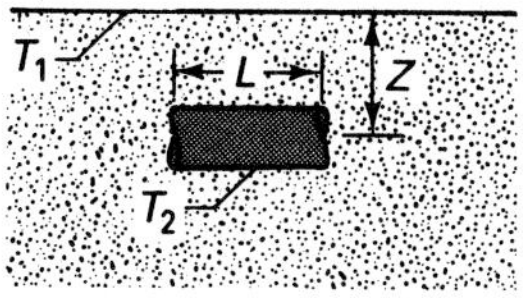	$\dfrac{\ln (L/r_1) - \ln [L/(2Z)]}{2\pi Lk}$
Two Plane Walls with Edge Section: inside dimension greater than δ. 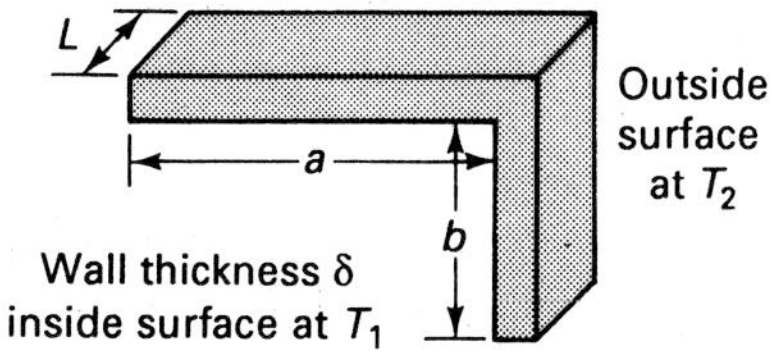	$\dfrac{1}{\left(\dfrac{aL}{\delta} + \dfrac{bL}{\delta} + 0.54L\right)k}$
Corner Section of Three Plane Walls: Inside dimensions greater than δ. 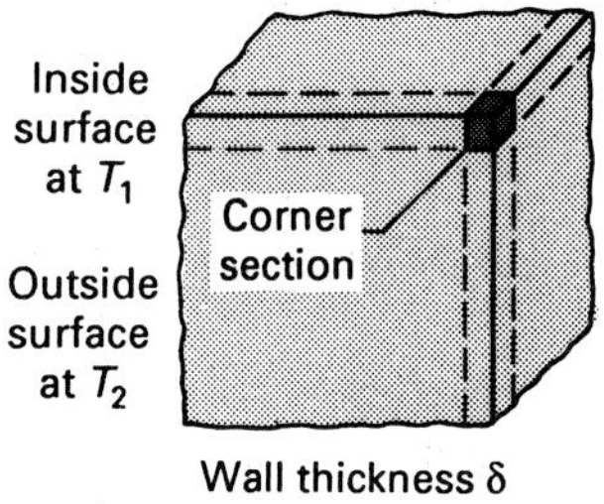	$\dfrac{1}{0.15\delta k}$

Source: Summarized from references 9 through 11 and Chap. 3.

EXAMPLE 3–11

Saturated steam at atmospheric pressure is passed through a long thin-walled horizontal pipe of 5 in. diameter, which is buried in the earth at a depth of 4 ft. Given an annual mean earth surface temperature of 51°F, estimate the mean rate of heat loss per unit length of pipe in the Chicago area for a soil thermal conductivity of 0.75 Btu/(h ft °F).

Solution

Objective Estimate the annual mean value of q'.

Schematic Steady-state indealization of thin-walled pipe buried in the earth.

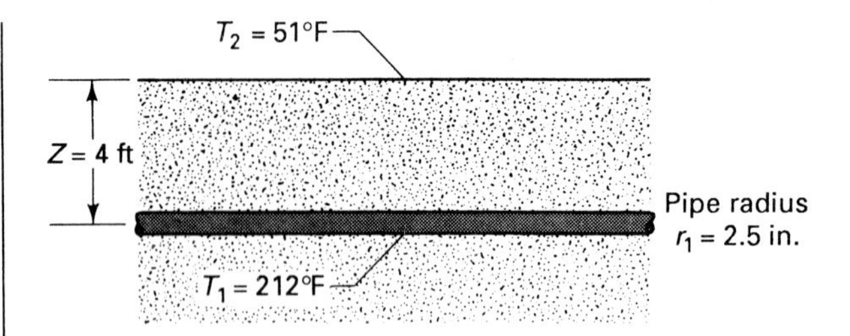

Assumptions/Conditions

steady-state
two-dimensional
uniform properties

Properties Earth: $k = 0.75$ Btu/(h ft °F).

Analysis According to the ASHRAE *Handbook of Fundamentals* [13], the annual mean earth temperature T_M is approximately constant for all depths up to about 200 ft, with the annual variation in daily mean surface temperature generally of the order of ±25°F in Chicago (see Table A–D–1). Therefore, although the temperature distribution within the earth changes over the course of a year, the mean rate of heat loss from the pipe can be approximated by assuming steady-state conditions with the earth surface temperature set equal to 51°F and the pipe temperature set equal to 212°F.

Referring to Table 3–2, we find that the thermal resistance R for this steady-state idealization of the actual process is given by

$$R = \frac{\cosh^{-1}(Z/r_1)}{2\pi Lk} = \frac{\ln [(Z + \sqrt{Z^2 - r_1^2})/r_1]}{2\pi Lk}$$

$$= \frac{\ln \{[4 + \sqrt{4^2 - (2.5/12)^2}]/(2.5/12)\}}{2\pi L[0.75 \text{ Btu/(h ft °F)}]} = \frac{0.774}{L} \frac{\text{h ft °F}}{\text{Btu}}$$

It follows that the rate of heat transfer becomes

$$q = \frac{T_1 - T_2}{R} = \frac{212°\text{F} - 51°\text{F}}{0.774/L} \frac{\text{Btu}}{\text{h ft °F}}$$

or

$$q' = \frac{q}{L} = 208 \text{ Btu/(h ft)}$$

3–9–2 Unsteady One-Dimensional Heat Transfer in Semi-Infinite Solids

Because of its importance in many practical applications, the unsteady temperature distribution and heat transfer in a semi-infinite solid with a change in conditions imposed at the surface has been extensively studied. Referring to Fig. 3–9 and assuming uniform properties, the energy equation for this type problem is given by

$$\frac{\partial^2 T}{\partial x^2} = \frac{1}{\alpha}\frac{\partial T}{\partial t} \tag{3–53}$$

Useful analytical solutions to this equation have been developed for four basic situations. The first three of these fundamental problems involve a uniform initial temperature distribution T_i, with an instantaneous change in the boundary condition imposed at $t = 0$. These three conditions include (1) a sudden step change in surface temperature T_s, (2) a sudden application of a constant heat flux q''_0, and (3) a sudden exposure of the surface to convection with T_F and h both constant.† The fourth fundamental problem involves the thermal response of a semi-infinite solid to a periodic change in the surface temperature. The initial-boundary conditions and solution results [i.e., temperature distribution $T(x,t)$, surface-heat flux $q''_s(t)$, and/or surface temperature $T_s(t)$] for these four basic cases are given in Table 3–4.

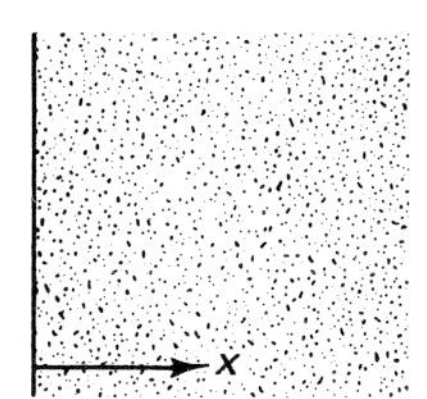

FIGURE 3–9
Semi-infinite solid with a change in boundary condition at $x = 0$.

It should be observed that these solutions involve the *error function* erf X, which is defined by the integral relation

$$\operatorname{erf} X = \frac{2}{\sqrt{\pi}}\int_0^X \exp(-\beta^2)\, d\beta \tag{3–54}$$

This important mathematical function is tabulated as a function of X in Appendix H–1. The *complementary error function* erfc X is related to the error function by

$$\operatorname{erfc} X = 1 - \operatorname{erf} X = \frac{2}{\sqrt{\pi}}\int_X^\infty \exp(-\beta^2)\, d\beta \tag{3–55}$$

The temperature distribution for a semi-infinite solid initially at T_i with surface suddenly exposed to convection (i.e., case iii) is shown in Fig. 3–10. It should be observed that the limiting curve associated with $h = \infty$ corresponds to a sudden step change in temperature (i.e., case i).

† $\bar{h}$ is represented by h for situations in which the coefficient of heat transfer is uniform over the surface.

TABLE 3–4 Solution results: Unsteady heat transfer in a semi-infinite solid with uniform properties.

Boundary condition	*Solution results*
Case i: Step change in surface temperature $T(x,0) = T_i$ $T(0,t) = T_0$ $T(\infty,t) = T_i$ 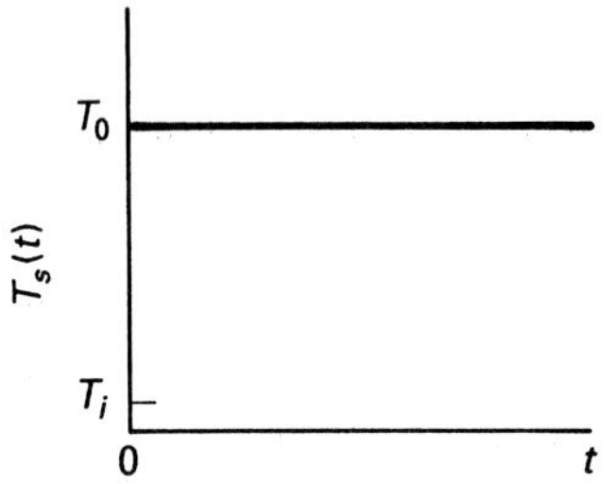	$\dfrac{T(x,t) - T_0}{T_i - T_0} = \operatorname{erf}\left(\dfrac{x}{2\sqrt{\alpha t}}\right)$ $q_s''(t) = \dfrac{k(T_0 - T_i)}{\sqrt{\pi \alpha t}}$ 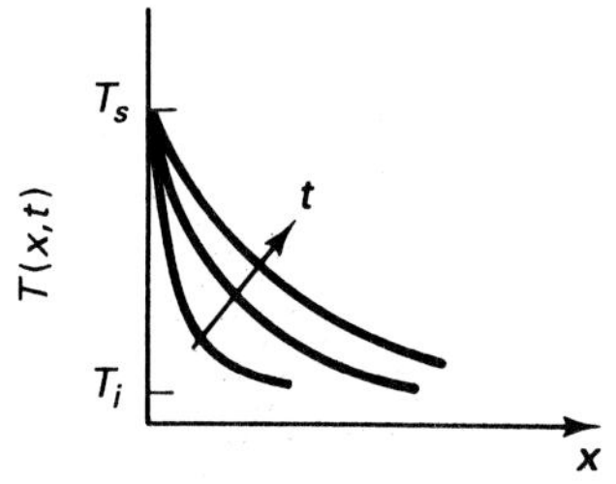
Case ii: Constant surface-heat flux $T(x,0) = T_i$ $q_s'' = -k\left.\dfrac{\partial T}{\partial x}\right\rvert_{x=0} = q_0''$ $T(\infty,t) = T_i$ 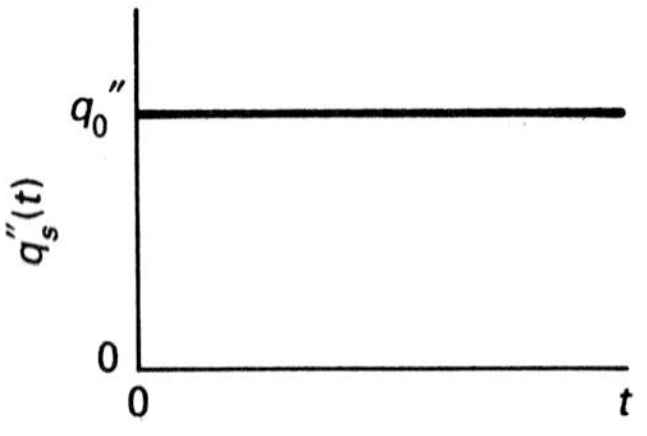	$T(x,t) - T_i = 2\dfrac{q_0''}{k}\sqrt{\dfrac{\alpha t}{\pi}}\exp\left(\dfrac{-x^2}{4\alpha t}\right) - \dfrac{q_0'' x}{k}\operatorname{erfc}\left(\dfrac{x}{2\sqrt{\alpha t}}\right)$ $T_s(t) - T_i = 2\dfrac{q_0''}{k}\sqrt{\dfrac{\alpha t}{\pi}}$ 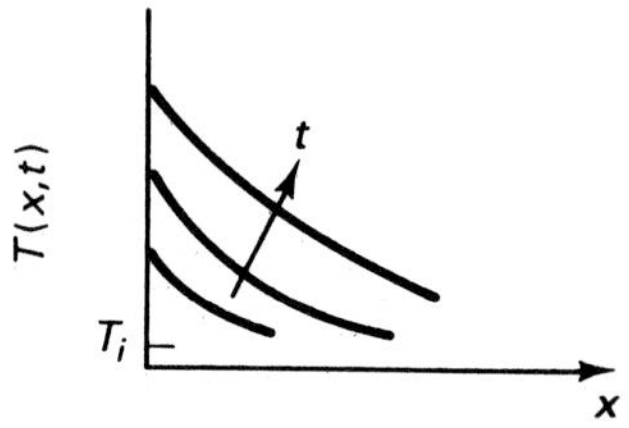

Continued on next page

TABLE 3-4 *(Continued)*

Boundary condition	*Solution results*
Case iii: Surface convection $T(x,0) = T_i$ $-k \left.\dfrac{\partial T}{\partial x}\right\vert_{x=0} = h[T_F - T(0,t)]$ $T(\infty,t) = T_i$	$\dfrac{T(x,t) - T_i}{T_F - T_i} = \operatorname{erfc}\left(\dfrac{x}{2\sqrt{\alpha t}}\right) - \left[\exp\left(\dfrac{hx}{k} + \dfrac{h^2\alpha t}{k^2}\right)\right]\left[\operatorname{erfc}\left(\dfrac{x}{2\sqrt{\alpha t}} + \dfrac{h\sqrt{\alpha t}}{k}\right)\right]$ $\dfrac{T_s(t) - T_i}{T_F - T_i} = 1 - \exp\left(\dfrac{h^2\alpha t}{k^2}\right)\operatorname{erfc}\left(\dfrac{h\sqrt{\alpha t}}{k}\right)$ $q_s''(t) = h(T_F - T_i)\exp\left(\dfrac{h^2\alpha t}{k^2}\right)\operatorname{erfc}\left(\dfrac{h\sqrt{\alpha t}}{k}\right)$ 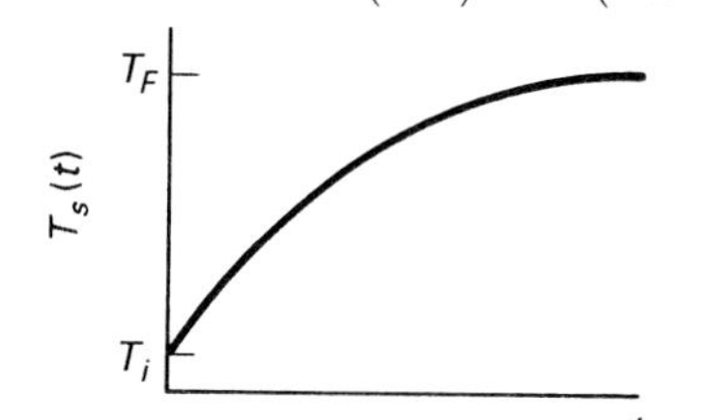 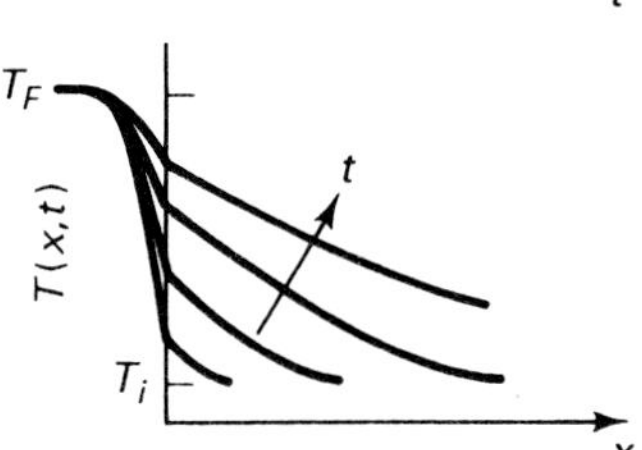
Case iv: Periodic surface temperature $T(0,t) = T_M - \Delta T_s \cos\left[\dfrac{2\pi}{\tau}(t - t_0)\right]$ $T(x,t) = T(x,t+n\tau);\ n = 0,1,2,3,\ldots$ 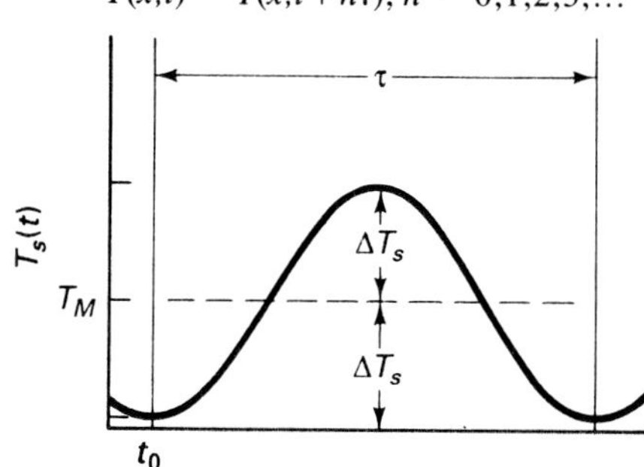 	$T(x,t) = T_M - \Delta T_s \exp\left(-x\sqrt{\dfrac{\pi}{\alpha\tau}}\right) \times \cos\left[\dfrac{2\pi}{\tau}(t - t_0) - x\sqrt{\dfrac{\pi}{\alpha\tau}}\right]$ $q_s''(t) = \dfrac{k\,\Delta T_s}{\sqrt{\alpha\tau/\pi}}\left\{-\cos\left[\dfrac{2\pi}{\tau}(t - t_0)\right] + \sin\left[\dfrac{2\pi}{\tau}(t - t_0)\right]\right\}$

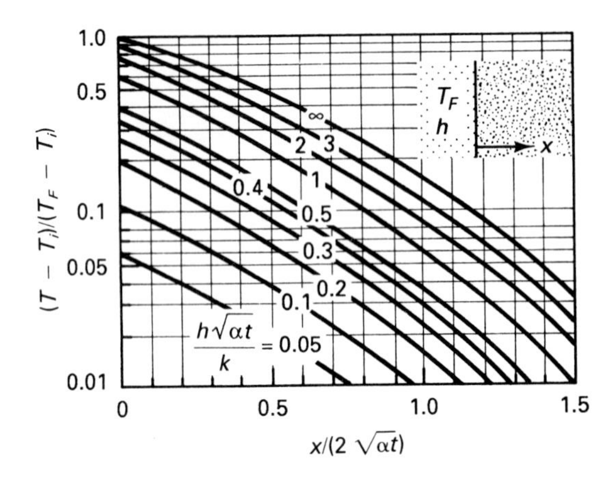

FIGURE 3–10
Chart for instantaneous temperature distribution in a semi-infinite solid with surface suddenly exposed to a fluid. (From Schneider [6]. Used with permission.)

EXAMPLE 3–12

Referring to Table 3–4, the temperature distribution in a semi-infinite solid initially at uniform temperature T_i with surface temperature suddenly changed to T_0 is

$$\frac{T(x,t) - T_0}{T_i - T_0} = \text{erf}\left(\frac{x}{2\sqrt{\alpha t}}\right)$$

for uniform properties. Use this relation to develop expressions for the instantaneous rate $q_s(t)$ and total accumulative Q heat transfer at the surface.

Solution

Objective Develop relations for $q_s(t)$ and Q.

Schematic Unsteady one-dimensional heat transfer in a semi-infinite solid.

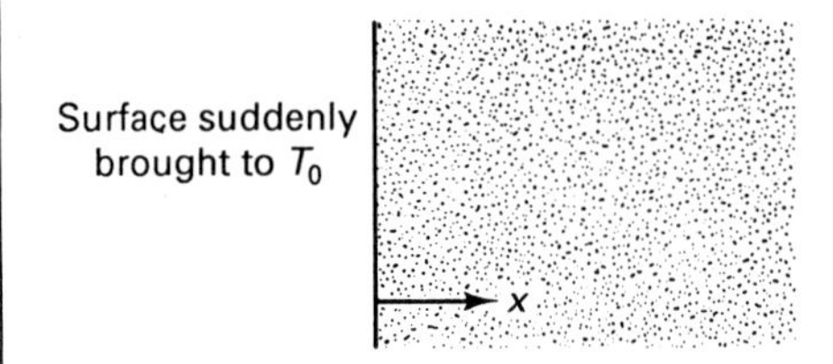

Assumptions/Conditions

unsteady-state
one spatial dimension
uniform properties

Analysis Making use of the defining equation for the error function given by Eq. (3–54), the temperature distribution is represented by

$$\frac{T(x,t) - T_0}{T_i - T_0} = \frac{2}{\sqrt{\pi}} \int_0^{x/(2\sqrt{\alpha t})} e^{-\beta^2} \, d\beta$$

Using this relation together with the Fourier law of conduction, the instantaneous rate of heat transfer within the semi-infinite solid is written as

$$q_x(t) = -kA \left.\frac{\partial T}{\partial x}\right|_x = -kA(T_i - T_0) \frac{2}{\sqrt{\pi}} \frac{\partial}{\partial x} \int_0^{x/(2\sqrt{\alpha t})} e^{-\beta^2} \, d\beta$$

The differentiation of this integral is accomplished by employing the Leibnitz rule (Appendix A), with the result

$$q_x(t) = -kA(T_i - T_0) \frac{2}{\sqrt{\pi}} \frac{e^{-x^2/(4\alpha t)}}{2\sqrt{\alpha t}} = \frac{kA(T_0 - T_i)}{\sqrt{\pi \alpha t}} e^{-x^2/(4\alpha t)}$$

Setting $x = 0$, we obtain

$$q_s(t) = \frac{kA(T_0 - T_i)}{\sqrt{\pi \alpha t}}$$

which is equivalent to the relation listed in Table 3–4 (case i) for $q_s''(t)$.

The total accumulative heat transfer Q is obtained by writing

$$Q = \int_0^t q_s(t) \, dt = \int_0^t \frac{kA(T_0 - T_i)}{\sqrt{\pi \alpha t}} \, dt = 2kA(T_0 - T_i) \sqrt{\frac{t}{\pi \alpha}}$$

EXAMPLE 3–13

One surface of a thick aluminum slab initially at 75°F is suddenly exposed to a fluid with $T_F = 250$°F and $h = 3000$ Btu/(h ft² °F). Determine the instantaneous surface temperature and heat flux after a time of 1 s.

Solution

Objective Determine T_s and q_s'' for $t = 1$ s.

Schematic Thick slab with one surface suddenly exposed to convection.

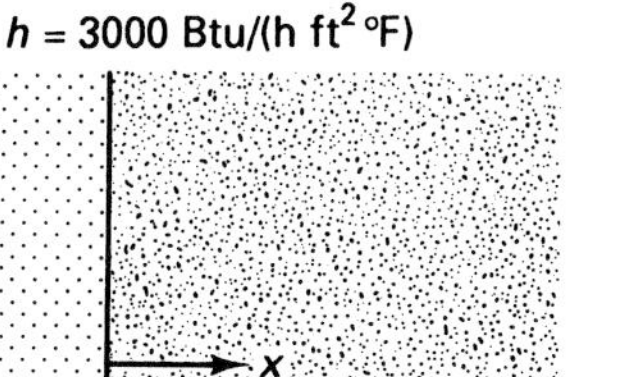

Assumptions/Conditions

unsteady-state
one spatial dimension
uniform properties

Properties Aluminum (Table A–C–1):

$$k = 236 \frac{\text{W}}{\text{m °C}} \frac{0.578 \text{ Btu/(h ft °F)}}{\text{W/(m °C)}} = 136 \text{ Btu/(h ft °F)}$$

$$\alpha = 9.75 \times 10^{-5} \frac{\text{m}^2}{\text{s}} \frac{10.8 \text{ ft}^2\text{/s}}{\text{m}^2\text{/s}} = 1.05 \times 10^{-3} \text{ ft}^2\text{/s}$$

Analysis Since the slab is thick and only one surface is exposed to convection, this system can be modeled as a semi-infinite solid. Referring to case iii of Table 3–4, the instantaneous surface temperature and heat flux are given by

$$\frac{T_s - T_i}{T_F - T_i} = 1 - \exp\left(\frac{h^2 \alpha t}{k^2}\right) \text{erfc}\left(\frac{h\sqrt{\alpha t}}{k}\right) \tag{a}$$

and

$$q_s'' = h(T_F - T_i) \exp\left(\frac{h^2 \alpha t}{k^2}\right) \text{erfc}\left(\frac{h\sqrt{\alpha t}}{k}\right) \tag{b}$$

Setting $t = 1$ s, we have

$$\frac{h\sqrt{\alpha t}}{k} = 3000 \frac{\text{Btu}}{\text{h ft}^2\text{°F}} \frac{\text{h ft °F}}{136 \text{ Btu}} \left[\frac{1.05 \times 10^{-3} \text{ ft}^2 (1 \text{ s})}{\text{s}}\right]^{1/2} = 0.715$$

$$\frac{h^2 \alpha t}{k^2} = 0.715^2 = 0.511$$

and, using Table A–H–1,

$$\text{erfc}\left(\frac{h\sqrt{\alpha t}}{k}\right) = 1 - \text{erf}(0.715) = 1 - 0.688 = 0.312$$

Substituting into Eqs. (a) and (b), we obtain

$$\frac{T_s - T_i}{T_F - T_i} = 1 - 0.312 \exp(0.511) = 0.480$$

or

$$T_s = (250\text{°F} - 75\text{°F})(0.480) + 75\text{°F} = 159\text{°F}$$

and

$$q_s'' = 3000 \frac{\text{Btu}}{\text{h ft}^2 \text{ °F}} (250\text{°F} - 75\text{°F})(0.312) \exp(0.511)$$

$$= 2.73 \times 10^5 \text{ Btu/(h ft}^2\text{)}$$

for $t = 1$ s.

It should be noted that Fig. 3–10 could be used instead of Eq. (a) to evaluate T_s. Following this practical but somewhat less accurate approach, the surface heat flux could be computed by use of the Newton law of cooling,

$$q_s'' = h(T_s - T_F)$$

EXAMPLE 3–14

The daily mean temperature distribution within the region near the surface of the earth at any time of the year is commonly approximated by the solution for a semi-infinite solid with a periodic boundary condition of the form

$$T_s(t) = T_M - \Delta T_s \cos\left[\frac{2\pi}{\tau}(t - t_0)\right] \qquad \text{at } x = 0 \tag{a}$$

where the period τ is set equal to 365 days, T_M is the annual mean earth temperature, ΔT_s is the amplitude of annual variation in surface soil temperature, and t_0 is the phase constant. The solution for the temperature distribution for this situation is given in Table 3–4 as (case iv)

$$T(x,t) = T_M - \Delta T_s \exp\left(-x\sqrt{\frac{\pi}{\alpha\tau}}\right)\cos\left[\frac{2\pi}{\tau}(t - t_0) - x\sqrt{\frac{\pi}{\alpha\tau}}\right] \tag{b}$$

Earth data for T_M, ΔT_s, and t_0 are listed in Appendix Table A–D–1 for selected cities in the U.S. Use this relation to estimate the depth of the *freezing line* (i.e., the minimum soil depth at which freezing will not occur over the course of a year under normal conditions) in the Chicago area, assuming damp, heavy soil.

Solution

Objective Approximate the freeze line depth x_{fl} in Chicago, Illinois.

Schematic Heat transfer near the surface of the earth.

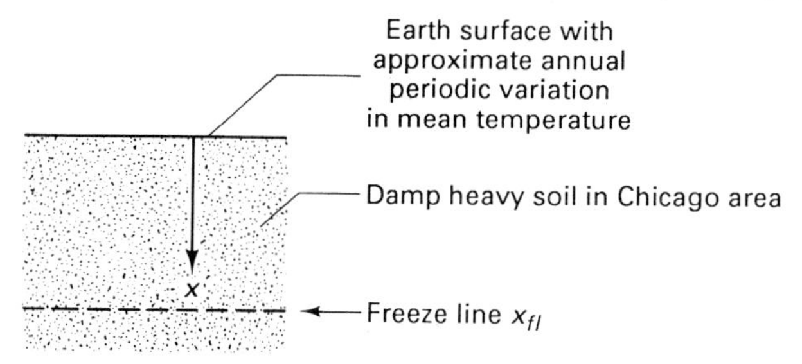

Assumptions/Conditions

unsteady-state
one spatial dimension
uniform properties

Properties Soil (damp heavy) (Table A–C–2): $k = 0.75$ Btu/(h ft °F), $\alpha = 0.6$ ft²/day.

Analysis The distribution in minimum temperature, which is associated with Eq. (b), is obtained by setting the cosine term equal to unity; that is,

$$T(x,t)_{\min} = T_M - \Delta T_s \exp\left(-x\sqrt{\frac{\pi}{\alpha\tau}}\right)$$

Rearranging, this expression is put into the form

$$x = \sqrt{\frac{\alpha\tau}{\pi}} \ln\left[\frac{\Delta T_s}{T_M - T(x,t)_{\min}}\right]$$

Setting $T(x,t)_{\min} = 32$°F and substituting for the various parameters and earth temperature data for Chicago (i.e., $T_M = 51$°F and $\Delta T_s = 25$°F), the depth of the freezing line becomes

$$x_{fl} = \sqrt{\frac{(0.6\ \text{ft}^2/\text{day})(365\ \text{day})}{\pi}} \ln\left(\frac{25°\text{F}}{51°\text{F} - 32°\text{F}}\right) = 2.29\ \text{ft}$$

Thus, under normal weather conditions the freezing line depth for damp heavy soil in the Chicago area is about 2.29 ft. Adding a 20% safety factor to allow for variations in weather, water pipes buried below 2.75 ft should be safe from freezing for this soil and location.

3–9–3 Unsteady One-Dimensional Heat Transfer in Flat Plates, Circular Cyclinders, and Spheres

We now turn our attention to unsteady one-dimensional heat transfer in the flat plate, circular cylinder, and spherical geometries pictured in Fig. 3–11. Solutions for the temperature distribution and heat transfer in these geometries have been developed for various boundary conditions.

Focusing attention on the case in which a flat plate of width $2L$ initially at uniform temperature T_i is suddenly subjected to convection, the mathematical formulation is given by (for uniform properties)

$$\frac{\partial^2 T}{\partial x^2} = \frac{1}{\alpha}\frac{\partial T}{\partial t} \tag{3–56}$$

$$T = T_i \qquad \text{at } t = 0$$

$$-k\frac{\partial T}{\partial x} = h(T - T_F) \qquad \text{at } x = L$$

and

$$\frac{\partial T}{\partial x} = 0 \qquad \text{at } x = 0$$

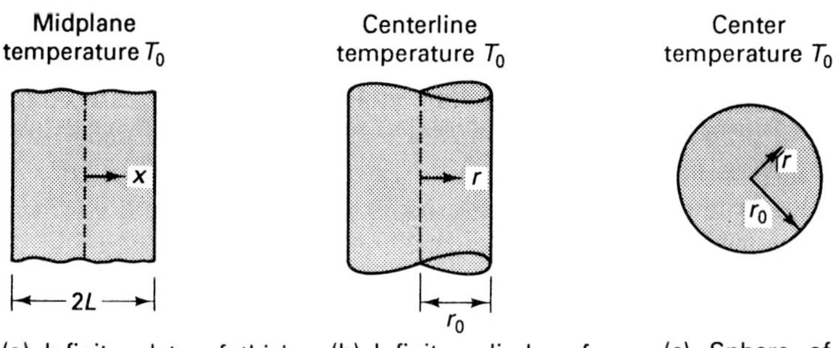

(a) Infinite plate of thickness $2L$; $\ell_0 = L$. (b) Infinite cylinder of radius r_0; $\ell_0 = r_0$. (c) Sphere of radius r_0; $\ell_0 = r_0$.

FIGURE 3–11 One-dimensional solids with a change in boundary conditions at surface; characteristic length ℓ_0.†

or

$$-k\frac{\partial T}{\partial x} = h(T_F - T) \qquad \text{at } x = -L$$

It should be observed that this formulation also applies to a flat plate of thickness L with surface at $x = 0$ insulated and surface at $x = L$ exposed to a convective fluid.

The exact solution to this system of equations for the case in which h is constant, which can be obtained by the separation-of-variables method introduced in Sec. 3–6–1, is given in Appendix E–2. The general solution is in the form of an infinite series, with the number of significant terms in the series dependent upon a dimensionless time known as the *Fourier number Fo* ($Fo = \alpha t/L^2 = \alpha t/\ell_0^2$). As pointed out by Heisler [14], for values of Fo greater than 0.2, which applies to 80% to 90% of the process period, the solution for the temperature distribution is approximated with an error of less than 1% by truncating the second- and higher-order terms in the series. The resulting approximate solution for dimensionless temperature Θ is

$$\Theta = \frac{T - T_F}{T_i - T_F} = \Theta_0 \cos\left(\gamma_1 \frac{x}{L}\right) \tag{3–57}$$

where the dimensionless midplane temperature Θ_0 is

$$\Theta_0 = \frac{T_0 - T_F}{T_i - T_F} = C_1 \exp(-\gamma_1^2 Fo) \tag{3–58}$$

and the coefficients γ_1 and C_1 are given in terms of Biot number Bi_0 ($Bi_0 = Bi = hL/k$) in Table 3–5. It follows that the surface temperature T_s and instantaneous rate of convection heat transfer q_c are given by

† Whereas $\ell = V/A_s$ is generally used as the characteristic length in the approximate lumped analysis approach of Chap. 2, ℓ_0 is used in the formal solution approach considered in this section. Note that ℓ and ℓ_0 are equivalent for the flat plate, but differ for the circular cylinder and sphere.

$$\Theta_s = \frac{T_s - T_F}{T_i - T_F} = \Theta_0 \cos \gamma_1 \tag{3–59}$$

and

$$q_c = hA_s(T_s - T_F) = hA_s(T_i - T_F)\,\Theta_0 \cos \gamma_1 \tag{3–60}$$

TABLE 3–5 Heisler relations: Unsteady one-dimensional convection cooling of a plane wall†

Coefficients C_1 and γ_1 as function of $Bi_0 = Bi = hL/k$			
Bi_0	γ_1 (rad)	C_1	Other solution relations
0.01	0.0998	1.0017	$\Theta_0 = C_1 \exp(-\gamma_1^2 Fo)$
0.02	0.1410	1.0033	
0.03	0.1732	1.0049	$\Theta = \Theta_0 \cos(\gamma_1 x/L)$
0.04	0.1987	1.0066	
0.05	0.2217	1.0082	$\Theta_s = \Theta_0 \cos \gamma_1$
0.06	0.2425	1.0098	$q_c = hA_s(T_i - T_F)\Theta_s$
0.07	0.2615	1.0114	
0.08	0.2791	1.0130	$\dfrac{Q}{Q_{max}} = 1 - \Theta_0 \dfrac{\sin \gamma_1}{\gamma_1}$
0.09	0.2956	1.0145	
0.10	0.3111	1.0160	
0.15	0.3779	1.0237	
0.20	0.4328	1.0311	
0.25	0.4801	1.0382	The coefficients C_1 and γ_1 are formally given in Appendix E by Eq. (E–2–3),
0.30	0.5218	1.0450	
0.4	0.5932	1.0580	$\gamma_1 \tan \gamma_1 = Bi_0$
0.5	0.6533	1.0701	
0.6	0.7051	1.0814	and Eq. (E–2–1),
0.7	0.7506	1.0919	
0.8	0.7910	1.1016	$C_1 = \dfrac{4 \sin \gamma_1}{2\gamma_1 + \sin(2\gamma_1)}$
0.9	0.8274	1.1107	
1.0	0.8603	1.1191	
2.0	1.0769	1.1795	
3.0	1.1925	1.2102	
4.0	1.2646	1.2287	
5.0	1.3138	1.2402	
6.0	1.3496	1.2479	
7.0	1.3766	1.2532	
8.0	1.3978	1.2570	
9.0	1.4149	1.2598	
10.0	1.4289	1.2620	
20.0	1.4961	1.2699	
30.0	1.5202	1.2717	
40.0	1.5325	1.2723	
50.0	1.5400	1.2727	
100.0	1.5552	1.2731	

† The Heisler relations are within 1% of the exact solution for $Fo > 0.2$.

To obtain an expression for the total accumulated heat transfer Q from the surface, we write†

$$Q = \Sigma E_o - \Sigma E_i = -\Delta E_s = -[E_s(t) - E_s(0)]$$
$$= -\int_V \rho c_v(T - T_i)\,dV = -\int_0^L \rho c_v A(T - T_i)\,dx \tag{3–61}$$
$$= \rho c_v A(T_i - T_F)\int_0^L (1 - \Theta)\,dx = \rho c_v V(T_i - T_F)\left(1 - \Theta_0 \frac{\sin\gamma_1}{\gamma_1}\right)$$

or

$$\frac{Q}{Q_{\max}} = 1 - \Theta_0 \frac{\sin\gamma_1}{\gamma_1} \tag{3–62}$$

where $Q_{\max} = \rho c_v V(T_i - T_F)$.

These solution results for dimensionless midplane temperature Θ_0, dimensionless temperature distribution Θ, and total accumulated heat transfer Q are represented by the convenient Heisler [14] charts shown in Fig. 3–12. With Fo and Bi_0 specified, Fig. 3–12(a) can be used to evaluate T_0; Fig. 3–12(b) can be used to evaluate the temperature T at any location off the midplane, such as the surface temperature $T = T_s$ at $x/L = 1$; and Fig. 3–12(c) can be used to evaluate Q.

The mathematical formulations for energy transfer from a circular cylinder and sphere, respectively, which are initially at uniform temperature T_i and suddenly exposed to a convecting fluid, are represented by (for uniform properties)

$$\frac{1}{r}\frac{\partial}{\partial r}\left(r\frac{\partial T}{\partial r}\right) = \frac{1}{\alpha}\frac{\partial T}{\partial t} \tag{3–63}$$

for a circular cylinder, and

$$\frac{1}{r^2}\frac{\partial}{\partial r}\left(r^2\frac{\partial T}{\partial r}\right) = \frac{1}{\alpha}\frac{\partial T}{\partial t} \tag{3–64}$$

for a sphere, with initial and boundary conditions of the form

$$T = T_i \qquad \text{at } t = 0$$
$$\frac{\partial T}{\partial r} = 0 \qquad \text{at } r = 0 \tag{3–65}$$
$$-k\frac{\partial T}{\partial r} = h(T - T_F) \qquad \text{at } r = r_0$$

† Q can also be represented by

$$Q = \int_0^t q_c\,dt$$

However, since Eq. (3–60) is restricted to the time domain for which $Fo > 0.2$, this alternative formulation is not useful in the present analysis.

for both geometries. The exact solution to these equations is given in Appendix E–2 for constant values of h. Approximate solutions, which are applicable for values of the Fourier number Fo ($Fo = \alpha t/r_0^2 = \alpha t/\ell_0^2$) greater than 0.2, are listed in Tables 3–6 and 3–7 and are represented graphically by the Heisler charts shown in Figs. 3–13 and 3–14. With Fo and the r_0 based Biot number Bi_0 specified, these solution results enable us to evaluate Θ_0, Θ, q_c, and Q.

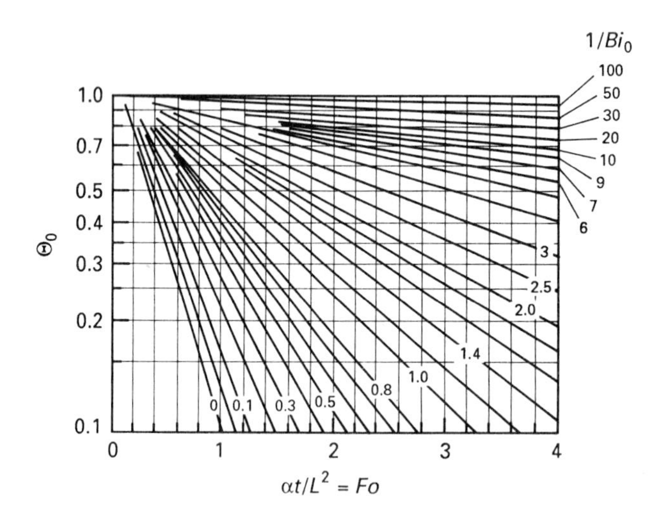

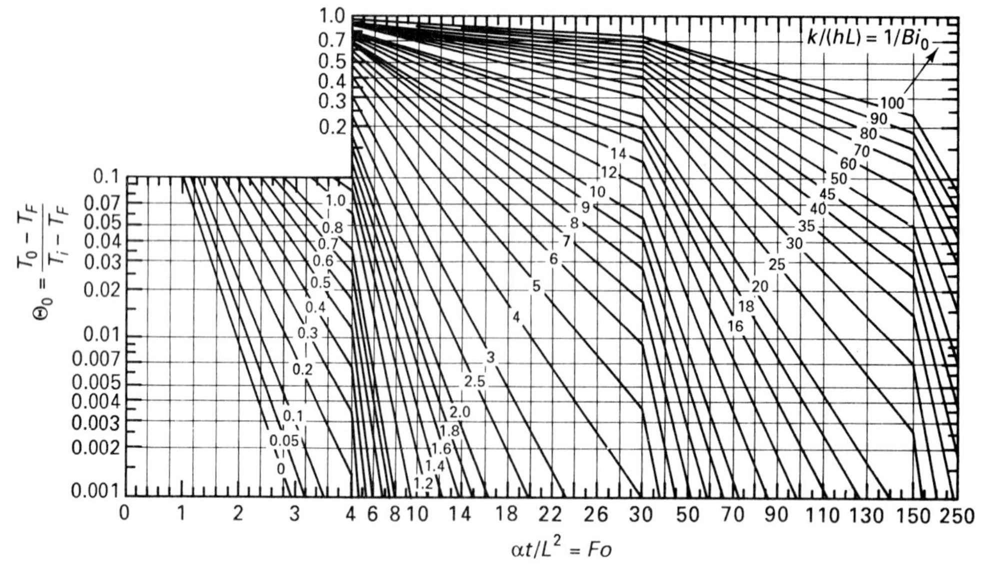

(a) Instantaneous midplane temperature T_0 (Heisler [14]).

FIGURE 3–12 Heisler charts: Unsteady one-dimensional convection cooling of a plane wall; $Bi_0 = Bi = hL/k$ and $Fo = \alpha t/L^2$.

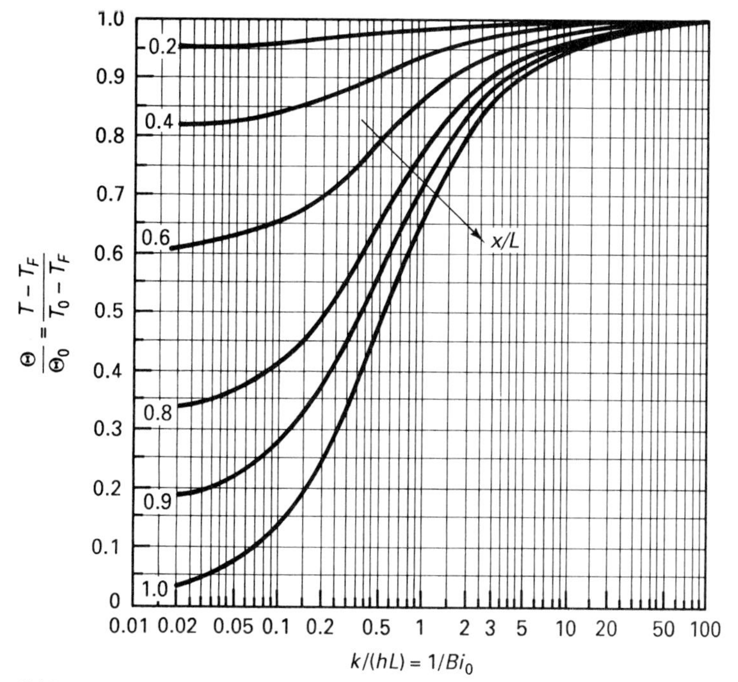

(b) Instantaneous temperature distribution for plane wall in terms of T_0 (Heisler [14]).

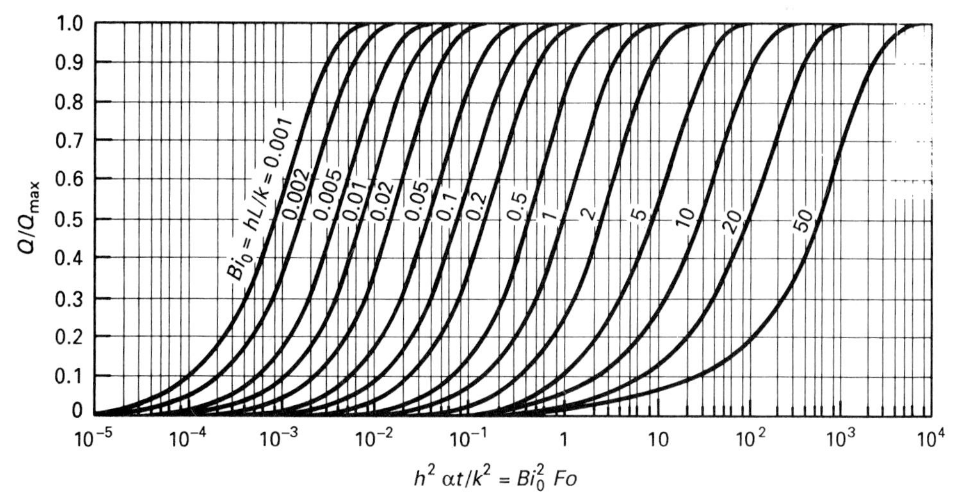

(c) Total accumulative heat transfer Q for plane wall. (From Grober and Grigull [15]. Used with permission.)

FIGURE 3–12 *(Continued)*

TABLE 3–6 Heisler relations: Unsteady one-dimensional convection cooling of an infinite circular cylinder†

Coefficients C_1 and γ_1 as function of $Bi_0 = hr_0/k$

Bi_0	γ_1 (rad)	C_1
0.01	0.1412	1.0025
0.02	0.1995	1.0050
0.03	0.2439	1.0075
0.04	0.2814	1.0099
0.05	0.3142	1.0124
0.06	0.3438	1.0148
0.07	0.3708	1.0173
0.08	0.3960	1.0197
0.09	0.4195	1.0222
0.10	0.4417	1.0246
0.15	0.5376	1.0365
0.20	0.6170	1.0483
0.25	0.6856	1.0598
0.30	0.7465	1.0712
0.4	0.8516	1.0932
0.5	0.9408	1.1143
0.6	1.0185	1.1346
0.7	1.0873	1.1539
0.8	1.1490	1.1725
0.9	1.2048	1.1902
1.0	1.2558	1.2071
2.0	1.5995	1.3384
3.0	1.7887	1.4191
4.0	1.9081	1.4698
5.0	1.9898	1.5029
6.0	2.0490	1.5253
7.0	2.0937	1.5411
8.0	2.1286	1.5526
9.0	2.1566	1.5611
10.0	2.1795	1.5677
20.0	2.2881	1.5919
30.0	2.3261	1.5973
40.0	2.3455	1.5993
50.0	2.3572	1.6002
100.0	2.3809	1.6015

Other solution relations

$$\Theta_0 = C_1 \exp(-\gamma_1^2 Fo)$$

$$\Theta = \Theta_0 J_0(\gamma_1 r/r_0)$$

$$\Theta_s = \Theta_0 J_0(\gamma_1)$$

$$q_c = hA_s(T_i - T_F)\Theta_s$$

$$\frac{Q}{Q_{max}} = 1 - 2\,\Theta_0 \frac{J_1(\gamma_1)}{\gamma_1}$$

J_0 – Bessel function of first kind, zero order

J_1 – Bessel function of first kind, first order

Representative values of J_0 and J_1 are tabulated in Table A–H–1–2.

The coefficients C_1 and γ_1 are formally given in Appendix E by Eq. (E–2–6),

$$\gamma_1 J_1(\gamma_1) = Bi_0 J_0(\gamma_1)$$

and Eq. (E–2–5),

$$C_1 = \frac{2}{\gamma_1} \frac{J_1(\gamma_1)}{J_0^2(\gamma_1) + J_1^2(\gamma_1)}$$

† The Heisler relations are within 1% of the exact solution for $Fo > 0.2$.

Notice in Figs. 3–12(b), 3–13(b), and 3–14(b) that the temperature T throughout the plane wall, circular cylinder, or sphere is only slightly dependent on location within the body for small values of the Biot number Bi_0 (or Bi). It follows that the unsteady lumped analysis developed in Chap. 2 can be used for such conditions; that is,

$$\Theta_0 = \frac{T_0 - T_F}{T_i - T_F} = \exp\left(-\frac{hA_s t}{\rho c_v V}\right) = \exp\left(-\frac{ht}{\rho c_v \ell}\right) \tag{3–66}$$

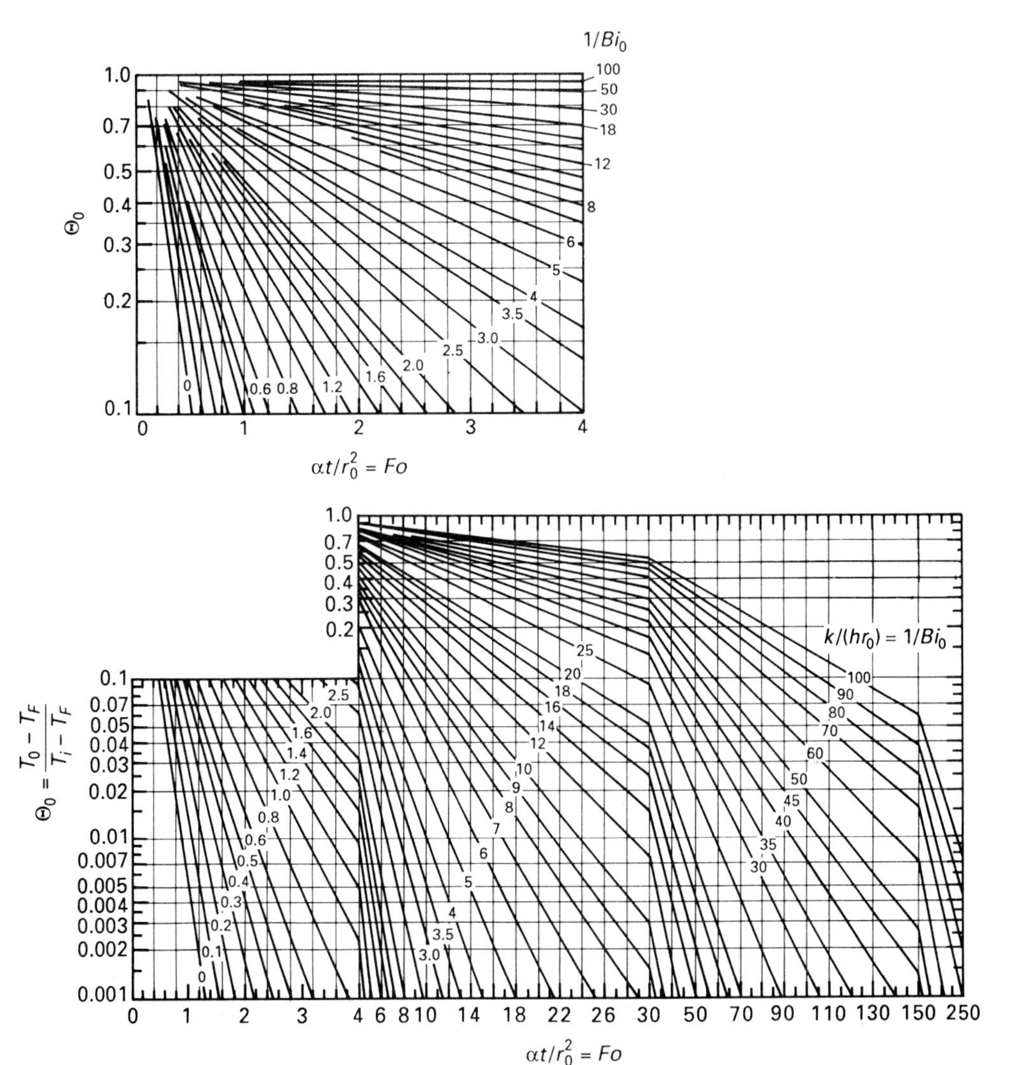

(a) Instantaneous centerline temperature T_0 (Heisler [14]).

FIGURE 3–13 Heisler charts: Unsteady one-dimensional convection cooling of an infinite cylinder; $Bi_0 = hr_0/k$ and $Fo = \alpha t/r_0^2$.

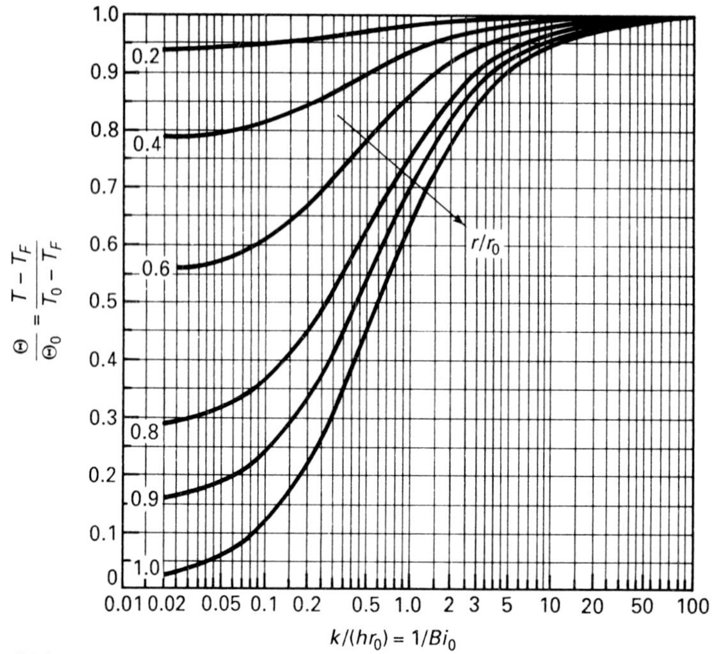

(b) Instantaneous temperature distribution for infinite cylinder in terms of T_0 (Heisler [14].

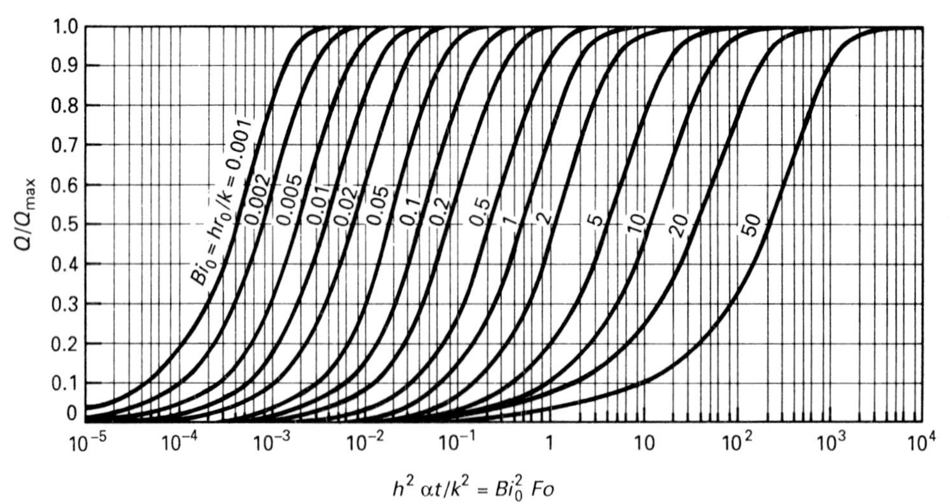

(c) Total accumulative heat transfer Q for infinite cylinder. (From Grober and Grigull [15]. Used with permission.)

FIGURE 3–13 *(Continued)*

TABLE 3–7 Heisler relations: Unsteady one-dimensional convection cooling of a sphere†

Coefficients C_1 and γ_1 as function of $Bi_0 = hr_0/k$

Bi_0	γ_1 (rad)	C_1
0.01	0.1730	1.0030
0.02	0.2445	1.0060
0.03	0.2989	1.0090
0.04	0.3450	1.0120
0.05	0.3852	1.0149
0.06	0.4217	1.0179
0.07	0.4550	1.0209
0.08	0.4860	1.0239
0.09	0.5150	1.0268
0.10	0.5423	1.0298
0.15	0.6608	1.0445
0.20	0.7593	1.0592
0.25	0.8448	1.0737
0.30	0.9208	1.0880
0.4	1.0528	1.1164
0.5	1.1656	1.1441
0.6	1.2644	1.1713
0.7	1.3525	1.1978
0.8	1.4320	1.2236
0.9	1.5044	1.2488
1.0	1.5708	1.2732
2.0	2.0288	1.4793
3.0	2.2889	1.6227
4.0	2.4556	1.7201
5.0	2.5704	1.7870
6.0	2.6537	1.8338
7.0	2.7165	1.8674
8.0	2.7654	1.8921
9.0	2.8044	1.9106
10.0	2.8363	1.9249
20.0	2.9857	1.9781
30.0	3.0372	1.9898
40.0	3.0632	1.9942
50.0	3.0788	1.9962
100.0	3.1102	1.9990

Other solution relations

$$\Theta_0 = C_1 \exp(-\gamma_1^2 Fo)$$

$$\Theta = \Theta_0 \frac{\sin(\gamma_1 r/r_0)}{\gamma_1 r/r_0}$$

$$\Theta_s = \Theta_0 \frac{\sin \gamma_1}{\gamma_1}$$

$$q_c = hA_s(T_i - T_F)\Theta_s$$

$$\frac{Q}{Q_{\max}} = 1 - 3\,\Theta_0 \frac{\sin \gamma_1 - \gamma_1 \cos \gamma_1}{\gamma_1^3}$$

The coefficients C_1 and γ_1 are formally given in Appendix E by Eq. (E–2–9),

$$1 - \gamma_1 \cot \gamma_1 = Bi_0$$

and Eq. (E–2–8),

$$C_1 = \frac{4(\sin \gamma_1 - \gamma_1 \cos \gamma_1)}{2\gamma_1 - \sin(2\gamma_1)}$$

† The Heisler relations are within 1% of the exact solution for $Fo > 0.2$.

from Eq. (2–143). The error in this equation is within about 10% for $Bi < 0.1$. It also follows that the instantaneous rate q_c and accumulative Q heat transfer for $Bi < 0.1$ can be approximated by Eq. (2–144),

$$\frac{q_c}{hA_s(T_i - T_F)} = \exp\left(-\frac{hA_s t}{\rho c_v V}\right) \tag{3–67}$$

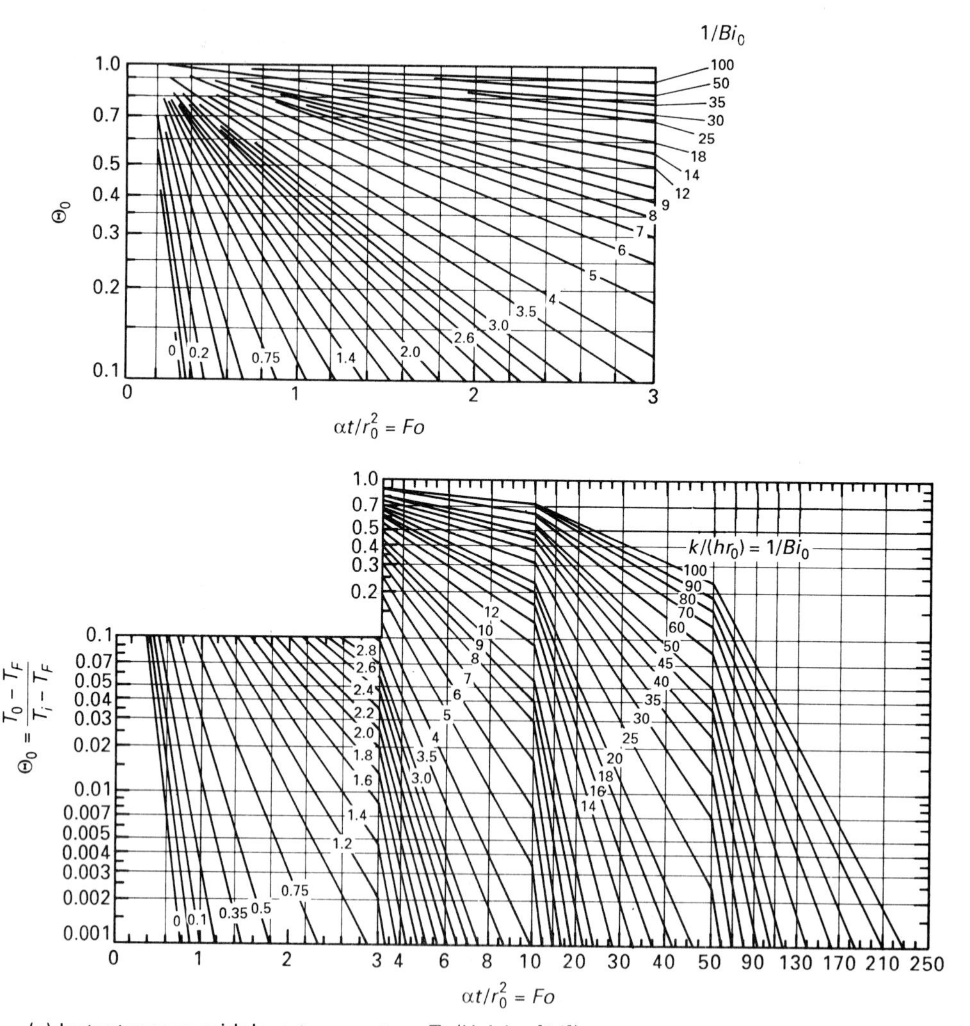

(a) Instantaneous midplane temperature T_0 (Heisler [14]).

FIGURE 3–14 Heisler charts: Unsteady one-dimensional convection cooling of a sphere; $Bi_0 = hr_0/k$ and $Fo = \alpha t/r_0^2$.

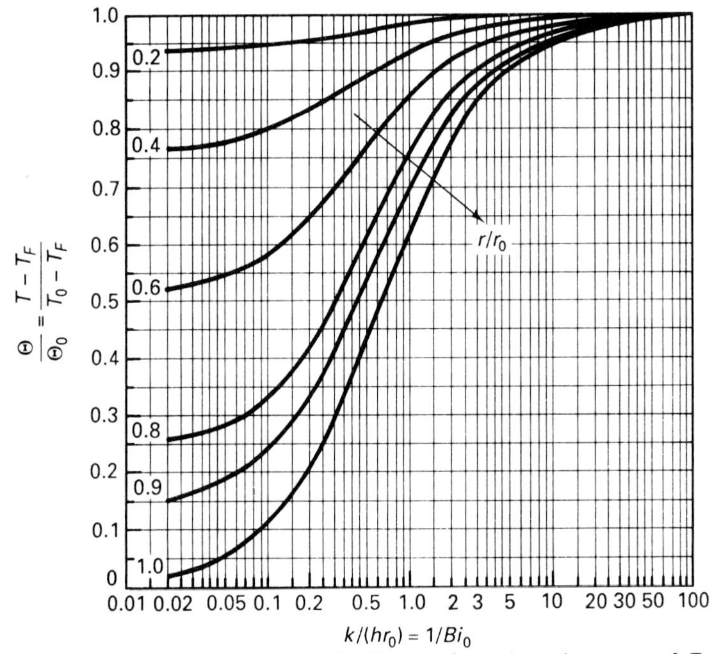

(b) Instantaneous temperature distribution for sphere in terms of T_0 (Heisler [14]).

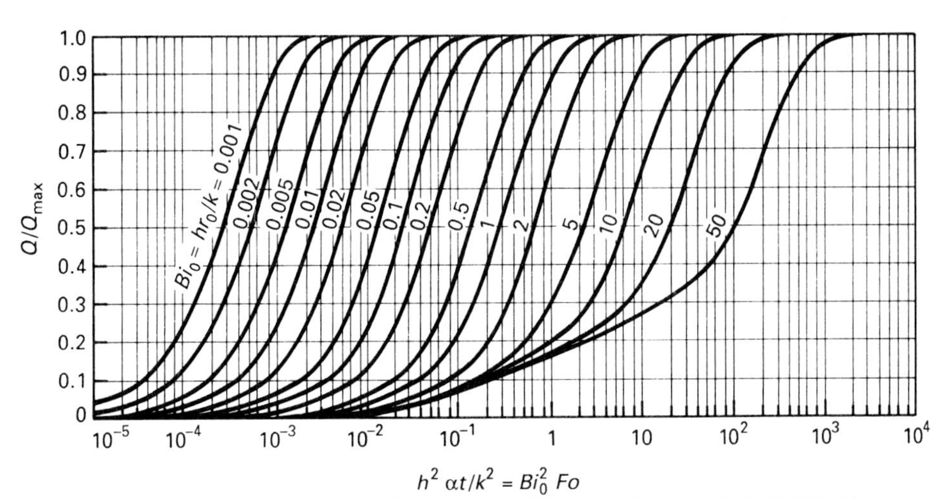

(c) Total accumulative heat transfer Q for sphere. (From Grober and Grigull [15]. Used with permission.)

FIGURE 3-14 *(Continued)*

and Eq. (2–147),

$$\frac{Q}{Q_{\max}} = 1 - \exp\left(-\frac{hA_s t}{\rho c_v V}\right) \tag{3-68}$$

These equations are written in terms of Fo and Bi_0 by simply expressing $hA_s t/(\rho c_v V)$ as

$$\frac{hA_s t}{\rho c_v V} = \frac{h\ell_0}{k}\frac{\alpha t}{\ell_0^2}\frac{\ell_0}{\ell} = Bi_0\,Fo\,\frac{\ell_0}{\ell} \tag{3-69}$$

Because the Heisler relations and charts given in Tables 3–5 to 3–7 and Figs. 3–12 to 3–14 are based on approximate solutions, which are restricted to situations for which the Fourier number Fo is not too small (i.e., $Fo = \alpha t/\ell_0^2 > 0.2$), these practical solution results can be used for all but the first 10% to 20% of an unsteady convective cooling or heating process. To obtain calculations for the small values of time t for which $Fo < 0.2$, the exact solution given in Appendix E–2 can be used. Alternatively, the solution for small values of time during which the temperature in the body interior (i.e., in the vicinity of the midplane, centerline, or center) is not significantly influenced by the change in surface condition can be approximated by the use of the solution results for a semi-infinite solid.

Finally, we should recognize that limiting results obtained from Figs. 3–12(a), 3–13(a), and 3–14(a) for large values of Biot number Bi_0 (i.e., $1/Bi_0 \simeq 0$) correspond to the case in which the surface is suddenly brought to a constant temperature $T_s = T_F$. Graphical solutions for the instantaneous rate q_s and total accumulative Q heat transfer at the surface of a plane wall, circular cylinder, and sphere for this important boundary condition are given in Fig. 3–15.

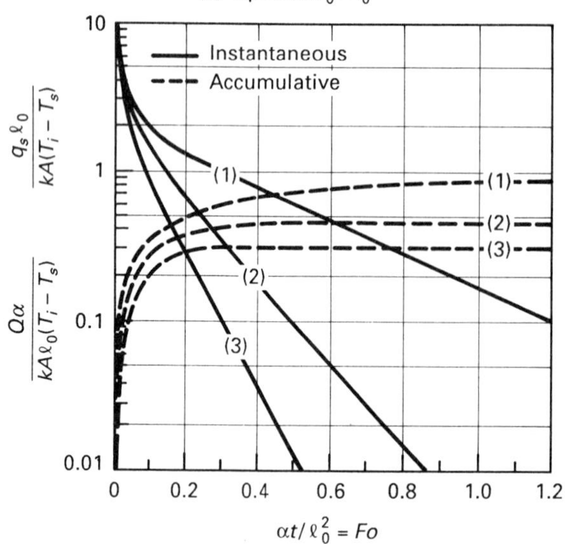

FIGURE 3–15 Unsteady one-dimensional heat transfer in plane wall, circular cylinder, and sphere with surface suddenly brought to a constant temperature T_0. (From Schneider [6]. Used with permission.)

EXAMPLE 3–15

A 2-in.-thick aluminum plate initially at 75°F is suddenly exposed to convection with $T_F = 250°\text{F}$ and $h = 3000\ \text{Btu/(h ft}^2\ °\text{F)}$. Determine the temperature at the surface after (1) 100 s, (2) 10 s, and (3) 1 s.

Solution

Objective Determine T_s for $t = 100$ s, $t = 10$ s, and $t = 1$ s.

Schematic Flat plate with both surfaces suddenly exposed to convection.

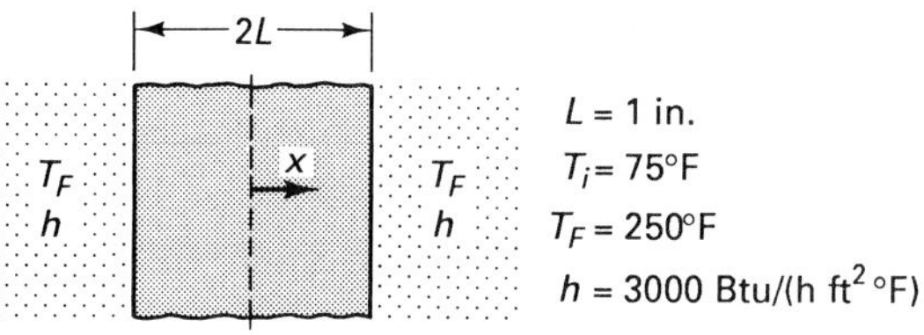

Assumptions/Conditions

unsteady-state
one spatial dimension
uniform properties

Properties Aluminum (Table A–C–1): $k = 136\ \text{Btu/(h ft °F)}$, $\alpha = 10.5 \times 10^{-4}$ ft²/s.

Analysis As a first step we compute the Biot number Bi_0.

$$Bi_0 = Bi = \frac{hL}{k} = 3000 \frac{\text{Btu}}{\text{h ft}^2\ °\text{F}} \frac{\text{ft}}{12} \frac{\text{h ft °F}}{136\ \text{Btu}} = 1.84 = \frac{1}{0.544}$$

Since $Bi \gg 0.1$, the approximate lumped analysis of Chap. 2 is inappropriate. However, the temperature distribution for this problem can be determined by use of the Heisler relations or charts given by Table 3–5 or Fig. 3–12, providing that the Fourier number Fo is not too small (i.e., $Fo \not< 0.2$). To compute Fo, we write

$$Fo = \frac{\alpha t}{L^2} = 10.5 \times 10^{-4} \frac{\text{ft}^2}{\text{s}} \frac{t}{(\text{ft}/12)^2} = 0.151\ t/\text{s}$$

such that

$$Fo = 0.151(100) = 15.1 \qquad \textit{case 1}$$

for $t = 100$ s,

$$Fo = 0.151(10) = 1.51 \qquad \textit{case 2}$$

for $t = 10$ s, and

$$Fo = 0.151(1) = 0.151 \qquad \textit{case 3}$$

for $t = 1$ s. Thus, the Heisler relations and charts can be used with confidence for cases 1 and 2, but provide limited accuracy for case 3.

Case 1: $t = 100$ s

Referring to the Heisler chart given by Fig. 3–12(a) and setting $Fo = 15.1$ and $1/Bi_0 = 0.544$, we find that the dimensionless midplane temperature Θ_0 is very small; that is,

$$\Theta_0 = \frac{T_0 - T_F}{T_i - T_F} \ll 0.001 \qquad T_0 \simeq T_F = 250°\text{F}$$

Using this result together with Fig. 3–12(b), we conclude that

$$T_s \simeq T_F = 250°\text{F}$$

Thus, the temperature throughout the entire plate is essentially at 250°F for $t = 100$ s. This result is confirmed by use of the Heisler relations, which indicate $\gamma_1 = 1.04$, $C_1 = 1.17$, $\Theta_0 = 1.05 \times 10^{-7}$, and $\Theta_s/\Theta_0 = 0.506$.

Case 2: $t = 10$ s

Setting $Fo = 1.51$ and following the approach taken in case 1, we obtain

$$\Theta_0 = \frac{T_0 - T_F}{T_i - T_F} = 0.23$$

or

$$T_0 = 0.23(75°\text{F} - 250°\text{F}) + 250°\text{F} = 210°\text{F}$$

from Fig. 3–12(a), and

$$\frac{\Theta_s}{\Theta_0} = \frac{T_s - T_F}{T_0 - T_F} = 0.5$$

or

$$T_s = 0.5(210°\text{F} - 250°\text{F}) + 250°\text{F} = 230°\text{F}$$

from Fig. 3–12(b). These values are in good agreement with the Heisler relations, which indicate $\gamma_1 = 1.04$, $C_1 = 1.17$, $\Theta_0 = 0.231$, and $\Theta_s/\Theta_0 = 0.506$.

Case 3: $t = 1$ s

Since $Fo < 0.2$ for case 3, we turn to the exact solution given by Eq. (E–2–1) in the Appendix, which indicates

$$\Theta_0 = \frac{T_0 - T_F}{T_i - T_F} = \sum_{n=1}^{\infty} C_n \exp\left(-\gamma_n^2 Fo\right) \cos \gamma_n$$

where

$$C_n = \frac{4 \sin \gamma_n}{2\gamma_n + \sin (2\gamma_n)} \tag{a}$$

from Eq. (E–2–2). Referring to Table A–E–2–1, the first three eigenvalues γ_n, which correspond to $Bi_0 = 1.84$, are

$$\gamma_1 = 1.04 \qquad \gamma_2 = 3.61 \qquad \gamma_3 = 6.55$$

Substituting these values into Eq. (a), we obtain

$$C_1 = 1.17 \qquad C_2 = -0.224 \qquad C_3 = 0.0784$$

Using these results and setting $Fo = 0.151$, Θ_0 becomes

$$\begin{aligned}\Theta_0 &= (1.17) \exp [-(1.04)^2(0.151)] \cos (1.04) \\ &\quad + (-0.224) \exp [-(3.61)^2(0.151)] \cos (3.61) \\ &\quad + (0.0784) \exp [-(6.55)^2(0.151)] \cos (6.55) + \cdots \\ &= 0.503 + 0.0279 + 0.000116 + \cdots \simeq 0.531\end{aligned} \tag{b}$$

It follows that

$$T_s = 0.531(75°\text{F} - 250°\text{F}) + 250°\text{F} = 157°\text{F}$$

Equation (b) indicates that the third- and higher-order terms in the exact solution can be neglected, but that the second term contributes about 5.3% to the solution. Thus, the error resulting from the use of the Heisler relations for this problem would be about 5.3%. Of course, we would expect larger errors in the Heisler relations for smaller values of Fo.

EXAMPLE 3–16

A brass sphere 50 cm in diameter initially at 80°C is placed in a cooling fluid with $T_F = 15°\text{C}$ and $h = 500\ \text{W/(m}^2\ °\text{C)}$. Determine the length of time required for the center of the sphere to cool to 30°C.

Solution

Objective Determine time t required for $T_0 = 30°\text{C}$.

Schematic Convection cooling of brass sphere.

Assumptions/Conditions

unsteady-state
one spatial dimension
uniform properties

Properties Brass (Table A–C–1): $k = 111$ W/(m °C), $\alpha = 3.41 \times 10^{-5}$ m^2/s.

Analysis First we calculate the Biot number Bi_0.

$$Bi_0 = \frac{hr_0}{k} = \frac{[500\ \text{W/(m}^2\ \text{°C)}](0.25\ \text{m})}{111\ \text{W/(m °C)}} = 1.126 = \frac{1}{0.888}$$

We also note that $Bi = (hr_0/3)/k = 0.375$. Because $Bi > 0.1$, we will utilize the Heisler relations/charts instead of the approximate lumped analysis approach developed in Chap. 2.

The dimensionless center temperature Θ_0 is

$$\Theta_0 = \frac{T_0 - T_F}{T_i - T_F} = \frac{30°\text{C} - 15°\text{C}}{80°\text{C} - 15°\text{C}} = 0.23$$

Referring to Fig. 3–14(a), we estimate $Fo = \alpha t/r_0^2 = 0.7$. For better accuracy, the Heisler relations listed in Table 3–7 can be used to obtain $\gamma_1 = 1.63$, $C_1 = 1.30$, and

$$\Theta_0 = C_1 \exp(-\gamma_1^2 Fo)$$

Solving for Fo, we obtain

$$Fo = \frac{1}{\gamma_1^2} \ln \frac{C_1}{\Theta_0} = \frac{1}{1.63^2} \ln \frac{1.30}{0.23} = 0.653$$

Using this more reliable value for Fo, the time t is given by

$$t = \frac{0.653(0.25\ \text{m})^2}{3.41 \times 10^{-5}\ \text{m}^2/\text{s}} = 1190\ \text{s}$$

As a point of interest, this value is 50% greater than the value indicated by Eq. (2–143), which is based on the lumped analysis approach.

3–9–4 Unsteady Two- and Three-Dimensional Heat-Transfer Systems

As we have seen, the temperature charts given by Figs. 3–10, 3–12, and 3–13 are applicable to unsteady one-dimensional heat transfer in semi-infinite solids, infinite plates, and infinite circular cylinders, as indicated in Table 3–8. Notice that the dimensionless temperature distribution Θ $[(T - T_F)/(T_i - T_F)]$ in these three unsteady one-dimensional systems is represented by $S(x,t)$, $P(x,t)$, and $C(r,t)$, respectively. These fundamental solutions can be combined in the form of products to obtain

TABLE 3–8 Unsteady one-dimensional temperature charts applicable to multidimensional systems

Geometry	*Temperature chart*	*Notation for dimensionless temperature distribution* $\Theta = \dfrac{T - T_F}{T_i - T_F} = 1 - \dfrac{T - T_i}{T_F - T_i}$
(i) Semi-infinite solid Uniform T_i T_F, h, x	Fig. 3–10	$S(x,t)$
(ii) Infinite plate Uniform T_i T_F, h, x, T_F, h, $2L$	Fig. 3–12	$P(x,t)$
(iii) Infinite cylinder Uniform T_i T_F, h, r, T_F, h, r_0	Fig. 3–13	$C(r,t)$

solutions for the unsteady temperature distribution in semi-infinite plates, short cylinders, and the other multidimensional systems shown in Fig. 3–16. For example, the temperature distribution is represented by

$$\Theta = \frac{T(x,y,t) - T_F}{T_i - T_F} = S(x,t)\,P(y,t) \tag{3–70}$$

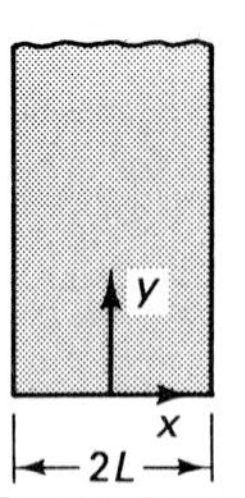

(a) Semi-infinite plate; $\Theta = P(x,t)S(y,t)$.

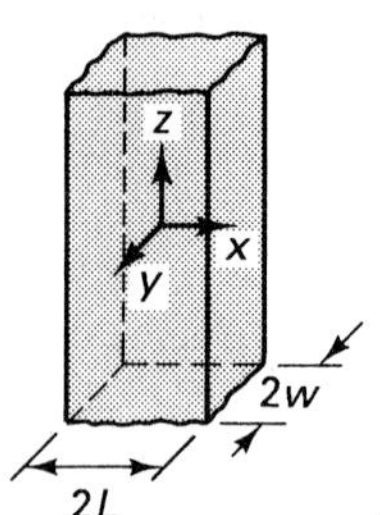

(b) Infinite rectangular bar; $\Theta = P(x,t)P(y,t)$.

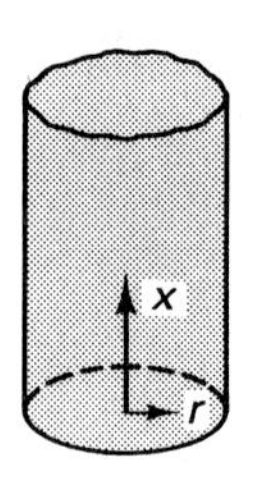

(c) Semi-infinite cylinder; $\Theta = S(x,t)C(r,t)$.

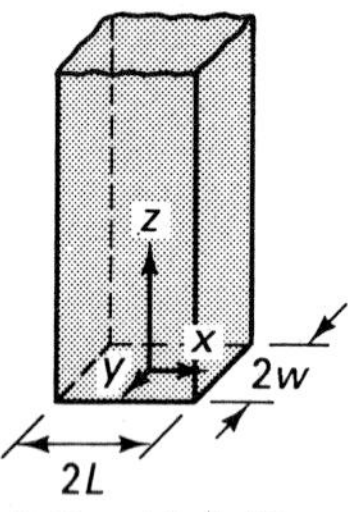

(d) Semi-infinite rectangular bar; $\Theta = P(x,t)P(y,t)S(z,t)$.

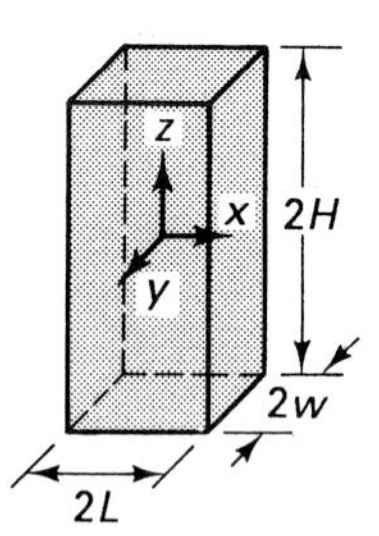

(e) Rectangular parallelpiped, $\Theta = P(x,t)P(y,t)P(z,t)$.

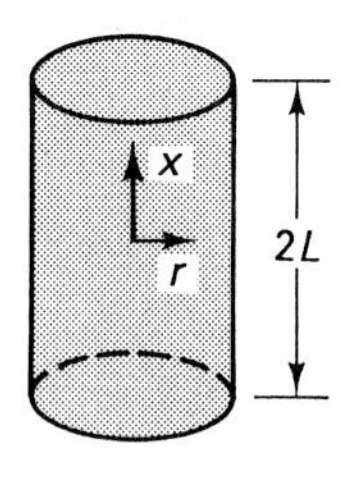

(f) Short cylinder, $\Theta = P(x,t)C(r,t)$.

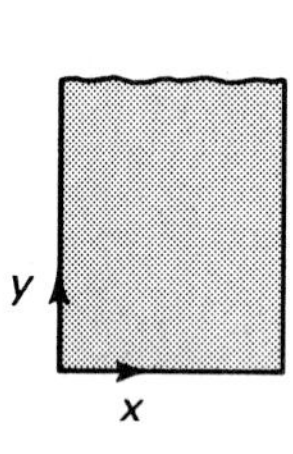

(g) One-quarter infinite solid; $\Theta = S(x,t)S(y,t)$.

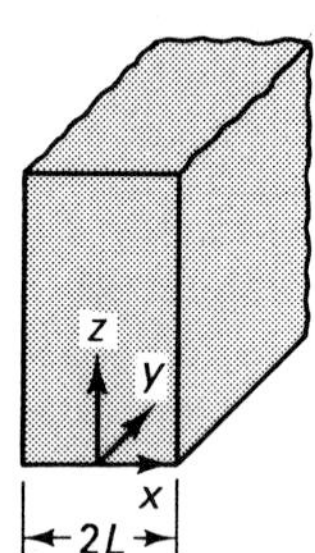

(h) One-quarter infinite plate; $\Theta = P(x,t)S(y,t)S(z,t)$.

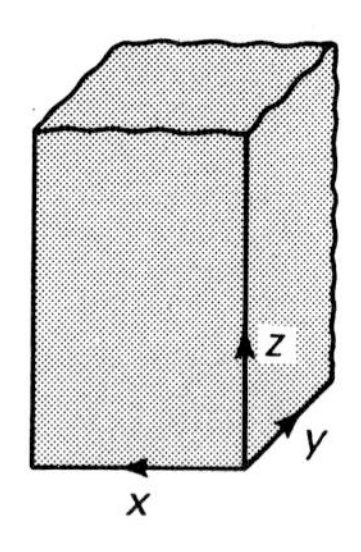

(i) One-eighth infinite solid; $\Theta = S(x,t)S(y,t)S(z,t)$.

FIGURE 3–16 Temperature distributions for unsteady multidimensional systems with uniform initial temperature T_i and convection boundary conditions expressed as products of unsteady one-dimensional solutions $S(\zeta,t)$, $P(\zeta,t)$, and $C(r,t)$; $\zeta = x, y, z,$ or r.

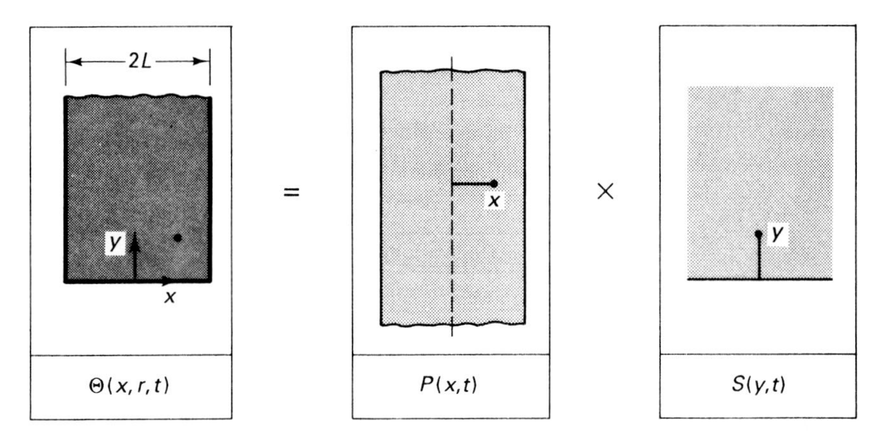

FIGURE 3–17 Product solution scheme for unsteady two-dimensional temperature distribution in a semi-infinite plate.

as shown in Fig. 3–17 for a semi-infinite plate, and

$$\Theta = \frac{T(x,r,t) - T_F}{T_i - T_F} = S(x,t)\, C(r,t) \tag{3–71}$$

as shown in Fig. 3–18 for a short cylinder. This practical solution approach is based on the separation-of-variables method, which is illustrated in Example 3–17.

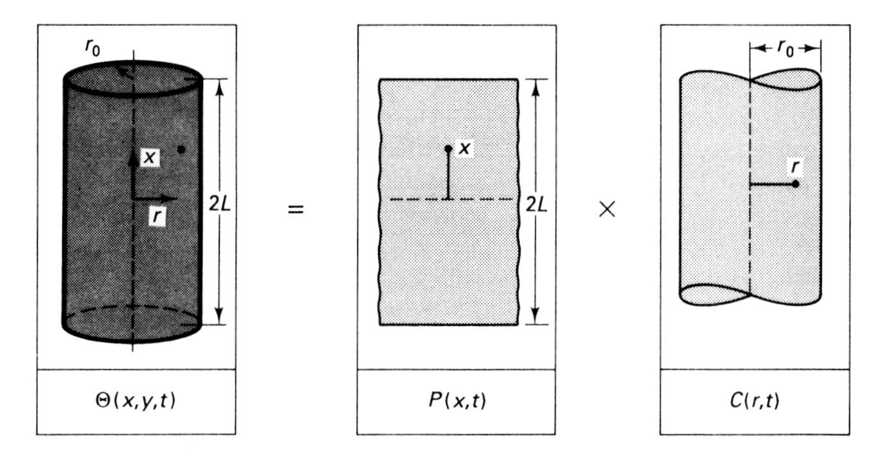

FIGURE 3–18 Product solution scheme for unsteady two-dimensional temperature distribution in a short cylinder.

EXAMPLE 3–17

The short circular cylinder shown in Fig. E3–17a, which is initially at a uniform temperature T_i, is suddenly exposed to a fluid with h and T_F specified. Demonstrate how the instantaneous temperature at any location within the cylinder can be determined by the use of Heisler charts.

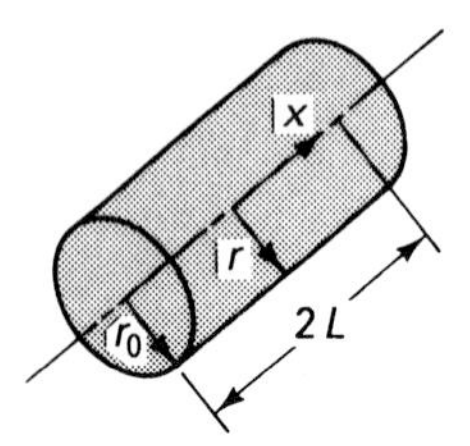

FIGURE E3–17a
Convective cooling or heating of short circular cylinder.

Solution

Objective Show how the Heisler charts can be used for this unsteady two-dimensional heat-transfer problem.

Assumptions/Conditions

unsteady-state
two spatial dimensions
uniform properties

Analysis The differential formulation for this multidimensional system takes the form [for $\Theta = (T - T_F)/(T_i - T_F)$]

$$\frac{1}{r}\frac{\partial}{\partial r}\left(r\frac{\partial\Theta}{\partial r}\right) + \frac{\partial^2\Theta}{\partial x^2} = \frac{1}{\alpha}\frac{\partial\Theta}{\partial t}$$

$$\Theta = 1 \qquad \text{at } t = 0$$

$$\frac{\partial\Theta}{\partial r} = 0 \qquad \text{at } r = 0 \qquad -k\frac{\partial\Theta}{\partial r} = h\Theta \qquad \text{at } r = r_0$$

$$k\frac{\partial\Theta}{\partial x} = h\Theta \qquad \text{at } x = -L \qquad -k\frac{\partial\Theta}{\partial x} = h\Theta \qquad \text{at } x = L$$

These equations can be reduced to two simpler problems by assuming the product solution

$$\Theta(x,r,t) = C(r,t)\,P(x,t)$$

Using this substitution, we obtain

$$\frac{1}{r}\frac{\partial}{\partial r}\left(r\frac{\partial C}{\partial r}\right) = \frac{1}{\alpha}\frac{\partial C}{\partial t} \qquad \frac{\partial^2 P}{\partial x^2} = \frac{1}{\alpha}\frac{\partial P}{\partial t}$$

$$C = 1 \quad \text{at } t = 0 \qquad P = 1 \quad \text{at } t = 0$$

$$\frac{\partial C}{\partial r} = 0 \quad \text{at } r = 0 \qquad k\frac{\partial P}{\partial x} = hP \quad \text{at } x = -L$$

$$-k\frac{\partial C}{\partial r} = hC \quad \text{at } r = r_0 \qquad -k\frac{\partial P}{\partial x} = hP \quad \text{at } x = L$$

The solutions to these two unsteady one-dimensional problems are represented by the Heisler charts given in Figs. 3–12 and 3–13. Therefore, we see that the solution to our multidimensional problem is equal to the product of the solutions to simpler unsteady one-dimensional problems. A simple geometric perspective of this practical result is represented by Fig. E3–17b, which shows a short circular cylinder resulting from the intersection of a flat plate and a circular cylinder.

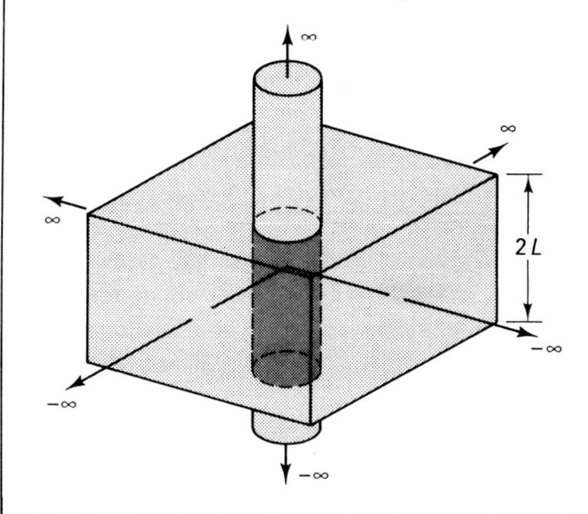

FIGURE E3–17b
Intersection of flat plate and circular cylinder.

EXAMPLE 3–18

In manufacturing a cylindrical stainless steel (AISI 302) disk 10 cm in diameter and 8 cm thick, the piece is quenched in an oil bath with $h = 400$ W/(m^2 K) from an initial temperature of 700 K to 300 K. Determine the minimum and maximum temperatures within the piece after a time of 10 minutes.

Solution

Objective Determine $T_{\min}$ and $T_{\max}$ for $t = 10$ min.

Schematic Convective cooling of short circular cylinder.

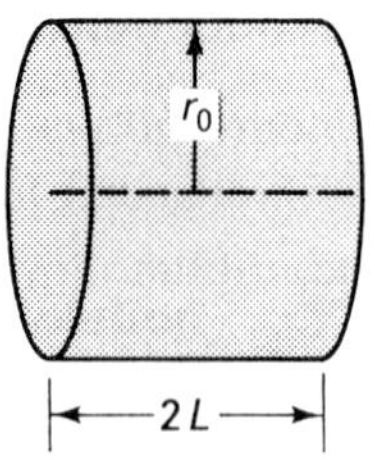

r_0

$2L$

$L = 0.04$ m
$r_0 = 0.05$ m
$T_i = 700$ K

Surrounding fluid
$T_F = 300$ K
$h = 400$ W/(m^2 K)

Assumptions/Conditions

unsteady-state
two spatial dimensions
uniform properties

Properties Stainless steel (AISI 302) at average temperature of 500 K (Table A–C–1): $\rho = 8050$ kg/m^3, $k = 18.6$ W/(m K), $c_v = 0.536$ kJ/(kg K); the thermal diffusivity α is calculated by writing

$$\alpha = \frac{18.6 \text{ W/(m K)}}{(8050 \text{ kg/m}^3)[536 \text{ J/(kg K)}]} = 4.31 \times 10^{-6} \text{ m}^2\text{/s}$$

Analysis As a first step, we calculate the value of Biot number Bi for the piece.

$$Bi = \frac{h\ell}{k} = \frac{hV}{kA_s} = \frac{h}{k}\frac{\pi r_0^2 (2L)}{2\pi r_0 (2L) + 2\pi r_0} = \frac{h}{k}\frac{r_0 L}{2L + r_0}$$

$$= \frac{400 \text{ W/(m}^2\text{ K)}}{18.6 \text{ W/(m K)}}\frac{0.05 \text{ m } (0.04 \text{ m})}{0.08 \text{ m} + 0.05 \text{ m}} = 0.331$$

Since the value of Bi is significantly greater than 0.1, we should account for the variation of temperature within the cylinder.

As we have seen, the dimensionless temperature distribution $\Theta(r,x,t)$ in a short cylinder such as this can be represented by the product of dimensionless temperature distributions in an infinite cylinder $C(r,t)$ and an infinite plate $P(x,t)$; that is (see Fig. 3–16f),

$$\Theta(r,x,t) = C(r,t)\, P(x,t)$$

where

$$C(r,t) = \frac{T(r,t) - T_F}{T_i - T_F} \qquad P(x,t) = \frac{T(x,t) - T_F}{T_i - T_F}$$

It is obvious that at any instant of time the temperature is a maximum at the center of the piece and is a minimum along the circumference at both ends. The calculations required to determine $T_{\min}$ and $T_{\max}$ for $t = 10$ min are summarized as follows:

Infinite Cylinder

$$\frac{\alpha t}{r_0^2} = \frac{(4.31 \times 10^{-6}\ \mathrm{m^2/s})(600\ \mathrm{s})}{(0.05\ \mathrm{m})^2} = 1.03$$

$$Bi_{0,C} = \frac{hr_0}{k} = \frac{[400\ \mathrm{W/(m^2\ K)}](0.05\ \mathrm{m})}{18.6\ \mathrm{W/(m\ K)}} = 1.08 = \frac{1}{0.930}$$

$$C(0,t) = \left.\frac{T_0 - T_F}{T_i - T_F}\right|_{\text{Cylinder}} = 0.26 \qquad \text{from Fig. 3–13a}$$

$$C(r_0,t) = 0.62(0.26) = 0.16 \qquad \text{from Fig. 3–13b}$$

Infinite Plate

$$\frac{\alpha t}{L^2} = \frac{(4.31 \times 10^{-6}\ \mathrm{m^2/s})(600\ \mathrm{s})}{(0.04\ \mathrm{m})^2} = 1.62$$

$$Bi_{0,P} = \frac{hL}{k} = \frac{[400\ \mathrm{W/(m^2\ K)}](0.04\ \mathrm{m})}{18.6\ \mathrm{W/(m\ K)}} = 0.860 = \frac{1}{1.16}$$

$$P(0,t) = \left.\frac{T_0 - T_F}{T_i - T_F}\right|_{\text{Plate}} = 0.35 \qquad \text{from Fig. 3–12a}$$

$$P(L,t) = 0.70(0.35) = 0.25 \qquad \text{from Fig. 3–12b}$$

Resultant Short Cylinder

Maximum temperature

$$\Theta_{\max} = C(0,t)\ P(0,t) = 0.26(0.35) = 0.091$$

$$T_{\max} = 0.091(700\ \mathrm{K} - 300\ \mathrm{K}) + 300\ \mathrm{K} = 336\ \mathrm{K}$$

Minimum temperature

$$\Theta_{\min} = C(r_0,t)\ P(L,t) = 0.16(0.25) = 0.040$$

$$T_{\min} = 0.040(700\ \mathrm{K} - 300\ \mathrm{K}) + 300\ \mathrm{K} = 316\ \mathrm{K}$$

3-10 SUMMARY

In this chapter we have introduced the differential formulation and related analytical, analogical, and graphical solution concepts that are commonly used in the analysis of basic steady and unsteady conduction heat transfer in isotropic materials involving two or more independent variables. These formulation and solution methods provide a basis for establishing an understanding of the physical nature of a problem. Furthermore, analytical methods are useful in establishing the effect of variations of the parameters on the solution, and in the development of criteria for limiting solution results.

Basic solution results in the form of analytic relations and charts are summarized in Sec. 3–9. These results provide the basis for a practical approach to analyzing many standard multidimensional conduction-heat-transfer problems involving convection and specified wall temperature or heat flux boundary conditions.

As we have seen, the analytical, analogical, and graphical solution methods all have their place in the science of heat transfer. However, numerical methods are generally required in the analysis of more complex problems. Therefore, special attention is given to basic numerical methods in the next chapter.

As indicated, our study has been restricted to isotropic media. The analysis of conduction heat transfer in anisotropic materials is introduced by Eckert and Drake [16].

CHAPTER 4

CONDUCTION HEAT TRANSFER: NUMERICAL APPROACH

4–1 INTRODUCTION

As discussed in Chap. 3, numerical methods provide the basis for analyzing the more complex problems for which other approaches are inadequate. Conduction-heat-transfer problems that fall into this category generally include those involving nonlinear boundary conditions, temperature-dependent properties, and complex geometries. Numerical methods are also commonly used in the analysis of systems involving radiation and convection.

The *finite-difference* method and the *finite-element* method are the two basic approaches generally used in numerical analysis. The finite-difference method is featured in this chapter and is thoroughly discussed in reference 1. The finite-element method is presented in some detail in references 1 and 2 and is introduced in Appendix F.

The finite-difference approach involves the use of (1) nodal networks, (2) finite-difference approximations for derivatives in space and time, (3) standard energy conservation formulation concepts, and (4) computer solution of systems of algebraic nodal equations. The nodal network and finite-difference approximations are introduced in Secs. 4–2 and 4–3, after which finite-difference formulation and solution concepts are introduced in the context of steady systems in Sec. 4–4 and unsteady systems in Sec. 4–5. The basic principles are introduced in the framework of two-dimensional rectangular systems with uniform properties and internal energy generation, and are extended to other multidimensional conduction-heat-transfer systems in several examples.

4–2 THE NODAL NETWORK

In the finite-difference approach to the analysis of conduction heat transfer in a rectangular solid such as is shown in Fig. 4–1, we designate a number of discrete *nodal points* at which the temperature is to be approximated. These nodal points are established by subdividing the entire system into subvolumes, with the distance between adjacent nodes represented by Δx or Δy. Each subvolume is treated as a lumped subsystem, with the temperature of a node assumed to represent the *average* temperature of its subvolume. The x and y location of each node within the *nodal network* is given by $(m - 1)\,\Delta x$ and $(n - 1)\,\Delta y$, respectively; the values of m and n take on integer values with m ranging from 1 to M, and n taking values from 1 to N.

The temperature at node (m,n) is designated by $T_{m,n}$ for steady-state processes. To extend the representation to unsteady systems, we simply designate the nodal temperature by $T^{\tau}_{m,n}$, where the time index τ takes on integer values 0, 1, 2, . . . , and is defined in terms of the time increment Δt by $t = \tau\,\Delta t$.

The mechanics of developing a nodal network is quite straightforward. We first sketch in horizontal and vertical construction lines that are Δy and Δx apart, with Δy commonly set equal to Δx. These construction lines also include the boundaries of the system. The nodes are then located at all intersections of the construction lines. The subvolumes are formed by sketching in horizontal and vertical lines that lie

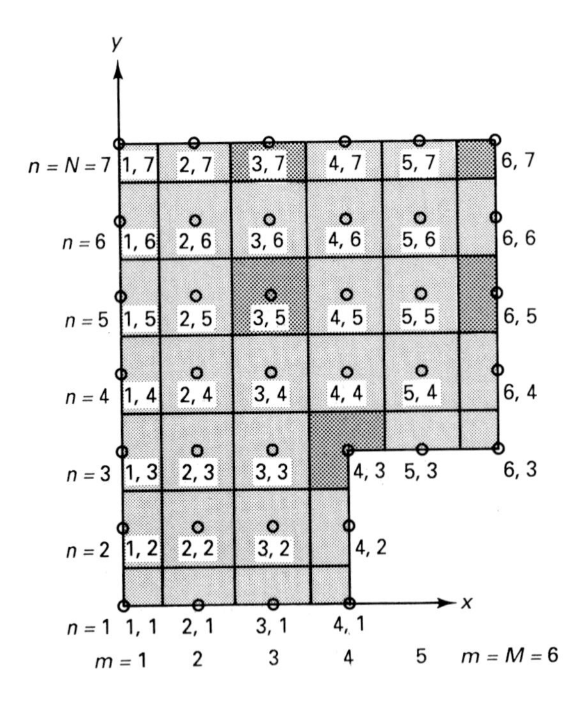

FIGURE 4–1
Representation of a rectangular plate by network of subvolumes and nodes. Shading indicates representative interior and exterior nodes. Plate thickness is δ.

halfway between adjacent construction lines. This procedure will produce the desired system of nodes and subvolumes. As we will see, the accuracy of a finite-difference representation is dependent upon the number of nodes, represented by Z_s, employed in the nodal network. Whereas a small number of nodes (*coarse grid*) can sometimes be used to obtain useful estimates, a relatively large number of nodes (*fine grid*) is generally required to develop accurate solutions.

4–3 FINITE-DIFFERENCE APPROXIMATIONS

The numerical finite-difference approach to analyzing conduction-heat-transfer processes features the use of finite-difference approximations for derivatives in temperature with respect to space and, in the case of unsteady processes, time t.

4–3–1 Spatial Derivatives

The formal defining equation for the partial derivative $\partial T/\partial x$ is given in Appendix A by Eq. (A–4),

$$\frac{\partial T}{\partial x} = \lim_{\Delta x \to 0} \frac{T(x+\Delta x,y) - T(x,y)}{\Delta x} \qquad \text{at } x,y \qquad (4\text{–}1)$$

By permitting the spatial increment Δx to remain finite, we obtain the well known finite-difference approximation

$$\frac{\partial T}{\partial x} = \frac{T(x+\Delta x,y) - T(x,y)}{\Delta x} \qquad \text{at } x,y \qquad (4\text{–}2)$$

Referring to Example 4–1, this equation can also be established by truncating second- and higher-order terms in the *Taylor series expansion*. Equation (4–2) is referred to as a *first-order* approximation because the order of the first term which is neglected in the Taylor series expansion is proportional to Δx. This first-order finite-difference approximation is expressed in terms of the nodal network for steady conditions by writing the *forward difference*,

$$\frac{\partial T}{\partial x} = \frac{T_{m+1,n} - T_{m,n}}{\Delta x} \qquad \text{at } m,n \qquad (4\text{–}3)$$

or the *backward difference*,

$$\frac{\partial T}{\partial x} = \frac{T_{m,n} - T_{m-1,n}}{\Delta x} \qquad \text{at } m,n \qquad (4\text{–}4)$$

both of which approximate the gradient *at the nodal point* (m,n) itself. The corresponding relations for the gradient in the y-direction are given by

$$\frac{\partial T}{\partial y} = \frac{T_{m,n+1} - T_{m,n}}{\Delta y} \qquad \text{at } m,n \qquad (4\text{–}5)$$

or

$$\frac{\partial T}{\partial y} = \frac{T_{m,n} - T_{m,n-1}}{\Delta y} \qquad \text{at } m,n \tag{4–6}$$

As shown in Example 4–1, the Taylor series expansion gives rise to a *second-order* finite-difference approximation for the gradient *at the interface between two subvolumes* of the form

$$\frac{\partial T}{\partial x} = \frac{T_{m+1,n} - T_{m,n}}{\Delta x} \qquad \text{at } m+1/2,n \tag{4–7}$$

and

$$\frac{\partial T}{\partial x} = \frac{T_{m,n} - T_{m-1,n}}{\Delta x} \qquad \text{at } m-1/2,n \tag{4–8}$$

The error in these more accurate central-difference approximations is proportional to Δx^2. Similar second-order approximations for gradients in the y-direction are given by

$$\frac{\partial T}{\partial y} = \frac{T_{m,n+1} - T_{m,n}}{\Delta y} \qquad \text{at } m,n+1/2 \tag{4–9}$$

and

$$\frac{\partial T}{\partial y} = \frac{T_{m,n} - T_{m,n-1}}{\Delta y} \qquad \text{at } m,n-1/2 \tag{4–10}$$

It should also be noted that a finite-difference approximation can be established for the second derivative $\partial^2 T/\partial x^2$ at a nodal point (m,n) by writing

$$\frac{\partial^2 T}{\partial x^2} = \frac{\partial T/\partial x_{m+1/2,n} - \partial T/\partial x_{m-1/2,n}}{\Delta x} \qquad \text{at } m,n \tag{4–11}$$

Substituting Eqs. (4–7) and (4–8) into this relation, we obtain

$$\frac{\partial^2 T}{\partial x^2} = \frac{T_{m+1,n} - 2\,T_{m,n} + T_{m-1,n}}{\Delta x^2} \qquad \text{at } m,n \tag{4–12}$$

which can be shown to have an error of the order of Δx^2. The corresponding second-order finite-difference approximation for $\partial^2 T/\partial y^2$ is written as

$$\frac{\partial^2 T}{\partial y^2} = \frac{T_{m,n+1} - 2\,T_{m,n} + T_{m,n-1}}{\Delta y^2} \qquad \text{at } m,n \tag{4–13}$$

As we have noted, the finite-difference approximations given by Eqs. (4–3)–(4–13) apply to steady-state systems. These relations are readily extended to unsteady systems by merely introducing the time index τ as a superscript for the nodal temperatures.

4–3–2 Time Derivatives

Following the approach introduced in the development of spatial derivatives, basic *first-order* approximations can be written for the time derivatives in terms of the *forward-time difference*,

$$\frac{\partial T}{\partial t} = \frac{T_{m,n}^{\tau+1} - T_{m,n}^{\tau}}{\Delta t} \qquad \text{at } m,n,\tau \tag{4–14}$$

or the *backward-time difference*,

$$\frac{\partial T}{\partial t} = \frac{T_{m,n}^{\tau} - T_{m,n}^{\tau-1}}{\Delta t} \qquad \text{at } m,n,\tau \tag{4–15}$$

Both of these finite-difference approximations are commonly used in the numerical solution of heat-transfer problems. However, care must be taken in the use of Eq. (4–14) to assure that the resulting systems of equations produce a stable solution. The issue of *stability* is considered in Sec. 4–5–2.

4–3–3 Summary

To summarize, finite-difference approximations have been presented in this section for spatial gradients in temperature at the subvolume interfaces and for spatial (first and second) and time derivatives at the individual nodes. Although these relations provide the basis for developing simple and accurate finite-difference solutions for many problems, it should be noted that more accurate higher-order finite-difference approximations can be developed for these derivatives by use of the Taylor series expansion. However, in addition to the question of accuracy, we must also concern ourselves with the issue of stability when analyzing unsteady conduction heat transfer problems.

EXAMPLE 4–1

Utilize the Taylor theorem [3–6] to evaluate the error of standard finite-difference approximations for $\partial T/\partial x$ in the context of steady two-dimensional conditions.

Solution

Objective Determine the error associated with basic finite-difference approximations for $\partial T/\partial x$.

Assumptions/Conditions

steady-state
two-dimensional

Analysis Based on the Taylor theorem, T at $x+\Delta x$ can be expanded in terms of T at x by

$$T(x+\Delta x,y) = T(x,y) + \Delta x \frac{\partial T}{\partial x} + \frac{1}{2!}\Delta x^2 \frac{\partial^2 T}{\partial x^2} + \frac{1}{3!}\Delta x^3 \frac{\partial^3 T}{\partial x^3} + \frac{1}{4!}\Delta x^4 \frac{\partial^4 T}{\partial x^4} + \cdots \quad \text{(a)}$$

Rearranging this equation, we obtain

$$\frac{\partial T}{\partial x} = \frac{T(x+\Delta x,y) - T(x,y)}{\Delta x} + O(\Delta x) \qquad \text{at } x,y \quad \text{(b)}$$

where $O(\Delta x)$ designates terms containing first- and higher-order powers of Δx. This equation provides the basis for the *forward-difference* approximation,

$$\frac{\partial T}{\partial x} = \frac{T_{m+1,n} - T_{m,n}}{\Delta x} \qquad \text{at } m,n \quad \text{(c)}$$

which has an error of the order of Δx. Hence, the error is reduced by approximately one-half by halving the increment Δx.

Similarly, $T(x-\Delta x,y)$ can be written as

$$T(x-\Delta x,y) = T(x,y) - \Delta x \frac{\partial T}{\partial x} + \frac{1}{2!}\Delta x^2 \frac{\partial^2 T}{\partial x^2} - \frac{1}{3!}\Delta x^3 \frac{\partial^3 T}{\partial x^3} + \frac{1}{4!}\Delta x^4 \frac{\partial^4 T}{\partial x^4} - \cdots \quad \text{(d)}$$

which leads to

$$\frac{\partial T}{\partial x} = \frac{T(x,y) - T(x-\Delta x,y)}{\Delta x} + O(\Delta x) \qquad \text{at } x,y \quad \text{(e)}$$

Thus, we conclude that the *backward-difference* approximation,

$$\frac{\partial T}{\partial x} = \frac{T_{m,n} - T_{m-1,n}}{\Delta x} \qquad \text{at } m,n \quad \text{(f)}$$

also has an error of the order of Δx.

A higher-order approximation can be developed for $\partial T/\partial x$ by subtracting Eq. (d) from Eq. (a); that is,

$$\frac{\partial T}{\partial x} = \frac{T(x+\Delta x,y) - T(x-\Delta x,y)}{2\,\Delta x} + O(\Delta x^2) \qquad \text{at } x,y \quad \text{(g)}$$

This equation gives rise to the *central-difference* approximation,

$$\frac{\partial T}{\partial x} = \frac{T_{m+1,n} - T_{m-1,n}}{2\,\Delta x} \qquad \text{at } m,n \quad \text{(h)}$$

which has an improved accuracy since the error is of the order of Δx^2.

Equation (h) provides the basis for writing finite-difference approximations for the temperature gradient at the interface between adjacent subvolumes given by Eq. (4–7),

$$\frac{\partial T}{\partial x} = \frac{T_{m+1,n} - T_{m,n}}{\Delta x} \qquad \text{at } m + 1/2, n \tag{i}$$

and Eq. (4–8),

$$\frac{\partial T}{\partial x} = \frac{T_{m,n} - T_{m-1,n}}{\Delta x} \qquad \text{at } m - 1/2, n \tag{j}$$

both of which have an error proportional to Δx^2.

EXAMPLE 4–2

Write second-order finite-difference approximations for the two-dimensional (x,y) Fourier law of conduction.

Solution

Objective Write second-order finite-difference approximations for Δq_x and Δq_y.

Schematic Finite-difference subvolume.

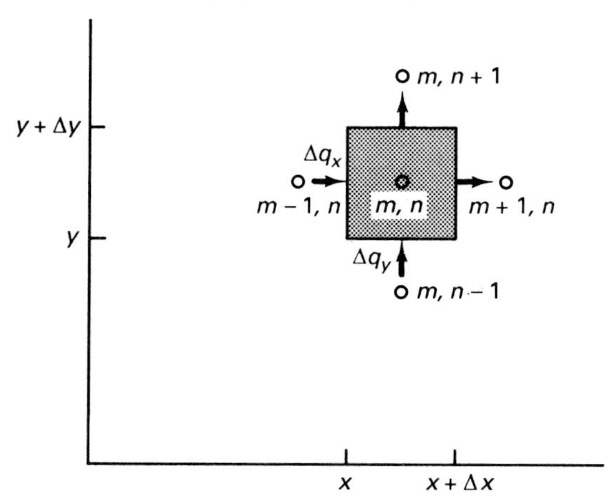

Assumptions/Conditions

two-dimensional

Analysis The two-dimensional form of the Fourier law of conduction can be written as

$$dq_x = -k\,dA_x \frac{\partial T}{\partial x} \qquad dq_y = -k\,dA_y \frac{\partial T}{\partial y}$$

Referring to the schematic, we see that the differential areas are approximated by $dA_x = \delta\ \Delta y$ and $dA_y = \delta\ \Delta x$, where $\Delta y = \Delta x$. The gradient $\partial T/\partial x$ at x can be approximated by the simple second-order difference (with respect to the face at x) given by Eq. (4–8),

$$\left.\frac{\partial T}{\partial x}\right|_x = \frac{T_{m,n} - T_{m-1,n}}{\Delta x}$$

Similarly, $\partial T/\partial y$ at y can be approximated by Eq. (4–10),

$$\left.\frac{\partial T}{\partial y}\right|_y = \frac{T_{m,n} - T_{m,n-1}}{\Delta y}$$

Thus, the steady two-dimensional Fourier law can be approximated by second-order finite-difference equations of the form

$$dq_x = \Delta q_x = -k\delta(T_{m,n} - T_{m-1,n})$$

and

$$dq_y = \Delta q_y = -k\delta(T_{m,n} - T_{m,n-1})$$

Similarly, second-order approximations can be written for dq_{x+dx} and dq_{y+dy} of the form

$$dq_{x+dx} = \Delta q_{x+\Delta x} = -k\delta(T_{m+1,n} - T_{m,n})$$

and

$$dq_{y+dy} = \Delta q_{y+\Delta y} = -k\delta(T_{m,n+1} - T_{m,n})$$

These relations will be used in the development of numerical formulations by means of an energy balance method in the sections that follow.

4–4 NUMERICAL ANALYSIS: STEADY SYSTEMS

Once the nodal network is established, finite-difference approximations can be used together with energy conservation principles to develop algebraic nodal equations for each of the Z_s nodes. Basic concepts pertaining to the formulation and solution of the nodal equations are introduced in this section for steady conduction-heat-transfer processes.

4–4–1 Finite-Difference Formulation

To develop nodal equations that represent the energy transfer within the system, we want to distinguish between *interior nodes* and *exterior nodes*.

Interior Nodal Equations

Nodal equations can be developed for a representative interior node (m,n) by (1) the discretization of the applicable differential energy equation, which we will refer to as the *discretization method*, or (2) the development of an energy balance for the corresponding control volume, which is known as the *control-volume finite-difference method* CVFDM (or the *energy-balance method* or the *finite-volume method*).

Following the discretization finite-difference method, we first write the applicable differential energy equation for steady two-dimensional conduction heat transfer with uniform properties and internal energy generation,

$$\frac{\partial^2 T}{\partial x^2} + \frac{\partial^2 T}{\partial y^2} + \frac{\dot{q}}{k} = 0 \tag{4–16}$$

[from Eq. (3–13)]. Substituting the second-order finite-difference approximations for $\partial^2 T/\partial x^2$ and $\partial^2 T/\partial y^2$ given by Eqs. (4–12) and (4–13), this equation takes the form

$$\frac{T_{m+1,n} - 2\,T_{m,n} + T_{m-1,n}}{\Delta x^2} + \frac{T_{m,n+1} - 2\,T_{m,n} + T_{m,n-1}}{\Delta y^2} + \frac{\dot{q}}{k} = 0 \tag{4–17}$$

or, with Δy set equal to Δx,

$$T_{m+1,n} + T_{m-1,n} + T_{m,n+1} + T_{m,n-1} - 4\,T_{m,n} + \frac{\dot{q}}{k}\Delta x^2 = 0 \tag{4–18}$$

which represents the nodal equation for any interior node (m,n). With m and n permitted to take on the integer values associated with each interior node, we obtain a system of algebraic equations that represent the finite-difference equivalent of the original partial differential equation, Eq. (4–16).

To introduce the alternative CVFDM for developing the nodal equations, we apply the first law of thermodynamics to an interior subvolume such as the one shown in Fig. 4–2, with the result

$$\Delta q_x + \Delta q_y + \dot{q}\,\Delta V = \Delta q_{x+\Delta x} + \Delta q_{y+\Delta y} + \cancel{\frac{\Delta E_s}{\Delta t}} \tag{4–19}$$

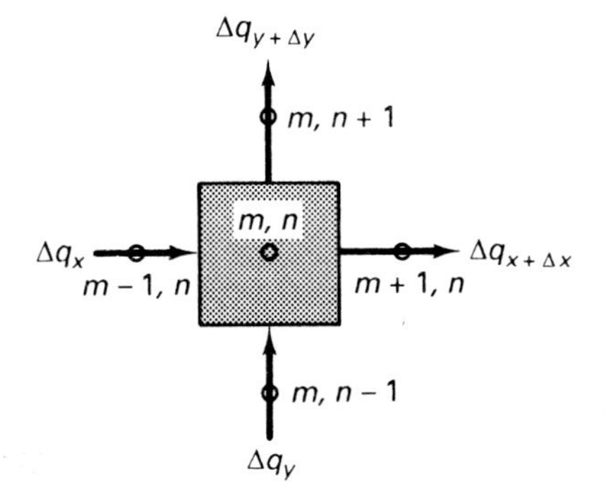

FIGURE 4–2
Representative interior node associated with rectangular plate of Fig. 4–1.

where $\Delta E_s/\Delta t = 0$ for steady-state conditions and $\Delta V = \delta \Delta x\ \Delta y$. Using the second-order finite-difference approximation for the Fourier law of conduction (see Example 4–2), Eq. (4–19) becomes

$$-k\delta\ \Delta y \frac{T_{m,n} - T_{m-1,n}}{\Delta x} - k\delta\ \Delta x \frac{T_{m,n} - T_{m,n-1}}{\Delta y} + \dot{q}\ \Delta V$$
$$= -k\delta\ \Delta y \frac{T_{m+1,n} - T_{m,n}}{\Delta x} - k\delta\ \Delta x \frac{T_{m,n+1} - T_{m,n}}{\Delta y} \qquad (4\text{–}20)$$

which gives rise to Eq. (4–17) and reduces to Eq. (4–18) for $\Delta y = \Delta x$.

Exterior Nodal Equations

Representative exterior nodes associated with the rectangular solid of Fig. 4–1 are shown in Fig. 4–3. In addition to being exposed to energy transfer at the boundaries, these nodes are characterized by subvolumes that are smaller in size than the interior

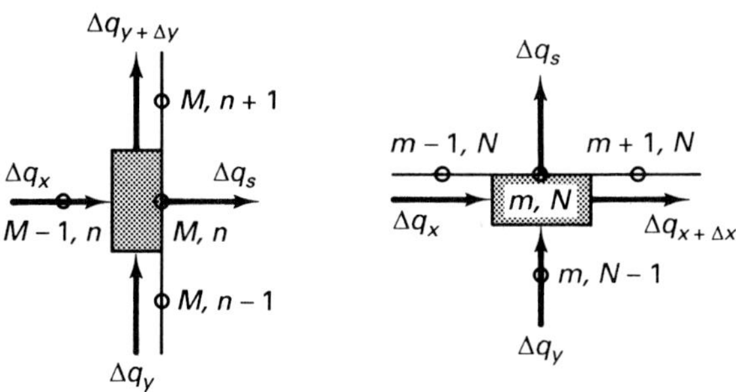

(a) Regular exterior node (M,n). (b) Regular exterior node (m,N).

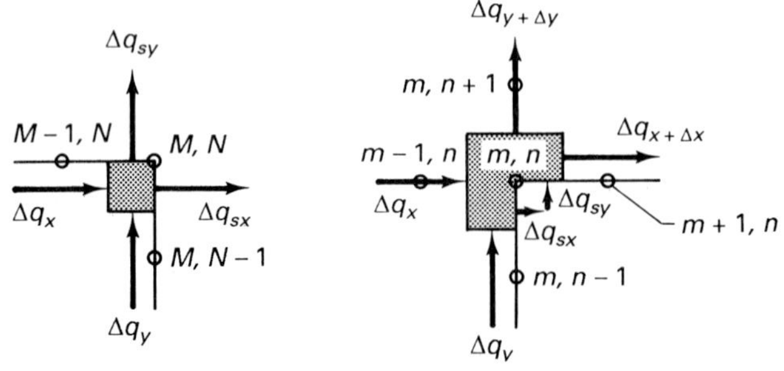

(c) Outer corner node (M,N) (d) Inner corner node (m,n).

FIGURE 4–3 Representative exterior nodes associated with rectangular plate of Fig. 4–1.

subvolumes. In order to characterize the energy transfer associated with exterior nodes such as these, we will make use of the CVFDM.

Referring to the regular exterior node shown in Fig. 4–3(a), we apply the first law of thermodynamics to obtain

$$\Delta q_x + \Delta q_y + \dot{q}\,\Delta V = \Delta q_s + \Delta q_{y+\Delta y} + \cancel{\frac{\Delta E_s}{\Delta t}} \tag{4–21}$$

where $\Delta E_s/\Delta t = 0$, $\Delta V = \delta\,\Delta x\,\Delta y/2$, and Δq_s is the rate of heat transfer out of the subvolume surface. Employing the second-order finite-difference approximation for the Fourier law of conduction, Eq. (4–21) gives rise to

$$-k\delta\,\Delta y\,\frac{T_{M,n} - T_{M-1,n}}{\Delta x} - k\delta\,\frac{\Delta x}{2}\,\frac{T_{M,n} - T_{M,n-1}}{\Delta y} + \dot{q}\,\Delta V = \Delta q_s - k\delta\,\frac{\Delta x}{2}\,\frac{T_{M,n+1} - T_{M,n}}{\Delta y} \tag{4–22}$$

Rearranging this equation with $\Delta q_s = \delta\,\Delta y\,q''_s$ and $\Delta y = \Delta x$, we write

$$2\,T_{M-1,n} + T_{M,n+1} + T_{M,n-1} - 4\,T_{M,n} + \left(\frac{\dot{q}}{k} - \frac{2}{\Delta x}\frac{q''_s}{k}\right)\Delta x^2 = 0 \tag{4–23}$$

Similar expressions can be developed for exterior nodes along the other surfaces and at the corners.

For convection, thermal radiation, or specified heat flux boundary conditions, the unknown exterior nodal temperatures satisfy equations such as Eq. (4–23) for steady-state conditions with q''_s $(= \Delta q_s/\Delta A_s)$ specified by

$$q''_s = h(T_{m,n} - T_F) \qquad \text{convection} \tag{4–24}$$

$$q''_s = \sigma F_{s-R}(T^4_{m,n} - T^4_R) \qquad \text{blackbody thermal radiation} \tag{4–25}$$

$$q''_s = f(x,y) \qquad \text{specified heat flux} \tag{4–26}$$

$$q''_s = 0 \qquad \text{insulated surface} \tag{4–27}$$

Once the surface temperature is calculated, Eqs. (4–24) and (4–25) provide the means by which the rate of heat transfer from the surface Δq_s can be determined for convection and blackbody thermal radiation boundary conditions. On the other hand, for specified wall-temperature boundary conditions, the exterior nodal temperatures are known *a priori*, with nodal equations such as Eq. (4–23) providing the basis for calculating the rate of heat transfer from the surface. For example, for an isothermal boundary condition at the surface $x = (M - 1)\,\Delta x$, the nodal temperatures along the surface are given by $T_{M,n} = T_{M,n+1} = T_{M,n-1} = T_0$, such that Eq. (4–23) reduces to

$$q''_s = \frac{k}{\Delta x}(T_{M-1,n} - T_0) + \frac{\Delta x}{2}\,\dot{q} \tag{4–28}$$

Thus, after $T_{M-1,n}$ has been determined, the rate of heat transfer across the surface can be calculated.

Nodal Equations: Summary

To recap, the full numerical finite-difference formulation for steady conditions consists of the Z_s algebraic equations produced at the interior and exterior nodes. Representative second-order nodal equations are given by Eq. (4–18) and Eqs. (4–23)–(4–27). Standard interior and exterior nodal equations are summarized in Table 4–1 for steady two-dimensional systems. Similar equations can be readily developed for steady one-dimensional and three-dimensional systems by use of the discretization finite-difference method and/or the CVFDM.

Accuracy

The differential formulation is generally the standard against which the numerical finite-difference formulation of a problem is compared. The error introduced by using finite-difference approximations of the derivatives decreases toward zero as the increments approach infinitesimal proportions. Consequently, the error of a finite-difference formulation for steady conduction heat transfer can be reduced by decreasing the volume ΔV ($= \delta \, \Delta x \, \Delta y$). For instance, the error associated with the second-order nodal equations listed in Table 4–1 is proportional to Δx^2. Thus, a 50% reduction in Δx would be expected to reduce the error by a factor of about four.

However, the use of smaller volumetric increments also brings about an increase in the number of subvolumes Z_s and algebraic equations required to represent a system. Of course, the length of time required to obtain a solution increases as the number of equations increases. Consequently, a compromise concerning element size must be made on the basis of the accuracy required and the cost of computation.

It should also be mentioned that improved accuracy can be achieved by the use of higher-order finite-difference approximations for derivatives.

The error associated with the finite-difference increment Δx (and Δt for unsteady systems) is commonly referred to as the *truncation error* or the *discretization error*. As we turn our attention to the development of solutions, it should also be noted that errors are introduced at each step in calculating nodal temperatures as a result of the use of a finite number of *significant figures*. Whereas the truncation error decreases with decreasing grid spacing, the *round-off error* increases. Fortunately, the number of significant figures that can be carried by modern digital computers is quite large, such that round-off error normally only becomes a factor in cases involving the use of extremely small increments (i.e., excessively large numbers of nodal equations).

The question remains: What grid size Δx must be utilized to achieve reasonable accuracy? The most reliable way to obtain an accurate solution is to actually solve the problem for a number of successively smaller increments in Δx. As long as the solution continues to converge, the round-off error can be assumed to be secondary. For example, with $L/\Delta x$ represented by M, solutions can be obtained for $M = 2, 3,$

TABLE 4–1 Summary of nodal equations for steady two-dimensional systems with internal heat generation

Type node	*Nodal equation*
1. Interior node 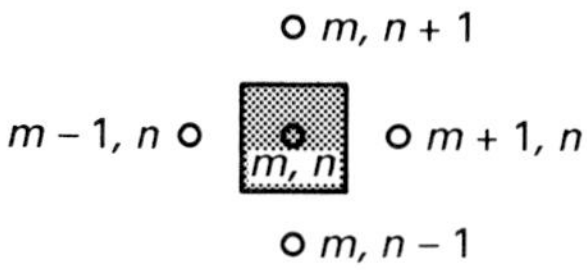	$T_{m+1,n} + T_{m-1,n} + T_{m,n+1} + T_{m,n-1} - 4T_{m,n} + \frac{\dot{q}}{k}\Delta x^2 = 0$ (1)
2. Exterior nodes	
a. General equations	
i. Regular exterior node (M,n) 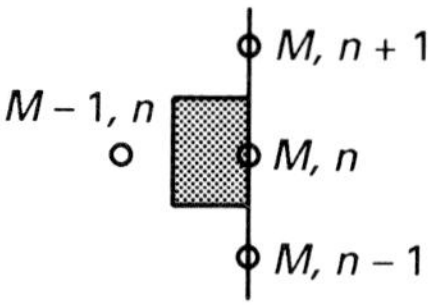	$2T_{M-1,n} + T_{M,n+1} + T_{M,n-1} - 4T_{M,n} + \frac{\dot{q}}{k}\Delta x^2 - 2\frac{q_s''}{k}\Delta x = 0$ (2i)
ii. Outer corner node (M,N) 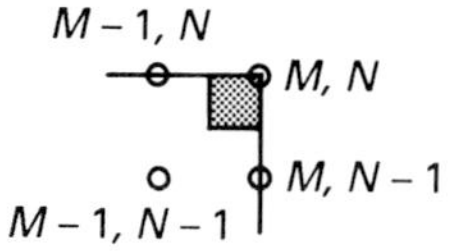	$2T_{M-1,N} + 2T_{M,N-1} - 4T_{M,N} + \frac{\dot{q}}{k}\Delta x^2 - 4\frac{q_s''}{k}\Delta x = 0$ (2ii)
iii. Inner corner node (m,n) 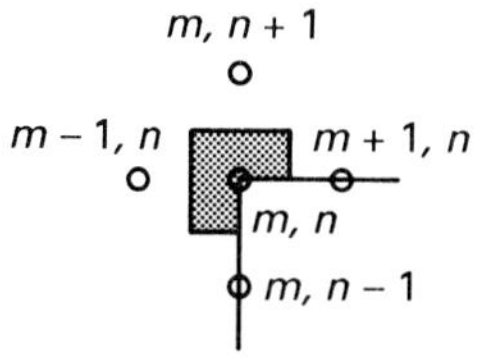	$\frac{2}{3}(T_{m+1,n} + 2T_{m-1,n} + 2T_{m,n+1} + T_{m,n-1} - 6T_{m,n}) + \frac{\dot{q}}{k}\Delta x^2 - \frac{4}{3}\frac{q_s''}{k}\Delta x = 0$ (2iii)
b. Boundary conditions†	
i. Specified temperature	$T_{m,n}$ known
ii. Specified heat flux	q_s'' known
iii. Insulated surface	$q_s'' = 0$
iv. Convection	$q_s'' = h(T_{m,n} - T_F)$
v. Blackbody thermal radiation	$q_s'' = \sigma F_{s-R}(T_{m,n}^4 - T_R^4)$

† These boundary conditions apply to each type of exterior node.

4, ... Since the finite-difference formulation for steady systems is stable, a value of M can be selected for which adequate convergence has been achieved, providing that the limiting point at which round-off error becomes significant has not been reached.

Alternatively, the accuracy of a finite-difference formulation can sometimes be estimated by using the formulation to obtain the numerical solution for simplified conditions or similar problems for which exact solutions are available. However, it must be emphasized that this approach does not guarantee the same solution accuracy for the problem being solved and for the similar problems with known solutions.

EXAMPLE 4–3

Develop a numerical finite-difference formulation for two-dimensional heat transfer in a square with 5-cm-long sides if three faces are maintained at 30°C and the other face exchanges blackbody thermal radiation with a surface at 500°C with $F_{s-R} = 0.25$. Utilize a grid spacing with $L/\Delta x = 3$. The thermal conductivity k and thickness δ of the square are assumed to be known. (The differential formulation for this problem is given in Example 3–1.)

Solution

Objective Develop the finite-difference nodal equations for energy transfer in this rectangular plate.

Schematic Finite-difference grid for rectangular plate.

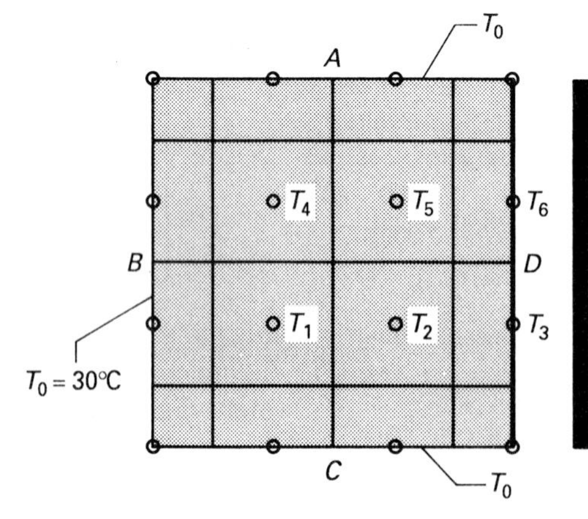

Assumptions/Conditions

steady-state
two-dimensional
uniform properties
blackbody thermal radiation

Analysis The finite-difference grid with $L/\Delta x = 3$ is shown in the schematic. This network involves the six unknown nodal temperatures T_1, T_2, T_3, T_4, T_5, and T_6. However, because of symmetry, we know that $T_4 = T_1$, $T_5 = T_2$, and $T_6 = T_3$. Therefore, we need only develop the nodal equations for T_1, T_2, and T_3. These nodal equations are developed by means of the CVFDM as follows:

Interior nodes at T_1 and T_2 $\Delta q_x + \Delta q_y = \Delta q_{x+\Delta x} + \Delta q_{y+\Delta y}$ (last term struck out)

$$-k\,\Delta y\,\delta\,\frac{T_{m,n} - T_{m-1,n}}{\Delta x} - k\,\Delta x\,\delta\,\frac{T_{m,n} - T_{m,n-1}}{\Delta y} = -k\,\Delta y\,\delta\,\frac{T_{m+1,n} - T_{m,n}}{\Delta x}$$

or

$$T_{m-1,n} + T_{m,n-1} + T_{m+1,n} = 3\,T_{m,n}$$

We therefore have linear equations of the form

$$T_0 + T_0 + T_2 = 3T_1 \tag{a}$$

for the node at T_1, and

$$T_1 + T_0 + T_3 = 3T_2 \tag{b}$$

for the node at T_2.

Exterior Node at T_3 $\Delta q_x + \Delta q_y = \Delta q_R$

$$-k\,\Delta y\,\delta\,\frac{T_3 - T_2}{\Delta x} - k\,\frac{\Delta x}{2}\,\delta\,\frac{T_3 - T_0}{\Delta y} = \sigma\,\Delta y\,\delta F_{s-R}(T_3^4 - T_R^4)$$

Rearranging, this nonlinear equation becomes

$$T_2 + \frac{1}{2}T_0 + \frac{\sigma\,\Delta y\,F_{s-R}T_R^4}{k} = \frac{3}{2}T_3 + \frac{\sigma\,\Delta y\,F_{s-R}T_3^4}{k} \tag{c}$$

Because Eqs. (a) through (c) can be solved for the temperatures T_1, T_2, and T_3, this completes our numerical finite-difference formulation for the temperature distribution with $L/\Delta x = 3$. Once these temperatures are known, calculations can be developed for the rate of heat transfer at any surface. To illustrate, the rate of heat transfer from radiating surface D is

$$q_R = 2\left[\sigma\,\Delta y\,\delta F_{s-R}(T_3^4 - T_R^4) + \sigma\,\frac{\Delta y}{2}\,\delta F_{s-R}(T_0^4 - T_R^4)\right]$$

To obtain the rate of heat transfer at the other surfaces, energy balances must be developed at each node along the surface of interest. For example, the rate of heat transfer at surface C is approximated by

$$q_C = -k\delta(T_1 - T_0 + T_2 - T_0) - \frac{k\delta}{2}(T_3 - T_0) + \sigma\delta\,\frac{\Delta y}{2}(T_0^4 - T_R^4)$$

As we shall see momentarily, the rather course grid spacing used in this illustration gives rise to an error of the order of 10%. To obtain higher levels of accuracy, a finer-grid mesh must be used.

4–4–2 Finite-Difference Solutions

The finite-difference formulation consists of nodal equations for each interior node and for those exterior nodes for which the temperature is unspecified. For the moment, we represent these equations for a steady two-dimensional system as follows:

$$i = 1 \qquad a_{11}T_1 + a_{12}T_2 + \cdots + a_{1j}T_j + \cdots + a_{1Z}T_Z = C_1 \tag{4–29–1}$$

$$i = 2 \qquad a_{21}T_1 + a_{22}T_2 + \cdots + a_{2j}T_j + \cdots + a_{2Z}T_Z = C_2 \tag{4–29–2}$$

$$\vdots$$

$$i = j \qquad a_{j1}T_1 + a_{j2}T_2 + \cdots + a_{jj}T_j + \cdots + a_{jZ}T_Z = C_j \tag{4–29–j}$$

$$\vdots$$

$$i = Z \qquad a_{Z1}T_1 + a_{Z2}T_2 + \cdots + a_{Zj}T_j + \cdots + a_{ZZ}T_Z = C_Z \tag{4–29–Z}$$

where Z is equal to Z_s minus the number of external nodes for which the temperature is specified, and i represents the row and j the column; the equations are generally ordered such that the matrix of elements a_{ij} are diagonally dominant (i.e., $|a_{jj}| > |a_{ij}|$ for $i = 1, 2, \ldots, Z, i \neq j$). This system of equations must be solved for the Z unknown nodal temperatures. Once the solution for the temperature distribution is determined, the rate of heat transfer within the body or at its surface can be obtained.

The mechanics involved in producing finite-difference calculations for the nodal temperatures and rate of heat transfer for linear systems are introduced in Examples 4–4 and 4–5 in the context of rather coarse grids, which result in small numbers of equations that are solved by hand. As would be expected, the accuracy of such coarse grid formulations is often inadequate. Since the number of equations required to accurately model multidimensional heat-transfer systems usually range from 10 or so to many thousands, more efficient computer solution techniques are generally used. Both iterative and direct numerical methods are available for the systematic solution of simultaneous algebraic equations. Several of the more simple and commonly used methods for solving linear and nonlinear systems of equations are discussed in the following sections.

EXAMPLE 4–4

A 1-cm-diameter 3-cm-long carbon (1%) steel fin transfers heat from the wall of a heat exchanger at 200°C to a fluid at 25°C with $\bar{h} = 120$ W/(m² °C). Develop a numerical finite-difference solution for the rate of heat transfer for the case in which the tip is insulated and thermal radiation effects are negligible. Utilize a grid spacing of $\Delta x = L/4$.

Solution

Objective Use the finite-difference approach with $\Delta x = L/4$ to develop an approximate solution for q_F.

Schematic Circular fin with insulated tip.

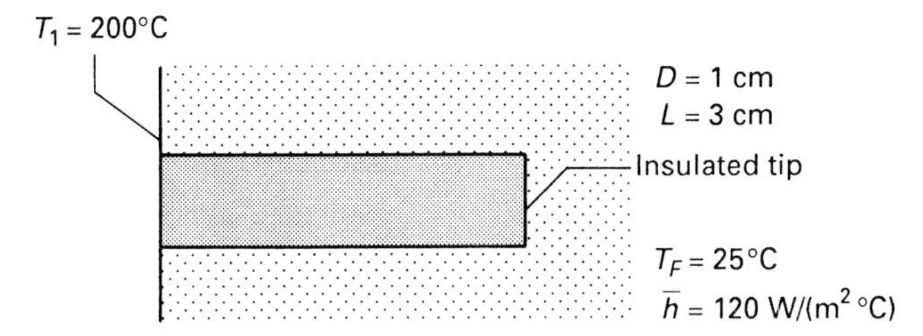

Assumptions/Conditions

steady-state
one-dimensional fin
uniform properties
negligible thermal radiation

Properties Carbon steel (1% C) at room temperature (Table A–C–1): $k = 43$ W/(m °C).

Analysis As shown in Example 2–12, the Biot number is equal to 0.00698, such that the heat transfer in this two-dimensional system can be approximated by a one-dimensional analysis.

A one-dimensional finite-difference grid is shown in Fig. E4–4 for $\Delta x = L/4$. Because T_1 is specified, this network only involves four unknown nodal temperatures. Therefore, four algebraic equations must be written to solve for T_2, T_3, T_4, and T_5. Following the CVFDM, an energy balance is developed for node (2) by writing

Node (2) $\quad q_x = q_{x+\Delta x} + \Delta q_c$

$$-kA\frac{T_2 - T_1}{\Delta x} = -kA\frac{T_3 - T_2}{\Delta x} + \bar{h}p\,\Delta x\,(T_2 - T_F)$$

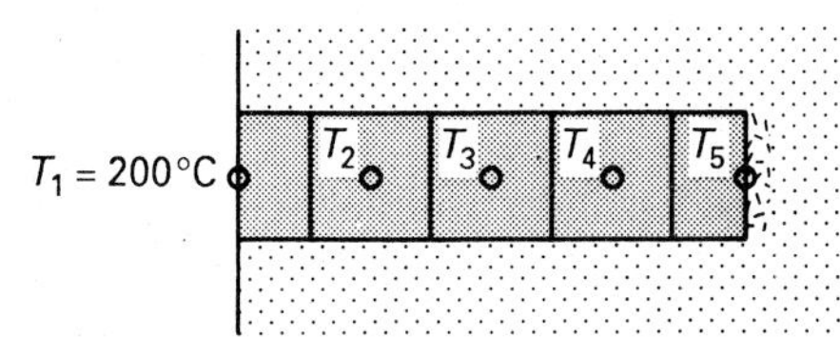

FIGURE E4–4
Finite-difference grid for fin with small Biot number; $\Delta x = L/4 = 0.0075$ m.

where $\bar{h}p\,\Delta x^2/(kA) = 0.0628$ for this problem. Rearranging, we obtain

$$T_1 - \left(2 + \frac{\bar{h}p\,\Delta x^2}{kA}\right) T_2 + T_3 + \frac{\bar{h}p\,\Delta x^2}{kA} T_F = 0$$

Similar equations are written for nodes (3) and (4).

Node (3) $$T_2 - \left(2 + \frac{\bar{h}p\,\Delta x^2}{kA}\right) T_3 + T_4 + \frac{\bar{h}p\,\Delta x^2}{kA} T_F = 0$$

Node (4) $$T_3 - \left(2 + \frac{\bar{h}p\,\Delta x^2}{kA}\right) T_4 + T_5 + \frac{\bar{h}p\,\Delta x^2}{kA} T_F = 0$$

The energy balance for node (5) takes the form

Node (5) $$q_x = \Delta q_c$$

$$-kA\frac{T_5 - T_4}{\Delta x} = \bar{h}p\frac{\Delta x}{2}(T_5 - T_F)$$

or

$$T_4 - \left(1 + \frac{\bar{h}p}{kA}\frac{\Delta x^2}{2}\right) T_5 + \frac{\bar{h}p}{kA}\frac{\Delta x^2}{2} T_F = 0$$

Substituting for the various parameters, our four nodal equations are summarized as follows:

$$200°\text{C} - 2.06T_2 + T_3 + 1.57\ °\text{C} = 0 \quad \text{(a)}$$

$$T_2 - 2.06T_3 + T_4 + 1.57\ °\text{C} = 0 \quad \text{(b)}$$

$$T_3 - 2.06T_4 + T_5 + 1.57\ °\text{C} = 0 \quad \text{(c)}$$

$$T_4 - 1.03T_5 + 0.785°\text{C} = 0 \quad \text{(d)}$$

These linear equations can be easily solved by standard algebraic manipulation. For example, Eqs. (a) and (b) can be combined to eliminate T_2 and Eqs. (c) and (d) can be coupled to remove T_5. The solution of the resulting two equations gives

$$T_3 = 155°\text{C} \qquad T_4 = 145°\text{C}$$

T_2 and T_5 are then found from Eqs. (a) and (d); that is,

$$T_2 = 173°\text{C} \qquad T_5 = 141°\text{C}$$

By comparing these calculations for the nodal temperatures with the exact analytical solution given by Eq. (2–116), we find a maximum error of only about 2%.

To obtain an approximation for the rate of heat transfer from the fin, we develop an energy balance at node (1) by writing

Node (1) $q_F = q_{\Delta x/2} + \Delta q_c$

$$q_F = -kA\frac{T_2 - T_1}{\Delta x} + \bar{h}p\frac{\Delta x}{2}(T_1 - T_F) = 14.6 \text{ W}$$

Referring back to Example 2–12, we find that the analytical solution for the rate of heat transfer q_F is

$$q_F = 15.1 \text{ W}$$

Thus, the error in our finite-difference calculations for q_F is about 3.3%.

To improve the accuracy of our solution, a finer grid can be utilized. To get a better feel for the effect of grid size on the accuracy of finite-difference solutions, it is suggested that this rather simple problem be solved for $\Delta x = L$, $\Delta x = L/2$, and $\Delta x = L/3$. A more accurate numerical finite-difference solution to this problem is developed in Example 4–6.

EXAMPLE 4–5

Develop an approximate numerical solution for the rate of heat transfer across surfaces A, B, C, and D of the rectangular plate shown in Fig. E4–5a by utilizing a grid with $\Delta x = L/3$. Also evaluate the accuracy of the solution.

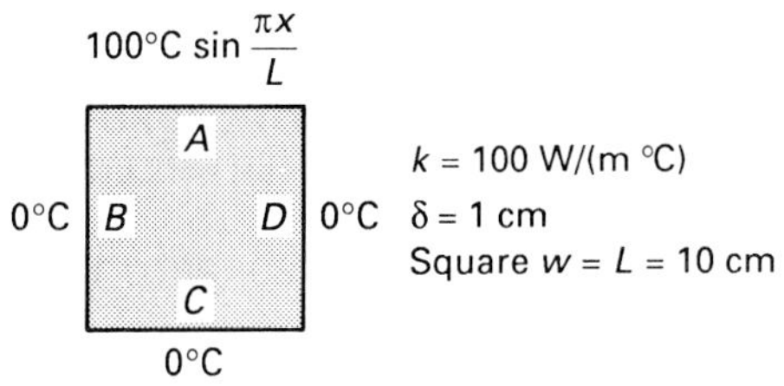

$100°\text{C} \sin\frac{\pi x}{L}$

A

0°C B D 0°C

C

0°C

$k = 100$ W/(m °C)
$\delta = 1$ cm
Square $w = L = 10$ cm

FIGURE E4–5a
Two-dimensional heat transfer in rectangular plate.

Solution

Objective Use the finite-difference approach with $\Delta x = L/3$ to develop an approximate solution for q_A, q_B, q_C, and q_D. Evaluate the accuracy of the calculations.

Assumptions/Conditions

steady-state
two-dimensional
uniform properties

Properties $k = 100$ W/(m °C).

Analysis The finite-difference grid for this problem is shown in Fig. E4–5b for $\Delta x = L/3$. This grid produces 16 nodes, but only four of the nodal temperatures are unknown; these are T_1, T_2, T_3, and T_4. Referring to Fig. E4–5b and utilizing Eq. (1) in Table 4–1, or by developing energy balances on each interior node, the nodal equations for this system are written as

Node (2,2) $\quad -4T_1 + T_2 + T_3 = 0$ (a)

Node (2,3) $\quad T_1 - 4T_2 + T_4 + 86.6°\text{C} = 0$ (b)

Node (3,2) $\quad T_1 - 4T_3 + T_4 = 0$ (c)

Node (3,3) $\quad + T_2 + T_3 - 4T_4 + 86.6°\text{C} = 0$ (d)

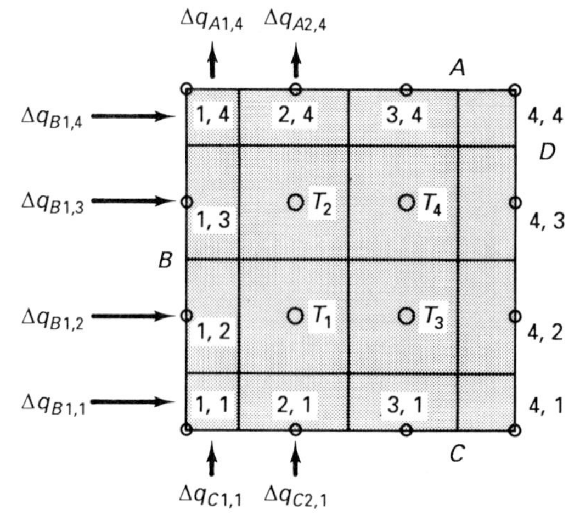

FIGURE E4–5b
Finite-difference grid for rectangular plate.

These four linear equations can be solved for the unknown nodal temperatures. However, our computational work can be eased by recognizing that system symmetry requires $T_3 = T_1$ and $T_4 = T_2$. Consequently, we really only have two unknown nodal temperatures to concern ourselves with. Utilizing this fact, Eqs. (a) and (b) become

Node (2,2) $\quad -3T_1 + T_2 = 0$ (e)

Node (2,3) $\quad T_1 - 3T_2 + 86.6°\text{C} = 0$ (f)

Equations (e) and (f) are easily solved to obtain

$$T_1 = \frac{1}{8}(86.6°\text{C}) = 10.8°\text{C} \qquad T_2 = \frac{3}{8}(86.6°\text{C}) = 32.4°\text{C}$$

A comparison of these predictions with the exact analytical solution developed in Sec. 3–6 reveals an error of 15.3% in T_1 and 8.06% in T_2.

Referring to Fig. E4–5b, the rate of heat transfer across the heating surface is

$$q_A = 2(\Delta q_{A1,4} + \Delta q_{A2,4})$$

To approximate the rate of heat transfer at nodes (1,4) and (2,4), energy balances are written for these nodes. Considering node (2,4) first, we write

$$\Delta q_{A2,4} = \Delta q_x + \Delta q_y = -k\delta \frac{\Delta y}{2} \frac{86.6^\circ\text{C} - 0^\circ\text{C}}{\Delta x} - k\delta\, \Delta x \frac{86.6^\circ\text{C} - T_2}{\Delta y}$$

Setting $T_2 = 32.4^\circ\text{C}$ and utilizing the values of k and δ for this problem (i.e., $k\delta = 1\ \text{W}/^\circ\text{C}$), this equation reduces to

$$\Delta q_{A2,4} = -k\delta \left(\frac{86.6^\circ\text{C} - 0^\circ\text{C}}{2} + 86.6^\circ\text{C} - 32.4^\circ\text{C} \right) = -97.5\ \text{W}$$

The negative sign indicates that heat is transferred into the surface at node (2,4).

The energy balance for the corner node (1,4) is written as

$$\Delta q_{B1,4} + \Delta q_y = \Delta q_{A1,4} + \Delta q_x$$

where

$$\Delta q_x = -k\delta \frac{\Delta y}{2} \frac{86.6^\circ\text{C} - 0^\circ\text{C}}{\Delta x} = -43.3\ \text{W}$$

$$\Delta q_y = -k\delta \frac{\Delta x}{2} \frac{0^\circ\text{C} - 0^\circ\text{C}}{\Delta y} = 0\ \text{W}$$

Thus, we have the result

$$\Delta q_{A1,4} = \Delta q_{B1,4} + 43.3\ \text{W} \tag{i}$$

where $\Delta q_{B1,4}$ is unknown. Because we have one equation but two unknowns, both $\Delta q_{A1,4}$ and $\Delta q_{B1,4}$ are indeterminant. About the best that we can do in this situation is to simply neglect the corner node, that is, assume that $\Delta q_{A1,4}$ is negligible.

Summing up, q_A is approximated by

$$q_A = 2(0\ \text{W} - 97.5\ \text{W}) = -195\ \text{W}$$

The exact solution for q_A is found from Eq. (3–44), which is developed in Sec. 3–6, to be -200 W. Thus, the error in q_A is only 2.5%. However, upon closer inspection we find that the exact solutions for $\Delta q_{A1,4}$ and $\Delta q_{A2,4}$ are -13.4 W and -86.6 W, respectively. The error in $\Delta q_{A2,4}$ is a substantial 12.6%, with the contribution of $\Delta q_{A1,4}$ and $\Delta q_{A4,4}$ being about 13% of q_A. These errors are of the order of the errors in the predictions for the temperature distribution. It just so happens that these errors compensate each other. We are not always so fortunate.

As seen in this example, the inherent indeterminant nature of the heat transfer through corner nodes for surfaces with specified temperatures is one of the main sources of error in this type of numerical finite-difference formulation. (This type of error does not occur for convection, thermal radiation, or specified heat flux boundary conditions.) This error can be reduced by decreasing the grid size. (This feat is sometimes accomplished by using variable grid spacing, with the smallest subvolumes utilized in the region where the accuracy is most critical.)

Following the pattern established in approximating q_A, the following calculations are obtained for q_B and q_C:

$$q_B = -86.5 \text{ W} \qquad q_C = -21.6 \text{ W}$$

Because of symmetry, we write

$$q_D = -q_B = 86.5 \text{ W}$$

To check our calculations, we perform an energy balance on the entire system.

$$q_A = q_B + q_C - q_D = (-86.5 - 21.6 - 86.5) \text{ W} = -194.6 \text{ W}$$

Because this value is within 0.3% of the direct calculation for q_A ($= -195$ W), we conclude that no calculation errors have been made. This small difference is a consequence of the use of three significant figures in our calculations and should not be confused with the actual error that results from the use of finite-difference approximations. Such an energy balance cannot be used to determine the accuracy of a numerical finite-difference analysis!

As indicated in Sec. 4–4–1, the error associated with our finite-difference approximations should be of the order of Δx^2. In checking back over the calculations for the temperatures and rates of heat transfer, we find that the average error is about 11%. Thus, we would expect that a grid spacing of $\Delta x/L = 0.1$ would be required to reduce the errors to the order of 1%. This is indeed the case, as is shown in Example 4–8.

Iterative Methods

The most efficient methods of developing solutions to conduction-heat-transfer problems requiring large numbers of nodes (50 or more) generally involve the use of iterative techniques. Iterative techniques have been devised for the simultaneous solution of systems of equations of the form of Eqs. (4–29–1) through (4–29–Z), which are similar to the simple iterative scheme introduced in Chap. 2. This approach involves the development of successive approximations for the unknowns that converge toward the exact solution.

The *Gauss-Seidel method* is one of the more popular iterative schemes for solving systems of linear algebraic equations. In this approach, Eqs. (4–29–1) through (4–29–Z) are rewritten as follows:

$$
\begin{aligned}
T_1 &= \frac{1}{a_{11}}[C_1 - (a_{12}T_2 + a_{13}T_3 + \cdots + a_{1j}T_j + \cdots + a_{1Z}T_Z)] \\
T_2 &= \frac{1}{a_{22}}[C_2 - (a_{21}T_1 + a_{23}T_3 + \cdots + a_{2j}T_j + \cdots + a_{2Z}T_Z)] \\
&\;\;\vdots \\
T_j &= \frac{1}{a_{jj}}[C_j - (a_{j1}T_1 + \cdots + a_{jj-1}T_{j-1} + a_{jj+1}T_{j+1} + \cdots + a_{jZ}T_Z)] \\
&\;\;\vdots \\
T_Z &= \frac{1}{a_{ZZ}}[C_Z - (a_{Z1}T_1 + a_{Z2}T_2 + \cdots + a_{Zj}T_j + \cdots + a_{ZZ-1}T_{Z-1})]
\end{aligned}
\tag{4–30}
$$

By introducing approximations for the unknown temperatures ($T_1 = T_1^{(1)}$, $T_2 = T_2^{(1)}$, $T_j = T_j^{(1)}$, etc.) into Eqs. (4–30–1) through (4–30–Z) new first-order calculations are developed for the unknowns, $T_1^{(2)}$, $T_2^{(2)}$, and so on. The first-order calculations are then substituted into Eqs. (4–30–1) through (4–30–Z) to produce second-order calculations. Assuming convergence, this procedure is continued until the kth iteration produces sufficient accuracy for all values of T_j; that is,

$$
\left| \frac{T_j^{(k+1)} - T_j^{(k)}}{T_j^{(k)}} \right| < \epsilon \tag{4–31}
$$

where superscript $k = 1, 2, \ldots$, and ϵ is a small number that determines the accuracy of the solution of Eqs. (4–29). In this connection, convergence of the Gauss-Seidel method is guaranteed if the equations are ordered such that the elements a_{ij} are diagonally dominant. (Formal convergence criteria for the Gauss-Seidel method is presented by Kreyszig [3].)

The mechanics involved in the use of the Gauss-Seidel iterative method for solving sets of linear equations is illustrated in Example 4–6. Notice that new temperature values are utilized in the iteration pattern as soon as they are calculated. This approach is adapted to digital computation in Examples 4–6 and 4–8. The related *successive-approximation method* for solving *nonlinear* equations is featured in Example 4–7. Although the Gauss-Seidel method is always convergent when applied to diagonally dominant equations, the convergence can be relatively slow. Iterative techniques that provide faster convergence are discussed in references 3 to 5. On the other hand, the use of the method of successive approximations for solving nonlinear equations requires that the equations be arranged in a form for which the iterative calculations will converge. This requirement is not always easily satisfied. Thus, the higher-order Newton-Raphson method is commonly used for solving nonlinear equations [3–5].

EXAMPLE 4–6

Develop a more accurate numerical finite-difference solution for the rate of heat transfer from the fin of Example 4–4 by employing the Gauss-Seidel iteration method.†

Solution

Objective Use the Gauss-Seidel iteration method to develop accurate finite-difference calculations for q_F.

Schematic Circular fin with insulated tip.

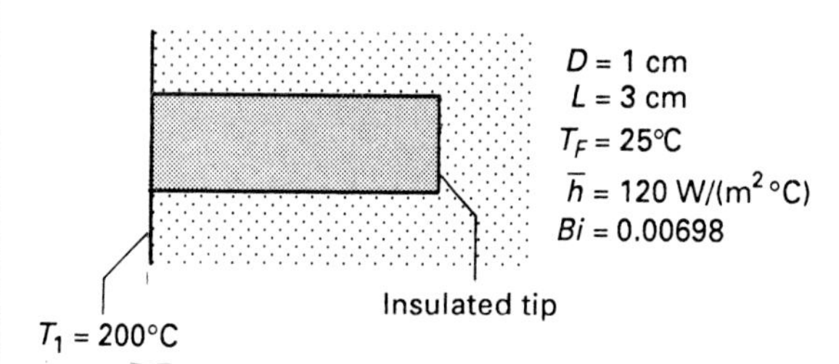

Assumptions/Conditions

- steady-state
- one-dimensional fin
- uniform properties
- negligible thermal radiation

Properties Carbon steel (1% C) at 200°C: $k = 43$ W/(m °C).

Analysis A finite-difference grid network is shown in Fig. E4–6a for M nodes. Following the CVFDM, we apply the first law of thermodynamics to each nodal control volume, with the result

$m = 2, 3, \ldots, M - 1 \qquad q_x = q_{x+\Delta x} + \Delta q_c$

$$T_{m-1} - \left(2 + \frac{\bar{h}p}{kA}\Delta x^2\right) T_m + T_{m+1} + \frac{\bar{h}p}{kA}\Delta x^2 T_F = 0$$

and

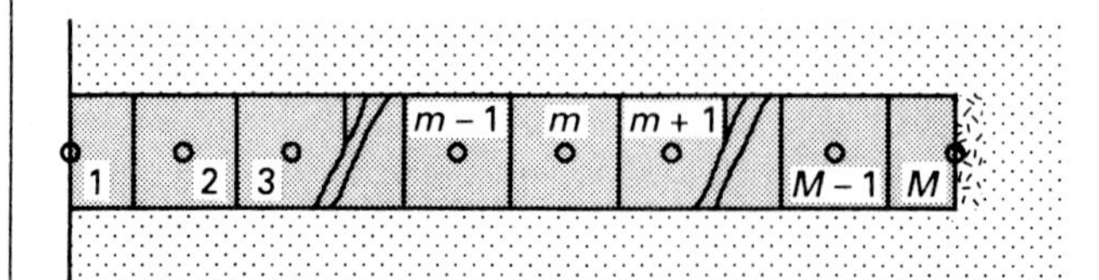

FIGURE E4–6a
Finite-difference grid for fin with small Biot number; $\Delta x = L/(M - 1)$.

† This problem is solved by the finite-element method in Appendix F.

$m = M \qquad q_x = \Delta q_c$

$$T_{M-1} - \left(1 + \frac{\bar{h}p}{kA}\frac{\Delta x^2}{2}\right) T_M + \left(\frac{\bar{h}p}{kA}\frac{\Delta x^2}{2}\right) T_F = 0$$

These $M-1$ equations are written in the Gauss-Seidel format as

$m = 2, 3, \ldots, M-1$

$$T_m = \frac{1}{2 + \bar{h}p\,\Delta x^2/(kA)}\left(T_{m-1} + T_{m+1} + \frac{\bar{h}p}{kA}\Delta x^2 T_F\right) \tag{a}$$

$m = M$

$$T_M = \frac{1}{1 + \bar{h}p\,\Delta x^2/(2kA)}\left(T_{M-1} + \frac{\bar{h}p}{kA}\frac{\Delta x^2}{2} T_F\right) \tag{b}$$

For example, for the $M = 5$ grid of Example 4–4, we have (to four significant figures)

$$T_2 = \frac{1}{2.063}(200°\text{C} + T_3 + 1.57°\text{C})$$

$$T_3 = \frac{1}{2.063}(T_2 + T_4 + 1.57°\text{C})$$

$$T_4 = \frac{1}{2.063}(T_3 + T_5 + 1.57°\text{C})$$

$$T_5 = \frac{1}{1.031}(T_4 + 0.785°\text{C})$$

Starting our iterative calculations with $T_m^{(1)} = 200°\text{C}$, first-round calculations are obtained as follows:

$$T_2^{(2)} = \frac{1}{2.063}(200°\text{C} + 200°\text{C} + 1.57°\text{C}) = 194.7°\text{C}$$

$$T_3^{(2)} = \frac{1}{2.063}(194.7°\text{C} + 200°\text{C} + 1.57°\text{C}) = 192.1°\text{C}$$

$$T_4^{(2)} = \frac{1}{2.063}(192.1°\text{C} + 200°\text{C} + 1.57°\text{C}) = 190.8°\text{C}$$

$$T_5^{(2)} = \frac{1}{1.031}(190.8°\text{C} + 0.785°\text{C}) = 185.8°\text{C}$$

A second iteration gives

$$T_2^{(3)} = \frac{1}{2.063}(200°\text{C} + 192.1°\text{C} + 1.57°\text{C}) = 190.8°\text{C}$$

$$T_3^{(3)} = \frac{1}{2.063}(190.8°\text{C} + 190.8°\text{C} + 1.57°\text{C}) = 185.7°\text{C}$$

$$T_4^{(3)} = \frac{1}{2.063}(185.7°\text{C} + 185.8°\text{C} + 1.57°\text{C}) = 180.8°\text{C}$$

$$T_5^{(3)} = \frac{1}{1.031}(180.8°\text{C} + 0.785°\text{C}) = 176.1°\text{C}$$

The iteration calculations for $T_2^{(k)}$, $T_3^{(k)}$, $T_4^{(k)}$, and $T_5^{(k)}$ converge within about 2% of the values obtained in Example 4–4 after 22 iterations. But because of the slow convergence, hand calculation is impractical for this problem. Therefore, in order to develop more accurate finite-difference solutions with larger values of M, numerical digital computation is now employed.

Utilizing BASIC computer language, Eqs. (a) and (b) are written as

```
T(J,2)=(T(J−1,1)+T(J+1,1)+CC1*TF)/(2+CC1)
```

for J $= m = 2, 3, \ldots, M-1$, and

```
T(M,2)=(T(M−1,1)+CC1/2*TF)/(1+CC1/2)
```

where CC1 = H*P*DX*DX/(K*A), H $= \bar{h}$, P $= p$, DX $= \Delta x$, K $= k$, and TF $= T_F$. The second index represents the new (index 2) and old (index 1) iteration values for the nodal temperatures. The rate of heat transfer q_F from the fin is obtained by performing an energy balance on node (1); q_F is expressed in BASIC by

```
QF= −K*A/DX*(T(2,2)−T(1,2))+H*P*DX/2*(T(1,2)−TF)        (j)
```

A simple flowchart and BASIC program are given in Figs. E4–6b and E4–6c, respectively, which solves the $M-1$ nodal equations by the Gauss-Seidel method. The program also calculates the rate of heat transfer from the fin. This BASIC program requires that the temperature at node M satisfy the specified iterative convergence criterion ϵ (designated by CHK in the program). (Logic that permits us to check and satisfy convergence for all nodal temperatures is developed in Example 4–8.)

Calculations for the rate of heat transfer from the fin are shown in Fig. E4–6d for $M = 3, 4, \ldots, 20$, and for $\epsilon = 10^{-2}, 10^{-3}, 10^{-4}$, and 10^{-5}. Note that the calculations are dependent upon both M and ϵ! The calculations for q_F essentially converge to approximately 15.1 W for $M \geq 6$ (i.e., $\Delta x < 0.006$ m) and $\epsilon = 10^{-5}$. This value is in agreement with the analytical solution developed in Example 2–12.

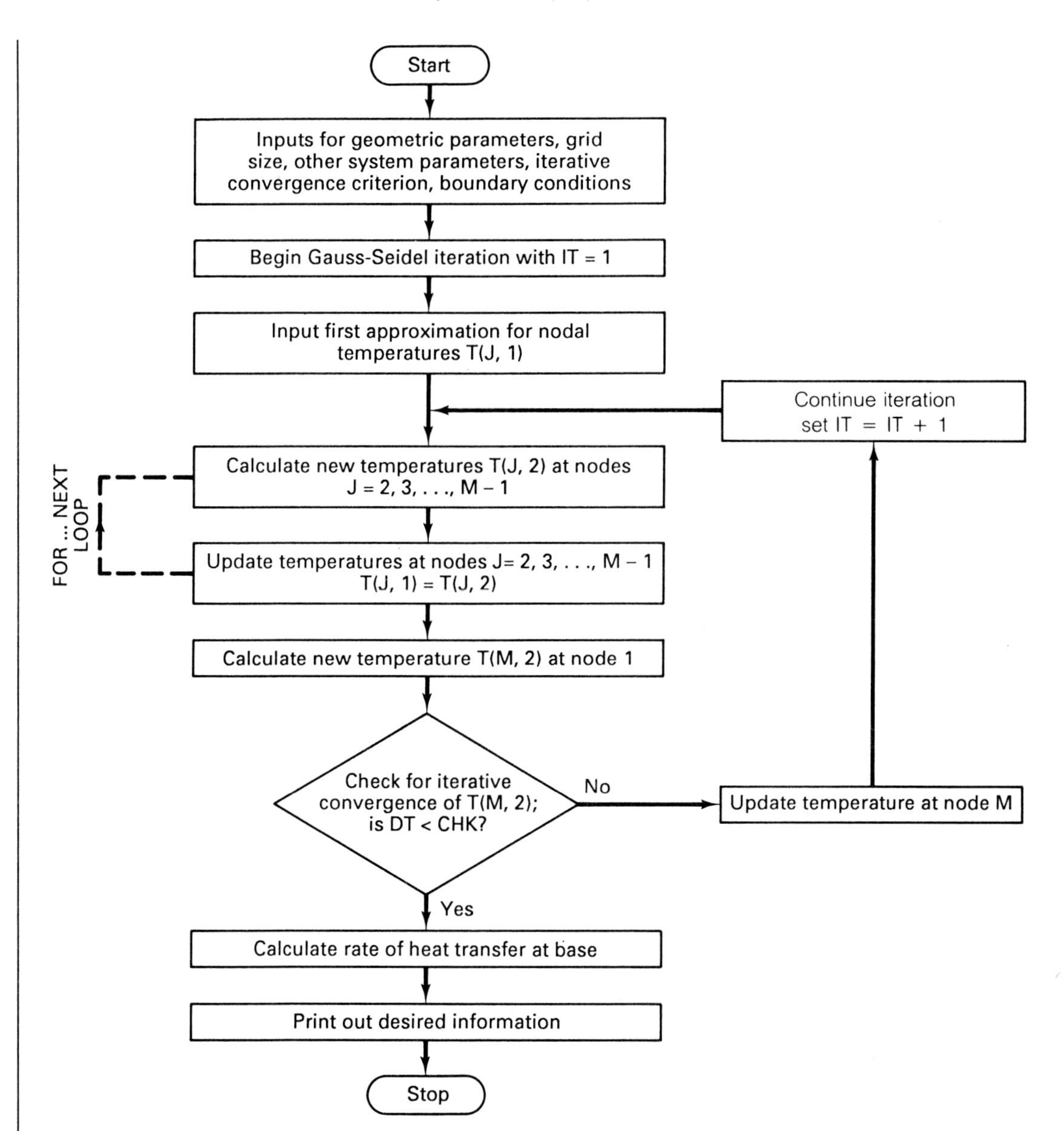

FIGURE E4–6b Program flowchart for Example 4–6.

```
                                                                  Comments
                                           Inputs for geometric parameters
   L=.03                                                                01
   D=.01                                                                02
   A=3.1416*D^2/4                                                       03
   P=3.1416*D                                                           04
                                                     Inputs for grid size
   FOR M=3 TO 20                                                        05
   M1=M-1                                                               06
   DX=L/M1                                                              07
                                                    Inputs for properties
   K=43                                                                 08
                                       Inputs for other system parameters
   H=120                                                                09
   CC1=H*P*DX^2/(K*A)                                                   10
   T1=200                                                               11
   TF=25                                                                12
                                Input for iterative convergence criterion
   CHK=.00001                                                           13
                                          Input for boundary conditions
   T(1,1)=T1                                                            14
   T(1,2)=T1                                                            15
                                    Begin Gauss Seidel iteration with IT = 1
   IT=1                                                                 16
                             Input first approximation for nodal temperatures
   FOR J=2 TO M                                                         17
   T(J,1)=T1                                                            18
                    Calculate new temperatures T(J,2) at nodes J = 2,3, . . . , M - 1
   NEXT J                                                               19
2  FOR J=2 TO M1                                                        20
   T(J,2)=(T(J-1,1)+T(J+1,1)+CC1*TF)/(2+CC1)                            21
                                   Update temperatures at J = 2,3, . . . , M - 1
   T(J,1)=T(J,2)                                                        22
                                 Calculate new temperature T(M,2) at node M
   NEXT J                                                               23
   T(M,2)=(T(M-1,1)+CC1/2*TF)/(1+CC1/2)                                 24
                                  Check for iterative convergence of T(M,2)
   DT=ABS((T(M,2)-T(M,1))/T(M,1))                                       25
   IF DT<CHK GOTO 4                                                     26
                                             Update temperature at node M
   T(M,1)=T(M,2)                                                        27
                                                       Continue iteration
   IT=IT+1                                                              28
   GOTO 2                                                               29
                                       Calculate rate of heat transfer at base
4  QF=-K*A/DX*(T(2,2)-T(1,2))+H*P*DX/2*(T(1,2)-TF)                      30
                                                Print out desired information
   LPRINT USING "####.##"; M, IT, T(M,2), QF                            31
   NEXT M                                                               32
   STOP                                                                 33
   END                                                                  34
```

FIGURE E4–6c BASIC program for Example 4–6.

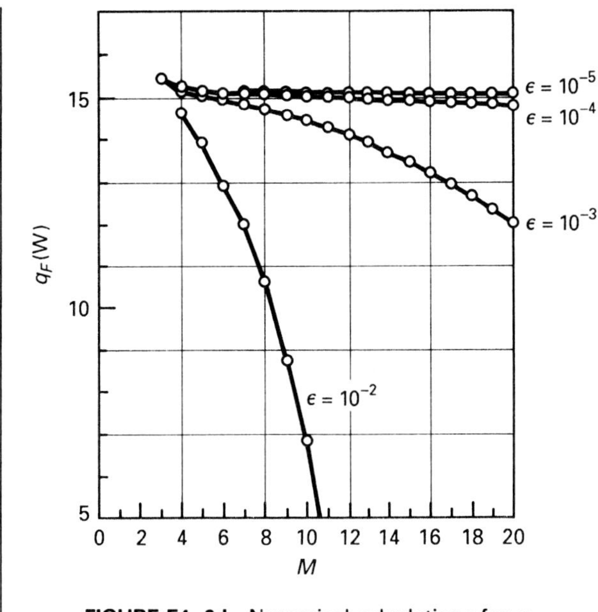

FIGURE E4–6d Numerical calculations for q_F.

EXAMPLE 4–7

Determine the rate of heat transfer and temperature distribution for the anodized aluminum fin shown in Fig. E4–7a. The base is at 80°C, the air and surrounding walls are at 25°C, and the coefficient of heat transfer due to natural convection cooling is 10 W/(m² °C).

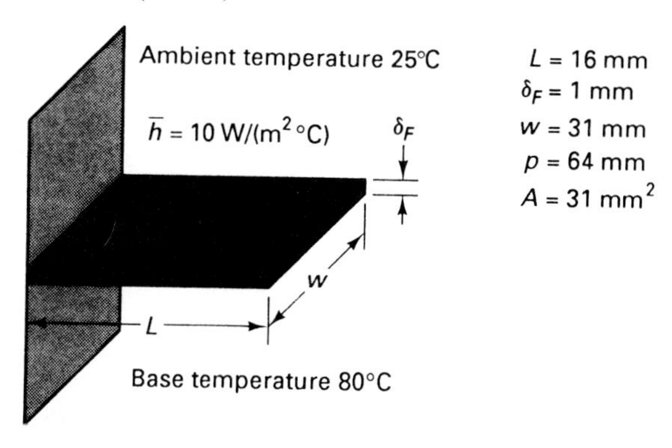

L = 16 mm
δ_F = 1 mm
w = 31 mm
p = 64 mm
A = 31 mm²

FIGURE E4–7a
Convecting and radiating fin.

Solution

Objective Calculate q_F and $T(x)$.

Assumptions/Conditions

steady-state
one-dimensional fin
uniform properties
convection and blackbody thermal radiation cooling
$F_{s-R} = 1$ from fin to surroundings

Properties Anodized aluminum (Example 2–13): $k = 200$ W/(m °C).

Analysis As shown in Example 2–14, a one-dimensional analysis is warranted for this problem because the Biot number is much less than 0.1. Because of the complexity of the problem, we will utilize the numerical approach.

Utilizing the finite-difference grid of Fig. E4–6a in Example 4–6, the nodal equations are given by

$m = 2, 3, \ldots, M - 1 \qquad q_x = q_{x+\Delta x} + \Delta q_c + \Delta q_R$

$$-kA\frac{T_m - T_{m-1}}{\Delta x} = -kA\frac{T_{m+1} - T_m}{\Delta x} + \bar{h}p\,\Delta x\,(T_m - T_F) + \sigma p\,\Delta x\,F_{s-R}(T_m^4 - T_R^4) \tag{a}$$

$m = M \qquad q_x = \Delta q_c + \Delta q_R + q_c + q_R$

$$-kA\frac{T_M - T_{M-1}}{\Delta x} = \bar{h}p\frac{\Delta x}{2}(T_M - T_F) + \sigma pF_{s-R}\frac{\Delta x}{2}(T_M^4 - T_R^4) + \bar{h}A(T_M - T_F) + \sigma AF_{s-R}(T_M^4 - T_R^4) \tag{b}$$

where q_c and q_R account for the heat transfer from the tip. Because of the nonlinear nature of these equations, we will employ the successive approximation method. This approach requires that Eqs. (a) and (b) be arranged in a form that will produce convergent iterative calculations. Therefore, following the experience obtained in Example 2–3, these equations are written as

$m = 2, 3, \ldots, M - 1$

$$T_m = \frac{T_{m-1} + T_{m+1} + CC_1\,T_F + CR_1\,T_R^4}{2 + CC_1 + CR_1\,T_m^3} \tag{c}$$

$m = M$

$$T_M = \frac{T_{M-1} + (CC_1/2 + CC_2)T_F + (CR_1/2 + CR_2)T_R^4}{1 + CC_1/2 + CC_2 + (CR_1/2 + CR_2)T_M^3} \tag{d}$$

where $CC_1 = \bar{h}p\,\Delta x^2/(kA)$, $CC_2 = \bar{h}\,\Delta x/k$, $CR_1 = \sigma pF_{s-R}\,\Delta x^2/(kA)$, and $CR_2 = \sigma F_{s-R}\,\Delta x/k$. Notice that T_m appears on both sides of these nonlinear equations.

Equations (c) and (d) are written in BASIC language as follows:

```
T(J,2) = (T(J-1,1) + T(J+1,1) + CC1*TF + CR1*TR^4)
         /(2 + CC1 + CR1*T(J,1)^3)                                  (e)
```

for J = 2, 3, . . . , M − 1, and

```
T(M,2) = (T(M-1,1) + (CC1/2 + CC2)*TF + (CR1/2 + CR2)*TR^4)
         /(1 + CC1/2 + CC2 + (CR1/2 + CR2)*T(M,1)^3)                (f)
```

for J = M, where CC1 = H*P*DX*DX/(K*A), CC2 = H*DX/K, CR1 = SIGMA*P*DX*DX*FSR/(K*A), and CR2 = SIGMA*DX*FSR/K. By applying the first law of thermodynamics to node (1), an expression is obtained for the rate of heat transfer q_F from the fin of the form

```
QF = -K*A/DX*(T(2,2) - T(1,2)) + H*P*DX/2*(T(1,2) - TF)
     + SIGMA*P*DX/2*FSR*(T(1,2)^4 - TR^4)                           (g)
```

The BASIC program presented in Example 4–6 is employed to solve these equations by utilizing the following inputs for geometric parameters, properties, and other system parameters:

Inputs for geometric parameters

L = 16 E−3

P = 64 E−3

A = 31 E−6

Inputs for properties

K = 200

Inputs for other system parameters

H = 10
SIGMA = 5.67 E−8
FSR = 1
CC1 = H*P*DX*DX/(K*A)
CC2 = H*DX/K
CR1 = SIGMA*P*DX*DX*FSR/(K*A)
CR2 = SIGMA*DX*FSR/K
T1 = 353
TF = 298
TR = 298

Of course, the equations for T(J,2), T(M,2), and QF must be specified in accordance with Eqs. (e), (f), and (g).

Calculations are shown in Fig. E4–7b for the rate of heat transfer from the fin for $M = 3, 4, \ldots, 20$ and $\epsilon = 10^{-5}$ and 10^{-7}. The calculations for q_F converge to approximately 1.02 W for $M \geq 3$ (i.e., $\Delta x \lesssim 0.006$ m) and $\epsilon = 10^{-7}$. To see the effect of radiation, the program is also run with the radiation terms eliminated. The calculations for q_F without radiation converge to about 0.575 W, which is consistent with the analytical solution given by Eq. (2–115). Thus, the thermal radiation contributes 43.6% to the total rate of heat transfer.

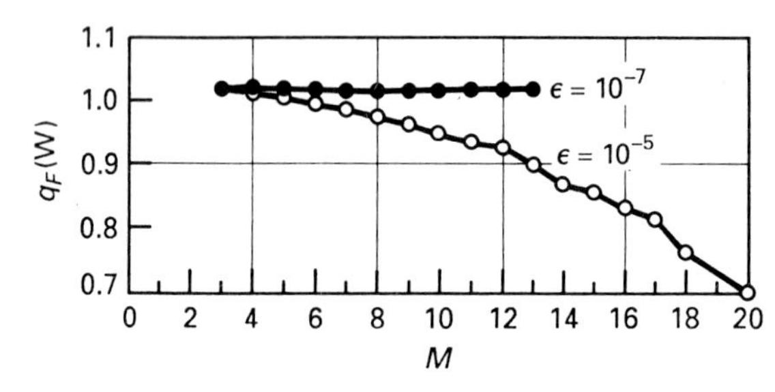

FIGURE E4–7b
Numerical calculations for q_F.

Assuming that no errors have been made in our analysis and program inputs, we can be reasonably sure of an accurate solution, since the calculations for q_F converge for increasing M and decreasing ϵ. However, to further reinforce our confidence in the solution accuracy, calculations are made for q_F for longer fin lengths to enable us to compare the results with the analytical solution of Example 2–14. The comparison of numerical calculations for q_F for fins of various lengths with the analytical result for a long fin is made in Fig. E4–7c. Notice that the numerical calculations approach the analytical solution of Example 2–14 for lengths of the order of 0.3 m and greater.

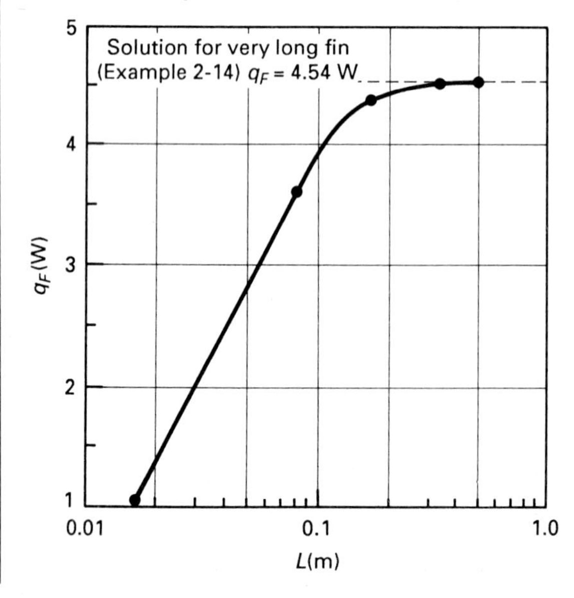

FIGURE E4–7c
Numerical calculations for q_F for various lengths.

The temperature distribution in the fin is shown in Table E4–7. The temperature drop along this short fin is seen to be very small, such that the fin efficiency η_F is quite high. To expand upon this point, we calculate the fin surface area A_F,

$$A_F = pL + A = 64 \text{ mm } (16 \text{ mm}) + 31 \text{ mm}^2$$

$$= 1060 \text{ mm}^2 = 0.00106 \text{ m}^2$$

and the maximum heat-transfer rate q_{max},

$$q_{max} = \bar{h}A_F(T_s - T_F) + \sigma A_F F_{s-R}(T_s^4 - T_R^4) = 0.583 \text{ W} + 0.459 \text{ W}$$

$$= 1.04 \text{ W}$$

It follows that

$$\eta_F = \frac{q_F}{q_{max}} = \frac{1.02 \text{ W}}{1.04 \text{ W}} = 0.981$$

TABLE E4–7 Numerical calculations for temperature distribution in fin

m	T_m (K)
1	353.0
2	352.7
3	352.5
4	352.3
5	352.1
6	351.9
7	351.8
8	351.74
9	351.68
10	351.66

EXAMPLE 4–8

Develop a BASIC program for determining the temperature distribution and rate of heat transfer across face A for the rectangular plate shown in Fig. E4–8a. Use the Gauss-Seidel iteration method and a grid network with $\Delta x = L/M$, where M can be any integer value. Then run the program to obtain an accurate numerical solution for (*a*) $T_1 = 0°\text{C}$ and $f(x) = 100°\text{C} \sin(\pi x/L)$, and (b) $T_1 = 0°\text{C}$ and $T_2 = 100°\text{C}$.

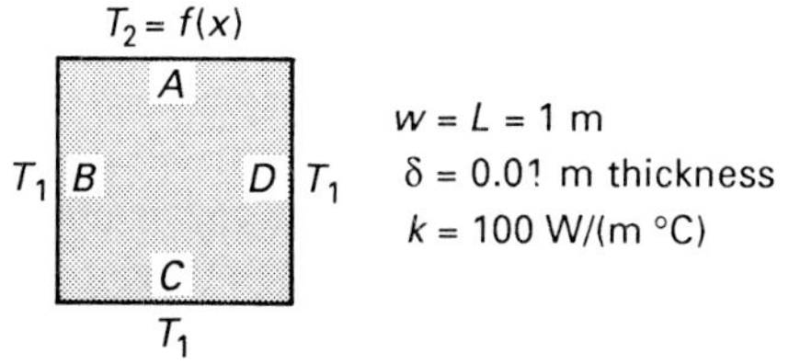

$T_2 = f(x)$

A

T_1 B

D T_1

C

T_1

$w = L = 1$ m

$\delta = 0.01$ m thickness

$k = 100$ W/(m °C)

FIGURE E4–8a
Two-dimensional heat transfer in square plate.

Solution

Objective Develop accurate finite-difference solutions for $T_{m,n}$ and q_A.

Assumptions/Conditions

steady-state
two-dimensional
uniform properties

Properties $k = 100$ W/(m °C).

Analysis A finite-difference grid with $M = 6$ is shown in Fig. E4–8b for this conduction-heat-transfer problem. The interior nodal equations take the general form

$$4\,T_{m,n} = T_{m+1,n} + T_{m-1,n} + T_{m,n+1} + T_{m,n-1}$$

with $m = 2, 3, \ldots, M-1$, and $n = 2, 3, \ldots, N-1$. Our formulation consists of $Z\,[= (M-2)(N-2)]$ such equations plus the boundary conditions. Using BASIC computer language, this system of equations is put into the interative form

T(J,I,2) = (T(J+1,I,1) + T(J−1,I,1) + T(J,I+1,1) + T(J,I−1,1))/4

where J = m = 2, 3, 4, . . . , $M-1$, and I = n = 2, 3, 4, . . . , $N-1$. The third index represents the new (index 2) and old (index 1) iteration values for the nodal temperatures. Using this format, the rate of heat transfer q_A is represented by

QA = K*D*(T(J,N−1,2) + T(J−1,N,2)/2
+ T(J+1,N,2)/2 − 2*T(J,N,2))

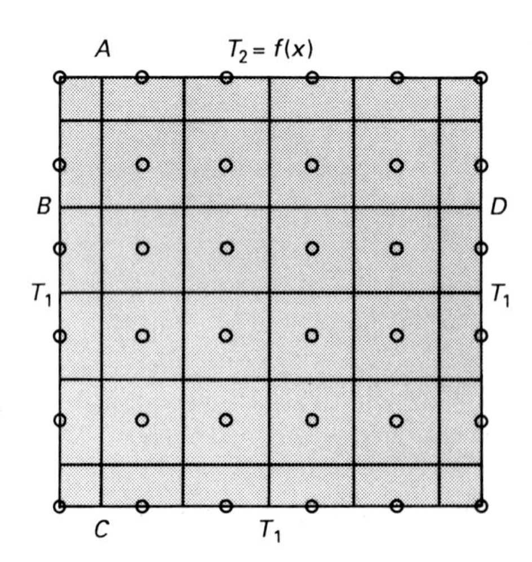

FIGURE E4–8b
Finite-difference grid for square plate.

A BASIC program, which employs the Gauss-Seidel iteration technique, is presented in Fig. A–G–1 in the Appendix. Whereas the program developed in Example 4–6 only checks for iterative convergence of one nodal temperature, the program developed for this example requires that all nodal temperatures satisfy the convergence criterion. The inputs for T_1, T_2, L, w, δ, and k are specified in accordance with the problem shown in Fig. E4–8a.

(a) $f(x) = 100°\text{C} \sin(\pi x/L)$, $T_1 = 0°\text{C}$: Calculations for q_A obtained by running this program on a digital computer are shown in Fig. E4–8c for values of $M = 2, 3, 4, \ldots, 20$ and for $\epsilon = 10^{-1}$ and 10^{-4}. Again we notice that the accuracy of the numerical solution depends upon both M and ϵ! The calculations for q_A are seen to converge to approximately -200.7 W for M equal to 9 with $\epsilon = 10^{-4}$. The exact solution given by Eq. (3–44) is -200 W, such that the error is only 0.35%.

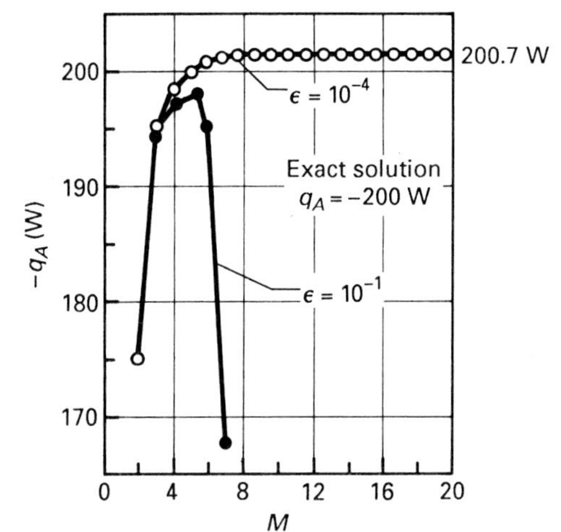

FIGURE E4–8c
Numerical calculations for q_A; $T_2 = 100 \sin(\pi x/L)$ °C.

Calculations are also easily obtained for q_B, q_C, and q_D. For example, the calculations for q_C converge to approximately -17.5 W at a value of M equal to 13. The exact solution for q_C is -17.3 W, such that the error is about 1.5%.

Calculations are shown in Table E4–8 for the nodal temperatures obtained for $M = 10$. The accuracy of these predictions is extremely good, with a minimum error of about 0.1%.

TABLE E4–8 Numerical calculations for temperature distribution in square plate†

0.0	34.20	64.28	86.60	98.48	98.48	86.6	64.28	34.20	0.0
0.0	24.16	45.41	61.18	69.57	69.57	61.1	45.41	24.16	0.0
0.0	17.03	32.01	43.13	49.04	49.04	43.1	32.01	17.03	0.0
0.0	11.96	22.48	30.28	34.44	34.44	30.2	22.48	11.96	0.0
0.0	8.33	15.65	21.09	23.98	23.98	21.0	15.65	8.33	0.0
0.0	5.70	10.72	14.44	16.42	16.42	14.4	10.72	5.70	0.0
0.0	3.77	7.08	9.54	10.84	10.84	9.5	7.08	3.77	0.0
0.0	2.28	4.29	5.78	6.58	6.58	5.7	4.29	2.28	0.0
0.0	1.08	2.02	2.73	3.10	3.10	2.7	2.02	1.08	0.0
0.0	0.0	0.0	0.0	0.0	0.0	0.0	0.0	0.0	0.0

† $T_2 = 100°\text{C} \sin(\pi x/L)$

(b) $T_2 = f(x) = 100°\text{C}$, $T_1 = 0°\text{C}$: The program presented in Fig. A–G–1 is easily run for $T_2 = 100°\text{C}$. Calculations produced for q_A and q_C for this condition are shown in Fig. E4–8d for $M = 2, 3, 4, \ldots, 20$ and $\epsilon = 10^{-4}$. For this problem, q_C converges to approximately -22.2 W, but convergence is nowhere in sight for q_A.

Referring back to Example 3–6, the exact solution for q_C is -22.1 W, but q_A is actually unbounded. Thus, our finite-difference solution is consistent with the exact analytical solution.

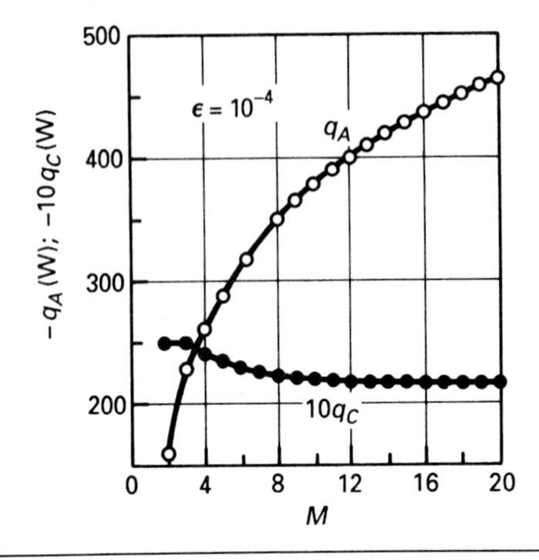

FIGURE E4–8d
Numerical calculations for q_A and q_C; T_2 = 100 °C.

Direct Methods

The most commonly used direct methods for solving systems of linear equations include matrix and Gaussian elimination methods [3–6].

In the Gaussian elimination method, the nodal equations, Eqs. (4–29–1) through (4–29–Z), are put into the form

$$a_{11}T_1 + a_{12}T_2 + \cdots + a_{1j}T_j + \cdots + a_{1Z}T_Z = C_1 \tag{4–32–1}$$

$$a'_{22}T_2 + \cdots + a'_{2j}T_j + \cdots + a'_{2Z}T_Z = C'_2 \tag{4–32–2}$$

$$\vdots$$

$$a'_{jj}T_j + \cdots + a'_{jZ}T_Z = C'_j \tag{4–32–j}$$

$$\vdots$$

$$a'_{ZZ}T_Z = C'_Z \tag{4–32–Z}$$

This arrangement is obtained by first dividing Eq. (4–29–1) by a_{11}. The resulting equation is used to algebraically eliminate T_1 from each of the remaining equations. For example, by combining Eqs. (4–29–1) and (4–29–2), we obtain

$$\begin{aligned} a'_{2j} &= a_{2j} - a_{1j}a_{21}/a_{11} \qquad j = 2, 3, \ldots, Z \\ C'_2 &= C_2 - C_1a_{21}/a_{11} \end{aligned} \tag{4–33}$$

The second equation is then used to eliminate T_2 from the remaining equations. This elimination procedure is continued until the final equation is obtained. This technique produces a solution for the unknown nodal temperature T_Z of the form

$$T_Z = \frac{C'_Z}{a'_{ZZ}} \tag{4–34}$$

The other unknowns can then be calculated in reverse order (i.e., $T_{Z-1}, T_{Z-2}, \ldots, T_j, \ldots, T_1$). This elimination procedure is simplified by the fact that a given nodal temperature T_j appears in no more than four equations for conduction problems involving two dimensions, such as x and y.

A number of standardized programs are available that employ matrix and Gaussian elimination methods. However, except for applications involving relatively few nodal equations, these direct methods are generally less efficient than iterative methods.

Unsteady Analysis Method

Because our next topic deals with the numerical solution of unsteady multidimensional conduction-heat-transfer problems, we merely mention here that steady-state problems are sometimes solved by means of unsteady analyses which are carried through to the steady-state limit.

4–5 NUMERICAL ANALYSIS: UNSTEADY SYSTEMS

Basic finite-difference formulation and solution concepts associated with the numerical analysis of unsteady conduction-heat-transfer processes are considered in this section.

4–5–1 Nodal Equations

To develop a numerical finite-difference formulation for unsteady conduction heat transfer in a rectangular solid with internal energy generation, the system is subdivided into Z_s subvolumes, with the temperature at each node designated by $T^{\tau}_{m,n}$. Using the CVFDM, which was introduced in the previous section, the first law of thermodynamics is applied to each interior and exterior node at an instant of time t. To illustrate, the development of an energy balance on an interior subvolume such as the one shown in Fig. 4–2 gives

$$\Delta q_x + \Delta q_y + \dot{q}\,\Delta V = \Delta q_{x+\Delta x} + \Delta q_{y+\Delta y} + \frac{\Delta E_s}{\Delta t} \tag{4–35}$$

where $\Delta E_s/\Delta t$ is no longer equal to zero and the heat-transfer rates are specified by unsteady finite-difference approximations for the Fourier law of conduction.

Based on our previous study, we know that $\Delta E_s/\Delta t$ can be expressed in terms of the specific heat and instantaneous temperature difference by

$$\frac{\Delta E_s}{\Delta t} = \rho c_v \, \Delta V \frac{\Delta T}{\Delta t} \tag{4–36}$$

where $\Delta T/\Delta t$ represents the finite-difference approximation for $\partial T/\partial t$. This relationship for $\Delta E_s/\Delta t$ together with the second-order finite-difference approximation for the Fourier law of conduction are substituted into Eq. (4–35), to obtain

$$\begin{aligned} -k\delta \, \Delta y \frac{T^{\tau}_{m,n} - T^{\tau}_{m-1,n}}{\Delta x} - k\delta \, \Delta x \frac{T^{\tau}_{m,n} - T^{\tau}_{m,n-1}}{\Delta y} + \dot{q} \, \Delta V \\ = -k\delta \, \Delta y \frac{T^{\tau}_{m+1,n} - T^{\tau}_{m,n}}{\Delta x} - k\delta \, \Delta x \frac{T^{\tau}_{m,n+1} - T^{\tau}_{m,n}}{\Delta y} + \rho c_v \, \Delta V \frac{\Delta T}{\Delta t} \end{aligned} \tag{4–37}$$

Setting $\Delta y = \Delta x$ and rearranging, this equation is put into the form

$$\frac{\alpha}{\Delta x^2}\left(T^{\tau}_{m+1,n} + T^{\tau}_{m-1,n} + T^{\tau}_{m,n+1} + T^{\tau}_{m,n-1} - 4\,T^{\tau}_{m,n} + \frac{\dot{q}}{k}\Delta x^2\right) = \frac{\Delta T}{\Delta t} \tag{4–38}$$

Similar nodal equations can be written for the exterior subvolumes. For example, the nodal equation for a regular exterior node at $m = M$ with surface heat flux q''_s is given by

$$\frac{\alpha}{\Delta x^2}\left(2\,T^{\tau}_{M-1,n} + T^{\tau}_{M,n+1} + T^{\tau}_{M,n-1} - 4\,T^{\tau}_{M,n} + \frac{\dot{q}}{k}\Delta x^2 - 2\frac{q''_s}{k}\Delta x\right) = \frac{\Delta T}{\Delta t} \tag{4–39}$$

It should be observed that these unsteady nodal equations and the corresponding relations for corner nodes can be obtained from the steady nodal equations given in Table 4–1 by merely replacing the right-hand side by $\Delta T/\Delta t$ and introducing the time index τ as a superscript for the nodal temperatures.

To complete the formulation, $\Delta T/\Delta t$ is usually expressed in terms of the *forward-time difference* [see Eq. (4–14)],

$$\frac{\Delta T}{\Delta t} = \frac{T^{\tau+1}_{m,n} - T^{\tau}_{m,n}}{\Delta t} \tag{4–40}$$

or the *backward-time difference* [see Eq. (4–15)],

$$\frac{\Delta T}{\Delta t} = \frac{T^{\tau}_{m,n} - T^{\tau-1}_{m,n}}{\Delta t} \tag{4–41}$$

both of which are *first-order* approximations.

The accuracy of the finite-difference formulations resulting from the use of Eqs. (4–38) and (4–39) and Eqs. (4–40) and (4–41) is controlled by component errors that are proportional to Δx^2 and Δt. Whereas accuracy requirements for limiting steady-state conditions can generally be used to establish an effective grid spacing

Δx, the time increment Δt normally must be determined by trial-and-error testing. However, in the case of the forward-time difference formulation, the selection of Δt is restricted by stability considerations.

Because the forward-time difference formulation produces algebraic equations that can be solved by straightforward explicit computational techniques, this approach will be emphasized in our study. Brief consideration will also be given to the alternative implicit finite-difference method, which is developed by use of the backward-time difference for $\Delta T/\Delta t$.

4–5–2 Explicit Method

Utilizing the forward-time difference, Eq. (4–38) takes the form

$$\frac{\alpha}{\Delta x^2}\left(T^{\tau}_{m+1,n} + T^{\tau}_{m-1,n} + T^{\tau}_{m,n+1} + T^{\tau}_{m,n-1} - 4\,T^{\tau}_{m,n} + \frac{\dot{q}}{k}\Delta x^2\right) = \frac{T^{\tau+1}_{m,n} - T^{\tau}_{m,n}}{\Delta t} \tag{4–42}$$

or

$$T^{\tau+1}_{m,n} = \frac{\alpha\,\Delta t}{\Delta x^2}\left(T^{\tau}_{m+1,n} + T^{\tau}_{m-1,n} + T^{\tau}_{m,n+1} + T^{\tau}_{m,n-1} + \frac{\dot{q}}{k}\Delta x^2\right) + \left(1 - 4\frac{\alpha\,\Delta t}{\Delta x^2}\right)T^{\tau}_{m,n} \tag{4–43}$$

This forward-time difference interior nodal equation expresses the nodal temperature $T^{\tau+1}_{m,n}$ at time $(\tau+1)\,\Delta t$ in terms of the nodal temperature distribution $T^{\tau}_{m,n}$, $T^{\tau}_{m+1,n}$, $T^{\tau}_{m-1,n}$, $T^{\tau}_{m,n+1}$, and $T^{\tau}_{m,n-1}$ at the earlier instant of time $\tau\,\Delta t$. Because the nodal temperature distribution is known at some initial instant of time, this type of equation is *explicit* in that the unknown nodal temperatures at the next instant of time can be calculated directly.

The explicit nodal equation for a regular exterior node at $m = M$ is given by

$$T^{\tau+1}_{M,n} = \frac{\alpha\,\Delta t}{\Delta x^2}\left(2\,T^{\tau}_{M-1,n} + T^{\tau}_{M,n+1} + T^{\tau}_{M,n-1} + \frac{\dot{q}}{k}\Delta x^2 - 2\frac{q''_s}{k}\Delta x\right) + \left(1 - 4\frac{\alpha\,\Delta t}{\Delta x^2}\right)T^{\tau}_{M,n} \tag{4–44}$$

Setting $q''_s = h(T^{\tau}_{m,n} - T_F)$ for the case of convection, this equation becomes

$$T^{\tau+1}_{M,n} = \frac{\alpha\,\Delta t}{\Delta x^2}\left(2\,T^{\tau}_{M-1,n} + T^{\tau}_{M,n+1} + T^{\tau}_{M,n-1} + \frac{\dot{q}}{k}\Delta x^2 + 2\frac{h\,\Delta x}{k}T_F\right) + \left[1 - \frac{\alpha\,\Delta t}{\Delta x^2}\left(4 + 2\frac{h\,\Delta x}{k}\right)\right]T^{\tau}_{M,n} \tag{4–45}$$

Equations (4–43) and (4–44) and other standard explicit nodal equations are listed in Table 4–2.

TABLE 4–2 Summary of explicit nodal equations for unsteady two-dimensional systems with internal heat generation

Type node	*Nodal equation*
1. Interior node 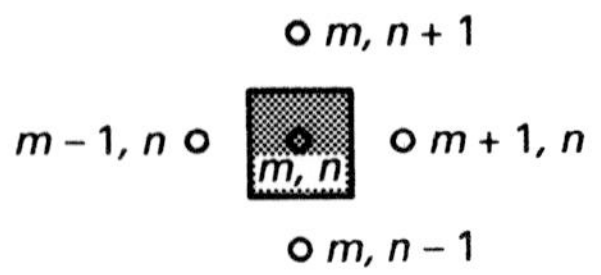	$$T_{m,n}^{\tau+1} = \frac{\alpha\,\Delta t}{\Delta x^2}\left(T_{m+1,n}^{\tau} + T_{m-1,n}^{\tau} + T_{m,n+1}^{\tau} + T_{m,n-1}^{\tau} + \frac{\dot{q}}{k}\Delta x^2\right) + \left(1 - 4\frac{\alpha\,\Delta t}{\Delta x^2}\right)T_{m,n}^{\tau} \qquad (1)$$
2. Exterior nodes	
a. General equations	
i. Regular exterior node (M,n) 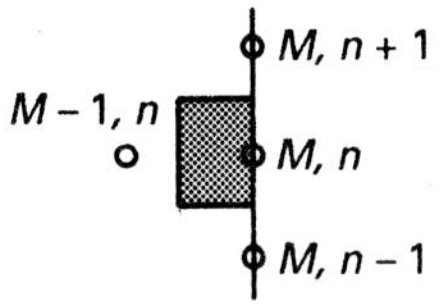	$$T_{M,n}^{\tau+1} = \frac{\alpha\,\Delta t}{\Delta x^2}\left(2\,T_{M-1,n}^{\tau} + T_{M,n+1}^{\tau} + T_{M,n-1}^{\tau} + \frac{\dot{q}}{k}\Delta x^2 - 2\frac{q_s''}{k}\Delta x\right) + \left(1 - 4\frac{\alpha\,\Delta t}{\Delta x^2}\right)T_{M,n}^{\tau} \qquad (2i)$$
ii. Outer corner node (M,N) 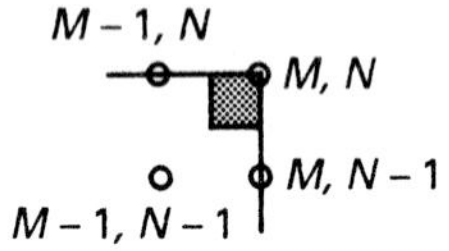	$$T_{M,N}^{\tau+1} = \frac{\alpha\,\Delta t}{\Delta x^2}\left(2\,T_{M-1,N}^{\tau} + 2\,T_{M,N-1}^{\tau} + \frac{\dot{q}}{k}\Delta x^2 - 4\frac{q_s''}{k}\Delta x\right) + \left(1 - 4\frac{\alpha\,\Delta t}{\Delta x^2}\right)T_{M,N}^{\tau} \qquad (2ii)$$
iii. Inner corner node (m,n) 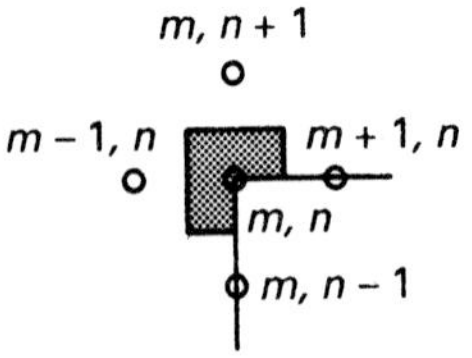	$$T_{m,n}^{\tau+1} = \frac{\alpha\,\Delta t}{\Delta x^2}\left[\frac{2}{3}\left(T_{m+1,n}^{\tau} + 2\,T_{m-1,n}^{\tau} + 2\,T_{m,n+1}^{\tau} + T_{m,n-1}^{\tau}\right) + \frac{\dot{q}}{k}\Delta x^2 - \frac{4}{3}\frac{q_s''}{k}\Delta x\right] + \left(1 - 4\frac{\alpha\,\Delta t}{\Delta x^2}\right)T_{m,n}^{\tau} \qquad (2iii)$$
b. Boundary conditions†	
i. Specified temperature	$T_{m,n}^{\tau}$ known
ii. Specified heat flux	q_s'' known
iii. Insulated surface	$q_s'' = 0$
iv. Convection	$q_s'' = h(T_{m,n}^{\tau} - T_F)$
v. Blackbody thermal radiation	$q_s'' = \sigma F_{s-R}[(T_{m,n}^{\tau})^4 - T_R^4]$

† These boundary conditions apply to each type of exterior node.

EXAMPLE 4-9

A plate of 3 mm thickness is initially at 50°C. One side of the plate is suddenly exposed to blackbody thermal radiation with $F_{s-R} = 0.5$ and $T_R = 500$°C. The other side of the plate is maintained at 50°C. Develop an explicit finite-difference formulation for this unsteady problem with $\Delta x = L/3$.

Solution

Objective Develop explicit finite-difference nodal equations for temperature using $\Delta x = L/3$.

Schematic Finite-difference grid for plane wall.

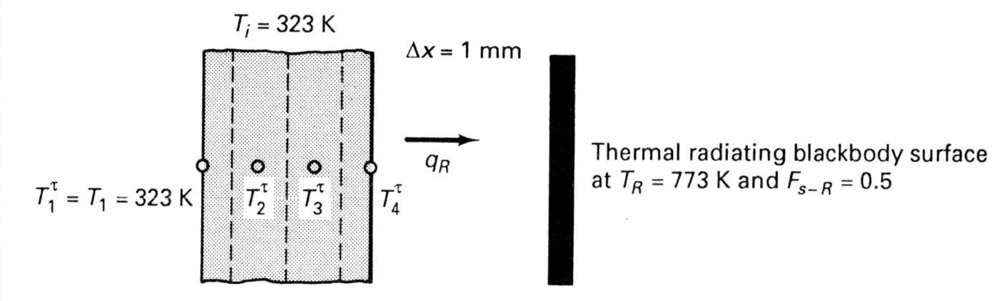

Assumptions/Conditions

unsteady-state
one spatial dimension
uniform properties
blackbody thermal radiation

Analysis The finite-difference grid network for this problem is shown in the schematic. Note that we have the three unknowns: T_2^τ, T_3^τ, and T_4^τ. Following the CVFDM, the energy balance on each node with unknown temperature is developed as follows, assuming uniform properties:

Interior Nodes m = 2 and 3 $\qquad q_x = q_{x+\Delta x} + \dfrac{\Delta E_s}{\Delta t}$

$$-kA\,\frac{T_m^\tau - T_{m-1}^\tau}{\Delta x} = -kA\,\frac{T_{m+1}^\tau - T_m^\tau}{\Delta x} + \rho c_v A\,\Delta x\,\frac{T_m^{\tau+1} - T_m^\tau}{\Delta t}$$

or

$$T_m^{\tau+1} - T_m^\tau = \frac{\alpha\,\Delta t}{\Delta x^2}(T_{m+1}^\tau + T_{m-1}^\tau - 2\,T_m^\tau)$$

Hence, we obtain

$$T_2^{\tau+1} - T_2^{\tau} = \frac{\alpha\,\Delta t}{\Delta x^2}(T_3^{\tau} + T_1^{\tau} - 2\,T_2^{\tau}) \tag{a}$$

and

$$T_3^{\tau+1} - T_3^{\tau} = \frac{\alpha\,\Delta t}{\Delta x^2}(T_4^{\tau} + T_2^{\tau} - 2\,T_3^{\tau}) \tag{b}$$

Exterior Node $m = M = 4$ $\qquad q_x = q_R + \dfrac{\Delta E_s}{\Delta t}$

$$-\,kA\,\frac{T_4^{\tau} - T_3^{\tau}}{\Delta x} = A q_R'' + \rho c_v A\,\frac{\Delta x}{2}\,\frac{T_4^{\tau+1} - T_4^{\tau}}{\Delta t}$$

or

$$T_4^{\tau+1} - T_4^{\tau} = \frac{2\alpha\,\Delta t}{\Delta x^2}\left(-\,\frac{\Delta x\,q_R''}{k} + T_3^{\tau} - T_4^{\tau}\right) \tag{c}$$

where $q_R'' = \sigma F_{s-R}[(T_4^{\tau})^4 - T_R^4]$. To complete the formulation, we write the initial condition as

$$T_2^0 = T_3^0 = T_4^0 = T_i$$

Utilizing this initial condition, Eqs. (a) through (c) can be solved for T_2^1, T_3^1, and T_4^1. These values of T_m^{τ} at $\tau = 1$ can, in turn, be utilized to obtain T_m^2. Following through in this fashion, calculations can be developed for T_m^{τ} for successively larger and larger values of τ.

Once T_m^{τ} is known, predictions can be developed for the rate of heat transfer. For example, the heat transfer flux at the radiating surface is simply

$$q_R'' = \sigma F_{s-R}[(T_4^{\tau})^4 - T_R^4]$$

To obtain the instantaneous rate of heat transfer at the other face, we merely perform an energy balance on exterior node (1); that is,

$$q_0 = q_{\Delta x/2} + \cancel{\frac{\Delta E_s}{\Delta t}} = -\,kA\,\frac{T_2^{\tau} - T_1^{\tau}}{\Delta x}$$

where $\Delta E_s/\Delta t$ is zero because $T_1^{\tau} = T_1$ is independent of time.

Stability Considerations

In connection with the specification of the finite-difference increments Δt and Δx (and Δy, which has been set equal to Δx), these increments must be selected such that the nodal calculations do not violate the physical requirement represented by the *second law of thermodynamics*. Otherwise, the resulting finite-difference "solution" will become unstable and blow up after a number of time steps have been taken. The physical basis for this instability is illustrated in Example 4–10. References 7 and 8 provide mathematical details pertaining to the stability and convergence of numerical solutions.

Focusing attention on systems with $\dot{q} = 0$, stability is assured by merely requiring that the coefficients associated with the term $T_{m,n}^{\tau}$ for all nodal equations within the nodal network be equal to or greater than zero. The resulting *stability criterion* for explicit finite-difference nodal equations takes the form

$$\frac{\alpha\,\Delta t}{\Delta x^2} \leq \frac{1}{X} \tag{4–46}$$

with X given in Table 4–3 for interior and representative exterior nodes and boundary conditions. With Δx specified, Δt can be *no larger* than $\Delta x^2/(\alpha X)$. This limitation prohibits the use of larger increments of time, regardless of accuracy and computational considerations.

Referring to Table 4–3, we observe that the minimum value of X for unsteady two-dimensional systems with $\dot{q} = 0$ is 4. However, since Eq. (4–46) must be satisfied for all nodes, X must be greater than 4 for convection and blackbody thermal radiation. In this connection, with $\alpha\,\Delta t/\Delta x^2$ set equal to $\frac{1}{4}$ for specified wall temperature or specified wall heat flux boundary conditions, the interior nodal equation given by Eq. (4–43) reduces to

$$T_{m,n}^{\tau+1} = \frac{1}{4}(T_{m+1,n}^{\tau} + T_{m-1,n}^{\tau} + T_{m,n+1}^{\tau} + T_{m,n-1}^{\tau}) \tag{4–47}$$

for $\dot{q} = 0$. Utilizing this value of $\alpha\,\Delta t/\Delta x^2$, we see that T_m^{τ} is equal to the arithmetic average of the four surrounding nodal temperatures at $t = \tau\,\Delta t$.

TABLE 4–3 Stability parameter for unsteady two-dimensional explicit nodal equations for $\dot{q} = 0$

Type node	X
Interior	4
Exterior	
Specified temperature	4
Specified heat flux	4
Convection	
Regular	$4 + 2\,h\,\Delta x/k$
Outer corner	$4 + 4\,h\,\Delta x/k$
Inner corner	$4 + \frac{4}{3}\,h\,\Delta x/k$
Blackbody thermal radiation	
Regular	$4 + 2\,\sigma F_{s-R}\,\Delta x\,(T_{m,n}^{\tau})^3/k$
Outer corner	$4 + 4\,\sigma F_{s-R}\,\Delta x\,(T_{m,n}^{\tau})^3/k$
Inner corner	$4 + \frac{4}{3}\,\sigma F_{s-R}\,\Delta x\,(T_{m,n}^{\tau})^3/k$

TABLE 4–4 Stability parameter for unsteady one-dimensional explicit nodal equations for $\dot{q} = 0$

Type node	X
Interior	2
Exterior	
Specified temperature	2
Specified heat flux	2
Convection	$2 + 2\,h\,\Delta x/k$
Blackbody thermal radiation	$2 + 2\,\sigma F_{s-R}\,\Delta x\,(T^{\tau}_{m,n})^3/k$

The stability criterion given by Eq. (4–46) also applies to explicit nodal equations for unsteady one-dimensional and three-dimensional systems with $\dot{q} = 0$; the values of X associated with one-dimensional nodal equations are listed in Table 4–4. With $\alpha\,\Delta t/\Delta x^2$ set equal to $\frac{1}{2}$ in the case of unsteady one-dimensional systems with $\dot{q} = 0$, the interior nodal equations are written as

$$T^{\tau+1}_{m} = \frac{1}{2}(T^{\tau}_{m+1} + T^{\tau}_{m-1}) \tag{4–48}$$

The fact that $T^{\tau+1}_{m}$ is equal to the average of the two adjacent nodal temperatures at the preceding increment of time provides the basis for a graphical approach to solving unsteady one-dimensional conduction-heat-transfer problems. This Schmidt graphical method is described in detail by Jacob [9].

Stability criterion can also be established for systems with $\dot{q} \neq 0$. In this regard, the values of X listed in Tables 4–3 and 4–4 provide a conservative criterion for $\dot{q} > 0$.

As a consequence of the stability requirements, the explicit computational approach developed above generally requires the use of relatively small increments in time. However, the fact that this finite-difference method produces direct calculations of future nodal temperatures sometimes compensates for the restriction in Δt.

EXAMPLE 4–10

Demonstrate that Eq. (4–43) violates the second law of thermodynamics for values of $\alpha\,\Delta t/\Delta x^2 > \frac{1}{4}$ and $\dot{q} = 0$.

Solution

Objective Show that the forward-time difference nodal equation,

$$T^{\tau+1}_{m,n} = \frac{\alpha\,\Delta t}{\Delta x^2}(T^{\tau}_{m+1,n} + T^{\tau}_{m-1,n} + T^{\tau}_{m,n+1} + T^{\tau}_{m,n-1}) + \left(1 - 4\frac{\alpha\,\Delta t}{\Delta x^2}\right)T^{\tau}_{m,n}$$

violates the second law of thermodynamics for $\alpha\,\Delta t/\Delta x^2 > \frac{1}{4}$.

Assumptions/Conditions

unsteady-state
two spatial dimensions
no internal energy generation

Analysis To demonstrate this point, we consider the case in which the sides of a square plate ($L \times L \times \delta$) initially at 110°F are suddenly brought to 100°F at time $t = t_1$. To simplify our test case, we utilize a grid with $\Delta x = L/2$, as shown in Fig. E4–10. At the instant $t = \tau\, \Delta t = 1\, \Delta t$, the nodal temperatures surrounding node (m,n) are given by

$$T^{\tau}_{m-1,n} = T^{\tau}_{m+1,n} = T^{\tau}_{m,n-1} = T^{\tau}_{m,n+1} = 100°\text{F}$$

but

$$T^{\tau}_{m,n} = 110°\text{F}$$

for $\tau = 1$. We now want to know the temperature at node (m,n) at the next increment of time. Therefore, we utilize Eq. (4–43) to obtain

$$T^{\tau+1}_{m,n} = \frac{\alpha\, \Delta t}{\Delta x^2} 400°\text{F} + \left(1 - 4\frac{\alpha\, \Delta t}{\Delta x^2}\right) 110°\text{F} \tag{a}$$

We select a value of Δt such that $\alpha\, \Delta t/\Delta x^2$ is greater than $\frac{1}{4}$, say

$$\frac{\alpha\, \Delta t}{\Delta x^2} = \frac{1}{3}$$

Substituting this value into Eq. (a), we have

$$T^{\tau+1}_{m,n} = \frac{400°\text{F}}{3} + \left(1 - \frac{4}{3}\right) 110°\text{F} = 96.7°\text{F}$$

Because $T^{\tau+1}_{m,n}$ is less than $T^{\tau}_{m,n}$, we conclude that heat has been conducted out of node (m,n) to the surrounding nodes. However, the fact that $T^{\tau+1}_{m,n}$ is actually less than 100°F indicates that heat has been transferred in directions of increasing temperature during part of the Δt time increment. This result is a violation of the second law of thermodynamics, which requires that heat cannot be transferred from a low-temperature system to a high-temperature system without the input of work.

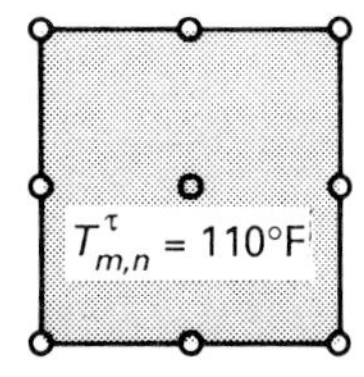

$T^{\tau}_{m-1,n} = T^{\tau}_{m+1,n} = T^{\tau}_{m,n-1} = T^{\tau}_{m,n+1} = 100°\text{F}$

FIGURE E4–10
Finite-difference grid for square plate; $\Delta x = L/2$ and $\tau = 1$.

By examining Eq. (a), it can be seen that the use of any time increment for which $\alpha\, \Delta t/\Delta x^2$ is greater than $\frac{1}{4}$ will bring about a violation of the second law. If, however, Δt is selected such that $\alpha\, \Delta t/\Delta x^2 \leq \frac{1}{4}$, $T^{\tau+1}_{m,n}$ does not fall below the surrounding nodal temperatures and no violation occurs.

Numerical Solutions

With the initial temperature distribution specified and with Δt and Δx specified in accordance with necessary stability and accuracy requirements, the explicit interior and exterior nodal equations provide the means by which the temperature distribution can be calculated at the next increment of time, 1 Δt. The temperature distribution at 1 Δt can then be used as an input to calculate the distribution at 2 Δt. This calculation procedure can be continued to obtain the temperature distribution over the number of time increments desired, or until the steady-state condition is approached. The use of the explicit numerical finite-difference method is illustrated in Example 4–11.

EXAMPLE 4–11

A plate [k = 50 W/(m °C), $\alpha = 2 \times 10^{-5}$ m^2/s] of 4-mm thickness is initially at 0°C. One side of the plate is then suddenly brought to a temperature of 100°C, with the other side maintained at 0°C. Develop an accurate explicit finite-difference solution for the rate of heat transfer from the hot surface.

Solution

Objective Use the explicit finite-difference approach to obtain an accurate solution for q_L.

Schematic Unsteady heat transfer in a plane wall.

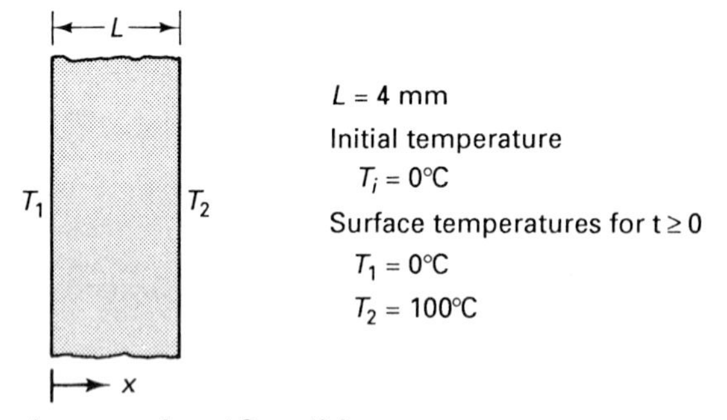

Assumptions/Conditions

unsteady-state
one spatial dimension
uniform properties

Properties k = 50 W/(m °C), $\alpha = 2 \times 10^{-5}$ m^2/s.

Analysis To get a feeling for the explicit solution technique, we first utilize the finite-difference grid with $\Delta x = L/4 = 1$ mm shown in Fig. E4–11a. Notice that this grid produces only three unknown nodal temperatures (T_2^{τ}, T_3^{τ}, and T_4^{τ}). Following the CVFDM, an explicit finite-difference energy balance is developed for one of these three interior nodes (m) by writing

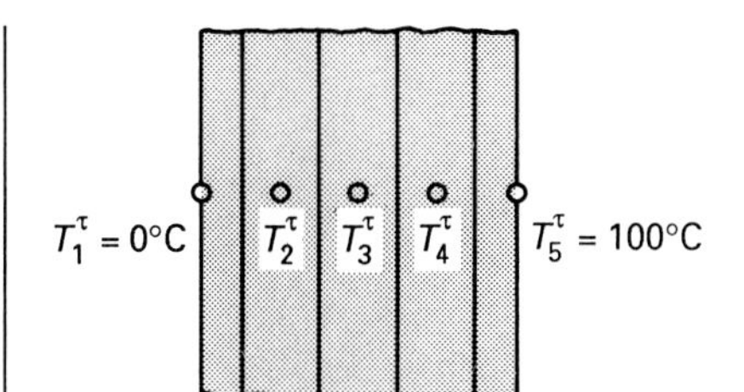

FIGURE E4–11a
Finite-difference grid for plane wall.

$$q_x = q_{x+\Delta x} + \frac{\Delta E_s}{\Delta t}$$

$$-kA\frac{T_m^\tau - T_{m-1}^\tau}{\Delta x} = -kA\frac{T_{m+1}^\tau - T_m^\tau}{\Delta x} + \rho c_v A\,\Delta x\frac{T_m^{\tau+1} - T_m^\tau}{\Delta t} \tag{a}$$

The solution for $T_m^{\tau+1}$ is

$$T_m^{\tau+1} = \frac{\alpha\,\Delta t}{\Delta x^2}(T_{m-1}^\tau + T_{m+1}^\tau) + \left(1 - 2\frac{\alpha\,\Delta t}{\Delta x^2}\right)T_m^\tau \tag{b}$$

To maintain stability, the coefficients associated with the T_m^τ term for each nodal equation must be equal to or greater than zero. Therefore, we require

$$1 - 2\alpha\frac{\Delta t}{\Delta x^2} \geq 0 \qquad \text{or} \qquad \frac{\alpha\,\Delta t}{\Delta x^2} \leq \frac{1}{2}$$

Setting $\alpha\,\Delta t/\Delta x^2 = \frac{1}{2}$, Eq. (b) becomes

$$T_m^{\tau+1} = \frac{1}{2}(T_{m-1}^\tau + T_{m+1}^\tau)$$

and Δt is given by

$$\Delta t = \frac{1}{2}\frac{\Delta x^2}{\alpha} = \frac{1}{2}\frac{(10^{-3}\ \text{m})^2}{2 \times 10^{-5}\ \text{m}^2/\text{s}} = 0.025\ \text{s}$$

To summarize, our three nodal equations are

$$T_2^{\tau+1} = \frac{1}{2}(T_1^\tau + T_3^\tau) = \frac{1}{2}(0°\text{C} + T_3^\tau) \tag{c}$$

$$T_3^{\tau+1} = \frac{1}{2}(T_2^\tau + T_4^\tau) \tag{d}$$

$$T_4^{\tau+1} = \frac{1}{2}(T_3^\tau + T_5^\tau) = \frac{1}{2}(T_3^\tau + 100°\text{C}) \tag{e}$$

The solution to these equations after the first increment of time ($t = \Delta t$) is

$$T_2^1 = \frac{1}{2}(0°\text{C} + 0°\text{C}) = 0°\text{C}$$

$$T_3^1 = \frac{1}{2}(0°\text{C} + 0°\text{C}) = 0°\text{C}$$

$$T_4^1 = \frac{1}{2}(0°\text{C} + 100°\text{C}) = 50°\text{C}$$

These results are then substituted back into Eqs. (c) through (e) to obtain predictions for T_2^2, T_3^2, and T_4^2. This procedure is continued as the solution is built up for increasing time steps. The predictions for T_m^τ are summarized in Table E4–11 for the number of time steps required to reach steady conditions. Note that the final steady-state profile is linear, which is consistent with the simple one-dimensional analysis developed in Chap. 2.

TABLE E4–11 Numerical calculations for T_m^τ

τ	T_2^τ	T_3^τ	T_4^τ
0	0.	0.	50.
1	0.	25.	50.
2	12.5	25.	62.5
3	12.5	37.5	62.5
4	18.8	37.5	68.8
5	18.8	43.8	68.8
6	21.9	43.8	71.9
7	21.9	46.9	71.9
8	23.4	46.9	73.4
9	23.4	48.4	73.4
10	24.2	48.4	74.2
11	24.2	49.2	74.2
12	24.6	49.2	74.6
13	24.6	49.6	74.6
14	24.8	49.6	74.8
15	24.8	49.8	74.8
16	24.9	49.8	74.9
17	24.9	49.9	74.9
18	25.0	49.9	75.0
19	25.0	50.0	75.0
20	25.0	50.0	75.0

Observing that no change occurs in the nodal temperatures (to three significant figures) for $\tau \geq 19$, the length of time required to reach steady-state conditions is approximated by

$$t_{ss} = \tau_{ss}\,\Delta t = 19(0.025\text{ s}) = 0.475\text{ s}$$

It should be mentioned that the calculations are generally stopped when the percent change in all nodal temperatures falls below a specified steady-state convergence criterion, SSC. For instance, with SSC set equal to 0.005, the calculations would be terminated at $\tau = 15$.

To determine the rate of heat transfer from the surface $x = L$ at time t, we apply the first law of thermodynamics to node (5), with the result

$$-kA\,\frac{T_5^\tau - T_4^\tau}{\Delta x} = q_L + \frac{\Delta E_s}{\Delta t}$$

or

$$q_L = -\frac{kA}{\Delta x}(T_5^\tau - T_4^\tau) - \rho c_v A\,\frac{\Delta x}{2}\left(\frac{T_5^{\tau+1} - T_5^\tau}{\Delta t}\right) \qquad \text{(f)}$$

where $T_5^{\tau+1} = T_5^\tau = 100°\text{C}$. Thus, to predict heat-transfer flux q_L'' at any instant t, the calculations for T_m^τ are substituted into Eq. (f). For example, at $t = 0.275$ s (i.e., $\tau = 11$), we obtain

$$q_L'' = \frac{k}{\Delta x}(74.2°\text{C} - 100°\text{C}) = -1.29\text{ MW/m}^2$$

As we have seen in Examples 4–6 through 4–8, one way to assure a reasonably accurate numerical finite-difference solution is to compare the solutions for smaller and smaller subvolume sizes. Because of the obvious computational involvement in producing solutions for smaller values of Δx and Δt, a simple BASIC program is developed for calculating the nodal temperatures and heat-transfer rate at the surface. The general nodal equation given by Eq. (b) is written in BASIC computer language as

```
T(J,2) = (T(J + 1,1) + T(J - 1,1))*S + (1 - 2 *S) *T(J,1)
```

where J $= m = 2, 3, 4, \ldots, M - 1$, and $S = \alpha\,\Delta t/\Delta x^2$. The second index designates the time stations $\tau + 1$ (index 2) and τ (index 1). The flowchart and BASIC program are given in Figs. E4–11b and E4–11c. In addition to calculating the instantaneous nodal temperatures for any specific grid space Δx and time increment Δt, this program calculates the heat flux at $x = L$. The instantaneous heat flux q_L'' is obtained from Eq. (f),

$$q_L'' = -\frac{k}{\Delta x}(T_5^\tau - T_4^\tau)$$

or

```
QFL = -K/DX*(T(M,2) - T(M - 1,2))
```

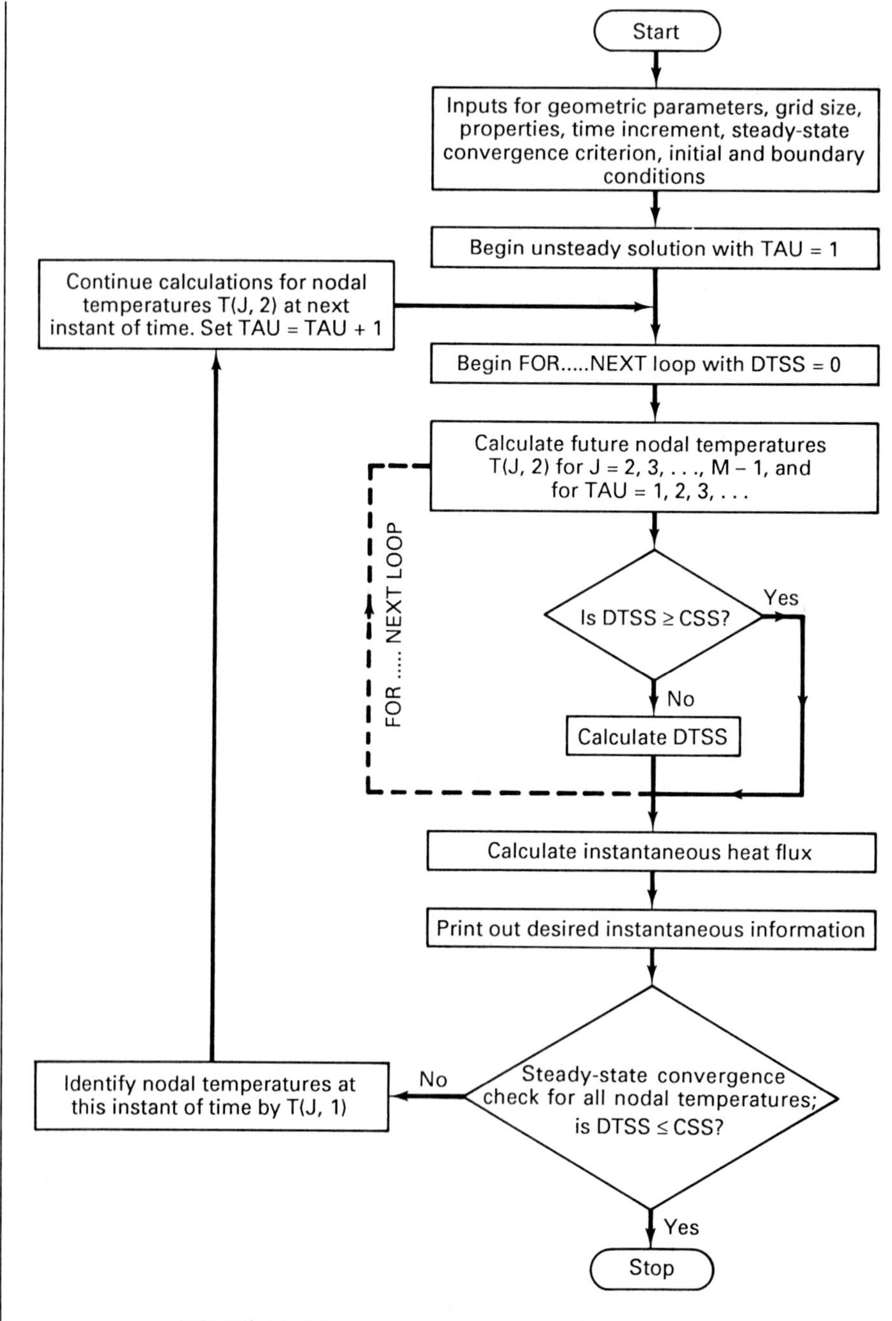

FIGURE E4–11b Program flowchart for Example 4–11.

```
                                                                   Comments
                                                Inputs for geometric parameters
   L=.004                                                                01
                                                           Inputs for grid size
   M=10                                                                  02
   M1=M-1                                                                03
   DX=L/M1                                                               04
                                                          Inputs for properties
   K=50                                                                  05
   ALPHA=.00002                                                          06
                                                       Input for time increment
   S=.5                                                                  07
   DTIME=S*DX^2/ALPHA                                                    08
                                    Input for steady-state convergence criterion
   CSS=.0001                                                             09
                                                  Inputs for initial conditions
   T1=0                                                                  10
   FOR J=1 TO M STEP 1                                                   11
                                                 Inputs for boundary conditions
   T(J,1)=T1                                                             12
   NEXT J                                                                13
   T(1,1)=T1                                                             14
   T(1,2)=T(1,1)                                                         15
   TM=100                                                                16
   T(M,1)=TM                                                             17
   T(M,2)=T(M,1)                                                         18
                                                Begin unsteady solution with τ=1
   TAU=1                                                                 19
                                                   Begin do loop with DTSS=0
2  DTSS=0                                                                20
                                       Calculate future nodal temperatures T(J,2)
   FOR J=2 TO M1                                                         21
   T(J,2)=(T(J+1,1)+T(J-1,1))*S+(1-2*S)*T(J,1)                           22
                                  Check for convergence to steady state for all nodes
   IF DTSS>CSS GOTO 3                                                    23
                                                                  Calculate DTSS
   DTSS=ABS ((T(J,2)-T(J,1))/T(J,2))                                     24
3  NEXT J                                                                25
                                               Calculate instantaneous heat flux
   QFL=-K/DX*(T(M,2)-T(M1,2))                                            26
                                      Print out desired instantaneous information
   LPRINT USING "#######.#";M, TAU, QFL                                  27
                                      Steady-state convergence check for all nodes
   IF DTSS<CSS GOTO 50                                                   28
                                     Identify current nodal temperatures by T(J,1)
   For J=2 TO M1                                                         29
   T(J,1)=T(J,2)                                                         30
   NEXT J                                                                31
```

FIGURE E4-11c BASIC program for Example 4-11.

```
                                  Calculate future nodal temperatures T(J,2)
      TAU=TAU+1                                                          32
      GOTO 2                                                             33
50    STOP                                                               34
      END                                                                35
```

Note: A dimension statement must be used for $M>10$. For example, for $10<M\leq100$ we may write **DIM** T(100,2).

FIGURE E4–11c (*Continued*)

Calculations obtained for q''_L by running this program are shown in Fig. E4–11d (to four significant figures) as a function of M and t with $\alpha\ \Delta t/\Delta x^2 = 0.5$. Note that as M increases, both Δx and Δt become smaller. Based on this result, we conclude that a reasonably accurate solution is obtained for $M \gtrsim 10$, but considerable error occurs for M much less than 10.

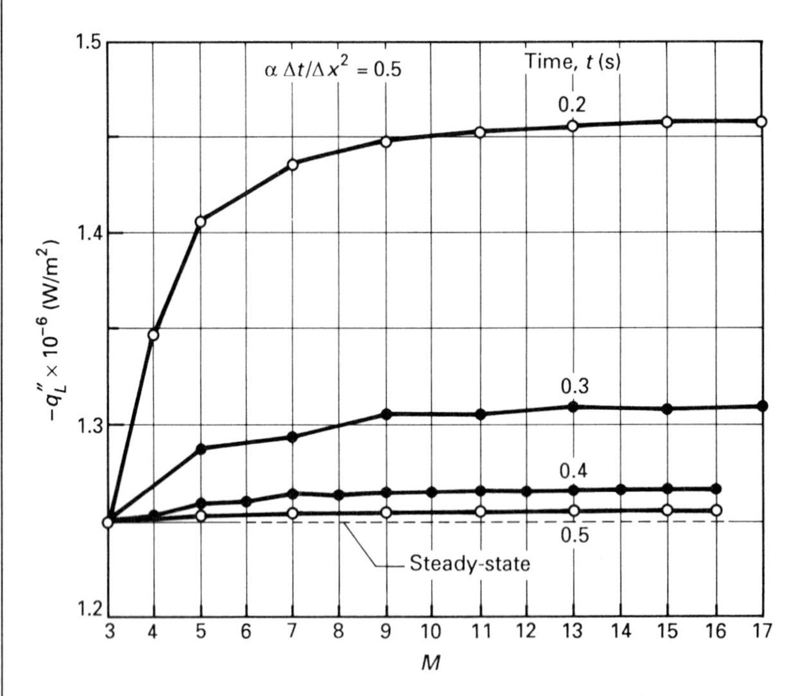

FIGURE E4–11d Numerical calculations for q''_L; $\alpha\ \Delta t/\Delta x^2 = 0.5$.

To see the effect of $\alpha\ \Delta t/\Delta x^2$ on our numerical solution, predictions are shown in Fig. E4–11e for the heat flux q''_L at surface $x = L$ versus τ for $\alpha\ \Delta t/\Delta x^2$ equal to 0.4, 0.5, and 0.6, with $M = 10$. For $\alpha\ \Delta t/\Delta x^2 \leq 0.5$, the solution is stable and converges to the proper steady-state value. However, for $\alpha\ \Delta t/\Delta x^2 = 0.6$, the numerical solution is seen to be unstable and divergent. This result underscores the importance of establishing the appropriate stability criteria for explicit finite-difference formulations of unsteady problems.

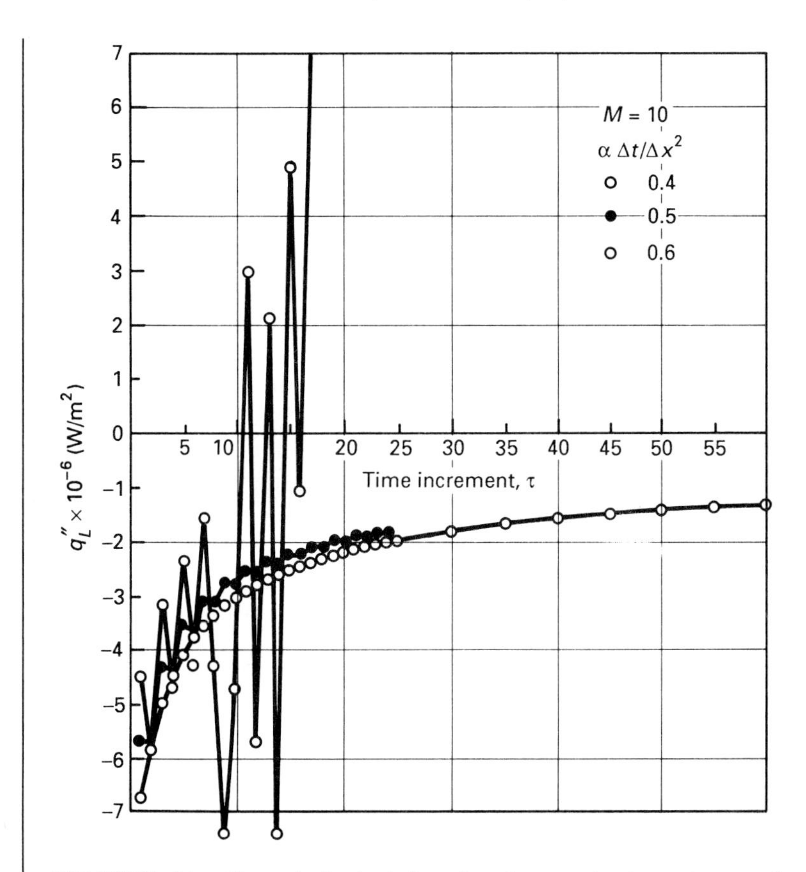

FIGURE E4–11e Numerical calculations for q_L''; several values of $\alpha\ \Delta t/\Delta x^2$.

This computer program can be modified in order to print out the temperature distribution at any value of τ. This task is suggested as an exercise.

4–5–3 Implicit Method

The alternative backward-time difference formulation is commonly used for problems in which the computational time becomes a factor. The substitution of the backward-time difference given by Eq. (4–41) into Eq. (4–38) gives rise to an interior nodal equation of the form

$$T^{\tau}_{m,n} - T^{\tau-1}_{m,n} = \frac{\alpha\ \Delta t}{\Delta x^2}\left(T^{\tau}_{m+1,n} + T^{\tau}_{m-1,n} + T^{\tau}_{m,n+1} + T^{\tau}_{m,n-1} - 4\,T^{\tau}_{m,n} + \frac{\dot{q}}{k}\,\Delta x^2\right) \tag{4–49}$$

or

$$T^{\tau+1}_{m,n} - T^{\tau}_{m,n} = \frac{\alpha\,\Delta t}{\Delta x^2}\left(T^{\tau+1}_{m+1,n} + T^{\tau+1}_{m-1,n} + T^{\tau+1}_{m,n+1} + T^{\tau+1}_{m,n-1} - 4\,T^{\tau+1}_{m,n} + \frac{\dot{q}}{k}\Delta x^2\right) \tag{4-50}$$

Similar equations can also be written for the exterior nodes. This backward-time-difference-type equation is implicit because the nodal temperature $T^{\tau+1}_{m,n}$ at any instant $(\tau + 1)\,\Delta t$ is given in terms of unknown nodal temperatures at that same instant of time as well as at the preceding instant $\tau\,\Delta t$. Thus, like the situation encountered in our solution of steady-state problems, the entire system of Z nodal equations must be solved simultaneously to develop calculations for $T^{\tau+1}_{m,n}$. However, this method has the advantage of being stable for all values of Δt. Consequently, larger increments in t can be utilized in this approach, such that less computer time is sometimes required than is necessary in the explicit method. Of course, it should be realized that the use of larger time increments results in larger discretization errors.

EXAMPLE 4–12

Develop an implicit finite-difference solution for the unsteady problem of Example 4–11.

Solution

Objective Use the implicit finite-difference approach to obtain an accurate solution for q_L.

Schematic Unsteady heat transfer in a plane wall.

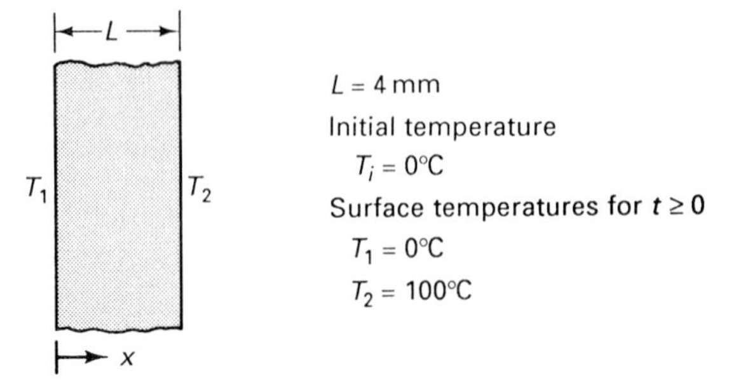

Assumptions/Conditions

unsteady-state
one spatial dimension
uniform properties

Properties $k = 50$ W/(m °C), $\alpha = 2 \times 10^{-5}$ m²/s.

Analysis The finite-difference grid for an implicit formulation is identical to the grid for an explicit formulation. Using the CVFDM, the first law of thermodynamics is applied to interior nodal control volume (m) to obtain

$$q_x = q_{x+\Delta x} + \frac{\Delta E_s}{\Delta t}$$

$$-kA\frac{T_m^\tau - T_{m-1}^\tau}{\Delta x} = -kA\frac{T_{m+1}^\tau - T_m^\tau}{\Delta x} + \rho c_v\, \Delta V \frac{T_m^\tau - T_m^{\tau-1}}{\Delta t}$$

$$T_{m-1}^\tau + T_{m+1}^\tau - \left(2 + \frac{\Delta x^2}{\alpha\, \Delta t}\right) T_m^\tau + \frac{\Delta x^2}{\alpha\, \Delta t} T_m^{\tau-1} = 0$$

or

$$T_{m-1}^{\tau+1} + T_{m+1}^{\tau+1} - \left(2 + \frac{\Delta x^2}{\alpha\, \Delta t}\right) T_m^{\tau+1} + \frac{\Delta x^2}{\alpha\, \Delta t} T_m^\tau = 0$$

where $\Delta E_s/\Delta t$ has been approximated by the backward difference, rather than the forward difference used in Example 4–11.

For this problem in which both surface temperatures are specified, m takes values of 2, 3, . . . , $M-1$, such that a total of $M-2$ unknown nodal temperatures occur at each time station $\tau = 1, 2, 3, \ldots$. However, unlike the explicit formulation of Example 4–11, these equations must be solved simultaneously for each value of τ. Therefore, the direct and indirect methods introduced in Sec. 4–4–2 can be utilized.

Because of its simplicity, the Gauss-Seidel iterative method is utilized. Our general nodal equation is therefore put into the form

$$T_m^{\tau+1} = \frac{1}{2 + \Delta x^2/(\alpha\, \Delta t)}\left(T_{m-1}^{\tau+1} + T_{m+1}^{\tau+1} + \frac{\Delta x^2}{\alpha\, \Delta t} T_m^\tau\right)$$

where $m = 2, 3, \ldots, M - 1$, and $\tau = 1, 2, 3, \ldots$. A finite-difference BASIC program is given in Fig. A–G–2, which solves this equation iteratively for $T_2^{\tau+1}$, $T_3^{\tau+1}$, $T_4^{\tau+1}$, . . . , $T_{M-1}^{\tau+1}$ at each increment of time $\tau = 1, 2, 3, \ldots$. This program is patterned after the BASIC program presented in Example 4–8. The solution for q_L'' is given by

$$q_L'' = -\frac{k}{\Delta x}(T_M^\tau - T_{M-1}^\tau)$$

Calculations produced by this program for the rate of heat transfer per unit area from the surface at $x = L$ with $\alpha\, \Delta t/\Delta x^2$ equal to 0.5, 1, and 2, and $M = 10$ are compared in Fig. E4–12 with the explicit calculations of Example 4–11. Although much larger increments in Δt can be used in the implicit approach, a price is paid in accuracy of the solution for small values of t.

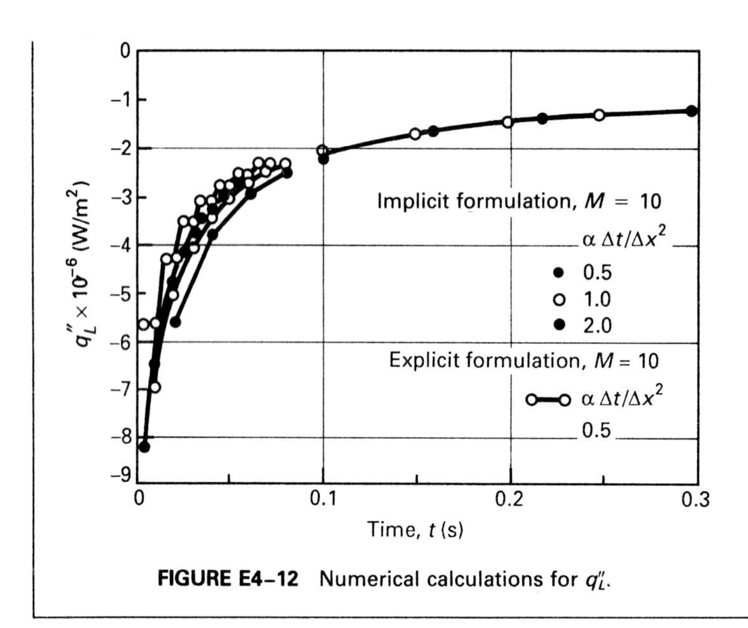

FIGURE E4-12 Numerical calculations for q_L''.

4-5-4 R/C Network Formulation

To complete our study of the numerical approach, we want to introduce the popular *R/C* network representation of the finite-difference formulations. To develop this perspective in the context of the CVFDM, we first consider the energy balance for the interior node shown in Fig. 4-4; that is,

$$\dot{q}\,\Delta V + \Delta q_x + \Delta q_y - \Delta q_{x+\Delta x} - \Delta q_{y+\Delta y} = \frac{\Delta E_s}{\Delta t} \tag{4-51}$$

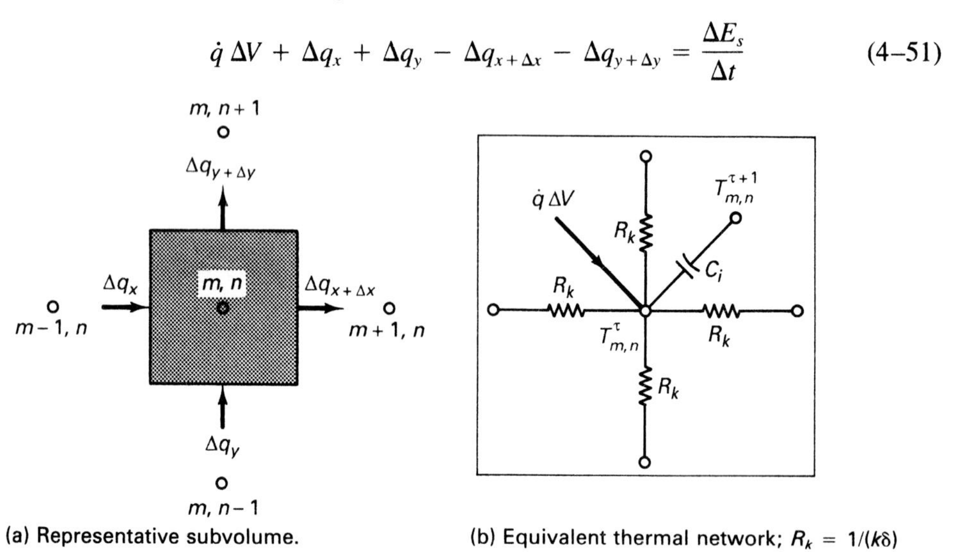

(a) Representative subvolume.

(b) Equivalent thermal network; $R_k = 1/(k\delta)$ and $C_i = \rho\,\Delta V\,c_v$.

FIGURE 4-4 Interior finite-difference node.

Utilizing the Fourier law of conduction, the heat-transfer rates can be expressed in terms of thermal resistances by writing

$$\Delta q_x = -k\delta\,\Delta y\,\frac{T^\tau_{m,n} - T^\tau_{m-1,n}}{\Delta x} = \frac{T^\tau_{m-1,n} - T^\tau_{m,n}}{1/(\delta k)}$$

$$\Delta q_{x+\Delta x} = -k\delta\,\Delta y\,(T^\tau_{m+1,n} - T^\tau_{m,n}) = \frac{T^\tau_{m,n} - T^\tau_{m+1,n}}{1/(\delta k)}$$

and

$$\Delta q_y = \frac{T^\tau_{m,n-1} - T^\tau_{m,n}}{1/(\delta k)} \qquad \Delta q_{y+\Delta y} = \frac{T^\tau_{m,n} - T^\tau_{m,n+1}}{1/(\delta k)}$$

Utilizing this result and a forward-difference approximation for $\partial T/\partial t$, Eq. (4–51) can be rewritten in the explicit form

$$\dot{q}\,\Delta V + \sum \frac{T^\tau_i - T^\tau_{m,n}}{R_i} = C_i\,\frac{T^{\tau+1}_{m,n} - T^\tau_{m,n}}{\Delta t} \tag{4–53}$$

where $R_i = R_k = 1/(\delta k)$, $C_i = \rho c_v\,\Delta V$, and T^τ_i represents the temperatures of the nodes that surround the (m,n) node. Equation (4–53) is represented in terms of an analogous electrical circuit in Fig. 4–4(b).

Similarly, the energy balance for the exterior node shown in Fig. 4–5(a) can be written as

$$\dot{q}\,\Delta V + \Delta q_x + \Delta q_y - \Delta q_s - \Delta q_{y+\Delta y} = \frac{\Delta E_s}{\Delta t} \tag{4–54}$$

where

$$\Delta q_x = \frac{T^\tau_{M-1,n} - T^\tau_{M,n}}{1/(\delta k)} \qquad \Delta q_y = \frac{T^\tau_{M,n-1} - T^\tau_{M,n}}{2/(\delta k)}$$
$$\Delta q_{y+\Delta y} = \frac{T^\tau_{M,n} - T^\tau_{M,n+1}}{2/(\delta k)} \tag{4–55}$$

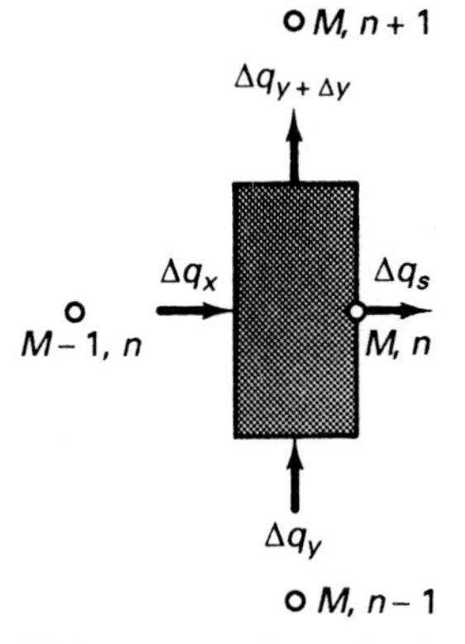

(a) Representative subvolume.

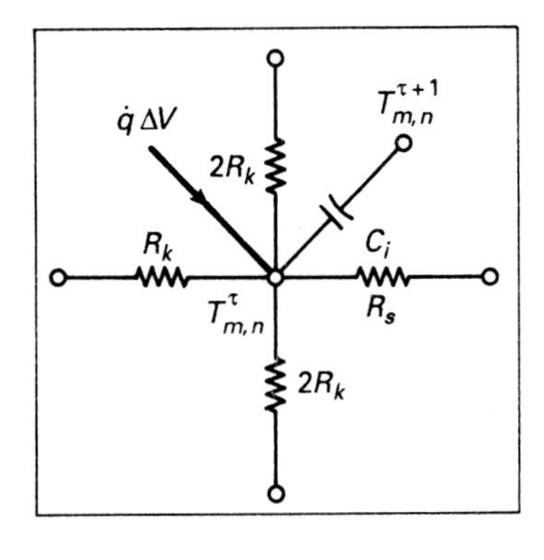

(b) Equivalent thermal network; $R_k = 1/(k\delta)$ and $C_i = \rho c_v\,\Delta V$.

FIGURE 4–5 Exterior finite-difference node.

The rate of heat transfer Δq_s from the surface by conduction, convection, or thermal radiation, can also be written in terms of a thermal resistance, such that Eq. (4–54) can be put into the form of Eq. (4–53). The thermal circuit for this exterior node is shown in Fig. 4–5(b). Notice that R_i is dependent upon the geometry of the subvolume and the boundary condition.

In summary, we find that Eq. (4–53) applies to both interior and exterior nodes and to both steady and unsteady multidimensional heat-transfer problems. In effect, this equation states that the summation of the currents flowing into the (m,n) node from the surroundings is equal to the flow of current into the capacitor. For steady-state conditions $T_{m,n}^{\tau+1} = T_{m,n}^{\tau}$, and no current flows to the capacitor. For this case, the summation of current flow into the (m,n) node is zero.

It should be mentioned that Eq. (4–53) must satisfy stability requirements which restrict Δt. For example, for $\dot{q} = 0$, Δt must be less than or equal to $C_i/\Sigma(1/R_i)$. If the restrictions imposed on Δt become too severe, an implicit formulation can be developed by merely utilizing a backward difference in $\partial T/\partial t$; that is,

$$\dot{q}\,\Delta V + \sum \frac{T_i^{\tau} - T_{m,n}^{\tau}}{R_i} = C_i \frac{T_{m,n}^{\tau} - T_{m,n}^{\tau-1}}{\Delta t} \tag{4–56}$$

Equations (4–53) and (4–56) provided the basis for building up electrical networks that are analogous to steady and unsteady multidimensional conduction-heat-transfer problems. To illustrate, a finite-difference grid and an electrical network are shown in Fig. 4–6 for unsteady conduction in a plane wall.

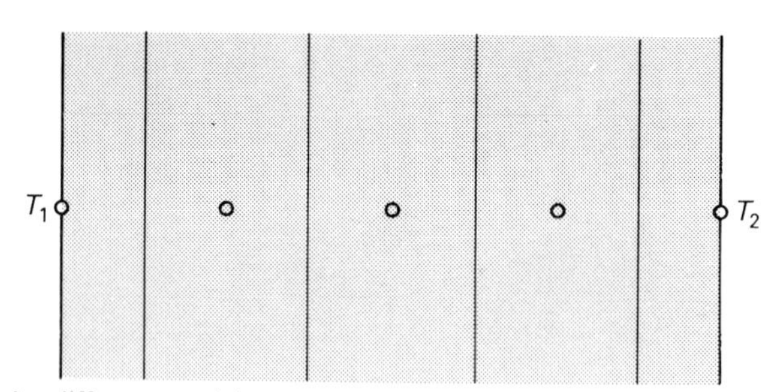

(a) Finite-difference grid.

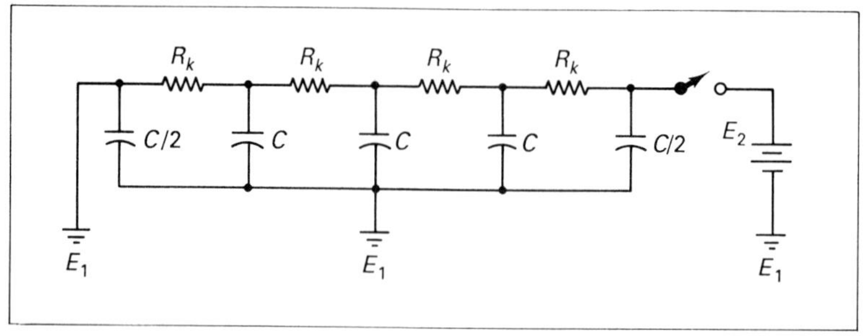

(b) R/C network; $R_k = \Delta x/(kA)$ and $C = \rho A c_v \Delta x$.

FIGURE 4–6 Unsteady conduction heat transfer in a plane wall.

The R/C network approach also lends itself to the analysis of three-dimensional rectangular, cylindrical, and spherical systems. The volume and resistance elements for the three standard coordinate systems are represented in Fig. 4–7 and Table 4–5. To generalize the notation, the node is represented by (l,m,n), with the plus and minus signs associated with the resistance subscripts used to designate location of the resistance relative to the node.

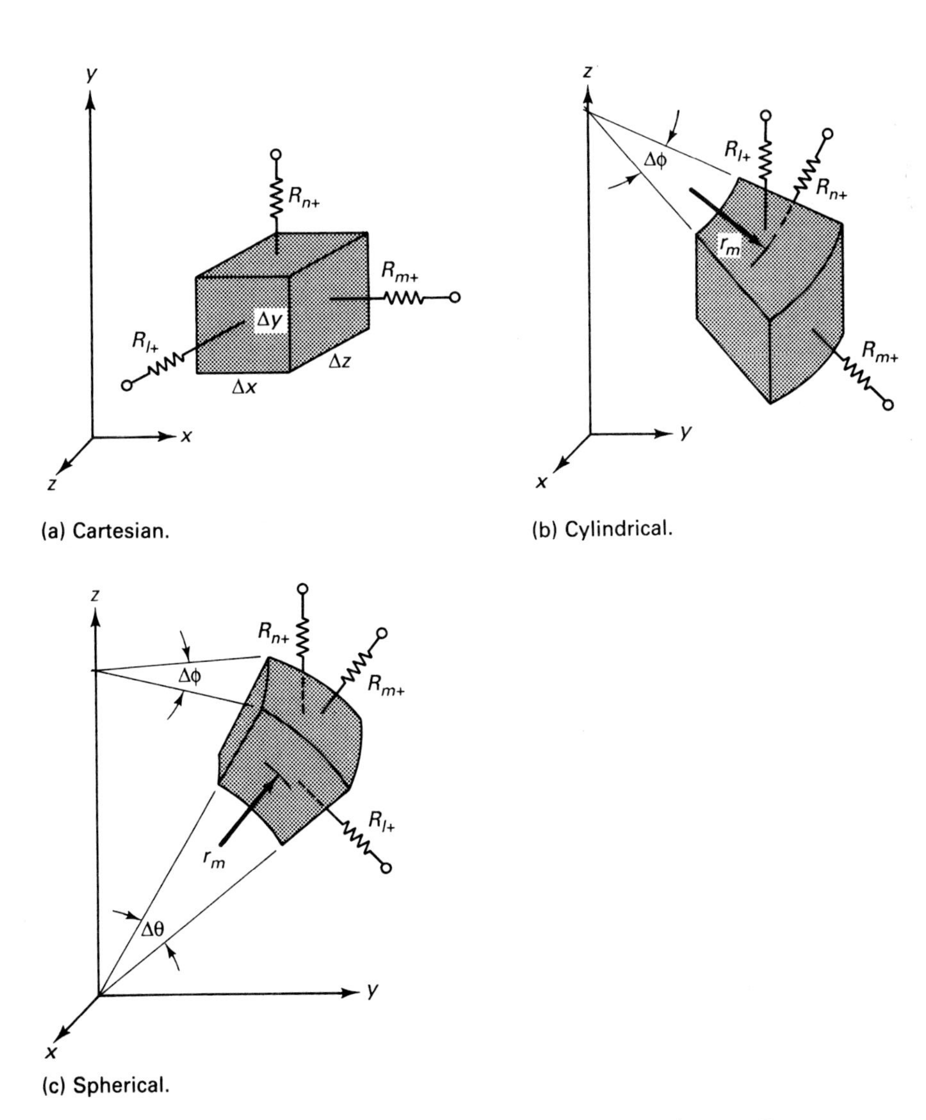

FIGURE 4–7 Finite-difference volume and resistance elements for coordinate systems.

TABLE 4–5 Internal element resistances and volumes for Cartesian, cylindrical, and spherical coordinate systems

	Cartesian	*Cylindrical*	*Spherical*
Volume element ΔV	$\Delta x\ \Delta y\ \Delta z$	$r_m\ \Delta r\ \Delta\phi\ \Delta z$	$r_m^2 \sin\theta\ \Delta r\ \Delta\phi\ \Delta\theta$
R_{m+}	$\dfrac{\Delta x}{\Delta y\ \Delta z\ k}$	$\dfrac{\Delta r}{(r_m + \Delta r/2)\ \Delta\phi\ \Delta z\ k}$	$\dfrac{\Delta r}{(r_m + \Delta r/2)^2 \sin\theta\ \Delta\phi\ \Delta\theta\ k}$
R_{m-}	$\dfrac{\Delta x}{\Delta y\ \Delta z\ k}$	$\dfrac{\Delta r}{(r_m - \Delta r/2)\ \Delta\phi\ \Delta z\ k}$	$\dfrac{\Delta r}{(r_m - \Delta r/2)^2 \sin\theta\ \Delta\phi\ \Delta\theta\ k}$
R_{n+}	$\dfrac{\Delta y}{\Delta x\ \Delta z\ k}$	$\dfrac{r_m\ \Delta\phi}{\Delta r\ \Delta z\ k}$	$\dfrac{\Delta\phi \sin\theta}{\Delta r\ \Delta\theta\ k}$
R_{n-}	$\dfrac{\Delta y}{\Delta x\ \Delta z\ k}$	$\dfrac{r_m\ \Delta\phi}{\Delta r\ \Delta z\ k}$	$\dfrac{\Delta\phi \sin\theta}{\Delta r\ \Delta\theta\ k}$
R_{l+}	$\dfrac{\Delta z}{\Delta x\ \Delta y\ k}$	$\dfrac{\Delta z}{r_m\ \Delta\phi\ \Delta r\ k}$	$\dfrac{\Delta\theta}{\sin(\theta + \Delta\theta/2)\ \Delta r\ \Delta\phi\ k}$
R_{l-}	$\dfrac{\Delta z}{\Delta x\ \Delta y\ k}$	$\dfrac{\Delta z}{r_m\ \Delta\phi\ \Delta r\ k}$	$\dfrac{\Delta\theta}{\sin(\theta - \Delta\theta/2)\ \Delta r\ \Delta\phi\ k}$
Nomenclature for increments	x,m	r,m	r,m
	y,n	ϕ,n	ϕ,n
	z,l	z,l	θ,l

4–6 SUMMARY

In this chapter we have considered the numerical finite-difference approach to the solution of conduction-heat-transfer problems. This approach involves the formulation of systems of nodal equations by use of nodal networks, finite-difference approximations for differential terms, and discretization or control-volume finite-difference methods. The nodal equations provide the basis for developing solutions for the nodal temperatures and heat fluxes. Since the grid space and time increment (for unsteady processes) are selected in accordance with accuracy requirements, finite-difference formulations generally result in fairly large numbers of nodal equations. In addition, in the case of explicit formulations for unsteady processes, stability criterion must also be satisfied.

Both iterative and direct solution approaches are available, with iterative techniques such as the Gauss-Seidel method generally providing the most effective solution approach. Because of the widespread use of digital computers, which can quickly solve the large numbers of algebraic equations often produced by finite-difference formulations, the numerical finite-difference solution approach is today a primary method of solution of multidimensional heat-transfer problems.

The finite-difference formulation concepts have been developed in this chapter in the context of conduction heat transfer in rectangular systems. Nodal equations are presented in Tables 4–1 and 4–2 for standard interior and exterior nodes associated with such simple geometries. In addition, *R*/*C* network finite-difference volume and resistance elements have been presented for standard three-dimensional Cartesian, cylindrical, and spherical systems. The formulation of finite-difference nodal equations for rectangular, cylindrical, and spherical geometries proves to be quite straightforward. Although finite-difference nodal equations can also be developed for complex geometries involving irregular boundaries, the finite-element method [1,2] is commonly selected for such applications. Two general finite-element formulation approaches include (1) discretization methods, which involve the use of variational calculus or the method of weighted residuals MWR, and (2) energy balance methods (also referred to as control-volume finite-element methods CVFEM). In this connection, a very attractive CVFEM, which is similar in nature to the control-volume finite-difference method CVFDM, has recently been developed by Baliga and Patankar [10], Schneider [11], and others. The MWR and the CVFEM are introduced in the context of one-dimensional heat-transfer systems in Appendix F.

The numerical concepts in this chapter are also widely used in the analysis of radiation and convection heat transfer. In connection with the solution of the fluid-flow and energy-transfer equations associated with convection systems, it should be noted that stability becomes an important issue when dealing with numerical approximations for terms such as $u\ \partial u/\partial x$ and $u\ \partial T/\partial x$, which appear in the fundamental conservation equations [4].

CHAPTER 5

RADIATION HEAT TRANSFER

5–1 INTRODUCTION

Radiation heat transfer is defined as the transfer of energy across a system boundary by means of an electromagnetic mechanism which is caused solely by a temperature difference. Some of the more basic and practical aspects of thermal radiation were introduced in Chap. 1. We now want to take a closer look at the physical mechanism, properties, and geometric factors associated with thermal radiation. In addition, modeling concepts are developed and design information is presented in this chapter which provide the basis for practical thermal analysis of radiation heat transfer. Practical thermal analyses will be developed for blackbody and diffuse nonblackbody surfaces, with special consideration given to unsteady systems, participating medium, combined mode processes, and solar radiation.

5–2 PHYSICAL MECHANISM

Whenever a charged particle undergoes acceleration, energy possessed by the particle is converted into a form of energy known as *electromagnetic radiation*. Electromagnetic radiation includes cosmic rays, gamma rays, X rays, ultraviolet radiation, visible light, infrared radiation, microwaves, broadcasting waves, and ultrasonic electrical waves. Electromagnetic radiation can be produced by various means, depending on the type of charged particles that are involved in the process. To illustrate, γ rays are produced by fission of nuclei or by radioactive disintegration, X rays by the bombardment of metals with high-energy electrons, microwaves by special types of electron tubes (klystrons, magnetrons, or traveling wave tubes), and radio waves by the excitation of certain crystals or by the flow of alternating current through electric conductors. It should also be noted that astronomical sources provide significant quantities of electromagnetic radiation that range from extremely short-wavelength cosmic rays to long-wavelength radio waves. Consequently, the reception and evaluation of X-ray, γ-ray, ultraviolet, visible, infrared, and radio-wave radiation are critical elements of modern astronomy.

Of particular interest to us is electromagnetic radiation that is produced by vibrational and rotational movements of atoms and molecules of a substance and/or by changes in the atomic energy levels of the least strongly bound electrons. Because the level of energy associated with the fluctuating motion of these small oscillators is indicated by temperature, the resulting electromagnetic radiation is referred to as *thermal radiation*. Thermal radiation heat transfer represents the exchange of thermal radiation between bodies at different temperature, with each body (1) converting internal energy into outflowing electromagnetic waves, and (2) absorbing incoming electromagnetic waves, which are converted into internal energy.

5-2-1 The Electromagnetic Spectrum

All the various types of electromagnetic waves are characterized by a frequency ν and a propagation velocity in free space (a vacuum or transparent medium) equal to the speed of light c. The speed of light c in a gas, liquid, or solid is related to the speed of light in a vacuum c_0 ($= 3 \times 10^8$ m/s) by the *index of refraction* $n = c_0/c$. (The index of refraction of air and most gases is essentially unity, but for liquids and solids such as water and glass it is of the order of 1.5.) The wavelength λ is defined in terms of ν and c by

$$\lambda = \frac{c}{\nu} \tag{5-1}$$

The wavelength λ is generally expressed in terms of the *micrometer* $\mu m = 10^{-6}$ m, *nanometer* nm $= 10^{-9}$ m, or *angstrom* Å $= 10^{-10}$ m. Whereas the propagation velocity and the wavelength of a radiant beam depend on the medium, the frequency ν depends only on the radiating source and is independent of the substance through which it is transmitted.

In addition to being described by continuous waves with characteristic wavelength λ (or frequency ν) and velocity c, electromagnetic radiation is also generally perceived as discrete packets of energy known as *quanta* or *photons*. This concept was first proposed in 1900 by Max Planck in the context of his *quantum theory*. Briefly, Planck related the *photon energy* e to the frequency ν by $e = h\nu = ch/\lambda$, where $h = 6.625 \times 10^{-34}$ J s is *Planck's constant*. Notice that according to this celebrated theory the energy associated with electromagnetic radiation is inversely proportional to wavelength. It should be noted that this perspective is compatible with properties of electromagnetic radiation that pertain to its absorption by matter and its ability to cause photobiologic effects.

The various types of electromagnetic radiation are characterized according to wavelength or frequency by the *electromagnetic spectrum*, which is shown in Fig. 5-1. Notice that the spectrum of electromagnetic radiation ranges from the very short-wavelength

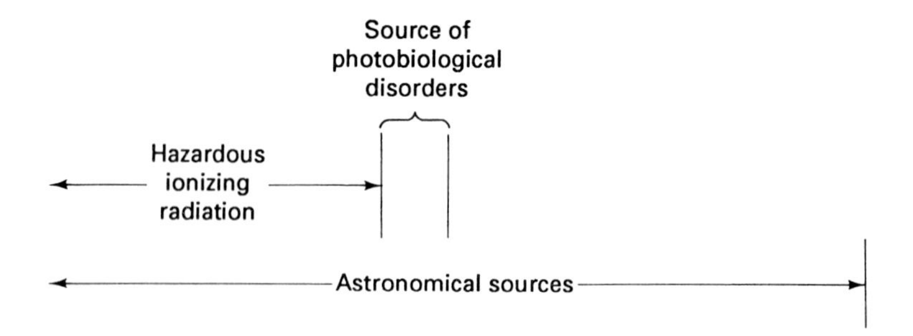

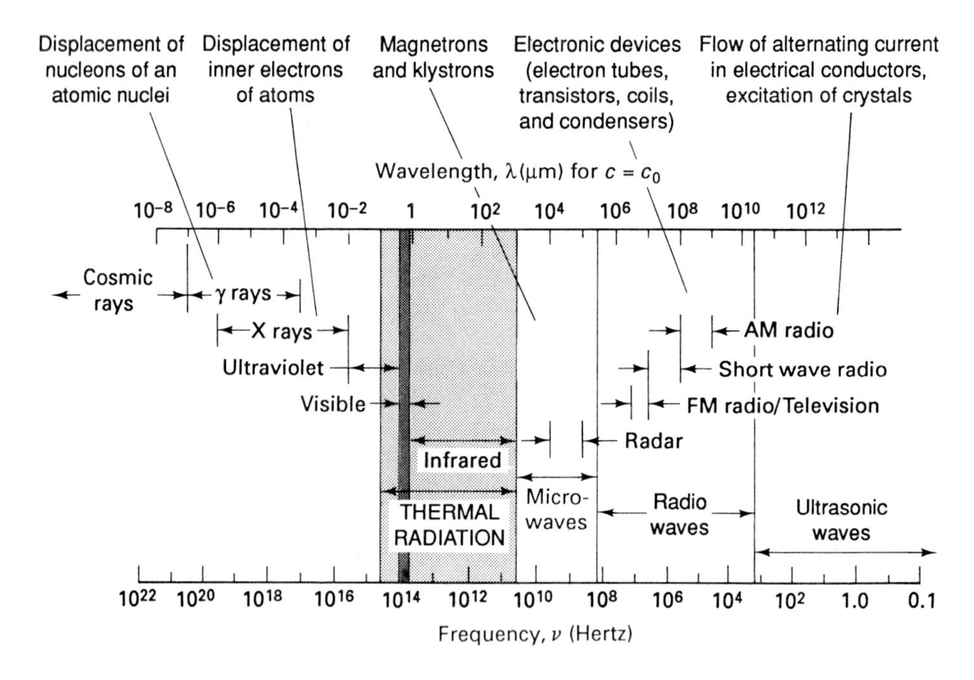

FIGURE 5–1 Electromagnetic frequency/wavelength spectrum.

cosmic-ray, γ-ray, and X-ray phenomena, through the intermediate-wavelength thermal radiation, to the long-wavelength microwaves, radio waves, and ultrasonic waves.

Thermal Radiation

For practical purposes, the thermal radiation wavelength band may be considered to extend from 0.1 to 1000 μm, which includes ultraviolet (0.1 to 0.38 μm), visible (0.38 to 0.76 μm), and infrared (0.76 to 1000 μm) regions. This part of the electromagnetic spectrum is focused upon in Fig. 5–2. Although the *solar radiation* band essentially lies in the heart of this thermal radiation region, astronomical sources provide thermal (as well as nonthermal) radiation with both shorter (UVC and vacuum UV) and longer (microwave and radio wave) wavelengths.

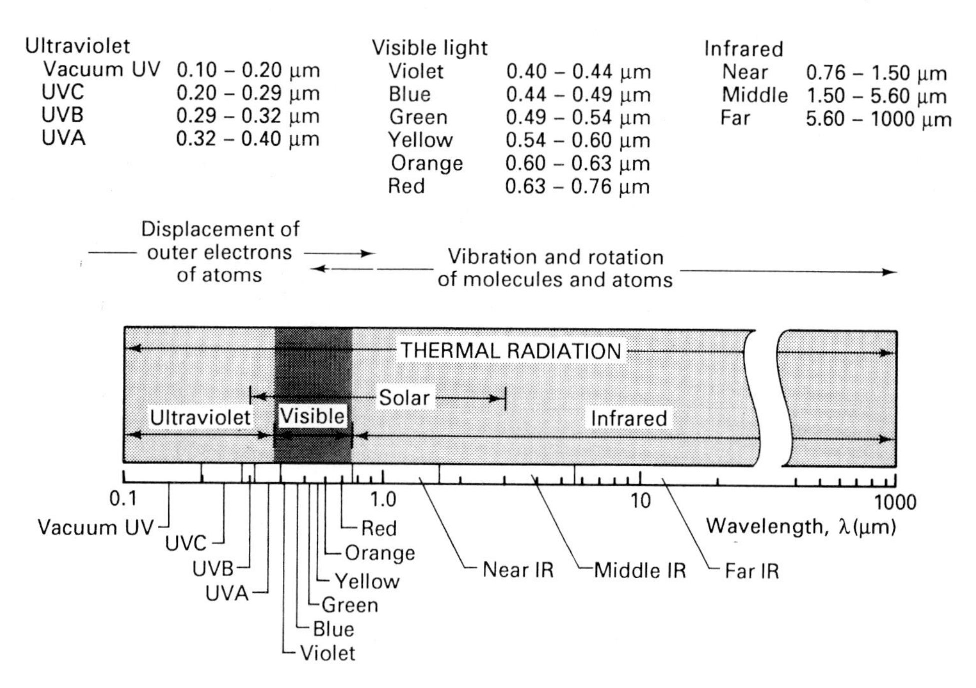

FIGURE 5-2 Electromagnetic spectrum for thermal radiation wavelength band ($c = c_0$).

***Ultraviolet Radiation* (*UV*)** The part of the ultraviolet radiation region which forms the low-wavelength boundary of the thermal radiation spectrum is mainly produced by changes in the atomic energy level that occurs when outer electrons of an atom are displaced. Ultraviolet radiation includes UVA (also called *black light*) (0.32–0.40 μm), UVB (also called *sunburn radiation*) (0.29–0.32 μm), UVC (0.20–0.29 μm), and vacuum UV (0.1–0.2 μm). The UVC (which causes sunburn and kills microorganisms) and vacuum UV components of solar radiation are completely absorbed by the ozone O_3 layer in the stratosphere. Furthermore, the ozone layer and atmosphere absorb large amounts of the UVB radiation, such that only about 2 to 3% of the radiation contained in terrestrial sunlight is in the UVA and UVB range. However, as portions of the ozone layer near the Antarctic and Arctic are destroyed by fluorochloromethanes (CFCs) from aerosols and refrigerants and other causes, significant increases in the amounts of UVB can be expected to reach the earth's surface. Whereas UVB is the primary natural cause of sunburn, strong evidence indicates that skin cancers such as malignant melanoma, which is sometimes lethal, and basal cell carcinoma result from long-term excessive exposure to UVB and UVA. It should also be noted that ultraviolet radiation can be produced by artificial light sources. For example, low-pressure mercury vapor lamps are commonly used to produce UVC radiation, which is used to destroy certain types of bacteria and to sterilize foodstuffs and medical equipment. In addition, fluorescent tubes are available that emit little or no UVB and shorter wavelength radiation. Such fluorescent UVA lamps are commonly used for artificial tanning. However, evidence concerning the relation between skin cancer and other photobiological disorders has caused dermatologists to issue a strong warning against the use of artificial UVA for cosmetic tanning.

Visible Radiation (Light) The term *light* is of course used to describe the visible portion of the electromagnetic spectrum (0.4–0.76 μm). Light consisting of narrow wavelength bands makes up the colors of the visible spectrum; that is, violet (0.40–0.44 μm), blue (0.44–0.49 μm), green (0.49–0.54 μm), yellow (0.54–0.60 μm), orange (0.60–0.63 μm), and red (0.63–0.76 μm). The summation of all visible wavelengths is referred to as *white light*. Visible and other electromagnetic radiation emanates naturally from the sun and stars and is produced artificially by devices such as fluorescent tubes consisting of a low-pressure mercury discharge source housed in a thin quartz glass tube coated on the inside with phosphor; incandescent lamps, which are usually designed with a coiled tungsten filament contained in an evacuated or inert-gas-filled glass envelope; lasers, carbon and xenon arcs, and low- or high-pressure mercury vapor lamps. The biological process of *photosynthesis* in which light is absorbed and transformed into chemical energy by green plants, algae, and certain bacteria ultimately provides the energy required by all living organisms.

Infrared Radiation (IR) Thermal radiation in the infrared region is primarily associated with molecules or lattice vibrations. All bodies at temperatures above absolute zero emit infrared radiation. In this connection, hot solid bodies generally emit far more infrared energy than visible and UV radiation. Other sources of infrared radiation include emissions of electronic discharges in gases and some lasers (e.g., CO_2, $\lambda = 10.6$ μm; neodymium, $\lambda = 1.06$ μm). Infrared radiation is very important in many temperature-sensing applications. Unlike visible and ultraviolet radiation, infrared (and microwave) radiation does not cause biological effects (either adverse or beneficial), except for producing heat, which makes possible the maintenance of temperatures that enable vital metabolic processes to occur.

Microwave Radiation

Although electromagnetic radiation produced by microwave tubes, known as *magnetrons*, is generally not classified as thermal radiation because of the method of generation, microwaves with wavelengths in the range 10^3 to 10^5 μm are reflected by metal, pass through glass and plastic, and are absorbed and converted into internal energy by food (water, sugar, and fat) molecules. As indicated in Chap. 1, this transport and absorption characteristic of microwaves is now widely used in microwave ovens, which are capable of efficiently cooking many foods.

5-3 THERMAL RADIATION PROPERTIES

The exchange of thermal radiation between surfaces is a function of (1) surface emissions properties; (2) surface absorption, reflection, and transmission properties; and (3) properties of the medium that lies within the path of the thermal radiation. Each of these issues is considered in this section.

5–3–1 Surface Emission Properties

The rate of thermal radiation emitted by a body is dependent upon the surface temperature T_s, the nature of the surface, and the electromagnetic radiation wavelength λ or frequency ν. The effect of each of these factors on the emission of thermal radiation must be considered.

Total Emissive Power

The effect of emitter surface temperature T_s on the rate of thermal radiation emitted by a body is seen by examining the *total emissive power* E, which was introduced in Chap. 1. Because the total emissive power represents the total rate of thermal radiation emitted per unit surface area (i.e., thermal radiation flux) over all wavelengths, it is sometimes designated by $E_{0\to\infty}$ instead of E. However, for convenience, we will retain the symbol E.

As indicated in Chap. 1, a surface that emits the maximum possible thermal radiation at any given temperature is called a *blackbody*. The total emissive power for thermal radiation in a vacuum from such ideal emitting blackbody surfaces is given by the *Stefan-Boltzmann law*,

$$E_b = \sigma T_s^4 \tag{5–2}$$

where the Stefan-Boltzmann constant σ is equal to 5.67×10^{-8} W/(m^2 K^4). Referring back to Sec. 1–2–2, we are reminded that the blackbody thermal radiation flux ranges from very significant levels for source temperatures of the order of 1000 K and above to quite small and often negligible quantities for normal environmental temperatures.

Although some surfaces and geometrical configurations approach ideal emitting conditions, perfect blackbody surfaces do not exist. (The nature of ideal blackbody thermal radiating surfaces will be explored further in Sec. 5–3–3.) The total emissive power of real nonblackbody surfaces is expressed in terms of E_b by

$$E = \epsilon E_b \tag{5–3}$$

where the *emissivity* ϵ ranges from zero to unity. (Because ϵ accounts for the thermal radiation emitted over all wavelengths into the entire hemispherical space above a surface, it is also commonly referred to as the total hemispherical emissivity.) It is important to note that ϵ is a property which is dependent only on the nature of the surface and its temperature T_s. The emissivity is given for common surfaces in Table 5–1, Fig. 5–3, and in Table A–C–7 of the Appendix; very comprehensive tabulations of radiation properties are available in *Thermophysical Properties of Matter* by Touloukian et al. [1–3] and in *Thermal Radiation Properties Survey* by Gubareff et al. [4]. Referring to Table 5–1, we observe that blackbody conditions are approached by surfaces coated with lampblack paint. On the other hand, metals have emissivities that range from very low values for polished surfaces to fairly high values for surfaces that have been oxidized or anodized. However, modifying terms such as polished, commercial finish, oxidized, anodized, and so on, which are used to describe the

TABLE 5-1 Emissivities of representative surfaces

Surface	*Emissivity*, ϵ	*Temperature*, T(K)
Aluminum		
Polished	0.04	500
Anodized	0.94	310
Brass		
Polished	0.07	320
Dull	0.22	320
Copper		
Polished	0.041	340
Slightly polished	0.12	320
Polished, lightly tarnished	0.05	320
Dull	0.15	320
Oxidized at 1030 K	0.50	590
Nickel, polished	0.09	270
Silver, polished	0.02	300
Stainless steel 18-8, polished	0.25	310
Tungsten, polished	0.33	3400
Asphalt	0.93	310
Glass, Pyrex	0.88	420
Parsons black paint	0.98	240
Lampblack paint	0.96	310

Source: Based on data primarily from Touloukian and DeWitt [1,2].

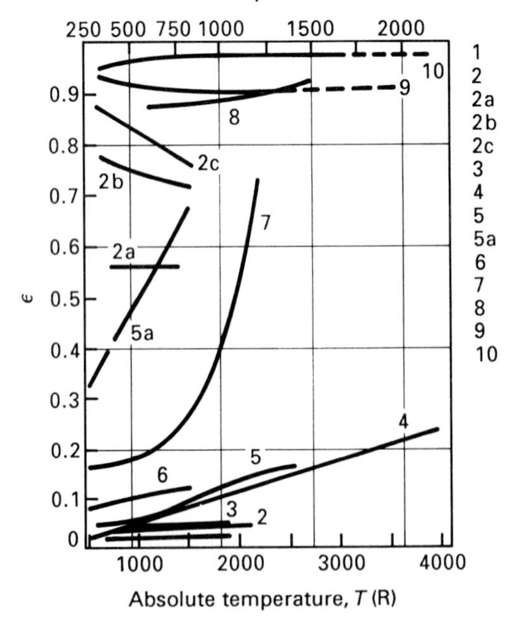

FIGURE 5-3 Dependence of emissivity ϵ of various surfaces on surface temperature. (Based on data from Touloukian and DeWitt [2] and Gubareff et al. [4].)

nature of a surface, are very subjective. As pointed out by Sparrow and Cess [5], because of the ambiguity of such terms, it is unwise to assume that radiative property values reported in literature apply with high precision to other similarly described materials.

Although the total emissive power E of real surfaces is less than E_b, E always increases with emitter surface temperature. This point is reinforced by Fig. 5–3, which shows the variation of ϵ with temperature. By multiplying ϵ by E_b $(= \sigma T_s^4)$, we find that E increases with increasing temperature, even for the materials for which ϵ itself decreases.

Subtotal Emissive Power

To see the effect of wavelength on thermal radiation, we consider the energy flux emitted from a surface over wavelengths from zero to λ. This *subtotal emissive power* $E_{0\to\lambda}$ is shown in Fig. 5–4 as a function of λ for a blackbody at several temperatures. $E_{b,0\to\lambda}$ increases from zero at small values of λ and approaches the total emissive power E_b as λ becomes large. Consistent with the electromagnetic spectrum shown in Figs. 5–1 and 5–2, we find that the significant contribution to the thermal radiation for these temperatures occurs within wavelengths of about 0.1 and 100 μm. For solar radiation, which has an effective blackbody source temperature of roughly 5800 K, the wavelength band essentially lies between 0.30 and 3.0 μm, with about 98% of the energy associated with $\lambda < 3.0$ μm. On the other hand, the wavelength range for a surface temperature of 400 K is mainly between 3.0 and 40 μm. As a matter of fact, less than 1% of the thermal radiation emitted at environmental temperatures below 400 K is contained in the part of the electromagnetic spectrum for which $\lambda <$ 3.0 μm.

A more general representation of the subtotal emissive power for a blackbody

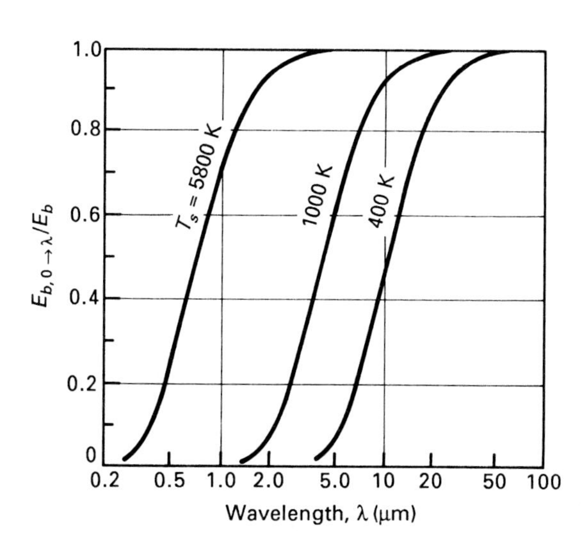

FIGURE 5–4
Subtotal emissive power $E_{b,0\to\lambda}$.

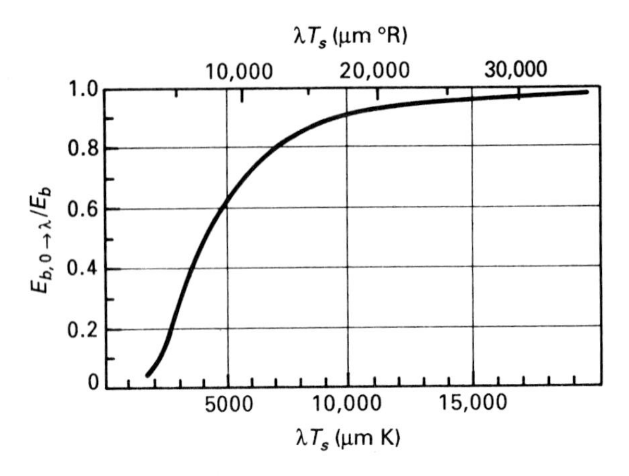

FIGURE 5–5
General representation of subtotal emissive power $E_{b,0\to\lambda}$.

that applies to the entire temperature range is given in Fig. 5–5 and Table A–H–2 of the Appendix in terms of $E_{b,0\to\lambda}/E_b$ versus λT_s.

Monochromatic Emissive Power

To complete the picture concerning the effect of wavelength on thermal radiation emitted from a surface, we introduced the *monochromatic emissive power* E_λ, which is defined as the thermal radiation flux emitted per unit wavelength $d\lambda$. This important thermal radiation emission property is related to the subtotal emissive power $E_{0\to\lambda}$ by

$$E_\lambda = \frac{dE_{0\to\lambda}}{d\lambda} \qquad \text{or} \qquad E_{0\to\lambda} = \int_0^\lambda E_\lambda \, d\lambda \tag{5–4,5}$$

As λ becomes large, it follows that

$$E = E_{0\to\infty} = \int_0^\infty E_\lambda \, d\lambda \tag{5–6}$$

Blackbody Thermal Radiation Theoretical predictions based on quantum theory were developed for the monochromatic emissive power for blackbody thermal radiation by M. Planck in 1901. The famous *Planck law* [6] is written as

$$E_{b\lambda} = \frac{C_1}{\lambda^5\{\exp\,[C_2/(\lambda T_s)] - 1\}} \tag{5–7}$$

where $C_1 = 3.743 \times 10^8$ W μm^4/m^2 and $C_2 = 1.439 \times 10^4$ μm K. This equation is shown in Fig. 5–6 for several temperatures, with λ taken as the independent variable. For each temperature, $E_{b\lambda}$ is seen to increase from zero at low wavelengths to a peak, and to then gradually fall back toward zero.

The peak in $E_{b\lambda}$ increases and shifts to shorter wavelengths as the temperature

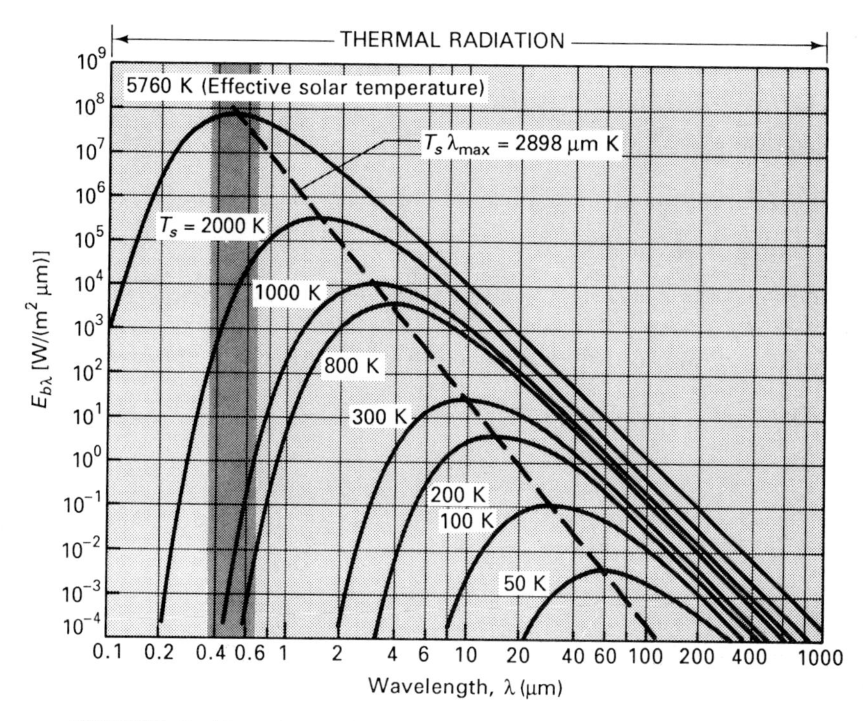

FIGURE 5-6 Monochromatic emissive power of a blackbody for several temperatures.

increases. The wavelength at which the peak occurs is given as a function of temperature by *Wien's displacement law*,

$$T_s \lambda_{\max} = 2898 \ \mu\text{m K} \tag{5-8}$$

This shift in $\lambda_{\max}$ and increase in $E_{b\lambda}$ with increasing temperature is responsible for the familiar change in color of heat-treated steel, which goes from a dull red at around 700°C to bright red, then to bright yellow, and finally becomes glowing white at approximately 1300°C. Referring to Fig. 5-6, we observe that little thermal radiation emitted by low-temperature blackbodies lies in the portion of the electromagnetic spectrum from 0.38 to 0.76 μm that is visible to the eye. However, as the temperature increases, more and more of the thermal radiation falls within the visible range, thus producing this array of color.

EXAMPLE 5-1

For practical purposes, the sun is generally considered to be a blackbody radiator with an effective temperature of about 5800 K. Determine the fraction of energy emitted by the sun that falls in the visible region of the electromagnetic spectrum.

Solution

Objective Determine the fraction of solar radiation that is visible.

Assumptions/Conditions

solar irradiation is equivalent to blackbody emission at 5800 K

Properties Sun at 5800 K: $\epsilon = 1$.

Analysis The wavelength band for solar radiation that is visible to the eye is 0.38 to 0.76 μm. Using Table A–H–2, we are able to determine the fraction of solar radiation with wavelength in the bands 0 to 0.38 μm and 0 to 0.76 μm; that is,

$$\frac{E_{b,0\rightarrow\lambda_1}}{E_b} = 0.102$$

for $\lambda_1 T_s = 0.38\ \mu\text{m}\ (5800\ \text{K}) = 2200\ \mu\text{m K}$, and

$$\frac{E_{b,0\rightarrow\lambda_2}}{E_b} = 0.55$$

for $\lambda_2 T_s = 0.76\ \mu\text{m}\ (5800\ \text{K}) = 4410\ \mu\text{m K}$. The difference between these two values gives the fraction of solar radiation falling in the visible spectrum; that is,

$$\frac{E_{b,\lambda_1\rightarrow\lambda_2}}{E_b} = 0.55 - 0.102 = 0.448$$

Therefore, approximately 44.8% of extraterrestrial solar radiation is visible to the human eye, with about 10.2% lying in the ultraviolet region and 45% in the infrared region. However, it should be noted that the amount of solar radiation actually reaching the surface of the earth is diminished by absorption within the stratosphere and atmosphere. Depending on environmental conditions, atmospheric pollution, latitude, season, time of day, and other factors, the total solar radiation reaching the earth's surface normally consists of approximately 60% IR, 37% visible light, and 3% UV.

Thermal Radiation Emitted from Real Surfaces The monochromatic emissive power E_λ of polished copper, anodized aluminum, and a blackbody are compared in Fig. 5–7. The ratio $E_\lambda/E_{b\lambda}$ at any given wavelength is known as the *monochromatic emissivity* ϵ_λ:

$$\epsilon_\lambda = \frac{E_\lambda}{E_{b\lambda}} \tag{5–9}$$

Like the emissivity ϵ, ϵ_λ is a property that is dependent on the surface alone. The monochromatic emissivities of these two real surfaces are shown in Fig. 5–8. Note

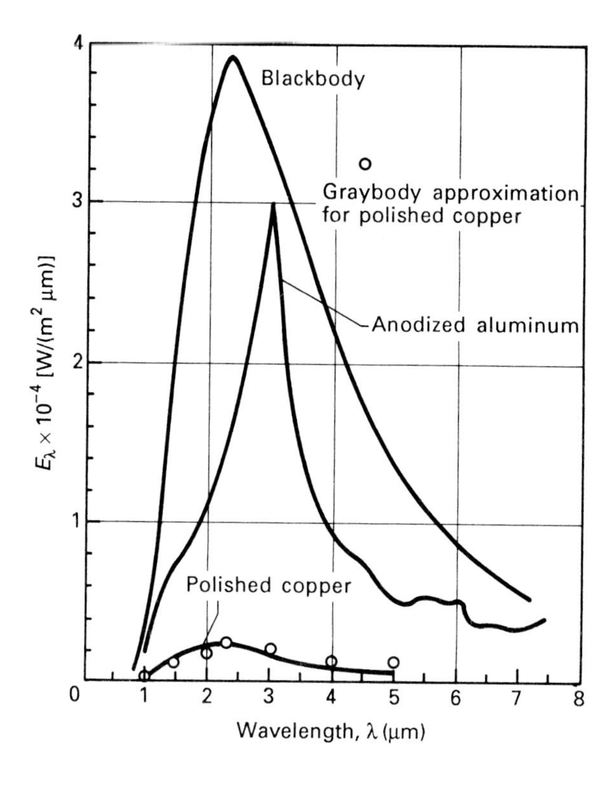

FIGURE 5-7
Monochromatic emissive power for blackbody and representative real surfaces at 1240 K.

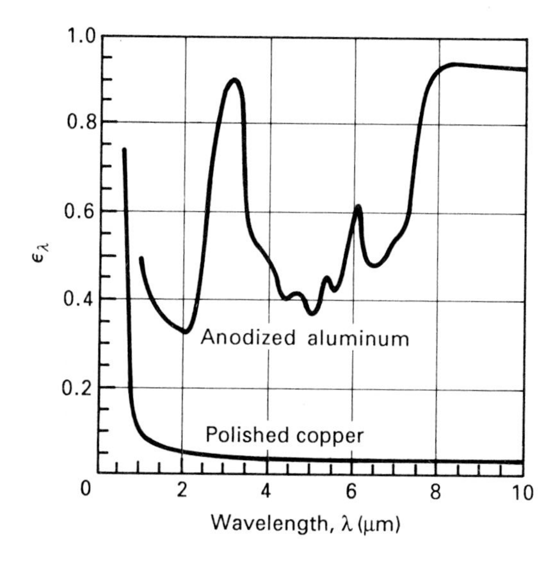

FIGURE 5-8
Monochromatic emissivity of polished copper and anodized aluminum (Dunkle et al. [7] and Seban [8]).

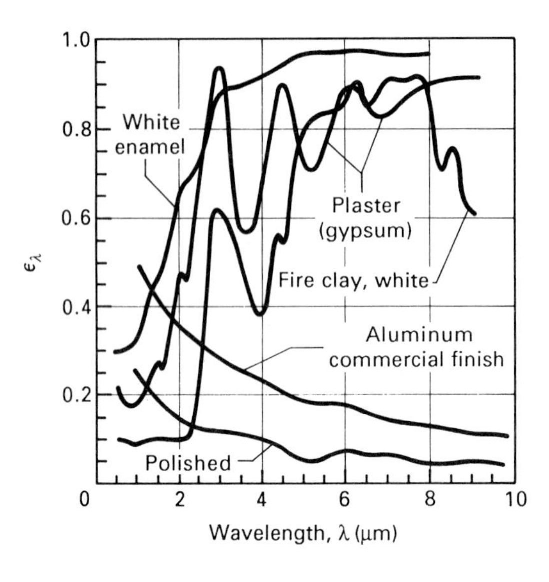

FIGURE 5–9
Monochromatic emissivity of several materials (Dunkle et al. [7] and Sieber [9]).

that ϵ_λ of these surfaces are less than unity and vary rather irregularly with λ. To at least some extent, all real surfaces exhibit these same characteristics. To reinforce this point, the monochromatic emissivities of several other materials at room temperature are shown in Fig. 5–9. An impressive listing of data for ϵ_λ is given in references 1–4 for many types of surfaces. Data in these references indicate that ϵ_λ is essentially independent of T_s for many substances. For example, ϵ_λ is nearly independent of surface temperature for metals such as polished copper, polished iron, and tungsten and for nonmetals such as carbon, Pyrex, and certain carbides. However, large changes occur in ϵ_λ with T_s for many nonmetallic substances, such as aluminum oxide (Al_2O_3).

The relationship between the emissivity ϵ and the monochromatic emissivity ϵ_λ is obtained by writing

$$\epsilon = \frac{E}{E_b} = \frac{\int_0^\infty E_\lambda \, d\lambda}{\int_0^\infty E_{b\lambda} \, d\lambda} = \frac{\int_0^\infty \epsilon_\lambda E_{b\lambda} \, d\lambda}{\int_0^\infty E_{b\lambda} \, d\lambda} \tag{5–10}$$

Note that for the case in which ϵ_λ is independent of λ, Eq. (5–10) reduces to

$$\epsilon = \epsilon_\lambda \tag{5–11}$$

Surfaces that satisfy this equation are known as *graybodies*. Referring to Fig. 5–8, we see that ϵ_λ of the polished copper surface exhibits approximate graybody behavior for wavelengths greater than about 2. A graybody approximation for the monochromatic emissive power $E_{g\lambda}$ for real surfaces such as this is given by

$$E_{g\lambda} = \epsilon E_{b\lambda} \tag{5–12}$$

This graybody approximation for the monochromatic emissive power of polished copper is shown in Fig. 5–7. Notice that $E_{g\lambda}$ follows the same wavelength-dependence pattern as $E_{b\lambda}$. The anodized aluminum and the other substances shown in Fig. 5–9 are observed to exhibit distinct nongraybody characteristics. Furthermore, the polished copper surface is nongray for wavelengths less than about unity.

EXAMPLE 5–2

The filament of an incandescent lamp operates at 2500 K. Assuming approximate graybody characteristics, determine the fraction of radiant energy emitted by the filament that falls in the visible spectrum.

Solution

Objective Determine the fraction of thermal radiation emitted by a graybody at 2500 K that produces light.

Assumptions/Conditions

graybody thermal radiation

Properties Lamp filament (graybody): $\epsilon_\lambda = \epsilon =$ constant.

Analysis Because the spectral distribution of radiant energy emitted by a graybody is the same as for a blackbody, we utilize Table A–H–2 to solve this problem. Following the pattern established in Example 5–1, we write

$$\frac{E_{b,0\rightarrow\lambda_1}}{E_b} = \frac{E_{g,0\rightarrow\lambda_1}}{E_g} = 0.00021$$

for $\lambda_1 T_s = 0.38\ \mu\text{m}\ (2500\ \text{K}) = 950\ \mu\text{m K}$, and

$$\frac{E_{b,0\rightarrow\lambda_2}}{E_b} = \frac{E_{g,0\rightarrow\lambda_2}}{E_g} = 0.0522$$

for $\lambda_2 T_s = 0.76\ \mu\text{m}\ (2500\ \text{K}) = 1900\ \mu\text{m K}$. Therefore, the fraction of energy that falls in the visible part of the spectrum is

$$\frac{E_{g,\lambda_1\rightarrow\lambda_2}}{E_g} = 0.0522 - 0.00021 = 0.052$$

For this operating temperature, only 5.2% of the radiant energy dissipated by the filament produces light, with most of the remaining energy producing infrared heating.

Thermal Radiation Intensity

Referring to Fig. 5–10, the *intensity I* is defined as the total rate of thermal radiation emitted per unit solid angle $d\omega$ and per unit area normal to the direction ϕ,θ. (The intensity is sometimes defined in the literature in terms of the total radiant energy

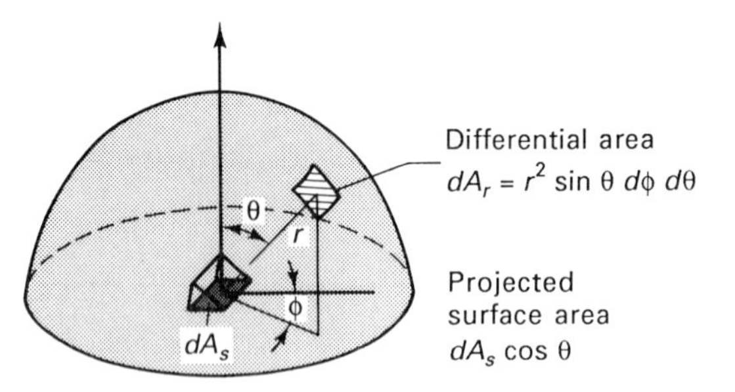

FIGURE 5–10
Geometric perspective for concept of intensity.

emitted, reflected, and transmitted from a surface.) For a surface area dA_s, the projected area is simply $dA_s \cos\theta$. The solid angle $d\omega$ is defined by

$$d\omega = \frac{dA_r}{r^2} \tag{5–13}$$

where the area dA_r of the hemispherical surface element is shown in Fig. 5–10. The intensity I is written in terms of the total emissive power E, which is the rate of energy emitted per unit surface area dA_s, as

$$I = \frac{dE}{\cos\theta \, d\omega} \tag{5–14}$$

Rearranging this equation and assuming no variations in the emissive properties with the azimuthal angle ϕ, we have

$$\frac{dE}{d\omega} = E_\theta = I \cos\theta \tag{5–15}$$

where E_θ is the directional emissive power.

An important aspect of ideal blackbody emitting surfaces is that the intensity I is the same in all directions. Surfaces that exhibit this characteristic are said to be *diffuse* emitters. Many real surfaces such as industrially rough surfaces approach diffuse conditions. Utilizing Eq. (5–15), I is equal to the directional emissive power normal to the surface E_0 for diffuse surfaces. For this case, Eq. (5–15) reduces to

$$E_\theta = E_0 \cos\theta \tag{5–16}$$

which is known as the *Lambert cosine law*.

The *directional emissivity* ϵ_θ,

$$\epsilon_\theta = \frac{E_\theta}{E_{b\theta}} = \frac{I}{I_b} \tag{5–17}$$

is shown in Fig. 5–11 for several real surfaces. Notice that the nonconductors are essentially diffuse for θ less than about 40 degrees but violate the Lambert cosine law for larger angles, with ϵ_θ falling to very small values. The metallic surfaces obey the Lambert cosine law over about the same range in θ, but ϵ_θ increases quite sharply before falling to zero at 90 degrees. Consistent with these findings, a hot metallic

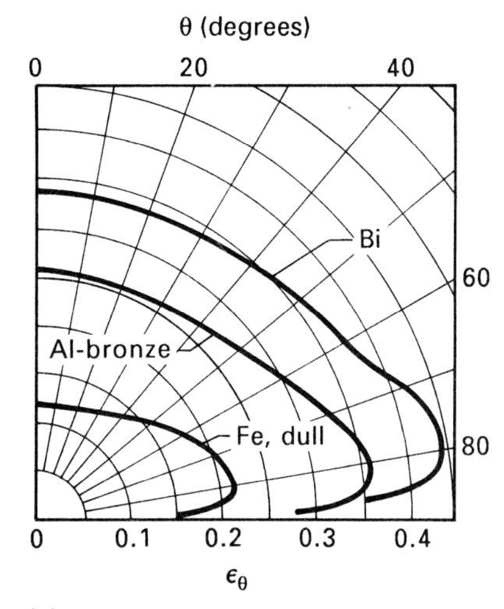

(a) Metal surfaces.

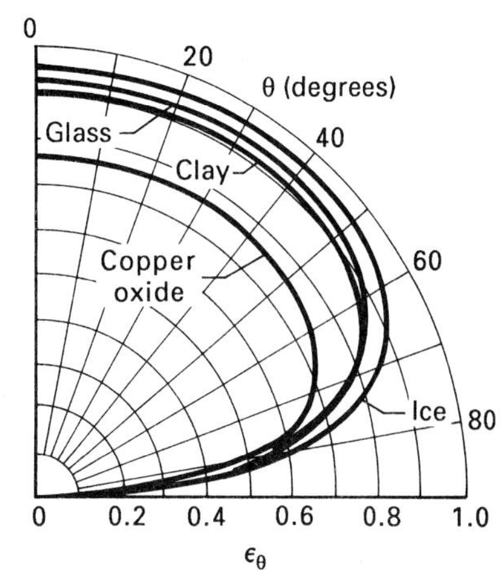

(b) Electric nonconducting surfaces.

FIGURE 5–11 Distribution of directional emissivity for several surfaces (Schmidt and Eckert [10]).

sphere will appear brighter near the base ($\theta \simeq 80$ degrees) than at the center $\theta \simeq 0$ degrees). The opposite holds true for a nonmetallic sphere. These directional effects are generally accounted for in design by utilizing the emissivity ϵ, which accounts for the radiant energy emitted into the entire hemispherical space above the surface. The emissivity ϵ is usually set equal to some fraction of ϵ_0. For example, $\epsilon/\epsilon_0 \simeq 1.2$ for bright metallic surfaces and $\epsilon/\epsilon_0 \simeq 0.96$ for nonconductors.

EXAMPLE 5–3

Develop an expression for the total emissive power E of a surface in terms of the intensity I.

Solution

Objective Express E in terms of I.

Analysis Based on Eqs. (5–13) and (5–14), we write

$$dE = I \frac{\cos\theta}{r^2} dA_r$$

Referring to Fig. 5–10, dA_r is equal to $(r\, d\theta)(r \sin\theta\, d\phi)$, such that dE becomes

$$dE = I \cos\theta \sin\theta\, d\phi\, d\theta$$

To obtain E, we integrate over the hemisphere (i.e., $0 \leqslant \phi \leqslant 2\pi$, $0 \leqslant \theta \leqslant \pi/2$).

$$E = \int_0^{\pi/2} \int_0^{2\pi} I \cos\theta \sin\theta\, d\phi\, d\theta$$

For a diffuse surface, I is uniform and is brought outside the integrals, such that

$$E = I \int_0^{\pi/2} \int_0^{2\pi} \cos\theta \sin\theta \, d\phi \, d\theta \tag{a}$$

By employing the double-angle trigonometric formula,

$$\sin(a + b) = \sin a \cos b + \cos a \sin b$$

we obtain

$$E = \pi I$$

5-3-2 Surface Irradiation Properties

Total Irradiation Properties

As illustrated in Fig. 5–12, thermal radiation incident upon a surface is absorbed, reflected, and transmitted through the body. The *absorptivity* α, *reflectivity* ρ, and *transmissivity* τ were defined in Chap. 1. These total hemispherical surface irradiation properties account for the fractions of incident thermal radiation flux G at *all* wavelengths over the entire hemisphere above a surface that are absorbed, reflected, and transmitted. The incoming thermal radiation flux G is called the *irradiation*. Referring to Fig. 5–12, the thermal irradiation received by the surface is distributed as follows:

Thermal radiation flux absorbed	αG
Thermal radiation flux reflected	ρG
Thermal radiation flux transmitted	τG
Total irradiation	G

The relationship between these surface irradiation properties is given by Eq. (1–13),

$$\alpha + \rho + \tau = 1 \tag{5–18}$$

Except for a few materials such as glass, rock salt, and other inorganic crystals, most solids are essentially opaque, with τ equal to zero.

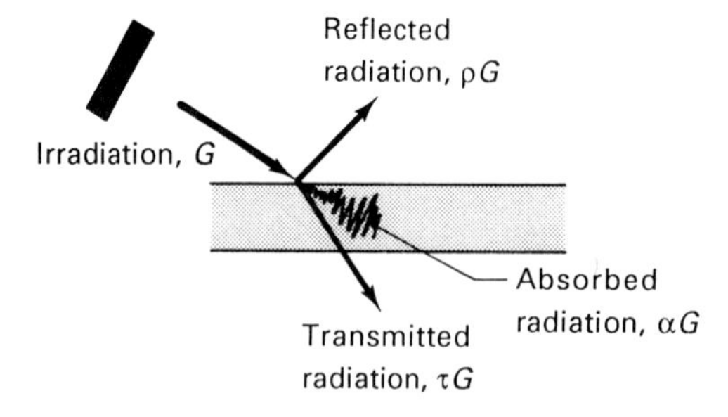

FIGURE 5–12
Absorption, reflection, and transmission of thermal radiation incident on a surface.

Although all real surfaces reflect and/or transmit at least some thermal radiation, the concept of an ideal blackbody surface which absorbs all incident irradiation (i.e., $\alpha = 1$, $\rho = 0$, $\tau = 0$) is extremely important. As we have already seen, it is for such ideal absorbing and emitting surfaces that the pioneering theoretical studies by Stefan, Boltzmann, Planck, and others were developed. Moreover, the radiative performance of ideal blackbody surfaces provides a standard against which the performance of real surfaces can be compared. The term "blackbody" stems from the observation that surfaces which absorb nearly all of the thermal radiation in the visible part of the electromagnetic spectrum are black in color as a result of the absence of reflected light. The eye is a very good indicator of reflected visible thermal radiation, but is totally insensitive to the reflection of thermal radiation outside this narrow wavelength spectrum. It just so happens that surfaces that appear black in color generally are also good absorbers of thermal radiation outside the visible range.

The total hemispherical absorption, reflection, and transmission properties are dependent upon the nature and temperature T_R of the emitting source and upon the character of the receiving surface. The importance of T_R and the type of surface are shown in Fig. 5-13. In this figure, the absorptivity α is shown as a function of emitter source temperature T_R for several common materials which are at room temperature. Notice that the white fire clay, which would be judged by the eye to be a poor absorber of thermal radiation, is actually a very good absorber ($\alpha \gtrsim 0.8$) of thermal radiation which is emitted from sources with temperatures below 500 K. However, white fire clay reflects most of the incoming solar radiation ($\alpha \simeq 0.1$) which is associated with an effective blackbody source temperature of approximately 5800 K.

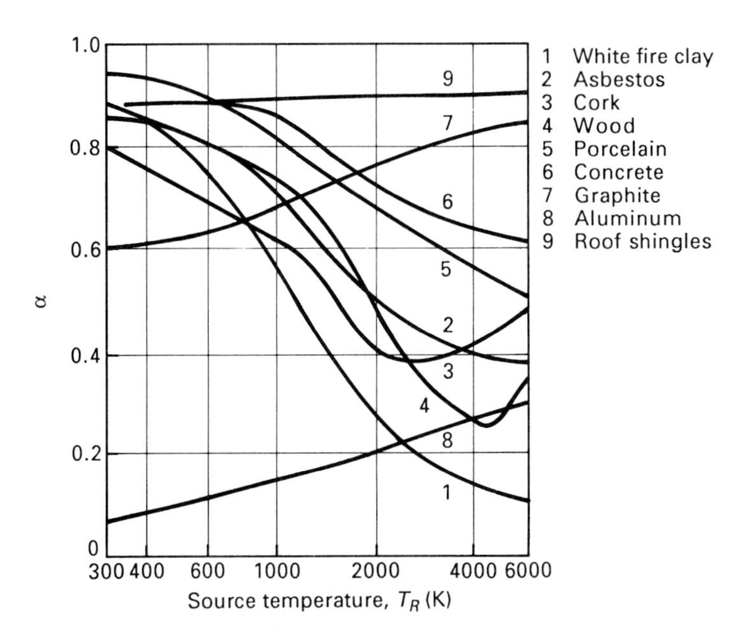

FIGURE 5-13 Dependence of absorptivity α on source temperature T_R of incident thermal radiation (Sieber [9]).

Data for α, ρ, and τ are available in references 1 through 4 and elsewhere for many surfaces. However, these data are generally restricted to situations involving irradiation from ideal emitting sources at one or two values of T_R. For example, considerable data are available for surfaces that receive solar radiation and radiation from blackbody sources at 300 K. But aside from the calculations by Sieber [9], which are shown in Fig. 5–13, relatively little information is available in the literature on the general effect of T_R on the surface irradiation properties. To circumvent this problem, α can sometimes be evaluated on the basis of tabulated data for ϵ by the use of *Kirchhoff's law* for thermal radiation. The limiting form of Kirchhoff's law indicates that α and ϵ are equal for thermal radiation exchange between a surface s and a *blackbody or graybody* R under conditions of *thermal equilibrium* (i.e., $T_R = T_s$). This form of Kirchhoff's law is developed in the context of a blackbody enclosure in Example 5–5. Other forms of this important law are considered in the next section.

EXAMPLE 5–4

No perfect blackbody surface has been found to exist. Referring to Table A–C–7 in the Appendix, we see that emissivities of the order of 0.95 to 0.98 are common for near-blackbodies such as surfaces coated with flat black paint. However, geometrical blackbodies can be constructed which perform even closer to the ideal. Show that blackbody conditions are approached by a small hole in the wall of a large cavity with opaque partially absorbing isothermal surface.

Solution

Objective Show that the radiation characteristics within a large cavity with opaque partially absorbing isothermal surface approach those of a blackbody.

Schematic Large cavity with uniform temperature T_R.

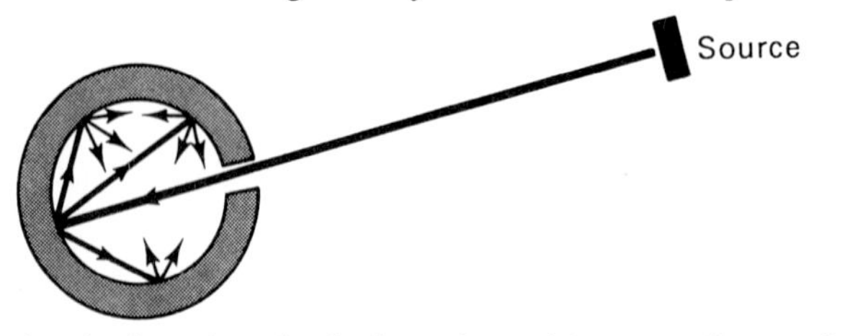

Analysis A spherical cavity with a small opening in its wall is shown in the schematic. If we trace the path of an incident ray of thermal radiation entering the cavity, we find that the ray is reflected within the interior of the cavity many times, with a part of the energy being absorbed each time. When the reflected ray eventually reaches the opening and escapes, its energy content is extremely small. Because nearly all of the thermal radiation entering the cavity is absorbed, the radiation leaving the hole approaches that of a blackbody at T_R. Furthermore, blackbody radiation is approximated within such a cavity, regardless of whether the surface is highly reflective or absorbing.

EXAMPLE 5-5

Referring to the large isothermal enclosure A_R containing a small body A_s shown in Fig. E5–5, develop a relation for α_s in terms of ϵ_s for the body.

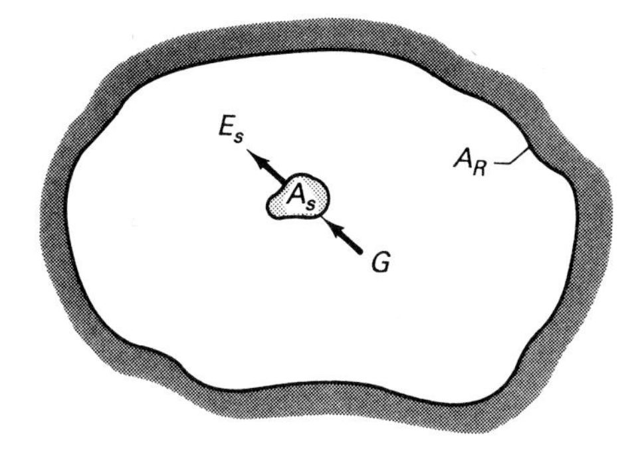

FIGURE E5–5
Small body contained in a large isothermal enclosure.

Solution

Objective Develop a relation between α_s and ϵ_s for a small body contained in a large isothermal enclosure.

Assumptions/Conditions

surface A_R is opaque
effect of surface A_s on irradiation from A_R is negligible

Analysis Assuming that the walls of the enclosure are opaque and that the body has no appreciable effect on the irradiation G, the enclosure forms a blackbody cavity with G represented by

$$G = E_b(T_R) = \sigma T_R^4$$

For steady-state conditions, thermal equilibrium must exist within the enclosure such that the temperature T_s of the body is equal to T_R. Thus, the net rate of energy transfer from the body must be equal to zero, and the energy balance becomes

$$\alpha_s G A_s - E_s(T_s) A_s = 0$$

or

$$\frac{\alpha_s G}{E_s(T_s)} = \frac{\alpha_s E_b(T_s)}{\epsilon_s E_b(T_s)} = \frac{\alpha_s}{\epsilon_s} = 1$$

Therefore, we conclude that $\alpha_s = \epsilon_s$, or simply

$$\alpha = \epsilon$$

for thermal radiation exchange between a surface s and a blackbody R under conditions of thermal equilibrium, which represents the limiting form of Kirchhoff's law.

Monochromatic Irradiation Properties

Spectral surface irradiation properties are now defined which account for the effect of wavelength λ. The *monochromatic absorptivity* α_λ is the fraction of incident thermal radiation with wavelength λ that is absorbed by a surface. Similarly, the *monochromatic reflectivity* ρ_λ and the *monochromatic transmissivity* τ_λ represent the fractions of incoming thermal radiation with wavelength λ that are reflected and transmitted, respectively. These important spectral surface irradiation properties are related by

$$\alpha_\lambda + \rho_\lambda + \tau_\lambda = 1 \tag{5-19}$$

The relationship between the monochromatic absorptivity α_λ of a receiving surface and its monochromatic emissivity ϵ_λ is given by Kirchhoff's law, which indicates

$$\alpha_\lambda = \epsilon_\lambda \tag{5-20}$$

for systems in which (1) the irradiation is diffuse, or (2) the surface is diffuse. The first of these conditions is approximately satisfied for many practical arrangements, and the second condition is approached for electrically nonconducting materials. This relation and its more general specular (i.e., directional) form, which is applicable to nondiffuse as well as diffuse thermal radiation, is developed by Planck [6] and Siegel and Howell [11].

In addition to the considerable amount of data available for ϵ_λ and α_λ, extensive tabulations of data have also been developed for ρ_λ and τ_λ. Again, references 1 through 4 are very good sources of data for these spectral surface irradiation properties.

The monochromatic transmissivity τ_λ is shown in Fig. 5–14(a) and (b) for ordinary window glass and type 2A diamond. It is observed that glass transmits thermal radiation quite well in the low-wavelength visible range of the electromagnetic

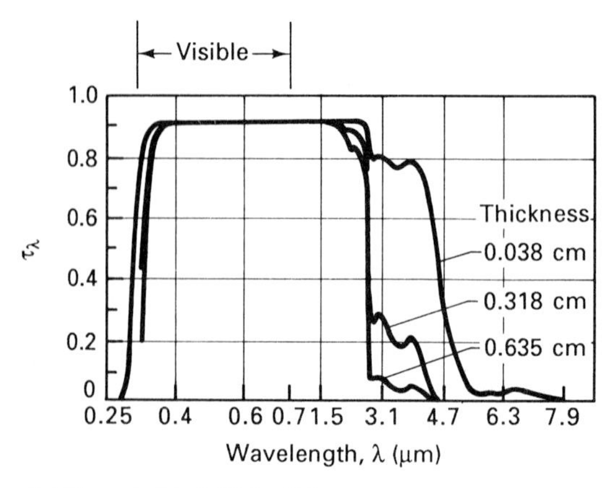

(a) Glass (with 0.02 Fe_2O_3) at room temperature (Dietz [12]).

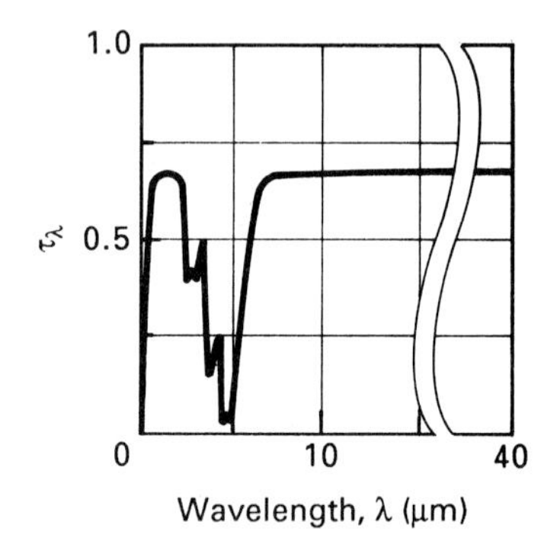

(b) Type 2A diamond at room temperature (Seal [13]).

FIGURE 5–14 Monochromatic transmissivity τ_λ.

spectrum. However, in the longer-wavelength infrared part of the spectrum ($\lambda \gtrsim 2.6$ μm), the glass is nearly opaque to thermal radiation, with most of the energy being absorbed and reflected. As we have already mentioned, this nongraybody characteristic of glass is responsible for the *greenhouse effect*, in which glass is transparent to short-wavelength irradiation from the sun but is essentially opaque to longer-wavelength thermal radiation emitted by the low-temperature interior of an enclosure. Some plastic films such as polyethylene have similar characteristics.

On the other hand, the type 2A diamond is transparent throughout much of the infrared region (6 μm $\lesssim \lambda \lesssim$ 40 μm), as well as in visible range. Because of its transparency across such a broad range of wavelengths and because of its great strength, type 2A diamond was utilized in the development of two 18.2-mm-diameter windows for the 1978 Pioneer Venus Space Mission. The function of the windows was to protect the infrared radiometer equipment from the extremely hostile Venusian environment. The conditions withstood by the diamond window included "a 4-month journey through the cold and vacuum of space, entry decelerations to 565*g*, searing heat (the Venusian surface is red hot), crushing pressure (100 times that of the earth's atmosphere), and a highly corrosive atmosphere containing sulfuric acid and other aggressive gases" [13].

Representative measurements are shown in Fig. 5-15 for the monochromatic reflectivity ρ_λ for several surfaces. These surfaces are essentially opaque, such that α_λ is equal to $1 - \rho_\lambda$. Notice that ρ_λ of the pure aluminum surface is quite high for all wavelengths. But ρ_λ for aluminum surfaces with certain coatings such as lead sulfide fall to much smaller values for wavelengths below 3 μm. This selective characteristic of certain types of metallic surfaces is very important in solar applications.

In order to express the absorptivity α in terms of α_λ (or ϵ_λ) we first introduce the *monochromatic irradiation* G_λ, which is defined as the irradiation flux per unit wavelength. The total irradiation G and the monochromatic irradiation G_λ are themselves related by

$$G = \int_0^\infty G_\lambda \, d\lambda \tag{5-21}$$

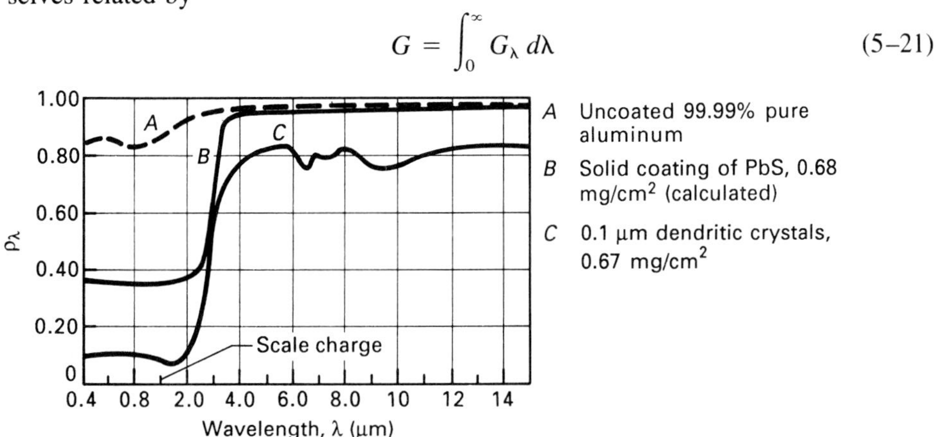

FIGURE 5-15 Monochromatic reflectivity ρ_λ for lead sulfide coatings on aluminum substrates (Williams et al. [14]).

The absorptivity is now formally defined by the equation

$$\alpha = \frac{\int_0^\infty \alpha_\lambda G_\lambda \, d\lambda}{\int_0^\infty G_\lambda \, d\lambda} \tag{5–22}$$

For the many practical situations in which $\alpha_\lambda = \epsilon_\lambda$, this expression becomes

$$\alpha = \frac{\int_0^\infty \epsilon_\lambda G_\lambda \, d\lambda}{\int_0^\infty G_\lambda \, d\lambda} \tag{5–23}$$

This equation provides a basis for the evaluation of α in terms of ϵ for several situations. First, for approximate graybody conditions, ϵ_λ is essentially independent of wavelength and equal to ϵ, such that Eq. (5–23) reduces to the *graybody* α *approximation*

$$\alpha = \epsilon \tag{5–24}$$

where both α and ϵ are evaluated at T_s. Although very few materials exist for which ϵ_λ is nearly constant for all wavelengths, some substances exhibit approximate graybody characteristics over significant parts of the spectrum. The key issue when using this graybody α approximation is that ϵ_λ (and α_λ) be essentially uniform in the wavelength range where there are appreciable amounts of both emitted and incident radiation. For example, referring to Fig. 5–8, this graybody α approximation could be used for a polished copper surface with T_s and T_R both less than approximately 1000 K, for which only 10% of the radiant energy is in the wavelength band 0 to 2 μm, but should not be used for situations in which T_R is much greater than 1000 K.

For nongraybody surfaces, the use of Eq. (5–23) in evaluating α requires that the monochromatic irradiation G_λ be specified. For situations involving blackbody or graybody radiation sources, G_λ in Eq. (5–23) can be replaced by $E_{b\lambda}$; that is,

$$\alpha = \frac{\int_0^\infty \epsilon_\lambda(T_s)\, E_{b\lambda}(T_R) \, d\lambda}{\int_0^\infty E_{b\lambda}(T_R) \, d\lambda} \tag{5–25}$$

By comparing Eqs. (5–10) and (5–25), we see that the absorptivity is approximately equal to the emissivity for an approximate blackbody or graybody source with T_R equal to T_s, which represents the limiting form of Kirchhoff's law. However, it should be noted that the rate of radiation heat transfer is zero for such isothermal conditions. It follows that the most practical consequence of the limiting form of Kirchhoff's law is its application to cases in which T_s and T_R are *not* equal. The first and most obvious extension of this law applies to systems in which the difference between T_s and T_R is small. For such cases, we simply utilize the *isothermal* α *approximation*

$$\alpha = \epsilon(T_s) \tag{5–26}$$

which is equivalent to the graybody α approximation given by Eq. (5–24). However, because of the strong dependence of the surface irradiation properties on the characteristics of the incident thermal radiation for nongraybody surfaces, the following *nonisothermal α approximation* is generally preferred:

$$\alpha = \epsilon(T_R) \tag{5–27}$$

This equation is obtained from Eq. (5–25) by merely setting $\epsilon_\lambda(T_s)$ equal to $\epsilon_\lambda(T_R)$. As we have already noted, this restriction on ϵ_λ is essentially satisfied by many metals and certain nonmetallic substances. [For thermal radiation from a blackbody or graybody source incident on a metallic surface with T_R low enough to exclude significant radiation in the near-infrared, visible, or ultraviolet ranges, some investigators prefer an α approximation of the form $\alpha = \epsilon(\sqrt{T_s T_R})$.]

For nongraybody surfaces for which $\epsilon_\lambda(T_s)$ is not approximately equal to $\epsilon_\lambda(T_R)$, Eq. (5–25) can be integrated numerically, provided that $\epsilon_\lambda(T_s)$ is known. This approach was used by Sieber [9] to obtain the absorptivities shown in Fig. 5–13.

To evaluate the graybody (or isothermal) and nonisothermal α approximations, we consider data for commerical-finish aluminum surfaces. Referring back to Fig. 5–9, we see that the aluminum surface exhibits nongraybody characteristics. A comparison of the data for ϵ shown in Fig. 5–3 with the values of α given in Fig. 5–13 reveals a very close agreement between the two. This result reinforces our confidence in the usefulness of the nonisothermal α approximation. On the other hand, the very fact that α for aluminum shown in Fig. 5–13 ranges from a value less than 0.1 for a source temperature T_R of 300 K to 0.3 for T_R equal to 6000 K alerts us to the limitation of the graybody or isothermal α approximation for this material. For that matter, except for applications involving small temperature differences, the use of the graybody α approximation would lead to serious error in the analysis of radiation from surfaces 1 through 8 listed in Fig. 5–13.

For situations in which most of the irradiation upon a surface originates from a single approximate blackbody or graybody source, the nonisothermal approximation is fairly easy to employ. For example, in the heating of a small metal ball in a furnace, α can readily be estimated by using Eq. (5–27). However, in applications involving multiple nongraybody surfaces and complex reflection patterns, the nonisothermal α approximation is less reliable and is not so easily administered, since the irradiation upon a surface originates from various sources. Consequently, the simple graybody α approximation is often utilized, at least as a first cut.

Directional Effects

Finally, mention should be made of the directional characteristics of irradiation surface properties. Two limiting types of reflection are illustrated in Fig. 5–16. The reflection is *diffuse* if the intensity of the reflected thermal radiation is constant for all angles of irradiation and reflection. On the other hand, if the angle of reflection is equal to the angle of incidence, the reflection is *specular*. Although real surfaces have neither diffuse nor specular irradiation surface properties, many common surfaces can be

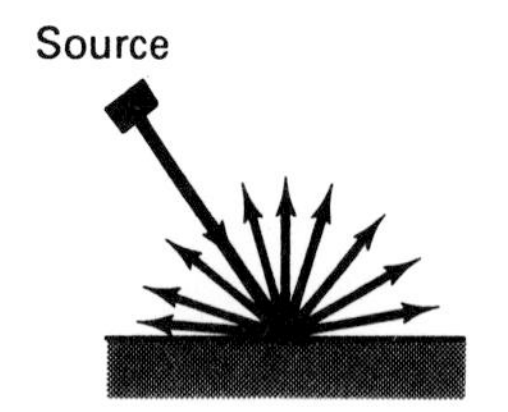

(a) Diffuse reflection.

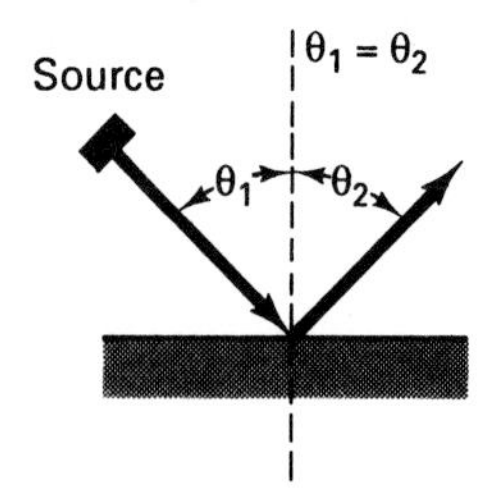

(b) Specular reflection.

FIGURE 5–16
Types of reflection.

placed in one or the other of these categories for purpose of design. For example, industrially rough surfaces essentially possess diffuse properties, with polished and smooth surfaces exhibiting near-specular characteristics.

EXAMPLE 5–6

Assuming that a glass plate transmits 90% of the incident thermal radiation in the wavelength range from 0.29 to 2.70 μm (see Fig. 5–14), is opaque outside this range, and reflects 5% for all wavelengths, determine the total transmissivity τ of the glass for solar radiation.

Solution

Objective Determine τ for this glass plate.

Schematic Transmission of solar radiation through a glass plate.

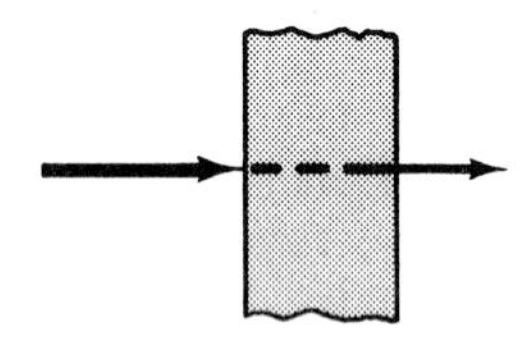

Properties

$$\tau_\lambda = 0.9 \qquad \text{for } 0.29\ \mu\text{m} < \lambda < 2.7\ \mu\text{m}$$
$$\tau_\lambda = 0 \qquad \text{for } \lambda \leqslant 0.29\ \mu\text{m},\ \lambda \geqslant 2.7\ \mu\text{m}$$

Assumptions/Conditions

solar irradiation is equivalent to blackbody emission at 5800 K

Analysis Assuming that the sun behaves as a blackbody radiator at 5800 K, we utilize Table A–H–2 to calculate the fraction of solar radiation that falls in the wavelength band from 0.29 to 2.70 μm. Using Table A–H–2, we obtain

$$\frac{E_{b,0\to\lambda_1}}{E_b} = 0.0254$$

for $\lambda_1 T_R = 0.29\ \mu\text{m}\ (5800\ \text{K}) = 1680\ \mu\text{m K}$, and

$$\frac{E_{b,0\rightarrow\lambda_2}}{E_b} = 0.972$$

for $\lambda_2 T_R = 2.70\ \mu\text{m}\ (5800\ \text{K}) = 15{,}700\ \mu\text{m K}$. Thus, the fraction of the solar radiation falling in the wavelength band for which τ_λ is 0.9 is

$$\frac{E_{b,\lambda_1\rightarrow\lambda_2}}{E_b} = 0.972 - 0.0254 = 0.947$$

It follows that the total transmissivity τ is

$$\tau = 0.947(0.9) = 0.852$$

To calculate the total absorptivity, we write

$$\alpha = 1 - \rho - \tau = 1 - 0.05 - 0.852 = 0.098$$

5–3–3 Thermal Radiation Properties of Gases

We are reminded that a vacuum provides the ideal medium for the transfer of thermal radiation from one surface to another. Thermal radiation passes through such evacuated spaces at the speed of light. Of course, in most practical situations at least some fluid, often in the form of a gas, lies in the path of the thermal radiation. Elementary gases such as oxygen (O_2), hydrogen (H_2), nitrogen (N_2), and dry air, which have symmetrical molecular structures, are essentially transparent to thermal radiation at low to moderate temperatures. Hence, the presence of such *nonparticipating gases* can generally be ignored. But, other polyatomic gases, such as water vapor (H_2O), carbon dioxide (CO_2), sulfur dioxide (SO_2), and various hydrocarbons absorb and emit significant amounts of thermal radiation. Gases such as these are known as *participating gases*. For example, water vapor contained at 1 atm pressure and 100°C between two plates 1 m apart would emit as much as 55% of the energy that would be emitted by a blackbody at 100°C with the same surface area as the plates. Carbon dioxide at the same temperature and pressure would emit about 20%. But carbon monoxide, which is diatomic, would only emit about 3% of the energy that would be emitted by an ideal radiator. Hence, the presence of participating gases between thermal radiating surfaces can have very important effects.

Thermal radiation emission and absorption characteristics of participating gases are generally much more complicated than for opaque solids. Whereas the emission and absorption of thermal radiation for opaque materials are surface phenomena, the thickness, shape, surface area, pressure, and temperature distribution can all affect thermal radiation in gases. Representative measurements for α_λ are shown in Fig. 5–17 for carbon dioxide in terms of the wave number $1/\lambda$. The absorbing spectrum of this important gas is seen to consist of distinct narrow bands. These absorbing and emitting patterns are typical of participating gases.

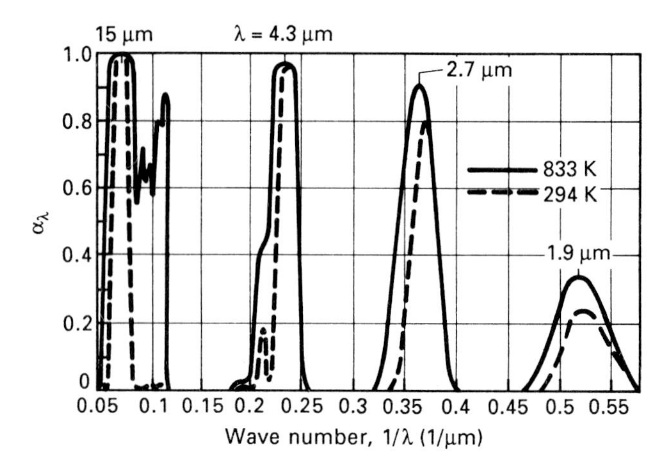

FIGURE 5–17 Monochromatic absorptivity α_λ for carbon dioxide (Edwards [15]).

Concerning the more practical information pertaining to the total properties, which account for all wavelengths and all directions in which thermal radiation passes, Hottel and Egbert [16] have developed the charts shown in Fig. 5–18(a) and (b) for water vapor and carbon dioxide total emissivities ϵ. Note that ϵ increases with increasing *mean beam length* L_e and partial pressure. The mean beam length, which accounts for all possible directions the thermal radiation may take, is listed in Table 5–2 for several standard systems. For situations in which L_e has not been evaluated, it is generally approximated by $L_e = 3.6V/A_s$. The data in Fig. 5–18(a) and (b) were obtained for a total pressure of 1 atm. Correction charts are available in Hottel [17] for systems under other total pressures. For systems involving combustion, both water vapor and carbon dioxide are present. For such situations, the total gas emissivity is simply equal to the sum of the emissivities for each component, minus a small correction factor that accounts for the mutual emission that takes place between the two gases. Corrections that account for this mutual emission factor are presented in Hottel [17]. This small difference can generally be neglected as a first approximation.

In regard to the total absorptivity α of participating gases, Hottel [17] has developed approximate graybody correlations for water vapor and carbon dioxide which take the forms

$$\alpha_w = \epsilon_w(T_s)\left(\frac{T_m}{T_s}\right)^{0.45} \qquad \alpha_c = \epsilon_c(T_s)\left(\frac{T_m}{T_s}\right)^{0.65} \tag{5–28a,b}$$

where T_m is the mean temperature of the gas and T_s the surface temperature. ϵ_w and ϵ_c are evaluated from Fig. 5–18 with the parameters P_wL_e and P_cL_e replaced by $P_wL_eT_s/T_m$ and $P_cL_eT_s/T_m$. For a mixture of water vapor and carbon dioxide, the absorptivity of the mixture is simply equal to the sum of α_w and α_c, minus a small correction.

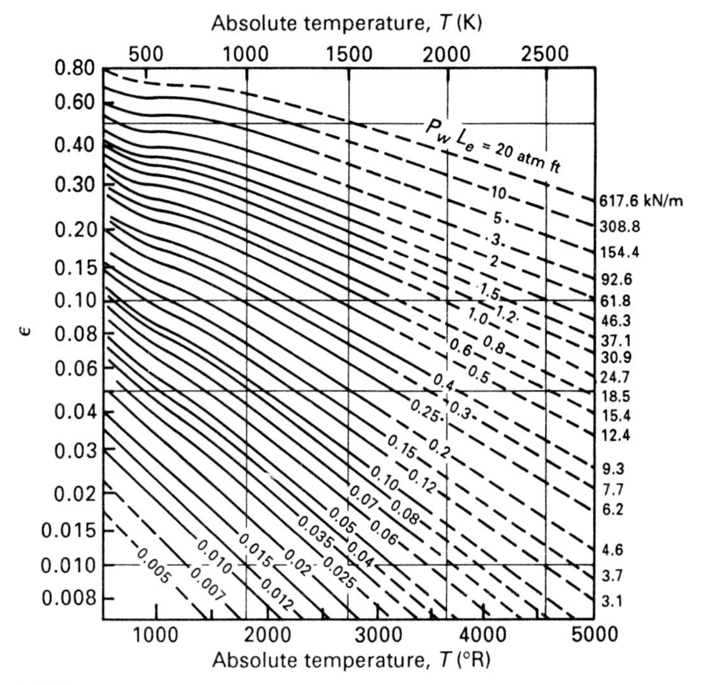

(a) Water vapor.

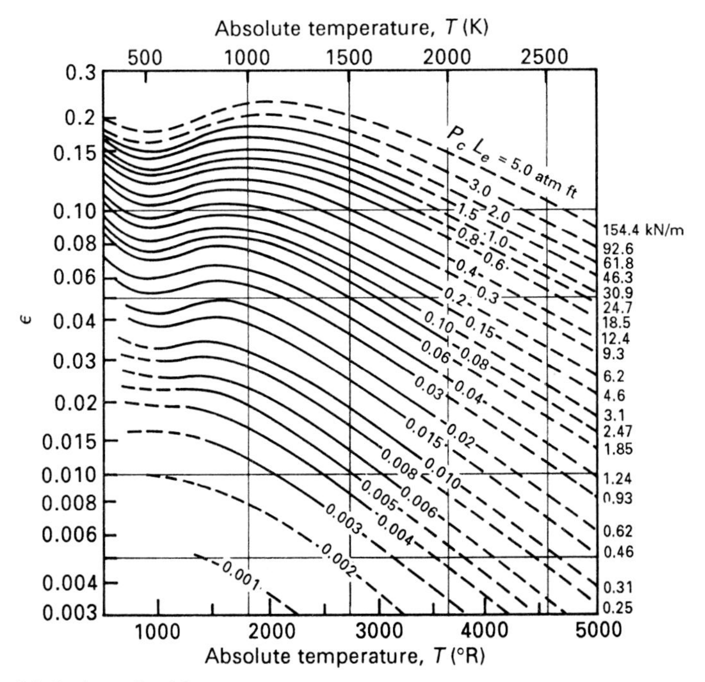

(b) Carbon dioxide.

FIGURE 5-18 Emissivities of gases at 1 atm total pressure (Hottel and Egbert [16]).

TABLE 5–2 Mean beam length L_e for radiation from entire gas volume

Gas volume	*Characteristic dimension*	L_e
Volume between two infinite planes	Separation distance L	$1.8L$
Cube; radiation to any face	Edge L	$0.60L$
Sphere	Diameter D	$0.65D$
Circular cylinder with $L = D$; radiation to entire surface	Diameter D	$0.60D$
Circular cylinder, with semi-infinite length; radiation to entire base	Diameter D	$0.65D$

Source: From Hottel [17] and Eckert and Drake [18].

It should be mentioned that in the high-temperature process of combustion which produces nonluminous products such as H_2O and CO_2, clouds of carbon particles radiate intensely at short wavelengths in the visible-light region of the electromagnetic spectrum. This visible thermal radiation which is emitted during combustion is what is referred to as the *flame*. The total emissivities of luminous flames range from values of the order of 0.2 for gaseous hydrocarbon fuels to almost unity for fuels such as oil which are burned under conditions of large carbon/hydrogen ratios. In engineering applications, the emission from luminous flames generally must be determined experimentally.

EXAMPLE 5–7

The walls of a cubical furnace 1 m on a side are maintained at 500 K. The products of combustion at 1 atm and 1500 K consist of 20% CO_2, 15% H_2O, and 65% N_2 (on molar basis). Assuming that the walls are essentially black, estimate the gas emissivity ϵ_m and absorptivity α_m.

Solution

Objective Approximate ϵ_m and α_m for the products of combustion within the furnace.

Schematic Cubical furnace with blackbody surfaces.

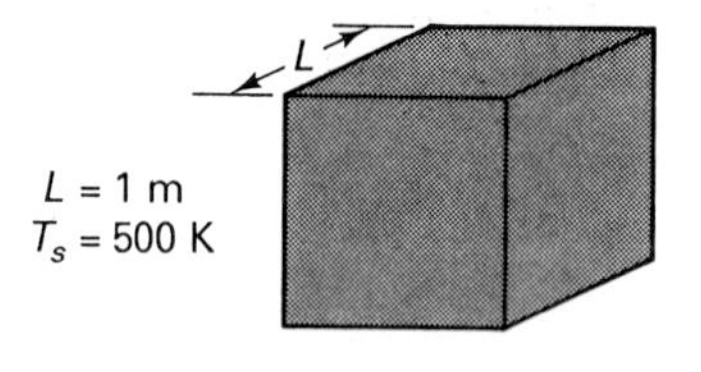

Mole fraction
of components
CO_2 20%
H_2O 15%
N_2 65%

$T_m = 1500$ K
$P_m = 1$ atm

Assumptions/Conditions

blackbody surface irradiation
enclosure contains participating gases H_2O and CO_2

Properties Walls: $\epsilon = 1$. The properties of the gases are considered in the analysis.

Analysis Assuming that the nitrogen is essentially nonparticipating, we focus our attention on the contributions of the water (w) and carbon-dioxide (c) to the thermal radiation properties of the mixture. The ideal gas law ($PV = mRT = n\bar{R}T$) indicates that the partial pressure of an ideal gas in a mixture is equal to the product of the total pressure of the mixture and the mole fraction of the gas. It follows that

$$P_w = 0.15(1 \text{ atm}) = 0.15 \text{ atm}$$
$$P_c = 0.2(1 \text{ atm}) = 0.2 \text{ atm}$$

Referring to Table 5–2, the mean equivalent beam length L_e is given by

$$L_e = 0.6L = 0.6(1 \text{ m}) = 0.6 \text{ m}$$

Thus, the pressure/length parameters are

$$P_w L_e = 0.15 \text{ atm } (0.6 \text{ m}) = 0.09 \text{ atm m} = 0.295 \text{ atm ft}$$
$$P_c L_e = 0.2 \text{ atm } (0.6 \text{ m}) = 0.12 \text{ atm m} = 0.393 \text{ atm ft}$$

Making use of Fig. 5–18, the emissivities ϵ_w and ϵ_c are approximated by

$$\epsilon_w = 0.08 \qquad \epsilon_c = 0.094$$

for a mean gas temperature of 1500 K. Thus, the total emissivity of the gas is approximated by

$$\epsilon_m = \epsilon_w + \epsilon_c = 0.08 + 0.094 = 0.174$$

By reference to Hottel [17], we find that the correction for ϵ_m, which accounts for the mutual emission of the two participating gases, is of the order of -10%. Therefore, as a first approximation, we are safe in assuming $\epsilon_m \simeq 0.174$.

To estimate the absorptivities, we calculate the modified pressure/length parameters.

$$P_w L_e \frac{T_s}{T_m} = 0.295 \text{ atm ft} \frac{500 \text{ K}}{1500 \text{ K}} = 0.0983 \text{ atm ft}$$

$$P_c L_e \frac{T_s}{T_m} = 0.393 \text{ atm ft} \frac{500 \text{ K}}{1500 \text{ K}} = 0.131 \text{ atm ft}$$

Evaluating the emissivities at T_s, we have

$$\epsilon_w(T_s) = 0.097 \qquad \epsilon_c(T_s) = 0.075$$

The absorptivities are now calculated by utilizing Eqs. (5–28a) and (5–28b).

$$\alpha_w = \epsilon_w(T_s)\left(\frac{T_m}{T_s}\right)^{0.45} = 0.097\left(\frac{1500\text{ K}}{500\text{ K}}\right)^{0.45} = 0.159$$

$$\alpha_c = \epsilon_c(T_s)\left(\frac{T_m}{T_s}\right)^{0.65} = 0.075\left(\frac{1500\text{ K}}{500\text{ K}}\right)^{0.65} = 0.153$$

Referring to Hottel [17], we find the correction for mutual absorption is less than 1%. It follows that the total absorptivity of the gas is about

$$\alpha_m = \alpha_w + \alpha_c = 0.159 + 0.153 = 0.312$$

5–4 RADIATION SHAPE FACTOR

As a final step toward our objective of developing practical predictions for radiation heat transfer, attention is turned to important geometric aspects of thermal radiation exchange between surfaces.

The *radiation shape factor* F_{s-R} introduced in Chap. 1 is defined as the fraction of thermal radiation leaving a diffuse surface A_s that passes through a nonparticipating medium to surface A_R. F_{s-R} is also commonly referred to as *view factor*, *configuration factor*, and *shape factor*. Because we will be dealing with systems involving more than one source, it is convenient to replace the subscript R by the source surface identification index j, where $j = 1, 2, 3, \ldots, N$. Thus, the radiation shape factor will generally be denoted by F_{s-j}, except for systems involving only two surfaces, for which case we will write F_{1-2}.

To illustrate, we consider the two simple systems shown in Fig. 5–19(a) and (b). The radiation shape factors F_{1-2} and F_{2-1} for the infinitely long parallel-plate system of Fig. 5–19(a) are both unity, because all the energy leaving either surface reaches the other surface. For the concentric-sphere arrangement shown in Fig. 5–19(b), F_{1-2} is also equal to unity. But, because only a fraction of the energy leaving the outer spherical surface A_2 reaches the inner surface A_1, F_{2-1} is less than unity.

We now turn our attention to fundamental principles that apply to radiation shape factors and design curves for several standard geometries.

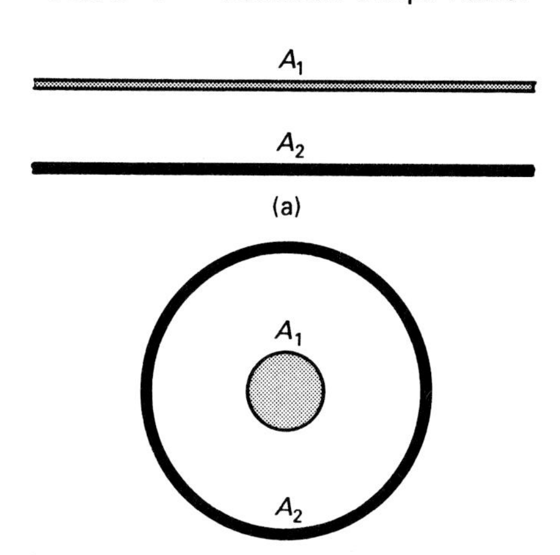

(a) Parallel plates; $F_{1-2} = 1$, $F_{2-1} = 1$.

(b) Concentric spheres; $F_{1-2} = 1$, $F_{2-1} < 1$.

FIGURE 5–19
Radiation systems.

5–4–1 Law of Reciprocity

The general relationship between F_{s-j} and F_{j-s} is given by the *reciprocity law*,

$$A_s F_{s-j} = A_j F_{j-s} \tag{5–29}$$

This relationship is developed in Sparrow and Cess [5] and Siegel and Howell [11] on the basis of geometric considerations. Based on this principle, we see that F_{2-1} for the spherical system shown in Fig. 5–19(b) is

$$F_{2-1} = \frac{A_1}{A_2} F_{1-2} = \left(\frac{r_1}{r_2}\right)^2 \tag{5–30}$$

Note that F_{2-1} approaches zero as r_1/r_2 decreases, and approaches unity as r_1 approaches r_2.

5–4–2 Summation Principles

Another important geometric concept pertains to the radiation shape factors from surface s to N surfaces forming an enclosure. This *first summation principle* is written as

$$\sum_{j=1}^{N} F_{s-j} = 1 \tag{5–31}$$

For example, for the three-surface enclosure shown in Fig. 5–20, we have

$$F_{1-2} + F_{1-3} = 1 \tag{5–32}$$

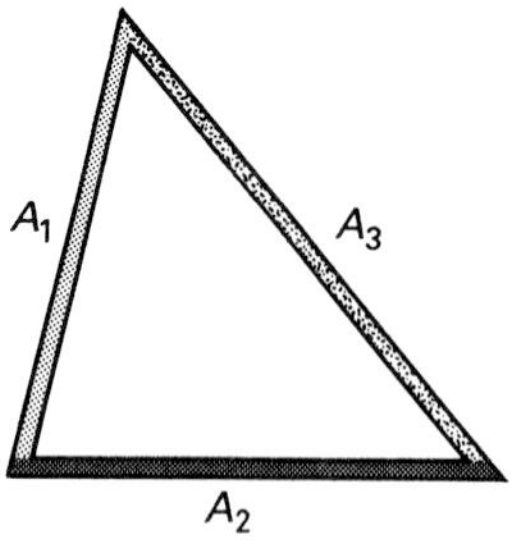

A_1 A_3 A_2

FIGURE 5–20
Trisurface enclosure.

This principle also requires that

$$F_{2-1} + F_{2-3} = 1 \qquad F_{3-1} + F_{3-2} = 1 \tag{5–33,34}$$

For systems involving concave surfaces, we must include the term F_{s-s}, which accounts for the fraction of thermal radiation leaving surface s that is directly incident upon itself. Thus, for the spherical system of Fig. 5–19(b), we write

$$F_{2-1} + F_{2-2} = 1 \tag{5–35}$$

A *second summation principle* states that the total radiation shape factor is equal to the sum of its parts. To illustrate, the radiation shape factor from surface A_1 to the combined surfaces of A_2 and A_3 in Fig. 5–20 is

$$F_{1-(2,3)} = F_{1-2} + F_{1-3} \tag{5–36}$$

By employing the law of reciprocity, this equation is put into the useful form

$$(A_2 + A_3)F_{(2,3)-1} = A_2F_{2-1} + A_3F_{3-1} \tag{5–37}$$

5–4–3 Design Curves

Relationships have been developed for radiation shape factors for a great many geometries. The theoretical approach to this task involves the use of the intensity concept for diffuse surfaces, as illustrated in Example 5–9.

Because the theoretical evaluation of radiation shape factors is generally quite involved, standard design curves are heavily relied upon in practice. Design equations for radiation shape factors are given in Tables 5–3 and 5–4 and Figs. 5–21, 5–22, and 5–23 for several arrangements that are commonly encountered. More comprehensive listings of radiation shape factors are given in references 11, 17, 20–22. As shown in Example 5–8, information on radiation shape factors for standard geometries such as those given in Figs. 5–21 and 5–22 can sometimes be extended to other geometrical arrangements of practical interest by utilizing the reciprocity and summation principles.

TABLE 5–3 Radiation shape factors for two-dimensional geometries

Geometry	*Relation*
Parallel Plates with Midlines Connected by Perpendicular $W_1 = w_1/L$, $W_2 = w_2/L$ 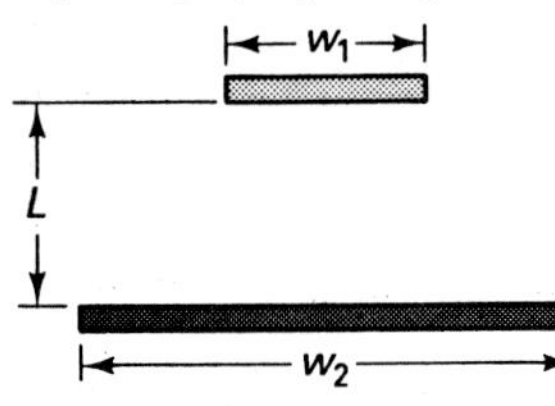	$F_{1-2} = \dfrac{[(W_1 + W_2)^2 + 4]^{1/2} - [(W_2 - W_1)^2 + 4]^{1/2}}{2W_1}$
Inclined Parallel Plates with Equal Width and a Common Edge 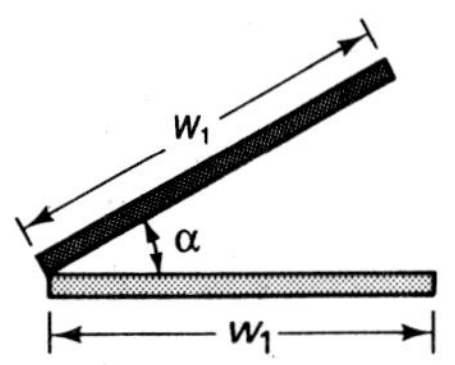	$F_{1-2} = 1 - \sin\left(\dfrac{\alpha}{2}\right)$
Perpendicular Plates with a Common Edge 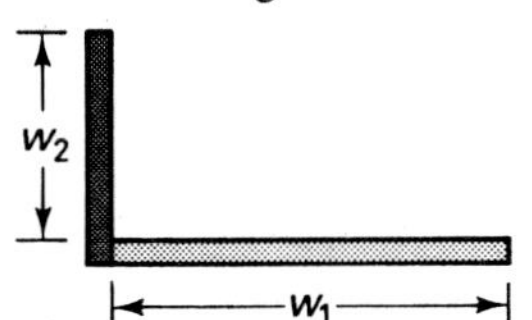	$F_{1-2} = \dfrac{1 + (w_2/w_1) - [1 + (w_2/w_1)^2]^{1/2}}{2}$
Three-Sided Enclosure 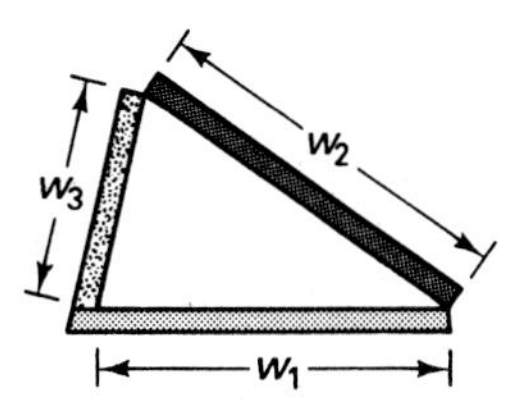	$F_{1-2} = \dfrac{w_1 + w_2 - w_3}{2w_1}$
Parallel Cylinders of Different Radius $R = r_2/r_1$, $S = s/r_1$, $C = 1 + R + S$ 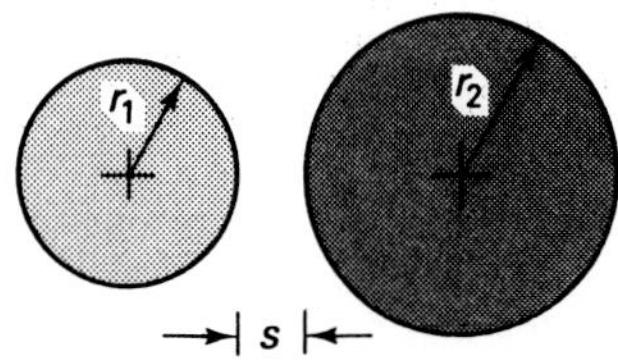	$F_{1-2} = \dfrac{1}{2\pi}\Big\{\pi + [C^2 - (R + 1)^2]^{1/2} - [C^2 - (R - 1)^2]^{1/2} + (R - 1)\cos^{-1}\left(\dfrac{R}{C} - \dfrac{1}{C}\right) - (R + 1)\cos^{-1}\left(\dfrac{R}{C} + \dfrac{1}{C}\right)\Big\}$

Sources: Howell [21] and Hamilton and Morgan [22].

TABLE 5–4 Radiation shape factors for three-dimensional geometries

Geometry	*Relation*
Aligned Parallel Rectangles $\bar{X} = X/Z$, $\bar{Y} = Y/Z$ (See Fig. 5–21) 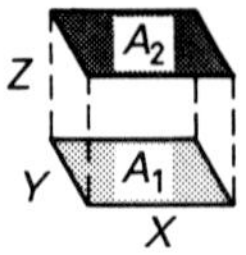	$F_{1-2} = \frac{2}{\pi \bar{X}\bar{Y}}\left\{\ln\left[\frac{(1+\bar{X}^2)(1+\bar{Y}^2)}{1+\bar{X}^2+\bar{Y}^2}\right]^{1/2} + \bar{X}(1+\bar{Y}^2)^{1/2}\tan^{-1}\left[\frac{\bar{X}}{(1+\bar{Y}^2)^{1/2}}\right] + \bar{Y}(1+\bar{X}^2)^{1/2}\tan^{-1}\left[\frac{\bar{Y}}{(1+\bar{X}^2)^{1/2}}\right] - \bar{X}\tan^{-1}\bar{X} - \bar{Y}\tan^{-1}\bar{Y}\right\}$
Perpendicular Rectangles with a Common Edge $H = Z/X$, $W = Y/X$ (See Fig. 5–22) 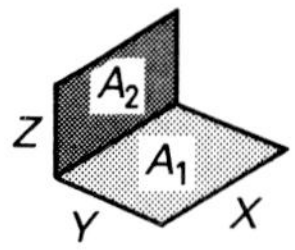	$F_{1-2} = \frac{1}{\pi W}\left(W\tan^{-1}\frac{1}{W} + H\tan^{-1}\frac{1}{H} - (H^2+W^2)^{1/2}\tan^{-1}\frac{1}{(H^2+W^2)^{1/2}} + \frac{1}{4}\ln\left\{\frac{(1+W^2)(1+H^2)}{1+W^2+H^2}\left[\frac{W^2(1+W^2+H^2)}{(1+W^2)(W^2+H^2)}\right]^{W^2} \times \left[\frac{H^2(1+H^2+W^2)}{(1+H^2)(H^2+W^2)}\right]^{H^2}\right\}\right)$
Coaxial Parallel Disks $R_1 = r_1/L$, $R_2 = r_2/L$ $S = 1 + (1+R_2^2)/R_1^2$ (See Fig. 5–23) 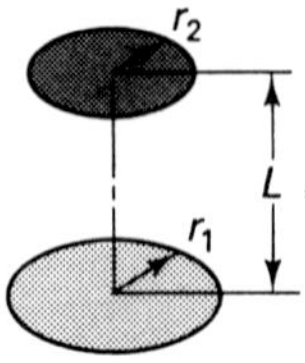	$F_{1-2} = \frac{1}{2}\{S - [S^2 - 4(r_2/r_1)^2]^{1/2}\}$

Sources: Mackey et al. [19], Howell [21], and Hamilton and Morgan [22].

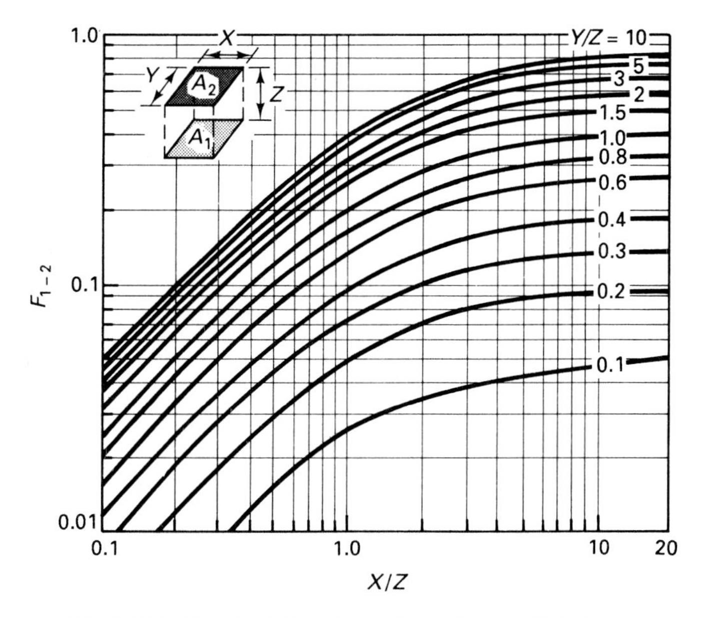

FIGURE 5-21 Radiation shape factor for parallel plates.

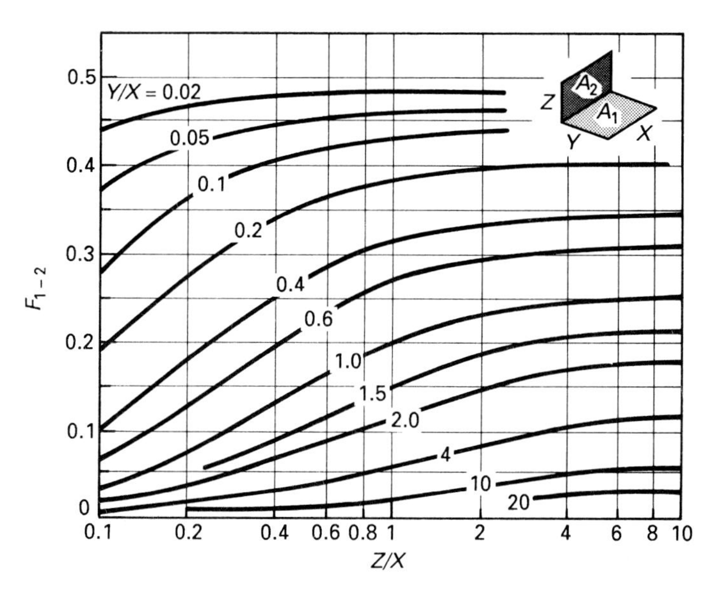

FIGURE 5-22 Radiation shape factor for perpendicular rectangles with a common edge.

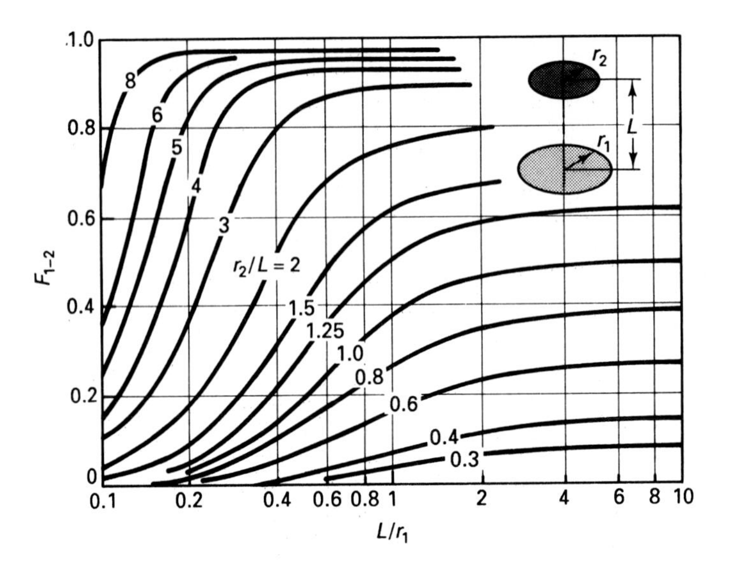

FIGURE 5–23 Radiation shape factor for coaxial parallel disks.

EXAMPLE 5–8

Determine the radiation shape factor for surfaces A_1 and A_2 shown in Fig. E5–8.

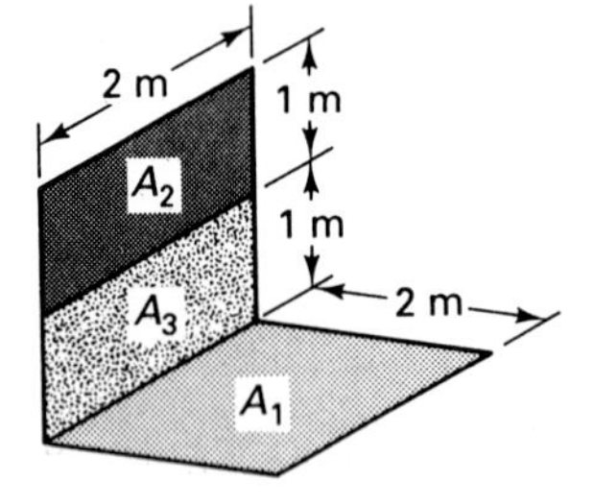

FIGURE E5–8
Trisurface system.

Solution

Objective Determine F_{1-2} and F_{2-1}.

Analysis According to the second summation principle, we write

$$F_{1-(2,3)} = F_{1-2} + F_{1-3}$$

or

$$F_{1-2} = F_{1-(2,3)} - F_{1-3}$$

Referring to Fig. 5–22, $F_{1-3} = 0.15$ and $F_{1-(2,3)} = 0.2$, such that

$$F_{1-2} = 0.2 - 0.15 = 0.05$$

Using the principle of reciprocity, F_{2-1} becomes

$$F_{2-1} = \frac{A_1}{A_2} F_{1-2} = \frac{2}{1}(0.05) = 0.1$$

EXAMPLE 5–9

Develop an expression for the radiation shape factor from the differential element dA_1 to the finite disk A_2 shown in Fig. E5–9.

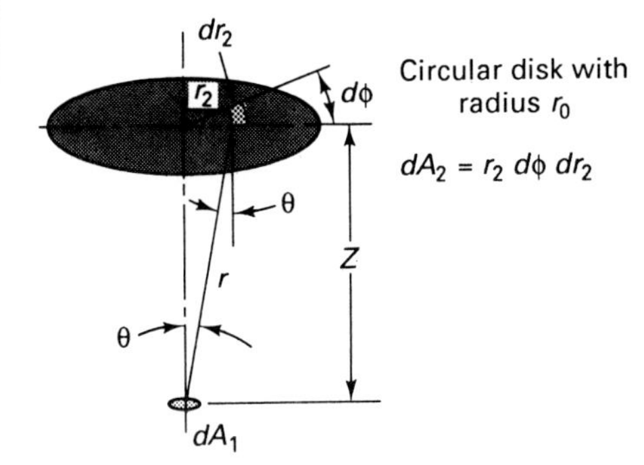

FIGURE E5–9
Parallel circular disks.

Solution

Objective Determine $F_{dA_1-A_2}$.

Assumptions/Conditions

diffuse surfaces

Analysis Referring to Fig. E5–9, the total rate of thermal radiation diffusely emitted by surface element dA_1 is $E_1\, dA_1$. To determine the portion of this energy that reaches surface A_2, we return to the definition of intensity given by Eq. (5–14),

$$I_1 = \frac{dE_1}{\cos\theta\, d\omega}$$

or

$$dE_1 = I_1 \cos\theta\, d\omega$$

where θ is the angle between the normal to surface dA_1 and the line r drawn between the area elements dA_1 and dA_2. The solid angle $d\omega$ is given by Eq. (5–13),

$$d\omega = \frac{dA_r}{r^2}$$

where dA_r is the differential area normal to the line r; that is,

$$dA_r = dA_2 \cos\theta$$

Thus, the thermal radiation flux emitted from dA_1 that reaches dA_2 is

$$dE_1 = I_1 \cos^2 \theta \frac{dA_2}{r^2}$$

where $I_1 = E_1/\pi$ for diffuse conditions (see Example 5–3).

The radiation shape factor $F_{dA_1-dA_2}$ is simply equal to the ratio dE_1/E_1; that is,

$$F_{dA_1-dA_2} = \frac{1}{\pi} \cos^2 \theta \frac{dA_2}{r^2}$$

To obtain $F_{dA_1-A_2}$, we must integrate over the entire surface A_2.

$$F_{dA_1-A_2} = \frac{1}{\pi} \int_{A_2} \cos^2 \theta \frac{dA_2}{r^2}$$

Noting that

$$dA_2 = r_2 \, d\phi \, dr_2$$

we obtain

$$F_{dA_1-A_2} = \frac{1}{\pi} \int_0^{r_0} \int_0^{2\pi} \cos^2 \theta \frac{r_2 \, d\phi \, dr_2}{r^2} \tag{a}$$

The quantities r and θ are expressed in terms of r_2 by

$$r^2 = Z^2 + r_2^2 \qquad \cos \theta = \frac{Z}{(Z^2 + r_2^2)^{1/2}}$$

Thus, Eq. (a) takes the form

$$F_{dA_1-A_2} = \frac{1}{\pi} \int_0^{r_0} \int_0^{2\pi} \frac{Z^2 \, r_2 \, d\phi \, dr_2}{(Z^2 + r_2^2)^2} = 2Z^2 \int_0^{r_0} \frac{r_2 \, dr_2}{(Z^2 + r_2^2)^2}$$

Setting $Z^2 + r_2^2 = \xi$, we continue the integration process,

$$F_{dA_1-A_2} = Z^2 \int_{Z^2}^{Z^2+r_0^2} \frac{d\xi}{\xi^2} = -\frac{Z^2}{\xi}\Bigg|_{Z^2}^{Z^2+r_0^2}$$

$$= Z^2\left(\frac{1}{Z^2} - \frac{1}{Z^2 + r_0^2}\right) = \frac{r_0^2}{Z^2 + r_0^2} \tag{b}$$

Notice that $F_{dA_1-A_2}$ appropriately approaches unity as the radius r_0 of the disk becomes large.

5–5 PRACTICAL THERMAL ANALYSIS OF RADIATION HEAT TRANSFER

Now that we have a basic understanding of the physical mechanism, properties, and geometric factors pertaining to the thermal radiation phenomenon, we are in a position to develop predictions for the rate of heat transfer (i.e., *net* rate of exchange of thermal radiation) that occurs in diffuse thermal radiating systems. Therefore, we turn our attention to the practical analysis of radiation heat transfer.

To illustrate the perspective that we will be taking, consider the radiating body in Fig. 5–24. The thermal radiation heat transfer q_R from surface A_s to other surrounding surfaces will be analyzed in sections to follow for cases in which the surface temperature T_s and thermal radiation properties are uniform over A_s. But we must recognize that the rate of thermal radiation heat transfer q_R from surface A_s must be in balance with changes in the rate of storage and the rates of energy transfer into and out of the body through other surfaces and by other means. That is, the energy transfer associated with this body must satisfy the first law of thermodynamics,

$$\Sigma \dot{E}_i = q_R + \frac{\Delta E_s}{\Delta t} \tag{5–38}$$

where $\Sigma \dot{E}_i$ is the net rate of energy transfer into the body, not including q_R. For steady-state conditions, $\Delta E_s/\Delta t$ is zero with T_s being independent of time, such that Eq. (5–38) reduces to

$$\Sigma \dot{E}_i = q_R \tag{5–39}$$

This equation simply indicates that the rate of thermal radiation heat transfer from surface A_s with steady surface temperature T_s must be replaced from an outside source, such as by the electrical generation of power within the body. If, on the other hand, a net rate of thermal radiation is received by surface A_s such that q_R is negative, this same rate of energy must be transferred from the body into a heat sink if steady-state conditions are to be maintained. For example, energy can be taken out of the back surface of the body by radiation or convection.

For unsteady conditions, T_s is time-dependent and q_R must satisfy Eq. (5–38). That is, q_R must be in balance with the rate of energy brought into a body $\Sigma \dot{E}_i$ and the rate of change in energy stored within the body $\Delta E_s/\Delta t$. In this regard, for bodies with negligible thermal resistance to conduction (i.e., small thermal radiation Biot number Bi_R), the temperature throughout the body is essentially uniform. For such

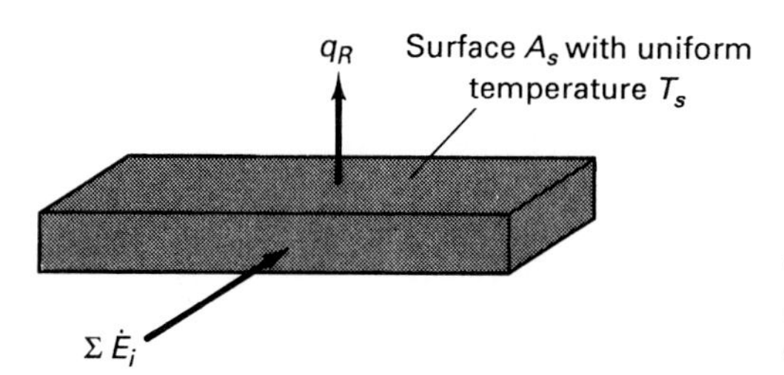

FIGURE 5–24
Thermal radiating body; $\Sigma \dot{E}_i$ represents the net rate of energy transfer into body, excluding radiation heat transfer q_R.

bodies, a lumped approximation can be made for $\Delta E_s/\Delta t$, with Eq. (5–38) taking the simpler form (for constant properties)

$$\Sigma \dot{E}_i = q_R + mc_v \frac{dT_s}{dt} \tag{5–40}$$

Keeping in mind that the energy transfer associated with the body of each radiating surface must satisfy the first law of thermodynamics, we now move on to the development of practical thermal analyses for the radiation heat transfer between two or more diffuse thermal radiating surfaces with uniform properties and uniform heating. To begin with, we focus on radiating surfaces in which conduction and convection are not significant. In addition to analyzing ideal blackbody and diffuse nonblackbody thermal radiating systems under steady-state conditions with a non-participating medium, consideration is given to systems involving unsteady conditions, participating medium, combined modes, and solar radiation. Approaches to analyzing more complex systems involving surfaces with nondiffuse characteristics, nonuniform properties, and nonuniform heating are introduced in references 5 and 11.

To simplify our notation, the subscript R for radiation heat transfer will be omitted throughout the remainder of this chapter, except for cases involving combined modes of heat transfer. Accordingly, the rate of radiation heat transfer from surface A_s to a second surface A_j will be designated by q_{s-j}. In addition, the total radiation-heat-transfer rate from surface A_s to N surfaces of an enclosure will be represented by q_s.

5–5–1 Blackbody Thermal Radiation

The analysis of radiation heat transfer in systems involving ideal blackbody surfaces is quite straightforward because the energy transfer is direct, with no reflections. In this section we will consider basic two-surface and multisurface blackbody systems. The analysis of these ideal systems provides a foundation for the practical analysis of real nonblackbody surfaces, which is considered in Sec. 5–5–3. In addition, the results of this analysis provide us with a new viewpoint concerning the analogy between the flow of electric current and radiation heat transfer.

Bisurface Systems

Representative blackbody systems consisting of two surfaces are shown in Fig. 5–25(a) and (b). Whereas no thermal radiation enters or leaves the closed system of Fig. 5–25(a), thermal radiation is propagated through the open spaces or "windows" of the radiatively open system of Fig. 5–25(b).

To determine the radiation heat transfer rate between two blackbody surfaces A_1 and A_2 in radiatively open or closed systems, we must recognize that the rate of thermal radiation leaving A_1 that reaches and is absorbed by A_2 is $A_1F_{1-2}E_{b1}$. Similarly, the rate of thermal radiation emitted by A_2 that is absorbed by A_1 is $A_2F_{2-1}E_{b2}$. It

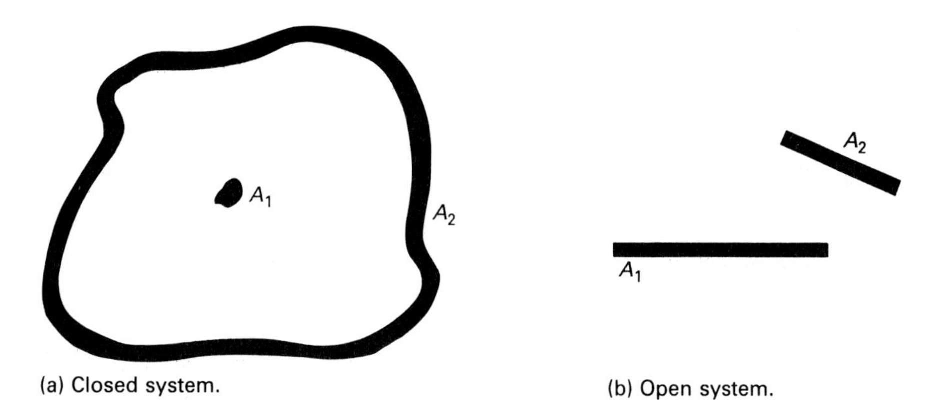

(a) Closed system. (b) Open system.

FIGURE 5–25 Bisurface thermal radiation blackbody systems.

follows that the rate of radiation heat transfer q_{1-2} from A_1 to A_2 in bisurface systems is equal to the net exchange of thermal radiation between A_1 and A_2; that is,

$$q_{1-2} = A_1F_{1-2}E_{b1} - A_2F_{2-1}E_{b2} \tag{5–41}$$

Utilizing the principle of reciprocity, which states that A_1F_{1-2} and A_2F_{2-1} are equal, this equation can be written as

$$q_{1-2} = A_1F_{1-2}(E_{b1} - E_{b2}) \tag{5–42}$$

Finally, introducing the Stefan-Boltzmann law, Eq. (5–2), we have

$$q_{1-2} = A_1F_{1-2}\sigma(T_1^4 - T_2^4) \tag{5–43}$$

Multisurface Systems

To determine the rate of radiation heat transfer from blackbody surface area A_s to N blackbody surfaces in radiatively closed or open systems such as those shown in Fig. 5–26(a) and (b), we must account for the exchange of thermal radiation between surface A_s and each of the N surfaces. The net rate of thermal radiation exchange between A_s and one of the surfaces A_j is simply

$$q_{s-j} = A_sF_{s-j}E_{bs} - A_jF_{j-s}E_{bj} = A_sF_{s-j}(E_{bs} - E_{bj}) \tag{5–44}$$

It follows that the total rate of radiation heat transfer q_s from A_s to all N blackbody surfaces is given by

$$q_s = \sum_{j=1}^{N} A_sF_{s-j}(E_{bs} - E_{bj}) \tag{5–45}$$

With E_{bs} and E_{bj} specified by the Stefan-Boltzmann law, we obtain

$$q_s = \sum_{j=1}^{N} A_sF_{s-j}\sigma(T_s^4 - T_j^4) \tag{5–46}$$

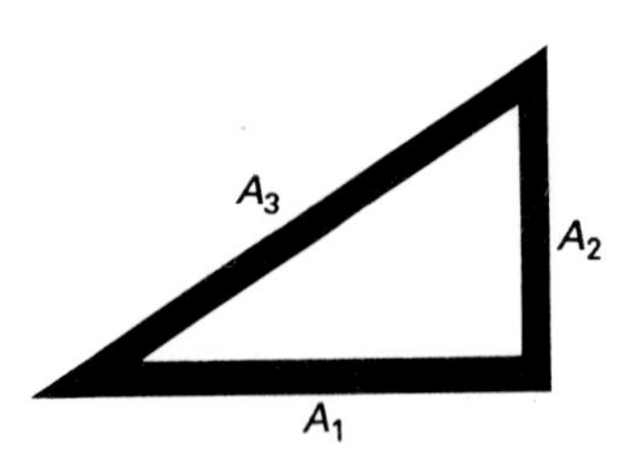

(a) Closed system.

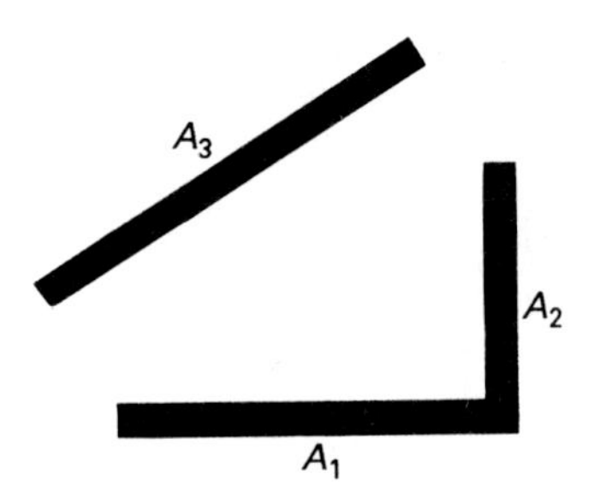

(b) Open system.

FIGURE 5–26 Trisurface blackbody thermal-radiation systems.

EXAMPLE 5–10

Demonstrate the validity of the principle of reciprocity by analyzing the radiation heat transfer between two blackbodies A_1 and A_2 which are at the same temperature.

Solution

Objective Deduce the principle of reciprocity, $A_1F_{1-2} = A_2F_{2-1}$.

Schematic Thermal radiation between two blackbodies.

Assumptions/Conditions

steady-state
blackbody thermal radiation
space between plates is evacuated

Properties $\epsilon_1 = 1, \epsilon_2 = 1$.

Analysis The rate of radiation heat transfer between two blackbodies is given by Eq. (5–41),

$$q_{1-2} = A_1F_{1-2}E_{b1} - A_2F_{2-1}E_{b2}$$

Of course, q_{1-2} must be zero and E_{b1} is equal to E_{b2}, since both surfaces are at the same temperature. It follows that

$$(A_1F_{1-2} - A_2F_{2-1})E_{b1} = 0$$

Because

$$E_{b1} = E_{b2} = \sigma T_1^4$$

the factor $(A_1F_{1-2} - A_2F_{2-1})$ must be zero. Thus, the reciprocity law

$$A_1F_{1-2} = A_2F_{2-1}$$

must be satisfied.

This same reasoning can be applied to any two blackbody surfaces A_s and A_j, with the result that

$$A_sF_{s-j} = A_jF_{j-s}$$

A more rigorous derivation of the principle of reciprocity is developed in references 5 and 11, which is based solely on geometric considerations.

Thermal Radiation Networks

As we have seen in Chaps. 1 through 3, the practical thermal analysis of many basic heat-transfer processes can be facilitated by use of the standard electrical analogy concept in which temperature T is taken to be analogous to electrical potential E_e. Based on this electrical analogy of the *first kind*, the radiation-heat-transfer resistance R_R between two blackbody surfaces is

$$R_R = \frac{1}{A_sF_{s-j}\sigma(T_s + T_j)(T_s^2 + T_j^2)} \tag{5-47}$$

This analogy concept is indeed useful for systems such as the ones shown in Fig. 5-25(a) and (b), which involve only two thermal radiating surfaces with known temperatures. However, because R_R is a function of temperature, the utility of this approach decreases as the number of thermal radiating surfaces with unknown temperature increases.

An alternative electrical analogy has been developed which more readily lends itself to the analysis of the more complex thermal radiation problems involving multiple surfaces and unknown surface temperatures. In this electrical analogy approach the total emissive power E is taken to be analogous to electrical voltage E_e. Equations (5-42) and (5-45) provide the basis for this powerful radiation-heat-transfer analysis tool. Simply put, the radiation-heat-transfer resistance R_{s-j} between surfaces A_s and A_j based on this electrical analogy of the *second kind* is

$$R_{s-j} = \frac{1}{A_sF_{s-j}} \tag{5-48}$$

To illustrate, analogous electrical circuits of the second kind (which we will refer to as *thermal radiation networks*) are utilized in Examples 5-11 through 5-13 for blackbody systems involving two, three, and four surfaces.

EXAMPLE 5–11

Referring to Fig. E5–11a, the blackbody surfaces A_1 and A_2 are at 27°C and 500°C, respectively. Determine the rate of radiation heat transfer between A_1 and A_2. Also sketch the thermal circuits of the first and second kinds.

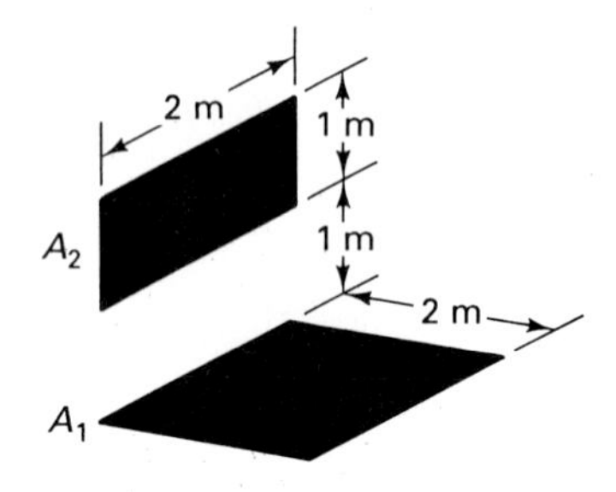

FIGURE E5–11a
Open thermal radiating blackbody system.

Solution

Objective Determine q_{1-2} and sketch the thermal radiation circuit and network.

Assumptions/Conditions

steady-state
blackbody thermal radiation
surrounding space is evacuated

Properties $\epsilon_1 = 1$, $\epsilon_2 = 1$.

Analysis The rate of radiation heat transfer is given by Eq. (5–42),

$$q_{1-2} = A_1 F_{1-2}(E_{b1} - E_{b2})$$

Referring back to Example 5–8, we see that $F_{1-2} = 0.05$. Calculating q_{1-2}, we obtain

$$q_{1-2} = 4 \text{ m}^2 (0.05)\left(5.67 \times 10^{-8} \frac{\text{W}}{\text{m}^2\text{ K}^4}\right)[(300 \text{ K})^4 - (773 \text{ K})^4]$$

$$= 4(0.05)(5.67)(3^4 - 7.73^4) \text{ W} = -3960 \text{ W}$$

Thermal circuits of the first and second kinds are shown in Figs. E5–11b and E5–11c. The thermal resistances R_R and R_{1-2} are given by

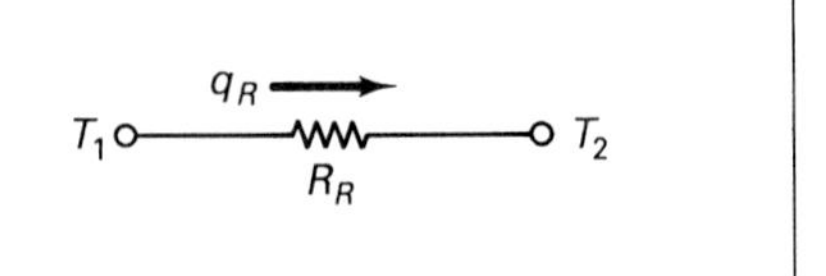

FIGURE E5–11b
Thermal circuit of first kind.

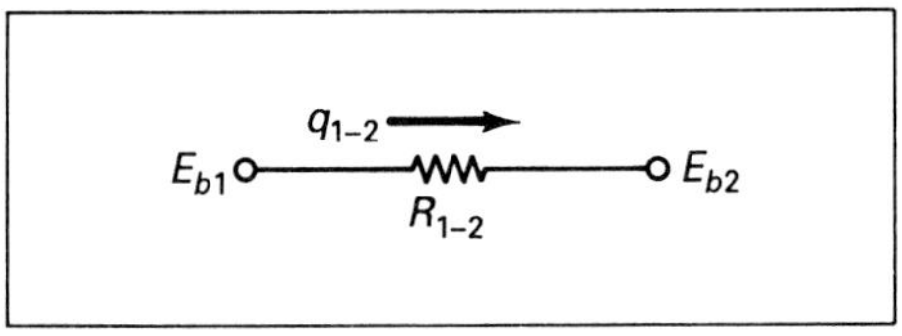

FIGURE E5-11c
Thermal radiation network.

$$R_R = \frac{1}{A_1 F_{1-2} \sigma (T_1 + T_2)(T_1^2 + T_2^2)} = 0.12 \text{ K/W}$$

and

$$R_{1-2} = \frac{1}{A_1 F_{1-2}} = 5/\text{m}^2$$

EXAMPLE 5-12

Utilize the thermal-radiation-network approach to evaluate the rate of radiation heat transfer between each of the three blackbody surfaces of the cylindrical enclosure shown in Fig. E5-12a.

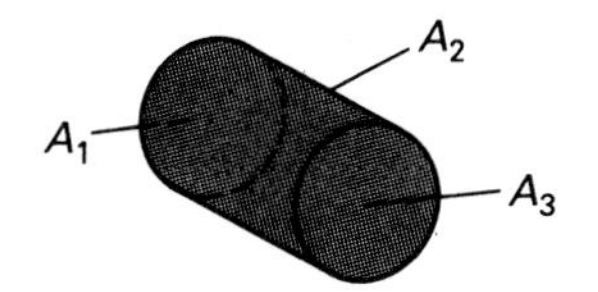

A_1 at $T_1 = 227°\text{C}$
A_2 at $T_2 = 200°\text{C}$
A_3 at $T_3 = 50°\text{C}$
$L = 1$ m, $D = 1$ m

FIGURE E5-12a
Cylindrical blackbody enclosure.

Solution

Objective Determine q_{1-2}, q_{1-3}, and q_{2-3}.

Assumptions/Conditions

steady-state
blackbody thermal radiation
enclosure is evacuated

Properties $\epsilon_1 = 1$, $\epsilon_2 = 1$, $\epsilon_3 = 1$.

Analysis The thermal radiation network is shown in Fig. E5-12b, where

$$E_{b1} = \sigma T_1^4 = 5.67 \times 10^{-8} \frac{\text{W}}{\text{m}^2\,\text{K}^4} (500 \text{ K})^4 = 3540 \text{ W/m}^2$$

$$E_{b2} = \sigma T_2^4 = \sigma(473 \text{ K})^4 = 2840 \text{ W/m}^2$$

$$E_{b3} = \sigma T_3^4 = \sigma(323 \text{ K})^4 = 617 \text{ W/m}^2$$

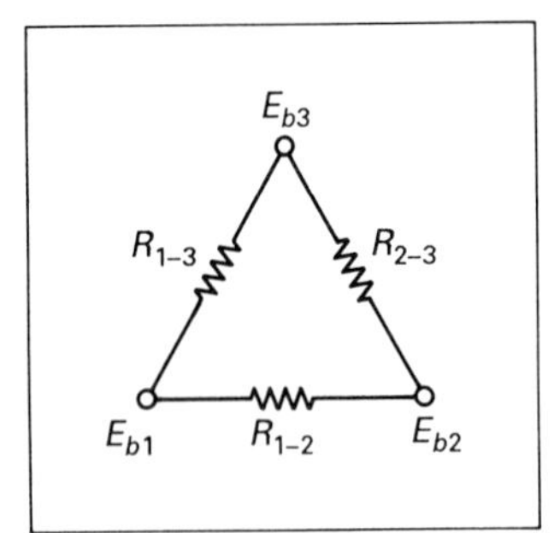

FIGURE E5–12b
Thermal radiation network.

Referring to Fig. 5–23, F_{1-3} is equal to 0.17. Utilizing the summation principle, we have

$$F_{1-2} = 1 - F_{1-3} = 0.83 \qquad F_{3-2} = 0.83$$

The thermal resistances are

$$R_{1-2} = \frac{1}{A_1 F_{1-2}} = \frac{1}{[\pi(1\text{ m})^2/4](0.83)} = 1.53/\text{m}^2$$

$$R_{1-3} = \frac{1}{A_1 F_{1-3}} = 7.47/\text{m}^2$$

$$R_{2-3} = \frac{1}{A_2 F_{2-3}} = \frac{1}{A_3 F_{3-2}} = 1.53/\text{m}^2$$

The rates of radiation heat transfer flowing in this network are now easily calculated.

$$q_{1-2} = \frac{E_{b1} - E_{b2}}{R_{1-2}} = \frac{(3540 - 2840)\text{ W/m}^2}{1.53/\text{m}^2} = 458\text{ W}$$

$$q_{1-3} = \frac{E_{b1} - E_{b3}}{R_{1-3}} = \frac{(3540 - 617)\text{ W/m}^2}{7.47/\text{m}^2} = 391\text{ W}$$

$$q_{2-3} = \frac{E_{b2} - E_{b3}}{R_{2-3}} = \frac{(2840 - 617)\text{ W/m}^2}{1.53/\text{m}^2} = 1450\text{ W}$$

The total rate of radiation heat transfer from each surface is

$$q_1 = q_{1-2} + q_{1-3} = 849\text{ W}$$

$$q_2 = q_{2-3} + q_{2-1} = q_{2-3} - q_{1-2} = 992\text{ W}$$

$$q_3 = q_{3-1} + q_{3-2} = -q_{1-3} - q_{2-3} = -1841\text{ W}$$

and we note that

$$q_1 + q_2 + q_3 = 0$$

which is consistent with requirements of the first law of thermodynamics.

EXAMPLE 5–13

The blackbody plates shown in Fig. E5–13a are located in a very large blackbody enclosure A_4 which is at 27°C. The back side of each plate is insulated, and 1 kW is electrically generated in plate 3. Sketch the thermal radiation network for this problem, and determine the temperature of surface A_3 and the rate of radiation heat transfer between A_1 and A_3.

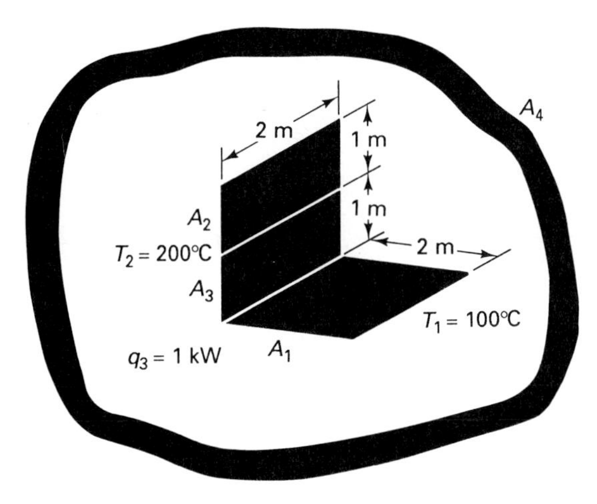

FIGURE E5–13a
Closed thermal radiating system.

Solution

Objective Sketch the thermal radiation network and determine T_3 and q_{1-3}.

Assumptions/Conditions

steady-state
blackbody thermal radiation
enclosure surface A_4 is very large
surrounding space is evacuated

Properties $\epsilon_1 = 1$, $\epsilon_2 = 1$, $\epsilon_3 = 1$, $\epsilon_4 = 1$.

Analysis Referring back to Example 5–8, we have

$$F_{1-2} = 0.05 \qquad F_{1-3} = 0.15$$

By reciprocity,

$$F_{3-1} = \frac{A_1}{A_3} F_{1-3} = 2(0.15) = 0.3$$

$$F_{2-1} = \frac{A_1}{A_2} F_{1-2} = 2(0.05) = 0.1$$

It is clear that $F_{2-3} = 0$. By employing the summation principle, we are able to write

$$F_{1-4} = 1 - F_{1-2} - F_{1-3} = 1 - 0.05 - 0.15 = 0.8$$

$$F_{2-4} = 1 - F_{2-1} - F_{2-3} = 1 - 0.1 - 0 = 0.9$$

$$F_{3-4} = 1 - F_{3-1} - F_{3-2} = 1 - 0.3 - 0 = 0.7$$

The thermal radiation network for this four-surface system is shown in Fig. E5–13b, where

$$R_{1-3} = \frac{1}{A_1 F_{1-3}} = \frac{1}{4\ \text{m}^2\ (0.15)} = 1.67/\text{m}^2 \qquad E_{b1} = \sigma(373\ \text{K})^4 = 1100\ \text{W/m}^2$$

$$R_{1-2} = \frac{1}{A_1 F_{1-2}} = \frac{1}{4\ \text{m}^2\ (0.05)} = 5/\text{m}^2 \qquad E_{b2} = \sigma(473\ \text{K})^4 = 2840\ \text{W/m}^2$$

$$R_{1-4} = \frac{1}{A_1 F_{1-4}} = \frac{1}{4\ \text{m}^2\ (0.8)} = 0.313/\text{m}^2 \qquad E_{b4} = \sigma(300\ \text{K})^4 = 459\ \text{W/m}^2$$

$$R_{2-3} = \frac{1}{A_2 F_{2-3}} = \frac{1}{2\ \text{m}^2\ (0)} = \infty/\text{m}^2$$

$$R_{3-4} = \frac{1}{A_3 F_{3-4}} = \frac{1}{2\ \text{m}^2\ (0.7)} = 0.714/\text{m}^2$$

$$R_{2-4} = \frac{1}{A_2 F_{2-4}} = \frac{1}{2\ \text{m}^2\ (0.9)} = 0.556/\text{m}^2$$

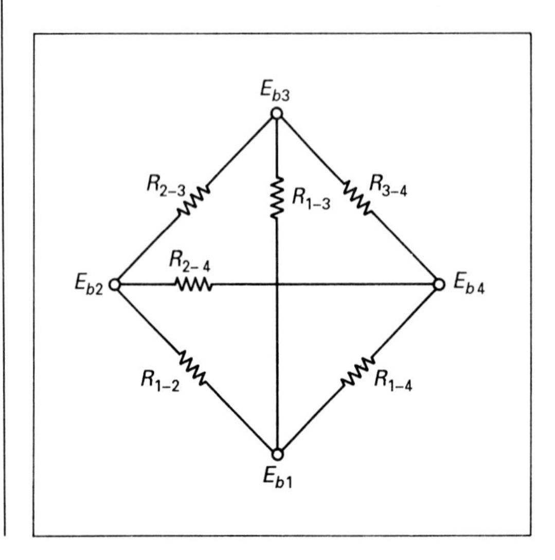

FIGURE E5–13b
Thermal radiation network.

To determine the unknown potential E_{b3}, we perform an energy balance at surface A_3.

$$q_3 = 1\text{ kW} = \frac{E_{b3} - E_{b1}}{R_{1-3}} + \frac{E_{b3} - E_{b2}}{R_{2-3}} + \frac{E_{b3} - E_{b4}}{R_{3-4}}$$

Making substitutions and solving for E_{b3}, we obtain

$$E_{b3} = \frac{1\text{ kW} + \dfrac{E_{b1}}{R_{1-3}} + \dfrac{E_{b2}}{R_{2-3}} + \dfrac{E_{b4}}{R_{3-4}}}{\dfrac{1}{R_{1-3}} + \dfrac{1}{R_{2-3}} + \dfrac{1}{R_{3-4}}}$$

$$= \frac{(1000 + 1100/1.67 + 2840/\infty + 459/0.714)\text{ W}}{(1/1.67 + 1/\infty + 1/0.714)\text{ m}^2} = 1150\text{ W/m}^2$$

To calculate T_3, we write

$$E_{b3} = \sigma T_3^4$$

$$T_3 = \left[\frac{1150\text{ W/m}^2}{5.67 \times 10^{-8}\text{ W/(m}^2\text{ K}^4)}\right]^{1/4} = 377\text{ K} = 104°\text{C}$$

The rate of radiation heat transfer between A_1 and A_3 is

$$q_{1-3} = \frac{E_{b1} - E_{b3}}{R_{1-3}} = \frac{(1100 - 1150)\text{ W/m}^2}{1.67/\text{m}^2} = -29.9\text{ W}$$

To check our solution, the summation $q_1 + q_2 + q_3 + q_4$ can be shown to be equal to zero.

5-5-2 Nonblackbody Thermal Radiation from Diffuse Opaque Surfaces

As we have seen, some real surfaces exhibit approximate blackbody characteristics, but most thermal radiation systems encountered in practice involve distinctly non-blackbody surfaces. Radiation heat transfer in common opaque nonblackbody systems is complicated by the occurrence of reflections. For systems involving more than one nonblackbody surface, the multiple reflection patterns of thermal radiation often become extremely complex. Fortunately, these complexities can be rather easily overcome for systems involving diffuse surfaces with uniform properties and heating conditions by the use of a practical thermal radiation network approach. Numerous real surface systems found in practice approximate these conditions. To develop this practical analysis approach for diffuse nonblackbody opaque systems, we first introduce the concept of radiosity.

Radiosity

Referring to Fig. 5–27, we define the *radiosity* J_s as the rate of thermal radiation emitted and reflected per unit surface area from A_s.

$$J_s = E_s + \rho_s G_s \tag{5–49}$$

Recall that the irradiation G_s is the thermal radiation flux incident on the surface A_s.

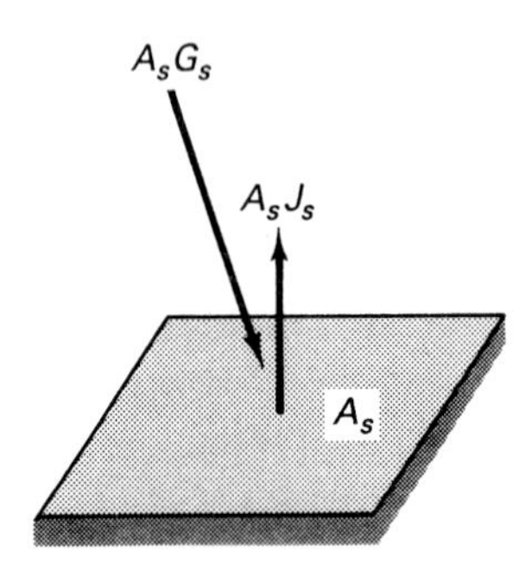

FIGURE 5–27
Thermal radiating surface A_s: concept of radiosity.

The total rate of radiation heat transfer from an opaque surface A_s is simply equal to the difference between the rate of thermal radiation leaving, $A_s J_s$, and the rate of thermal radiation coming in, $A_s G_s$; that is,

$$q_s = A_s(J_s - G_s) \tag{5–50}$$

To eliminate the irradiation G_s, we combine Eqs. (5–49) and (5–50).

$$q_s = A_s\left[J_s - \frac{1}{\rho_s}(J_s - E_s)\right] = \frac{A_s}{\rho_s}[E_s - (1 - \rho_s)J_s] \tag{5–51}$$

Since the body is opaque,

$$\rho_s = 1 - \alpha_s \tag{5–52}$$

and we have

$$q_s = \frac{A_s\alpha_s}{\rho_s}\left(\frac{\epsilon_s}{\alpha_s}E_{bs} - J_s\right) = \frac{A_s\alpha_s}{1 - \alpha_s}\left(\frac{\epsilon_s}{\alpha_s}E_{bs} - J_s\right) \tag{5–53}$$

Based on this equation, the total rate of radiation heat transfer q_s can be represented by a thermal radiation network element as shown in Fig. 5–28; the node potentials are $(\epsilon_s/\alpha_s)E_{bs}$ and J_s, and the thermal resistance R_s is

$$R_s = \frac{1 - \alpha_s}{A_s\alpha_s} = \frac{\rho_s}{A_s\alpha_s} \tag{5–54}$$

As indicated in Sec. 5–3, the total absorptivity α_s of a surface is dependent upon the sources from which the irradiation G_s originates. Because of the reflection

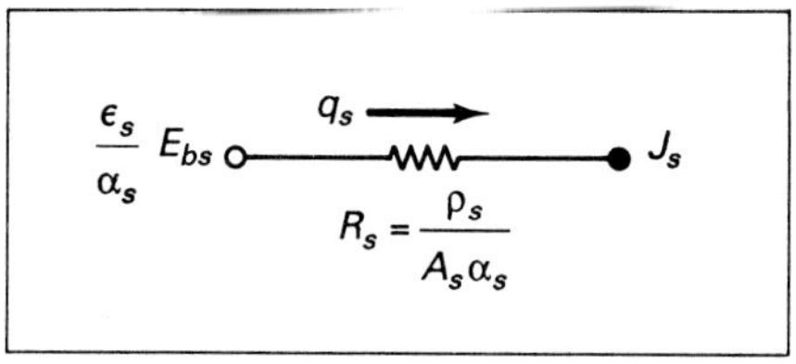

FIGURE 5-28
Thermal radiation network element for q_s.

patterns associated with nonblackbody surface systems, special care must be taken in the evaluation of α_s for nongraybody conditions, as illustrated in several examples in the next section. For systems with surfaces that exhibit approximate graybody characteristics, α_s is independent of the source of the various components of incoming irradiation, and α_s can be set equal to ϵ_s. For this limiting condition, we write

$$q_s = \frac{A_s\epsilon_s}{1 - \epsilon_s}(E_{bs} - J_s) \tag{5-55}$$

The thermal radiation network element for this simple situation is shown in Fig. 5-29. The node potentials for this case are simply E_{bs} and J_s, and R_s is given by

$$R_s = \frac{1 - \epsilon_s}{A_s\epsilon_s} = \frac{\rho_s}{A_s\epsilon_s} \tag{5-56}$$

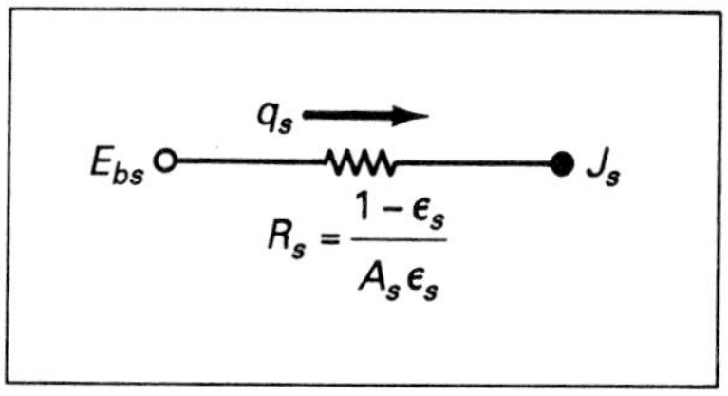

FIGURE 5-29
Thermal radiation network element for q_s: graybody conditions.

Equation (5-53) and its limiting form Eq. (5-55), together with the corresponding thermal radiation network elements given in Figs. 5-28 and 5-29, provide the key building block for the development of a simple practical analysis of radiation heat transfer. However, before we turn our attention to the completion of our practical analysis approach for bisurface and multisurface systems, it should be mentioned that another approach to handling Eqs. (5-49) and (5-50) that is favored by some is to eliminate the radiosity J_s. This alternative method is introduced in Example 5-14.

EXAMPLE 5-14

The radiation heat transfer q_s associated with graybody surfaces is given in terms of E_{bs} and radiosity J_s by Eq. (5-55). Develop an alternative formulation for q_s by eliminating J_s instead of G_s in Eqs. (5-49) and (5-50).

Solution

Objective Develop an alternative relationship for q_s for graybody surfaces.

Assumptions/Conditions

diffuse opaque graybody surfaces

Analysis Starting with Eqs. (5–49) and (5–50),

$$J_s = E_s + \rho_s G_s \qquad q_s = A_s(J_s - G_s)$$

we eliminate J_s, to obtain

$$q_s = A_s(E_s + \rho_s G_s - G_s) = A_s(E_s - \alpha_s G_s) \tag{a}$$

The irradiation G_s falling on surface A_s is expressed in terms of the radiosities of the surrounding surfaces by

$$A_s G_s = \sum_{j=1}^{N} A_j F_{j-s} J_j$$

Utilizing the principle of reciprocity, this equation takes the form

$$A_s G_s = \sum_{j=1}^{N} A_s F_{s-j} J_j$$

or

$$G_s = \sum_{j=1}^{N} F_{s-j} J_j \tag{b}$$

The total rate of thermal radiation absorbed by surface A_s is

$$A_s \alpha_s G_s = \sum_{j=1}^{N} \alpha_{sj} A_s F_{s-j} J_j \tag{c}$$

where α_{sj} is the absorptivity component associated with the J_j source. Substituting this result into Eq. (a), we have

$$q_s = A_s \left(E_s - \sum_{j=1}^{N} \alpha_{sj} F_{s-j} J_j \right) \tag{d}$$

Although this equation does not conveniently lend itself to electrical analogy representation, as does Eq. (5–53), it does provide a basis for the analytical or numerical solution of more complex thermal radiation systems. For the simple limiting case of graybody conditions, Eq. (d) reduces to

$$q_s = A_s \epsilon_s \left(E_{bs} - \sum_{j=1}^{N} F_{s-j} J_j \right) \tag{e}$$

As in the case of Eq. (5–53), Eqs. (d) and (e) apply to systems with uniform emission, reflection, and irradiation over each surface.

As a point of interest, it is noted that the total absorptivity α_s at surface A_s can be obtained by combining Eqs. (b) and (c); that is,

$$\alpha_s = \frac{\sum_{j=1}^{N} \alpha_{sj} F_{s-j} J_j}{\sum_{j=1}^{N} F_{s-j} J_j} \tag{f}$$

Bisurface Systems

We consider radiation heat transfer between two opaque diffuse surfaces A_1 and A_2. For uniform radiosities, the rate of thermal radiation leaving A_1 that reaches A_2 is $A_1F_{1-2}J_1$, and from A_2 to A_1 is $A_2F_{2-1}J_2$. Therefore, the thermal radiation-heat-transfer rate q_{1-2} from surface A_1 to A_2 is

$$q_{1-2} = A_1F_{1-2}J_1 - A_2F_{2-1}J_2 \tag{5-57}$$

or, since $A_1F_{1-2} = A_2F_{2-1}$,

$$q_{1-2} = A_1F_{1-2}(J_1 - J_2) \tag{5-58}$$

This equation provides the basis for the thermal radiation network element shown in Fig. 5-30.

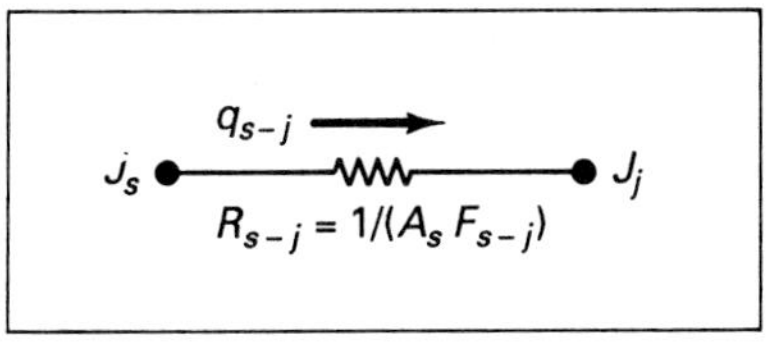

q_{s-j}
J_s J_j
$R_{s-j} = 1/(A_s F_{s-j})$

FIGURE 5-30
Thermal radiation network element between potentials J_s and J_j.

To close our two-surface system analysis, we couple the nodes J_1 and J_2 to the surface potentials $(\epsilon_1/\alpha_1)E_{b1}$ and $(\epsilon_2/\alpha_2)E_{b2}$ by using the thermal radiation network element given in Fig. 5-28. The resulting thermal radiation network is shown in Fig. 5-31. Because Eq. (5-58) was developed for surfaces with uniform radiosities, this thermal radiation network only applies to systems with uniform emission, reflection, and irradiation over each surface. The requirement that the reflected radiation flux be uniform is strictly satisfied only for symmetrical systems, such as infinite parallel plates or concentric spheres, with each surface having a uniform temperature and uniform thermal radiation properties. However, this convenient thermal radiation network approach is often applied to nonsymmetrical systems as a first approximation.

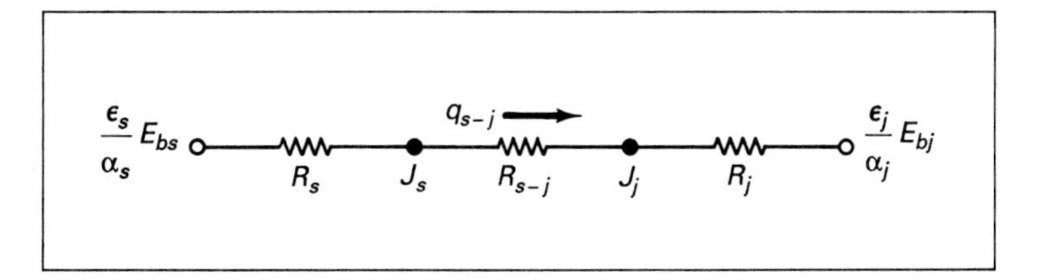

FIGURE 5-31 Thermal radiation network for bisurface system.

The rate of radiation heat transfer q_{1-2} associated with this thermal radiation network is given by

$$q_{1-2} = \frac{(\epsilon_1/\alpha_1)E_{b1} - (\epsilon_2/\alpha_2)E_{b2}}{R_1 + R_{1-2} + R_2} = \frac{(\epsilon_1/\alpha_1)E_{b1} - (\epsilon_2/\alpha_2)E_{b2}}{\dfrac{1-\alpha_1}{A_1\alpha_1} + \dfrac{1}{A_1F_{1-2}} + \dfrac{1-\alpha_2}{A_2\alpha_2}} \quad (5\text{–}59)$$

For blackbody conditions, R_1 and R_2 are both zero, such that the thermal radiation network reduces to the one shown in Fig. E5–11c, and Eq. (5–59) reduces to Eq. (5–42). For approximate graybody conditions, Eq. (5–59) simplifies to

$$q_{1-2} = \frac{E_{b1} - E_{b2}}{\dfrac{1-\epsilon_1}{A_1\epsilon_1} + \dfrac{1}{A_1F_{1-2}} + \dfrac{1-\epsilon_2}{A_2\epsilon_2}} \quad (5\text{–}60)$$

EXAMPLE 5–15

Consider exchange of radiation heat transfer between the small surface A_1 and an enclosure A_2 shown in Fig. E5–15a. Demonstrate that the enclosure approximates blackbody conditions as A_2 becomes large.

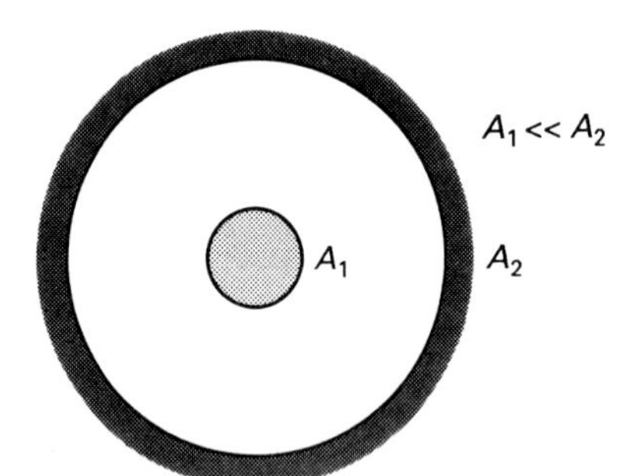

FIGURE E5–15a
Closed thermal radiating system.

Solution

Objective Show that blackbody conditions are approached within this enclosure as A_2 becomes large or as A_1 becomes small.

Assumptions/Conditions

steady-state
diffuse opaque surfaces
enclosure is evacuated

Analysis The thermal radiation network for this system is shown in Fig. E5–15b, where

$$R_1 = \frac{\rho_1}{A_1\alpha_1} \qquad R_{1-2} = \frac{1}{A_1F_{1-2}} \qquad R_2 = \frac{\rho_2}{A_2\alpha_2}$$

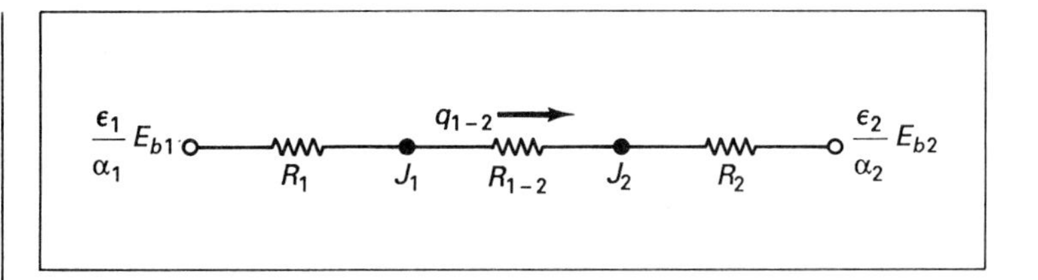

FIGURE E5–15b Thermal radiation network.

For very large values of A_2, most of the energy incident upon A_2 originates from itself, unless the body is a perfect reflector. Consequently, the nonisothermal α approximation reduces to the graybody α approximation with ϵ_2/α_2 equal to unity. Further, the resistance R_2 approaches zero as A_2 increases. For these conditions, the radiosity J_2 is essentially equal to E_{b2}, such that blackbody conditions are approached at the surface of the enclosure.

Incidentally, the thermal radiation network for this system implicitly accounts for thermal radiation from surface A_2 to itself.

EXAMPLE 5–16

Develop an expression for the net rate of exchange of thermal radiation between two diffuse nonblackbody infinite parallel plates A_1 and A_2 without the use of radiosity.

Solution

Objective Develop a relation for q_{1-2} for this parallel plate system.

Schematic Nonblackbody thermal radiation between infinite parallel plates.

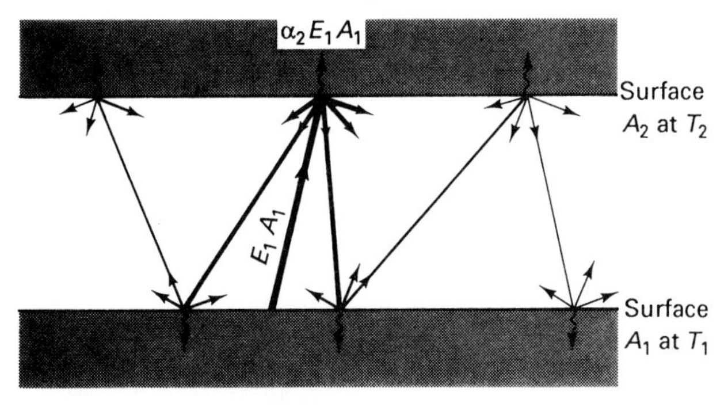

Assumptions/Conditions

steady-state
diffuse nonblackbody surfaces
space between plates is evacuated

Analysis The reflection and absorption pattern of thermal radiation originally emitted from surface A_1 is shown in the schematic. The energy rate content of this thermal radiation before absorption and reflection is E_1A_1. Of this amount of radiation power, the quantity $\alpha_2E_1A_1$ is absorbed by A_2 and $\rho_2E_1A_1$ is reflected back toward A_1. Of this reflected energy rate, the quantity $\rho_1(\rho_2E_1A_1)$ is rereflected toward surface A_2. Surface A_2 absorbs $\alpha_2(\rho_1\rho_2E_1A_1)$ of this rate of energy and reflects the remainder, $\rho_2(\rho_1\rho_2E_1A_1)$. The total rate of thermal radiation originating from surface A_1 that is absorbed by A_2 is written as

$$\alpha_2E_1A_1[1 + \rho_1\rho_2 + (\rho_1\rho_2)^2 + \cdots + (\rho_1\rho_2)^n + \cdots]$$

or

$$\alpha_2E_1A_1 \sum_{n=0}^{\infty} (\rho_1\rho_2)^n$$

In a similar fashion, we can show that the total rate of thermal radiation originating from surface A_2 that is absorbed by A_1 is

$$\alpha_1E_2A_2 \sum_{n=0}^{\infty} (\rho_1\rho_2)^n$$

It follows that the net rate of thermal radiation exchange between A_1 and A_2 becomes

$$q_{1-2} = (\alpha_2E_1A_1 - \alpha_1E_2A_2)\left[\sum_{n=0}^{\infty} (\rho_1\rho_2)^n\right]$$

where $A_1 = A_2$.

Referring to reference 23, the series $\Sigma_{n=0}^{\infty} x^n$ converges to $1/(1 - x)$ for x less than unity. Thus, the rate of radiation heat transfer is

$$q_{1-2} = \frac{A_1(\alpha_2E_1 - \alpha_1E_2)}{1 - \rho_1\rho_2}$$

Introducing the identities $E_1 = \epsilon_1E_{b1}$, $E_2 = \epsilon_2E_{b2}$, $\rho_1 = 1 - \alpha_1$, and $\rho_2 = 1 - \alpha_2$, we obtain

$$q_{1-2} = \frac{A_1\left(\dfrac{\epsilon_1}{\alpha_1}E_{b1} - \dfrac{\epsilon_2}{\alpha_2}E_{b2}\right)}{\dfrac{1}{\alpha_1} + \dfrac{1}{\alpha_2} - 1}$$

which is equivalent to Eq. (5–59) for $A_1 = A_2$ and $F_{1-2} = 1$. For graybody conditions, we have

$$q_{1-2} = \frac{A_1(E_{b1} - E_{b2})}{\dfrac{1}{\epsilon_1} + \dfrac{1}{\epsilon_2} - 1}$$

This equation is equivalent to Eq. (5–60) for this infinite parallel-plate geometry.

The fact that this direct approach is consistent with the radiosity-based network approach should reinforce our confidence in the more efficient network method. Further, because this direct method becomes extremely unwieldly for multisurface systems, the concept of radiosity is the key to developing practical thermal analyses for radiation processes.

EXAMPLE 5–17

Surface A_2 of the system shown in Fig. E5–17a is a graybody with emissivity of 0.56 and surface A_1 is a blackbody. Determine the rate of radiation heat transfer between A_1 and A_2 if $T_1 = 27°C$ and $T_2 = 500°C$.

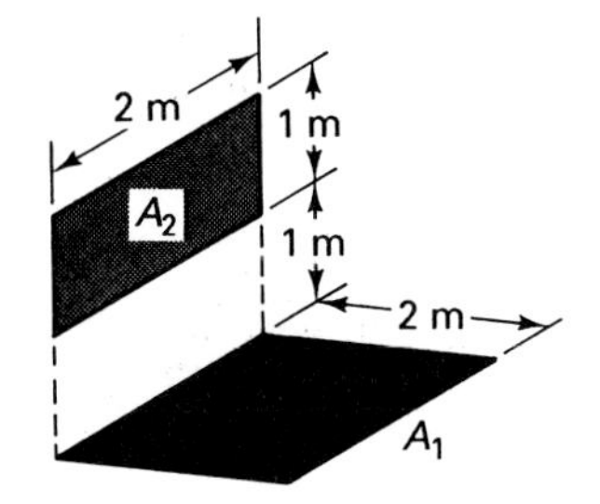

FIGURE E5–17a
Open thermal radiating system.

Solution

Objective Determine q_{2-1}.

Assumptions/Conditions

steady-state
blackbody and diffuse opaque graybody surfaces
surrounding space is evacuated

Properties $\epsilon_1 = 1$, $\epsilon_2 = 0.56$. Because surface A_2 is a graybody, we also have

$$\alpha_2 = \epsilon_2 = 0.56$$

Analysis The radiation shape factor is given in Example 5–11 as

$$F_{1-2} = 0.05$$

Assuming diffuse radiation, the approximate thermal radiation network for this nonsymmetrical bisurface system is shown in Fig. E5–17b, where $E_{b1} = 459 \text{ W/m}^2$, $E_{b2} = 20{,}200 \text{ W/m}^2$, and

$$R_{1-2} = \frac{1}{A_1 F_{1-2}} = 5/\text{m}^2$$

$$R_2 = \frac{\rho_2}{A_2 \alpha_2} = \frac{1 - 0.56}{2 \text{ m}^2 (0.56)} = 0.393/\text{m}^2$$

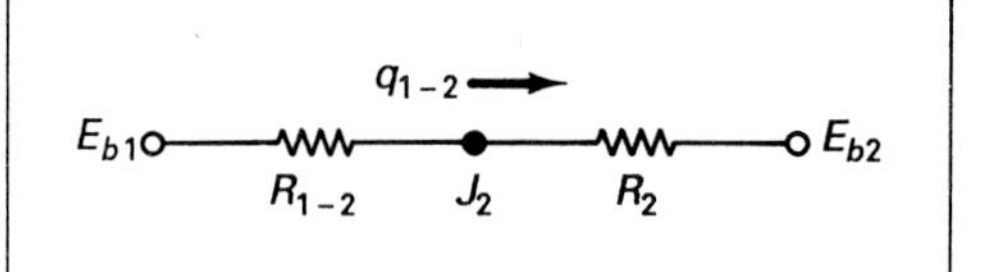

FIGURE E5–17b
Thermal radiation network.

It follows that the rate of radiation heat transfer q_{2-1} is approximately

$$q_{2-1} = -q_{1-2} = \frac{(20{,}200 - 459)\ \mathrm{W/m^2}}{(5 + 0.393)/\mathrm{m^2}} = 3660\ \mathrm{W}$$

This result lies about 7.5% below the value of 3960 W obtained in Example 5–11 for blackbody radiation from both surfaces.

To improve the accuracy of our analysis for this nonsymmetrical system, surface A_2 can be subdivided into smaller areas.

EXAMPLE 5–18

Surface A_2 of the system shown in Fig. E5–17a is made of oxidized nickel and surface A_1 is a blackbody. Determine the rate of radiation heat transfer between A_1 and A_2 if $T_1 = 27°\mathrm{C}$ and $T_2 = 500°\mathrm{C}$.

Solution

Objective Determine q_{2-1}.

Schematic Open thermal radiating system.

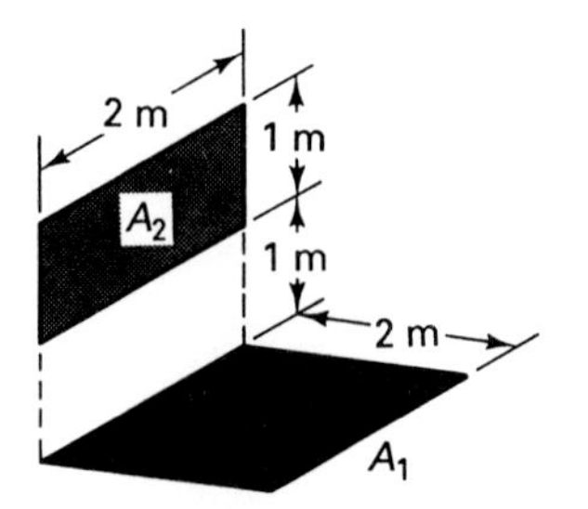

Assumptions/Conditions

steady-state
blackbody and diffuse opaque graybody surfaces
surrounding space is evacuated

Properties

$\epsilon_1 = 1.$

Oxidized nickel (Fig. 4–3): $\epsilon = 0.63$ at 500°F, $\epsilon = 0.32$ at 27°F.

Therefore, we take

$$\epsilon_2 = 0.63$$

Analysis Noting that all the thermal radiation from A_1 to A_2 is emitted by blackbody A_1, which is at a temperature of 27°C, we employ the nonisothermal α approximation to obtain

$$\alpha_2 = 0.32$$

As we have already seen, $F_{1-2} = 0.05$.

The thermal radiation network for this system is shown in Fig. E5-18 with $(\epsilon_2/\alpha_2)E_{b2}$ and R_2 given by

$$\frac{\epsilon_2}{\alpha_2} E_{b2} = \frac{0.63}{0.32}\left(20{,}200 \frac{\text{W}}{\text{m}^2}\right) = 39{,}800 \text{ W/m}^2$$

$$R_2 = \frac{\rho_2}{A_2\alpha_2} = \frac{1 - 0.32}{2 \text{ m}^2 (0.32)} = 1.06/\text{m}^2$$

The rate of radiation heat transfer is written as

$$q_{2-1} = \frac{(\epsilon_2/\alpha_2)\, E_{b2} - E_{b1}}{R_{1-2} + R_2} = \frac{(39{,}800 - 459) \text{ W/m}^2}{(5 + 1.06)/\text{m}^2} = 6490 \text{ W}$$

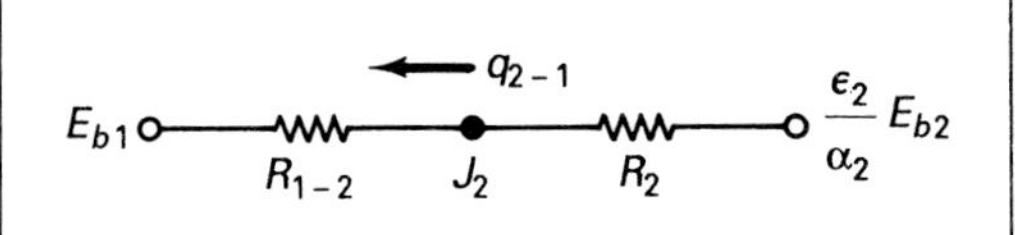

FIGURE E5-18
Thermal radiation network.

As a point of interest, the graybody α approximation gives

$$\alpha_2 = 0.63 \qquad R_2 = \frac{\rho_2}{A_2\alpha_2} = \frac{1 - 0.63}{2 \text{ m}^2 (0.63)} = 0.294/\text{m}^2$$

The resulting prediction for the rate of radiation heat transfer is

$$q_{2-1} = \frac{E_{b2} - E_{b1}}{R_{1-2} + R_2} = \frac{(20{,}200 - 459) \text{ W/m}^2}{(5 + 0.294)/\text{m}^2} = 3730 \text{ W}$$

which is a substantial 40% below the value obtained by utilizing the preferred nonisothermal α approximation.

As in Example 5-17, the accuracy of our analysis for this nonsymmetrical system can be improved by breaking surfaces A_1 and A_2 into smaller areas.

Multisurface Systems

Following the pattern of our analysis of bisurface systems, the net rate of thermal radiation between opaque diffuse surfaces A_s and A_j of a multisurface system with uniform radiosities is

$$q_{s-j} = A_s F_{s-j} J_s - A_j F_{j-s} J_j \qquad (5\text{–}61)$$

or, based on the principle of reciprocity,

$$q_{s-j} = \frac{J_s - J_j}{R_{s-j}} \qquad R_{s-j} = \frac{1}{A_s F_{s-j}} \qquad (5\text{–}62,63)$$

Thus, we have a general thermal radiation network element such as the one shown in Fig. 5–30 for each surface combination.

The radiosity J_s associated with each surface is linked to its surface potential $(\epsilon_s/\alpha_s)E_{bs}$ by means of the element shown in Fig. 5–28. To illustrate, thermal radiation networks are shown in Figs. 5–32 and 5–33 for representative three-surface and four-surface diffuse systems. For blackbody surfaces, R_s is zero, and these thermal radiation

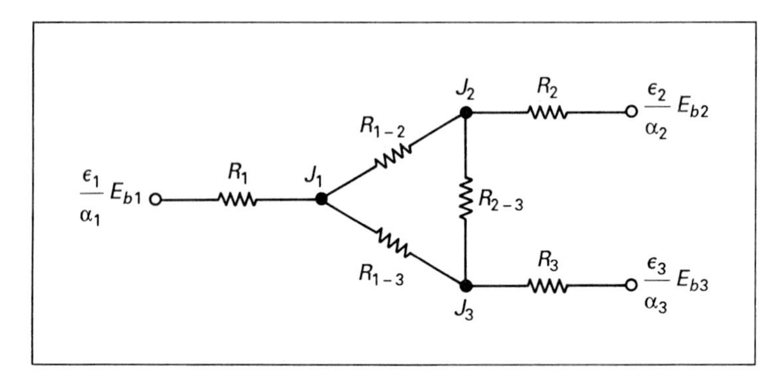

FIGURE 5–32 Thermal radiation network for trisurface system.

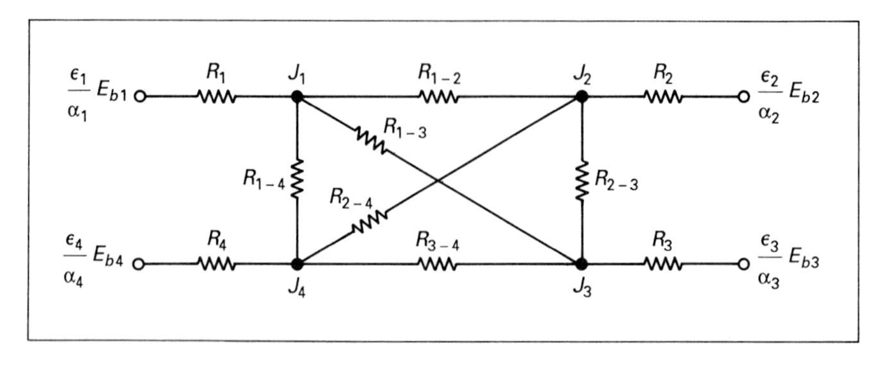

FIGURE 5–33 Thermal radiation network for four-surface system.

networks reduce to the networks shown in Examples 5–12 and 5–13. And, for graybody conditions, α_s is set equal to ϵ_s. As in the case of bisurface systems, these multisurface thermal radiation networks strictly only apply to symmetrical systems, but are commonly used as a first estimate for nonsymmetrical systems.

Referring back to the development of Eq. (5–53),

$$q_s = \frac{A_s\alpha_s}{1 - \alpha_s}\left(\frac{\epsilon_s}{\alpha_s} E_{bs} - J_s\right) \tag{5–53}$$

which is the basis for the network element of Fig. 5–28, we are again reminded that the rate of radiation heat transfer q_s under steady-state conditions must be in balance with the net rate of energy entering the body, $\sum \dot{E}_i$. Certain situations occur in practice in which the only significant energy transfer to or from a body is caused by thermal radiation. Two rather special cases of practical importance that fall into this category are thermal radiation shields and reradiating surfaces.

Thermal Radiation Shields As suggested in Chap. 1, the heat transfer through a wall can be greatly reduced by utilizing a composite construction that consists of layers of highly reflective materials separated by evacuated spaces. The thin plates or shells utilized in such superinsulative composite walls are known as *thermal radiation shields*. Referring to Fig. 5–34(a), we see that under steady-state conditions the rate of radiation heat transfer q_{1-s} from enclosure surface A_1 to the single thermal radiation shield is equal to the rate of radiation heat transfer q_{s-2} from the shield to enclosure surface A_2. Although such highly reflective surfaces have specular characteristics, the total radiation in this symmetrical system and in an infinite parallel-plate system can be treated as diffuse. A thermal radiation network is shown for this

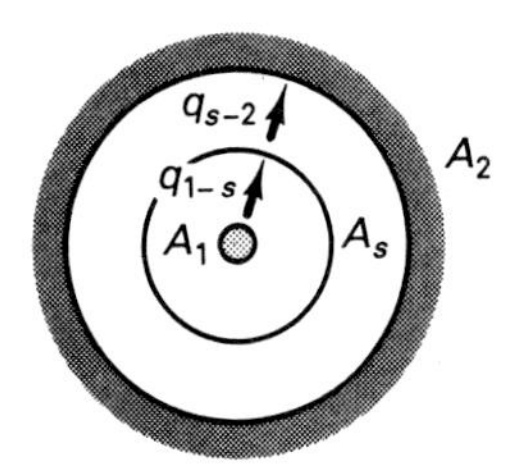

(a) Concentric sphere system.

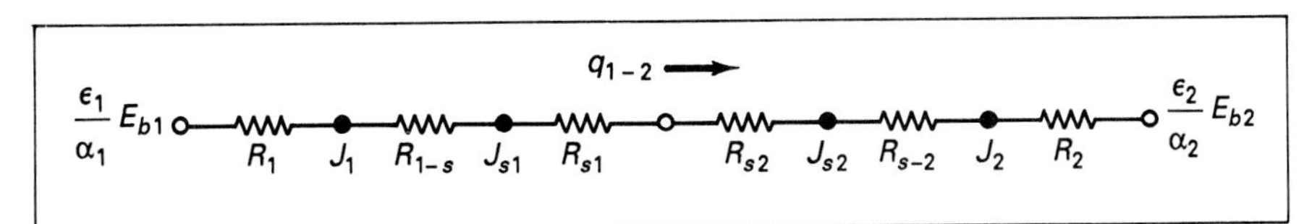

(b) Thermal radiation network.

FIGURE 5–34 Thermal radiation shield.

arrangement in Fig. 5–34(b). Note that the thermal radiation shield is assumed to be very thin with negligible resistance to conduction. Based on this thermal radiation network, the rate of radiation heat transfer through the shield from surface A_1 to A_2 is

$$q_{1-2} = \frac{(\epsilon_1/\alpha_1)E_{b1} - (\epsilon_2/\alpha_2)E_{b2}}{\dfrac{\rho_1}{A_1\alpha_1} + \dfrac{1}{A_1F_{1-s}} + \dfrac{\rho_{s1}}{A_s\alpha_{s1}} + \dfrac{\rho_{s2}}{A_s\alpha_{s2}} + \dfrac{1}{A_sF_{s-2}} + \dfrac{\rho_2}{A_2\alpha_2}} \tag{5–64}$$

with F_{1-s} and F_{s-2} both equal to unity. Because the shield surfaces are highly reflective, the thermal radiation reaching surface A_1 primarily originates from surface A_1 itself. Therefore, the nonisothermal α approximation essentially reduces to the graybody α approximation (i.e., $\epsilon_1/\alpha_1 \simeq 1$). The same is true for surface A_2, with $\epsilon_2/\alpha_2 \simeq 1$. As a final simplification, the radiation properties of the shield can generally be taken as the average of its two surfaces; that is,

$$\alpha_s = \frac{\alpha_{s1} + \alpha_{s2}}{2} \qquad \rho_s = \frac{\rho_{s1} + \rho_{s2}}{2} = 1 - \alpha_s \tag{5–65a,b}$$

Introducing these simplifications, q_{1-2} is given by

$$q_{1-2} = \frac{E_{b1} - E_{b2}}{\dfrac{1-\epsilon_1}{A_1\epsilon_1} + \dfrac{1}{A_1} + \dfrac{1-\epsilon_2}{A_2\epsilon_2} + \dfrac{2-\epsilon_s}{A_s\epsilon_s}} \tag{5-66}$$

For N shields, it is a simple matter to show that q_{1-2} becomes

$$q_{1-2} = \frac{E_{b1} - E_{b2}}{\dfrac{1-\epsilon_1}{A_1\epsilon_1} + \dfrac{1}{A_1} + \dfrac{1-\epsilon_2}{A_2\epsilon_2} + \displaystyle\sum_{j=1}^{N} \frac{1}{A_{sj}}\frac{2-\epsilon_{sj}}{\epsilon_{sj}}} \tag{5–67}$$

For a parallel-plate arrangement with each reflective shield having approximately the same radiation properties, this equation reduces to

$$q_{1-2} = \frac{A_1(E_{b1} - E_{b2})}{\dfrac{1-\epsilon_1}{\epsilon_1} + 1 + \dfrac{1-\epsilon_2}{\epsilon_2} + \dfrac{2-\epsilon_s}{\epsilon_s}N} = \frac{A_1(E_{b1} - E_{b2})}{\dfrac{1}{\epsilon_1} + \dfrac{1}{\epsilon_2} - 1 + \dfrac{2-\epsilon_s}{\epsilon_s}N} \tag{5–68}$$

Taking one step further, for the case in which the radiation properties of all the surfaces are approximately equal, we have

$$q_{1-2} = \frac{\epsilon A_1(E_{b1} - E_{b2})}{(2-\epsilon)(1+N)} \tag{5–69}$$

An examination of Eqs. (5–67) through (5–69) reveals that the rate of radiation heat transfer becomes small as (1) the reflectivity of the shield increases toward unity (i.e., as α_s or ϵ_s falls toward zero), and (2) the number of shields increases.

EXAMPLE 5–19

A blackbody plate at 0°F exchanges radiation with a parallel stainless steel 301 plate at 1500°F. Determine the percent reduction in heat transfer if a thin polished aluminum plate with emissivity and absorptivity approximately equal to 0.08 is placed between these two large plates.

Solution

Objective Determine the percent change in q_{1-2} that is caused by the use of the radiation shield A_s.

Schematic Thermal radiation exchange between plates with radiation shield.

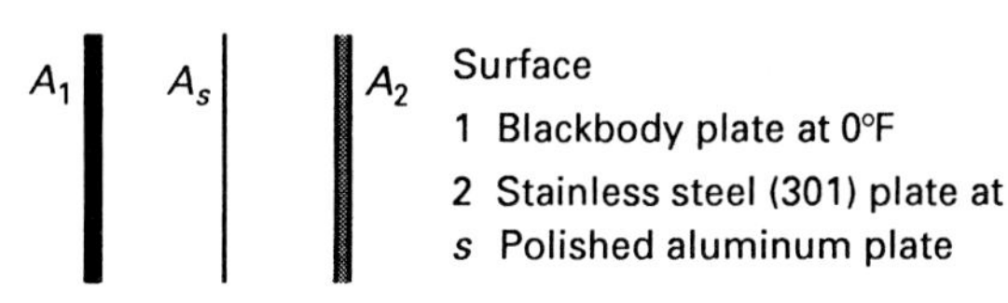

Assumptions/Conditions

steady-state
blackbody and diffuse opaque graybody surfaces
space between plates is evacuated

Properties

$\epsilon_1 = 1.$
Polished aluminum: $\epsilon_s = \alpha_s = 0.08.$
Stainless steel (301) at 1500°F (Fig. 4–3): $\epsilon_2 = 0.5.$

Analysis For the case in which no radiation shield is used, we employ the non-isothermal α approximation,

$$\alpha_2 = \epsilon_2(T_1) \simeq 0.16$$

The thermal radiation network for the parallel plates without the shield is shown in Fig. E5–19a, where

$$E_{b1} = \sigma T_1^4 = 0.171 \times 10^{-8} \frac{\text{Btu}}{\text{h ft}^2\ {}^\circ\text{R}^4} (460^\circ\text{R})^4 = 76.6 \text{ Btu/(h ft}^2)$$

$$E_{b2} = \sigma T_2^4 = \sigma(1960^\circ\text{R})^4 = 2.52 \times 10^4 \text{ Btu/(h ft}^2)$$

$$\frac{\epsilon_2}{\alpha_2} E_{b2} = \frac{0.5}{0.16}\left(2.52 \times 10^4 \frac{\text{Btu}}{\text{h ft}^2}\right) = 7.88 \times 10^4 \text{ Btu/(h ft}^2)$$

$$R_{1-2} = \frac{1}{A_1 F_{1-2}} = \frac{1}{A_1}$$

$$R_2 = \frac{\rho_2}{A_2\alpha_2} = \frac{1 - 0.16}{A_1(0.16)} = \frac{5.25}{A_1}$$

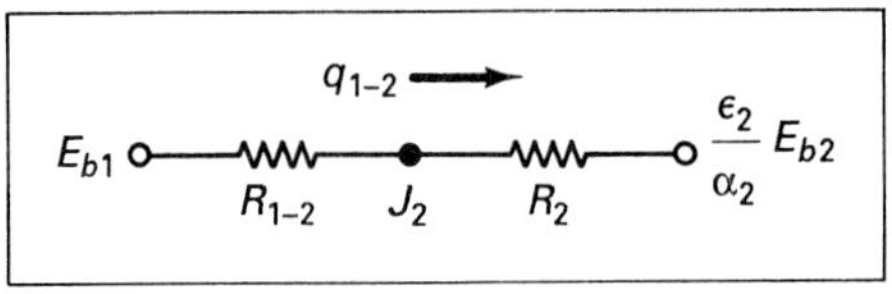

FIGURE E5–19a Thermal radiation network for parallel plates.

Calculating the rate of radiation heat transfer q_{2-1}, we obtain

$$q_{2-1} = \frac{(\epsilon_2/\alpha_2)E_{b2} - E_{b1}}{R_{1-2} + R_2} = \frac{(7.88 \times 10^4 - 76.6) \text{ Btu/(h ft}^2)}{(1 + 5.25)/A_1}$$

$$q''_{2-1} = 1.26 \times 10^4 \text{ Btu/(h ft}^2)$$

The thermal radiation network for the situation in which a thin plate lies between A_1 and A_2 is shown in Fig. E5–19b, where

$$R_{1-s} = \frac{1}{A_1 F_{1-s}} = \frac{1}{A_1}$$

$$R_{s1} = R_{s2} = \frac{\rho_s}{A_s \alpha_s} = \frac{1 - 0.08}{A_s(0.08)} = \frac{11.5}{A_1}$$

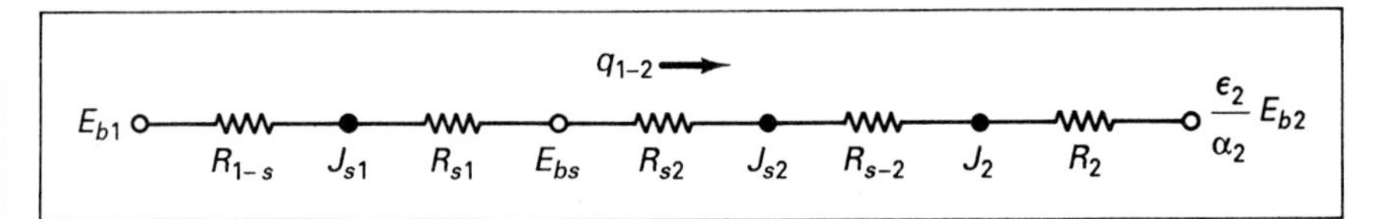

FIGURE E5–19b Thermal radiation network for parallel plates with shield.

As we have already noted, the emissivity of the stainless steel plate at 1500°F is approximately 0.5. Because most of the radiation incident upon A_2 actually originates at surface A_2 itself, we evaluate α_2 at a source temperature of 1500°F, as a first estimate; that is,

$$\alpha_2 = \epsilon_2(T_2) = 0.5$$

It follows that

$$R_2 = \frac{\rho_2}{A_2\alpha_2} = \frac{1 - 0.5}{A_1(0.5)} = \frac{1}{A_1}$$

The rate of radiation heat transfer is then

$$q_{2-1} = \frac{E_{b2} - E_{b1}}{R_{1-s} + R_{s1} + R_{s2} + R_{s-2} + R_2} = \frac{(2.52 \times 10^4 - 76.6) \text{ Btu/(h ft}^2)}{(1 + 11.5 + 11.5 + 1 + 1)/A_1}$$

$$q''_{2-1} = 966 \text{ Btu/(h ft}^2)$$

Thus, based on this approximate analysis, we have a reduction in the radiation heat transfer of about 92%.

To determine the approximate temperature of the radiation shield, we write

$$q_{2-1} = \frac{E_{bs} - E_{b1}}{R_{1-s} + R_{s1}}$$

$$E_{bs} = E_{b1} + (R_{1-s} + R_{s1})q_{2-1} = [76.6 + (1 + 11.5)966] \text{ Btu/(h ft}^2)$$

$$\sigma T_s^4 = 12{,}200 \text{ Btu/(h ft}^2) \qquad T_s = 1630°\text{R} = 1170°\text{F}$$

Our analysis can be refined by evaluating the absorptivity at surface A_2 at a temperature which better represents the actual source of radiation. For example, the radiosity J_{s2} represents the radiant energy incident upon A_2. Therefore, we could evaluate α_2 at the temperature of a blackbody with total emissive power equal to J_{s2}.

Reradiating Surfaces In the furnace arrangement of Fig. 5–35(a), the plate on the insulated floor is heated to a steady-state temperature which is totally governed by the thermal radiation from the walls at T_2 and ceiling at T_1. Under steady-state conditions, the total rate of radiation heat transfer q_s from the plate to its enclosure

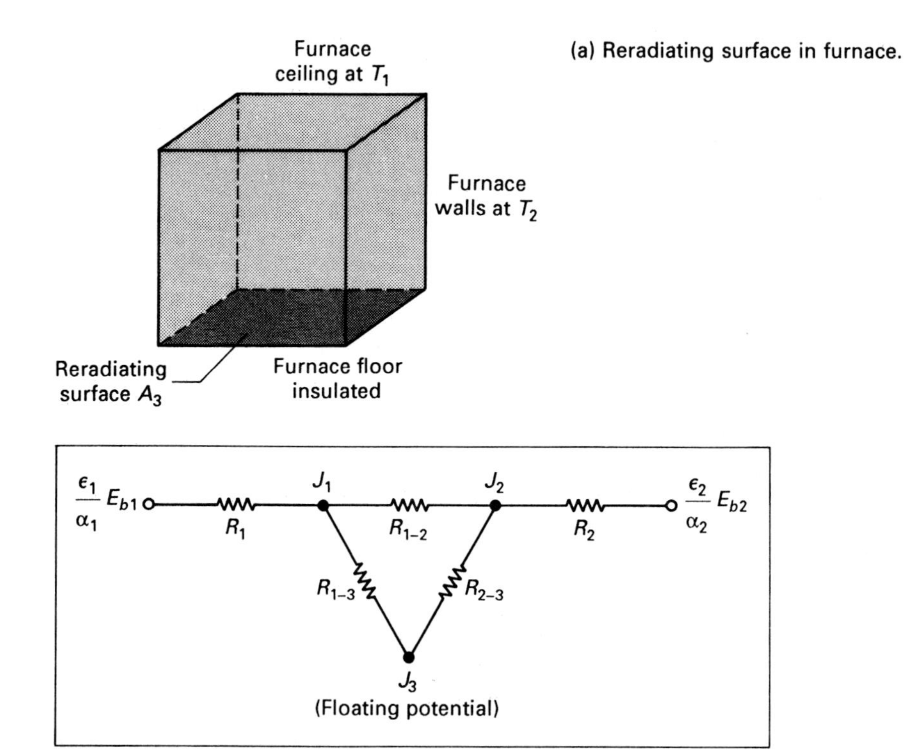

(a) Reradiating surface in furnace.

(b) Thermal radiation network.

FIGURE 5–35 Reradiation.

is clearly zero. Bodies such as this with q_s equal to zero are said to have *reradiating surfaces*.

For the usual case in which the reflectivity ρ_s of an irradiating surface is not equal to unity, the thermal resistance R_s given by Eq. (5–54) or Eq. (5–56) is finite. Thus, with q_s set equal to zero in Eq. (5–55), we conclude that for irradiating surfaces with diffuse characteristics, the floating potential E_{bs} must be equal to the radiosity J_s. This conclusion gives rise to the thermal radiation network shown in Fig. 5–35(b) for the system in Fig. 5–35(a). Solving this network for q_{1-2}, we obtain

$$q_{1-2} = \frac{(\epsilon_1/\alpha_1)E_{b1} - (\epsilon_2/\alpha_2)E_{b2}}{\dfrac{\rho_1}{A_1\alpha_1} + \dfrac{\rho_2}{A_2\alpha_2} + R_R} \tag{5–70}$$

where

$$R_R = \left[A_1F_{1-2} + \frac{1}{1/(A_1F_{1-3}) + 1/(A_2F_{2-3})}\right]^{-1} \tag{5–71}$$

Incidentally, for polished metallic surfaces such as copper (refer back to Fig. 5–8), the absorptivity α_s and emissivity ϵ_s are very small, with ρ_s being almost equal to unity. The thermal resistance R_s for such highly reflective surfaces is very large. If we neglect directional aspects, the thermal radiation network for a near-perfect reflector with specified temperature will be exactly the same as for a reradiating surface with floating potential. But, in reality, important nondiffuse specular characteristics of polished surfaces enter into the picture for many geometric arrangements that compromise the accuracy of Eqs. (5–62) and (5–63).

EXAMPLE 5–20

Outline the steps required for determining the temperature of the reradiating plate surface A_3 shown in Fig. 5–35(a) if the temperature T_1 of the ceiling is 1000°C and the temperature T_2 of the walls is 500°C. Radiation from the interior of the furnace exhibits graybody conditions with emissivity equal to 0.8 and the plate is made of commercial aluminum.

Solution

Objective Summarize the steps required for determining T_3.

Schematic Reradiating surface in furnace.

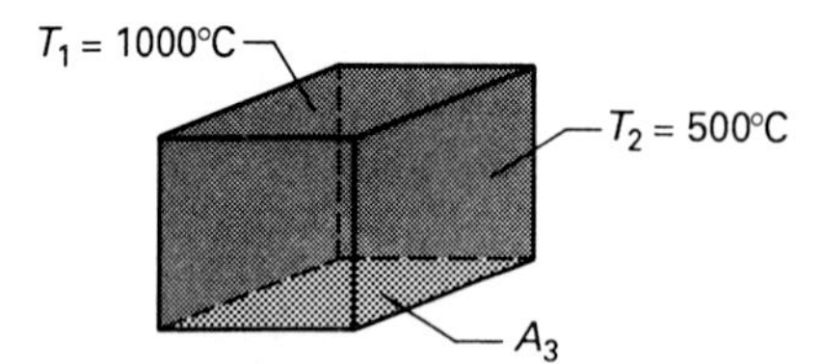

Assumptions/Conditions

steady-state
diffuse opaque graybody surfaces
surface A_3 is reradiating
furnace is evacuated

Properties $\epsilon_1 = 0.8$, $\epsilon_2 = 0.8$.

Analysis First, the radiation properties ϵ_3 and α_3 of surface A_3 can be obtained from Fig. 5–3 and by utilizing either the nonisothermal or graybody α approximation. According to the graybody α approximation, $\alpha_1 = \epsilon_1 = 0.8$ and $\alpha_2 = \epsilon_2 = 0.8$.

Second, the radiation shape factors can be evaluated by utilizing Fig. 5–21 and the principles of reciprocity and summation.

Third, referring to Fig. 5–35, which represents the approximate thermal radiation network for this problem, q_{1-2} can be calculated from Eq. (5–70), J_1 and J_2 can be calculated by writing

$$q_1 = q_{1-2} = \frac{(\epsilon_1/\alpha_1)E_{b1} - J_1}{R_1}$$

$$q_1 = -q_2 = \frac{J_2 - (\epsilon_2/\alpha_2)E_{b2}}{R_2}$$

q_{1-3} can be calculated from

$$q_{1-3} = q_1 - \frac{J_1 - J_2}{R_{1-2}}$$

and E_{b3} ($= J_3$) can be obtained from

$$q_{1-3} = \frac{J_1 - E_{b3}}{R_{1-3}} = \frac{J_1 - \sigma T_3^4}{R_{1-3}}$$

Radiation Factor

In looking back over the analyses developed for diffuse thermal radiation between opaque surfaces, we observe that the rate of radiation heat transfer q_{s-j} between two surfaces A_s and A_j can be expressed in the compact form

$$q_{s-j} = A_s\mathscr{F}_{s-j}\left(\frac{\epsilon_s}{\alpha_s}E_{bs} - \frac{\epsilon_j}{\alpha_j}E_{bj}\right) \tag{5–72}$$

We will refer to $\mathscr{F}_{s-j}$ as the *radiation factor*. For approximate graybody conditions, Eq. (5–72) reduces to

$$q_{s-j} = A_s\mathscr{F}_{s-j}(E_{bs} - E_{bj}) = A_s\mathscr{F}_{s-j}\sigma(T_s^4 - T_j^4) \tag{5–73}$$

For example, for thermal radiation exchange in a bisurface system, $\mathscr{F}_{1-2}$ is given by

$$\mathscr{F}_{1-2} = \frac{1}{\dfrac{\rho_1}{\alpha_1} + \dfrac{1}{F_{1-2}} + \dfrac{A_1}{A_2}\dfrac{\rho_2}{\alpha_2}} \tag{5–74}$$

This equation reduces to

$$\mathscr{F}_{1-2} = \frac{1}{\dfrac{1-\epsilon_1}{\epsilon_1} + \dfrac{1}{F_{1-2}} + \dfrac{A_1}{A_2}\dfrac{1-\epsilon_2}{\epsilon_2}} \tag{5–75}$$

for approximate graybody conditions, and to

$$\mathscr{F}_{1-2} = F_{1-2} \tag{5–76}$$

for the limiting blackbody case.

For purposes of design, $\mathscr{F}_{s-j}$ is listed in Table 5–5 for several practical arrangements. For the more complex geometries for which $\mathscr{F}_{s-j}$ is not tabulated, the thermal radiation network should be solved systematically by numerical or analytical techniques. The systematic solution of thermal radiation problems is discussed in the following section.

TABLE 5–5 Radiation factor $\mathscr{F}_{1-2}$

	$\mathscr{F}_{1-2}$	
System	*General*	*Graybody conditions*
Surfaces A_1 and A_2	$\left(\dfrac{\rho_1}{\alpha_1} + \dfrac{1}{F_{1-2}} + \dfrac{A_1}{A_2}\dfrac{\rho_2}{\alpha_2}\right)^{-1}$	$\left(\dfrac{\epsilon_1 - 1}{\epsilon_1} + \dfrac{1}{F_{1-2}} + \dfrac{A_1}{A_2}\dfrac{\epsilon_2 - 1}{\epsilon_2}\right)^{-1}$
Infinite parallel plates	$\left(\dfrac{\rho_1}{\alpha_1} + 1 + \dfrac{\rho_2}{\alpha_2}\right)^{-1}$	$\left(\dfrac{1}{\epsilon_1} + \dfrac{1}{\epsilon_2} - 1\right)^{-1}$
Concentric cylinders or spheres	$\left(\dfrac{\rho_1}{\alpha_1} + 1 + \dfrac{D_1}{D_2}\dfrac{\rho_2}{\alpha_2}\right)^{-1}$	$\left(\dfrac{1-\epsilon_1}{\epsilon_1} + 1 + \dfrac{A_1}{A_2}\dfrac{1-\epsilon_2}{\epsilon_2}\right)^{-1}$
Small body A_1 inside large body A_2	$\left(\dfrac{\rho_1}{\alpha_1} + 1\right)^{-1}$	ϵ_1
Blackbody surfaces A_1 and A_2	F_{1-2}	F_{1-2}
Surfaces A_1 and A_2 with with one radiation shield		$\left(\dfrac{1-\epsilon_1}{\epsilon_1} + 1 + \dfrac{A_1}{A_2}\dfrac{1-\epsilon_2}{\epsilon_2} + \dfrac{A_1}{A_s}\dfrac{2-\epsilon_s}{\epsilon_s}\right)^{-1}$
Parallel plates with N radiation shields; $\epsilon_{sj} = \epsilon_s$		$\left(\dfrac{1-\epsilon_1}{\epsilon_1} + 1 + \dfrac{1-\epsilon_2}{\epsilon_2} + \dfrac{2-\epsilon_s}{\epsilon_s}N\right)^{-1}$

Systematic Solution Approach

To develop a systematic solution for the radiation heat transfer in a thermal radiation network, an energy balance can be made on each J node. For the moment, we focus our attention on systems with opaque diffuse surfaces. Referring to the J_s node shown in Fig. 5–36, we have our building-block equation,

$$q_s = \frac{A_s\alpha_s}{\rho_s}\left(\frac{\epsilon_s}{\alpha_s}E_{bs} - J_s\right) \tag{5–53}$$

where $\rho_s = 1 - \alpha_s$. Based on the first law of thermodynamics, q_s must also satisfy

$$q_s = \sum_{j=1}^{N} A_s F_{s-j}(J_s - J_j) \tag{5–77}$$

To obtain a nodal equation for J_s for the case in which the surface temperature T_s is known, we combine Eqs. (5–53) and (5–77), with the result

$$\frac{\alpha_s}{\rho_s}J_s + \sum_{j=1}^{N} F_{s-j}(J_s - J_j) = \frac{\epsilon_s}{\rho_s}E_{bs} \tag{5–78}$$

On the other hand, if the rate of radiation heat transfer q_s from a surface is specified with T_s being unknown, then Eq. (5–77) serves as the nodal equation for J_s. A nodal equation of one of these types can be written for each of the N surfaces. For example, for the three-surface system of Fig. 5–37, we write

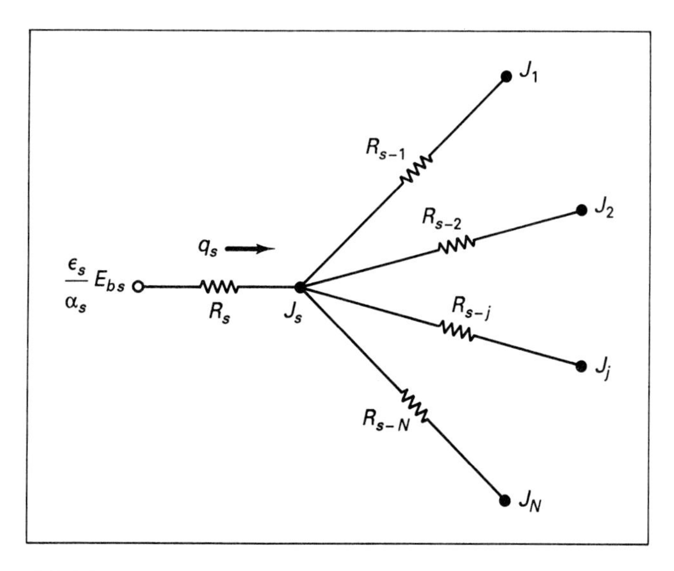

FIGURE 5–36 Segment of thermal radiation network involving J_s node.

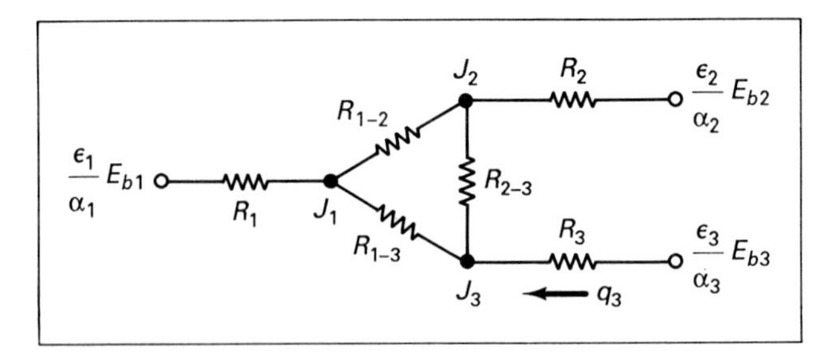

FIGURE 5–37 Thermal radiation network for trisurface system with T_1, T_2, and q_3 specified.

$$\left(\frac{\alpha_1}{\rho_1} + F_{1-2} + F_{1-3}\right)J_1 - F_{1-2}J_2 - F_{1-3}J_3 = \frac{\epsilon_1}{\rho_1}E_{b1} \qquad (5\text{–}79\text{–}1)$$

$$-F_{2-1}J_1 + \left(\frac{\alpha_2}{\rho_2} + F_{2-1} + F_{2-3}\right)J_2 - F_{2-3}J_3 = \frac{\epsilon_2}{\rho_2}E_{b2} \qquad (5\text{–}79\text{–}2)$$

$$-F_{3-1}J_1 - F_{3-2}J_2 + (F_{3-1} + F_{3-2})J_3 = \frac{q_3}{A_3} \qquad (5\text{–}79\text{–}3)$$

In general, for N surfaces the nodal equations are written as

$$a_{11}J_1 + a_{12}J_2 + \cdots + a_{1j}J_j + \cdots + a_{1N}J_N = C_1 \qquad (5\text{–}80\text{–}1)$$

$$a_{21}J_1 + a_{22}J_2 + \cdots + a_{2j}J_j + \cdots + a_{2N}J_N = C_2 \qquad (5\text{–}80\text{–}2)$$

$$\vdots$$

$$a_{j1}J_1 + a_{j2}J_2 + \cdots + a_{jj}J_j + \cdots + a_{jN}J_N = C_j \qquad (5\text{–}80\text{–j})$$

$$\vdots$$

$$a_{N1}J_1 + a_{N2}J_2 + \cdots + a_{Nj}J_j + \cdots + a_{NN}J_N = C_N \qquad (5\text{–}80\text{–N})$$

The N unknowns $J_1, J_2, \ldots, J_j, \ldots, J_s, \ldots, J_N$ in these N equations can be solved by the analytical or numerical techniques which were introduced in Chap. 3. For example, the Gauss-Seidel approach can be used to obtain a hand calculation or computer solution. Or, hand calculation and computer solutions can be affected by powerful matrix methods. For situations in which the graybody α approximation can be utilized, the solutions are quite straightforward. However, for nongraybody conditions, α_s is expressed in terms of the unknown radiosities J_j, such that this approach requires iteration. For this situation, α_s can be specified by Eq. (f) in Example 5–14.

Once the radiosities have been determined, the unknown radiation-heat-transfer rate q_s or temperature T_s can be determined for each surface by employing Eq. (5–53).

As an alternative approach, the equation developed in Example 5–14 can be used. But here again, because of the dependence of the various components of absorptivity α_{sj} on the radiosities J_j, this approach also involves iteration, unless graybody conditions are assumed.

EXAMPLE 5–21

The following information is available for the very long three-surface graybody enclosure shown in Fig. E5–21:

$$T_1 = 200°\text{C} \qquad \epsilon_1 = 0.2$$

$$T_2 = 27°\text{C} \qquad \epsilon_2 = 0.7$$

$$q_3'' = 1\ \text{kW/m}^2 \qquad \epsilon_3 = 0.5$$

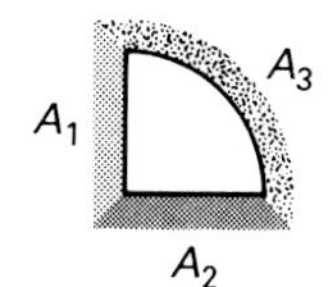

Radius $r_0 = 1$ m
Length very long

FIGURE E5–21
Trisurface graybody enclosure.

Determine the radiation-heat-transfer fluxes from surfaces A_1 and A_2 and the temperature of surface A_3.

Solution

Objective Determine q_1'', q_2'', and T_3.

Assumptions/Conditions

steady-state
diffuse opaque graybody surfaces
enclosure is evacuated

Properties $\epsilon_1 = 0.2$, $\epsilon_2 = 0.7$, $\epsilon_3 = 0.5$.

Analysis The approximate thermal radiation network for this nonsymmetrical system is given in Fig. 5–37.

We first consider the geometric aspects of the problem. As seen in Prob. 5–15, F_{1-2} is equal to 0.293. Based on the reciprocity and summation principles, we write

$$F_{2-1} = \frac{A_1}{A_2} F_{1-2} = 1(0.293) = 0.293$$

$$F_{1-3} = 1 - F_{1-2} = 1 - 0.293 = 0.707$$

$$F_{3-1} = \frac{A_1}{A_3} F_{1-3} = \frac{4}{2\pi}(0.707) = 0.45$$

$$F_{3-2} = \frac{A_2}{A_3} F_{2-3} = 0.45$$

Because A_3 is concave, we also have

$$F_{3-3} = 1 - F_{3-1} - F_{3-2} = 0.1$$

The total emissive powers E_{b1} and E_{b2} are written as

$$E_{b1} = \sigma T_1^4 = 5.67 \times 10^{-8} \frac{\text{W}}{\text{m}^2\ \text{K}^4}(473\ \text{K})^4 = 2.84\ \text{kW/m}^2$$

$$E_{b2} = \sigma T_2^4 = \sigma(300\ \text{K})^4 = 0.459\ \text{kW/m}^2$$

And, because we are dealing with opaque graybody surfaces, the absorptivities and reflectivities are given by

$$\alpha_1 = \epsilon_1 = 0.2 \qquad \rho_1 = 1 - \alpha_1 = 0.8$$

$$\alpha_2 = \epsilon_2 = 0.7 \qquad \rho_2 = 1 - \alpha_2 = 0.3$$

$$\alpha_3 = \epsilon_3 = 0.5 \qquad \rho_3 = 1 - \alpha_3 = 0.5$$

The nodal equations at the three radiosity nodes are given by Eqs. (5–79–1), (5–79–2), and (5–79–3). Substituting for the various potentials, radiation shape factors, and radiation properties, we obtain

$$\left(\frac{0.2}{0.8} + 0.293 + 0.707\right)J_1 - 0.293J_2 - 0.707J_3 = \frac{0.2}{0.8}\left(2.84\ \frac{\text{kW}}{\text{m}^2}\right)$$

$$-\ 0.293J_1 + \left(\frac{0.7}{0.3} + 0.293 + 0.707\right)J_2 - 0.707J_3 = \frac{0.7}{0.3}\left(0.459\ \frac{\text{kW}}{\text{m}^2}\right)$$

$$-\ 0.45J_1 - 0.45J_2 + (0.45 + 0.45)J_3 = 1\ \text{kW/m}^2$$

or

$$1.25J_1 - 0.293J_2 - 0.707J_3 = 0.71\ \text{kW/m}^2$$

$$-\ 0.293J_1 + 3.33J_2 - 0.707J_3 = 1.07\ \text{kW/m}^2$$

$$-0.45J_1 - 0.45J_2 + 0.9J_3 = 1\ \text{kW/m}^2$$

These three equations are easily solved for J_1, J_2, and J_3 by elimination, iteration, or other means. The radiosity values are found to be

$$J_1 = 2.51 \text{ kW/m}^2 \qquad J_2 = 1.17 \text{ kW/m}^2 \qquad J_3 = 2.95 \text{ kW/m}^2$$

With the radiosity values known, calculations are obtained for q_1'' by utilizing Eq. (5–53).

$$q_s = \frac{A_s \alpha_s}{\rho_s}\left(\frac{\epsilon_s}{\alpha_s} E_{bs} - J_s\right)$$

$$q_1'' = \frac{2840 - 2510}{0.8/0.2} \frac{\text{W}}{\text{m}^2} = 82.5 \text{ W/m}^2$$

$$q_2'' = \frac{459 - 1170}{0.3/0.7} \frac{\text{W}}{\text{m}^2} = -1660 \text{ W/m}^2$$

Because q_3'' is known, Eq. (5–53) is rearranged to obtain the surface temperature T_3.

$$E_{b3} = \frac{\alpha_3}{\epsilon_3}\left(q_3 \frac{\rho_3}{A_3 \alpha_3} + J_3\right) = \frac{0.5}{0.5}\left(1000 \frac{0.5}{0.5} + 2950\right) \frac{\text{W}}{\text{m}^2} = 3950 \text{ W/m}^2$$

$$T_3 = \left(\frac{E_{b3}}{\sigma}\right)^{1/4} = \left[\frac{3950 \text{ W/m}^2}{5.67 \times 10^{-8} \text{ W/(m}^2\text{ K}^4)}\right]^{1/4} = 514 \text{ K} = 241°\text{C}$$

To check our solution, we calculate the rates of heat transfer per unit length from each surface.

$$\frac{q_1}{L} = q_1'' r_0 = 82.5 \frac{\text{W}}{\text{m}^2} (1 \text{ m}) = 82.5 \text{ W/m}$$

$$\frac{q_2}{L} = -1660 \frac{\text{W}}{\text{m}^2} (1 \text{ m}) = -1660 \text{ W/m}$$

$$\frac{q_3}{L} = 1000 \frac{\text{W}}{\text{m}^2}\left(\frac{2\pi}{4} \text{ m}\right) = 1570 \text{ W/m}$$

Summing these three values, we find that the system is in balance to within three significant figures.

5–5–3 Unsteady Thermal Radiation Systems

The principles pertaining to thermal radiation presented in this chapter apply for unsteady as well as steady conditions. However, for unsteady systems, we must account for the rate of change in stored energy within bodies with unsteady temperature and we must be aware of changes in the thermal radiation properties that may be brought about by the change in temperature with time.

As suggested earlier in this chapter, practical lumped analyses can sometimes be developed for unsteady thermal radiating systems in which the radiation Biot numbers Bi_R ($\equiv \bar{h}_R \ell / k$) of the bodies with unsteady temperature are of the order of 0.1 and less. The analysis of such systems follows the pattern established in Sec. 2–7 for unsteady one-dimensional convection heat transfer.

To illustrate, we consider the situation in which a small metallic ball initially at T_i with surface area A_1 is placed in an oven with steady surface temperature T_2. Because the unsteady temperature T_1 within the ball is essentially uniform at any instant t for small values of Bi_R, the lumped differential form of the first law is written as (for constant properties)

$$q_{2-1} = \frac{dU}{dt} = \rho V c_v \frac{dT_1}{dt} \tag{5–81}$$

For approximate graybody conditions, q_{2-1} is given by Eq. (5–73),

$$q_{2-1} = A_1 \mathscr{F}_{1-2} \sigma (T_2^4 - T_1^4) \tag{5–82}$$

where $\mathscr{F}_{1-2}$ is given by Eq. (5–74),

$$\mathscr{F}_{1-2} = \frac{1}{\dfrac{\rho_1}{\alpha_1} + \dfrac{1}{F_{1-2}} + \dfrac{A_1}{A_2}\dfrac{\rho_2}{\alpha_2}} \tag{5–74}$$

Setting F_{1-2} equal to unity and assuming that the variation in radiation properties α_1 and α_2 is small as the temperature T_1 changes from T_i to T_2, $\mathscr{F}_{1-2}$ is taken as constant. Hence, our lumped differential formulation becomes

$$\frac{dT_1}{dt} + \frac{\sigma A_1 \mathscr{F}_{1-2}}{\rho V c_v} (T_1^4 - T_2^4) = 0 \tag{5–83}$$

$$T_1 = T_i \qquad \text{at } t = 0 \tag{5–84}$$

The solution of this nonlinear ordinary differential equation is developed in Example 5–22.

For unsteady systems in which the thermal radiation Biot number is not small or for nonsymmetrical conditions, a multidimensional analysis must be developed which accounts for the effects of conduction heat transfer. The topic of mixed-mode thermal radiation is discussed in Sec. 5–5–5.

EXAMPLE 5–22

The walls of a furnace are kept at 1230°C. A thin copper plate (1 m by 1 m by 1 mm) initially at 50°C is then placed in the oven with all surfaces exposed to radiation from the furnace walls. All surfaces can be considered to be black. Determine the temperature of the plate and the heat transfer flux as a function of time. How long does it take for the plate to reach steady state?

Solution

Objective Determine the instantaneous plate temperature T_1, heat transfer rate q''_R, and time t_{ss} required to reach steady state.

Schematic Plate heated in an oven: all blackbody surfaces.

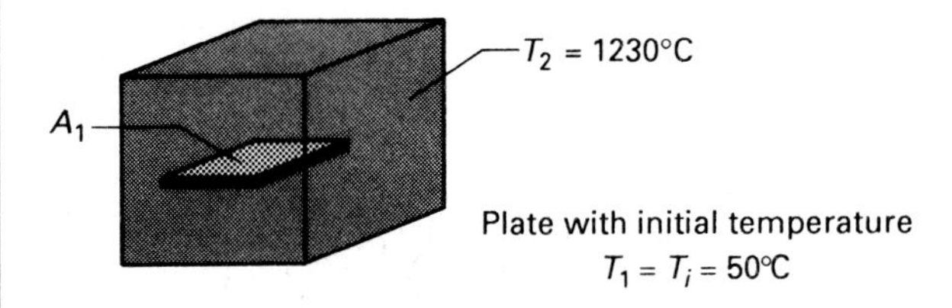

Assumptions/Conditions

unsteady-state
blackbody surfaces
enclosure is evacuated

Properties All surfaces: $\epsilon = 1$.

Analysis Assuming that the thermal radiation Biot number is less than 0.1, the differential formulation for this problem is given by Eqs. (5–83) and (5–84).

$$q_{1-2} + \frac{dU}{dt} = 0$$

$$\frac{dT_1}{dt} + \frac{\sigma A_1 \mathcal{F}_{1-2}}{\rho V c_v}(T_1^4 - T_2^4) = 0$$

$$T_1 = T_i = 323 \text{ K}$$

where $\mathcal{F}_{1-2} = F_{1-2} = 1$ for blackbody radiation.

Separating the variables in this nonlinear first-order differential equation, we write

$$\frac{dT_1}{T_1^4 - T_2^4} = \frac{dT_1}{(T_1^2 - T_2^2)(T_1^2 + T_2^2)} = -\frac{\sigma A_1 \mathcal{F}_{1-2}}{\rho V c_v}\,dt$$

The left-hand side of this relationship can be broken into partial fractions as follows:

$$\frac{dT_1}{(T_1^2 - T_2^2)(T_1^2 + T_2^2)} = \left(\frac{C_1}{T_1^2 - T_2^2} + \frac{C_2}{T_1^2 + T_2^2}\right) dT_1$$

or

$$1 = C_1 T_1^2 + C_1 T_2^2 + C_2 T_1^2 - C_2 T_2^2$$

where C_1 and C_2 must be evaluated. By equating the coefficients associated with T_1, we obtain

$$0 = C_1 + C_2$$

We also have

$$1 = (C_1 - C_2)T_2^2$$

Hence, C_1 and C_2 are given by

$$C_1 = -C_2 \qquad C_1 = \frac{1}{2T_2^2}$$

Therefore, our differential equation takes the simpler form

$$\frac{dT_1}{2T_2^2(T_1^2 - T_2^2)} - \frac{dT_1}{2T_2^2(T_1^2 + T_2^2)} = -\frac{\sigma A_1 \mathscr{F}_{1-2}}{\rho V c_v}\,dt$$

This equation can be integrated with the help of integration tables. The result is

$$\frac{1}{4T_2^3}\ln\left|\frac{T_1 - T_2}{T_1 + T_2}\right| - \frac{1}{2T_2^3}\tan^{-1}\frac{T_1}{T_2} = -\frac{\sigma A_1 \mathscr{F}_{1-2}}{\rho V c_v}\,t + C_3$$

Applying the initial condition, we have

$$C_3 = \frac{1}{2T_2^3}\left(\frac{1}{2}\ln\left|\frac{T_i - T_2}{T_i + T_2}\right| - \tan^{-1}\frac{T_i}{T_2}\right)$$

The solution now can be written as

$$\frac{1}{2}\left(\ln\left|\frac{T_1 - T_2}{T_1 + T_2}\right| - \ln\left|\frac{T_i - T_2}{T_i + T_2}\right|\right) - \left(\tan^{-1}\frac{T_1}{T_2} - \tan^{-1}\frac{T_i}{T_2}\right) = -2T_2^3\frac{\sigma A_1 \mathscr{F}_{1-2}}{\rho V c_v}\,t \tag{a}$$

Setting $\mathscr{F}_{1-2} = 1$, $A_1 = 2\ \text{m}^2$, $V = 10^{-3}\ \text{m}^3$, $T_2 = 1500$ K, $\rho = 8950$ kg/m^3, and $c_v = 0.383$ kJ/(kg °C), Eq. (a) reduces to the form

$$t = \frac{\text{s}}{0.223}\left[\left(\tan^{-1}\frac{T_1}{1500\ \text{K}} - 0.212\right) - \frac{1}{2}\left(\ln\left|\frac{T_1 - 1500\ \text{K}}{T_1 + 1500\ \text{K}}\right| + 0.438\right)\right] \tag{b}$$

This equation is solved for t by taking T_1 as the independent variable. The resulting predictions for T_1 are shown in Fig. E5–22.

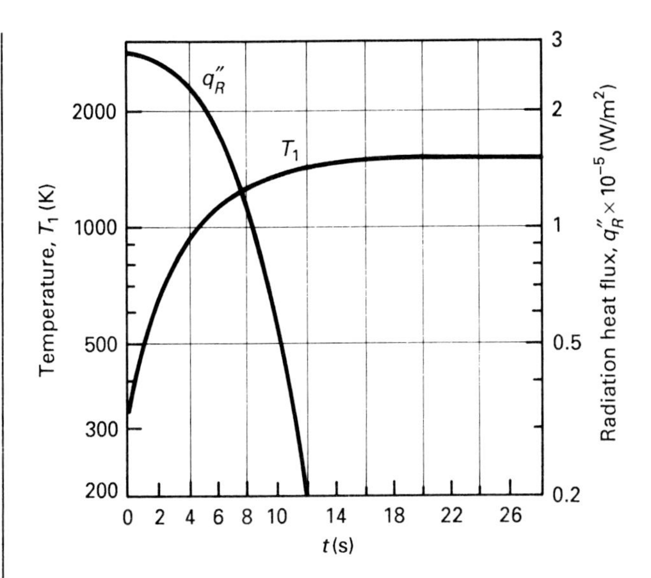

FIGURE E5-22 Calculations for T_1 and q_R''.

To estimate the time required for steady state to occur, we set T_1 equal to 99% of T_R (i.e., $T_1 = 1490$ K). For this value of T_1, Eq. (b) gives

$$t_{ss} = 14.4 \text{ s}$$

The rate of radiation heat transfer at any instant of time t is obtained by writing

$$q_R = \sigma A_1 F_{1-2}(T_1^4 - T_2^4)$$

where T_1 is taken from Fig. E5-22. Calculations for q_R'' are shown in Fig. E5-22.

5-5-4 Transparent Medium

Because of the significance of thermal radiation in transparent gases and glass panes, we now develop practical analyses for radiation heat transfer through participating gaseous and solid mediums. Attention is first given to gases which emit, absorb, and transmit thermal radiation, but which exhibit negligible reflection and scattering characteristics. Then consideration is turned to thermal radiation through transparent solids such as glass.

Participating Gases

To introduce some of the basic concepts involved in the analysis of thermal radiation in participating gases, we restrict ourselves to simplistic but very practical approximate gray gas systems. The basic approach in which participating gases are treated as graybodies was introduced and developed by Hottel in the 1930s in the context of high-gas-temperature systems with highly oxidized or contaminated walls with high emissivity and absorptivity. The gas emissivities given in Fig. 5–18(a) and (b) for water vapor and carbon dioxide were obtained by Hottel and Egbert [16,17] for such systems. Consequently, the gray gas techniques are most reliable for enclosures with blackbody or near-blackbody surfaces. For this reason, we will focus our attention on blackbody and gray near-blackbody enclosures.

Blackbody Enclosures We first consider gray gas/blackbody enclosure systems.

Gray Gas/Single-Surface Enclosure For thermal radiation between an isothermal blackbody enclosure to the gas medium, q_{s-m} is simply equal to the difference between (1) the thermal radiation emitted by the enclosure which is absorbed by the gas, $\alpha_m(A_s F_{s-m} E_{bs})$, and (2) the thermal radiation emitted by the gas, $A_m F_{m-s} \epsilon_m E_{bm}$; that is,

$$q_{s-m} = \alpha_m(A_s F_{s-m} E_{bs}) - A_m F_{m-s} \epsilon_m E_{bm} \tag{5–85}$$

Utilizing reciprocity, it follows that

$$q_{s-m} = A_s F_{s-m}(\alpha_m E_{bs} - \epsilon_m E_{bm}) \tag{5–86}$$

where $F_{s-m} = 1$. Referring back to Sec. 5–3, the emissivity ϵ_m of the gas is evaluated at the mean gas temperature T_m and the influence of the source temperature T_s on the gas absorptivity α_m can be accounted for by utilizing approximate graybody correlations of the form of Eqs. (5–28a) and (5–28b).

EXAMPLE 5–23

The walls of a cubical furnace 1 m on a side are maintained at 227°C. The products of combustion at 1 atm and 1230°C consist of 20% CO_2, 15% H_2O, and 65% N_2. Assuming that the walls are essentially black, determine the radiation heat transfer between the gas and the walls.

Solution

Objective Determine q_{m-s}.

Schematic Gas contained in cubical furnace with blackbody surfaces.

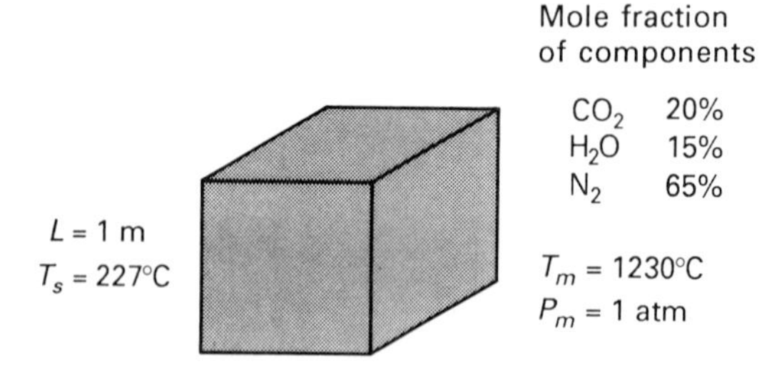

Assumptions/Conditions

steady-state

blackbody surfaces

enclosure contains participating gases H_2O and CO_2

Properties Walls: $\epsilon = 1$.

Analysis The gas emissivity and absorptivity are shown in Example 5–7 to be approximately

$$\epsilon_m = 0.174 \qquad \alpha_m = 0.312$$

The total emissive powers E_{bs} and E_{bm} are given by

$$E_{bs} = \sigma T_s^4 = 5.67 \times 10^{-8} \frac{\text{W}}{\text{m}^2\,\text{K}^4}(500\text{ K})^4 = 3.54 \times 10^3\ \text{W/m}^2$$

$$E_{bm} = \sigma T_m^4 = \sigma(1500\text{ K})^4 = 2.87 \times 10^5\ \text{W/m}^2$$

The radiation heat transfer between the burning gases and the surface of the furnace are calculated by utilizing Eq. (5–86).

$$q_{m-s} = -q_{s-m} = A_s F_{s-m}(\epsilon_m E_{bm} - \alpha_m E_{bs})$$

$$= 6\ \text{m}^2\,(1)[0.174(2.87 \times 10^5) - 0.312(3.54 \times 10^3)]\frac{\text{W}}{\text{m}^2} = 293\ \text{kW}$$

Gray Gas/Multisurface Enclosure For multisurface blackbody enclosures, we have equations similar to Eq. (5–85) for each surface. For example, for two surfaces A_s and A_j we have

$$q_{s-m} = A_s F_{s-m}(\alpha_{ms} E_{bs} - \epsilon_m E_{bm}) \qquad (5\text{–}87)$$

and

$$q_{j-m} = A_j F_{j-m}(\alpha_{mj} E_{bj} - \epsilon_m E_{bm}) \qquad (5\text{–}88)$$

where the second subscript on the absorptivity of the gas medium designates the source of the irradiation.

In addition, an expression must be developed for the rate of thermal radiation q_{s-j} transmitted through the medium from surface s to surface j. The rate of thermal radiation leaving surface A_s that reaches A_j is $A_s F_{s-j} E_{bs} \tau_{ms}$ and the rate from surface A_j to A_s is $A_j F_{j-s} E_{bj} \tau_{mj}$. Therefore, the rate of thermal radiation from A_s to A_j is

$$q_{s-j} = A_s F_{s-j}(E_{bs}\tau_{ms} - E_{bj}\tau_{mj}) \qquad (5\text{–}89)$$

If the surface temperatures T_s and T_j are not too different, τ_{ms} and τ_{mj} can be represented by an average value τ_m, such that Eq. (5–89) becomes

$$q_{s-j} = A_s F_{s-j}\tau_m(E_{bs} - E_{bj}) \qquad (5\text{–}90)$$

Following through, α_{ms} and α_{mj} in Eqs. (5–87) and (5–88) can be set equal to α_m. Otherwise, Eq. (5–89) can be left in its present form and the distinction between α_{ms} and α_{mj} in Eqs. (5–87) and (5–88) can be retained.

Equations (5–87) through (5–90) provide the basis for the thermal radiation network representation of multisurface blackbody systems involving a participating gray gas. For example, the parallel-plate system in Fig. 5–38(a) is represented by the network shown in Fig. 5–38(b) for the case in which $\tau_{m1} \simeq \tau_{m2} \simeq \tau_m$ and $\alpha_{m1} \simeq \alpha_{m2} \simeq \alpha_m$. Based on this thermal network, we see that the total radiation-heat-transfer rate q_m from the medium is

$$\begin{aligned} q_m &= \frac{(\epsilon_m/\alpha_m)E_{bm} - E_{b1}}{R_{1-m}} + \frac{(\epsilon_m/\alpha_m)E_{bm} - E_{b2}}{R_{2-m}} \\ &= A_1F_{1-m}\alpha_m\left(\frac{\epsilon_m}{\alpha_m}E_{bm} - E_{b1}\right) + A_2F_{2-m}\alpha_m\left(\frac{\epsilon_m}{\alpha_m}E_{bm} - E_{b2}\right) \end{aligned} \tag{5–91}$$

The rate q_m must also satisfy the first law of thermodynamics,

$$\Sigma\, \dot{E}_i = q_m + \frac{\Delta E_s}{\Delta t} \tag{5–92}$$

which, for steady-state conditions, reduces to

$$q_m = \Sigma\, \dot{E}_i \tag{5–93}$$

If no energy is transferred to the medium from external sources (i.e., $\Sigma\, \dot{E}_i = 0$), then q_m is zero under steady-state conditions and the node E_{bm} becomes a simple floating point. Under these passive equilibrium conditions, the same rate of energy

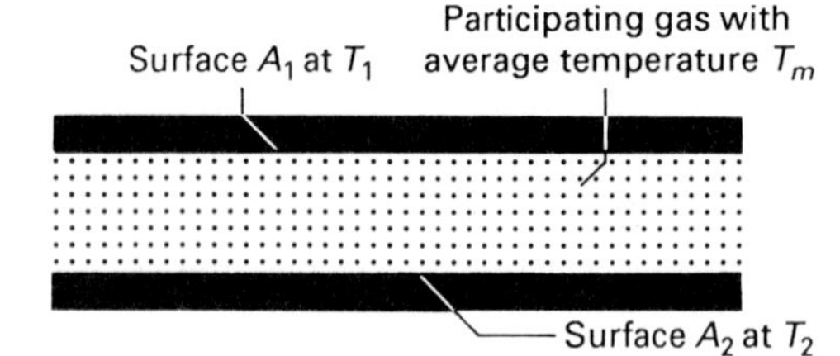

(a) Parallel-plate system.

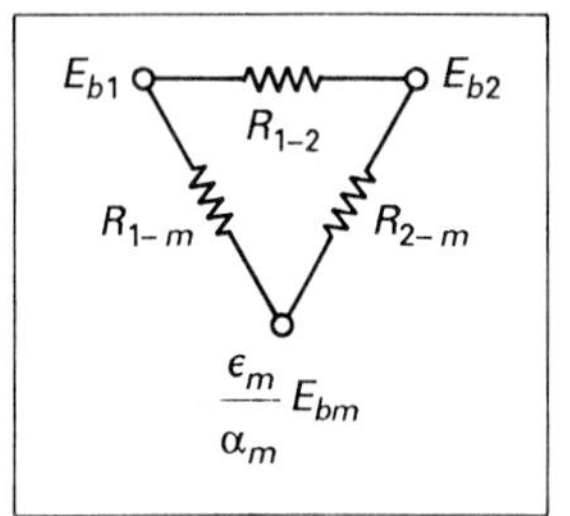

$$R_{1-2} = \frac{1}{A_1F_{1-2}\tau_m}$$

$$R_{1-m} = \frac{1}{A_1F_{1-m}\alpha_m}$$

$$R_{2-m} = \frac{1}{A_2F_{2-m}\alpha_m}$$

(b) Thermal radiation network.

FIGURE 5–38
Radiation heat transfer in a bisurface blackbody system with participating gas.

is emitted by the medium as is absorbed. The solution to the thermal network for this simple case gives rise to

$$q_{1-2} = A_1 \left[F_{1-2}\tau_m + \frac{\alpha_m}{\dfrac{1}{F_{1-m}} + \dfrac{A_1}{A_2 F_{2-m}}} \right] (E_{b1} - E_{b2}) \tag{5–94}$$

Notice that for a nonparticipating gas with $\alpha_m = 0$, this equation reduces to Eq. (5–42),

$$q_{1-2} = A_1 F_{1-2}(E_{b1} - E_{b2}) \tag{5–42}$$

Near-Blackbody Enclosures For situations in which the emissivity of the wall of a single enclosure is of the order of 0.8 and larger, Hottel [17] has shown that the net rate of radiation heat transfer can be approximated by multiplying Eq. (5–86) by the factor $(\epsilon_s + 1)/2$.

$$q_{s-m} = A_s F_{s-m}(\alpha_m E_{bs} - \epsilon_m E_{bm}) \frac{\epsilon_s + 1}{2} \tag{5–95}$$

But it should be emphasized that this approximation is only valid for near-blackbody surfaces.

For enclosures with low emittance surfaces, the gray gas method is not appropriate. For problems of this type, the radiation properties of the gas must be obtained experimentally, or more comprehensive analyses must be developed that account for the band-absorption characteristics of the gas. Higher-order analyses are discussed in references 5, 11, and 24.

Transparent Solids

As indicated in the preceding sections, the dependence of medium irradiation properties on the source is sometimes important in dealing with participating gases. This factor is very critical in important applications involving the transfer of solar radiation through a glass medium into an enclosure. Therefore, this effect will be accounted for as we consider the practical solution approach for radiation in transparent solids.

Blackbody Bisurface System The analysis of thermal radiation between two blackbody surfaces A_1 and A_2 that are separated by a nonreflecting transparent solid medium is identical to the analysis developed in the previous section for a gray gas/multisurface blackbody enclosure. The basic equations for a parallel-plate bisurface system are taken from Eqs. (5–87) through (5–89); that is,

$$q_{1-m} = A_1 F_{1-m}(\alpha_{m1} E_{b1} - \epsilon_m E_{bm}) \tag{5–96}$$

$$q_{2-m} = A_2 F_{2-m}(\alpha_{m2} E_{b2} - \epsilon_m E_{bm}) \tag{5–97}$$

$$q_{1-2} = A_1 F_{1-2}(E_{b1}\tau_{m1} - E_{b2}\tau_{m2}) \tag{5–98}$$

where

$$\alpha_{m1} + \tau_{m1} = 1 \qquad \alpha_{m2} + \tau_{m2} = 1 \tag{5–99,100}$$

In order to develop a thermal radiation network for the important case in which τ_{m1} and τ_{m2} are significantly different, we rearrange Eqs. (5–96) through (5–98) as follows:

$$q_{1-m} = A_1F_{1-m}\frac{\alpha_{m1}E_{b1} - \epsilon_m E_{bm}}{E_{b1} - E_{bm}}(E_{b1} - E_{bm}) \tag{5–101}$$

$$q_{2-m} = A_2F_{2-m}\frac{\alpha_{m2}E_{b2} - \epsilon_m E_{bm}}{E_{b2} - E_{bm}}(E_{b2} - E_{bm}) \tag{5–102}$$

$$q_{1-2} = A_1F_{1-2}\frac{E_{b1}\tau_{m1} - E_{b2}\tau_{m2}}{E_{b1} - E_{b2}}(E_{b1} - E_{b2}) \tag{5–103}$$

These equations provide the basis for the thermal radiation network shown in Fig. 5–39.

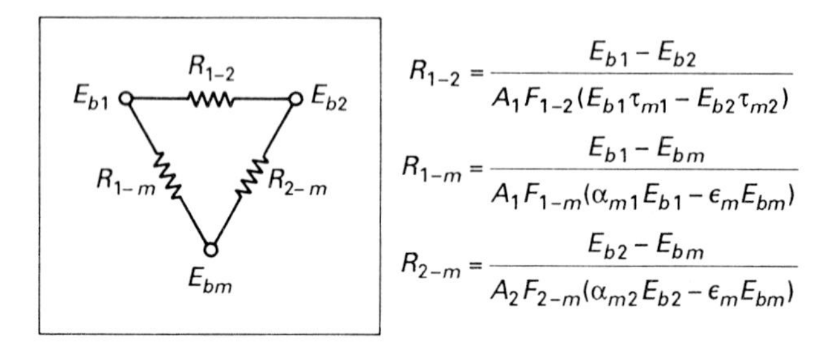

FIGURE 5–39 Thermal radiation network for blackbody bisurface system with transparent solid medium.

The absorptivities and transmissivities can be approximated by the method developed in Example 5–13. Of course, for situations in which the surface temperatures T_1 and T_2 are of the same order of magnitude, $\tau_{m1} \simeq \tau_{m2}$ (and $\alpha_{m1} \simeq \alpha_{m2}$) such that this network reduces to the thermal radiation network shown in Fig. 5–38(b). However, for important applications involving the transfer of solar or high-temperature radiation through glass into enclosures, say from A_1 to A_2, we have a maximum difference between τ_{m1} and τ_{m2}, with the glass being essentially transparent to the high-temperature radiation ($\tau_{m1} \simeq 1$, $\alpha_{m1} \simeq 0$) and nearly opaque to the energy emitted within the enclosure ($\tau_{m2} \simeq 0$, $\alpha_{m2} \simeq 1$). By referring to Fig. 5–39 and by reexamining Eqs. (5–96) through (5–98), we see that this combination of glass properties allows a large net rate of thermal radiation to be transmitted through the glass.

By performing an energy balance on the E_{bm} node, we have (neglecting convection)

$$q_{1-m} = q_{m-2} \tag{5–104}$$

That is, the rate of radiation heat transfer between A_1 and the glass is equal to the net rate of radiant exchange between the glass and A_2. It follows that the total rate of radiation heat transfer from A_1 is given by

$$q_1 = q_{1-2} + q_{1-m} \tag{5–105}$$

The rate of radiation heat transfer q_{1-2} transmitted directly through the glass can be evaluated by using Eq. (5–103). To evaluate q_{1-m}, we must solve for $\epsilon_m E_{bm}$. This is done by utilizing Eq. (5–104). Substituting for q_{1-m} and q_{m-2} in this equation, we obtain

$$A_1F_{1-m}(\alpha_{m1}E_{b1} - \epsilon_m E_{bm}) = A_2F_{2-m}(\epsilon_m E_{bm} - \alpha_{m2}E_{b2})$$

$$\epsilon_m E_{bm} = \frac{A_1F_{1-m}\alpha_{m1}E_{b1} + A_2F_{2-m}\alpha_{m2}E_{b2}}{A_1F_{1-m} + A_2F_{2-m}} \tag{5–106}$$

Employing the principle of reciprocity and recognizing that $F_{m-1} = 1$ and $F_{m-2} = 1$, Eq. (5–106) reduces to

$$\epsilon_m E_{bm} = \frac{\alpha_{m1}E_{b1} + \alpha_{m2}E_{b2}}{2} \tag{5–107}$$

The substitution of this expression for $\epsilon_m E_{bm}$ into Eq. (5–101) or Eq. (5–102) gives

$$\begin{aligned} q_{1-m} = q_{m-2} &= A_mF_{m-1}\left(\alpha_{m1}E_{b1} - \frac{\alpha_{m1}E_{b1} + \alpha_{m2}E_{b2}}{2}\right) \\ &= A_m\left(\frac{\alpha_{m1}E_{b1} - \alpha_{m2}E_{b2}}{2}\right) \end{aligned} \tag{5–108}$$

Returning to Eq. (5–105), q_1 becomes

$$q_1 = A_1F_{1-2}(E_{b1}\tau_{m1} - E_{b2}\tau_{m2}) + A_m\left(\frac{\alpha_{m1}E_{b1} - \alpha_{m2}E_{b2}}{2}\right) \tag{5–109}$$

Based on geometric considerations, F_{1-2} can be set equal to F_{1-m}, such that $A_1F_{1-2} = A_1F_{1-m} = A_mF_{m-1} = A_m$. Thus, our final expression for q_1 takes the form

$$\begin{aligned} q_1 &= A_m\left(E_{b1}\tau_{m1} - E_{b2}\tau_{m2} + \frac{\alpha_{m1}E_{b1} - \alpha_{m2}E_{b2}}{2}\right) \\ &= \frac{A_m}{2}\left[E_{b1}(1 + \tau_{m1}) - E_{b2}(1 + \tau_{m2})\right] \end{aligned} \tag{5–110}$$

This rate of thermal radiation heat transfer from the high-temperature source is equal to the total rate of thermal radiation received by A_2, $-q_2$.

EXAMPLE 5–24

Determine the rate of radiation heat transfer from the interior of a furnace with surface area of 1 m^2 at 2000°C through a 0.1-m^2 glass plate to the interior of a room with a 10-m^2 surface area at 27°C. The glass properties are specified as $\tau_\lambda = 0.90$ for

$0.29\ \mu m < \lambda < 2.7\ \mu m$, $\tau_\lambda = 0$ outside this range, and $\rho_\lambda = 0$ for all wavelengths. Assume that the walls of the furnace and room can be approximated as blackbodies.

Solution

Objective Determine q_{1-2}.

Schematic Furnace with glass window.

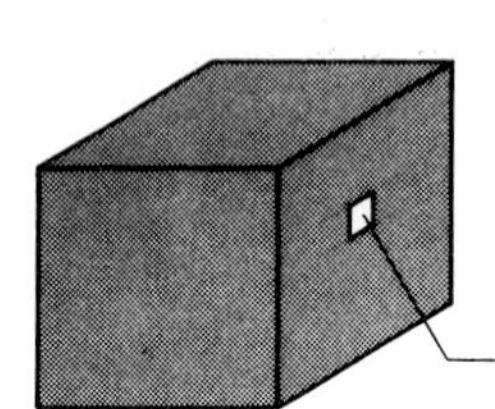

Assumptions/Conditions

steady-state
glass window in blackbody enclosure
enclosure is evacuated

Properties

Furnace and room walls: $\epsilon = 1$.
Glass:

$$\tau_\lambda = 0.9 \qquad \text{for } 0.29\ \mu m < \lambda < 2.7\ \mu m$$
$$\tau_\lambda = 0 \qquad \text{for } \lambda \leq 0.29\ \mu m,\ \lambda \geq 2.7\ \mu m$$
$$\rho_\lambda = 0 \qquad \text{for all } \lambda$$

Analysis Following the pattern established in Example 5–6, we obtain $E_{b,\lambda_1 \to \lambda_2}/E_b$ for both source temperatures in the wavelength range $0.29\ \mu m < \lambda < 2.7\ \mu m$. For the 2270 K temperature source, the corresponding λT range is $659\ \mu m\ K < \lambda T < 6140\ \mu m\ K$. Referring to Table A–H–2, we find

$$\frac{E_{b,\lambda_1 \to \lambda_2}}{E_b} = 0.748 - 0.17 \times 10^{-7} \simeq 0.748$$

It follows that

$$\tau_{m1} = 0.9(0.748) = 0.673$$

For the 300 K source, the λT range is from 87 to 810 μm K and

$$\frac{E_{b,\lambda_1 \to \lambda_2}}{E_b} \simeq 0.738 \times 10^{-4}$$

such that τ_{m2} is essentially zero.

To obtain the total rate of radiation heat transfer we employ Eq. (5–110).

$$q_1 = A_m \left(E_{b1}\tau_{m1} - E_{b2}\tau_{m2} + \frac{\alpha_{m1}E_{b1} - \alpha_{m2}E_{b2}}{2} \right)$$

where $\tau_{m1} = 0.673$, $\tau_{m2} = 0$, $\alpha_{m1} = 0.327$, $\alpha_{m2} = 1$, and

$$E_{b1} = \sigma T_1^4 = 5.67 \times 10^{-8} \frac{\text{W}}{\text{m}^2\,\text{K}^4} (2270\ \text{K})^4 = 1.51 \times 10^6\ \text{W/m}^2$$

$$E_{b2} = \sigma T_2^4 = \sigma(300\ \text{K})^4 = 459\ \text{W/m}^2$$

Substituting into Eq. (5–110), we calculate

$$q_1 = 0.1\ \text{m}^2 \left[1.51 \times 10^6\,(0.673) - 0 + \frac{0.327(1.51 \times 10^6) - 1(459)}{2} \right] \frac{\text{W}}{\text{m}^2}$$

$$= 1.26 \times 10^5\ \text{W}$$

For purpose of comparison, the radiation-heat-transfer rate is calculated for an opening with no glass plate.

$$q_{1-2} = A_m(E_{b1} - E_{b2}) = 0.1\ \text{m}^2\,(1.51 \times 10^6 - 459)\frac{\text{W}}{\text{m}^2} = 1.51 \times 10^5\ \text{W}$$

Thus, we find that the glass plate reduces the rate of radiation heat transfer by about 16%.

Effects of Diffuse Nonblackbody Surfaces and Reflecting Medium Of course, the transfer of thermal radiation through a transparent solid medium generally occurs in the context of nonblackbody surfaces. In addition, reflection by the surfaces of the medium is sometimes significant. For example, the reflectivity of glass is usually of the order of 0.1. These complexities are approximately accounted for in Examples 5–25 and 5–26 by utilizing the radiosity concept.

EXAMPLE 5–25

Develop a thermal radiation network for diffuse nonblackbody parallel plates which are separated by a nonreflecting transparent solid medium.

Solution

Objective Develop a thermal radiation network for this parallel plate system.

Schematic Parallel plates separated by transparent plate.

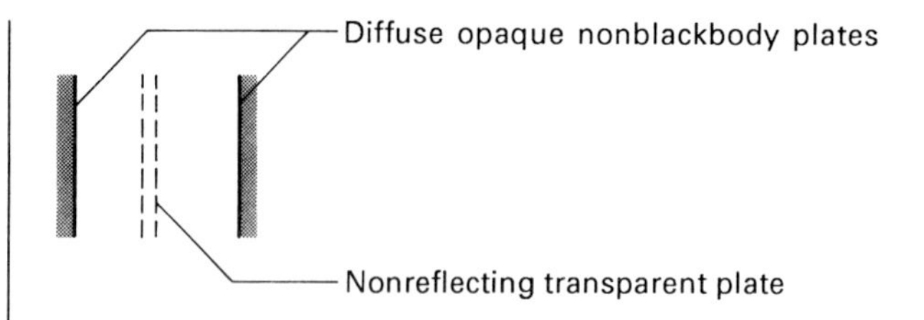

Assumptions/Conditions

steady-state

enclosed spaces are evacuated

Analysis To analyze radiation heat transfer between two diffuse surfaces A_1 and A_2 which are separated by a transparent solid medium, we utilize the concepts that have been introduced in the previous two sections. First, the net rate of radiation heat transfer from each surface is expressed in terms of radiosity by relationships of the form of Eq. (5–53); that is,

$$q_1 = \frac{A_1\alpha_1}{\rho_1}\left(\frac{\epsilon_1}{\alpha_1}E_{b_1} - J_1\right) \qquad q_2 = \frac{A_2\alpha_2}{\rho_2}\left(\frac{\epsilon_2}{\alpha_2}E_{b_2} - J_2\right)$$

A second set of equations is written for the net rate of radiation heat transfer between each surface and the medium A_m. By recognizing that q_{1-m} must be equal to the difference between (1) the energy leaving A_1 that is absorbed by the medium, $\alpha_{m1}(A_1F_{1-m}J_1)$, and (2) the thermal radiation which is emitted by the medium and reaches A_1, $A_mF_{m-1}\epsilon_mE_{bm}$, we have

$$q_{1-m} = \alpha_{m1}(A_1F_{1-m}J_1) - A_mF_{m-1}\epsilon_mE_{bm} = A_1F_{1-m}(\alpha_{m1}J_1 - \epsilon_mE_{bm})$$

Similarly, q_{2-m} takes the form

$$q_{2-m} = A_2F_{2-m}(\alpha_{m2}J_2 - \epsilon_mE_{bm})$$

Finally, the rate of radiation heat transfer between A_1 and A_2 is equal to the difference between (1) the rate of thermal radiation leaving A_1 that reaches A_2, $A_1F_{1-2}J_1\tau_{m1}$, and (2) the rate from A_2 to A_1, $A_2F_{2-1}J_2\tau_{m2}$; that is,

$$q_{1-2} = A_1F_{1-2}(J_1\tau_{m1} - J_2\tau_{m2}) \tag{a}$$

These equations provide the basis for the thermal radiation network shown in Fig. E5–25, where

$$R_{1-2} = \frac{J_1 - J_2}{A_1F_{1-2}(J_1\tau_{m1} - J_2\tau_{m2})}$$

$$R_{1-m} = \frac{J_1 - E_{bm}}{A_1F_{1-m}(\alpha_{m1}J_1 - \epsilon_mE_{bm})}$$

$$R_{2-m} = \frac{J_2 - E_{bm}}{A_2F_{2-m}(\alpha_{m2}J_2 - \epsilon_mE_{bm})}$$

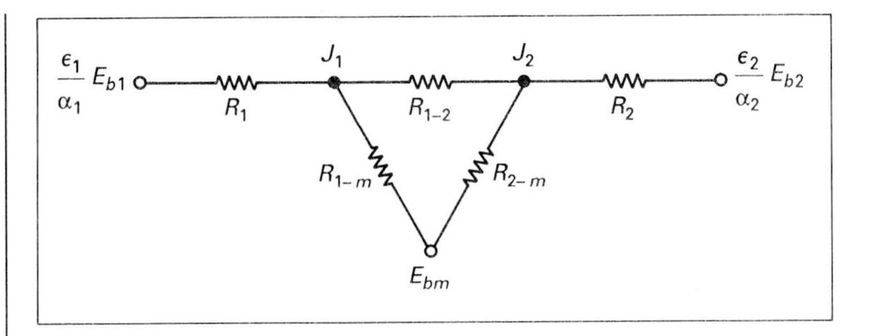

FIGURE E5–25 Thermal radiation network.

Note that this network reduces to the one shown in Fig. 5–39 for blackbody surfaces A_1 and A_2.

EXAMPLE 5–26

Develop a thermal radiation network for two diffuse nonblackbody parallel plates which are separated by a reflecting transparent solid medium.

Solution

Objective Develop a thermal radiation network for this parallel plate system.

Schematic Parallel plates separated by reflecting/transparent plate.

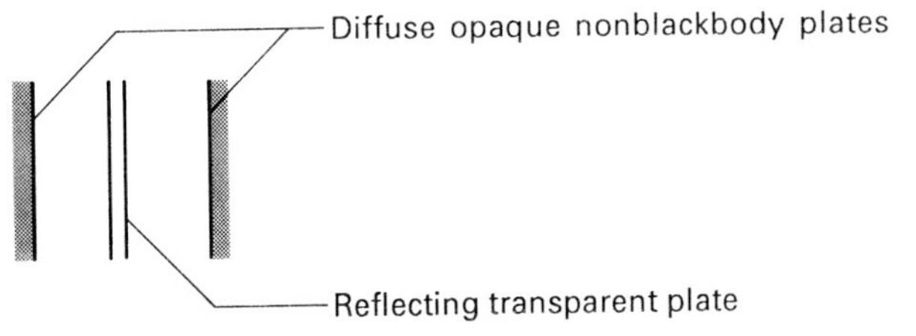

Assumptions/Conditions

steady-state
enclosed spaces are evacuated

Analysis To account for reflection from either side of the solid medium in a bisurface system, we utilize the radiosity concept. The radiosity J_{m1} from one side of the solid is written in terms of the thermal radiation emitted and reflected as

$$J_{m1} = \epsilon_m E_{bm} + \rho_{m1} G_{m1} \tag{a}$$

where G_{m1} is the irradiation from the surface A_1. It is important to note that the thermal radiation transmitted through the solid medium is *not* included in our defining equation for radiosity, but rather is treated separately. The net rate of radiation heat transfer from one side of the solid medium q_{m1} (not including energy transmitted through the solid) is expressed by

$$q_{m1} = A_m(\epsilon_m E_{bm} - \alpha_{m1} G_{m1}) \tag{b}$$

Combining Eqs. (a) and (b) to eliminate G_{m1}, we have

$$q_{m1} = A_m\left[\epsilon_m E_{bm} - \frac{\alpha_{m1}}{\rho_{m1}}(J_{m1} - \epsilon_m E_{bm})\right] = \frac{A_m}{\rho_{m1}}[E_{bm}\epsilon_m(\rho_{m1} + \alpha_{m1}) - \alpha_{m1}J_{m1}]$$

$$= \frac{A_m}{\rho_{m1}}(1 - \tau_{m1})\left(\epsilon_m E_{bm} - \frac{\alpha_{m1}J_{m1}}{1 - \tau_{m1}}\right)$$

Similarly, an equation can be written for q_{m2} of the form

$$q_{m2} = \frac{A_m}{\rho_{m2}}(1 - \tau_{m2})\left(\epsilon_m E_{bm} - \frac{\alpha_{m2}J_{m2}}{1 - \tau_{m2}}\right)$$

To obtain an expression for the net rate of thermal radiation from surface A_1 to the solid medium excluding the energy transmitted through the medium, we take the difference between (1) the rate of nontransmitted thermal radiation leaving A_1 that reaches the medium, $A_1F_{1-m}J_1(1 - \tau_{m1})$, and (2) the rate of nontransmitted thermal radiation leaving the medium that reaches A_1, $A_mF_{m-1}J_{m1}$. That is,

$$q_{1-m} = A_1F_{1-m}J_1(1 - \tau_{m1}) - A_mF_{m-1}J_{m1} = A_1F_{1-m}[J_1(1 - \tau_{m1}) - J_{m1}]$$

$$= A_1F_{1-m}(1 - \tau_{m1})\left(J_1 - \frac{J_{m1}}{1 - \tau_{m1}}\right)$$

Similarly, we write

$$q_{2-m} = A_2F_{2-m}(1 - \tau_{m2})\left(J_2 - \frac{J_{m2}}{1 - \tau_{m2}}\right)$$

To account for the thermal radiation transmitted through the medium, we utilize Eq. (a) in Example 5–25,

$$q_{1-2} = A_1F_{1-2}(J_1\tau_{m1} - J_2\tau_{m2})$$

Finally, the rate of radiation heat transfer from each surface is given by use of Eq. (5–53).

$$q_1 = \frac{A_1\alpha_1}{\rho_1}\left(\frac{\epsilon_1}{\alpha_1}E_{b1} - J_1\right) \qquad q_2 = \frac{A_2\alpha_2}{\rho_2}\left(\frac{\epsilon_2}{\alpha_2}E_{b2} - J_2\right)$$

These expressions permit us to construct the thermal radiation network shown in Fig. E5–26, where

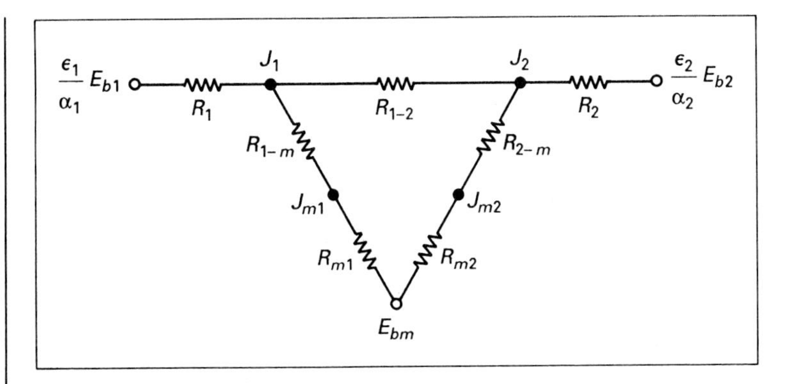

FIGURE E5-26 Thermal radiation network.

$$R_{1-2} = \frac{J_1 - J_2}{A_1 F_{1-2}(J_1 \tau_{m1} - J_2 \tau_{m2})}$$

$$R_{1-m} = \frac{J_1 - J_{m1}}{A_1 F_{1-m}(1 - \tau_{m1}) \left(J_1 - \dfrac{J_{m1}}{1 - \tau_{m1}} \right)}$$

$$R_{m1} = \frac{J_{m1} - E_{bm}}{\dfrac{A_m}{\rho_m}(1 - \tau_{m1}) \left(\epsilon_m E_{bm} - \dfrac{\alpha_{m1} J_{m1}}{1 - \tau_{m1}} \right)}$$

with similar relations for R_{2-m} and R_{m2}.

5-5-5 Thermal Radiation Systems with Combined Modes

Most thermal radiation problems encountered in practice involve at least one of the other two modes of heat transfer. For example, radiation heat transfer in a furnace includes conduction heat transfer through the walls and within bodies being heated as well as convection heat transfer within and without the furnace. Another example is the transfer of heat through superinsulative walls, which involves all three modes.

Practical analyses were developed in Chaps. 1 through 4 for mixed-mode blackbody radiation systems by utilizing the thermal resistance R_R. This practical approach is extended to more realistic one-dimensional nonblackbody combined-mode systems in Examples 5–27 and 5–28. For more complex multidimensional mixed-mode systems, which involve nonuniform surface temperatures, numerical solution techniques are required.

EXAMPLE 5–27

A lightly oxidized thin copper plate with a 1-m^2 surface area is mounted horizontally outdoors with the lower surface insulated. Determine the temperature of the plate on a clear winter night for which the air temperature is 1°C and $\bar{h}$ is 10 W/(m^2 °C). (*Note*: According to Duffie and Beckman [25], the sky can be considered as a blackbody radiator with temperature given by

$$T_{sky} = T_{air} - A$$

where $A = 20°C$ in the winter and $A = 6°C$ in the summer.)

Solution

Objective Determine the plate temperature T_s.

Schematic Thin copper plate (lightly oxidized).

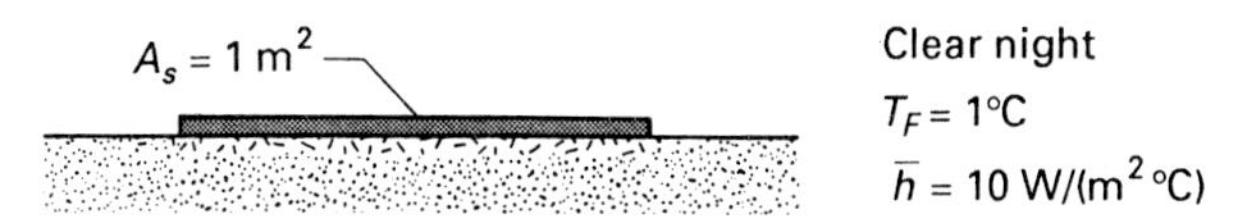

Assumptions/Conditions

steady-state
diffuse opaque graybody plate
sky considered to be a blackbody radiator

Properties

Copper, lightly oxidized (Fig. 5–3): $\epsilon_s = 0.56$.

Sky: $\epsilon_{sky} = 1$.

Analysis Focusing attention on the radiation heat transfer, the thermal radiation network is given in Fig. E5–27a, where

$$T_R = T_{air} - 20°C = 254\ K$$

$$E_{bR} = \sigma T_R^4 \qquad R_{s-R} = \frac{1}{A_s F_{s-R}} = \frac{1}{A_s}$$

Utilizing the nonisothermal or graybody approximation, we have

$$\alpha_s = \epsilon_s = 0.56$$

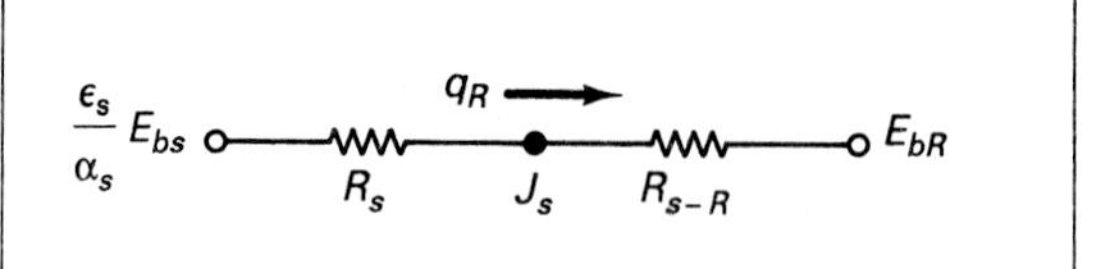

FIGURE E5–27a
Thermal radiation network.

Therefore, R_s is given by

$$R_s = \frac{\rho_s}{A_s\alpha_s} = \frac{1 - 0.56}{A_s(0.56)} = \frac{0.786}{A_s}$$

The rate of radiation heat transfer is now expressed in terms of the unknown surface potential T_s by

$$q_R = \frac{(\epsilon_s/\alpha_s)E_{bs} - E_{bR}}{R_s + R_{s-R}} = \frac{E_{bs} - E_{bR}}{(0.786 + 1)/A_s} = \frac{\sigma A_s(T_s^4 - T_R^4)}{1.79} \tag{a}$$

where $T_R = 254$ K.

Turning to the overall problem, a mixed-mode thermal circuit is shown in Fig. E5–27b. We obtain R_R by rearranging Eq. (a) as follows:

$$q_R = \left[\frac{\sigma A_s}{1.79}(T_s + T_R)(T_s^2 + T_R^2)\right](T_s - T_R) = \frac{T_s - T_R}{R_R}$$

$$R_R = \frac{1.79}{\sigma A_s(T_s + T_R)(T_s^2 + T_R^2)}$$

Of course, R_c is simply equal to $1/(\bar{h}A_s)$.

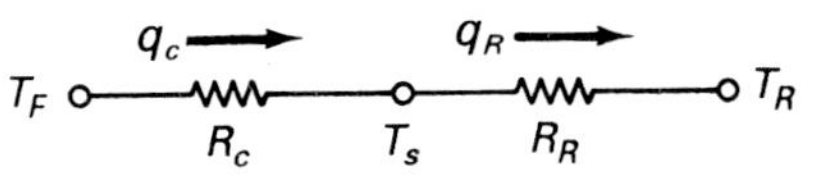

FIGURE E5–27b
Thermal circuit.

To determine the unknown surface temperature T_s, an energy balance is performed on the T_s node.

$$q_c = q_R \qquad \bar{h}A_s(T_F - T_s) = \frac{\sigma A_s}{1.79}(T_s^4 - T_R^4)$$

$$T_s = T_F - \frac{\sigma}{1.79\bar{h}}(T_s^4 - T_R^4) = 274 \text{ K} - \frac{3.17 \times 10^{-9}}{\text{K}^3}[T_s^4 - (254 \text{ K})^4]$$

$$= 287 \text{ K} - \frac{3.17 \times 10^{-9}}{\text{K}^3}T_s^4$$

This equation is solved by iteration, with the result that

$$T_s \simeq 270 \text{ K} = -3°\text{C}$$

Thus, we find that although the air temperature is 1°C, the surface temperature is below freezing because of radiation to the sky.

EXAMPLE 5–28

Determine the rate of heat transfer through a superinsulative spherical wall which consists of an inner surface at $-40°\text{C}$ with a radius 1.01 m, a 1-mm evacuated space, and a 10-cm-thick layer of insulation [$k = 0.60$ W/(m °C)]. The surrounding fluid is at 35°C with $\bar{h} = 150$ W/(m^2 °C). The inner radiative surface is constructed of a highly reflective material with $\rho = 0.9$ while the other surface is a blackbody.

Solution

Objective Determine q_{1-2}.

Schematic Superinsulative spherical wall.

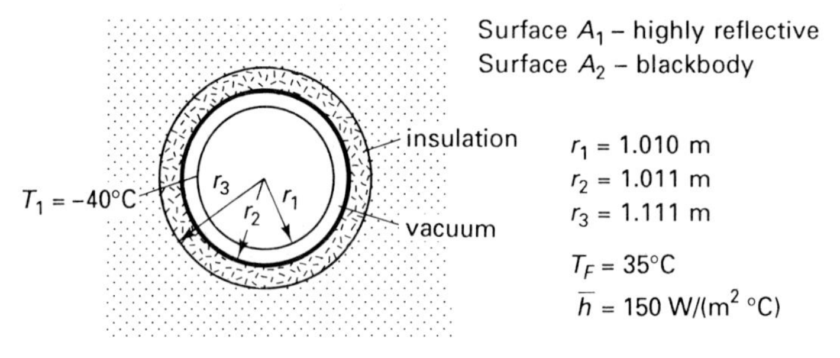

Assumptions/Conditions

steady-state
one-dimensional

Properties

Insulation: $k = 0.6$ W/(m °C).
Surfaces: $\rho_1 = 0.9$, $\alpha_2 = 1$.

Analysis The thermal circuit for this mixed-mode radiation, conduction, and convection system is shown in Fig. E5–28a. The thermal resistances for conduction and convection are given by

$$R_k = \frac{r_3 - r_2}{4\pi r_3 r_2 k} = \frac{(1.111 - 1.011)\text{ m}}{4\pi(1.111\text{ m})(1.011\text{ m})[0.6\text{ W/(m °C)}]} = 0.0118\text{ °C/W}$$

$$R_c = \frac{1}{\bar{h}A_s} = \frac{1}{[150\text{ W/(m}^2\text{ °C)}](4\pi)(1.111\text{ m})^2} = 4.30 \times 10^{-4}\text{ °C/W}$$

q_{1-2} →
$T_1 = 233$ K — R_R — T_2 — R_k — T_3 — R_c — $T_F = 308$ K

FIGURE E5–28a Thermal circuit.

To determine the resistance R_R for thermal radiation, we utilize the thermal radiation network shown in Fig. E5–28b, where

$$E_{b2} = \sigma T_2^4 \qquad R_1 = \frac{\rho_1}{A_1\alpha_1} = \frac{0.9}{A_1(0.1)} = \frac{9}{A_1} \qquad R_{1-2} = \frac{1}{A_1F_{1-2}} = \frac{1}{A_1}$$

and, assuming that $\epsilon_1 \simeq \alpha_1$,

$$\frac{\epsilon_1}{\alpha_1}E_{b1} \simeq E_{b1} = \sigma T_1^4$$

It follows that the rate of radiation heat transfer q_{1-2} can be written as

$$q_{1-2} = \frac{E_{b1} - E_{b2}}{9/A_1 + 1/A_1} = \frac{A_1(E_{b1} - E_{b2})}{10}$$

Rearranging this expression, we have

$$q_{1-2} = \frac{[A_1\sigma(T_1 + T_2)(T_1^2 + T_2^2)](T_1 - T_2)}{10}$$

or

$$q_{1-2} = \frac{T_1 - T_2}{R_R} \tag{a}$$

where $A_1 = 4\pi(1.01\text{ m})^2$, $T_1 = 233$ K, and

$$R_R = \frac{10}{A_1\sigma(T_1 + T_2)(T_1^2 + T_2^2)}$$

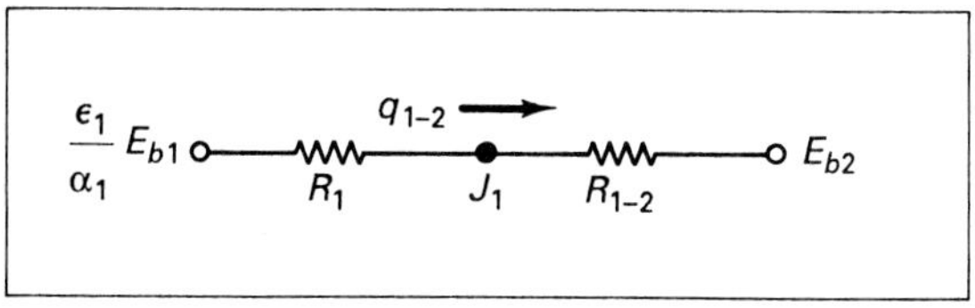

FIGURE E5–28b Thermal radiation network.

Returning to the circuit in Fig. E5–28a, we perform an energy balance on the T_2 node, with the result that

$$\frac{233\text{ K} - T_2}{R_R} = \frac{T_2 - 308\text{ K}}{(1.18 \times 10^{-2} + 4.3 \times 10^{-4})\text{ K/W}}$$

Rearranging, we obtain

$$T_2 = \frac{233\text{ K}/R_R + 2.52 \times 10^4\text{/W}}{81.8\text{ W/K} + 1/R_R}$$

This equation is solved for T_2 by iteration. Starting with $T_2 = 290$ K,

$$R_R = 0.191\text{ K/W} \qquad T_2 = 304\text{ K}$$

A second iteration gives

$$R_R = 0.174 \text{ K/W} \qquad T_2 \simeq 303 \text{ K}$$

Substituting this result into Eq. (a), we obtain

$$q_{1-2} = \frac{233 \text{ K} - 303 \text{ K}}{0.174 \text{ K/W}} = -402 \text{ W}$$

As a point of interest, we note that for blackbody radiation at surface A_1, the resistance R_R reduces to

$$R_R = \frac{1}{A_1\sigma(T_1 + T_2)(T_1^2 + T_2^2)}$$

which is a factor of 10 less than for the case in which a highly reflective surface is used. For the blackbody situation, we find

$$T_2 = 270 \text{ K} \qquad q_R = -2.22 \times 10^4 \text{ W}$$

This heat loss is greater by a factor of 55 than for the case in which one reflective surface is employed. This result gives a good indication of why highly reflective surfaces are used in superinsulative walls.

5-6 SOLAR RADIATION

5-6-1 The Solar Resource

The sun is an essentially spherical body ($r_0 \simeq 0.695 \times 10^6$ km) of extremely high-temperature matter. Within its inner core ($r \lessapprox 0.23r_0$), temperatures of the order of 8×10^6 K to 40×10^6 K are maintained by a continuous fusion process in which mass is converted into energy. This fusion process produces X-ray and γ-ray electromagnetic radiation that emanates from the high-density core. In the low-density gaseous region between $0.7r_0$ and r_0, energy is also transported by convection. The temperature is believed to drop from about 130,000 K to 5000 K across this convective zone. The *photosphere*, which makes up the outer layer of the convection zone, is essentially opaque and well defined. Evaluations of thermal radiation received from the sun, ranging from visible to long wavelength radio waves, indicate that three gaseous layers lie outside the photosphere, with temperatures increasing from about 5000 K within the inner layer to as high as 10^6 K in the outer layer.

Solar radiation consists of energy emitted by the various layers, with the major contributions being provided by the photosphere. Although the radiation emitted by the sun originates from the various temperature zones, for practical purposes the sun can be considered as a blackbody radiator at an effective temperature of about 5800 K. A more accurate value for the effective temperature of the sun according to

Thekaekara [26] is 5762 K. Based on Thekaekara's estimate, the effective total emissive power of the sun is calculated to be about

$$E_{\text{sun}} = \sigma T_{\text{sun}}^4 = 5.67 \times 10^{-8} \frac{\text{W}}{\text{m}^2\,\text{K}^4} (5762\ \text{K})^4 = 6.25 \times 10^7\ \text{W/m}^2$$

The approximate monochromatic emissive power $E_{b\lambda}$ of the sun calculated by the use of the Planck law, Eq. (5–7), is shown in Fig. 5–6.

Of course, only a small fraction of this enormous solar radiation flux reaches the outer fringes of the earth's atmosphere, and an even smaller portion reaches the surface of the earth itself. Based on direct measurements of solar irradiation in the outer reaches of earth's atmosphere, which have been obtained by high-altitude aircraft, balloons, and spacecraft, the monochromatic extraterrestrial solar irradiation $G_{s\lambda}$ reaching the earth's atmosphere for the mean earth–sun distance of about 1.50×10^8 km is given by the standard NASA curve shown in Fig. 5–40. This solar irradiation consists of very short wavelength γ rays, X rays, and ultraviolet rays, as well as thermal radiation. The total extraterrestrial solar irradiation G_s reaching the earth's atmosphere for the mean earth–sun distance is approximately 1350 W/m^2. Strictly speaking, G_s varies by about $\pm 3\%$ as a result of changes in the earth–sun distance and the conditions on the sun. However, for practical purposes G_s can be taken as a constant. For this reason G_s is generally referred to as the *solar constant*.

The solar irradiation flux is further attenuated by the atmosphere before it reaches the surface of the earth. The γ-ray and X-ray components of the solar irradiation are absorbed in the ionosphere by nitrogen, oxygen, and other materials, and nearly all of the UVC and much of the UVB is absorbed by the ozone layer in the stratosphere.

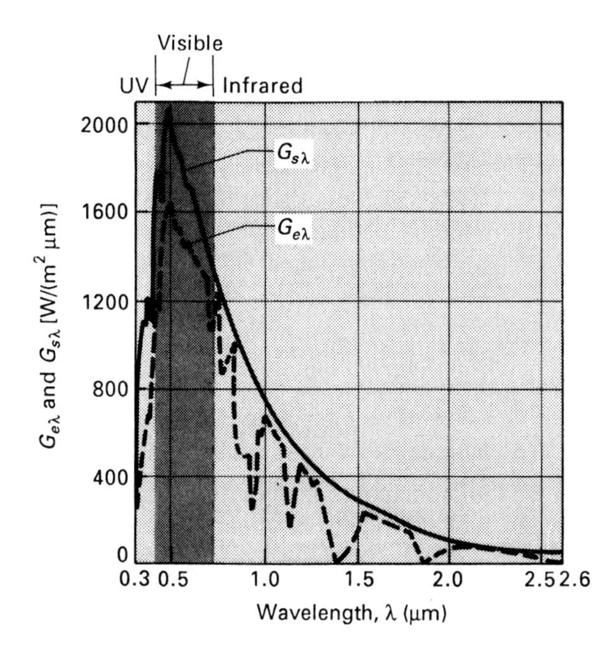

FIGURE 5–40
Representative monochromatic extraterrestrial solar irradiation $G_{s\lambda}$ and monochromatic solar irradiation $G_{e\lambda}$ (Thekaekara [26]).

Absorption also contributes to the attenuation of longer wavelength radiation, with O_2 and O_3 affecting visible light, and water vapor and CO_2 affecting the infrared. Of course, atmospheric pollutants also contribute to the absorption of solar radiation. In addition to absorption, molecular scattering in the atmosphere results in the redirection of much of the incoming solar radiation. According to the work of J. W. S. Rayleigh, the degree of scattering is inversely proportional to λ^4. This phenomenon is known as the *Rayleigh effect*. It follows that the influence of scatter increases as wavelength decreases according to the progression IR $<$ Visible $<$ UVA $<$ UVB $<$ UVC. In this connection, it has been estimated that as much as 50% of the UVB radiation reaching the atmosphere is scattered to space [27]. It follows that only about 3.5% of the visible light and even less infrared radiation entering the atmosphere is scattered to space.

The geometric factors associated with irradiation can be handled with the aid of celestial geometry. However, the lack of meteorological information generally makes it impractical to base predictions for the terrestrial solar resource at any location on extraterrestrial solar irradiation. Rather, the terrestrial solar irradiation is usually measured over a reasonable length of time at the location of interest. Data of this type have been collected at a number of stations across the United States and Europe for many years. Numerous other stations are being set up to measure the solar resource throughout the world. Such data are provided on hourly, daily, and annual bases, with the annual solar irradiation data being of the most practical value in the design of solar systems. Representative experimental measurements for the monochromatic solar irradiation $G_{e\lambda}$ reaching the earth's surface for a very clear atmosphere are shown in Fig. 5–40. The irregularities in $G_{e\lambda}$ are caused by absorption of water vapor, carbon dioxide, and oxygen. Notice that the solar radiation reaching the surface of the earth primarily consists of thermal radiation with wavelength ranging from 0.3 to 2.5 μm.

To illustrate the importance of weather conditions and time of day, the total terrestrial solar irradiation G_e measured in the springtime at Madison, Wisconsin, is shown in Fig. 5–41 for both clear and cloudy days.

Concerning the location of a receiving surface, the solar irradiation is a function of the geometrical relation between the surface and the sun, which is continuously changing on both a daily and annual basis. One of the most important practical aspects involving location is the total number of hours of daytime per year. Another important factor pertains to the orientation of the surface itself. Of course, the maximum irradiation flux on a receiving surface at a given location on the earth is obtained by continual adjustment of the surface orientation to maintain normal solar incidence throughout the daylight hours. But such adjustment requires rotation about two axes, which is generally not practical, except for high-temperature power generation stations. Based on the analysis of much experimental data, researchers in the area of solar radiation have recommended that for maximum annual collection, a receiver should be oriented toward the equator with a slope approximately equal to the latitude ϕ. For best winter collection the slope should be about $\phi + 10$ degrees, and for optional summer irradiation the slope should be approximately $\phi - 10$ degrees.

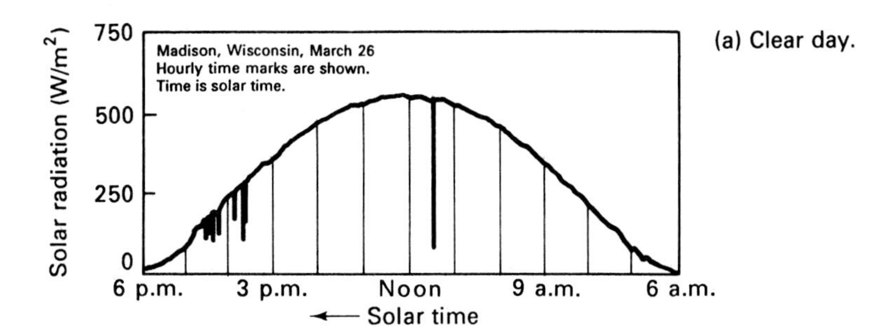

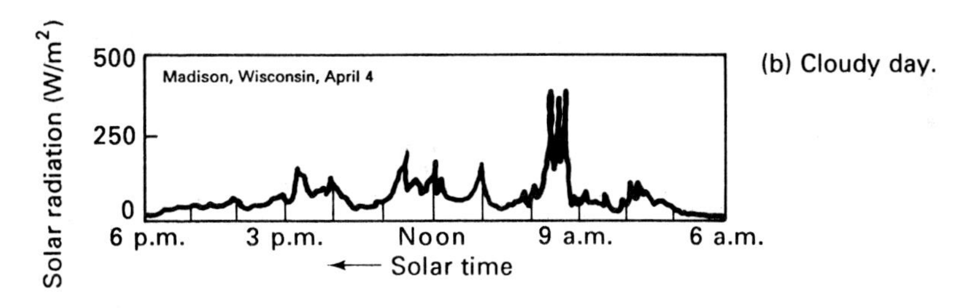

FIGURE 5-41 Total solar irradiation on a horizontal surface vs. time. (From Duffie and Beckman [25]. Used with permission.)

The solar radiation reaching the surface of the earth is absorbed and reflected. Some of the energy that is absorbed is immediately reemitted, with the remainder being stored thermally in the land masses and oceans, chemically in vegetation through the process of photosynthesis, and mechanically in form of wind and eventually reemitted from water vapor as it condenses in the upper atmosphere.

EXAMPLE 5-29

Assuming that the sun radiates as a blackbody at 5760 K, develop a prediction for the extraterrestrial irradiation G_s on the atmosphere of the earth.

Solution

Objective Calculate the extraterrestrial irradiation G_s.

Schematic Earth-sun geometric relationship (not to scale).

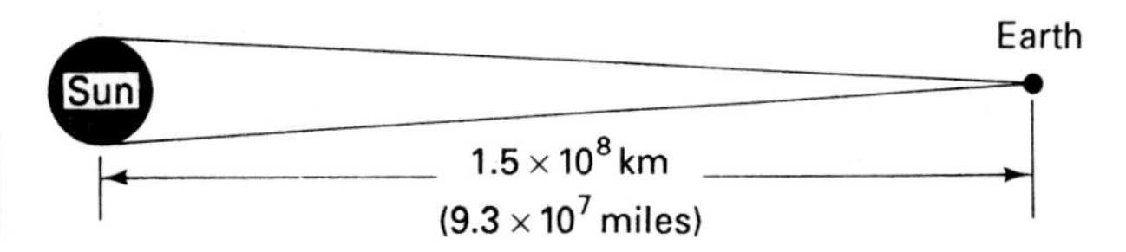

Assumptions/Conditions

solar irradiation is equivalent to blackbody emission at 5760 K

Properties Sun at 5760 K: $\epsilon = 1$.

Analysis Referring to the geometric relationship between the earth and the sun shown in the schematic, the solar irradiation G_s reaching the earth's outer atmosphere is given by

$$G_s \, dA_e = A_{\text{sun}} F_{A_{\text{sun}} - dA_e} E_{\text{sun}} \tag{a}$$

where the differential area dA_e is normal to the line r drawn between the earth and the sun. Based on reciprocity, we write

$$G_s = F_{dA_e - A_{\text{sun}}} E_{\text{sun}} \tag{b}$$

By visualizing the differential area dA_e and sun as two parallel disks, $F_{dA_e - A_{\text{sun}}}$ can be approximated by Eq. (b) of Example 5–9,

$$F_{dA_e - A_{\text{sun}}} = \frac{r_0^2}{Z^2 + r_0^2} \tag{c}$$

where the distance Z between the earth and sun is about 1.5×10^8 km and the radius of the sun r_0 is about 0.695×10^6 km. Combining Eqs. (b) and (c) and employing the Stefan-Boltzmann law, we have

$$G_s = \frac{\sigma T_{\text{sun}}^4}{(Z/r_0)^2 + 1} = \frac{[5.67 \times 10^{-8} \text{ W/(m}^2 \text{ K}^4)](5760 \text{ K})^4}{[(1.5 \times 10^8)/(0.695 \times 10^6)]^2 + 1} = 1340 \text{ W/m}^2$$

This value of the radiation constant G_s lies between the standard value of 1353 W/m^2 proposed in 1971 by Thekaekara and Drummond [28] and an earlier standard of 1322 W/m^2.

5–6–2 Solar-Energy Systems

A variety of solar-energy systems have been devised to satisfy applications ranging from the heating of water, the heating and cooling of residential and commercial buildings, and the operation of desalination plants, to the production of power on a commercial basis and the energizing of electronic equipment in remote locations on land, sea, and in space. Several basic types of solar-energy systems include solar-thermal systems, ocean thermal systems, wind systems, photovoltaic devices, as well as others. We will restrict our attention to solar-thermal systems.

Solar-Thermal Systems

All solar systems that receive, collect, store, and utilize thermal radiation directly from the sun fall under the heading of solar thermal. These systems consist of two

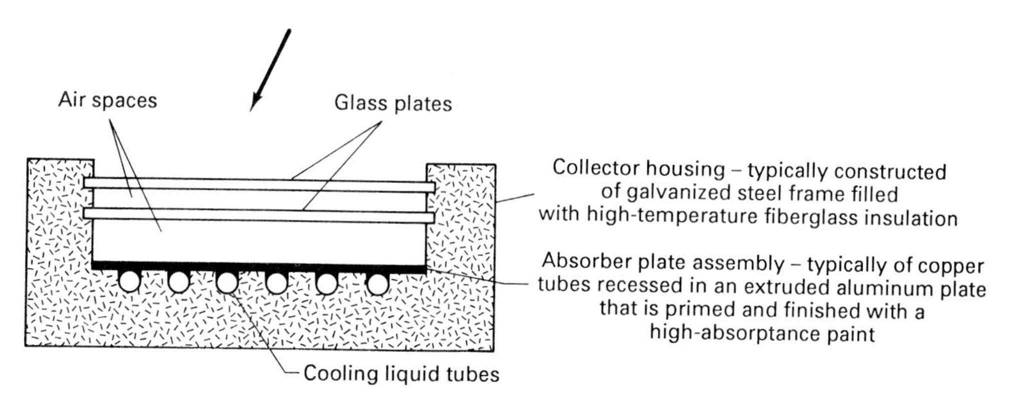

FIGURE 5–42 Flat plate collector cross-section.

basic components: a collector and a storage unit. Of course, these solar components must be interfaced with conversion devices (i.e., air-conditioning units and engines), loads, auxiliary energy supplies, and controls to obtain a total energy system.

The solar collector is actually a thermal radiation heat exchanger in that such devices transfer solar energy to a fluid. The cross section of a typical flat-plate collector arrangement is illustrated in Fig. 5–42. The simple flat-plate collector consists of a near-black solar energy absorbing surface, which collects and transfers the solar energy to a fluid; a transparent cover, which minimizes radiation and convection losses to the atmosphere; and back insulation, which minimizes conduction losses. Flat-plate collectors are utilized in applications requiring moderate temperatures up to about 100°C above ambient. Applications involving the use of flat-plate collectors include water and environmental heating, refrigeration, and seawater desalination.

It should be noted that concentrating collectors are capable of operating at much higher temperatures than flat-plate collectors. Whereas single-axis systems can operate at temperatures up to 315°C, systems with double-axis focusing are able to attain temperatures as high as 3600°C.

Because of the intermittent nature of solar radiation, the storage unit is an essential part of most solar systems. Simply put, the procedure for storing energy in thermal solar-energy systems consists of heating a fluid or solid which is confined in a well-insulated space. For example, hot water is often circulated from the collector into a heavily insulated tank. For systems that utilize air as the working fluid, the hot air is sometimes passed from the collector into a porous bed of stone, which receives and stores the energy. However, simple methods such as these are generally restricted to storage periods of only a few days, at best, because of imperfect thermal insulation and size limitations. Other, more sophisticated methods, which involve materials that store energy by change of phase from solid to liquid, have been found to reduce the storage volume and extend the length of time energy can be stored.

5-7 SUMMARY

As we have seen, the physical mechanism of thermal radiation heat transfer involves the transport of energy between bodies at different temperature by means of electromagnetic waves that are associated with the body temperature, that travel at the speed of light, and that are characterized by wavelength or frequency. The thermal radiation spectrum essentially encompasses ultraviolet radiation (0.1–0.38 μm), visible light (0.38–0.76 μm), and infrared radiation (0.76–1000 μm). Basic concepts pertaining to thermal emission and irradiation properties, in addition to geometric factors, were introduced, and practical thermal analyses were developed for thermal radiation transfer in ideal or near-ideal systems involving diffuse surfaces with uniform and known thermal radiation properties ϵ, α, ρ, and τ, uniform heating conditions (including uniform emission, reflection, and emission), and nonparticipating or gray gases. The use of this basic approach in analyzing problems involving solar radiation, combined modes, and unsteady conditions was also introduced. As we have seen, the concept of *radiosity* is the key to the practical analysis of radiation heat transfer. Either the thermal radiation network approach or the alternative approach of Example 5–14 can be used.

The radiosity concept and other approaches, such as the Monte Carlo method, are used in the analysis of nonideal systems involving surfaces with nondiffuse characteristics, complex spectral properties, and nonuniform heating. Introductions to the analysis of these types of nonideal systems, as well as systems involving multidimensional combined modes and participating gases, are presented by Siegel and Howell [11] and Sparrow and Cess [5]. With regard to solar radiation, which was briefly touched upon in Sec. 5–6, this timely topic is more thoroughly introduced in references 25, 29, and 30.

CHAPTER 6

CONVECTION HEAT TRANSFER: INTRODUCTION

To introduce the subject of convection heat transfer, we consider the main features that distinguish the various convection systems from one another, the concept of the boundary layer, the approaches to the analysis of convection heat transfer that are available to us, and characteristics of internal flow.

6–1 CHARACTERIZING FACTORS ASSOCIATED WITH CONVECTION SYSTEMS

Convection flow systems can be categorized according to several basic factors. These several categories are now introduced.

6–1–1 Forced and Natural Convection

Forced convection and natural convection are two basic mechanisms by which fluid motion can be produced. *Forced convection* represents fluid flow that is caused by mechanical devices such as pumps, fans, or compressors. Examples of forced convection systems include the forced-draft air cooler illustrated in Fig. 6–1, as well as gas turbines, condensers, and evaporators in steam power plants and in refrigeration units, and oil and gas pipelines, to name only a few.

Natural convection refers to fluid motion that is caused by temperature- (or concentration-) induced density gradients within the fluid. Natural convection (or free-convection) flow of air over a steam pipe is represented in Fig. 6–2. Notice that the less-dense air near the steam pipe rises while the heavier cool air falls. Other familiar examples of thermal driven natural convection flow include circulation through fireplaces and the cooling of electronic devices. Concentration driven natural con-

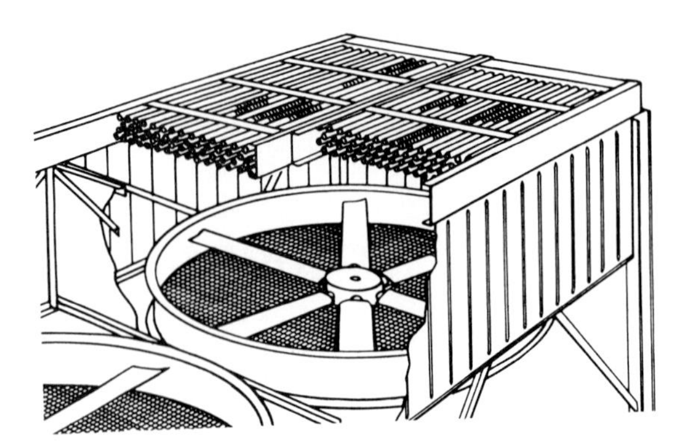

FIGURE 6–1
Cutaway view of forced-draft air cooler. (Courtesy of Yuba Heat Transfer Corporation.)

vection occurs in interfacial mass transfer processes such as evaporation from a vertical porous wet surface.

In practice, many convection-heat-transfer systems involve both the forced and natural convection mechanisms. The topic of combined natural and forced convection is considered in Chap. 9.

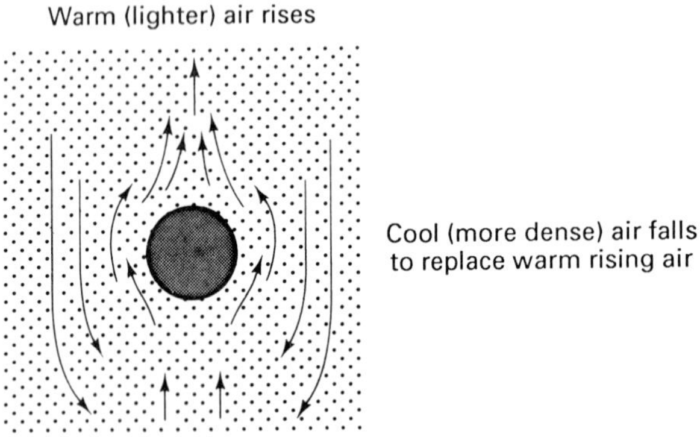

Warm (lighter) air rises

Cool (more dense) air falls to replace warm rising air

FIGURE 6–2
Natural convection flow of air over a heated steam pipe.

6–1–2 Internal and External Flow

Examples of practical internal and external convection flow systems are illustrated in Figs. 6–3 through 6–5. Flow in tubes, channels, annuli, and heat exchangers are examples of internal flows. External flows involve such geometries as flat plates, wing foils, cylinders, and so forth.

It is important to note that the velocity and temperature distributions are generally functions of axial location x in both external and internal flow fields. Such fields are said to be hydrodynamically and thermally developing. However, under certain conditions the velocity and temperature distributions are geometrically similar in the streamwise x-direction. Situations for which *similar flow* fields occur include flow in

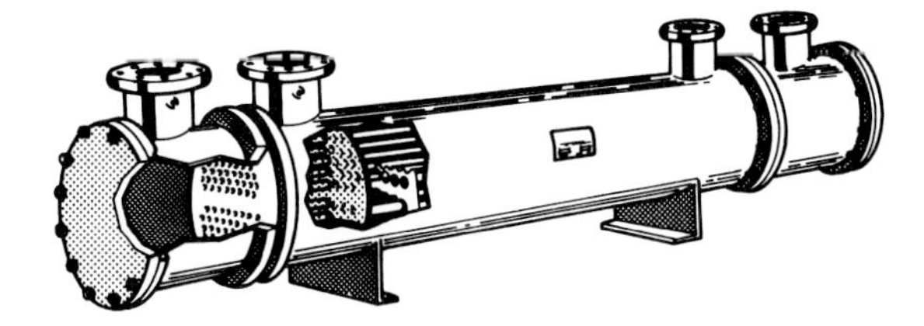

FIGURE 6–3 Typical industrial shell-and-tube heat exchanger—single pass construction on shell side and tube side. (Courtesy of Enerquip.)

FIGURE 6–4 Fin-tube forced-air duct heater. (Courtesy of INDEECO—Industrial Engineering and Equipment Company.)

FIGURE 6–5 External flow over wing foil. (Courtesy of H. Werle, ONERA, Paris.)

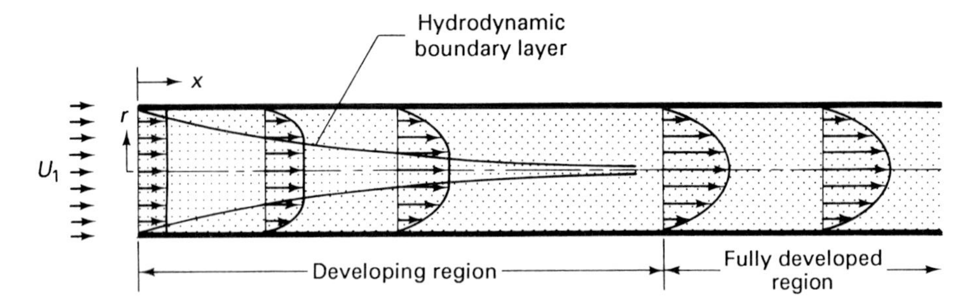

FIGURE 6–6 Typical axial velocity distributions for laminar tube flow.

a tube in the region well downstream from the entrance, flow over a flat plate with uniform free-stream velocity, flow over wedge-shaped bodies, and natural-convection flow over a vertical isothermal surface. To illustrate, Fig. 6–6 shows representative distributions in axial velocity u for tube flow. The velocity profiles are seen to be developing in the entrance region and are fully developed (unchanging) in the region downstream. Further consideration is given to fully developed internal flows in Secs. 6–4 and 7–2–1 and to similar external flows in Secs. 7–2–2 and 7–2–3 and Appendix I.

6–1–3 Time and Space Dimensions

As in the analysis of conduction-heat-transfer systems, convection-heat-transfer processes are categorized according to the time and space dimensions that the temperature and velocity distributions are dependent upon. In our study, attention will be focused upon steady one- and two-dimensional processes such as occur in a circular tube with uniform wall temperature or uniform wall heat flux. However, one should be aware of the importance of more complex multidimensional processes involving unsteady operation and nonuniform heating. For example, unsteady effects are important in the start-up of a boiler and in pulsating flows, and variations in the surface temperature or heat flux around the perimeter of a tube could be important if the tube is radiantly heated on only one side.

6–1–4 Laminar and Turbulent Flow

Laminar forced or natural convection flows exist when individual elements of fluid follow smooth streamline paths, whereas the flow is considered turbulent when the movement of elements of fluid is unsteady and random in nature. This important distinction between laminar and turbulent flow is demonstrated in Fig. 6–7, in which the path followed by fluid is marked by dye. In both cases, the main flow is from left to right. It should be noted that the random fluctuations associated with turbulent flow are superimposed upon the main flow.

One of the simplest ways of experimentally determining whether the flow is laminar or turbulent is to utilize very small electrical heating probes, such as those

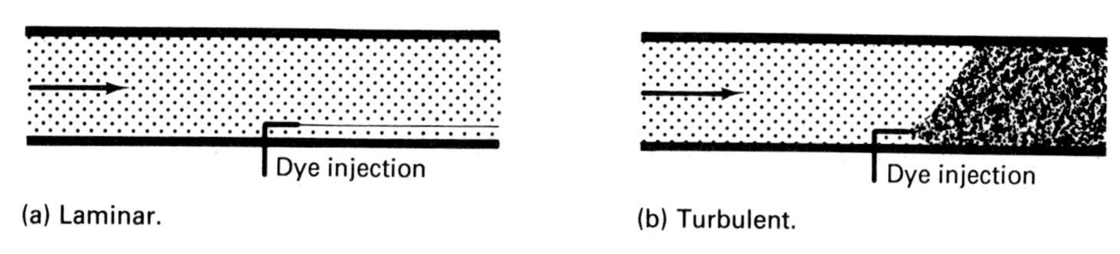

(a) Laminar. (b) Turbulent.

FIGURE 6–7 Typical dye streak patterns for channel or tube flow.

shown in Fig. 6–8, which can be mounted within a flow stream or flush with the surface of a wall. These *anemometer* probes are maintained at an essentially constant temperature by controlling the instantaneous electrical current flow. The instantaneous bridge voltage for a flush-mounted probe is shown in Fig. 6–9 for laminar and turbulent channel flow of liquid mercury. The unsteady character of the turbulent condition is clearly indicated by this signal. Similar signals are obtained from standard probes mounted in the flow stream. When properly calibrated, the signal from a standard anemometer probe can be used to determine the velocity at the location of the probe.

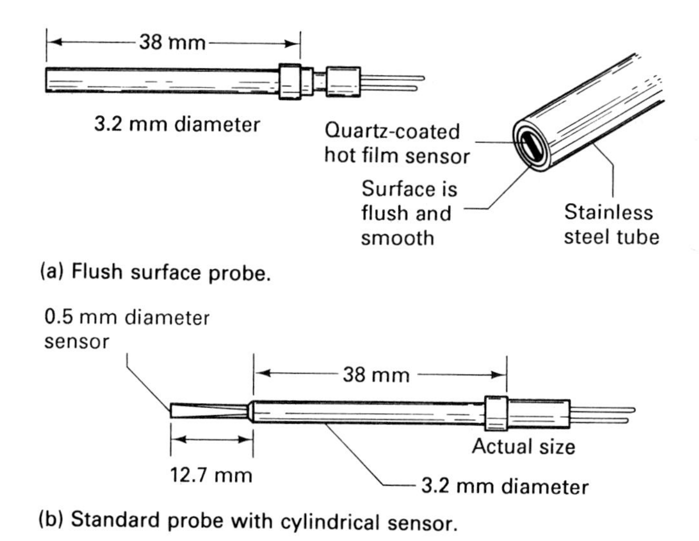

(a) Flush surface probe.

(b) Standard probe with cylindrical sensor.

FIGURE 6–8 Anemometer probes. (Courtesy of TSI Inc.)

6–1–5 Boundary Conditions

The two simplest thermal boundary conditions encountered in convection heat transfer are the uniform wall-heat-flux and wall-temperature conditions. A more general situation is also frequently encountered in heat exchangers involving convection between two fluids separated by a wall. For situations such as this, the temperatures or heat

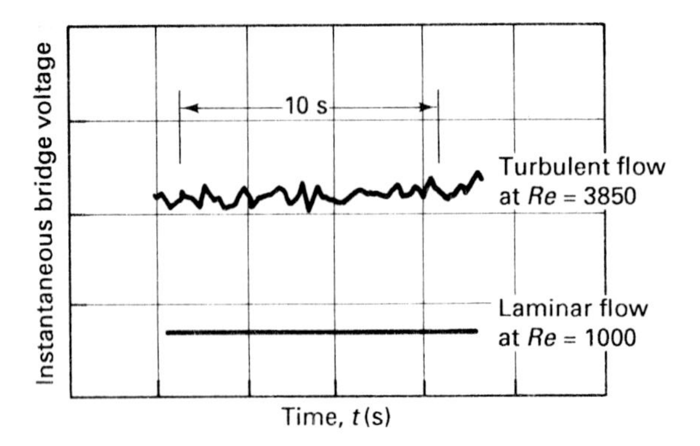

FIGURE 6–9
Signals from flush-mounted anemometer probe for channel flow of liquid mercury; Reynolds number *Re* is defined by Eq. (6–28).

fluxes at the surfaces are not known *a priori*. These three thermal boundary conditions are illustrated in Fig. 6–10. In this connection, a uniform wall-heat-flux condition can be experimentally achieved by the generation of energy in the wall itself by the flow of electric current. A uniform wall-temperature boundary condition is approximated for the case in which heat is convected from a saturated fluid at constant temperature and very high coefficient of heat transfer through a thin metallic wall into the fluid of interest. As mentioned in Sec. 6–1–3, more complex boundary conditions are sometimes encountered in practice, which involve variations in wall temperature or heat flux around the system perimeter.

For momentum transfer, we can use the nonslip condition $u = 0$ at a stationary wall for most fluids. Of course, if the wall itself possesses an axial velocity u_0, then $u = u_0$ at the wall. For nonporous walls, the transverse velocity v is zero at the wall. But for transpired flows $v_0 \neq 0$ since fluid actually passes through the wall.

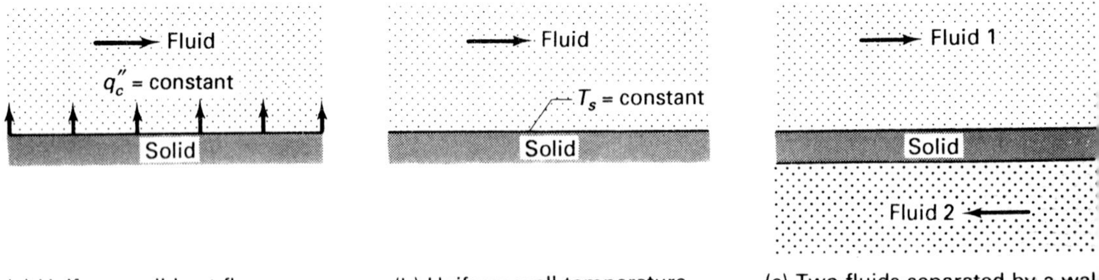

FIGURE 6–10 Three common types of convection boundary conditions.

6–1–6 Type of Fluid

The thermophysical properties of a fluid are in general dependent on its chemical composition, temperature, pressure, and phase. Focusing attention on chemically homogeneous fluids, the most prominent thermophysical properties associated with forced and natural convection heat transfer include density ρ, viscosity μ, specific enthalpy i, specific heat c_P, and thermal conductivity k.

Our study of convection heat transfer will be restricted to processes in which the effects of compressibility are small—that is, on processes in which pressure-induced changes in the density can be neglected. Liquids usually can be treated as incompressible. In addition, compressibility effects can generally be neglected for gas flow processes that operate in the low-Mach-number subsonic range. However, compressibility effects must be accounted for in processes involving highly compressed liquids and supersonic gas flow in which significant pressure related changes in density occur.

In regard to the properties μ and k, many of the fluids encountered in practice can be classified as being Newtonian and isotropic. Such fluids satisfy the following fluid-stress and conduction-heat-transfer laws for two-dimensional (x, r or x, y) conditions:

$$\tau = -\mu \frac{\partial u}{\partial r} \qquad \text{or} \qquad \tau = \mu \frac{\partial u}{\partial y} \tag{6–1a,b}$$

$$q_r'' = -k \frac{\partial T}{\partial r} \qquad \text{or} \qquad q_y'' = -k \frac{\partial T}{\partial y} \tag{6–2a,b}$$

Whereas most fluids are essentially isotropic, some important fluids such as blood exhibit distinct non-Newtonian characteristics. In our study of the fundamentals of convection heat transfer, we will concentrate entirely on Newtonian and isotropic fluids.

The *specific enthalpy i,* defined by

$$i = e + \frac{P}{\rho} \tag{6–3}$$

accounts for the internal energy (e) and flow energy (P/ρ) associated with mass entering or exiting a control volume. Referring to Fig. 6–11, the rate of enthalpy $d\dot{H}$ entering a differential control volume is expressed in terms of i and the mass flow rate $d\dot{m}$ by†

$$d\dot{H} = i \, d\dot{m} \tag{6–4}$$

With dm expressed in terms of density ρ, velocity u, and cross-sectional area dA by

$$d\dot{m} = \rho u \, dA \tag{6–5}$$

(a) Mass flow rate. (b) Rate of enthalpy.

FIGURE 6–11 Mass flow rate $d\dot{m}$ and rate of enthalpy $d\dot{H}$ for fluid with velocity u and density ρ entering a differential control volume with cross-sectional area dA.

† The rate of momentum $d\dot{M}$ associated with mass flow rate $d\dot{m}$ is given by $d\dot{M} = u \, d\dot{m}$.

Eq. (6–4) becomes

$$d\dot{H} = i\rho u \, dA \tag{6–6}$$

Information pertaining to the variation in i with pressure P and temperature T is given in the *Steam Tables* [1] and other references for many common substances. The specific enthalpy is expressed in terms of quality X for two-phase fluids by

$$i = Xi_{fg} + i_f = Xi_g + (1 - X)i_f \tag{6–7}$$

where i_{fg} $(= i_g - i_f)$ is the *latent heat of vaporization*. The specific enthalpy of saturated liquid i_f and that of saturated vapor i_g are tabulated as a function of T or P. As a matter of practicality, e and i are generally expressed in terms of specific heats c_v and c_P for single-phase fluids by Eqs. (1–4) and (1–5),

$$c_v = \left.\frac{\partial e}{\partial T}\right|_v \qquad c_P = \left.\frac{\partial i}{\partial T}\right|_P \tag{6–8,9}$$

in general, or

$$c_v = \frac{de}{dT} \qquad c_P = \frac{di}{dT} \tag{6–10,11}$$

for ideal gases, or

$$c_v = c_P = \frac{di}{dT} \tag{6–12}$$

for incompressible liquids. It should be noted that the specific enthalpy i for ideal fluids (i.e., ideal gases and incompressible liquids) can be expressed in terms of c_P as

$$i = \int_{T_R}^{T} c_P \, dT + i_R \tag{6–13}$$

or, for uniform c_P,

$$i = c_P(T - T_R) + i_R \tag{6–14}$$

where i_R is the value of the specific enthalpy at the reference temperature T_R. In this connection, it is the value of relative specific enthalpy $i - i_R$ that is tabulated in the *Steam Tables* [1].

The dimensionless *Prandtl number Pr*,

$$Pr = \frac{\mu c_P}{k} \tag{6–15}$$

is a key parameter in characterizing convection in single-phase fluids. The Prandtl numbers of several common fluids are given in Tables A–C–3 through A–C–5 in the Appendix. Referring to Fig. 6–12, the Prandtl number ranges from very small values for liquid metals to very large values for highly viscous liquids such as oil. The Prandtl number for gases is generally of the order of unity.

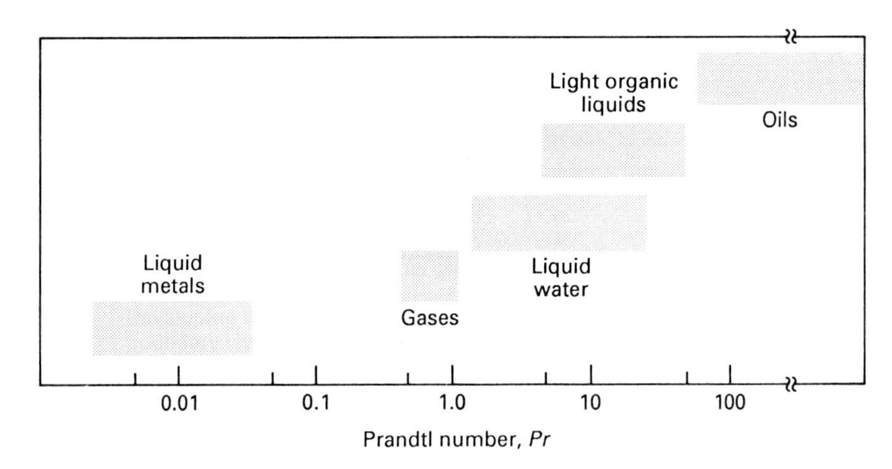

FIGURE 6-12 Prandtl-number spectrum of common type fluids.

The performance of convection processes is also dependent upon whether the fluid exists as a single-phase gas or liquid, or as a multiple-phase substance. Of course, boiling and condensation are the most common types of multiple-phase convection processes. Both of these two-phase convection heat transfer processes are of great technological importance.

Convection heat transfer processes involving chemically nonhomogeneous fluids with interfacial mass transfer include evaporation such as occurs in body perspiration, natural- and forced-draft cooling towers, drying, humidification, and psychrometry; ablation cooling; transpiration cooling; and combustion. The analysis of convection heat transfer processes of this type requires the modeling of species diffusion mass transfer, which is caused on a molecular scale by concentration gradients of two or more of the components. Although the topic of convection heat and mass transfer in chemically nonhomogeneous systems can be quite complex and is a specialization in the field of chemical engineering, practical solution approaches have been developed which are in common use in the engineering profession. The practical solution approach to convection heat and mass transfer is introduced in the *mass transfer supplement*.

6-1-7 Other Factors

Numerous other factors are encountered in convection heat transfer that characterize the particular problem being considered. For example, for situations in which the temperature variation is not large, the properties are generally assumed to be uniform. However, for cases in which the properties vary noticeably with temperature, this temperature dependence must be accounted for. Further, factors such as kinetic energy, energy dissipation, body forces, axial conduction, thermal radiation, and three-dimensional effects can be neglected under certain conditions, but are sometimes quite important. In analyzing any convection-heat-transfer problem, care must be taken to ensure that all the significant factors are accounted for.

6–2 THE BOUNDARY LAYER

For flow over surfaces such as the flat plate shown in Fig. 6–13, the relative velocity is brought to zero at the wall. As the fluid proceeds downstream, the influence of the surface on the velocity reaches further and further into the fluid because of viscous shearing forces. For most practical applications, the region in which the viscous effects are significant is confined to a distinct layer near the surface known as the *hydrodynamic boundary layer* (HBL). In practice, the thickness of the HBL, represented by δ, is generally designated as the distance from the wall at which the velocity differs by some small percentage ϵ (say 1%) from the free-stream or centerline velocity.

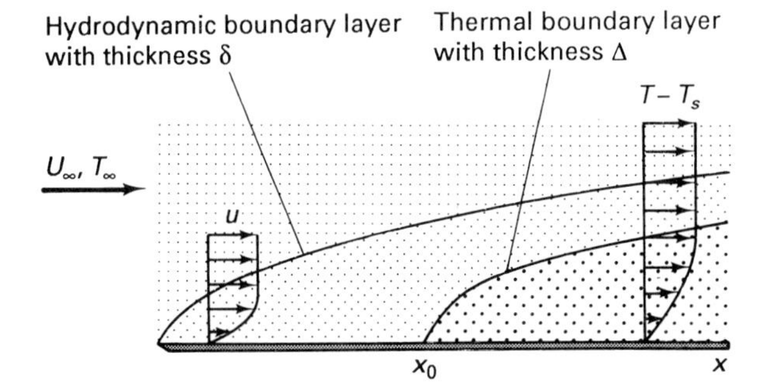

FIGURE 6–13 Development of hydrodynamic and thermal boundary layers for flow over a flat plate with heating in the region $x \geq x_0$.

Similar to the development of the hydrodynamic boundary layer, as fluid flows over a body the temperature of the fluid in contact with the wall is brought to the surface temperature T_s. As the fluid moves downstream from the point at which heating or cooling is initiated, the effect of the wall on the temperature of the fluid penetrates deeper into the fluid, causing the development of a thermal boundary layer (TBL) such as illustrated in Fig. 6–13. The thickness of the TBL is represented by Δ.

The concept of the boundary layer was introduced in 1904 by Prandtl [2] in the context of basic flows of the type shown in Figs. 6–6 and 6–13 for which the boundary layers are very thin or are confined by the system geometry. The approximations proposed by Prandtl for boundary-layer flows of this type provided the impetus for early developments in boundary-layer theory. For future reference, these classic *boundary-layer approximations* are given in the context of two-dimensional flow with x-component velocity u and y-component velocity v by

$$u \gg v \qquad \frac{\partial u}{\partial y} \gg \frac{\partial u}{\partial x}, \frac{\partial v}{\partial y}, \frac{\partial v}{\partial x} \qquad \frac{\partial P}{\partial y} \approx 0 \qquad \frac{\partial P}{\partial x} \approx \frac{dP}{dx} \qquad \text{for the HBL} \tag{6–16}$$

and

$$\frac{\partial T}{\partial y} \gg \frac{\partial T}{\partial x} \qquad \text{for the TBL} \tag{6–17}$$

Prandtl's pioneering work is generally recognized as one of the most important achievements in the modern developments of viscous fluid flow and convection heat transfer.

As illustrated by Fig. 6–13, for external forced convection flow the boundary layers continue to grow as x increases, with the velocity and temperature at the outer edge of the boundary layer being equal to the free-stream values U_∞ and T_∞. Thus, the reference velocity U_F and reference temperature T_F for external forced-convection flow are generally set equal to U_∞ and T_∞, respectively. (The distinguishing features of natural convection boundary layers are considered in Chaps. 7 and 9.) On the other hand, for internal flows such as the one shown in Fig. 6–6, the thickness of the boundary layers that form around the perimeter of the confining surface grow as x increases, until coming together, with the reference velocity and temperature generally specified in terms of bulk-stream values. The bulk-stream characteristics associated with internal flow are introduced in Sec. 6–4.

Referring back to the defining relation for the Prandtl number Pr, Eq. (6–15), by multiplying the numerator and denominator by density, we obtain

$$Pr = \frac{\mu}{\rho}\frac{\rho c_P}{k} = \frac{\nu}{\alpha} = \frac{\text{kinematic viscosity}}{\text{thermal diffusivity}} \tag{6–18}$$

Written in this form, the Prandtl number can be seen to represent the *relative effectiveness of molecular transport of momentum and energy within the hydrodynamic and thermal boundary layers*.

The hydrodynamic characteristics associated with forced-convection boundary layer flow are generally expressed in terms of the *Reynolds number* Re_L,

$$Re_L = \frac{U_F \mathbf{L}}{\nu} \tag{6–19}$$

where $\mathbf{L}$ is a reference length (such as D, D_H, x, or L). The Reynolds number actually represents the *ratio between the inertia and viscous forces* acting upon the fluid. This important dimensionless parameter provides an index that indicates whether the flow is likely to be laminar or turbulent. For example, with $U_F = U_\infty$ and $\mathbf{L} = x$ for forced convection flow over a flat plate with uniform free-stream velocity, the flow is normally laminar for $Re_x \lesssim 5 \times 10^5$ and turbulent for $Re_x \gtrsim 5 \times 10^5$.

6–3 APPROACHES TO THE ANALYSIS OF CONVECTION HEAT TRANSFER

The engineering analysis of convection-heat-transfer processes can be achieved by means of theoretical or practical approaches, with the practical approach sometimes being supplemented by dimensional analysis.

6–3–1 Theoretical Analysis

Convection heat transfer is generally more complex than conduction in a solid or in a stationary fluid because of the superimposed effects of fluid motion. Aside from this complicating factor, the basic concepts involved in the treatment of heat transfer from a solid/fluid interface to a fluid are the same as the treatment of the heat transfer within a solid. Similar to the analysis of conduction heat transfer, the complete theoretical solution of a convection-heat-transfer problem requires the development of a mathematical formulation that represents the energy transport within the fluid itself. In addition, because energy is transported by fluid movement, a mathematical formulation must be developed for the fluid motion. These formulations involve the application of the fundamental conservation principles for mass, momentum, and energy in the context of differential control volumes and generally take the form of systems of partial differential equations.

Strictly speaking, it is the formulation and solution of such equations of motion and energy that provide the means of obtaining predictions for the velocity and temperature distributions within a fluid and the wall shear stress or pressure drop and convection heat transfer. The formulation and solution of the equations of motion and energy will be presented for several convection processes in Chap. 7, which pertains to the theoretical analysis of convection.

6–3–2 Practical Analysis

The engineer is often primarily interested in knowing the pressure drop or drag and the rate of heat transfer (or the surface temperature) and is not always concerned with the details of the velocity and temperature distributions within the fluid. Therefore, a much simpler lumped-analysis approach has been developed that provides a means for obtaining practical engineering calculations. This practical analysis approach involves the use of coefficients of friction or drag and heat transfer and application of the fundamental conservation principles to lumped and lumped/differential control volumes, which generally result in much simpler mathematical formulations than are obtained in the theoretical approach. For situations in which relations for the coefficients of friction and heat transfer are available, the practical lumped-analysis approach will be seen to produce calculations for the wall shear stress (or pressure drop) and the rate of heat transfer (or the surface temperature) for prescribed fluid-flow/heating conditions. However, the velocity and temperature distributions within the fluid cannot be obtained by the lumped approach. If the velocity and temperature profiles are required or if information pertaining to the coefficients of friction and heat transfer for the specific problem of interest is not available, the full theoretical analysis referred to above must be developed or experimental measurements and empirical correlations must be obtained.

Coefficients of Friction and Heat Transfer

The *local* wall-shear stress τ_s and wall-heat flux q''_c are traditionally expressed in terms of coefficients of friction and heat transfer.

The *local Fanning friction factor* f_x is defined by†

$$\tau_s = \rho U_F^2 \frac{f_x}{2} \tag{6–20}$$

where U_F is a reference velocity that depends on the geometry. The *mean Fanning friction factor* $\overline{f}$ over an area of surface A_s is defined as

$$\overline{f} = \frac{1}{A_s} \int_{A_s} f_x \, dA_s \tag{6–21}$$

Both f_x and $\overline{f}$ are dimensionless coefficients. Related dimensionless coefficients that account for entrance, exit, and component pressure losses in internal flow systems are introduced in Chaps. 8 and 11.

The *local coefficient of heat transfer* h_x is defined by the *general Newton law of cooling*, Eq. (1–21),

$$q_c'' = \frac{dq_c}{dA_s} = h_x(T_s - T_F) \tag{6–22}$$

[The heat flux from the wall is given in terms of the distribution in temperature T by Eq. (6–2b) at $y = 0$.] The mean coefficient of heat transfer $\overline{h}$ is defined in terms of h_x by

$$\overline{h} = \frac{1}{A_s} \int_{A_s} h_x \, dA_s \tag{6–23}$$

The coefficients of heat transfer [with dimensions W/(m^2 °C) or Btu/(h ft^2 °F)] are usually expressed in terms of the dimensionless Nusselt number Nu_L (and $\overline{Nu_L}$),

$$Nu_L = \frac{h_x L}{k} \qquad \overline{Nu_L} = \frac{\overline{h} L}{k} \tag{6–24a,b}$$

or dimensionless *Stanton number St* (and $\overline{St}$),

$$St = \frac{Nu_L}{Re_L \, Pr} = \frac{h_x}{\rho c_P U_F} \qquad \overline{St} = \frac{\overline{Nu_L}}{Re_L \, Pr} = \frac{\overline{h}}{\rho c_P U_F} \tag{6–25a,b}$$

where ρ, c_P, and k are properties of the fluid. It should be noted that the Nusselt number represents the *ratio of convection heat transfer for fluid in motion to conduction heat transfer for a motionless layer of fluid*. On the other hand, the Stanton number, which combines the Nusselt number, Reynolds number, and Prandtl number, indicates the *relative magnitude of the actual convection heat flux and the enthalpy energy flux capacity of the fluid flow*.

As we shall see in the next several chapters, theoretical and empirical relations are available for coefficients of friction and heat transfer for many standard convection

† The Fanning friction factor is also commonly represented by C_f. The product $4f$ is referred to as the *Darcy friction factor* λ, $\lambda = 4f$.

processes. These relations are generally expressed in terms of dimensionless parameters effectively representing the geometry, flow, and fluid characteristics. For example, the key independent dimensionless parameters for forced convection include the Nusselt number (or Stanton number), the Prandtl number Pr, and the Reynolds number Re_L.

As an aside, it should be mentioned that Eq. (6–22) actually appears to have first been proposed by Fourier (1768–1830), many years after the death of Newton (1643–1727) [4]. However, it was Newton's earlier work with the concept of temperature that laid the groundwork for the development of the concept of heat transfer in the late eighteenth century and the eventual framing of Eq. (6–22). Strictly speaking, the heat flux q_c'' is not directly proportional to the temperature difference $T_s - T_F$ for all convection heat-transfer processes. For example, q_c'' is a nonlinear function of $T_s - T_F$ for natural convection, boiling and condensation, and forced convection with large temperature differences, such that h_x is a function of $T_s - T_F$. However, h_x is independent (or at least approximately so) of $T_s - T_F$ for many practical forced-convection processes involving mild to moderate temperature differences, such that Eq. (6–22) effectively separates the key variables q_c'' and $T_s - T_F$. As we shall see, the coefficient of heat transfer has been assimilated into the modern practical engineering approach to the analysis of heat exchangers and other convection processes. As we get further into the study of convection heat transfer, additional benefits and occasional liabilities associated with the use of coefficients of friction f_x and heat transfer h_x will be uncovered. Mostly for the better, but sometimes for the worse, these coefficients of friction and heat transfer "have been assimilated irrevocably into the engineering literature and will no doubt be with us forever."†

6–3–3 Dimensional Analysis

For cases that must be dealt with empirically, dimensional analysis is commonly used to determine the pertinent dimensionless parameters. This approach is based on the principle of dimensional continuity, which requires that all terms in every equation be dimensionally consistent. To properly apply the dimensional-analysis approach to convection, we must know what dimensional parameters the coefficients of friction and heat transfer depend upon. These parameters can generally be established on the basis of a thorough physical understanding of the problem.

The formal theoretical principle used in dimensional analysis is known as the *Buckingham π theorem* [5]. The Buckingham π theorem indicates that the number N of *independent* dimensionless groups π_i which are associated with a physical phenomenon, is equal to the total number of significant dimensional parameters I minus the number of fundamental dimensions J, which are required to define the dimensions of all the I parameters, with the relationship among the various dimensionless groups taking the form

$$\pi_1 = \text{fn}\,(\pi_2, \pi_3, \ldots, \pi_i, \ldots, \pi_N) \tag{6–26}$$

† This statement is a generalization of an argument made by White [3] pertaining to coefficients of friction.

The mechanics involved in the general dimensional analysis approach are discussed by Buckingham [5], Bridgeman [6], and Langhaar [7], and introductions to the use of this tool in analyzing convection-heat-transfer and fluid-mechanics processes are given by Kreith and Bohn [8], Lienhard [9], and Fox and McDonald [10]. The use of this approach is illustrated in Example 6–1 in the context of a basic forced-convection system.

EXAMPLE 6–1

Assuming that the coefficient of heat transfer $\bar{h}$ for forced convection flow in a tube with uniform wall temperature is dependent upon the geometric, hydrodynamic, and thermal parameters listed in Table E6–1, utilize the dimensional-analysis approach to determine the key dimensionless groups for this problem.

TABLE E6–1 Forced convection parameters—flow in a tube

Dimensional parameter	*Symbol*	*Dimensions*
Geometric		
Axial distance	x	$[L]$
Diameter	D	$[L]$
Hydrodynamic		
Entrance velocity	U_1	$[L/t]$
Density	ρ	$[m/L^3]$
Viscosity	μ	$[m/(tL)]$
Thermal		
Thermal conductivity	k	$[mL/(t^3\ T)]$
Specific heat	c_P	$[L^2/(t^2\ T)]$
Coefficient of heat transfer	$\bar{h}$	$[m/(t^3\ T)]$

Solution

Objective Use the dimensional-analysis approach to obtain the primary dimensionless groups pertaining to the coefficient of heat transfer $\bar{h}$ for forced convection flow in a tube with uniform wall temperature T_0.

Schematic Flow in a tube with uniform wall temperature T_0.

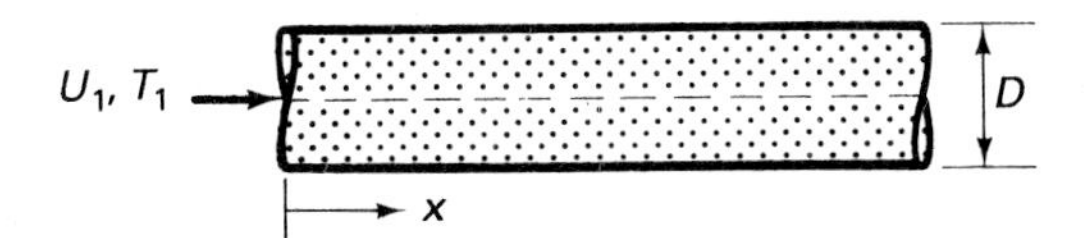

Assumptions/Conditions

forced convection
uniform properties

standard conditions (i.e., steady-state; two-dimensional; ideal, Newtonian, isotropic, single-phase, and chemically homogeneous fluid; incompressible flow; nonporous walls; and negligible buoyancy, kinetic energy, viscous dissipation, external body forces, axial conduction, and thermal radiation)

Properties Assume that fluid properties ρ, μ, k, and c_P are specified.

Analysis Referring to Table E6–1, we see that this problem involves eight dimensional parameters ($I = 8$) and four fundamental dimensions ($J = 4$). Therefore, according to the Buckingham π theorem,

$$N = I - J = 8 - 4 = 4$$

which indicates four dimensionless groups. Following the procedure outlined by various authors [5–9], to determine the dimensionless groups π_i (where $i = 1, 2, 3, 4$), we write

$$\pi_i = U_1^a \mu^b \rho^c k^d c_P^e \bar{h}^f D^g x^h$$

Substituting for the dimensions of each parameter and noting that the π_i groups are dimensionless, we have

$$1 = \left[\frac{L}{t}\right]^a \left[\frac{m}{Lt}\right]^b \left[\frac{m}{L^3}\right]^c \left[\frac{mL}{t^3T}\right]^d \left[\frac{L^2}{t^2T}\right]^e \left[\frac{m}{t^3T}\right]^f [L]^g[L]^h$$

Because the summation of the exponents of each fundamental dimension must be equal to zero, we conclude that

$$\begin{aligned} b + c + d + f &= 0 \quad \text{for mass } m & \text{(a)} \\ a - b - 3c + d + 2e + g + h &= 0 \quad \text{for length } L & \text{(b)} \\ -a - b - 3d - 2e - 3f &= 0 \quad \text{for time } t & \text{(c)} \\ -d - e - f &= 0 \quad \text{for temperature } T & \text{(d)} \end{aligned}$$

These four equations must be satisfied for each dimensionless group. But, because we have eight unknowns, the values of four of the exponents must be specified for each of the four dimensionless groups π_1, π_2, π_3, and π_4. (The fact that these dimensionless groups must be independent simply means that no group can be expressed as the product of any combination of the other groups.) To help in selecting these four exponents for each dimensionless group, we will purposefully seek one thermal group (π_1) involving $\bar{h}$ and D, one thermophysical property group (π_2), one hydrodynamic group (π_3), and one geometric group (π_4).

To obtain dimensionless thermal group π_1, we set f equal to unity, and the hydrodynamic and geometric exponents a, b, and h equal to zero. Solving Eqs. (a) through (d) with these inputs, we obtain $c = 0$, $d = -1$, $e = 0$, and $g = 1$. Therefore, this dimensionless thermal group is

$$\pi_1 = \frac{\bar{h}D}{k}$$

which is the mean *Nusselt number* $\overline{Nu}$.

To formulate a thermophysical property group, a, f, g, and h are set equal to zero. Equations (a) through (d) then give rise to $b = e$, $c = 0$, and $d = -e$. Setting the common exponent e equal to unity, we obtain

$$\pi_2 = \frac{\mu c_P}{k}$$

which is recognized as the *Prandtl number Pr*.

To obtain a dimensionless hydrodynamic group, the thermal exponents d, e, and f and the geometric exponent h are set equal to zero. It follows that $a = 1$, $b = -1$, $c = 1$, and $g = 1$, such that

$$\pi_3 = \frac{\rho U_1 D}{\mu}$$

which is the *Reynolds number Re*.

Finally, to develop a dimensionless geometric group, we leave g unspecified with $h = 1$, $b = 0$, $e = 0$, and $f = 0$. Substituting these values into Eqs. (a) through (d), we obtain $a = 0$, $c = 0$, $d = 0$, and $g = -1$. Hence, we have the natural dimensionless geometric grouping

$$\pi_4 = \frac{x}{D}$$

Based on these results, we can expect correlations for the coefficient of heat transfer associated with forced convection in smooth tubes to take the general form

$$\frac{\bar{h}D}{k} = \overline{Nu} = \text{fn}\,(Re, Pr, x/D) \tag{e}$$

And this is indeed the case, as we shall see in Chap. 8, which presents design equations for the coefficient of heat transfer.

It should be noted that the actual functional relationship between $\overline{Nu}$ and Re, Pr, and x/D, which is dependent upon the specific operating conditions, can often be determined on the basis of experimental data. In this connection, Eq. (e) reduces the data correlation problem from one involving the eight variables U_1, μ, ρ, k, c_P, $\bar{h}$, D, and x to one with only the four dimensionless groups, $\overline{Nu}$, Re, Pr, and x/D. To obtain an empirical correlation for $\bar{h}$ in terms of U_1, μ, ρ, k, c_P, D, and x, it would be necessary to vary each of these seven parameters at least three or four times. Assuming that we obtained four data points for each of these seven variables, we would require 4^7, or about 16,400 individual measurements! On the other hand, to obtain a correlation for $\overline{Nu}$ in terms of Re, Pr, and x/D with as much information would only require 4^3, or 64, data points. Thus, although dimensional analysis cannot be used to determine the functional relationship between a parameter such as $\bar{h}$ and the other system parameters, this approach can be utilized to greatly simplify the correlation of experimental data to produce empirical correlations for design.

It should be noted that the dimensionless parameters that characterize a convection heat-transfer process can also often be obtained by the theoretical approach introduced in Sec. 6–3–1. This more comprehensive approach will be developed in Chap. 7 for several basic convection processes.

Because of their common use and significance, basic dimensionless parameters that are used to characterize convection-heat-transfer processes are summarized for forced convection, natural convection, and boiling and condensation in Tables 8–9, 9–8, and 10–4. In this connection, Arpaci and Larsen [11] provide an excellent discussion of the interpretation of dimensionless groups for convection heat transfer.

6–4 CHARACTERISTICS OF INTERNAL FLOW

To establish a framework for the theoretical and practical analysis of internal flow, we now introduce the key bulk-stream characteristics and discuss the nature of developing and fully developed flow in the context of uniform property flow in passages with uniform cross-sectional area A and impermeable walls, such as the circular tube shown in Fig. 6–14. (The concepts presented in this section are extended to variable property flow in Appendix L and to flow through tube banks and heat-exchanger cores in Chaps. 8 and 11.) For arrangements of this type, flow enters with a uniform velocity U_1, the mass flow rate $\dot{m}$ is constant, and heating is initiated at some distance x_0 from the entrance.

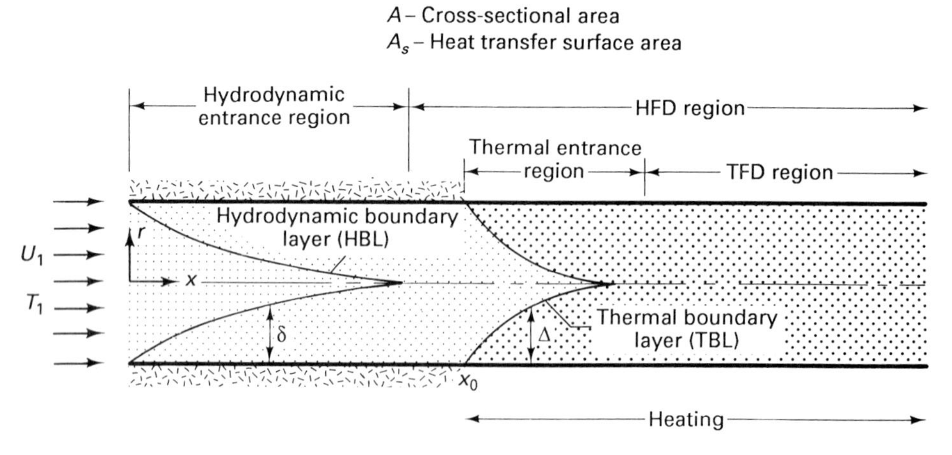

FIGURE 6–14 Hydrodynamic and thermal entrance regions and fully developed regions associated with flow in a circular tube.

The reference length L for internal flows is generally set equal to the *hydraulic diameter*† D_H, which is defined by

$$D_H = 4\frac{A}{p_w} \tag{6–27}$$

† The hydraulic radius r_H also commonly appears in the literature; r_H is related to D_H by $D_H = 4r_H$.

where p_w is the wetted perimeter. For tubular passages with p_w equal to the heat-transfer surface perimeter p, Eq. (6–27) reduces to

$$\frac{D_H}{L} = 4\frac{A}{pL} = 4\frac{A}{A_s} \tag{6–28}$$

where A_s ($= pL$) is the heat-transfer surface area. The hydraulic diameters of several common passages are given as follows:

Circular Tube

$$D_H = 4\frac{\pi D^2/4}{\pi D} = D \tag{6–29}$$

Circular Annulus

$$D_H = \frac{4\pi(D_o^2 - D_i^2)/4}{\pi(D_o + D_i)} = D_o - D_i \tag{6–30}$$

Parallel Plates
(width w, depth Z)

$$D_H = 4\frac{wZ}{2Z} = 2w \tag{6–31}$$

6–4–1 Hydrodynamic Entrance and Fully Developed Regions

As indicated in Fig. 6–14, the hydrodynamic boundary layer thickness δ for flow in a tube grows as x increases, until it reaches the centerline. Beyond the axial location at which this occurs, the growth of the boundary layer is constrained and the shape of the velocity profile becomes independent of x; that is,

$$\frac{\partial u}{\partial x} = 0 \tag{6–32}$$

for constant mass flow rate $\dot{m}$ and constant properties. In this region, the flow is *hydrodynamically fully developed* (HFD). The region upstream of the HFD region is known as the *hydrodynamic entrance region.*

Internal flows are generally characterized in terms of the *bulk-stream mass flux G*,

$$G = \frac{\dot{m}}{A} = \frac{1}{A}\int_{\dot{m}} d\dot{m} = \frac{1}{A}\int_A \rho u \, dA = \frac{\rho}{A}\int_A u \, dA \tag{6–33}$$

or the *bulk-stream velocity* U_b,

$$U_b = \frac{\dot{m}}{\rho A} = \frac{G}{\rho} = \frac{1}{A}\int_A u \, dA \tag{6–34}$$

for uniform density. For uniform property flow in systems with constant mass flow rate, U_b is constant and is equal to the entering velocity U_1.

Setting $L = D_H$ and $U_F = U_b$ in Eq. (6–19) and designating the Reynolds number for internal flow by Re rather than Re_D, we write

$$Re = \frac{U_b D_H}{\nu} \qquad Re = \frac{G D_H}{\mu} \qquad Re = \frac{\dot{m} D_H}{\mu A} \tag{6–35a,b,c}$$

Using any of these three equivalent relations for Re, the flow is generally laminar for values of Re less than about 2000, and the flow is turbulent for values of Re greater than this value. In the turbulent region, the flow is usually fully turbulent for $Re \gtrsim 10^4$ and transitional turbulent for $2000 \lesssim Re \lesssim 10^4$.

The Fanning friction factor for internal flow is defined by Eq. (6–20) with $U_F = U_b$; that is,

$$\tau_s = \rho U_b^2 \frac{f_x}{2} \tag{6–36}$$

The mean coefficient of friction over the length 0 to x for two-dimensional flow is given by

$$\bar{f} = \frac{1}{x} \int_0^x f_x \, dx \tag{6–37}$$

It should be noted that the hydrodynamic entrance region is characterized by a local friction factor f_x that varies with x. However, f_x is independent of x in the HFD region for systems with uniform mass flow rate $\dot{m}$ and uniform properties. To see this point, we couple the Newtonian shear law, Eq. (6–1), with the defining equation for f_x, Eq. (6–36), and the defining equation for HFD flow, Eq. (6–32), as follows:

$$\begin{aligned} \tau_s &= \rho U_b^2 \frac{f_x}{2} = \mu \frac{\partial u}{\partial y}\bigg|_0 \\ \frac{d\tau_s}{dx} &= \frac{\rho U_b^2}{2} \frac{df_x}{dx} = \mu \frac{\partial}{\partial y}\left(\frac{\partial u}{\partial x}\right)\bigg|_0 = 0 \end{aligned} \tag{6–38}$$

Hence, τ_s is constant and

$$f_x = f = \text{constant} \tag{6–39}$$

in the HFD region. The symbol f will be consistently used in our study to designate the Fanning friction factor for HFD flow.

6–4–2 Thermal Entrance and Fully Developed Regions

Because the fluid is heated or cooled as it flows downstream, its enthalpy or energy content changes with x. To characterize this change, we define the bulk-stream enthalpy rate $\dot{H}_b$ as the total rate of enthalpy transferred through the cross-sectional area A at any axial location x. Referring to Fig. 6–15, $\dot{H}_b$ is expressed in terms of the distribution in specific enthalpy i at x by

$$\dot{H}_b = \int_{\dot{m}} i \, d\dot{m} = \int_A i\rho u \, dA = \rho \int_A iu \, dA \tag{6–40}$$

for uniform density.

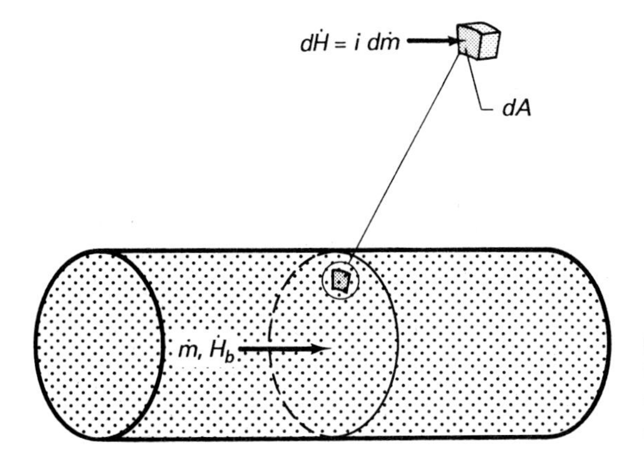

FIGURE 6–15
Representation of bulk-stream enthalpy rate $\dot{H}_b$ for flow in a tube with mass flow rate $\dot{m}$.

To express $\dot{H}_b$ in terms of the temperature distribution T over the cross section, we make use of Eq. (6–14) for ideal fluids; that is,

$$i = c_P(T - T_R) + i_R \tag{6–41}$$

for uniform specific heat c_P, where the subscript R represents the reference state. Combining this equation with Eq. (6–40), we have

$$\dot{H}_b = \rho \int_A [c_P(T - T_R) + i_R]u\, dA = \rho c_P \int_A uT\, dA + \dot{m}(i_R - c_P T_R) \tag{6–42}$$

The local average thermal energy state of the fluid over the entire cross section A is characterized by the *bulk-stream temperature* T_b, which is formally defined for uniform property flow as the average value of T for which Eq. (6–42) is satisfied; that is,

$$\dot{H}_b = \rho c_P \int_A uT_b\, dA + \dot{m}(i_R - c_P T_R) = \dot{m}[c_P(T_b - T_R) + i_R] \tag{6–43}$$

We are now in a position to establish the gradient $d\dot{H}_b/dT_b$, which is essential to the practical thermal analysis approach. Based on Eq. (6–43), $d\dot{H}_b/dT_b$ is given by

$$\frac{d\dot{H}_b}{dT_b} = \frac{d}{dT_b}\{\dot{m}[c_P(T_b - T_R) + i_R]\} \tag{6–44}$$

or, since $\dot{m}$ (= GA) is independent of T_b, and since T_R, i_R, and c_P are constant,†

$$d\dot{H}_b = \dot{m}c_P\, dT_b \tag{6–45}$$

Finally, to express T_b in terms of T, we equate Eqs. (6–42) and (6–43), with the result

$$\dot{m}[c_P(T_b - T_R) + i_R] = \rho c_P \int_A uT\, dA + \dot{m}(i_R - c_P T_R)$$

† As shown in Appendix L, Eq. (6–45) also applies to variable property flow.

or

$$T_b = \frac{\rho}{\dot{m}} \int_A uT\,dA = \frac{1}{AU_b} \int_A uT\,dA \tag{6–46}$$

which is the defining equation for the bulk-stream temperature associated with uniform property flow. Since T_b represents the temperature that would be measured if the fluid flowing through the cross-sectional area were collected and mixed in a cup, this parameter is also commonly referred to as the *mixing cup temperature*. Note that whereas T_b is a function of x for heating or cooling, its counterpart U_b is constant for flow in passages with uniform properties, cross-sectional area, and mass flow rate.

The thickness Δ of the TBL grows with respect to x until it becomes constrained by the system geometry (see Fig. 6–14). Beyond the axial location at which Δ becomes a constant, the shape of the temperature profile becomes independent of x for uniform wall temperature, uniform wall heat flux, and certain other boundary conditions; that is,†

$$\frac{\partial}{\partial x}\left(\frac{T - T_s}{T_b - T_s}\right) = 0 \tag{6–47}$$

which is the defining equation for *thermal fully developed flow* (TFD).

The coefficient of heat transfer h_x for internal flow is defined by Eq. (6–22) with $T_F = T_b$.

$$q_c'' = \frac{dq_c}{dA_s} = h_x(T_s - T_b) \tag{6–48}$$

The mean coefficient of heat transfer over the length 0 to x is defined by

$$\bar{h} = \frac{1}{x} \int_0^x h_x\,dx \tag{6–49}$$

As will be seen, $\bar{h}$ is primarily used in the analysis of problems involving uniform wall-temperature conditions and in heat-exchanger applications.

Whereas the coefficients h_x and $\bar{h}$ are functions of x in the thermal entrance region, the TFD region is characterized by a coefficient h_x that is independent of x for uniform mass flow rate and uniform properties. To demonstrate this point, we combine the Fourier law of conduction, Eq. (6–2b), with the general Newton law of cooling, Eq. (6–48), to obtain

$$q_c'' = h_x(T_s - T_b) = -k \left.\frac{\partial T}{\partial y}\right|_0 \tag{6–50}$$

† For situations in which the wall temperature varies around the perimeter, T_s is replaced by the peripheral mean wall temperature.

where k is the thermal conductivity of the fluid and $\partial T/\partial y|_0$ is the temperature gradient within the fluid at $y = 0$. Rearranging this equation and introducing the defining equation for TFD flow, Eq. (6–47), we have

$$h_x = -k\frac{\partial}{\partial y}\left(\frac{T}{T_s - T_b}\right)\bigg|_0 = k\frac{\partial}{\partial y}\left(\frac{T - T_s}{T_b - T_s}\right)\bigg|_0$$
$$\frac{dh_x}{dx} = k\frac{\partial}{\partial y}\left[\frac{\partial}{\partial x}\left(\frac{T - T_s}{T_b - T_s}\right)\right]\bigg|_0 = 0 \tag{6–51}$$

Hence, we see that

$$h_x = h = \text{constant} \tag{6–52}$$

where h specifically designates the coefficient of heat transfer for TFD conditions.

As indicated earlier, the coefficient of heat transfer is generally expressed in terms of the Nusselt number or Stanton number. To distinguish between h, h_x, and $\bar{h}$, when employing these dimensionless parameters, we will use the following notation for internal flow:

$$Nu = \frac{hD_H}{k} \qquad Nu_x = \frac{h_xD_H}{k} \qquad \overline{Nu} = \frac{\bar{h}D_H}{k}$$

$$St = \frac{h}{\rho c_P U_b} \qquad St_x = \frac{h_x}{\rho c_P U_b} \qquad \overline{St} = \frac{\bar{h}}{\rho c_P U_b}$$

6–5 SUMMARY

In this chapter we discussed features that characterize convection-heat-transfer processes and the concept of the boundary layer. We also gave brief consideration to the theoretical and practical approaches to analyzing convection heat transfer, and we introduced important bulk-stream characteristics for internal flows.

The theoretical approach to the analysis of convection heat transfer involves the development of mathematical formulations for the fluid flow and energy transfer, which generally take the form of systems of partial differential equations. Representative convection-heat-transfer processes are analyzed in Chap. 7 and Appendixes I to K, which deal with the theory of convection. This treatment provides a theoretical basis for some of the relations for coefficients of friction and heat transfer used in practical engineering analysis, and also provides a framework for dealing with more complex problems.

The practical analysis approach involves the use of relations for coefficients of friction and heat transfer together with fundamental conservation principles. This approach generally results in simple ordinary differential or algebraic equations and offers a means of determining the pressure drop or drag and the heat-transfer performance. Practical engineering analyses are developed for basic forced convection, natural convection, and boiling and condensation processes in Chaps. 8, 9, and 10 and Appendixes L and M. This material provides the basis for the evaluation and design of heat exchangers, which is the topic of Chap. 11 and Appendixes N and O.

The theoretical and practical approaches to the analysis of convection heat transfer involve the use of a number of important dimensionless parameters. The interpretation of the primary dimensionless groups associated with the analysis of forced convection, natural convection, and boiling and condensation is summarized in Tables 8–9, 9–8, and 10–4.

Finally, our study of convection heat transfer will be restricted to systems involving fluids which are ideal (i.e., ideal gases and incompressible liquids). References 12 and 13 deal with the important topic of compressible flow. Heat and mass transfer in chemically nonhomogeneous fluids is considered in the *mass transfer supplement*.

CHAPTER 7

CONVECTION HEAT TRANSFER: THEORETICAL ANALYSIS

7–1 INTRODUCTION

In this chapter we introduce the theoretical treatment of convection heat transfer for boundary layer flows. As indicated in Chap. 6, the complete theoretical analysis of convection-heat-transfer problems requires the use of physical laws in the development of mathematical formulations and solutions for the fluid flow and energy transfer within the fluid. The development of differential, integral, or numerical formulations involves (1) the application of the fundamental principles pertaining to mass, momentum, and energy to a control volume within the flow field; and (2) the use of the particular laws pertaining to fluid shear stress and heat flux. These physical laws, which were presented in Chap. 1, are summarized in Table 7–1. Once the mathematical formulations are developed, solutions are obtained for the velocity and temperature distributions and the friction factor and Nusselt number.

In this introductory study, attention is given to developing the fundamentals of laminar and turbulent-flow theory. The treatment of basic convection-heat-transfer theory will be presented in the context of several classic steady one- and two-dimensional internal and external flows and will feature the use of simple analytical and numerical solution techniques. Our study of the theory of laminar and turbulent flow will be followed by a brief introduction to the use of numerical methods in the analysis of thermal boundary layer flow and complex convection-heat-transfer processes.

7–2 LAMINAR FLOW THEORY

The general approach to analyzing laminar convection processes involves the development of (1) mathematical formulations for continuity, momentum, and energy transfer within the fluid; (2) solutions for the velocity profiles u and v, and temperature distribution T within the fluid; and (3) predictions for the wall shear stress, wall heat

TABLE 7–1 Fundamental and particular laws: Summary

1. *Fundamental laws*
   a. Conservation of mass (continuity)
   Rate of creation of mass = 0

$$\Sigma \dot{m}_o - \Sigma \dot{m}_i + \frac{\Delta m_s}{\Delta t} = 0 \qquad \text{(i)}$$

   b. Momentum principle (Newton's second law of motion relative to direction x)
   Rate of creation of momentum (RCM_x) = Sum of forces (ΣF_x)

$$\Sigma \dot{M}_{o,x} - \Sigma \dot{M}_{i,x} + \frac{\Delta M_{s,x}}{\Delta t} = \Sigma F_x \qquad \text{(ii)}$$

   c. Conservation of energy (first law of thermodynamics)
   Rate of creation of energy = 0

$$\Sigma \dot{E}_o - \Sigma \dot{E}_i + \frac{\Delta E_s}{\Delta t} = 0 \qquad \text{(iii)}$$

2. *Particular laws*
   a. Newton law of viscosity, x direction†

$$\tau = \mu \frac{\partial u}{\partial y} \qquad \text{rectangular systems} \qquad \text{(iv)}$$

$$\tau = \mu \frac{\partial u}{\partial y} = -\mu \frac{\partial u}{\partial r} \qquad \text{cylindrical internal flow system} \qquad \text{(v)}$$

   b. Fourier law of conduction
   Axial x direction

$$q''_x = -k \frac{\partial T}{\partial x} \qquad \text{(vi)}$$

   Direction perpendicular to surface

$$q''_y = -k \frac{\partial T}{\partial y} \qquad \text{rectangular system} \qquad \text{(vii)}$$

$$q''_y = -k \frac{\partial T}{\partial y} = k \frac{\partial T}{\partial r} \qquad \text{cylindrical internal flow system (note that } q''_r = -q''_y\text{)} \qquad \text{(viii)}$$

† Equations (iv) and (v) are boundary layer approximations that neglect the contribution of $\partial v/\partial x$. Equations (iv) and (v) actually take the more general forms $\tau = \mu(\partial u/\partial y + \partial v/\partial x)$ and $\tau = -\mu(\partial u/\partial r - \partial v/\partial x)$, respectively.

flux (or wall temperature), and/or coefficients of friction and heat transfer. In our study of this fundamental topic, both differential and integral approaches will be used, with consideration given to the major forced convection/natural convection and internal flow/external flow categories. The basic systems to be studied include two-dimensional flow in tubes and flow over flat plates. The concepts introduced in the study of these classical convection-heat-transfer processes provide a foundation for the theoretical analysis of the more complex problems that are generally encountered in practice.

7-2-1 Flow in Tubes

The problem of convection heat transfer for steady two-dimensional laminar flow in a tube with uniform heat flux is considered in this section. This basic problem is illustrated in Fig. 7-1. Although emphasis is placed on circular tubes, the concepts introduced in this section apply to other internal flow systems with uniform cross-sectional area, such as annuli, channels, and parallel plates.

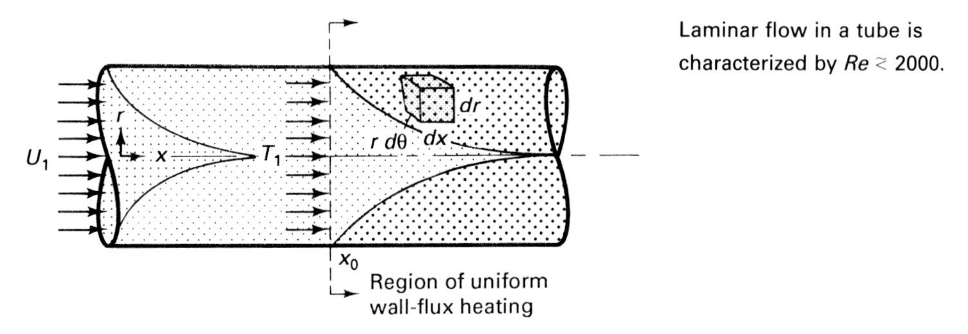

FIGURE 7-1 Laminar flow in a circular tube with uniform wall-flux heating for $x \geq x_0$.

As mentioned in Chap. 6, hydrodynamic and thermal boundary layers develop in the entrance region for internal flow systems, with the flow becoming fully developed in the region downstream where the hydrodynamic and thermal boundary layers are independent of x. Recall that fully developed conditions are defined by

$$\frac{\partial u}{\partial x} = 0 \qquad \text{HFD} \tag{7-1}$$

$$\frac{\partial}{\partial x}\left(\frac{T - T_s}{T_b - T_s}\right) = 0 \qquad \text{TFD} \tag{7-2}$$

For these conditions, both the friction factor and the coefficient of heat transfer are independent of x (i.e., $f_x = f$ and $h_x = h$).

In our analysis of this problem, we first develop differential formulations for continuity, momentum, and energy transfer.

Mathematical Formulation

To develop the differential formulation for continuity, momentum, and energy within the entrance and fully developed region, the fundamental laws are applied to the differential volume of fluid $r\,d\theta\,dr\,dx$ shown in Fig. 7-1. The steps required in the development of the differential formulation follow the guidelines established in Chaps. 2 and 3 for conduction heat transfer.

Continuity Applying the principle of conservation of mass to the control volume shown in Figs. 7–1 and 7–2, we obtain

$$\Sigma\, \dot{m}_o - \Sigma\, \dot{m}_i + \frac{\Delta m_s}{\Delta t} = 0$$

$$d\dot{m}_{x+dx} + d\dot{m}_{r+dr} - d\dot{m}_x - d\dot{m}_r = 0 \qquad (7\text{–}3)$$

where $\Delta m_s/\Delta t = 0$ for steady flow, $d\dot{m}_x = \rho u r\, d\theta\, dr$ and $d\dot{m}_r = \rho v r\, d\theta\, dx$. Utilizing the definition of the partial derivative, this expression takes the form

$$\frac{\partial}{\partial x}(d\dot{m}_x)\, dx + \frac{\partial}{\partial r}(d\dot{m}_r)\, dr = 0 \qquad (7\text{–}4)$$

or

$$\frac{\partial}{\partial x}(\rho u) + \frac{1}{r}\frac{\partial}{\partial r}(\rho r v) = 0 \qquad (7\text{–}5)$$

For uniform property flow, the continuity equation becomes

$$\frac{\partial u}{\partial x} + \frac{1}{r}\frac{\partial}{\partial r}(r v) = 0 \qquad (7\text{–}6)$$

The formulation for continuity is completed by writing the boundary condition for v,

$$v = 0 \qquad \text{at } r = r_0 \qquad (7\text{–}7)$$

(The x boundary condition for u will be taken care of in the formulation for momentum transfer, which follows.)

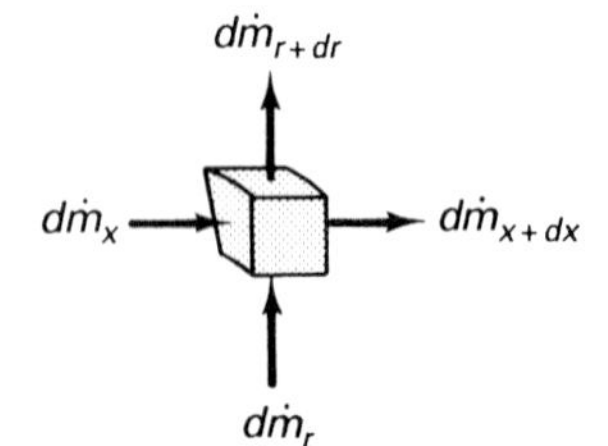

FIGURE 7–2
Conservation of mass relative to differential cylindrical control volume; $dA_x = r\, d\theta\, dr$, $dA_r = r\, d\theta\, dx$, $dV = r\, d\theta\, dr\, dx$.

In the HFD region where $\partial u/\partial x = 0$, the continuity equation for incompressible flow reduces to

$$\frac{d}{dr}(r v) = 0 \qquad (7\text{–}8)$$

By coupling this simple equation with the boundary condition, we find that $v = 0$ in the HFD region.

Momentum Transfer To develop the momentum equation, we apply Newton's second law of motion to the element shown in Figs. 7–1 and 7–3; that is,

$$\Sigma \dot{M}_{o,x} - \Sigma \dot{M}_{i,x} + \frac{\Delta M_{s,x}}{\Delta t} = \Sigma F_x \tag{7–9}$$

The sum of forces is

$$\begin{aligned}\Sigma F_x &= (\tau\, dA)_r + (P\, dA)_x - (\tau\, dA)_{r+dr} - (P\, dA)_{x+dx} \\ &= (\tau r\, d\theta\, dx)|_r + (Pr\, d\theta\, dr)|_x - (\tau r\, d\theta\, dx)|_{r+dr} - (Pr\, d\theta\, dr)|_{x+dx} \\ &= -\frac{\partial}{\partial r}(r\tau)\, d\theta\, dx\, dr - \frac{\partial P}{\partial x} r\, d\theta\, dr\, dx = -\frac{1}{r}\frac{\partial}{\partial r}(r\tau)\, dV - \frac{\partial P}{\partial x} dV\end{aligned} \tag{7–10}$$

where $dV = r\, d\theta\, dr\, dx$, assuming negligible normal viscous stresses, buoyant forces, and other body forces. The rate of creation of axial momentum is represented by RCM_x,

$$\begin{aligned}RCM_x &= d\dot{M}_{x+dx} + d\dot{M}_{r+dr} - d\dot{M}_x - d\dot{M}_r \\ &= \frac{\partial}{\partial x}(d\dot{M}_x)\, dx + \frac{\partial}{\partial r}(d\dot{M}_r)\, dr\end{aligned} \tag{7–11}$$

where $d\dot{M}_x = u\, d\dot{m}_x$ and $d\dot{M}_r = u\, d\dot{m}_r$. Substituting for $d\dot{M}_x$ and $d\dot{M}_r$, RCM_x becomes

$$\begin{aligned}RCM_x &= \frac{\partial}{\partial x}(u\, d\dot{m}_x)\, dx + \frac{\partial}{\partial r}(u\, d\dot{m}_r)\, dr \\ &= \frac{\partial}{\partial x}(u\rho u)\, r\, d\theta\, dr\, dx + \frac{\partial}{\partial r}(u\rho v r)\, d\theta\, dx\, dr \\ &= \frac{\partial}{\partial x}(u\rho u)\, dV + \frac{1}{r}\frac{\partial}{\partial r}(u\rho v r)\, dV\end{aligned} \tag{7–12}$$

Making use of the continuity equation, Eq. (7–5), this expression reduces to

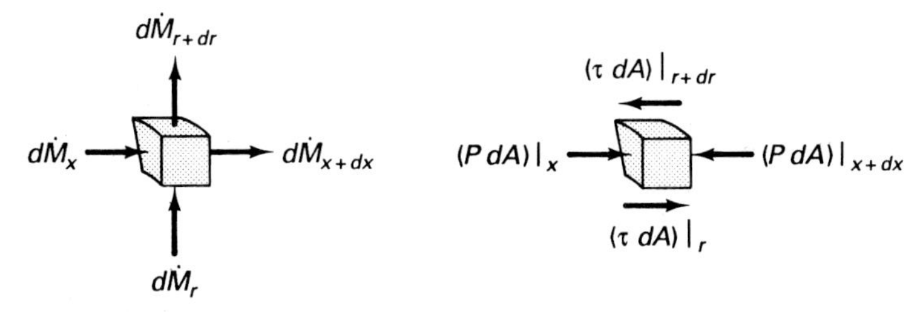

(a) Momentum transfer. (b) Forces.

FIGURE 7–3 Momentum principle relative to differential cylindrical control volume; x-direction.

$$RCM_x = u\left[\frac{\partial}{\partial x}(\rho u) + \frac{1}{r}\frac{\partial}{\partial r}(\rho v r)\right] dV + \left(\rho u \frac{\partial u}{\partial x} + \rho v \frac{\partial u}{\partial r}\right) dV$$
$$= \rho\left(u\frac{\partial u}{\partial x} + v\frac{\partial u}{\partial r}\right) dV \tag{7–13}$$

Substituting Eqs. (7–10) and (7–13) into Eq. (7–9), we obtain

$$\rho\left(u\frac{\partial u}{\partial x} + v\frac{\partial u}{\partial r}\right) = -\frac{1}{r}\frac{\partial}{\partial r}(r\tau) - \frac{\partial P}{\partial x} \tag{7–14}$$

Utilizing Newton's law of viscosity [Eq. (v) in Table 7–1] and assuming that P is essentially a function of x alone, the momentum equation is written as

$$\rho\left(u\frac{\partial u}{\partial x} + v\frac{\partial u}{\partial r}\right) = \frac{1}{r}\frac{\partial}{\partial r}\left(\mu r\frac{\partial u}{\partial r}\right) - \frac{dP}{dx} \tag{7–15}$$

or, for constant properties,

$$u\frac{\partial u}{\partial x} + v\frac{\partial u}{\partial r} = \frac{\nu}{r}\frac{\partial}{\partial r}\left(r\frac{\partial u}{\partial r}\right) - \frac{1}{\rho}\frac{dP}{dx} \tag{7–16}$$

The differential momentum equation is coupled with boundary conditions for u of the form

$$u = U_1 \qquad \text{at } x = 0 \tag{7–17}$$

for a uniform distribution at the entrance,

$$u = 0 \qquad \text{at } r = r_0 \tag{7–18}$$

for no slip at the wall, and

$$\frac{\partial u}{\partial r} = 0 \qquad \text{at } r = 0 \tag{7–19}$$

because of symmetry.

For HFD flow, $\partial u/\partial x = 0$ and $v = 0$ (based on continuity) such that our differential formulation for momentum transfer in a uniform property fluid is given by

$$\frac{\nu}{r}\frac{d}{dr}\left(r\frac{du}{dr}\right) - \frac{1}{\rho}\frac{dP}{dx} = 0 \tag{7–20}$$

and Eqs. (7–18) and (7–19). Referring back to Eq. (7–13), we see that the rate of creation of momentum RCM_x for flow with uniform density is equal to zero for HFD conditions, such that the momentum equation actually represents the force balance $\sum F_x = 0$.

EXAMPLE 7–1

Show that the shear stress varies linearly with r or y for fully developed laminar flow in a circular tube.

Solution

Objective Show that the relation for τ is linear for HFD flow in a circular tube.

Schematic Laminar HFD flow in a circular tube.

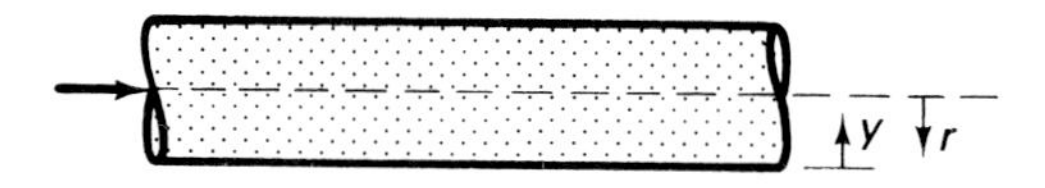

Assumptions/Conditions

forced convection
uniform properties
steady-state
Newtonian fluid
nonporous walls
negligible buoyancy and external body forces

Analysis To see the behavior of τ for laminar HFD flow in a circular tube, Eq. (7–20) is first put into the form [see Eq. (7–14)]

$$\frac{1}{r}\frac{d}{dr}(r\tau) + \frac{dP}{dx} = 0 \tag{a}$$

where

$$\tau = \mu\frac{du}{dy} = -\mu\frac{du}{dr}$$

for Newtonian fluids. Noting that $r\tau = 0$ at $r = 0$, Eq. (a) is integrated to obtain

$$r\tau = -\frac{r^2}{2}\frac{dP}{dx} \qquad \text{or} \qquad \tau = -\frac{r}{2}\frac{dP}{dx}$$

Setting $\tau = \tau_s = \tau_0$ at $r = r_0$ to eliminate dP/dx, the solution becomes

$$\frac{\tau}{\tau_0} = \frac{r}{r_0} = 1 - \frac{y}{r_0}$$

Thus, the distribution in τ is indeed linear.

Energy Transfer In analyzing the energy transfer for this problem, we must account for significant components of the molecular conduction heat transfer and energy transported by fluid motion (i.e., advection). To simplify the analysis, we will neglect the effects of kinetic energy, viscous dissipation, body forces, axial conduction, and thermal radiation. (Kinetic energy and/or viscous dissipation are important for viscous fluids such as oil and in high-speed aerodynamic problems, the effects of axial conduction are sometimes significant for liquid metals, and thermal radiation must be accounted for in the analysis of systems involving participating gases and high temperatures.)

Applying the first law of thermodynamics to the control volume dV shown in Fig. 7–4, we write

$$\Sigma \dot{E}_o - \Sigma \dot{E}_i + \cancel{\frac{\Delta E_s}{\Delta t}} = 0$$

$$\underbrace{d\dot{H}_{x+dx} + d\dot{H}_{r+dr} - d\dot{H}_x - d\dot{H}_r}_{\textit{advection}} + \underbrace{dq_{r+dr} - dq_r}_{\textit{molecular conduction}} = 0 \tag{7–21}$$

where $\Delta E_s/\Delta t = 0$ for steady-state conditions, $d\dot{H}_x = i\, d\dot{m}_x$, $d\dot{H}_r = i\, d\dot{m}_r$ and i is the specific enthalpy.

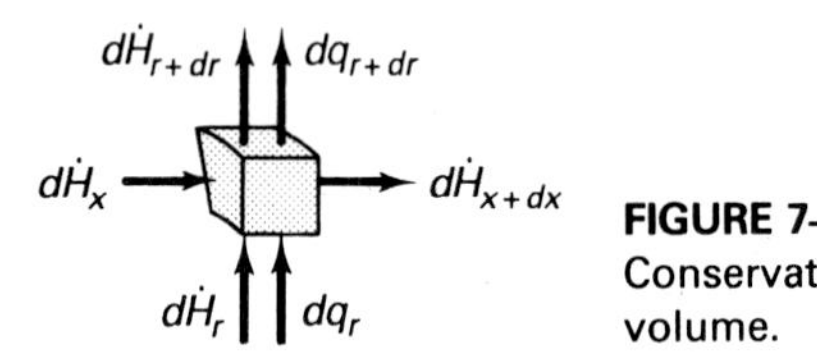

FIGURE 7–4
Conservation of energy relative to differential cylindrical control volume.

Utilizing the definition of the partial derivative, it follows that

$$\frac{\partial}{\partial x}(d\dot{H}_x)\, dx + \frac{\partial}{\partial r}(d\dot{H}_r)\, dr + \frac{\partial}{\partial r}(dq_r)\, dr = 0 \tag{7–22}$$

Introducing the Fourier law of conduction [Eq. (viii) in Table 7–1] and substituting for $d\dot{H}_x$ and $d\dot{H}_r$, this equation takes the form

$$\frac{\partial}{\partial x}(\rho u i) + \frac{1}{r}\frac{\partial}{\partial r}(\rho r v i) = \frac{1}{r}\frac{\partial}{\partial r}\left(rk\frac{\partial T}{\partial r}\right) \tag{7–23}$$

or, after making use of the continuity equation, Eq. (7–6),

$$\rho\left(u\frac{\partial i}{\partial x} + v\frac{\partial i}{\partial r}\right) = \frac{1}{r}\frac{\partial}{\partial r}\left(rk\frac{\partial T}{\partial r}\right) \tag{7–24}$$

With i expressed in terms of c_P, we obtain

$$\rho c_P \left(u \frac{\partial T}{\partial x} + v \frac{\partial T}{\partial r} \right) = \frac{1}{r} \frac{\partial}{\partial r} \left(rk \frac{\partial T}{\partial r} \right) \tag{7–25}$$

For uniform thermal conductivity, this equation reduces to

$$u \frac{\partial T}{\partial x} + v \frac{\partial T}{\partial r} = \frac{\alpha}{r} \frac{\partial}{\partial r} \left(r \frac{\partial T}{\partial r} \right) \tag{7–26}$$

Assuming that heating is initiated at x_0, the thermal boundary conditions are

$$T = T_1 \qquad \text{at } x = x_0 \tag{7–27}$$

$$\frac{\partial T}{\partial r} = 0 \qquad \text{at } r = 0 \tag{7–28}$$

and

$$k \frac{\partial T}{\partial r} = q_0'' \qquad \text{at } r = r_0 \tag{7–29}$$

for uniform wall-flux heating of the fluid.

For HFD flow, $v = 0$ and the energy equation for uniform properties becomes

$$u \frac{\partial T}{\partial x} = \frac{\alpha}{r} \frac{\partial}{\partial r} \left(r \frac{\partial T}{\partial r} \right) \tag{7–30}$$

For TFD conditions, $\partial T/\partial x$ is prescribed in accordance with the defining equation for TFD flow given by Eq. (7–2). This point will be expanded upon momentarily.

EXAMPLE 7–2

Use a lumped analysis to express dT_b/dx and T_b in terms of q_0'' for flow in a tube with uniform wall-flux heating.

Solution

Objective Develop expressions for dT_b/dx and T_b using a simple lumped analysis.

Schematic Flow in a tube with uniform wall-flux heating.

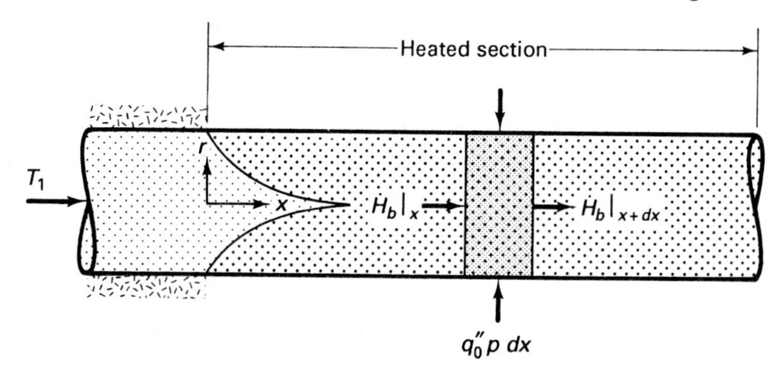

Assumptions/Conditions

forced convection
uniform properties
ideal and single-phase fluid
steady-state
no interfacial mass transfer
negligible buoyancy
negligible kinetic energy, viscous dissipation, external body forces, axial conduction, and thermal radiation

Analysis Applying the first law of thermodynamics to the lumped/differential control volume $A\,dx$ shown in the schematic, we obtain

$$\Sigma\,\dot{E}_o - \Sigma\,\dot{E}_i + \cancel{\frac{\Delta E_s}{\Delta t}} = 0$$

$$\dot{H}_b\big|_{x+dx} - \dot{H}_b\big|_x - q_0''p\,dx = 0$$

or

$$q_0''p = \frac{d\dot{H}_b}{dx} \tag{a}$$

Using Eq. (6–45) to express $d\dot{H}_b$ in terms of T_b and c_P for ideal fluids, this equation gives

$$q_0''p = \dot{m}c_P\frac{dT_b}{dx} \tag{b}$$

or, since $\dot{m} = \rho A U_b$,

$$\frac{dT_b}{dx} = \frac{q_0''p}{\dot{m}c_P} = \frac{4q_0''}{\rho c_P U_b D} \tag{c}$$

To obtain an expression for T_b, we integrate Eq. (c) over the length 0 to x, with the result

$$T_b = T_1 + \frac{p}{\dot{m}}\int_0^x \frac{q_0''}{c_P}\,dx \tag{d}$$

which reduces to the linear relation

$$T_b = T_1 + \frac{q_0''px}{\dot{m}c_P} = T_1 + \frac{4q_0''x}{\rho c_P U_b D} \tag{e}$$

for uniform wall flux and uniform specific heat.

It should be noted that Eqs. (a)–(e) apply over the entire length of the heated section, whether or not the flow is fully developed and whether the flow is laminar or turbulent. This simple lumped formulation perspective is featured in Chaps. 8–11, which deal with the practical analysis of convection.

Solution

To solve for the rate of convection heat transfer for laminar flow in a tube, the fluid flow and energy equations must be solved for the velocity and temperature distributions. Once these distributions are known, the friction factor and coefficient of heat transfer can be determined. For problems involving variable property developing flow, the more general continuity, momentum, and energy equations given by Eqs. (7–5), (7–15), and (7–25) must be solved simultaneously. For the somewhat simpler case of uniform property developing flow, Eqs. (7–6) and (7–16) are first solved for the velocity distribution, after which Eq. (7–26) is solved for the temperature profile. Numerical and approximate analytical solutions are available in the literature for these type problems in which the flow is developing [1–12].

Attention is now turned to the development of solutions for the simpler case involving uniform property fully developed laminar flow. An analytical solution is first developed for the fluid-flow aspects of the problem, after which predictions are developed for the temperature distribution and coefficient of heat transfer.

Fluid Flow—HFD Region For HFD flow, $v = 0$ and the momentum equation and boundary conditions are given by

$$\frac{\nu}{r}\frac{d}{dr}\left(r\frac{du}{dr}\right) - \frac{1}{\rho}\frac{dP}{dx} = 0 \tag{7–20}$$

and

$$u = 0 \quad \text{at } r = r_0 \qquad \frac{du}{dr} = 0 \quad \text{at } r = 0 \tag{7–18,19}$$

To solve this problem, we first develop a solution for the dimensionless velocity distribution.

Dimensionless Velocity Distribution Separating the variables in Eq. (7–20) and integrating, we have

$$r\frac{du}{dr} = \frac{1}{\mu}\frac{dP}{dx}\frac{r^2}{2} + C_1 \tag{7–31}$$

where C_1 is set equal to zero in accordance with Eq. (7–19). Separating the variables and integrating once again, the distribution in velocity u becomes

$$u = \frac{1}{\mu}\frac{dP}{dx}\frac{r^2}{4} + C_2 \tag{7–32}$$

Utilizing the no slip condition at the wall, C_2 is given by

$$C_2 = -\frac{1}{\mu}\frac{dP}{dx}\frac{r_0^2}{4} \tag{7–33}$$

such that our solution for u is

$$u = -\frac{1}{4\mu}\frac{dP}{dx}r_0^2\left[1 - \left(\frac{r}{r_0}\right)^2\right] \tag{7–34}$$

Alternatively, by setting u equal to the centerline velocity U_c at r equal to zero, we have the somewhat more convenient expression,

$$\frac{u}{U_c} = 1 - \left(\frac{r}{r_0}\right)^2 \tag{7–35}$$

Although we now have a theoretical expression for the velocity profile, our solution is incomplete because U_c and dP/dx are as yet unknown.

Bulk-Stream Velocity and Friction Factor In order to express these unknown parameters in terms of the bulk-stream velocity U_b, which is equal to U_1 (see Fig. 7–1), we couple Eq. (7–35) with the defining equation for U_b, Eq. (6–34),

$$U_b = \frac{1}{A}\int_A u\,dA \tag{7–36}$$

as follows:

$$\begin{aligned} U_b &= \frac{1}{\pi r_0^2}\int_0^{r_0} u\,2\pi r\,dr = \frac{2}{r_0^2}\int_0^{r_0} ur\,dr \\ &= \frac{2}{r_0^2}\int_0^{r_0} rU_c\left[1 - \left(\frac{r}{r_0}\right)^2\right]dr = \frac{U_c}{2} = -\frac{1}{8\mu}\frac{dP}{dx}r_0^2 \end{aligned} \tag{7–37}$$

The substitution of this result back into Eq. (7–35) gives

$$u = 2U_b\left[1 - \left(\frac{r}{r_0}\right)^2\right] \tag{7–38}$$

where $U_b = U_1$. This expression is shown in Fig. 7–5 to agree very well with experimental data.

Now that the velocity distribution is fully specified, Newton's law of viscous shear is used to obtain an expression for the wall shear stress $\tau_s = \tau_0$.

$$\tau_0 = \mu\left.\frac{du}{dy}\right|_0 = -\mu\left.\frac{du}{dr}\right|_{r_0} = \mu 2U_b\left.\left(\frac{2r}{r_0^2}\right)\right|_{r_0} = 4\frac{\mu U_b}{r_0} \tag{7–39}$$

It follows that the Fanning friction factor f becomes

$$f = \frac{\tau_0}{\rho U_b^2/2} = \frac{8\mu U_b}{\rho U_b^2 r_0} = 16\frac{\nu}{U_b D} = \frac{16}{Re} \tag{7–40}$$

This equation is compared with experimental data in Fig. 7–6. The agreement between theory and experiment is exceptional for Reynolds numbers below the transitional value of approximately 2000.

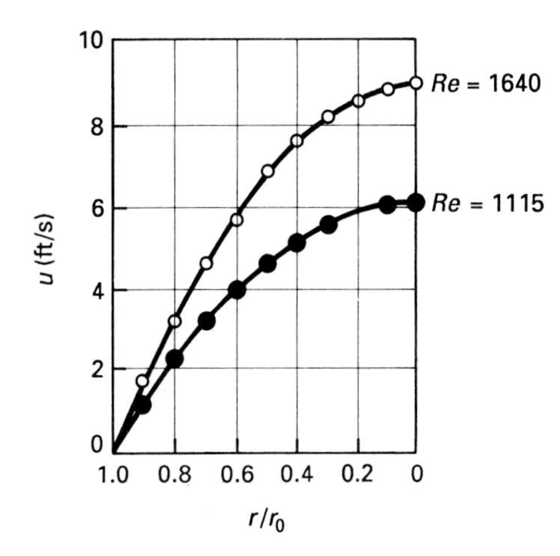

FIGURE 7–5
Comparison of Eq. (7–38) with experimental data for u—HFD laminar flow in circular tube. (Data from Senecal [13] for flow of air in a 0.75-in.-I.D. tube.)

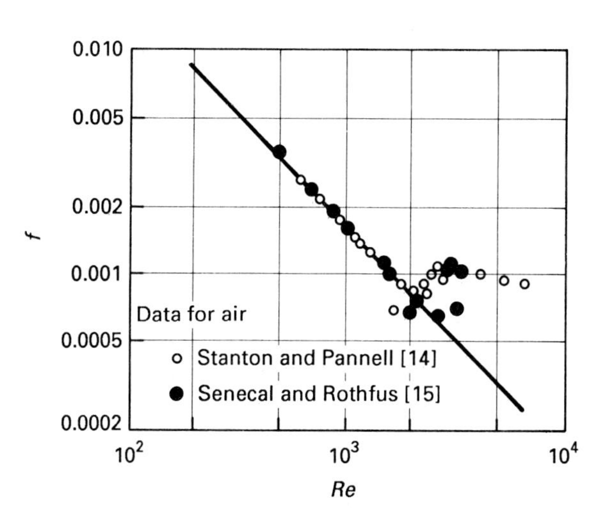

FIGURE 7–6
Comparison of Eq. (7–40) with experimental data for f—HFD laminar flow in circular tubes.

Energy Transfer—TFD Region With the velocity distribution specified by Eq. (7–38) for HFD conditions, the energy equation becomes

$$2U_b\left[1 - \left(\frac{r}{r_0}\right)^2\right]\frac{\partial T}{\partial x} = \frac{\alpha}{r}\frac{\partial}{\partial r}\left(r\frac{\partial T}{\partial r}\right) \qquad (7\text{–}41)$$

For the TFD region, the term $\partial T/\partial x$ can be specified in terms of T, T_s, and T_b by utilizing the defining equation for TFD conditions [Eq. (6–47)],

$$\frac{\partial}{\partial x}\left(\frac{T - T_s}{T_b - T_s}\right) = 0 \qquad (7\text{–}2)$$

and by recognizing that the local coefficient of heat transfer h is independent of x. For a uniform wall-heat-flux boundary condition, it follows from the general Newton law of cooling,

$$q_c'' = q_0'' = h(T_s - T_b) \tag{7–42}$$

that $T_s - T_b$ is independent of x. Thus,

$$\frac{dT_b}{dx} = \frac{dT_s}{dx} \tag{7–43}$$

such that Eq. (7–2) gives

$$\frac{\partial T}{\partial x} = \frac{dT_s}{dx} = \frac{dT_b}{dx} \tag{7–44}$$

Hence, the energy equation takes the one-dimensional form

$$2U_b\left[1 - \left(\frac{r}{r_0}\right)^2\right]\frac{dT_b}{dx} = \frac{\alpha}{r}\frac{d}{dr}\left(r\frac{dT}{dr}\right) \tag{7–45}$$

for uniform wall-flux heating.

Referring to the simple lumped analysis developed in Example 7–2, dT_b/dx is expressed in terms of q_0'' by

$$\frac{dT_b}{dx} = \frac{q_0''p}{\dot{m}c_P} = \frac{4q_0''}{\rho c_P U_b D} \tag{7–46}$$

Substituting this relation into Eq. (7–45), the energy equation takes the form

$$\frac{4q_0''}{k}\left(\frac{r}{r_0}\right)\left[1 - \left(\frac{r}{r_0}\right)^2\right] = \frac{d}{dr}\left(r\frac{dT}{dr}\right) \tag{7–47}$$

This equation is coupled with the r boundary conditions

$$\frac{\partial T}{\partial r} = 0 \qquad \text{at } r = 0 \tag{7–28}$$

$$k\frac{\partial T}{\partial r} = q_0'' \qquad \text{at } r = r_0 \tag{7–29}$$

Following the pattern established in the solution of the fluid-flow problem, we first develop a solution for the dimensionless temperature distribution.

Dimensionless Temperature Distribution A first integration of Eq. (7–47) from 0 to r and the use of Eq. (7–28) gives

$$r\frac{dT}{dr} = \frac{4q_0''}{r_0 k}\left(\frac{r^2}{2} - \frac{r^4}{4r_0^2}\right) \tag{7–48}$$

[This same result is obtained by integrating from r_0 to r with the condition at the wall prescribed by Eq. (7–29).] A second integration yields

$$T = \frac{4q_0''}{r_0 k}\left(\frac{r^2}{4} - \frac{r^4}{16r_0^2}\right) + C_1 \tag{7–49}$$

As just indicated, both boundary conditions satisfy Eq. (7–48), but neither can be utilized to obtain C_1 in Eq. (7–49). We circumvent this anomaly by setting T equal to T_s at $r = r_0$. However, it is important to note that T_s is still an unknown function of x that eventually must be evaluated in terms of the specified input q_0''. The use of this intermediate step gives

$$C_1 = T_s - \frac{4q_0''}{r_0 k}\frac{3}{16}r_0^2 \tag{7–50}$$

such that the temperature profile becomes

$$T - T_s = -\frac{q_0'' r_0}{k}\left[\frac{3}{4} - \left(\frac{r}{r_0}\right)^2 + \frac{1}{4}\left(\frac{r}{r_0}\right)^4\right] \tag{7–51}$$

To put this expression into a more convenient dimensionless format, T is set equal to T_c at r equal to zero, with the result

$$\frac{T - T_s}{T_c - T_s} = 1 - \frac{4}{3}\left(\frac{r}{r_0}\right)^2 + \frac{1}{3}\left(\frac{r}{r_0}\right)^4 \tag{7–52}$$

where $T_c - T_s = -\frac{3}{4}q_0'' r_0/k$. Whereas the HFD velocity distribution is a second-order polynomial, the temperature distribution is seen to be fourth order.

Bulk-Stream Temperature and Nusselt Number In order to express T_s in terms of T_b, we utilize the defining expression for T_b given by Eq. (6–46). This equation reduces to the following form for a circular-tube geometry:

$$T_b = \frac{2}{r_0^2 U_b}\int_0^{r_0} uTr\,dr \tag{7–53}$$

The substitution of the expressions for u and T given by Eqs. (7–38) and (7–51) into this equation results in

$$T_b = T_s - \frac{11}{24}\frac{q_0'' r_0}{k} \tag{7–54}$$

Referring to Example 7–2, because T_b is also given by Eq. (E7–2e), which is based on the simple lumped analysis approach, Eq. (7–54) specifies T_s.

Eliminating T_s in Eq. (7–51) by the use of Eq. (7–54), we have a final expression for the temperature distribution, which can be written as

$$T - T_b = -\frac{q_0'' r_0}{k}\left[\frac{7}{24} - \left(\frac{r}{r_0}\right)^2 + \frac{1}{4}\left(\frac{r}{r_0}\right)^4\right] \tag{7–55}$$

To obtain an expression for the Nusselt number Nu, we merely rearrange Eq. (7–54), with the result

$$h = \frac{q_0''}{T_s - T_b} = \frac{24}{11}\frac{k}{r_0} \tag{7–56}$$

or

$$Nu = \frac{hD}{k} = \frac{48}{11} = 4.36 \tag{7–57}$$

for uniform wall-heat flux. This equation is consistent with the limiting calculations developed by Sellars et al. [5] for thermal developing flow.

Referring to Example 7–3, we also note that $Nu = 3.66$ for the case of fully developed laminar flow in a circular tube with *uniform wall temperature*. The fact that the solution for uniform wall temperature is 16% lower than the value given by Eq. (7–57) for uniform wall-heat flux indicates the significance of the form of the thermal boundary condition for laminar flow.

As mentioned earlier, the effects of viscous dissipation and kinetic energy should be taken into account for applications involving high-speed flow or viscous fluids. To quantify this point, we note that the relative significance of viscous dissipation and kinetic energy increases with increasing values of the product $Pr\ Ec$, where Pr is the Prandtl number and $Ec = U_F^2/[c_P(T_s - T_F)]$ is the *Eckert number*. Viscous dissipation and kinetic energy effects can be safely neglected for $Pr\ Ec \ll 1$, but should be accounted for when this criterion is not satisfied. Notice that the Eckert number indicates *the significance of the kinetic energy of the flow relative to the enthalpy difference across the boundary layer.*

It should also be noted that the effects of axial conduction generally can be neglected for situations in which the *Peclet number* $Pe = Re\ Pr$ is not small; that is for $Pe \gtrsim 100$. The Peclet number represents the *ratio of enthalpy flow rate to heat conduction rate*. Since Pe is proportional to the Prandtl number Pr, we should not be surprised to find that problems in which axial conduction is significant often involve liquid metals.

EXAMPLE 7–3

Write the mathematical formulation for heat transfer associated with hydrodynamic fully developed laminar flow in a circular tube with uniform wall temperature.

Solution

Objective Write the energy equation and boundary conditions for HFD laminar flow in a circular tube with $T_s = T_0$.

Schematic HFD laminar flow in a circular tube: uniform wall temperature T_0.

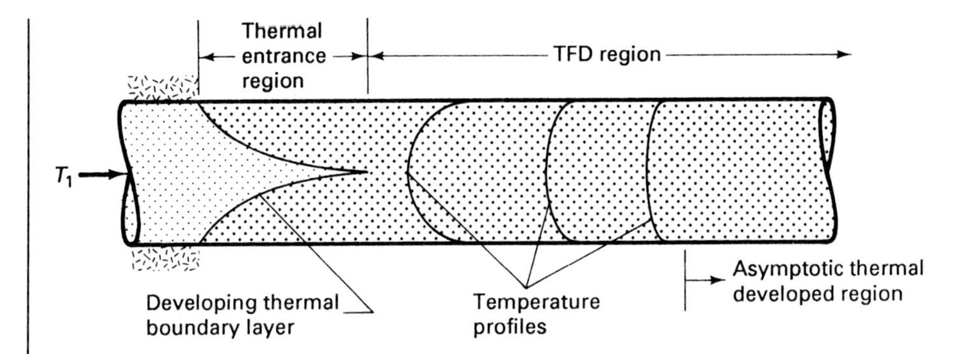

Assumptions/Conditions

forced convection

uniform properties

standard conditions (i.e., steady-state; two-dimensional; ideal, Newtonian, isotropic and single-phase fluid; incompressible flow; no interfacial mass transfer; and negligible buoyancy, kinetic energy, viscous dissipation, external body forces, axial conduction, and thermal radiation)

Analysis The differential formulation for energy transfer in the thermal entrance region is represented by Eq. (7–30),

$$u\frac{\partial T}{\partial x} = \frac{\alpha}{r}\frac{\partial}{\partial r}\left(r\frac{\partial T}{\partial r}\right)$$

or, with u given by Eq. (7–38), by

$$2U_b\left[1 - \left(\frac{r}{r_0}\right)^2\right]\frac{\partial T}{\partial x} = \frac{\alpha}{r}\frac{\partial}{\partial r}\left(r\frac{\partial T}{\partial r}\right) \tag{a}$$

with boundary conditions of the form

$$T = T_0 \qquad \text{at } r = r_0 \qquad \frac{\partial T}{\partial x} = 0 \qquad \text{at } r = 0$$

The temperature gradient $\partial T/\partial x$ for TFD flow is expressed in terms of T, T_0, and T_b by expanding Eq. (7–2); that is

$$\frac{\partial}{\partial x}\left(\frac{T - T_0}{T_b - T_0}\right) = \frac{T_b - T_0}{(T_b - T_0)^2}\left(\frac{\partial T}{\partial x} - \frac{dT_0}{dx}\right) - \frac{T - T_0}{(T_b - T_0)^2}\left(\frac{dT_b}{dx} - \frac{dT_0}{dx}\right) = 0$$

which reduces to

$$\frac{\partial T}{\partial x} = \frac{T - T_0}{T_b - T_0}\frac{dT_b}{dx}$$

Using this relation, Eq. (a) becomes

$$2U_b\left[1 - \left(\frac{r}{r_0}\right)^2\right]\frac{dT_b}{dx}\left(\frac{T - T_0}{T_b - T_0}\right) = \frac{\alpha}{r}\frac{\partial}{\partial r}\left(r\frac{\partial T}{\partial r}\right) \quad \text{(b)}$$

Although Eq. (b) does not lend itself to simple analytical treatment, solutions have been developed by an iterative method involving the use of successive approximations for the temperature distribution [11]. The resulting solution for Nusselt number Nu is given by

$$Nu = 3.66$$

for uniform wall temperature conditions.

It should be noted that for this uniform wall-temperature problem the heat flux approaches zero and the bulk-stream temperature T_b approaches T_0 as x increases. This region in which T_b no longer changes with x is characterized by $\partial T/\partial x = 0$ and is referred to as *asymptotic thermal developed flow*. Asymptotic thermal developed flows of practical importance commonly occur in annular and Couette flow systems.

EXAMPLE 7–4

Develop relations for the friction factor, Nusselt number, and distributions in T_b and T_s for fully developed laminar flow between parallel plates with uniform wall-flux heating.

Solution

Objective Develop relations for f, Nu, T_b, and T_s for TFD laminar flow between parallel plates with $q_c'' = q_0''$.

Schematic TFD laminar flow between parallel plates with uniform wall-heat flux.

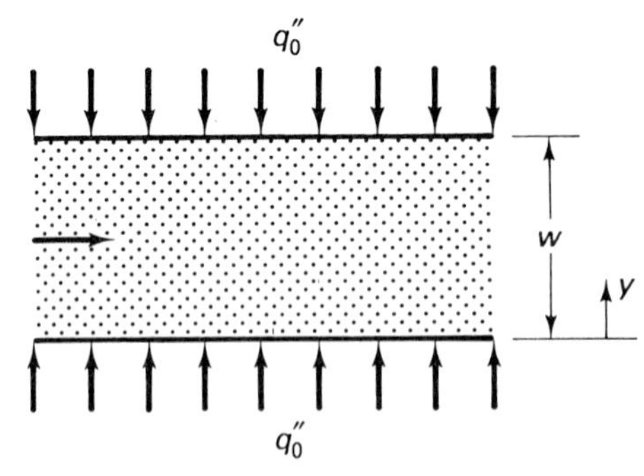

Assumptions/Conditions

forced convection
uniform properties
standard conditions

Analysis The mathematical formulation for TFD flow between parallel plates is represented by

$$\nu \frac{d^2u}{dy^2} - \frac{1}{\rho}\frac{dP}{dx} = 0 \tag{a}$$

and

$$\rho c_P u \frac{\partial T}{\partial x} = k \frac{\partial^2 T}{\partial y^2} \tag{b}$$

for negligible viscous dissipation, kinetic energy, and other standard conditions, where

$$u = 0 \quad \text{at } y = 0 \qquad \frac{du}{dy} = 0 \quad \text{at } y = \frac{w}{2} \tag{c,d}$$

$$-k\frac{\partial T}{\partial y} = q_0'' \quad \text{at } y = 0 \qquad \frac{\partial T}{\partial y} = 0 \quad \text{at } y = \frac{w}{2} \tag{e,f}$$

and

$$\frac{\partial T}{\partial x} = \frac{dT_s}{dx} = \frac{dT_b}{dx} \tag{g}$$

for uniform wall-heat flux. Using the lumped/differential formulation approach introduced in Example 7–2, we are also able to write

$$\frac{dT_b}{dx} = \frac{q_0''p}{\dot{m}c_P} = \frac{q_0''2Z}{\rho c_P U_b Z w} = \frac{2q_0''}{\rho c_P U_b w} \tag{h}$$

The velocity distribution is readily obtained by integrating Eq. (a) twice; that is,

$$\frac{du}{dy} = \frac{1}{\mu}\frac{dP}{dx}y + C_1$$

$$u = \frac{1}{\mu}\frac{dP}{dx}\frac{y^2}{2} + C_1 y + C_2$$

Using Eqs. (c) and (d) to evaluate C_1 and C_2, our solution for u takes the form

$$u = \frac{1}{\mu}\frac{dP}{dx}\left(\frac{y^2}{2} - \frac{yw}{2}\right)$$

To determine the bulk-stream velocity, we write

$$U_b = \frac{1}{w}\int_0^w u\,dy = \frac{1}{w}\int_0^w \frac{1}{\mu}\frac{dP}{dx}\left(\frac{y^2}{2} - \frac{yw}{2}\right)dy$$

$$= \frac{1}{w\mu}\frac{dP}{dx}\left(\frac{y^3}{6} - \frac{y^2w}{4}\right)\Bigg|_0^w = -\frac{w^2}{12}\frac{1}{w}\frac{dP}{dx}$$

Using this result to express dP/dx in terms of U_b, the velocity distribution becomes

$$u = 6U_b\left[\frac{y}{w} - \left(\frac{y}{w}\right)^2\right] \tag{i}$$

Expressions are obtained for the wall-shear stress τ_0 and friction factor f by writing

$$\tau_0 = \mu\frac{du}{dy}\bigg|_0 = \mu 6U_b\left[\frac{1}{w} - 2\frac{y}{w^2}\right]\bigg|_0 = \frac{6\mu U_b}{w} = \rho U_b^2\frac{f}{2}$$

and

$$f = 12\frac{\nu}{U_b w}$$

The hydraulic diameter D_H for flow between parallel plates is given by

$$D_H = \frac{4A}{p} = \frac{4wZ}{2Z} = 2w$$

Thus, the friction factor is expressed in terms of Reynolds number Re by

$$f = 24\frac{\nu}{U_b D_H} = \frac{24}{Re}$$

Substituting Eq. (i) for u into Eq. (b) and using Eqs. (g) and (h) to represent $\partial T/\partial x$, the energy equation for TFD flow with uniform wall-heat flux reduces to a simple second-order differential equation of the form

$$\frac{d^2T}{dy^2} = \frac{12}{w}\frac{q_0''}{k}\left[\frac{y}{w} - \left(\frac{y}{w}\right)^2\right]$$

This equation is integrated twice to obtain

$$\frac{dT}{dy} = \frac{12}{w}\frac{q_0''}{k}\left(\frac{y^2}{2w} - \frac{y^3}{3w^2}\right) + C_1$$

$$T = \frac{12}{w}\frac{q_0''}{k}\left(\frac{y^3}{6w} - \frac{y^4}{12w^2}\right) + C_1 y + C_2$$

Using Eq. (f) to evaluate C_1 and representing the unspecified wall temperature at $y = 0$ by T_s, the solution for T is put into the form

$$T - T_s = \frac{q_0''w}{k}\left[-\frac{y}{w} + 2\left(\frac{y}{w}\right)^3 - \left(\frac{y}{w}\right)^4\right]$$

This relation is combined with Eq. (i) for u and the defining relation for bulk-stream temperature T_b, Eq. (6–46), to obtain

$$T_b = \frac{1}{U_b w}\int_0^w uT\,dy = \frac{1}{U_b w}\int_0^w 6U_b\left[\frac{y}{w} - \left(\frac{y}{w}\right)^2\right]$$

$$\times\left\{T_s + \frac{q_0''w}{k}\left[2\left(\frac{y}{w}\right)^3 - \left(\frac{y}{w}\right)^4 - \frac{y}{w}\right]\right\}dy = T_s - \frac{17}{70}\frac{q_0''w}{k}$$

Rearranging, the Nusselt number Nu becomes

$$Nu = \frac{hD_H}{k} = \frac{q_0''}{T_s - T_b}\frac{2w}{k} = 2\frac{70}{17} = 8.24$$

To express T_b in terms of x, we simply integrate Eq. (h), with T_b set equal to T_1 at $x = 0$, with the result

$$T_b = T_1 + \frac{2q_0''}{\rho c_P U_b}\frac{x}{w} \tag{j}$$

Using this relation, the wall temperature T_s becomes

$$T_s = T_b + \frac{17}{70}\frac{q_0''w}{k} = T_1 + \frac{2q_0''}{\rho c_P U_b}\frac{x}{w} + \frac{17}{70}\frac{q_0''w}{k} \tag{k}$$

in the thermal fully developed region.

As we have noted, this solution is restricted to applications for which viscous dissipation and kinetic energy are negligible. The analysis is generalized to account for these effects in Appendix I and Example 7–5.

EXAMPLE 7–5

Develop relations for the Nusselt number Nu and distributions in T_b and T_s for thermal fully developed laminar flow of a highly viscous fluid between parallel plates with uniform wall-flux heating.

Solution

Objective Develop relations for Nu, T_b, and T_s for TFD laminar flow of a viscous fluid between parallel plates with $q_c'' = q_0''$.

Schematic TFD laminar flow between parallel plates with uniform wall-heat flux: highly viscous fluid.

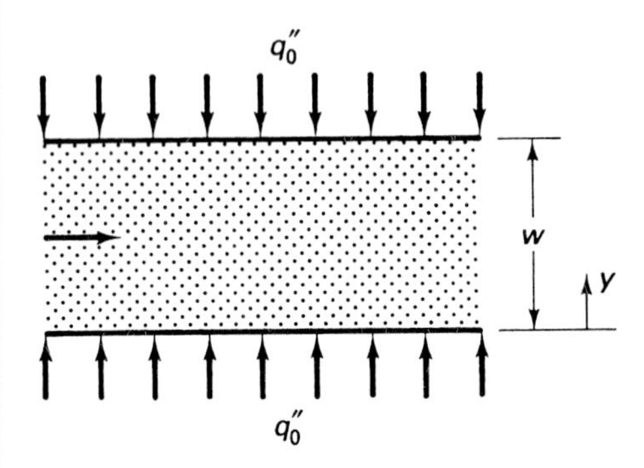

Assumptions/Conditions

forced convection

uniform properties

standard conditions, except for significant viscous dissipation effects

Analysis As shown in Appendix I–1, the energy equation for fully developed laminar flow between parallel plates takes the form

$$\rho c_P u \frac{\partial T}{\partial x} = k \frac{\partial^2 T}{\partial y^2} + \mu \left(\frac{du}{dy}\right)^2 \tag{a}$$

where the term $\mu(du/dy)^2$ represents the *energy dissipation* per unit volume, the velocity u is given by Eq. (E7–4i),

$$u = 6U_b \left[\frac{y}{w} - \left(\frac{y}{w}\right)^2\right]$$

and $\partial T/\partial x = dT_b/dx$ for uniform wall-heat flux.

Referring to Fig. E7–5, the lumped/differential formulation for this case is given by

$$\dot{H}_b|_{x+dx} - \dot{H}_b|_x - q_0''p\, dx - d\dot{E}_\tau = 0$$

or

$$q_0''p\, dx + d\dot{E}_\tau = \frac{d\dot{H}_b}{dx}\, dx = \dot{m}c_P \frac{dT_b}{dx}\, dx \tag{b}$$

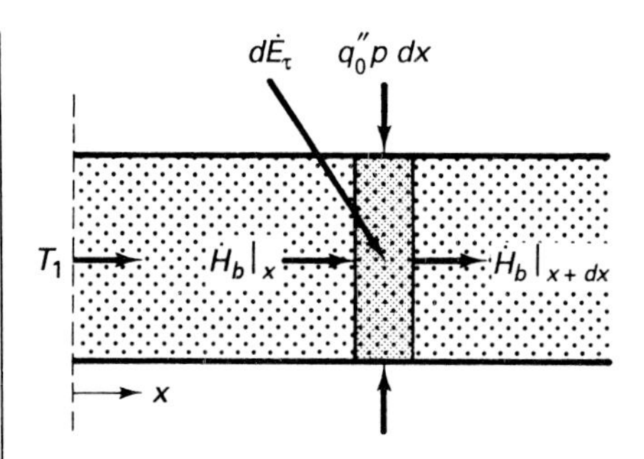

Z–Depth of plates

FIGURE E7–5
Lumped/differential formulation for energy transfer, including viscous dissipation.

where $d\dot{E}_\tau$ represents the rate of viscous energy dissipation within the lumped/differential element. With $d\dot{E}_\tau$ expressed in terms of the energy dissipation per unit volume $\mu(du/dy)^2$, we obtain

$$d\dot{E}_\tau = \int \mu\left(\frac{du}{dy}\right)^2 dV = Z\,dx \int_0^w 36\,\frac{\mu U_b^2}{w^2}\left[1 - 4\frac{y}{w} + 4\left(\frac{y}{w}\right)^2\right] dy$$

$$= 12\,\frac{\mu U_b^2}{w} Z\,dx$$

Substituting for $d\dot{E}_\tau$, $\dot{m}$, and the perimeter p, Eq. (b) becomes

$$2\,q_0'' + 12\,\frac{\mu U_b^2}{w} = \rho c_P U_b w\,\frac{dT_b}{dx}$$

or

$$\frac{dT_b}{dx} = \frac{1}{\rho c_P U_b w}\left(2q_0'' + 12\,\frac{\mu U_b^2}{w}\right) \tag{c}$$

Combining these results with Eq. (a), we have

$$\frac{6}{w}\left[\frac{y}{w} - \left(\frac{y}{w}\right)^2\right]\left(2q_0'' + 12\,\frac{\mu U_b^2}{w}\right)$$
$$= k\,\frac{d^2T}{dy^2} + 36\,\frac{\mu U_b^2}{w^2}\left[1 - 4\frac{y}{w} + 4\left(\frac{y}{w}\right)^2\right]$$

or

$$\frac{d^2T}{dy^2} = \frac{12}{w}\frac{q_0''}{k}\left[\frac{y}{w} - \left(\frac{y}{w}\right)^2\right] + 36\,\frac{\mu U_b^2}{kw^2}\left[-1 + 6\frac{y}{w} - 6\left(\frac{y}{w}\right)^2\right]$$

with the accompanying boundary conditions given by Eqs. (E7–4e) and (E7–4f). Following the approach used in Example 7–4, the solution for T is given by

$$T - T_s = \frac{q_0''w}{k}\left[-\frac{y}{w} + 2\left(\frac{y}{w}\right)^3 - \left(\frac{y}{w}\right)^4\right]$$
$$+ \frac{\mu U_b^2}{k}\left[-18\left(\frac{y}{w}\right)^2 + 36\left(\frac{y}{w}\right)^3 - 18\left(\frac{y}{w}\right)^4\right]$$

Introducing the defining relation for bulk-stream temperature T_b, we obtain

$$T_b = \frac{1}{U_b w}\int_0^w uT\,dy = T_s - \frac{17}{70}\frac{q_0''w}{k} - \frac{54}{70}\frac{\mu U_b^2}{k}$$

or

$$T_s - T_b = \frac{17}{70}\frac{q_0''w}{k} + 0.771\frac{\mu U_b^2}{k} \tag{d}$$

and†

$$Nu = \frac{q_0''}{T_s - T_b}\frac{2w}{k} = \frac{140}{17}\left[1 - 0.771\frac{\mu U_b^2}{k(T_s - T_b)}\right]$$
$$= 8.24\,(1 - 0.771\,Pr\,Ec)$$

where $Ec = U_b^2/[c_P(T_s - T_b)]$ is the Eckert number. Notice that the effect of viscous dissipation is to increase the temperature difference $T_s - T_b$ and to decrease Nu, with the effect on Nu being less than 1% for this application when $Pr\,Ec < 0.013$. For a representative temperature difference $T_s - T_b$ equal to 25°C, this criterion is satisfied for air [$Pr = 0.70$, $c_P = 1.01$ kJ/(kg °C)] with $U_b < 21.6$ m/s, for water [$Pr = 5.0$, $c_P = 4.18$ kJ/(kg °C)] with $U_b < 16.5$ m/s, for oil [$Pr = 100$, $c_P = 2.4$ kJ/(kg °C)] with $U_b < 2.79$ m/s, and for oil [$Pr = 1000$, $c_P = 2.0$ kJ/(kg °C)] with $U_b < 0.805$ m/s. Thus, we would expect the effects of viscous dissipation to be negligible for many applications involving the flow of air or water, but often to be significant for flow of viscous fluids such as oil.

To complete the analysis, we integrate Eq. (c) with T_b set equal to T_1 at $x = 0$, with the result

$$T_b = T_1 + \left(2q_0'' + 12\frac{\mu U_b^2}{w}\right)\frac{x}{\rho c_P U_b w} \tag{e}$$

This result is combined with Eq. (d) to obtain a relation for T_s as a function of x of the form

$$T_s = T_1 + \left(2q_0'' + 12\frac{\mu U_b^2}{w}\right)\frac{x}{\rho c_P U_b w} + \frac{17}{70}\frac{q_0''w}{k} + 0.771\frac{\mu U_b^2}{k} \tag{f}$$

which is applicable to the thermal fully developed region. Notice that Eqs. (e) and (f) reduce to Eqs. (E7–4j) and (E7–4k) for negligible viscous dissipation.

† This solution can also be written in the form $Nu = 8.24/(1 + 6.35\,Br)$, where $Br = \mu U_b^2/(q_0''2w)$ is known as the *Brinkman number* [12]. We also note that $Br = Ec\,Pr/Nu$.

7-2-2 Boundary Layer Flow over Plane Surfaces

We now turn our attention to the problem illustrated in Fig. 7-7 of convection heat transfer associated with steady two-dimensional laminar boundary layer flow over a plane surface that is heated or cooled in the region $x \geqslant x_0$. We have already seen in Chap. 6 that hydrodynamic and thermal boundary layers develop over the surface, with δ and Δ continually increasing with x.

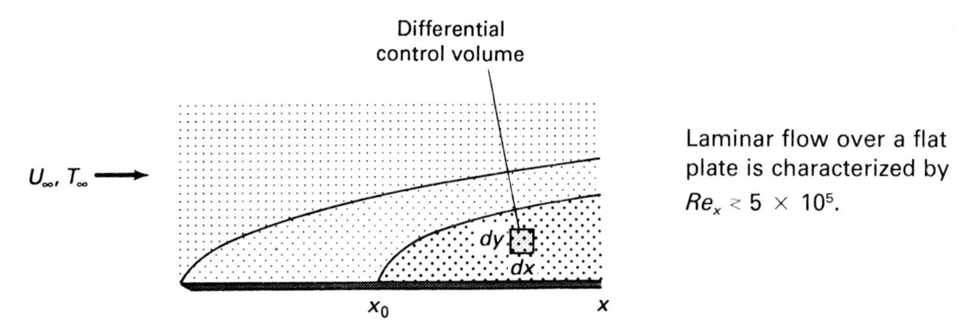

FIGURE 7-7 Hydrodynamic and thermal boundary layers for flow over a flat plate with heating in the region $x \geq x_0$.

Whereas the free-stream temperature T_∞ is independent of x for boundary layer flow applications such as this, the free-stream velocity U_∞ may be uniform or may be a function of x, depending on the arrangement. For the case of parallel flow over a flat plate such as is shown in Fig. 7-7, U_∞ is uniform. On the other hand, U_∞ follows the power law $U_\infty = Cx^m$ for flow over the wedge-shape body shown in Fig. 7-8. A large number of nonuniform distributions are encountered in practice.

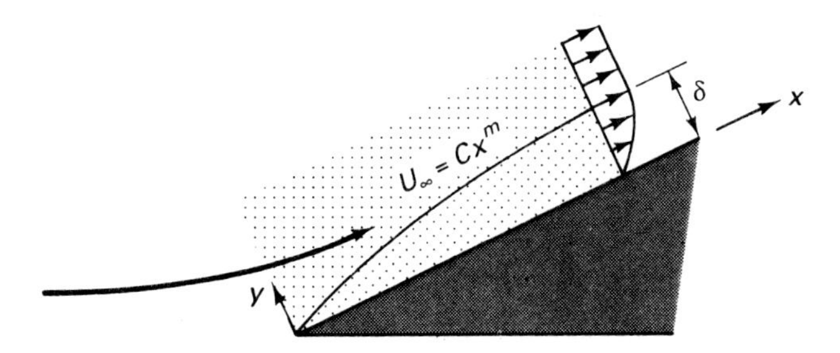

FIGURE 7-8 Flow over a wedge-shape body.

Differential Formulation

Following the approach developed in the preceding section, the application of the principles of conservation pertaining to mass, momentum, and energy to the differential rectangular control volume shown in Fig. 7-7 leads to a system of boundary layer equations for uniform properties given by (see Appendix I-1)

$$\frac{\partial u}{\partial x} + \frac{\partial v}{\partial y} = 0 \tag{7–58}$$

$$u\frac{\partial u}{\partial x} + v\frac{\partial u}{\partial y} = \nu\frac{\partial^2 u}{\partial y^2} - \frac{1}{\rho}\frac{dP}{dx} \tag{7–59}$$

for negligible gravitational and other body force effects, and

$$u\frac{\partial T}{\partial x} + v\frac{\partial T}{\partial y} = \alpha\frac{\partial^2 T}{\partial y^2} \tag{7–60}$$

for negligible kinetic energy, viscous dissipation, body forces, axial conduction, and thermal radiation effects. These equations are similar to Eqs. (7–6), (7–16), and (7–26) and reduce from the more general equations developed in Appendix I–1 for viscous and high-speed flow. The seven accompanying boundary conditions are

$$u = U_\infty \quad \text{at } x = 0 \qquad u = U_\infty \quad \text{as } y \to \infty \tag{7–61,62}$$

$$u = 0 \quad \text{at } y = 0 \qquad v = 0 \quad \text{at } y = 0 \tag{7–63,64}$$

$$T = T_\infty \quad \text{at } x = x_0 \qquad T = T_\infty \quad \text{as } y \to \infty \tag{7–65,66}$$

and

$$T = T_s \qquad \text{at } y = 0 \tag{7–67a}$$

for specified wall temperature, or

$$-k\frac{\partial T}{\partial y} = q_c'' \qquad \text{at } y = 0 \tag{7–67b}$$

for specified wall-heat flux. By satisfying Eq. (7–59) in the region outside the boundary layer, we are able to express the pressure gradient in terms of the free-stream velocity U_∞ by

$$-\frac{dP}{dx} = \rho U_\infty \frac{dU_\infty}{dx} \tag{7–68}$$

It should also be noted that $v \neq 0$ at $y = 0$ for flow over a porous plate with blowing or suction. The important problem of transpired boundary layer flow is addressed in references 11 and 16–18.

Whereas the differential formulation for boundary layer flow generally requires the use of somewhat involved numerical solution techniques, an alternative integral formulation perspective can be used that often lends itself to simpler solution techniques.

Integral Formulation

The integral formulation can be developed by either of two methods. The *indirect method* involves the integration of the differential equations. This approach will be introduced in the context of natural convection flow in Sec. 7–2–3. In the *direct approach* to the development of the integral equations, we apply the fundamental laws pertaining to mass, momentum, and energy transfer to the lumped/differential fluid volumes $\delta\, dx\, dz$ and $\Delta\, dx\, dz$. The direct approach is now used to develop the integral formulation for laminar boundary layer flow over plane surfaces.

Continuity Applying the principle of conservation of mass to the control volume $\delta\, dx\, dz$ shown in Fig. 7–9, we obtain

$$\Sigma\, \dot{m}_o - \Sigma\, \dot{m}_i + \frac{\Delta m_s}{\Delta t} = 0$$

$$\dot{m}_{x+dx} - \dot{m}_x - d\dot{m}_\delta = 0 \tag{7–69}$$

where $\Delta m_s/\Delta t = 0$ for steady flow, and the mass flow through the wall is zero $[v(x,0) = 0]$; $d\dot{m}_\delta$ is the rate of mass transfer from the free stream into the control volume. The total mass-flow rate $\dot{m}_x$ is related to $d\dot{m}_x\ (= \rho u\, dz\, dy)$ by

$$\dot{m}_x = \int d\dot{m}_x = \int_0^\delta (\rho u\, dz)\, dy \tag{7–70}$$

By utilizing the definition of the derivative, Eq. (7–69) reduces to

$$d\dot{m}_\delta = \frac{d\dot{m}_x}{dx}\, dx = \left(\frac{d}{dx}\int d\dot{m}_x\right) dx = \left[\frac{d}{dx}\int_0^\delta (\rho u\, dz)\, dy\right] dx \tag{7–71}$$

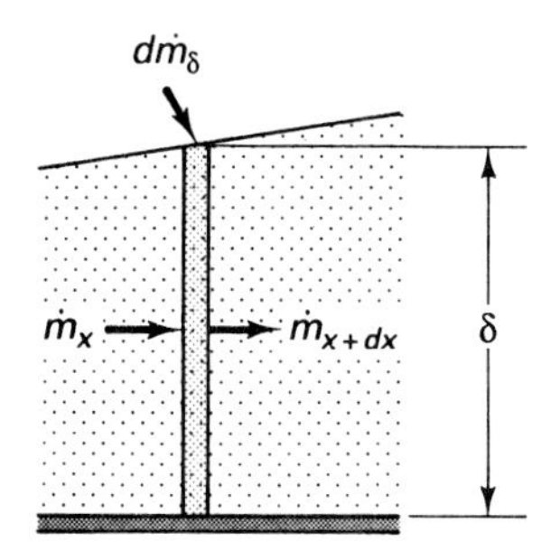

FIGURE 7–9
Conservation of mass relative to lumped/differential rectangular control volume; $dV = \delta\, dx\, dz$.

Similarly, the differential mass flow rate $d\dot{m}_\Delta$ from the free stream into the thermal boundary layer volume $\Delta\, dx\, dz$ can be represented by

$$d\dot{m}_\Delta = \left(\frac{d}{dx}\int d\dot{m}_x\right) dx = \left[\frac{d}{dx}\int_0^\Delta (\rho u\, dz)\, dy\right] dx \tag{7–72}$$

Momentum Transfer Applying the momentum principle to the lumped/differential volume $\delta\, dx\, dz$ shown in Fig. 7–10, we obtain

$$RCM_x = \Sigma\, F_x \tag{7–73}$$

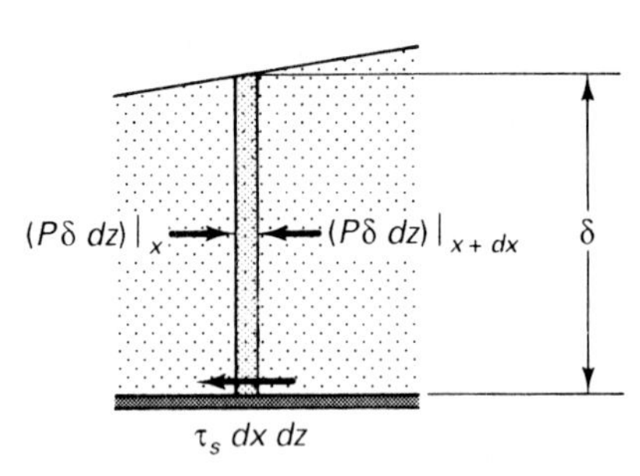

(a) Forces.

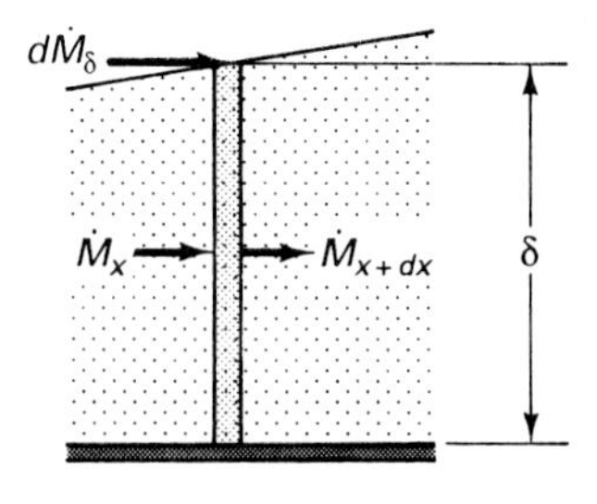

(b) Momentum transfer.

FIGURE 7–10 Momentum principle (x-direction) relative to lumped/differential rectangular control volume; $dV = \delta\, dx\, dz$.

where

$$\Sigma F_x = -\tau_s\, dx\, dz + (P\delta\, dz)|_x - (P\delta\, dz)|_{x+dx} \tag{7–74}$$

and

$$RCM_x = \dot{M}_o - \dot{M}_i + \frac{\Delta M_s}{\Delta t} = \dot{M}_{x+dx} - \dot{M}_x - d\dot{M}_\delta \tag{7–75}$$

with $\Delta M_s/\Delta t = 0$ for steady flow, $d\dot{M}_\delta = U_\infty\, d\dot{m}_\delta$, and

$$\dot{M}_x = \int u\, d\dot{m}_x = \int_0^\delta u(\rho u\, dz)\, dy \tag{7–76}$$

Utilizing the definition of the derivative, Eq. (7–73) reduces to

$$\frac{d\dot{M}_x}{dx}\, dx - d\dot{M}_\delta = -\tau_s\, dx\, dz - \frac{dP}{dx}\,\delta\, dx\, dz \tag{7–77}$$

The substitution of Eq. (7–76) and the defining relationship for $d\dot{M}_\delta$ into this expression gives

$$\left(\frac{d}{dx}\int u\, d\dot{m}_x\right) dx - U_\infty\, d\dot{m}_\delta = -\tau_s\, dx\, dz - \frac{dP}{dx}\,\delta\, dx\, dz \tag{7–78}$$

Substituting for $d\dot{m}_x$ and $d\dot{m}_\delta$, Eq. (7–78) becomes

$$\frac{d}{dx}\int_0^\delta u^2\, dy - U_\infty \frac{d}{dx}\int_0^\delta u\, dy = -\frac{\tau_s}{\rho} - \frac{\delta}{\rho}\frac{dP}{dx} \tag{7–79}$$

for uniform property flow. Using Eq. (7–68) to eliminate $-dP/dx$ in favor of $\rho U_\infty\, dU_\infty/dx$ and rearranging, we obtain

$$U_\infty \frac{d}{dx}\int_0^\delta u\, dy - \frac{d}{dx}\int_0^\delta u^2\, dy + \delta U_\infty \frac{dU_\infty}{dx} = \frac{\tau_s}{\rho} \tag{7–80}$$

With U_∞ taken inside the differential and with τ_s represented by Newton's law of viscosity, Eq. (7–80) takes the form

$$\frac{d}{dx}\int_0^\delta u(U_\infty - u)\,dy + \frac{dU_\infty}{dx}\int_0^\delta (U_\infty - u)\,dy = \nu \left.\frac{\partial u}{\partial y}\right|_0 = \frac{\tau_s}{\rho} \tag{7–81}$$

which represents the *integral momentum equation.*

For the case of uniform free-stream velocity flow, $dU_\infty/dx = 0$ such that Eq. (7–81) reduces to

$$\frac{d}{dx}\int_0^\delta u(U_\infty - u)\,dy = \nu \left.\frac{\partial u}{\partial y}\right|_0 = \frac{\tau_s}{\rho} \tag{7–82}$$

The boundary conditions that accompany the integral momentum equation are given by Eqs. (7–61)–(7–64). Notice that the boundary condition for v given by Eq. (7–64) has already been satisfied in the development of the integral continuity equation, which itself has been incorporated into the integral momentum equation.

It should be noted that higher-order integral equations for fluid flow can be developed by multiplying the differential continuity and momentum equations by weighing functions and integrating across the boundary layer. For example, the *integral mechanical energy equation* can be obtained by multiplying the momentum equation through by u, rearranging and coupling this equation with the continuity equation, and integrating across the boundary layer. Higher-order integral equations such as the mechanical energy equation are used in the development of the most modern multiple-parameter integral methods.

Energy Transfer The application of the first law of thermodynamics to the lumped volume $\Delta\, dx\, dz$ shown in Fig. 7–11 gives (for negligible potential and kinetic energy, viscous dissipation, body force, axial conduction, and thermal radiation effects)

$$\Sigma \dot{E}_o - \Sigma \dot{E}_i + \cancel{\frac{\Delta E_s}{\Delta t}} = 0$$

$$\dot{H}_{x+dx} - \dot{H}_x - d\dot{H}_\Delta - dq_c = 0 \tag{7–83}$$

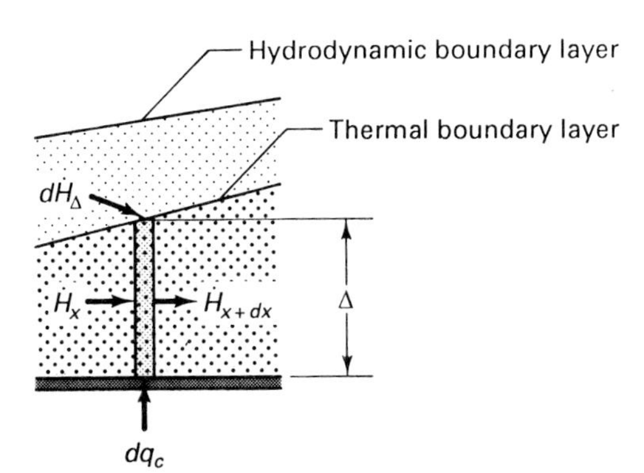

FIGURE 7–11
Conservation of energy relative to lumped/differential rectangular control volume; $dV = \Delta\, dx\, dz$.

where $\Delta E_s/\Delta t = 0$ for steady-state conditions, $d\dot{H}_\Delta = i_\infty \, d\dot{m}_\Delta$, and

$$\dot{H}_x = \int d\dot{H}_x = \int i \, d\dot{m}_x = \int_0^\Delta i(\rho u \, dz) \, dy \tag{7–84}$$

Utilizing the definition of the derivative, Eq. (7–83) is written as

$$d\dot{H}_\Delta + dq_c = \frac{d\dot{H}_x}{dx} dx \tag{7–85}$$

With $\dot{H}_x$ given by Eq. (7–84) and with $d\dot{m}_\Delta$ specified on the basis of continuity by Eq. (7–72), we have

$$i_\infty \left(\frac{d}{dx} \int d\dot{m}_x \right) dx + dq_c = \left(\frac{d}{dx} \int i \, d\dot{m}_x \right) dx \tag{7–86}$$

or

$$\frac{d}{dx} \int (i - i_\infty) \, d\dot{m}_x = \frac{dq_c}{dx} \tag{7–87}$$

Replacing $d\dot{m}_x$ by $(\rho u \, dz) \, dy$ and setting $i - i_\infty = c_P(T - T_\infty)$ for flow of ideal fluids with uniform properties, the *integral energy equation* becomes

$$\rho c_P \frac{d}{dx} \int_0^\Delta u(T - T_\infty) \, dy = \frac{dq_c}{dx \, dz} = q_c'' \tag{7–88}$$

The integral energy equation is coupled with the thermal boundary conditions given by Eqs. (7–65) through (7–67). For the case in which a specified wall-heat flux is maintained, q_c'' is simply replaced by the specified input. On the other hand, for the case in which the wall temperature is specified, the Fourier law of conduction is utilized, with Eq. (7–88) taking the form

$$\rho c_P \frac{d}{dx} \int_0^\Delta u(T - T_\infty) \, dy = -k \left. \frac{\partial T}{\partial y} \right|_0 = q_c'' \tag{7–89}$$

As in the case of the fluid flow, higher-order integral equations can be developed for energy transfer. However, such higher-order integral energy equations are not presently in common use.

Solution

Because of the nature of the differential formulation given by Eqs. (7–58)–(7–60) and the accompanying boundary conditions, the development of exact solutions for laminar boundary layer flow usually requires the use of numerical-solution techniques [19–21]. The numerical-solution approaches developed in the literature generally involve the use of transformations that provide a basis for achieving efficient numerical calculations. A particularly important feature of the transformation approach is that the transformed equations reduce to ordinary differential equations for the special class of flows for which the velocity and temperature distributions are geometrically similar in the streamwise direction. Boundary layer flow over a flat plate with uniform free-stream velocity and uniform wall temperature or uniform wall-heat flux, and flow over wedge-shape bodies are included in this special category of *similar flows*. Similarity solutions are developed in Appendix I–2 and references 11 and 16–18.

The simpler approximate integral approach involves the solution of ordinary rather than partial differential equations for nonsimilar as well as similar steady two-dimensional boundary layers. In addition, the integral approach provides a frame of reference for the development of the classic transformations used in the numerical-transformation approach. Because of the relative simplicity and versatility of the integral approach, this very popular and practical method will be featured in this section. Because we are dealing with uniform property flow, the fluid-flow aspects of the problem are treated first, after which the energy transfer will be handled. The practical approach presented in this section is extended to the more general case involving nonuniform free-stream velocity and transpiration in references 22 and 23.

We now turn our attention to the classic problem of flow over a heated or cooled flat plate with uniform free-stream velocity. Referring to Eq. (7–68), boundary layer flow with uniform free-stream velocity is characterized by a zero axial pressure gradient; that is,

$$\frac{dP}{dx} = -\rho U_\infty \frac{dU_\infty}{dx} = 0 \tag{7–90}$$

Integral Solution—Fluid Flow The integral momentum equation for boundary layer flow over a flat plate with uniform free-stream velocity,

$$\frac{d}{dx}\int_0^\delta u(U_\infty - u)\,dy = \nu \left.\frac{\partial u}{\partial y}\right|_0 \tag{7–82}$$

involves the one unknown u. The boundary layer thickness δ is of course dependent upon u. A simple approximate solution to this integral form of the momentum equation can be obtained by treating δ as the unknown, with the velocity profile u being approximated in terms of δ on the basis of the physics of the problem.

Dimensionless Velocity Distribution Referring to Fig. 7–12, which depicts the laminar hydrodynamic boundary layer, approximate velocity profiles are easily sketched in that satisfy the boundary conditions

$$u = U_\infty \qquad \text{as } y \to \infty \qquad \text{or at } y \simeq \delta \tag{7–62}$$

$$u = 0 \qquad \text{at } y = 0 \tag{7–63}$$

and the physical requirement

$$\frac{\partial u}{\partial y} = 0 \qquad \text{as } y \to \infty \qquad \text{or at } y \simeq \delta \tag{7–91}$$

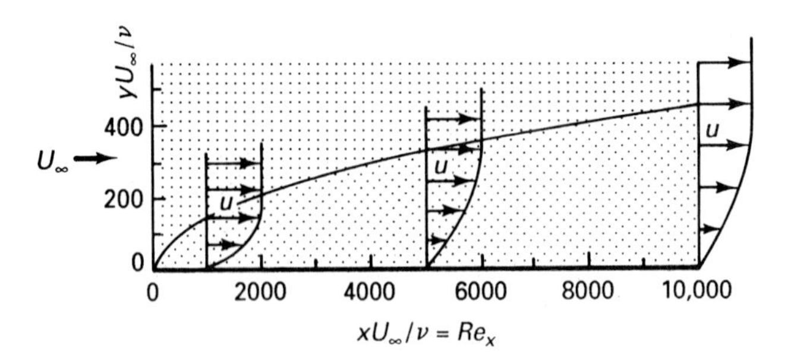

FIGURE 7–12 Representative velocity distribution for laminar boundary layer flow over a flat plate with uniform free-stream velocity.

Many analytical expressions are available that meet these minimal requirements. One example is the simple second-order polynomial approximation

$$\frac{u}{U_\infty} = 2\frac{y}{\delta} - \left(\frac{y}{\delta}\right)^2 \qquad \text{for } y \leqslant \delta \tag{7–92a}$$

$$= 1 \qquad \text{for } y \geqslant \delta \tag{7–92b}$$

Improvements in our approximation for u can be made by satisfying one or two higher-order requirements, such as

$$\frac{\partial^2 u}{\partial y^2} = 0 \qquad \text{at } y = 0 \tag{7–93}$$

which is indicated by the differential momentum equation, Eq. (7–59) with $dP/dx = 0$, and

$$\frac{\partial^2 u}{\partial y^2} = 0 \qquad \text{at } y = \delta \tag{7–94}$$

in accordance with physical requirements. We can also show that the differential equation must satisfy

$$\frac{\partial^3 u}{\partial y^3} = 0 \qquad \text{at } y = 0 \qquad \text{and} \qquad \text{as } y \to \infty \text{ or at } y \simeq \delta \tag{7–95}$$

plus higher-order derivatives at the outer edge of the boundary layer. The exact solution for u satisfies all these requirements. However, it should be noted that the fulfillment of all these constraints at $y = 0$ and at $y = \delta$ does not guarantee that u satisfies the fundamental differential formulation [Eqs. (7–58) and (7–59)] in the region $0 < y < \delta$.

As already indicated, the second-order profile given by Eq. (7–92) satisfies the primary conditions given by Eqs. (7–62), (7–63), and (7–91). It even satisfies the requirements given by Eq. (7–95), but it fails to satisfy the specifications given by Eqs. (7–93) and (7–94). On the basis of previous studies, more accurate approximations for u have been developed by the use of third-, fourth-, and fifth-order polynomials. The simplest and most convenient of these higher-order approximations is obtained on the basis of the third-order polynomial†

$$u = a_0 + a_1 y + a_2 y^2 + a_3 y^3 \tag{7–96}$$

Since we now have four coefficients, Eqs. (7–62), (7–63), (7–91), and (7–93) can be satisfied. The coupling of Eq. (7–96) with these conditions gives $a_0 = 0$, $2a_2 = 0$, $a_0 + a_1\delta + a_2\delta^2 + a_3\delta^3 = U_\infty$, and $a_1 + 2a_2\delta + 3a_3\delta^2 = 0$. It follows that $a_1 = 3U_\infty/(2\delta)$, and $a_3 = -U_\infty/(2\delta^3)$, such that Eq. (7–96) becomes

$$\frac{u}{U_\infty} = \frac{3}{2}\frac{y}{\delta} - \frac{1}{2}\left(\frac{y}{\delta}\right)^3 \qquad \text{for } y \leqslant \delta \tag{7–97a}$$

$$= 1 \qquad \text{for } y \geqslant \delta \tag{7–97b}$$

The velocity profiles shown in Fig. 7–12 are based on this equation. Introducing Newton's law of viscosity, we are able to express the wall-shear stress τ_s in terms of δ by

$$\tau_s = \mu \left.\frac{\partial u}{\partial y}\right|_0 = \frac{3}{2}\frac{\mu U_\infty}{\delta} \tag{7–98}$$

As we shall see, this third-order polynomial gives rise to approximate solutions that are within about 3% of the exact solution. Comparable results can be achieved by the use of a third-order polynomial based on the primary constraints together with Eq. (7–94) instead of Eq. (7–93). This level of accuracy is also achieved by the use of a fourth-order polynomial, which satisfies both Eqs. (7–93) and (7–94), and a fifth-order polynomial, which satisfies one of the conditions represented by Eq. (7–95). However, the use of higher-order conditions at the boundaries to establish approximations for the velocity distribution generally proves to be ineffective. Alternative modern methods for achieving more accurate and reliable approximations involve the use of one or more higher-order integral equations. For example, a fourth-order polynomial can be developed with the five coefficients specified in accordance with Eqs. (7–62), (7–63), (7–91), and (7–93) and the integral mechanical energy equation. The use of this two-parameter approach results in solutions that are within 1% of the exact solution. However, this approach is somewhat more involved than the simple one-parameter approach featured in our study. An attractive multiple-parameter approach of a different kind, which involves the use of approximations for the viscous stress in terms of u, has been developed by Dorodnitsyn [24], Holt [25], and Fletcher [26].

Boundary Layer Thickness δ and Friction Factor Now that reasonable approximations are available for the velocity profile u in terms of δ, we return to the integral momentum equation in order to develop predictions for δ. Utilizing our third-order polynomial approximation, the integral momentum equation becomes

† An equivalent approach, which provides a basis for extending the analysis to laminar transpired flow and turbulent flow, involves the use of polynomial approximations for shear stress τ in terms of y [22].

$$U_\infty^2 \frac{d}{dx} \int_0^\delta \left[\frac{3}{2}\left(\frac{y}{\delta}\right) - \frac{1}{2}\left(\frac{y}{\delta}\right)^3\right]\left[1 - \frac{3}{2}\frac{y}{\delta} + \frac{1}{2}\left(\frac{y}{\delta}\right)^3\right] dy = \frac{3}{2}\frac{\nu U_\infty}{\delta} \tag{7–99}$$

Performing the integration called for in this equation, we obtain

$$\delta \frac{d\delta}{dx} = \frac{140}{13}\frac{\nu}{U_\infty} \tag{7–100}$$

This differential equation in δ is coupled with the requirement

$$\delta = 0 \qquad \text{at } x = 0 \tag{7–101}$$

Equation (7–100) is readily integrated to give

$$\delta^2 = \frac{280}{13}\frac{\nu x}{U_\infty} \qquad \delta = 4.64\sqrt{\frac{\nu x}{U_\infty}} \tag{7–102,103}$$

This equation was actually used in the construction of Figs. 7–7 and 7–12.

Substituting this result for δ into Eq. (7–97a), we obtain an expression for the velocity distribution of the form

$$\begin{aligned}\frac{u}{U_\infty} &= \frac{3}{2}\left[\sqrt{\frac{13}{280}}\frac{y}{\sqrt{x\nu/U_\infty}}\right] - \frac{1}{2}\left[\sqrt{\frac{13}{280}}\frac{y}{\sqrt{x\nu/U_\infty}}\right]^3 \qquad \text{for } y \leq \delta \\ &= 0.323\eta - 0.005\eta^3 \qquad \text{for } \eta \leq 4.64\end{aligned} \tag{7–104}$$

where $\eta = y/\sqrt{x\nu/U_\infty} = \sqrt{280/13}\, y/\delta$. This equation is compared with experimental data and the numerical solution in Fig. 7–13. With u expressed in terms of η (or

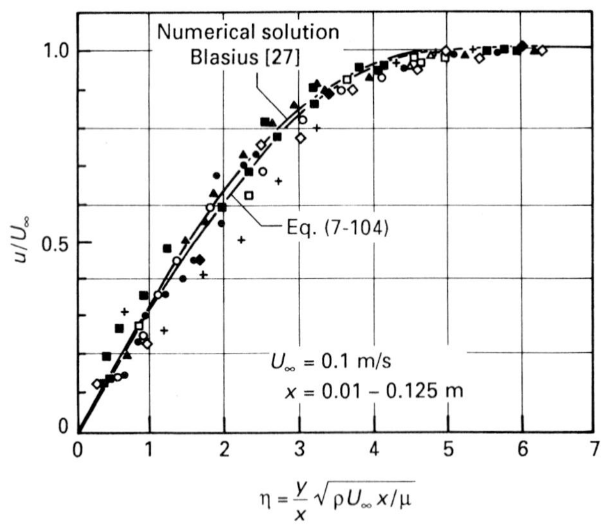

FIGURE 7–13 Velocity distribution for laminar boundary layer flow over a flat plate with uniform free-stream velocity. (Data from Hansen [2].)

y/δ), we find that the velocity profiles are geometrically similar at all values of x. Consequently, η is known as a *similarity coordinate*.

To obtain the friction factor f_x, Eqs. (7–98) and (7–102) are combined, with the result

$$f_x = \frac{\tau_s}{\rho U_\infty^2/2} = 3\frac{\nu}{U_\infty \delta} = 3\sqrt{\frac{13}{280}}\sqrt{\frac{\nu}{U_\infty x}} = \frac{0.646}{Re_x^{1/2}} \tag{7–105}$$

This expression is only 2.7% below the exact solution,

$$f_x = \frac{0.664}{Re_x^{1/2}} \tag{7–106}$$

which was first developed by Blasius [27] and is shown to be in good agreement with experimental data for laminar flow in Fig. 7–14.

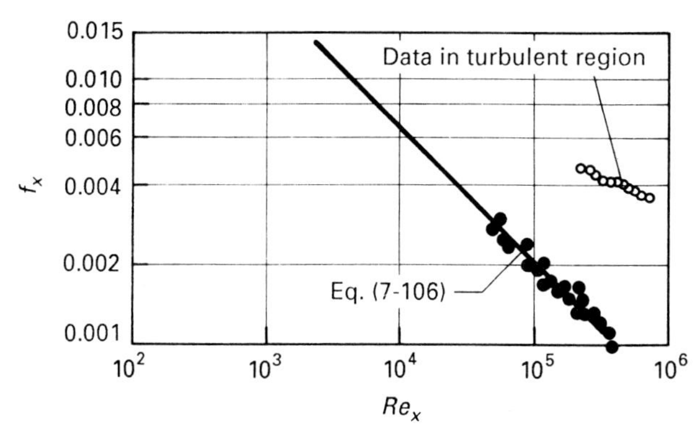

FIGURE 7–14 Friction factor f_x—laminar boundary layer flow over a flat plate with uniform free-stream velocity.

Integral Solution—Energy Transfer Consideration is now turned to the development of approximate integral solutions for several basic types of thermal boundary conditions.

To develop an approximate solution for the thermal boundary layer thickness Δ, which appears in the integral energy equation,

$$\rho c_P \frac{d}{dx}\int_0^\Delta u(T - T_\infty)\,dy = -k\frac{\partial T}{\partial y}\bigg|_0 = q_c'' \tag{7–89}$$

we must develop an approximation for the temperature distribution.

Dimensionless Temperature Distribution Representative temperature distributions for flow over a flat plate with *uniform wall temperature* are illustrated in Fig. 7–15 for cases in which the thermal boundary layer thickness Δ is greater than, nearly equal to, and less than δ.† Following the pattern of our fluid-flow analysis,

† Concerning the result shown in Fig. 7–15 for $Pr = 0.72$, the integral analysis to follow indicates that $\Delta < \delta$ for uniform wall-heat flux, but $\Delta > \delta$ for uniform wall temperature. This point illustrates the significant effect of the form of the thermal boundary condition on the solution for laminar flow.

T is approximated by the third-order polynomial

$$T = b_0 + b_1 y + b_2 y^2 + b_3 y^3 \tag{7–107}$$

where the four coefficients are selected in accordance with the following four conditions, which are analogous to those used in the fluid flow analysis:

$$T = T_\infty \qquad \frac{\partial T}{\partial y} = 0 \qquad \text{as } y \to \infty \text{ or at } y \simeq \Delta \tag{7–66,108}$$

$$T = T_s \qquad \frac{\partial^2 T}{\partial y^2} = 0 \qquad \text{at } y = 0 \tag{7–67a,109}$$

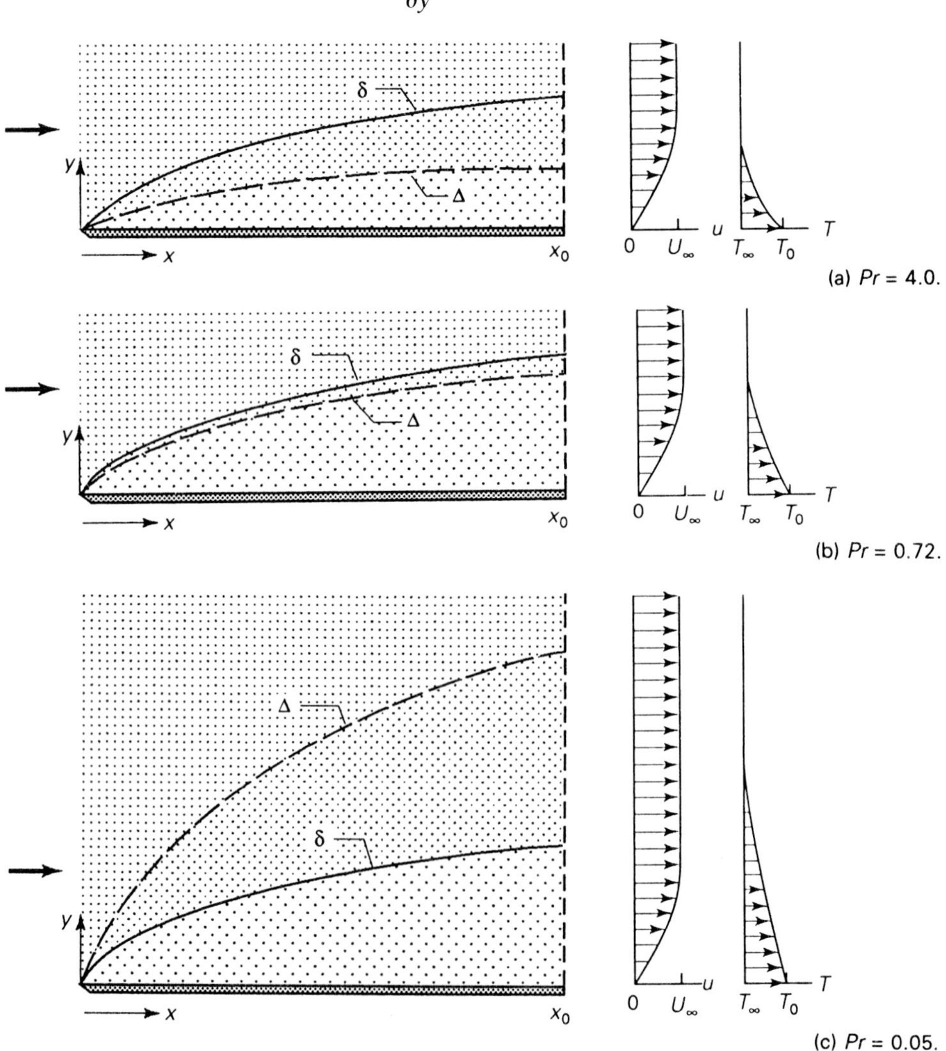

FIGURE 7–15 Laminar forced convection flow over a flat plate with uniform free-stream velocity and uniform wall-heat flux. (Velocity and temperature profiles shown at $x = x_0$.)

[Equation (7–109) is obtained by evaluating the differential energy equation, Eq. (7–60), at the wall where u and v are both zero.] Following through with the evaluation of the coefficients b_n, we obtain

$$\frac{T - T_s}{T_\infty - T_s} = \frac{3}{2}\frac{y}{\Delta} - \frac{1}{2}\left(\frac{y}{\Delta}\right)^3 \qquad \text{for } y \leqslant \Delta \tag{7–110a}$$

$$= 1 \qquad \text{for } y \geqslant \Delta \tag{7–110b}$$

which is of the same form as Eq. (7–97) for the dimensionless velocity distribution, except for the important distinction between Δ and δ. The relationship between wall-heat flux q''_c, wall temperature T_s, and the thermal boundary layer thickness Δ is developed by applying the Fourier law of conduction; that is,

$$q''_c = -k\frac{\partial T}{\partial y}\bigg|_0 = \frac{3}{2}\frac{k(T_s - T_\infty)}{\Delta} \tag{7–111}$$

Using Eq. (7–102) for δ, this equation gives rise to an expression for Nusselt number Nu_x of the form

$$Nu_x = \frac{q''_c}{T_s - T_\infty}\frac{x}{k} = \frac{3}{2}\frac{x/\delta}{r} = \frac{3}{2}\sqrt{\frac{13}{280}}\frac{Re_x^{1/2}}{r} = 0.323\frac{Re_x^{1/2}}{r} \tag{7–112}$$

where $r = \Delta/\delta$ is the boundary-layer-thickness ratio.

Thermal Boundary Layer Thickness Δ and Nusselt Number To evaluate the integral appearing in the integral energy equation, T is approximated by Eq. (7–110), with u represented by Eq. (7–97a) in the region $0 \leqslant y \leqslant \delta$ and, for conditions in which Δ is greater than δ, by Eq. (7–97b) [i.e., $u = U_\infty$] in the region $\delta < y \leqslant \Delta$. Following through with this integration, we obtain

$$\int_0^\Delta u(T - T_\infty)\,dy$$

$$= \int_0^\Delta \left\{U_\infty(T_s - T_\infty)\left[\frac{3}{2}\frac{y}{\delta} - \frac{1}{2}\left(\frac{y}{\delta}\right)^3\right]\left[1 - \frac{3}{2}\frac{y}{\Delta} + \frac{1}{2}\left(\frac{y}{\Delta}\right)^3\right]\right\} dy \tag{7–113a}$$

$$= \frac{3}{2}U_\infty(T_s - T_\infty)\delta\left[\frac{1}{10}\left(\frac{\Delta}{\delta}\right)^2 - \frac{1}{140}\left(\frac{\Delta}{\delta}\right)^4\right] = \frac{3}{20}U_\infty(T_s - T_\infty)\delta\left(r^2 - \frac{r^4}{14}\right)$$

for $\Delta \leqslant \delta$, and

$$\int_0^\Delta u(T - T_\infty)\,dy = \int_0^\delta u(T - T_\infty)\,dy + \int_\delta^\Delta U_\infty(T - T_\infty)\,dy$$

$$= \frac{3}{2}U_\infty(T_s - T_\infty)\,\delta\left\{\frac{1}{4}\left[\frac{\Delta}{\delta} - 1 + \frac{2}{5}\frac{\delta}{\Delta} - \frac{1}{35}\left(\frac{\delta}{\Delta}\right)^3\right]\right\} \tag{7–113b}$$

for $\Delta \geqslant \delta$. To consolidate, we introduce the boundary-layer-thickness ratio $r = \Delta/\delta$ and write

$$\int_0^\Delta u(T - T_\infty)\,dy = U_\infty(T_s - T_\infty)\delta\, f(r) \tag{7–114}$$

where

$$f(r) = \frac{3}{2}\left(\frac{r^2}{10} - \frac{r^4}{140}\right) \qquad \text{for } r \leqslant 1 \text{ or } f(r) \leqslant 0.139 \tag{7–115a}$$

and

$$f(r) = \frac{3}{8}\left(r - 1 + \frac{2}{5}r^{-1} - \frac{r^{-3}}{35}\right) \qquad \text{for } r \geqslant 1 \text{ or } f(r) \geqslant 0.139 \tag{7–115b}$$

Using this result together with Eq. (7–111), the integral energy equation,

$$\rho c_P \frac{d}{dx}\int_0^\Delta u(T - T_\infty)\,dy = q_c'' \tag{7–89}$$

is put into the form

$$\frac{d}{dx}\left[U_\infty(T_s - T_\infty)\delta\, f(r)\right] = \frac{3}{2}\frac{\alpha(T_s - T_\infty)}{\delta r} \tag{7–116a}$$

in terms of wall temperature T_s, or

$$\frac{d}{dx}\left[\frac{U_\infty\, q_c''\, \delta^2\, F(r)}{\alpha}\right] = q_c'' \tag{7–116b}$$

in terms of wall-heat flux q_c'', where

$$F(r) = \frac{2}{3}r\,f(r) = \frac{r^3}{10} - \frac{r^5}{140} = \frac{r^3}{10}\left(1 - \frac{r^2}{14}\right) \qquad \text{for } r \leqslant 1 \text{ or } F(r) \leqslant 0.0929 \tag{7–117a}$$

and

$$F(r) = \frac{2}{3}r\,f(r) = \frac{1}{4}\left(r^2 - r + \frac{2}{5} - \frac{r^{-2}}{35}\right) \qquad \text{for } r \geqslant 1 \text{ or } F(r) \geqslant 0.0929 \tag{7–117b}$$

Setting $\Delta = 0$ at the location at which heating or cooling is initiated, the distribution in r with x can be evaluated by solving Eq. (7–116a) for specified wall temperature or Eq. (7–116b) for specified wall-heat flux. With δ given by Eq. (7–103) for uniform free-stream velocity, the solution for r can then be used to determine the temperature distribution and Nusselt number. (The solution is extended to non-uniform free-stream velocity and transpiration in references 22 and 23.)

Before proceeding with the solution for representative distributions in T_s and q_c'', the function $F(r)$ should be examined. Calculations are easily obtained for $F(r)$ by taking r as the independent variable. This function is shown in Fig. 7–16. Because we are interested in obtaining simple explicit solutions, Eq. (7–117a) is approximated by the *cubic equation*,

$$F(r) = \frac{r^3}{10} \qquad \text{for } r \leq 1 \text{ or } F(r) \leq 0.1 \tag{7-118a}$$

and Eq. (7–117b) is approximated by the *quadratic equation*,

$$F(r) = \frac{1}{4}\left(r^2 - r + \frac{2}{5}\right) \qquad \text{for } r \geq 1 \text{ or } F(r) \geq 0.1 \tag{7-118b}$$

It follows that r can be approximated by

$$r = [10\, F(r)]^{1/3} \qquad \text{for } r \leq 1 \text{ or } F(r) \leq 0.1 \tag{7-119a}$$

and

$$r = \frac{1}{2}\left\{1 + \sqrt{1 - 4\left[\frac{2}{5} - 4\, F(r)\right]}\right\}$$

$$= \frac{1}{2}[1 + \sqrt{16\, F(r) - 0.6}] \qquad \text{for } r \geq 1 \text{ or } F(r) \geq 0.1 \tag{7-119b}$$

Equations (7–118a) and (7–118b) are shown in Fig. 7–16. The maximum error in

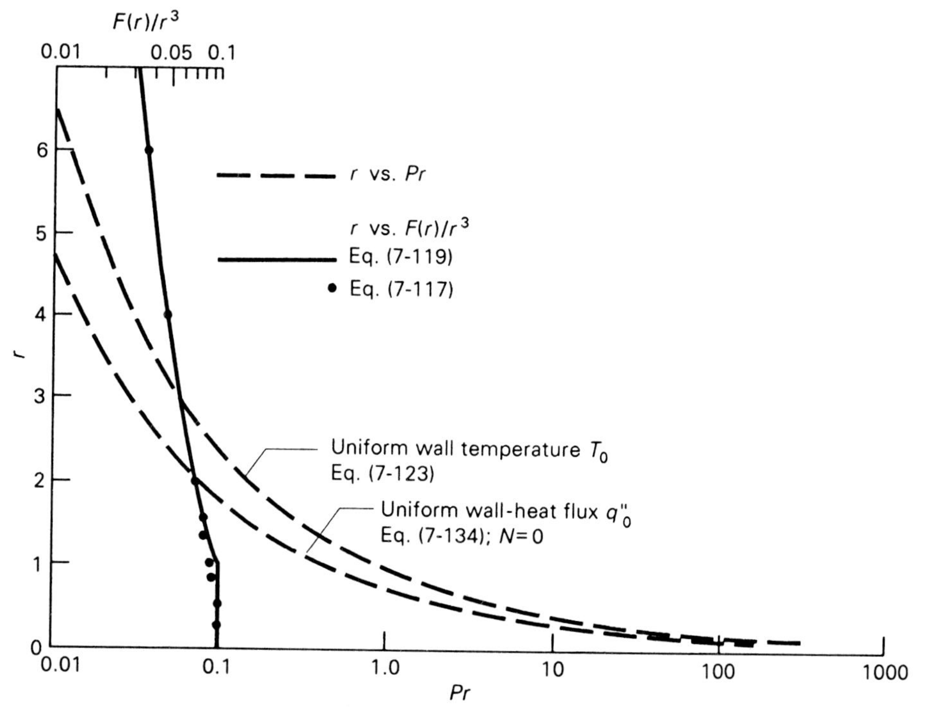

FIGURE 7–16 Distribution in r for uniform free-stream velocity.

these approximations for r relative to Eq. (7–117) is 2.4% at $r = 1$. Substituting these approximations into Eq. (7–112), we obtain

$$Nu_x = \frac{0.323\, Re_x^{1/2}}{[10\, F(r)]^{1/3}} \qquad \text{for } r \leqslant 1 \text{ or } F(r) \leqslant 0.1 \tag{7–120a}$$

and

$$Nu_x = \frac{0.646\, Re_x^{1/2}}{1 + \sqrt{16\, F(r) - 0.6}} \qquad \text{for } r \geqslant 1 \text{ or } F(r) \geqslant 0.1 \tag{7–120b}$$

To complete the solution, $F(r)$ is evaluated by solving the integral energy equation for thermal boundary conditions of interest.

Uniform Wall Temperature The solution to Eq. (7–116a) for arbitrary wall temperature generally requires simple numerical computations. However, a closed-form analytical solution can be obtained for the case of uniform wall temperature if one recognizes that r is uniform for this geometrically similar thermal boundary layer. This important point is shown in Appendix I–2 and will be demonstrated momentarily in the context of the integral solution for power-law wall-flux heating. Following through with the solution for $T_s = T_0$ and uniform free-stream velocity, Eq. (7–116a) reduces to

$$\delta \frac{d\delta}{dx} = \frac{3}{2} \frac{\alpha}{U_\infty r\, f(r)} = \frac{\alpha}{U_\infty\, F(r)} \tag{7–121}$$

Combining this relation with Eq. (7–100), we obtain

$$F(r) = \frac{13}{140} \frac{\alpha}{\nu} = \frac{13}{140\, Pr} \tag{7–122}$$

It follows that

$$r = \frac{0.976}{Pr^{1/3}} \qquad \text{for } r \leqslant 1 \text{ or } Pr \geqslant 0.929 \tag{7–123a}$$

from Eq. (7–119a), and

$$r = \frac{1}{2}\left(1 + \sqrt{\frac{1.49}{Pr} - 0.6}\right) \qquad \text{for } r \geqslant 1 \text{ or } Pr \leqslant 0.929 \tag{7–123b}$$

from Eq. (7–119b). Thus, we find that $\Delta \leqslant \delta$ for $Pr \geqslant 0.929$ and $\Delta \geqslant \delta$ for $Pr \leqslant 0.929$. Calculations for r are shown as a function of Pr in Fig. 7–16.

To see the nature of the temperature distribution, Eq. (7–110a) is put into the form

$$\frac{T - T_s}{T_\infty - T_s} = \frac{3}{2} \frac{y/\delta}{r} - \frac{1}{2}\left(\frac{y/\delta}{r}\right)^3 \qquad \text{for } 0 \leqslant y \leqslant \Delta \tag{7–124}$$

or, using Eq. (7–103) for δ,

$$\frac{T - T_s}{T_\infty - T_s} = 0.323\,\frac{\eta}{r} - 0.005\left(\frac{\eta}{r}\right)^3 \qquad \text{for } 0 \leqslant \eta \leqslant 4.64r \qquad (7\text{–}125)$$

where $\eta = y/\sqrt{\nu x/U_\infty}$ and r is approximated by Eq. (7–123). Calculations for the distributions in temperature and velocity are compared in Fig. 7–17 for several values of Pr. These results reinforce our earlier conclusion that the thermal boundary layer lies within the hydrodynamic boundary layer for large values of Pr, but extends beyond the hydrodynamic boundary layer for small values of Pr.

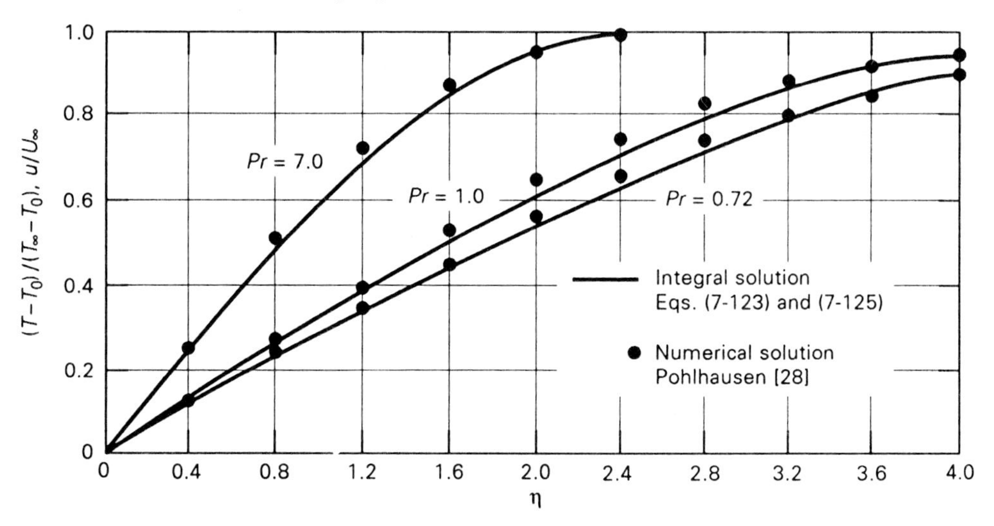

FIGURE 7–17 Solutions for temperature and velocity profiles for laminar boundary layer flow over a flat plate with uniform wall temperature and uniform free-stream velocity; velocity and temperature profiles coincide for $Pr = 1$.

It also follows that the relation for Nu_x given by Eq. (7–120) becomes

$$Nu_x = 0.331\,Re_x^{1/2}\,Pr^{1/3} \qquad \begin{array}{l}\text{for } r \leqslant 1\\ \text{or } Pr \geqslant 0.928\end{array} \qquad (7\text{–}126a)$$

and

$$Nu_x = \frac{0.646\,Re_x^{1/2}}{1 + \sqrt{1.49/Pr - 0.6}} \qquad \begin{array}{l}\text{for } r \geqslant 1\\ \text{or } Pr \leqslant 0.928\end{array} \qquad (7\text{–}126b)$$

This equation is compared with numerical solution results in Fig. 7–18 in terms of $Nu_x/\sqrt{Re_x}$ versus Pr. The difference in the approximate integral and numerical solutions for uniform wall temperature range from 2.4% for high Pr to 6% for liquid metals ($Pr < 0.05$).

In this connection, the numerical solution results for uniform wall temperature are correlated to within 1% by

$$Nu_x = 0.332\,Re_x^{1/2}\,Pr^{1/3} \qquad \text{for } 0.6 \leqslant Pr \leqslant 10 \qquad (7\text{–}127a)$$

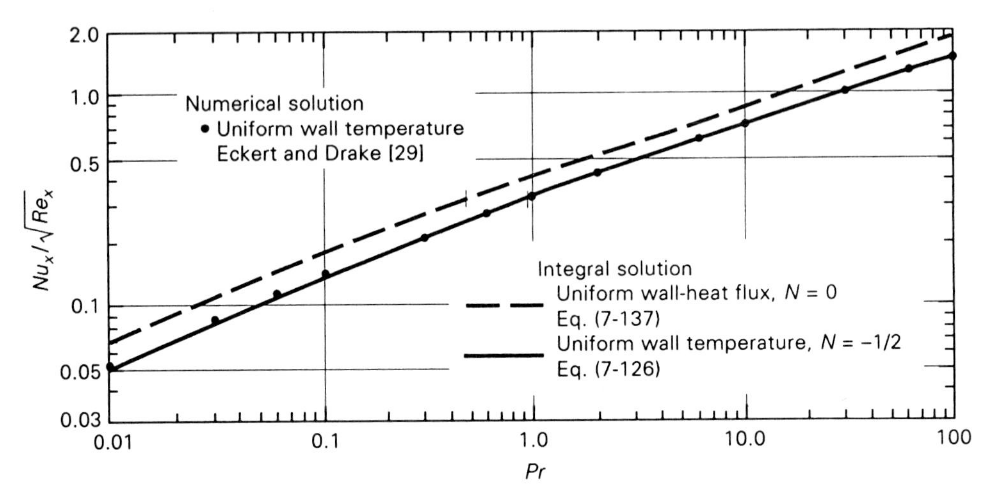

FIGURE 7–18 Nusselt number for uniform free-stream velocity and standard thermal boundary conditions.

and

$$Nu_x = 0.339\, Re_x^{1/2}\, Pr^{1/3} \qquad \text{for } 10 \leqslant Pr \tag{7–127b}$$

with the limiting solution for very low values of Pr given by

$$Nu_x = 0.565\, Re_x^{1/2}\, Pr^{1/2} = 0.565\, Pe_x^{1/2} \tag{7–128}$$

where $Pe_x = Re_x\, Pr$ is the local Peclet number. However, it should be noted that the integral and numerical solutions given above for low values of Pr do not account for the effect of *axial conduction*, which becomes significant at low flow rates. Based on theoretical considerations and experience, we restrict their use to conditions in which the Peclet number is greater than 100.

Specified Wall-Heat Flux For the case of arbitrarily specified wall-heat flux, Eq. (7–116b) is readily integrated to obtain

$$\frac{U_\infty q_c''\, \delta^2}{\alpha} F(r) = \int_0^x q_c''\, dx \tag{7–129}$$

or

$$F(r) = \frac{\alpha}{U_\infty \delta^2} \frac{1}{q_c''} \int_0^x q_c''\, dx \tag{7–130}$$

Using Eq. (7–102) to eliminate δ^2, we have

$$F(r) = \frac{13}{280\, Pr} \frac{1}{x q_c''} \int_0^x q_c''\, dx \tag{7–131}$$

To provide a specific example, we consider the case in which the heat flux is specified by a power law of the form

$$q_c'' = q_0'' x^N \qquad (7\text{–}132)$$

where q_0'' and N are constants. Substituting this input for q_c'' into Eq. (7–131), we obtain

$$F(r) = \frac{13}{280\ Pr}\frac{1}{xq_0''x^N}\int_0^x q_0''x^N\,dx = \frac{13}{280\ Pr\ (N+1)} \qquad (7\text{–}133)$$

such that r becomes

$$r = \frac{0.774}{(N+1)^{1/3}\ Pr^{1/3}} \qquad \begin{array}{l}\text{for } r \leqslant 1\\ \text{or } Pr \geqslant 0.464/(N+1)\end{array} \qquad (7\text{–}134a)$$

from Eq. (7–119a), and

$$r = \frac{1}{2}\left[1 + \sqrt{\frac{0.743}{(N+1)\ Pr}} - 0.6\right] \qquad \begin{array}{l}\text{for } r \geqslant 1\\ \text{or } Pr \leqslant 0.464/(N+1)\end{array} \qquad (7\text{–}134b)$$

from Eq. (7–119b). This equation is shown in Fig. 7–16 for uniform wall-heat flux ($N = 0$). Notice that for this case $\Delta \leqslant \delta$ for $Pr \geqslant 0.464$ and $\Delta \geqslant \delta$ for $Pr \leqslant 0.464$.

The fact that r is independent of x for power-law wall-flux heating indicates that the temperature distribution associated with this boundary condition is geometrically similar at all values of x. To see the nature of the distribution in wall temperature for such geometrically similar thermal boundary layers, Eqs. (7–103) and (7–132) are substituted into Eq. (7–111), with the result

$$T_s - T_\infty = \frac{2}{3}\frac{q_c''}{k}\delta r = 3.09\left(\frac{\nu x}{U_\infty}\right)^{1/2} x^N \frac{q_0''}{k} r = K x^{N+1/2} \qquad (7\text{–}135)$$

where $K = 3.09\sqrt{\nu/U_\infty}\, q_0''/k$ and r is constant. Thus, the wall temperature also follows a power-law distribution. In particular, we observe that power-law wall-flux heating with $N = -1/2$ corresponds to a uniform wall-temperature distribution, for which case Eq. (7–133) reduces to Eq. (7–122).

Substituting Eq. (7–133) into Eq. (7–120), the solution for Nu_x, which is associated with power-law distributions in wall-heat flux and wall temperature, becomes†

$$Nu_x = 0.417\ Re_x^{1/2}\ Pr^{1/3}\ (N+1)^{1/3} \qquad \begin{array}{l}\text{for } r \leqslant 1\\ \text{or } Pr \geqslant 0.464/(N+1)\end{array} \qquad (7\text{–}136a)$$

† The use of fourth-order polynomial approximations results in solutions of the form

$$Nu_x = 0.450\ Re_x^{1/2}\ Pr^{1/3}\ (N+1)^{1/3} \qquad \begin{array}{l}\text{for } r \leqslant 1\\ \text{or } Pr \geqslant 0.440/(N+1)\end{array}$$

and

$$Nu_x = \frac{0.686\ Re_x^{1/2}}{1 + \sqrt{\dfrac{0.783}{(N+1)\ Pr} - \dfrac{7}{9}}} \qquad \begin{array}{l}\text{for } r \geqslant 1\\ \text{or } Pr \leqslant 0.440/(N+1)\end{array}$$

which is somewhat more accurate than Eq. (7–136) for uniform wall-heat flux and low Pr, but less accurate for uniform wall temperature and high Pr.

and

$$Nu_x = \frac{0.646\, Re_x^{1/2}}{1 + \sqrt{\dfrac{0.743}{(N+1)\, Pr} - 0.6}} \qquad \begin{array}{l}\text{for } r \geqslant 1 \\ \text{or } Pr \leqslant 0.464/(N+1)\end{array} \qquad (7\text{–}136b)$$

These equations reduce to Eq. (7–126) for uniform wall temperature ($N = -1/2$), and

$$Nu_x = 0.417\, Re_x^{1/2}\, Pr^{1/3} \qquad \begin{array}{l}\text{for } r \leqslant 1 \\ \text{or } Pr \geqslant 0.464\end{array} \qquad (7\text{–}137a)$$

and

$$Nu_x = \frac{0.646\, Re_x^{1/2}}{1 + \sqrt{0.743/Pr - 0.6}} \qquad \begin{array}{l}\text{for } r \geqslant 1 \\ \text{or } Pr \leqslant 0.464\end{array} \qquad (7\text{–}137b)$$

for uniform wall-flux heating ($N = 0$). Equations (7–137a) and (7–137b) are shown in Fig. 7–18.

For the case of uniform wall-heat flux, the integral solution is within 5.6% of the numerical solution results reported by Levy [30]†; the exact solution is correlated by

$$Nu_x = 0.442\, Re_x^{1/2}\, Pr^{1/3} \qquad (7\text{–}138)$$

for moderate values of Pr. Comparing this equation with Eq. (7–126a), we find that Nu_x is about 33% greater for uniform wall-heat flux than for uniform wall temperature.

The integral approach developed in this section is readily applied to other distributions in wall-heat flux and can be adapted to arbitrary distributions in wall temperature by the use of simple numerical techniques. In addition, the method can be applied to situations involving nonuniform free-stream velocity, transpiration, wall curvature, natural convection, turbulent flow, and other factors.

EXAMPLE 7–6

Use the integral solution approach to develop approximate analytical solutions for the Nusselt number associated with laminar boundary layer flow over a flat plate with uniform free-stream velocity and unheated starting length followed by (a) uniform wall-heat flux and (b) uniform wall temperature.

Solution

Objective Develop approximate integral solutions for Nu_x.

† The fourth-order integral solution for uniform wall-heat flux is given by

$$Nu_x = 0.450\, Re_x^{1/2}\, Pr^{1/3} \qquad \text{for } Pr \geqslant 0.44$$

which is within 2% of the exact solution for moderate values of Pr.

Schematic Laminar flow over a flat plate with uniform free-stream velocity and heated surface condition specified by (a) uniform wall-heat flux for $x \geqslant x_0$ and (b) uniform wall temperature for $x \geqslant x_0$.

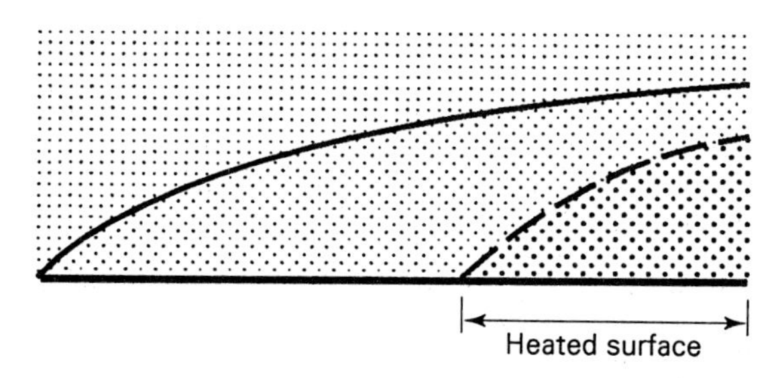

Assumptions/Conditions

forced convection
uniform properties
standard conditions

Analysis Referring to the preceding formulation, the use of third-order polynomial approximations for velocity and temperature permits us to express the integral energy equation in the form of Eq. (7–116a),

$$\frac{d}{dx}\left[U_\infty(T_s - T_\infty)\delta\,\frac{F(r)}{r}\right] = \frac{\alpha(T_s - T_\infty)}{\delta r} \qquad \text{(a)}$$

for specified wall temperature, and Eq. (7–116b),

$$\frac{d}{dx}\left[\frac{U_\infty q_c''\delta^2\,F(r)}{\alpha}\right] = q_c'' \qquad \text{(b)}$$

for specified wall-heat flux, with r approximated by Eq. (7–119),

$$r = [10\,F(r)]^{1/3} \qquad \text{for } F(r) \leqslant 0.1 \qquad \text{(c)}$$

and

$$r = \frac{1}{2}\,[1 + \sqrt{16\,F(r) - 0.6}] \qquad \text{for } F(r) \geqslant 0.1 \qquad \text{(d)}$$

and the Nusselt number Nu_x given by Eq. (7–120),

$$Nu_x = \frac{0.323\,Re_x^{1/2}}{r} = \frac{0.323\,Re_x^{1/2}}{[10\,F(r)]^{1/3}} \qquad \text{for } F(r) \leqslant 0.1 \qquad \text{(e)}$$

and

$$Nu_x = \frac{0.646\,Re_x^{1/2}}{1 + \sqrt{16\,F(r) - 0.6}} \qquad \text{for } F(r) \geqslant 0.1 \qquad \text{(f)}$$

(a) Step wall-heat flux The solution to Eq. (b) for arbitrarily specified heat flux is given by Eq. (7–131),

$$F(r) = \frac{13}{280\,Pr}\frac{1}{xq_c''}\int_0^x q_c''\,dx$$

For step wall-flux heating,

$$q_c'' = 0 \qquad \text{for } x < x_0 \qquad q_c'' = q_0'' \qquad \text{for } x \geqslant x_0$$

this equation reduces to

$$F(r) = \frac{13}{280}\frac{1 - x_0/x}{Pr}$$

Substituting this result into Eqs. (e) and (f), the solution for Nu_x becomes

$$Nu_x = \frac{0.417\,Re_x^{1/2}\,Pr^{1/3}}{(1 - x_0/x)^{1/3}} \qquad \text{for } Pr \geqslant 0.464(1 - x_0/x) \qquad \text{(g)}$$

and

$$Nu_x = \frac{0.646\,Re_x^{1/2}}{1 + \sqrt{0.743(1 - x_0/x)/Pr - 0.6}} \qquad \text{for } Pr \leqslant 0.464(1 - x_0/x)$$

These equations reduce to Eqs. (7–137a) and (7–137b) for $x_0 = 0$ or for large values of x/x_0.

(b) Step wall temperature As previously mentioned, the solution of the integral energy equation for specified wall temperature generally requires the use of numerical-solution methods. However, a useful approximate solution can be developed for the important case of step wall-temperature heating. For this particular case, Eq. (a) reduces to

$$r\frac{d}{dx}\left[\delta\frac{F(r)}{r}\right] = \frac{\alpha}{U_\infty\delta}$$

where $r = 0$ at $x = x_0$. Using Eq. (c) to eliminate $F(r)$ in favor of r over the region for which $r \leqslant 1$, we obtain†

$$r\frac{d}{dx}(\delta r^2) = 10\frac{\alpha}{U_\infty\delta} \qquad r\frac{dr^2}{dx} + \frac{r^3}{\delta}\frac{d\delta}{dx} = \frac{10\alpha}{U_\infty\delta^2}$$

Rearranging further and substituting Eq. (7–103) for δ, this equation becomes

$$\frac{2}{3}\frac{dr^3}{dx} + \frac{r^3}{2x} = \frac{13}{28\,Pr}\frac{1}{x}$$

† For situations in which $r \geqslant 1$, the use of Eq. (d) results in a nonlinear differential equation that requires numerical treatment.

The homogeneous solution of this linear equation is obtained by writing

$$\frac{dr_H^3}{r_H^3} = -\frac{3}{4}\frac{dx}{x}$$

followed by

$$\ln\frac{r_H^3}{C} = -\frac{3}{4}\ln x = \ln x^{-3/4} \qquad r_H^3 = Cx^{-3/4}$$

Combining this with the particular solution [$r_p = 13/(14\ Pr)$] and setting $r = 0$ at $x = x_0$, the solution for r becomes

$$r = \left(\frac{13}{14\ Pr}\right)^{1/3}\left[1 - \left(\frac{x_0}{x}\right)^{3/4}\right]^{1/3} \qquad \begin{array}{l}\text{for } r \leqslant 1\\ \text{or } Pr \geqslant 0.928[1 - (x_0/x)^{3/4}]\end{array}$$

Substituting this result into Eq. (e), we have

$$Nu_x = \frac{0.331\ Re_x^{1/2}\ Pr^{1/3}}{[1 - (x_0/x)^{3/4}]^{1/3}} \qquad \text{for } Pr \geqslant 0.928[1 - (x_0/x)^{3/4}] \qquad \text{(h)}$$

which reduces to Eq. (7–126a) for $x_0 = 0$ or large values of x/x_0. Because of the linearity of the energy equation, Eq. (h) can be used together with the principle of superposition to develop approximate analytical solutions for certain nonuniform wall-temperature distributions. Although this equation is theoretically restricted by the criterion $Pr \geqslant 0.928\ [1 - (x_0/x)^{3/4}]$, the practical range of applicability is represented by $Pr \geqslant 0.6\ [1 - (x_0/x)^{3/4}]$.

7–2–3 Natural Convection on Vertical Surfaces

We now turn our attention to the fundamental problem of natural convection flow over a vertical heated or cooled surface that is immersed in an extensive quiescent fluid with ambient temperature T_∞ and density ρ_∞. According to the well-known *Archimedes principle*, a body with density ρ immersed in a fluid with density ρ_∞ experiences an *upward buoyancy force* equal to the weight of the displaced fluid. The net effect of the body weight and the buoyancy force will cause the body to rise if ρ is *less* than ρ_∞ or to sink if ρ is *greater* than ρ_∞. It is the buoyancy force f_B, which is caused by temperature-induced density variations in the region near a heated or cooled surface, that provides the driving mechanism for natural convection. Consequently, a heated vertical surface placed in a quiescent fluid will produce a region of warm, less dense, upward-moving fluid near the wall. Similarly, a region of cool, more dense, downward-moving fluid will be sustained near a cooled vertical surface.

Representative distributions in temperature and velocity associated with these two situations are shown in Fig. 7–19. Comparing these profiles with those shown in Fig. 7–15 for forced convection, we observe two important differences. The one

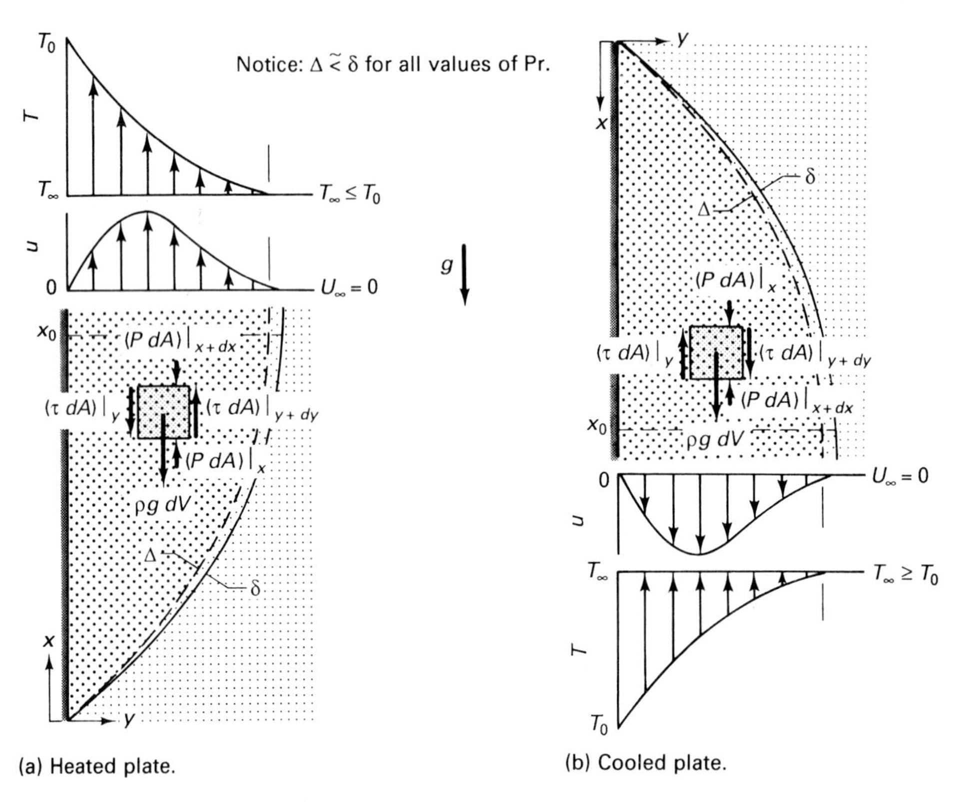

(a) Heated plate. (b) Cooled plate.

FIGURE 7–19 Laminar natural convection flow over a vertical plate at uniform temperature T_0 in a quiescent fluid at temperature T_∞. (Velocity and temperature profiles are shown at $x = x_0$.)

very obvious difference is in the nature of the velocity profile. Very much unlike the velocity distribution for forced convection, for this natural convection flow u increases from zero at the wall to a maximum value, and then decreases to zero at the outer edge of the boundary layer. On the other hand, we observe that the temperature distributions are quite similar, except for the somewhat subtle fact that the thermal boundary layer generally lies within the hydrodynamic boundary layer (i.e., $r < 1$), *regardless* of the value of the Prandtl number *Pr*.† The fact that Δ is less than or of the order of δ for small values of *Pr* is a consequence of the motion-producing buoyancy forces acting within the entire thermal boundary layer. For moderate to large values of *Pr*, the thermal boundary layer lies well within the hydrodynamic boundary layer, as in the case of forced convection, since $\nu >> \alpha$. In this case, the buoyancy force produces a strong fluid shear layer that drives the outer flow field. Thus, for buoyant flows such as this, motion is generally produced across or beyond the entire thermal boundary layer.

† Strictly speaking, Δ/δ does slightly exceed unity for fluids with very low values of *Pr*. For example, $\Delta/\delta \simeq 1.11$ for $Pr = 0.01$ [31].

Mathematical Formulation

The presence of a significant imbalance in the density-related gravitational body force acting on the fluid is the feature that distinguishes natural convection from forced convection. Because the effects of the forces acting on the fluid are accounted for by the momentum equation for standard boundary layer flows, it follows that the continuity and energy equations for natural convection with moderate temperature differences are of the same form as for uniform property forced convection; that is,

$$\frac{\partial u}{\partial x} + \frac{\partial v}{\partial y} = 0 \tag{7–139}$$

$$u\frac{\partial T}{\partial x} + v\frac{\partial T}{\partial y} = \alpha \frac{\partial^2 T}{\partial y^2} \tag{7–140}$$

To develop the momentum equation for natural convection, we simply include the gravitation body force $\rho g_x \, dV$ in the force balance, as shown in Fig. 7–19. To simplify the formulation we will focus attention on the heated plate [Fig. 7–19(a)] for which $g_x = -g$. {The resulting formulation is adapted to the cooled plate [Fig. 7–19(b)] by merely changing the sign associated with terms involving the gravitational acceleration g.}

Applying Newton's second law of motion to the differential element shown in Fig. 7–19(a), we write

$$RCM_x = \Sigma F_x \tag{7–141}$$

where

$$\begin{aligned} \Sigma F_x &= (\tau \, dx \, dz)|_{y+dy} - (\tau \, dx \, dz)|_y + (P \, dy \, dz)|_x - (P \, dy \, dz)|_{x+dx} - \rho g \, dx \, dy \, dz \\ &= \left(\frac{\partial \tau}{\partial y} - \frac{\partial P}{\partial x} - \rho g\right) dV \end{aligned} \tag{7–142}$$

and

$$\begin{aligned} RCM_x &= d\dot{M}_{x+dx} + d\dot{M}_{y+dy} - d\dot{M}_x - d\dot{M}_y \\ &= \frac{\partial}{\partial x}(d\dot{M}_x)\, dx + \frac{\partial}{\partial y}(d\dot{M}_y)\, dy \end{aligned} \tag{7–143}$$

Substituting $d\dot{M}_x = u \, d\dot{m}_x$ and $d\dot{M}_y = u \, d\dot{m}_y$ and using the continuity equation, Eq. (7–139), RCM_x becomes

$$\begin{aligned} RCM_x &= \left[\rho\left(u\frac{\partial u}{\partial x} + v\frac{\partial u}{\partial y}\right) + \rho u \left(\frac{\partial u}{\partial x} + \frac{\partial v}{\partial y}\right)\right] dV \\ &= \rho\left(u\frac{\partial u}{\partial x} + v\frac{\partial u}{\partial y}\right) dV \end{aligned} \tag{7–144}$$

Substituting these results for RCM_x and ΣF_x into Eq. (7–141), the momentum equation becomes

$$\rho\left(u\frac{\partial u}{\partial x}+v\frac{\partial u}{\partial y}\right)=\mu\frac{\partial^2 u}{\partial y^2}-\frac{dP}{dx}-\rho g \tag{7–145}$$

for uniform viscosity. Except for the gravitational force term $-\rho g$, this equation is identical to the momentum equation for forced convection. As $y \to \infty$, Eq. (7–145) reduces to

$$-\frac{dP}{dx}=\rho_\infty g \tag{7–146}$$

Substituting this result back into Eq. (7–145), the momentum equation takes the form

$$\rho\left(u\frac{\partial u}{\partial x}+v\frac{\partial u}{\partial y}\right)=\mu\frac{\partial^2 u}{\partial y^2}+g(\rho_\infty-\rho) \tag{7–147}$$

where the term $g(\rho_\infty - \rho)$ represents the driving buoyancy force f_B. (Notice that f_B is the difference between the two body forces $-\rho g$ and $-\rho_\infty g$.) The effect of buoyancy is determined by accounting for the dependence of the gravitational term ρg on temperature. For isothermal conditions, ρg is constant ($=\rho_\infty g$) and no flow occurs.

In this formulation for moderate temperature differences, the properties associated with the nonbuoyancy terms are generally evaluated at the free-stream temperature T_∞ or film temperature $T_f = (T_s + T_\infty)/2$. The dependence of ρ on T is often expressed in terms of the *coefficient of thermal expansion* β, which is defined by

$$\beta=\frac{1}{V}\frac{\partial V}{\partial T}\bigg|_P=-\frac{1}{\rho}\frac{\partial \rho}{\partial T}\bigg|_P \tag{7–148}$$

For ideal gases, β is expressed in terms of the absolute temperature T by $\beta = 1/T$. This property is listed in Table A–C–3 for common liquids. Expanding the defining equation for β, we obtain (for P = constant)

$$\int_{\rho_\infty}^{\rho} d\rho \simeq -\int_{T_\infty}^{T}\rho\beta\, dT = -\overline{\rho\beta}(T-T_\infty) \tag{7–149}$$

or

$$\rho-\rho_\infty=-\overline{\rho\beta}(T-T_\infty) \tag{7–150}$$

where

$$\overline{\rho\beta}=\frac{1}{T-T_\infty}\int_{T_\infty}^{T}\rho\beta\, dT \tag{7–151}$$

Substituting this result for ρ into Eq. (7–147), we obtain

$$\rho\left(u\frac{\partial u}{\partial x}+v\frac{\partial u}{\partial y}\right)=\mu\frac{\partial^2 u}{\partial y^2}+g\,\overline{\rho\beta}(T-T_\infty) \tag{7–152}$$

For situations in which the temperature differences are moderate, the term $\overline{\rho\beta}$ can be approximated by $\rho\beta$, where ρ and β are evaluated at T_∞ or T_f, such that Eq. (7–152) reduces to

$$u\frac{\partial u}{\partial x} + v\frac{\partial u}{\partial y} = \nu\frac{\partial^2 u}{\partial y^2} + g\beta(T - T_\infty) \tag{7–153}$$

The evaluation of the fluid properties at the free-stream or film temperature and the use of $\overline{\rho\beta} = \rho\beta$ is known as the *Boussinesq approximation* [32]. In effect, this simplification involves the assumption of uniform properties, except for variable density in the buoyancy term.

To complete the differential formulation, we write the boundary conditions that accompany the continuity, momentum, and energy equations as follows:

$$u = 0 \quad \text{at } x = 0 \qquad u = 0 \quad \text{as } y \to \infty \tag{7–154,155}$$

$$u = 0 \quad \text{at } y = 0 \qquad v = 0 \quad \text{at } y = 0 \tag{7–156,157}$$

$$T = T_\infty \quad \text{at } x = x_0 \qquad T = T_\infty \quad \text{as } y \to \infty \tag{7–158,159}$$

and

$$T = T_s \quad \text{or} \quad -k\frac{\partial T}{\partial y} = q_c'' \quad \text{at } y = 0 \tag{7–160a,b}$$

As in the case of forced convection, practical approximate integral methods can be effectively used in the analysis of natural convection boundary layer flow. Following the indirect approach, to develop the integral momentum equation for laminar natural convection boundary layer flow we integrate the differential momentum equation across the boundary layer. First, Eq. (7–153) is rewritten as follows:

$$\frac{\partial u^2}{\partial x} - u\frac{\partial u}{\partial x} + \frac{\partial}{\partial y}(uv) - u\frac{\partial v}{\partial y} = \nu\frac{\partial^2 u}{\partial y^2} + g\beta(T - T_\infty) \tag{7–161}$$

Utilizing the continuity equation, Eq. (7–139), this expression reduces to the form

$$\frac{\partial u^2}{\partial x} + \frac{\partial}{\partial y}(uv) = \nu\frac{\partial^2 u}{\partial y^2} + g\beta(T - T_\infty) \tag{7–162}$$

Integrating each term of this equation with respect to y, we obtain

$$\int_0^y \frac{\partial u^2}{\partial x}\,dy + \int_0^y \frac{\partial}{\partial y}(uv)\,dy = \int_0^y \nu\frac{\partial^2 u}{\partial y^2}\,dy + \int_0^y g\beta(T - T_\infty)\,dy \tag{7–163}$$

or

$$\frac{\partial}{\partial x}\int_0^y u^2\,dy + (uv)\Big|_0^y = \nu\frac{\partial u}{\partial y}\Big|_0^y + \int_0^y g\beta(T - T_\infty)\,dy \tag{7–164}$$

Setting y equal to δ and noting that $u = 0$ at $y = \delta$, the integral momentum equation reduces to

$$\frac{d}{dx}\int_0^{\delta} u^2\,dy = -\nu\frac{\partial u}{\partial y}\bigg|_0 + \int_0^{\delta} g\beta(T - T_\infty)\,dy \tag{7–165}$$

The integral formulation is completed by writing the integral energy equation, which is identical in form to Eq. (7–89); that is,

$$\rho c_P \frac{d}{dx}\int_0^{\Delta} u(T - T_\infty)\,dy = q_c'' = -k\frac{\partial T}{\partial y}\bigg|_0 \tag{7–166}$$

Solution

Because the momentum equation for natural convection involves distributions in both velocity u and temperature T, the development of a solution to this problem generally requires that the fluid flow and energy equations be solved simultaneously. As in the case of forced convection, both numerical/transformation and approximate integral methods are commonly used to solve these equations [31]†. In this connection, results based on the transformation analysis approach (see Appendix I–2) indicate that the hydrodynamic and thermal boundary layers are geometrically similar for laminar natural convection flow over a vertical surface with uniform wall-heat flux, uniform wall-temperature, and power-law distributions in T_s and q_c''. This is also found to be the case for forced convection. To develop a practical solution to the problem at hand, we again turn to the approximate integral method.

Integral Solution—Fluid Flow and Energy Transfer The integral momentum and energy equations for natural convection flow over a vertical surface with mild to moderate temperature differences are represented by Eqs. (7–165) and (7–166). Following the approach used in the development of integral solutions for forced convection, the integral equations can be solved by the use of approximations for u and T in terms of the boundary layer thicknesses δ and Δ.

Dimensionless Velocity and Temperature Distributions To maintain consistency with the analysis developed for forced convection in the preceding section, we will approximate the distributions in velocity u and temperature T by third-order polynomials; that is,

$$u = a_0 + a_1 y + a_2 y^2 + a_3 y^3 \tag{7–167}$$

and

$$T = b_0 + b_1 y + b_2 y^2 + b_3 y^3 \tag{7–168}$$

To evaluate these eight coefficients, the boundary conditions for u and T given by Eqs. (7–154) through (7–160) are coupled with the physical requirements

† An approximate solution for laminar film condensation on a vertical surface, which represents a natural convection flow process involving two phases, is developed in Appendix J.

$$\frac{\partial u}{\partial y} = 0 \qquad \text{at } y = \delta \qquad \frac{\partial T}{\partial y} = 0 \qquad \text{at } y = \Delta \tag{7–169,170}$$

Two additional conditions are written by satisfying Eqs. (7–153) and (7–140) at the wall.

$$\frac{\partial^2 u}{\partial y^2} = -\frac{g\beta(T_s - T_\infty)}{\nu} \qquad \frac{\partial^2 T}{\partial y^2} = 0 \qquad \text{at } y = 0 \tag{7–171,172}$$

The evaluation of the coefficients a_n and b_n in accordance with these eight conditions gives

$$u = \frac{U_x}{4}\left[\frac{y}{\delta} - 2\left(\frac{y}{\delta}\right)^2 + \left(\frac{y}{\delta}\right)^3\right] \tag{7–173}$$

where $U_x = g\beta(T_s - T_\infty)\,\delta^2/\nu$, and

$$\frac{T - T_s}{T_\infty - T_s} = \frac{3}{2}\frac{y}{\Delta} - \frac{1}{2}\left(\frac{y}{\Delta}\right)^3 \qquad \text{for } y \le \Delta \tag{7–174a}$$

$$= 1 \qquad \text{for } y \ge \Delta \tag{7–174b}$$

which is identical to the approximation given by Eq. (7–110) for forced convection. It also follows that

$$q_c'' = -k\left.\frac{\partial T}{\partial y}\right|_0 = \frac{3}{2}\frac{k\,(T_s - T_\infty)}{\Delta} \tag{7–175}$$

and

$$Nu_x = \frac{q_c''}{T_s - T_\infty}\frac{x}{k} = \frac{3}{2}\frac{x}{\Delta} \tag{7–176}$$

Boundary Layer Thickness and Nusselt Number Following the approach taken in the analysis of forced convection boundary layer flow, these equations can be substituted into the integral momentum and energy equations, Eqs. (7–165) and (7–166), to produce two equations in the unknowns δ and Δ. However, because both Eqs. (7–165) and (7–166) involve u, T, δ, and Δ, the resulting expressions appear to be somewhat intimidating. Consequently, analysts have developed a variety of approximate approaches in order to avoid difficulties. These approaches generally involve the use of the simplifying assumption $\delta \simeq \Delta$, which proves to be quite reasonable for natural convection flow of fluids with moderate to small values of Prandtl number.

The simplest approach of this type is a one-parameter method in which the single unknown Δ is obtained by the solution of either the integral momentum equation or the integral energy equation. To develop this practical approach, Eqs. (7–173) and (7–174a) are used to evaluate the integral appearing in the integral energy equation, with the result

$$\int_0^\Delta u(T - T_\infty)\,dy = \int_0^\Delta \frac{U_x}{4}(T_s - T_\infty)\left[\frac{y}{\Delta} - 2\left(\frac{y}{\Delta}\right)^2 + \left(\frac{y}{\Delta}\right)^3\right] \times \left[1 - \frac{3}{2}\frac{y}{\Delta} + \frac{1}{2}\left(\frac{y}{\Delta}\right)^3\right] dy = \frac{U_x(T_s - T_\infty)\Delta}{105} = \frac{1}{105}\frac{g\beta(T_s - T_\infty)^2\Delta^3}{\nu} \tag{7-177}$$

Substituting this result into the integral energy equation, Eq. (7–166), and using Eq. (7–175) to express $T_s - T_\infty$ in terms of q_c'', we obtain

$$\frac{1}{105}\frac{d}{dx}\left[\frac{g\beta(T_s - T_\infty)^2\Delta^3}{\nu}\right] = \frac{3}{2}\frac{\alpha}{\Delta}(T_s - T_\infty) \tag{7-178a}$$

for specified wall temperature T_s, or

$$\frac{(2/3)^2}{105}\frac{d}{dx}\left(\frac{g\beta q_c''^2\Delta^5}{k\alpha\nu}\right) = q_c'' \tag{7-178b}$$

for specified wall-heat flux q_c''. With Δ set equal to zero at $x = 0$, these equations can be solved for the distribution in Δ for arbitrarily specified inputs for T_s or q_c''. Whereas simple numerical approaches are required for arbitrarily specified wall temperature, analytical solutions are readily developed for uniform wall temperature and arbitrarily specified wall-heat flux.

Uniform Wall Temperature For uniform wall temperature, Eq. (7–178a) is put into the form

$$\Delta\frac{d\Delta^3}{dx} = 3\Delta^3\frac{d\Delta}{dx} = \frac{3}{4}\frac{d\Delta^4}{dx} = \frac{315}{2}\frac{\alpha\nu}{g\beta(T_0 - T_\infty)} \tag{7-179}$$

Setting $\Delta = 0$ at $x = 0$, the solution to this equation is

$$\Delta^4 = \frac{4}{3}\frac{315}{2}\frac{\alpha\nu x}{g\beta(T_0 - T_\infty)} \tag{7-180}$$

or

$$\frac{\Delta}{x} = 3.81\left[\frac{\nu^2}{g\beta x^3(T_0 - T_\infty)}\frac{\alpha}{\nu}\right]^{1/4} = \frac{3.81}{(Gr_x\,Pr)^{1/4}} = \frac{3.81}{Ra_x^{1/4}} \tag{7-181}$$

where the *Grashof number* Gr_x and *Rayleigh number* Ra_x are defined by

$$Gr_x = \frac{g\beta(T_0 - T_\infty)x^3}{\nu^2} \tag{7-182}$$

and

$$Ra_x = Gr_x\,Pr = \frac{g\beta(T_0 - T_\infty)x^3}{\alpha\nu} \tag{7-183}$$

with properties evaluated at the free-stream temperature T_∞. The form of this equation suggests that the Rayleigh number Ra_x is the primary characterizing parameter for this natural convection process. The Rayleigh number represents the *ratio of buoyancy forces to change in viscous forces in the boundary layer.* Whereas the Grashof number is comparable to the Reynolds number for forced convection (since both parameters pertain to momentum transfer), the Rayleigh number characterizes the coupling through buoyancy of momentum and energy. It so happens that Ra_L is the key parameter for most natural convection flows. In this connection, experimental evidence indicates that natural convection flow over a vertical surface is generally laminar for $Ra_x \lesssim 10^9$ and turbulent for $Ra_x \gtrsim 10^9$.

To establish the range of conditions for which this solution is appropriate, the integral momentum equation can be solved with the assumption $\delta = \Delta$. This step leads to a solution for Δ of the form $\Delta/x = 3.60/Gr_x^{1/4}$. This relation is equivalent to Eq. (7–181) for $Pr = 1.25$ and is within about 10% for $0.8 \lesssim Pr \lesssim 1.8$. Therefore, the analysis should be reasonably accurate over this important range of Prandtl numbers, which includes common gases. But we would expect this simple result to degenerate for values of Pr much outside this range.

With δ and Δ approximated by Eq. (7–181), the velocity and temperature profiles can be obtained from Eqs. (7–173) and (7–174a). To compare the solution with numerical results available in the literature, Eq. (7–173) is put into the form

$$\frac{ux}{2\nu\, Ra_x^{1/2}} = \frac{1.81}{Pr}\left[\frac{y}{\Delta} - 2\left(\frac{y}{\Delta}\right)^2 + \left(\frac{y}{\Delta}\right)^3\right] \tag{7–184}$$

with y/Δ represented by

$$\frac{y}{\Delta} = \frac{y}{x}\frac{x}{\Delta} = \frac{1}{3.81}\left(\frac{y}{x} Ra_x^{1/4}\right) \tag{7–185}$$

Calculations obtained from Eqs. (7–174), (7–184), and (7–185) are shown together with numerical solution results obtained by Ostrach [33] and experimental data for air (with $Pr = 0.73$) in Fig. 7–20. The approximate integral solutions for the distributions in velocity and temperature are observed to be quite respectable.

In connection with this result, it should be noted that the form of the integral solution for Δ/x given by Eq. (7–181) and the form of Eqs. (7–174) and (7–184) indicate that the distributions in dimensionless velocity $ux/(\nu\, Ra_x^{1/2})$ and temperature $(T - T_s)/(T_\infty - T_s)$ are geometrically similar when expressed in terms of $(y/x)Ra_x^{1/4}$. This result provides a basis for similarity variables that are commonly employed in numerical approaches (see Appendix I–2).

To obtain an expression for Nusselt number Nu_x, we substitute Eq. (7–181) for Δ/x into Eq. (7–176), with the result

$$Nu_x = 0.394\, Ra_x^{1/4} \tag{7–186}$$

As shown in Chap. 9, this equation gives rise to a relation for mean Nusselt number $\overline{Nu}_L$ that is in good agreement with experimental data for air. However, we expect the solution to be unreliable for fluids with values of Pr much different from unity. To check this point, Eq. (7–186) is compared with numerical solution results for $Nu_x/Ra_x^{1/4}$ versus Pr in Fig. 7–21. This figure indicates that the simple one-parameter integral solution is within 5% of the numerical solution for Pr between 0.6 and 1.4, but degenerates for values of Pr outside this range.

Whereas this one-parameter method is incapable of fully accounting for the effect of Pr on natural convection flow, two-parameter methods that satisfy both the integral

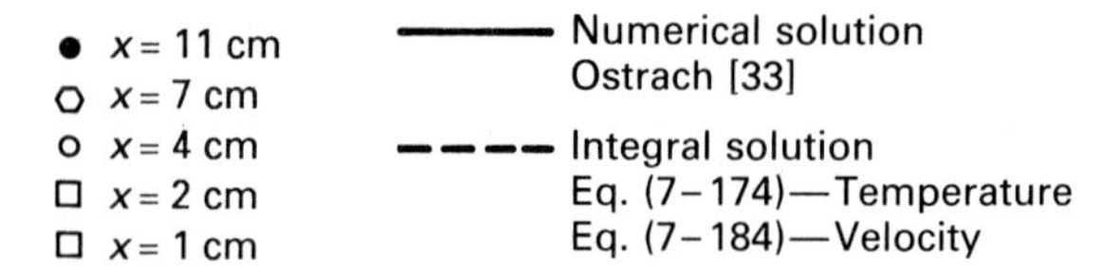

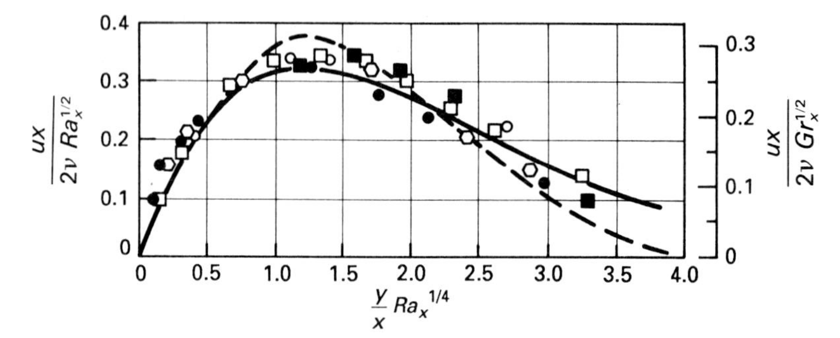

(a) Velocity distribution.

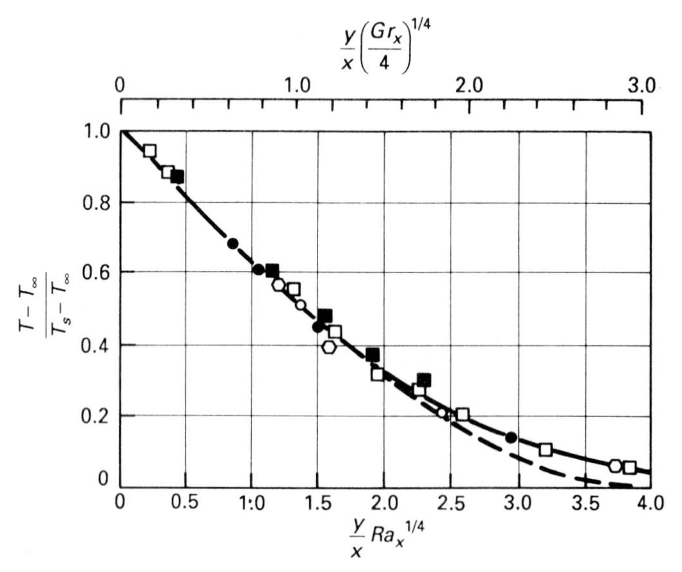

(b) Temperature distribution.

FIGURE 7–20 Distributions in (a) velocity and (b) temperature for laminar natural convection flow of air ($Pr = 0.73$) over a vertical plate with uniform wall temperature.

momentum equation and integral energy equation perform quite satisfactorily. Two-parameter methods are more involved but prove to be manageable (see Appendix K). Calculations for the Nusselt number obtained on the basis of a two-parameter method are shown in Fig. 7-21 to be in quite good agreement with the numerical calculations. The error in the calculations for Nu_x ranges from about 2% for large Pr to less than 0.5% for moderate Pr, but reaches an order of 15% for small values of Pr. In this connection, the numerical solution is represented to within 2% over the entire range of Pr by the following correlation by Ede [35]:

$$\frac{Nu_x}{Ra_x^{1/4}} = \frac{3}{4}\left(\frac{Pr}{2.5 + 5\sqrt{Pr} + 5\,Pr}\right)^{1/4} \tag{7-187}$$

This equation is also shown in Fig. 7-21.

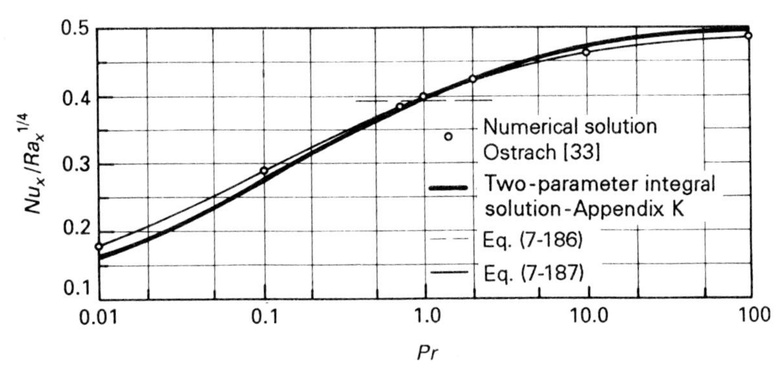

FIGURE 7-21 Nusselt number for laminar natural convection flow over a vertical plate with uniform wall temperature.

Specified Wall-Heat Flux To develop solutions for cases in which the wall-heat flux q''_c is specified, we simply integrate Eq. (7-178b), with the result

$$\frac{g\beta q''^2_c\Delta^5}{k\alpha\nu} = 105\left(\frac{3}{2}\right)^2 \int_0^x q''_c\,dx \tag{7-188}$$

or

$$\left(\frac{\Delta}{x}\right)^5 = 105\left(\frac{3}{2}\right)^2 \frac{k\nu^2\alpha}{g\beta q''_c x^4 \nu}\frac{1}{xq''_c}\int_0^x q''_c\,dx$$

$$= \frac{236}{Gr^*_x\,Pr}\frac{1}{xq''_c}\int_0^x q''_c\,dx \tag{7-189}$$

and

$$\frac{\Delta}{x} = \frac{2.98}{Ra_x^{*1/5}}\left(\frac{1}{xq''_c}\int_0^x q''_c\,dx\right)^{1/5} \tag{7-190}$$

where the *flux Grashof number* Gr_x^* and *flux Rayleigh number* Ra_x^* are defined by

$$Gr_x^* = \frac{g\beta q_c'' x^4}{k\nu^2} \tag{7–191}$$

$$Ra_x^* = Gr_x^*\ Pr = \frac{g\beta q_c'' x^4}{\alpha\nu} \tag{7–192}$$

An inspection of these defining relations indicates

$$Gr_x^* = Gr_x\ Nu_x \qquad Ra_x^* = Ra_x\ Nu_x \tag{7–193,194}$$

Notice that, with q_c'' known, Gr_x^* and Ra_x^* are readily calculated.

For the interesting case in which q_c'' is specified by the power law,

$$q_c'' = q_0'' x^N \tag{7–195}$$

we obtain

$$\frac{\Delta}{x} = \frac{2.98}{Ra_x^{*1/5}}\left(\frac{1}{xq_0''x^N}\int_0^x q_0''x^N\,dx\right)^{1/5} = \frac{2.98}{(N+1)^{1/5}\,Ra_x^{*1/5}} \tag{7–196}$$

To see the distribution in $T_s - T_\infty$ that is associated with this power-law distribution in q_c'', we put Eq. (7–175) into the form

$$\begin{aligned} T_s - T_\infty &= \frac{2}{3}\frac{q_c''\Delta}{k} = \frac{2}{3}\frac{q_0''}{k}x^{N+1}\frac{\Delta}{x} \\ &= \frac{2}{3}\frac{2.98}{(N+1)^{1/5}}\frac{q_0''\,x^{N+1}}{k\,Ra_x^{*1/5}} = Kx^{N+1/5} \end{aligned} \tag{7–197}$$

where

$$K = \frac{1.98}{(N+1)^{1/5}}\frac{q_0''^{4/5}}{k^{4/5}}\left(\frac{\nu\alpha}{g\beta}\right)^{1/5} \tag{7–198}$$

such that the wall temperature also follows a power-law distribution. Notice that $N = -1/5$ corresponds to the case of uniform wall temperature for laminar natural convection flow.

To obtain an expression for Nusselt number Nu_x, we substitute Eq. (7–196) into Eq. (7–176).

$$Nu_x = \frac{3}{2}\frac{(N+1)^{1/5}}{2.98}Ra_x^{*1/5} = 0.503(N+1)^{1/5}\,Ra_x^{*1/5} \tag{7–199}$$

This equation reduces to

$$Nu_x = 0.503\ Ra_x^{*1/5} \tag{7–200}$$

for uniform wall-heat flux ($N = 0$). This result is within 5% of the numerical solution by Ostrach [33] for *Pr* between 0.3 and 0.85. For the case of uniform wall temperature ($N = -1/5$), Eq. (7–199) gives

$$Nu_x = 0.481\ Ra_x^{*1/5} \tag{7–201}$$

To compare this result with the solution for uniform wall temperature given by Eq. (7–186), we write

$$Nu_x = 0.481\ (Ra_x\ Nu_x)^{1/5} \tag{7–202}$$

or

$$Nu_x = 0.4\ Ra_x^{1/4} \tag{7–203}$$

Equations (7–186) and (7–203) are within 1.5%.

7–3 TURBULENT FLOW THEORY

As indicated in Chap. 6, turbulent flow is characterized by nonstreamline unsteady random motion of fluid elements within the flow stream. This chaotic churning action is brought about by minute disturbances within the system, which become unstable for values of the Reynolds number above a certain critical value. For Reynolds numbers below this value, disturbances still occur but are stabilized or dampened by the viscous properties of the fluid, such that laminar conditions prevail.

In the following sections, attention is given to general characteristics of turbulent convection processes and to the classical approach to modeling turbulence. In addition, brief consideration is given to the solution of basic turbulent convection processes.

7–3–1 Characteristics of Turbulent Convection Processes

The Unsteady Nature of Turbulent Flow

Extensive experimental flow visualization and anemometer studies have been conducted over the past few years that provide us with a qualitative description of the mechanism associated with turbulent convection processes. First, these studies demonstrate the unsteady character of the entire flow field, including the region very close to the wall. This point is illustrated by Fig. 7–22, in which fluctuations in the instantaneous axial velocity u at a point very close to the wall are shown for turbulent flow over a flat plate.

A fairly realistic description of the turbulent transport mechanism is shown in Fig. 7–23, which pictures a burst process that involves the intermittent exchange of fluid between the turbulent core and wall region. As mentioned by Kays and Crawford [11], relatively large elements of low-velocity fluid adjacent to the wall surface periodically lift off the surface and eventually break up in the turbulent core. Of course,

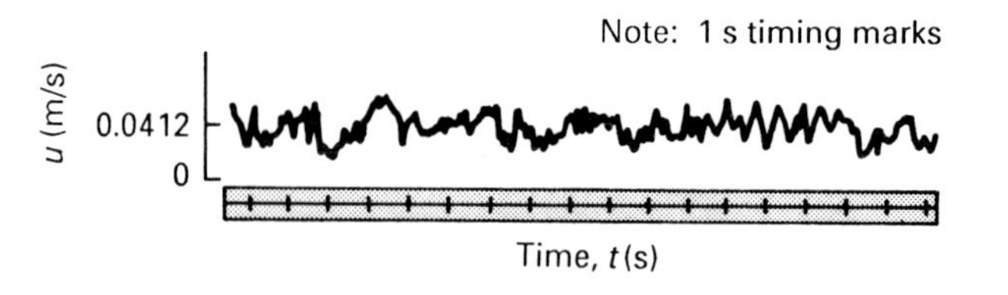

Turbulent flow over a flat plate is characterized by $Re_x \gtrsim 5 \times 10^5$.

FIGURE 7–22 Measurements of instantaneous axial velocity u (at $y = 0.00127$ m; $y^+ = 8$) for fully turbulent boundary layer flow over a flat plate with $U_\infty = 0.131$ m/s (Runstadler et al. [36]).

continuity considerations require that the fluid ejected from the wall region by the burst or lift-off process must be replaced by inrushing fluid with properties, such as axial velocity and temperature, which are associated with the turbulent core.

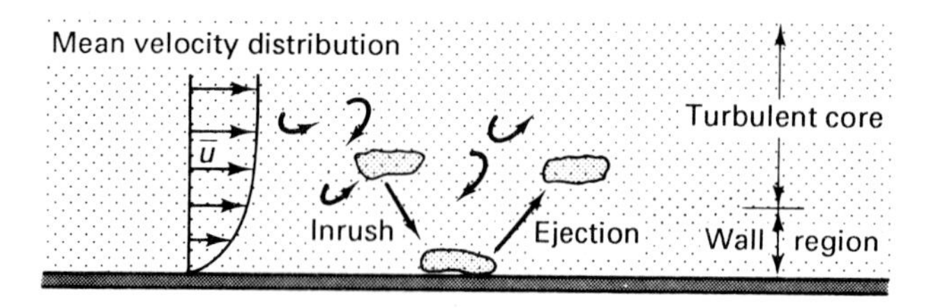

FIGURE 7–23 Turbulent burst process.

Because turbulence is characterized by fluctuating transport properties, the time-variant velocity profile u is often expressed in terms of mean $\bar{u}$ and fluctuating u' values; that is,

$$u = \bar{u} + u' \tag{7–204}$$

where

$$\bar{u} = \frac{1}{t}\int_0^t u\,dt \tag{7–205}$$

and

$$\overline{u'} = \frac{1}{t}\int_0^t u'\,dt = 0 \tag{7–206}$$

The flow is said to be steady on a time-average basis if $\partial\bar{u}/\partial t = 0$. u, $\bar{u}$, and u' are shown in Fig. 7–24 for a situation in which the flow undergoes a change from laminar to turbulent conditions. We see that $u' = 0$ for laminar flow and $u' \neq 0$ for turbulent flow. For developing flows, we have $\partial\bar{u}/\partial x \neq 0$, whereas HFD internal flow is defined for turbulent conditions by

$$\frac{\partial\bar{u}}{\partial x} = 0 \tag{7–207}$$

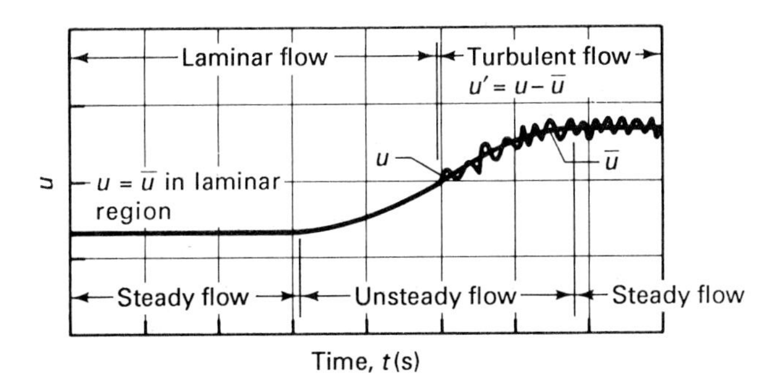

FIGURE 7-24 Representative instantaneous axial velocity u for process that undergoes a change from laminar to turbulent flow.

The characteristics v, T, and P are also expressed in terms of mean and fluctuating values. In regard to the temperature profile, TFD conditions are defined for turbulent flow in terms of the time-average temperature distribution $\bar{T}$ by

$$\frac{\partial}{\partial x}\left(\frac{\bar{T} - T_s}{T_b - T_s}\right) = 0 \tag{7-208}$$

Unlike the problem of laminar flow, predictions for the instantaneous unsteady transport properties for turbulent flow cannot be obtained. However, calculations can be developed for the mean transport characteristics for many basic turbulent flow processes by the use of existing theoretical approaches and certain empirical inputs.

The fundamental and particular laws utilized in the analysis of laminar convection processes also apply to turbulent flow. Hence, the principles of continuity, momentum, and energy, together with the Newton law of viscosity and the Fourier law of conduction, will provide the backbone for our study of turbulent flow. But, it is the unsteady and time-average forms of these principles that must be used in the analysis of turbulence.

The Consequences of Turbulent Flow

The practical approach developed in Chap. 8 applies to turbulent flow as well as to laminar flow. However, the appropriate coefficients of friction and heat transfer for turbulent flow must be utilized in these rather simple analyses. Based on the physical picture of turbulence portrayed by Fig. 7-23, we expect that the macroscopic exchange of fluid between the wall and turbulent core regions will bring about a larger transport of momentum and heat than for laminar flow. The coefficients of friction and heat transfer given in Chap. 8 for laminar and turbulent flow substantiate this qualitative observation. This point is reinforced by Figs. 7-6 and 7-14 in which experimental measurements for the friction factors associated with laminar and turbulent HFD flow

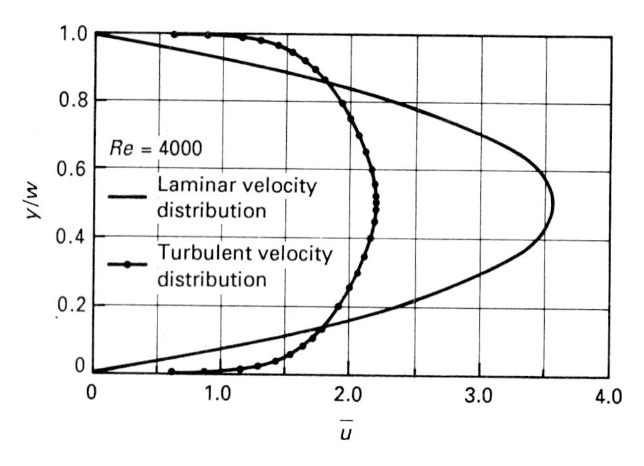

Turbulent flow in a tube channel is characterized by $Re \gtrsim Re_c$, with Re_c being greater than 2000 for very carefully controlled conditions.

FIGURE 7–25 Velocity distributions for both laminar and (transitional) turbulent HFD channel flow. (From Knudsen and Katz [37]. Used with permission.)

in circular tubes and flow over flat plates are shown. Notice the increase in friction factor as the flow transcends from laminar to turbulent conditions.

The consequence of this unsteady exchange of fluid between the wall and core regions is also vividly portrayed by the velocity profiles shown in Fig. 7–25 for channel flow. Although the Reynolds number is equal to 4000 for both these profiles, one is laminar and the other is (transitional) turbulent. Notice that much larger wall-region velocities are found in turbulent flow than in laminar flow. Hence, the transport of momentum from the bulk stream to the wall region is indeed most effective for turbulent flow.

Similarly, convection heat and mass transfer are much more effective for turbulent flow than for laminar flow. Therefore, industrial convection heat and mass transfer equipment is often operated in the turbulent flow regime in order to achieve high transfer rates.

7–3–2 Classical Approach to Modeling Turbulence

The key modeling concepts involved in the classical analysis of turbulent convection processes are introduced in this section in the context of the differential formulation approach for forced convection over plane surfaces. In addition, the very useful integral formulation will be presented.

Differential Formulation

Fluid Flow In the classical approach to turbulence, the continuity and momentum equations are first written in terms of the instantaneous unsteady velocity distribution, after which equations are developed for the time-average velocity distribution.

Instantaneous Equations Applying the principles of continuity and momentum to an unsteady fluctuating uniform property flow field, we have

Continuity

$$\frac{\partial u}{\partial x} + \frac{\partial v}{\partial y} = 0 \tag{7–209}$$

Momentum (x-direction)

$$\rho\left(\frac{\partial u}{\partial t} + u\frac{\partial u}{\partial x} + v\frac{\partial u}{\partial y}\right) = \frac{\partial}{\partial y}\left(\mu\frac{\partial u}{\partial y}\right) - \frac{\partial P}{\partial x} \tag{7–210}$$

The development of these equations follows step by step the formulation of the differential continuity and momentum equations for unsteady laminar boundary layer flow.

Because of the complex random unsteady nature of turbulence, Eqs. (7–209) and (7–210) cannot be solved for the instantaneous fluctuating profiles u and v. However, these equations provide a theoretical basis for the development of time-average equations which, when coupled with certain empirical inputs, can be solved for the important mean velocity distributions $\bar{u}$ and $\bar{v}$. Once we know $\bar{u}$ and $\bar{v}$, the mean wall-shear stress $\bar{\tau}_s$ can be determined from

$$\bar{\tau}_s = \mu\left.\frac{\partial \bar{u}}{\partial y}\right|_0 \tag{7–211}$$

and calculations can eventually be obtained for the mean temperature distribution $\bar{T}$ and Nusselt number by solving the time-average energy equation. (For variable property conditions, the time average fluid flow and energy equations must be solved simultaneously.)

Time-Average Equations To develop the time-average continuity equation, u and v in Eq. (7–209) are replaced by their mean and fluctuation components $u = \bar{u} + u'$, $v = \bar{v} + v'$; that is,

$$\frac{\partial \bar{u}}{\partial x} + \frac{\partial u'}{\partial x} + \frac{\partial \bar{v}}{\partial y} + \frac{\partial v'}{\partial y} = 0 \tag{7–212}$$

Then, taking the time average of each term, we write

$$\overline{\frac{\partial \bar{u}}{\partial x}} + \overline{\frac{\partial u'}{\partial x}} + \overline{\frac{\partial \bar{v}}{\partial y}} + \overline{\frac{\partial v'}{\partial y}} = 0 \tag{7–213}$$

or simply,

$$\frac{\partial \bar{u}}{\partial x} + \frac{\partial \bar{v}}{\partial y} = 0 \tag{7–214}$$

which is identical in form to Eq. (7–58) for laminar flows.

To develop the time-average momentum equation we first rewrite Eq. (7–210) as

$$\rho\left[\frac{\partial u}{\partial t}+\frac{\partial u^2}{\partial x}+\frac{\partial}{\partial y}(uv)-u\left(\frac{\partial u}{\partial x}+\frac{\partial v}{\partial y}\right)\right]=\frac{\partial}{\partial y}\left(\mu\frac{\partial u}{\partial y}\right)-\frac{\partial P}{\partial x} \tag{7–215}$$

By introducing the mean and fluctuating properties into this equation and taking the time average of each term, we obtain (for $\partial\overline{u}/\partial t = 0$)

$$\rho\left[\frac{\partial\overline{u}^2}{\partial x}+\frac{\partial}{\partial y}(\overline{\overline{u}\,\overline{v}})+\frac{\partial\overline{u'^2}}{\partial x}+\frac{\partial}{\partial y}(\overline{u'v'})\right]=\frac{\partial}{\partial y}\left(\mu\frac{\partial\overline{u}}{\partial y}\right)-\frac{\partial\overline{P}}{\partial x} \tag{7–216}$$

Although the fluctuation terms on the left-hand side of this equation actually represent the transport of momentum by turbulent eddy motion, this equation is traditionally coupled with the continuity equation and put into the form

$$\rho\left(\overline{u}\frac{\partial\overline{u}}{\partial x}+\overline{v}\frac{\partial\overline{u}}{\partial y}\right)=\frac{\partial}{\partial y}\left(\mu\frac{\partial\overline{u}}{\partial y}-\rho\,\overline{u'v'}\right)-\frac{\partial}{\partial x}(\overline{P}+\rho\,\overline{u'^2}) \tag{7–217}$$

where $\overline{\partial u'^2}/\partial x$ is generally assumed to be negligible on the basis of experience. Because the term $\mu\ \partial\overline{u}/\partial y$ represents the mean shear stress associated with molecular transport, the term $-\rho\ \overline{u'v'}$ has come to be called the *apparent turbulent shear stress* $\overline{\tau}_t$; that is,

$$\overline{\tau}_t = -\rho\,\overline{u'v'} \tag{7–218}$$

$\overline{\tau}_t$ is also referred to as the *Reynolds stress* in honor of the man who first introduced this concept [38]. The actual mean shear stress, $\mu\ \partial\overline{u}/\partial y$, is designated by $\overline{\tau}_y$. Based on this convention, the two terms $\overline{\tau}$ and $\overline{\tau}_y$ taken together can be thought of as an *apparent total shear stress* which we shall denote by $\overline{\tau}$.

$$\overline{\tau}=\overline{\tau}_y+\overline{\tau}_t=\mu\frac{\partial\overline{u}}{\partial y}-\rho\,\overline{u'v'} \tag{7–219}$$

Based on these definitions for $\overline{\tau}$ and $\overline{\tau}_t$ and assuming negligible pressure gradients in the y direction, the time-average momentum equation, Eq. (7–217), takes the compact and useful form

$$\rho\left(\overline{u}\frac{\partial\overline{u}}{\partial x}+\overline{v}\frac{\partial\overline{u}}{\partial y}\right)=\frac{\partial\overline{\tau}}{\partial y}-\frac{d\overline{P}}{dx} \tag{7–220}$$

With the Reynolds stress term equal to zero for laminar conditions, this equation is seen to reduce to the momentum equation for laminar boundary layer flow, Eq. (7–59).

The differential formulation for turbulent fluid flow is completed by writing the boundary conditions and by specifying the turbulence parameter $\overline{\tau}_t$. The boundary conditions are similar in form to the boundary conditions for laminar flow. For example, we have the no-slip condition at the wall

$$\overline{u}=0 \qquad \text{at} \quad y=0 \tag{7–221}$$

Specification of Reynolds Stress The distinctive feature of these turbulent fluid-flow equations is the fact that the Reynolds stress $\bar{\tau}_t$ must be specified by the use of theoretical and/or empirical inputs. Although $\bar{\tau}_t$ can be evaluated on the basis of experimental measurements for u' and v', the usual analytical approach, known as the mean-field or Boussinesq [39] method, is to relate this turbulence parameter to the local mean axial velocity $\bar{u}$ by

$$\bar{\tau}_t = \mu_t \frac{\partial \bar{u}}{\partial y} \qquad (7\text{–}222)$$

where the eddy viscosity μ_t must be specified; ν_t ($= \mu_t/\rho$) is the *eddy diffusivity* or *eddy kinematic viscosity*. (The symbol ϵ_M is often used instead of ν_t in the turbulence literature.) Notice the similarity between this equation and Newton's law of viscosity, $\tau_y = \mu\ \partial u/\partial y$ or $\bar{\tau}_y = \mu\ \partial \bar{u}/\partial y$. However, unlike μ which is a property, μ_t is a turbulence characteristic that varies across the flow field. With the Reynolds stress $\bar{\tau}_t$ expressed in terms of the eddy viscosity μ_t, the apparent total shear stress becomes

$$\bar{\tau} = \bar{\tau}_y + \bar{\tau}_t = (\mu + \mu_t) \frac{\partial \bar{u}}{\partial y} \qquad (7\text{–}223)$$

or

$$\frac{\bar{\tau}}{\rho} = (\nu + \nu_t) \frac{\partial \bar{u}}{\partial y} \qquad (7\text{–}224)$$

The differential time-average momentum equation can be written in terms of μ_t as

$$\rho \left(\bar{u} \frac{\partial \bar{u}}{\partial x} + \bar{v} \frac{\partial \bar{u}}{\partial y} \right) = \frac{\partial}{\partial y} \left[(\mu + \mu_t) \frac{\partial \bar{u}}{\partial y} \right] - \frac{d\bar{P}}{dx} \qquad (7\text{–}225)$$

Representative measurements for ν_t are shown in Fig. 7–26 for fully turbulent flow in terms of the dimensionless distances y^+ and y/δ; the dimensionless distance y^+ is given by

$$y^+ = \frac{U^* y}{\nu} \qquad (7\text{–}226)$$

The *friction velocity* U^* is defined by

$$U^* = \sqrt{\frac{\bar{\tau}_s}{\rho}} \qquad (7\text{–}227)$$

and δ represents the channel half-width, tube radius, or boundary layer thickness. Although these data are for channel flow, data for fully turbulent flow in tubes and concentric-circular tube annuli and over surfaces for which no abrupt changes occur follow the same pattern, such that the same correlations for ν_t can generally be applied to basic fully turbulent flow fields with mild to moderate pressure gradients and transpiration rates.

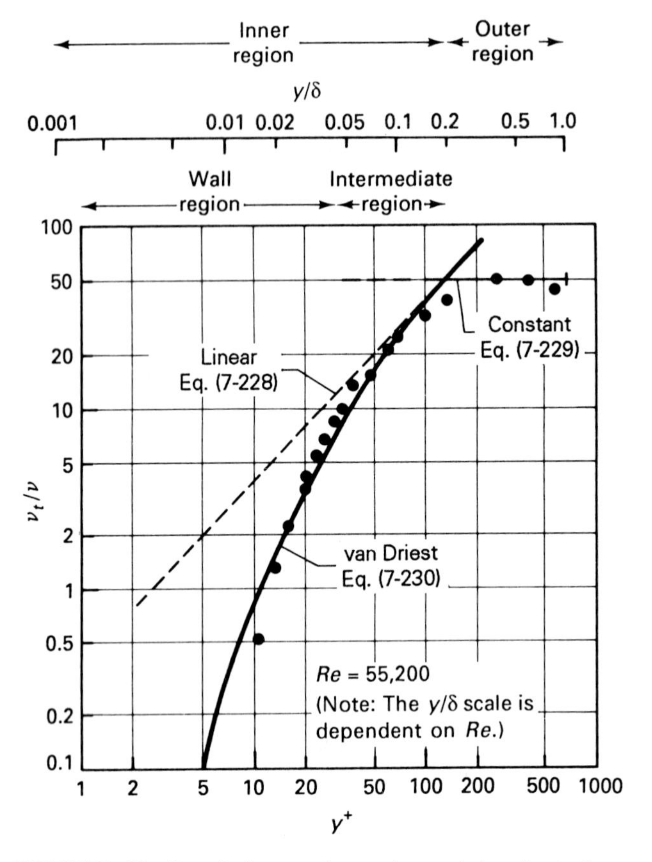

FIGURE 7–26 Correlations and experimental data for turbulent eddy diffusivity ν_t. (Data for channel flow by Hussain and Reynolds [40].)

For practical purposes, a fully turbulent flow field can be broken into the following zones:

Inner region	$y/\delta \lessapprox y_1/\delta$
Wall region	$y^+ \lessapprox 50$
Intermediate region	$y^+ \gtrapprox 50$
Outer region	$y/\delta \gtrapprox y_1/\delta$

where y_1/δ generally lies between 0.15 and 0.2. The domain outside the wall region is also known as the turbulent core. Referring to Fig. 7–26, we approximate ν_t/ν in the intermediate and outer regions (i.e., in the turbulent core) by expressions of the form

$$\frac{\nu_t}{\nu} = \kappa y^+ \qquad \begin{array}{l}\text{Intermediate region} \\ y^+ \gtrapprox 50, \quad y/\delta \lessapprox y_1/\delta\end{array} \tag{7–228}$$

and

$$\frac{\nu_t}{\nu} = \alpha_1 \delta^+ \qquad \begin{array}{l}\text{Outer region}\\ y/\delta \gtrsim y_1/\delta\end{array} \tag{7–229}$$

where the empirical constant κ is approximately 0.41 for fully turbulent flow and α_1 ($= \kappa y_1/\delta$) is usually between 0.06 and 0.08. (The alternative *mixing length* approach to characterizing the turbulent core is introduced in Example 7–7.) In the wall region, ν_t/ν is clearly nonlinear.

One of the objectives in the analysis of turbulence is to minimize the number of empirical inputs by employing reasonable theoretical models of the actual turbulent transport process. These efforts have been most productive in treating the region adjacent to the wall. Among the various mathematical models for the behavior of ν_t within the wall region for fully turbulent flow, the formulation by van Driest [41] is the most popular. This classical analysis gives rise to an expression for ν_t/ν within the entire inner region of the form

$$\frac{\nu_t}{\nu} = (\kappa y^+)^2 \left[1 - \exp\left(-\frac{y^+}{a^+}\right)\right]^2 \frac{du^+}{dy^+} \tag{7–230}$$

where $u^+ = \bar{u}/U^*$ and a^+ is a damping parameter that is approximately equal to 25 for mild pressure gradients. This equation reduces to Eq. (7–228) as y^+ approaches a value of the order of 50. Equation (7–230) is shown to correlate the experimental data in Fig. 7–26 quite well within both the wall and intermediate regions. This basic approach has been extended to boundary layer flows with transpiration and moderate to strong pressure gradients by Kays and Moffat [42], Cebeci and Smith [43], and others. Other empirical correlations for ν_t within the inner region have been developed by Rotta [44], Reichardt [45], Deissler [46], and Spalding [47].

EXAMPLE 7–7

A popular alternative mean-field representation for $\bar{\tau}_t$ was proposed by Prandtl [48] in 1910 of the form

$$\bar{\tau}_t = \rho \ell^2 \left(\frac{\partial \bar{u}}{\partial y}\right)^2 \tag{a}$$

where the *mixing length* ℓ is imagined to be the small transverse distance that fluid elements move during a turbulent fluctuation.

Prandtl developed the mixing length concept on the basis of rather simple intuitive reasoning pertaining to the behavior of fluid elements with fluctuating velocity components u' and v'. Referring to Fig. E7–7, because of diffusion and viscous, pressure, and impact forces, momentum transfer occurs all along the flight path of a fluid element. Therefore, the element actually begins to lose its identity at the moment its transverse flight begins. However, to model this very complex turbulent transport mechanism, Prandtl envisioned an idealized process in which the turbulent eddies retain their identity over the entire flight path, with complete mixing occurring abruptly

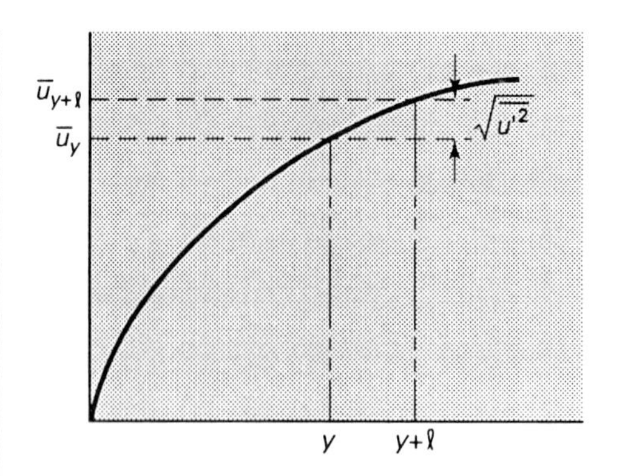

(b) Representative distribution in mean velocity.

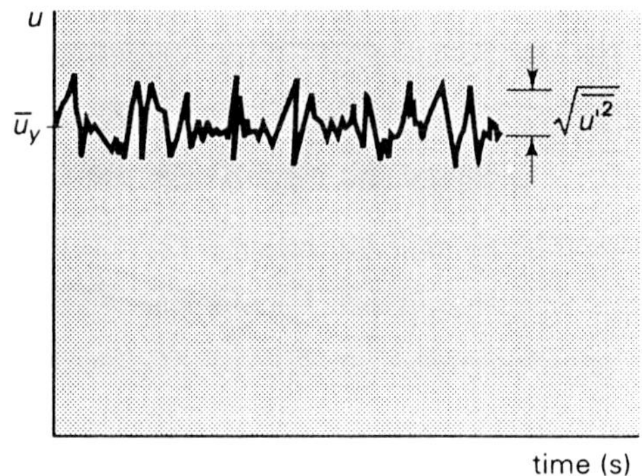

(c) Representative instantaneous velocity at distance y from wall.

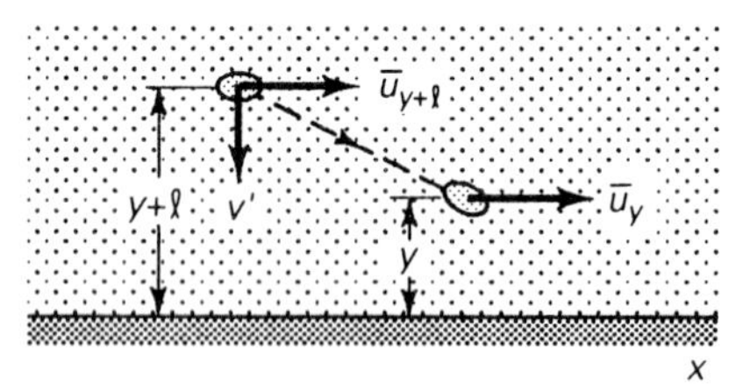

(a) Velocities associated with movement of individual element of fluid.

FIGURE E7–7 Perspective for Prandtl mixing length.

at a mean distance ℓ. Thus, an element of fluid moving from $y + \ell$ to y is assumed to arrive at y with mean velocity $\bar{u}_{y+\ell}$. It is the arrival of such elements of fluid that is envisaged to result in the fluctuation in the measured instantaneous velocity at y. To follow through with this notion, the mean fluctuation $\sqrt{\overline{u'^2}}$ in signal measured at y is said to be the result of the arrival of elements of fluid from locations $y \pm \ell$, where the mean velocity $\bar{u}_{y\pm\ell}$ differs from $\bar{u}_y$ by the rms value $\sqrt{\overline{u'^2}}$. Assuming that the distance ℓ is small, this perspective translates into the simple relation

$$\frac{\sqrt{\overline{u'^2}}}{\ell} = \frac{|\bar{u}_{y\pm\ell} - \bar{u}_y|}{|(y \pm \ell) - y|} = \left|\frac{\Delta \bar{u}}{\Delta y}\right| \simeq \left|\frac{\partial \bar{u}}{\partial y}\right| \qquad \text{or} \qquad \sqrt{\overline{u'^2}} \simeq \ell \left|\frac{\partial \bar{u}}{\partial y}\right| \tag{b,c}$$

Prandtl also assumed that v' would be of the same order of magnitude as u', such that $\sqrt{\overline{v'^2}}$ can be represented by

$$\sqrt{\overline{v'^2}} \simeq \left|\frac{\partial \bar{u}}{\partial y}\right| \tag{d}$$

With $\overline{u'v'}$ approximated by the product $\sqrt{\overline{u'^2}}\,\sqrt{\overline{v'^2}}$, Prandtl obtained his famous mixing length equation for Reynolds stress; that is,

$$\bar{\tau}_t = -\rho\,\overline{u'v'} = -\rho\,\sqrt{\overline{u'^2}}\,\sqrt{\overline{v'^2}} = \rho\ell^2\left(\frac{\partial \bar{u}}{\partial y}\right)^2$$

Based on experimental evidence, the mixing length within the turbulent core is commonly approximated by

$$\ell^+ = \kappa y^+ \qquad \begin{array}{l}\text{Intermediate region}\\ y^+ \geqslant 50, \quad y/\delta \leqslant y_1/\delta\end{array} \tag{e}$$

and

$$\ell^+ = \alpha_1 \delta^+ \qquad \begin{array}{l}\text{Outer region}\\ y/\delta \geqslant y_1/\delta\end{array} \tag{f}$$

where $\ell^+ = \ell U^*/\nu$. Use Eq. (7–230) to develop an equivalent relationship for ℓ^+ within the wall region.

Solution

Objective Develop a relationship for mixing length ℓ that is equivalent to Eq. (7–230) for ν_t within the wall region.

Assumptions/Conditions

single phase

Prandtl's assumptions are summarized as follows:

turbulent eddies retain identity over entire flight path

complete mixing occurs abruptly at a mean distance ℓ

ℓ is small

v' is of the same order of magnitude as u'.

Analysis To relate ℓ to ν_t, we compare Eq. (a) with Eq. (7–222), with the result

$$\bar{\tau}_t = \mu_t \frac{\partial \bar{u}}{\partial y} = \rho\ell^2\left(\frac{\partial \bar{u}}{\partial y}\right)^2$$

$$\nu_t = \ell^2 \frac{\partial \bar{u}}{\partial y} \qquad \text{or} \qquad \frac{\nu_t}{\nu} = \ell^{+2}\frac{\partial u^+}{\partial y^+} \tag{g,h}$$

Comparing Eq. (h) with Eq. (7–230), we obtain

$$\ell^+ = \kappa y^+\left[1 - \exp\left(-\frac{y^+}{a^+}\right)\right] \tag{i}$$

The bracketed term is generally referred to as the damping factor. Notice that as y^+ increases the damping factor approaches unity and Eq. (i) reduces to Eq. (e), which applies in the intermediate region of the turbulent core. Equation (i) is used in many modern numerical codes for analyzing turbulent boundary layer flow.

Energy Transfer Following the pattern established in the development of the fluid-flow equations, the time-average energy equation for standard conditions can be written as

$$\rho c_P\left(\bar{u}\frac{\partial \bar{T}}{\partial x} + \bar{v}\frac{\partial \bar{T}}{\partial y}\right) = -\frac{\partial \overline{q''}}{\partial y} \tag{7–231}$$

where the *apparent total heat flux* $\overline{q''}$ is defined in terms of the mean heat flux $\overline{q''_y}$ and the *apparent turbulent heat flux* $\overline{q''_t}$ by

$$\overline{q''} = \overline{q''_y} + \overline{q''_t} \tag{7–232}$$

with

$$\overline{q''_y} = -k\frac{\partial \bar{T}}{\partial y} \qquad \text{mean heat flux} \tag{7–233}$$

$$\overline{q''_t} = \rho c_P \overline{v'T'} \qquad \text{apparent turbulent heat flux} \tag{7–234}$$

Notice that Eq. (7–231) reduces to the energy equation for laminar flow, Eq. (7–60), as the turbulent fluctuation term approaches zero.

The time-average energy equation must, of course, be coupled with appropriate boundary conditions. As an example, for turbulent flow over a surface with uniform wall-flux heating, we have

$$-k\frac{\partial \bar{T}}{\partial y} = \overline{q''_0} \qquad \text{at } y = 0 \tag{7–235}$$

Finally, our formulation is closed by specifying the turbulence parameter $\overline{q''_t}$. The apparent turbulent heat flux is generally expressed in terms of $\bar{T}$ by

$$\overline{q''_t} = -k_t\frac{\partial \bar{T}}{\partial y} \qquad \text{or} \qquad \frac{\overline{q''_t}}{\rho c_P} = -\alpha_t\frac{\partial \bar{T}}{\partial y} \tag{7–236,237}$$

where k_t is the *eddy thermal conductivity* and α_t is the *eddy thermal diffusivity*. (The symbol ϵ_H is often used in place of α_t in the turbulence literature.) It follows that the apparent total heat flux is represented by

$$\overline{q''} = -(k + k_t)\frac{\partial \bar{T}}{\partial y} \tag{7–238}$$

or

$$\frac{\overline{q''}}{\rho c_P} = -(\alpha + \alpha_t)\frac{\partial \bar{T}}{\partial y} \tag{7–239}$$

With $\overline{q''_t}$ expressed in this mean-field format, the differential time-average energy equation takes the form

$$\rho c_P\left(\bar{u}\frac{\partial \bar{T}}{\partial x} + \bar{v}\frac{\partial \bar{T}}{\partial y}\right) = \frac{\partial}{\partial y}\left[(k + k_t)\frac{\partial \bar{T}}{\partial y}\right] \tag{7–240}$$

Although ν_t and α_t are not fluid properties, they are often expressed in the form of a dimensionless ratio called the *turbulent Prandtl number* Pr_t,

$$Pr_t = \frac{c_P \mu_t}{k_t} = \frac{\nu_t}{\alpha_t} \tag{7–241}$$

As early as 1874, Reynolds postulated that Pr_t should be approximately unity because $\overline{q_t''}$ and $\overline{\tau}_t$ both result from the same mechanism; that is, $\alpha_t \simeq \nu_t$ or $Pr_t \simeq 1$. This simple assumption has been utilized with quite good results for basic fully turbulent flow of moderate-to-high-Prandtl-number fluids. However, recent experimental evidence for moderate Pr fluids indicates that Pr_t is greater than unity in at least a portion of the wall region, falls to a value of the order of 0.9 in the intermediate region out to about $y/\delta = 0.4$, and eventually falls to very small values in the outermost part of the boundary layer [49,50]. The net effect of this Pr_t variation with y^+ is evidently balanced out across the boundary layer, such that the $Pr_t = 1$ approximation provides reasonable engineering predictions. However, the turbulent Prandtl number cannot be assumed to be unity for low-Prandtl-number liquid metals. Formulations have been developed for Pr_t for liquid metals by Ázer and Chao [51], Jenkins [52], and others.

Integral Formulation

Following the pattern established in Sec. 7–2–3, we integrate the differential time-average continuity, momentum, and energy equations to obtain the integral formulation for turbulent developing flow. The resulting time-average integral momentum and energy equations take the form

$$\frac{d}{dx}\int_0^\delta \overline{u}(U_\infty - \overline{u})\,dy + \frac{dU_\infty}{dx}\int_0^\delta (U_\infty - \overline{u})\,dy = \frac{\overline{\tau}_s}{\rho} \tag{7–242}$$

or, for uniform free-stream velocity,

$$\frac{d}{dx}\int_0^\delta \overline{u}(U_\infty - \overline{u})\,dy = \frac{\overline{\tau}_s}{\rho} \tag{7–243}$$

and

$$\frac{d}{dx}\int_0^\Delta \overline{u}(\overline{T} - T_\infty)\,dy = \frac{\overline{q_c''}}{\rho c_P} \tag{7–244}$$

for basic steady-flow processes with uniform properties.

The boundary conditions for the integral formulation are identical to those utilized in the differential formulation. Notice that these integral equations do *not* involve the turbulent fluctuation terms $\overline{\tau}_t$ and $\overline{q_t''}$! Although the integral formulation for turbulent flow is of the same form as for laminar flow, it should be observed that these modeling equations for turbulence involve $\overline{u}$, $\overline{T}$, $\overline{\tau}_s$, and $\overline{q_c''}$ rather than u, T, τ_s, and q_c''. Thus, the turbulent fluctuating momentum and energy transport mechanism that the parameters $\overline{\tau}_t$ and $\overline{q_c''}$ represent still manages to play its role in the formulation.

In this regard, the rather well-behaved nature of the profiles for laminar flow can be attributed to the fact that the momentum and energy transfer mechanisms operate evenhandedly across the entire flow field. On the other hand, the fact that the mean fluctuation momentum and energy transfer (or apparent turbulent shear stress and heat flux, $\bar{\tau}_t$ and $\overline{q_t''}$) associated with turbulent flow vary with y in a nonlinear fashion produces turbulent profiles that possess quite distinctive and different characteristics in the wall region, the intermediate region, and the outer region. Nevertheless, approximate relations can be developed for the distributions in mean velocity and temperature, which can be used in the formulation of reliable integral solutions for the local mean wall-shear stress and heat transfer. This basic approach to the analysis of turbulence will be introduced in the context of fully turbulent boundary layer flow over a flat plate with uniform free-stream velocity.

7–3–3 Other Approaches to Modeling Turbulence

The algebraic eddy viscosity (mixing length) representation for Reynolds stress τ_t presented in Sec. 7–3–2 applies to uniform property fully turbulent near-equilibrium† boundary layer flows with pressure gradient and transpiration. Practical algebraic inputs of this kind have also been developed for situations involving factors such as curved surfaces, variable properties, buoyancy, high speeds, transitional turbulence, and rough surfaces. Because of its relative simplicity, this approach is used in many of the numerical codes that have been developed for basic engineering applications.

However, it must be pointed out that the use of such algebraic eddy viscosity turbulence models has produced mixed results when applied to some of the more complex processes involving strong nonequilibrium boundary layers, large free-stream turbulence, large surface curvature, strong recirculation, and certain other factors. This point is particularly well documented by the 1981 *Stanford Conference on Complex Turbulent Flows* [53] in which many numerical and integral solution techniques were tested against a broad range of flows. It is for complex turbulent flows that our lack of reliable information pertaining to the distribution in μ_t (or ℓ) and k_t (or Pr_t) is most critical.

In an attempt to overcome some of the inadequacies of the algebraic eddy viscosity turbulence modeling approach, extensive study has been directed toward the

† Equilibrium boundary layers represent the category of turbulent boundary layer flows for which the distributions in mean velocity and temperature within the outer region are essentially goemetrically similar. (The same relationships among the conditions acting at the boundaries that produce exact similar boundary layers for laminar flows give rise to equilibrium boundary layers for turbulent flow.) Because of the nature of turbulent boundary layers, the velocity and temperature distributions within the outer region are generally approximately similar for situations with no abrupt changes in the boundary conditions. Such near-equilibrium conditions occur in many boundary-layer applications. However, strong nonequilibrium conditions exist in the vicinity of abrupt changes in the free-stream velocity, transpiration velocity, or thermal boundary conditions, with equilibrium or near-equilibrium conditions generally being approached within a very short distance (i.e., $\Delta x \simeq 5\delta$).

development of higher-order analyses with μ_t expressed in terms of turbulence quantities, such as turbulent kinetic energy K and/or turbulent dissipation ε, which involve the solution of one or more partial differential transport equations for turbulence quantities. One- and two-equation models of this type have achieved considerable success [54–61]. Other approaches to modeling turbulence that have been introduced in recent years and are in various stages of development include lag equation methods [11,62], the Reynolds stress method [63–67], the algebraic stress method [60,68–71], the two-fluid method [72,73], the large eddy simulation method [74,75], and the surface renewal method [76–79]. The turbulence models most commonly featured in modern general-purpose computational fluid dynamics codes include eddy viscosity, $K-\varepsilon$, and algebraic stress models.

Although no single method for modeling turbulence has been devised that is universally applicable to the wide range of conditions encountered in practice, the methods that are presently available to the engineer provide a remarkable capability.

7–3–4 Solutions

The time-average differential equations that represent the transport of mass, momentum, and energy for turbulent convection processes can be solved by numerical and integral solution techniques. These methods of solution follow the patterns established in the analysis of laminar flow. To demonstrate the mechanics involved in developing solutions for basic turbulent convection-heat-transfer processes, consideration will be given to fully turbulent TFD internal flow and fully turbulent boundary layer flow over a flat plate, after which attention will be turned to the famous Reynolds analogy.

Fully Turbulent Flow in Tubes

To minimize the mathematical details, the analysis of convection heat transfer for fully developed, fully turbulent internal flow is presented in the context of the parallel-plate system shown in Fig. 7–27. Experimental measurements for friction factor and heat transfer coefficient indicate that momentum transfer and convection heat transfer are fairly insensitive to the geometry for fully turbulent flow in internal flow systems

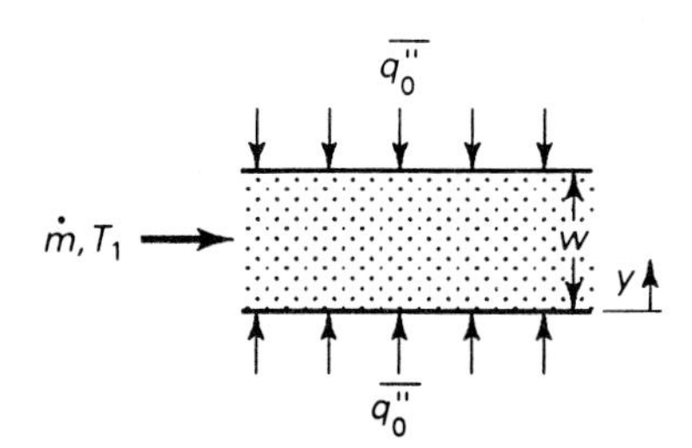

FIGURE 7–27
Fully developed turbulent flow between parallel plates with uniform wall-flux heating.

with uniform cross-sectional area. Hence, calculations for coefficients of friction and heat transfer that are obtained for fully turbulent flow between parallel plates can also be used for the standard circular tube geometry, as well as other geometries with uniform cross section.† As indicated in Chap. 6, for internal flow in tubes the flow is generally classified as transitional turbulent for $2000 \lesssim Re \lesssim 10^4$ and fully turbulent for $Re \gtrsim 10^4$.

Fluid Flow—HFD Region Our primary objectives in this section are to develop calculations for the mean velocity profile $\bar{u}$ and the mean wall-shear stress $\bar{\tau}_0$ or friction factor f. This information will then be utilized in the following section to develop solutions for the mean temperature distribution and the Nusselt number.

For HFD turbulent flow between parallel plates, the classical time-average momentum equation reduces to

$$\frac{d}{dy}\left[(\mu + \mu_t)\frac{d\bar{u}}{dy}\right] - \frac{d\bar{P}}{dx} = 0 \tag{7–245}$$

The boundary conditions are represented by

$$\bar{u} = 0 \quad \text{at } y = 0 \qquad \frac{d\bar{u}}{dy} = 0 \quad \text{at } y = \frac{w}{2} \tag{7–246,247}$$

The formulation is closed by specifying the turbulent viscosity μ_t. As indicated in the previous section, ν_t can be approximated by the van Driest equation,

$$\frac{\nu_t}{\nu} = (\kappa y^+)^2\left[1 - \exp\left(-\frac{y^+}{a^+}\right)\right]^2 \frac{du^+}{dy^+} \tag{7–230}$$

within the wall region,

$$\frac{\nu_t}{\nu} = \kappa y^+ \tag{7–228}$$

in the intermediate region, and

$$\frac{\nu_t}{\nu} = \alpha_1 \delta^+ \tag{7–229}$$

in the outer region, where $\kappa = 0.41$, $\alpha_1 = 0.08$, $a^+ = 25$, and $\delta = w/2$ for HFD flow between parallel plates.

Dimensionless Mean Velocity Distribution Equation (7–245) is separated and integrated to obtain

$$(\mu + \mu_t)\frac{d\bar{u}}{dy} = \frac{d\bar{P}}{dx}y + C_1 \tag{7–248}$$

† See Sec. 8–2–1 for an exception pertaining to annular flow.

Using the boundary condition given by Eq. (7–247), we have

$$(\mu + \mu_t)\frac{d\bar{u}}{dy} = \frac{d\bar{P}}{dx}\left(y - \frac{w}{2}\right) \tag{7–249}$$

A somewhat more convenient form of this equation can be obtained by eliminating the pressure gradient term in favor of the mean wall-shear stress $\bar{\tau}_0$. This is accomplished by writing

$$\frac{d\bar{P}}{dx} = -\frac{2}{w}(\mu + \mu_t)\left.\frac{d\bar{u}}{dy}\right|_0 = -\frac{2}{w}\bar{\tau}_0 \tag{7–250}$$

The coupling of this result with Eq. (7–249) gives

$$(\mu + \mu_t)\frac{d\bar{u}}{dy} = \bar{\tau}_0\left(1 - \frac{y}{w/2}\right) \tag{7–251}$$

or, after introducing the dimensionless parameters u^+ and y^+,

$$\left(1 + \frac{\nu_t}{\nu}\right)\frac{du^+}{dy^+} = 1 - \frac{y}{w/2} \tag{7–252}$$

where y/w is also equal to y^+/w^+.† To obtain a relation for u^+, we again separate the variables and integrate, with the result

$$u^+ = \int_0^{y^+} \frac{1 - y/(w/2)}{1 + \nu_t/\nu}\,dy^+ \tag{7–253}$$

Although this equation can be integrated numerically as it stands, the use of several practical approximations simplifies our computational task.

Focusing attention on the inner region $[y/\delta = y/(w/2) < 0.2]$, where the $y/(w/2)$ term is small, Eq. (7–253) becomes

$$u^+ = \int_0^{y^+} \frac{1}{1 + \nu_t/\nu}\,dy^+ \qquad \begin{array}{l}\text{Inner region}\\ y/(w/2) < 0.2\end{array} \tag{7–254}$$

The numerical integration of this equation in the wall region where ν_t/ν can be approximated by the van Driest equation, Eq. (7–230), is discussed in Example 7–8. Calculations obtained using the van Driest equation are shown to compare very favorably with experimental data for u^+ in Fig. 7–28.

In the innermost part of the wall region, known as the *viscous sublayer*, where ν_t is much less than ν, Eq. (7–254) reduces to the linear equation

$$u^+ = y^+ \tag{7–255}$$

† We note that Eq. (7–251) also takes the alternative dimensionless form

$$\bar{\tau} = \bar{\tau}_0\left(1 - \frac{y}{w/2}\right) \tag{7–251a}$$

This equation indicates a linear distribution in $\bar{\tau}$ across the flow field, with $\bar{\tau} \simeq \bar{\tau}_0$ in the inner region where $y/(w/2)$ is small.

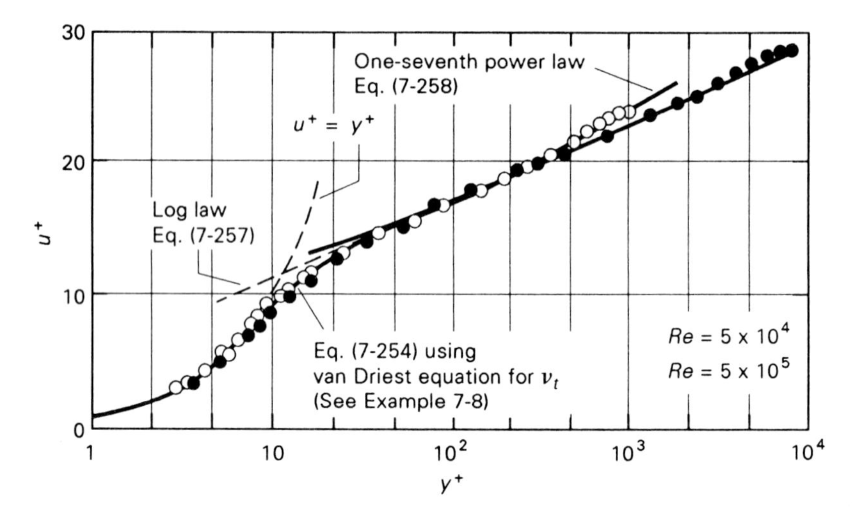

FIGURE 7–28 Dimensionless mean velocity distributions: fully turbulent HFD flow in circular tubes. (Data from Lindgren [80].)

This limiting solution for mean velocity near the wall is in good agreement with the experimental data in Fig. 7–28.

In the intermediate region where ν_t/ν is approximately equal to κy^+, Eq. (7–254) indicates

$$u^+ = \int \frac{1}{1 + \kappa y^+} dy^+ + C = \frac{1}{\kappa} \ln \left(y^+ + \frac{1}{\kappa} \right) + C \tag{7–256}$$

where C is a constant of integration. Since ν_t is much greater than ν in this region, this equation may be approximated by the famous logarithmic law,

$$u^+ = \frac{1}{\kappa} \ln y^+ + C \tag{7–257}$$

With κ set equal to 0.41, the empirical constant C is found to be equal to 5. It should be noted that the constant C and the constant a^+ in the van Driest equation are related. That is, the specification of one predetermines the other. The logarithmic law is compared with experimental data in Fig. 7–28. The agreement between Eq. (7–257) and the data is exceptional within the intermediate region. This equation is even in reasonable agreement with the data in the outer region, which is due to compensating errors in the approximations for $\bar{\tau}$ and ν_t/ν in this region. But, as expected, Eq. (7–257) fails to represent the data in the wall region.

As shown in Fig. 7–28, this logarithmic equation can be approximated by

$$u^+ = 8.7 y^{+1/7} \tag{7–258}$$

between $y^+ = 30$ and 500. This simple one-seventh power law has been found to corrclate experimental data fairly well in the intermediate and outer regions for Reynolds numbers in the range 10^4 to 10^5.

Because the theoretical basis for Eq. (7-257) and others developed from Eq. (7-254) is restricted to the inner region, we will refer to such equations as *laws of the inner region*. (These types of equations are frequently referred to as *laws of the wall* in the turbulence literature.) It is our good fortune that these equations also represent the experimental data reasonably well in the outer region. However, the development of more accurate theoretically based expressions for u^+ in the outer region requires that Eq. (7-253) or the underlying differential equations be solved.

We now turn our attention to the development of calculations for the friction factor f. Once f is known, the mean velocity distribution $\bar{u}$ can be obtained by utilizing the defining equation for u^+; that is,

$$\bar{u} = u^+ U^* = u^+ \sqrt{\frac{\bar{\tau}_0}{\rho}} = u^+ U_b \sqrt{\frac{f}{2}} \tag{7-259}$$

EXAMPLE 7-8

Develop an expression for u^+ within the entire inner region for fully turbulent HFD flow between parallel plates by utilizing the van Driest equation for ν_t.

Solution

Objective Use the van Driest equation for ν_t to develop an inner law for u^+ for fully turbulent HFD flow between parallel plates.

Schematic Fully turbulent HFD flow between parallel plates.

Assumptions/Conditions

forced convection
uniform properties
standard conditions

Analysis The van Driest expression for ν_t within the inner region is given by Eq. (7-230),

$$\frac{\nu_t}{\nu} = (\kappa y^+)^2 \left[1 - \exp\left(-\frac{y^+}{a^+}\right)\right]^2 \frac{du^+}{dy^+} \tag{7-230}$$

The relationship between ν_t and du^+/dy^+ can be obtained from the form of the momentum equation that is applicable within the inner region,

$$\left(1 + \frac{\nu_t}{\nu}\right)\frac{du^+}{dy^+} = 1 \qquad \text{or} \qquad \frac{du^+}{dy^+} = \frac{1}{1 + \nu_t/\nu} \tag{a}$$

Hence, Eq. (7–230) can be put into the form

$$\frac{\nu_t}{\nu} = (\kappa y^+)^2 \left[1 - \exp\left(-\frac{y^+}{a^+}\right)\right]^2 \frac{1}{1 + \nu_t/\nu}$$

or

$$\left(\frac{\nu_t}{\nu}\right)^2 + \frac{\nu_t}{\nu} - (\kappa y^+)^2 \left[1 - \exp\left(-\frac{y^+}{a^+}\right)\right]^2 = 0$$

The solution to this quadratic equation is

$$\frac{\nu_t}{\nu} = \frac{-1 + \{1 + 4(\kappa y^+)^2[1 - \exp(-y^+/a^+)]^2\}^{1/2}}{2}$$

We are now ready to obtain an expression for u^+. Substituting this expression for ν_t into Eq. (a) and integrating, we have

$$u^+ = \int_0^{y^+} \frac{1}{1 + \dfrac{-1 + \{1 + 4(\kappa y^+)^2[1 - \exp(-y^+/a^+)]^2\}^{1/2}}{2}}\, dy^+ \tag{b}$$

Calculations obtained by the numerical integration of this equation are shown in Fig. 7–28 to be in excellent agreement with experimental data. A simple BASIC program for performing this numerical integration is presented in Fig. A–G–3 (see Example 7–11).

EXAMPLE 7–9

Develop solutions for the dimensionless velocity profile u^+ in the inner region for fully turbulent HFD flow in a circular tube.

Solution

Objective Develop an inner law for u^+ for fully turbulent HFD flow in a circular tube.

Schematic Fully turbulent HFD flow in a circular tube.

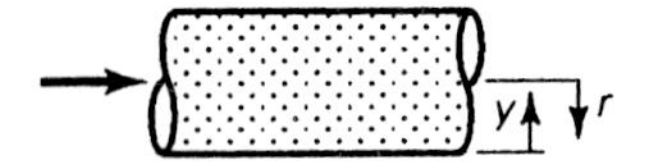

Assumptions/Conditions

forced convection
uniform properties
standard conditions

Analysis The time-average differential momentum equation for turbulent flow in a circular tube can be represented by

$$\rho\left(\bar{u}\frac{\partial \bar{u}}{\partial x} + \bar{v}\frac{\partial \bar{u}}{\partial r}\right) = -\frac{1}{r}\frac{\partial}{\partial r}(r\bar{\tau}) - \frac{d\bar{P}}{dx}$$

For HFD flow, this equation reduces to

$$\frac{d}{dr}(r\bar{\tau}) = -r\frac{d\bar{P}}{dx}$$

Integrating once, we obtain

$$r\bar{\tau} = -\frac{r^2}{2}\frac{d\bar{P}}{dx} + C_1$$

Because $\bar{\tau} = 0$ at $r = 0$, $C_1 = 0$. Hence, we have

$$\bar{\tau} = -\frac{r}{2}\frac{d\bar{P}}{dx} \tag{a}$$

Setting $\bar{\tau} = \bar{\tau}_0$ at $r = r_0$, we see that

$$\frac{d\bar{P}}{dx} = -\frac{2\bar{\tau}_0}{r_0}$$

such that Eq. (a) also takes the form

$$\bar{\tau} = \bar{\tau}_0\frac{r}{r_0}$$

With r replaced by $r_0 - y$, this equation takes the form

$$\bar{\tau} = \bar{\tau}_0\left(1 - \frac{y}{r_0}\right)$$

This equation for turbulent flow in a circular tube is seen to be of the same form as Eq. (7–251) for turbulent flow between parallel plates, except for the fact that $w/2$ is replacing by r_0. Therefore, the expressions for u^+ that were developed for HFD flow between parallel plates also apply to HFD tube flow, with $w/2$ being replaced by r_0. It follows that u^+ is approximated by Eq. (E7–8b) for the wall region and by Eq. (7–257) for the region $y^+ > 50$ and $y/\delta < 0.2$. Equation (7–257) can also be utilized as a first approximation in the outer region.

Bulk-Stream Velocity and Friction Factor As in the analysis of laminar internal flow, the final step in obtaining calculations for the velocity distribution $\bar{u}$ and the friction factor f involves the evaluation of the bulk-stream velocity U_b. By definition, U_b for turbulent flow between parallel plates is given by

$$U_b = \frac{1}{w} \int_0^w \bar{u}\, dy \tag{7–260}$$

With $\bar{u}$ expressed in terms of u^+, we obtain

$$U_b = \frac{U^*}{w^+} \int_0^{w^+} u^+\, dy^+ \tag{7–261}$$

where $w^+ = U^* w/\nu$. Because $U^* = U_b \sqrt{f/2}$, we have

$$\sqrt{\frac{2}{f}} = \frac{2}{w^+} \int_0^{w^+/2} u^+\, dy^+ \tag{7–262}$$

Therefore, with u^+ known, calculations can be obtained for f.

As a first approximation, we will use the simple one-seventh power law given by Eq. (7–258) throughout the entire flow field. Although this equation fails to represent the data for u^+ in the wall region, the volume-flow rate in this region is negligibly small when compared to the volume-flow rate in the intermediate and outer region. Hence, the error introduced by this approximation is small. Substituting the one-seventh power law into Eq. (7–262), we have

$$\sqrt{\frac{2}{f}} = \frac{2}{w^+} \int_0^{w^+/2} 8.7 y^{+1/7}\, dy^+ = \frac{8.7}{8/7} \left(\frac{w^+}{2} \right)^{1/7} \tag{7–263}$$

where

$$w^+ = \frac{U^* w}{\nu} = \frac{U_b w}{\nu} \sqrt{\frac{f}{2}} = \frac{Re}{2} \sqrt{\frac{f}{2}} \tag{7–264}$$

Solving for f, we obtain

$$f = 0.081\, Re^{-1/4} \tag{7–265}$$

This equation lies only 2.5% above the well-known Blasius equation,

$$f = 0.079\, Re^{-1/4} \tag{7–266}$$

which was obtained on the basis of an empirical curve fit to data for HFD turbulent flow in circular tubes and between parallel plates in the Reynolds number range from 5000 to 10^5.

To obtain calculations for f for larger values of Re, the logarithmic correlation for u^+ can be used. As shown in Example 7–10, the use of this more accurate correlation for u^+ gives rise to an implicit expression for f of the form

$$\sqrt{\frac{2}{f}} = C - \frac{1}{\kappa} + \frac{1}{\kappa} \ln \left(\frac{Re}{4} \sqrt{\frac{f}{2}} \right) \tag{7–267}$$

for flow between parallel plates. Setting $\kappa = 0.41$ and $C = 5$, we obtain

$$\sqrt{\frac{2}{f}} = 2.44 \ln\left(Re\sqrt{\frac{f}{2}}\right) - 0.82 \tag{7-268}$$

Rearranging this equation, an explicit expression is obtained for Re in terms of f.

$$Re = 1.4\sqrt{\frac{2}{f}} \exp\left(0.41\sqrt{\frac{2}{f}}\right) \tag{7-269}$$

An expression is also developed in Example 7-10 for turbulent HFD flow in a circular tube, which takes the form

$$\sqrt{\frac{2}{f}} = 2.44 \ln\left(Re\sqrt{\frac{f}{2}}\right) - 0.351 \tag{7-270}$$

This expression was first derived by Prandtl in 1935. However, Prandtl adjusted the constants on the basis of experimental measurements by Nikuradse [81] to obtain better agreement with the data in the low-Reynolds-number range. The final Prandtl-Nikuradse equation is given by

$$\sqrt{\frac{2}{f}} = 2.46 \ln\left(Re\sqrt{\frac{f}{2}}\right) + 0.292 \tag{7-271}$$

This equation is shown to be in excellent agreement with the experimental data in Fig. 7-29 over the entire turbulent flow range. The equation for flow between parallel plates, Eq. (7-268), is approximately 20% above Eq. (7-271). However, this difference is on the same order of the scatter in the experimental data.

For values of $Re > 10^4$, Eq. (7-271) can be approximated by a simple explicit correlation for f given by

$$f = 0.046\,Re^{-0.2} \qquad \text{for } Re \gtrsim 10^4 \tag{7-272}$$

This simple equation is shown in Fig. 7-29 to compare very well with experimental data.

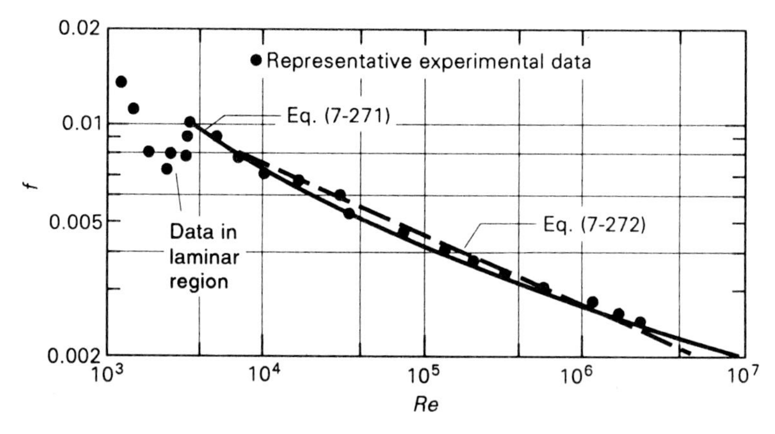

FIGURE 7-29 Friction factor f: HFD flow in circular tubes.

EXAMPLE 7–10

Utilize the logarithmic inner law to develop expressions for the Fanning friction factor for fully turbulent HFD flow (a) between parallel plates and (b) in circular tubes.

Solution

Objective Use the log law,

$$u^+ = \frac{1}{\kappa}\ln y^+ + C$$

to develop expressions for f.

Schematic Fully turbulent HFD flow.

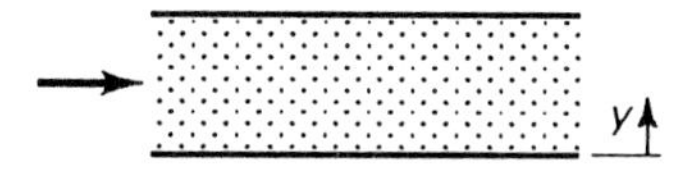

(a) Flow between parallel plates.

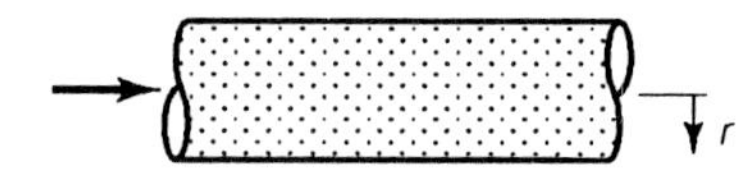

(b) Flow in a circular tube.

Assumptions/Conditions

forced convection
uniform properties
standard conditions

Analysis

(a) Fully Turbulent HFD Flow between Parallel Plates For flow between parallel plates, f is given by Eq. (7–262),

$$\sqrt{\frac{2}{f}} = \frac{2}{w^+}\int_0^{w^+/2} u^+\,dy^+ \tag{7–262}$$

Utilizing the logarithmic inner law for u^+, we have

$$\sqrt{\frac{2}{f}} = \frac{2}{w^+}\int_0^{w^+/2}\left(C + \frac{1}{\kappa}\ln y^+\right)dy^+ \tag{7–267}$$

$$= \frac{2}{w^+}\left[Cy^+ + \frac{1}{\kappa}(y^+\ln y^+ - y^+)\right]\Bigg|_0^{w^+/2} = C - \frac{1}{\kappa} + \frac{1}{\kappa}\ln\left(\frac{Re}{4}\sqrt{\frac{f}{2}}\right)$$

With $C = 5$ and $\kappa = 0.41$, we have

$$\sqrt{\frac{2}{f}} = 2.56 + 2.44\ln\left(\frac{Re}{4}\sqrt{\frac{f}{2}}\right) = 2.44\ln\left(Re\sqrt{\frac{f}{2}}\right) - 0.82 \tag{7–268}$$

(b) Fully Turbulent HFD Flow in a Circular Tube For a circular-tube geometry, U_b is defined by

$$U_b = \frac{2}{r_0^2}\int_0^{r_0} r\bar{u}\,dr = \frac{2}{r_0^2}\int_0^{r_0} (r_0 - y)\bar{u}\,dy$$

Introducing the friction velocity U^*, this equation takes the form

$$\sqrt{\frac{2}{f}} = \frac{2}{r_0^+}\int_0^{r_0^+}\left(1 - \frac{y^+}{r_0^+}\right) u^+\,dy^+$$

With u^+ specified by the logarithmic law, we have

$$\sqrt{\frac{2}{f}} = \frac{2}{r_0^+}\int_0^{r_0^+}\left(C + \frac{1}{\kappa}\ln y^+ - C\,\frac{y^+}{r_0^+} - \frac{y^+}{\kappa r_0^+}\ln y^+\right) dy^+$$

$$= \frac{2}{r_0^+}\left[C r_0^+ + \frac{1}{\kappa}(r_0^+ \ln r_0^+ - r_0^+) - \frac{C}{2} r_0^+ - \frac{1}{\kappa r_0^+}\left(\frac{r_0^{+2}}{2}\ln r_0^+ - \frac{r_0^{+2}}{4}\right)\right]$$

$$= \frac{1}{\kappa}\ln r_0^+ + C - \frac{3}{2\kappa} = 2.44\ln\left(\frac{Re}{2}\sqrt{\frac{f}{2}}\right) + 1.34$$

$$= 2.44\ln\left(Re\sqrt{\frac{f}{2}}\right) - 0.351 \qquad (7\text{–}270)$$

Energy Transfer—TFD Region We now seek to develop calculations for the mean temperature distribution $\overline{T}$ and the Nusselt number Nu for fully turbulent TFD flow between parallel plates with uniform wall-flux heating. Our analysis will be seen to involve steps that are similar to those that were utilized in the solution of the laminar flow problem and that are analogous to those that were employed in the analysis of the turbulent momentum transfer.

The classical differential formulation for the mean energy transfer for turbulent TFD flow between parallel plates is given by

$$\rho c_P \bar{u}\,\frac{\partial \overline{T}}{\partial x} = \frac{\partial}{\partial y}\left[(k + k_t)\,\frac{\partial \overline{T}}{\partial y}\right] \qquad (7\text{–}273)$$

and

$$\frac{\partial \overline{T}}{\partial y} = 0 \quad \text{at } y = \frac{w}{2} \qquad -k\,\frac{\partial \overline{T}}{\partial y} = \overline{q_0''} \quad \text{at } y = 0 \qquad (7\text{–}274,275)$$

for a uniform wall-heat flux. For this uniform wall-heat-flux boundary condition,

$\partial\overline{T}/\partial x$ is evaluated by utilizing the defining equation for TFD flow, Eq. (7–208), and a lumped energy balance, with the result

$$\frac{\partial\overline{T}}{\partial x} = \frac{dT_b}{dx} = \frac{dT_s}{dx} = \frac{\overline{q_0''}p}{\dot{m}c_P} = \frac{4\overline{q_0''}}{\rho c_P U_b D_H} \tag{7–276}$$

where $D_H = 2w$ for flow between parallel plates. Hence, the energy equation takes the simpler form

$$\frac{d}{dy}\left[(k + k_t)\frac{d\overline{T}}{dy}\right] = \rho c_P \overline{u}\frac{dT_b}{dx} = \frac{4\overline{q_0''}}{D_H}\frac{\overline{u}}{U_b} \tag{7–277}$$

As indicated in Sec. 7–2, the turbulent thermal conductivity k_t can be expressed in terms of ν_t by

$$\alpha_t = \frac{k_t}{\rho c_P} = \frac{\nu_t}{Pr_t} \tag{7–278}$$

where the turbulent Prandtl number Pr_t is approximately equal to unity for fluids with moderate values of Prandtl number.

Dimensionless Mean Temperature Distribution Recognizing that

$$-\left[(k + k_t)\frac{d\overline{T}}{dy}\right]\bigg|_0 = -k\frac{d\overline{T}}{dy}\bigg|_0 = \overline{q_0''} \tag{7–279}$$

Eq. (7–277) is separated and integrated to obtain

$$(k + k_t)\frac{d\overline{T}}{dy} + \overline{q_0''} = \frac{4\overline{q_0''}}{U_b D_H}\int_0^y \overline{u}\, dy \tag{7–280}$$

or

$$-(k + k_t)\frac{d\overline{T}}{dy} = \overline{q_0''}\left(1 - \frac{4}{U_b D_H}\int_0^y \overline{u}\, dy\right) \tag{7–281}$$

This equation is put into the dimensionless form

$$\left(\frac{\alpha}{\nu} + \frac{\alpha_t}{\nu}\right)\frac{dT^+}{dy^+} = 1 - \frac{4}{Re}\int_0^{y^+} u^+\, dy^+ \tag{7–282}$$

where T^+ is defined in terms of the unknown wall temperature T_s by

$$T^+ = \frac{(T_s - \overline{T})\rho c_P U^*}{\overline{q_0''}} \tag{7–283}$$

and the integral $\int_0^{y^+} u^+\, dy^+$ can be evaluated by specifying u^+ in accordance with the fluid-flow analysis.

Separating the variables and integrating once again, we write

$$T^+ = \int_0^{y^+}\left[\frac{1 - (4/Re)\int_0^{y^+} u^+\, dy^+}{1/Pr + (\nu_t/\nu)/Pr_t}\right] dy^+ \tag{7–284}$$

With ν_t, Pr_t, and u^+ specified, this equation can be numerically integrated. However, to simplify the problem, we break this integral into two parts. In the inner region $[y/(w/2) \lesssim 0.2]$, where the integral $\int_0^{y+} u^+ \, dy^+$ is small, we write

$$T^+ = \int_0^{y+} \frac{1}{1/Pr + (\nu_t/\nu)/Pr_t} dy^+ \qquad \begin{array}{l}\text{Inner region}\\ y/(w/2) \lesssim 0.2\end{array} \tag{7–285}$$

In the outer region where the velocity profile is quite flat, we approximate $\bar{u}$ by U_b, with the result that

$$\frac{4}{Re}\int_0^{y+} u^+ \, dy^+ = \frac{4}{2wU_b/\nu} \frac{U_b}{U^*} \frac{yU^*}{\nu} = 2\frac{y}{w} \tag{7–286}$$

and

$$T^+ = \int \frac{1 - y/(w/2)}{1/Pr + (\nu_t/\nu)/Pr_t} dy^+ + B \qquad \begin{array}{l}\text{Outer region}\\ y/(w/2) \gtrsim 0.2\end{array} \tag{7–287}$$

where B is a constant of integration. With Pr and Pr_t set equal to unity, Eqs. (7–285) and (7–287) are seen to be equivalent to the relationships for u^+ given by Eqs. (7–254) and (7–253), respectively.

Concerning the wall region, Eq. (7–285) has been solved numerically by investigators for various inputs for ν_t/ν and Pr_t. To illustrate, the numerical integration of this equation with Pr_t set equal to unity and with ν_t/ν given by the van Driest equation, Eq. (7–230), is considered in Example 7–11. Calculations obtained in Example 7–11 for T^+ are shown in Fig. 7–30 to be in excellent agreement with experimental data for fluids with moderate values of Prandtl number. In the viscous sublayer where ν_t is much less than ν, Eq. (7–285) gives rise to the limiting

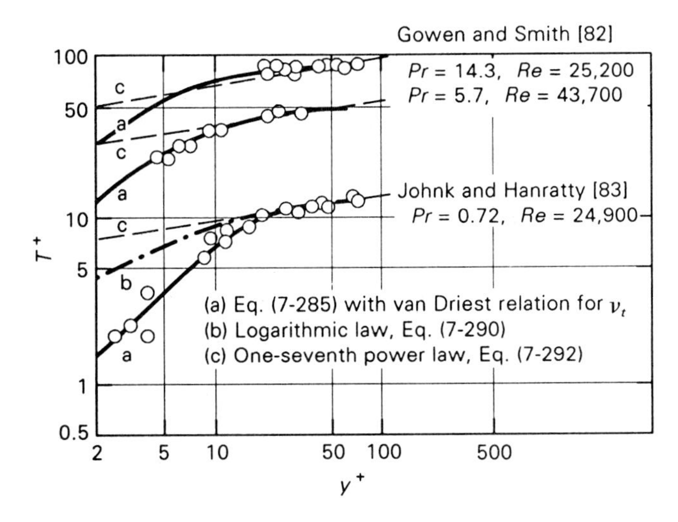

FIGURE 7–30 Dimensionless mean temperature distributions: fully turbulent TFD flow in circular tubes.

result

$$T^+ = Pr\, y^+ \qquad \text{for } y^+ < 5 \tag{7–288}$$

This equation is consistent with the experimental data and with the van Driest calculations shown in Fig. 7–30.

In the intermediate part of the inner region where ν_t/ν is approximately equal to κy^+, Eq. (7–285) indicates (for $Pr_t = 1$)

$$T^+ = \int \frac{1}{1 + \kappa y^+}\, dy^+ + B = \frac{1}{\kappa} \ln \left(y^+ + \frac{1}{\kappa\, Pr} \right) + B \tag{7–289}$$

where the constant B is a function of Prandtl number Pr. Since α_t is much greater than α in this region for fluids other than liquid metals, this equation reduces to the more familiar form

$$T^+ = \frac{1}{\kappa} \ln y^+ + B \tag{7–290}$$

As shown in Example 7–11, the numerical integration of Eq. (7–285) with ν_t/ν properly specified within the wall region provides us with the relationship between B and Pr. The constant B is given in Table 7–2 for representative values of Pr. White [16] has proposed a useful correlation for B of the form

$$B = 12.7\, Pr^{2/3} - 7.7 \tag{7–291}$$

This correlation is shown in Table 7–2 to be quite reasonable for values of Pr on the order of unity and greater.

Equation (7–290) is compared with experimental data in Fig. 7–30 for turbulent TFD flow in circular tubes. The agreement between this logarithmic inner law and the data is seen to be quite good throughout the intermediate region. As in the case for the u^+ correlations, we find that our inner law for T^+ can also be used to approximate the data in the outer region.

TABLE 7–2 Calculations for B and B_1

	B		
Pr	Example 7–11	Eq. (7–291)	B_1
0.5	0.217	0.689	5.71
0.72	2.74	2.93	7.15
1.0	5.50	5.50	8.73
2.0	13.4	13.2	13.2
3.0	20.0	19.7	17.0
5.0	31.2	30.9	23.4
5.7	34.7	34.5	25.4
7.5	43.3	43.1	30.3
10.0	54.1	53.6	36.5
14.3	71.2	70.8	46.3

Equation (7–290) can also be approximated by a simple one-seventh power law of the form

$$T^+ = B_1 y^{+1/7} \tag{7–292}$$

for moderate values of Pr. The coefficient B_1 is obtained by equating Eqs. (7–290) and (7–292) at some value of y^+ within the intermediate region. For example, if we equate these expressions at $y^+ = 50$, then $B_1 = (9.78 + B)/1.75$. Calculations for B_1 based on this relationship are given in Table 7–2 for several values of Pr. This one-seventh power law is shown in Fig. 7–30 to correlate the data for T^+ quite well in the intermediate region. Similar to the restriction imposed on the one-seventh power law for u^+, Eq. (7–292) is best suited for values of Reynolds number between 10^4 and 10^5.

Concerning the theoretical treatment of the outer region, Eq. (7–287) can be integrated numerically. However, as indicated above, the dimensionless temperature distribution within the outer region can be approximated by our simple inner laws.

Next, we develop calculations for the Nusselt number. With Nu and f known, calculations can be obtained for the mean temperature distribution $\overline{T}$ by utilizing the defining equation for T^+; that is,

$$\frac{\overline{T} - T_s}{T_b - T_s} = \frac{\overline{q_0''}}{T_s - T_b} \frac{T^+}{\rho c_P U^*} = \frac{Nu\ T^+}{Re\ Pr\ \sqrt{f/2}} \tag{7–293}$$

EXAMPLE 7–11

Develop calculations for T^+ within the inner region for fully turbulent TFD flow between parallel plates. Utilize the van Driest equation for ν_t and the Reynolds approximation $\alpha_t = \nu_t$ (i.e., $Pr_t = 1$).

Solution

Objective Develop an inner law for T^+ for fully turbulent TFD flow.

Schematic Fully turbulent TFD flow between parallel plates.

Assumptions/Conditions

forced convection
uniform properties
standard conditions

Analysis The solution for T^+ within the inner region is given by Eq. (7–285),

$$T^+ = \int_0^{y^+} \frac{1}{1/Pr + (\nu_t/\nu)/Pr_t}\, dy^+ \tag{7–285}$$

We have seen in Example 7–8 that van Driest's equation for ν_t can be written as

$$\frac{\nu_t}{\nu} = \frac{-1 + \{1 + 4(\kappa y^+)^2\,[1 - \exp(-y^+/a^+)]^2\}^{1/2}}{2} \tag{a}$$

Setting Pr_t equal to unity and utilizing the van Driest equation for ν_t, we have

$$T^+ = \int_0^{y^+} \frac{1}{\dfrac{1}{Pr} + \dfrac{-1 + \{1 + 4(\kappa y^+)^2\,[1 - \exp(-y^+/a^+)]^2\}^{1/2}}{2}}\, dy^+ \tag{b}$$

A BASIC program designed to integrate Eq. (b) numerically is presented in Fig. A–G–3. Calculations for T^+ obtained by running this program on a digital computer are presented in Fig. 7–30 for various values of Prandtl number with $\kappa = 0.41$ and $a^+ = 25$. The calculations for each value of Pr approach the limiting logarithmic relation given by Eq. (7–290),

$$T^+ = \frac{1}{\kappa} \ln y^+ + B \tag{7–290}$$

as y^+ increases. To determine the value of B, we set T^+ in Eq. (b) equal to Eq. (7–290) for values of $y^+ > 100$. This simple step is also included in the BASIC program. Calculations for B obtained by running this program are given in Table 7–2.

By comparing Eq. (E7–8b) with Eq. (b) above, we see that the velocity profile u^+ is identical to the temperature distribution T^+ for $Pr = 1$ and $Pr_t = 1$.

Bulk-Stream Temperature and Nusselt Number In the classical approach, calculations are obtained for the Nusselt number by utilizing the defining equations for the bulk-stream temperature T_b. (It should be recalled that this same approach was utilized in our analysis of the laminar flow problem.) For turbulent flow between parallel plates, T_b is defined by

$$T_b = \frac{1}{wU_b} \int_0^{w} \bar{u}\bar{T}\, dy = \frac{2}{wU_b} \int_0^{w/2} \bar{u}\bar{T}\, dy \tag{7–294}$$

or, in terms of T^+,

$$T_b^+ = \frac{T_s - T_b}{q_0''/(\rho c_P U^*)} = \frac{2U^*}{w^+ U_b} \int_0^{w^+/2} u^+ T^+\, dy^+ \tag{7–295}$$

Solving for $\overline{q''_0}/(T_s - T_b)$, we obtain

$$\frac{\overline{q''_0}}{T_s - T_b} = \frac{\rho c_P U_b}{(2/w^+)\int_0^{w^+/2} u^+ T^+ \, dy^+} \tag{7-296}$$

The Nusselt number is therefore given by

$$Nu = \frac{\overline{q''_0}}{T_s - T_b}\frac{2w}{k} = \frac{Re\ Pr}{(2/w^+)\int_0^{w^+/2} u^+ T^+ \, dy^+} \tag{7-297}$$

Using the solutions for u^+ and T^+ that have just been developed, Eq. (7-297) can be integrated by analytical or numerical methods. Perhaps the simplest approach to this problem is to approximate u^+ and T^+ by the one-seventh power laws. Another simple approach is to approximate $\bar{u}$ by U_b and to approximate T^+ by the logarithmic law, Eq. (7-290). Following this approach, we write

$$\begin{aligned} Nu &= \frac{Re\ Pr}{\dfrac{U_b}{U^*}\dfrac{2}{w^+}\displaystyle\int_0^{w^+/2}\left(\frac{1}{\kappa}\ln y^+ + B\right)dy^+} \\ &= \frac{\sqrt{f/2}\ Re\ Pr}{\dfrac{2}{w^+}\left[\dfrac{1}{\kappa}(y^+ \ln y^+ - y^+) + By^+\right]\Bigg|_0^{w^+/2}} \\ &= \frac{\sqrt{f/2}\ Re\ Pr}{\dfrac{1}{\kappa}\left[\ln\left(\dfrac{Re}{4}\sqrt{\dfrac{f}{2}}\right) - 1\right] + B} \end{aligned} \tag{7-298}$$

where B is given in Table 7-2. To put this equation into a more manageable form, we utilize the relationship between f and Re given by Eq. (7-267); that is,

$$\sqrt{\frac{2}{f}} = C - \frac{1}{\kappa} + \frac{1}{\kappa}\ln\left(\frac{Re}{4}\sqrt{\frac{f}{2}}\right) \tag{7-267}$$

By coupling Eqs. (7-298) and (7-267) to eliminate the logarithmic term, we obtain

$$Nu = \frac{(f/2)\ Re\ Pr}{1 + \sqrt{f/2}\,(B - C)} \tag{7-299}$$

Substituting Eq. (7-291) for B and setting $C = 5$, this equation reduces to

$$Nu = \frac{(f/2)\ Re\ Pr}{1 + 12.7\sqrt{f/2}\,(Pr^{2/3} - 1)} \tag{7-300}$$

Equation (7-300) is in excellent agreement with experimental data for a broad range in Pr. To illustrate, Eq. (7-300) is compared with experimental data for air

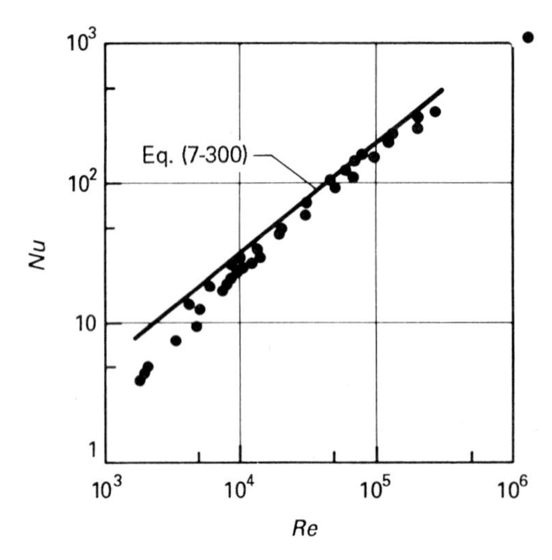

FIGURE 7–31
Comparison of calculations for Nusselt number with representative experimental data for TFD turbulent flow of air in circular tubes.

in Fig. 7–31. This same equation was obtained by White [16] for tube flow. Furthermore, Eq. (7–300) is almost identical to the popular Petukhov-Kirillov correlation [84] given by

$$Nu = \frac{(f/2)\,Re\,Pr}{1.07 + 12.7\,\sqrt{f/2}\,(Pr^{2/3} - 1)} \tag{7–301}$$

Fully Turbulent Boundary Layer Flow over Plane Surfaces

We now consider turbulent boundary layer flow over a flat plate with uniform free-stream velocity and uniform wall-flux heating maintained over the surface. The development of the hydrodynamic boundary layer for this case is illustrated in Fig. 7–32. Strictly speaking, unless the boundary layer is artificially disturbed (tripped), the transition from laminar to turbulent flow generally occurs near the location at which $Re_x = 5 \times 10^5$, depending on surface roughness and certain other factors. Notice the large increase in the boundary layer thickness as the transition from laminar to turbulent flow occurs. For example, at $Re_x = 10^7$ the turbulent boundary layer is approximately 10 times as thick as the boundary layer would be if it were laminar.

To develop a convenient closed-form solution to this important problem, we will employ the integral approach.

Fluid Flow The classical differential formulation for the velocity field associated with turbulent boundary layer flow is given by Eqs. (7–214) and (7–225),

$$\frac{\partial \bar{u}}{\partial x} + \frac{\partial \bar{v}}{\partial y} = 0 \tag{7–302}$$

$$\rho\left(\bar{u}\,\frac{\partial \bar{u}}{\partial x} + \bar{v}\,\frac{\partial \bar{u}}{\partial y}\right) = \frac{\partial}{\partial y}\left[(\mu + \mu_t)\,\frac{\partial \bar{u}}{\partial y}\right] - \cancel{\frac{dP}{dx}} \tag{7–303}$$

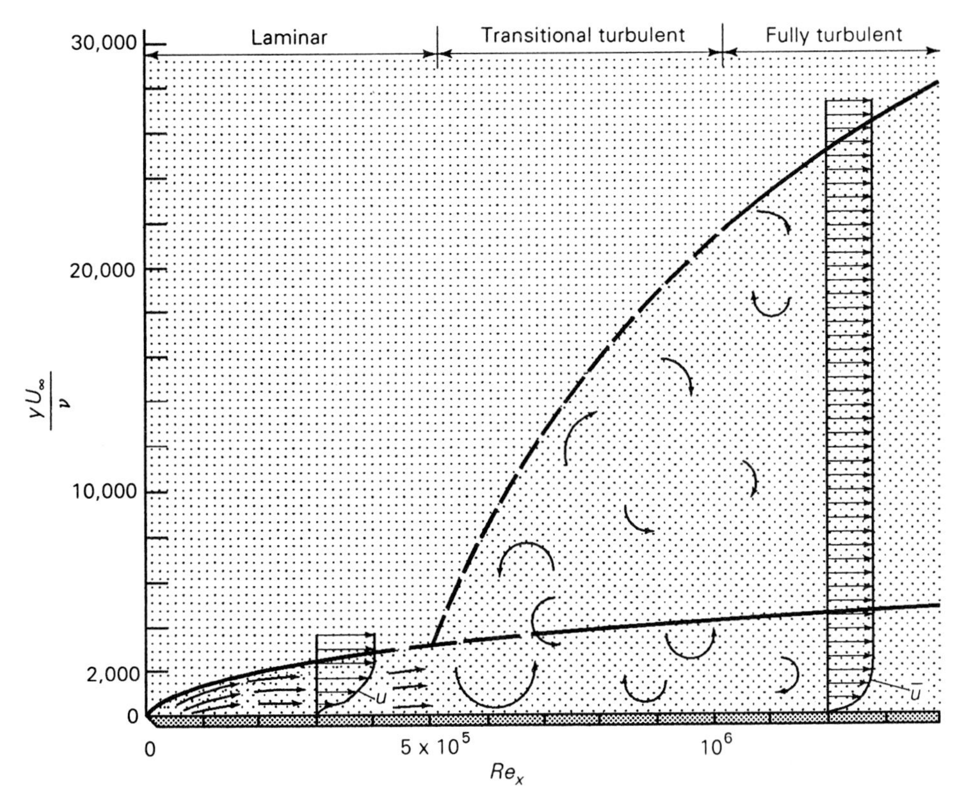

FIGURE 7–32 Laminar to turbulent transition of boundary layer for flow over a flat plate with uniform free-stream velocity.

where $d\overline{P}/dx = 0$ for uniform free-stream velocity, and

$$\overline{u} = U_\infty \qquad \text{at } x = 0 \tag{7–304}$$

$$\overline{u} = 0, \overline{v} = 0 \qquad \text{at } y = 0 \tag{7–305,306}$$

$$\overline{u} = U_\infty \qquad \text{as } y \to \infty \text{ or at } y = \delta \tag{7–307}$$

The formulation is completed by the specification of eddy viscosity μ_t or mixing length ℓ. As indicated in Sec. 7–3–3, the eddy diffusivity ν_t can be approximated by the van Driest equation,

$$\frac{\nu_t}{\nu} = (\kappa y^+)^2 \left[1 - \exp\left(-\frac{y^+}{a^+}\right)\right]^2 \frac{du^+}{dy^+} \tag{7–230}$$

within the wall region,

$$\frac{\nu_t}{\nu} = \kappa y^+ \tag{7-228}$$

in the intermediate region, and

$$\frac{\nu_t}{\nu} = \alpha_1 \delta^+ \tag{7-229}$$

in the outer region,† where $\kappa = 0.41$, $\alpha_1 = 0.08$, $a^+ = 25$, and δ continues to grow for developing external flow. These approximations are utilized in many of the modern differential/numerical approaches to analyzing turbulent boundary layer flow processes.

These equations provide the basis for developing analytical and numerical solutions for the mean velocity distribution and mean wall-shear stress for turbulent boundary layer flow. Our emphasis will be placed on the development of simplified analytical solutions for u^+ and f_x for fully turbulent boundary layer flow over a flat plate with uniform free-stream velocity.

Dimensionless Mean Velocity Distribution Following the path taken in the classical analysis of fully turbulent HFD flow between parallel plates, we integrate Eq. (7–303) to obtain

$$(\mu + \mu_t)\frac{\partial \overline{u}}{\partial y} = \overline{\tau}_0 + \int_0^y \rho\left(\overline{u}\frac{\partial \overline{u}}{\partial x} + \overline{v}\frac{\partial \overline{u}}{\partial y}\right) dy \tag{7-308}$$

or

$$\overline{\tau} = \overline{\tau}_0 + \int_0^y \rho\left(\overline{u}\frac{\partial \overline{u}}{\partial x} + \overline{v}\frac{\partial \overline{u}}{\partial y}\right) dy \tag{7-309}$$

Whereas $\overline{\tau}$ was found to be a simple linear function of y for turbulent fully developed internal flow, Eq. (7–309) indicates that the relationship between $\overline{\tau}$ and y is nonlinear for developing external boundary layer flow. However, in the inner region ($y/\delta < 0.2$) where $\overline{\tau} \simeq \overline{\tau}_0$, Eq. (7–309) takes the simple form

$$(\mu + \mu_t)\frac{d\overline{u}}{dy} = \overline{\tau}_0 \qquad \text{or} \qquad \left(1 + \frac{\nu_t}{\nu}\right)\frac{du^+}{dy^+} = 1 \tag{7-310,311}$$

such that the solution for u^+ becomes

$$u^+ = \int_0^{y^+} \frac{1}{1 + \nu_t/\nu}\, dy^+ \qquad \begin{matrix}\text{Inner region}\\ y/\delta \lesssim 0.2\end{matrix} \tag{7-312}$$

This equation is identical to Eq. (7–254) for the inner region of a turbulent HFD internal flow. Therefore, the solutions for u^+ that were developed in Sec.

† An alternative empirical relation that is commonly used to characterize the outer region is given in terms of mixing length ℓ by $\ell^+ = \alpha_0 \delta^+$.

7–3–4 for the inner region of internal flows also apply to this external-flow problem. With ν_t approximated by the van Driest equation, Eq. (7–230), Eq. (7–312) can be used to calculate u^+ throughout the inner region. In the intermediate region where $\nu_t/\nu = \kappa y^+$, u^+ can be approximated by the logarithmic law, Eq. (7–257),

$$u^+ = \frac{1}{\kappa} \ln y^+ + C \qquad \begin{array}{l}\text{Intermediate region} \\ y^+ \gtrsim 50,\ y/\delta \lesssim 0.2\end{array} \tag{7–313}$$

where $\kappa = 0.41$ and $C = 5$, or by the one-seventh power law, Eq. (7–258),

$$u^+ = 8.7 y^{+1/7} \qquad \begin{array}{l}\text{Intermediate region} \\ y^+ \gtrsim 50,\ y/\delta \lesssim 0.2\end{array} \tag{7–314}$$

which is best suited for values of Re_x in the range 10^5 to 10^7. Equations (7–312) and (7–313) are shown in Fig. 7–33 to be in quite adequate agreement with experimental data.

Although these inner laws can be used as a first approximation for u^+ throughout the turbulent core, the development of a formal solution to Eq. (7–308) that is applicable to the outer region requires that the convective effects represented by the terms $\bar{u}\,\partial\bar{u}/\partial x + \bar{v}\,\partial\bar{u}/\partial y$ be accounted for. Numerical approaches are available in the literature [19–21,43] that involve the solution of the underlying differential equations, Eqs. (7–302) and (7–303). Solutions have also been developed by the use of polynomial approximations for $\bar{\tau}$ that are applicable across the entire bound-

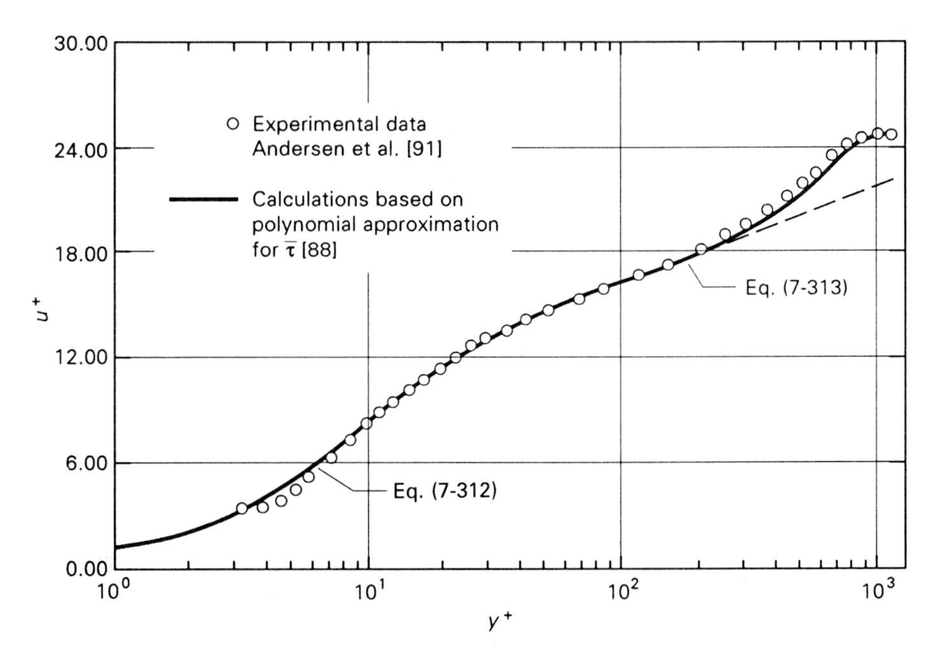

FIGURE 7–33 Dimensionless mean velocity distribution for fully turbulent boundary layer flow over a flat plate with uniform free-stream velocity.

ary layer [88,89]. Calculations obtained on the basis of this $\bar{\tau}$-*approximation* approach are shown in Fig. 7–33. Alternatively, the distribution in u^+ within the turbulent core can be represented by Coles' [90] empirical *law of the wake*, which takes the form

$$u^+ = C + \frac{1}{\kappa} \ln y^+ + \frac{1}{\kappa} \sin^2 \frac{\pi y}{2\delta} \tag{7–315}$$

for uniform free-stream velocity.

It is important to note that the simple inner laws given by Eqs. (7–313) and (7–314) and the law of the wake given by Eq. (7–315) are restricted to turbulent boundary layer flows with uniform free-stream velocity and no transpiration. In this connection, the numerical approach and the $\bar{\tau}$-approximation approach have been generalized to account for nonuniform free-stream velocity and transpiration. A generalized form of Coles' law of the wake is also available, which is reported to be applicable to nonuniform free-stream velocity flow over nonporous surfaces.

As in the case of internal turbulent HFD flow, relations for u^+ must be coupled with calculations for the friction factor in order to obtain the distribution in mean velocity $\bar{u}$ for external turbulent boundary layer flow.

Integral Solution for Friction Factor Calculations can be developed for the friction factor f_x for turbulent boundary layer flow by coupling relations for u^+ with the integral momentum equation. To develop calculations for f_x based on the integral approach, the integral momentum equation,

$$\frac{d}{dx} \int_0^{\delta} \bar{u}(U_\infty - \bar{u})\, dy = \frac{\bar{\tau}_s}{\rho} \tag{7–243}$$

is written in terms of the dimensionless parameters u^+ and y^+; that is,

$$\nu \frac{d}{dx} \int_0^{\delta^+} u^+(U_\infty - u^+ U^*)\, dy^+ = U_\infty^2 \frac{f_x}{2}$$

or

$$\frac{d}{dRe_x} \left[\sqrt{\frac{f_x}{2}} \int_0^{\delta^+} u^+(U_\infty^+ - u^+)\, dy^+ \right] = \frac{f_x}{2} \tag{7–316}$$

where $U_\infty^+ = U_\infty/U^* = \sqrt{2/f_x}$. By specifying u^+, Eq. (7–316) can be solved for f_x or δ^+.

To demonstrate, u^+ is approximated by the simple one-seventh power law, Eq. (7–314). Substituting this correlation for u^+ into the integral momentum equation, we obtain

$$\frac{d}{dRe_x} \left[\sqrt{\frac{f_x}{2}} \int_0^{\delta^+} 8.7 y^{+1/7} (8.7\delta^{+1/7} - 8.7 y^{+1/7})\, dy^+ \right] = \frac{f_x}{2}$$

$$\frac{d}{dRe_x} \left[\sqrt{\frac{f_x}{2}} (8.7)^2\, \delta^{+9/7} \left(\frac{1}{8/7} - \frac{1}{9/7} \right) \right] = \frac{f_x}{2} \tag{7–317}$$

This gives us one equation in the two unknowns f_x and δ. A second independent relationship between these two unknowns is obtained by setting $\bar{u} = U_\infty$ at $y = \delta$ in our power-law correlation; that is,

$$\sqrt{\frac{2}{f_x}} = 8.7\delta^{+1/7} \tag{7-318}$$

Utilizing this expression, Eq. (7-317) reduces to

$$\frac{7}{72(8.7)^7}\frac{d}{dRe_x}\left(\frac{2}{f_x}\right)^4 = \frac{f_x}{2} \tag{7-319}$$

Separating the variables and rearranging, we have

$$\frac{2}{f_x}\,d\left(\frac{2}{f_x}\right)^4 = \frac{72}{7}(8.7)^7\,dRe_x \tag{7-320}$$

or

$$4\left(\frac{2}{f_x}\right)^4 d\left(\frac{2}{f_x}\right) = \frac{72}{7}(8.7)^7\,dRe_x \tag{7-321}$$

As a first approximation, this equation is assumed to apply over the whole plate length with $2/f_x = 0$ at $Re_x = 0$. Utilizing this boundary condition, the solution for f_x is

$$\frac{4}{5}\left(\frac{2}{f_x}\right)^5 = \frac{72}{7}(8.7)^7\,Re_x \qquad \text{or} \qquad f_x = 0.0581\,Re_x^{-0.2} \tag{7-322,323}$$

This expression lies only 1.5% below a commonly used empirical correlation given by [11]

$$f_x = 0.0592\,Re_x^{-0.2} \tag{7-324}$$

Equation (7-324) is shown to be in good agreement with experimental data in Fig. 7-34 for values of Re_x from transition up to about 10^7.

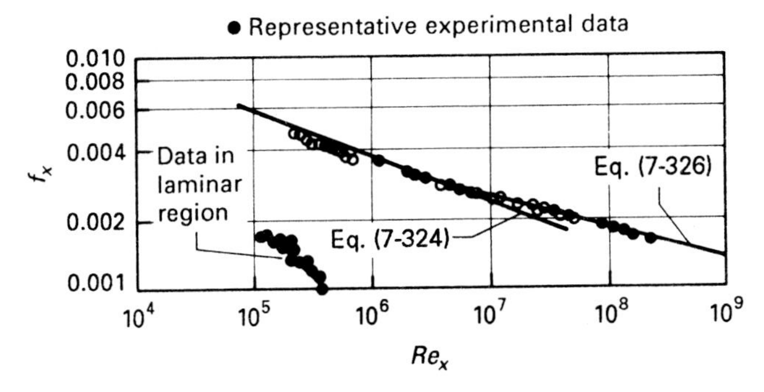

FIGURE 7-34 Friction factor f_x—turbulent boundary layer flow over a flat plate with uniform free-stream velocity.

Integral soutions for f_x have been developed that correlate the data over a broader Reynolds number range by utilizing more accurate inputs for u^+, such as the logarithmic inner law, the Coles' law of the wake, and calculations based on the $\overline{\tau}$-approximation method. For example, the use of the logarithmic inner law gives

$$Re_x = \frac{e^{-\kappa C}}{\kappa^3}\left[e^{\kappa\sqrt{2/f_x}}\left(\kappa^2\frac{2}{f_x} - 4\kappa\sqrt{\frac{2}{f_x}} + 6\right) - 6 - 2\kappa\sqrt{\frac{2}{f_x}}\right] \tag{7–325}$$

This equation represents the data quite well over the entire Reynolds number range for turbulent flow. White [16] has developed an explicit correlation for f_x that is in very good agreement with Eq. (7–325), which takes the form

$$f_x = \frac{0.455}{\ln^2(0.06\,Re_x)} \tag{7–326}$$

This equation is shown to be in good agreement with experimental data in Fig. 7–34.

The integral approach introduced in this section for turbulent boundary layer flow is generalized to account for nonuniform free-stream velocity and transpiration in reference 89.

Energy Transfer The classical differential formulation for energy transfer associated with fully turbulent boundary layer flow over a flat plate with uniform heating maintained over the plate is given by Eq. (7–240) and accompanying boundary conditions; that is,

$$\rho c_P\left(\overline{u}\frac{\partial\overline{T}}{\partial x} + \overline{v}\frac{\partial\overline{T}}{\partial y}\right) = \frac{\partial}{\partial y}\left[(k + k_t)\frac{\partial\overline{T}}{\partial y}\right] \tag{7–327}$$

$$\overline{T} = T_\infty \quad \text{at } x = 0 \qquad \overline{T} = T_\infty \quad \text{as } y \to 0 \tag{7–328,329}$$

and

$$-k\frac{\partial\overline{T}}{\partial y} = \overline{q_c''} \quad \text{at } y = 0 \tag{7–330}$$

for specified wall-heat flux. The formulation is closed by the specification of k_t or α_t. For fluids with moderate values of Prandtl number, we set Pr_t equal to unity, as a first approximation; that is,

$$\alpha_t = \frac{k_t}{\rho c_P} = \nu_t \tag{7–331}$$

Dimensionless Mean Temperature Distribution Integrating Eq. (7–327), we obtain

$$(\alpha + \alpha_t)\frac{\partial\overline{T}}{\partial y} = \int_0^y\left(\overline{u}\frac{\partial\overline{T}}{\partial x} + \overline{v}\frac{\partial\overline{T}}{\partial y}\right)dy - \frac{\overline{q_c''}}{\rho c_P} \tag{7–332}$$

In the inner region in which the integral on the right side of this equation is small, this equation reduces to

$$-(\alpha + \alpha_t)\frac{d\overline{T}}{dy} = \frac{\overline{q''_c}}{\rho c_P} \qquad \begin{array}{l}\text{Inner region}\\ y/\delta \lesssim 0.2\end{array} \tag{7–333}$$

By expressing $\overline{T}$ in terms of the dimensionless temperature profile T^+, we obtain

$$\left(\frac{\alpha}{\nu} + \frac{\alpha_t}{\nu}\right)\frac{dT^+}{dy^+} = 1 \qquad \begin{array}{l}\text{Inner region}\\ y/\delta \lesssim 0.2\end{array} \tag{7–334}$$

or

$$T^+ = \int_0^{y^+} \frac{1}{1/Pr + (\nu_t/\nu)/Pr_t}\, dy^+ \tag{7–335}$$

which is identical to our result within the inner region for fully turbulent TFD flow between parallel plates, Eq. (7–285). As we have already seen, the solution to this equation in the intermediate region takes the form

$$T^+ = \frac{1}{\kappa}\ln y^+ + B \qquad \begin{array}{l}\text{Intermediate region}\\ y^+ \gtrsim 50,\ y/\delta \lesssim 0.2\end{array} \tag{7–336}$$

for fluids other than liquid metals, where B can be approximated in terms of Pr by Eq. (7–291). This logarithmic equation can be approximated by the one-seventh power law

$$T^+ = B_1 y^{+1/7} \tag{7–337}$$

where $B_1 = (9.78 + B)/1.75$. In the wall region, T^+ can be obtained by the numerical integration of Eq. (7–335) with ν_t specified by equations, such as the van Driest expression, that apply in the zone $y^+ \lesssim 50$. It is the solution of Eq. (7–335) within this region that provides us with calculations for B. Calculations are given in Table 7–2 for B and B_1.

Equations (7–335), (7–336), and (7–337) are compared with an empirical correlation of experimental data for turbulent boundary layer flow of air in Fig. 7–35. These expressions are seen to be in good agreement with the correlation within the turbulent core.

The inner laws for T^+ are generally utilized to approximate the dimensionless temperature distribution in the outer region. To obtain more accurate calculations within this region, the underlying differential equations can be solved numerically or approximations can be used for the distribution in $\overline{q''}$ that apply across the entire boundary layer.

Integral Solution for Nusselt Number Integral solutions are developed in references 92 and 93 for turbulent thermal boundary layer flow. To develop a simple integral solution for the local mean Nusselt number for fully turbulent boundary

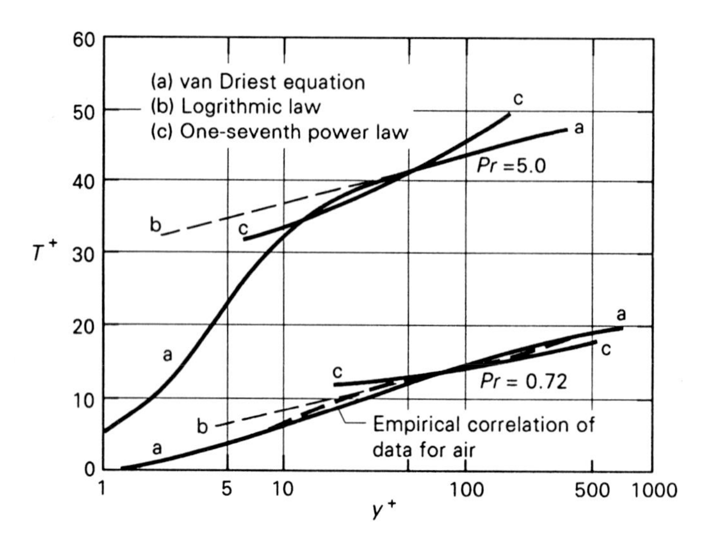

FIGURE 7–35 Dimensionless temperature distribution for fully turbulent boundary layer flow over flat plates.

layer flow, we start with the integral energy equation for turbulent flow,

$$\frac{d}{dx}\int_0^{\Delta} \overline{u}(\overline{T} - T_\infty)\,dy = \frac{\overline{q_c''}}{\rho c_P} \tag{7–244}$$

Following the approach that was utilized in our integral solution for f_x, this equation is written in terms of u^+ and T^+ as follows:

$$\nu \frac{d}{dx}\int_0^{\Delta^+} u^+\left[\frac{(\overline{T} - T_0) - (T_\infty - T_0)}{\overline{q_c''}/(\rho c_P U^*)}\right]\frac{\overline{q_c''}}{\rho c_P U^*}\,dy^+ = \frac{\overline{q_c''}}{\rho c_P}$$

or

$$\frac{d}{dRe_x}\left[\frac{\overline{q_c''}}{\sqrt{f_x/2}}\int_0^{\Delta^+} u^+(T_\infty^+ - T^+)\,dy^+\right] = \overline{q_c''} \tag{7–338}$$

Setting $\Delta^+ = 0$ at $x = 0$, the solution to Eq. (7–338) for arbitrarily specified wall-heat flux $\overline{q_c''}$ is given by

$$\sqrt{\frac{2}{f_x}}\int_0^{\Delta^+} u^+(T_\infty^+ - T^+)\,dy^+ = \frac{1}{\overline{q_c''}}\int_0^{Re_x} \overline{q_c''}\,dRe_x$$

or

$$\int_0^{\Delta^+} u^+(T_\infty^+ - T^+)\,dy^+ = I_\Delta \tag{7–339}$$

where

$$I_\Delta = \sqrt{\frac{f_x}{2}}\frac{1}{\overline{q_c''}}\int_0^{Re_x} \overline{q_c''}\, dRe_x \tag{7-340}$$

With u^+ and T^+ specified, this equation provides us with calculations for Δ^+ in terms of Re_x. Then, setting $\overline{T} = T_\infty$ at $y = \Delta$, our defining equation for T^+ gives

$$T_\infty^+ = (T_s - T_\infty)\frac{\rho c_P U^*}{\overline{q_c''}} \tag{7-341}$$

or

$$Nu_x = \frac{\overline{q_c''}}{T_s - T_\infty}\frac{x}{k} = \frac{\sqrt{f_x/2}\, Re_x\, Pr}{T_\infty^+} \tag{7-342}$$

To illustrate, with u^+ and T^+ approximated by the logarithmic laws, Eqs. (7–313) and (7–336), Eq. (7–339) gives rise to

$$\begin{aligned} I_\Delta &= \int_0^{\Delta^+}\left(\frac{1}{\kappa}\ln y^+ + C\right)\left(\frac{1}{\kappa}\ln \Delta^+ - \frac{1}{\kappa}\ln y^+\right) dy^+ \\ &= \Delta^+\left(\frac{1}{\kappa^2}\ln \Delta^+ + \frac{C}{\kappa} - \frac{2}{\kappa^2}\right) \end{aligned} \tag{7-343}$$

Setting $\kappa = 0.41$ and $C = 5$, this equation reduces to

$$I_\Delta = 5.95\Delta^+ \ln \Delta^+ + 0.297\Delta^+ \tag{7-344}$$

which is readily solved for I_Δ in terms of Δ^+. The resulting calculations are correlated to within 1% by

$$I_\Delta = 17.6\Delta^{+9/8} \quad \text{or} \quad \Delta^+ = \left(\frac{I_\Delta}{17.6}\right)^{8/9} = 0.0781 I_\Delta^{8/9} \tag{7-345,346}$$

Using this result together with Eqs. (7–291), (7–336), and (7–342), we obtain

$$\begin{aligned} Nu_x &= \frac{\sqrt{f_x/2}\, Re_x\, Pr}{(1/\kappa)\ln \Delta^+ + B} = \frac{\sqrt{f_x/2}\, Re_x\, Pr}{(1/0.41)\ln (0.0781 I_\Delta^{8/9}) + 12.7\, Pr^{2/3} - 7.7} \\ &= \frac{\sqrt{f_x/2}\, Re_x\, Pr}{2.17 \ln I_\Delta + 12.7\, Pr^{2/3} - 13.9} \end{aligned} \tag{7-347}$$

where I_Δ is specified in terms of $\overline{q_c''}$ and Re_x by Eq. (7–340). Setting $\overline{q_c''} = \overline{q_0''}$ for the case of uniform wall-flux heating, Eq. (7–347) becomes

$$Nu_x = \frac{\sqrt{f_x/2}\, Re_x\, Pr}{2.17 \ln (\sqrt{f_x/2}\, Re_x) + 12.7\, Pr^{2/3} - 13.9} \tag{7-348}$$

This equation is shown in Fig. 7–36 to lie approximately 10% above experimental data for turbulent boundary layer flow of air over a flat plate with uniform wall-

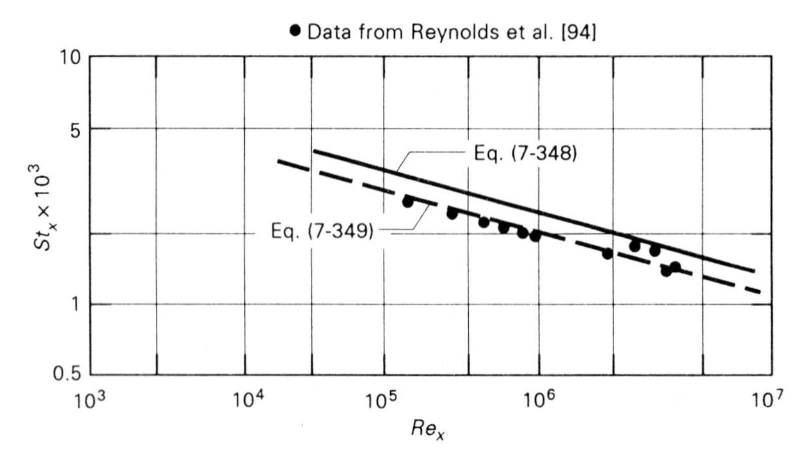

FIGURE 7–36 Comparison of relations for Stanton number with experimental data for turbulent flow of air over a flat plate with uniform free-stream velocity and uniform wall temperature.

temperature heating. This result reinforces our earlier conclusion that the Nusselt number (or Stanton number) is less sensitive to the form of the thermal boundary condition for turbulent flow than for laminar flow. In this connection, these data for uniform wall temperature are in good agreement with the following correlation developed by White [16]:

$$Nu_x = \frac{(f_x/2)\, Re_x\, Pr}{1 + 12.7\sqrt{f_x/2}\,(Pr^{2/3} - 1)} \qquad \text{for } 0.5 \lesssim Pr \tag{7–349}$$

The Reynolds Analogy

The *Reynolds analogy* provides a practical relation between heat and momentum transfer for turbulent boundary layer flow over isothermal surfaces of fluids with Prandtl number of the order of unity. To develop this relation we first combine Eqs. (7–223) and (7–238), with the result

$$\frac{\overline{q''}}{\overline{\tau}} = -\frac{(k + k_t)\,\partial\overline{T}/\partial y}{(\mu + \mu_t)\,\partial\overline{u}/\partial y} = -\frac{k + k_t}{\mu + \mu_t}\frac{d\overline{T}}{d\overline{u}} \tag{7–350}$$

or

$$\frac{\overline{q''}}{\overline{\tau}} = -c_P\left(\frac{k/k_t + 1}{Pr\, k/k_t + Pr_t}\right)\frac{d\overline{T}}{d\overline{u}} \tag{7–351}$$

Since $k_t >> k$ throughout most of the boundary layer, this equation can be approximated by

$$\frac{\overline{q''}}{\overline{\tau}} = -\frac{c_P}{Pr_t}\frac{d\overline{T}}{d\overline{u}} \tag{7–352}$$

Furthermore, since the eddy diffusivity ν_t and eddy thermal diffusivity α_t both result from the same mechanism of transverse fluctuation, $\alpha_t \simeq \nu_t$ such that $Pr_t \simeq 1$ and Eq. (7–352) reduces to

$$\frac{q''}{\overline{\tau}} = -c_P \frac{dT}{d\overline{u}} \tag{7–353}$$

Finally, assuming that the ratio $\overline{q''}/\overline{\tau}$ is essentially constant across the boundary layer, Eq. (7–353) is integrated to obtain

$$\frac{\overline{q_c''}}{\overline{\tau}_s} = -c_P \frac{T_F - T_s}{U_F} \tag{7–354}$$

or

$$h_x = \frac{\overline{q_c''}}{T_s - T_F} = \frac{\overline{\tau}_s c_P}{U_F} \tag{7–355}$$

and

$$Nu_L = \frac{h_x L}{k} = \frac{f_x}{2} Re_L \, Pr \tag{7–356}$$

which is a general form of the Reynolds analogy. Equation (7–356) can be combined with relations for friction factor to determine the Nusselt number for standard situations involving fluids with moderate values of Prandtl number.

To illustrate, with f approximated by Eq. (7–272),

$$f = 0.046 \, Re^{-0.2} \tag{7–357}$$

for fully turbulent HFD flow in a circular tube with $L = D$, Nu becomes

$$Nu = \frac{hD}{k} = 0.023 \, Re^{0.8} \, Pr \tag{7–358}$$

Similarly, using Eq. (7–324),

$$f_x = 0.0592 \, Re_x^{-0.2} \tag{7–359}$$

for fully turbulent flow over a flat plate with $L = x$, we obtain

$$Nu_x = \frac{h_x x}{k} = 0.0296 \, Re_x^{0.8} \, Pr \tag{7–360}$$

These relations prove to be quite reasonable for fluids with values of Pr approximately equal to unity. However, because of the underlying assumptions, the accuracy of the Reynolds analogy decreases as the difference $|Pr - 1|$ increases. For example, the error in the Reynolds analogy reaches 10% for $0.85 \lesssim Pr \lesssim 1.2$ and 20% for $0.27 \lesssim Pr \lesssim 1.3$. Recognizing this deficiency and making use of experimental information, Colburn [95] proposed an analogy of the form

$$Nu_L = \frac{f_x}{2} Re_L \, Pr^{1/3} \tag{7–361}$$

The Colburn analogy proves to be in good agreement with experimental data for fluids with moderate values of Pr. Referring back to the theoretical analysis for flow in tubes and flow over flat plates presented earlier in this section, we see that the deficiencies of the Reynolds analogy can now be dealt with by the specification of the turbulent viscosity μ_t or mixing length ℓ and turbulent Prandtl number Pr_t across the boundary layer.

7-4 COMPUTER ANALYSIS OF CONVECTION HEAT TRANSFER: INTRODUCTION

Numerical methods are commonly employed in the analysis of convection heat transfer. Convection problems for which numerical methods can or must be employed include fully developed flow in tubes with rectangular (or other noncircular) cross section, two- or three-dimensional external or internal boundary layer flows, and complex two- or three-dimensional flows that do not satisfy the boundary-layer approximations. The complexity of these approaches ranges from simple techniques for performing numerical integration, to basic methods for solving ordinary differential equations, to basic explicit or implicit methods for solving the boundary layer equations, to comprehensive methods for solving complex convection flows.

In dealing with convection heat transfer, it is the advective terms that appear in the energy equation and the fact that the fluid flow must be characterized by continuity and momentum equations that distinguish the analysis from those developed in Chap. 4 for conduction. In addition to special problems associated with the numerical treatment of the advective terms, the analysis of two- and three-dimensional convection processes is further complicated by (1) the linkage of the energy equation to the continuity and momentum equations, (2) the nonlinear nature of the resulting nodal equations for energy and momentum transfer, and (3), in the case of turbulent flow, the need for appropriate turbulence modeling. Approaches to dealing with these complications are well established for basic boundary layer flows and appear to be reasonably well in hand for many complex flow processes.

The use of numerical methods to develop algebraic models (algorithms) for fluid flow and convection heat-transfer problems is commonly referred to as *computational fluid dynamics* CFD. The primary CFD approaches are the finite-difference method (i.e., discretization or control-volume type) and the finite-element method. Whereas the finite-difference method is the most familiar and commonly used numerical approach for solving convection heat-transfer problems, the finite-element method is becoming more widely used. General-purpose finite-difference and finite-element codes are presently commercially available that are capable of simulating convection heat-transfer processes involving one, two, and three dimensions, steady and unsteady conditions, laminar and turbulent flow, regular and irregular geometries, incompressible and compressible fluids, single- and multiphase flow, chemical reaction, and combustion. As pointed out by Pepper and Baker [96], finite-difference methods are relatively simple to formulate, can be readily generalized from one to two or three dimensions, can be adapted with reasonable accuracy to irregular boundaries by the use of boundary-fitted coordinates, and require considerably less computational work than finite-element methods (for equivalent nodes). The primary advantage of the finite-element method is its generality and capability of handling complex boundary shapes. The formulation of finite-difference and finite-element methods for convection heat transfer is introduced in references 97 and 98.

In connection with the numerical treatment of the advective terms, both finite-difference and finite-element methods are susceptible to destabilizing computational errors (known as *dispersion errors*) that can arise from the numerical approximation of large spatial gradients. Thus, an advection stability criterion generally must be imposed on the grid spacing for any particular numerical method or steps must be taken to eliminate the problem without unduly compromising the accuracy of the solution. Otherwise, artificial oscillations or wiggles will appear in the calculations that obscure the actual solution. The special problem of achieving stable but accurate numerical solutions for convection heat transfer is considered in references 97 and 99.

7–5 SUMMARY

For purposes of analysis, the two primary categories in convection-heat-transfer systems include *laminar flow* and *turbulent flow* conditions. Whereas the mathematical formulations for laminar flow systems are generally quite straightforward, the solution of the resulting equations can be quite complex. For the simplest problems, as typified by the fully developed tube flow problem introduced in Sec. 7–2–1, analytical solutions can be readily obtained. Analytical solutions to a number of simple laminar convection problems are presented in textbooks by Kays and Crawford [11], White [16], Burmeister [17], and others. Somewhat more complex laminar flow problems involving hydrodynamic and/or thermal development, such as the forced and natural convection boundary layer flows considered in Secs. 7–2–2 and 7–2–3, can be fairly easily solved by means of the integral approach. Analytical solutions can even be developed for some of these types of problems by introducing similarity coordinates. These problems of intermediate complexity are also sometimes solved numerically. The most complex problems involving compressibility effects, variable properties, axial conduction, strong adverse and favorable pressure gradients, mass transfer through a wall, unsteady conditions, and other complications are generally handled by numerical methods.

The analysis of turbulent convection heat transfer involves the conceptual problem of how to mathematically model the complex turbulent transport process itself. The classical mean-field approach with eddy viscosity μ_t and eddy thermal conductivity k_t represented by algebraic relations proves to be adequate for basic boundary layer flows, with higher-order methods required for more complex problems. Once μ_t (or mixing length ℓ) and k_t (or turbulent Prandtl number Pr_t) are known, the solution of turbulent convection heat-transfer problems follows the same basic pattern as for laminar flow. Analytical techniques can be utilized for the simplest problems. But numerical and integral techniques are generally called on to handle the more complex problems, with numerical approaches becoming more and more dominant.

The finite-difference method is the simplest and most commonly used numerical approach to analyzing convection heat transfer. The finite-element method is also becoming more widely used. Several powerful general-purpose finite-difference and finite-element codes are presently commercially available for analyzing a broad range of convection processes. However, as pointed out by Taborek and Palen [100], computers can be employed to achieve superior results providing that the software is considered strictly a tool, subject to engineering guidance and careful result evaluation. The effective use of computer programs for solving convection heat-transfer problems requires an understanding of the theory of convection and basic numerical concepts.

CHAPTER 8

CONVECTION HEAT TRANSFER: PRACTICAL ANALYSIS—FORCED CONVECTION

8–1 INTRODUCTION

As we have seen, convection processes are categorized according to the geometry of the heat-transfer surface. Three basic categories of forced-convection systems include (1) internal flow in circular tubes, annuli, and noncircular tubular passages with uniform cross-sectional area; (2) external flow over surfaces such as flat plates, cylinders, and spheres; and (3) flow across tube banks. The practical thermal and hydraulic analysis of each of these types of forced-convection processes will be considered in this chapter, with emphasis given to ideal fluids.

8–2 INTERNAL FLOW

As indicated in Chap. 6, the practical analysis approach involves the use of coefficients of friction, pressure drop, and heat transfer. Relations for the convection coefficients that are used in the practical analysis of flow through tubular passages are considered in Sec. 8–2–1. To complete the presentation, the practical thermal and hydraulic analysis approaches, which are used in the evaluation and design of heat-transfer equipment, are developed in Secs. 8–2–2 and 8–2–3 (and Appendix M), with special attention given to variable property flow and fin tubes. The concepts presented in these sections provide a frame of reference for the practical analysis of flow through tube banks and heat-exchanger cores, which is considered in Sec. 8–4 and Chap. 11.

8–2–1 Coefficients of Friction, Pressure Drop, and Heat Transfer

A representative heat-transfer core consisting of a tubular passage connected to inlet and exit headers is illustrated in Fig. 8–1. The *contraction ratio* A/A_1 is designated by σ. To characterize appropriately the hydrodynamic and thermal performance of

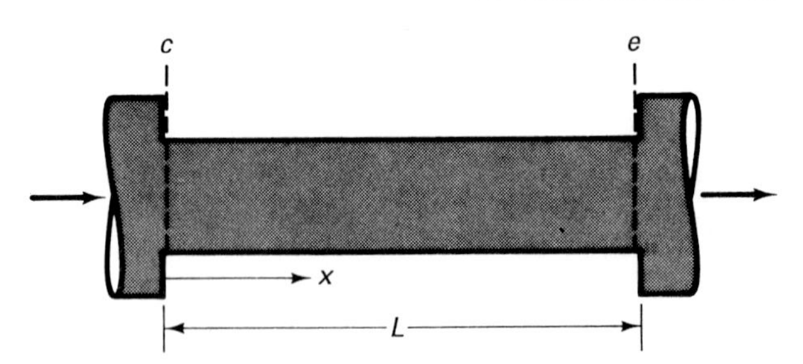

FIGURE 8–1 Passage with abrupt contraction entrance and abrupt expansion exit.

heat-transfer cores such as this, we must account for the effects of flow in the developing and fully developed regions and at the exit.

The coefficient of heat transfer (h, h_x, $\overline{h}$), which is generally expressed in terms of the Nusselt number,

$$Nu = \frac{hD_H}{k} \qquad Nu_x = \frac{h_x D_H}{k} \qquad \overline{Nu} = \frac{\overline{h}D_H}{k} \tag{8–1}$$

and the friction factor (f, f_x, $\overline{f}$) provide important practical information pertaining to developing and fully developed internal flows.

Because the pressure drop is the primary hydrodynamic dependent variable of interest in the evaluation and design of heat-transfer cores, coefficients for pressure drop are also in common use. The pressure drop over a region extending from the entrance at $x = 0$ (where the pressure is represented by P_c) to x is sometimes represented by

$$P_c - P = \frac{px}{A}\frac{f_{\text{app}}}{2}\rho U_b^2 = \frac{2x}{D_H} f_{\text{app}}\, \rho U_b^2 \tag{8–2}$$

where f_{app} is referred to as the *apparent friction factor*. This approach was introduced in 1942 by Langhaar [1]. An alternative approach to characterizing the pressure drop, which is generally used in modern heat exchanger design methods, involves a relationship of the form

$$P_c - P = \frac{\rho U_b^2}{2}\left(K_c + \frac{4x}{D_H} f\right) \tag{8–3}$$

where K_c is the *entrance-loss coefficient* and f is the friction factor for fully developed flow. For heat-transfer cores with exit headers, we must also account for the pressure loss ΔP_{loss} associated with the irreversible free expansion and momentum changes following an abrupt expansion. This pressure loss is generally expressed in terms of the *expansion-loss coefficient* K_e by

$$\Delta P_{\text{loss}} = K_e \frac{\rho U_b^2}{2} \tag{8–4}$$

The practical thermal and hydraulic analysis approaches considered in Secs. 8–2–2 and 8–2–3 feature the use of the heat-transfer coefficients h, h_x, and $\bar{h}$ and the hydrodynamic coefficients f, K_c, and K_e. We now turn our attention to relations that are available in the literature for these and related coefficients. We shall refer to these theoretical and empirical relations as *convection correlations*. Extensive surveys of convection correlations are available in *Engineering Sciences Data* [2], *Handbook of Heat Transfer* [3], *Compact Heat Exchangers* [4], and *Heat Exchanger Design Handbook* [5].

Convection Correlations

Convection correlations for Nusselt number, friction factor, and entrance/expansion loss coefficients appearing in the literature for flow in tubular passages are generally expressed in terms of the Reynolds number ($Re = U_b D_H/\nu = G D_H/\mu$), Prandtl number ($Pr = \mu c_P/k$), dimensionless distance x/D_H, contraction ratio ($\sigma = A/A_1$), geometry, and thermal boundary conditions by relations of the form

$$\textit{Nusselt number} = \text{fn}\,(Re, Pr, x/D_H, \textit{geometry, thermal boundary conditions})$$

$$f = \text{fn}\,(Re, x/D_H, \textit{geometry})$$

$$K_c = \text{fn}\,(Re, x/D_H, \sigma, \textit{geometry}) \tag{8–5}$$

$$K_e = \text{fn}\,(Re, L/D_H, \sigma, \textit{geometry})$$

and are categorized according to whether the flow is laminar or turbulent. The basis for the theoretical relations presented in this section is developed in Chap. 7.

Laminar Flow For practical purposes, hydrodynamic and thermal fully developed flow can be said to occur for [6]

$$\frac{x}{D_H} \gtrsim 0.05\,Re \qquad \text{HFD} \tag{8–6a}$$

$$\frac{x}{D_H} \gtrsim 0.05\,Re\,Pr \qquad \text{TFD} \tag{8–6b}$$

For HFD and TFD laminar flow with uniform properties, f and Nu are given by

$$f = \frac{C_1}{Re} \qquad Nu = \frac{hD_H}{k} = C_2 \tag{8–7,8}$$

where the constants C_1 and C_2 are dependent upon geometry, and C_2 is dependent upon the thermal boundary conditions. Representative values of C_1 and C_2 are given

TABLE 8–1 HFD and TFD laminar flow: Coefficients for Eqs. (8–7) and (8–8)

		Nusselt Number $Nu = C_2$			
Geometry	*Friction factor* $f\,Re = C_1$	*Uniform wall temperature*		*Uniform wall heat flux*	
Square tube	14.2	2.98		3.61	
Circular tube	16.0	3.66		4.36	
Infinite parallel plates	24.0	7.54		8.24	
Circular annulus					
D_i/D_o		Nu_i[a]	Nu_o[b]	Nu_{ii}[c]	Nu_{oo}[d]
0	16.0	∞	3.66	∞	4.364
0.05	21.57	17.46	4.06	17.81	4.792
0.1	22.34	11.56	4.11	11.91	4.834
0.2	23.09			8.499	4.833
0.25		7.37	4.23		
0.4	23.68			6.583	4.979
0.5		5.74	4.43		
0.6	23.90			5.912	5.099
0.8	23.98			5.58	5.24
1.0	24.0	4.86	4.86	5.385	5.385

[a] Nu_i–uniform temperature at inner surface; outer surface insulated.
[b] Nu_o–uniform temperature at outer surface; inner surface insulated.
[c] Nu_{ii}–uniform heat flux at inner surface; outer surface insulated.
[d] Nu_{oo}–uniform heat flux at outer surface; inner surface insulated.

in Table 8–1 and in Fig. 8–2. Note that $C_1 = 16$ for flow in a circular tube, with $C_2 = 3.66$ for uniform wall-temperature heating, and $C_2 = 4.36$ for uniform wall-flux heating. Thus, for this particular geometry, the Nusselt number for uniform wall-heat flux is nearly 20% greater than for uniform wall temperature.

It should be noted that because the convection coefficients vary around the perimeter for rectangular passages and other noncircular tubes, f and h are taken as averages over the perimeter.

In the case of fully developed laminar flow in annuli, Table 8–1 and Fig. 8–2 provide correlations for Fanning friction factor f (averaged over both surfaces) and Nusselt number as a function of the diameter ratio D_i/D_o. The correlations for Nusselt number represented by Nu_{ii}, Nu_i, Nu_{oo}, and Nu_o pertain to four basic boundary conditions for which one of the two surfaces is insulated; these include Nu_{ii}—uniform flux at inner surface; Nu_i—uniform temperature at inner surface; Nu_{oo}—uniform flux at outer surface; and Nu_o—uniform temperature at outer surface. Correlations are available in references 3 to 6 that express the Nusselt numbers at the inner and outer surfaces in terms of these basic solutions for situations in which both surfaces are heated or cooled.

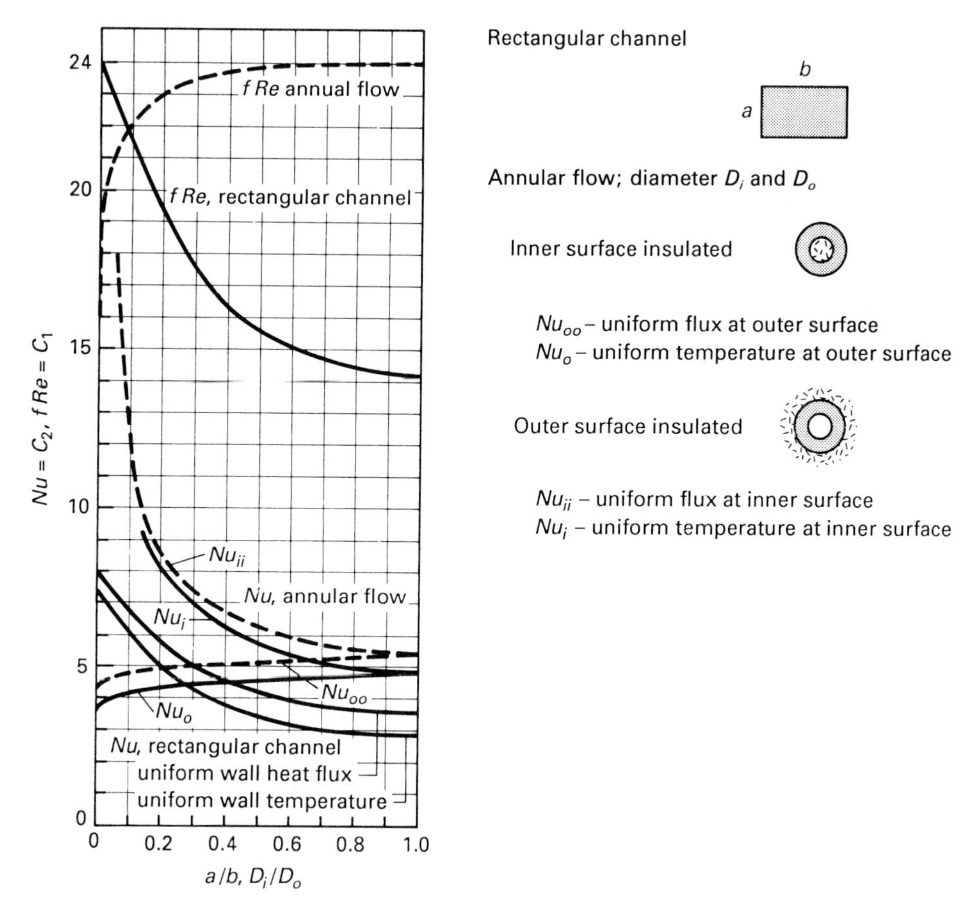

FIGURE 8–2 Nusselt number and Fanning friction factor for laminar fully developed flow (Kays and Clark [7] and Lundberg et al. [8]).

Turning to hydrodynamic developing flow, for the case in which the frontal area A_1 is much greater than the cross-sectional area A of the tubular passage (i.e., $\sigma \simeq 0$), the velocity distribution within the passage develops from an essentially uniform profile at $x = 0$ to a fully developed profile. This problem has been studied for the case of laminar flow in a circular tube by Langhaar [1], Sparrow et al. [9–11], and others. Solution results obtained by Langhaar for f_x, $\bar{f}$, and f_{app} are shown in Fig. 8–3. This solution indicates that f_x asymptotically falls toward the HFD value f as $(x/D)/Re$ approaches a value of 0.05. This result provides the basis for the criterion for HFD conditions given by Eq. (8–6). The entrance-loss coefficient K_c for this case is given in Fig. 8–4. Entrance-loss coefficients have also been obtained for flow in circular tubes with nonzero values of the contraction ratio σ. Correlations are shown in Fig. 8–5 for K_c as a function of σ and Re for flow in one or more circular tubes.

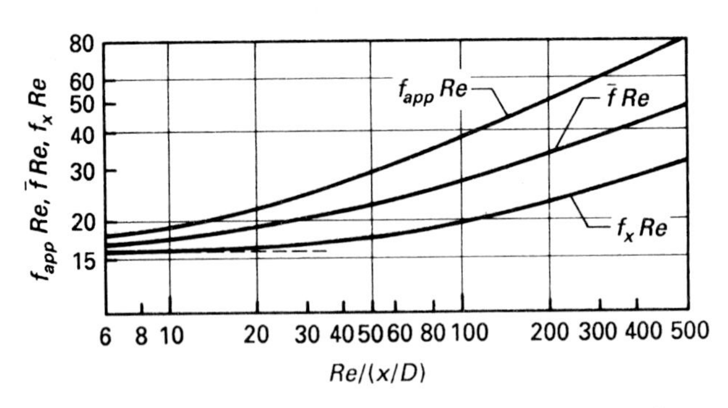

FIGURE 8–3 Calculations for coefficients of friction by Langhaar [1]—hydrodynamic developing laminar flow in a circular tube.

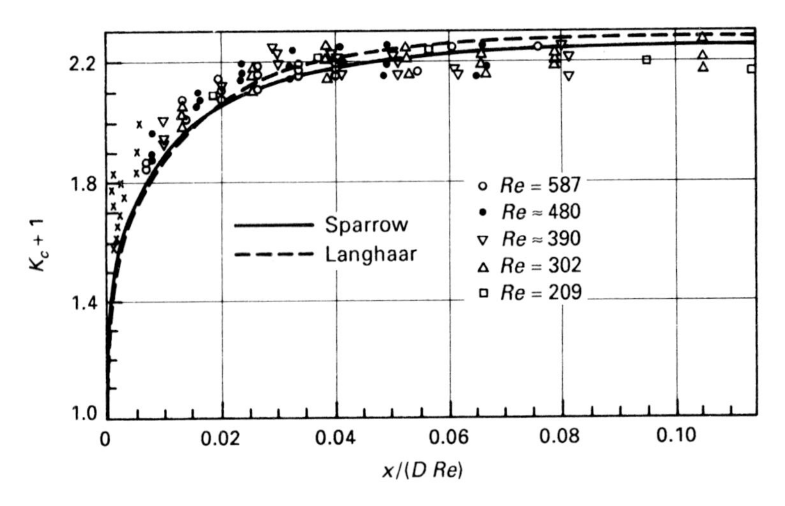

FIGURE 8–4 Entrance-loss coefficient K_c for laminar flow in a circular tube with $\sigma = 0$ (McComas and Eckert [12]).

This figure also shows correlations for the expansion-loss coefficient K_e. Entrance-loss and expansion-loss coefficients associated with other cross-sectional geometries are given in references 4 and 12 through 14.

Theoretical and empirical correlations are available in the literature for the coefficient of heat transfer for thermal developing laminar flow in circular tubes and other geometries. Typical of the design equations that are available is the following popular correlation by Hausen [15] for HFD flow in a circular tube with uniform wall-temperature heating:

$$\overline{Nu} = 3.66 + \frac{0.0668 \dfrac{Re\,Pr}{x/D}}{1 + 0.04\left(\dfrac{Re\,Pr}{x/D}\right)^{2/3}} \tag{8–9}$$

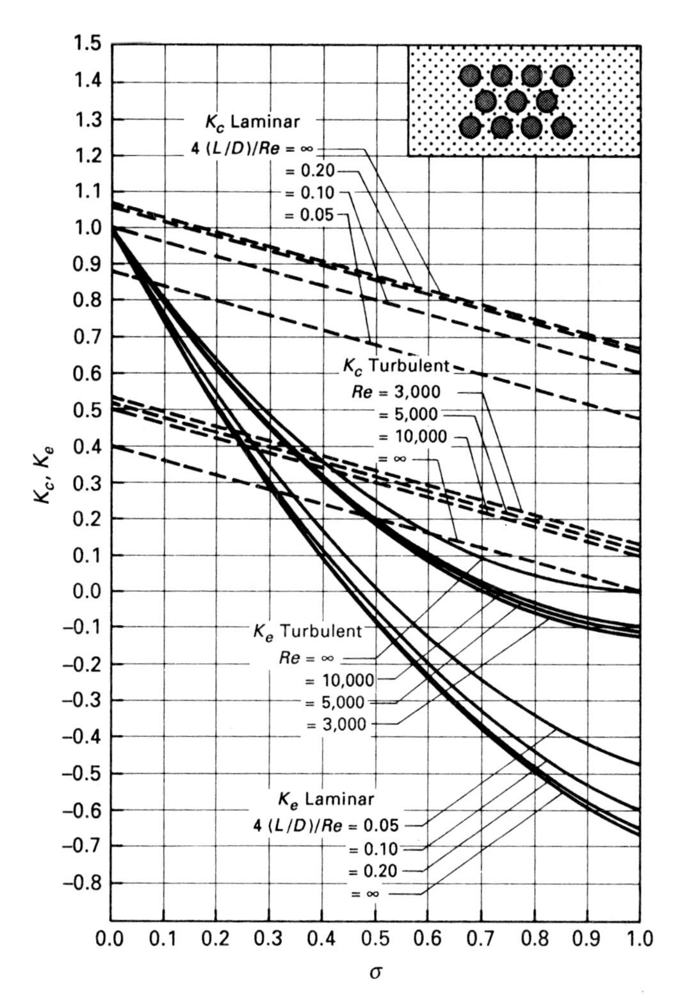

FIGURE 8–5 Entrance-loss and expansion-loss coefficients for a single or multiple-circular tube heated core with abrupt-contraction entrance and abrupt-expansion exit. (From Kays and London [4]. Used with permission.)

For small values of x/D, this equation reduces to

$$\overline{Nu} = 1.67 \left(\frac{Re\, Pr}{x/D} \right)^{1/3} \qquad \text{for} \quad \frac{x/D}{Re\, Pr} \lesssim 0.01 \tag{8–10}$$

These equations are shown in Fig. 8–6. For thermal developing HFD laminar flow with uniform wall-flux heating in short tubes, the local Nusselt number Nu_x can be approximated by [16]

$$Nu_x = 1.30 \left(\frac{Re\, Pr}{x/D} \right)^{1/3} \qquad \text{for} \quad \frac{x/D}{Re\, Pr} \lesssim 0.01 \tag{8–11}$$

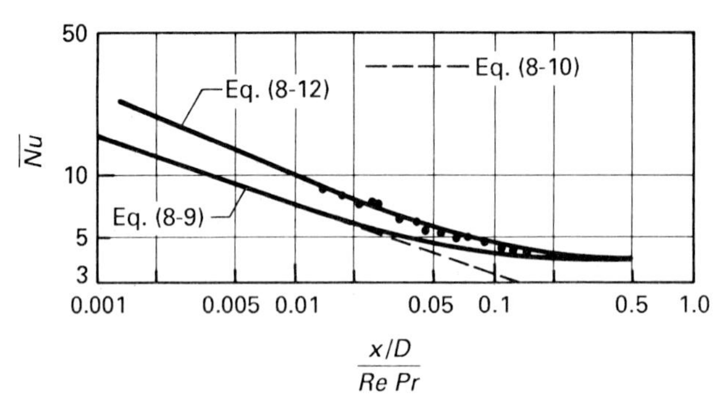

FIGURE 8–6 Correlations and experimental data for thermal developing flow with uniform wall-temperature heating. (Data for air ($Pr = 0.72$) from Kays [17].)

Correlations are also available for combined hydrodynamic and thermal developing laminar flows. For example, Kays [17] has developed a correlation for developing laminar flow of air in a circular pipe with uniform wall temperature of the form

$$\overline{Nu} = 3.66 + \frac{0.104 \dfrac{Re\ Pr}{x/D}}{1 + 0.016 \left(\dfrac{Re\ Pr}{x/D}\right)^{0.8}} \tag{8–12}$$

This equation is compared with Eq. (8–9) and with experimental data for developing flow in Fig. 8–6. Note that the Nusselt number is larger for hydrodynamic developing flow than for HFD flow. Another correlation for developing flow with constant wall temperature, which has been widely used for fluids with $Pr \gtrsim 0.5$, is given by [18]

$$\overline{Nu} = 1.86 \left(\frac{Re\ Pr}{x/D}\right)^{1/3} \qquad \text{for } \frac{x/D}{Re\ Pr} \lesssim 0.01 \tag{8–13}$$

It should be noted that Eqs. (8–9) to (8–13) do not account for the effects of axial conduction. Consequently, these equations are generally restricted to situations in which the *Peclet number* $Pe = Re\ Pr$, which represents the *ratio of enthalpy-flow rate to heat-conduction rate*, is not small; that is, $Pe \gtrsim 100$. The parameter $Re\ Pr/(x/D) = Pe/(x/D)$ appearing in these equations is the *Graetz number* Gz. Equations (8–10), (8–11), and (8–13) are most accurate for values of $Gz \gtrsim 100$ (i.e., $x/D \lesssim 0.01\ Re\ Pr$). However, for preliminary design calculations involving fluids with $Pe > 100$, these equations can be extended to $Gz \gtrsim 20$ (i.e., $x/D \lesssim 0.05\ Re\ Pr$), and Nu can be approximated by the TFD value for $Gz \lesssim 20$ (i.e., $x/D \gtrsim 0.05\ Re\ Pr$).

Turbulent Flow Fully turbulent flow generally occurs for $Re \gtrsim 10^4$, with the flow being transitional turbulent in the range $2000 \lesssim Re \lesssim 10^4$.

Hydrodynamic and thermal developed turbulent flow are established over much shorter distances than for laminar flow. For example, HFD conditions occur for x/D greater than a value of about 10.† TFD conditions occur over about the same distance for gases such as air, with the length of the thermal entrance region decreasing as the Prandtl number Pr increases, and vice versa.

The Fanning friction factor for fully developed and fully turbulent flow in smooth tubes is well represented by the Prandtl/Nikuradse equation [20],

$$\sqrt{\frac{2}{f}} = 2.46 \ln\left(Re\sqrt{\frac{f}{2}}\right) + 0.292 \tag{8–14}$$

This implicit equation has been shown to be very closely approximated by the explicit formula [21]

$$f = (1.58 \ln Re - 3.28)^{-2} \tag{8–15}$$

Equation (8–15) is compared with experimental data in Fig. 8–7.

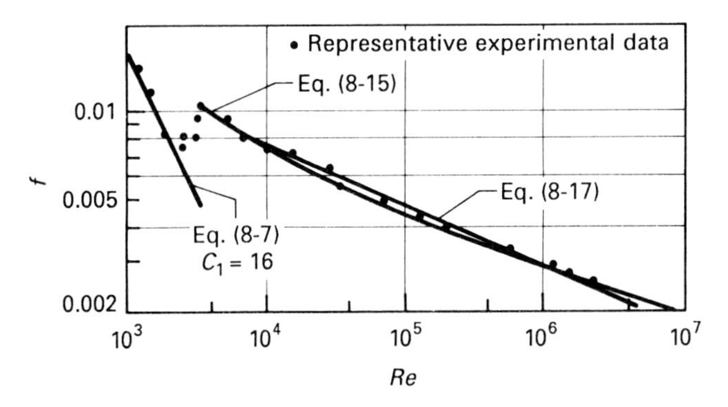

FIGURE 8–7 Correlations and experimental data for friction factor f—HFD flow in circular tubes.

For the Nusselt number, the following correlation by Petukhov and Kirillov [22] and White [23] is recommended:

$$Nu = \frac{(f/2)\,Re\,Pr}{1.07 + 12.7\sqrt{f/2}\,(Pr^{2/3} - 1)} \tag{8–16}$$

This equation is reported to have an accuracy of 5 to 6% in the ranges $10^4 \lesssim Re \lesssim 5 \times 10^6$ and $0.5 \lesssim Pr \lesssim 200$, and 10% for the same Reynolds number range and for $200 \lesssim Pr \lesssim 2000$ [24]. Equation (8–16) is shown in Fig. 8–8, with f given by Eq. (8–15). Notice that the Nusselt number is much larger for turbulent flow than for laminar flow.

† Shah and Bhatti [19] suggest the following criterion for the turbulent hydrodynamic entrance length:

$$\frac{x}{D} \gtrsim 1.36\,Re^{0.25} \qquad \text{for } Re > 10^4$$

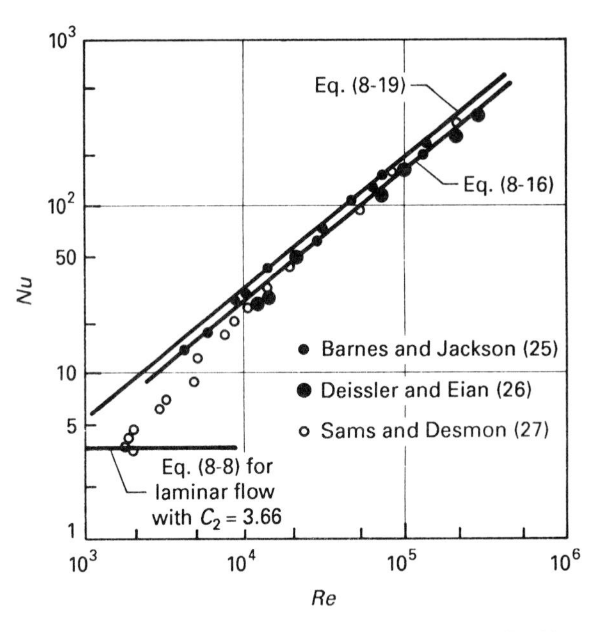

FIGURE 8–8 Correlations and experimental data for Nusselt number—TFD turbulent flow of air.

Simpler but somewhat less accurate correlations for the Fanning friction factor f and the Nusselt number Nu for fully developed and fully turbulent flow ($Re \gtrsim 10^4$) in smooth tubes are given by Eq. (7–272),

$$f = 0.046\, Re^{-0.2} \tag{8–17}$$

and

$$Nu = \frac{f}{2} Re\, Pr^n \qquad \text{or} \qquad Nu = 0.023\, Re^{0.8}\, Pr^n \tag{8–18,19}$$

where $n = 0.5$ for $0.5 \lesssim Pr \lesssim 5.0$ and $n = 1/3$ for $Pr \gtrsim 5.0$. The earliest correlations of the form of Eq. (8–19) were developed by Dittus and Boelter [28] in 1930 ($n = 0.4$ for heating and $n = 0.3$ for cooling) and Colburn [29] in 1933 ($n = 1/3$). With n set equal to 1/3, Eq. (8–18) is known as the *Colburn analogy*. Equations (8–17) and (8–19) are compared with experimental data in Figs. 8–7 and 8–8.

The convection correlations introduced in this section for fully developed flow in circular tubes can be applied with reasonable accuracy to noncircular tubes by expressing the Reynolds number Re and Nusselt number Nu in terms of the hydraulic diameter D_H. However, as in the case of laminar flow processes, f and h for turbulent flow in noncircular cross sections are taken as averages over the perimeter. In this connection, for TFD fully turbulent conditions, Nu is essentially independent of the form of the thermal boundary conditions (relative to the streamwise direction, x), except for low-Prandtl number fluids such as liquid metals. Hence, the convection correlations can be used for uniform wall-flux or uniform wall-temperature conditions.

In the case of fully developed annular flow, the Fanning friction factor and Nusselt number are dependent upon the diameter ratio D_i/D_o, as well as on Re and Pr. However, f (averaged over both surfaces) and Nu_{oo} and Nu_o (for moderate to high values of Pr) are only weakly dependent upon D_i/D_o, such that the convection correlations for f and Nu associated with fully developed flow in circular tubes usually can be employed as a reasonable first approximation. On the other hand, the difference between Nu_{ii} and Nu ranges from small values of the order of 1% for $0.8 \lesssim D_i/D_o$ and moderate values up to about 10% for $0.5 \lesssim D_i/D_o \lesssim 0.8$, to well over 10% for $D_i/D_o \lesssim 0.5$. To illustrate, calculations published by Kays and Leung [30] for Nu/Nu_{ii} and Nu/Nu_{oo} are shown as a function of D_i/D_o and Pr with $Re = 10^5$ in Fig. 8–9. Whereas Nu/Nu_{oo} is generally no smaller than about 0.95 for $D_i/D_o \gtrsim 0.2$, Nu/Nu_{ii} falls significantly below 0.95 for moderate Pr and $D_i/D_o \lesssim 0.5$. Although the ratios Nu/Nu_{ii} and Nu/Nu_{oo} are slightly dependent on Re, Fig. 8–9 can be used to estimate these values for fully turbulent conditions.

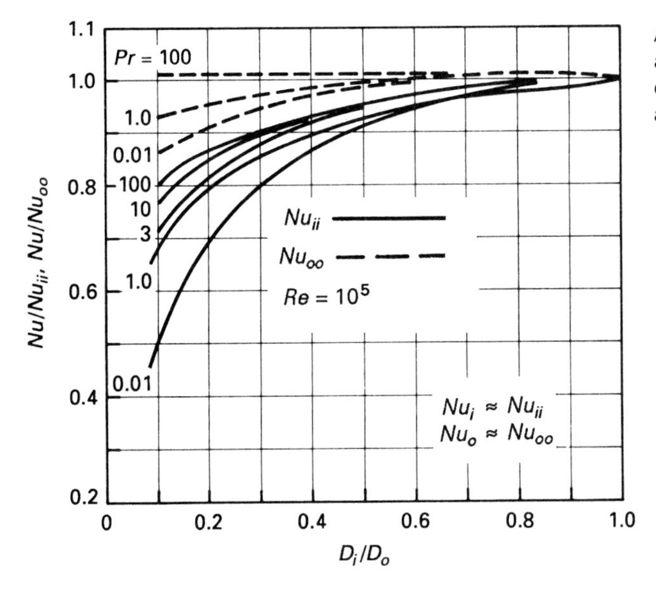

Annular flow; diameters D_i and D_o (see Fig. 8–2 for definitions of Nu_i, Nu_{ii}, Nu_o and Nu_{oo}).

FIGURE 8–9 Calculations for Nu_{ii} and Nu_{oo} associated with fully developed turbulent annular flow; Nu represents the Nusselt number for fully developed flow between parallel plates or in a circular tube (Kays and Leung [30]).

Experimental and theoretical studies have been conducted for HFD transitional turbulent flow by Patel and Head [31] and others. These studies indicate that the correlations for f and Nu begin to separate from the correlations for fully turbulent flow at $Re \lesssim 10^4$ and approach the correlations for laminar conditions as Re falls toward a value of the order of 2000. This trend in the Nusselt number is clearly seen in Figs. 8–7 and 8–8. Because the range in Reynolds number generally encountered in heat-exchanger applications extends from 500 to 15,000, Kays and London [4] provide performance curves that account for transitional turbulence effects. A correlation for the Nusselt number Nu for transitional turbulent internal flows proposed by Taborek [32] features the simple relation

$$Nu = C_{tr}\, Nu_{L2} + (1 - C_{tr})\, Nu_{T8} \tag{8–20a}$$

$$C_{tr} = \frac{Re_T - Re}{Re_T - Re_L} = 1.33 - \frac{Re}{6000} \tag{8–20b}$$

which prorates Nu between the Nusselt number Nu_{L2} for laminar flow at $Re = 2000$ and the Nusselt number Nu_{T8} for fully turbulent flow at $Re = 8000$.†

The friction factor and Nusselt number are dependent on x/D_H, Re and the entrance configuration for developing turbulent flow. Performance curves based on experimental data for air are given by Kays and London [4] for several types of tubes with abrupt-contraction entrances. A convenient correlation for thermal developing flow in short smooth tubes is given by [3]‡

$$\overline{Nu} = Nu\left(1 + \frac{C}{x/D}\right) \qquad \text{for } \frac{x_c}{D} \lesssim \frac{x}{D} \lesssim 60 \tag{8–21a}$$

and

$$\overline{Nu} = Nu\,\frac{1.11\, Re^{0.2}}{(x/D)^{0.8}} \qquad \text{for } x \lesssim x_c \tag{8–21b}$$

where $x_c/D = 0.625\, Re^{0.25}$, $C = 1.4$ for the case in which HFD conditions exist at the entrance and $C = 6$ when no hydrodynamic calming section is used. This correlation is sometimes used in preliminary design calculations for heat exchangers.

To provide a practical means of determining the pressure drop ΔP for developing turbulent flow in heat-exchanger cores with abrupt entrance contraction and exit expansion, Kays [14] has developed relations for the entrance-loss coefficient K_c and the expansion-loss coefficient K_e for a number of standard cross sections. Correlations for K_c and K_e, which are applicable to cores consisting of single or multiple-circular tubes, are given in Fig. 8–5.

† This simple approach is extended to the calculation of friction factor f for transitional turbulent flow by writing

$$f = C_{tr}\, f_{L2} + (1 - C_{tr})\, f_{T4} \qquad C_{tr} = \frac{Re_T - Re}{Re_T - Re_L} = 2 - \frac{Re}{2000}$$

where the friction factors represented by f_{L2} and f_{T4} are evaluated at $Re = 2000$ and $Re = 4000$, respectively. Following this approach, f is approximated by standard fully turbulent correlations for $Re \geq 4000$. As an alternative approach to characterizing HFD transitional turbulent flow, f and Nu are sometimes simply approximated by Eq. (7–266) and Eq. (8–16) or Eq. (8–18).

‡ Although this equation was developed for turbulent flow of air, it is also often used as a first approximation for other fluids, except for liquid metals.

For liquid metals ($0.002 \lesssim Pr \lesssim 0.05$), large amounts of scatter exist in the published data for Nusselt number. A correlation developed by Subbotin et al. [33] and Seban and Shimazaki [34], which is commonly used for design purposes for TFD flows, takes the form

$$Nu = 5.0 + 0.025\,(Re\,Pr)^{0.8} \tag{8–22}$$

where $Re\,Pr \gtrsim 100$ and $L/D \gtrsim 30$. This equation represents the mean of most of the data in the literature for uniform wall-heat flux. Other correlations commonly used for liquid metals include those developed by Skupinski et al. [35] and Labarsky and Kaufman [36].

EXAMPLE 8–1

Determine the mass flux, bulk-stream velocity, Reynolds number, friction factor and coefficient of heat transfer for fully developed flow of air in a 10-cm-diameter circular tube for mass-flow rates of 0.0025 kg/s, 0.0075 kg/s, and 0.025 kg/s. Assume uniform wall-temperature heating and approximate uniform properties at 27°C and atmospheric pressure.

Solution

Objective Determine G, U_b, Re, f, and h.

Schematic Fully developed flow of air in a circular tube: uniform wall temperature.

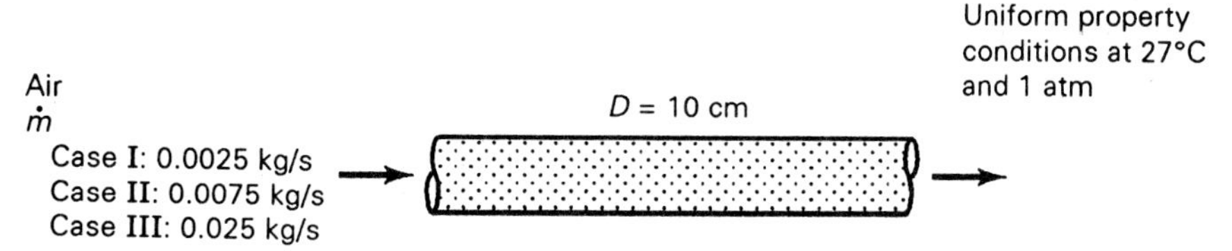

Assumptions/Conditions

forced convection
uniform properties
standard conditions

Properties Air at $T = 27°C = 300$ K (Table A–C–5): $\rho = 1.18$ kg/m^3, $\mu = 1.85 \times 10^{-5}$ kg/(m s), $k = 0.0262$ W/(m °C), $Pr = 0.708$.

Analysis The mass flux G, bulk-stream velocity U_b, and Reynolds number Re for these three cases are obtained from the defining relations

$$G = \frac{\dot{m}}{A} \qquad U_b = \frac{\dot{m}}{\rho A} = \frac{G}{\rho} \qquad Re = \frac{U_b D_H}{\nu} = \frac{G D_H}{\mu} = \frac{\dot{m} D_H}{\mu A}$$

which indicate

$$G_{\mathrm{I}} = \frac{\dot{m}_{\mathrm{I}}}{A} = \frac{0.0025 \text{ kg/s}}{\pi(0.1 \text{ m})^2/4} = 0.318 \text{ kg/(m}^2 \text{ s)}$$

$$U_{b,\mathrm{I}} = \frac{G_{\mathrm{I}}}{\rho} = \frac{0.318 \text{ kg/(m}^2 \text{ s)}}{1.18 \text{ kg/m}^3} = 0.269 \text{ m/s}$$

$$Re_{\mathrm{I}} = \frac{G_{\mathrm{I}} D}{\mu} = \frac{[0.318 \text{ kg/(m}^2 \text{ s)}](0.1 \text{ m})}{1.85 \times 10^{-5} \text{ kg/(m s)}} = 1720$$

for case I,

$$G_{\mathrm{II}} = 0.955 \text{ kg/(m}^2 \text{ s)} \qquad U_{b,\mathrm{II}} = 0.809 \text{ m/s} \qquad Re_{\mathrm{II}} = 5160$$

for case II, and

$$G_{\mathrm{III}} = 3.18 \text{ kg/(m}^2 \text{ s)} \qquad U_{b,\mathrm{III}} = 2.7 \text{ m/s} \qquad Re_{\mathrm{III}} = 17{,}200$$

for case III. Hence, the flow is laminar for case I, transitional turbulent for case II, and fully turbulent for case III.

Case I—laminar flow For laminar fully developed flow, the friction factor f_{I} is given by Eq. (8–7),

$$f_{\mathrm{I}} = \frac{16}{Re_{\mathrm{I}}} = \frac{16}{1720} = 0.0093$$

which requires $x/D \gtrsim 0.05\ Re = 86$, and the Nusselt number is given by Eq. (8–8),

$$Nu_{\mathrm{I}} = 3.66$$

which requires $x/D \gtrsim 0.05\ Re\ Pr = 60.9$. It follows that

$$h_{\mathrm{I}} = Nu_{\mathrm{I}} \frac{k}{D} = 3.66 \, \frac{0.0262 \text{ W/(m °C)}}{0.1 \text{ m}} = 0.96 \text{ W/(m}^2 \text{ °C)}$$

Case II—transitional turbulent flow For transitional turbulent flow with $Re = 5160$, we approximate the friction factor f_{II} by use of Eq. (8–15) (since $Re > 4000$),

$$f_{\mathrm{II}} = (1.58 \ln 5160 - 3.28)^{-2} = 0.00956$$

To evaluate the Nusselt number Nu_{II}, we write

$$Nu_{\mathrm{II}} = C_{tr}\, Nu_{L2} + (1 - C_{tr})\, Nu_{T8} \tag{a}$$

from Eq. (8–20), where

$$C_{tr} = 1.33 - \frac{5160}{6000} = 0.470 \qquad f_{T8} = (1.58 \ln 8000 - 3.28)^{-2} = 0.00839$$

$$Nu_{T8} = \frac{(0.00839/2)(8000)(0.708)}{1.07 + 12.7\sqrt{0.00839/2}\,(0.708^{2/3} - 1)} = 26.4$$

Substituting into Eq. (a), we obtain

$$Nu_{\text{II}} = 0.470(3.66) + (1 - 0.47)(26.4) = 15.7$$

and

$$h_{\text{II}} = 15.7 \frac{0.0262 \text{ W/(m °C)}}{0.1 \text{ m}} = 4.11 \text{ W/(m}^2 \text{ °C)}$$

Case III—fully turbulent flow For fully turbulent flow with $x/D \gtrsim 10$, we write

$$f_{\text{III}} = (1.58 \ln 17{,}200 - 3.28)^{-2} = 0.0068$$

from Eq. (8–15),

$$Nu_{\text{III}} = \frac{(0.0068/2)(17{,}200)(0.708)}{1.07 + 12.7\sqrt{0.0068/2}\,(0.708^{2/3} - 1)} = 64.6$$

from Eq. (8–16), and

$$h_{\text{III}} = 64.6 \frac{0.0262 \text{ W/(m °C)}}{0.1 \text{ m}} = 16.9 \text{ W/(m}^2 \text{ °C)}$$

EXAMPLE 8–2

Air enters a 10-cm-diameter circular tube at 27°C and 1.25 atm with a bulk-stream velocity of 0.2 m/s. The tube length is 5 m and the wall is maintained at 35°C. Determine the mass-flow rate, mass flux, Reynolds number and mean coefficient of heat transfer, assuming uniform property conditions.

Solution

Objective Determine $\dot{m}$, G, Re, and $\bar{h}$ over the length L.

Schematic Air flow in a circular tube: uniform wall temperature.

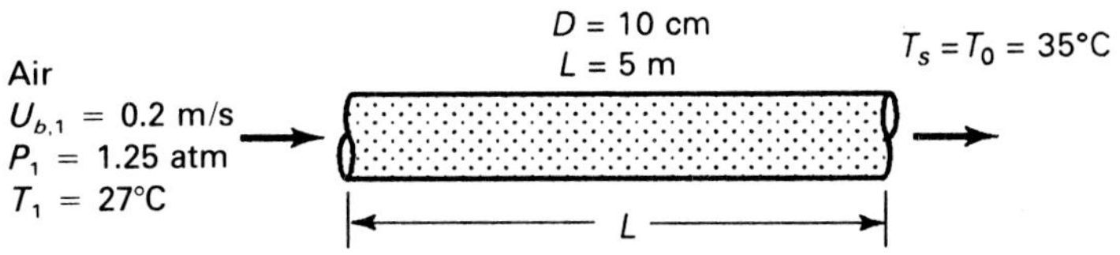

Air
$U_{b,1} = 0.2$ m/s
$P_1 = 1.25$ atm
$T_1 = 27$°C
$D = 10$ cm
$L = 5$ m
$T_s = T_0 = 35$°C
L

Assumptions/Conditions

forced convection

hydrodynamic and thermal developing flow occurs in the entrance region

uniform properties

ideal gas behavior and other standard conditions

Properties Because the values of μ, k, and Pr for air are fairly insensitive to pressure, these properties can simply be obtained from Table A–C–5; the properties at 300 K and 1 atm are $\mu = 1.85 \times 10^{-5}$ kg/(m s), $k = 0.0262$ W/(m °C), $Pr = 0.708$, and $\rho = 1.18$ kg/m³. Using the ideal gas law to account for the effect of pressure on density, we obtain

$$\rho = 1.25\left(1.18\,\frac{\text{kg}}{\text{m}^3}\right) = 1.48\ \text{kg/m}^3$$

Analysis To calculate $\dot{m}$, G, and Re we write

$$\dot{m} = (\rho A U_b)_1 = 1.48\,\frac{\text{kg}}{\text{m}^3}\,\frac{\pi(0.1\ \text{m})^2}{4}\left(0.2\,\frac{\text{m}}{\text{s}}\right) = 0.00232\ \text{kg/s}$$

$$G = \rho U_b = 1.48\,\frac{\text{kg}}{\text{m}^3}\left(0.2\,\frac{\text{m}}{\text{s}}\right) = 0.296\ \text{kg/(m}^2\ \text{s)}$$

$$Re = \frac{GD}{\mu} = \frac{[0.296\ \text{kg/(m}^2\ \text{s)}](0.1\ \text{m})}{1.85 \times 10^{-5}\ \text{kg/(m s)}} = 1600$$

This value of Re indicates that the flow is laminar.

Assuming that combined hydrodynamic and thermal developing flow occurs in the entrance region, the mean Nusselt number over the length of the tube can be approximated by use of Eq. (8–12),

$$\overline{Nu} = 3.66 + \frac{0.104\ Re\ Pr/(L/D)}{1 + 0.016\,[Re\ Pr/(L/D)]^{0.8}}$$

where

$$\frac{L/D}{Re\ Pr} = \frac{5\ \text{m}/(0.1\ \text{m})}{1600(0.708)} = 0.0441$$

Substituting for the term $(L/D)/(Re\ Pr)$, we obtain

$$\overline{Nu} = 3.66 + 2.32 = 5.98$$

or

$$\bar{h} = \overline{Nu}\,\frac{k}{D} = 5.98\,\frac{0.0262\ \text{W/(m °C)}}{0.1\ \text{m}} = 1.57\ \text{W/(m}^2\ \text{°C)}$$

Notice that this result for $\overline{Nu}$ is 39% greater than the value for TFD flow, $Nu = 3.66$, which occurs for $(x/D)/(Re\ Pr) \gtrsim 0.05$ (i.e., $x \gtrsim 5.66$ m).

Effects of Property Variation

The significance of property variation caused by temperature should always be kept in mind when analyzing convection-heat-transfer processes. The two primary effects that are brought about by temperature-induced property change are (1) direct effects caused by temperature variations (a) over the cross section and (b) in the direction of flow; and (2) indirect effects involving the creation of buoyancy forces by variations in the density (or concentration) with temperature. The second of these two effects pertains to combined forced and natural convection, which is discussed in Chap. 9.

The effects of property variation on the local momentum and heat-transfer characteristics that result from the variation in temperature over the cross-section are generally accounted for by utilizing correction factors for the friction factor and Nusselt number. For liquids, where variations in the viscosity are generally the main consideration, these correction factors take the form

$$\frac{f}{f_{cp}} = \left(\frac{\mu_s}{\mu_b}\right)^m \qquad \frac{Nu}{Nu_{cp}} = \left(\frac{\mu_s}{\mu_b}\right)^n \tag{8–23,24}$$

where the subscript *cp* represents the constant property characteristics, μ_s is evaluated at the wall temperature T_s, the other properties are evaluated at the bulk-stream temperature T_b, and the exponents m and n are dependent upon the conditions. On the other hand, the variations in density and thermal conductivity as well as viscosity are usually the most important factors for gases. These three properties are related to the *absolute* temperature for gases in such a way as to permit the use of correction factors of the form

$$\frac{f}{f_{cp}} = \left(\frac{T_s}{T_b}\right)^m \qquad \frac{Nu}{Nu_{cp}} = \left(\frac{T_s}{T_b}\right)^n \tag{8–25,26}$$

over the range $0.33 \lessapprox T_s/T_b \lessapprox 3.0$. Representative values of m and n are given in Table 8–2 for liquid and gas flow with various conditions. The relationships for friction factor and Nusselt number that are presented throughout this chapter for ideal isothermal conditions may be adjusted to account for mild to moderate property variations by merely multiplying by the appropriate viscosity or temperature-ratio correction factor.

The variation in properties over the length of a tube is often more pronounced than the variation over the cross section. The effect of property variation in the direction of flow on the coefficients of friction and heat transfer are generally accounted for in design work by evaluating the properties at the arithmetic average of the inlet and outlet bulk-stream temperatures, with the Reynolds number defined by Eq. (6–35b). According to Kays and London [4], this approach is adequate for cases involving gas flow with the absolute temperature variation along the tube being less than 2 to 1. When this approach is used, the resulting average value of T_b can also be used to represent T_b in Eqs. (8–23) to (8–26), which are used to correct for local property variation effects over the cross section. For situations in which the temperature variation in the flow direction is large, Kays and London [4] suggest that analyses

TABLE 8–2 Variable property conditions—coefficients associated with Eqs. (8–23)–(8-26)

Fluid	*Condition*	*m*	*n*	*References*
Liquid	Laminar			
	Cooling	0.50	−0.14	18,26
	Heating	0.58	−0.14	18,26
	Turbulent[a]			
	Cooling	0.25	−0.25	6,24
	Heating	0.25	−0.11	6,24,37
Gas	Laminar	1.0	0	6
	Turbulent			
	Cooling	−0.1	0	6
	Heating[b]	−0.1	−0.5	6

[a] The *Nu* correction for turbulent flow of liquids is recommended by Petukhov [24] for the ranges $0.08 \lesssim \mu_s/\mu_b \lesssim 40$; $2 \lesssim Pr \lesssim 140$; and $10^4 \lesssim Re \lesssim 1.25 \times 10^5$.
[b] Sleicher and Rouse [38] recommend

$$n = 0.3 - \left(\log \frac{T_s}{T_b}\right)^{0.25}$$

be developed for separate sections of the tube or heat exchanger over which the temperature variation is not excessive.

Enhancement of Heat Transfer

The correlations for heat transfer and friction presented earlier in this section only apply to passages with plain smooth surfaces. The heat transfer in industrial applications is often enhanced by the use of fins, rough surfaces, and other techniques. A review of the various methods of enhancement has been developed by Bergles and associates [39,40]. For example, heat transfer for turbulent tube flow has been increased by as much as 400% by the use of rough surfaces. However, large increases in the friction factor have also been found to occur, such that care must be taken to utilize surfaces that are efficient as far as both heat transfer and pressure drops are concerned.

Thermal and hydraulic analyses of systems employing enhancement features such as rough surfaces or fins require the use of special design correlations. Empirical correlations for friction and heat transfer are generally provided by the commercial manufacturers of finned tubes or rough surfaces.

For tubes with rough surfaces characterized by roughness elements with average height ϵ, f is commonly evaluated by use of the well-known *Moody chart*†, which

† The Moody friction factor chart is traditionally presented in terms of the Moody (or Darcy-Weissback) friction factor λ, $\lambda = -(dP/dx)D_H/(\rho U_b^2/2)$, where dP/dx represents the pressure gradient for HFD flow. It should be noted that $\lambda = 4f$. Unfortunately, the symbol f is sometimes used in the literature to represent the Moody friction factor as well as the Fanning friction factor. Therefore, care must be taken to distinguish between these two types of friction factors.

is shown in Fig. 8–10. Typical values of the *roughness* ϵ are given in Table 8–3. However, Fig. 8–10 indicates that the primary characterizing factor for flow in rough tubes is the *relative roughness* ϵ/D (and Reynolds number Re) rather than the average roughness height ϵ itself.

TABLE 8–3 Representative values of roughness ϵ

Pipe or tube material	ϵ (μm)
Commercial steel or wrought iron	46
Cast iron	250
Riveted steel	900–9000
Asphalted cast iron	120
Galvanized surface	150
Drawn tubing	1.5
Concrete	300–3000
Wood stave	180–900

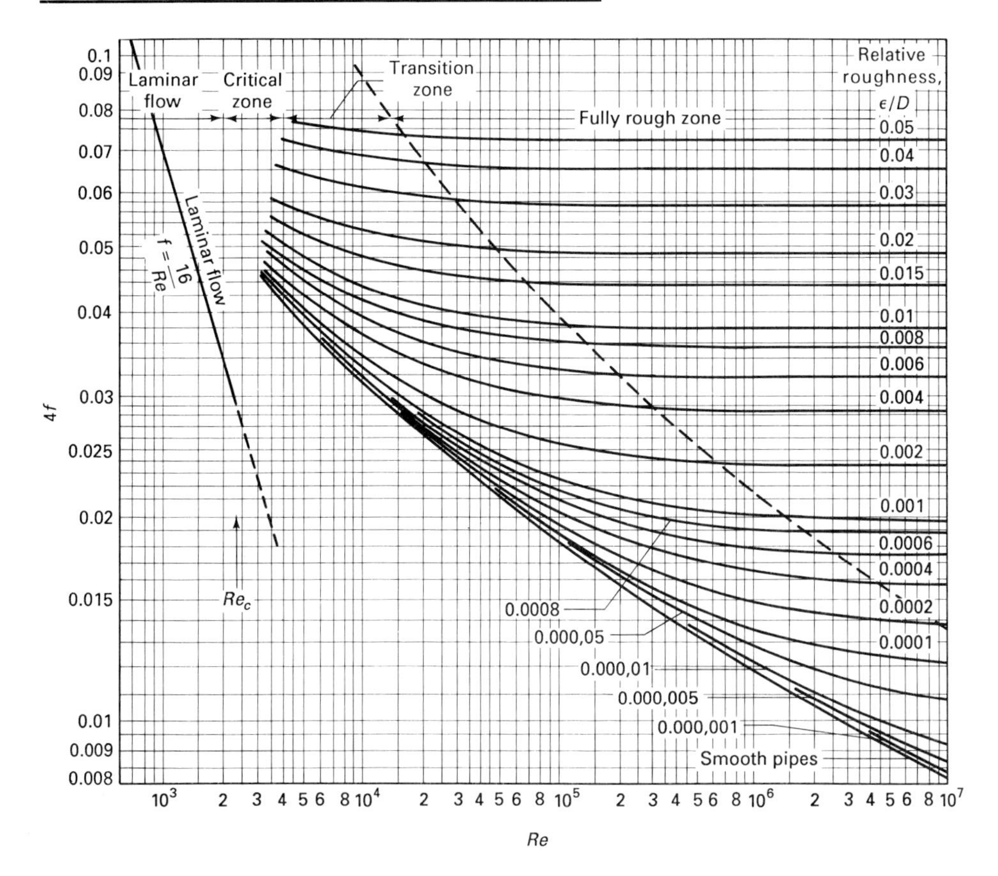

FIGURE 8–10 The Moody friction factor chart for rough surfaces (Moody [41]).

Other Factors

The coefficients of friction and heat transfer are also sometimes influenced by other factors such as unsteady operation, nonuniform distributions in wall temperature or heat flux, viscous dissipation, phase change (see Chap. 10), and mass transfer through the wall. As indicated earlier, when the convection process is complicated by factors such as these, their influence on the coefficients should be accounted for. Consequently, one must be prepared to search the literature for appropriate correlations, or even to undertake or commission an experimental or theoretical study.

EXAMPLE 8–3

Show the effect of temperature-induced property variation on the shape of the velocity distribution for laminar flow of (a) gas in a heated tube and (b) liquid in a cooled tube.

Solution

Objective Sketch velocity distributions for uniform and variable property conditions.

Assumptions/Conditions

forced convection
variable properties
standard conditions

Analysis Referring to Tables A–C–3 through A–C–5, we observe that the viscosity is directly proportional to temperature for gases and inversely proportional for liquids. As a result, when a gas is heated or when a liquid is cooled, the fluid near the wall is more viscous than the fluid further away from the wall. Consequently, we would expect the actual velocity distribution for these conditions to be lower relative to the isothermal velocity profile in the region near the wall and larger in the region away from the wall. This perspective is shown in Fig. E8–3 for laminar tube flow. Of course, the opposite result would be expected for the cooling of a gas or heating of a liquid.

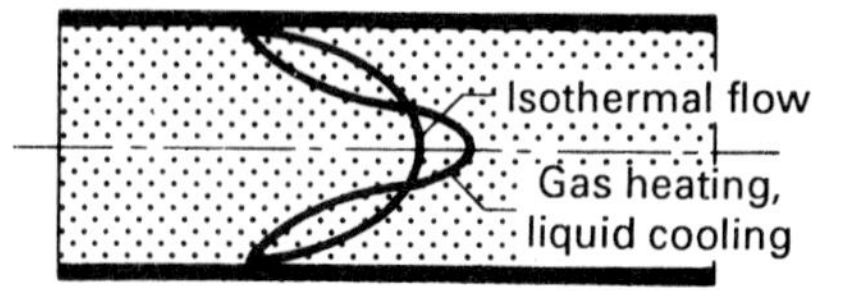

FIGURE E8–3
Influence of property variation on velocity distribution for laminar tube flow.

EXAMPLE 8–4

Air at 11.4°C and 1 atm pressure enters a 10-cm-diameter tube with a mass-flow rate of 0.0185 kg/s. The tube wall is maintained at 175°C and the outlet temperature is 42.6°C. Determine the friction factor and the coefficient of heat transfer for approximate fully developed conditions.

Solution

Objective Determine f and h for fully developed conditions.

Schematic Fully developed flow of air in a circular tube: uniform wall temperature.

$D = 10$ cm $\quad T_s \doteq T_0 = 175°C$

Air
$\dot{m} = 0.0185$ kg/s
$P_1 = 1$ atm
$T_1 = 11.4°C$

$T_2 = 42.6°C$

Assumptions/Conditions

forced convection
moderate property variation†
standard conditions

Properties Air at 27°C (Table A–C–5): $\rho = 1.18$ kg/m^3, $\mu = 1.85 \times 10^{-5}$ kg/(m s), $k = 0.0262$ W/(m °C), $Pr = 0.708$.

Analysis To account (approximately) for the effect of property variations on the coefficients of friction and heat transfer, we will evaluate the fluid properties at the arithmetic average bulk-stream temperature and we will employ the correction factors given by Eqs. (8–25) and (8–26),

$$\frac{f}{f_{cp}} = \left(\frac{T_0}{T_b}\right)^m \qquad \frac{Nu}{Nu_{cp}} = \left(\frac{T_0}{T_b}\right)^n$$

To provide a basis for evaluating the friction factor and coefficient of heat transfer, we compute the mass flux G and Reynolds number Re by writing

$$G = \frac{\dot{m}}{A} = \frac{0.0185 \text{ kg/s}}{\pi(0.1 \text{ m})^2/4} = 2.36 \text{ kg/(m}^2\text{ s)}$$

† The term *moderate property variation* is used in this chapter to indicate situations in which a uniform property analysis is used, with the effect of property variation accounted for by the use of the average fluid temperature and correction factors such as those given by Eqs. (8–23) to (8–26).

and

$$Re = \frac{GD}{\mu} = \frac{[2.36\ \text{kg/(m}^2\ \text{s)}](0.1\ \text{m})}{1.85 \times 10^{-5}\ \text{kg/(m s)}} = 12{,}800$$

such that the flow is fully turbulent. (Note that the average bulk-stream velocity $U_b = G/\rho$ is equal to 2 m/s.) Using Eqs. (8–15) and (8–16) to evaluate f_{cp} and Nu_{cp}, we obtain

$$f_{cp} = (1.58 \ln 12{,}800 - 3.28)^{-2} = 0.00735$$

and

$$Nu_{cp} = \frac{(0.00735/2)(12{,}800)(0.708)}{1.07 + 12.7\sqrt{0.00735/2}\,(0.708^{2/3} - 1)} = 36.5$$

Referring to Table 8–2, we find that $m = -0.1$ and $n = -0.5$ for turbulent flow and heating. Setting $T_0 = 448$ K and $T_b = 300$ K, we have

$$f = 0.00735\left(\frac{448\ \text{K}}{300\ \text{K}}\right)^{-0.1} = 0.00735(0.96) = 0.00706$$

and

$$Nu = 36.5\left(\frac{448\ \text{K}}{300\ \text{K}}\right)^{-0.5} = 36.5(0.818) = 29.9$$

Therefore, the coefficient of heat transfer is

$$h = Nu\frac{k}{D} = 29.9\,\frac{0.0262\ \text{W/(m °C)}}{0.1\ \text{m}} = 7.83\ \text{W/(m}^2\ \text{°C)}$$

8–2–2 Practical Thermal Analysis

Practical lumped analyses are developed in this section for the overall heat-transfer performance of steady internal flow in passages with uniform cross-sectional area, as illustrated in Fig. 8–11. The inlet bulk-stream temperature, fluid properties, mass

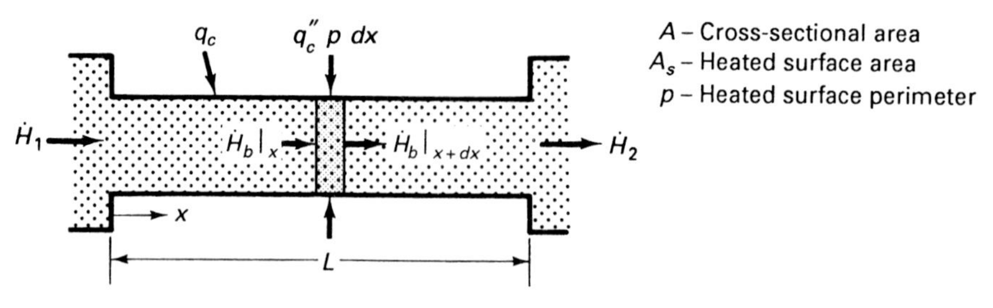

FIGURE 8–11 Bulk-stream enthalpy and heat-transfer rate associated with flow through lumped volume AL and lumped/differential volume $A\,dx$ within a tubular passage.

flow rate, and system dimensions (or outlet bulk-stream temperature T_2) are assumed to be known. The main objective of these analyses is to predict the overall rate of heat transfer q_c and the outlet temperature T_2 (or the system dimensions) in terms of the specified system parameters and boundary conditions of interest. In addition, we are generally interested in knowing the distribution in bulk-stream temperature T_b and the wall temperature T_s (for cases in which the heat flux is specified).

To develop expressions for the two primary parameters q_c and T_2, two independent equations are developed by applying the first law of thermodynamics, Eq. (1–1),

$$\Sigma \dot{E}_o - \Sigma \dot{E}_i + \cancel{\frac{\Delta E_s}{\Delta t}} = 0 \tag{8–27}$$

to (1) the lumped volume AL, and (2) the lumped/differential volume $A\,dx$ (see Fig. 8–11). Applying the first law of thermodynamics to the fluid volume AL, we obtain

$$\dot{H}_2 - \dot{H}_1 - q_c = 0 \tag{8–28}$$

or

$$q_c = \dot{H}_2 - \dot{H}_1 = \int_{\dot{H}_1}^{\dot{H}_2} d\dot{H}_b \tag{8–29}$$

Using Eq. (6–45) to express $d\dot{H}_b$ in terms of T_b and c_P for ideal fluids, this equation takes the form

$$q_c = \dot{m} \int_{T_1}^{T_2} c_P\, dT_b \tag{8–30}$$

for uniform $\dot{m}$, which reduces to

$$q_c = \dot{m} c_P (T_2 - T_1) \tag{8–31}$$

for uniform specific heat. Similarly, the application of the first law of thermodynamics to the fluid volume $A\,dx$ gives

$$\dot{H}_b\big|_{x+dx} - \dot{H}_b\big|_x - q_c''\, p\, dx = 0 \tag{8–32}$$

or

$$q_c'' p = \frac{d\dot{H}_b}{dx} \tag{8–33}$$

where p is the perimeter of the heated surface. Substituting for $d\dot{H}_b$, this equation becomes

$$q_c'' p = \dot{m} c_P \frac{dT_b}{dx} \tag{8–34}$$

The boundary condition that accompanies this first-order ordinary differential equation is given by

$$T_b = T_1 \qquad \text{at } x = 0 \tag{8–35}$$

Attention is now turned to the solution of these equations for specified wall-heat flux and uniform wall-temperature boundary conditions. These solutions are developed in the context of basic uniform property conditions, after which consideration is given to the effects of variable properties.

Specified Wall-Heat Flux

For cases in which the wall heat flux q_c'' is specified as a function of x, the overall rate of heat transfer q_c is simply

$$q_c = \int_{A_s} q_c'' \, dA_s = p \int_0^L q_c'' \, dx \tag{8–36}$$

This equation reduces to

$$q_c = q_0'' A_s = q_0'' pL \tag{8–37}$$

for uniform wall-heat flux (i.e., $q_c'' = q_0''$). With q_c known, the outlet bulk-stream temperature T_2 for uniform property flow is obtained from Eq. (8–31),

$$T_2 = T_1 + \frac{q_c}{\dot{m} c_P} \tag{8–38}$$

To obtain an expression for T_b, we integrate Eq. (8–34) over the length 0 to x, with the result

$$T_b - T_1 = \frac{p}{\dot{m} c_P} \int_0^x q_c'' \, dx \tag{8–39}$$

For uniform wall-flux heating, this equation yields

$$T_b - T_1 = \frac{q_0'' p x}{\dot{m} c_P} = \frac{4 q_0'' D_H}{k} \frac{p}{p_w} \frac{x/D_H}{Re\,Pr} \tag{8–40}$$

This expression is shown in Fig. 8–12 for flow in a circular tube (with $p = p_w = \pi D$). Notice that the bulk-stream temperature T_b is linear over the length of the tube. In addition, because T_b is independent of the coefficient of heat transfer h_x, the predictions for T_b are the same for laminar and turbulent flow.

An expression is obtained for the unknown surface temperature T_s by combining Eq. (8–39) with the defining equation for the coefficient of heat transfer (the general Newton law of cooling), Eq. (6–48); that is,

$$q_c'' = h_x (T_s - T_b) \tag{8–41}$$

$$T_s - T_1 = \frac{q_c''}{h_x} + \frac{p}{\dot{m} c_P} \int_0^x q_c'' \, dx \tag{8–42}$$

For a uniform wall-heat flux, our prediction for T_s takes the form

$$T_s - T_1 = \frac{4 q_0'' D_H}{k} \left(\frac{1}{4\,Nu_x} + \frac{x/D_H}{Re\,Pr} \frac{p}{p_w} \right) \tag{8–43}$$

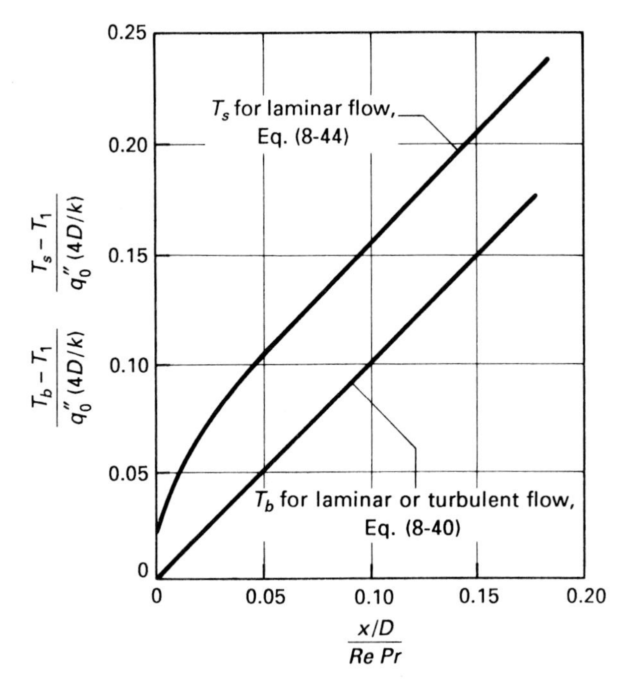

FIGURE 8–12 Bulk-stream and wall temperatures for flow in a circular tube with uniform wall-heat flux.

Note that T_s is a function of the coefficient of heat transfer. Accordingly, to obtain calculations for T_s, h_x or Nu_x must be specified. For example, for HFD laminar thermal developing flow in a circular tube, Nu_x can be approximated by Eq. (8–11) in the entrance region and by Eq. (8–8) ($Nu = 4.36$) in the TFD region, such that Eq. (8–43) becomes

$$\frac{T_s - T_1}{4q_0''D/k} = 0.192\left(\frac{x/D}{Re\;Pr}\right)^{1/3} + \frac{x/D}{Re\;Pr} \qquad \text{for } \frac{x/D}{Re\;Pr} \lesssim 0.05 \qquad (8\text{-}44a)$$

$$= 0.0573 + \frac{x/D}{Re\;Pr} \qquad \text{for } \frac{x/D}{Re\;Pr} \gtrsim 0.05 \qquad (8\text{-}44b)$$

This equation is shown in Fig. 8–12. The wall temperature T_s is seen to be nonlinear in the thermal entrance region, but is linear in the TFD region. Similar calculations can be made for turbulent flow. Because the Nusselt number for turbulent flow is much larger than for laminar flow, we conclude on the basis of Eq. (8–43) that the wall will be cooler for turbulent flow.

EXAMPLE 8–5

Air at 27°C and atmospheric pressure enters a 10-cm-diameter tube of 1-m length. The mass-flow rate is 0.0185 kg/s and the heat flux from the surface of the tube is specified by

$$q''_c = q''_0 \exp\left(-\frac{x}{L}\right)$$

where $q''_0 = 0.541$ kW/m². Determine the outlet temperature T_2, the local bulk-stream temperature T_b, and the surface temperature T_s at the outlet.

Solution

Objective Determine T_2, the distribution in T_b, and T_s at $x = L$.

Schematic Air flow in a circular tube: nonuniform wall-heat flux.

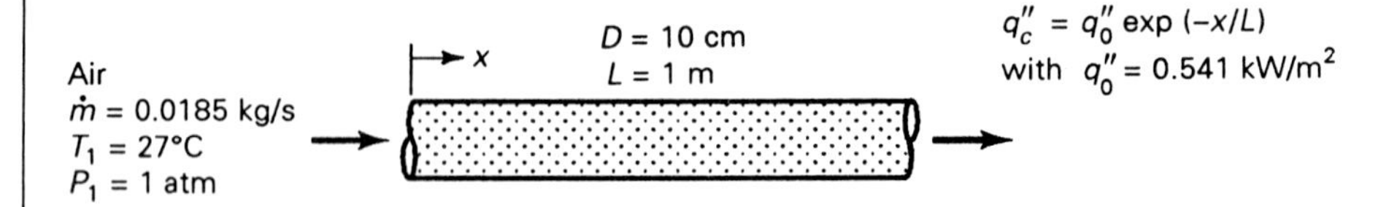

Assumptions/Conditions

forced convection
uniform properties
no hydrodynamic calming section
standard conditions

Properties Air at 27°C (Table A–C–5): $\rho = 1.18$ kg/m³, $\mu = 1.85 \times 10^{-5}$ kg/(m s), $c_P = 1.01$ kJ/(kg °C), $k = 0.0262$ W/(m °C), $Pr = 0.708$.

Analysis To obtain the outlet temperature for approximate uniform property conditions, we utilize Eq. (8–31), which was obtained by performing an energy balance on the lumped control volume AL; that is,

$$q_c = \dot{m}c_P(T_2 - T_1)$$

where

$$q_c = p\int_0^L q''_c\, dx = pq''_0 \int_0^L e^{-x/L}\, dx$$

$$= pLq''_0(1 - e^{-1}) = \pi(0.1 \text{ m})(1 \text{ m})\left(0.541 \frac{\text{kW}}{\text{m}^2}\right)(1 - e^{-1}) = 0.107 \text{ kW}$$

Following through with the calculation for T_2, we write

$$T_2 = \frac{q_c}{\dot{m}c_P} + T_1 = \frac{0.107 \text{ kW}}{(0.0185 \text{ kg/s})[1.01 \text{ kJ/(kg °C)}]} + 27°\text{C} = 32.7°\text{C}$$

Referring to Table A–C–5, we see that the variations in properties of air over the temperature range from 27°C to 32.7°C are very small. Consequently, our use of a uniform property analysis with the properties evaluated at 27°C is quite reasonable.

The bulk-stream temperature T_b is obtained from Eq. (8–39), which was developed by applying the first law of thermodynamics to the lumped/differential element $A\,dx$.

$$T_b - T_1 = \frac{p}{\dot{m}c_P}\int_0^x q_c''\,dx = \frac{pLq_0''}{\dot{m}c_P}\int_0^x e^{-x/L}\,\frac{dx}{L}$$

$$= \frac{\pi(0.1 \text{ m})(1 \text{ m})(0.541 \text{ kW/m}^2)}{(0.0185 \text{ kg/s})[1.01 \text{ kJ/(kg °C)}]}(1 - e^{-x/L})$$

$$T_b = 27°\text{C} + 9.10°\text{C}\,(1 - e^{-x/L}) \qquad \text{(a)}$$

This equation is shown in Fig. E8–5. For the purpose of comparison, T_b is also shown for a uniform heat flux of 0.541 kW/m².

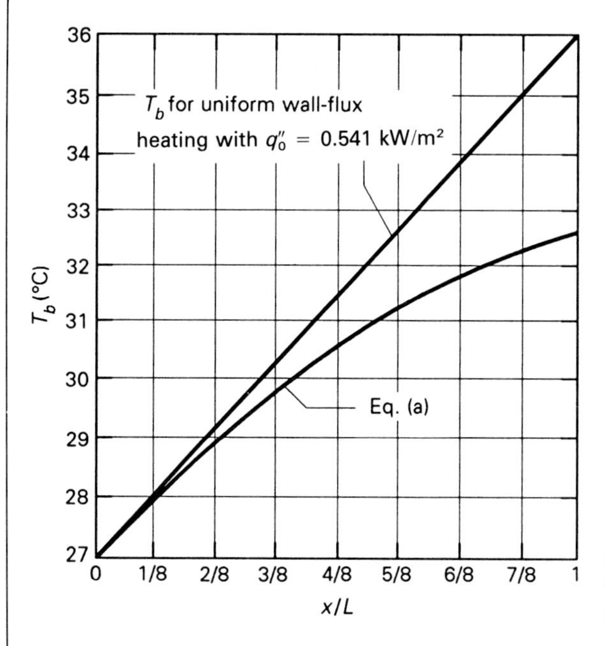

FIGURE E8–5
Calculations for bulk-stream temperature T_b.

To estimate the local wall temperature T_s at $x = L$, we utilize the general Newton law of cooling given by Eq. (8–41),

$$q_c'' = h_x(T_s - T_b)$$

Referring to Example 8–4, the flow is turbulent with $Re = 12{,}800$ and $Nu_{cp} = 36.5$, such that

$$h_{cp} = 36.5 \frac{0.0262 \text{ W/(m °C)}}{0.1 \text{ m}} = 9.56 \text{ W/(m}^2 \text{ °C)}$$

Assuming no calming section and $h \simeq h_{cp}$, Eq. (8–21a) is used to estimate h_x at $x = L$, with the result

$$h_L = h\left(1 + \frac{C}{L/D}\right) = 9.56 \frac{\text{W}}{\text{m}^2 \text{ °C}}\left(1 + \frac{6}{10}\right) = 15.3 \text{ W/(m}^2 \text{ °C)}$$

It follows that the surface temperature at $x = L$ can be approximated by

$$T_{s,L} = \frac{q''_{c,L}}{h_L} + T_2 = \frac{(0.541 \times 10^3 \text{ W/m}^2)(e^{-1})}{15.3 \text{ W/(m}^2 \text{ °C)}} + 32.7\text{°C} = 45.7\text{°C}$$

Due to the relatively small surface-to-bulk-stream temperature difference, we conclude that the approximation $h \simeq h_{cp}$ should be quite reasonable.

Specified Wall Temperature

For specified wall-temperature heating, the heat flux q''_c is expressed in terms of the general Newton law of cooling, Eq. (8–41). Using this representation for q''_c, Eq. (8–34) becomes

$$h_x p(T_s - T_b) = \dot{m}c_P \frac{dT_b}{dx} \qquad (8\text{–}45)$$

We now consider the solution to this equation and Eq. (8–31),

$$q_c = \dot{m}c_P(T_2 - T_1) \qquad (8\text{–}31)$$

for uniform wall temperature and arbitrarily specified wall temperature conditions.

Solution for Uniform Wall Temperature For a uniform wall temperature boundary condition ($T_s = T_0$), Eq. (8–45) can be written as

$$\frac{d(T_b - T_0)}{T_b - T_0} = -\frac{ph_x}{\dot{m}c_P}\,dx \qquad (8\text{–}46)$$

which is integrated to obtain

$$\ln \frac{T_b - T_0}{T_1 - T_0} = -\frac{p}{\dot{m}c_P}\int_0^x h_x\,dx = -\frac{\bar{h}px}{\dot{m}c_P} \qquad (8\text{–}47)$$

where the mean coefficient of heat transfer $\bar{h}$ is defined by Eq. (6–49),

$$\bar{h} = \frac{1}{x}\int_0^x h_x\,dx \qquad (8\text{–}48)$$

Thus, the solution for T_b becomes

$$\frac{T_b - T_0}{T_1 - T_0} = \exp\left(-\frac{\bar{h}px}{\dot{m}c_P}\right) \tag{8–49}$$

or

$$\frac{T_b - T_1}{T_0 - T_1} = 1 - \exp\left(-\frac{\bar{h}px}{\dot{m}c_P}\right) \tag{8–50}$$

As shown in Fig. 8–13, this equation indicates that T_b asymptotically approaches T_0 as x increases.

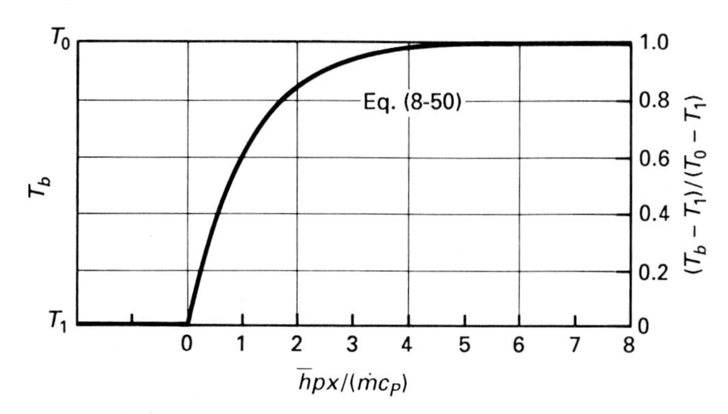

FIGURE 8–13 Bulk-stream temperature for flow with uniform wall temperature.

To obtain a relation for the outlet bulk-stream temperature T_2 we set x equal to L in Eq. (8–50), with the result

$$\frac{T_2 - T_1}{T_0 - T_1} = 1 - \exp\left(-\frac{\bar{h}A_s}{\dot{m}c_P}\right) \tag{8–51}$$

where $A_s = pL$. Finally, the coupling of Eqs. (8–31) and (8–51) gives rise to an expression for q_c of the form

$$q_c = \dot{m}c_P(T_0 - T_1)\left[1 - \exp\left(-\frac{\bar{h}A_s}{\dot{m}c_P}\right)\right] \tag{8–52}$$

In the limit, as the dimensionless parameter $\bar{h}A_s/(\dot{m}c_P)$ increases, the outlet temperature T_2 approaches T_0 and the total rate of heat transfer q_c approaches a maximum value of $\dot{m}c_P(T_0 - T_1)$. This thermodynamic limit occurs at a value of $\bar{h}A_s/(\dot{m}c_P)$ of the order of 4 to 5.

To obtain calculations for q_c, T_2, and T_b for a given flow condition, we must utilize appropriate correlations for the mean coefficient of heat transfer.

Although the theoretical aspects of the analysis for uniform wall-temperature heating are complete, the predictions for q_c are now put into the effectiveness, efficiency, and log mean temperature difference formats, all of which are in common use. In addition, a simple working relationship is developed for the heating–pumping power ratio $q_c/\dot{W}_p$, which can be used as a criterion for estimating the relative significance of the energy required to pump the fluid.

Effectiveness The *effectiveness* ϵ of a convection-heat-transfer process is defined by

$$q_c = \epsilon q_{max} \tag{8–53}$$

where the maximum possible rate of heat transfer q_{max} for uniform wall-temperature heating is

$$q_{max} = \dot{m}c_P(T_0 - T_1) \tag{8–54}$$

Introducing Eq. (8–31), we see that the effectiveness also represents the temperature ratio $(T_2 - T_1)/(T_0 - T_1)$. Referring to Eq. (8–51), ϵ is written as

$$\epsilon = \frac{T_2 - T_1}{T_0 - T_1} = 1 - \exp\left(-\frac{\bar{h}A_s}{\dot{m}c_P}\right) \tag{8–55}$$

This equation is shown in Fig. 8–14 in terms of $\bar{h}A_s/(\dot{m}c_P)$.

The dimensionless parameter $hA_s/(\dot{m}c_P)$ is sometimes referred to as the *number of transfer units NTU*. The *NTU* provides a comparison between the thermal capacity $\bar{h}A_s$ of the system and the capacity rate $\dot{m}c_P$, and provides an indication of the physical size of the system. The larger the *NTU*, the larger the surface area A_s and the closer

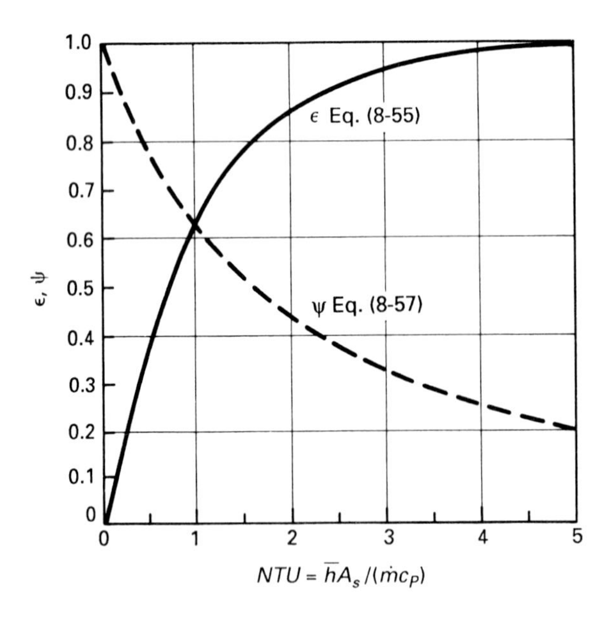

FIGURE 8–14
Effectiveness and efficiency for uniform wall-temperature heating.

the heat-transfer system comes to approaching its thermodynamic limit. However, because ϵ increases very gradually with NTU for $\epsilon \gtrsim 0.8$, the value of NTU is generally maintained in the range 0.5 to 2.5 in order to achieve effective use of the surface area.

Efficiency The significance of the magnitude of NTU is reinforced by putting Eq. (8–52) into the form

$$q_c = \dot{m}c_P(T_0 - T_1)\epsilon = \bar{h}A_s(T_0 - T_1)\frac{\epsilon}{\bar{h}A_s/(\dot{m}c_P)} = \bar{h}A_s(T_0 - T_1)\psi \tag{8–56}$$

where the *efficiency* ψ is defined by

$$\psi = \frac{\epsilon}{NTU} = \frac{1 - \exp(-NTU)}{NTU} = \frac{1 - \exp[-\bar{h}A_s/(\dot{m}c_P)]}{\bar{h}A_s/(\dot{m}c_P)} \tag{8–57}$$

As shown in Fig. 8–14, ψ is in the range 1 to 0.5 for $NTU \lesssim 1.5$, but is less than 0.5 for $NTU \gtrsim 1.5$ and approaches zero as NTU becomes large. Because ψ is proportional to the total heat flux q_c/A_s, ψ provides a clear indication of the effectiveness of the surface in transferring heat for specified values of $\bar{h}$ and $T_0 - T_1$.

Log Mean Temperature Difference (LMTD) To put the calculations for q_c into the $LMTD$ form, we replace the capacity rate $\dot{m}c_P$ in Eq. (8–31) by utilizing Eq. (8–51); that is,

$$\dot{m}c_P = -\frac{\bar{h}A_s}{\ln[(T_2 - T_0)/(T_1 - T_0)]} = \frac{\bar{h}A_s}{\ln[(T_0 - T_1)/(T_0 - T_2)]} \tag{8–58}$$

This substitution gives rise to the famous $LMTD$ equation for heat transfer,

$$q_c = \bar{h}A_s\, LMTD \tag{8–59}$$

where

$$LMTD = \frac{T_2 - T_1}{\ln[(T_0 - T_1)/(T_0 - T_2)]} = \frac{\Delta T_1 - \Delta T_2}{\ln(\Delta T_1/\Delta T_2)} \tag{8–60}$$

and $\Delta T_1 = T_0 - T_1$ and $\Delta T_2 = T_0 - T_2$. Comparing Eqs. (8–56) and (8–59), we find that $LMTD$ and ψ are related by

$$LMTD = (T_0 - T_1)\psi \tag{8–61}$$

The $LMTD$ format is most useful for making design calculations in order to size a heat-transfer system (i.e., to determine the necessary thermal capacity $\bar{h}A_s$) when the inlet, outlet, and surface temperatures are specified. However, for situations in which the outlet temperature is not known, use of the $LMTD$ approach to evaluate T_2 involves time-consuming iterations. For such cases, the more expedient approach generally is to simply utilize Eqs. (8–51) and (8–52) directly, or to use the effectiveness or efficiency concepts.

EXAMPLE 8–6

Water at 20°C and 1 atm enters a 10-cm-diameter tube of 10-m length. The mass-flow rate of the water is 10 kg/s and the wall temperature is 80°C. Determine the outlet temperature of the water.

Solution

Objective Determine T_2.

Schematic Flow of water in a circular tube: uniform wall temperature.

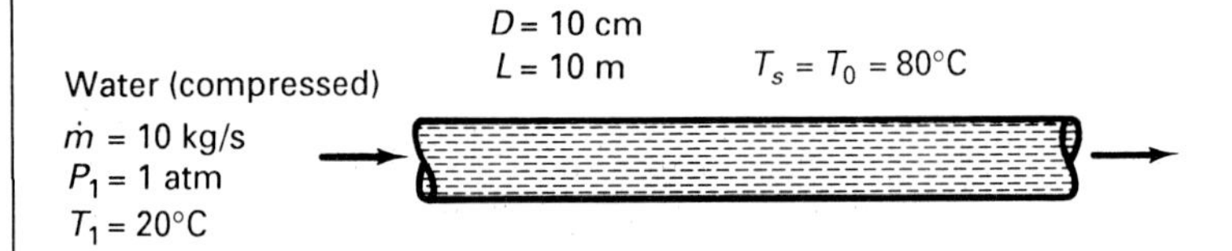

Assumptions/Conditions

forced convection
negligible entrance effects on $\bar{h}$
moderate property variation
standard conditions

Properties

Water at $T = 20°\text{C} = 293$ K (Table A–C–3); $\rho = 1/v_f = 1000$ kg/m^3, $\mu = 0.00101$ kg/(m s), $c_P = 4.18$ kJ/(kg °C), $k = 0.603$ W/(m °C), $Pr = 7.01$.

Water at $T = 80°\text{C} = 353$ K: $\mu = 3.54 \times 10^{-4}$ kg/(m s).

Analysis Because the outlet temperature is not known, we will utilize the effectiveness approach. Calculating q_{max}, we obtain

$$q_{max} = \dot{m}c_P(T_0 - T_1) = 10\,\frac{\text{kg}}{\text{s}}\left(4.18\,\frac{\text{kJ}}{\text{kg °C}}\right)(80°\text{C} - 20°\text{C}) = 2510 \text{ kW}$$

Thus, the actual rate of heat transfer is given by

$$q_c = \epsilon q_{max} = \epsilon 2510 \text{ kW} \tag{a}$$

where the effectiveness ϵ is given by Eq. (8–55) or Fig. 8–14. To obtain ϵ, we must evaluate the number of transfer units $NTU = \bar{h}A_s/(\dot{m}c_P)$, which necessitates the calculation of $\bar{h}$.

The Reynolds number Re is computed by writing

$$G = \frac{\dot{m}}{A} = \frac{10 \text{ kg/s}}{\pi(0.1 \text{ m})^2/4} = 1270 \text{ kg/(m}^2\text{ s)}$$

or $U_b = G/\rho = 1.27$ m/s, and

$$Re = \frac{GD}{\mu} = \frac{[1270 \text{ kg/(m}^2\text{ s)}](0.1 \text{ m})}{0.00101 \text{ kg/(m s)}} = 126{,}000$$

such that the flow is turbulent. Utilizing Eqs. (8–15) and (8–16) to approximate Nu for isothermal conditions, we have

$$f_{cP} = (1.58 \ln 126{,}000 - 3.28)^{-2} = 0.00429$$

$$Nu_{cP} = \frac{(0.00429/2)(126{,}000)(7.01)}{1.07 + 12.7\sqrt{0.00429/2}\,(7.01^{2/3} - 1)} = 719$$

To correct for nonuniform viscosity, we use Eq. (8–24).

$$Nu = Nu_{cP}\left(\frac{\mu_0}{\mu_b}\right)^{-0.11} = 719\left(\frac{3.54 \times 10^{-4}}{1.01 \times 10^{-3}}\right)^{-0.11} = 807$$

Since $x/D = 100$, we approximate $\bar{h}$ by h to obtain

$$\bar{h} = Nu\frac{k}{D} = 807\,\frac{0.603 \text{ W/(m °C)}}{0.1 \text{ m}} = 4860 \text{ W/(m}^2\text{ °C)}$$

The number of transfer units $\bar{h}A_s/(\dot{m}c_P)$ is then found to be

$$\frac{\bar{h}A_s}{\dot{m}c_P} = \frac{[4.86 \text{ kW/(m}^2\text{ °C)}][\pi(0.1 \text{ m})(10 \text{ m})]}{(10 \text{ kg/s})[4.18 \text{ kJ/(kg °C)}]} = 0.365$$

Using Eq. (8–55), we calculate the effectiveness ϵ.

$$\epsilon = 1 - \exp\left(-\frac{\bar{h}A_s}{\dot{m}c_P}\right) = 0.305$$

Returning to Eq. (a), q_c is

$$q_c = 0.305(2510 \text{ kW}) = 766 \text{ kW}$$

To obtain the outlet temperature T_2, we substitute into Eq. (8–31), with the result

$$q_c = \dot{m}c_P(T_2 - T_1)$$

$$T_2 = \frac{766 \text{ kW}}{(10 \text{ kg/s})[4.18 \text{ kJ/(kg °C)}]} + 20\text{°C} = 18.3\text{°C} + 20\text{°C} = 38.3\text{°C}$$

Referring to Table A–C–3, we find that the specific heat of water at 38.3°C is equal to the value at 20°C to within three significant figures. Therefore, our assumption of uniform specific heat is quite adequate. However, we note that the viscosity μ changes considerably over this temperature range. Therefore, our analysis can be refined by approximating the properties at the arithmetic average of the inlet and outlet temperatures. This refinement is suggested as an exercise.

EXAMPLE 8–7

Air at 1.2 atm flowing at a rate of 0.1 kg/s is to be cooled from 400 K to 300 K in a 8.4-cm-diameter tube with uniform wall temperature of 250 K. Determine the length of the tube.

Solution

Objective Determine the length L required to bring the outlet temperature T_2 to 300 K.

Schematic Air flow in a circular tube: uniform wall temperature.

Air
$\dot{m}$ = 0.1 kg/s
P_1 = 1.2 atm
T_1 = 400 K
D = 8.4 cm
$T_s = T_0$ = 250 K
T_2 = 300 K

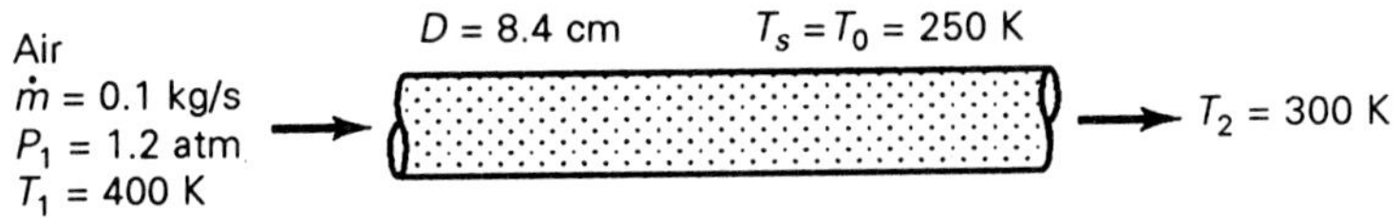

Assumptions/Conditions

forced convection
negligible entrance effects on $\bar{h}$
moderate property variation
standard conditions

Properties Air at the average inlet and outlet temperature of 350 K and 1.2 atm (Table A–C–5): $\mu = 2.08 \times 10^{-5}$ kg/(m s), $c_P = 1.01$ kJ/(kg °C), $k = 0.03$ W/(m °C), $Pr = 0.697$; $\rho = 1.03$ kg/m³ based on the ideal gas law.

Analysis Noting that the specific heat is essentially uniform over the temperature range from 300 K to 400 K, we utilize Eq. (8–31) to calculate q_c.

$$q_c = \dot{m}c_P(T_2 - T_1) = 0.1\,\frac{\text{kg}}{\text{s}}\left(1.01\,\frac{\text{kJ}}{\text{kg K}}\right)(300\text{ K} - 400\text{ K}) = -10.1\text{ kW}$$

Because the inlet and outlet temperatures are known, we can easily utilize the *LMTD* relationship

$$q_c = \bar{h}A_s\, LMTD \tag{8–59}$$

where

$$LMTD = \frac{\Delta T_1 - \Delta T_2}{\ln(\Delta T_1/\Delta T_2)} = \frac{(250\text{ K} - 400\text{ K}) - (250\text{ K} - 300\text{K})}{\ln[(250\text{ K} - 400\text{ K})/(250\text{ K} - 300\text{ K})]} = -91\text{ K}$$

To obtain $\bar{h}$, we calculate the Reynolds number Re by writing

$$G = \frac{\dot{m}}{A} = \frac{0.1\text{ kg/s}}{\pi(0.084\text{ m})^2/4} = 18\text{ kg/(m}^2\text{ s)}$$

or $U_b = G/\rho = 17.5$ m/s, and

$$Re = \frac{GD}{\mu} = \frac{[18\ \text{kg/(m}^2\ \text{s)}](0.084\ \text{m})}{2.08 \times 10^{-5}\ \text{kg/(m s)}} = 72{,}700$$

Thus, the flow is turbulent. Utilizing Eqs. (8–15) and (8–16) to calculate Nu for uniform property conditions, we have

$$f_{cp} = (1.58 \ln 72{,}700 - 3.28)^{-2} = 0.00482$$

$$Nu_{cp} = \frac{(0.00482/2)(72{,}700)(0.697)}{1.07 + 12.7\sqrt{0.00482/2}\,(0.697^{2/3} - 1)} = 130$$

Noting that $n = 0$ in Table 8–2, we conclude that the property effect associated with the wall-to-bulk-stream temperature difference indicated by Eq. (8–26) is negligible. Thus,

$$Nu = Nu_{cp} = 130$$

Approximating $\bar{h}$ by h, we obtain

$$\bar{h} = Nu\frac{k}{D} = 130\frac{0.03\ \text{W/(m °C)}}{0.084\ \text{m}} = 46.4\ \text{W/(m}^2\ \text{°C)}$$

Setting $q_c = -10.1$ W, $LMTD = -91$ K $= -91$°C, and $\bar{h} = 46.4$ W/(m² °C), the length can now be calculated from Eq. (8–59).

$$A_s = \frac{q_c}{\bar{h}\ LMTD} = \frac{-10.1\ \text{kW}}{[46.4\ \text{W/(m}^2\ \text{°C)}](-91\text{°C})} = 2.39\ \text{m}^2$$

$$L = \frac{2.39\ \text{m}^2}{\pi(0.084\ \text{m})} = 9.06\ \text{m}$$

Because the resulting value of L/D (= 108) is fairly large, the approximation used for $\bar{h}$ (i.e., $\bar{h} \simeq h$) is judged to be reasonable.

EXAMPLE 8–8

A 1-cm-diameter tube of 1-m length with inlet contraction and exit expansion represented by $\sigma = 0.7$ is used to heat air at atmospheric pressure and 12°C. Determine the total rate of heat transfer and the outlet temperature for a wall temperature of 175°C, mass-flow rate of 0.00925 kg/s, and uniform inlet conditions.

Solution

Objective Determine q_c and T_2.

Schematic Air flow in a heated tubular core: uniform wall temperature.

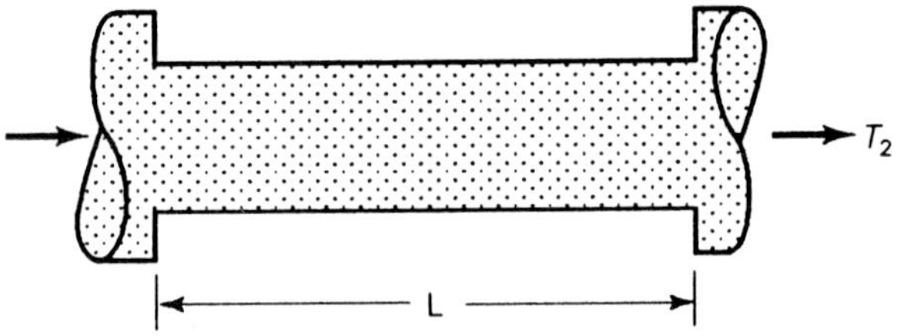

Air
$\dot{m}$ = 0.00925 kg/s
P_1 = 1 atm
T_1 = 12°C

T_2

L

T_0 = 175°C
σ = 0.7
D = 1 cm
L = 1 m

Assumptions/Conditions

forced convection
moderate property variation
negligible entrance effects on $\bar{h}$
standard conditions

Properties Because the outlet temperature T_2 is not known, we will first develop an analysis with the bulk-stream properties evaluated at a convenient temperature of the order of T_1, after which the analysis will be refined by evaluating the properties at the approximate average value of the inlet and exit temperatures.

Air at T = 300 K (Table A–C–5): ρ = 1.18 kg/m^3, μ = 1.85 × 10^{-5} kg/(m s), c_P = 1.01 kJ/(kg °C), k = 0.0262 W/(m °C), and Pr = 0.708.

Analysis The Reynolds number Re is computed by writing

$$G = \frac{\dot{m}}{A} = \frac{0.00925 \text{ kg/s}}{\pi(0.01 \text{ m})^2/4} = 118 \text{ kg/(m}^2\text{ s)}$$

or $U_b = G/\rho$ = 100 m/s = 224 mile/h (indicating low-Mach number subsonic flow),

$$Re = \frac{GD_H}{\mu} = \frac{[118 \text{ kg/(m}^2\text{ s)}](0.01 \text{ m})}{1.85 \times 10^{-5} \text{ kg/(m s)}} = 63{,}800$$

It follows that the flow is fully turbulent, such that Nu can be approximated by Eq. (8–16),

$$Nu = \frac{(f/2)\, Re\, Pr}{1.07 + 12.7\sqrt{f/2}\,(Pr^{2/3} - 1)}$$

with f given by Eq. (8–15),

$$f = (1.58 \ln Re - 3.28)^{-2}$$

Substituting for Re and setting $\overline{Nu} = Nu$ since L/D is fairly large, we obtain f = 0.00496 and $\overline{Nu}$ = 119. To approximately correct for the effect of property variation over the cross section on Nu, we use Eq. (8–26); that is,

$$Nu = Nu_{cp}\left(\frac{T_s}{T_b}\right)^n = 119\left(\frac{448}{300}\right)^{-0.5} = 97.4$$

Therefore, the mean coefficient of heat transfer $\bar{h}$ becomes

$$\bar{h} = h = Nu\frac{k}{D} = 97.4\frac{0.0262\ \text{W/(m °C)}}{0.01\ \text{m}} = 255\ \text{W/(m}^2\ \text{°C)}$$

Because the specific heat c_P is nearly uniform over the temperature range occurring in this problem, the exit temperature T_2 is given by Eq. (8–51),

$$T_2 - T_1 = (T_0 - T_1)\left[1 - \exp\left(-\frac{\bar{h}A_s}{\dot{m}c_P}\right)\right]$$

Substituting for $\bar{h}$ and the other parameters, our first approximation for the number of transfer units is $NTU = \bar{h}A_s/(\dot{m}c_P) = 0.857$. Thus,

$$T_2 = 12°\text{C} + (175°\text{C} - 12°\text{C})[1 - \exp(-0.857)] = 106°\text{C}$$

It follows that the average of the inlet and exit bulk-stream temperatures T_m is approximately 59°C, such that the analysis can be refined by taking the properties as $\rho = 1.07$ kg/m³, $\mu = 1.99 \times 10^{-5}$ kg/(m s), $c_P = 1.01$ kJ/(kg °C), $k = 0.0287$ W/(m °C), and $Pr = 0.701$. Using these property values, we obtain $Re = 59{,}300$, $f_{cp} = 0.00504$, $Nu_{cp} = 112$, $Nu = 91.7$, $\bar{h} = 263$ W/(m² °C), $NTU = 0.885$, and $T_2 = 108$°C. Substituting into Eq. (8–31),

$$q_c = \dot{m}c_P(T_2 - T_1)$$

the rate of heat transfer q_c is 0.897 kW.

EXAMPLE 8–9

Air with a mass-flow rate of 0.00925 kg/s and entering conditions given as $P_1 = 1$ atm and $T_1 = 12$°C is heated by a core of five 1-cm-diameter tubes of 1-m length with a contraction ratio σ of 0.7. Determine the total rate of heat transfer and the outlet temperature for the case in which the wall temperature is 175°C.

Solution

Objective Determine q_c and T_2.

Schematic Air flow in a core of five tubes: uniform wall temperature.

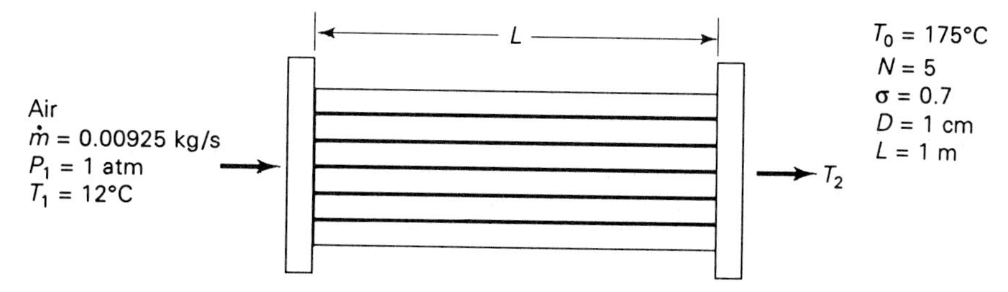

Assumptions/Conditions

forced convection
negligible entrance effects on $\bar{h}$
moderate property variation
standard conditions

Properties Following the approach taken in Example 8–8, the bulk-stream properties will first be evaluated at a convenient value of 300 K, which is on the order of T_1. The properties of air at 300 K are listed in Example 8–8.

Analysis The hydraulic diameter for this multiple-tube core is given by

$$D_H = \frac{4A}{p} = \frac{4N\pi D^2/4}{N\pi D} = D = 0.01 \text{ m}$$

The mass flux G is

$$G = \frac{\dot{m}}{A} = \frac{0.00925 \text{ kg/s}}{5\pi(0.01 \text{ m})^2/4} = 23.6 \text{ kg/(m}^2 \text{ s)}$$

or $U_b = G/\rho = 20$ m/s. It follows that our first-order approximation for the Reynolds number Re becomes

$$Re = \frac{GD_H}{\mu} = \frac{[23.6 \text{ kg/(m}^2 \text{ s)}](0.01 \text{ m})}{1.85 \times 10^{-5} \text{ kg/(m s)}} = 12{,}700$$

Using Eqs. (8–15) and (8–16) to approximate f and Nu for uniform property conditions, we obtain $f_{cp} = 0.00737$ and $Nu_{cp} = 36.4$. The correction for property variation over the cross section obtained by use of Eq. (8–26) gives $Nu = 29.8$ and $h = 78$ W/(m² °C). Using these results together with the approximation $\bar{h} \simeq h$, Eq. (8–51) indicates $NTU = 1.31$ and an outlet bulk-stream temperature T_2 of 131°C. Thus, our first-order approximate solution for $T_m = (T_1 + T_2)/2$ is 71.5°C = 344 K.

To refine our analysis, the properties are evaluated at a temperature of 350°C, which is convenient and sufficiently close to 344 K; that is, $\rho = 0.998$ kg/m³, $\mu = 2.08 \times 10^{-5}$ kg/(m s), $k = 0.03$ W/(m °C), $c_P = 1.01$ kJ/(kg °C), and $Pr = 0.697$. Following through with the calculations, we obtain $Re = 11{,}300$, $f_{cp} = 0.0076$, $Nu_{cp} = 33.2$, $Nu = 29.1$, $\bar{h} = 87.3$ W/(m² °C), $T_2 = 137$°C, and using Eq. (8–31), $q_c = 2.35$ kW.

Solution for Nonuniform Wall Temperature Whereas an analytical solution was easily developed for the situation in which the wall temperature is uniform, analytical solutions generally cannot be obtained for nonuniform wall-temperature heating. However, Eqs. (8–31) and (8–45) can be put into a practical finite-difference form that accounts for variation in T_s. The numerical solution of these equations for nonuniform wall temperature is illustrated in Example 8–10.

EXAMPLE 8–10

Develop a one-dimensional numerical formulation for convection heat transfer in a tube with nonuniform wall temperature T_s. Then solve Example 8–6 for the case in which T_s is given by

$$T_s = 20°\text{C} + 80°\text{C}\left[1 - \cos\left(\frac{4\pi x}{L}\right)\right] \tag{a}$$

Solution

Objective (1) Develop a one-dimensional finite-difference formulation for tube flow with nonuniform wall temperature, and (2) determine T_2 for water with $\dot{m} = 10$ kg/s, $P_1 = 1$ atm, $T_1 = 20°\text{C}$, $D = 10$ cm, $L = 10$ m, and T_s specified by Eq. (a).

Schematic Flow in a circular tube: nonuniform wall temperature.

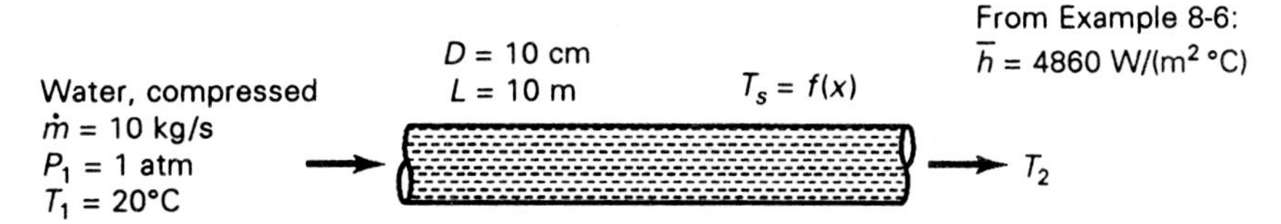

Assumptions/Conditions

forced convection
negligible entrance effects on $\bar{h}$
$\bar{h}$ approximately independent of boundary conditions
uniform properties
standard conditions

Properties Water at 20°C (Table A–C–3): $\rho = 1000$ kg/m^3, $\mu = 0.00101$ kg/(m s), $c_P = 4.18$ kJ/(kg °C), $k = 0.603$ W/(m °C), $Pr = 7.01$.

Analysis Using the practical thermal analysis approach, the mathematical formulation is represented by Eqs. (8–31) and (8–45); that is,

$$q_c = \dot{m}c_P(T_2 - T_1) \tag{b}$$

$$ph_x(T_s - T_b) = \dot{m}c_P\frac{dT_b}{dx} \tag{c}$$

where

$$T_b = T_1 \qquad \text{at } x = 0 \tag{d}$$

To develop a solution to Eqs. (b) and (c) for the case in which T_s is nonuniform, these equations are written in finite-difference format as follows:

$$q_c = \dot{m}c_P(T_{b,M} - T_{b,1}) \tag{e}$$

and

$$ph_m(T_{s,m} - T_{b,m}) = \frac{\dot{m}}{\Delta x} c_P(T_{b,m+1} - T_{b,m}) \tag{f}$$

or

$$T_{b,m+1} = \frac{ph_m \, \Delta x}{\dot{m}c_P} T_{s,m} + \left(1 - \frac{ph_m \, \Delta x}{\dot{m}c_P}\right) T_{b,m} \tag{g}$$

where $T_{b,1} = T_1$ and m ranges from 1 to M as shown in Fig. E8–10a.

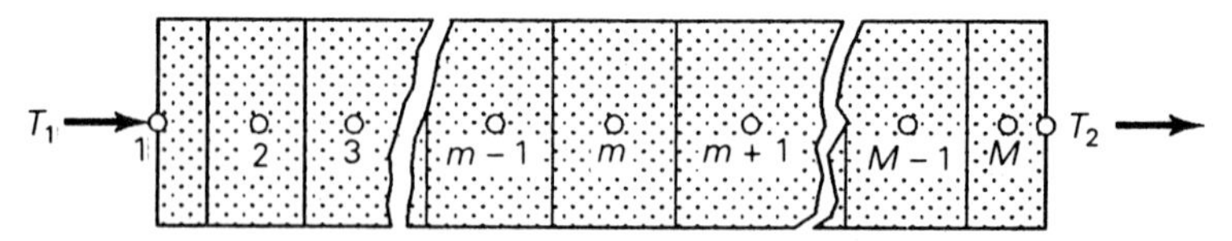

FIGURE E8–10a Finite-difference grid for tube flow: convection with T_s and $\bar{h}$ specified.

To develop numerical calculations for the case in which $h \simeq \bar{h}$ and T_s is given by Eq. (a), Eqs. (e) and (g) are written in BASIC (or FORTRAN) as follows:

```
T(J+1)=(P*H/(MDOT*CP))*DX*TS
        +T(J)*(1-(P*H/(MDOT*CP)*DX))          (h)

QC=MDOT*CP*(T(M)-T(1))                         (i)
```

where $D = 0.1$ m, $L = 10$ m, $h = \text{H} = 4860$ W/(m^2 °C), $p = \text{P} = \pi D$, $\dot{m} = \text{MDOT} = 10$ kg/s, and $c_P = \text{CP} = 4.18$ kJ/(kg °C). A BASIC program is given in Appendix A–G–4, which solves these equations for any specified wall temperature. Calculations are shown for T_b in Fig. E8–10b for T_s specified by Eq. (a). The solution is found to converge for values of M on the order of 100. The outlet temperature T_2 is 44.3°C and the total rate of heat transfer q_c is 1020 kW.

To serve as a check and for purpose of comparison, the program is also run with T_s set equal to 80°C. The calculations for this uniform wall-temperature condition are also shown in Fig. E8–10b. The outlet temperature is 38.3°C and the rate of heat transfer is 763 kW. These results are consistent with the solution developed in Example 8–6.

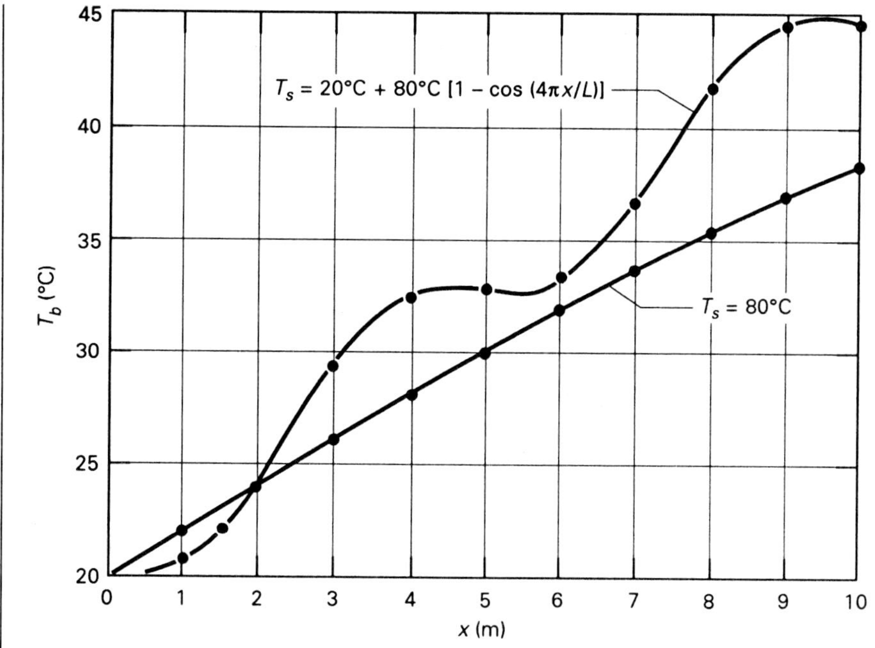

FIGURE E8–10b Calculations for bulk-stream temperature T_b.

Variable Property Flow

For small to moderate variations in specific heat c_P over the length of a passage, a uniform property analysis can generally be used with c_P approximated by the average of the values occurring at the inlet and exit. However, for situations in which significant changes in c_P occur, the effect of property variation should be more formally accounted for.

The mathematical formulation obtained by means of the practical thermal analysis approach for variable property flow is given by Eq. (8–30),

$$q_c = \dot{H}_2 - \dot{H}_1 = \dot{m} \int_{T_1}^{T_2} c_P \, dT_b \tag{8–30}$$

and Eq. (8–34),

$$q_c''p = \dot{m}c_P \frac{dT_b}{dx} \tag{8–34}$$

with $T_b = T_1$ at $x = 0$, where c_P is a specified function of T_b.

As illustrated in Example 8–11, these equations are readily solved for the case of specified wall-heat-flux conditions, with the solutions being of an analytical or numerical character, depending on whether the integrals $\int q_c'' \, dx$ and $\int c_P \, dT_b$ can be evaluated analytically or require numerical quadrature. On the other hand, for specified wall-temperature conditions, Eq. (8–34) is put into the form of Eq. (8–45),

$$h_x p(T_s - T_b) = \dot{m} c_P \frac{dT_b}{dx} \tag{8–45}$$

As in the case of nonuniform wall-temperature heating, the solution of this equation for variable c_P generally requires the use of numerical methods similar to the approach used in Example 8–10.

EXAMPLE 8–11

Engine oil flowing at a rate of 0.25 kg/s is to be heated from 20°C to 140°C in a 10-cm-diameter tube with uniform wall-heat flux maintained over a 10-m length. Determine the required heat flux and the distribution in bulk-stream temperature.

Solution

Objective Determine the required value of q_0'' and the distribution in T_b.

Schematic Flow of engine oil in a tube: uniform wall-flux heating.

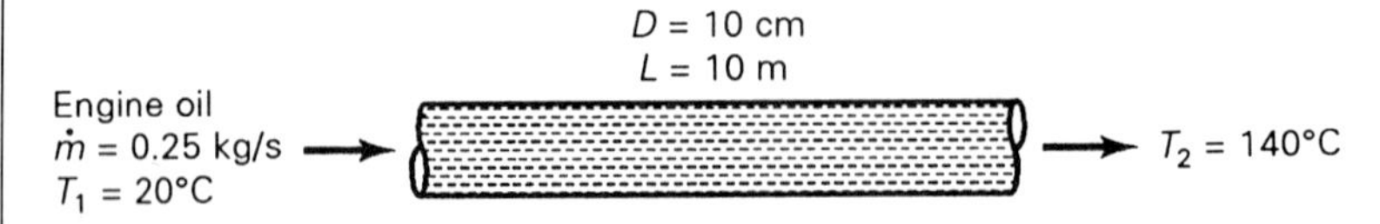

Assumptions/Conditions

forced convection
variable c_P
standard conditions

Properties Referring to Table A–C–3, we find that the specific heat of engine oil varies by about 27% for temperature ranging from 20°C to 140°C. By plotting c_P versus temperature, the variation is found to be well approximated by a linear equation of the form

$$c_P = a + bT$$

using the absolute temperature scale, where $a = 0.61$ kJ/(kg K) and $b = 0.00433$ kJ/(kg K^2).

Analysis To determine the heat-transfer rate, this correlation for c_P is substituted into Eq. (8–30), with the result

$$q_c = \dot{m} \int_{T_1}^{T_2} c_P \, dT_b = \dot{m} \int_{T_1}^{T_2} (a + bT_b)\, dT_b = \dot{m}\left(aT_b + \frac{b}{2} T_b^2 \right)\Bigg|_{T_1}^{T_2}$$

$$= 0.25 \frac{\text{kg}}{\text{s}} \left\{ \left(0.61 \frac{\text{kJ}}{\text{kg K}} \right) (413 \text{ K} - 293 \text{ K}) + \frac{0.00433}{2} \frac{\text{kJ}}{\text{kg K}^2} [(413 \text{ K})^2 - (293 \text{ K})^2] \right\} = 64.2 \text{ kW}$$

and

$$q_0'' = \frac{q_c}{\pi D L} = \frac{64.2 \text{ kW}}{\pi(0.1 \text{ m})(10 \text{ m})} = 20.4 \text{ kW/m}^2$$

The local bulk-stream temperature T_b is represented by Eq. (8–34),

$$q_c'' p = q_0'' p = \dot{m} c_P \frac{dT_b}{dx}$$

Separating and integrating, we have

$$\int_0^x q_0'' p \, dx = \dot{m} \int_{T_1}^{T_b} c_P \, dT_b = \dot{m} \int_{T_1}^{T_b} (a + bT_b)\, dT_b$$

$$q_0'' p x = \dot{m} \left(aT_b + \frac{b}{2} T_b^2 \right)\Bigg|_{T_1}^{T_b} = \dot{m} \left[a(T_b - T_1) + \frac{b}{2} (T_b^2 - T_1^2) \right]$$

or

$$\frac{b}{2} T_b^2 + aT_b - \left(\frac{q_0'' p x}{\dot{m}} + aT_1 + \frac{b}{2} T_1^2 \right) = 0$$

The solution to this quadratic equation is

$$T_b = -\frac{a}{b} + \left[\left(\frac{a}{b} \right)^2 + 2 \left(\frac{q_0'' p x}{b \dot{m}} + \frac{a}{b} T_1 + \frac{T_1^2}{2} \right) \right]^{1/2}$$

$$= -141 \text{ K} + \left(1.88 \times 10^5 + 1.19 \times 10^5 \frac{x}{L} \right)^{1/2} \text{ K} \qquad \text{(a)}$$

which is shown in Fig. E8–11.

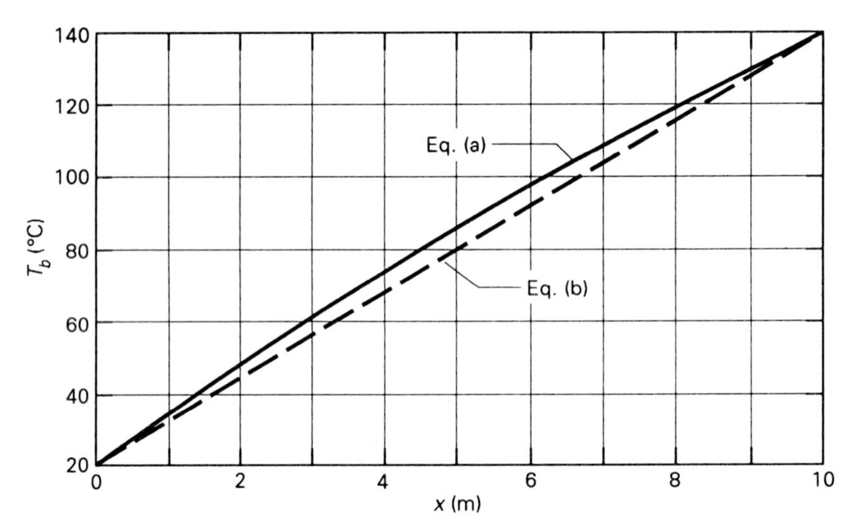

FIGURE E8–11 Calculations for bulk-stream temperature T_b.

It should be noted that for this problem in which the specific heat varies linearly with temperature, the use of a uniform property analysis with c_P evaluated at the arithmetic mean of the inlet and outlet temperatures produces reasonable accuracy. For example, a uniform property analysis for T_b gives rise to Eq. (8–40),

$$T_b - T_1 = \frac{q_0'' px}{\dot{m}c_P} = \frac{q_c}{\dot{m}c_P}\frac{x}{L}$$

where $c_P = 2.14$ kJ/(kg °C) and $q_c = 64.2$ kW. Substituting into this equation, we obtain

$$T_b = 20°\text{C} + 120°\text{C}\,\frac{x}{L} \qquad \text{(b)}$$

This equation is shown in Fig. E8–11 to be within 4.4% of Eq. (a).

Fin Tubes

Fins are often used to enhance the heat-transfer performance for flow of gases or viscous fluids that are characterized by low coefficients of heat transfer. The cross section of a typical annular passage with longitudinal fins is shown in Fig. 8–15.

The geometric characteristics of fin tubes include fin *type* (i.e., longitudinal, circumferential, internal, external, etc.), fin *dimensions* (e.g., height H_F and thickness δ_F), number of fins per tube N_F, and tube diameter(s). With the basic geometric characteristics specified, calculations can be made for the cross-sectional area A (sometimes referred to as the *net flow area*, wetted perimeter p_w, and total area of finned surface A_s. The total surface area A_s and total fin surface area A_{fins} are represented

FIGURE 8–15
Longitudinal fin-tube double pipe, $N_F = 36$.

by†

$$A_s = N_F(A_F + A_p) = N_F A_o \tag{8–62}$$

and

$$A_{\text{fins}} = N_F A_F \tag{8–63}$$

where A_F is the surface area of an individual fin, A_p is the area of the associated prime surface, and $A_o = A_F + A_p$. This information is generally expressed in terms of the *finned area fraction*.

$$\frac{A_{\text{fins}}}{A_s} = \frac{A_F}{A_o} \tag{8–64}$$

the perimeter of the finned surface p,

$$p = \frac{A_s}{L} \tag{8–65}$$

and the hydraulic diameter D_H,

$$D_H = \frac{4A}{p_w} \tag{8–66}$$

Notice that the wetted perimeter p_w and finned surface perimeter p are related by $p = p_w$ for an internally finned tube, and

$$p_w = \pi D_{i,s} + p \tag{8–67}$$

for an annulus with externally finned tubes. To illustrate, geometric characteristics and relations are given in Table 8–4 for standard longitudinal fin tubes of the type shown in Fig. 8–15.

As indicated in Sec. 8–2–1, empirical correlations for friction factor f and the Nusselt number Nu (assuming $Nu_i \simeq Nu$) have been established for standard fin-tube units available in the market. Representative correlations for flow in a longitudinal

† The characteristic lengths and areas are expressed in terms of the fin dimensions H_F and δ_F and number of fins per tube N_F in Table 8–4.

TABLE 8–4 Geometric dimensions of representative longitudinal fin-tube double pipes.†

Outer pipe		Inner pipe								
IPS (in.)	$D_{i,s}$ (mm)	d_o (mm)	d_i (mm)	N_F	H_F (mm)	A (mm^2)	$A_{s,o}/L$ (m^2/m)	D_H (mm)	A_F/A_o	A_o/A_i
1 1/2	40.9	12.7	9.40	8	12.7	1100	0.243	11.8	0.865	8.23
1 1/2	40.9	12.7	9.40	12	12.7	1050	0.345	8.89	0.915	11.7
2	52.5	25.4	19.9	16	12.7	1480	0.486	9.08	0.865	7.79
2	52.5	25.4	19.9	20	12.7	1430	0.588	7.61	0.894	9.42
2 1/2	62.7	26.7	21.0	12	12.7	2390	0.389	16.4	0.812	5.90
2 1/2	62.7	26.7	21.0	16	12.7	2350	0.490	13.7	0.858	7.45
3	77.9	25.4	19.9	16	25.4	3900	0.893	13.7	0.926	14.3
3	77.9	25.4	19.9	20	25.4	3810	1.10	11.4	0.943	17.6
3	77.9	48.3	40.9	24	12.7	2670	0.761	10.6	0.829	5.93
3	77.9	48.3	40.9	28	12.7	2620	0.863	9.48	0.853	6.72
3 1/2	90.1	48.3	40.9	24	19.0	4140	1.07	12.3	0.878	8.30
3 1/2	90.1	48.3	40.9	28	19.0	4080	1.22	10.9	0.896	9.48
4	102	48.3	40.9	28	25.4	5750	1.57	12.1	0.920	12.3
4	102	48.3	40.9	36	25.4	5570	1.98	9.68	0.940	15.4

Geometric relations‡

$$A_s = A_o N_F = [(2H_F + \delta_F)N_F + \pi d_o - N_F\delta_F]L \qquad A_{\text{fins}} = A_F N_F = [(2H_F + \delta_F)N_F]L$$

$$\frac{A_{\text{fins}}}{A_s} = \frac{A_F}{A_o} = \frac{(2H_F + \delta_F)N_F}{(2H_F + \delta_F)N_F + \pi d_o - N_F\delta_F} \qquad p = \frac{A_s}{L} = (2H_F + \delta_F)N_F + \pi d_o - N_F\delta_F$$

$$\frac{A_o}{A_i} = \frac{A_s}{\pi d_i} = \frac{A_s/L}{\pi d_i} \qquad A = \frac{\pi}{4}(D_{i,s}^2 - d_o^2) - H_F\delta_F N_F$$

$$p_w = \pi D_{i,s} + p \qquad D_H = \frac{4A}{p_w}$$

† Schedule 40 shell STD (standard pressure units). Fin thickness, δ_F = 0.035 in. = 0.889 mm. Other values for δ_F commonly found in practice include 0.610 mm (0.024 in.) and 1.27 mm (0.05 in.).

‡ These geometric relations were used to calculate the values for A, $A_{s,o}/L$, D_H, A_F/A_o and A_o/A_i listed in this table. Alternative relations are sometimes used which are based on an approximation for the root width of the fin channel by $R_{w,F} = \pi(d_o + \delta_F)/N_F$.

fin-tube annulus are shown in Fig. 8–16. These correlations for f and Nu can be approximated by Eq. (8–15),

$$f = (1.58 \ln Re - 3.28)^{-2} \tag{8–15}$$

and Eq. (8–16),

$$Nu = \frac{(f/2)\,Re\,Pr}{1.07 + 12.7\sqrt{f/2}\,(Pr^{2/3} - 1)} \tag{8–16}$$

for fully turbulent flow with Re above about 6000. The correlations represented in Fig. 8–16 are adapted to other values of Prandtl number and to fin tubes with cut-and-twist enhancement in reference 43.

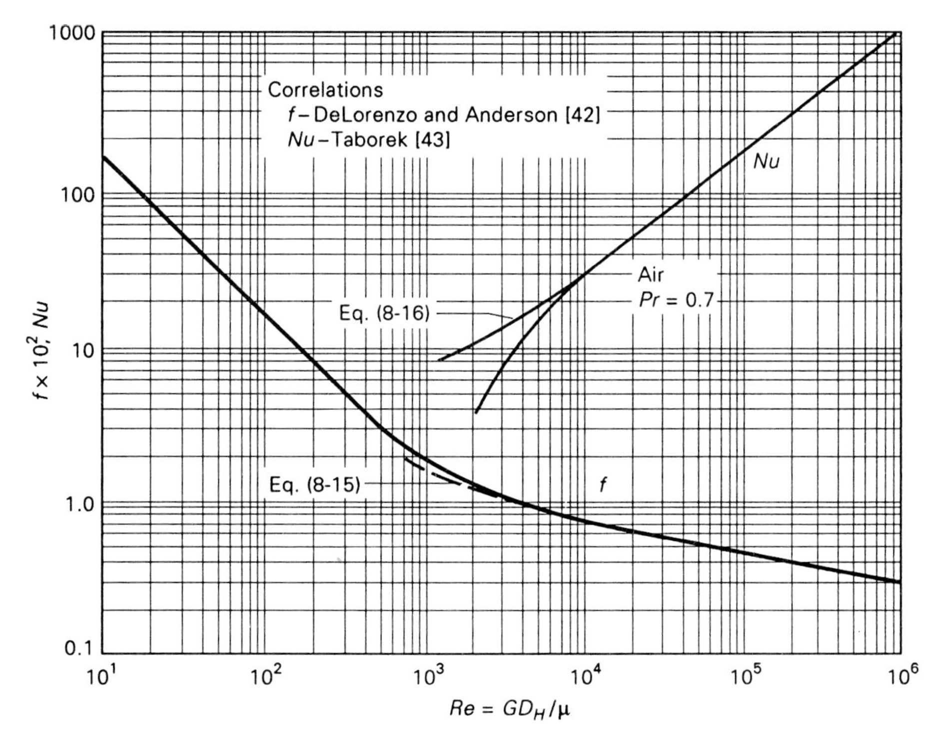

FIGURE 8–16 Correlations for Fanning friction factor f and Nusselt number Nu for flow in longitudinal fin-tube annulus. (Assume $Nu_i \simeq Nu$.)

To extend the practical thermal analysis approach to flow in a fin-tube annulus, we must account for the difference in base temperature and fin-surface temperature. Referring back to Sec. 2–6–4, this is done by expressing the local heat flux q_c'' in the form of Eq. (2–132),

$$q_c'' = \eta_o h_x (T_s - T_b) \tag{8–68}$$

where the *net surface efficiency* η_o is defined by Eq. (2–127),

$$\eta_o = 1 - \frac{A_F}{A_o}(1 - \eta_F) \tag{8–69}$$

and η_F is the *fin efficiency*. Curves for η_F are shown in Fig. 2–17 for longitudinal and other type fins. Combining Eq. (8–69) with Eq. (8–34), we obtain

$$\eta_o h_x p (T_s - T_b) = \dot{m} c_P \frac{dT_b}{dx} \tag{8–70}$$

The solution to this equation for uniform wall temperature T_0 and uniform properties is given by

$$\frac{T_2 - T_1}{T_0 - T_1} = 1 - \exp\left(\frac{-\overline{\eta_o h}\, A_s}{\dot{m}c_P}\right) \tag{8–71}$$

Combining this result with Eq. (8–31),

$$q_c = \dot{m}c_P(T_2 - T_1) \tag{8–31}$$

we obtain

$$q_c = \dot{m}c_P(T_0 - T_1)\left[1 - \exp\left(\frac{-\overline{\eta_o h}\, A_s}{\dot{m}c_P}\right)\right] \tag{8–72}$$

or

$$q_c = \overline{\eta_o h} A_s\, LMTD \tag{8–73}$$

where $LMTD$ is defined by Eq. (8–60), and

$$\overline{\eta_o h} = \frac{1}{L}\int_0^L \eta_o h_x\, dx = \frac{1}{A_s}\int_{A_s} \eta_o h_x\, dA_s \tag{8–74}$$

In practice, η_o is generally taken as a constant, such that $\overline{\eta_o h} \simeq \eta_o \overline{h}$.

EXAMPLE 8–12

Air at 1.2 atm flowing at a rate of 0.1 kg/s is to be cooled from 400 K to 300 K in the longitudinal fin-tube annulus shown in Fig. E8–12. The outer surface is insulated and the inner surface is maintained at a uniform temperature of 250 K. Determine the length of the unit.

$N_F = 36$
$H_F = 1$ in. $= 25.4$ mm
$\delta_F = 0.035$ in. $= 0.889$ mm
$D_{i,s} = 102$ mm
$d_o = 48.3$ mm
$d_i = 40.9$ mm

FIGURE E8–12 Longitudinal fin-tube annulus.

Solution

Objective Determine the length L required to bring the outlet temperature T_2 to 300 K.

Schematic Air flow in a longitudinal fin-tube annulus: uniform wall temperature at inner surface, outer surface insulated.

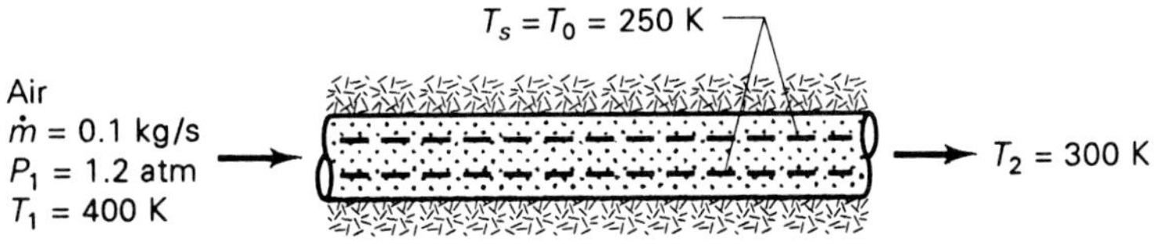

Air
$\dot{m}$ = 0.1 kg/s
P_1 = 1.2 atm
T_1 = 400 K

$T_s = T_0$ = 250 K

T_2 = 300 K

Assumptions/Conditions

forced convection
negligible entrance effects on $\bar{h}$
negligible heat transfer through fin tip
moderate property variation
standard conditions

Properties

Air at T_m = 350 K and P = 1.2 atm (Table A–C–5): $\mu = 2.08 \times 10^{-5}$ kg/(m s), c_P = 1.01 kJ/(kg °C), k = 0.03 W/(m °C), Pr = 0.697, ρ = 1.03 kg/m^3 (based on ideal gas law).

Carbon steel (0.5% C) [Table A–C–1(a)]: k = 54 W/(m °C).

Analysis Aside from the fact that we are dealing with flow in a longitudinal fin-tube annulus, the analysis follows the pattern of the analysis developed in Example 8–7 for the flow of air in a circular tube with uniform wall temperature. Thus, we are immediately able to write

$$q_c = \dot{m}c_P(T_2 - T_1) = 0.1\frac{\text{kg}}{\text{s}}\left(1.01\frac{\text{kJ}}{\text{kg K}}\right)(300\text{ K} - 400\text{ K}) = -10.1\text{ kW}$$

from Eq. (8–31), and

$$q_c = \eta_o\bar{h}A_s\ LMTD \tag{a}$$

from Eq. (8–73), where η_o is the net surface efficiency and

$$LMTD = \frac{\Delta T_1 - \Delta T_2}{\ln(\Delta T_1/\Delta T_2)}$$

$$= \frac{(250\text{ K} - 400\text{ K}) - (250\text{ K} - 300\text{ K})}{\ln[(250\text{ K} - 400\text{ K})/(250\text{ K} - 300\text{ K})]} = -91\text{ K}$$

To provide a basis for evaluating $\bar{h}$ and η_o, we turn to Table 8–4 which indicates the following geometric dimensions for this fin-tube application:

$$A = 5570 \text{ mm}^2 \qquad A_s/L = 1.98 \text{ m}$$

$$D_H = 9.68 \text{ mm} \qquad A_F/A_o = 0.94$$

Each of these values is readily calculated using the geometric relations listed in Table 8–4.

These inputs for cross-sectional area A and hydraulic diameter D_H are used to compute the Reynolds number Re by writing

$$G = \frac{\dot{m}}{A} = \frac{0.1 \text{ kg/s}}{0.00557 \text{ m}^2} = 18 \text{ kg/(m}^2 \text{ s)}$$

or $U_b = G/\rho = 17.5$ m/s, and

$$Re = \frac{GD_H}{\mu} = \frac{[18 \text{ kg/(m}^2 \text{ s)}](0.00968 \text{ m})}{[2.08 \times 10^{-5} \text{ kg/(m s)}]} = 8380$$

This value of Re indicates essentially fully turbulent conditions, such that f and Nu can be approximated by Eqs. (8–15) and (8–16); that is,

$$f_{cp} = (1.58 \ln 8380 - 3.28)^{-2} = 0.00827$$

$$Nu_{cp} = \frac{(0.00827/2)(8380)(0.697)}{1.07 + 12.7 \sqrt{0.00827/2}\,(0.697^{2/3} - 1)} = 27$$

and since $n = 0$ from Table 8–2,

$$Nu = Nu_{cp} = 27$$

It follows that

$$\bar{h} = h = 27 \frac{0.03 \text{ W/(m °C)}}{0.00968 \text{ m}} = 83.7 \text{ W/(m}^2 \text{ °C)}$$

The net surface efficiency η_o is expressed in terms of the fin efficiency η_F by Eq. (8–69),

$$\eta_o = 1 - \frac{A_F}{A_o}(1 - \eta_F) \tag{8–69}$$

To evaluate η_F, we assume negligible heat transfer from the fin tip and write†

$$\eta_F = \frac{\sqrt{\bar{h}p^*k_FA^*}}{\bar{h}A_F} \tanh (mH_F) = \frac{\tanh (mH_F)}{mH_F} = \frac{\exp (2mH_F) - 1}{mH_F[\exp (2mH_F) + 1]}$$

† To distinguish the geometric characteristics of a fin from those of the annular-flow passage, we designate $A^* = L\delta_F$ for the cross-sectional area of a fin, and $p^* = 2L$ for the perimeter of a fin. Thus, the fin parameter m is represented by $m^2 = \bar{h}p^*/(k_FA^*)$.

from Eq. (2–124) and Table 2–2, where

$$(mH_F)^2 = \frac{\bar{h}p^*H_F^2}{k_F A^*} = \frac{2\bar{h}H_F^2}{k_F \delta_F} = \frac{2[83.7 \text{ W/(m}^2\ ^\circ\text{C)}](0.0254 \text{ m})^2}{[54 \text{ W/(m }^\circ\text{C)}](0.000889 \text{ m})} = 2.25$$

$$mH_F = 1.5$$

Thus, we obtain

$$\eta_F = \frac{\tanh 1.5}{1.5} = 0.603$$

and, upon substitution into Eq. (8–69),

$$\eta_o = 1 - 0.94(1 - 0.603) = 0.627$$

The fin-side surface area A_s can now be calculated from Eq. (a).

$$A_s = \frac{q_c}{\eta_o \bar{h}\ LMTD} = \frac{-10.1 \text{ kW}}{0.627[83.7 \text{ W/(m}^2\ ^\circ\text{C)}](-91^\circ\text{C})} = 2.11 \text{ m}^2$$

Setting $A_s/L = 1.98$ m from Table 8–4, the unit length is obtained by writing

$$L = \frac{2.11 \text{ m}^2}{1.98 \text{ m}} = 1.07 \text{ m}$$

To see the relative effectiveness of the longitudinal fin-tube surface for this application, we note that use of a circular tube with essentially the same cross-sectional area requires a length of 9.06 m (see Example 8–7), which is a factor of approximately 8.5. Because the fin surface is relatively inexpensive compared to the prime tube surface, we conclude that the use of finned surfaces can result in cost-effective designs.

8–2–3 Hydraulic Considerations

The pressure drop ΔP, which must be overcome by pumps or blowers for flow through heat-transfer cores such as the tubular passage shown in Fig. 8–17, can be an important design consideration. The total pressure drop ΔP_{1-2} $(= P_1 - P_2)$ across a core such as this includes an inlet core pressure drop ΔP_{1-c} $(= P_1 - P_c)$ caused by the abrupt contraction, a pressure drop ΔP_{c-e} $(= P_c - P_e)$ due to friction and momentum rate change within the core, and a core exit pressure rise ΔP_{2-e} $(= P_2 - P_e)$ resulting from the increase in cross-sectional area at the outlet.

We now turn our attention to the practical hydraulic analysis approach to establishing the pressure drop ΔP_{1-2} and the related pumping power $\dot{W}_p$ for steady incompressible flow through tubular-heating cores. Although the flow will be treated as incompressible [i.e., $\rho \neq \text{fn}(P)$], we will account for changes in bulk-stream density caused by differences in temperature T_b over the length of the heated core.

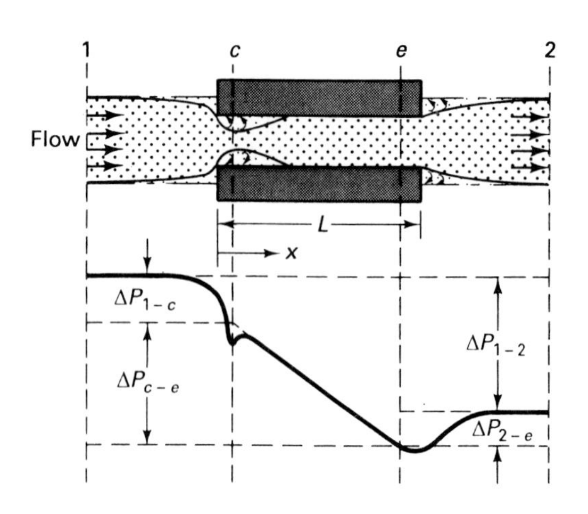

FIGURE 8–17
Entrance, core, and exit components of pressure drop for flow through a tubular passage. (From Shah [44]. Used with permission.)

Relations are developed in Appendix M for the total pressure drop ΔP_{1-2} in terms of f, K_c, K_e, and σ for both uniform and variable property flows. The total pressure drop for the case of uniform property flow is given by†

$$\Delta P_{1-2} = \frac{\rho U_b^2}{2}\left(K_c + \frac{4L}{D_H}f + K_e\right) \tag{8–75}$$

Notice that this equation reduces to

$$\Delta P_{1-2} = \frac{\rho U_b^2}{2}\frac{4L}{D_H}f \tag{8–76}$$

for negligible entrance and exit losses. For situations involving gas flow with large temperature differences, ΔP_{1-2} is represented by

$$\Delta P_{1-2} = \frac{G^2 v_1}{2}\left[(1 - \sigma^2 + K_c) + \frac{4L}{D_H}f\frac{v_m}{v_1} + 2\left(\frac{v_2}{v_1} - 1\right) - (1 - \sigma^2 - K_e)\frac{v_2}{v_1}\right] \tag{8–77}$$

which accounts for the variation in specific volume and bulk-stream velocity over the length of the core; the mean specific volume v_m is generally approximated by the arithmetic average of the inlet and exit values.

† Equation (8–75) is obtained from the more general form (Appendix M)

$$\Delta P_{1-2} = \frac{\rho U_b^2}{2}\left[(1 - \sigma^2 + K_c) + \frac{4L}{D_H}f - (1 - \sigma^2 - K_e)\right]$$

For the case in which $K_e = 0$ and $\sigma = 1$ at the exit, ΔP_{1-2} becomes

$$\Delta P_{1-2} = \frac{\rho U_b^2}{2}\left(1 - \sigma^2 + K_c + \frac{4L}{D_H}f\right)$$

The pumping power $\dot{W}_p$, which is required to overcome a pressure drop ΔP, is

$$\dot{W}_p = \frac{\dot{m}\,\Delta P}{\rho} \qquad (8\text{–}78)$$

for uniform density, or

$$\dot{W}_p = \dot{m}\,\Delta P\,v_m \qquad (8\text{–}79)$$

for nonuniform properties.

To examine the nature of the pumping power, we consider HFD flow with uniform properties. Using Eq. (8–76) to represent ΔP for HFD flow, we obtain

$$\dot{W}_p = \frac{(\rho A U_b)[(\rho U_b^2/2)(4L/D_H)f]}{\rho} = \frac{\mu^3}{\rho^2}\frac{p_w L}{D_H^3}\frac{f}{2}Re^3 \qquad (8\text{–}80)$$

With the friction factor represented by Eq. (8–7) for laminar flow and by Eq. (8–17) for turbulent flow, Eq. (8–80) reduces to

$$\dot{W}_p = \frac{C_1}{2}\frac{\mu^3}{\rho^2}\frac{p_w L}{D_H^3}Re^2 = \frac{C_1}{8}\frac{\mu}{\rho^2}\frac{\dot{m}^2 p_w^2 L}{A^3} \qquad (8\text{–}81)$$

for laminar flow, and

$$\dot{W}_p = 0.023\frac{\mu^3}{\rho^2}\frac{p_w L}{D_H^3}Re^{2.8} = \frac{0.023}{4^{0.2}}\frac{\mu^{0.2}}{\rho^2}\frac{\dot{m}^{2.8} p_w^{1.2} L}{A^3} \qquad (8\text{–}82)$$

for turbulent flow. These equations indicate that, among other things, the pumping power is inversely proportional to the *square* of the fluid density for both laminar and turbulent flow. Therefore, the pumping power is generally small for liquids, but can be quite significant for gases and other low-density fluids. In this connection, it is quite possible to expend as much mechanical energy in pumping low-density fluids as is transferred as heat. Because mechanical energy is generally valued at 4 to 10 times as much as its equivalent in heat transfer for most thermal-power systems [4], the hydraulic considerations are often critical in the development of economical heat-exchanger designs. As shown in Appendix M, the pumping–heating power ratio $\dot{W}_p/q_c$ is approximately proportional to the square of the bulk-stream velocity, U_b^2 {i.e., $[\dot{m}/(\rho A)]^2$}. Thus, for applications in which the pumping power is likely to be important, the designer can increase the cross-sectional area in the heat-exchanger core by increasing the number of flow passages or the hydraulic diameter.

EXAMPLE 8–13

Water at 27°C and 1.2 atm is heated in a core consisting of a 2-cm-diameter tube with 3-m length and 0.5 contraction ratio. The mass-flow rate is 0.25 kg/s and the tube wall is maintained at 50°C. Determine the heat-transfer rate, the pressure drop, and the pumping power.

Solution

Objective Determine q_c, ΔP_{1-2}, and $\dot{W}_p$.

Schematic Flow of water in a heated tubular core: uniform wall temperature.

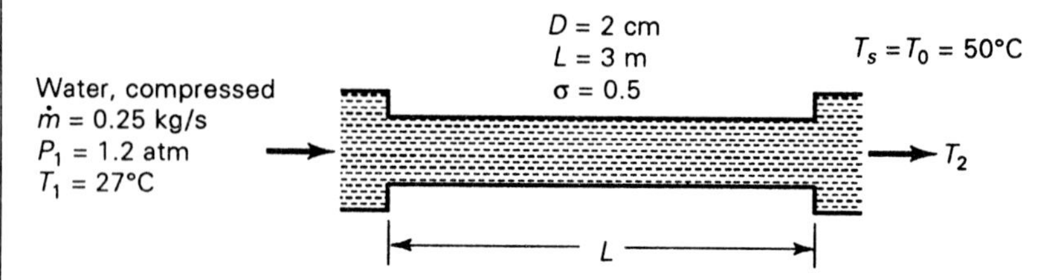

Assumptions/Conditions

forced convection
negligible entrance effects on $\bar{h}$
moderate property variation
standard conditions

Properties

Water at $T = 27°C = 300$ K (Table A–C–3): $\rho = 1/v_f = 997$ kg/m³, $\mu = 8.55 \times 10^{-4}$ kg/(m s), $c_P = 4.18$ kJ/(kg °C), $k = 0.613$ W/(m °C), $Pr = 5.83$.

Water at $T = 50°C = 323$ K: $\mu = 5.48 \times 10^{-4}$ kg/(m s).

Analysis To compute the Reynolds number Re we write

$$G = \frac{\dot{m}}{A} = \frac{0.25 \text{ kg/s}}{\pi(0.02 \text{ m})^2/4} = 796 \text{ kg/(m}^2\text{ s)}$$

or $U_b = G/\rho = 0.798$ m/s, and

$$Re = \frac{GD}{\mu} = \frac{[796 \text{ kg/(m}^2\text{ s)}](0.02 \text{ m})}{8.55 \times 10^{-4} \text{ kg/(m s)}} = 18{,}600$$

such that the flow is turbulent.

Since the tube length-to-diameter ratio L/D (= 150) is quite large, we will approximate the mean coefficient of heat transfer $\bar{h}$ by the value for TFD flow, h. Using Eqs. (8–15) and (8–16) to compute f and Nu for uniform property conditions, we obtain

$$f_{cp} = [1.58 \ln 18{,}600 - 3.28]^{-2} = 0.00666$$

and

$$Nu_{cp} = \frac{(0.00666/2)(18{,}600)(5.83)}{1.07 + 12.7\sqrt{0.00666/2}\,(5.83^{2/3} - 1)} = 133$$

To correct for the effect of property variation on friction factor and Nusselt number, we write

$$f = f_{cp}\left(\frac{\mu_0}{\mu_b}\right)^{0.25} = 0.00666\left(\frac{5.48 \times 10^{-4}}{8.55 \times 10^{-4}}\right)^{0.25} = 0.00596$$

from Eq. (8–23), and

$$Nu = Nu_{cp}\left(\frac{\mu_0}{\mu_b}\right)^{-0.11} = 133\left(\frac{5.48 \times 10^{-4}}{8.55 \times 10^{-4}}\right)^{-0.11} = 140$$

from Eq. (8–24). It follows that

$$\bar{h} = h = Nu\frac{k}{D} = 140\,\frac{0.613\ \text{W/(m °C)}}{0.02\ \text{m}} = 4290\ \text{W/(m}^2\ \text{°C)}$$

The rate of heat transfer q_c can now be computed by use of Eq. (8–52),

$$q_c = \dot{m}c_P(T_0 - T_1)\left[1 - \exp\left(-\frac{\bar{h}A_s}{\dot{m}c_P}\right)\right]$$

where the number of transfer units NTU is

$$NTU = \frac{\bar{h}A_s}{\dot{m}c_P} = \frac{[4290\ \text{W/(m}^2\ \text{°C)}][\pi(0.02\ \text{m})(3\ \text{m})]}{(0.25\ \text{kg/s})[4.18\ \text{kJ/(kg °C)}]} = 0.774$$

Substituting for NTU and other parameters, we obtain

$$q_c = 0.25\,\frac{\text{kg}}{\text{s}}\left(4.18\,\frac{\text{kJ}}{\text{kg °C}}\right)(50\text{°C} - 27\text{°C})[1 - \exp(-0.774)] = 13\ \text{kW}$$

It also follows that the outlet bulk-stream temperature T_2 becomes

$$T_2 = 27\text{°C} + \frac{13\ \text{kW}}{(0.25\ \text{kg/s})[4.18\ \text{kJ/(kg °C)}]} = 39.4\text{°C}$$

from Eq. (8–31).

Turning to the hydraulic analysis, assuming that the change in density over the length of the core is small, the total pressure drop is represented by Eq. (8–75),

$$\Delta P_{1-2} = \frac{\rho U_b^2}{2}\left(K_c + \frac{4L}{D}f + K_e\right) = \frac{G^2}{2\rho}\left(K_c + \frac{4L}{D}f + K_e\right) \qquad (8\text{–}75)$$

Using Fig. 8–5 to evaluate the entrance-loss coefficient K_c and the expansion-loss coefficient K_e for $\sigma = 0.5$ and $Re = 18{,}600$, we find that $K_c = 0.28$ and $K_e = 0.21$. Substituting into Eq. (8–75), the pressure drop becomes

$$\Delta P_{1-2} = \frac{[796\ \text{kg/(m}^2\ \text{s)}]^2}{2(997\ \text{kg/m}^3)}\left[0.28 + \frac{4(3\ \text{m})(0.00596)}{0.02\ \text{m}} + 0.21\right]$$

$$= 318\,\frac{\text{kg}}{\text{m s}^2}(0.28 + 3.58 + 0.21) = 1290\ \text{Pa} = 1.29\ \text{kPa}$$

where 1 pascal (Pa) = 1 N/m². Notice that the entrance and expansion losses account for about 12% of the total pressure drop.

The pumping power associated with flow through the core is computed by writing

$$\dot{W}_p = \frac{\dot{m}\Delta P_{1-2}}{\rho} = \frac{(0.25 \text{ kg/s})[1290 \text{ kg/(m s}^2)]}{997 \text{ kg/m}^3} = 0.323 \text{ W}$$

from Eq. (8–78). Thus, the pumping–heating power ratio $\dot{W}_p/q_c$ (= 0.323 W/13 kW = 2.48×10^{-5}) is very small, which is generally the case for applications involving the flow of liquids.

Finally, we note that the analysis could be refined by evaluating the properties at the average inlet and outlet bulk-stream temperatures. However, because the temperature rise over the length of the core is only 12.4°C, the change in calculations resulting from this correction would be within about 5%.

EXAMPLE 8–14

Example 8–8 deals with air flowing in a heated tube of 1-cm diameter and 1-m length, with inlet contraction and exit expansion represented by σ = 0.7, uniform entering bulk-stream conditions given by P_1 = 1 atm and T_1 = 12°C, wall temperature of 175°C, and mass-flow rate of 0.00925 kg/s. Determine the pressure drop and pumping power.

Solution

Objective Determine ΔP and $\dot{W}_p$ over the length L.

Schematic Air flow in a heated tubular core: uniform wall temperature.

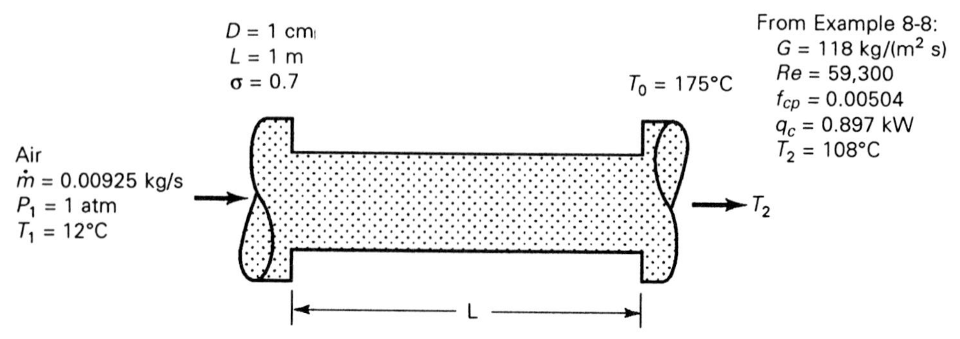

Assumptions/Conditions

forced convection
moderate property variation
standard conditions

Properties The hydrodynamic properties evaluated at the average bulk-stream temperature (i.e., $T_m = 60°C = 333$ K) are (Table A–C–5) $\rho = 1.06$ kg/m^3 and $\mu = 2 \times 10^{-5}$ kg/(m s). Except for density, which varies by about 25% over the length of the tube, the axial variation in viscosity and other properties is less than 10%. Therefore, we will account for the effect of the variation in density and bulk-stream velocity over the length of the tube. The density at the inlet is given by $\rho_1 = 1.24$ kg/m^3.

Analysis Referring to Example 8–8, $G = 118$ kg/(m^2 s) and $f_{cp} = 0.00504$. Correcting for property variation over the cross section on the friction factor, Eq. (8–25) gives

$$f = 0.00504 \left(\frac{448 \text{ K}}{333 \text{ K}}\right)^{-0.1} = 0.00489$$

To account for the effect of variation in density and bulk-stream velocity over the length of the core on the pressure drop, we use Eq. (8–77),

$$\Delta P = \frac{G^2 v_1}{2} \left[(1 - \sigma^2 + K_c) + \frac{4L}{D} f \frac{v_m}{v_1} + 2\left(\frac{v_2}{v_1} - 1\right) - (1 - \sigma^2 - K_e)\frac{v_2}{v_1} \right]$$

To estimate the entrance- and expansion-loss coefficients, we use Fig. 8–5 which indicates $K_c = 0.13$ and $K_e = 0.1$. Substituting into Eq. (8–77) and using the ideal gas law to approximate the specific-volume ratios, the pressure drop becomes

$$\Delta P_{1-2} = \frac{[118 \text{ kg/(m}^2 \text{ s)}]^2}{2(1.24 \text{ kg/m}^3)} \left[(1 - 0.7^2 + 0.13) + \frac{4(1 \text{ m})(0.00489)}{0.01 \text{ m}} \frac{333 \text{ K}}{285 \text{ K}} + 2\left(\frac{318 \text{ K}}{285 \text{ K}} - 1\right) - (1 - 0.7^2 - 0.1)\frac{381 \text{ K}}{285 \text{ K}} \right]$$

$$= 5610(0.64 + 2.29 + 0.674 - 0.41) \text{ N/m}^2 = 17.9 \text{ kPa}$$

It should be noted that use of the uniform property relation for pressure drop represented by Eq. (8–75) gives $\Delta P_{1-2} = 14.4$ kPa. Thus, the effects of variation in density and bulk-stream velocity over the length of the core accounts for about 20% of the total pressure drop.

The pumping power is computed by writing

$$\dot{W}_p = \frac{\dot{m}\,\Delta P_{1-2}}{\rho_m} = \frac{(0.00925 \text{ kg/s})(17{,}900 \text{ N/m}^2)}{1.06 \text{ kg/m}^3} = 156 \text{ W} = 0.156 \text{ kW}$$

The total heat-transfer rate for this application is shown in Example 8–8 to be 0.897 kW. Thus, the pumping–heating power ratio is 0.156. This example demonstrates the significance of pumping power when considering the overall energy requirements and performance of internal flow heat-transfer processes involving gas flow with relatively high mass flux. For applications such as this, the pumping–heating power ratio can be reduced by the use of heat-exchanger cores with larger cross-sectional area. To illustrate, calculations for the five-tube-core exchanger of Example 8–9 indicate $f = 0.00738$, $K_c = 0.23$, $K_e = 0.03$, $\Delta P_{1-2} = 1.03$ kPa, $\dot{W}_p = 9.28$ W, and $\dot{W}_p/q_c = 0.00395$.

8–3 EXTERNAL FLOW

We now turn our attention to the practical analysis of external flow over surfaces such as flat plates, cylinders, and spheres. As indicated in Sec. 6–2, external flows are characterized by the development of boundary layers that continue to grow in the streamwise direction. The flow is generally laminar for small values of x, but develops into turbulence beyond some critical value of x $(= x_c)$ for large enough flow rates. These flows are generally characterized in engineering analysis by setting $U_F = U_\infty$ and $T_F = T_\infty$. For uniform property analyses, the properties for external flows are generally evaluated at the free-stream temperature T_∞.

Because of significant differences between external longitudinal flow over flat plates and cylinders and flow across cylinders and spheres, these two basic situations will be discussed separately.

8–3–1 Flow over Flat Plates and Cylinders

Boundary layer flow over a flat plate heated in the region $x > x_0$ is shown in Fig. 8–18. An external flow such as this is characterized by uniform free-stream velocity U_∞ and zero pressure gradient. The reference length L for this type flow is commonly set equal to x, such that the Reynolds number Re_x becomes

$$Re_x = \frac{U_\infty x}{\nu} \tag{8–70}$$

The flow is usually laminar for $Re_x \lesssim 10^5$, transitional turbulent for $10^5 \lesssim Re_x \lesssim 5 \times 10^5$, and fully turbulent for $5 \times 10^5 \lesssim Re_x$.

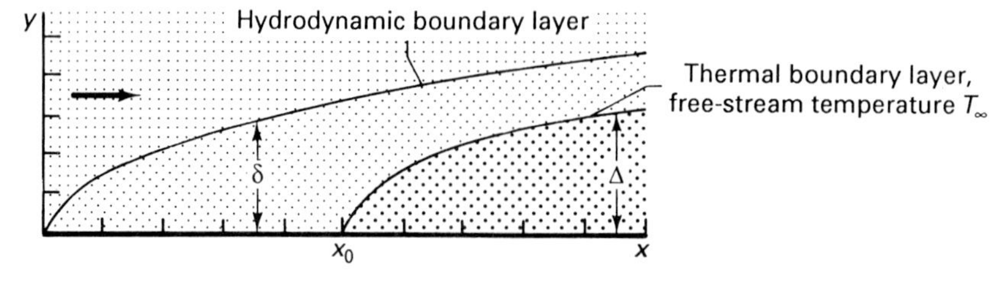

FIGURE 8–18 Hydrodynamic and thermal boundary layer flow over a flat plate with uniform free-stream velocity and heating in the region $x \geq x_0$.

Coefficients of Friction and Heat Transfer

The definitions for the Fanning friction factor f_x and coefficient of heat transfer h_x for flow over flat plates and cylinders are given by Eqs. (6–20) and (6–22) with $U_F = U_\infty$ and $T_F = T_\infty$; that is,

$$\tau_s = \rho U_\infty^2 \frac{f_x}{2} \qquad q_c'' = h_x(T_s - T_\infty) \tag{8–84,85}$$

The standard definitions for $\bar{f}$ and $\bar{h}$ given by Eqs. (6–21) and (6–23) also apply. The coefficients h_x and $\bar{h}$ for external flows are usually expressed in terms of the Nusselt number ($Nu_x = h_x x/k$, $\overline{Nu} = \bar{h}L/k$) or Stanton number [$St_x = h_x/(\rho c_P U_\infty)$, $\overline{St} = \bar{h}/(\rho c_P U_\infty)$].

Laminar Flow The exact solution for the friction factor f_x for laminar boundary layer flow with uniform free-stream velocity is given by

$$f_x = \frac{0.664}{Re_x^{1/2}} \tag{8–86}$$

after Blasius [45]. This equation is compared with experimental data in Fig. 8–19. To obtain an expression for the mean friction factor $\bar{f}$ over the length L, this equation is integrated in accordance with the defining equation, Eq. (6–21), with the result

$$\bar{f} = \frac{1}{L}\int_0^L f_x\,dx = \frac{0.664}{L}\sqrt{\frac{\nu}{U_\infty}}\int_0^L x^{-1/2}\,dx = 2 f_L = \frac{1.328}{Re_L^{1/2}} \tag{8–87}$$

Convenient correlations accurate to within 1% have been developed for the local Nusselt number for laminar boundary layer flow with uniform wall temperature over the entire surface ($x_0 = 0$), which take the form (Chap. 7)

$$Nu_x = 0.332\,Re_x^{1/2}\,Pr^{1/3} \qquad \text{for } 0.6 \lessapprox Pr \lessapprox 10 \tag{8–88}$$

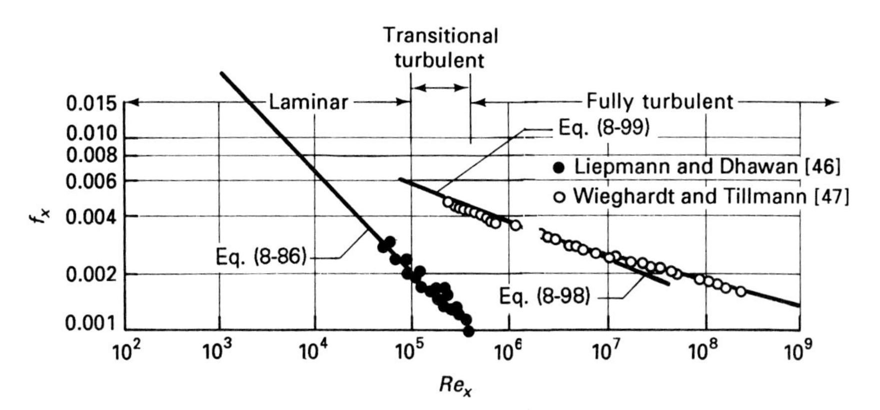

FIGURE 8–19 Correlations for f_x—laminar and turbulent boundary layer flow.

and

$$Nu_x = 0.339\, Re_x^{1/2}\, Pr^{1/3} \qquad \text{for } 10 \lesssim Pr \tag{8–89}$$

Equation (8–88) is accurate to within 2 to 3% for the wider range $0.3 \lesssim Pr \lesssim 10{,}000$. This equation is commonly expressed in the form of the Colburn analogy [29],

$$Nu_x = \frac{f_x}{2} Re_x\, Pr^{1/3} \tag{8–90}$$

For lower values of Pr, the Nusselt number can be approximated to within 3% by Eq. (7–126b),

$$Nu_x = \frac{0.646\, Re_x^{1/2}}{1 + (1.49/Pr - 0.6)^{1/2}} \qquad \text{for } Pr \lesssim 0.928,\quad Pe \gtrsim 100 \tag{8–91}$$

which is applicable to conditions for which the Peclet number Pe is greater than 100.

For the important case of laminar boundary layer flow with uniform wall-flux heating, the Nusselt number can be approximated by (Chap. 7)

$$Nu_x = 0.442\, Re_x^{1/2}\, Pr^{1/3} \qquad \text{for } 0.3 \lesssim Pr \lesssim 5 \tag{8–92}$$

$$Nu_x = 0.45\, Re_x^{1/2}\, Pr^{1/3} \qquad \text{for } 5 \lesssim Pr \tag{8–93}$$

and

$$Nu_x = \frac{0.646\, Re_x^{1/2}}{1 + (0.743/Pr - 0.6)^{1/2}} \qquad \text{for } Pr \lesssim 0.464,\quad Pe \gtrsim 100 \tag{8–94}$$

Notice that the Nusselt number for laminar boundary layer flow with uniform wall-heat flux is about 33% greater than for uniform wall temperature.

Relations have also been developed for the coefficient of heat transfer for nonuniform thermal boundary conditions. For example, for the important case in which the heated section is preceded by an unheated starting length, the Nusselt number can be represented by (see Example 7–6)

$$Nu_x = \frac{Nu_x|_{x0=0}}{[1 - (x_0/x)^{3/4}]^{1/3}} \qquad \text{for } Pr \gtrsim 0.6\,[1 - (x_0/x)^{3/4}] \tag{8–95}$$

for a step wall temperature (i.e., $T_s = T_\infty$ for $x < x_0$ and $T_s = T_0$ for $x \geq x_0$), and

$$Nu_x = \frac{Nu_x|_{x0=0}}{(1 - x_0/x)^{1/3}} \qquad \text{for } Pr \gtrsim 0.464\,(1 - x_0/x) \tag{8–96}$$

for a step wall-heat flux (i.e., $T_s = T_\infty$ for $x < x_0$ and $q''_c = q''_0$ for $x \geq x_0$), where $Nu_x|_{x0=0}$ represents the solution for no unheated starting length. More general relations are developed in Chap. 7 for cases in which the distribution in heat flux is arbitrarily specified. However, numerical methods generally must be used for cases in which the wall temperature is arbitrarily specified.

The relations for laminar boundary layer flow with uniform free-stream velocity indicate that the Nusselt number Nu_x is proportional to $Re_x^{1/2}$ [i.e., $Nu_x = Re_x^{1/2}$ fn (Pr)]. Using this result together with the defining equation for the mean coefficient of heat transfer, Eq. (6–23), we obtain

$$\bar{h} = \frac{1}{L}\int_0^L h_x\,dx = \frac{1}{L}\int_0^L Nu_x \frac{k}{x}\,dx$$
$$\overline{Nu}_L = \frac{\bar{h}L}{k} = \int_0^L \frac{Nu_x}{x}\,dx = \sqrt{\frac{U_\infty}{\nu}}\int_0^L x^{-1/2}\,dx\ \text{fn}(Pr) = 2\,Re_L^{1/2}\,\text{fn}(Pr) = 2\,Nu_{x=L} \tag{8–97}$$

Thus, we find that both the mean coefficient of heat transfer $\bar{h}$ and the mean friction factor $\bar{f}$ are *twice* the local coefficients h_L and f_L for this important type of laminar boundary layer flow.

Turbulent Flow The friction factor for turbulent boundary layer flow with uniform free-stream velocity can be approximated to within about 15% by Eq. (7–324),

$$f_x = 0.0592\,Re_x^{-0.2} \tag{8–98}$$

A somewhat less convenient but more accurate expression for turbulent flow recommended by White [23] takes the form

$$f_x = \frac{0.455}{\ln^2(0.06\,Re_x)} \tag{8–99}$$

These equations are also shown together with experimental data and the laminar flow results in Fig. 8–19. The local Fanning friction factor has not been universally characterized for transitional turbulent boundary layer flow. Therefore, correlations such as Eqs. (8–98) and (8–99) are generally utilized in both the transitional and fully turbulent regions.

One of the more simple and useful correlations for Nusselt number for fully turbulent boundary layer flow with uniform free-stream velocity is given by [6]

$$Nu_x = \frac{f_x}{2}Re_x\,Pr^n \qquad \text{or} \qquad Nu_x = 0.0296\,Re_x^{0.8}\,Pr^n \tag{8–100,101}$$

where $n = 0.5$ for $0.5 \lessapprox Pr \lessapprox 5.0$ and $n = \frac{1}{3}$ for $Pr \gtrapprox 5.0$. [Note the similarity between this correlation and the correlation for TFD fully turbulent tube flow given by Eqs. (8–18) and (8–19).] For improved accuracy, the following relationship developed by White [23] is recommended for $0.5 \lessapprox Pr$:

$$Nu_x = \frac{(f_x/2)\,Re_x\,Pr}{1 + 12.7\sqrt{f_x/2}\,(Pr^{2/3} - 1)} \tag{8–102}$$

For all practical purposes, this equation is equivalent to the Petukhov-Kirillov correlation, Eq. (8–16), which was recommended for fully turbulent TFD tube flow. Notice that Eq. (8–102) reduces to Eq. (8–100) as the Prandtl number approaches unity. As in the case of internal flow, the effect of the thermal boundary condition on turbulent external boundary layer flow is generally small, except for abrupt changes or low values of Pe, such that Eqs. (8–100)–(8–102) can be used for uniform wall temperature or uniform wall-heat flux.†

Practical integral solutions are developed in Chap. 7 for turbulent boundary layer flow with arbitrarily specified wall-heat flux. For the case of an unheated starting length followed by uniform wall-heat flux, Nu_x can be approximated by [from Eq. (7–347)]

$$Nu_x = \frac{\sqrt{f_x/2}\, Re_x\, Pr}{2.17 \ln [(1 - x_0/x)\sqrt{f_x/2}\, Re_x] + 12.7\, Pr^{2/3} - 13.9} \qquad \text{for } 0.5 \lesssim Pr \qquad (8\text{–}103)$$

which reduces to Eq. (7–348),

$$Nu_x = \frac{\sqrt{f_x/2}\, Re_x\, Pr}{2.17 \ln (\sqrt{f_x/2}\, Re_x) + 12.7\, Pr^{2/3} - 13.9} \qquad \text{for } 0.5 \lesssim Pr \qquad (8\text{–}104)$$

for uniform wall-heat flux. This equation lies approximately 5% above Eq. (8–102). Because the heat transfer is less sensitive to the exact form of the boundary condition for turbulent flow than for laminar flow, Eq. (8–103) can also be used for step wall-temperature heating.

For the general case in which transition occurs at a point sufficiently removed from the leading or trailing edges, the mean coefficients $\bar{f}$ and $\bar{h}$ over the length L are affected by conditions within both the laminar and turbulent zones. Under these circumstances, the mean coefficients $\bar{\Lambda}$ (i.e., $\bar{f}/2$, $\bar{h}$) are expressed in terms of the local coefficients Λ (i.e., $f_x/2$, h_x) by

$$\bar{\Lambda} = \frac{1}{L}\left[\int_0^{x_c} \Lambda_{\text{lam}}\, dx + \int_{x_c}^{L} \Lambda_{\text{turb}}\, dx\right] \qquad (8\text{–}105)$$

Using Eqs. (8–86) and (8–98) to represent f_x in the laminar and turbulent boundary layers, Eq. (8–105) gives

$$\begin{aligned}\frac{\bar{f}}{2} &= \frac{1}{L}\left[0.332\sqrt{\frac{\nu}{U_\infty}}\int_0^{x_c} x^{-0.5}\, dx + 0.0296\left(\frac{\nu}{U_\infty}\right)^{0.2}\int_{x_c}^{L} x^{-0.2}\, dx\right] \\ &= \frac{1}{L}\left[2(0.332)\sqrt{\frac{\nu}{U_\infty}}\, x_c^{0.5} + \frac{0.0296}{0.8}\left(\frac{\nu}{U_\infty}\right)^{0.2}(L^{0.8} - x_c^{0.8})\right] \\ &= 0.037\, Re_L^{-0.2} + (0.664\, Re_{x_c}^{-0.5} - 0.037\, Re_{x_c}^{-0.2})\frac{x_c}{L}\end{aligned} \qquad (8\text{–}106)$$

† The Nusselt number for turbulent boundary layer flow with uniform wall-heat flux is sometimes approximated by $Nu_x = 0.0308\, Re_x^{0.8}\, Pr^{1/3}$, which is only 4% above Eq. (8–101) for $n = \frac{1}{3}$.

Similarly, using Eqs. (8–88) and (8–101) to represent the local coefficient h_x, the mean coefficient of heat transfer $\bar{h}$ is represented in terms of Nusselt number by

$$\overline{Nu}_L = [0.037\, Re_L^{0.8} + (0.664\, Re_{x_c}^{0.5} - 0.037\, Re_{x_c}^{0.8})]\, Pr^{1/3} \qquad (8\text{–}107)$$

With Re_{x_c} set equal to a typical value of 5×10^5, these equations become

$$\frac{\bar{f}}{2} = 0.037\, Re_L^{-0.2} - 0.00174\,\frac{x_c}{L} \qquad (8\text{–}108)$$

and

$$\overline{Nu}_L = (0.037\, Re_L^{0.8} - 871)\, Pr^{1/3} \qquad (8\text{–}109)$$

Notice that for cases in which the boundary layer is tripped at the leading edge or in which L is much greater than x_c, Eqs. (8–106) and (8–107) reduce to

$$\bar{f} = 0.074\, Re_L^{-0.2} \qquad \overline{Nu}_L = 0.037\, Re_L^{0.8}\, Pr^{1/3} \qquad (8\text{–}110,111)$$

Further Considerations Correction factors of the type introduced for tube flow are often utilized to account for the effect of property variations on the friction factor and coefficient of heat transfer. As a matter of fact, Eqs. (8–23) and (8–24) and the values of m and n given in Table 8–2 are recommended by Kays and Crawford [6] for boundary layer flow of liquids, with the reference temperature set equal to the free-stream temperature T_∞. Similarly, Eqs. (8–25) and (8–26) are sometimes used for gases, with m and n being dependent upon the situation. Alternatively, Eckert [48] has suggested that the effect of property variation for boundary layer flow of gases be accounted for by evaluating the properties at the film temperature $T_f = (T_s + T_\infty)/2$.

As in the case of internal flow, the heat transfer is often enhanced for flow over flat plates and cylinders by the use of rough surfaces or fins. But for such surfaces, the manufacturers' design correlations for $\bar{h}$ and $\bar{f}$ must be employed.

Once again, we should be reminded that the coefficients of friction and heat transfer are also sometimes influenced by other factors, such as nonuniform free-stream velocity, curved surfaces, viscous dissipation, phase change, and unsteady operation. When the convection process is complicated by factors such as these, their influence on the coefficients should be accounted for.

Practical Analyses

The coefficients of friction and heat transfer can be used to determine the total force F_L acting on a plate and the total rate of heat transfer q_c or the distribution in wall temperature T_s.

Momentum Transfer Based on the definition for wall-shear stress τ_s, the differential shear force dF_x, which is caused by fluid flowing over the surface of a flat plate or cylinder at any axial location x, is given by

$$dF_x = \tau_s \, dA_s = \tau_s p \, dx \tag{8–112}$$

It follows that the total shear force F_L acting on the body becomes

$$F_L = \int_0^L \tau_s p \, dx \tag{8–113}$$

Introducing the Fanning friction factor, we obtain

$$F_L = \int_0^L \rho U_\infty^2 \frac{f_x}{2} p \, dx = \rho \frac{U_\infty^2}{2} p \int_0^L f_x \, dx = \rho U_\infty^2 \frac{\bar{f}}{2} A_s \tag{8–114}$$

for uniform free-stream velocity and uniform properties.

Heat Transfer To determine the total convection-heat-transfer rate from the surface of a flat plate or cylinder of length L, we first express the rate of heat transfer from a differential surface area dA_s ($= p \, dx$) in terms of the local heat flux q_c''.

$$dq_c = q_c'' \, dA_s = q_c'' p \, dx \tag{8–115}$$

Thus, the total rate of convection heat transfer over the entire surface

$$q_c = p \int_0^L q_c'' \, dx \tag{8–116}$$

where the evaluation of the integral depends on the form of the thermal boundary condition.

Specified Wall-Heat Flux For the case in which the wall-heat flux is specified, Eq. (8–116) can be integrated directly to obtain q_c. For example, for a uniform wall-heat flux ($q_c'' = q_0''$), q_c is simply

$$q_c = q_0'' \, pL = q_0'' A_s \tag{8–117}$$

To obtain the local wall temperature, we introduce the coefficient of heat transfer h_x by utilizing the general Newton law of cooling, Eq. (8–85),

$$q_c'' = h_x(T_s - T_\infty) \tag{8–118}$$

Solving for T_s, we have

$$T_s - T_\infty = \frac{q_c''}{h_x} = q_c'' \frac{x/k}{Nu_x} \tag{8–119}$$

To calculate T_s, Eq. (8–119) must be coupled with the appropriate correlations for Nu_x, as illustrated in Example 8–15. Notice that the wall operates at much lower temperatures for turbulent flow than for laminar flow (assuming negligible viscous dissipation effects).

Specified Wall Temperature With the wall temperature specified as a function of x, h_x is immediately introduced into Eq. (8–116) by utilizing the general Newton law of cooling; that is,

$$q_c = p \int_0^L h_x(T_s - T_\infty)\, dx \tag{8-120}$$

For a uniform wall-temperature boundary condition ($T_s = T_0$), Eq. (8–120) gives rise to an expression for q_c in terms of the mean coefficient of heat transfer $\bar{h}$ of the form

$$q_c = (T_0 - T_\infty)p \int_0^L h_x\, dx = \bar{h}A_s(T_0 - T_\infty) \tag{8-121}$$

This equation is recognized as the simplified form of the Newton law of cooling introduced in Chap. 1.

EXAMPLE 8–15

Air at 27°C and 1 atm flows at a rate of 3 m/s over a plate of 1-m length. The plate is heated electrically, with a uniform wall-heat flux of 100 W/m^2. Determine the approximate temperature along the plate.

Solution

Objective Determine the distribution in wall temperature T_s.

Schematic Air flow over a flat plate: uniform wall-heat flux.

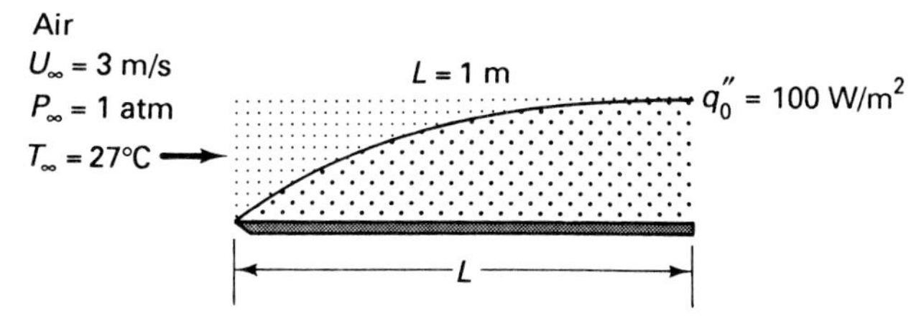

Assumptions/Conditions

forced convection
uniform properties
standard conditions

Properties Air at 27°C (Table A–C–5): $\rho = 1.18$ kg/m^3, $\mu = 1.85 \times 10^{-5}$ kg/(m s), $\nu = 1.57 \times 10^{-5}$ m^2/s, $k = 0.0262$ W/(m °C), $Pr = 0.708$.

Analysis The local surface temperature T_s of the plate is given by

$$T_s = T_\infty + \frac{q_c''}{k}\frac{x}{Nu_x} \tag{a}$$

from Eq. (8–119), where Nu_x is dependent on location x and on whether the flow is laminar or turbulent. To see whether the flow becomes turbulent, we calculate Re_L.

$$Re_L = \frac{LU_\infty}{\nu} = \frac{1 \text{ m } (3 \text{ m/s})}{1.57 \times 10^{-5} \text{ m}^2/\text{s}} = 1.91 \times 10^5$$

Because Re_L is only slightly greater than 10^5, the flow may or may not be turbulent.

Using Eq. (8–92) to approximate Nu_x for laminar flow and uniform property conditions, T_s becomes

$$T_s = T_\infty + \frac{q_c''\nu}{0.442\, kU_\infty} \frac{Re_x^{1/2}}{Pr^{1/3}}$$

$$= 27°\text{C} + \frac{(100 \text{ W/m}^2)(1.57 \times 10^{-5} \text{ m}^2/\text{s})}{0.442[0.0262 \text{ W/(m °C)}](3 \text{ m/s})(0.708)^{1/3}} Re_x^{1/2} \quad \text{(b)}$$

$$= 27°\text{C} + 0.0507\, Re_x^{1/2} \text{ °C}$$

This equation is shown in Fig. E8–15. Notice that T_s increases with x by the factor $x^{1/2}$.

If the flow is tripped such that turbulence occurs over most of the plate, Nu_x can be approximated by Eq. (8–101). For this situation T_s is given by

$$T_s = T_\infty + \frac{q_c''}{0.0296} \frac{\nu}{kU_\infty} \frac{Re_x^{0.2}}{Pr^{0.5}} = 27°\text{C} + 0.802\, Re_x^{0.2} \text{ °C} \quad \text{(c)}$$

for uniform property conditions. This equation is also shown in Fig. E8–15.

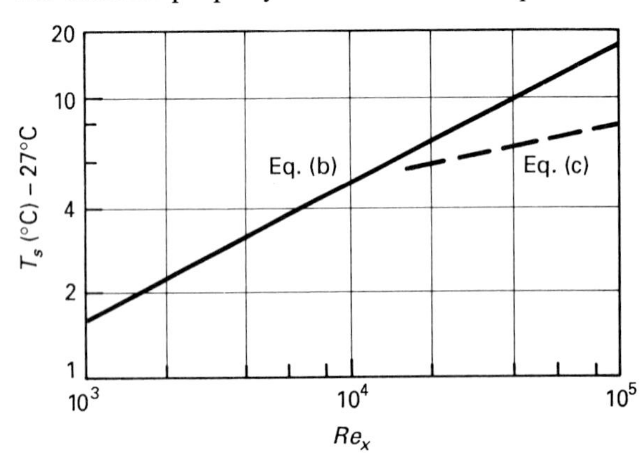

FIGURE E8–15 Calculations for wall temperature.

Because the calculations for wall temperature increase by only about 20°C, we would expect the property variation effects to be small. Referring to Eq. (a), by evaluating k at the film temperature and using a correction of the form of Eq. (8–26) for Nu_x, we find that the maximum wall temperature decreases by about 3% for laminar flow and increases by less than 1% for turbulent flow.

8–3–2 Flow Across Cylinders and Spheres

Referring to Fig. 8–20, flow normal to the axis of a circular cylinder and over a sphere is characterized by a diversion of the flow at the *forward stagnation point* where the fluid is brought to rest, and the variation of the free-stream velocity U_∞ and pressure P_∞ and growth of boundary layers in the streamwise direction x along the surface of the body. The boundary layer develops under the influence of a *favorable pressure gradient* ($dP/dx < 0$) and *accelerating* free-stream velocity ($dU_\infty/dx > 0$) over the forward part of the body until the point of minimum pressure P_∞ and maximum velocity U_∞. Beyond this point the boundary layer develops under the complicating influence of *adverse pressure gradient* ($dP/dx > 0$) and *decelerating* flow ($dU_\infty/dx < 0$) which generally produces a region of flow reversal followed by an increase in the growth of the boundary layer and the development of a wake flow involving irregular vortex type motion which persists well downstream of the body. The location beyond which the flow reversal occurs, known as the *separation point*, is characterized by $\partial u/\partial y|_0 = 0$ (i.e., $\tau_s = 0$), where the coordinate y is normal to the surface at all points along the surface. The specific features of the flow are strongly dependent upon the value of the Reynolds number and on whether the flow is laminar or turbulent. Descriptions of these details are provided by Schlichting [49] and White [23].

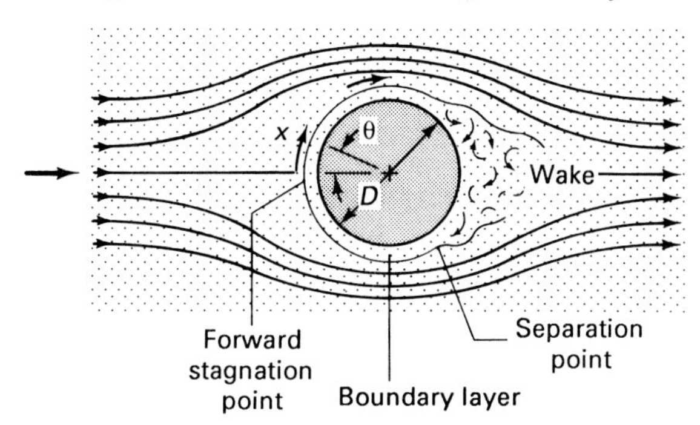

FIGURE 8–20 Boundary layer flow over a circular cylinder in crossflow.

Although a local Reynolds number is sometimes written for flow normal to a cylinder or sphere in terms of the distance along the curved surface, a more convenient overall Reynolds number is given by

$$Re_{D_N} = \frac{U_\infty D_N}{\nu} \tag{8–122}$$

where the reference length D_N is the height of the body as seen by the onrushing fluid. For example, for circular cylinders and spheres, D_N is simply equal to the diameter D.

Coefficients of Drag and Heat Transfer

The net force acting on a body in crossflow is the result of both the total wall-shear stress and overall pressure drop. Because it is this drag force F_D that is generally needed in design, a *drag coefficient* is defined as

$$F_D = \frac{\rho U_\infty^2}{2} C_D A_N \tag{8–123}$$

where A_N is the area of the body normal to the direction of flow. Figure 8–21 shows the coefficients of drag for flow across cylinders and spheres.

Although empirical correlations are available for the local coefficient of heat transfer h_θ for flow across cylinders and spheres, practical considerations have resulted in the use of a mean coefficient $\bar{h}$, which is defined by

$$\bar{h} = \frac{1}{2\pi} \int_0^{2\pi} h_\theta \, d\theta \tag{8–124}$$

Accordingly, for standard uniform wall-temperature heating, the rate of heat transfer q_c is given by the simple Newton law of cooling,

$$q_c = \bar{h} A_s (T_s - T_\infty) \tag{8–125}$$

Numerous empirical correlations are available in the literature for the mean Nusselt number $\overline{Nu}_D$ for flow across cylinders. One of the more recently developed correlations for gas or liquid flow over a circular cylinder with uniform wall temperature is given by [50]

$$\overline{Nu}_D = (0.4\, Re_D^{0.5} + 0.06\, Re_D^{2/3})\, Pr^{0.4} \left(\frac{\mu_\infty}{\mu_0}\right)^{0.25} \tag{8–126}$$

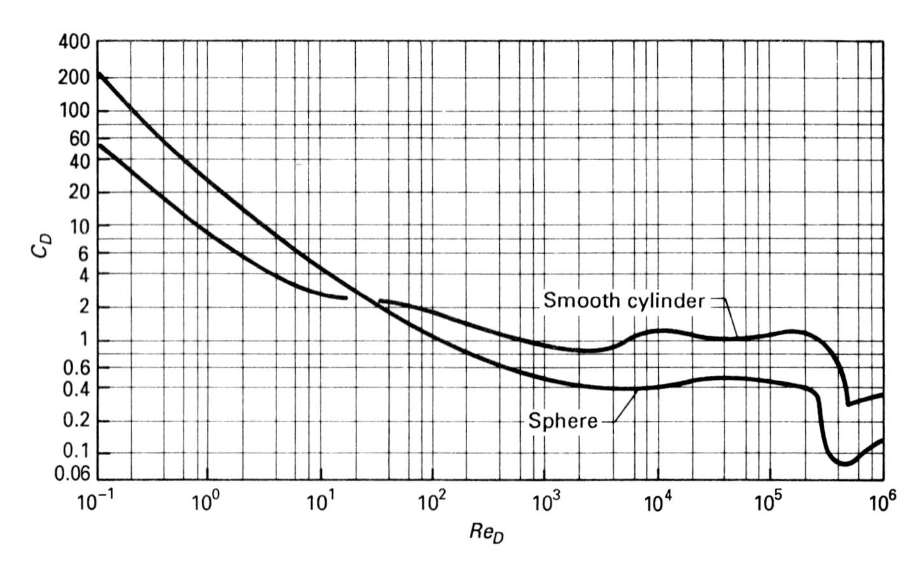

FIGURE 8–21 Representative drag coefficients for circular cylinder in crossflow and for flow over a sphere. (From Schlichting [49]. Adapted with permission.)

where $\overline{Nu}_D = \bar{h}D/k$, $10 \lesssim Re_D \lesssim 10^5$, $0.67 \lesssim Pr \lesssim 300$, $0.25 \lesssim \mu_\infty/\mu_0 \lesssim 5.2$, and the properties other than μ_0 are evaluated at T_∞. This correlation has been reported by Whitaker [51] to lie within ±25% of the experimental data.

A similar correlation has been recommended by Whitaker [50] for flow over a sphere which takes the form

$$\overline{Nu}_D = 2 + (0.4\,Re_D^{0.5} + 0.06\,Re_D^{2/3})\,Pr^{0.4}\left(\frac{\mu_\infty}{\mu_0}\right)^{0.25} \tag{8–127}$$

where $3.5 \lesssim Re_D \lesssim 7.6 \times 10^4$, $0.71 \lesssim Pr \lesssim 380$, $1 \lesssim \mu_\infty/\mu_0 \lesssim 3.2$, and all properties except μ_0 are evaluated at T_∞. For higher Reynolds numbers, the following correlation by Achenbach [52] is recommended for gases:

$$\overline{Nu}_D = 430 + a\,Re_D + b\,Re_D^2 + c\,Re_D^3 \tag{8–128}$$

where $a = 0.005$, $b = 2.5 \times 10^{-11}$, $c = -3.1 \times 10^{-17}$, $4 \times 10^5 \lesssim Re_D \lesssim 5 \times 10^6$, $Pr \simeq 0.71$, and the properties are evaluated at the film temperature T_f.

Correlations are available for flow of liquid metals over circular cylinders and spheres in references 53 and 54 and for crossflow over noncircular and nonspherical bodies in references 2, 3, and 55. In addition, design correlations are available for finned tubes which are widely employed in industry.

Practical Analyses

The use of Eqs. (8–123) and (8–125) directly give the total drag and rate of convection heat transfer from bodies with uniform surface temperature in crossflow.

EXAMPLE 8–16

Air at atmospheric pressure and 35°C flows across a 1-cm-diameter wire at a velocity of 100 m/s. The power generated by electric current per unit length of the wire is 2000 W/m. Estimate the mean surface temperature of the wire.

Solution

Objective Estimate the mean surface temperature T_0.

Schematic Air flow across a wire with uniform energy generation.

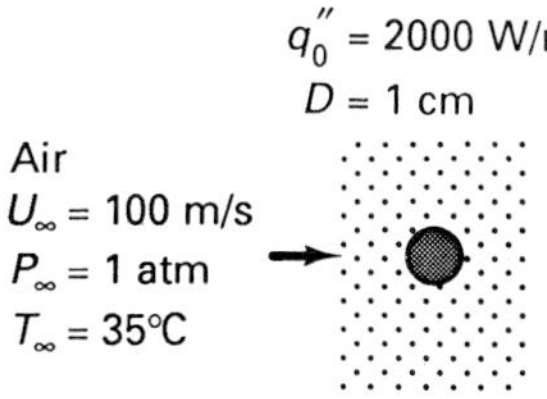

$q_0'' = 2000$ W/m
$D = 1$ cm

Air
$U_\infty = 100$ m/s
$P_\infty = 1$ atm
$T_\infty = 35$°C

Assumptions/Conditions

forced convection
moderate property variation
standard conditions

Properties Air at 35°C: $\rho = 1.16\ \text{kg/m}^3$, $\mu = 1.88 \times 10^{-5}\ \text{kg/(m s)}$, $\nu = 1.62 \times 10^{-5}\ \text{m}^2\text{/s}$, $k = 0.0268\ \text{W/(m °C)}$, $Pr = 0.706$.

Analysis Although the heat-transfer correlations found in the literature were developed for uniform wall-temperature heating, these correlations can be used to obtain an approximate solution for the surface temperature for this uniform heat-flux condition. Calculating the Reynolds number Re_D, we have

$$Re_D = \frac{U_\infty D}{\nu} = \frac{(100\ \text{m/s})(0.01\ \text{m})}{1.62 \times 10^{-5}\ \text{m}^2\text{/s}} = 61{,}700$$

To obtain the mean Nusselt number $\overline{Nu}_D$, we use Eq. (8–126),

$$\overline{Nu}_D = (0.4\, Re_D^{0.5} + 0.06\, Re_D^{2/3})\, Pr^{0.4} \left(\frac{\mu_\infty}{\mu_0}\right)^{0.25}$$

Approximating μ_∞/μ_0 by unity, we obtain

$$\overline{Nu}_D = 168$$

or

$$\bar{h} = \overline{Nu}_D \frac{k}{D} = 168 \frac{0.0268\ \text{W/(m °C)}}{0.01\ \text{m}} = 450\ \text{W/(m}^2\ \text{°C)}$$

An estimate can now be made for the surface temperature by using the Newton law of cooling.

$$T_0 - T_\infty = \frac{q_c}{\bar{h}A_s} = \frac{q_c}{L}\frac{1}{\bar{h}p} = \frac{2000\ \text{W/m}}{[450\ \text{W/(m}^2\ \text{°C)}]\pi(0.01\ \text{m})}$$

$$T_0 = 141\text{°C} + 35\text{°C} = 176\text{°C}$$

To refine the calculation for wall temperature, μ_0 in Eq. (8–126) can be evaluated at 176°C. This step gives rise to $T_0 = 186$°C, which represents a 5.7% change. Further refinement in the value of μ_0 is not justified because of the approximate nature of the heat-transfer correlation.

8–4 FLOW ACROSS TUBE BANKS

As we shall see in Chap. 11, flow across tube bundles is extremely important in heat-exchanger applications. To introduce this subject, we consider the basic geometric and bulk-stream characteristics associated with flow through unfinned tube banks in Sec. 8–4–1, present the practical thermal and hydraulic analysis approaches in Secs. 8–4–2 and 8–4–3, and consider the analysis of fin-tube banks in Sec. 8–4–4.

8–4–1 Geometric and Bulk-Stream Characteristics

The flow and heat-transfer characteristics for unfinned tube banks are dependent upon the tube diameter D, the longitudinal pitch S_L and transverse pitch S_T measured between tube centers, number of tube rows crossed N_L and tube columns N_T, arrangement of the tubes in the bank, the flow rates, and the fluid. Standard in-line and staggered tube-bank arrangements generally used by manufacturers are shown in Fig. 8–22.

Following the convention used by Kays and London [4], the tube bank length L represents an equivalent flow length measured from the leading edge of the first tube row to the leading edge of a tube row that would follow the last tube row, were another tube row present; that is,

$$L = N_L S_L \tag{8–129}$$

The frontal area A_1 and volume V of the tube bank are represented by

$$A_1 = N_T S_T Z \tag{8–130}$$

and

$$V = LA_1 = NS_L S_T Z \tag{8–131}$$

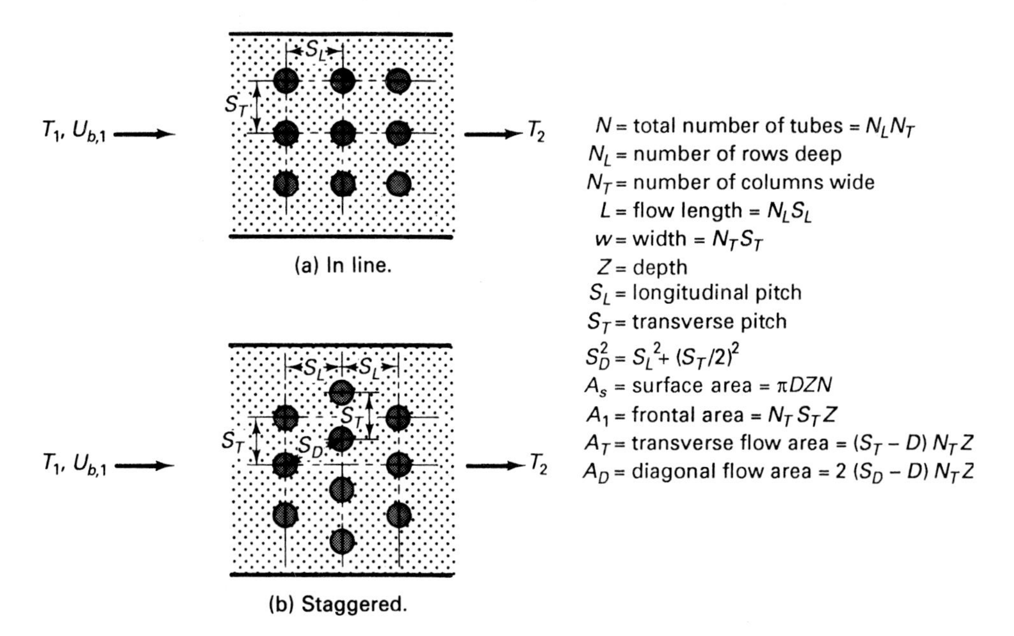

FIGURE 8–22 Tube-bank arrangements.

and the total surface area A_s from which heat is convected is given by

$$A_s = \pi DNZ \tag{8-132}$$

where Z is the depth. In practice the surface area A_s is often expressed in terms of the *surface-area density* β,

$$\beta = \frac{A_s}{V} = \frac{\pi D}{S_L S_T} \tag{8-133}$$

Figure 8–23 shows representative flow patterns for water flowing over a tube bank. The flow exhibits complex features such as separation, recirculation (i.e., flow reversal), and vortex motion that are characteristic of external flow across cylinders. However, because the flow is confined by this multiple-tube geometry, it is classified as internal flow for purpose of analysis.

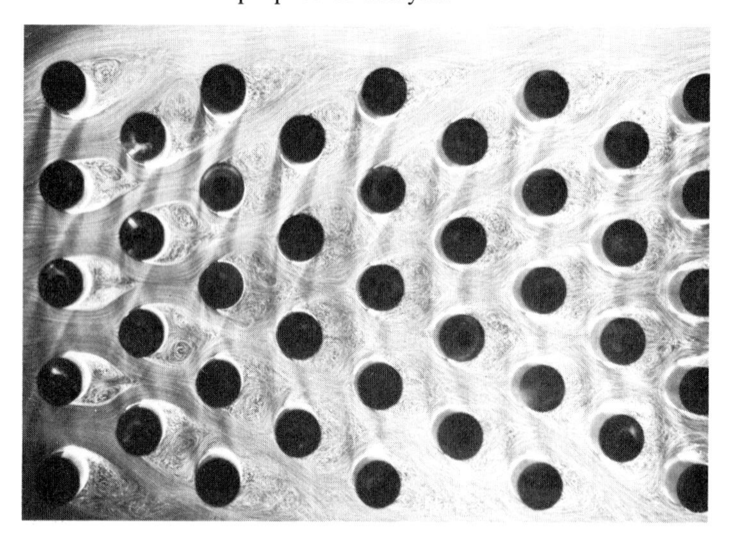

FIGURE 8–23 Flow pattern for water flowing across staggered array of tubes. (Courtesy of H. Werle, ONERA, Paris.)

It follows that bulk-stream characteristics are used in the development of the practical lumped analysis approach for flow across tube banks. However, because this particular application involves N separate heating surfaces and multiple-flow passages with a nonuniform cross-sectional flow area A, the actual distributions in bulk-stream velocity $U_{b,\mathrm{act}}$ and bulk-stream temperature $T_{b,\mathrm{act}}$ within the tube bank are not easily characterized. To overcome this complexity, the tube-bank core is modeled by a number of uniform and continuous flow channels, as illustrated in Fig. 8–24. In this approach, the cross-sectional area is set equal to the *minimum free-flow area* A_{min} available for fluid flow within the tube bank, and the *effective heat-transfer surface perimeter* p is given by

$$p = \frac{A_s}{L} = \frac{\pi DNZ}{L} = \frac{\pi DN_T Z}{S_L} \tag{8-134}$$

which is equal to the *effective wetted perimeter* p_w.

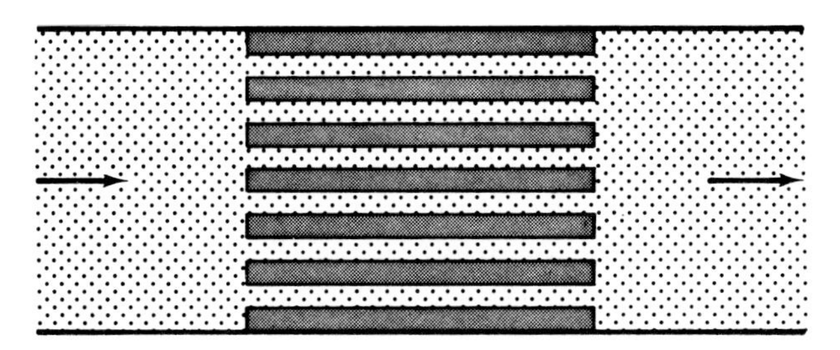

FIGURE 8–24 Tube-bank core model for practical lumped analysis; G and U_b based on the minimum free-flow area A_{min} in the core. (From Kays and London [4]. Used with permission.)

Referring to Fig. 8–22, $A_{min} = (S_T - D)N_T Z$ for in-line arrangements. For staggered arrays, the minimum free-flow area may occur between adjacent tubes in a row or between diagonally opposed tubes. Hence, A_{min} is the smaller of the two values $(S_T - D)N_T Z$ and $2(S_D - D)N_T Z$, where the factor 2 is a consequence of the fluid dividing as it moves from the transverse plane through the diagonal planes. These relations for A_{min} are represented by

$$A_{min} = C_A N_T D Z \tag{8–135}$$

where the *area coefficient* C_A is

$$C_A = \frac{S_T}{D} - 1 \qquad \text{for in-line arrays, or staggered arrays with } S_D \geqslant (S_T + D)/2 \tag{8–136a}$$

$$C_A = 2\left(\frac{S_D}{D} - 1\right) \qquad \text{for staggered arrays with } S_D < (S_T + D)/2 \tag{8–136b}$$

and the *diagonal pitch* S_D is

$$S_D = \left[S_L^2 + \left(\frac{S_T}{2}\right)^2\right]^{1/2} \tag{8–137}$$

It follows that the contraction ratio σ is given by

$$\sigma = \frac{A_{min}}{A_1} = \frac{C_A D}{S_T} \tag{8–138}$$

To specify the hydraulic diameter D_H for flow across tube banks, it is customary to write

$$D_H = \frac{4A_{min}}{p_w} = \frac{4C_A S_L}{\pi} \tag{8–139}$$

or

$$\frac{D_H}{4L} = \frac{A_{min}}{A_s} = \frac{C_A}{\pi N_L} \tag{8–140}$$

The hydraulic diameter can also be expressed in terms of β and σ by

$$D_H = 4\frac{A_{\min}}{A_s}\frac{V}{A_1} = 4\frac{\sigma}{\beta} \tag{8–141}$$

The bulk-stream hydrodynamic characteristics within the core model are represented by U_b and G and satisfy the following relation for uniform property conditions:†

$$\dot{m} = \rho A_{\min} U_b = A_{\min} G \tag{8–142}$$

The principle of conservation of mass also permits us to write‡

$$\dot{m} = \rho A_1 U_{b,1} = A_1 G_1 \tag{8–143}$$

It follows that U_b is related to the entering bulk-stream velocity $U_{b,1}$ by

$$U_b = \frac{A_1}{A_{\min}} U_{b,1} = \frac{U_{b,1}}{\sigma} = \frac{S_T}{C_A D} U_{b,1} \tag{8–144}$$

Similarly, we see that

$$G = \frac{A_1}{A_{\min}} G_1 = \frac{G_1}{\sigma} = \frac{S_T}{C_A D} G_1 \tag{8–145}$$

It also follows that the Reynolds number Re is defined by

$$Re = \frac{U_b D_H}{\nu} = \frac{G D_H}{\mu} \tag{8–146}$$

Based on this definition for the Reynolds number, the flow has been reported to be laminar for Re less than a value of the order of 200, and the flow is transitional turbulent for $200 \lessapprox Re \lessapprox 6000$.

To complete the picture, the bulk-stream thermal characteristics within the core model are represented by $\dot{H}_b$ and T_b, which satisfy Eq. (6–45),

$$d\dot{H}_b = \dot{m} c_P \, dT_b \tag{8–147}$$

8–4–2 Practical Thermal Analysis

To adapt the practical thermal analysis approach developed in Sec. 8–2–2 to flow across tube banks, we define the local coefficient of heat transfer h_x by

$$q''_c = h_x(T_s - T_b) \tag{8–148}$$

† The properties ρ, μ, and ν are replaced by the mean values ρ_b, μ_b, and ν_b and c_P is taken as a function of T_b for a variable property analysis. The bulk-stream hydrodynamic characteristics within the core model are also commonly represented by $U_{b,\max}$ and $G_{\max}$.

‡ To avoid confusion between bulk-stream velocity U_b and the overall coefficient of heat transfer U which is defined in Chap. 11, the entering bulk-stream velocity is now designated by $U_{b,1}$.

where q_c'' represents an *effective local heat flux* transferred across the core model (Fig. 8–24) surface area $p\ dx$. Using this definition for h_x, the relations developed in Sec. 8–2–2 for flow in tubes with specified wall-heat flux and uniform wall-temperature heating are applicable to flow through tube banks and heat-exchanger cores.

For example, the total heat-transfer rate q_c and exit bulk-stream temperature T_2 for the case of uniform wall temperature with uniform properties are given by Eq. (8–31),

$$q_c = \dot{m}c_P(T_2 - T_1) \tag{8–149}$$

or Eq. (8–52),

$$q_c = \dot{m}c_P(T_0 - T_1)\left[1 - \exp\left(-\frac{\bar{h}A_s}{\dot{m}c_P}\right)\right] \tag{8–150}$$

and Eq. (8–51)

$$\frac{T_2 - T_1}{T_0 - T_1} = 1 - \exp\left(-\frac{\bar{h}A_s}{\dot{m}c_P}\right) \tag{8–151}$$

where $\bar{h}$ is the mean coefficient of heat transfer. As we have seen, these equations can be combined to obtain

$$q_c = \bar{h}A_s\, LMTD \tag{8–152}$$

where the log mean temperature difference $LMTD$ is defined by Eq. (8–60),

$$LMTD = \frac{T_2 - T_1}{\ln\left[(T_0 - T_1)/(T_0 - T_2)\right]} = \frac{\Delta T_1 - \Delta T_2}{\ln\left(\Delta T_1/\Delta T_2\right)} \tag{8–153}$$

Correlations have been developed for the mean coefficient of heat transfer $\bar{h}$ on the basis of experimental data for flow over tube banks with approximate uniform wall temperature by Kays and London [4], Whitaker [50], Grimison [56], Zukauskas [57,58], Bergelin et al. [59], the Babcock and Wilcox Co. [60], and others. Data obtained by Kays and London [4] for a specific tube bank are shown by Fig. 8–25 in terms of $\overline{St}$ [$= \bar{h}/(\rho c_P U_b)$]. A general correlation developed by Kays and London [4] for staggered tube-bank arrangements of infinite extent is shown in Fig. 8–26. This correlation is used in conjunction with Fig. 8–27 to account for the effect on $\bar{h}$ of a finite number of tube rows in a bank. Although the experimental data on which these curves are based were obtained for air, these correlations can also be used with reasonable confidence for liquids with moderately high values of Prandtl number.

The correlation developed for $\bar{h}$ by Zukauskas [57] is of the form

$$\overline{Nu}_D = C\, Re_D^m\, Pr^n \left(\frac{Pr}{Pr_s}\right)^{0.25} F_c \tag{8–154}$$

where the constants C, m, and n are given in Table 8–5, and the tube-row correction factor F_c is given in Fig. 8–28.

It should be noted that the solutions developed in Sec. 8–2–2 and other references for arbitrary wall flux, arbitrary wall temperature, and variable property internal flows

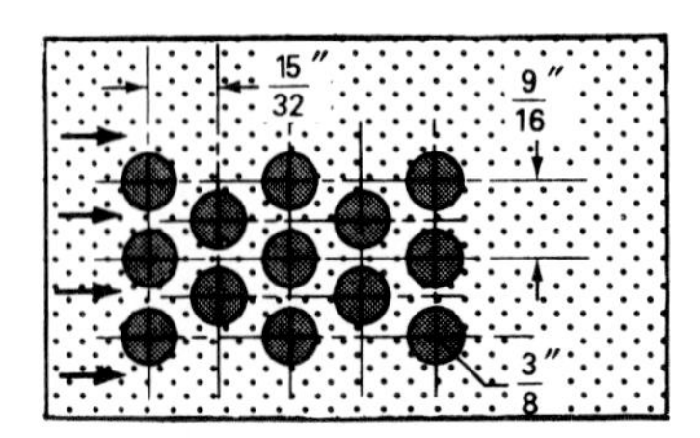

$D = 0.375$ in. $= 9.525$ mm
$D_H = 0.0249$ ft $= 7.569$ mm
$\sigma = 0.333$
$\beta = 53.6\ ft^2/ft^3 = 175.2\ m^2/m^3$
Note: Minimum free-flow area is in spaces transverse to flow.

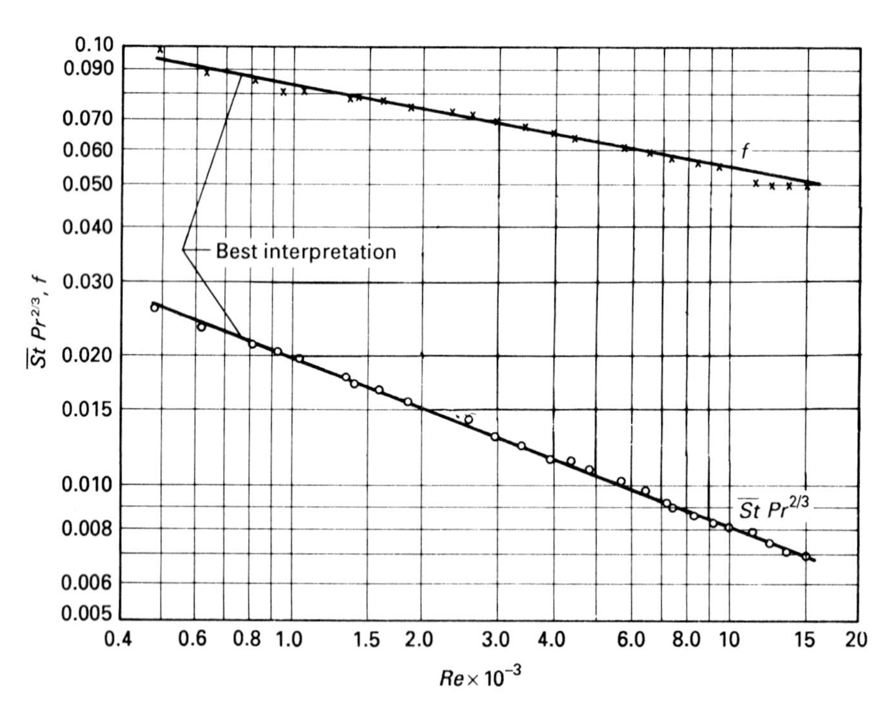

FIGURE 8–25 Flow normal to a staggered tube bank; surface S1.50–1.25. (From Kays and London [4]. Used with permission.)

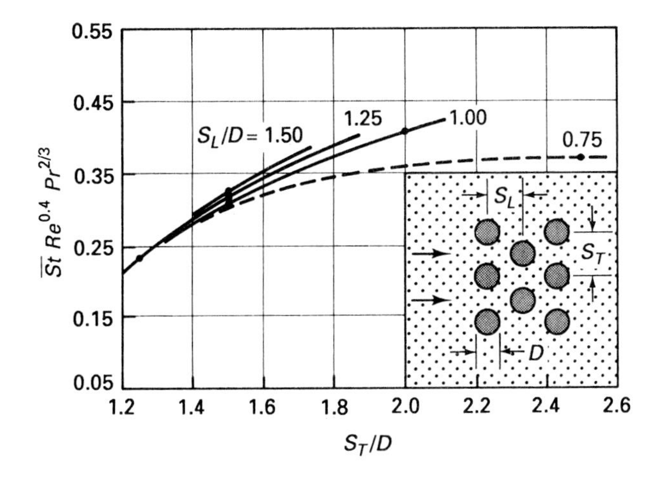

FIGURE 8–26 Mean Stanton number correlation—gas flow normal to an infinite bank of staggered circular tubes; $300 \le Re \le 15{,}000$. (From Kays and London [4]. Used with permission.)

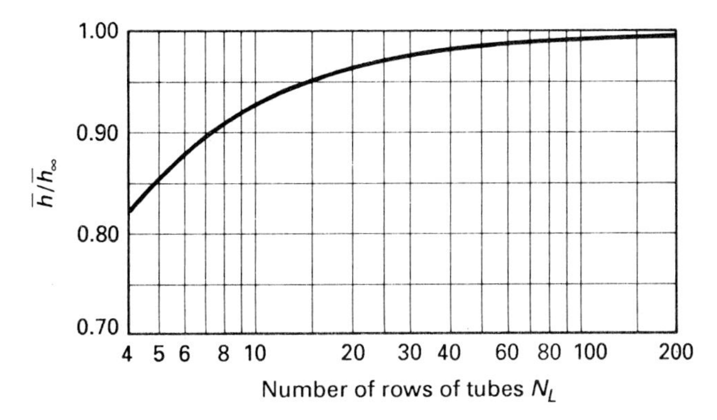

FIGURE 8–27 Overall influence of row-to-row variations on the heat transfer coefficient for tube banks. (From Kays and London [4]. Used with permission.)

TABLE 8–5 Constants of Eq. (8–154) for flow across a tube bank

Configuration	*Range of Re_D*	C	m	n
In line	10^0–10^2	0.9	0.4	0.36
	10^2–10^3	0.52	0.5	0.36
	10^3–2×10^5	0.27	0.63	0.36
	2×10^5–2×10^6	0.033	0.8	0.4
Staggered				
	10^0–5×10^2	1.04	0.4	0.36
	5×10^2–10^3	0.71	0.5	0.36
	10^3–2×10^5	$0.35(S_T/S_L)^{0.2}$	0.6	0.36
	2×10^5–2×10^6	$0.031(S_T/S_L)^{0.2}$	0.8	0.36

Source: From Zukauskas [57].

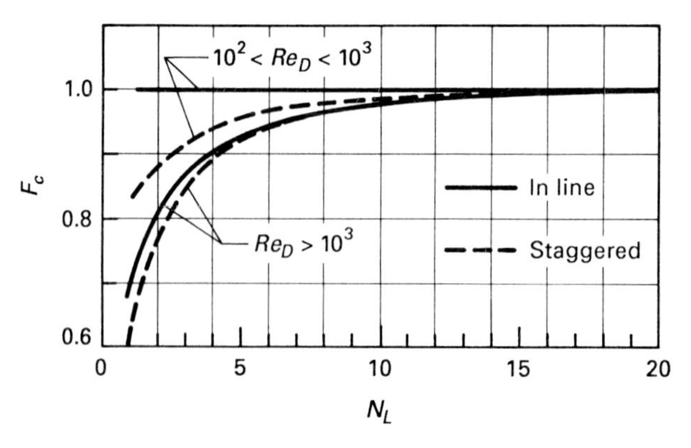

FIGURE 8–28 A correction factor to account for the tube-row effect for heat transfer in flow normal to bare tube banks (Zukauskas [57]).

also apply to flow across tube banks. However, because of the complex nature of the actual process, little information pertaining to the distribution in the local coefficient of heat transfer h_x, which appears in these solutions, is available in the literature. Therefore, to develop first-order solutions for cases involving such arbitrary conditions, h_x is generally approximated by correlations for $\bar{h}$ which are based on uniform wall temperature and uniform properties.

8–4–3 Hydraulic Considerations

The practical lumped approach to the analysis of flow through tube banks and heat-exchanger cores developed by Kays and London [4] involves the use of a Fanning friction factor f defined by

$$\tau_s = \rho U_b^2 \frac{f}{2} \tag{8–155}$$

where τ_s represents an effective local wall-shear stress acting on the core-model (see Fig. 8–24) surface area $p\,dx$ that accounts for the effects of viscous shear and form drag within the actual core, assuming fully developed conditions. In this general approach, ΔP_{1-2} is given in terms of the entrance-loss and expansion-loss coefficients K_c and K_e by Eq. (8–75),

$$\Delta P_{1-2} = \frac{\rho U_b^2}{2}\left(K_c + \frac{4L}{D_H} f + K_e\right) \tag{8–156}$$

for uniform property conditions. For specific applications involving tube banks, K_c and K_e have traditionally been set equal to zero, such that ΔP_{1-2} is represented by

$$\Delta P_{1-2} = \frac{4L}{D_H}\frac{f}{2}\rho U_b^2 = \frac{\pi N_L}{C_A}\frac{f}{2}\rho U_b^2 \tag{8–157}$$

with the entrance and exit losses accounted for by f. For situations in which significant changes in density occur over the length of the tube bank, ΔP_{1-2} is obtained from Eq. (8–77) (with $K_1 = K_2 = 0$); that is,

$$\Delta P_{1-2} = \frac{G^2 v_1}{2}\left[(1 + \sigma^2)\left(\frac{v_2}{v_1} - 1\right) + \frac{4L}{D_H} f \frac{v_m}{v_1}\right] \tag{8–158}$$

where $4L/D_H = \pi N_L/C_A$.

Experimental data for f are presented by Kays and London [4] for a number of in-line and staggered tube-bank arrangements. Direct test data obtained for a specific tube bank (staggered arrangement with $D = 0.009525$ m, $S_L = 0.01191$ m, $S_T = 0.01429$ m) are shown in Fig. 8–25. Correlations developed on the basis of data for all of the tube banks with staggered array considered by Kays and London are shown in Fig. 8–29. Similar correlations have been developed by Grimison [56] for both in-line and staggered tube-bank arrangements.

In this connection, it is important to note that somewhat different approaches to the correlation of pressure drop for flow across tube banks appear in the literature [57–63]. For example, the correlation developed by Zukauskas [57] can be written as

$$\Delta P_{1-2} = 4N_L \frac{f_i}{2} \rho U_b^2 \tag{8–159}$$

Correlations developed by Zukauskas [57] for f_i are shown in terms of $4f_i/\chi$ versus Re_D by Fig. 8–30 where $Re_D = U_b D/\nu$; χ is a correction factor which is generally a function of Re_D and both S_L/D and S_T/D. By comparing Eqs. (8–157) and (8–159),

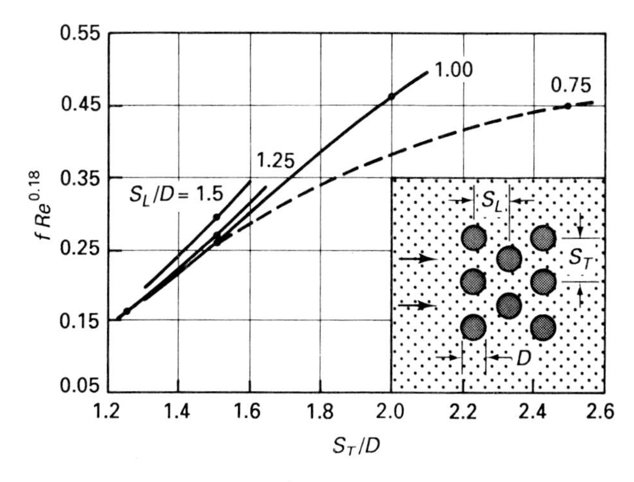

FIGURE 8–29 Friction factor correlation—flow normal to an infinite bank of staggered circular tubes; $300 \leq Re \leq 15{,}000$. (From Kays and London [4]. Used with permission.)

we observe that f_i and f are related by

$$f = \frac{4C_A f_i}{\pi} \tag{8–160}$$

It follows that the term $\pi N_L f/C_A$ in Eqs. (8–156) to (8–158) can be replaced by $4N_L f_i$.

With ΔP known, the pumping power can be obtained by use of Eq. (8–78),

$$\dot{W}_p = \frac{\dot{m}\,\Delta P}{\rho} \tag{8–161}$$

for uniform property flow, or Eq. (8–79),

$$\dot{W}_p = \dot{m}\,\Delta P\, v_m \tag{8–162}$$

for variable property conditions.

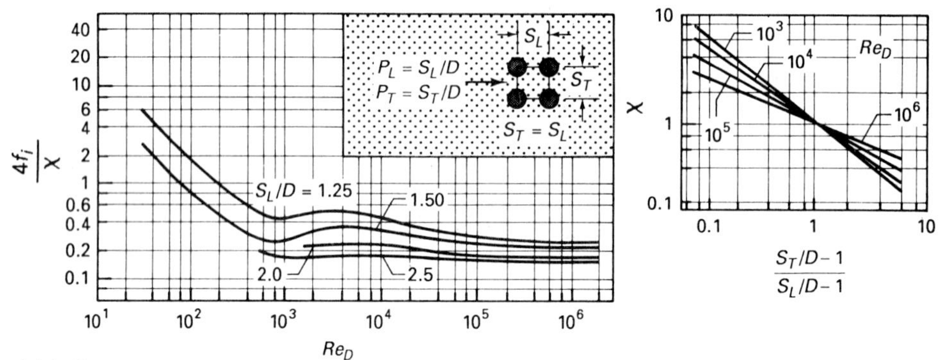

(a) In-line arrangement.

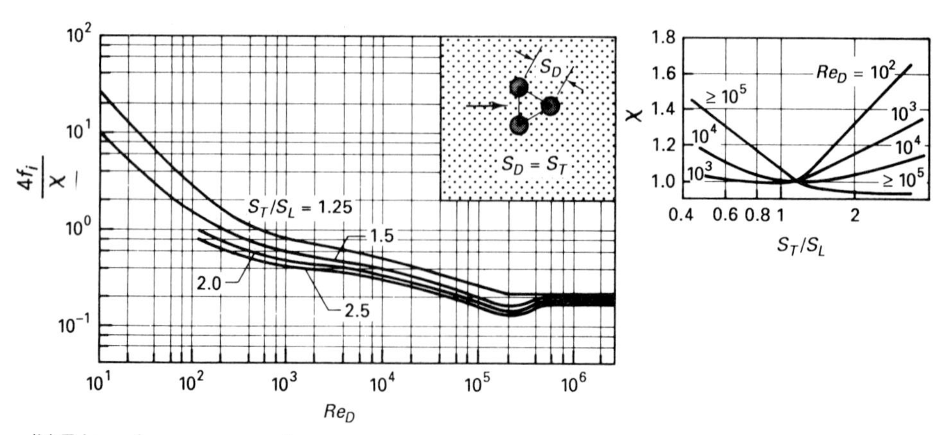

(b) Triangular arrangement.

FIGURE 8–30 Friction factor f_i and correction factor χ for flow across tube banks (Zukauskas [57]).

EXAMPLE 8–17

The basic geometric parameters listed for the tube-bank surface S1.50–1.25 shown in Fig. 8–24 are

$$D = 0.375 \text{ in.} \quad S_L = 15/32 \text{ in.} \quad S_T = 9/16 \text{ in.}$$

Confirm the values for hydraulic diameter D_H, contraction ratio σ, and surface-area density β, which are given in the figure.

Solution

Objective Show that $D_H = 0.0249$ ft, $\sigma = 0.333$, and $\beta = 53.6 \text{ ft}^{-1}$.

Schematic Tube-bank surface S1.50–1.25.

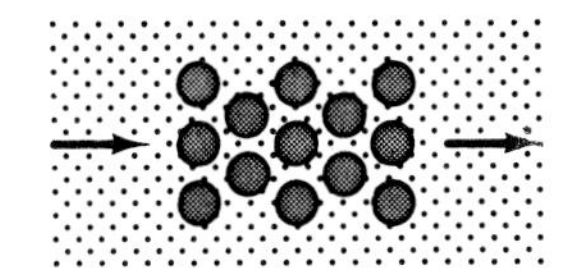

$D = 0.375$ in.
$S_L = 15/32$ in.
$S_T = 9/16$ in.

Analysis To provide a basis for evaluating D_H, σ, and β for this staggered arrangement, we compute

$$S_D = \left[S_L^2 + \left(\frac{S_T}{2}\right)^2\right]^{1/2} = \left[\left(\frac{15}{32}\text{ in.}\right)^2 + \left(\frac{9/16}{2}\text{ in.}\right)^2\right]^{1/2} = 0.547 \text{ in.}$$

and

$$\frac{S_T + D}{2} = \frac{9/16 \text{ in.} + 0.375 \text{ in.}}{2} = 0.469 \text{ in.}$$

Because $S_D > (S_T + D)/2$, $A_{\min}$ is taken as the area in spaces transverse to the flow; that is,

$$C_A = \frac{S_T}{D} - 1 = \frac{9/16}{0.375} - 1 = 0.5$$

Thus, using Eq. (8–135), $A_{\min}$ is given by

$$\frac{A_{\min}}{N_T Z} = C_A D = 0.5(0.375 \text{ in.}) = 0.188 \text{ in.} = 0.0156 \text{ ft} = 0.00476 \text{ m}$$

We are now able to compute D_H from Eq. (8–139),

$$D_H = \frac{4A_{\min}}{p_w} = \frac{4C_A S_L}{\pi} = \frac{4(0.5)(15/32 \text{ in.})}{\pi}$$

$$= 0.298 \text{ in.} = 0.0249 \text{ ft} = 0.00758 \text{ m}$$

σ from Eq. (8–138),

$$\sigma = \frac{A_{\min}}{A_1} = \frac{C_A D}{S_T} = \frac{0.5(0.375\text{ in.})}{0.562} = 1/3$$

and β from Eq. (8–133),

$$\beta = \frac{A_s}{V} = \frac{\pi D}{S_L S_T} = \frac{\pi(0.375\text{ in.})}{(15/32\text{ in.})(9/16\text{ in.})} = 4.47/\text{in.} = 53.6/\text{ft} = 176/\text{m}$$

These results for D_H, σ, and β are consistent with the values given by Fig. 8–25.

EXAMPLE 8–18

A tube bank is used in a commercial heat-pump unit to heat air at 1 atm and 15°C. The tube bank consists of eighty 9.52-mm-diameter tubes in a staggered array with 10 rows, 2.5-m depth, 1.19-cm longitudinal pitch, and 1.43-cm transverse pitch. The air enters with a velocity of 6.8 m/s and the tubes are maintained at 65°C. Determine the temperature of the exiting air, total rate of heat transfer, pressure drop, and pumping power.

Solution

Objective Determine T_2, q_c, ΔP, and $\dot{W}_p$.

Schematic Air flow over a tube bank: uniform wall temperature.

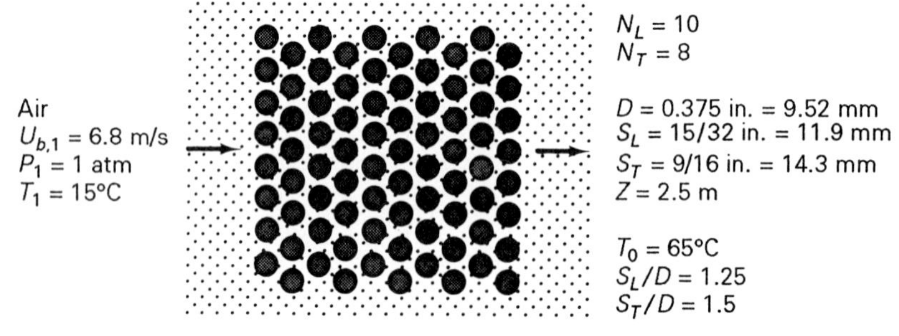

Assumptions/Conditions

forced convection
moderate property variation
standard conditions

Properties

Air at 15°C (Table A–C–5): $\rho = 1.23\text{ kg/m}^3$, $\mu = 1.76 \times 10^{-5}\text{ kg/(m s)}$, $c_P = 1.01\text{ kJ/(kg °C)}$, $k = 0.0253\text{ W/(m °C)}$, $Pr = 0.711$.

Air at 65°C: $Pr = 0.7$.

Analysis The entrance area A_1 and core surface area A_s are given by

$$A_1 = N_T S_T Z = 8(0.0143 \text{ m})(2.5 \text{ m}) = 0.286 \text{ m}^2$$

and

$$A_s = \pi DNZ = \pi(0.00952 \text{ m})(80)(2.5 \text{ m}) = 5.98 \text{ m}^2$$

This particular arrangement is identical to the tube-bank surface S1.50–1.25, which was considered in Example 8–17. Therefore, we also have $C_A = 0.5$, $D_H = 0.00758 \text{ m} = 0.758 \text{ cm}$, and $\sigma = 1/3$. It follows that

$$A_{\text{min}} = C_A N_T DZ = 0.5(8)(0.00952 \text{ m})(2.5 \text{ m}) = 0.0953 \text{ m}^2$$

Turning to the hydraulic aspects, the mass-flow rate $\dot{m}$ is

$$\dot{m} = \rho A_1 U_{b,1} = 1.23 \frac{\text{kg}}{\text{m}^3} (0.286 \text{ m}^2)\left(6.8 \frac{\text{m}}{\text{s}}\right) = 2.39 \text{ kg/s}$$

and the core-model-mass flux G is

$$G = \frac{\dot{m}}{A_{\text{min}}} = \frac{2.39 \text{ kg/s}}{0.0953 \text{ m}^2} = 25.1 \text{ kg/(m}^2 \text{ s)}$$

or $U_{b,c} = G/\rho_1 = 20.4$ m/s.

These results can be used to compute the Reynolds number Re (and/or Re_D) and to evaluate the friction factor f and coefficient of heat transfer $\bar{h}$. The Reynolds number Re is

$$Re = \frac{GD_H}{\mu} = \frac{[25.1 \text{ kg/(m}^2 \text{ s)}](0.00758 \text{ m})}{1.76 \times 10^{-5} \text{ kg/(m s)}} = 10{,}800$$

Using Figs. 8–29 and 8–26, we obtain

$$f = 0.27\, Re^{-0.18} = 0.27(10{,}800)^{-0.18} = 0.0507$$

and

$$\overline{St}_\infty Pr^{2/3} = 0.32\, Re^{-0.4} = 0.00779$$

or, since $\overline{St} = \overline{Nu}/(Re\, Pr)$,

$$\overline{Nu}_\infty = \frac{\bar{h}_\infty D_H}{k} = 0.32\, Re^{0.6} Pr^{1/3} = 0.32(10{,}800)^{0.6}(0.711)^{1/3} = 75.1$$

which gives

$$\bar{h}_\infty = \overline{Nu}_\infty \frac{k}{D_H} = 75.1 \frac{0.0253 \text{ W/(m}^2 \text{ °C)}}{0.00758 \text{ m}} = 251 \text{ W/(m}^2 \text{ °C)}$$

Using Fig. 8–27 to correct for the number of rows, $\bar{h}$ becomes

$$\bar{h} = 0.93\, \bar{h}_\infty = 0.93\left(251 \frac{\text{W}}{\text{m}^2 \text{ °C}}\right) = 233 \text{ W/(m}^2 \text{ °C)}$$

Finally, to approximately correct for the property variation over the cross section, we use Eqs. (8–25) and (8–26); that is,

$$\frac{f}{f_{cp}} = \left(\frac{T_0}{T_b}\right)^m = \left(\frac{338}{288}\right)^{-0.1} = 0.984$$

$$f = 0.0507(0.984) = 0.0499$$

and

$$\frac{\bar{h}}{\bar{h}_{cp}} = \frac{\overline{Nu}}{\overline{Nu}_{cp}} = \left(\frac{T_0}{T_b}\right)^n = \left(\frac{338}{288}\right)^{-0.5} = 0.923$$

$$\bar{h} = 233 \frac{\text{W}}{\text{m}^2\ {}^\circ\text{C}}(0.923) = 215\ \text{W/(m}^2\ {}^\circ\text{C)}$$

The outlet temperature T_2 and heat-transfer rate are given by Eqs. (8–151),

$$\frac{T_2 - T_1}{T_0 - T_1} = 1 - \exp\left(-\frac{\bar{h}A_s}{\dot{m}c_P}\right) \tag{8–151}$$

and Eq. (8–149),

$$q_c = \dot{m}c_P(T_2 - T_1) \tag{8–149}$$

To evaluate T_2, we first compute the number of transfer units NTU.

$$NTU = \frac{\bar{h}A_s}{\dot{m}c_P} = \frac{[215\ \text{W/(m}^2\ {}^\circ\text{C)}](5.98\ \text{m}^2)}{(2.39\ \text{kg/s})[1.01\ \text{kJ/(kg}\ {}^\circ\text{C)}]} = 0.533$$

Substituting this value into Eq. (8–151), T_2 becomes

$$T_2 = 15^\circ\text{C} + (65^\circ\text{C} - 15^\circ\text{C})[1 - \exp(-0.533)] = 35.7^\circ\text{C}$$

Using Eq. (8–149), q_c is

$$q_c = 2.39\frac{\text{kg}}{\text{s}}\left(1.01\frac{\text{kJ}}{\text{kg}\ {}^\circ\text{C}}\right)(35.7^\circ\text{C} - 15^\circ\text{C}) = 50\ \text{kW}$$

[This result is also obtained by use of the *LMTD* equation for q_c, Eq. (8–152).]

This calculation for T_2 indicates a moderate variation in fluid properties over the length of the tube bank. The refinement of the analysis by use of average fluid properties (based on this value of T_2) is suggested as an exercise.

To continue with the analysis, because the variation in density over the length of the core is within about 6.5%, the pressure drop can be approximated by Eq. (8–157) by setting $\rho = \rho_1$ and $U_b = U_{b,c}$; that is,†

† Using Eq. (8–158) to account for the effect of the change in density and bulk-stream velocity over the length of the core, the calculation for pressure drop becomes $\Delta P_{1-2} = 0.851$ kPa, which is only 5.8% larger than the value obtained by use of Eq. (8–157).

$$\Delta P_{1-2} = \frac{\pi N_L}{C_A}\frac{f}{2}\rho U_b^2 = \frac{\pi(10)}{0.5}\frac{0.0499}{2}\left(1.23\,\frac{\text{kg}}{\text{m}^3}\right)(20.4\text{ m})^2$$

$$= 802\text{ N/m}^2 = 0.802\text{ kPa}$$

Using Eq. (8–161), the pumping power is

$$\dot{W}_p = \frac{\dot{m}\,\Delta P_{1-2}}{\rho} = \frac{(2.39\text{ kg/s})(802\text{ N/m}^2)}{1.23\text{ kg/m}^3} = 1560\text{ W} = 1.56\text{ kW}$$

Comparing q_c and $\dot{W}_p$, we find that the pumping–heating power ratio is only 0.032. However, with the velocity increased by a factor of 2, the pumping–heating power ratio increases by a factor of about 4.

In connection with the correlations for f and $\overline{St}$ given in Figs. 8–29 and 8–26, Kays and London [4] established these relations on the basis of experimental data obtained for a number of tube-bank arrangements. Direct data taken for the particular arrangement considered in this example are shown in Fig. 8–25. Using this figure, $f = 0.053$ and $\overline{St}_\infty\, Pr^{2/3} = 0.008$, which are within 4% and 1% of the results obtained using Figs. 8–29 and 8–26.

To use the alternative correlations developed by Zukauskas [57], we compute the Reynolds number Re_D.

$$Re_D = \frac{GD}{\mu} = \frac{D}{D_H}Re = \frac{0.00952\text{ m}}{0.00758\text{ m}}\,10{,}800 = 13{,}600$$

Using this input together with Figs. 8–30 and 8–28 and Eq. (8–154), we obtain

$$F_c = 0.975 \qquad \frac{4f_i}{\chi} = 0.38 \qquad \chi = 1.0$$

and

$$\overline{Nu}_D = C\,Re_D^m\,Pr^n\left(\frac{Pr}{Pr_0}\right)^{0.25}F_c = 0.35\left(\frac{S_T}{S_L}\right)^{0.2}Re_D^{0.6}\,Pr^{0.36}\left(\frac{Pr}{Pr_0}\right)^{0.25}F_c$$

$$= 0.35\left(\frac{1.43}{1.19}\right)^{0.2}(13{,}600)^{0.6}\,(0.711)^{0.36}\left(\frac{0.711}{0.700}\right)^{0.25}(0.975) = 94.5$$

Thus,

$$f = \frac{4C_Af_i}{\pi} = \frac{4(0.5)[0.38(1.0)/4]}{\pi} = 0.0605$$

and

$$\bar{h} = 94.5\,\frac{0.0253\text{ W/(m °C)}}{0.00952\text{ m}} = 241\text{ W/(m}^2\text{ °C)}$$

These results are within about 19% and 10% of the values obtained using the correlations of Kays and London [4].

EXAMPLE 8–19

A tube bank with tubes maintained at approximately 58°F is to be used to cool air at atmospheric pressure from 260°F to 80°F. The entering velocity is 10 ft/s and $\frac{3}{8}$-in.-diameter tubes are to be placed in a staggered array with 10 columns, $\frac{3}{8}$ in. longitudinal pitch, and 0.75 in. transverse pitch. Determine the number of tubes that are required, the pumping power per unit depth, and the pumping–heating power ratio.

Solution

Objective Determine the required number of rows N_L, the pumping power per unit depth $\dot{W}_p/Z$, and the pumping–heating power ratio $\dot{W}_p/q_c$.

Schematic† Air flow over a tube bank: uniform wall temperature.

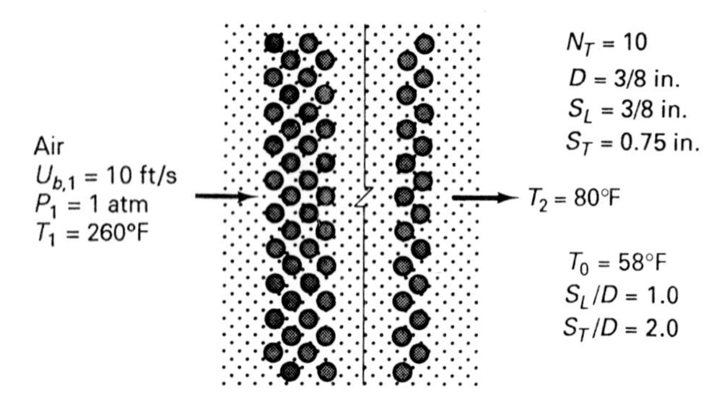

Assumptions/Conditions

forced convection
moderate property variation
standard conditions

Properties Air at average bulk-stream temperature T_m = 170°F = 630°R = 350 K (Table A–C–5):

$$\rho = 0.998 \frac{\text{kg}}{\text{m}^3} \frac{0.0624\ \text{lb}_\text{m}/\text{ft}^3}{\text{kg/m}^3} = 0.0623\ \text{lb}_\text{m}/\text{ft}^3$$

$$\mu = 2.08 \times 10^{-5} \frac{\text{kg}}{\text{m s}} \frac{0.672\ \text{lb}_\text{m}/(\text{ft s})}{\text{kg}/(\text{m s})} = 1.40 \times 10^{-5}\ \text{lb}_\text{m}/(\text{ft s})$$

† This particular geometric arrangement is denoted as surface S2.00–1.00 in the work of Kays and London [4].

$$c_P = 1.01\,\frac{\text{kJ}}{\text{kg K}}\,\frac{0.239\ \text{Btu/(lb}_\text{m}\ ^\circ\text{F)}}{\text{kJ/(kg K)}} = 0.241\ \text{Btu/(lb}_\text{m}\ ^\circ\text{F)}$$

$$k = 0.030\,\frac{\text{W}}{\text{m}^\circ\ \text{C}}\,\frac{0.578\ \text{Btu/(h ft }^\circ\text{F)}}{\text{W/(m }^\circ\text{C)}} = 0.0173\ \text{Btu/(h ft }^\circ\text{F)}$$

$$Pr = 0.697$$

The density at the inlet and exit temperatures varies from 0.0551 $\text{lb}_\text{m}/\text{ft}^3$ (0.883 kg/m^3) to about 0.0735 $\text{lb}_\text{m}/\text{ft}^3$ (1.18 kg/m^3), which represents a 34% increase. Therefore, the effects of density variation over the length of the tube bank should be considered.

Analysis The entrance area A_1 and surface area A_s are expressed in terms of the tube bank depth Z by

$$A_1 = N_T S_T Z = 10\left(\frac{0.75}{12}\ \text{ft}\right)Z = 0.625Z\ \text{ft}$$

and

$$A_s = \pi D N Z = \pi\left(\frac{0.375}{12}\ \text{ft}\right)(10)N_L Z = 0.982 N_L Z\ \text{ft} \tag{a}$$

The minimum free-flow area A_{min} is determined by writing

$$S_D = \left[S_L^2 + \left(\frac{S_T}{2}\right)^2\right]^{1/2} = \left[(0.375\ \text{in.})^2 + \left(\frac{0.75\ \text{in.}}{2}\right)^2\right]^{1/2} = 0.53\ \text{in.}$$

and

$$\frac{S_T + D}{2} = \frac{0.75\ \text{in.} + 0.375\ \text{in.}}{2} = 0.562\ \text{in.}$$

In this case $S_D < (S_T + D)/2$, such that C_A is given by Eq. (8–136b),

$$C_A = 2\left(\frac{S_D}{D} - 1\right) = 2\left(\frac{0.53\ \text{in.}}{0.375\ \text{in.}} - 1\right) = 0.827$$

Using this result, we obtain

$$A_{\text{min}} = C_A N_T D Z = 0.827(10)\left(\frac{0.375}{12}\ \text{ft}\right)Z = 0.258Z\ \text{ft}$$

$$D_H = \frac{4C_A S_L}{\pi} = \frac{4(0.827)(0.375\ \text{in.})}{\pi} = 0.395\ \text{in.} = 0.0329\ \text{ft}$$

$$\sigma = \frac{A_{\text{min}}}{A_1} = \frac{0.258Z\ \text{ft}}{0.625Z\ \text{ft}} = 0.413$$

The mass-flow rate $\dot{m}$ and core-model-mass flux G are computed by writing

$$\dot{m} = \rho_1 A_1 U_{b,1} = 0.0551\,\frac{\text{lb}_\text{m}}{\text{ft}^3}\,(0.625Z\ \text{ft})\left(10\,\frac{\text{ft}}{\text{s}}\right) = 0.344Z\ \text{lb}_\text{m}/(\text{ft s})$$

and

$$G = \frac{\dot{m}}{A_{min}} = \frac{0.344Z \text{ lb}_m/(\text{ft s})}{0.258Z \text{ ft}} = 1.33 \text{ lb}_m/(\text{ft}^2 \text{ s})$$

With the viscosity evaluated at T_m, the Reynolds number Re becomes

$$Re = \frac{GD_H}{\mu} = \frac{[1.33 \text{ lb}_m/(\text{ft}^2 \text{ s})](0.0329 \text{ ft})}{1.4 \times 10^{-5} \text{ lb}_m/(\text{ft s})} = 3130$$

Using this result together with Figs. 8–29 and 8–26, f and $\overline{h}_\infty$ become

$$f = 0.46\, Re^{-0.18} = 0.108$$

and

$$\frac{\overline{h}_\infty D_H}{k} = \overline{Nu}_\infty = 0.46\, Re^{0.6}\, Pr^{1/3} = 51$$

or

$$\overline{h}_\infty = \overline{Nu}_\infty \frac{k}{D_H} = 51 \frac{0.0173 \text{ Btu/(h ft °F)}}{0.0329 \text{ ft}} = 26.8 \text{ Btu/(h ft}^2 \text{ °F)}$$

Since N_T is unknown, the correction for the effect of the number of rows on $\overline{h}$ will be made later. Correcting for the effect of property variation over the cross section, we obtain

$$f = f_{cp}\left(\frac{518\text{°R}}{630\text{°R}}\right)^{-0.1} = 0.108(1.02) = 0.11$$

and, since $n = 0$,

$$\overline{h} = \overline{h}_{cp} = 26.8 \text{ Btu/(h ft}^2 \text{ °F)}$$

The rate of heat transfer q_c is given by Eq. (8–149),

$$q_c = \dot{m}c_P(T_2 - T_1) = 0.344Z \frac{\text{lb}_m}{\text{ft s}}\left(0.241 \frac{\text{Btu}}{\text{lb}_m \text{ °F}}\right)(80\text{°F} - 260\text{°F})$$

$$= -14.9Z \text{ Btu/(ft s)} = -51.5Z \text{ kW/m}$$

and Eq. (8–152),

$$q_c = \overline{h}A_s\, LMTD$$

where $LMTD$ is given by Eq. (8–153),

$$LMTD = \frac{T_2 - T_1}{\ln [(T_0 - T_1)/(T_0 - T_2)]} = \frac{80\text{°F} - 260\text{°F}}{\ln [(58\text{°F} - 260\text{°F})/(58\text{°F} - 80\text{°F})]}$$

$$= -81.2\text{°F}$$

Combining these relations and solving for A_s, we have

$$A_s = \frac{q_c}{\bar{h}\ LMTD} = \frac{-14.9Z\ \text{Btu/(ft s)}}{[26.8\ \text{Btu/(h ft}^2\ ^\circ\text{F)}](-81.2^\circ\text{F})} = 24.6Z\ \text{ft}$$

Substituting this result into Eq. (a), N_L is given by

$$N_L = \frac{A_s}{0.982Z\ \text{ft}} = \frac{24.6Z\ \text{ft}}{0.982Z\ \text{ft}} = 25.1$$

Using this result together with Fig. 8–27, the analysis is refined by writing

$$\bar{h} = 0.97\bar{h}_\infty = 0.97\left(26.8\ \frac{\text{Btu}}{\text{h ft}^2\ ^\circ\text{F}}\right) = 26\ \text{Btu/(h ft}^2\ ^\circ\text{F)}$$

$$A_s = \frac{-14.9Z\ \text{Btu/(ft s)}}{[26\ \text{Btu/(h ft}^2\ ^\circ\text{F)}](-81.2^\circ\text{F})} = 25.4Z\ \text{ft}$$

and

$$N_L = \frac{25.4Z\ \text{ft}}{0.982Z\ \text{ft}} = 25.9$$

Rounding off, we conclude that 26 tube rows must be used in this application. Thus, a total of 260 tubes is required to satisfy the heating requirement. Because of this large number of tubes, alternative means of accomplishing this task should be considered. The use of finned tubes and compact heat exchangers for applications such as this with large heating requirements is considered in Sec. 8–4–4 and Chap. 11.

Turning to the hydraulic analysis, to account for the effects of density variation over the length of the tube bank on the pressure drop, we use Eq. (8–158),†

$$\Delta P_{1-2} = \frac{G^2 v_1}{2}\left[(1 + \sigma^2)\left(\frac{v^2}{v_1} - 1\right) + \frac{\pi N_L}{C_A} f \frac{v_m}{v_1}\right]$$

$$= \left(1.33\ \frac{\text{lb}_\text{m}}{\text{ft}^2\ \text{s}}\right)^2 \frac{1}{2(0.0551\ \text{lb}_\text{m}/\text{ft}^3)}\left[(1 + 0.413^2)\left(\frac{0.0551\ \text{lb}_\text{m}/\text{ft}^3}{0.0735\ \text{lb}_\text{m}/\text{ft}^3} - 1\right)\right.$$
$$\left. + \frac{\pi(26)(0.108)}{0.827}\ \frac{0.0551\ \text{lb}_\text{m}/\text{ft}^3}{0.0623\ \text{lb}_\text{m}/\text{ft}^3}\right] = (-4.7 + 151)\frac{\text{lb}_\text{m}}{\text{ft s}^2} \qquad \text{(b)}$$

$$= 151\ \frac{\text{lb}_\text{m}}{\text{ft s}^2}\ \frac{\text{lb}_\text{f}\ \text{s}^2}{32.2\ \text{lb}_\text{m}\ \text{ft}} = 4.69\ \text{lb}_\text{f}/\text{ft}^2 = 0.0326\ \text{psi} = 224\ \text{Pa}$$

This pressure drop is in the appropriate range for conventional fans that develop low static pressure.

† Note that $v_2/v_1 = (1/\rho_2)/(1/\rho_1) = \rho_1/\rho_2$ can also be represented in terms of the absolute temperatures by $v_2/v_1 = T_2/T_1$. Similarly, v_m/v_1 can be set equal to T_m/T_1.

This result is substituted into Eq. (8–162) to obtain the pumping power $\dot{W}_p$.

$$\dot{W}_p = \dot{m}\,\Delta P\,v_m = 0.344Z\frac{\text{lb}_\text{m}}{\text{ft s}}\left(4.69\frac{\text{lb}_\text{f}}{\text{ft}^2}\right)\left(\frac{1}{0.0623\ \text{lb}_\text{m}/\text{ft}^3}\right)$$

$$= 25.9Z\ \text{ft lb}_\text{f}/(\text{ft s}) = 0.0333Z\ \text{Btu}/(\text{ft s})$$

Thus, the pumping–heating power ratio is

$$\frac{\dot{W}_p}{q_c} = \frac{0.0333Z\ \text{Btu}/(\text{ft s})}{-14.9Z\ \text{Btu}/(\text{ft s})} = -0.00223$$

Concerning the result of the hydraulic analysis, it is interesting to note that despite the large density variation over the length of the tube bank, the effect of acceleration represented by the first term in Eq. (b) is only about 3% of the core effect. This is often the case for applications involving tube banks and compact heat exchangers, such that the simpler uniform property relation given by Eq. (8–157) (with ρ set equal to ρ_m) often suffices.

8–4–4 Fin-Tube Banks

Fin tubes are commonly used for applications requiring large heating rates. A representative arrangement with staggered circular fin tubes is illustrated in Fig. 8–31.

The geometric parameters that characterize fin-tube banks include the fin *type* and *dimensions* (e.g., fin height H_F, fin thickness δ_F, and fin pitch S_F for circumferential fins†), as well as tube diameter D, tube pitches S_L and S_T, number of tube rows N_L and columns N_T, and tube arrangement. The relations for length L, frontal area A_1, and volume V given by Eqs. (8–129),

$$L = N_L S_L \tag{8–163}$$

Eq. (8–130),

$$A_1 = N_T S_T Z \tag{8–164}$$

and Eq. (8–131),

$$V = LA_1 = NS_L S_T Z \tag{8–165}$$

are common to tube banks with finned as well as unfinned tubes. In addition, we are able to represent the total surface area A_s and total fin surface area A_{fins} by

$$A_s = N[N_F(A_F + A_p)] = N(N_F A_o) \quad \text{and} \quad A_{\text{fins}} = N(N_F A_F) \tag{8–166,167}$$

where N_F is the number of fins per tube.

† The fin pitch S_F for a circumferential fin is defined in terms of the number of fins per tube N_F by $S_F = Z/N_F$.

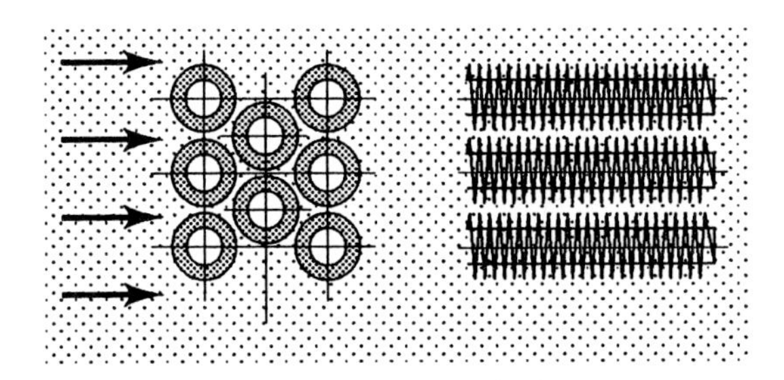

FIGURE 8–31 Fin-tube bank. (From Kays and London [4]. Used with permission.)

Continuing with the geometric considerations presented in Sec. 8–4–1, a tube-bank core model of the form illustrated by Fig. 8–24 is employed, with the effective perimeter represented by $p = A_s/L$ and the characteristic cross-sectional area taken as the minimum free-flow area A_{min}. With the fin type and basic dimensions specified, calculations can be made for A_{min}, total surface area A_s, fin surface area A_F, and prime surface area A_p. This information is generally provided in terms of the contraction ratio $\sigma = A_{min}/A_1$, the surface-area density $\beta = A_s/V$, and the finned area fraction,

$$\frac{A_{\text{fins}}}{A_s} = \frac{A_F}{A_o} \tag{8–168}$$

Knowing σ and β, the hydraulic diameter D_H is given by Eq. (8–141),

$$D_H = \frac{4A_{min}}{p_w} = 4\frac{\sigma}{\beta} \tag{8–169}$$

where $p_w = p$.

The relations for bulk-stream characteristics given by Eqs. (8–142) to (8–147) for tube banks with plain tubes also apply for fin tubes. In addition, the practical hydraulic analysis approach presented in Sec. 8–4–3 and a slightly generalized form of the practical thermal analysis approach presented in Sec. 8–4–2 apply for tube banks with fin tubes.

Following the practical thermal analysis approach introduced in Sec. 8–2–2, the effective local heat flux q_c'' for fin tubes is expressed in terms of the net surface efficiency η_o by Eq. (8–68),

$$q_c'' = \eta_o h_x (T_s - T_b) \tag{8–170}$$

where η_o is given by Eq. (8–69),

$$\eta_o = 1 - \frac{A_F}{A_o}(1 - \eta_F) \tag{8–171}$$

The fin efficiency η_F is given by Fig. 2–17 for circumferential and other type fins. Using Eq. (8–170), the solution for uniform wall-temperature heating with uniform properties is given by Eq. (8–71),

$$\frac{T_2 - T_1}{T_0 - T_1} = 1 - \exp\left(\frac{-\overline{\eta_o h} A_s}{\dot{m} c_P}\right) \tag{8–172}$$

and

$$q_c = \dot{m} c_P (T_2 - T_1) \tag{8–173a}$$

$$= \dot{m} c_P (T_0 - T_1)\left[1 - \exp\left(\frac{-\overline{\eta_o h} A_s}{\dot{m} c_P}\right)\right] \tag{8–173b}$$

or

$$q_c = \overline{\eta_o h} A_s \, LMTD \tag{8–174}$$

where *LMTD* is defined by Eq. (8–60). In practice, η_o is generally taken as a constant, such that $\overline{\eta_o h} \simeq \eta_o \bar{h}$. Correlations of experimental data for $\overline{St}$ (staggered array)† are given by Kays and London [4], Gray and Webb [64], Webb [65], Briggs and Young [66], Robinson and Briggs [67], and Rabas et al. [68].‡ Representative data are shown in Fig. 8–32 for a tube bank with circumferential fin tubes placed in a staggered array. The mean coefficient of heat transfer associated with flow across staggered fin tubes has been reported to approach an asymptotic value after only about four tube rows [19]. Therefore, no correction factor for number of tube rows is required.

The practical hydraulic analysis approach involves the specification of the friction factor f, use of Eq. (8–157),

$$\Delta P_{1-2} = \frac{4L}{D_H} \frac{f}{2} \rho U_b^2 \tag{8–175}$$

or Eq. (8–158),

$$\Delta P_{1-2} = \frac{G^2 v_1}{2}\left[(1 + \sigma^2)\left(\frac{v_2}{v_1} - 1\right) + \frac{4L}{D_H} f \frac{v_m}{v_1}\right] \tag{8–176}$$

for pressure drop, and use of Eq. (8–161),

$$\dot{W}_p = \frac{\dot{m}\, \Delta P}{\rho} \tag{8–177}$$

or Eq. (8–162),

$$\dot{W}_p = \dot{m}\, \Delta P \, v_m \tag{8–178}$$

for pumping power. Experimental data and correlations for f are given by Kays and

† The heat transfer coefficient for in-line arrangements is substantially lower than that for staggered configurations because of the flow bypass through the gap if fins are not touching. Therefore, no correlations are available for in-line fin-tube banks [19].

‡ However, much of the heat-transfer data for fin-tubes has been retained as proprietary information by heat-exchanger manufacturers.

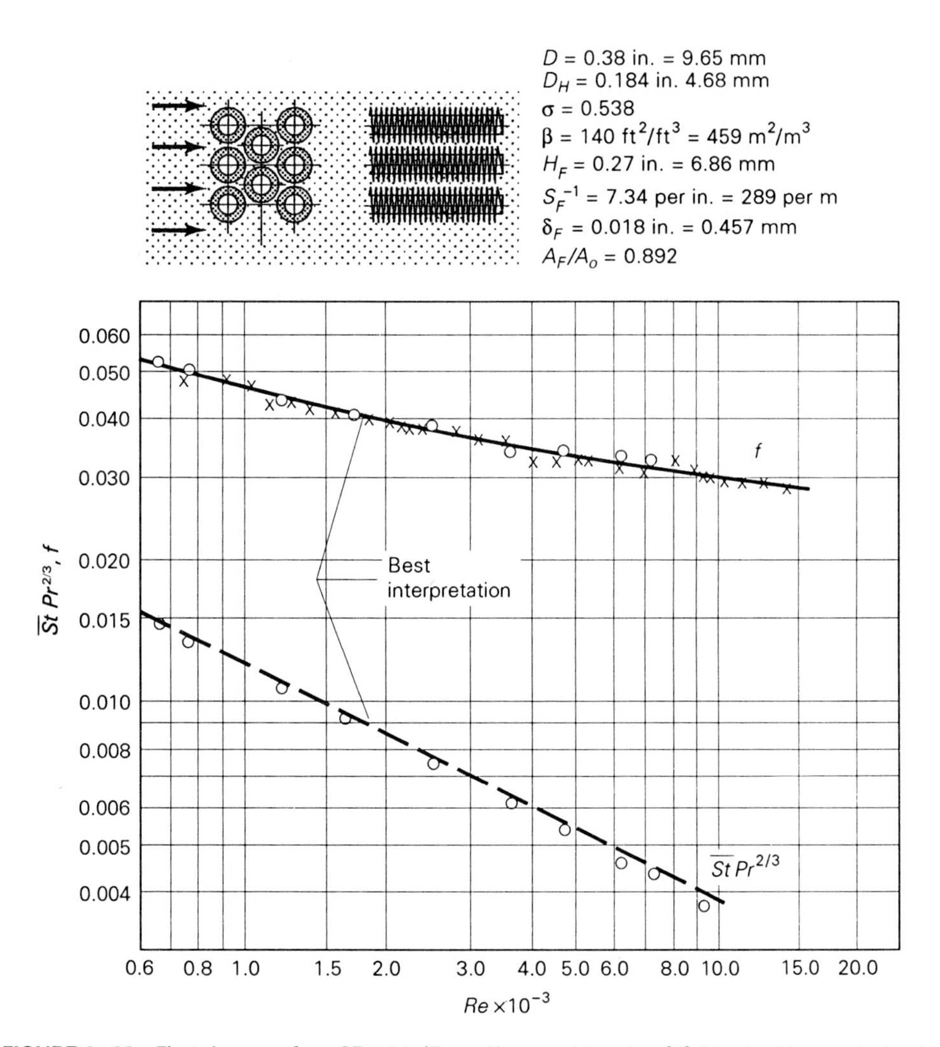

FIGURE 8-32 Fin tubes; surface CF-7.34. (From Kays and London [4]. Used with permission.)

London [4], Gray and Webb [64], and Webb [65] for tube banks with several tube geometries. Representative data and a correlation for f are shown in Fig. 6–32.

EXAMPLE 8–20

The number of plain tubes required in the application considered in Example 8–19 was found to be 260. Determine the number of fin tubes required for the case in which surface CF 7.34 is used instead of S 2.00–1.00.

Solution

Objective Determine the required number of rows N_L of fin tubes.

Schematic Air flow over fin-tube bank: uniform wall temperature.

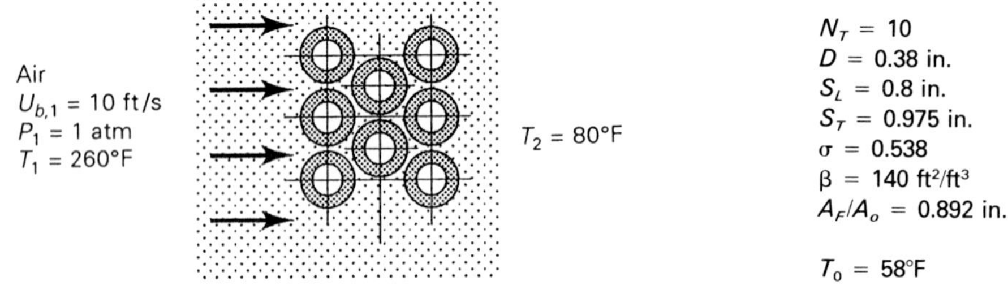

Assumptions/Conditions

forced convection
negligible thermal contact resistance between fin and tube wall
moderate property variation
standard conditions

Properties

Air (Example 8–19): $\rho = 0.0623$ lb$_m$/ft^3, $\mu = 1.4 \times 10^{-5}$ lb$_m$/(ft s), $c_P = 0.241$ Btu/(lb$_m$ °F), $k = 0.0173$ Btu/(h ft °F), and $Pr = 0.697$ at average bulk-stream temperature $T_m = 170$°F; $\rho = 0.0551$ lb$_m$/ft^3 at entrance bulk-stream temperature $T_1 = 260$°F.

Aluminum at 58°F [Table A–C–1(a)]: $k = 236$ W/(m °C) $= 136$ Btu/(h ft °F).

Analysis The frontal area A_1 and volume V are given by

$$A_1 = N_T S_T Z = 10 \frac{0.975 \text{ ft}}{12} Z = 0.812Z \text{ ft}$$

and

$$V = LA_1 = NS_L S_T Z = 10N_L(0.8 \text{ in.})(0.975 \text{ in.})Z$$

$$= 7.8N_L Z \text{ in.}^2 = 0.0542N_L Z \text{ ft}^2$$

With the surface-area density β specified, A_s is represented by

$$A_s = \beta V = \frac{140}{\text{ft}} 0.0542 N_L Z \text{ ft}^2 = 7.59 N_L Z \text{ ft} \tag{a}$$

Similarly, the minimum free-flow area A_{min} is specified in terms of the input for σ by

$$A_{\text{min}} = \sigma A_1 = 0.538(0.812Z \text{ ft}) = 0.437Z \text{ ft}$$

and the hydraulic diameter D_H is given by

$$D_H = \frac{4A_{\text{min}}}{p_w} = 4\frac{\sigma}{\beta} = 4\frac{0.538}{140/\text{ft}} = 0.0154 \text{ ft} = 0.184 \text{ in.}$$

We are now able to compute the mass-flow rate $\dot{m}$, mass flux G, and Reynolds number Re as follows:

$$\dot{m} = (\rho A U_b)_1 = \left(0.0551 \frac{\text{lb}_\text{m}}{\text{ft}^3}\right)(0.812Z \text{ ft})\left(10\frac{\text{ft}}{\text{s}}\right) = 0.448Z \text{ lb}_\text{m}/(\text{ft s})$$

$$G = \frac{\dot{m}}{A_{\text{min}}} = \frac{0.448Z \text{ lb}_\text{m}/(\text{ft s})}{0.437Z \text{ ft}} = 1.03 \text{ lb}_\text{m}/(\text{ft}^2 \text{ s})$$

$$Re = \frac{GD_H}{\mu} = \frac{[1.03 \text{ lb}_\text{m}/(\text{ft}^2 \text{ s})](0.0154 \text{ ft})}{1.4 \times 10^{-5} \text{ lb}_\text{m}/(\text{ft s})} = 1130$$

Using this result for Re together with Fig. 8–32, the mean Stanton number is approximated by

$$\overline{St}\, Pr^{2/3} = 0.011$$

such that $\bar{h}$ becomes

$$\bar{h} = 0.011 \frac{k}{D_H} Re\, Pr^{1/3} = 0.011 \frac{0.0173 \text{ Btu/(h ft °F)}}{0.0154 \text{ ft}} (1130)(0.7)^{1/3}$$

$$= 12.4 \text{ Btu/(h ft}^2 \text{ °F)}$$

Referring to Table 8–2, $n = 0$ for laminar or turbulent flow of a gas being cooled, such that $\bar{h}$ need not be corrected for property variation over the cross section.

The rate of heat transfer is represented by Eq. (8–173a),

$$q_c = \dot{m} c_P (T_2 - T_1) = -19.4Z \text{ Btu/(ft s)}$$

and Eq. (8–174),

$$q_c = \overline{\eta_o h}\, A_s\, LMTD$$

where $\overline{\eta_o h} \simeq \eta_o \bar{h}$ and $LMTD = -81.2$°F, and the net surface efficiency η_o is given by Eq. (8–171),

$$\eta_o = 1 - \frac{A_F}{A_o}(1 - \eta_F)$$

The fin efficiency η_F is obtained from Fig. 2–17b. Setting $\bar{h}$ = 13.6 Btu/(h ft² °F), k_F = 136 Btu/(h ft °F) (for aluminum), δ_F = 0.018 in., $L = r_2 - r_1 = H_F$ = 0.27 in., $L_c = L + \delta_F/2$ = 0.279 in., $r_{2c} = r_2 + \delta_F/2$ = 0.469 in., $A_p = L_c\delta_F$ = 0.00502 in.², $L_c^{3/2}\,[\bar{h}/(k_F A_p)]^{1/2}$ = 0.181, r_{2c}/r_1 = 2.47, we find that $\eta_F \simeq 0.94$, such that

$$\eta_o = 1 - 0.892(1 - 0.94) = 0.946$$

It follows that

$$A_s = \frac{q_c}{\eta_o \bar{h}\,LMTD} = \frac{-19.4Z\ \text{Btu/(ft s)}}{0.946[12.4\ \text{Btu/(h ft}^2\ ^\circ\text{F)}](-81.2^\circ\text{F})} = 73.3Z\ \text{ft}$$

This result is substituted into Eq. (a) to obtain

$$N_L = \frac{A_s}{7.59Z\ \text{ft}} = \frac{73.3Z\ \text{ft}}{7.59Z\ \text{ft}} = 9.66$$

which indicates approximately 10 rows or 100 fin tubes. Thus, we find that a factor of 2.6 fewer fin tubes than plain tubes (260) is required for this application.

8–5 SUMMARY

The practical engineering analysis of internal flow convection-heat-transfer processes involves the use of correlations for friction factor f, entrance- and expansion-loss coefficients K_c and K_e, and the coefficient of heat transfer h in conjunction with lumped/differential formulations for momentum and energy transfer. In this chapter we have presented correlations for f, K_c, K_e, and h for standard single-phase incompressible internal flows which are commonly encountered in practice. Correlations for laminar and turbulent flow presented in this chapter are summarized in Table 8–6. In addition, mathematical formulations for the fluid flow and heat transfer have been developed for uniform and variable property flows. The resulting practical thermal and hydraulic analysis approach applies to internal flows with both uniform and nonuniform cross-sectional area, such that the heat transfer, pressure drop, and pumping power can be computed for heat-transfer cores involving single or multiple tubular passages as well as flow across tube banks. Relations for the key geometric and flow characteristics developed in Sec. 8–4 for flow across tube banks are summarized in Table 8–7. Because of its importance, the practical analysis approach for internal flows is summarized in Table 8–8. As we shall see in Chap. 11, this lumped analysis approach provides the framework for the evaluation and design of heat exchangers. Consideration has been given to fin tubes and fin-tube banks in Secs. 8–2–2 and 8–4–4.

The practical approach to the analysis of external flow features the use of defining relations and correlations for friction factor f_x and the coefficient of heat transfer h_x. Representative correlations for friction factor f_x and Nusselt number Nu_x for laminar and turbulent flow over surfaces with various heating conditions are listed in Table 8–9.

Finally, the primary dimensionless parameters used in the analysis of forced convection are summarized in Table 8–10.

TABLE 8–6 Summary of convection correlations for internal flow

Condition	*Correlation*	*Identification*
LAMINAR FLOW†		
HFD	$f = C_1/Re$	Eq. (8–7), Table 8–1 Fig. 8–2
TFD	$Nu = C_2$	Eq. (8–8), Table 8–1 Fig. 8–2
Thermal entry $T_s = T_0$	$\overline{Nu} = 3.66 + \dfrac{0.0668 \dfrac{Re\,Pr}{x/D}}{1 + 0.04\left(\dfrac{Re\,Pr}{x/D}\right)^{2/3}}$	Eq. (8–9) Hausen [15]
$T_s = T_0$ $\dfrac{x/D}{Re\,Pr} \leqslant 0.01$	$\overline{Nu} = 1.67\left(\dfrac{Re\,Pr}{x/D}\right)^{1/3}$	Eq. (8–10) Limiting form of Eq. (8–9)
$q_c'' = q_0''$ $\dfrac{x/D}{Re\,Pr} \leqslant 0.01$	$(Nu)_x = 1.30\left(\dfrac{Re\,Pr}{x/D}\right)^{1/3}$	Eq. (8–11) Sellars et al. [16]
Hydrodynamic and thermal entry	Performance curves for f, $\bar{f}$, f_{app} Performance curves for K_c and K_e	Fig. 8–2 Langhaar [1] Figs. 8–4, 8–5 Kays and London [4]
Gas $T_s = T_0$	$\overline{Nu} = 3.66 + \dfrac{0.104 \dfrac{Re\,Pr}{x/D}}{1 + 0.016\left(\dfrac{Re\,Pr}{x/D}\right)^{0.8}}$	Eq. (8–12) Kays [14]
$\dfrac{x/D}{Re\,Pr} \leqslant 0.01$	$\overline{Nu} = 1.86\left(\dfrac{Re\,Pr}{x/D}\right)^{1/3}$	Eq. (8–13) Seider and Tate [18]
TRANSITIONAL TURBULENT FLOW	$f = C_{tr} f_{L2} + (1 - C_{tr}) f_{T4}$ $C_{tr} = 2 - Re/2000$ $Nu = C_{tr}\, Nu_{L2} + (1 - C_{tr})\, Nu_{T8}$ $C_{tr} = 1.33 - Re/6000$	Eq. (8–20) Taborek [32]

TABLE 8–6 *(Continued)*

Condition	*Correlation*	*Identification*
TURBULENT FLOW		
HFD		
$4000 \lesssim Re$	$f = (1.58 \ln Re - 3.28)^{-2}$	Eq. (8–15)
$10^4 \lesssim Re$	$f = 0.046\, Re^{-0.2}$	Eq. (8–17)
TFD $10^4 \leqslant Re \leqslant 5 \times 10^6$ $0.5 \leqslant Pr \leqslant 200$	$Nu = \dfrac{(f/2)\, Re\, Pr}{1.07 + 12.7\sqrt{f/2}\,(Pr^{2/3} - 1)}$	Eq. (8–16)
$10^4 \leqslant Re$	$Nu = 0.023\, Re^{0.8}\, Pr^n$ $n = 0.5$ for $0.5 \lesssim Pr \lesssim 5.0$ $n = 1/3$ for $0.5 \lesssim Pr$	Eq. (8–19) (related correlations by Dittus and Boelter [28] and Colburn [29]
Thermal Entry		
$x_c/D \leqslant x/D \leqslant 60$	$\overline{Nu} = Nu\left(1 + \dfrac{1.4}{x/D}\right)$	Eq. (8–21a)
$x/D \leqslant x_c/D = 0.625\, Re^{0.25}$	$\overline{Nu} = Nu\dfrac{1.11\, Re^{0.2}}{(x/D)^{0.8}}$	Eq. (8–21b) *Handbook of Heat Transfer* [3]
TFD, Liquid metals $Re\, Pr \geqslant 100$ $L/D \geqslant 30$	$Nu = 5 + 0.025\,(Re\, Pr)^{0.8}$	Eq. (8–22) Subbotin et al. [33]
Hydrodynamic and thermal entry	Performance curves for K_c and K_e	Eq. 8–5 Kays and London [4]
$x_c/D \leqslant x/D \leqslant 60$	$\overline{Nu} = Nu\left(1 + \dfrac{6}{x/D}\right)$	Eq. (8–21a) *Handbook of Heat Transfer* [3]
$x/D \leqslant x_c/D = 0.625\, Re^{0.25}$	See Eq. (8–21b)	
Rough surfaces HFD	Moody friction factor chart	Fig. 8–10 Moody [43]
	$Nu = \dfrac{(f/2)\, Re\, Pr}{1.07 + 12.7\sqrt{f/2}\,(Pr^{2/3} - 1)}$ with f obtained from Moody chart	Eq. (8–16)
VARIABLE PROPERTY FLOW		
Gas	$\dfrac{f}{f_{cp}} = \left(\dfrac{T_s}{T_b}\right)^m \quad \dfrac{Nu}{Nu_{cp}} = \left(\dfrac{T_s}{T_b}\right)^n$	Eq. (8–25, 26) m, n from Table 8–2
Liquid	$\dfrac{f}{f_{cp}} = \left(\dfrac{\mu_s}{\mu_b}\right)^m \quad \dfrac{Nu}{Nu_{cp}} = \left(\dfrac{\mu_s}{\mu_b}\right)^n$	Eqs. (8–23, 24) m, n from Table 8–2
TUBE BANKS	Performance curves for f and $\overline{St}$	Kays and London [4] Zukauskas [58]

† Correlations for Nusselt number are restricted to $Pr \geqslant 0.5$.

TABLE 8–7 Flow across tube banks: Relations for geometric and flow characteristics†

Characteristic	*Relation*	*Equation*
Derived Geometric Characteristics		
Length	$L = N_L S_L$	(8–129)
Frontal area	$A_1 = N_T S_T Z$	(8–130)
Volume	$V = A_1 L = N S_L S_T Z$	(8–131)
Surface area	$A_s = \pi D N Z$	(8–132)
Surface-area density	$\beta = \dfrac{A_s}{V} = \dfrac{\pi D}{S_L S_T}$	(8–133)
Heated perimeter	$p = \dfrac{A_s}{L} = \dfrac{\pi D N_T Z}{S_L}$	(8–134)
Wetted perimeter	$p_w = p$	
Minimum free-flow area	$A_{\min} = C_A N_T D Z$	(8–135)
Area Coefficient		
In-line array	$C_A = \dfrac{S_T}{D} - 1$	(8–136a)
Staggered array	$C_A = \dfrac{S_T}{D} - 1 \quad \text{for } S_D \geqslant \dfrac{S_T + D}{2}$	(8–136a)
	$C_A = 2\left(\dfrac{S_D}{D} - 1\right) \quad \text{for } S_D < \dfrac{S_T + D}{2}$	(8–136b)
	$S_D = \left[S_L^2 + \left(\dfrac{S_T}{2}\right)^2\right]^{1/2}$	(8–137)
Contraction ratio	$\sigma = \dfrac{A_{\min}}{A_1} = \dfrac{C_A D}{S_T}$	(8–138)
Hydraulic diameter	$D_H = \dfrac{4A_{\min}}{p_w} = \dfrac{4C_A S_L}{\pi}$	(8–139)
	$= 4\dfrac{\sigma}{\beta}$	(8–141)
Flow Characteristics		
Mass-flow rate	$\dot{m} = \rho A_{\min} U_b = A_{\min} G$	(8–142)
	$= \rho A_1 U_{b,1} = A_1 G_1$	(8–143)
Bulk-stream velocity	$U_b = \dfrac{U_{b,1}}{\sigma} = \dfrac{S_T}{C_A D} U_{b,1}$	(8–144)
Bulk-stream mass flux	$G = \dfrac{G_1}{\sigma} = \dfrac{S_T}{C_A D} G_1$	(8–145)
Reynolds number	$Re = \dfrac{U_b D_H}{\nu} = \dfrac{G D_H}{\mu}$	(8–146)

† Basic geometric dimensions: D = tube diameter, S_L = longitudinal pitch, S_T = transverse pitch, Z = depth.

TABLE 8–8 Summary of the practical thermal analysis approach: Internal flow†

Control volume	*Variables and relations*	*Energy equation* $\sum \dot{E}_o - \sum \dot{E}_i + \frac{\Delta E_s}{\Delta t} = 0$
Lumped/differential $dV = A\,dx$ 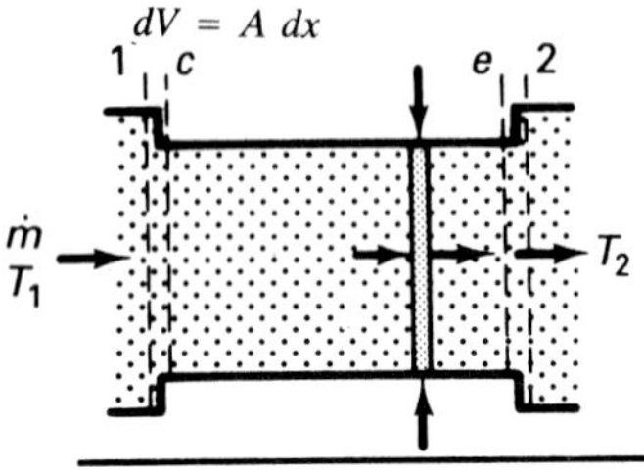 1 c e 2 $\dot{m}$ T_1 T_2	Independent variables $\dot{m}$ T_1, T_s or q'' Dependent variables $T_b(x)$ $q_s''(x)$ or $T_s(x)$ Bulk-stream relation $d\dot{H}_b = \dot{m}c_P\,dT_b$	Lumped/differential formulation $q_c'' p = \frac{d\dot{H}_b}{dx} = \dot{m}c_P \frac{dT_b}{dx}$ Lumped formulation $q_c'' = \dot{H}_2 - \dot{H}_1 = \dot{m}c_P(T_2 - T_1)$
Coefficients	*Dimensionless parameters*	*Practical solution results*
Newton law of cooling (Coefficient of heat transfer h_x) $q_c'' = h_x(T_s - T_b)$ Fanning friction factor f_x $\tau_s = \rho U_b^2 \frac{f_x}{2}$ Entrance-loss coefficient K_c $P_c - P = \frac{\rho U_b^2}{2}\left(\frac{4x}{D_H} f + K_c\right)$ Expansion-loss coefficient K_e $\Delta P_{\text{loss}} = K_e \frac{\rho U_b^2}{2}$	Reynolds number $Re = \frac{GD_H}{\mu} = \frac{U_b D_H}{\nu}$ Prandtl number $Pr = \frac{\mu c_P}{k} = \frac{\nu}{\alpha}$ Nusselt number $Nu = \frac{hD_H}{k}$ Stanton number $St = \frac{Nu}{Re\,Pr}$	Solution results given in Sec. 8–2–2 for (1) specified wall-heat flux, (2) specified wall temperature, (3) variable properties. Solution for uniform wall-temperature heating given by $q_c = \dot{m}c_P(T_0 - T_1)\left[1 - \exp\left(-\frac{\bar{h}A_s}{\dot{m}c_P}\right)\right]$ which is commonly represented in terms of effectiveness ϵ, efficiency ψ, or *LMTD* by $q_c = \epsilon \dot{m}c_P(T_0 - T_1)$ $q_c = \psi \bar{h}A_s(T_0 - T_1)$ $q_c = \bar{h}A_s\,LMTD$ Hydraulic: $\Delta P_{1-2} = \frac{\rho U_b^2}{2}\left(K_c + \frac{4L}{D_H} f + K_e\right)$ $\dot{W}_p = \frac{\dot{m}\,\Delta P}{\rho}$ [See Eqs. (8–77) and (8–79) for variable density.]

† See Tables 8–6 and 8–9 for correlations.

TABLE 8–9 Summary of convection correlations for external flow—flat plate with uniform free-stream velocity

Condition	*Correlation*	*Identification*
LAMINAR FLOW		
Uniform wall temperature	$f_x = 0.664\, Re_x^{-1/2}$	Eq. (8–86)
$10 \lesssim Pr$	$Nu_x = 0.339\, Re_x^{1/2}\, Pr^{1/3}$	Eq. (8–89)
$0.6 \lesssim Pr \lesssim 10$	$Nu_x = 0.332\, Re_x^{1/2}\, Pr^{1/3}$	Eq. (8–88)
$Pr \lesssim 0.928$ $Pe \gtrsim 100$	$Nu_x = \dfrac{0.646\, Re_x^{1/2}}{1 + (1.49/Pr - 0.6)^{1/2}}$	Eq. (8–91)
Uniform Wall-Heat Flux		
$5 \lesssim Pr$	$Nu_x = 0.45\, Re_x^{1/2}\, Pr^{1/3}$	Eq. (8–93)
$0.3 \lesssim Pr \lesssim 5$	$Nu_x = 0.442\, Re_x^{1/2}\, Pr^{1/3}$	Eq. (8–92)
$Pr \lesssim 0.464$ $Pe \gtrsim 100$	$Nu_x = \dfrac{0.646\, Re_x^{1/2}}{1 + (0.743/Pr - 0.6)^{1/2}}$	Eq. (8–94)
Step Wall Temperature		
$Pr \gtrsim 0.6[1 - (x_0/x)^{3/4}]$	$Nu_x = \dfrac{Nu_x\vert_{x0=0}}{[1 - (x_0/x)^{3/4}]^{1/3}}$	Eq. (8–95)
Step Wall-Heat Flux		
$Pr \gtrsim 0.464(1 - x_0/x)$	$Nu_x = \dfrac{Nu_x\vert_{x0=0}}{(1 - x_0/x)^{1/3}}$	Eq. (8–96)
Mean Coefficients	$\bar{f} = 2 f_L \quad \overline{Nu}_L = 2\, Nu_{x=L}$	
TURBULENT FLOW		
	$f_x = 0.0592\, Re_x^{-0.2}$	Eq. (8–98)
	$f_x = \dfrac{0.455}{\ln^2 (0.06\, Re_x)}$	Eq. (8–99)
Uniform Wall Temperature		
$0.5 \lesssim Pr \lesssim 5.0$	$Nu_x = 0.0296\, Re_x^{0.8}\, Pr^{0.5}$	Eq. (8–101)
$5.0 \lesssim Pr$	$Nu_x = 0.0296\, Re_x^{0.8}\, Pr^{1/3}$	Eq. (8–101)
$0.5 \lesssim Pr$	$Nu_x = \dfrac{(f_x/2)\, Re_x\, Pr}{1.0 + 12.7\sqrt{f_x/2}\,(Pr^{2/3} - 1)}$	Eq. (8–102)
Uniform Wall-Heat Flux		
$0.5 \lesssim Pr$	$Nu_x = \dfrac{\sqrt{f_x/2}\, Re_x\, Pr}{2.17 \ln(\sqrt{f_x/2}\, Re_x) + B_0}$ $B_0 = 12.7\, Pr^{2/3} - 13.9$	Eq. (8–104)
Step Wall-Heat Flux		
$0.5 \lesssim Pr$	$Nu_x = \dfrac{\sqrt{f_x/2}\, Re_x\, Pr}{2.17 \ln[(1 - x_0/x)\sqrt{f_x/2}\, Re_x] + B_0}$	Eq. (8–103)
Mixed	$\dfrac{\bar{f}}{2} = 0.037\, Re_L^{-0.2} + (0.664\, Re_{x_c}^{-0.5} - 0.037\, Re_{x_c}^{-0.2})\dfrac{x_c}{L}$	Eq. (8–106)
$Pr \gtrsim 0.5$	$\overline{Nu}_L = [0.037\, Re_L^{0.8} + (0.664\, Re_{x_c}^{0.5} - 0.037\, Re_{x_c}^{0.8})]\, Pr^{1/3}$	Eq. (8–107)
$x_0 = 0$ or x/x_0 large	$\bar{f} = 0.074\, Re_L^{-0.2}$	Eq. (8–110)
$Pr \gtrsim 0.5$	$\overline{Nu}_L = 0.037\, Re_L^{0.8}\, Pr^{1/3}$	Eq. (8–111)

TABLE 8–10 Dimensionless parameters for convection heat transfer: Forced convection

Dimensionless group	*Symbol*	*Definition*	*Interpretation*
Eckert number	Ec	$\dfrac{U_F^2}{c_P(T_s - T_F)}$	$\dfrac{\text{kinetic energy of flow}}{\text{enthalpy difference across boundary layer}}$
Friction factor	f_x	$\dfrac{\tau_s}{\rho U_F^2/2}$	$\dfrac{\text{shear force}}{\text{inertia force}}$
Graetz number	Gz	$\dfrac{Re\,Pr}{x/D}$	$\dfrac{\text{enthalpy flow rate}}{\text{axial heat conduction}}$
Nusselt number	Nu_L or Nu	$\dfrac{hL}{k}$	$\dfrac{\text{convection for fluid in motion}}{\text{conduction for motionless fluid layer}}$
Prandtl number	Pr	$\dfrac{\nu}{\alpha}$	relative effectiveness of molecular transport of momentum and energy within the boundary layer
Peclet number	Pe	$Re_L\,Pr$	$\dfrac{\text{enthalpy flow rate}}{\text{heat conduction rate}}$
Reynolds number	Re_L or Re	$\dfrac{U_F L}{\nu}$	$\dfrac{\text{inertia forces}}{\text{viscous forces}}$
Stanton number	St	$\dfrac{Nu_L}{Re_L\,Pr}$	$\dfrac{\text{actual convection heat flux}}{\text{enthalpy energy flux capacity}}$

CHAPTER 9

CONVECTION HEAT TRANSFER: PRACTICAL ANALYSIS—NATURAL CONVECTION

9–1 INTRODUCTION

As indicated in Chap. 6, flow caused by temperature (or concentration)†-induced density gradients within the fluid is known as *natural convection* (or *free convection*). The most familiar natural-convection flow fields occur as a result of the influence of gravity on fluids in which density gradients have been thermally established. For example, when a vertical cold plate is placed in warm stationary fluid, the temperature of the fluid near the wall will be decreased by conduction heat transfer. As the temperature of the fluid falls, its density will, of course, increase. This difference in density results in the downward flow of the heavier cold fluid near the plate and upward flow of the lighter warm fluid. Similarly, the placement of a hot plate in a cool motionless fluid will result in the upward flow of the light warm fluid near the plate. Examples of natural convection in gravitational force fields include the cooling of electrical devices such as power transistors and transformers, the heating or cooling of building walls on windless days, and the heating of a pan of water.

Another very important type of natural convection flow field occurs in the presence of centrifugal forces that are also proportional to fluid density. This type of natural convection is commonly used to cool rotating components such as turbine blades.

An instrument known as the *Mach–Zehnder interferometer* is often utilized to study natural convection flows. This optical instrument produces interference fringes that are the result of changes in the index of refraction caused by small density differences within the fluid. Consequently, lines of constant temperature can be

† Natural convection caused by concentration-induced density gradients involves interfacial mass transfer which is considered in the *mass transfer supplement*.

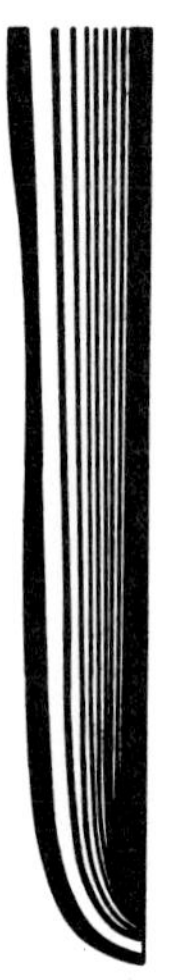
(a) Laminar flow.

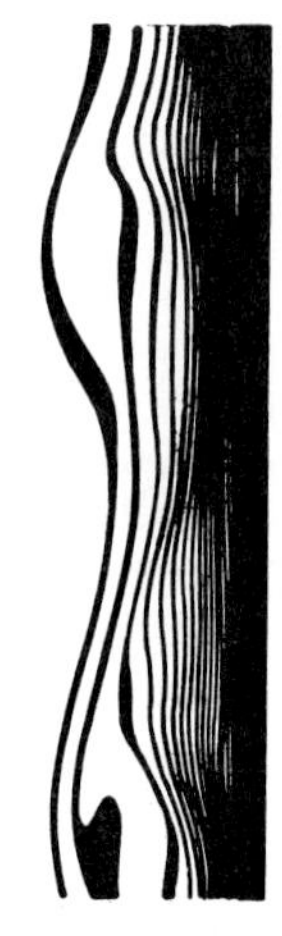
(b) Turbulent flow.

FIGURE 9–1
Interferograms for natural convection flow of air over a vertical flat plate. (Courtesy of E. R. G. Eckert and E. E. Soehngen.)

determined by the use of this instrument. For example, Fig. 9–1 shows photographs of the fringe pattern for natural convection flow over a vertical heated flat plate. This figure clearly indicates the instantaneous flow pattern for this geometry. Because of the streamline pattern observed in Fig. 9–1(a), the flow can be assumed to be laminar over this part of the plate. On the other hand, the pattern observed in Fig. 9–1(b) is clearly irregular, which is characteristic of turbulent flow.

As suggested in Chap. 6, the same classifications found in forced convection pertaining to the geometry of the fluid-solid interface, the nature of the path followed by individual elements of fluid, the type of boundary conditions, and so on, also apply to natural convection systems. Consideration is now given to the characterizing parameters used in the practical analysis of natural convection processes, after which specific attention is given to external natural convection flow, internal natural convection flow, natural convection flow in enclosed spaces, and combined natural and forced convection.

9–2 CHARACTERIZING PARAMETERS FOR NATURAL CONVECTION

Because no pumping or blowing is involved in natural convection flow, we will focus attention on the practical thermal analysis approach, which features the use of local and mean coefficients of heat transfer. Theoretical considerations pertaining to natural convection boundary layers are introduced in Sec. 7–2–3 of Chap. 7.

For systems in which the local coefficient of heat transfer h_x can be conveniently obtained, h_x is defined by the general Newton law of cooling,

$$q_c'' = \frac{dq_c}{dA_s} = h_x(T_s - T_F) \tag{9–1}$$

and $\bar{h}$ is defined in terms of h_x by

$$\bar{h} = \frac{1}{A_s}\int_{A_s} h_x \, dA_s \tag{9–2}$$

For more complex systems in which the use of a local coefficient is not practical, $\bar{h}$ is defined by

$$q_c = \bar{h}A_s(\overline{T_s - T_F}) \tag{9–3}$$

where the mean temperature difference $\overline{T_s - T_F}$ depends upon the system geometry and thermal boundary conditions. The local and mean coefficients of heat transfer for natural convection systems are generally expressed in terms of the Nusselt number Nu_L. For example, for flow over a flat plate of length L, the local Nusselt number is represented by $Nu_x = h_x x/k$ and the mean Nusselt number by $\overline{Nu}_L = \bar{h}L/k$.

The Nusselt number Nu_L is generally expressed in terms of the Rayleigh number Ra_L and the Prandtl number Pr for natural convection processes; that is,

$$Nu_L = \text{fn}\,(Ra_L, Pr) \tag{9–4}$$

The *Rayleigh number* Ra_L is defined by

$$Ra_L = \frac{g\beta(T_s - T_F)\boldsymbol{L}^3}{\alpha\nu} \tag{9–5}$$

where the reference fluid temperature T_F and the reference length $\boldsymbol{L}$ are dependent upon the system geometry. The *coefficient of thermal expansion* β is given in Table A–C–3 for several liquids. For ideal gases β can be shown to be equal to $1/T$, where T is the absolute temperature of the gas. However, for practical purposes β can generally be taken as a constant for moderate temperature differences. Because the Rayleigh number characterizes the coupling through buoyancy of momentum to energy, this dimensionless parameter is the primary variable in natural convection flows. As indicated in Chap. 7, Ra_L represents the product of Pr and the well known *Grashof number* Gr_L,

$$Gr_L = \frac{g\beta(T_s - T_F)\boldsymbol{L}^3}{\nu^2} = \frac{Ra_L}{Pr} \tag{9–6}$$

which represents the relative significance of buoyant and viscous effects.†

9–3 EXTERNAL NATURAL CONVECTION FLOW

Our attention is first focused on flow over a vertical flat plate with $U_\infty = 0$, which is the classic example of natural convection heat transfer. Consideration will then be

† General defining relations for Ra_L and Gr_L, which are applicable to temperature and concentration (including phase change) driven natural convection flows, are given by

$$Ra_L = Gr_L\,Pr = \frac{g(\rho_F - \rho_s)\boldsymbol{L}^3}{\rho\alpha\nu} \qquad Gr_L = \frac{g(\rho_F - \rho_s)\boldsymbol{L}^3}{\rho\nu^2}$$

given to a general class of external flow processes, which includes cylindrical, spherical, and rectangular solid geometries.

The reference temperature T_F for external natural convection flow is equal to the free-stream temperature T_∞, which is uniform for nonstratified conditions. Using the Boussinesq approximation, β and the other fluid properties are evaluated at the film temperature T_f or free-stream temperature T_∞.

9–3–1 Flow over a Vertical Flat Plate

Similar to the situation in forced-convection boundary layer flow over a flat plate, hydrodynamic and thermal boundary layers develop along the wall of a vertical flat plate in a natural convection flow field. As illustrated in Fig. 9–2 for natural convection flow over a heated plate, the thicknesses of the boundary layers increase with x. The flow is laminar toward the front of the plate, but develops into turbulence at a point downstream at which the local Rayleigh number Ra_x,

$$Ra_x = \frac{g\beta(T_s - T_\infty)x^3}{\alpha\nu} \tag{9–7}$$

is equal to about 10^9.

For natural convection flow over a vertical flat plate, Eq. (9–1) applies; that is, the local heat flux q_c'' is represented by

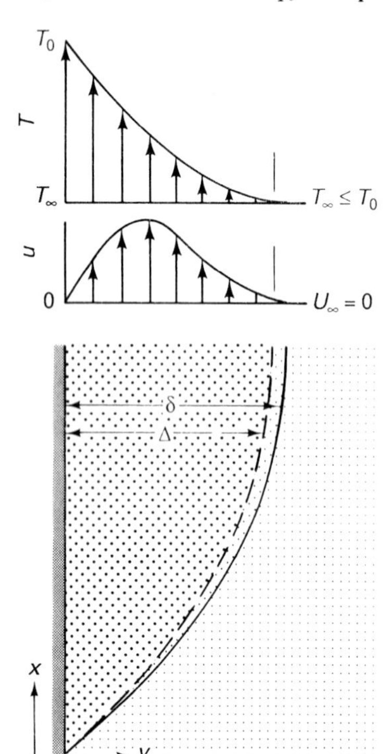

FIGURE 9–2
Representative temperature and velocity profiles for laminar natural convection flow over a heated vertical flat plate.

$$q_c'' = h_x(T_s - T_\infty) \tag{9–8}$$

where h_x is generally expressed in terms of the Nusselt number Nu_x,

$$Nu_x = \frac{q_c''}{T_s - T_\infty}\frac{x}{k} = \frac{h_x x}{k} \tag{9–9}$$

Experimental and theoretical studies have been conducted for uniform wall temperature, uniform wall-heat flux, and other boundary conditions. To provide a basis for the practical thermal analysis of natural convection flow over vertical surfaces, we will focus attention on uniform wall-temperature and uniform wall-heat-flux conditions.

Uniform Wall Temperature

For uniform wall-temperature heating ($T_s = T_0$ = constant), the total heat-transfer rate q_c over a plate with surface area A_s is obtained from Eq. (9–8) by writing

$$q_c = \int_{A_s} h_x(T_s - T_\infty)\, dA_s = (T_0 - T_\infty)\int_{A_s} h_x\, dA_s = \bar{h}A_s(T_0 - T_\infty) \tag{9–10}$$

where $\bar{h}$ is defined by Eq. (9–2).

Theoretical solutions based on boundary layer theory (Chap. 7) for laminar natural convection flow over a vertical plate with uniform wall temperature have been developed by Ostrach [1] and others. Figure 9–3 shows calculations obtained by Ostrach for Nusselt number Nu_x in terms of $Nu_x/Ra_x^{1/4}$ versus Pr. These results are represented to within 1% by correlations developed by Ede [2] and Churchill and Usagi [3]. The correlation by Ede takes the form

$$\frac{Nu_x}{Ra_x^{1/4}} = \frac{3}{4}\left(\frac{Pr}{2.5 + 5.0\sqrt{Pr} + 5.0\,Pr}\right)^{1/4} = \text{fn}_1\,(Pr) \tag{9–11}$$

where the right-hand side is represented by $\text{fn}_1\,(Pr)$ and the local Rayleigh number Ra_x is

$$Ra_x = Gr_x\,Pr = \frac{g\beta(T_0 - T_\infty)x^3}{\alpha\nu} \tag{9–12}$$

Equation (9–11) is shown in Fig. 9–3. Using this result, an expression is obtained for the mean Nusselt number $\overline{Nu}_L$ by writing

$$\begin{aligned}
\bar{h} &= \frac{1}{L}\int_0^L h_x\, dx = \frac{1}{L}\int_0^L Nu_x\frac{k}{x}\, dx = \frac{k}{L}\int_0^L \text{fn}_1\,(Pr)\frac{Ra_x^{1/4}}{x}\, dx \\
&= \frac{k}{L}\text{fn}_1\,(Pr)\left[\frac{g\beta(T_0 - T_\infty)}{\alpha\nu}\right]^{1/4}\int_0^L x^{-1/4}\, dx \qquad (9\text{–}13) \\
&= \frac{k}{L}\text{fn}_1\,(Pr)\left[\frac{g\beta(T_0 - T_\infty)}{\alpha\nu}\right]^{1/4}\frac{L^{3/4}}{3/4}
\end{aligned}$$

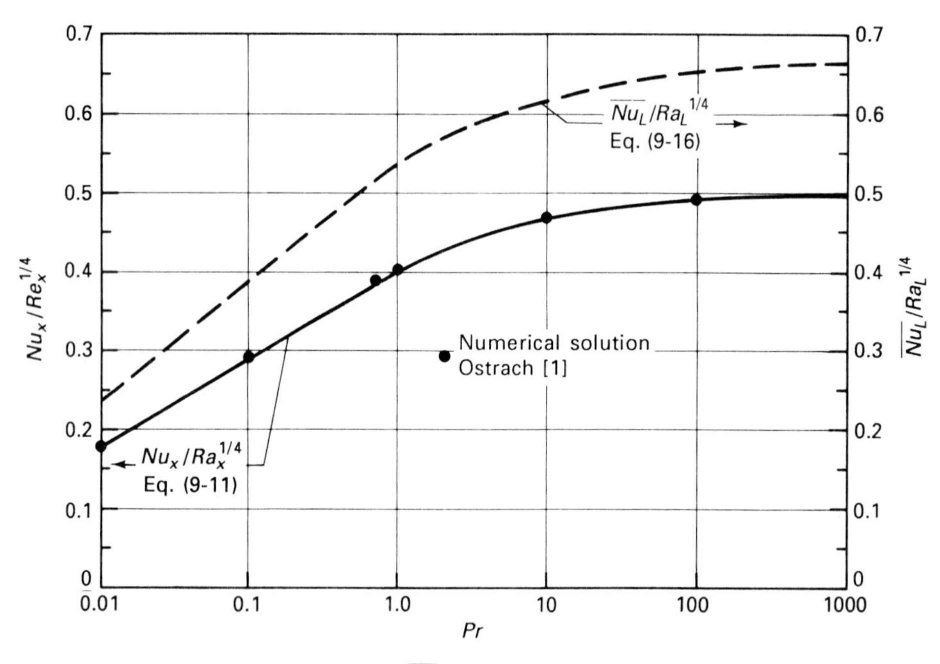

FIGURE 9–3 Nusselt number Nu_x (and $\overline{Nu_L}$) in terms of Rayleigh number Ra_x (and Ra_L) and Prandtl number Pr for vertical surface with uniform wall temperature.

and

$$\overline{Nu_L} = \frac{\bar{h}L}{k} = \frac{4}{3}\,\text{fn}_1\,(Pr)\,Ra_L^{1/4} \tag{9–14}$$

or

$$\overline{Nu_L} = \frac{4}{3} Nu_{x=L} \tag{9–15}$$

Using Eq. (9–10) to represent Nu_x, we obtain

$$\frac{\overline{Nu_L}}{Ra_L^{1/4}} = \frac{4}{3}\,\text{fn}_1\,(Pr) = \left(\frac{Pr}{2.5 + 5.0\sqrt{Pr} + 5.0\,Pr}\right)^{1/4} \tag{9–16}$$

This relation, which approximates the exact solution obtained by Ostrach, is compared in Fig. 9–4 with experimental data for gases with values of Prandtl number Pr of the order of unity. The agreement is seen to be quite good in the range $10^5 \lesssim Ra_L \lesssim 10^9$. Equation (9–16) is also shown in Fig. 9–3.

Because of limitations in approximations employed in boundary layer theory, Eqs. (9–11) and (9–16) are restricted to laminar flow with $10^4 \lesssim Ra_x \lesssim 10^9$. However, empirical correlations are available in the literature, which apply in the region for which $Ra_x < 10^4$ as well as in the turbulent zone where $Ra_x > 10^9$. A general

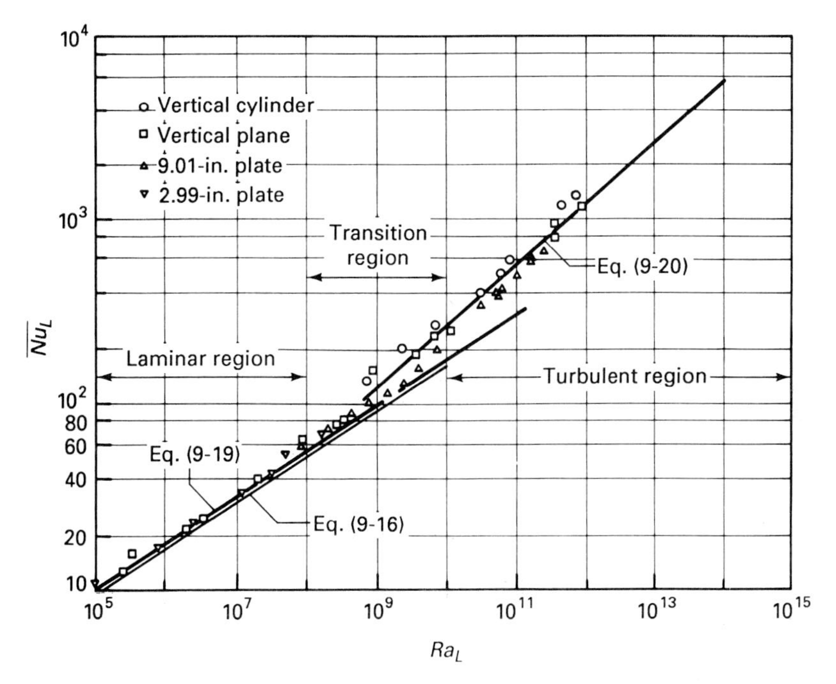

FIGURE 9–4 Experimental data and correlations for natural convection on a vertical flat plate. (Data assembled by Eckert and Jackson [4].)

correlation developed for $\overline{Nu}_L$ by Churchill and Chu [5] that covers the entire range of Ra_L is of the form

$$\overline{Nu}_L = \left\{0.825 + \frac{0.387\, Ra_L^{1/6}}{[1 + (0.492/Pr)^{9/16}]^{8/27}}\right\}^2 \tag{9–17}$$

For somewhat better accuracy within the laminar flow region, Churchill and Chu [5] suggest the alternative correlation

$$\overline{Nu}_L = 0.68 + \frac{0.670\, Ra_L^{1/4}}{[1 + (0.492/Pr)^{9/16}]^{4/9}} \qquad \text{for } 0 < Ra_L \lesssim 10^9 \tag{9–18}$$

Other useful relations for uniform wall-temperature heating of gases are given by

$$\overline{Nu}_L = 0.525\, Ra_L^{1/4} \qquad \text{for } 10^4 \lesssim Ra_L \lesssim 10^9 \tag{9–19}$$

for laminar flow [from Eq. (7–164)],† and

$$\overline{Nu}_L = 0.13\, Ra_L^{1/3} \qquad \text{for } 10^9 \lesssim Ra_L \tag{9–20}$$

for turbulent flow [7]. These simple correlations are shown to be in good agreement with experimental data in Fig. 9–4.

† Another popular correlation for laminar flow is given by [6]

$$\overline{Nu}_L = 0.59\, Ra_L^{1/4} \qquad \text{for } 10^4 \lesssim Ra_L \lesssim 10^9$$

EXAMPLE 9–1

The power amplifier shown in Fig. E9–1 is mounted vertically in air at 25°C. The case is made of anodized aluminum with a surface area of about 3800 mm^2 and a height of 40 mm. Determine the mean coefficient of heat transfer for natural convection cooling, assuming a uniform case temperature of 125°C. Also estimate the power dissipation from the unit.

FIGURE E9–1
Power amplifier.

Solution

Objective Determine the mean coefficient of heat transfer $\bar{h}$ and the total rate of heat transfer q.

Schematic Vertical flat plate representation of power amplifier.

$T_0 = 125°C$

L, q

$A_s = 3.8 \times 10^{-3}\ m^2$
$L = 0.04$ m
$T_\infty = 25°C$
$T_R = 25°C$

The power amplifier is mounted vertically with the back surface fastened to a support, which we will assume to be insulated. Because of its small thickness, we will represent the cooling surface of the device by a flat plate.

Assumptions/Conditions

natural convection and blackbody thermal radiation cooling

Boussinesq approximation with β and other fluid properties evaluated at T_f

standard conditions, except for significance of buoyancy and thermal radiation

Properties Air at film temperature $T_f = 75°C = 348$ K (Table A–C–5): $\rho = 0.994 \text{ kg/m}^3$, $\nu = 2.06 \times 10^{-5} \text{ m}^2/\text{s}$, $k = 0.0299$ W/(m °C), $Pr = 0.697$. The coefficient of thermal expansion β is approximated by

$$\beta = \frac{1}{T_f} = \frac{1}{348 \text{ K}} = 2.87 \times 10^{-3}/\text{K}$$

Analysis To determine the mean Nusselt number $\overline{Nu}_L$, we first compute the Grashof number Gr_L and Rayleigh number Ra_L.

$$Gr_L = \frac{g\beta(T_0 - T_\infty)L^3}{\nu^2}$$

$$= \frac{(9.81 \text{ m/s}^2)(2.87 \times 10^{-3}/\text{K})(125°\text{C} - 25°\text{C})(0.04 \text{ m})^3}{(2.06 \times 10^{-5} \text{ m}^2/\text{s})^2} = 4.25 \times 10^5$$

$$Ra_L = Gr_L\, Pr = 4.25 \times 10^5\,(0.697) = 2.96 \times 10^5$$

Because the Rayleigh number Ra_L is well below 10^9, the flow is judged to be laminar. Based on Eq. (9–16) the mean Nusselt number Nu_L is given by

$$\overline{Nu}_L = \left(\frac{Pr}{2.5 + 5\sqrt{Pr} + 5\,Pr}\right)^{1/4} Ra_L^{1/4}$$

$$= \left[\frac{0.697}{2.5 + 5\sqrt{0.697} + 5(0.697)}\right]^{1/4} (2.96 \times 10^5)^{1/4} = 11.9$$

such that the mean coefficient of heat transfer becomes

$$\bar{h} = \overline{Nu}_L \frac{k}{L} = 11.9 \frac{0.0299 \text{ W/(m °C)}}{0.04 \text{ m}} = 8.90 \text{ W/(m}^2 \text{ °C)}$$

The total rate of heat transfer q from the surface is

$$q = q_c + q_R$$

where q_c is given by Eq. (9–10). Because the plate is made of black anodized aluminum, we will utilize the blackbody approximation for q_R given by Eq. (1–18), with T_R set equal to the ambient temperature. Substituting for q_c and q_R, we have

$$q = \bar{h}A_s(T_0 - T_\infty) + \sigma A_s F_{s-R}(T_0^4 - T_R^4)$$

$$= 8.90 \frac{\text{W}}{\text{m}^2 \text{ °C}} (3.8 \times 10^{-3} \text{ m}^2)(125°\text{C} - 25°\text{C})$$

$$+ 5.67 \times 10^{-8} \frac{\text{W}}{\text{m}^2\,\text{K}^4}(3.8 \times 10^{-3}\,\text{m}^2)(1)[(398\,\text{K})^4 - (298\,\text{K})^4]$$

$$= 3.38\,\text{W} + 3.71\,\text{W} = 7.09\,\text{W}$$

The natural-convection heat transfer accounts for about 48% of the total power dissipated by the power amplifier.

Uniform Wall-Heat Flux

Theoretical solutions have been obtained by Ostrach [1] for the distribution in wall temperature T_s for uniform wall-flux heating (i.e., $q_c'' = q_0'' =$ constant). These results for T_s are generally expressed in terms of the local Nusselt number Nu_x versus the *local flux Rayleigh number* Ra_x^*, where

$$Nu_x = \frac{q_0''}{T_s - T_\infty}\frac{x}{k} \tag{9–21}$$

$$Ra_x^* = Nu_x\,Ra_x = Gr_x^*\,Pr = \frac{g\beta q_0'' x^4}{k\alpha\nu} \tag{9–22}$$

and the *local flux Grashof number* Gr_x^* is defined by

$$Gr_x^* = Nu_x\,Gr_x = \frac{g\beta q_0'' x^4}{k\nu^2} \tag{9–23}$$

The solution for Nu_x is shown as a function of Ra_x^* in Fig. 9–5 for $10^4 \lesssim Ra_x \lesssim 10^9$. Fujii and Fujii [8] and Churchill and Ozoe [9] have developed correlations for uniform wall-heat flux. The correlation by Fujii and Fujii is given by

$$\frac{Nu_x}{Ra_x^{*\,1/5}} = \left(\frac{Pr}{4 + 9\sqrt{Pr} + 10\,Pr}\right)^{1/5} = \text{fn}_2\,(Pr) \tag{9–24}$$

This correlation is within about 1% of the exact solution results shown in Fig. 9–5, and is valid in the range $10^4 \lesssim Ra_x \lesssim 10^9$. As illustrated in Example 9–3, correlations of this type for Nu_x versus Ra_x^* provide an efficient and practical means of calculating the distribution of $T_s - T_\infty$ for a specified wall-heat flux q_0''.

Combining Eqs. (9–21) and (9–24), we obtain an expression for the distribution in dimensionless wall temperature of the form

$$\frac{T_s - T_\infty}{q_0''}\frac{k}{L} = \frac{(x/L)^{1/5}}{Ra_L^{*\,1/5}\,\text{fn}_2\,(Pr)} \tag{9–25}$$

Thus, we see that the distribution in $T_s - T_\infty$ for laminar natural convection in the region away from the leading edge is proportional to $(x/L)^{1/5}$.

To compare the solution results for uniform wall-heat flux and uniform wall temperature, the calculations for Nu_x versus Ra_x^* are expressed in terms of Nu_x versus

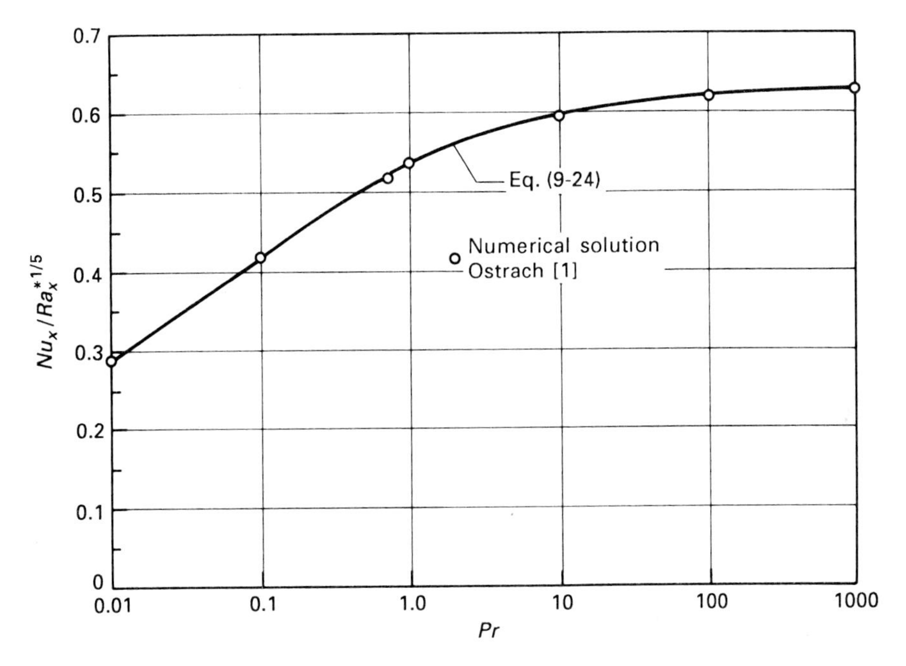

FIGURE 9–5 Nusselt number Nu_x in terms of flux Rayleigh number Ra_x^* and Prandtl number Pr for vertical surface with uniform wall-heat flux.

Ra_x and shown in Fig. 9–6. Notice that the calculations for uniform wall-heat flux lie approximately 15% above the results for uniform wall temperature. Equation (9–24) is expressed in terms of Nu_x and Ra_x by writing

$$\frac{Nu_x}{(Nu_x\,Ra_x)^{1/5}} = \left(\frac{Pr}{4 + 9\sqrt{Pr} + 10\,Pr}\right)^{1/5} = \text{fn}_2\,(Pr) \tag{9–26}$$

or

$$\frac{Nu_x}{Ra_x^{1/4}} = \left(\frac{Pr}{4 + 9\sqrt{Pr} + 10\,Pr}\right)^{1/4} = [\text{fn}_2\,(Pr)]^{5/4} \tag{9–27}$$

This equation is also shown in Fig. 9–6. In this connection, we also observe that

$$\frac{Nu_x}{Ra_x^{1/4}} = \frac{Nu_x\,Ra_x}{Ra_x^{5/4}} = \frac{Ra_x^*}{Ra_x^{5/4}} \tag{9–28}$$

It follows that the representation of natural convection correlations in terms of Ra_x^* versus Ra_x provides an equivalent and viable alternative to the use of Nu_x versus Ra_x for specified wall temperature and Nu_x versus Ra_x^* for specified wall-heat flux.

Practical relations for natural convection flow of gases over a vertical plate with uniform wall-heat flux have been developed of the forms

$$Nu_x = 0.424\,Ra_x^{1/4} \qquad \text{or} \qquad Nu_x = 0.503\,Ra_x^{*1/5} \tag{9–29a,b}$$

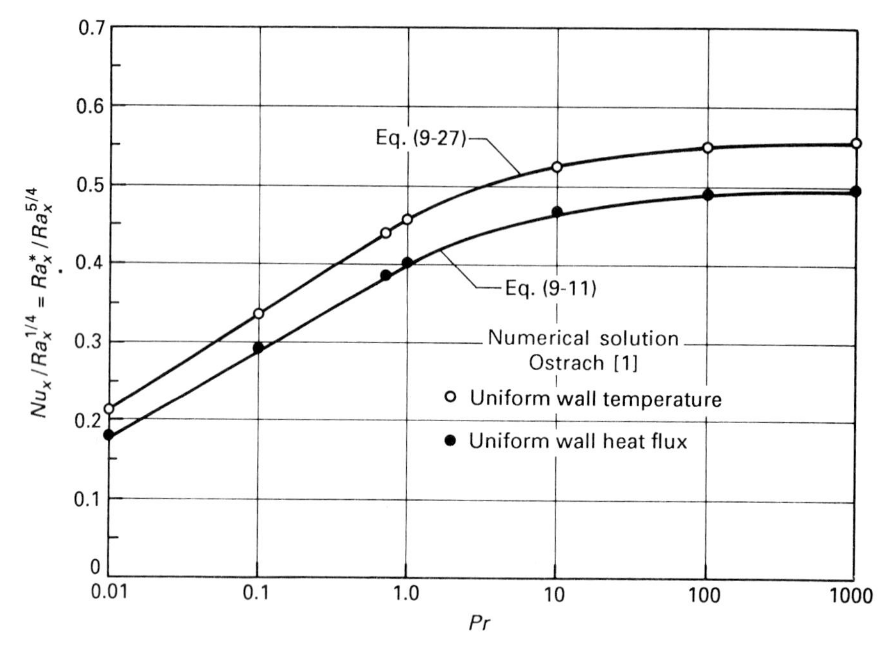

FIGURE 9–6 Comparison of solution and correlations for natural convection flow over vertical surfaces with uniform wall temperature and uniform wall-heat-flux conditions.

for laminar flow (see Chap. 7), and

$$Nu_x = 0.0942\, Ra_x^{1/3} \qquad \text{or} \qquad Nu_x = 0.170\, Ra_x^{*1/4} \tag{9–30a,b}$$

for turbulent flow [10].

Hydraulic Considerations

Solutions and correlations are also available in the literature for the friction factor f_x for natural convection flow over vertical plates. However, since no power is expended in mechanical pumping for natural convection flow, further hydraulic considerations are generally not required.

EXAMPLE 9–2

A vertical plate 10 cm high and 5 cm wide is heated electrically with the heat flux over the surface being uniformly equal to 1.11 kW/m^2. Determine the temperature distribution over the plate if both sides are cooled by natural convection with an ambient air temperature of 38°C.

Solution

Objective Determine the distribution in surface temperature T_s.

Schematic Vertical flat plate: natural convection heating of air.

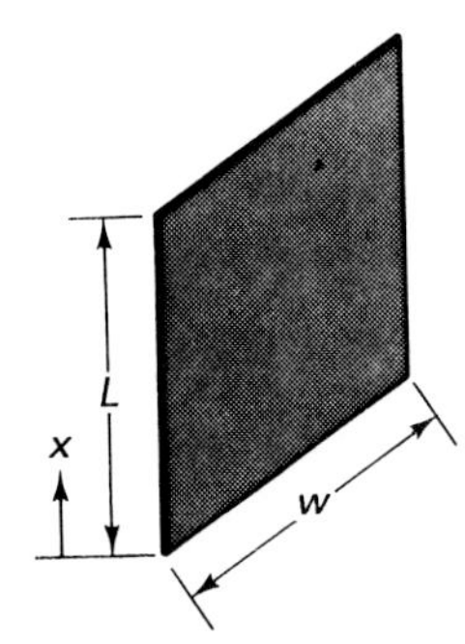

$T_F = 38°C$
$q_0'' = 1.11\ kW/m^2$
$L = 10\ cm$
$w = 5\ cm$

Assumptions/Conditions

natural convection
Boussinesq approximation with β and other fluid properties evaluated at T_∞
standard conditions, except for significance of buoyancy

Properties Because the surface temperature T_s is unknown, we will evaluate the properties of the air at $T = 38°C = 311$ K. Referring to Table A–C–5, we write $\rho = 1.12\ kg/m^3$, $\nu = 1.67 \times 10^{-5}\ m^2/s$, $k = 0.0266$ W/(m °C), $Pr = 0.706$, and

$$\beta = \frac{1}{T_\infty} = \frac{1}{311\ K} = 3.22 \times 10^{-3}/K$$

Analysis Because the heat flux is specified in this problem, we first write relations for the flux Grashof number Gr_x^* and the flux Rayleigh number Ra_x^*.

$$Gr_x^* = \frac{g\beta q_0'' x^4}{k\nu^2} = \frac{(9.81\ m/s^2)(3.22 \times 10^{-3}/K)(1110\ W/m^2)x^4}{[0.0266\ W/(m\ °C)](1.67 \times 10^{-5}\ m^2/s)^2}$$

$$= (4.73 \times 10^{12})\left(\frac{x}{m}\right)^4$$

$$Ra_x^* = Gr_x^*\, Pr = 3.34 \times 10^{12}\left(\frac{x}{m}\right)^4$$

Assuming that the flow is laminar, we combine this relation for Ra_x^* with the correlation for Nu_x given by Eq. (9–24) (or Fig. 9–6) with the result

$$Nu_x = \frac{q_0''}{T_s - T_\infty}\frac{x}{k} = \left(\frac{Pr}{4 + 9\sqrt{Pr} + 10\,Pr}\right)^{1/5} Ra_x^{*\,1/5}$$

$$= 0.519\, Ra_x^{*\,1/5} = 166\left(\frac{x}{m}\right)^{4/5}$$

or

$$T_s - T_\infty = \frac{q_0''}{166}\frac{x}{k}\left(\frac{m}{x}\right)^{4/5} = \frac{(1110\ W/m^2)/166}{0.0266\ W/(m\ °C)}\, x^{1/5}\, m^{4/5} = 251\left(\frac{x}{m}\right)^{1/5}\ °C$$

The maximum temperature is at the top of the plate ($x = 0.1$ m).

$$T_{s,\max} - T_\infty = 251\left(\frac{0.1\text{ m}}{\text{m}}\right)^{1/5} °\text{C} = 158°\text{C}$$

$$T_{s,\max} = 158°\text{C} + 38°\text{C} = 196°\text{C}$$

It follows that the maximum Grashof number and Rayleigh number become

$$Gr_L = \frac{g\beta(T_s - T_\infty)L^3}{\nu^2} = \frac{(9.81\text{ m/s}^2)(3.22 \times 10^{-3}/\text{K})(158°\text{C})(0.1\text{ m})^3}{(1.67 \times 10^{-5}\text{ m}^2/\text{s})^2}$$
$$= 1.79 \times 10^7$$

and

$$Ra_L = Gr_L\,Pr = 1.79 \times 10^7\,(0.706) = 1.26 \times 10^7$$

Because $Ra_L < 10^9$, the flow is indeed laminar.

The solution can be refined by evaluating the properties at a film temperature, with T_s equal to the temperature at $x = L/2$.

9–3–2 General External Natural Convection Flows

The rate of heat transfer associated with many standard external natural convection flows encountered in practice can be represented directly by Eq. (9–3) with T_F set equal to T_∞; that is,

$$q_c = \bar{h}A_s(\bar{T}_s - T_\infty) \qquad \text{or} \qquad q_c'' = \bar{h}(\bar{T}_s - T_\infty) \qquad (9\text{–}31\text{a,b})$$

where q_c'' ($= q_c/A$) is the mean heat flux over the surface and $\bar{T}_s$ is the mean surface temperature. The mean coefficient of heat transfer $\bar{h}$ is generally represented by correlations for mean Nusselt number $\overline{Nu}_L$,

$$\overline{Nu}_L = \frac{\bar{h}L}{k} \qquad (9\text{–}32)$$

in terms of the Rayleigh number Ra_L,

$$Ra_L = Gr_L\,Pr = \frac{g\beta(\bar{T}_s - T_\infty)L^3}{\nu\alpha} \qquad (9\text{–}33)$$

where β and other fluid properties are evaluated at the free-stream or film temperature. The characteristic length L used in the defining relations for $\overline{Nu}_L$ and Ra_L is generally taken to approximate the length of travel of the fluid in the boundary layer.

Churchill [11] has developed a general convection correlation that is applicable to a variety of natural convection flows for which the primary buoyant driving force is directed tangential to the surface. This correlation is given by

$$\overline{Nu}_L = (a + 0.331\, b\, Ra_L^{1/6})^2 \tag{9-34}$$

where

$$b = \frac{1.17}{[1 + (0.5/Pr)^{9/16}]^{8/27}} \tag{9-35}$$

The coefficient a and the characteristic length $\boldsymbol{L}$ are listed in Table 9-1 for various geometries. This correlation has been reported to be in good agreement with experimental data for both isothermal and uniformly heated surfaces over the entire range of Ra_L and all Pr. Alternative natural convection correlations for these geometries are given in references 6, 7, 12, 13, and others.

Referring to Fig. 9-7, the driving force for natural convection flow over a horizontal plate is entirely due to buoyant forces acting in the direction normal to the surface. This normal buoyant force results in direct vertical motion and in an axial pressure gradient that indirectly drives the flow along the plate. Natural convection correlations have been developed for isothermal horizontal plates of the general form

$$\overline{Nu}_L = b\, Ra_L^m \tag{9-36}$$

The characteristic length $\boldsymbol{L}$ and coefficients b and m are given in Table 9-2 for two basic arrangements. Numerous other convection correlations are available in the literature for horizontal plates that account for various vertical edge conditions.

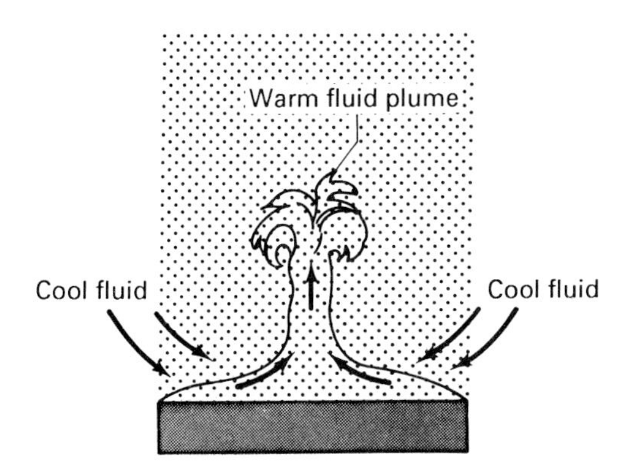

FIGURE 9-7
Illustration of natural convection flow over a hot horizontal surface facing up. (Adapted from Gebhart et al. [17]. Used with permission.)

Applications involving the geometries indicated in Tables 9-1 and 9-2 include steam and water pipes, air ducts, room walls, floors and ceilings, and cryogenic containers. For applications involving specified heat-flux conditions, the correlations for $\overline{Nu}_L$ given in this section can be written and plotted in terms of the *flux* Rayleigh number Ra_L^* in order to compute the temperature difference $\overline{T}_s - T_\infty$ without iteration. Further information is available in the literature on topics such as natural convection associated with rotating bodies and finned surfaces. For comprehensive reviews of external natural convection flow, papers by Churchill [11] and Gebhart [15,16] and the reference text by Gebhart and associates [17] are recommended.

TABLE 9–1 Natural convection: Vertical or inclined surfaces—characteristic length L and coefficients for Eq. (9–34)

Geometry	L	a	*Comments*
Vertical Surfaces			
Plate 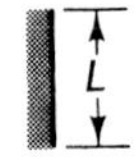	L	0.825	
Cylinder 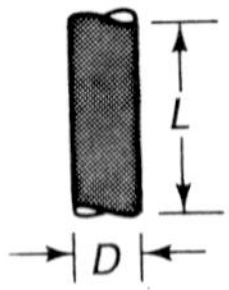	L	0.825	$D/L \gtrsim 35/Gr_L^{1/4}$ See [12,13] for smaller values of D/L
Inclined Surface Heated surface facing down 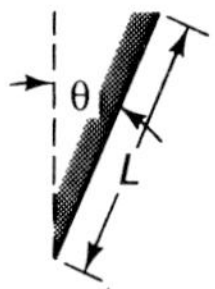	L	0.825	Replace g by $g \cos \theta$ in Ra_L See [14] for heated surface up or cooled surface down
Cooled surface facing up 			
Horizontal Cylinder 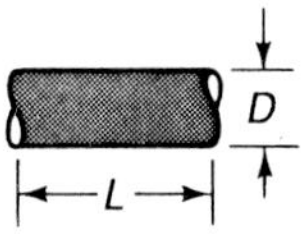	πD	1.06	$L \gg D$
Sphere 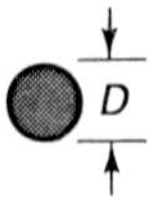	$\frac{\pi}{2} D$	1.77	
Vertical cone 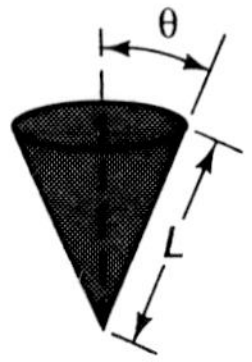	$\frac{4}{5} L$	0.735	Replace g by $g \cos \theta$ in Ra_L
Other Surfaces			
Inclined disk	$\frac{9}{11} D$	0.748	Replace g by $g \cos \theta$ in Ra_L
Spherelike surface with area A_s and volume V	$3\pi V/A_s$	$A_s^{3/2}/(6V)$	

Source: Churchill [11].

TABLE 9-2 Natural convection: Horizontal surfaces—characteristic length L and coefficients for Eq. (9-36)

Horizontal plane surfaces with surface area A_s and perimeter p	Ra_L	b	m	*References and comments*
$L = A_s/p$				[18,19]
Hot surface up (or cold surface down)	10^4–10^7 10^7–10^{11}	0.54 0.15	$\frac{1}{4}$ $\frac{1}{3}$	Laminar Turbulent
Hot surface down (or cold surface up)	10^5–10^{11}	0.27	$\frac{1}{4}$	Laminar

EXAMPLE 9-3

Estimate the mean coefficient of heat transfer for the power amplifier of Example 9-1 if it is mounted horizontally.

Solution

Objective Determine the effect on $\bar{h}$ of mounting a surface horizontally rather than vertically.

Schematic Horizontal flat plate representation of power amplifier.

$A_s = 3800\ \text{mm}^2$ $T_\infty = 25°\text{C}$ From Example 9-1
$L = 40\ \text{mm}$ $T_0 = 125°\text{C}$ $Ra_L = 2.96 \times 10^5$

L

Assumptions/Conditions

natural convection
Boussinesq approximation with β and other fluid properties evaluated at T_f
standard conditions, except for significance of buoyancy

Analysis The mean Nusselt number $\overline{Nu}_L$ for horizontally mounted surfaces can be obtained from Eq. (9–36) with the characteristic length L and coefficients given in Table 9–2; that is,

$$\overline{Nu}_L = b\,Ra_L^m \qquad L = \frac{A_s}{p}$$

and $m = \frac{1}{4}$ or $m = \frac{1}{3}$, depending on the conditions. To determine the perimeter p, we write

$$w = \frac{A_s}{L} = \frac{3800 \text{ mm}^2}{40 \text{ mm}} = 95 \text{ mm}$$

$$p = 2(w + L) = 2(95 \text{ mm} + 40 \text{ mm}) = 270 \text{ mm}$$

$$L = \frac{A_s}{p} = \frac{3800 \text{ mm}^2}{270 \text{ mm}} = 14.1 \text{ mm} = 0.0141 \text{ m}$$

Using this value for L and the result from Example 9–1 for Ra_L, we obtain

$$Ra_L = Ra_L\left(\frac{L}{L}\right)^3 = 2.96 \times 10^5\left(\frac{14.1}{40}\right)^3 = 1.3 \times 10^4$$

Thus the flow is laminar.

For the case in which the device is mounted with the hot surface up, we obtain

$$\overline{Nu}_L = 0.54\,Ra_L^{1/4} = 0.54(1.3 \times 10^4)^{1/4} = 5.77$$

and

$$\bar{h} = \overline{Nu}_L\frac{k}{L} = 5.77\,\frac{0.0299 \text{ W/(m °C)}}{0.0141 \text{ m}} = 12.2 \text{ W/(m}^2\text{ °C)}$$

On the other hand, with the hot surface facing down, we obtain

$$\overline{Nu}_L = 0.27\,Ra_L^{1/4} = 2.88$$

and

$$\bar{h} = 2.88\,\frac{0.0299 \text{ W/(m °C)}}{0.0141 \text{ m}} = 6.11 \text{ W/(m}^2\text{ °C)}$$

Clearly, one should avoid mounting the power amplifier with the hot surface facing down, if at all possible.

EXAMPLE 9–4

Determine the rate of heat transfer by natural convection from the surface of the cabinet shown in Fig. E9–4. The back of the cabinet is mounted on a vertical wall. Its surface temperature is 125°C and the ambient temperature is 25°C.

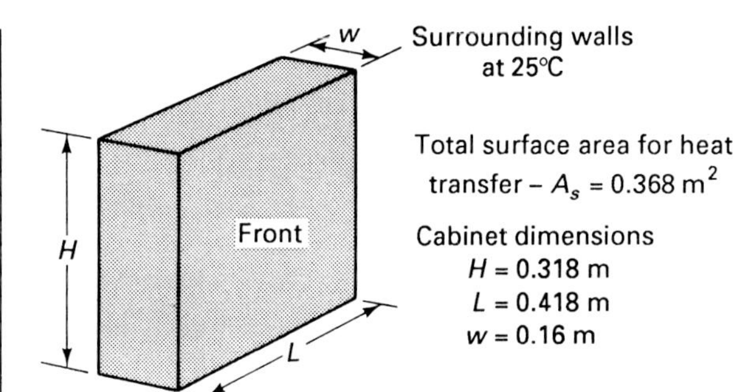

FIGURE E9–4
Cabinet mounted on a vertical wall.

Solution

Objective Determine the rate of convection heat transfer q_c.

Assumptions/Conditions

natural convection
Boussinesq approximation with β and other fluid properties evaluated at T_f
standard conditions, except for significance of buoyancy

Properties Air at $T_f = 75°\text{C} = 348$ K (Table A–C–5): $\rho = 0.994$ kg/m³, $\nu = 2.06 \times 10^{-5}$ m²/s, $k = 0.0299$ W/(m °C), $Pr = 0.697$, and

$$\beta = \frac{1}{T_f} = \frac{1}{348\text{ K}} = 2.87 \times 10^{-3}/\text{K}$$

Analysis As a first approximation, we will treat each surface independently. To provide a convenient means of computing the coefficient of heat transfer for each vertical and horizontal surface, the Grashof and Rayleigh numbers are represented in terms of $\boldsymbol{L}$ by

$$Gr_L = \frac{(9.81\text{ m/s}^2)(2.87 \times 10^{-3}/\text{K})(125°\text{C} - 25°\text{C})}{(2.06 \times 10^{-5}\text{ m}^2/\text{s})^2} \boldsymbol{L}^3 = 6.63 \times 10^9 \left(\frac{\boldsymbol{L}}{\text{m}}\right)^3$$

$$Ra_L = Gr_L\, Pr = 4.62 \times 10^9 \left(\frac{\boldsymbol{L}}{\text{m}}\right)^3$$

The characteristic length $\boldsymbol{L}$ and surface areas for the vertical and horizontal surfaces are specified as follows:

Vertical

$$A_{s,wH} = 0.16\text{ m }(0.318\text{ m}) = 0.0509\text{ m}^2$$

$$A_{s,LH} = 0.418\text{ m }(0.318\text{ m}) = 0.133\text{ m}^2$$

$$\boldsymbol{L}_v = H = 0.318\text{ m}$$

Horizontal

$$A_{s,wL} = 0.16 \text{ m } (0.418 \text{ m}) = 0.0669 \text{ m}^2$$

$$p = 2(0.16 \text{ m})(0.418 \text{ m}) = 0.578 \text{ m}$$

$$L_h = \frac{A_{s,wL}}{p} = \frac{0.0669 \text{ m}^2}{0.578 \text{ m}} = 0.116 \text{ m}$$

We are now in a position to evaluate Ra_L, $\overline{Nu}_L$, $\bar{h}$, and q_c for each surface.

Vertical

$$Ra_{L,v} = 4.62 \times 10^9 (0.318)^3 = 1.49 \times 10^8$$

$$\overline{Nu}_{L,v} = \left(\frac{Pr}{2.5 + 5\sqrt{Pr} + 5\,Pr}\right)^{1/4} Ra_{L,v}^{1/4} = 56.5 \qquad \text{from Eq. (9–16)}$$

$$\bar{h}_v = 56.5 \frac{0.0299 \text{ W/(m °C)}}{0.318 \text{ m}} = 5.27 \text{ W/(m}^2\text{ °C)}$$

$$q_{c,v} = \bar{h} A_{s,v}(T_0 - T_\infty)$$

$$= 5.27 \frac{\text{W}}{\text{m}^2\text{ °C}} [2(0.0509 \text{ m}^2) + 0.133 \text{ m}^2](125\text{°C} - 25\text{°C}) = 124 \text{ W}$$

Horizontal

$$Ra_{L,h} = 4.62 \times 10^9 (0.116)^3 = 7.21 \times 10^6$$

Top

$$\overline{Nu}_{L,\text{top}} = 0.54\, Ra_{L,h}^{1/4} = 28 \qquad \text{from Eq. (9–36), Table 9–2}$$

$$\bar{h}_{\text{top}} = 28 \frac{0.0299 \text{ W/(m °C)}}{0.116 \text{ m}} = 7.22 \text{ W/(m}^2\text{ °C)}$$

$$q_{c,\text{top}} = 7.22 \frac{\text{W}}{\text{m}^2\text{ °C}} (0.116 \text{ m}^2)(125\text{°C} - 25\text{°C}) = 83.8 \text{ W}$$

Bottom

$$\overline{Nu}_{L,\text{bottom}} = 0.27\, Ra_{L,h}^{1/4} = 14 \qquad \text{from Eq. (9–36), Table 9–2}$$

$$\bar{h}_{\text{bottom}} = 14 \frac{0.0299 \text{ W/(m °C)}}{0.116 \text{ m}} = 3.61 \text{ W/(m}^2\text{ °C)}$$

$$q_{c,\text{bottom}} = 3.61 \frac{\text{W}}{\text{m}^2\text{ °C}} (0.116 \text{ m}^2)(125\text{°C} - 25\text{°C}) = 41.9 \text{ W}$$

Summing up the contributions of the vertical, top, and bottom surfaces, the total rate of heat transfer is

$$q_c = q_{c,v} + q_{c,\text{top}} + q_{c,\text{bottom}} = 124 \text{ W} + 83.8 \text{ W} + 41.9 \text{ W} = 250 \text{ W}$$

As an aside, the mean coefficient of heat transfer over the entire cabinet can be obtained by writing

$$\bar{h} = \frac{q_c}{A_s(T_0 - T_\infty)} = \frac{250 \text{ W}}{0.368 \text{ m}^2 (125°\text{C} - 25°\text{C})} = 6.79 \text{ W/(m}^2 \text{ °C)}$$

9–4 INTERNAL NATURAL CONVECTION FLOW

Internal natural convection flows are often encountered in heat-transfer applications involving fin units, fireplaces, and solar heating units, as well as other systems. To illustrate, the schematic of a natural circulation solar water heater is shown in Fig. 9–8. With the storage tank located above the solar collector, water circulates by natural convection when solar energy is captured by the collector.

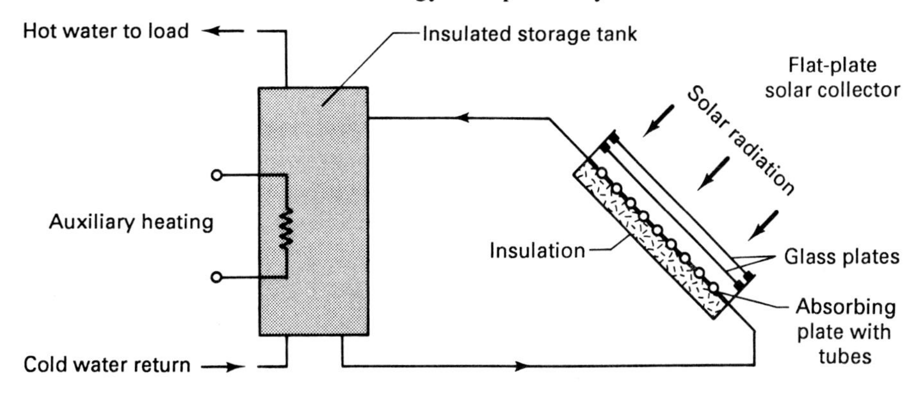

FIGURE 9–8 Schematic of a natural-convection solar heating system.

Internal natural convection flow systems are complicated by the fact that the bulk-stream temperature T_b generally varies with axial location x. The flow pattern depends upon the relative values of the wall temperatures and the temperature of entering fluid T_1. For example, for flow between vertical parallel plates with uniform surface temperatures T_0 and T_w, fluid rises in the vicinity of the warmer wall and falls in the region of the cooler surface for $T_0 > T_1 > T_w$. On the other hand, with T_0 and T_w equal and greater than T_1, fluid rises throughout the entire system. These two situations are illustrated in Fig. 9–9(a) and (b) for laminar flow conditions. Such parallel-plate geometries are often used to approximate cooling fins in transformers, radiators, and other industrial devices, and in natural-circulation solar flat-plate collectors.

Correlations have been developed on the basis of Eq. (9–3) for internal natural convection flow with uniform wall temperatures T_0 and T_w, where† $\overline{T_s - T_F} = \bar{T}_s - T_1 = (T_0 + T_w)/2 - T_1$; that is,

$$q_c = \bar{h}A_s(\bar{T}_s - T_1) \tag{9–37}$$

† Correlations are also found in the literature with $\overline{T_s - T_F}$ set equal to $T_0 - T_w$.

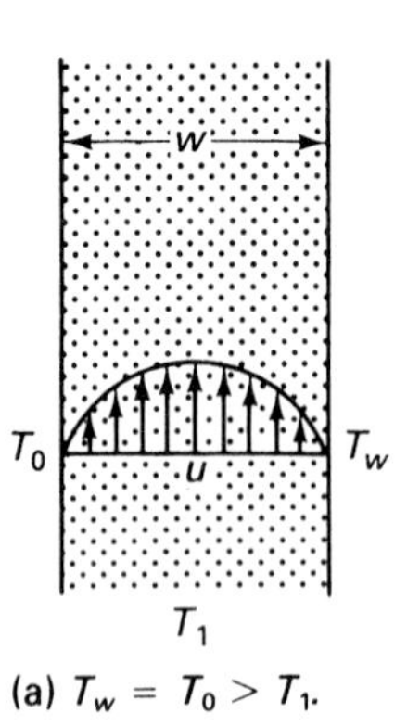

(a) $T_w = T_0 > T_1$.

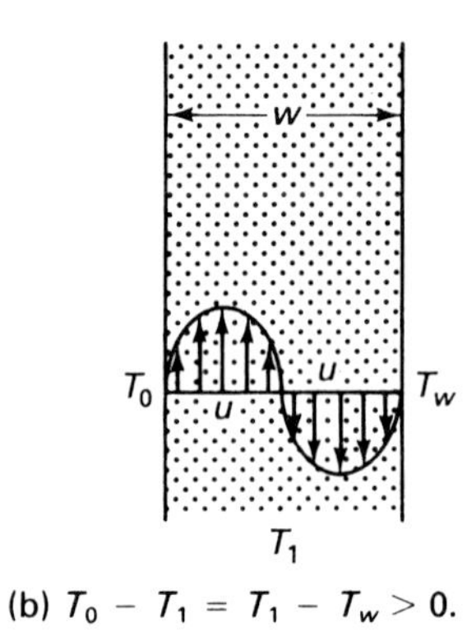

(b) $T_0 - T_1 = T_1 - T_w > 0$.

FIGURE 9–9
Representative internal natural-convection flow patterns for parallel-plate geometry.

Heat-transfer correlations for other cases involving internal natural convection flow are available in the literature. For example, correlations have been developed for natural convection flow between inclined parallel plates by Tabor [20] and by Dropkin and Somerscales [21].

9–5 NATURAL CONVECTION FLOW IN ENCLOSED SPACES

A typical enclosed natural-convection flow system and flow pattern is shown in Fig. 9–10. This type of natural circulation is particularly important in the cooling of electronic devices. In addition, the development of natural-convection flow in enclosures is a factor in the use air gaps to insulate building walls and cryogenic chambers, and in the design of solar flat-plate collectors.

Correlations have been developed for natural-convection heat transfer in enclosures with surfaces of length H maintained at uniform temperatures T_1 and T_2. These correlations for $\bar{h}$ are based on Eq. (9–3) with $\overline{T_s - T_F} = T_1 - T_2$; that is,

$$q_c = \bar{h}A_s(T_1 - T_2) \tag{9–38}$$

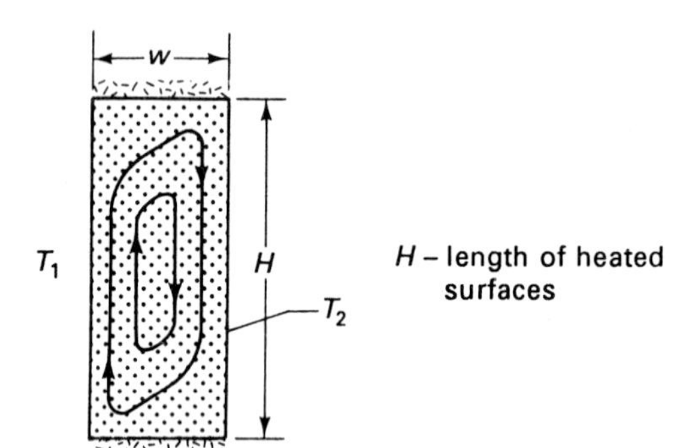

FIGURE 9–10
Representative natural-convection flow pattern for an enclosed vertical space; $L = w$, $T_1 > T_2$.

TABLE 9–3 Natural convection: Characteristics for enclosed spaces with uniform wall temperatures T_1 and T_2

Geometry	L	A_s	$R_c = L/(\overline{Nu_L}\, kA_s)$
Rectangular enclosure; width w, length of heated surfaces H	w	$A_1 = A_2$	$\dfrac{w}{A_1 k \overline{Nu_L}}$
Vertical concentric annuli; length H, end surfaces insulated	$r_2 - r_1$	$\dfrac{A_2 - A_1}{\ln (A_2/A_1)}$	$\dfrac{\ln (r_2/r_1)}{2\pi H k \overline{Nu_L}}$
Concentric spheres	$r_2 - r_1$	$\sqrt{A_1 A_2}$	$\dfrac{r_2 - r_1}{4\pi r_1 r_2 k \overline{Nu_L}}$

The mean coefficient $\bar{h}$ is generally expressed in terms of Ra_L, L, and H by an empirical expression of the form

$$\frac{\bar{h}L}{k} = \overline{Nu_L} = C\, Ra_L^m \left(\frac{H}{L}\right)^n \tag{9–39}$$

where $Ra_L = g\beta L^3(T_1 - T_2)/(\alpha\nu)$; L and A_s are dependent upon the geometry (see Table 9–3). The properties are generally evaluated at the average temperature $(T_1 + T_2)/2$ for internal flows. The coefficients C, m, and n are given in Table 9–4 for vertical rectangular and annular enclosures (with the enclosing end surfaces insulated), and concentric spheres with H/L set equal to unity). The works by Raithby and Hollands [22]

TABLE 9–4 Natural convection: Convection correlations for enclosed spaces—coefficients for Eq. (9–39)†

Geometry	*Fluid*	Ra_L	Pr	C	m	n	H/L
Rectangular enclosures with vertical surfaces heated, and vertical concentric annuli	Gas	$<2 \times 10^3$	0.5–2	1	0	0	—
		2×10^3–2×10^5	0.5–2	0.197	$\frac{1}{4}$	$-\frac{1}{9}$	11–42
	Liquid	2×10^5–10^7	0.5–2	0.073	$\frac{1}{3}$	$-\frac{1}{9}$	11–42
		10^4–10^7	$1 - 2 \times 10^4$	C_1	$\frac{1}{4}$	-0.3	10–40
		10^6–10^9	1–20	0.046	$\frac{1}{3}$	0	1–40
	Gas or liquid	$<10^{10}$	$<10^5$	C_2	0.28	$-\frac{1}{4}$	2–10
		$>\dfrac{(0.2 + Pr)\, 10^3}{Pr}$	10^{-3}–10^5	C_3	0.29	0	1–2
Concentric spheres, set $H/L = 1$	Gas or liquid	10^2–10^9	0.7–4×10^3	0.228	0.226	—	—

† Vertical surfaces heated and cooled. $C_1 = 0.42\, Pr^{0.012}$, $C_2 = 0.22\, [Pr/(0.2 + Pr)]^{0.28}$, $C_3 = 0.18\, [Pr/(0.2 + Pr)]^{0.29}$.

Sources: Jakob [24,25], Graff and Held [26], Emery et al. [27,28], Catton [29], and Scanlan et al. [30,31].

and Kuehn and Goldstein [23] provide additional useful correlations for concentric annuli and spheres.

In regard to natural convection in rectangular enclosures with horizontal heated surfaces, the flow patterns depend upon whether the hotter plate is on the top or the bottom. In the first case, the low-density fluid lies above the heavier fluid, such that no buoyancy effects occur. For this situation, we have pure conduction heat transfer in the fluid, such that $\overline{Nu}_w = 1$.

For the second case, the heavier fluid is on the top. For values of Ra_w less than a critical value, the buoyancy forces are not large enough to cause the fluid to turn over. The critical value of Ra_w for gases is about 1700. For this case, we have stability with no natural convection currents. The flow pattern for values of Ra_w between 1700 and 3.2×10^5 is laminar and takes the form of cells of circulating fluid. For larger values of Ra_w, the flow becomes turbulent and the cellular pattern no longer exists. An empirical equation of the form of Eq. (9–39) with $n = 0$ is recommended for natural convection in horizontal rectangular enclosures heated from below; that is,

$$\overline{Nu}_w = C\, Ra_w^m \qquad (9\text{–}40)$$

The values of C and m and the ranges in Ra_w are shown in Table 9–5.

TABLE 9–5 Natural convection: Convection correlations for horizontal rectangular enclosures heated from below—coefficients for Eq. (9–40)

Fluid	Ra_w	*Pr*	*C*	*m*	*References*
Gas	$< 1.7 \times 10^3$		1	0	[24–26]
	1.7×10^3–7.0×10^3	0.5–2	0.059	0.4	[32–34]
	7.0×10^3–3.2×10^5	0.5–2	0.212	$\frac{1}{4}$	
	$3.2 \times 10^5 <$	0.5–2	0.061	$\frac{1}{3}$	
Liquid	$< 1.7 \times 10^3$		1	0	[25,33–37]
	1.7×10^3–6.0×10^3	1–5000	0.012	0.6	
	6.0×10^3–3.7×10^4	1–5000	0.375	0.2	
	3.7×10^4–10^8	1–20	0.13	0.3	
	$10^8 <$	1–20	0.057	$\frac{1}{3}$	

For natural convection in inclined enclosures with heating from below such as the one shown in Fig. 9–11, studies by Hollands et al. [22,35,38], Catton et al. [39], and Ayyaswamy and Catton [40] indicate that the Nusselt number correlations for vertical enclosures can be utilized up to a critical tilt angle θ^*, with the gravitational acceleration g replaced by its directional component along the surface of the plate, $g \cos\theta$; θ^* is shown in Table 9–6 as a function of the aspect ratio H/w. Utilizing this substitution for $7 \times 10^3 < Ra_w < 3.2 \times 10^5$, we have

$$\overline{Nu}_w = \overline{Nu}_w\big|_{\theta=0} (\cos\theta)^{1/4} \qquad (9\text{–}41)$$

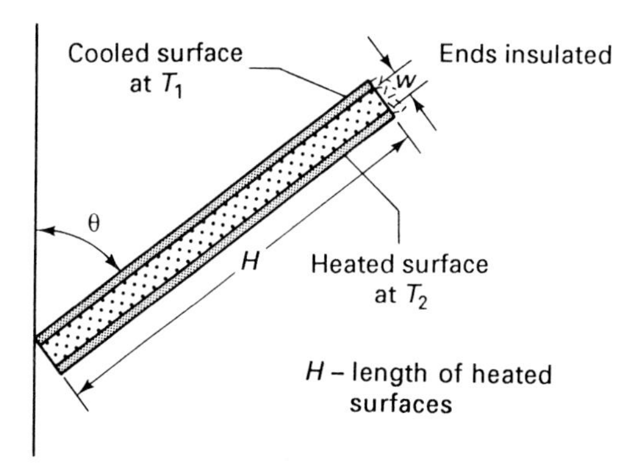

FIGURE 9–11
Natural convection in an inclined enclosed rectangular space.

This equation has been reported to correlate experimental data in the region $0 \lesssim \theta \lesssim \theta^*$ for all aspect ratios. For tilt angles between θ^* and $\pi/2$ and large aspect ratios ($H/w \gtrsim 12$), Hollands et al. [41] recommend a correlation of the form

$$\overline{Nu}_w = 1 + 1.44\left[1 - \frac{1708}{Ra_w \sin\theta}\right]^{\bullet}\left\{1 - \frac{1708\,[\sin 1.8(\pi/2 - \theta)]^{1.6}}{Ra_w \sin\theta}\right\} + \left[\left(\frac{Ra_w \sin\theta}{5830}\right)^{1/3} - 1\right]^{\bullet} \tag{9–42}$$

where the notation $[\ ]^{\bullet}$ indicates that the quantity in brackets is to be set equal to zero when its value is less than zero. For situations in which the upper surface of an inclined rectangular enclosure is heated and the lower surface is cooled, Catton [29] recommends the following correlation by Arnold et al. [42] for all aspect ratios:

$$\overline{Nu}_w = 1 + (\overline{Nu}_w|_{\theta=0} - 1)\cos\theta \tag{9–43}$$

It should be noted that the rate of heat transfer q_c is also sometimes expressed in terms of a thermal resistance for natural convection in enclosures as

$$q_c = \frac{T_1 - T_2}{R_c} \tag{9–44}$$

where $R_c = L/(\overline{Nu}_L\, kA_s)$; specific expressions for R_c are given in Table 9–3. Because of the similarity between these expressions for R_c and the expressions developed in

TABLE 9–6 Critical tilt angle θ^* for inclined rectangular cavities

H/w	1	3	6	12	>12
θ^*	65°	37°	30°	23°	20°

Source: From Hollands et al. [41].

Chap. 2 for R_k for one-dimensional conduction heat transfer, the product $\overline{Nu}_L\, k$ is sometimes referred to as the *apparent thermal conductivity* k_e; that is,

$$k_e = \overline{Nu}_L\, k \tag{9–45}$$

For $\overline{Nu}_L = 1$ or $k_e = k$, we have pure conduction heat transfer through the fluid with no natural convection effect.

For more information on the topic of natural convection in enclosures, review articles by Catton [29] and Ostrach [43] are suggested.

EXAMPLE 9–5

Two concentric blackbody spheres with 2-cm and 5-cm radii are separated by air at 0.5 atm. The surface temperatures are equal to 154°C and 0°C. Determine the total rate of heat transfer in this system.

Solution

Objective Determine the total rate of heat transfer q by natural convection and radiation.

Schematic Air contained between concentric blackbody spheres.

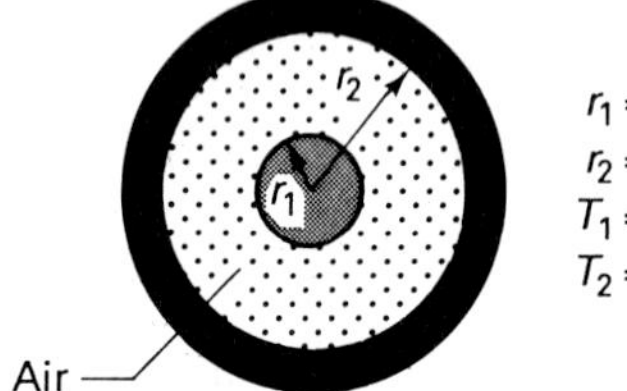

Assumptions/Conditions

natural convection and blackbody thermal radiation

Boussinesq approximation with β and other fluid properties evaluated at average temperature

standard conditions, except for significance of buoyancy and blackbody thermal radiation

Properties Air at average temperature $(T_1 + T_2)/2 = 77°\text{C} = 350$ K (Table A–C–5): $\rho = 0.998\ \text{kg/m}^3$, $\nu = 2.08 \times 10^{-5}\ \text{m}^2/\text{s}$, $k = 0.03$ W/(m °C), $Pr = 0.697$, and

$$\beta = \frac{1}{350\ \text{K}} = 2.86 \times 10^{-3}/\text{K}$$

Analysis The total rate of heat transfer is given by

$$q = q_c + q_R$$

The rate of heat transfer by natural convection is written as

$$q_c = \bar{h} A_s (T_1 - T_2)$$

where

$$A_s = \sqrt{A_1 A_2} = 4\pi r_1 r_2 = 4\pi(0.02\text{ m})(0.05\text{ m}) = 0.0126\text{ m}^2$$

$$L = r_2 - r_1 = 0.05\text{ m} - 0.02\text{ m} = 0.03\text{ m}$$

from Table 9-3,

$$\overline{Nu}_L = 0.228\, Ra_L^{0.226} \tag{9-39}$$

and

$$Gr_L = \frac{g\beta(T_1 - T_2)L^3}{\nu^2}$$

$$= \frac{(9.81\text{ m/s}^2)(2.86 \times 10^{-3}/\text{K})(154°\text{C} - 0°\text{C})(0.03\text{ m})^3}{(2.08 \times 10^{-5}\text{ m}^2/\text{s})^2} = 2.7 \times 10^5$$

$$Ra_L = Gr_L\, Pr = 1.88 \times 10^5$$

Substituting this value of Ra_L into Eq. (9-39), we obtain

$$\overline{Nu}_L = 0.228(1.88 \times 10^5)^{0.226} = 3.55$$

and

$$\bar{h} = 3.55\,\frac{0.03\text{ W/(m °C)}}{0.03\text{ m}} = 3.55\text{ W/(m}^2\text{ °C)}$$

Thus, we obtain

$$q_c = 3.55\,\frac{\text{W}}{\text{m}^2\text{ °C}}(0.0126\text{ m}^2)(154°\text{C} - 0°\text{C}) = 6.89\text{ W}$$

The rate of radiation heat transfer is given by Eq. (1-18),

$$q_R = \sigma A_1 F_{1-2}(T_1^4 - T_2^4)$$

$$= 5.67 \times 10^{-8}\,\frac{\text{W}}{\text{m}^2\text{ K}^4}[4\pi(0.02\text{ m})^2][(427\text{ K})^4 - (273\text{ K})^4] = 7.89\text{ W}$$

Thus, the total heat-exchange rate is

$$q = q_c + q_R = 6.89\text{ W} + 7.89\text{ W} = 14.8\text{ W}$$

The natural convection accounts for about 47% of this total.

9-6 COMBINED NATURAL AND FORCED CONVECTION

Strictly speaking, natural convection occurs in any nonisothermal forced convection system in which gravitational or centrifugal force fields are present. For example, natural convection superimposed upon forced convection in vertical tubes brings about an increase in heat-transfer rate for laminar upward flow and a decrease for laminar downward flow. That is, the heat transfer is enhanced for laminar flow in vertical tubes when the buoyancy forces are in the direction of flow. On the other hand, the opposite has been found to be true for turbulent flow in vertical tubes. In fact, severe deteriorations in the rate of heat transfer have been reported for turbulent upward vertical flow of high-pressure supercritical fluids in nuclear reactors and fossil-fired steam generators [44,45].

As a guide in determining the significance of natural convection in forced flow fields, the buoyancy forces are generally small and can be neglected for situations in which $Gr_L \ll Re_L^2$. However, for cases in which Gr_L and Re_L^2 are of the same order or magnitude, both natural convection and forced convection are usually significant. This point is reinforced by the theoretical and experimental results shown in Fig.

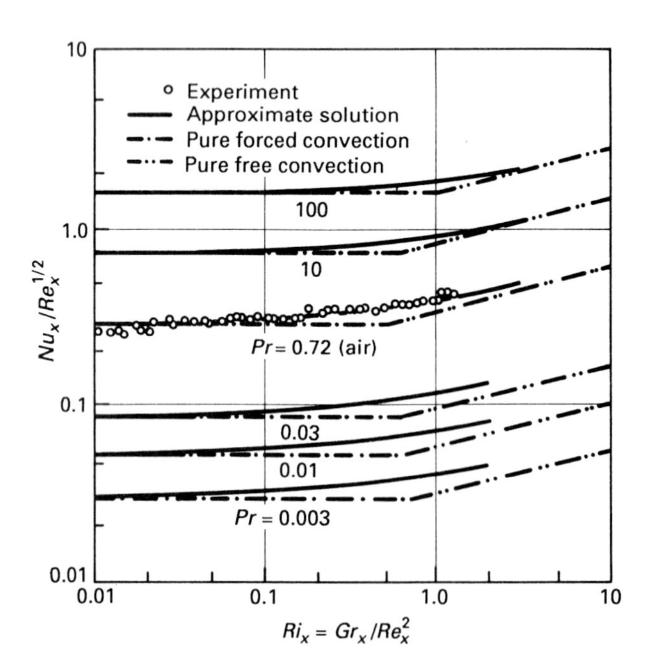

FIGURE 9-12 Local Nusselt number Nu_x for combined natural and forced convection from an isothermal vertical plate (Lloyd and Sparrow [46]).

9–12 for upward laminar forced convection flow over a vertical heated flat plate with uniform wall temperature. Notice that the theoretical predictions and data for air approach the limiting solution for pure forced convection as the *Richardson number* Ri_x,

$$Ri_x = \frac{Gr_x}{Re_x^2} \tag{9–46}$$

falls toward a value of the order of 0.02. The natural convection effect is of the order of 10% for Ri_x equal to about 0.225. The theoretical calculations and experimental data are also seen to approach limiting curves for pure natural convection as Ri_x approaches a value of the order of 10.

Several dimensionless number criteria have been developed to establish the limits between forced and combined convection-heat-transfer regimes. For example, Metais and Eckert [47] published preliminary limit criterion charts in 1964 for vertical and horizontal tube flow of moderate-Prandtl-number fluids. Natural convection limit criteria have also been developed for liquid metals [48] and for supercritical fluids [49].

For information concerning Nusselt number correlations for combined forced and natural convection, one can refer to references 50–52.

9–7 SUMMARY

As we have seen, the heat-transfer performance of natural convection flows is generally characterized in terms of the Rayleigh number Ra_L or flux Rayleigh number Ra_L^*. Practical natural convection correlations have been presented in this chapter for Nusselt number Nu_L for a number of arrangements involving external flow, internal flow, and flow in enclosures. These natural convection correlations and related dimensionless parameters are summarized in Tables 9–7 and 9–8. With the Rayleigh number known for specified wall temperature, these correlations are readily used to obtain the heat-transfer rate. Similarly, with the flux Rayleigh number known for specified wall-heat flux, the correlation for Nusselt number in terms of flux Rayleigh number provides a straightforward means of obtaining the wall temperature.

Brief consideration has also been given to the important topic of combined natural and forced convection, which becomes important in certain forced-convection flows.

Numerous other important natural convection flows such as thermal plumes, buoyant jets, and flow in stratified ambient media are treated in the literature. The reference textbook by Gebhart et al. [17] provides timely in-depth coverage of a wide range of natural convection flows.

TABLE 9–7 Summary of convection correlations for natural convection†

Condition	*Correlation*	*Identification*
Vertical surface		
Isothermal		
Laminar flow $10^4 < Ra_L < 10^9$	$\frac{Nu_x}{Ra_x^{1/4}} = \frac{3}{4}\left(\frac{Pr}{2.5 + 5\sqrt{Pr} + 5\,Pr}\right)^{1/4}$	Eq. (9–11) Ede [2]
	$\frac{\overline{Nu}_L}{Ra_L^{1/4}} = \left(\frac{Pr}{2.5 + 5\sqrt{Pr} + 5\,Pr}\right)^{1/4}$	Eq. (9–16)
Laminar and turbulent flow, all Ra_L	$\overline{Nu}_L = \left\{0.825 + \frac{0.387\,Ra_L^{1/6}}{[1 + (0.492/Pr)^{9/16}]^{8/27}}\right\}^2$	Eq. (9–17) Churchill & Chu [5]
Laminar flow $0 < Ra_L < 10^9$	$\overline{Nu}_L = 0.68 + \frac{0.670\,Ra_L^{1/4}}{[1 + (0.492/Pr)^{9/16}]^{4/9}}$	Eq. (9–18) Churchill & Chu [5]
Gas Laminar flow $10^4 < Ra_L < 10^9$	$\overline{Nu}_L = 0.525\,Ra_L^{1/4}$	Eq. (9–19) McAdams [7]
Gas Turbulent flow $10^9 < Ra_L$	$\overline{Nu}_L = 0.13\,Ra_L^{1/3}$	Eq. (9–20) McAdams [7]
Uniform wall-heat flux		
Laminar flow $10^4 < Ra_x < 10^9$	$\frac{Nu_x}{Ra_x^{*1/5}} = \left(\frac{Pr}{4 + 9\sqrt{Pr} + 10\,Pr}\right)^{1/5}$	Eq. (9–24) Fujii & Fujii [8]
	$\frac{Nu_x}{Ra_x^{1/4}} = \left(\frac{Pr}{4 + 9\sqrt{Pr} + 10\,Pr}\right)^{1/4}$	Eq. (9–27) Equivalent to Eq. (9–24)
Gas Laminar flow	$Nu_x = 0.424\,Ra_x^{1/4}$	Eq. (9–29a) Chap. 7
	$Nu_x = 0.503\,Ra_x^{*1/5}$	Eq. (9–29b) Equivalent to Eq. (9–29a)
Gas Turbulent flow	$Nu_x = 0.0942\,Ra_x^{1/3}$	Eq. (9–30a) Vliet & Liu [10]
	$Nu_x = 0.170\,Ra_x^{*1/4}$	Eq. (9–30b) Equivalent to Eq. (9–30a)
General external surfaces—Isothermal or uniform heat flux		
Vertical plates, cylinders, inclined surfaces, horizontal cylinders, spheres, vertical cones, and other surfaces all Ra_L, Pr	$\overline{Nu}_L = (a + 0.331\,b\,Ra_L^{1/6})^2$	Eq. (9–34) a, $\mathbf{L}$—Table 9–1 b—Eq. (9–35) Churchill [11]

† $Gr_L = g\beta(T_s - T_\infty)\mathbf{L}^3/\nu^2$, $Ra_L = Gr_L\,Pr = g\beta(T_s - T_\infty)\mathbf{L}^3/(\alpha\nu)$, $Ra_L^* = Nu_L\,Ra_L = g\beta q_0''\,\mathbf{L}^4/(k\alpha\nu)$, $Gr_L^* = g\beta q_0''\,\mathbf{L}^4/(k\nu^2)$.

TABLE 9–7 *(Continued)*

Condition	*Correlation*	*Identification*
Horizontal isothermal plates	$\overline{Nu}_L = b\,Ra_L^m$	Eq. (9–36) b, m, L—Table 9–2 Goldstein et al. [18] Lloyd & Moran [19]
Enclosed spaces Rectangular enclosures, vertical concentric annuli, concentric spheres	$\overline{Nu}_L = C\,Ra_L^m \left(\frac{H}{L}\right)^n$	$[q_c = \bar{h}A_s(T_1 - T_2)]$ Eq. (9–39) L, A_s—Table 9–3 C, m, n, and Ra_L range—Table 9–4 various references
Horizontal plates heated from below	$\overline{Nu}_w = C\,Ra_w^m$	Eq. (9–40) C, m—Table 9–5 various references
Inclined rectangular space Heated from below cooled from above $0 < \theta < \theta^*$	$\overline{Nu}_w = (\overline{Nu}_w\|_{\theta=0})(\cos\theta)^{1/4}$	Eq. (9–41) θ^*—Table 9–6 Hollands et al. [22,35,38]
$\theta^* < \theta < \pi/2$	See Eq. (9–42)	$H/w > 12$ Hollands et al. [41]
Heated from above cooled from below	$\overline{Nu}_w = 1 + (\overline{Nu}_w\|_{\theta=0} - 1)\cos\theta$	Eq. (9–43) Arnold et al. [42]

TABLE 9–8 Dimensionless parameters for convection heat transfer: Natural convection

Dimensionless group	*Symbol*	*Definition*	*Interpretation*
Grashof number	Gr_L	$\dfrac{g\beta(T_s - T_\infty)L^3}{\nu^2}$	Relative significance of buoyant and viscous effects for single phase flow
Nusselt number†	Nu_L	$\dfrac{hL}{k}$	$\dfrac{\text{convection for fluid in motion}}{\text{conduction for motionless fluid layer}}$
Prandtl number†	Pr	$\dfrac{\nu}{\alpha}$	relative effectiveness of momentum and energy transfer by molecular diffusion within the boundary layer
Rayleigh number	Ra_L	$Gr_L\ Pr$	$\dfrac{\text{buoyancy forces}}{\text{change in shear forces}}$
Reynolds number†	Re_L	$\dfrac{U_F L}{\nu}$	$\dfrac{\text{inertia forces}}{\text{viscous forces}}$
Flux Rayleigh number	Ra_L^*	$Nu_L\ Ra_L$	Dimensionless heat flux for natural convection
Richardson number	Ri_x	$\dfrac{Gr_x}{Re_x^2}$	Buoyant force/inertia force; Relative significance of natural convection and forced convection

† Dimensionless parameters also used in analysis of forced convection.

CHAPTER 10

CONVECTION HEAT TRANSFER: PRACTICAL ANALYSIS— BOILING AND CONDENSATION

10–1 INTRODUCTION

The study of two- and three-phase substances is one of the main topics covered in introductory courses in thermodynamics. Based on our background in thermodynamics, we know that a phase change occurs when a single-phase substance is brought to the saturation state. For example, when the temperature of a subcooled liquid is raised to the saturation temperature T_{sat}, vaporization or *boiling* occurs. On the other hand, when the temperature of a superheated vapor is lowered to T_{sat}, *condensation* occurs. The thermodynamic study of boiling and condensation as well as melting and freezing is developed in the context of idealistic equilibrium conditions. In practice, however, heat-transfer processes involving phase change occur under nonequilibrium conditions in which the difference between the wall-surface temperature T_s and T_{sat} is not zero. The study of boiling and condensation heat transfer deals with such nonequilibrium liquid-to-vapor and vapor-to-liquid phase-change processes.

Familiar engineering applications of boiling and condensation heat transfer occur in power or refrigeration cycles. In addition, because of the large heat fluxes that can be accomplished in processes that involve phase change, boiling and condensation processes are also used in compact heat exchangers.

It should be mentioned that two-phase heat-transfer processes such as boiling and condensation are considerably more involved than are single-phase convection-heat-transfer processes because of the complicating effects of factors such as property variation, surface tension, latent heat of vaporization, and surface conditions. However, because we will be dealing with the practical thermal analysis of the simplest types of boiling and condensation-heat-transfer processes, the complexities that will be encountered in our introductory treatment of this topic will be minimized.

The general characterizing parameters associated with two-phase heat-transfer processes are introduced in the following section, after which specific attention is given to the phenomena of boiling and condensation in Secs. 10–3 and 10–4.

10–2 CHARACTERIZING PARAMETERS FOR TWO-PHASE HEAT-TRANSFER PROCESSES

To characterize two-phase heat-transfer processes, we generally must account for (1) the body force $g(\rho_f - \rho_g)$, which accompanies the large differences in density between the liquid and vapor states; (2) surface tension σ, for situations involving a liquid-vapor interface with finite curvature or temperature (or concentration) -induced variations in σ over the interface;† (3) the latent heat of vaporization relative to the difference between the surface and saturation temperatures $i_{fg}/\Delta T = i_{fg}/(T_s - T_{sat})$; (4) the thermophysical properties of the fluid ρ, μ, c_P, and k (liquid f or vapor g); and (5) a characteristic length $\boldsymbol{L}$.‡ Assuming that the two-phase coefficient of heat transfer $\bar{h} = q_c''/(T_s - T_{sat})$ is dependent upon these parameters, we write

$$\bar{h} = \frac{q_c''}{T_s - T_{sat}} = \text{fn}\,[g(\rho_f - \rho_g), \sigma, i_{fg}/(T_s - T_{sat}), \rho, \mu, c_P, k, \boldsymbol{L}] \tag{10–1}$$

Noting that this relation involves *nine* independent parameters and *four* fundamental dimensions, the dimensional analysis approach indicates *five* independent dimensionless groups and a dimensionless equation of the form

$$\frac{q_c''}{T_s - T_{sat}}\frac{\boldsymbol{L}}{k} = \text{fn}\left[\frac{\mu c_P}{k}, \frac{g(\rho_f - \rho_g)\boldsymbol{L}^3}{\rho\nu^2}, \frac{c_P(T_s - T_{sat})}{i_{fg}}, \frac{g(\rho_f - \rho_g)\boldsymbol{L}^2}{\sigma}\right] \tag{10–2}$$

or

$$\overline{Nu}_L = \text{fn}\,[Pr, Gr_L, Ja, Bo_L] \tag{10–3}$$

where

$$\overline{Nu}_L = \frac{q_c''}{T_s - T_{sat}}\frac{\boldsymbol{L}}{k} = \frac{\bar{h}\boldsymbol{L}}{k} \tag{10–4}$$

is the Nusselt number, Pr is the Prandtl number,

$$Gr_L = \frac{g(\rho_f - \rho_g)\boldsymbol{L}^3}{\rho\nu^2} = \frac{\rho g(\rho_f - \rho_g)\boldsymbol{L}^3}{\mu^2} \tag{10–5}$$

is a generalized form of the *Grashof number*, which represents *the relative significance of buoyant and viscous effects for two-phase flow*,

$$Ja = \frac{c_P(T_s - T_{sat})}{i_{fg}} \tag{10–6}$$

† The surface tension σ, which results from a net inward force of attraction on molecules at the interface toward molecules within the liquid phase, is defined as the work done in extending the surface of a liquid one unit area. The units commonly used for σ are N/m, kg/s^2 or dyne/cm.

‡ As discussed in Sec. 10–4–2, additional parameters are required for dropwise condensation.

is the *Jakob number*, which represents *the ratio of the maximum sensible energy absorbed to the latent energy absorbed*, and

$$Bo_L = \frac{g(\rho_f - \rho_g)L^2}{\sigma} \qquad (10\text{–}7)$$

is the *Bond number*, which represents the *ratio of gravitational forces to surface tension forces*. The surface tension σ is a function of both the fluid and temperature, as shown in Tables A–C–3 and A–C–6 of the Appendix. The enthalpy of vaporization i_{fg} is listed in Tables A–C–3(a) and A–C–3(c) for water and several other common fluids. Both σ and i_{fg} decrease as the saturation pressure increases.

These and several other dimensionless parameters will be used in the following sections of this chapter to characterize boiling and condensation heat-transfer processes. As we will see, the correlations for boiling and condensation involve the use of properties of saturated liquid (ρ_f, μ_f, and $c_{P,f}$) or saturated vapor (ρ_g, μ_g, and $c_{P,g}$) in the definition of Gr_L and Ja. Unless otherwise indicated, the Nusselt number Nu_L will be expressed in terms of k_f. In this connection, $Gr_{L,f}$, Ja_f, Bo_L, and Pr_g all experience increases and Pr_f decreases with increasing saturation temperature T_{sat} (or saturation pressure P_{sat}).

10–3 BOILING HEAT TRANSFER

Boiling heat transfer occurs when a fluid is exposed to a surface with temperature sufficiently greater than the saturation temperature T_{sat}.

The two types of boiling that are found in practice are *pool boiling* and *flow boiling*. Pool boiling occurs in the absence of bulk fluid flow, such as in the boiling of a pan of water. This type of boiling can also be easily produced by submerging an electrically heated wire or a steam-heated tube in a liquid, as shown in Figs. 10–1 and 10–2. Common examples of flow boiling processes include fossil fuel or nuclear steam generators and refrigerant evaporators.

Pool and flow boiling processes are further classified according to whether the bulk-liquid temperature is equal to or less than the saturation temperature. If the bulk temperature of the liquid is below T_{sat}, the process is referred to as *subcooled* boiling. Otherwise, the process is known as *saturated* boiling.

10–3–1 Pool Boiling

Boiling Curve for Pool Boiling

For a given fluid, operating pressure, and heating surface, the heat flux q_c'' that occurs during pool boiling is dependent upon the excess temperature $\Delta T_e = T_s - T_{sat}$. To illustrate this point, q_c'' is plotted in terms of ΔT_e in Fig. 10–3 for pool boiling of water. This general boiling curve applies to pool boiling of water at any subcritical

FIGURE 10–1 Nucleate pool boiling on an electrically heated wire. (Courtesy of Professor E. Hahne.)

pressure. Scales are also shown on this figure for pool boiling of water at atmospheric pressure [1,2]. Similarly shaped pool boiling curves have been developed for many other fluids and operating conditions. Although the general shape of the boiling curve is widely agreed upon, considerable controversy exists concerning the quantitative aspects of these curves, especially for fluids other than water [3].

The temperature of the fluid that is in contact with the wall is T_s. Thus, although the bulk-liquid temperature is always less than or equal to T_{sat}, the fluid very near the heating surface is superheated, with the degree of superheat being equal to the excess temperature ΔT_e. As we move along the boiling curve from left to right, the degree of superheat of fluid in contact with the heating surface increases.

A boiling curve such as this exhibits rather distinct regimes, which include natural convection, nucleate boiling, transition boiling, and film boiling.

Natural Convection For small values of excess temperature (i.e., $\Delta T_e < \Delta T_{e,A}$) the fluid motion occurs by means of natural convection. The natural convection circulates the slightly superheated liquid from the vicinity of the heating surface to the surface of the pool, where evaporation occurs. No boiling per se occurs in the natural convection regime, such that the heat transfer can be computed by the methods introduced in Chap. 9.

FIGURE 10–2
Film pool boiling of methanol on a vertical steam-heated copper tube. (Courtesy of Professor J. W. Westwater.)

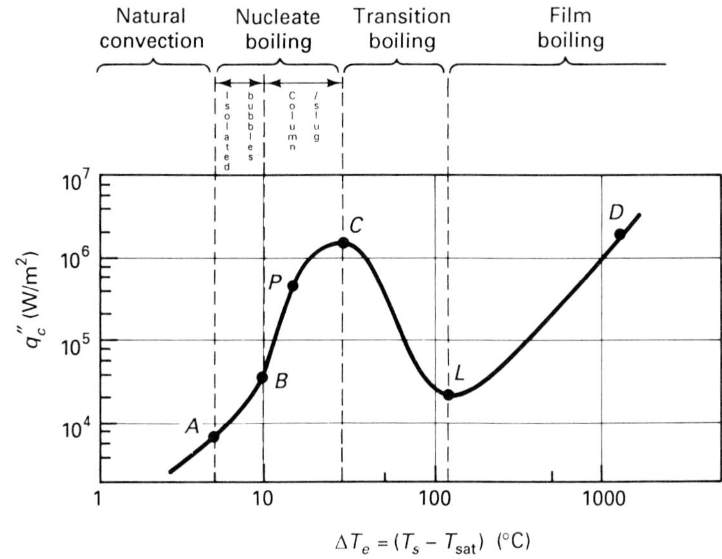

FIGURE 10–3 Representative boiling curve for pool boiling of water at one atmosphere.

Nucleate Boiling For excess temperatures between $\Delta T_{e,A}$ and $\Delta T_{e,C}$ liquid is transformed into vapor nuclei (i.e., bubbles) at various preferential sites on the heating surface. The nucleate boiling regime is generally characterized according to the behavior of the bubbles.

In the region between the *onset of nucleate boiling* ONB (point A) and point B, the nucleation sites produce isolated bubbles that eventually break away from the heating surface and rise toward the surface of the pool. The bubble growth and separation result in liquid being circulated and drawn toward the heating surface. It is this intermittent entrainment of cool liquid onto the hot surface which is primarily responsible for the increase in heat flux q''_c in this *isolated bubble-nucleate boiling regime*, with the contribution of evaporation (i.e., energy carried by vapor bubbles) being secondary.

In the region between point B and point C where the heat flux is a maximum, the number of active nucleation sites becomes very large, such that the bubbles coalesce, escape from the surface in the form of jets or columns, and merge into slugs of vapor. The large heat fluxes in this *column/slug-nucleate boiling regime* are produced by the combined effects of liquid entrainment and evaporation. Whereas the effects of evaporation increase with ΔT_e, the contribution of the liquid entrainment reaches a maximum at point P. Beyond this point the inflow of cooler liquid toward the heating surface is inhibited by the escaping vapor, with the result that the gradient in q''_c with respect to ΔT_e decreases until the point is reached at which the surface is no longer continually wetted and the heat flux is a maximum. The *maximum heat flux* q''_{max} is also commonly referred to as the *critical heat flux* or the *peak heat flux*.

Transition Boiling A dramatic decrease in heat flux occurs in the region immediately beyond the nucleate boiling regime. In this transition boiling regime a vapor film is intermittently built up and partially destroyed, such that partial nucleate boiling continues to occur. However, operation in the transition boiling regime does not occur in practice unless it is possible to control the excess temperature ΔT_e, such as in a tube heated by steam.

Film Boiling A continuous stable vapor film covers the heating surface in the film boiling regime. The location L at which the heat flux is a minimum is known as the *Leidenfrost point*. This point is named for J. G. Leidenfrost, who observed in 1756 that water globules on a very hot metal surface boil away very slowly as they "dance about." Because the vapor has a much lower thermal conductivity than the liquid, very large temperature differences occur across the vapor film. In this high-temperature region of the boiling curve the heat flux is seen to once again increase with increasing excess temperature. This increase in q''_c can be credited to the large driving potential $T_s - T_{sat}$ and to the effects of thermal radiation. The surface temperature T_s continues to increase with q''_c until the melting point of the material is reached and *burnout* occurs.

Several regimes of pool boiling have been captured by camera for boiling on a copper rod immersed in a fluid as shown in Fig. 10–4. The fin is heated at the wall to the right, such that the excess temperature is a minimum at the left end of

FIGURE 10–4 Several regimes of pool boiling of isopropanol on a copper fin. (Courtesy of Professor J. W. Westwater.)

the rod and increases as we move toward the wall. Both nucleate boiling regimes are seen to be active over the left one-third of the rod, with transition and film boiling occurring over the remainder of the rod.

The boiling curve shown in Fig. 10–3 can be produced by gradually increasing the surface temperature T_s. By slowly increasing T_s we can reach and maintain steady-state boiling at any point along this curve. However, for the case in which the electrical power input $\dot{W}$ to the heater is slowly raised, stable operation between the points C and D cannot be maintained. For this situation, increasing the heat flux just beyond the maximum heat flux, point C, results in a rapid often destructive transition to film boiling at point D, where the excess temperature may be sufficiently high to cause meltdown of the boiling surface. This point is expanded upon in the following example.

EXAMPLE 10–1

Utilize the lumped-analysis approach to explain the rapid unstable transition from point C to point D on the boiling curve for the case in which the electrical power input to the heater is gradually increased.

Solution

Objective Explain the sudden unstable increase in T_s for the case in which power $\dot{W}$ to a boiler element is gradually increased.

Schematic Boiling of liquid by electric heater.

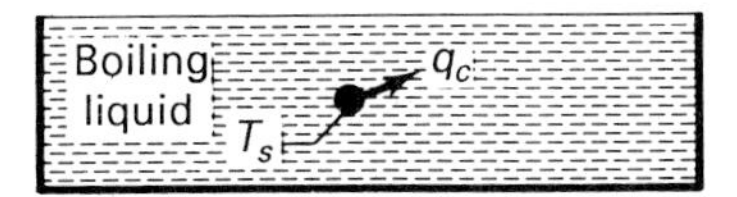

$\dot{W}$ – Power input
E_s – Energy storage

Assumptions/Conditions

pool boiling

standard conditions, except for unsteady conditions, two-phase fluid, significance of buoyancy and significance of thermal radiation in film boiling regime

Analysis To aid in our understanding of this point, we write the lumped energy balance for a boiler element with small Biot number (i.e., negligible resistance to conduction) as follows:

$$\Sigma \dot{E}_i = \Sigma \dot{E}_o + \frac{\Delta E_s}{\Delta t} \qquad \dot{W} = q_c'' A_s + \rho V c_v \frac{dT_s}{dt}$$

Based on this equation, we conclude that the surface temperature T_s will increase with increase in power $\dot{W}$ to the heater. The boiling heat flux q_c'' responds to any change in T_s in accordance with the boiling curve. Putting this information together, in the natural convection and nucleate boiling regimes where q_c'' increases with increasing T_s, a small increase in $\dot{W}$ will produce an increase in T_s, which in turn will cause q_c'' to increase until a new steady-state surface temperature is reached and $q_c'' A_s$ is equal to $\dot{W}$. However, if after reaching the peak heat flux at point D, we again increase $\dot{W}$ by a very small amount, an unstable sequence of events is set in motion in which the initial increase in T_s produces a *decrease* in q_c''! This degradation in the heat flux q_c'' then stimulates the rate of increase in surface temperature dT_s/dt. As T_s quickly increase with time at this fixed value of $\dot{W}$, q_c'' will change roughly in accordance with the boiling curve until burnout occurs or equilibrium is reached near point D. At this point, the rate of heat generation within the heater $\dot{W}$ is again in balance with the rate of heat transfer from the surface $q_c'' A_s$. Because point C on the boiling curve is associated with this sometimes destructive event, it is often referred to as the *burnout point*.

Convection Correlations for Pool Boiling

Following the pattern established in the analysis of single-phase heat-transfer processes, the rate of heat transfer associated with the nucleate, transition, and film boiling regimes of pool boiling is given in terms of the *mean coefficient of boiling heat transfer* $\bar{h}_B$ as†

$$q_c'' = \bar{h}_B (T_s - T_{sat}) \qquad (10\text{–}8)$$

Note that it is the excess temperature $T_s - T_{sat}$ that is utilized in this defining equation. Unlike the situation encountered in forced convection in which $\bar{h}$ is often independent of $T_s - T_F$, $\bar{h}_B$ is always a function of $T_s - T_{sat}$. Consequently, some experimentalists report their pool boiling heat-transfer data directly in terms of q_c'' instead of $\bar{h}_B$.

† The standard coefficient of heat transfer $\bar{h}$ is generally utilized in the correlation of data in the natural convection regime.

Despite the fact that considerable controversy exists concerning the accuracy of available correlations for pool boiling, various design curves have been used to adequately, although perhaps not optimally, design boiling-heat-transfer equipment. Because of the burnout threat in many boiling-heat-transfer applications, the safety of the design comes first and the question of efficiency comes second.

Empirical design correlations for q_c'' or $\bar{h}_B$ have been developed for nucleate and film boiling regimes of pool boiling. Some data even exist for the transition boiling regime [4], but general correlations are not presently available for this zone. It should be noted that the geometrical orientation of the surface is relatively unimportant for pool boiling. Therefore, the same correlations are generally used for horizontal, vertical, or inclined surfaces.

Experimental curves for nucleate pool boiling of water on a 0.024-in.-diameter platinum wire are shown in Fig. 10–5 for various saturation pressures. The points at which the maximum heat flux occurs are also shown. The experimental evidence indicates that the heat flux is approximately proportional to ΔT_e^3 throughout most of the nucleate boiling regime. The data have been correlated by Rohsenow [6], Fritz [7], and many others. The popular correlation by Rohsenow takes the form

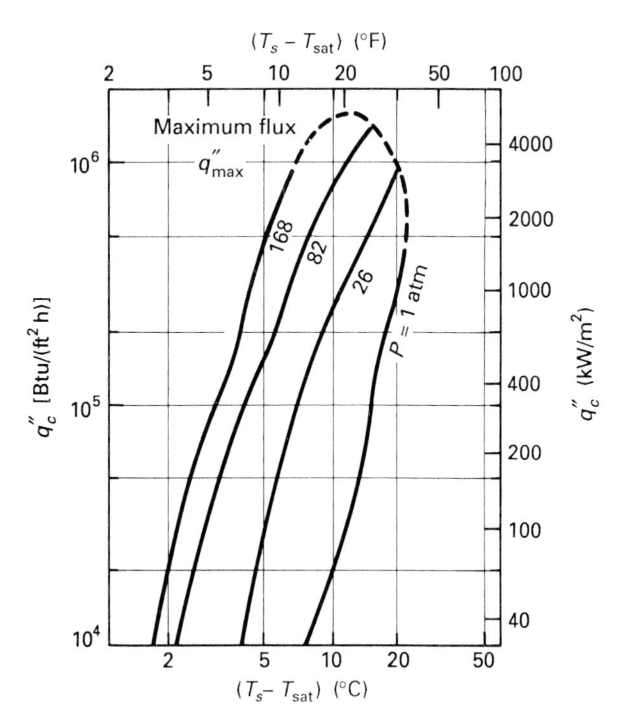

FIGURE 10–5 Nucleate pool boiling of water at various pressures on electrically heated platinum wire (Addoms [5]).

$$\frac{c_{P,f}(T_s - T_{sat})}{i_{fg}\, Pr_f^s} = C_{sf}\left[\frac{q_c''}{\mu_f i_{fg}}\sqrt{\frac{\sigma}{g(\rho_f - \rho_g)}}\right]^{1/3} \qquad (10\text{–}9a)$$

or equivalently

$$q_c'' = \mu_f i_{fg}\left[\frac{g(\rho_f - \rho_g)}{\sigma}\right]^{1/2}\left[\frac{c_{P,f}(T_s - T_{sat})}{C_{sf} i_{fg}\, Pr_f^s}\right]^3 \qquad (10\text{–}9b)$$

where C_{sf} and s are empirical constants. Table 10–1 gives values of C_{sf} and s for various clean surface-fluid combinations. For example, C_{sf} is equal to 0.013 for a copper-water interface. The Prandtl number exponent s is generally equal to unity for water and 1.7 for other fluids. The uncertainty associated with Rohsenow's correlation is about 25% for ΔT_e, but is as large as 100% for q_c''. Thus, ΔT_e can be computed with far more confidence than q_c''.

Equation (10–9) can also be written in the dimensionless form

$$\overline{Nu_L} = \frac{Ja_f^2\, Bo_L^{1/2}\, Pr_f^{1-3s}}{C_{sf}} \qquad (10\text{–}9c)$$

Due to the nature of this equation and the variations in Pr_f, Ja_f, and Bo_L with saturation pressure, we conclude that the nucleate boiling heat flux can be enhanced by increasing P_{sat} or T_{sat}.

TABLE 10–1 Values of the empirical constants C_{sf} and s in Eq. (10–9) for various liquid-surface combinations

Liquid-surface combination	C_{sf}	s
Water-copper		
polished	0.0130	1.0
scored	0.0068	1.0
Water-stainless steel		
mechanically polished	0.0130	1.0
ground and polished	0.0060	1.0
chemically etched	0.0130	1.0
Water-brass	0.0060	1.0
Water-nickel	0.0060	1.0
Water-platinum	0.0130	1.0
Benzene-chromium	0.0100	1.7
Carbon tetrachloride-copper	0.0130	1.7
Ethyl alcohol-chromium	0.0027	1.7
Isopropyl alcohol-copper	0.00250	1.7
n-Butyl alcohol-copper	0.00300	1.7
n-Pentane-copper		
polished	0.0154	1.7
lapped	0.0049	1.7
n-Pentane-chromium	0.015	1.7

Source: From references 5 and 8 through 11.

It should be emphasized that these and other correlations for nucleate pool boiling only apply to that part of the boiling curve for which ΔT_e is less than the maximum excess temperature $\Delta T_{e,\max}$. The maximum heat flux attainable with nucleate boiling is generally quite crucial to the designer because of the importance of avoiding the burnout dangers of the film boiling regime, particularly in high-performance constant-heat flux systems. The maximum flux for saturated nucleate boiling can be approximated by relations developed by Zuber [12], Rohsenow and Griffith [13], Kutateladze [14], Lienhard and Dhir [15], and others. The equation developed by Zuber on the basis of a hydrodynamic stability analysis is written as

$$q''_{\max} = \frac{\pi}{24} \rho_g i_{fg} \left[\frac{\sigma g(\rho_f - \rho_g)}{\rho_g^2} \right]^{1/4} \left(\frac{\rho_f + \rho_g}{\rho_f} \right)^{1/2} \tag{10–10}$$

Since ρ_g is generally much smaller than ρ_f, this equation is commonly approximated by

$$q''_{\max} = \frac{\pi}{24} \rho_g i_{fg} \left[\frac{\sigma g(\rho_f - \rho_g)}{\rho_g^2} \right]^{1/4} \tag{10–11}$$

Although these equations can be utilized as a first approximation for most fluid-surface combinations with relatively large characteristic length, a correction factor must be applied for applications in which the characteristic length of the surface is small relative to the mean bubble diameter [15,16].

According to Eq. (10–11), $q''_{\max}$ can be increased by increasing the gravitational field or by selecting a fluid with a large value of i_{fg}. For example, in the selection of a fluid, i_{fg} is higher for water than other common liquids, such that water will sustain the highest maximum heat flux. Further, because the fluid properties are functions of pressure, $q''_{\max}$ can be increased to a maximum for any given fluid by optimizing the pressure. The variation of $q''_{\max}$ with pressure is indicated in Fig. 10–5 for water. The optimum pressure for water is about 100 atm and the maximum heat flux is approximately 3.8×10^6 W/m^2. As an aside, with Eq. (10–11) expressed in dimensionless form (see Table 10–3), we note that $Nu_{L,\max}\, Ja_f = q''_{\max} L c_{P,f}/(i_{fg} k_f)$ is proportional to $Gr_{L,f}^{1/2} Pr_f / Bo_L^{1/4}$.

Zuber [12] has also proposed an equation for the *minimum heat flux* $q''_{\min}$ associated with the Leidenfrost point, which is given by

$$q''_{\min} = 0.09 \rho_g i_{fg} \left[\frac{\sigma g(\rho_f - \rho_g)}{(\rho_f + \rho_g)^2} \right]^{1/4} \tag{10–12}$$

The uncertainty of this relation is about 50% for fluids at moderate pressure, but is larger at high pressures. According to this correlation, $q''_{\min} L c_{P,f}/(i_{fg} k_f)$ is proportional to $Gr_{L,f}^{1/2} Pr_f / Bo_L^{1/4}$.

Correlations for film boiling on the surface of a horizontal tube or sphere of diameter D have been developed that take the form

$$\overline{h_B} = C\left[\frac{g(\rho_f - \rho_g)i'_{fg}k_g^3}{\nu_g(T_s - T_{sat})D}\right]^{1/4} \tag{10–13}$$

where $C = 0.62$ for the tube [17] and $C = 0.67$ for the sphere [15], and the vapor properties are evaluated at the film temperature $(T_s + T_{sat})/2$. The *corrected latent heat* i'_{fg} is defined as

$$i'_{fg} = i_{fg} + 0.4c_{P,g}(T_s - T_{sat}) = i_{fg}(1 + 0.4\,Ja_g) \tag{10–14}$$

according to Bromley [17]. Equation (10–13) is expressed conveniently in terms of dimensionless parameters by

$$\overline{Nu_D} = C\left[Gr_{D,g}\,Pr_g\left(\frac{1}{Ja_g} + 0.4\right)\right]^{1/4} \tag{10–15}$$

where the product $Gr_{D,g}\,Pr_g$ is the Rayleigh number $Ra_{D,g}$. This equation indicates that the film-boiling heat flux should increase with increasing values of P_{sat} or T_{sat}.

For temperatures much above 300°C, the effect of thermal radiation across the vapor film becomes significant. The most practical approach to accounting for thermal radiation involves the use of an effective thermal radiation coefficient $\overline{h_R}$,

$$\overline{h_R} = \frac{\epsilon_s\sigma(T_s^4 - T_{sat}^4)}{T_s - T_{sat}} \tag{10–16}$$

where ϵ_s is the emissivity of the surface and σ now represents the Stefan-Boltzmann constant. Using this relation, Bromley [17] proposed a simple correlation of the form

$$\overline{h_{B,R}} = \overline{h_B} + \frac{3}{4}\overline{h_R} \qquad \text{for } \overline{h_R} < \overline{h_B} \tag{10–17}$$

where $\overline{h_B}$ is given by Eq. (10–13). For situations in which the restriction $\overline{h_R} < \overline{h_B}$ is not satisfied, Bromley recommended the more involved relation

$$\overline{h_{B,R}} = \overline{h_B}\left(\frac{\overline{h_B}}{\overline{h_{B,R}}}\right)^{1/3} + \overline{h_R} \tag{10–18}$$

which requires the use of iterative solution methods.

EXAMPLE 10–2

Water at atmospheric conditions is boiled by an electrically heated 5-mm-diameter platinum wire. Determine the maximum wall temperature and heat flux that can be attained in the safe nucleate boiling regime.

Solution

Objective Determine T_{max} and q''_{max}.

Schematic Boiling of water by heated wire.

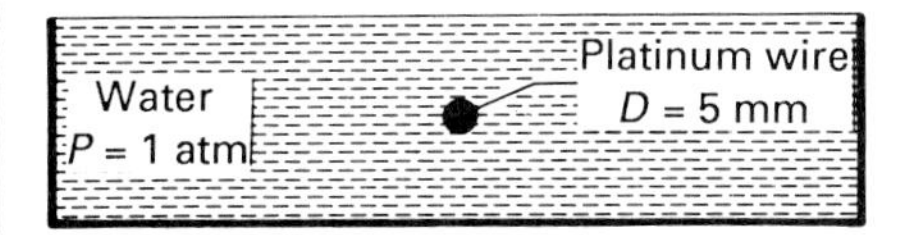

Assumptions/Conditions

pool boiling
standard conditions, except for two-phase fluid and significance of buoyancy

Properties Saturated water at 100°C (Table A–C–3): $\rho_f = 1/v_f = 958\ \text{kg/m}^3$, $\rho_g = 1/v_g = 0.595\ \text{kg/m}^3$, $i_{fg} = 2260\ \text{kJ/kg}$, $c_{P,f} = 4.22\ \text{kJ/(kg °C)}$, $\mu_f = 2.79 \times 10^{-4}\ \text{kg/(m s)}$, $Pr_f = 1.76$, $\sigma = 5.89 \times 10^{-2}\ \text{N/m}$.

Analysis Substituting into the Zuber correlation for $q''_{\max}$, Eq. (10–10), becomes

$$q''_{\max} = \frac{\pi}{24}\rho_g i_{fg}\left[\frac{\sigma g(\rho_f - \rho_g)}{\rho_g^2}\right]^{1/4}\left(\frac{\rho_f + \rho_g}{\rho_f}\right)^{1/2} = \frac{\pi}{24}\left(0.595\ \frac{\text{kg}}{\text{m}^3}\right)\left(2260\ \frac{\text{kJ}}{\text{kg}}\right)$$

$$\times\left[\frac{(5.88 \times 10^{-2}\ \text{N/m})(9.81\ \text{m/s}^2)(958 - 0.595)\ \text{kg/m}^3}{(0.595\ \text{kg/m}^3)^2}\right]^{1/4}\left(\frac{958 + 0.595}{958}\right)^{1/2}$$

$$= 1110\ \text{kW/m}^2$$

Combining this result with the Rohsenow correlation, Eq. (10–9), $T_{\max}$ is given by

$$T_{\max} - T_{\text{sat}} = i_{fg}\,Pr_f^s\,\frac{C_{sf}}{c_{P,f}}\left[\frac{q''_{\max}}{\mu_f i_{fg}}\sqrt{\frac{\sigma}{g(\rho_f - \rho_g)}}\right]^{1/3}$$

where $C_{sf} = 0.013$ and $s = 1.0$. Following through with the calculation, we obtain

$$T_{\max} - T_{\text{sat}} = 20.1°\text{C} \qquad \text{or} \qquad T_{\max} = 120°\text{C}$$

These calculations are observed to be consistent with the results shown in Fig. 10–5.

EXAMPLE 10–3

Saturated water at 134°F is boiled by a 3.28-ft-long 0.816-in.-diameter copper tube with surface temperature equal to 150°F. Determine the rate of heat transfer and the rate of evaporation.

Solution

Objective Determine q''_c and the rate of evaporation $\dot{m}$.

Schematic Boiling of water by heated tube.

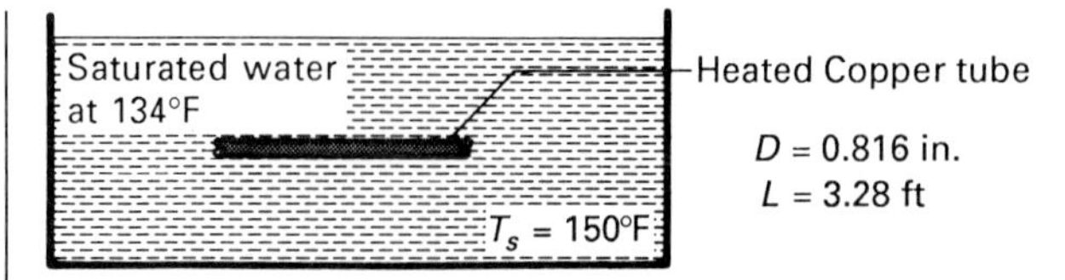

Assumptions/Conditions

pool boiling
standard conditions, except for two-phase fluid and significance of buoyancy

Properties Water at $T = 134°\text{F} = 330$ K (Table A–C–3): $P_{\text{sat}} = 0.172$ bar, $Pr_f = 3.15$,

$$\rho_f = \frac{1}{0.00102}\frac{\text{kg}}{\text{m}^3}\frac{0.0624\ \text{lb}_\text{m}/\text{ft}^3}{\text{kg/m}^3} = 61.2\ \text{lb}_\text{m}/\text{ft}^3$$

$$\rho_g = \frac{1}{8.82}\frac{\text{kg}}{\text{m}^3}\frac{0.0624\ \text{lb}_\text{m}/\text{ft}^3}{\text{kg/m}^3} = 0.00707\ \text{lb}_\text{m}/\text{ft}^3$$

$$i_{fg} = 2370\frac{\text{kJ}}{\text{kg}}\frac{0.4299\ \text{Btu/lb}_\text{m}}{\text{kJ/kg}} = 1020\ \text{Btu/lb}_\text{m}$$

$$c_{P,f} = 4.18\frac{\text{kJ}}{\text{kg K}}\frac{0.2388\ \text{Btu/(lb}_\text{m}\ °\text{F)}}{\text{kJ/(kg K)}} = 0.998\ \text{Btu/(lb}_\text{m}\ °\text{F)}$$

$$\mu_f = 4.89 \times 10^{-4}\frac{\text{kg}}{\text{m s}}\frac{0.6720\ \text{lb}_\text{m}/(\text{ft s})}{\text{kg/(m s)}} = 3.29 \times 10^{-4}\ \text{lb}_\text{m}/(\text{ft s})$$

$$\sigma = 6.67 \times 10^{-2}\frac{\text{N}}{\text{m}}\left(0.2248\frac{\text{lb}_\text{f}}{\text{N}}\right)\left(\frac{\text{m}}{3.281\ \text{ft}}\right) = 4.56 \times 10^{-3}\ \text{lb}_\text{f}/\text{ft}$$

Analysis The excess temperature is

$$T_s - T_{\text{sat}} = 150°\text{F} - 134°\text{F} = 16°\text{F}$$

Referring to Fig. 10–5 and noting that the pressure is about 0.17 atm, we conclude that this boiling process is in the nucleate boiling regime. It follows that Eq. (10–9b) can be used to compute the heat flux.

$$q_c'' = \mu_f i_{fg}\left[\frac{g(\rho_f - \rho_g)}{\sigma}\right]^{1/2}\left[\frac{c_{P,f}(T_s - T_{\text{sat}})}{i_{fg}\,Pr_f^s\,C_{sf}}\right]^3$$

$$= 3.29 \times 10^{-4}\frac{\text{lb}_\text{m}}{\text{ft s}}\left(1020\frac{\text{Btu}}{\text{lb}_\text{m}}\right)\left[\frac{(32.2\ \text{ft/s}^2)(61.2\ \text{lb}_\text{m}/\text{ft}^3)}{4.56 \times 10^{-3}\ \text{lb}_\text{f}/\text{ft}}\right]^{1/2}$$

$$\times\left(\frac{\text{lb}_\text{f}\ \text{s}^2}{32.2\ \text{ft lb}_\text{m}}\right)^{1/2}\left\{\frac{[0.998\ \text{Btu/(lb}_\text{m}\ °\text{F)}](16°\text{F})}{(1020\ \text{Btu/lb}_\text{m})(3.15)(0.013)}\right\}^3$$

$$= 2.17\ \text{Btu/(s ft}^2) = 7820\ \text{Btu/(h ft}^2) = 24.6\ \text{kW/m}^2$$

Thus, the total rate of heat transfer is

$$q_c = q_c''\pi DL = 2.17 \frac{\text{Btu}}{\text{s ft}^2} \pi \left(\frac{0.816 \text{ ft}}{12}\right) (3.28 \text{ ft}) = 1.52 \text{ Btu/s} = 1.60 \text{ kW}$$

To calculate the rate of evaporation $\dot{m}$, we write

$$\dot{m} = \frac{q_c}{i_{fg}} = \frac{1.52 \text{ Btu/s}}{1020 \text{ Btu/lb}_\text{m}} = 1.49 \times 10^{-3} \text{ lb}_\text{m}\text{/s}$$

Enhancement of Pool Boiling

The heat-transfer correlations given in the previous section apply to smooth surfaces. Many methods have been utilized for producing improvements in pool boiling heat transfer. Perhaps the most common approaches involve the use of rough surfaces and extended surfaces. Several other methods include surface treatment, liquid additives, and mechanical agitation or vibration. A survey of various methods of heat-transfer enhancement is provided by Webb [18]. The main idea in using rough or extended surfaces is to increase the number of nucleation sites. The effect of enhancement techniques such as these on the boiling heat flux is shown in Fig. 10–6. The Thermoexcel-E surface shown in Fig. 10–7 is representative of machined rough surfaces that are available in the marketplace. The use of properly designed fins or rough surfaces such as Thermoexcel-E has also been reported to substantially increase the critical heat flux and delay the onset of transition and film boiling [18,19].

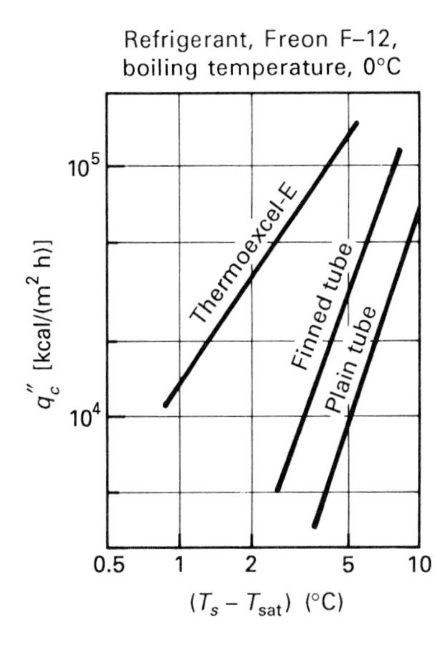

FIGURE 10–6
Nucleate pool boiling heat-transfer performance of several surfaces (Arai et al. [19]).

(a) Plan view.

(b) Cross section.

FIGURE 10–7
Thermoexcel-E surface for enhanced boiling. (Courtesy of Hitachi, Ltd.)

10–3–2 Flow Boiling

Physical Description of Flow Boiling

Flow boiling occurs in internal and external flow forced and natural convection systems. External flow boiling over a cylinder or plate is much like pool boiling because the free-stream temperature and quality are essentially constant with respect to the flow direction. But internal flow boiling is more complex because of changes in the bulk-stream temperature and quality with respect to x.

In internal flow boiling processes, liquid (or a liquid-vapor mixture) enters a heating chamber such as the tube shown in Fig. 10–8. Beyond the axial location at which the boiling process is initiated (point a), the quality of the liquid-vapor mixture increases with distance until total vaporization occurs. Thus, for the case in which a subcooled liquid enters the heating chamber, the bulk-fluid temperature increases to a value somewhat less than T_{sat}, at which point nucleate boiling is initiated. The bulk temperature then quickly reaches T_{sat} and the quality goes from zero to unity. After

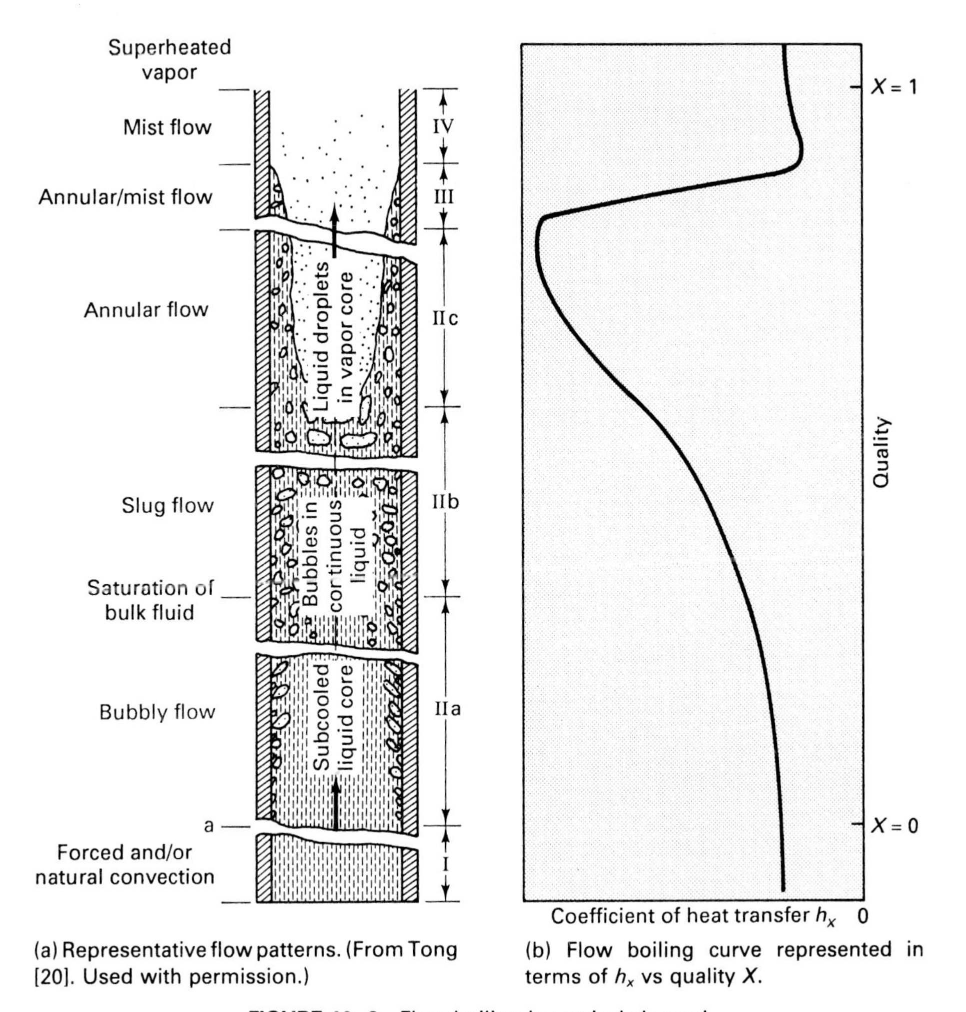

(a) Representative flow patterns. (From Tong [20]. Used with permission.)

(b) Flow boiling curve represented in terms of h_x vs quality X.

FIGURE 10–8 Flow boiling in vertical channel.

all the liquid has been vaporized, the entire bulk fluid becomes superheated. Strictly speaking, T_{sat} varies slightly with x because of pressure losses within the chamber. However, the assumption that T_{sat} is essentially constant is generally quite adequate.

Boiling Curve for Flow Boiling

A representative boiling curve for internal flow boiling in heated tubes or channels is shown in Fig. 10–9. With the excess temperature ΔT_e represented by $T_s - T_b$ for internal flow and by $T_s - T_{sat}$ for external flow, the boiling curves for flow boiling and pool boiling are of the same general form. Points on the flow boiling curve

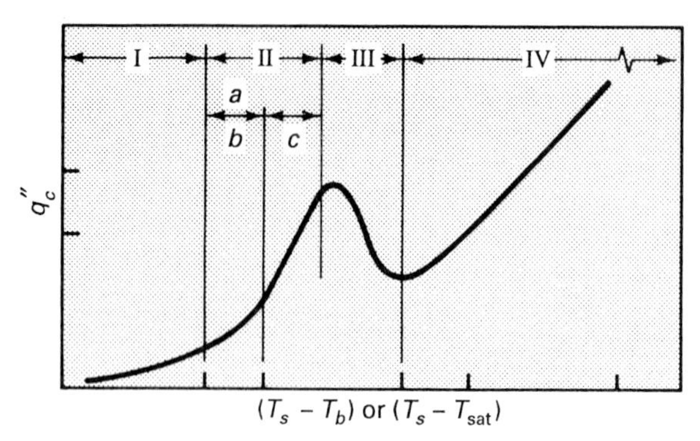

FIGURE 10–9
Representative boiling curve for internal flow boiling.

represent the excess temperature ΔT_e at various axial locations x on the surface versus the local heat flux q''_c. The actual numerical values of q''_c and the temperature difference are a function of pressure, geometry, mean velocity, thermal boundary conditions, and degree of subcooling. In this connection, flow boiling curves are also commonly presented in terms of the local coefficient of heat transfer h_x and quality X, as illustrated in Fig. 10–8. It should be noted that the local coefficient of heat transfer h_x is defined in the traditional sense; that is,

$$dq_c = h_x \, dA_s \, (T_s - T_F) \tag{10–19}$$

where $T_F = T_b$ for internal flow and $T_F = T_\infty$ for external flow.

The same basic regimes found in pool boiling also occur in flow boiling: forced and/or natural convection (I); nucleate boiling (II); transition boiling (III); and film boiling (IV). However, for internal flow boiling processes, because the fluid is totally confined within the tube or channel with no free interface from which vapor can escape, rather distinctive two-phase flow patterns are associated with each of these regimes. Take, for example, the situation in which a subcooled liquid enters a heated vertical tube in which T_s is greater than T_{sat}. Forced and/or natural convection occurs in region I where $T_s - T_b$ is small. But, as in the case of pool boiling, no actual boiling takes place in this zone. When the temperature difference $T_s - T_b$ is sufficiently large to produce regime II nucleate boiling, three basic flow patterns are produced.

First, bubbles form at nucleation sites but do not carry far into the main stream because the local bulk temperature T_b is still below T_{sat}. However, as T_b approaches T_{sat}, more and more bubbles appear in the bulk stream. Because of the presence of discrete bubbles in the flow stream, regime IIa is sometimes called the *bubbly flow regime*. The bubbles that occur throughout the bulk stream then begin to coalesce into slugs of vapor. Although the quality is quite low in region IIb, the percent of the volume that is occupied by vapor is as large as 50%. Consequently, the bulk flow rate increases. This portion of the nucleate boiling zone is sometimes called the *slug flow regime*. In the third part of the nucleate boiling region, the tube wall is covered by a thin liquid film with vapor and liquid droplets flowing in the center of the tube. Although nucleate boiling occurs in this region, vapor is also believed to be generated by vaporization at the liquid-vapor interface. Large increases in the heat flux continue to occur in this zone, until the heat flux reaches a peak, which is commonly referred to as the *departure from nucleate boiling* (DNB). The coefficients of heat transfer in the vicinity of the DNB can be 20 to 50 times as large as the coefficient for single-phase flow. This part of the nucleate boiling regime (IIc) is also known as the *annular flow regime*.

As in the case of pool boiling, the heat flux in flow boiling processes experiences a drastic decrease in regime III. In this transition boiling region, a wall-drying process occurs in which a vapor film with large thermal resistance gradually develops on the surface. Region III is also sometimes known as the *annular/mist regime*.

In the *film boiling regime*, the wall is completely dried out with liquid droplets being dispersed throughout the bulk of the fluid. Heat is transferred from the wall to the vapor and then from the vapor to the liquid until complete vaporization occurs. Regime IV is also known as the *mist flow regime*.

As in the case of pool boiling, the coefficients of heat transfer are quite high in the nucleate flow boiling regimes. Hence, the wall temperature is usually not too much larger than the bulk-fluid temperature or saturated temperature and the walls are not in danger of meltdown or burnout. However, because of the greatly reduced coefficients of heat transfer in the regimes III and IV, these regions are avoided in certain applications. For example, fossil-fuel-fired boilers or nuclear reactors are usually operated in the nucleate boiling regimes to avoid the onset of DNB. The coefficient of heat transfer attained in these systems is sufficiently high to permit satisfactory operation as long as fouling is prevented by proper water treatment. On the other hand, steam generators for pressurized-water-reactor systems (i.e., water-to-boiling-water heat exchangers) and certain types of process heat exchangers are sometimes operated in regimes III and IV. In these cases, the temperature of the heat source is within the safe operating range of the equipment. In once-through boilers in which superheated steam is produced, the wall-heat flux is generally reduced in the section of the heating chamber where regimes III and IV occur.

The effect of fluid enthalpy (or quality) and heat flux on the wall temperature T_s and bulk-fluid temperature T_b and on the location of the DNB point are indicated in Fig. 10–10 for an upward flow of water in a vertical uniformly heated tube. Notice that T_b increases until the fluid reaches the saturation temperature. T_b then drops slightly in the region where steam is being generated because of a small pressure drop. In the single-phase superheat region, T_b once again increases. Except for the low-heat-flux case, the behavior in surface temperature T_s is quite another matter. For moderate heat flux, T_s remains somewhat above T_b until the DNB point associated with regime III is reached in the region of high steam quality, and a degradation in heat transfer occurs. In this nucleate/film boiling regime, T_s experiences a rather pronounced increase to a mild peak, and then decreases as the steam quality approaches 100%. T_s once again gradually increases in the superheat region. For higher heat fluxes, the DNB point and regime III boiling is reached at lower steam quality and the peak in T_s is higher. As the heat flux increases, the peak in T_s increases and moves in the direction of lower and lower bulk-stream enthalpy, until the metal melts or the tube or channel ruptures.

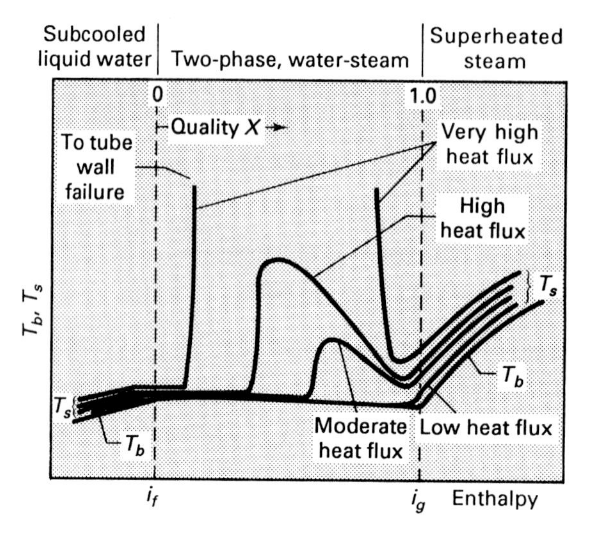

FIGURE 10–10 Bulk fluid and tube-wall temperature for flow boiling of water. (From *Steam—Its Generation and Use.* Copyright 1978 by Babcock and Wilcox. Used with permission.)

EXAMPLE 10–4

Sketch a boiling curve for external forced convection flow boiling on an electrically heated wire.

Solution

Objective Show the effect of U_∞ for flow boiling on an electrically heated wire.

Schematic Flow boiling on an electrically heated wire.

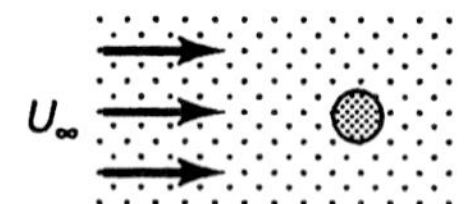

Assumptions/Conditions

flow boiling

standard conditions, except for two-phase fluid and significance of buoyancy

Analysis For cases in which the free-stream velocity is zero, natural convection and pool boiling control. Therefore, we start with the pool boiling curve shown in Fig. E10–4a. As U_∞ increases, the coefficient of heat transfer in the single-phase natural convection region increases. This effect would be expected to carry over to the nucleate boiling regimes. Thus, boiling curves are sketched in for low, moderate, and high values of U_∞, which should account for the qualitative effects of forced convection.

To reinforce this result, experimental data reported by Lung et al. [21] are shown in Fig. E10–4b. Our external flow boiling curves are in general agreement

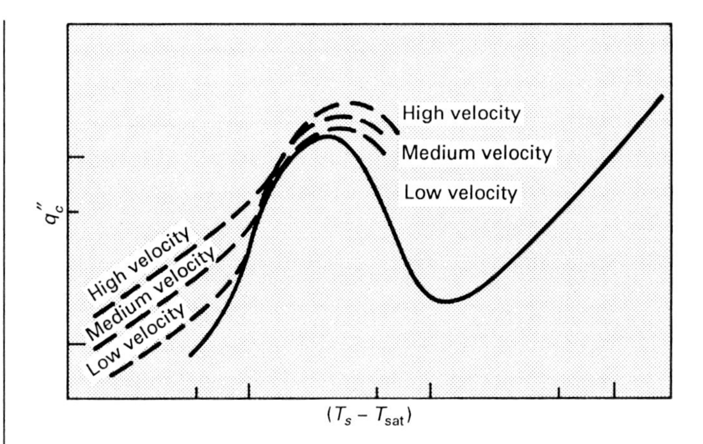

FIGURE E10-4a Boiling curve for external flow boiling.

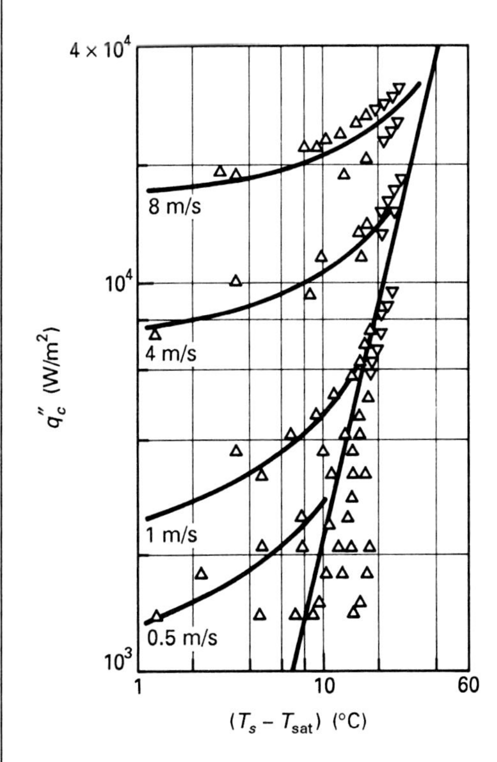

FIGURE E10-4b Experimental data for flow boiling.

with these data. Note that the heat-transfer process is essentially controlled by boiling for larger values of $T_s - T_{sat}$ and by forced convection for smaller values. Experimental and theoretical studies also indicate that the peak heat flux is markedly increased for forced convection.

Convection Correlations for Flow Boiling

Some success has been achieved in developing correlations for the coefficient of heat transfer and the peak heat flux for flow boiling processes. In particular, fairly simple correlations are available for nucleate boiling, in which case the total heat flux q_c'' is artificially broken into boiling- and nonboiling-convection components. This idea is reflected in the following equation:

$$q_c'' = q_c''|_{\text{boiling}} + q_c''|_{\text{nonboiling}} \tag{10–20}$$

where $q_c''|_{\text{boiling}}$ is approximated by the correlations for two-phase nucleate pool boiling and $q_c''|_{\text{nonboiling}}$ is obtained from the standard correlations for single-phase convection. In regimes IIb and IIc, where the heat-transfer process is essentially governed by the boiling mechanism, Eq. (10–20) can be approximated by

$$q_c'' = q_c''|_{\text{boiling}} \tag{10–21}$$

In this region, the heat-transfer flux is essentially independent of the velocity and forced or natural convection effects. These assumptions are consistent with the data shown in Example 10–4.

Empirical burnout or peak heat-flux correlations have been developed for various flow boiling systems. A comprehensive survey of the work done in this area has been presented by Hewitt [22].

Empirical correlations have also been developed for flow boiling heat transfer associated with regimes III and IV. However, the quantitative treatment of this two-phase flow subject is complex and lies beyond the scope of our introductory study. Several references on the subject of two-phase flow include the works by Tong [20], Hsu and Graham [23], Collier [24], and Butterworth and Hewitt [25].

Enhancement of Flow Boiling

Techniques similar to those used in pool boiling are used for improving flow boiling characteristics [18]. One of the more successful methods involves the use of internally ribbed tubes such as the one shown in Fig. 10–11. This type tube has been reported to suppress the onset of DNB and to permit operation at higher heat fluxes than is possible with smooth tubes [25,26]. However, because of the expense of manufacturing internally ribbed tubes, the Babcock and Wilcox Company recommends their use only in high-pressure systems (above 150 atm).

FIGURE 10–11 Single-lead ribbed tube. (From *Steam—Its Generation and Use*. Copyright 1978 by Babcock and Wilcox. Used with permission.)

10-4 CONDENSATION HEAT TRANSFER

Condensation heat transfer occurs when a vapor is exposed to a surface with temperature T_s less than the saturation temperature T_{sat}. The condensate that is formed flows down the surface under the influence of gravity. If the liquid does not wet the surface, the process is called *dropwise condensation*. Under these conditions, droplets form on the surface as shown in Fig. 10-12. These drops generally flow over the surface along random paths. For those liquids that wet the surface, the condensate forms a film that builds up as it flows down the surface. This *film condensation* process is illustrated in Fig. 10-13. Because of the low velocities associated with the film for laminar flow conditions, the energy transfer across the film is primarily due to conduction, such that the temperature distribution is essentially linear. The existence of such a film, whether in laminar or turbulent flow, increases the resistance to heat transfer, with the result that dropwise condensation is as much as 10 times more effective than film condensation. However, most surfaces become wetted when exposed to a condensing vapor over an extended length of time. Consequently, film condensation is generally encountered in industrial applications and is usually planned for in design work. In this connection, the rate of heat transfer is generally lower when a vapor exists in the presence of a noncondensable gas. Therefore, noncondensable gases are sometimes vented to improve the efficiency of the condensation-heat-transfer process.

Although our attention will be restricted to the surface condensation processes mentioned above, it should be noted that condensation can also occur in the absence of solid boundaries. Examples of these other types of condensation include the formation of fog and the exposure of a vapor to a cool liquid.

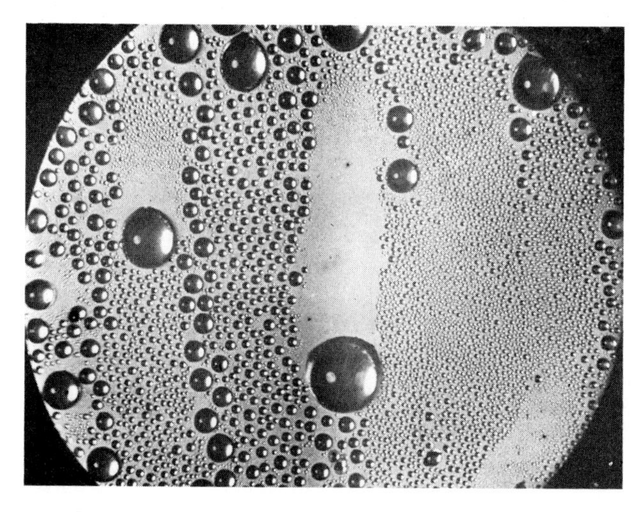

FIGURE 10-12 Dropwise condensation of steam under ideal conditions. (Courtesy of Professor M. N. Ozisik.)

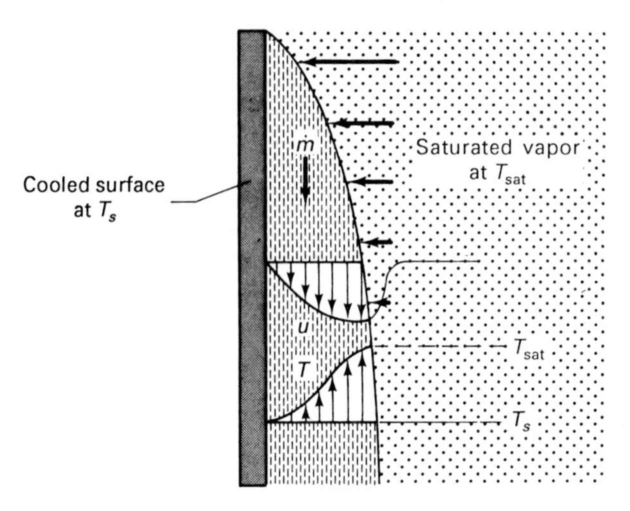

FIGURE 10–13 Film of condensate on vertical surface.

10–4–1 Film Condensation

According to the work of Shafrin and Zisman [27], a liquid will wet the surface if the surface tension σ of the fluid is less than a *critical surface tension* σ_{cr}, which is characteristic of the surface alone. Aside from certain fluid-surface combinations, such as those involving mercury (σ_{Hg} = 0.47 N/m at 100°C) or special materials such as Teflon (σ_{cr} = 0.018 N/m), σ is usually less than σ_{cr}, such that film condensation is most common.

We now turn to the practical analysis of film condensation, which involves the use of a liquid film Reynolds number Re_f and convection correlations for various standard geometries.

Reynolds Number for Film Condensation

As in single-phase forced-convection processes, the heat transfer in film condensation is dependent upon whether the flow within the film is laminar or turbulent. To provide a criterion by which to characterize the flow, a Reynolds number Re_f for condensate liquid film flow has been defined, which takes the form

$$Re_f = \frac{D_H U_b}{\nu_f} = \frac{4A}{p}\frac{U_b}{\nu_f} \tag{10–22}$$

where U_b is the condensate bulk velocity and $A = p\delta_f$; δ_f is the local thickness of the film and p is the perimeter. Based on the principle of continuity, U_b can be expressed in terms of the condensate mass flow rate $\dot{m}$ as

$$\dot{m} = \rho_f A U_b \tag{10–23}$$

Thus, Re_f can be written as

$$Re_f = \frac{4\dot{m}}{p\mu_f} \tag{10–24}$$

Notice that the value of Re_f increases as the film grows. For vertical tubes and plates, the critical Reynolds number is approximately 1800.

The properties of the liquid film are generally evaluated at the film temperature $(T_s + T_{sat})/2$.

Convection Correlations

Consistent with the traditional approach to correlating heat transfer data for single-phase convection processes, the local rate of heat transfer for film condensation is generally expressed in terms of a *coefficient of condensation heat transfer* h_C.

$$dq_c = h_C\, dA_s\, (T_{sat} - T_s) \tag{10–25}$$

For a uniform wall-temperature boundary condition, the total rate of heat transfer over the entire surface is given by

$$q_c = \overline{h_C} A_s (T_{sat} - T_s) \tag{10–26}$$

where the mean coefficient of condensation heat transfer is defined by

$$\overline{h_C} = \frac{1}{L}\int_0^L h_C\, dx \tag{10–27}$$

A useful relationship between $\overline{h_C}$ and Re_f can be developed by equating the rate of convection heat transfer from the film to the wall q_c, to the rate of energy transfer associated with phase change $\dot{m} i_{fg}$, and sensible heat, which is proportional to $\dot{m} c_{P,f}\,(T_{sat} - T_s)$ (see Appendix J); that is,

$$q_c = \dot{m} i_{fg} + C_1 \dot{m} c_{P,f} (T_{sat} - T_s) = \dot{m}\,[i_{fg}(1 + C_1\, Ja_f)] = \dot{m} i'_{fg} \tag{10–28}$$

where i'_{fg} is a modified latent heat similar in form to Eq. (10–14),

$$i'_{fg} = i_{fg}(1 + C_1\, Ja_f) \tag{10–29}$$

and $C_1 = 0.68$ according to the analysis of Rohsenow [28]. It follows that

$$\overline{h_C} = \frac{q_c}{A_s(T_{sat} - T_s)} = \frac{\dot{m} i'_{fg}}{A_s(T_{sat} - T_s)} \tag{10–30}$$

and

$$Re_f = \frac{4\dot{m}}{p\mu_f} = \frac{4q_c}{p\mu_f\, i'_{fg}} = \frac{4\overline{h_C} A_s (T_{sat} - T_s)}{p\mu_f\, i'_{fg}} \tag{10–31}$$

Thus, with $\overline{h_C}$ known on the basis of empirical or theoretical correlations, Re_f can be calculated. Several design correlations are now presented for the coefficient of condensation heat transfer associated with external film condensation. Because film condensation involves wetted surfaces with no liquid drop effects, the surface tension σ does not appear in the correlations. Referring to the dimensional analysis result introduced in Sec. 10–2, we should expect correlations involving the four dimensionless parameters Nu_L, Pr, Gr_L, and Ja.

For information on the more complex topic of film condensation inside tubes and channels, the review by Collier [24] is recommended.

Vertical Surface Referring to Appendix J, Nusselt [29] and Rohsenow [28] developed a theoretical expression for laminar film condensation on an isothermal surface, which can be represented by

$$h_C = 0.707 \left[\frac{\rho_f g(\rho_f - \rho_g) k_f^3 i'_{fg}}{\mu_f (T_{sat} - T_s) x} \right]^{1/4} \tag{10–32}$$

This relation has been shown to be applicable for $Pr \gtrsim 0.5$ and $Ja_f \lesssim 1.0$, with high accuracy for $Re_f \lesssim 30$ and with errors within 20% for $Re_f \lesssim 1800$. Based on this famous result for the local coefficient of condensation heat transfer, an expression can be written for the mean coefficient $\overline{h_C}$ of the form

$$\overline{h_C} = 0.943 \left[\frac{\rho_f g(\rho_f - \rho_g) k_f^3 i'_{fg}}{\mu_f (T_{sat} - T_s) L} \right]^{1/4} \tag{10–33a}$$

or

$$\overline{Nu_L} = 0.943 \left[Gr_{L,f} Pr_f \left(\frac{1}{Ja_f} + 0.68 \right) \right]^{1/4} \tag{10–33b}$$

By coupling Eqs. (10–31) and (10–33), the relation for Re_f associated with laminar flow becomes

$$Re_f = 3.77 \left[\frac{L(T_{sat} - T_s)}{\mu_f i'_{fg}} \right]^{3/4} \left[\frac{\rho_f g(\rho_f - \rho_g) k_f^3}{\mu_f^2} \right]^{1/4} \tag{10–34a}$$

or

$$Re_f = 3.77 \left[Gr_{L,f}^{1/4} Pr_f^{-3/4} \left(\frac{1}{Ja_f} + 0.68 \right)^{-3/4} \right] \tag{10–34b}$$

This equation in either form can be used to calculate Re_f. If Re_f is less than 1800, h_C and $\overline{h_C}$ can then be calculated by the use of Eqs. (10–32) and (10–33). Taking this one step further, we couple Eqs. (10–33) and (10–34) to obtain

$$\overline{h_C} = 1.47\, Re_f^{-1/3} \left[\frac{\rho_f g(\rho_f - \rho_g) k_f^3}{\mu_f^2} \right]^{1/3} \tag{10–35a}$$

or

$$\overline{Nu_L} = 1.47\, Re_f^{-1/3}\, Gr_{L,f}^{1/3} \tag{10–35b}$$

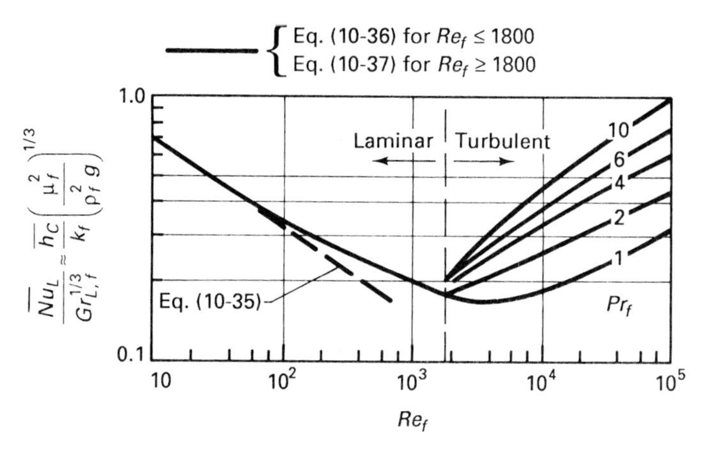

FIGURE 10–14 Correlations for film condensation heat transfer on a vertical surface.

This expression is shown in Fig. 10–14. It should be noted that since $\rho_f >> \rho_g$, the ordinate in this figure is generally expressed in terms of $\overline{h_C}/k_f\,[\mu_f^2/(\rho_f^2 g)]^{1/3}$, which is essentially equivalent to the more exact term $\overline{Nu_L}/Gr_{L,f}^{1/3}$.

For improved accuracy, which accounts for the effects of waves that form on the condensate film in the range $30 \lessapprox Re_f \lessapprox 1800$ and for turbulent flow conditions that occur for $Re_f \gtrapprox 1800$, the following correlations have been recommended for vertical surfaces:†

$$\overline{h_C} = \frac{Re_f}{1.08\,Re_f^{1.22} - 5.2}\left[\frac{\rho_f g(\rho_f - \rho_g)k_f^3}{\mu_f^2}\right]^{1/3} \tag{10–36a}$$

or

$$\overline{Nu_L} = \frac{Re_f}{1.08\,Re_f^{1.22} - 5.2}\,Gr_{L,f}^{1/3} \tag{1036b}$$

for $30 \lessapprox Re_f \lessapprox 1800$ by Kutateladze [30], and

$$\overline{h_C} = \frac{Re_f}{8750 + 58\,Pr_f^{-0.5}\,(Re_f^{0.75} - 253)}\left[\frac{\rho_f g(\rho_f - \rho_g)k_f^3}{\mu_f^2}\right]^{1/3} \tag{10–37a}$$

or

$$\overline{Nu_L} = \frac{Re_f}{8750 + 58\,Pr_f^{-0.5}\,(Re_f^{0.75} - 253)}\,Gr_{L,f}^{1/3} \tag{10–37b}$$

for $Re_f \gtrapprox 1800$ by Labuntsov [31]. These correlations are also shown in Fig. 10–14.

Equations (10–32) through (10–36) can also be used for film condensation on the outside of vertical tubes if the tube diameter is large compared to the film thickness δ. It should also be noted that because ρ_f is generally so much larger than ρ_g, the

† See Table 10–3 for explicit relations for Re_f.

term $\rho_f - \rho_g$ appearing in condensation-heat-transfer correlations can be approximated by ρ_f for many situations.

Inclined Surface For plates that are inclined by an angle θ with the vertical, the equations above can be utilized as a first approximation for film condensation on the upper surface by replacing the gravitational acceleration by its component parallel to the surface, $g \cos \theta$. However, these equations cannot be extended to inclined tubes.

Horizontal Tube or Sphere For laminar film condensation on the outside of a horizontal tube or sphere, $\overline{h_C}$ can be approximated by correlations of the same general form as Eq. (10–33); that is,

$$\overline{h_C} = C\left[\frac{\rho_f g(\rho_f - \rho_g)k_f^3 i'_{fg}}{\mu_f(T_{sat} - T_s)D}\right]^{1/4} \tag{10–38a}$$

or

$$\overline{Nu_D} = C\left[Gr_{D,f}Pr_f\left(\frac{1}{Ja_f} + 0.68\right)\right]^{1/4} \tag{10–38b}$$

where $C = 0.729$ for the cylinder and $C = 0.815$ for the sphere [16].

For the case of condensation on the outside of a bank of horizontal tubes, the heat flux to the lower tubes is reduced due to condensate falling from above tubes. A conservative estimate can be made for laminar condensation on a vertical row of N tubes by simply replacing D in the correlation for a single tube by ND.

EXAMPLE 10–5

Based on the fact that the temperature distribution within a laminar condensed liquid film on a vertical surface is essentially linear, develop an approximate expression for the thickness of the film in terms of h_C.

Solution

Objective Develop an approximation for δ_f in terms of h_C.

Schematic Condensed liquid film.

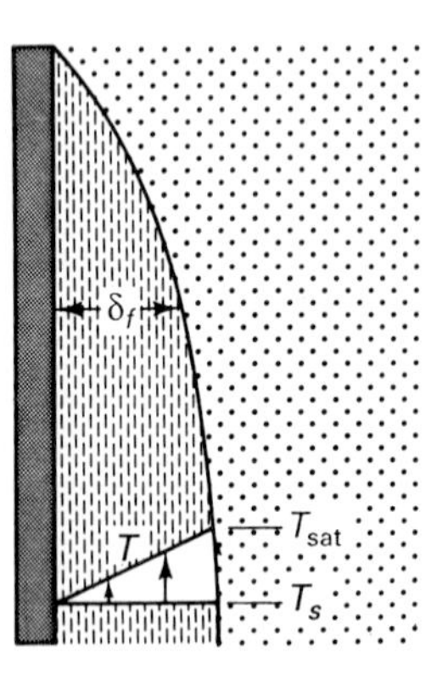

Assumptions/Conditions

condensation
wetted surface
laminar flow
uniform properties within liquid phase
standard conditions, except for two-phase fluid and significance of buoyancy

Analysis The condensate film thickness δ_f can be estimated by performing a simple thermal analysis. Assuming as a first approximation that the temperature distribution across the film is linear, we write

$$dq_c = dq_y|_0$$

$$h_C \, dA_s \, (T_s - T_{sat}) = -k_f \, dA_s \left.\frac{\partial T}{\partial y}\right|_0 = -k_f \, dA_s \frac{T_{sat} - T_s}{\delta_f}$$

or

$$\delta_f = \frac{k_f}{h_C}$$

Thus, we can approximate δ_f once relations are available for h_C.

EXAMPLE 10–6

Steam at a saturation temperature of 57°C condenses on the surface of a 1-m-long 2-cm-diameter vertical tube with a surface temperature of 48°C. Assuming film condensation, determine the total rate of heat transfer and the total mass flow rate of condensate over the entire tube length.

Solution

Objective Determine q_c and $\dot{m}$.

Schematic Film condensation on a vertical tube.

Steam at T_{sat} = 57°C

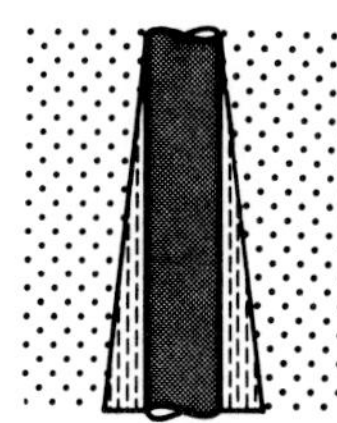

D = 2 cm
L = 1 m
T_s = 48°C

Assumptions/Conditions

condensation
wetted surface
uniform properties within each phase
standard conditions, except for two-phase fluid and significance of buoyancy
$\rho_g \ll \rho_f$

Properties Water at a film temperature of 325 K (Table A–C–3): $\rho_f = 1/v_f = 987$ kg/m^3, $i_{fg} = 2380$ kJ/kg, $k_f = 0.645$ W/(m °C), $c_{P,f} = 4.18$ kJ/(kg °C), $\mu_f = 5.28 \times 10^{-4}$ kg/(m s).

Analysis Assuming for the moment that the flow is laminar and that the flat plate correlations can be used, the film Reynolds number Re_f is given by Eq. (10–34),

$$Re_f = 3.77 \left[\frac{L(T_{\text{sat}} - T_s)}{\mu_f i'_{fg}}\right]^{3/4} \left[\frac{\rho_f g(\rho_f - \rho_g)k_f^3}{\mu_f^2}\right]^{1/4}$$

where

$$Ja_f = \frac{c_{P,f}(T_{\text{sat}} - T_s)}{i_{fg}} = \frac{[4.18 \text{ kJ/(kg °C)}](57°\text{C} - 48°\text{C})}{2380 \text{ kJ/kg}} = 0.0158$$

and

$$i'_{fg} = i_{fg}(1 + C_1 Ja_f) = 2380 \frac{\text{kJ}}{\text{kg}} [1 + 0.68\,(0.0158)] = 2410 \text{ kJ/kg}$$

Noting that $\rho_f \gg \rho_g$, we write

$$Re_f = 3.77 \left\{\frac{1 \text{ m}\,(57°\text{C} - 48°\text{C})}{[5.28 \times 10^{-4} \text{ kg/(m s)}](2410 \text{ kJ/kg})}\right\}^{3/4} \times \left\{\frac{(9.81 \text{ m/s}^2)(987 \text{ kg/m}^3)^2[0.645 \text{ W/(m °C)}]^3}{[5.28 \times 10^{-4} \text{ kg/(m s)}]^2}\right\}^{1/4} = 161$$

For this value of Re_f the flow is laminar and the wave effect should be relatively small, such that Eq. (10–34) is judged to be reasonable.

The condensate mass flow rate is immediately calculated from Eq. (10–24).

$$\dot{m} = \frac{p\mu_f Re_f}{4} = \frac{\pi(0.02 \text{ m})[5.28 \times 10^{-4} \text{ kg/(m s)}](161)}{4} = 1.34 \times 10^{-3} \text{ kg/s}$$

To calculate $\bar{h}_C$, we utilize Eq. (10–35a).

$$\bar{h}_C = 1.47\, Re_f^{-1/3} \left[\frac{g\rho_f(\rho_f - \rho_g)k_f^3}{\mu_f^2}\right]^{1/3}$$

$$= 1.47(161)^{-1/3}\left\{\frac{(9.81\ \text{m/s}^2)(987\ \text{kg/m}^3)^2[0.645\ \text{W/(m °C)}]^3}{[5.28\times 10^{-4}\ \text{kg/(m s)}]^2}\right\}^{1/3}$$

$$= 5700\ \text{W/(m}^2\ \text{°C)}$$

The total rate of heat transfer over the tube is

$$q_c = \bar{h}_C A_s(T_{sat} - T_s) = 5700\ \frac{\text{W}}{\text{m}^2\ \text{°C}}\ \pi(1\ \text{m})(0.02\ \text{m})(57\text{°C} - 48\text{°C})$$

$$= 3.22\ \text{kW}$$

As seen in Prob. 10–29 the local film thickness δ_f for this situation is approximated by

$$\delta_f = 1.52 \times 10^{-4}\ (x/L)^{1/4}\ \text{m}$$

Because δ_f is much less than the tube diameter D, we are justified in utilizing the flat-plate correlation.

Referring back to Example 10–3 on boiling heat transfer, we see that for the same saturation temperature and temperature difference of 9°C ($\simeq$16°F), the heat-transfer rate from the vertical surface is approximately 2 times larger for this laminar filmwise condensation than for the boiling. However, at high pressures, boiling heat transfer is much more effective than laminar film condensation.

EXAMPLE 10–7

Referring back to Example 10–6 in which 1.34×10^{-3} kg/s of condensate was produced on a 1-m-long vertical 2-cm-diameter tube, determine the length of a horizontal 2-cm-diameter tube which is required to produce the same condensate mass flow rate.

Solution

Objective Determine the required length L.

Schematic Film condensation on a horizontal tube.

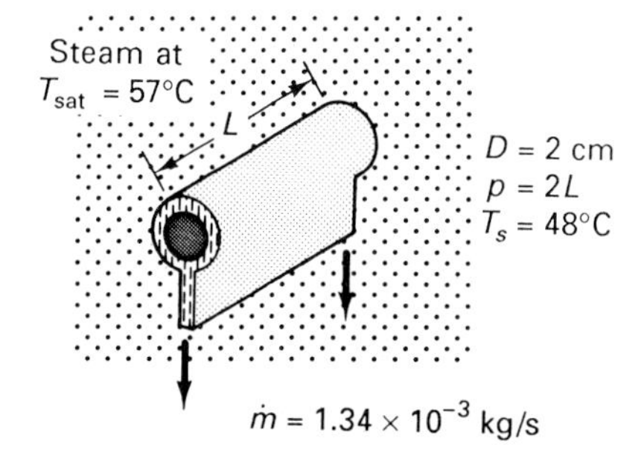

Assumptions/Conditions

condensation
wetted surface
uniform properties within each phase
standard conditions, except for two-phase fluid and significance of buoyancy
$\rho_g << \rho_f$

Properties Water at a film temperature of 325 K (Table A–C–3): $\rho_f = 1/v_f = 987$ kg/m^3, $i_{fg} = 2380$ kJ/kg, $k_f = 0.645$ W/(m °C), $c_{P,f} = 4.18$ kJ/(kg °C), $\mu_f = 5.28 \times 10^{-4}$ kg/(m s).

Analysis First, by comparing Eqs. (10–33) and (10–38), we have

$$\frac{\bar{h}_{C,\text{horiz}}}{\bar{h}_{C,\text{vert}}} = \frac{0.729}{0.943}\left(\frac{L}{D}\right)^{1/4} = 0.773\left(\frac{L}{D}\right)^{1/4}$$

Thus, we see immediately that laminar film condensation is more effective for horizontal tubes if the length L is greater than $2.8D$.

Because $\dot{m}$ is known, we will utilize Eq. (10–30) to solve for the length.

$$L = \frac{A_s}{\pi D} = \frac{\dot{m} i'_{fg}}{\pi D \bar{h}_C (T_{\text{sat}} - T_s)} \tag{a}$$

First, assuming laminar flow and setting $i'_{fg} = i_{fg}(1 + C_1\, Ja_f) = 2410$ kJ/(kg °C), Eq. (10–38) is used to calculate $\bar{h}_C$.

$$\bar{h}_C = 0.729\left\{\frac{(9.81\text{ m/s}^2)(987\text{ kg/m}^3)^2[0.645\text{ W/(m °C)}]^3(2410\text{ kJ/kg})}{[5.28 \times 10^{-4}\text{ kg/(m s)}](57°\text{C} - 48°\text{C})(0.02\text{ m})}\right\}^{1/4}$$

$$= 1.17 \times 10^4\text{ W/(m}^2\text{ °C)}$$

Notice that this coefficient is over twice as large as the coefficient obtained for the vertical tube! Substituting into Eq. (a), we have

$$L = \frac{(1.34 \times 10^{-3}\text{ kg/s})(2410\text{ kJ/kg})}{\pi(0.02\text{ m})[1.17 \times 10^4\text{ W/(m}^2\text{ °C)}](57°\text{C} - 48°\text{C})} = 0.488\text{ m}$$

To determine whether the flow is actually laminar, we calculate the film Reynolds number. Utilizing Eq. (10–24),

$$Re_f = \frac{4\dot{m}}{p\mu_f}$$

with $p = 2L$ for this horizontal tube geometry, we obtain

$$Re_f = \frac{4(1.34 \times 10^{-3}\text{ kg/s})}{2(0.488\text{ m})[5.28 \times 10^{-4}\text{ kg/(m s)}]} = 10.4$$

Thus, the flow is laminar and Eq. (10–38) does indeed apply.

To conclude, we are able to use a tube that is less than half as long as was required for the vertical arrangement. Consequently, horizontal tube arrangements are often used in condenser design. However, it should be noted that $\bar{h}_C$ can be greater for turbulent film condensation on vertical tubes than for laminar condensation on horizontal tubes.

Enhancement of Film Condensation

The key to improving film condensation heat transfer for a specified fluid, surface temperature T_s, and saturation temperature T_{sat} is to decrease the liquid film thickness, which acts as a thermal resistance, and to stimulate turbulence. To accomplish these objectives, rough surfaces and finned tubes have been designed that improve the condensate mass flow $\dot{m}$. Figure 10–15 shows the heat-transfer coefficient for film condensation on horizontal smooth, finned, and rough surfaces. The high heat fluxes obtained by the Thermoexcel-C saw tooth surface have been attributed to a superior condensate dropping ability [32].

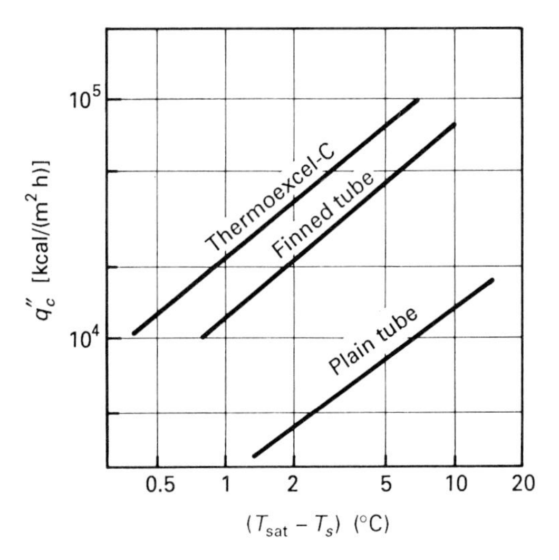

FIGURE 10–15 Condensation-heat-transfer performance of several surfaces (Arai et al. [19]).

For more information on the enhancement of film condensation for both internal and external flow systems, the review article by Webb [18] is suggested.

10–4–2 Dropwise Condensation

To accomplish dropwise condensation, the fluid surface tension σ must be greater than the critical surface tension σ_{cr}. Representative values of σ_{cr} are listed in Table 10–2 for several surfaces. Referring to Table A–C–6, we conclude that dropwise

TABLE 10–2 Critical surface tensions for representative solid materials

Material	$\sigma_{cr} \times 10^3$ N/m
Nylon	46
Polyvinyl chloride	39
Polystyrene	33
Polyethylene	31
Kel-F	31
Teflon	18
Platinum with monolayer of	
perfluorobutyric acid	10
perfluorolauric acid	6

Source: Sharfin and Zisman [27].

condensation can be achieved on all of these surfaces for water at saturation temperatures below about 160°C.

Experimental data for dropwise condensation of steam are shown in Fig. 10–16. Note the very high coefficients of heat transfer that are achieved. As a matter of fact, dropwise condensation is one of the most effective of all known heat-transfer mechanisms. Consequently, efforts are continuously being made to harness this dy-

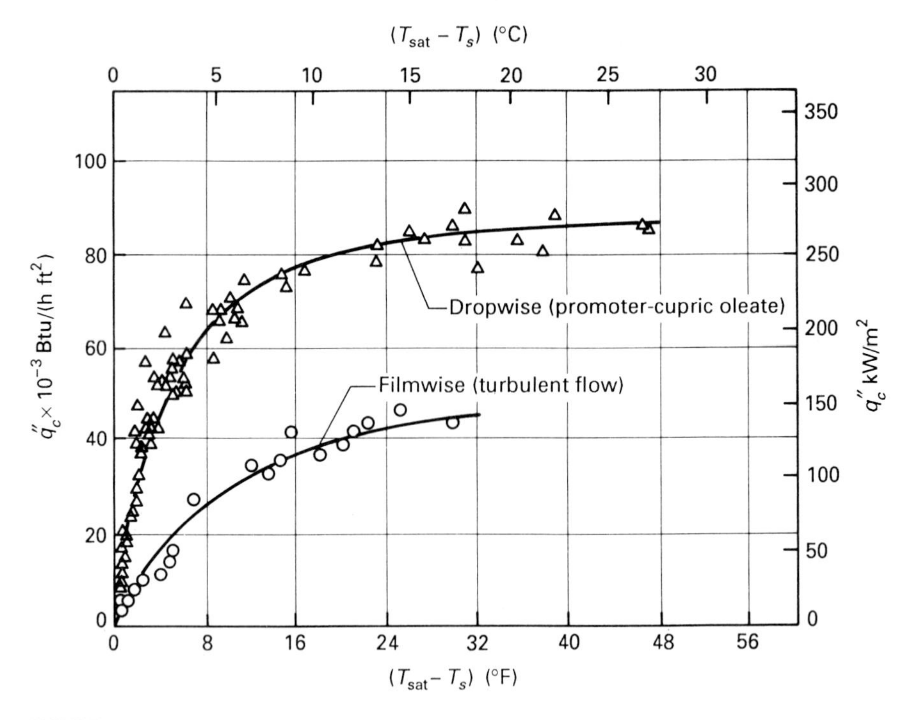

FIGURE 10–16 Heat flux for condensation of steam at atmospheric pressure on a vertical copper surface (Welch and Westwater [33]).

namic mechanism for industrial use. Various methods have been utilized to attempt to achieve consistent dropwise condensation, such as the use of surface coatings, which reduce the value of σ_{cr}. Surface coatings and promoters that are sometimes used include Teflon, noble metals, silicones, various waxes, fatty acids (e.g., oleic and stearic), and long-chain hydrocarbons. However, these and most other nonwetting agents do not retain their effectiveness for long lengths of time because of fouling, oxidation, and removal. As an exception, fluoroplastic Emralon 300™ has shown promise as a coating, with a lifetime expectancy of as much as 20 years anticipated [34].

Experimental evidence indicates that the dropwise condensation heat flux can be correlated in terms of $d\sigma/dT$, T_{sat}, the properties of the saturated liquid, and other parameters featured in Eq. (10–1); that is,

$$q_c'' = \text{fn}\,[g(\rho_f - \rho_g), \sigma, T_s - T_{sat}, i_{fg}, \rho_f, \mu_f, c_{P,f}, k_f, \boldsymbol{L}, |d\sigma/dT|, T_{sat}] \qquad (10\text{–}39)$$

This equation involves *twelve* parameters and *five* fundamental dimensions. Using dimensional analysis, an equation involving *seven* dimensionless groups is obtained, which is represented by

$$\overline{Nu}_L = \text{fn}\,[Pr_f, Ja_f, Bo_L, S_{L,f}, Ma_{L,f}, T_{sat}/\Delta T_e] \qquad (10\text{–}40)$$

where Nu_L is the Nusselt number, Pr_f is the Prandtl number, Ja_f is the Jacob number, Bo_L is the Bond number,

$$S_{L,f} = \frac{\sigma}{L\rho_f\, i_{fg}} \qquad (10\text{–}41)$$

is a *surface energy parameter* that represents the *ratio of surface energy per unit area to the energy flux required to overcome this energy*, and

$$Ma_{L,f} = \frac{L|d\sigma/dT|\, T_{sat}}{\mu_f\, \nu_f} \qquad (10\text{–}42)$$

is the *modified Marangoni number*, which represents the *ratio of the imbalance in surface tension forces to liquid tangential viscous forces*.

A correlation for dropwise condensation of steam and ethylene glycol that involves six of these seven dimensionless parameters has been proposed by Peterson and Westwater [35], which is represented by

$$\frac{2\sigma T_{sat}\,\bar{h}_C}{i_{fg}\rho_f k_f(T_{sat} - T_s)} = 1.46 \times 10^{-6} \left[\frac{k_f(T_{sat} - T_s)}{\mu_f i_{fg}}\right]^{-1.63} Pr_f^{1/2} \left[\frac{2\sigma|d\sigma/dT|T_{sat}}{\mu_f^2\, i_{fg}}\right]^{1.16} \qquad (10\text{–}43a)$$

or

$$\overline{Nu}_L = 1.63 \times 10^{-6}\, Pr_f^{2.13}\, Ja_f^{-1.63}\, S_{L,f}^{0.16}\, Ma_{L,f}^{1.16}\, \frac{\Delta T_e}{T_{sat}} \qquad (10\text{–}43b)$$

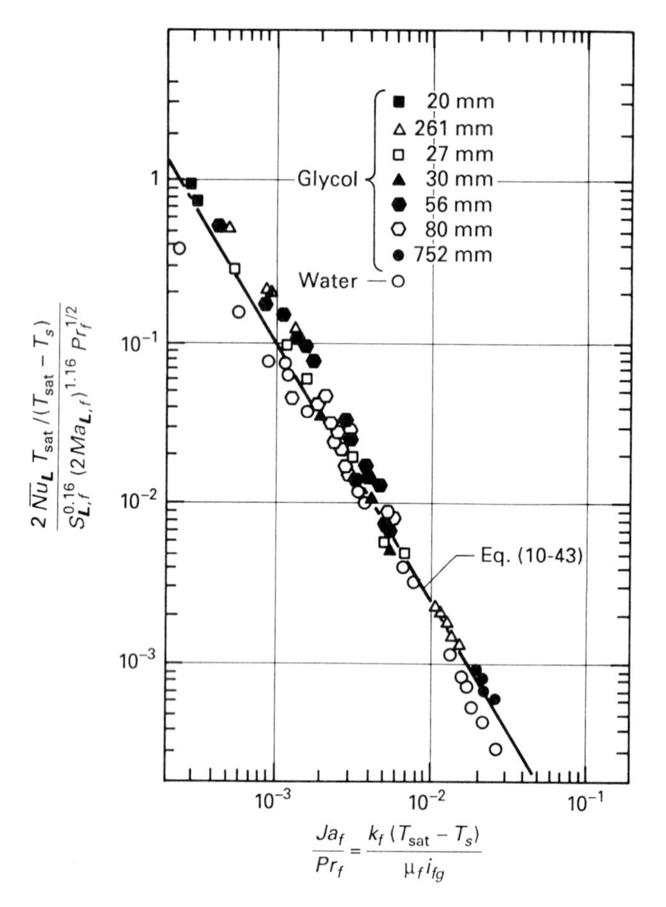

FIGURE 10–17 Convection correlation for dropwise condensation on vertical surface (Peterson and Westwater [35]).

for $1.65 \lesssim Pr_f \lesssim 23.6$ and $7.8 \times 10^{-4} \lesssim (2\ Ma_{L,f}\, S_{L,f}) \lesssim 2.65 \times 10^{-2}$. This correlation is compared with experimental data in Fig. 10–17. By inspecting this correlation, we note that the modified Marangoni number $Ma_{L,f}$ accounts for the effect of surface tension, with the Bond number Bo_L not appearing, and that the characteristic length L actually cancels out.

EXAMPLE 10–8

Steam at a saturation temperature of 57°C condenses on the surface of a 1-m-long 2-cm-diameter vertical tube with a surface temperature of 48°C. Assuming that the surface has been coated with an effective nonwetting agent such that dropwise condensation is achieved, determine the total rate of heat transfer over the length of the tube.

Solution

Objective Determine q_c, assuming dropwise condensation.

Schematic Dropwise condensation on a vertical tube.

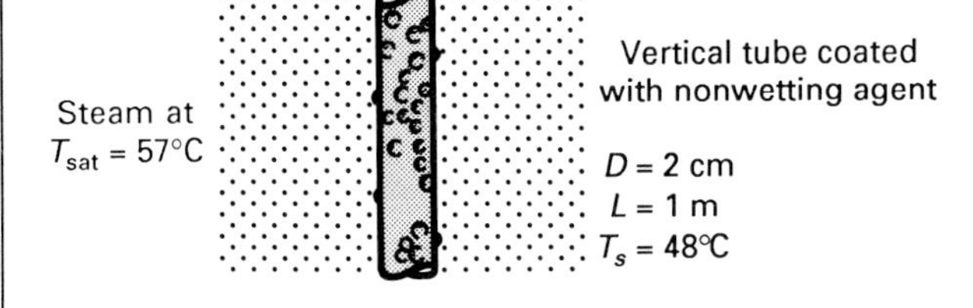

Assumptions/Conditions

condensation
nonwetted surface
uniform properties within each phase
standard conditions, except for two-phase fluid and significance of buoyancy
$\rho_g \ll \rho_f$

Properties Water at T_{sat} = 57°C = 330 K (Table A–C–3): $\rho_f = 1/v_f = 984$ kg/m^3, $i_{fg} = 2370$ kJ/kg, $k_f = 0.650$ W/(m °C), $c_{P,f} = 4.18$ kJ/(kg °C), $\mu_f = 4.89 \times 10^{-4}$ kg/(m s), $Pr_f = 3.15$, $\sigma = 0.0666$ N/m, and

$$\frac{d\sigma}{dT} = \frac{1}{2}\left(\frac{0.0658 - 0.0666}{5} + \frac{0.0666 - 0.0675}{5}\right)$$

$$= -1.7 \times 10^{-4} \text{ N/(m K)}$$

Analysis To make use of the Peterson and Westwater correlation for dropwise condensation, Eq. (10–43), we compute the pertinent dimensionless parameters as follows:

Jakob number

$$Ja_f = \frac{c_{P,f}(T_{sat} - T_s)}{i_{fg}} = \frac{[4.18 \text{ kJ/(kg °C)}]\,(57°\text{C} - 48°\text{C})}{2370 \text{ kJ/kg}} = 0.0159$$

Surface-energy parameter

$$S_{L,f} = \frac{\sigma}{L\rho_f\, i_{fg}} = \frac{0.0666 \text{ N/m}}{1 \text{ m }(984 \text{ kg/m}^3)(2370 \text{ kJ/kg})} = 2.86 \times 10^{-11}$$

Modified Marangoni number

$$Ma_{L,f} = \frac{\rho_f L|d\sigma/dT|T_{\text{sat}}}{\mu_f^2} = \frac{(984 \text{ kg/m}^3)(1 \text{ m})[1.7 \times 10^{-4} \text{ N/(m K)}](330 \text{ K})}{[4.89 \times 10^{-4} \text{ kg/(m s)}]^2}$$

$$= 2.31 \times 10^8$$

Since Pr_f and the product $S_{L,f}\,Ma_{L,f} = 2.86 \times 10^{-11}(2.31 \times 10^8) = 0.00661$ both fall in the appropriate range, we substitute into Eq. (10–43) to obtain

$$\overline{Nu_L} = 1.63 \times 10^{-6}\,(3.15)^{2.13}\,(0.0159)^{-1.63}\,(2.86 \times 10^{-11})^{0.16}$$

$$\times\ (2.31 \times 10^8)^{1.16}\left(\frac{9}{330}\right) = 45{,}200$$

It follows that

$$\bar{h}_C = \frac{\overline{Nu_L}}{L/k_f} = \frac{45{,}200}{1 \text{ m}/[0.65 \text{ W/(m °C)}]} = 29{,}400 \text{ W/(m}^2\text{ °C)}$$

and

$$q_c = \bar{h}_C A_s (T_{\text{sat}} - T_s)$$

$$= 29{,}400 \frac{\text{W}}{\text{m}^2\text{ °C}}[\pi(1 \text{ m})(0.02 \text{ m})](57\text{°C} - 48\text{°C}) = 16.6 \text{ kW}$$

Comparing this result with the value obtained for laminar film condensation in Example 10–6, we find an increase in $\bar{h}_C$ or q_c by a factor of just over 5.

EXAMPLE 10–9

The *heat pipe* is a device that utilizes evaporation and condensation to transfer heat extremely effectively. A basic heat pipe system is shown in Fig. E10–9a. A heat pipe consists of a closed pipe lined with a wicking material. A condensible fluid is contained within the pipe, with gas filling the hollow core and liquid permeating the wicking material. When one end of the pipe is exposed to a temperature above the

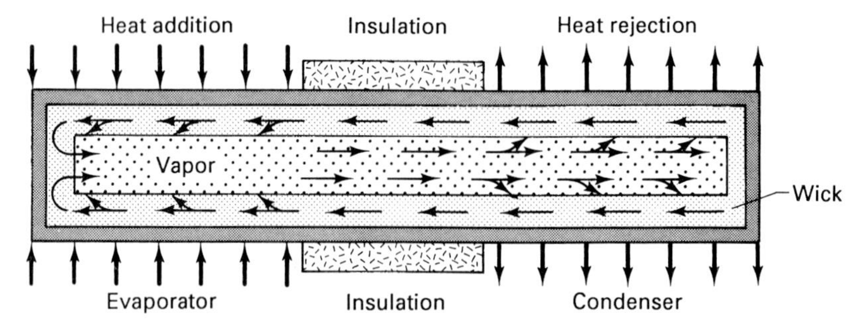

FIGURE E10–9a Basic heat pipe.

saturation temperature T_{sat} of the fluid, and the other end is exposed to a temperature below T_{sat}, the fluid within the pipe circulates as shown in Fig. E10–9a. This circulation is caused by (1) the evaporation of liquid in the heated portion of the wick, (2) the condensation of fluid back into the cooled section of the wick, and (3) capillary liquid flow within the wick.

Because of the effectiveness of the vaporization and condensation mechanisms and the capillary pumping action, properly designed heat pipes are capable of transferring tens and even hundreds of times as much heat as can be transferred in solid metal bars of equal size. To illustrate the temperature distribution and heat flux in a sodium–stainless steel heat pipe is shown in Fig. E10–9b. Let us compare the heat transfer in this system with conduction heat transfer in a solid stainless steel 1-in.-diameter 18.5-in.-long rod.

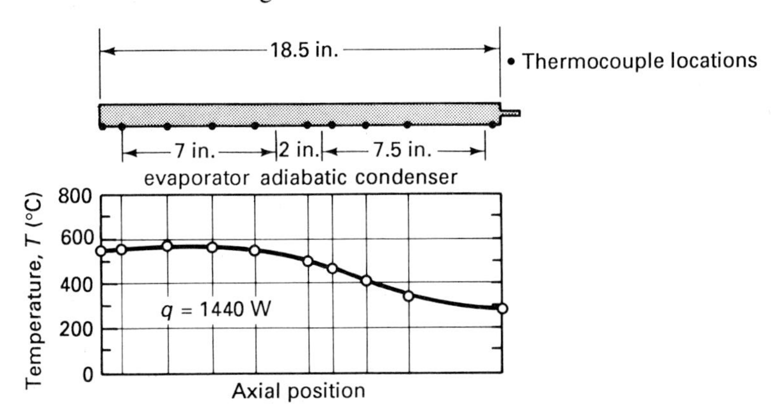

FIGURE E10–9b Axial temperature distributions for sodium heat pipe (Dzakowic et al. [36]).

Solution

Objective For purpose of comparison, calculate q in a solid rod with $T_1 = 550°\text{C}$ and $T_2 = 300°\text{C}$.

Properties Stainless steel (AISI 302) at 400°C (Table A–C–1): $k = 17.3$ W/(m °C).

Analysis Referring to Fig. E10–9b, the heat pipe transfers 1440 W across a temperature difference of approximately 250°C. The rate of heat transfer in a stainless steel bar with this temperature drop is

$$q = \frac{kA}{L}(T_1 - T_2) = 17.3 \frac{\text{W}}{\text{m °C}} \frac{\pi(1 \text{ in.})^2}{4(18.5 \text{ in.})}(250°\text{C})$$

$$= 184 \frac{\text{W in.}}{100 \text{ cm}} \frac{2.54 \text{ cm}}{\text{in.}} = 4.66 \text{ W}$$

Comparing this rate of heat transfer with the 1440 W transferred in the heat pipe, we have a factor of over 300.

10–5 SUMMARY

In this chapter we have dealt with issues pertaining to the identification of the basic physical features associated with boiling and condensation, and with the presentation of characterizing parameters and convection correlations that can be used in the practical analysis approach. The convection correlations and dimensionless parameters featured in this chapter are summarized in Tables 10–3 and 10–4.

As we have seen, large heat fluxes can be established by boiling and condensation processes. However, we have also learned of inherent dangers associated with boiling heat transfer applications involving direct heating by electrical, fossil-fuel, or nuclear energy sources. In such cases, care must be taken to avoid operation beyond the safe nucleate boiling regimes. In the case of condensation heat transfer, we generally expect and design for film condensation, since the surface tension σ of most fluids is lower than the critical surface tension σ_{cr}. However, some success has been achieved in the development of surface coatings and promoters that make possible the decrease in σ_{cr} and the establishment of dropwise condensation, which results in particularly high heat fluxes.

TABLE 10–3 Summary of convection correlations for boiling and condensation

Condition	*Correlation*	*Identification*
Nucleate boiling		
Heat flux	$q''_c = \mu_f i_{fg} \left[\frac{g(\rho_f - \rho_g)}{\sigma}\right]^{1/2} \left[\frac{c_{P,f}(T_s - T_{sat})}{C_{sf}\, i_{fg}\, Pr_f^s}\right]^3$	Eq. (10–9b) C_{sf}, s—Table 10–1 Rohsenow [6]
	or	
	$\overline{Nu}_L = Ja_f^2\, Bo_L^{1/2}\, Pr_f^{1-3s}\, C_{sf}^{-1}$	Eq. (10–9c)
Maximum heat flux	$q''_{max} = \frac{\pi}{24} \rho_g i_{fg} \left[\frac{\sigma g(\rho_f - \rho_g)}{\rho_g^2}\right]^{1/4} \left(\frac{\rho_f + \rho_g}{\rho_f}\right)^{1/2}$	Eq. (10–10) or Eq. (10–11) Zuber [12]
	or	
	$\frac{q''_{max}}{i_{fg}} \frac{L c_{P,f}}{k_f} = \frac{\pi}{24} \frac{Gr_{L,f}^{1/2} Pr_f}{Bo_L^{1/4}} \left[\frac{\rho_g(\rho_f + \rho_g)}{\rho_f^2}\right]^{1/2}$	
Minimum heat flux	$q''_{min} = 0.09\, \rho_g i_{fg} \left[\frac{\sigma g(\rho_f - \rho_g)}{(\rho_f + \rho_g)^2}\right]^{1/4}$	Eq. (10–12) Zuber [12]
	or	
	$\frac{q''_{min}}{i_{fg}} \frac{L\, c_{P,f}}{k_f} = 0.09 \frac{Gr_{L,f}^{1/2} Pr_f}{Bo_L^{1/4}} \left[\frac{\rho_g^2}{\rho_f(\rho_f + \rho_g)}\right]^{1/2}$	

TABLE 10-3 *(Continued)*

Condition	*Correlation*	*Identification*
Film Boiling horizontal tube $C = 0.62$ sphere $C = 0.67$	$\bar{h}_B = C\left[\dfrac{g(\rho_f - \rho_g)\, i_{fg}\,(1 + 0.4\, Ja_g)\, k_g^3}{\nu_g(T_s - T_{sat})D}\right]^{1/4}$ or	Eq. (10–13) Zuber [12]
Properties of vapor evaluated at film temperature	$\overline{Nu}_D = C\left[Gr_{D,g}\, Pr_g\left(\dfrac{1}{Ja_g} + 0.4\right)\right]^{1/4}$	Eq. (10–15)
with radiation	$\bar{h}_R = \dfrac{\epsilon_s\sigma(T_s^4 - T_{sat}^4)}{T_s - T_{sat}}$	Eq. (10–16)
$\bar{h}_R < \bar{h}_B$	$\bar{h}_{B,R} = \bar{h}_B + \dfrac{3}{4}\bar{h}_R$	Eq. (10–17)
	$\bar{h}_{B,R} = \bar{h}_B\left(\dfrac{\bar{h}_B}{\bar{h}_{B,R}}\right)^{1/3} + \bar{h}_R$	Eq. (10–18)
Film condensation Vertical surface† Laminar flow	$h_C = 0.707\left[\dfrac{\rho_f g(\rho_f - \rho_g)k_f^3 i_{fg}(1 + 0.68\, Ja_f)}{\mu_f(T_{sat} - T_s)x}\right]^{1/4}$	Eq. (10–32) Nusselt [29]
	$\bar{h}_C = 0.943\left[\dfrac{\rho_f g(\rho_f - \rho_g)k_f^3 i_{fg}(1 + 0.68\, Ja_f)}{\mu_f(T_{sat} - T_s)L}\right]^{1/4}$ or	Eq. (10–33a)
	$\overline{Nu}_L = 0.943\left[Gr_{L,f}\, Pr_f\left(\dfrac{1}{Ja_f} + 0.68\right)\right]^{1/4}$ or	Eq. (10–33b)
	$\bar{h}_C = 1.47\, Re_f^{-1/3}\left[\dfrac{\rho_f g(\rho_f - \rho_g)k_f^3}{\mu_f^2}\right]^{1/3}$ or	Eq. (10–35a)
	$\overline{Nu}_L = 1.47\, Re_f^{-1/3}\, Gr_{L,f}^{1/3}$	Eq. (10–35b)
$30 < Re_f < 1800$	$\bar{h}_C = \dfrac{Re_f}{1.08\, Re_f^{1.22} - 5.2}\left[\dfrac{\rho_f g(\rho_f - \rho_g)k_f^3}{\mu_f^2}\right]^{1/3}$ or	Eq. (10–36a) Kutateladze [30]
	$\overline{Nu}_L = \dfrac{Re_f}{1.08\, Re_f^{1.22} - 5.2}\, Gr_{L,f}^{1/3}$ or	Eq. (10–36b)
	$Re_f = \left\{\dfrac{1}{1.08}\left[5.2 + \dfrac{4L(T_{sat} - T_s)}{\mu_f\, i'_{fg}}\, C_F^{1/3}\right]\right\}^{1/1.22}$	

TABLE 10–3 (*Continued*)

Condition	*Correlation*	*Identification*
Turbulent flow $Re_f > 1800$	$\bar{h}_C = \dfrac{Re_f\, C_F^{1/3}}{8750 + 58\, Pr_f^{-1/2}\,(Re_f^{3/4} - 253)}$	Eq. (10–37a) Labuntsov [31]
	or	
	$\overline{Nu}_L = \dfrac{Re_f}{8750 + 58\, Pr_f^{-1/2}\,(Re_f^{3/4} - 253)}\, Gr_{L,f}^{1/3}$	Eq. (10–37b)
	or	
	$Re_f = \left\{253 + \dfrac{Pr_f^{1/2}}{58}\left[\dfrac{4L(T_{\text{sat}} - T_s)C_F^{1/3}}{\mu_f i'_{fg}} - 8750\right]\right\}^{4/3}$	
Inclined plates	Use correlations for vertical surfaces with g replaced by $g \cos\theta$	
Horizontal tubes $C = 0.729$	$\bar{h}_C = C\left[\dfrac{\rho_f g(\rho_f - \rho_g)k_f^3\,(1 + 0.68\, Ja_f)}{\mu_f(T_{\text{sat}} - T_s)D}\right]^{1/4}$	Eq. (10–38a) Lienhard et al. [16]
and	or	
Spheres $C = 0.815$	$\overline{Nu}_D = C\left[Gr_{D,f}Pr_f\left(\dfrac{1}{Ja_f} + 0.68\right)\right]^{1/4}$	Eq. (10–38b)
Dropwise condensation	$\overline{Nu}_L = 1.63 \times 10^{-6}\, Pr_f^{2.13}\, Ja_f^{-1.63}\, S_{L,f}^{0.16}\, Ma_{L,f}^{1.16}\, \dfrac{\Delta T_e}{T_{\text{sat}}}$	Eq. (10–43b) Peterson and Westwater [35]

† $C_F = \rho_f g(\rho_f - \rho_g)k_f^3/\mu_f^2$.

TABLE 10-4 Dimensionless parameters for convection heat transfer: Boiling and condensation

Dimensionless group	*Symbol*	*Definition*	*Interpretation*
Bond number	Bo_L	$\frac{g(\rho_f - \rho_g)L^2}{\sigma}$	$\frac{\text{gravitational forces}}{\text{surface tension forces}}$
Jakob number	Ja	$\frac{c_P(T_s - T_{sat})}{i_{fg}}$	$\frac{\text{maximum sensible energy absorbed}}{\text{latent heat absorbed}}$
Grashof number	Gr_L	$\frac{g(\rho_f - \rho_g)L^3}{\rho\nu^2}$	Relative significance of buoyant and viscous effects for two-phase flow
Nusselt number†	Nu_L	$\frac{hL}{k}$	$\frac{\text{convection for fluid in motion}}{\text{conduction for motionless fluid layer}}$
Prandtl number†	Pr	$\frac{\nu}{\alpha}$	relative effectiveness of momentum and energy transfer by molecular diffusion within the boundary layer
Reynolds number (for condensation film)	Re_f	$\frac{4\dot{m}}{p\mu_f}$	$\frac{\text{inertia forces}}{\text{viscous forces}}$
Surface energy parameter	$S_{L,f}$	$\frac{\sigma}{L\rho_f i_{fg}}$	$\frac{\text{surface energy per unit area}}{\text{energy flux to overcome surface energy}}$
Modified Marangoni number	$Ma_{L,f}$	$\frac{L\lvert d\sigma/dT\rvert T_{sat}}{\mu_f \nu_f}$	$\frac{\text{imbalance in surface tension forces}}{\text{liquid tangential viscous forces}}$

† Dimensionless parameters also used in analysis of forced convection.

CHAPTER 11

CONVECTION HEAT TRANSFER: PRACTICAL ANALYSIS— HEAT EXCHANGERS

11–1 INTRODUCTION

Heat exchangers are devices that transfer heat between fluids at different temperatures. Heat exchangers are used in a wide range of industrial and commercial applications. Examples are found in the power, air-conditioning, refrigeration, cryogenics, heat recovery, process, automotive, aircraft, marine, and manufacturing industries, as well as in many products available in the marketplace.

In this chapter we will describe the more common types of heat exchangers that are commercially available, discuss the important functions of evaluation and design, and utilize the principles presented in Chaps. 6 through 10 to develop the practical approach to heat exchanger analysis, which features the effectiveness, *LMTD*, and efficiency methods. With this background, specific attention is given to the analysis of double-pipe, shell-and-tube, and crossflow heat exchangers and multipass heat exchanger networks. Finally, to complete our study, brief consideration is given to the computer analysis of heat exchangers.

11–2 TYPES OF HEAT EXCHANGERS

As indicated by Shah [1], a heat exchanger consists of (1) active heat exchanging elements such as a core or a matrix containing the heat transfer surface, and (2) passive fluid distribution elements such as headers, manifolds, tanks, inlet and outlet nozzles or pipes, and seals. The surface of the exchanger core, which is in direct contact with fluids and through which heat is transferred, is referred to as the *heat transfer surface*. The various types of heat exchangers that are commercially available may be categorized according to the geometric configuration of the heat transfer surface and flow arrangement as well as certain other considerations.

11–2–1 Double-Pipe Heat Exchangers

A fin tube *double-pipe* heat exchanger module is shown in Fig. 11–1. In the hairpin unit shown in this figure, two double-pipes are joined at one end by a U-bend and a return bend housing on the shell side. As illustrated in Fig. 11–2, double-pipe heat exchangers involve combinations of tube and annular flow, with the tube and annular flows being in the same direction for *parallel-flow* (concurrent) arrangements and in opposing directions for *counterflow*. As we shall see, heat transfer is most effectively achieved by counterflow. However, this is at the expense of higher wall temperatures. Thus, counterflow operation is normally selected, unless the higher wall temperatures associated with this arrangement become a factor.

Whereas double-pipe heat exchangers were originally used as small-size classical counterflow heat exchangers, present applications involve the use of double-pipe hairpin units with multiple plain or longitudinal finned tubes. Nevertheless, because of the relatively small surface area and cross-sectional flow area, double-pipe heat exchangers are generally reserved for applications that require low to moderate heat

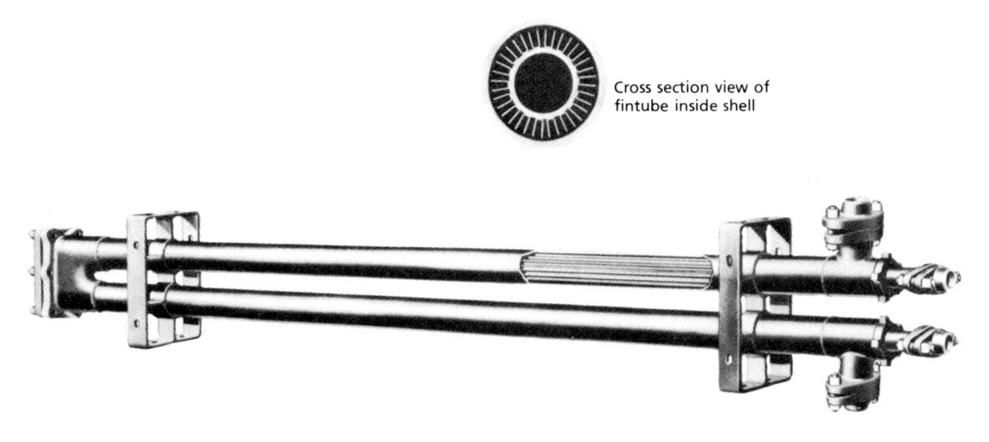

FIGURE 11–1 Double-pipe hairpin heat exchanger module. (Courtesy of Brown Fintube Company, Houston, Texas.)

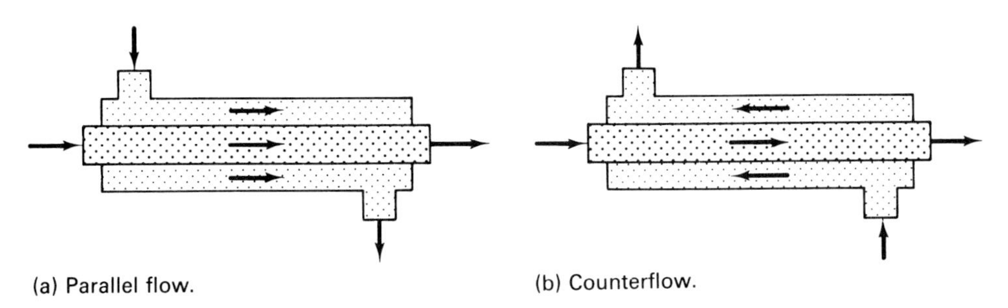

(a) Parallel flow. (b) Counterflow.

FIGURE 11–2 Flow arrangements in double-pipe heat exchangers.

transfer rates. As a rule of thumb, applications for which double-pipe heat exchangers are economical normally involve heat transfer surface areas less than 50 m^2 and heat transfer rates less than 1000 kW. As pointed out by Taborek [2], situations in which double-pipe heat exchangers are selected include (1) applications requiring pure counterflow due to large temperature cross,† (2) applications involving a high pressure stream, (3) applications with uncertainty of performance, because the flexibility of modular construction permits the addition of units, (4) applications with widely different flow rates, because better utilization of available pressure drop is often possible through series–parallel arrangements, and (5) applications with low heat transfer coefficient for one stream, because of the availability of longitudinal finned tubes which provide a large increase in heat transfer surface area.

In applications requiring the transfer of high rates of heat, other kinds of heat exchangers are utilized. Two types that provide sufficiently large surface areas are shell-and-tube heat exchangers and crossflow heat exchangers.

11–2–2 Shell-and-Tube Heat Exchangers

Typical shell-and-tube heat exchangers are shown in Figs. 6–3 and 11–3. Shell-and-tube heat exchangers are classified according to the shell type and the number of tube passes and certain other geometric characteristics. The basic components of shell-and-tube heat exchangers are indicated in Fig. 11–3.

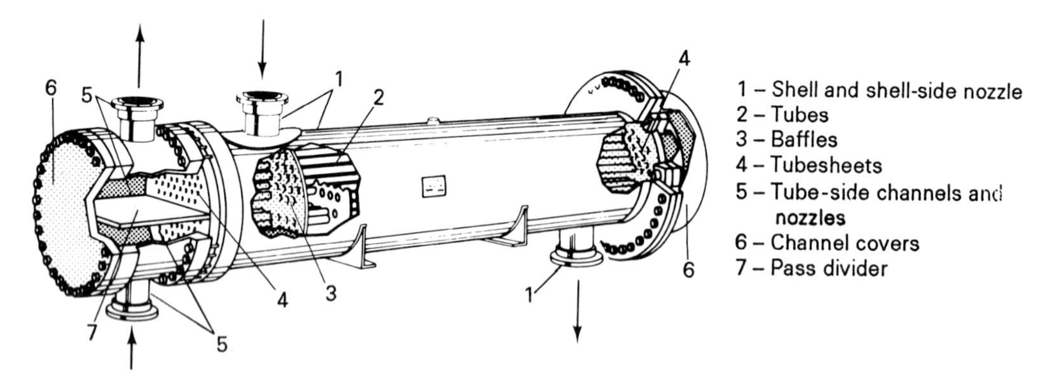

FIGURE 11–3 Typical shell-and-tube heat exchanger: TEMA-E shell with two tube passes and fixed tubesheets. (Courtesy of Enerquip.)

The standard shell type shown in Fig. 11–3 is TEMA E, with entry and outlet nozzles at the opposite ends.‡ Other types of shells are sketched in Fig. 11–4. The selection of shell type to a large extent depends on the shell-side pressure drop requirements.

† The term temperature cross indicates that the outlet temperature of the cold fluid is higher than the outlet temperature of the hot fluid.

‡ TEMA stands for Tubular Exchanger Manufacturers Association.

	Shell types	Applications	Limitations
E	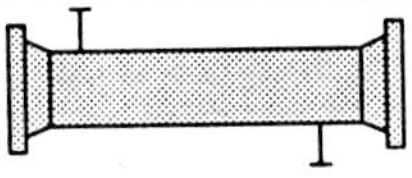 One-pass shell	Normally most practical arrangement.†	
F	 Two-pass shell with longitudinal baffle	Because of economic considerations, commonly used in applications that would require two E shells in series.	Requires careful design and construction because of tendencies for thermal leakage. Presents problems for removing or replacing tube bundle. Results in higher pressure drops.‡
G	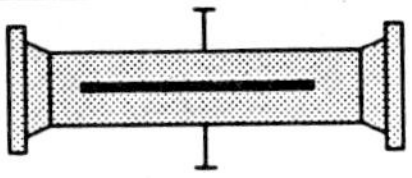 Split flow	Used primarily for reboilers and occassionally for no-phase change flow.	Relatively higher thermal efficiency than E shell.‡
H	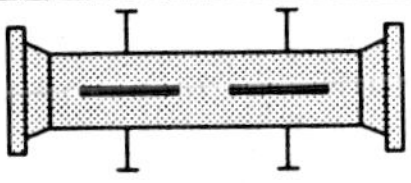 Double split flow	Used in place of G shell for applications with long tube lengths.	Relatively higher thermal efficiency than E shell.‡
J	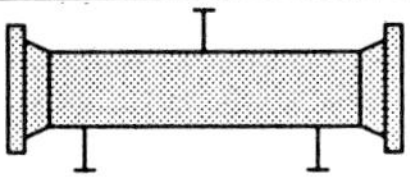 Divided flow	Used for applications requiring low pressure drop.	Relatively higher thermal efficiency than E shell.
K	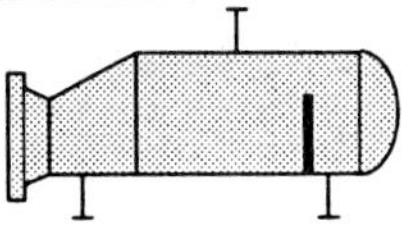 Kettle-type reboiler	Used as a vaporizer.	Sometimes characterized by low circulation in bundle, which can increase fouling.
X	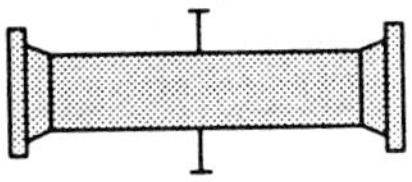 Crossflow	Used for gases with low pressure drop requirement and for condensing vapors at low pressure (i.e., vacuum condensation).	Flow maldistribution can be a problem.

† The E-shell permits the use of all tube bundle types, baffles types, and any tube-pass arrangement and is the only type shell that can be positioned vertically. The nozzles are located on each end of the shell, either on the same or opposite side.

‡ Flow leakage around longitudinal baffle can occur which is detrimental to performance. This problem can be alleviated by welding the horizontal baffle to the shell (if removable bundle is not required) or by restricting the pressure drop to within 5 to 7 psi.

FIGURE 11–4 Shell types. (Adapted from TEMA [3] and Taborek [4]. Used with permission.)

The number of tubes N used in shell-and-tube heat exchangers varies from as few as 20 or so to over 1000, with the number of tube passes N_{tp} ranging from 1 to 16. Depending upon the type tube bundle, the tubes are held in place by one or two *tubesheets* that are mounted at one or both ends of the bundle. The tubes are fastened to the tubesheet by either expanding the tubes into grooves cut into the holes in the tubesheet or welding.

Tube bundles are classified according to how the tubesheets and tubes are arranged, with the bundle being either permanently enclosed in the shell or removable.† Of the nonremovable bundle types, the **U**-*tube* (which requires only one tubesheet) is the least expensive and accommodates large thermal expansion. *Fixed tubesheet* construction is also relatively inexpensive compared to removable bundle types, but is restricted to processes involving small to moderate temperature differences due to limited provision for thermal expansion. If fouling is no concern or cleaning can be accomplished by chemical means and conditions permit two or more (even) tube passes, the **U**-tube is the prime choice. Alternatively, the fixed tubesheet bundle is preferred for applications requiring brush cleaning and/or a single tube pass, providing that thermal expansion is not large. The more expensive *floating head* designs which permit the removal of the tube bundle are normally used in situations in which the tubes must be removed for cleaning. In addition to being capable of handling large thermal expansion, floating head construction can be arranged for either single or multiple tube passes. Floating head tube bundle types include (a) *externally sealed* or *packed* (TEMA W), which allows near maximum tube count and is the least expensive; (b) *split ring* (TEMA S), which is of intermediate cost and requires extensive disassembly for bundle removal; and (c) pull through (TEMA T), which permits the use of fewer tubes and is most expensive but requires minimum disassembly.

For fixed tubesheet and floating head type bundles, the length L of a shell-and-tube heat exchanger is considered to be the length of the straight tubing. In the case of **U**-tube arrangements, the length L is taken as the distance from the tubesheet to the tangent of the tube bend.

As we shall see, the use of multiple tube passes results in some loss in thermal performance due to the nature of the relative flow patterns of the two fluids. A simple mass balance indicates that, for a given total tube-side mass-flow rate, the tube-side bulk-stream velocity $U_{b,t}$ is directly proportional to the number of tube passes. In this connection, for large shell-and-tube heat exchangers with many tubes, it is often necessary to utilize multiple tube passes in order to achieve sufficiently high values of $U_{b,t}$. Otherwise, the low velocities associated with a single-pass arrangement may result in poor heat-transfer performance and be conducive to excessive fouling.

Baffles are generally used to increase the heat-transfer performance in commercial and shell-and-tube exchangers by diverting the flow on the shell side across the tubes. The use of baffles also provides a means of strengthening the mechanical structure of the exchanger. The most common arrangement features the use of segmental baffles, as shown in Fig. 11–5. Segmental baffles are characterized by the ratio of the height of the cutout segment to the inside diameter $D_{i,s}$ of the shell. The optimum heating-to-pumping power performance has been found to occur for a baffle cut B_c of about 20 to 25% [6]. The *baffle spacing*, also known as *baffle pitch*, is

† The following description of tube bundle types is based on "Characteristics of Constructional Elements" by Taborek [4].

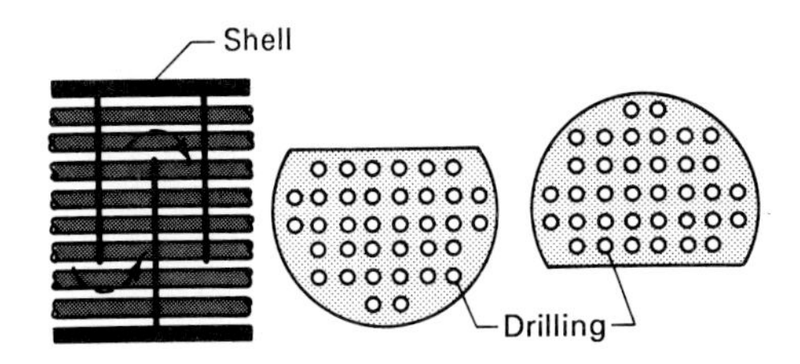

FIGURE 11–5
Segmental baffles (25% cut) used in shell-and-tube heat exchangers. (From Kern [5]. Used with permission.)

designated by L_b. The baffle spacing is generally maintained between 0.2 to 1 times $D_{i,s}$ except for large-diameter shells which necessitate restrictions on the maximum unsupported length for mechanical reasons.† The best shell-side performance is normally achieved with baffle spacing between 0.4 and 0.6 of $D_{i,s}$ [4].

Segmental baffles are normally used unless pressure drop or mechanical considerations necessitate the use of other arrangements. Alternative baffle types include double- or triple-segmental baffles, no-tubes-in-window segmental baffles, disk-and-doughnut baffles, and rodbaffles, which consist of vertical and horizontal rods that provide support and containment for the tubes. These baffle configurations are illustrated in Fig. 11–6. Notice that, whereas the segmental and disc-and-doughnut type baffles direct the flow back and forth across the tube bank, with the greatest tendency for crossflow associated with the use of single-segmental baffles, the rodbaffle arrangement results in shell-side flow that is essentially parallel to the tubes. Relative to the standard single-segmental baffle type shown in Fig. 11–5, the pressure drop is reduced by a factor of about 2 for double-segmental and disc-and-doughnut baffles and by a factor of about 3 for triple-segmental baffles, with rodbaffles providing even larger reductions.

Three standard types of tube arrangements used in shell-and-tube heat exchangers are shown in Fig. 11–7.‡ The triangular and square arrangements are similar to the staggered and in-line tube bank arrays introduced in Sec. 8–4. These arrangements are characterized by tube pitch S_t, tube outside diameter d_o, and shell inside diameter $D_{i,s}$. As pointed out by Kern and Kraus [8], these various layouts permit some selectivity on the shell side between heat transfer and allowable pressure drop. In applications in which the pressure drop is not a primary consideration, such as high-pressure operation with liquids, it may be desirable to pack as much surface as possible into a given size shell. On the other hand, a less dense arrangement that provides lower resistance to the flow may be required for gases at moderate pressure. In this connection, the triangular arrangement has the highest tube density and produces a good heat transfer-to-pumping power ratio. The square and rotated square arrangements permit only about 85% of the tube density, but offer the advantage of convenient

† According to TEMA, the maximum baffle pitches for class A tubes (steel and steel alloys) and class B tubes (aluminum and copper alloys) are 60 and 52 in. for 3/4-in. O.D. tubes and 73 and 63 in. for 1-in. O.D. tubes. General relations for $L_{b,\max}$ are given in the *Heat Exchanger Design Handbook* [6].

‡ A 60° staggered layout involving triangular pitch with flow parallel to the base is also sometimes used. However, this arrangement produces a lower heat transfer-to-pressure drop ratio and is therefore not generally recommended [6].

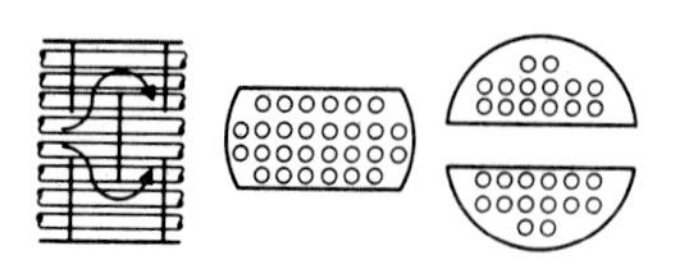

(a) Double-segmental baffle.

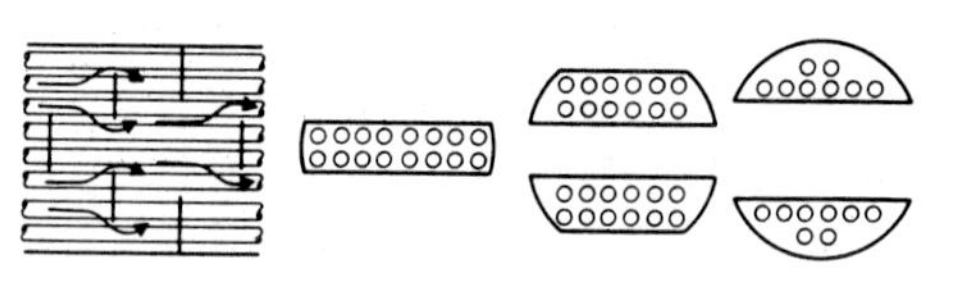

(b) Triple-segmental baffle.

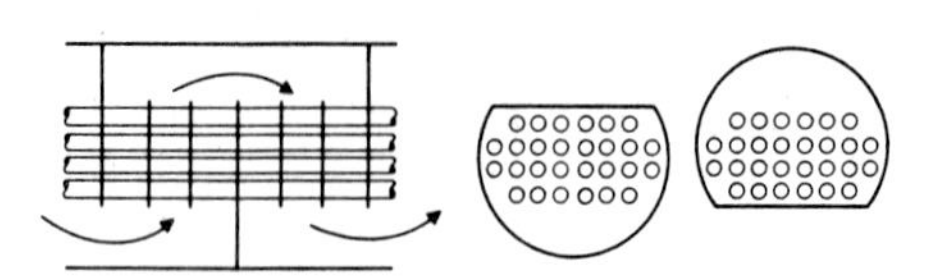

(c) No-tubes-in-window segmental baffle.

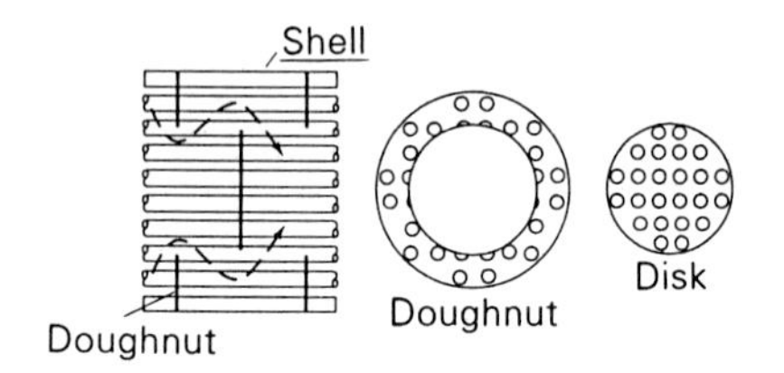

(d) Disk-and-doughnut baffles.

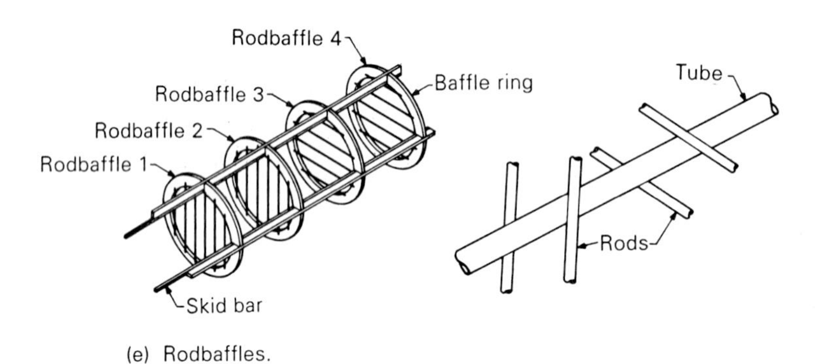

(e) Rodbaffles.

FIGURE 11–6 Other baffle types. (From Shah and Mueller [7]. Used with permission.)

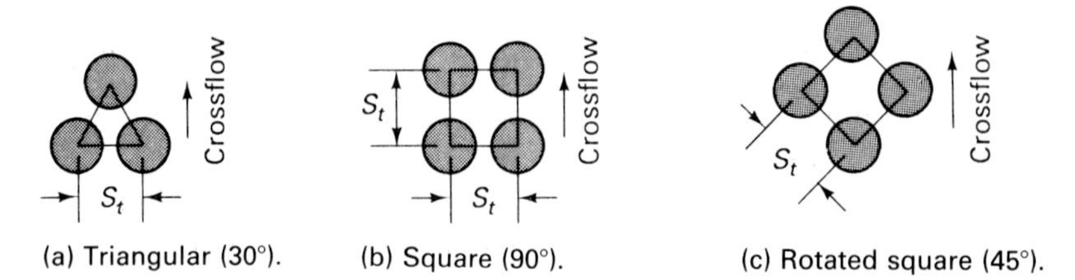

(a) Triangular (30°). (b) Square (90°). (c) Rotated square (45°).

FIGURE 11–7 Typical tube arrangements used for shell-and-tube heat exchangers.

mechanical cleaning. The square layout provides the highest heat transfer-to-pressure drop ratio for turbulent flow, but the worst for laminar flow [7]. Standards for minimum tube pitch have been established to assure that the structural integrity of the tube-sheets is not unduly compromised.

Typical tube outside diameters d_o used in shell-and-tube heat exchangers include 5/8, 3/4, 7/8, and 1 in. The number of tubes that can be used depends on the shell inside diameter $D_{i,s}$, tube diameter d_o, pitch S_t, layout, and tube passes N_{tp}. The maximum number of tubes of a given diameter and layout that can be accommodated within a shell is known as the *tube count*. Representative tube counts (N) for 3/4-in. diameter tubes are given in Table A–O–1.†

Because of their versatility, shell-and-tube heat exchangers are the most common type exchanger used in industry. However, other type heat exchangers are used in applications in which space and weight are a premium, such as in automobiles and aircraft.

11–2–3 Crossflow Heat Exchanger

Crossflow heat exchangers consist of a number of interconnected tubes or passageways that are separated by one or more channels, as illustrated in Fig. 11–8. Heat exchangers of this type are often used when one or both of the fluids is gas. In this connection, heat exchangers designed for the purpose of cooling a hot fluid stream by passing air in crossflow over one or more plane or finned tubes are known as *air coolers*.

Crossflow heat exchangers and solution results are categorized according to whether the fluid streams can be considered to be *mixed* or *unmixed*. Mixing of a stream implies that its flow is unrestricted in the lateral direction (i.e., in the direction of the other stream). Convesely, unmixed flow implies that the flow of a stream is restricted in the lateral direction. The three possible combinations of mixed and unmixed streams are represented schematically in Fig. 11–9. As we will see in Sec. 11–5, the heat transfer performance of crossflow heat exchangers is better for unmixed streams than for mixed streams. Although the categorization of a stream is sometimes clear cut, as in the cases shown in Fig. 11–8(d)–(f), the actual flow is better described as being partially mixed in many applications. For example, the degree of mixing for crossflow over the high-fin tubes shown in Fig. 11–8(b),(c) ranges from mixed for wide spacing to unmixed for close spacing. However, in spite of this reality, solution approaches available in the literature require that the stream be designated by one extreme or the other (i.e., mixed or unmixed).

The plate heat exchanger illustrated in Fig. 11–10 is another important type of unmixed crossflow arrangement. Plate heat exchangers are built of thin plates that are clamped together to form passages. As shown in Fig. 11–10, the plates are fitted with gaskets, which are used to seal the unit and to direct the two liquids counter-currently through the narrow passages between alternate pairs of heat transfer plates. Heat exchangers of this type are easily dismantled for cleaning, inspection, or alteration

† Analytical approaches to estimating tube count are presented in references 6 and 9.

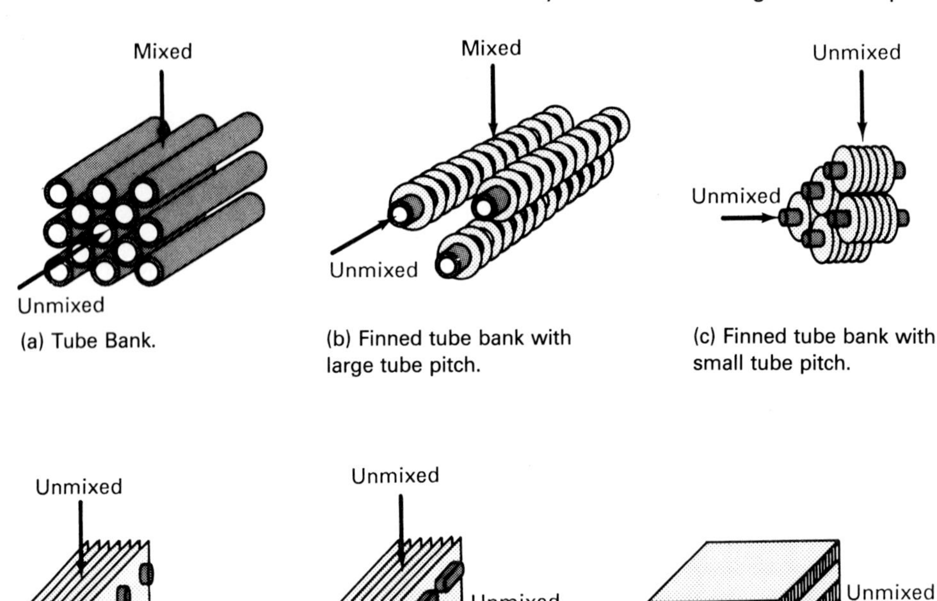

FIGURE 11–8 Crossflow heat exchangers. (From Kern and Kraus [8] and Kays and London [10]. Used with permission.)

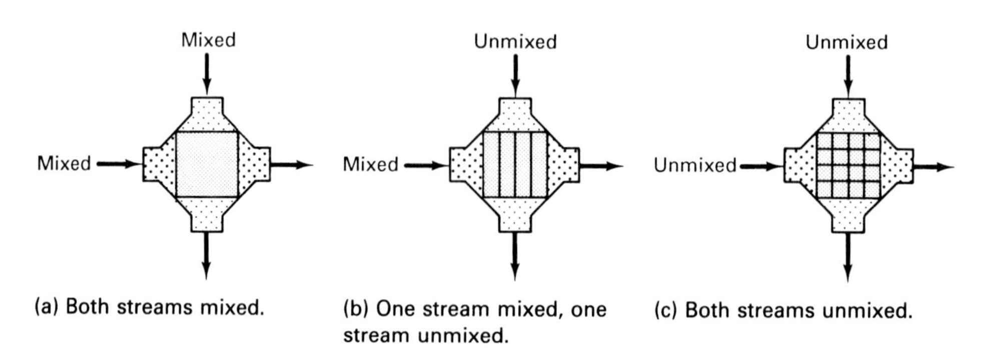

FIGURE 11–9 Mixed and unmixed categories of crossflow heat exchangers.

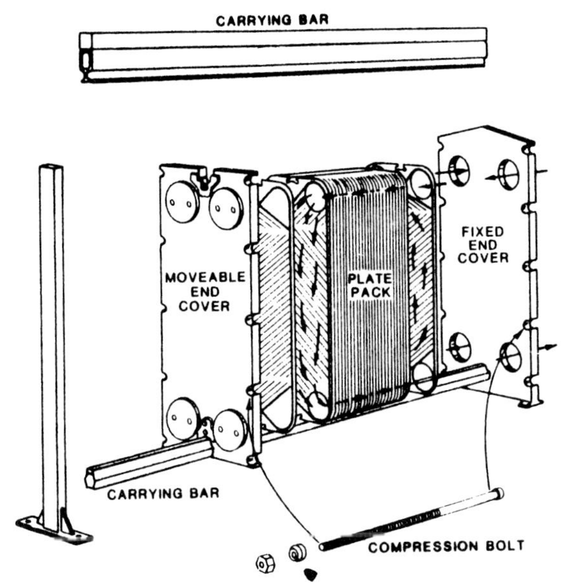

FIGURE 11-10
Gasketed plate heat exchanger—exploded view. (From *Heat Exchanger Design Handbook* [11]. Used with permission.)

and are widely used in the dairy, beverage, food, pharmaceutical, and synthetic rubber industries [1]. However, because of limitation in gasket material, plate heat exchangers are generally restricted to pressures below 10 to 15 atm and temperatures below 150°C.

11-2-4 Regenerators

In the types of heat exchangers described above, two fluids are separated by a thin wall through which heat is transferred.† These direct transfer heat exchangers are commonly called *recuperators*. Another type of heat exchanger known as the *regenerator* involves the alternate use of the same flow passages by both fluids. In such storage-type heat exchangers, energy is first transferred by convection from one fluid and stored in the matrix wall, and later transferred from the matrix to the other fluid. Thus, energy is periodically stored and rejected by the matrix wall. Examples of storage-type heat exchangers include the rotary regenerator and the fixed dual-bed regenerator shown in Fig. 11-11. Representative matrix surfaces are shown in Fig. 11-12. In addition to providing a very compact heat transfer surface, the surface of a regenerator matrix tends to be self-cleaning because of the periodic flow reversals and lack of permanent flow-stagnation regions. This self-cleaning feature has been demonstrated by the Ljungstrom air preheaters used in TVA power plants, which burn various grades of coal [10].

† In certain types of heat exchangers two immiscible fluids are in direct contact. Water cooling towers are an example of such *direct contact* heat exchangers.

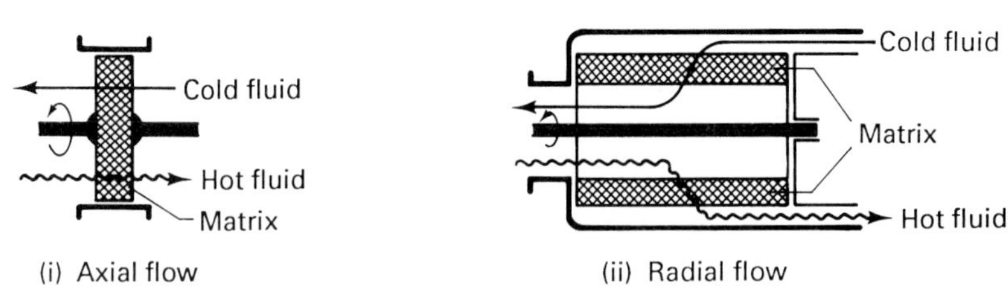

(a) Rotary regenerators.

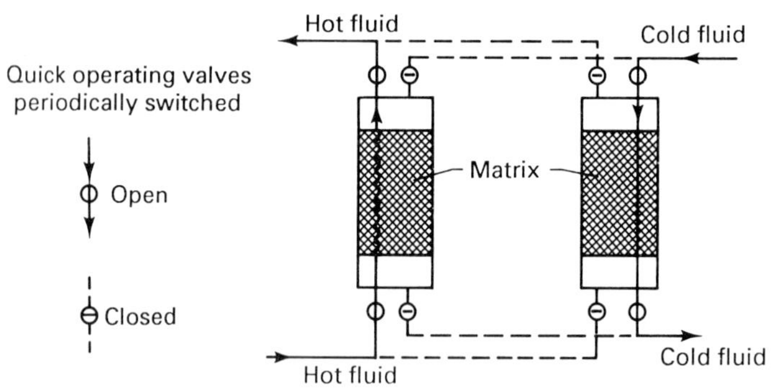

(b) Fixed-dual bed regenerator.

FIGURE 11–11 Regenerators. (From Kays and London [10]. Used with permission.)

11–2–5 Heat Exchanger Performance/Size

High-performance heat exchangers are characterized by high heating–pumping power ratios and small volumes. Various methods are routinely used to enhance the performance and reduce the size of heat exchangers. For example, the rate of heat transfer to fluids flowing in tubular heat exchangers is often enhanced by the use of longitudinal, circumferential or spiraled fins, spines, ribs or grooves. Several types of finned tubes are shown in Fig. 11–13. Similarly, tube-fin and plate-fin heat exchangers are commonly constructed of corrugated (wavy) or interrupted (strip, louvered or perforated) fins. Louvered fins are formed by cutting small strips in the fin and lifting the strips above the base fin surface.

Low-fin tubes of the type shown in Fig. 11–14 are in common use in shell-and-tube heat exchangers. Convection correlations for plain tubes are generally adapted to low-fin tubes by replacing d_o by an *equivalent diameter* d_{req} [6,8]. This characteristic length represents the tube root diameter d_r plus an additional increment approximately equal to the thickness of a cylinder equivalent to the fin metal. Table A–O–3 lists d_o, d_r, d_{req}, the outside (finned) surface area per unit length $A_{s,o}/L$, outside-to-inside surface area ratio $A_{s,o}/A_{s,i}$, and other dimensions for several standard low-fin tubes with 19 fins per inch. These low-fin tubes provide a surface area that is approximately 2.5 times larger than the surface area of plain tubes with the same nominal O.D. Because the fins do not protrude beyond the original tubes from which they are

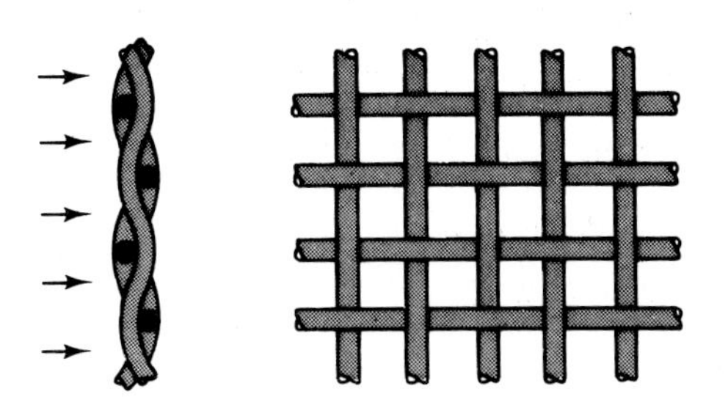

(a) Woven screen matrix.

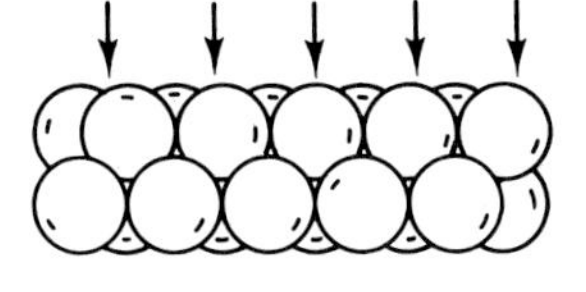

(b) Sphere bed matrix.

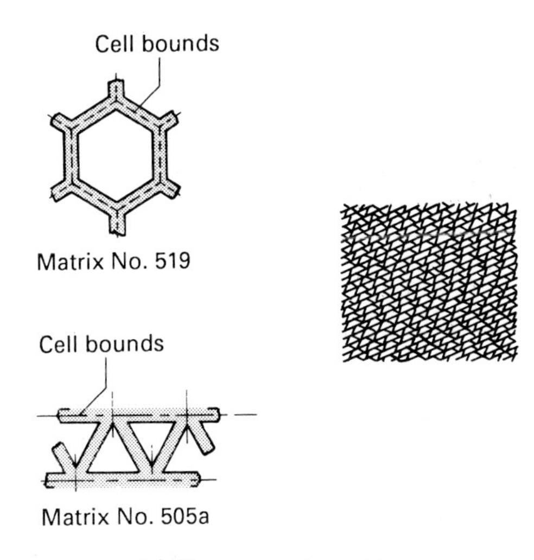

(c) Glass ceramic matrix.

FIGURE 11-12 Regenerator matrix surfaces. (From Kays and London [10]. Used with permission.)

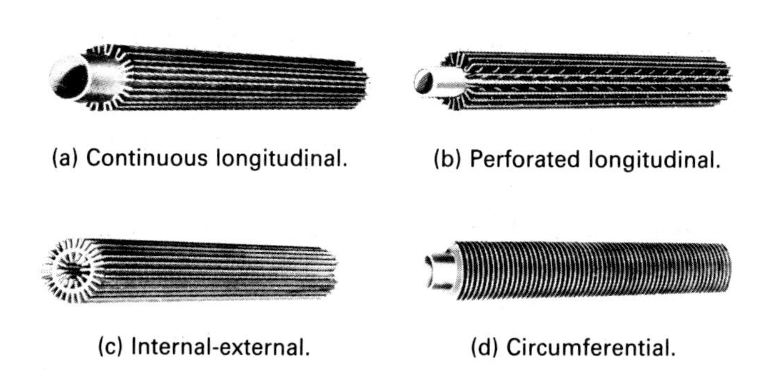

(a) Continuous longitudinal.

(b) Perforated longitudinal.

(c) Internal-external.

(d) Circumferential.

FIGURE 11-13 Fintubes. (Courtesy of Brown Fintube Company, Houston, Texas.)

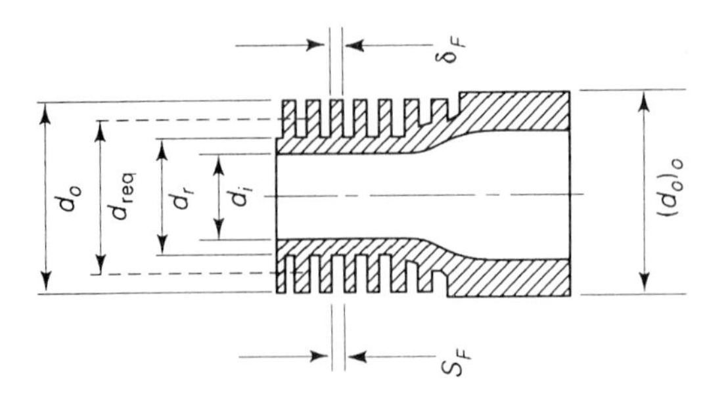

$(d_o)_o$	outside diameter of plain end
d_o	diameter over fin
d_r	root diameter
d_i	inside diameter
d_{req}	equivalent diameter
δ_F	fin thickness
S_F	distance between fins
H_F	fin height; $H_F = (d_o - d_r)/2$
X_F	wall thickness; $X_F = (d_r - d_i)/2$
N_F	Number of fins per unit length; $N_F = 1/(S_F + \delta_F)$

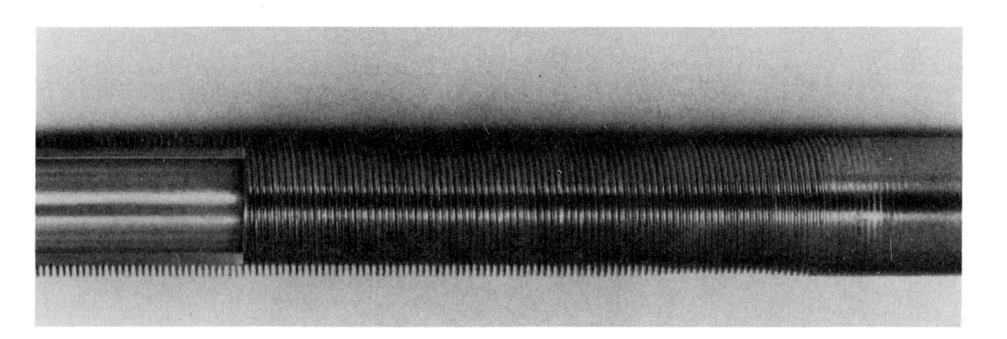

FIGURE 11–14 Typical low-fin tubing used in shell-and-tube heat exchanges. (Courtesy of Wolverine Tube Company.)

extruded, low-fin tubes can be used interchangeably and as replacements for plain tubes in shell-and-tube heat exchangers.

Other methods of enhancing the heat-transfer performance of tubular heat exchangers involve the use of various types of inserts that promote mixing. The *turbulator* tube insert shown in Fig. 11–15 has been reported to provide significant increases in heat-transfer rate and discourages the buildup of fouling deposits on the heat-transfer surface. Because they produce quite good heat transfer-to-pumping power ratios, twisted tape inserts such as this are commonly used to improve the performance of existing heat exchangers. Other devices such as inserts with wire loops or fluted spheres mounted on connecting rods, coiled tubes, and mechanical aids that stir the fluid or scrape the surface are also found in commercial practice.

Heat exchangers with *surface-area density* β (i.e., ratio of surface area to volume A_s/V) greater than about 700 m^2/m^3 are referred to as *compact heat exchangers*. Because of their smaller size and weight, compact heat exchangers are prevalent in aircraft, vehicular, and marine systems. Two compact crossflow heat exchangers designed for use in aircraft are shown in Fig. 11–16. A spectrum of surface-area density prepared by Shah [1] for representative types of heat exchangers is shown in Fig. 11–17. A typical shell-and-tube heat exchanger with 25.4 mm (1 in.) diameter tubes, which is commonly used in power plant condensers, will have a value of β of about

FIGURE 11–15 Turbulator tube insert. (Courtesy of Brown Fintube Company, Houston, Texas.)

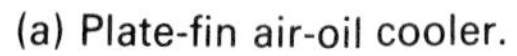

(a) Plate-fin air-oil cooler.

(b) Air-oil temperature regulator.

FIGURE 11–16 Compact heat exchangers for use in aircraft. (Courtesy of AiResearch Division—Allied Signal Aerospace Company.)

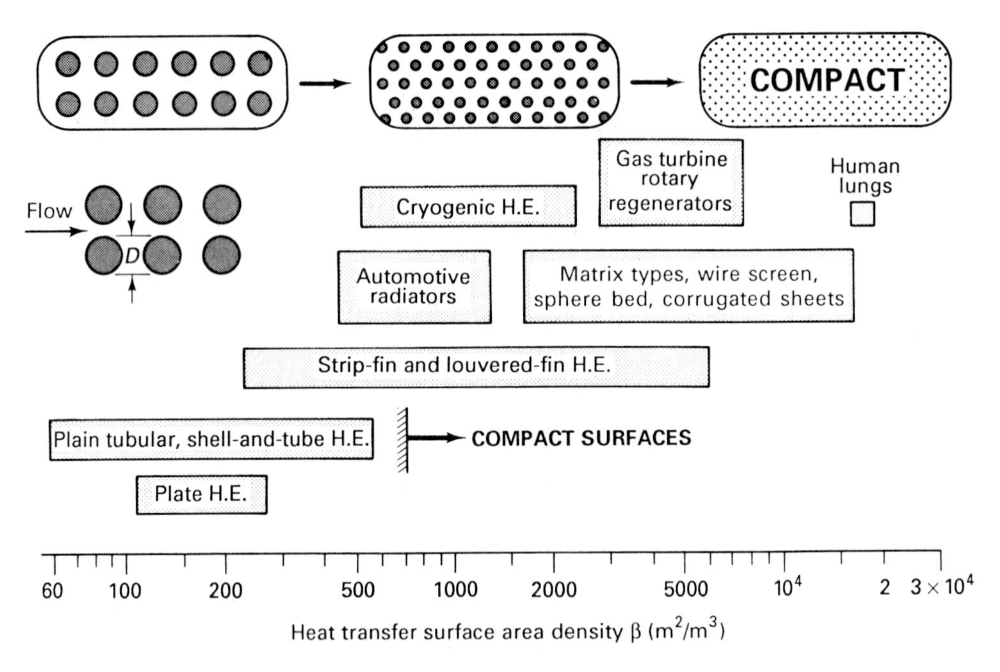

FIGURE 11–17 Surface-area density spectrum of heat exchangers. (From Shah [1]. Used with permission.)

130 m²/m³, such that it would be considered to be noncompact.† On the other hand, modern automobile radiators with 5.5 fins/cm (14 fins/in.) qualify as compact heat exchangers since they generally have an area density of the order of 1100 m²/m³, which is approximately equivalent to 3-mm-diameter tubes. Interestingly, the human lungs are an amazingly compact heat and mass transfer system, having a surface-area density of about 17,500 m²/m³, which is equivalent to about 0.19 mm tubes.

11–3 EVALUATION AND DESIGN

As suggested in Chap. 1, the objective of the thermal and hydraulic evaluation function is to determine the total rate of heat transfer, outlet temperatures, and pressure drop that can be produced by an existing system under given operating conditions (i.e., fluids, mass-flow rates, and inlet temperatures). The evaluation or rating of heat-exchanger performance provides the basis of (1) prescribing changes in the operation, and (2) determining when an existing unit must be cleaned, overhauled, modified, or replaced.

In its broadest sense, heat exchanger design encompasses (1) component selection, (2) thermal and hydraulic design, (3) mechanical and metallurgical design, (4)

† The relation between D_H and β used by Shah [1] to approximately represent typical heat exchanger configurations is given by $\beta = 3333/D_H$.

architectural design, and (5) operation and maintenance considerations, all of which should be performed in the context of the overall plant or system, with interaction between the requirements of each of these individual design facets.

The objective of the thermal and hydraulic design function is to determine the surface area required to transfer a specified rate of heat with acceptable pressure drop for given fluids, mass-flow rates, and terminal (i.e., inlet and exit) temperatures. The economic thermal-hydraulic design of a heat exchanger for no phase change fluids normally requires the full utilization of the available pressure drop and high heat transfer-to-pumping power ratio (i.e., effective pressure drop-to-heat transfer conversion).

The primary purpose of the mechanical and metallurgical design function is to provide sufficiently strong and durable enclosures to contain and separate the fluids flowing through the heat exchanger. The mechanical design provides specifications of the required thickness for the walls of components such as shells, tubes, tubesheets, end covers, flanges, and nozzles. From the metallurgical standpoint, the materials to be used for construction must be selected so that they (1) provide the necessary strength to contain the design pressures and (2) are compatible with the working fluids and other exchanger construction materials [12].

In practice, the design of heat exchangers generally involves iterative calculations that provide optimum solution results (based on cost, size, or weight), flow-induced vibration analysis, design details, a TEMA specification sheet, maximum allowable working pressure, bill of materials, drawings, and cost estimates. An overview of heat exchanger design methodology is provided by Taborek [6] and Shah [13].†

The primary design documents in the United States and many other countries are the ASME Boiler and Pressure Vessel Code, Sec. VIII, Div. I [14] and TEMA [3]. The ASME code is an extensive volume of mechanical design regulations that protects against hazardous failures and generally has the force of law. On the other hand, TEMA is a set of supplementary thermal/hydraulic and mechanical design standards for shell-and-tube heat exchangers that pertains to nonsafety matters such as economy, operational features, and maintenance.

The purpose of our study is to provide an introduction to the thermal and hydraulic analysis of heat exchangers. Practical thermal and hydraulic analyses will be developed momentarily for double-pipe, shell-and-tube, and crossflow heat exchangers and multipass heat exchanger networks. However, as a preliminary to the development of the practical thermal analysis of heat exchangers, we must first consider the concept of the overall coefficient of heat transfer.

11-4 OVERALL COEFFICIENT OF HEAT TRANSFER

Unlike the single fluid processes studied in the previous chapters, the surface temperatures or heat fluxes are not specified for heat exchangers. This complication is overcome by use of the overall coefficient of heat transfer.

† The subject of flow-induced vibrations is introduced in references 7 and 11.

11–4–1 Local Overall Coefficient of Heat Transfer

Referring to the counterflow arrangement shown in Fig. 11–18, the differential rate of heat transfer dq across a wall of thickness δ, length dx, and uniform perimeters p_c and p_h can be written in terms of the local bulk-stream temperatures T_h and T_c and local thermal resistance R' as†

$$dq = \frac{T_h - T_c}{R'_h + R'_k + R'_c} = \frac{T_h - T_c}{1/(h\,dA_s)_h + R'_k + 1/(h\,dA_s)_c} \tag{11–1}$$

where h_h and h_c represent the local coefficient of heat transfer of the hot and cold streams, respectively, the heat-transfer surface areas are represented by $dA_{s,h} = p_h\,dx$ and $dA_{s,c} = p_c\,dx$; $R'_k = \delta/(k\,dA_s)$ for plane walls and $R'_k = [\ln(d_o/d_i)]/(2\pi k\,dx)$ for cylindrical tube walls with inside and outside diameters d_i and d_o. For counterflow (or parallel-flow) arrangements such as this with uniform inlet temperatures $T_{h,i}$ and $T_{c,i}$, the distributions in T_h and T_c are generally functions of x. The defining equation for the *local overall coefficient of heat transfer U* is

$$dq = U\,dA_s\,(T_h - T_c) \tag{11–2}$$

where the reference surface area dA_s is normally set equal to $dA_{s,h}$ or $dA_{s,c}$. Combining Eqs. (11–1) and (11–2), we obtain an expression for the local overall coefficient of heat transfer of the form

$$\frac{1}{U\,dA_s} = \frac{1}{(U\,dA_s)_h} = \frac{1}{(U\,dA_s)_c} = \frac{1}{(h\,dA_s)_h} + R'_k + \frac{1}{(h\,dA_s)_c} \tag{11–3}$$

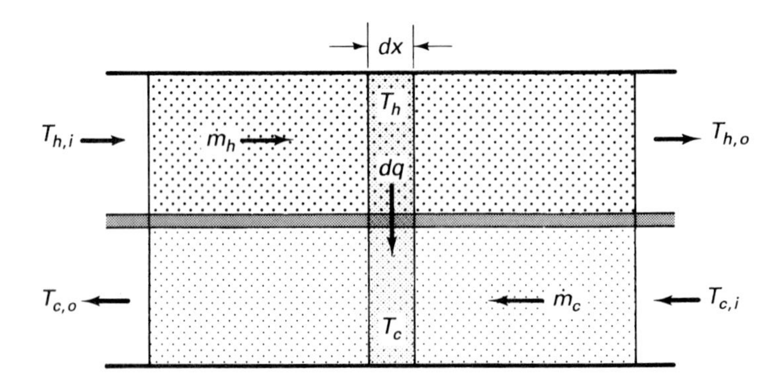

FIGURE 11–18 Sketch for rate of heat transfer dq between two counterflowing fluids separated by a wall.

† Alternative representations commonly found in the literature for shell-and-tube and double-pipe heat exchangers are of the form

$$dq = \frac{T_s - T_t}{R'_s + R'_k + R'_t} \quad \text{or} \quad dq = \frac{T_o - T_i}{R'_o + R'_k + R'_i}$$

where the shell or annular side is represented by s or o and the tube side by t or i. To avoid confusion with T_s, the wall temperature will be represented by T_w in this chapter.

Since the perimeters are uniform, this equation can also be written as†

$$\frac{1}{UA_s} = \frac{1}{(UA_s)_h} = \frac{1}{(UA_s)_c} = \frac{1}{(hA_s)_h} + R_k + \frac{1}{(hA_s)_c} \tag{11–4}$$

where $A_s = pL$, $A_{s,h} = p_h L$, $A_{s,c} = p_c L$, and L is the length of the heat exchanger wall; $R_k = \delta/(kA_s)$ for plane walls, and $R_k = [\ln(d_o/d_i)]/(2\pi kL)$ for cylindrical tube walls. From these equations we see that U_h and U_c are geometrically related by $UA_s = U_h A_{s,h} = U_c A_{s,c}$. We also observe that U is nonuniform for situations in which h_h and/or h_c are nonuniform.

The local coefficients of heat transfer h_h and h_c are generally represented by the use of correlations for Stanton number St_h and St_c (or Nusselt number Nu_h and Nu_c) in terms of hydraulic diameter D_H and Reynolds number Re of the type introduced in Chaps. 8 to 10 for internal flow through passages with smooth, finned, rough, or otherwise modified surfaces.

11–4–2 Mean Overall Coefficient of Heat Transfer

Similar to the defining equation for $\bar{h}$, the *mean overall coefficient of heat transfer* $\bar{U}$ for a wall with surface area A_s is given by

$$\bar{U} = \frac{1}{A_s}\int_{A_s} U\,dA_s \tag{11–5}$$

As we will see, this formal defining relation for $\bar{U}$ sometimes occurs in the analysis of heat exchangers. It should also be noted that $\bar{U}$ is commonly interpreted less formally as the average value of the local overall coefficient of heat transfer that can be used in ideal solutions for which U is assumed to be constant throughout the heat exchanger.

The mean overall coefficient of heat transfer $\bar{U}$ can be evaluated by the use of Eq. (11–5) together with general (design) correlations for the coefficients of heat transfer of the hot and cold fluids. For the two limiting cases in which either the hot-side or cold-side thermal resistance controls, $\bar{U}$ can be approximated by $\bar{h}_h$ or $\bar{h}_c$, respectively (for clean unfinned surfaces). On the other hand, for cases in which these thermal resistances are of the same order of magnitude, both h_h and h_c must be used in the evaluation of $\bar{U}$. However, in practice U is often treated as a constant, such that $\bar{U} \simeq U$. In applications for which this is permissible, $\bar{U}$ is sometimes calculated by evaluating the coefficients h_h and h_c at the midpoint of the heat exchanger. $\bar{U}$ is

† This equation commonly appears in the literature as

$$\frac{1}{U_h} = \frac{1}{h_h} + \frac{\delta}{k(A_k/A_{s,h})} + \frac{1}{h_c(A_{s,c}/A_{s,h})}$$

based on surface area $A_{s,h}$, or

$$\frac{1}{U_c} = \frac{1}{h_h(A_{s,h}/A_{s,c})} + \frac{\delta}{k(A_k/A_{s,c})} + \frac{1}{h_{s,c}}$$

based on surface area $A_{s,c}$, where A_k is the effective area of the wall; $A_k = A_s$ for plane walls and $A_k = (d_o - d_i)L/\ln(d_o/d_i)$ for cylindrical tube walls.

also commonly approximated by setting h_h and h_c in Eq. (11–4) equal to the mean values $\bar{h}_h$ and $\bar{h}_c$ at the outlets; that is,

$$\frac{1}{\bar{U}A_s} = \frac{1}{(\bar{h}A_s)_h} + R_k + \frac{1}{(\bar{h}A_s)_c} \tag{11–6}$$

Whereas design work involves the use of Eq. (11–5) or relations such as Eq. (11–6) together with generalized correlations for h_h and h_c or $\bar{h}_h$ and $\bar{h}_c$ to calculate $\bar{U}$, the evaluation function is commonly undertaken with the aid of specific (rating) correlations for $\bar{U}$, which are established on the basis of the analysis of previously recorded performance data (i.e., terminal temperatures, fluid properties, and mass-flow rates) for the unit in question. The use of the practical thermal analysis approach in calculating $\bar{U}$ will be considered in Sec. 11–5–1.

Representative values of $\bar{U}$ are given in Table 11–1 for various situations found in practice.

TABLE 11–1 Overall heat-transfer coefficients—representative values

	$\bar{U}$	
Application	Btu/(h ft^2 °F)	W/(m^2 °C)
Steam condenser	200–10^3	10^3–6×10^3
Freon-12 condenser with water coolant	50–200	300–10^3
Water-to-water	100–300	600–2×10^3
Water-to-oil	20–60	100–400
Water-to-gasoline	60–90	300–500
Steam-to-light fuel oil	30–60	200–400
Steam-to-heavy fuel oil	10–30	60–200
Steam-to-gasoline	50–200	300–10^3
Finned-tube heat exchanger; water in tubes, air over tubes	5–10† 70–150‡	30–60† 420–840‡
Finned-tube heat exchanger; steam in tubes, air over tubes	5–50† 70–700‡	30–300† 420–4200‡

† Based on area of finned surface.
‡ Based on area of unfinned surface.
Source: Based on [15].

True Mean Temperature Difference

Referring to the defining equation for the local overall coefficient of heat transfer U,

$$dq = U(T_h - T_c)\, dA_s \tag{11–2}$$

we note that

$$q = \int_{A_s} U(T_h - T_c)\, dA_s \tag{11–7}$$

It follows that q can be expressed in terms of $\overline{U}$ by writing

$$q = \overline{U}A_s \, \Delta T_m \tag{11–8}$$

where the *true mean temperature difference* ΔT_m is defined by

$$\Delta T_m = \frac{1}{\overline{U}A_s} \int_{A_s} U(T_h - T_c) \, dA_s \tag{11–9}$$

This relation reduces to

$$\Delta T_m = \frac{1}{A_s} \int_{A_s} (T_h - T_c) \, dA_s \tag{11–10}$$

for uniform overall coefficient of heat transfer (i.e., $\overline{U} = U$). As we shall see, with ΔT_m expressed in terms of the effectiveness P (or ϵ), efficiency ψ, or $LMTD$, Eq. (11–8) provides the basis for the practical thermal analysis of heat exchangers.

11–4–3 Effect of Fins

Finned tubes are commonly used in double-pipe and shell-and-tube heat exchangers when the ratio of the tube-side to shell-side coefficients of heat transfer is greater than about 2 to 1 [6].

To characterize the heat transfer in heat exchangers with finned surfaces, the local thermal resistances to convection are approximated in terms of the *net surface efficiency* η_o by Eq. (2–133), such that Eq. (11–2) leads to the more general relation

$$\frac{1}{U \, dA_s} = \frac{1}{(\eta_o h \, dA_s)_h} + (R'_{tc})_h + R'_k + (R'_{tc})_c + \frac{1}{(\eta_o h \, dA_s)_c} \tag{11–11}$$

where h_h and h_c now represent the local *mean* coefficients of heat transfer, R'_{tc} accounts for the thermal contact resistance between fin and wall, η_o is expressed in terms of *fin efficiency* η_F, fin surface area A_F and prime surface area A_p by Eq. (2–127),

$$\eta_o = 1 - \frac{A_F}{A_o}(1 - \eta_F) \tag{11–12}$$

and $A_o = A_F + A_p$; A_F/A_o is the *finned area fraction*. As mentioned in Chap. 2, η_o is equal to unity for unfinned surfaces but is less than unity for finned surfaces. With the finned surface represented by an effective unfinned surface with equivalent surface area A_s and uniform effective perimeter $p = A_s/L$, Eq. (11–11) takes the more practical form [see Eqs. (2–130) and (2–133)]

$$\frac{1}{UA_s} = \frac{1}{(\eta_o h A_s)_h} + (R_{tc})_h + R_k + (R_{tc})_c + \frac{1}{(\eta_o h A_s)_c} \tag{11–13}$$

The coefficients of heat transfer h_h and h_c are usually specified by the use of design correlations that express Stanton number (St_h, St_c) in terms of Reynolds number

Re $[= (GD_H/\mu)_h, (GD_H/\mu)_c]$ for various fin tube geometries. The hydraulic diameter D_H is represented by Eq. (6–27),

$$D_H = \frac{4A}{p_w} \tag{11–14a}$$

for flow in tubes or annuli, or Eq. (8–139), which indicates

$$D_H = \frac{4A_{\min}}{p} = \frac{4A_{\min}}{A_s/L} = 4\frac{\sigma}{\beta} \tag{11–14b}$$

for crossflow over tube banks with $p_w = p$, where the minimum free-flow area $A_{\min}$ and surface area A_s (or the contraction ratio σ and surface-area density β) are specified by the supplier.

11–4–4 Effect of Fouling

It should be noted that the overall coefficient of heat transfer is often reduced by fouling deposits of various forms (sedimentation, crystalline, bacterial, etc.) that accumulate on the heat-exchanger walls. For example, Fig. 11–19 shows severe fouling

FIGURE 11–19 Ash deposit fouling on secondary superheater tubes. (From *Steam—Its Generation and Use*. Copyright 1978 by Babcock and Wilcox. Used with permission.)

on the super-heater tubes of a boiler. The effect of such deposits is usually accounted for by the *fouling factor* F_f, which is defined in terms of the coefficient of heat transfer h by

$$F_f = \frac{1}{h_f} - \frac{1}{h} \tag{11–15}$$

where h_f represents the coefficient of heat transfer after fouling has occurred. It follows that the overall coefficient of heat transfer U for a wall with fouling on both surfaces is represented by

$$\frac{1}{UA_s} = \frac{1}{(hA_s)_h} + \left(\frac{F_f}{A_s}\right)_h + R_k + \left(\frac{F_f}{A_s}\right)_c + \left(\frac{1}{hA_s}\right)_c \tag{11–16}$$

for unfinned surfaces, and

$$\frac{1}{UA_s} = \frac{1}{(\eta_o hA_s)_h} + (R_{tc})_h + \left(\frac{F_f}{\eta_o A_s}\right)_h + R_k + \left(\frac{F_f}{\eta_o A_s}\right)_c + (R_{tc})_c + \frac{1}{(\eta_o hA_s)_c} \tag{11–17}$$

for finned surfaces. Representative fouling factors are given in Table 11–2. A review of the literature on fouling in heat exchangers is provided in references 16 to 18.

The primary parameters upon which the rate of fouling deposition generally depends include the bulk-stream velocity and the difference between the surface temperature and the bulk-stream temperature. Because of the relation between fouling and bulk-stream velocity, heat exchangers are normally designed to operate at velocities above a specified minimum value, which is typically on the order of 1 m/s.

TABLE 11–2 Fouling factors—representative values

	F_f	
Type fluid	(h ft² °F/Btu)	(m² °C/kW)
Seawater:		
Below 50°C	0.5×10^{-3}	0.09
Above 50°C	1×10^{-3}	0.2
Treated boiler feedwater above 50°C	1×10^{-3}	0.2
Fuel oil	5×10^{-3}	0.9
Quenching oil	4×10^{-3}	0.7
Alcohol vapors	0.5×10^{-3}	0.09
Steam, non-oil-bearing	0.5×10^{-3}	0.09
Industrial air	2×10^{-3}	0.4
Refrigerating liquid	1×10^{-3}	0.2

Source: Based on [3].

As an additional point of practical interest, the use of fin surfaces usually results in a reduction in the effect of fouling deposition. This positive effect is in part the result of the lower temperature difference between the fin surface and the bulk stream. Furthermore, because of the increased surface area of fin surfaces, much more fouling is required on the extended surface than on the bare surface to achieve equal reduction in performance. It follows that in some applications plant maintenance time can be significantly reduced by the use of fins [19].

Corrosion in heat exchangers is another problem that is encountered in practice, especially in the chemical industry. To overcome this problem, glass and plastic tubes and shell or shell linings are often used. For example, the single-pass shell-and-tube heat exchanger shown in Fig. 11–20 is constructed of flexible Teflon tubes and a Teflon-lined carbon steel shell. Although the thermal conductivity of Teflon and other plastic materials is lower than that of metals and glass, its resistance to corrosion and thermal or mechanical shock and other characteristics give it an advantage over other corrosive-resistant materials for some applications. It should also be noted that the excellent thermal conduction properties of borosilicate materials make the use of glass quite attractive in many applications involving corrosive chemicals.

As pointed out by Taborek [20], various design and operational provisions (such as maintaining high bulk-stream velocity, restricting the wall temperature, specifying

FIGURE 11–20 Shell-and-tube heat exchanger with Teflon tubes and Teflon-lined carbon steel shell. (Courtesy of AMETEK Haveg Division.)

corrosion-resistant materials, and maintaining water quality control) can be made, but ultimately most exchangers must be cleaned. The ability to perform cleaning efficiently is often a decisive factor in exchanger-type selection.

11–4–5 Effect of Baffles

As indicated in Sec. 11–2–2, one purpose of baffles is to increase the heat transfer by directing the flow stream on the shell-side back and forth across the tubes. According to TEMA standards, the spacing L_b for segmental baffles normally should lie within 0.2 $D_{i,s}$ to $D_{i,s}$, with the overall coefficient of heat transfer generally being higher for smaller spacing. However, because of clearances that are inherent to the construction of shell-and-tube heat exchangers, significant fractions of the fluid flows through leakage and bypass areas, which are much less effective in transferring heat. Except for situations in which the shell-side coefficient of heat transfer $\bar{h}_s$ is very large (such that the effect on $\bar{U}$ is negligible), the specification of $\bar{h}_s$ in design work requires that the contribution of the main crossflow stream and the leakage and bypass streams as well as the flow through baffle windows be taken into account. Approaches to the evaluation of $\bar{h}_s$ for shell-and-tube heat exchangers with segmental baffles are considered in Sec. 11–7.

11–4–6 Effect of Multiple Tube Passes

The typical situation encountered in shell-and-tube heat exchangers involves two (or more) tube passes, as illustrated in Fig. 11–21. The complicating factors that enter into this picture include the actual nonuniformity in the local overall coefficient of heat transfer U and shell-side and tube-side bulk-stream temperatures T_s and T_t over any cross section (as well as in the axial direction) and the combination of axial counterflow and parallel flow. For effectively baffled shell-side flow, U, T_s, and the

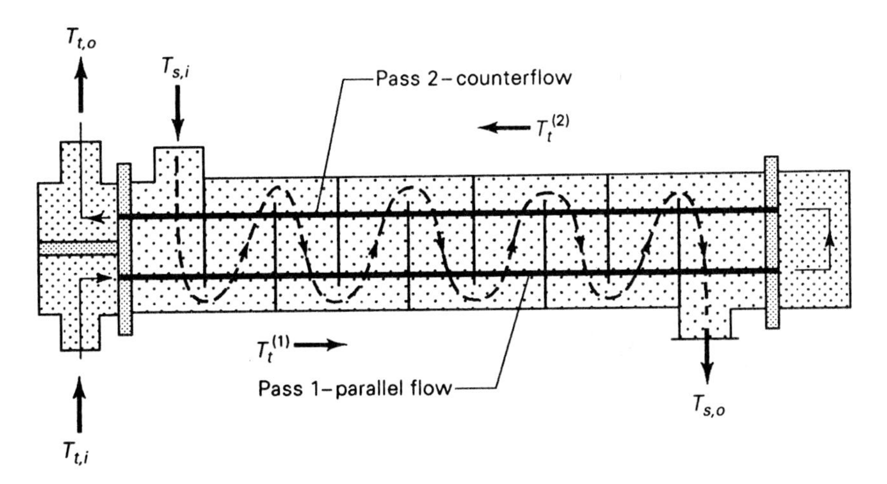

FIGURE 11–21 1-2 TEMA E shell-and-tube arrangement.

tubular bulk-stream temperature $T_t^{(i)}$ for a given pass i are generally assumed to be uniform over the cross section. In addition, U is commonly assumed to be constant over the entire length of the exchanger. These approximations will be used in Sec. 11–7 to adapt the defining equation for U given by Eq. (11–2) to well-baffled shell-and-tube heat exchangers with multiple tube passes.

11–4–7 Effect of Crossflow

Equations (11–1) to (11–4) also apply to more complex situations involving crossflow with streamwise directions x for the hot fluid (or cold fluid) and z for the cold fluid (or hot fluid), as illustrated in Fig. 11–22. Whereas the bulk-stream temperatures T_h and T_c are both functions of x alone (i.e., one dimensional) for simple counterflow and parallel-flow arrangements, the dimensions of T_h and T_c depend on whether streams in the crossflow are mixed or unmixed. The dimensions of T_h and T_c are indicated in Table 11–3 for mixed and unmixed crossflow arrangements. Because both x and z are involved in the specification of T_h and T_c for any combination of mixed and unmixed crossflow streams, the problem is two dimensional rather than one dimensional. It follows that U can be approximately uniform or can be a function of either or both x and z for heat exchangers with crossflow.

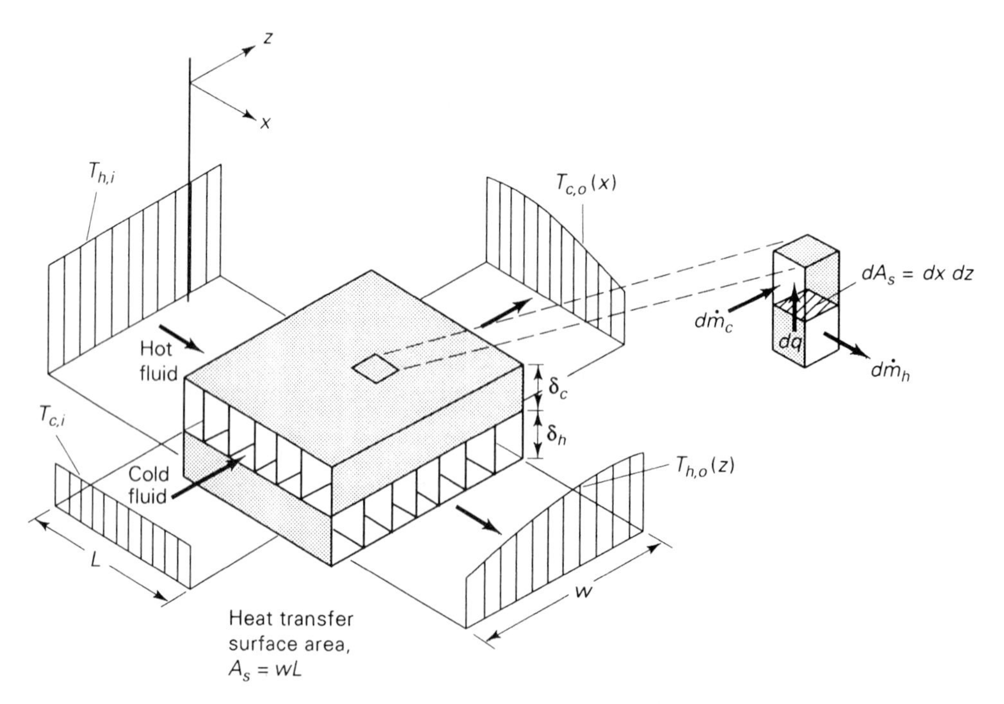

FIGURE 11–22 Sketch for rate of heat transfer dq and distributions in terminal temperatures for unmixed–unmixed crossflow arrangement.

TABLE 11–3 Dimensions of T_h and T_c for mixed and unmixed crossflow arrangements

Stream	*Streamwise direction*	*Dimensions*	
		Mixed	*Unmixed*
Hot	x	$T_h(x)$	$T_h(x,z)$
Cold	z	$T_c(z)$	$T_c(x,z)$

11–4–8 Effect of Two-Phase Flow

The bulk-stream temperature distributions associated with two-phase flow through a boiler with small pressure drop are illustrated in Fig. 11–23(a). For flow processes such as this, the bulk-stream temperature of the two-phase fluid (T_c for boiling, T_h for condensation) corresponds to the saturation temperature T_{sat}. As we have seen, boiling and condensation are commonly characterized by large values of the mean coefficient of heat transfer ($\overline{h}_c$ for boiling, $\overline{h}_h$ for condensation), such that the mean overall coefficient of heat transfer $\overline{U}$ is sometimes controlled by the thermal resistance on the single-phase side; that is (for a thin-wall tube with negligible fouling),

$$\frac{1}{\overline{U}A_s} = \frac{1}{(\overline{h}A_s)_h} + \frac{1}{(\overline{h}A_s)_c} \tag{11–18}$$

which reduces to

$$\overline{U}_h = \overline{h}_h \tag{11–19a}$$

for boiling heat transfer with $\overline{h}_c \gg \overline{h}_h$ and $A_s = A_{s,h}$, or

$$\overline{U}_c = \overline{h}_c \tag{11–19b}$$

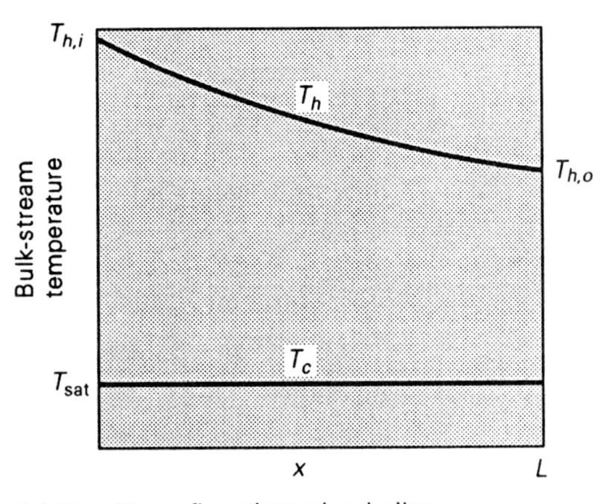

(a) Two-Phase flow through a boiler.

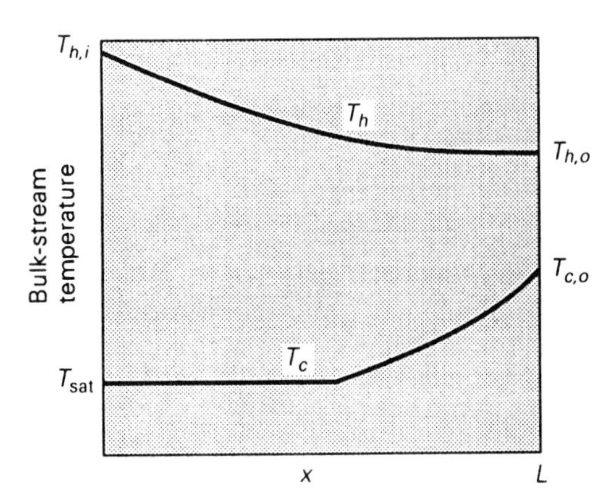

(b) Two-phase flow followed by superheated flow through a boiler.

FIGURE 11–23 Distributions in bulk-stream temperatures.

for condensation with $\bar{h}_h \gg \bar{h}_c$ and $A_s = A_{s,c}$. Whereas this is likely to be the case for boiling or condensation of steam, the convective resistance for organic vapors is generally on the same order of magnitude as the resistance associated with single-phase flow of water at low turbulent Reynolds number. For situations such as this in which the convective resistance on the two-phase side must be accounted for, the mean coefficient of heat transfer for boiling or condensation is obtained by the use of correlations of the type presented in Chap. 10. Referring back to this chapter, we are reminded that the convection correlations for boiling and condensation involve the surface temperature T_s. Because T_s generally changes significantly over the length of the heat exchanger, the analysis of problems of this type usually involves nonuniform distributions in U and requires iteration.

Whereas the bulk-stream temperature of a boiling (or condensing) fluid is uniform as long as the fluid is saturated and the pressure drop is small, the temperature increases (or decreases) sharply when the fluid becomes superheated (or subcooled), as illustrated in Fig. 11–23(b). For this case, $\bar{U}$ may be approximated by Eq. (11–19a) [or Eq. (11–19b)] in the two-phase zone, but is given by Eq. (11–18) in the superheated (or subcooled) region, which generally results in much lower values of $\bar{U}$ because of the lower values of $\bar{h}_c$ (or $\bar{h}_h$).

Finally, when dealing with two-phase flow, we must keep in mind the influence of pressure drop on the saturation temperature. For situations in which the pressure drop within a boiler, evaporator, or condenser is significant, the decrease in saturation temperature should be accounted for.

EXAMPLE 11–1

Freon F-12 at −20°C and 0.4 MPa flows in the annulus of a small double-pipe heat exchanger at a rate of 0.53 kg/s. Hot water at 98°C and 0.2 MPa passes through the tube with a mass-flow rate of 0.07 kg/s. The heat exchanger is constructed of thin-walled copper tubing with 1-cm inside diameter, 2-cm outside diameter, and 3.5-m length. Determine the approximate mean overall coefficient of heat transfer.

Solution

Objective Determine $\bar{U}$.

Schematic Double-pipe heat exchanger: counterflow arrangement.

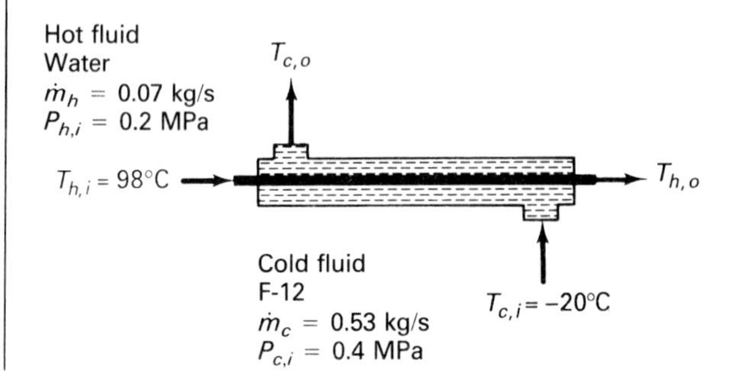

Assumptions/Conditions

negligible conduction resistance R_k
negligible fouling
negligible entrance effects
uniform properties
standard conditions†

Properties Assuming that both fluids are compressed, the properties are approximated by use of Table A–C–3 for saturated liquid.

Hot water at $T = 98°C = 371$ K: $\rho = 1/v_f = 961$ kg/m³, $\mu = 2.86 \times 10^{-4}$ kg/(m s), $c_P = 4.21$ kJ/(kg °C), $k = 0.68$ W/(m °C), $Pr = 1.78$.

Freon F-12 at −20°C:‡ $\rho = 1460$ kg/m³, $\nu = 2.35 \times 10^{-7}$ m²/s, $\mu = \rho\nu = 3.43 \times 10^{-4}$ kg/(m s), $c_P = 0.907$ kJ/(kg °C), $k \simeq 0.071$ W/(m °C), $Pr \simeq 4.4$.

Analysis The Reynolds number of the hot water flowing in the tube is calculated first by writing

$$D_{H,h} = d_i = d_o = 1 \text{ cm} \qquad A_h = \frac{\pi}{4}(0.01 \text{ m})^2 = 7.85 \times 10^{-5} \text{ m}^2$$

$$G_h = \frac{\dot{m}_h}{A_h} = \frac{0.07 \text{ kg/s}}{7.85 \times 10^{-5} \text{ m}^2} = 892 \text{ kg/(m}^2 \text{ s)} \qquad U_{b,h} = \frac{G_h}{\rho_h} = 0.928 \text{ m/s}$$

$$Re_h = \left(\frac{GD_H}{\mu}\right)_h = \frac{[892 \text{ kg/(m}^2 \text{ s)}](0.01 \text{ m})}{2.86 \times 10^{-4} \text{ kg/(m s)}} = 31{,}200$$

Thus, the flow is fully turbulent. Because L/D (= 350) is much greater than 10, we assume that the flow is fully developed with negligible entrance effects. To approximate the coefficient of heat transfer h_h, we utilize Eqs. (8–15) and (8–16).

$$f_h = (1.58 \ln 31{,}200 - 3.28)^{-2} = 0.00585$$

$$Nu_h = \frac{(0.00585/2)(31{,}200)(1.78)}{1.07 + 12.7\sqrt{0.00585/2}\,(1.78^{2/3} - 1)} = 117$$

$$h_h = Nu_h\left(\frac{k}{D_H}\right)_h = 117\,\frac{0.68 \text{ W/(m °C)}}{0.01 \text{ m}} = 7960 \text{ W/(m}^2 \text{ °C)}$$

For the Freon flowing in the annulus, we write

$$D_{H,c} = D_{i,s} - d_o = 1 \text{ cm}$$

$$A_c = \frac{\pi}{4}[(0.02 \text{ m})^2 - (0.01 \text{ m})^2] = 2.36 \times 10^{-4} \text{ m}^2$$

† The term *standard conditions* is used in this chapter to refer to the conditions listed in Example 7–3 plus negligible heat loss to the surroundings.

‡ k and Pr are given to within two significant figures.

$$G_c = \frac{\dot{m}_c}{A_c} = \frac{0.53 \text{ kg/s}}{2.36 \times 10^{-4} \text{ m}^2} = 2250 \text{ kg/(m}^2 \text{ s)} \qquad U_{b,c} = \frac{G_c}{\rho_c} = 1.54 \text{ m/s}$$

$$Re_c = \left(\frac{GD_H}{\mu}\right)_c = \frac{[2250 \text{ kg/(m}^2 \text{ s)}](0.01 \text{ m})}{3.43 \times 10^{-4} \text{ kg/(m s)}} = 65{,}600$$

Referring to Fig. 8–9 and noting that $d_o/D_{i,s} = D_i/D_o = 0.666$, we find that $Nu_c = Nu_{ii} \approx Nu/0.97$. Thus, for this application Nu_c can be set equal to Nu as a reasonable conservative approximation. Using Eqs. (8–15) and (8–16) to calculate Nu, we obtain

$$f_c = (1.58 \ln 65{,}600 - 3.28)^{-2} = 0.00493$$

$$Nu_c = \frac{(0.00493/2)(65{,}600)(4.4)}{1.07 + 12.7\sqrt{0.00493/2}\,(4.4^{2/3} - 1)} = 334$$

$$h_c = Nu_c\left(\frac{k}{D_H}\right)_c = 334\,\frac{0.071 \text{ W/(m °C)}}{0.01 \text{ m}} = 2370 \text{ W/(m}^2 \text{ °C)}$$

The mean overall coefficient of heat transfer $\overline{U}$ is now approximated by using Eq. (11–6) with $A_s = A_{s,h} \simeq A_{s,c}$, $\overline{h}_h \simeq h_h$, $\overline{h}_c \simeq h_c$, and $R_k \simeq 0$,

$$\frac{1}{\overline{U}} = \frac{1}{h_h} + \frac{1}{h_c} = \left(\frac{1}{7960} + \frac{1}{2370}\right)\frac{\text{m}^2 \text{ °C}}{\text{W}}$$

$$\overline{U} = 1830 \text{ W/(m}^2 \text{ °C)}$$

If the outlet temperatures and wall temperatures are known, our calculations for h_h and h_c can be refined by evaluating the properties of each fluid at the arithmetic average of its inlet and outlet temperatures and by utilizing the correction factors introduced in Chap. 8.

It should be noted that the use of the somewhat less conservative approximation for Nu_c indicated by Fig. 8–9 gives rise to $Nu_c = Nu/0.97 = 344$, $h_c = 2440$ W/(m² °C), and $\overline{U} = 1870$ W/(m² °C). Thus, this refinement results in a change of only 2% in the calculation for $\overline{U}$.

EXAMPLE 11–2

A parallel-flow double-pipe heat exchanger is constructed of 0.113-in.-thick steel (0.5% carbon) tubing with 0.824-in.-I.D. inner tube and 2-in.-I.D. outer tube. The tube-side and shell-side local coefficients of heat transfer for a certain application are reported to be approximated by 200 Btu/(h ft² °F) and 1000 Btu/(h ft² °F), respectively, and the tube-side fouling factor is 5.67×10^{-4} (h ft² °F)/Btu. Calculate the local overall coefficient of heat transfer.

Solution

Objective Determine U.

Schematic Double-pipe heat exchanger: parallel-flow arrangement.

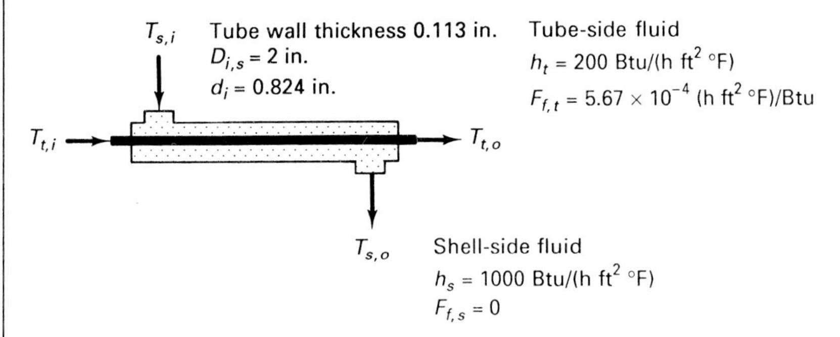

Assumptions/Conditions

uniform properties
standard conditions

Properties Carbon steel (0.5% C) (Table A–C–1):

$$k \simeq 54 \frac{\text{W}}{\text{m °C}} \frac{0.578 \text{ Btu/(h ft °F)}}{\text{W/(m °C)}} = 31.2 \text{ Btu/(h ft °F)}$$

Analysis The local overall coefficient of heat transfer for this situation is represented by Eq. (11–16),

$$\frac{1}{UA_s} = \frac{1}{(hA_s)_t} + \left(\frac{F_f}{A_s}\right)_t + R_k + \left(\frac{F_f}{A_s}\right)_s + \frac{1}{(hA_s)_s}$$

where the subscripts t and s are used in place of h and c, and

$$R_k = \frac{\ln (d_o/d_i)}{2\pi k L}$$

Setting $A_s = A_{s,t} = \pi d_i L$, the local overall coefficient of heat transfer relative to the tube-side surface area is given as

$$\frac{1}{U_t} = \frac{1}{h_t} + F_{f,t} + \frac{d_i}{2k} \ln\left(\frac{d_o}{d_i}\right) + F_{f,s}\frac{d_i}{d_o} + \frac{1}{h_s}\frac{d_i}{d_o}$$

Substituting for the various parameters, we obtain

$$\frac{1}{U_t} = \left[\frac{1}{200} + 5.67 \times 10^{-4} + \frac{0.824/12}{2(31.2)} \ln\left(\frac{1.05}{0.824}\right) + 0 + \frac{1}{1000}\left(\frac{0.824}{1.05}\right)\right] \frac{\text{h ft}^2 \text{ °F}}{\text{Btu}}$$

$$= (0.005 + 0.000567 + 0.000267 + 0.000785)\,\frac{\text{h ft}^2\ ^\circ\text{F}}{\text{Btu}}$$

$$= 0.00664 \text{ h ft}^2\ ^\circ\text{F/Btu}$$

or

$$U_t = 151 \text{ Btu/(h ft}^2\ ^\circ\text{F)}$$

EXAMPLE 11–3

Air at 15°C and 1 atm with entering velocity of 6.8 m/s is to be heated in a crossflow heat exchanger with eighty 9.52-mm-diameter 2.5-m-long thin-wall tubes. The tubes are placed 10 tubes deep in a staggered array with longitudinal and transverse pitches of 1.19 cm and 1.43 cm, respectively. Warm exhaust gas at 65°C enters the tubes with a mass-flow rate of 0.25 kg/s. Determine the mean overall coefficient of heat transfer for this application, assuming that the properties of the exhaust gas are approximately the same as for air.

Solution

Objective Determine $\overline{U}$.

Schematic Crossflow heat exchanger: mixed–unmixed arrangement.

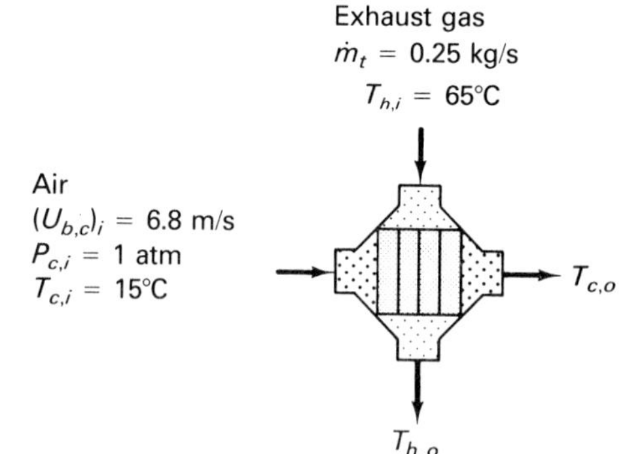

$N = 80$, $N_L = 10$, staggered array
Thin-wall tubes, $D = 9.52$ mm
$S_L = 1.19$ cm, $S_T = 1.43$ cm
$Z = 2.5$ m

From Example 8–18:
$\overline{h}_c = 233$ W/(m² °C)

Assumptions/Conditions

negligible conduction resistance R_k
negligible fouling
negligible entrance effects
uniform properties
standard conditions

Properties Exhaust gas at $T = 65$°C $= 338$ K (Table A–C–3 for air); $\rho = 1.04$ kg/m³, $\mu = 2.02 \times 10^{-5}$ kg/(m s), $c_P = 1.01$ kJ/(kg °C), $k = 0.0291$ W/(m °C), $Pr = 0.7$.

Analysis Referring back to Example 8–18, the mean coefficient of heat transfer $\bar{h}_c$ for flow of air over the tube bank is approximately 233 W/(m² °C) (neglecting variable property effects).

To determine the mean coefficient of heat transfer $\bar{h}_h$ for the tubular flow of warm exhaust gas, we must calculate the Reynolds number Re_h. The mass flux G_h is given by

$$G_h = \frac{\dot{m}_h}{A_h} = \frac{0.25 \text{ kg/s}}{80\pi(0.00952 \text{ m})^2/4} = 43.9 \text{ kg/(m}^2\text{ s)}$$

or $U_{b,h} = (G/\rho)_h = 42.2$ m/s, such that Re_h becomes

$$Re_h = \left(\frac{GD}{\mu}\right)_h = \frac{[43.9 \text{ kg/(m}^2\text{ s)}](0.00952 \text{ m})}{2.02 \times 10^{-5} \text{ kg/(m s)}} = 20{,}700$$

Thus, the flow is turbulent and the Nusselt number Nu_h can be obtained by writing

$$f_h = (1.58 \ln 20{,}700 - 3.28)^{-2} = 0.00648$$

$$Nu_h = \frac{(0.00648/2)(20{,}700)(0.7)}{1.07 + 12.7\sqrt{0.00648/2}\,(0.7^{2/3} - 1)} = 51.2$$

Noting that $Z/D \gg 10$, $\bar{h}_h$ is approximated by

$$\bar{h}_h \simeq h_h = \left(Nu\frac{k}{D}\right)_h = 51.2\frac{0.0291 \text{ W/(m °C)}}{0.00952 \text{ m}} = 156 \text{ W/(m}^2\text{ °C)}$$

Assuming that the thermal resistance of the tube wall and fouling are small, the mean coefficient of heat transfer is approximated by writing

$$\frac{1}{\bar{U}} = \frac{1}{\bar{h}_h} + \frac{1}{\bar{h}_c} = \left(\frac{1}{156} + \frac{1}{233}\right)\frac{\text{m}^2\text{ °C}}{\text{W}}$$

such that

$$\bar{U} = 93.4 \text{ W/(m}^2\text{ °C)}$$

As in Example 11–1, this approximation for $\bar{U}$ can be refined to account for property variation with temperature once the outlet temperatures and the wall temperatures are known.

11–5 HEAT EXCHANGER ANALYSIS

The situations that must be dealt with in the practical analysis of direct transfer heat exchangers (recuperators) range from relatively simple unfinned double-pipe modules involving basic counterflow or parallel flow with uniform cross-sectional area to more complex finned-tube double-pipe, baffled shell-and-tube, and crossflow arrangements with combinations of mixed and unmixed crossflow, counterflow and parallel flow, finned and unfinned tubes, and uniform and nonuniform cross-sectional areas.

To develop a general approach that is applicable to this variety of flow arrangements, we model the heat exchanger core by uniform and continuous flow channels through which the fluids pass in the appropriate relative counterflow, parallel flow, and crossflow paths. Following the approach in Sec. 8–4, the cross-sectional area of the core model is represented by the minimum free-flow area A_{min} for flow passages with nonuniform cross section, such as crossflow over tube banks or flow over finned tubes.

To provide a general framework for analysis, the conditions of the thoroughly mixed entering and exiting streams will primarily be specified by the use of subscripts with reference to the hot (h) and cold (c) fluids according to the schedule listed in Table 11–4. An alternative perspective that is commonly used in the literature is based on the shell or mixed (s) and tube or unmixed (t) sides, as is also indicated in Table 11–4.

TABLE 11–4 Subscripts used to designate mixed entering and exiting conditions of the fluid stream†

	Flow direction		*Station*		
Reference	*inlet*	*outlet*	1	2	*n*
hot fluid	h,i	h,o	$h,1$	$h,2$	h,n
cold fluid	c,i	c,o	$c,1$	$c,2$	c,n
shell or mixed	s,i	s,o	$s,1$	$s,2$	s,n
tube or unmixed	t,i	t,o	$t,1$	$t,2$	t,n

† Other conventions are sometimes used in the literature. Examples include in for i, out for o, i for t, and o for s. In addition, capital T and lower case t are sometimes used to differentiate between the hot and cold (or shell and tubular) fluids.

The general aspects associated with the practical thermal and hydraulic analysis of direct transfer heat exchangers are developed in this section, with specific attention given to steady operation of double-pipe, shell-and-tube, and crossflow heat exchangers in Secs. 11–6 through 11–8. The practical analysis of regenerators and unsteady operation of recuperators is presented in references 7, 10, and 11.

11–5–1 Practical Thermal Analysis

The practical thermal analysis of heat exchangers features the representation of the heat-transfer rate in terms of the local overall coefficient of heat transfer U by Eq. (11–2),†

$$dq = U\,dA_s\,(T_h - T_c) \qquad (11\text{–}2)$$

and, referring to Fig. 11–24, is based on lumped and lumped/differential formulations of the first law of thermodynamics of the general forms

$$q = \dot{H}_{h,i} - \dot{H}_{h,o} \qquad q = \dot{H}_{c,o} - \dot{H}_{c,i} \qquad (11\text{–}20,21)$$

$$dq = -d\dot{H}_h \qquad dq = d\dot{H}_c \qquad (11\text{–}22,23)$$

† As indicated in Sec. 11–7 and Appendix N–5, Eqs. (11–2), (11–22), and (11–23) are put into a somewhat different form for shell-and-tube heat exchangers.

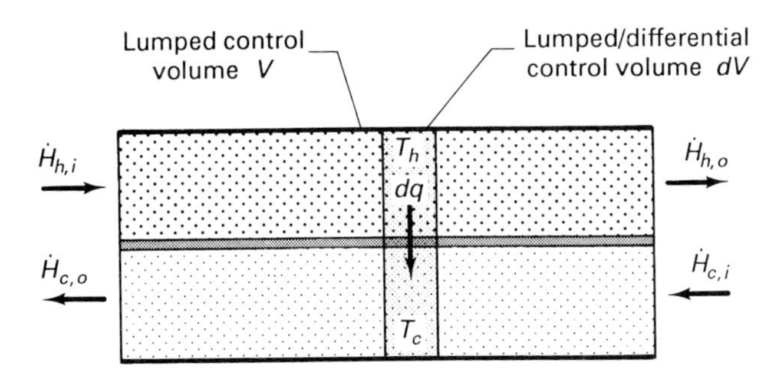

FIGURE 11–24 General perspective for lumped and lumped/differential formulation presented in context of a basic counterflow arrangement.

which represent the total and differential rates of heat transfer from the hot fluid [Eqs. (11–20) and (11–22)] and to the cold fluid [Eqs. (11–21) and (11–23)]. The solution of Eqs. (11–2) and (11–20) to (11–23) depends on the nature of the distributions in bulk-stream temperatures T_h and T_c and on whether the fluids are single-phase or two-phase. Whereas the bulk-stream temperatures are both functions of x alone for counterflow and/or parallel-flow in double-pipe and well mixed shell-and-tube heat exchanger units, two independent variables (e.g., x and z) are required to describe the distributions in T_h and T_c for flow in crossflow heat exchangers. Solution results for double-pipe, shell-and-tube, and crossflow heat exchangers are considered in Secs. 11–6 through 11–8.

For single-phase fluids, Eqs. (11–20) to (11–23) become

$$q = (\dot{m}c_P)_h(T_{h,i} - T_{h,o}) \qquad q = (\dot{m}c_P)_c(T_{c,o} - T_{c,i}) \tag{11–24,25}$$

for uniform specific heats, and

$$dq = -(\dot{m}c_P)_h\, dT_h \qquad dq = (\dot{m}c_P)_c\, dT_c \tag{11–26,27}$$

for one-dimensional distributions in bulk-stream temperature $T_h(x)$ and $T_c(x)$, such as occur in double-pipe or well baffled shell-and-tube heat exchangers, or

$$dq = -(d\dot{m}\; c_P)_h\, dT_h \qquad dq = (d\dot{m}\; c_P)_c\, dT_c \tag{11–28,29}$$

for two-dimensional systems involving crossflow (see Sec. 11–8–1). The products $(\dot{m}c_P)_h$ and $(\dot{m}c_P)_c$ are known as the *capacity rates* and are commonly designated by C_h and C_c, respectively. These equations provide the basis for relating the total rate of heat transfer q and the outlet temperatures $T_{h,o}$ and $T_{c,o}$ to the overall coefficient of heat transfer, surface area A_s, and other system parameters (i.e., fluid properties, mass-flow rates $\dot{m}_h$ and $\dot{m}_c$, inlet fluid temperatures $T_{h,i}$ and $T_{c,i}$, and geometry and flow arrangement).

We now want to consider the primary methods that are available for analyzing energy transfer in heat exchangers, criteria for economical use of heat exchanger surface area, general limiting solution results, and the specification of the mean overall coefficient of heat transfer.

Heat Transfer Analysis Methods

The most commonly used methods for analyzing the thermal performance of heat exchangers include the effectiveness method, the *LMTD* method, and the efficiency method†, all of which give rise to the same solution result for any given problem.

Effectiveness Method To generalize the concept of effectiveness, which was introduced in Chap. 8, we define the *effectiveness* P_I by

$$q = P_I\, q_{\max,I} \tag{11–30}$$

where subscript I refers to the one of the two fluids that is specified as the reference stream and $q_{\max,I}$ is the thermodynamic maximum possible rate of heat transfer to the reference fluid; the nonreference fluid is indicated by subscript II. As we shall see, two approaches are available in the literature for specifying the reference fluid. These two approaches will be considered momentarily.

In the context of the notation used in Fig. 11–24, $q_{\max,I}$ is represented by

$$q_{\max,I} = C_I(T_{h,i} - T_{c,i}) = C_I\, \Delta T_{\mathrm{MAX}} \tag{11–31}$$

where $C_I = (\dot{m}c_P)_I$ and $\Delta T_{\mathrm{MAX}} = T_{h,i} - T_{c,i}$ is the thermodynamic limit on the maximum absolute change in temperature of *either* fluid. Combining Eqs. (11–30) and (11–31), we obtain

$$q = P_I C_I (T_{h,i} - T_{c,i}) = P_I C_I\, \Delta T_{\mathrm{MAX}} \tag{11–32}$$

such that P_I can be expressed in terms of C_I by

$$P_I = \frac{q}{q_{\max,I}} = \frac{q}{C_I(T_{h,i} - T_{c,i})} = \frac{q}{C_I\, \Delta T_{\mathrm{MAX}}} \tag{11–33}$$

With q represented by Eqs. (11–24) and (11–25), Eq. (11–33) takes the form

$$P_I = \frac{C_h(T_{h,i} - T_{h,o})}{C_I(T_{h,i} - T_{c,i})} = \frac{C_c(T_{c,o} - T_{c,i})}{C_I(T_{h,i} - T_{c,i})} \tag{11–34}$$

Thus, we see that the effectiveness P_I represents the ratio of the absolute change in temperature of the reference stream I to the thermodynamic limit ΔT_{MAX}. The values of P_I range between zero and unity.

As shown in the context of the analysis of double-pipe heat exchangers (Sec. 11–6), the effectiveness P_I can be expressed in terms of the *capacitance ratio* R_I,

$$R_I = \frac{C_I}{C_{II}} \tag{11–35}$$

and the number of transfer units NTU_I,

$$NTU_I = \frac{\overline{U}A_s}{C_I} \tag{11–36}$$

† Also commonly referred to as the ψ-P method or the θ method.

which provides a dimensionless index of the size of the heat exchanger; the term $\overline{U}A_s$ represents the heat exchanger thermal capacity. In general, P_I is a function of R_I, NTU_I, and the *geometry* and *flow arrangement*; that is,

$$P_I = P_I(NTU_I, R_I) \quad \text{or} \quad NTU_I = NTU_I(P_I, R_I) \tag{11–37a,b}$$

with the solution results generally expressed in the form of analytical relations and/or charts for double-pipe, shell-and-tube, and crossflow heat exchangers. The form of the functional relations represented by Eq. (11–37) is independent of which fluid is specified as the reference stream I for *symmetrical* exchangers, which include double-pipe, well-mixed 1-2 E-type shell-and-tube, unmixed–unmixed- and mixed–mixed-crossflow arrangements operating under certain conditions (see Appendix N–1). Other exchangers are *nonsymmetrical* in the sense that the form of the functional relations represented by Eq. (11–37) depends on which fluid is taken as the reference stream I.

As an additional point of significance, the true mean temperature difference ΔT_m is expressed in terms of the effectiveness P_I by combining Eqs. (11–8) and (11–32), with the result

$$\Delta T_m = \frac{P_I}{\overline{U}A_s/C_I}(T_{h,i} - T_{c,i}) = \psi(T_{h,i} - T_{c,i}) \tag{11–38}$$

where the heat exchanger *efficiency* ψ is defined by

$$\psi = \frac{P_I}{\overline{U}A_s/C_I} = \frac{P_I}{NTU_I} \tag{11–39}$$

It follows from Eq. (11–38) that the efficiency ψ is also equal to the ratio of the true mean temperature difference ΔT_m to the maximum possible ΔT_{MAX}; that is,

$$\psi = \frac{\Delta T_m}{\Delta T_{\text{MAX}}} = \frac{\Delta T_m}{T_{h,i} - T_{c,i}} \tag{11–40}$$

The effectiveness method employing charts or analytical relations proves to be very useful in both the evaluation and design of heat exchangers. Although the effectiveness method provides a basis for solving basic evaluation and design problems (with $\overline{U}$ specified) without iteration, iterative solution techniques are generally required for comprehensive design problems for which $\overline{U}$ must be determined by use of convection correlations, as will be illustrated in several examples. The effectiveness method is most commonly used by automotive, aircraft, air-conditioning, refrigeration, and other industries that employ compact heat exchangers [7]. The use of the effectiveness method is illustrated in Examples 11–4, 11–5, 11–10, 11–11, 11–13, 11–17, and 11–19 to 11–23.

The relations for effectiveness P_I presented in this section provide the basis for two formats for the effectiveness method that are currently in use for the design and evaluation of heat exchangers, which we will refer to as the *P-NTU*$_I$ method and the ϵ-*NTU* method. We now turn our attention to the defining relations associated with these two perspectives. The effectiveness relations are presented in the context of matrix formalism in Appendix N–2.

P-NTU$_I$ Method The most straightforward approach to specifying the reference stream I is to merely select one fluid or the other, regardless of the relative magnitudes of the capacitance rates of the two fluids. The P-NTU_I method is associated with this approach to designating the reference stream.

To illustrate, with the fluids designated by tube or unmixed t and shell or mixed s, the defining relations for the primary parameters P_I, R_I, and NTU_I and the related parameter ψ_I associated with the P-NTU_I method become

$$P_t = \frac{T_{t,o} - T_{t,i}}{T_{s,i} - T_{t,i}} = \frac{1}{R_t}\frac{T_{s,i} - T_{s,o}}{T_{s,i} - T_{t,i}} \tag{11–41}$$

$$R_t = \frac{C_t}{C_s} = \frac{T_{s,i} - T_{s,o}}{T_{t,o} - T_{t,i}} \tag{11–42}$$

$$NTU_t = \frac{\overline{U}A_s}{C_t} \qquad \psi_t = \frac{P_t}{NTU_t} \tag{11–43,44}$$

for $C_I = C_t$, or

$$P_s = \frac{T_{s,i} - T_{s,o}}{T_{s,i} - T_{t,i}} = \frac{1}{R_s}\frac{T_{t,o} - T_{t,i}}{T_{s,i} - T_{t,i}} \tag{11–45}$$

$$R_s = \frac{C_s}{C_t} = \frac{T_{t,o} - T_{t,i}}{T_{s,i} - T_{s,o}} \tag{11–46}$$

$$NTU_s = \frac{\overline{U}A_s}{C_s} \qquad \psi_s = \frac{P_s}{NTU_s} \tag{11–47,48}$$

for $C_I = C_s$. The relationships for and between these two sets of defining equations for P_I, R_I, NTU_I, and ψ_I are summarized in Table 11–5. For symmetrical exchangers, the functional relations for $P_t(NTU_t,R_t)$ and $P_s(NTU_s,R_s)$ are identical. Although the functional relations for $P_t(NTU_t,R_t)$ and $P_s(NTU_s,R_s)$ are not identical for nonsymmetrical exchangers, it so happens that they are generally *approximately* equal, with the difference being within about 2.5% for $\psi > 0.35$.† Thus, the analytical or graphical functional relations may be used interchangeably with either fluid taken as the reference stream for symmetrical exchangers (with no restrictions) and for nonsymmetrical exchangers (with the error normally being within about 2.5% for $\psi > 0.35$).‡

† The condition $\psi > 0.35$ is generally satisfied for shell-and-tube heat exchangers operating with NTU_I in the normal range between 0.5 and 1.5 to 2.0. However, ψ often falls well below 0.35 for compact crossflow exchangers, which commonly operate at much higher values of NTU_I. For such cases, the error in assuming interchangeability for a nonsymmetrical exchanger can be over 10%.

‡ As an exception, certain solution results for specified nonuniform distribution in U are clearly not interchangeable in either the exact or approximate sense.

TABLE 11–5 Interrelations for P_I, R_I, NTU_I, and ψ_I

Effectiveness P_I	*Capacitance ratio* R_I	NTU_I	*Efficiency* ψ_I
$P_t = R_s P_s$ $P_s = R_t P_t$	$R_t = 1/R_s$ $R_s = 1/R_t$	$NTU_t = R_s\, NTU_s$ $NTU_s = R_t\, NTU_t$	$\psi_t = \psi_s$ $\psi_s = \psi_t$

Supplemental relations:

General
$$R_t = \frac{C_t}{C_s} = \frac{P_s}{P_t} = \frac{NTU_s}{NTU_t} = \frac{1}{R_s} = \frac{\Delta T_s}{\Delta T_t} = \frac{T_{s,i} - T_{s,o}}{T_{t,o} - T_{t,i}}$$

Symmetrical exchanger: Functional relation $P_s(NTU_s,R_s)$ = functional relation $P_t(NTU_t,R_t)$

Nonsymmetrical exchanger: Functional relation $P_s(NTU_s,R_s) \simeq$ functional relation $P_t(NTU_t,R_t)$ with error generally less than 2.5% for $\psi > 0.35$.

To simplify the presentation and to maintain consistency within much of the literature, we will designate the parameters P, R, and ψ without the reference stream subscript I. Using this convention, solutions are represented by functional relations of the form

$$P = P(NTU_I,R) \qquad \text{or} \qquad NTU_I = NTU_I(P,R) \tag{11–49a,b}$$

In this connection, closed-form expressions of the form $P(NTU_I,R)$ are available for many standard-type heat exchangers [see Table A–N–1–1(a)]. On the other hand, closed-form expressions of the form $NTU_I(P,R)$ are only available for basic heat exchanger types such as double-pipe, well-baffled 1-2 TEMA-E shell-and-tube, and crossflow with one fluid mixed and one fluid unmixed [see Table A–N–1–1(b)].

Finally, we simply note that the P–NTU_I method is commonly employed in the current heat transfer literature.

ϵ–NTU Method Following the suggestion of Kays and London [10], the ϵ–NTU method is identified with the specification of the fluid with minimum capacity rate as the reference stream I; that is,

$$C_I = (\dot{m}c_P)_{\min} = C_{\min} \tag{11–50}$$

To distinguish this approach from the P–NTU_I method, we designate the effectiveness P_I by ϵ,

$$\epsilon = \frac{q}{q_{\max}} = \frac{q}{C_{\min}(T_{h,i} - T_{c,i})} = \frac{q}{C_{\min}\,\Delta T_{\text{MAX}}} \tag{11–51}$$

the capacitance ratio R_I by C^*,†

$$C^* = \frac{C_{\min}}{C_{\max}} \tag{11–52}$$

the number of transfer units NTU_I by NTU,

† Whereas R generally ranges from 0 to ∞, C^* varies from 0 to 1.

$$NTU = \frac{\overline{U}A_s}{C_{\min}} \tag{11–53}$$

and the efficiency ψ by

$$\psi = \frac{\epsilon}{NTU} \tag{11–54}$$

In connection with the defining equation for ϵ given by Eq. (11–51), it should be noted that $q_{\max}$ now corresponds to the thermodynamic limiting case in which the outlet temperature of the fluid with minimum capacity rate approaches the inlet temperature of the other fluid.†

To express ϵ in terms of the terminal temperatures, we combine Eq. (11–51) with Eqs. (11–24) and (11–25), with the result

$$\epsilon = \frac{C_h}{C_{\min}} \frac{T_{h,i} - T_{h,o}}{T_{h,i} - T_{c,i}} = \frac{C_c}{C_{\min}} \frac{T_{c,o} - T_{c,i}}{T_{h,i} - T_{c,i}} \tag{11–55}$$

which indicates

$$\epsilon = \frac{T_{h,i} - T_{h,o}}{T_{h,i} - T_{c,i}} = \frac{1}{C^*} \frac{T_{c,o} - T_{c,i}}{T_{h,i} - T_{c,i}} \tag{11–56}$$

for $C_{\min} = C_h \le C_c$, or

$$\epsilon = \frac{T_{c,o} - T_{c,i}}{T_{h,i} - T_{c,i}} = \frac{1}{C^*} \frac{T_{h,i} - T_{h,o}}{T_{h,i} - T_{c,i}} \tag{11–57}$$

for $C_{\min} = C_c \le C_h$.

Relations are available in the literature for ϵ as a function of C^*, NTU, and the geometry and flow arrangements; that is,

$$\epsilon = \epsilon(NTU, C^*) \qquad \text{or} \qquad NTU = NTU(\epsilon, C^*) \tag{11–58a,b}$$

Closed-form expressions for $\epsilon(NTU,C^*)$ and $NTU(\epsilon,C^*)$ are listed in Tables A–N–1–2(a) and (b) for standard-type heat exchangers.

As pointed out by Taborek [6] and Shah and Mueller [7], the disadvantage of the ϵ–NTU method is that the analyst must keep track of which fluid corresponds to $C_{\min}$, a situation which may be changing in the course of iterative design calculations. Aside from the inconvenience of the need for tracking the $C_{\min}$ fluid, the capabilities of the ϵ–NTU method are comparable to the P–NTU_t method.

LMTD Method Practical solution results for heat exchangers are also commonly expressed in the form

$$q = \overline{U}A_s(F\ LMTD) \tag{11–59}$$

† Except for pure counterflow, crossflow with both fluids unmixed, and certain other arrangements for which $\epsilon \to 1$ as $NTU \to \infty$ (see Table A–N–1–2), $q_{\max}$ is not physically attainable for $C^* \neq 0$.

where *LMTD* represents the log mean temperature difference associated with an equivalent counterflow single-pass arrangement with the same hot and cold stream entering and exiting temperatures,

$$LMTD = \frac{(T_{h,i} - T_{c,o}) - (T_{h,o} - T_{c,i})}{\ln[(T_{h,i} - T_{c,o})/(T_{h,o} - T_{c,i})]} = \frac{\Delta T_1 - \Delta T_2}{\ln(\Delta T_1/\Delta T_2)} \qquad (11\text{–}60)$$

where $\Delta T_1 = T_{h,i} - T_{c,o}$, $\Delta T_2 = T_{h,o} - T_{c,i}$, and F is the *LMTD correction factor* (with values between unity and zero) that accounts for the extent to which the actual process differs from ideal counterflow. Comparing Eqs. (11–8) and (11–59), we see that F is related to the true mean temperature difference ΔT_m by

$$F = \frac{\Delta T_m}{LMTD} \qquad (11\text{–}61)$$

Thus, F is a ratio comparing the true mean temperature difference ΔT_m to the ideal temperature difference, which is represented by *LMTD*. For this reason, F is considered to be a physically meaningful design parameter.

The *LMTD* correction factor F is traditionally represented by the functional relation

$$F = F(P,R) \qquad (11\text{–}62)$$

for a specific geometry and flow arrangement, where the effectiveness P and capacitance ratio R are defined by Eqs. (11–41) and (11–42) or Eqs. (11–45) and (11–46). *F-P* design charts are available for many of the types of heat exchangers employed in industry. However, it should be noted that explicit expressions of the standard form $F(P,R)$ are only available for a few basic heat exchanger types, such as those listed in Table A–N–1–1(b). On the other hand, closed-form solutions represented by the functional relation

$$F = F(NTU_I,R) \qquad (11\text{–}63)$$

are available for the heat exchanger types listed in Table A–N–1–1(a), where NTU_I is defined by Eq. (11–36) in a manner consistent with the definition of P and R.

The *LMTD* method employing standard *F-P* charts provides a convenient method for solving the basic heat exchanger design problem, but involves iterative calculations when used in the evaluation of an existing heat exchanger with unspecified outlet temperatures, whether or not $\overline{U}$ is specified. On the other hand, the closed-form expressions represented by $F(NTU_I,R)$ provide a basis for achieving noniterative (albeit somewhat involved) solutions for basic evaluation problems. The *LMTD* method has been traditionally most widely used by process, power, and petrochemical industries that operate shell-and-tube, plate, and other noncompact heat exchangers [7]. The use of the *LMTD* method is illustrated in Examples 11–6 to 11–9, 11–12, 11–14 to 11–16, and 11–18.

Efficiency Method Introducing the heat exchanger *efficiency* ψ defined by Eq. (11–39),

$$\psi = \frac{P}{NTU_I} = \frac{P}{\overline{U}A_s/C_I} \tag{11–64}$$

(also see Sec. 8–2–2), the relation for q given by Eq. (11–32) becomes

$$q = PC_I(T_{h,i} - T_{c,i}) = \psi \overline{U}A_s(T_{h,i} - T_{c,i}) \tag{11–65}$$

Notice that ψ represents the efficiency of a heat exchanger relative to the limiting situation for which the inlet temperature difference $T_{h,i} - T_{c,i} = \Delta T_{\text{MAX}}$ is maintained over the entire surface area A_s. It follows that ψ falls from unity at $A_s = 0$ toward zero as A_s increases. Referring to Eq. (11–40),

$$\psi = \frac{\Delta T_m}{\Delta T_{\text{MAX}}} = \frac{\Delta T_m}{T_{h,i} - T_{c,i}} \tag{11–40}$$

we are also reminded that ψ is equal to the ratio of the true mean temperature difference ΔT_m to the inlet temperature difference $T_{h,i} - T_{c,i}$.

Clearly, solutions of the form $P(NTU_I,R)$ for a specific geometry and flow arrangement can be used to express ψ in terms of NTU_I and R; that is,

$$\psi = \frac{P}{NTU_I} = \frac{P(NTU_I,R)}{NTU_I} = \psi(NTU_I,R) \tag{11–66}$$

Closed-form solutions of this kind are given in Table A–N–1–1(a). Similarly, solutions of the form $NTU_I(P,R)$ can be used to obtain

$$\psi = \frac{P}{NTU_I(P,R)} = \psi(P,R) \tag{11–67}$$

Explicit relations of this form are given in Table A–N–1–1(b).

The efficiency method combines all the variables of the effectiveness and *LMTD* methods in such a way as to eliminate several inconveniences of these methods for hand or computer calculations. This method features the use of ψ-P design curves, known as Mueller's charts, with ψ as the ordinate and P as the abscissa, which incorporate curves for constant values of R and $1/NTU_I$ [21,22]. Whereas F is sometimes incorporated into the ψ-P design curves, Taborek [6] suggests the use of dual design graphs that consist of the ψ-P chart and the F-P chart. The efficiency method is illustrated in Examples 11–12 and 11–18 (see Appendix N–9).

Overview The effectiveness, *LMTD*, and efficiency methods are all in common use for evaluating and designing heat exchangers. The key relations used in these practical analysis methods are summarized in Table 11–6.

To establish the relationship between P (or ϵ), F, and ψ, we should first take note of the key defined (or observed) interrelations listed in Table 11–7. Notice that the slope of straight lines drawn through the origin is equal to ψ on a P-NTU_I or ϵ-NTU chart and $1/NTU_I$ on a ψ-P chart.

TABLE 11-6 Summary of relations employed in the effectiveness, *LMTD*, and efficiency methods

Characteristic	Method: Effectiveness ϵ-*NTU*	Effectiveness *P*-NTU_I	*LMTD*†	Efficiency
Heat-transfer rate q	$q = \epsilon C_{\min}(T_{h,i} - T_{c,i})$	$q = PC_I(T_{h,i} - T_{c,i})$	$q = \overline{U}A_s(F\ LMTD)$	$q = \psi \overline{U}A_s(T_{h,i} - T_{c,i})$
Functional relations				
Closed form	$\epsilon = \epsilon(NTU, C^*)$ $NTU = NTU(\epsilon, C^*)$	$P = P(NTU_I, R)$ $NTU_I = NTU_I(P,R)$	$F = F(NTU_I, R)$ $F = F(P,R)$	$\psi = \psi(NTU_I, R)$ $\psi = \psi(P,R)$
Chart form	ϵ vs. NTU, C^*	P vs. NTU_I, R	F vs. P, R	ψ vs. P, R, NTU_I, F
Dimensionless variables	$NTU = \overline{U}A_s/C_{\min}$ $C^* = C_{\min}/C_{\max}$	$NTU_I = \overline{U}A_s/C_I$ $R = C_I/C_{II}$	$NTU_I = \overline{U}A_s/C_I$ $R = C_I/C_{II}$ $P = \dfrac{q}{C_I(T_{h,i} - T_{c,i})}$ $LMTD = \dfrac{\Delta T_1 - \Delta T_2}{\ln(\Delta T_1/\Delta T_2)}$	$NTU_I = \overline{U}A_s/C_I$ $R = C_I/C_{II}$ $P = \dfrac{q}{C_I(T_{h,i} - T_{c,i})}$
Other variables			$\Delta T_1 = T_{h,i} - T_{c,o}$ $\Delta T_2 = T_{h,o} - T_{c,i}$	
True mean temperature difference	$\Delta T_m = \dfrac{\epsilon}{NTU}(T_{h,i} - T_{c,i})$	$\Delta T_m = \dfrac{P}{NTU_I}(T_{h,i} - T_{c,i})$	$\Delta T_m = F\ LMTD$	$\Delta T_m = \psi(T_{h,i} - T_{c,i})$

† $\Delta T_1 = T_{h,i} - T_{c,i}$ and $\Delta T_2 = T_{h,o} - T_{c,o}$ for parallel-flow double-pipe exchanger. NTU_I is not generally employed in the standard *LMTD/F-P* chart approach.

The *LMTD* correction factor F is expressed in terms of the effectiveness P or ϵ by writing (for counterflow)

$$q = \overline{U}A_s(F\ LMTD) = PC_I(T_{h,i} - T_{c,i}) \tag{11-68}$$

from Eqs. (11-32) and (11-59), and

$$F = \frac{P}{\overline{U}A_s/C_I} \frac{T_{h,i} - T_{c,i}}{\Delta T_1 - \Delta T_2} \ln \frac{\Delta T_1}{\Delta T_2} \tag{11-69}$$

Using the relations for $\Delta T_1/\Delta T_2$ and $(\Delta T_1 - \Delta T_2)/(T_{h,i} - T_{c,i})$ listed in Table 11-7, we obtain†

$$F = \frac{1}{NTU_I\,(1 - R)} \ln \frac{1 - PR}{1 - P} \tag{11-70}$$

or

$$F = \frac{1}{NTU\,(1 - C^*)} \ln \frac{1 - \epsilon C^*}{1 - \epsilon} \tag{11-71}$$

† See Table A-N-1-1 for solution results for $R = C^* = 1$.

TABLE 11–7 Relationships between dimensionless parameters employed in the effectiveness, *LMTD*, and efficiency methods

Type	*Relationships*	
Defined or observed	$P = \epsilon \dfrac{C_{min}}{C_I}$ $\quad R = \dfrac{C_I}{C_{II}}$ $\quad NTU_I = NTU \dfrac{C_{min}}{C_I}$	
	$P = \epsilon \quad R = C^* \quad NTU_I = NTU$	for $C_{min} = C_I$
	$P = \epsilon C^* \quad R = \dfrac{1}{C^*} \quad NTU_I = NTU\, C^* \quad NTU = NTU_I\, R$	for $C_{min} = C_{II}$
	$\psi = \dfrac{\epsilon}{NTU} = \dfrac{P}{NTU_I}$	
	$\dfrac{d\epsilon}{dNTU} = \psi \qquad \dfrac{dP}{dNTU_I} = \psi$	for ψ constant
	$\dfrac{d\psi}{dP} = \dfrac{1}{NTU_I}$	for NTU_I constant
	$\dfrac{\Delta T_1}{\Delta T_2} = \dfrac{T_{h,i} - T_{c,o}}{T_{h,o} - T_{c,i}} = \dfrac{1 - P}{1 - PR}$	
	$\dfrac{\Delta T_1}{\Delta T_2} = \dfrac{T_{h,i} - T_{c,o}}{T_{h,o} - T_{c,i}} = \dfrac{1 - \epsilon C_{min}/C_c}{1 - \epsilon C_{min}/C_h}$	
	$\dfrac{\Delta T_1 - \Delta T_2}{T_{h,i} - T_{c,i}} = -P(1 - R)$	
	$\dfrac{\Delta T_1 - \Delta T_2}{T_{h,i} - T_{c,i}} = -\epsilon \dfrac{C_{min}}{C_c}\left(1 - \dfrac{C_c}{C_h}\right)$	
Derived	$F = \dfrac{1}{NTU\,(1 - C^*)} \ln \dfrac{1 - \epsilon C^*}{1 - \epsilon}$	
	$= \dfrac{\epsilon}{NTU\,(1 - \epsilon)}$	for $C^* = 1$
	$F = \dfrac{1}{NTU_I\,(1 - R)} \ln \dfrac{1 - PR}{1 - P}$	
	$= \dfrac{P}{NTU_I\,(1 - P)}$	for $R = 1$
	$\psi = \dfrac{FP(1 - R)}{\ln\,[(1 - PR)/(1 - P)]}$	
	$= F(1 - P)$	for $R = 1$

To relate ψ and F, we write†

$$q = \overline{U}A_s(F\;LMTD) = \psi \overline{U}A_s(T_{h,i} - T_{c,i}) \tag{11–72}$$

from Eqs. (11–59) and (11–65), and

$$\psi = F \frac{LMTD}{T_{h,i} - T_{c,i}} = F \frac{\Delta T_1 - \Delta T_2}{(T_{h,i} - T_{c,i}) \ln (\Delta T_1/\Delta T_2)}$$

$$= \frac{FP(1 - R)}{\ln\,[(1 - PR)/(1 - P)]} \tag{11–73}$$

† See Table A–N–1–1 for solution results for $R = C^* = 1$.

These relationships between F and P (or ϵ) and ψ and F are summarized in Table 11–7.

As we have indicated, the functional relations for $P(NTU_I,R)$ and $NTU_I(P,R)$ are interchangeable for symmetrical exchangers (with no restrictions) and are approximately interchangeable for nonsymmetrical exchangers (with the error being within about 2.5% for $\psi \gtrsim 0.35$). Because of the relation between the various parameters, this exact or approximate interchangeability also applies to the ϵ-NTU method, the $LMTD$ method, and the effectiveness method.

As indicated earlier, the effectiveness, $LMTD$, and efficiency methods are the primary approaches used in the evaluation and design of heat exchangers. The steps required in the use of these methods are summarized in Table 11–8 for evaluation and design problems. Except for the fact that use of the $LMTD/F$-P chart method for solving basic evaluation problems involves iteration [see Table 11–8(b)], these methods are equivalent. We are also reminded that the chart or computer solution of comprehensive design problems for which $\overline{U}$ is unspecified generally requires iteration.

Minimum Performance Criteria

The effect of increasing the surface area A_s of a heat exchanger is to increase the heat transfer q toward a maximum as the exit fluid temperatures approach limiting values. Thus, the larger the exchanger surface area A_s, the lower the total heat flux, with q/A_s approaching zero as A_s increases. It follows that the economical design of heat exchangers requires that each unit be sized according to criteria that assure that the exchanger surface area is effectively utilized.

As indicated in the previous section, the efficiency ψ, which falls from unity at $A_s = 0$ toward zero as A_s increases, represents the efficiency of a heat exchanger relative to the limiting situation for which the inlet temperature difference $T_{h,i} - T_{c,i}$ is maintained over the entire surface area A_s. Because the efficiency is proportional to the total heat flux q/A_s, ψ provides a useful measure of the overall performance of the heat-exchanger surface. To achieve the economical use of heat-exchanger surface area, ψ should be maintained above a minimum value ψ_{MIN}, with the optimal value of ψ depending on the equipment and operating costs and other factors. As a rule of thumb, ψ_{MIN} can be set equal to 0.35 for double-pipe and shell-and-tube heat exchangers. As we shall see, this minimum performance criterion generally results in values of F_{MIN} between 0.75 and unity and corresponding values of NTU_{MAX} in the range 1.25 to 2.75. However, this criterion is too restrictive for units in series and crossflow exchangers where smaller values of ψ_{MIN} or F_{MIN} can sometimes be tolerated and economically justified [6]. This is particularly true of compact heat exchangers, which are commonly operated at high values of NTU_I. In this connection, Kovarik [23] has recently shown that high values of NTU_I (which correspond to low values of ψ) are optimal for crossflow exchangers only if the incremental costs of heat-transfer surface area A_s and frictional losses are low.

An alternative approach featured in the *Heat Exchanger Design Handbook* [6] to establishing a rule of thumb criterion for the effective use of surface area for

TABLE 11–8(a) Summary of analysis steps for evaluation and design using effectiveness, general *LMTD*, and efficiency methods

Evaluation (rating) problem: Given—A_s, C_h, C_c, geometry, flow arrangement, and inlet temperatures $T_{h,i}$ and $T_{c,i}$

Method	
Effectiveness (ϵ-*NTU*)	*Effectiveness* (*P-NTU$_I$*)
1. Calculate $\overline{U}$ preferably using specific rating correlations. 2. Determine $C_{\min}$ and calculate NTU and C^*. 3. Use the appropriate ϵ-NTU chart or $\epsilon(NTU,C^*)$ functional relation to evaluate ϵ. 4. Calculate q from $q = \epsilon C_{\min}(T_{h,i} - T_{c,i})$. 5. Calculate the outlet temperatures from $T_{h,o} = T_{h,i} - \dfrac{q}{C_h}$ $T_{c,o} = T_{c,i} + \dfrac{q}{C_c}$	1. Calculate $\overline{U}$ preferably using specific rating correlations. 2. Specify C_I and calculate NTU_I and R. 3. Use the appropriate P-NTU_I chart or $P(NTU_I,R)$ functional relation to evaluate P. 4. Calculate q from $q = PC_I(T_{h,i} - T_{c,i})$. 5. Calculate the outlet temperatures from $T_{h,o} = T_{h,i} - \dfrac{q}{C_h}$ $T_{c,o} = T_{c,i} + \dfrac{q}{C_c}$

Design (sizing) problem: Given—C_h, C_c, geometry and flow arrangement, and terminal temperatures $T_{h,i}$, $T_{c,i}$, $T_{h,o}$, $T_{c,o}$

Method	
Effectiveness (ϵ-*NTU*)	*Effectiveness* (*P-NTU$_I$*)
1. Calculate $\overline{U}$ using convection correlations.[a] 2. Determine $C_{\min}$ and calculate ϵ and C^*. 3. Use the appropriate ϵ-NTU chart or $NTU(\epsilon,C^*)$ functional relation to evaluate NTU. 4. Calculate A_s from $A_s = NTU\dfrac{C_{\min}}{\overline{U}}$	1. Calculate $\overline{U}$ using convection correlations.[a] 2. Specify C_I and calculate P and R. 3. Use the appropriate P-NTU_I chart or $NTU_I(P,R)$ functional relation to evaluate NTU_I. 4. Calculate A_s from $A_s = NTU_I\dfrac{C_I}{\overline{U}}$

[a] Because $\overline{U}$ is a function of the system dimensions, the calculations for $\overline{U}$ generally requires iteration.

noncompact heat exchangers is to require that $T_{c,o}$ does not exceed $T_{h,o}$.† The relation that expresses the locus of points for which $T_{h,o} = T_{c,o}$ is represented in terms of the effectiveness P and the heat capacity ratio R by

$$q_o = C_h(T_{h,i} - T_{h,o}) = C_h(T_{h,i} - T_{c,o}) = C_c(T_{c,o} - T_{c,i}) \tag{11–74}$$

† The location within a heat exchanger at which $T_c = T_h$ is known as the *temperature cross*. By setting $T_{c,o} = T_{h,o}$, the temperature cross occurs at the heat-exchanger outlet.

TABLE 11–8(a) *(Continued)*

Determine—Heat-transfer rate q and outlet temperatures $T_{h,o}$ and $T_{c,o}$

General LMTD[b]	*Efficiency* (ψ-P)
1. Calculate $\overline{U}$ preferably using specific rating correlations.	1. Calculate $\overline{U}$ preferably using specific rating correlations.
2. Specify C_I and calculate NTU_I and R.	2. Specify C_I and calculate NTU_I and R.
3. Use the appropriate ψ-P chart or $F(NTU_I,R)$ functional relation to evaluate F.	3. Use the appropriate ψ-P chart or $\psi(NTU_I,R)$ functional relation to evaluate ψ.[d]
4. Evaluate P by use of the appropriate ψ-P chart or Eq. (11–70), which is put into the form (for counterflow)[c] $P = \dfrac{1 - \exp[F(NTU_I)(1 - R)]}{R - \exp[F(NTU_I)(1 - R)]}$	4. Calculate q from $q = \psi\overline{U}A_s(T_{h,i} - T_{c,i})$.
5. Knowing P and R, calculate $T_{c,o}$, $T_{h,o}$, and $LMTD$.	5. Calculate the outlet temperatures from $T_{h,o} = T_{h,i} - \dfrac{q}{C_h}$, $T_{c,o} = T_{c,i} + \dfrac{q}{C_c}$
6. Calculate q from $q = \overline{U}A_s(F\,LMTD)$.	

Determine—A_s

General LMTD	*Efficiency* (ψ-P)
1. Calculate $\overline{U}$ using convection correlations.[a]	1. Calculate $\overline{U}$ using convection correlations.[a]
2. Specify C_I and calculate P and R. Also calculate $LMTD$ and q.	2. Specify C_I and calculate P and R.
3. Use the appropriate F-P chart or $F(P,R)$ functional relation to evaluate F.	3. Use the appropriate ψ-P chart or $\psi(P,R)$ functional relation to evaluate ψ (or NTU_I).
4. Calculate A_s from $A_s = \dfrac{q}{\overline{U}(F\,LMTD)}$	4. Calculate A_s from $A_s = \dfrac{q}{\psi\overline{U}(T_{h,i} - T_{c,i})} = NTU_I\dfrac{C_I}{\overline{U}}$

[b] Steps for evaluation using the standard $LMTD$ method are summarized in Table 11–8(b).

[c] See Eq. (11–137) for parallel flow in double-pipe heat exchanger.

[d] $\psi(P,R)$ design curves incorporate the straight line relation $1/NTU_I = d\psi/dP$, such that ψ can be readily obtained for specified values of NTU_I and R.

With the cold fluid designated as the reference stream, Eq. (11–74) indicates

$$R_o = \frac{C_c}{C_h} = \frac{T_{h,i} - T_{c,o}}{T_{c,o} - T_{c,i}} = \frac{T_{h,i} - T_{c,i} + T_{c,i} - T_{c,o}}{T_{c,o} - T_{c,i}}$$
$$= \frac{1}{P} - 1 = \frac{1 - P}{P} \quad \text{or} \quad P_o = \frac{1}{1 + R} \tag{11–75,76}$$

Notice that this result is also obtained from Eq. (11–74) by designating the hot fluid as the reference stream and recognizing that $T_{h,o} = T_{c,o}$. To express Eq. (11–76) in

TABLE 11–8(b) Summary of analysis steps for evaluation using standard *LMTD* method†

Evaluation (rating) problem:
Given—A_s, C_h, C_c, geometry, flow arrangement, and inlet temperatures $T_{h,i}$ and $T_{c,i}$
Determine—Heat-transfer rate q and outlet temperatures $T_{h,o}$ and $T_{c,o}$

Steps	*Steps*
1. Calculate $\overline{U}$ preferably using specific rating correlations. 2. Specify C_I and calculate R. 3. Set first approximation ($i = 1$) for outlet temperature. 4. Calculate $LMTD^{(i)}$ and $P^{(i)}$ and obtain first approximation for $F^{(i)}$ from the appropriate *F-P* chart.	5. Determine $q^{(i)}$ from $q^{(i)} = \overline{U}A_s\,(F^{(i)}\,LMTD^{(i)})$. 6. Calculate new outlet temperatures to compare with previous approximations. 7. Repeat steps 4 to 6 continuously updating approximations until satisfactory convergence is obtained.

† The iterative nature of this solution is due to the fact that the standard *F-P* charts do not incorporate lines of constant NTU_I. This limitation is eliminated by the ψ-*P* curves, which indicate the relation between F, P, R, and NTU_I, as well as ψ.

terms of ϵ and C^*, we write

$$P_o = \frac{\epsilon_o C_{\min}}{C_I} = \frac{1}{1 + C_I/C_{II}} \tag{11–77}$$

which indicates

$$\epsilon_o = \frac{1}{C_{\min}/C_I + C_{\min}/C_{II}} = \frac{1}{1 + C^*} \tag{11–78}$$

Thus, we require $P \leq P_o$ or $\epsilon \leq \epsilon_o$. Alternatively, this criterion can be represented by $F \geq F_o$, $\psi \geq \psi_o$, or $NTU_I \leq NTU_{I,o}$, which provide equivalent restrictions.† As we shall see, this minimum performance criterion is generally compatible with the ψ_{MIN} criterion. Fore xample, the outlet temperature criterion normally results in values of ψ_o in the range from 0.3 to 0.4 for shell-and-tube heat exchangers. However, under certain conditions involving small or large values of R, this criterion results in values of ψ_o that fall well below 0.3, thus suggesting the possibility of poor exchanger-surface-area performance.

The ψ_{MIN} and F_o criteria provide simple indicators that alert the designer to situations in which operation may be uneconomical.

The actual economic optimization analysis associated with achieving a given heat-transfer rate q in a heat exchanger involves a consideration of the costs related to (1) size of the heat-transfer surface area (capital cost), (2) the pumping or fan power (operating cost), and (3) the cost of the thermal energy supplied to the exchanger [23]. The capital cost, which includes the purchase price of the equipment and expenses related to maintenance, financing, taxes, and insurance, is normally proportional to surface area A_s. The operating cost, which accounts for the expense of achieving and maintaining the flow through both sides of the exchanger, is proportional to the power

† F_o is generally represented by $F_{\min}$ in the literature.

used on both sides. Finally, the cost of the thermal energy supplied to the heat exchanger is simply proportional to the product of the heat-transfer rate q and the unit cost of thermal energy. Following the analysis of Kovarik [23], which is summarized in Appendix N–3, it turns out that the optimal design of a heat exchanger is not influenced by the unit cost of thermal energy. The implication of this conclusion is extremely significant since the possible dependence of heat exchanger design on the cost of fuel required to supply the necessary thermal energy could create significant uncertainties in times of unstable fuel cost.

Limiting Solution Results for Heat Exchanger with One Isothermal Stream ($C^* = R = 0$)

Recognizing that Eqs. (11–24) and (11–25) can be represented by

$$q = C_{\min}\,\Delta T_{\max} = C_{\max}\,\Delta T_{\min} \tag{11–79}$$

where $\Delta T_{\max}$ and $\Delta T_{\min}$ represent the change in temperature of the fluid associated with the $C_{\min}$ stream and the $C_{\max}$ stream, respectively, we conclude that

$$C^* = \frac{C_{\min}}{C_{\max}} = \frac{\Delta T_{\min}}{\Delta T_{\max}} \simeq 0 \tag{11–80}$$

for applications with one stream characterized by approximately uniform bulk-stream temperature (i.e., $\Delta T_{\min} \simeq 0$) or large capacity rate (i.e., $1/C_{\max} \simeq 0$). This limiting situation commonly occurs for applications involving two-phase flow (boiling or condensation) or one strong stream with very high capacity rate relative to the weak stream. Because the temperature of the $C_{\max}$ fluid stream is uniform for $C^* = 0$, the $C_{\min}$ fluid stream experiences simple uniform wall-temperature heating, such that a generalized form of the basic solution developed in Sec. 8–2–3 applies. That is, setting $\dot{m}c_P = C_{\min}$, $T_0 - T_1 = T_{h,i} - T_{c,i}$ and replacing $\bar{h}$ by $\overline{U}$ to account for resistances due to the wall, fins, and fouling, we obtain [from Eq. (8–52)]

$$q = C_{\min}(T_{h,i} - T_{c,i})[1 - \exp(-NTU)] \tag{11–81}$$

where $NTU = \overline{U}A_s/C_{\min}$. As we shall see, all the various solution results for q associated with double-pipe, shell-and-tube, and crossflow heat exchangers reduce to Eq. (11–81) as $C^* \to 0$.

Equation (11–81) is expressed in terms of the effectiveness ϵ by

$$q = \epsilon C_{\min}(T_{h,i} - T_{c,i}) \tag{11–82}$$

where

$$\epsilon = 1 - \exp(-NTU) \qquad \text{for } C^* = 0 \tag{11–83a}$$

or equivalently

$$NTU = -\ln(1 - \epsilon) \qquad \text{for } C^* = 0 \tag{11–83b}$$

This relation for ϵ is shown in Fig. 11–25.

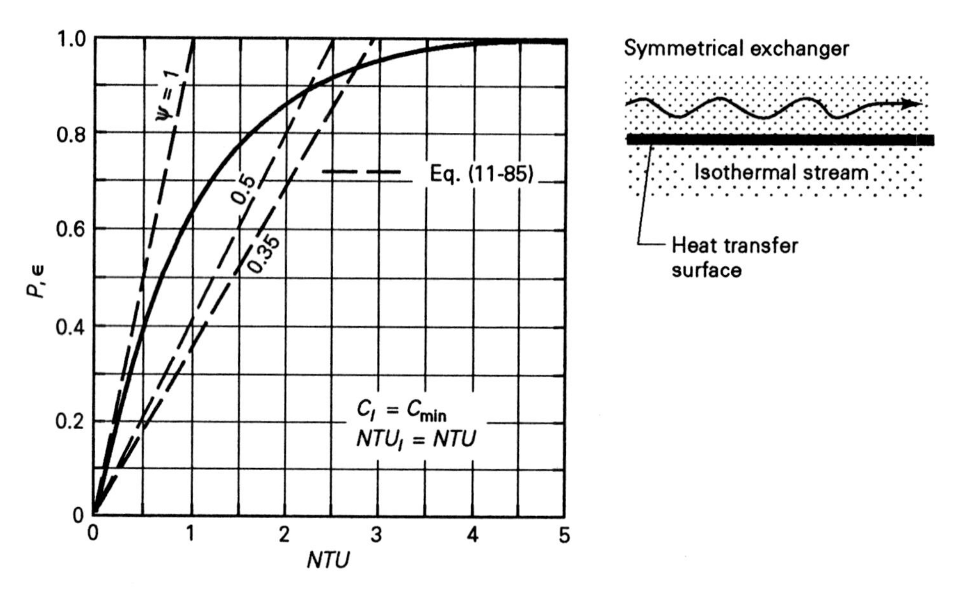

FIGURE 11–25 Limiting solution results for heat exchanger with one isothermal stream ($C^* = R = 0$): effectiveness P or ϵ.

This important limiting result is expressed in terms of the effectiveness P for the corresponding case in which $R = 0$ (and $C_I = C_{\text{min}}$) by writing

$$P = 1 - \exp(-NTU_I) \qquad \text{for } R = 0 \tag{11–84a}$$

or

$$NTU_I = NTU = -\ln(1 - P) \qquad \text{for } R = 0 \tag{11–84b}$$

Notice that with the C_{min} fluid taken as the reference stream, P and ϵ are equivalent.

In connection with Fig. 11–25, the locus of points corresponding to constant values of efficiency ψ is deduced from the defining equation which indicates

$$\frac{d\epsilon}{dNTU} = \psi = \text{constant} \tag{11–85}$$

This straight line relation is shown in Fig. 11–25 for several values of ψ. In this connection, with ψ_{MIN} set equal to 0.35, we would not want to operate a heat exchanger with $NTU > 2.65$ under the limiting condition for which $C^* = 0$.†

To express the solution for this limiting case in terms of the efficiency ψ, we simply write

$$q = \psi \bar{U} A_s (T_{h,i} - T_{c,i}) \tag{11–86}$$

where

$$\psi = \frac{\epsilon}{NTU} = \frac{1 - \exp(-NTU)}{NTU} = \frac{\epsilon}{\ln[1/(1 - \epsilon)]} \tag{11–87}$$

† Notice that the outlet temperature criterion for limiting the size of heat exchangers breaks down for applications with small values of C^*.

or

$$\psi = \frac{P}{NTU_I} = \frac{1 - \exp(-NTU_I)}{NTU_I} = \frac{P}{\ln[1/(1 - P)]} \tag{11–88}$$

where $C_I = C_{\min}$ and $NTU_I = NTU$. Equation (11–87) is shown in terms of ψ versus ϵ in Fig. 11–26 (see Fig. 8–14 for ψ versus NTU). This figure clearly shows the decrease in efficiency ψ with increasing values of ϵ or NTU. Notice that NTU is represented by the straight line relation

$$\frac{d\psi}{d\epsilon} = \frac{1}{NTU} = \text{constant} \tag{11–89}$$

in accordance with the defining equation for ψ.

Finally, we note that the solution given by Eq. (11–81) is expressed in terms of $LMTD$ by [from Eq. (8–59)]

$$q = \overline{U}A_s\, LMTD \tag{11–90}$$

where $\Delta T_1 = T_{h,i} - T_{c,o}$ and $\Delta T_2 = T_{h,o} - T_{c,i}$, such that $F = 1$ for $C^* = R = 0$.

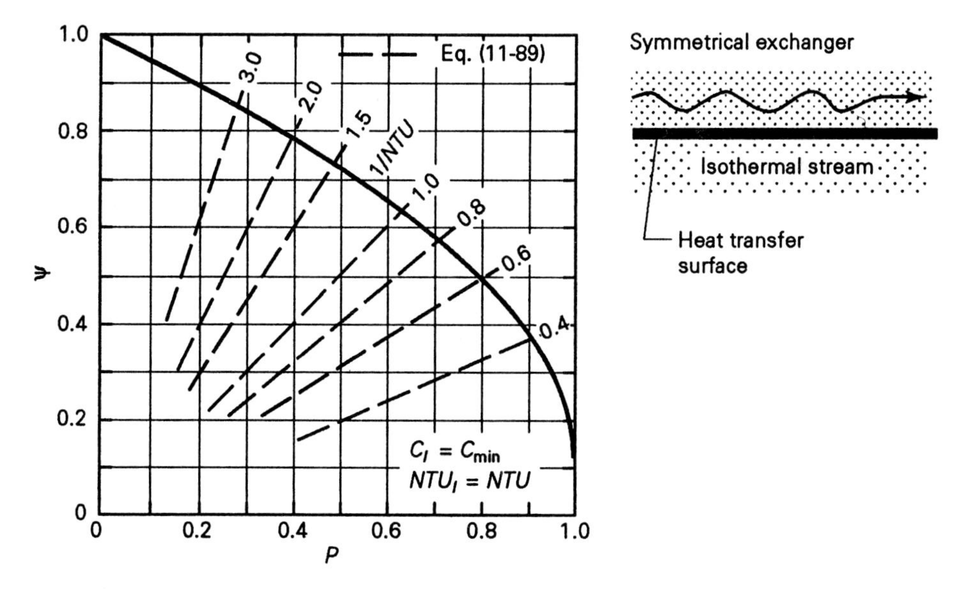

FIGURE 11–26 Limiting solution results for heat exchanger with one isothermal stream ($C^* = R = 0$): efficiency ψ.

Evaluation of $\overline{U}$

In design work the mean overall coefficient of heat transfer $\overline{U}$ is generally evaluated by the use of design correlations for the mean (or local) coefficients of heat transfer $\bar{h}_h$ and $\bar{h}_c$ (or h_h and h_c) represented by Nusselt number $\overline{Nu}$ [$(\bar{h}D_H/k)_h$, $(\bar{h}D_H/k)_c$] or

Stanton number $\overline{St}$ $[\bar{h}_h/(\rho c_P U_b)_h, \bar{h}_c/(\rho c_P U_b)_c]$ in terms of Reynolds number Re $[(GD_H/\mu)_h, (GD_H/\mu)_c]$ and Prandtl number Pr $[(\mu c_P/k)_h, (\mu c_P/k)_c]$, and inputs for resistance associated with fouling (and fins), using the relations presented in Sec. 11–4. The nature of the correlations for $\bar{h}_h$ and $\bar{h}_c$ is dependent on the geometry, flow arrangement, and on whether the flow is laminar or turbulent. Correlations of the type introduced in Chap. 8 for $\bar{h}$ can generally be used for double-pipe and crossflow heat exchangers. However, the evaluation of $\bar{h}$ for baffled-shell-side flow requires special attention. This point will be considered in Sec. 7.

In the thermal analysis of existing heat exchangers, correlations can be readily established for $\bar{U}$ on the basis of recorded measurements for the terminal temperatures and flow rates for approximate uniform property conditions by the use of solutions for effectiveness P or ϵ, the *LMTD* correction factor F, or the efficiency ψ. For example, using the P-NTU_I method, P can be computed from Eq. (11–41) or Eq. (11–45) and R can be obtained from Eq. (11–42) or Eq. (11–46). Knowing P and R, the $P(NTU_I,R)$ functional relation that corresponds to the geometry and flow arrangement of interest can be used to evaluate the number of transfer units NTU_I, after which $\bar{U}$ can be computed by putting Eq. (11–36) into the form

$$\bar{U} = \frac{C_I}{A_s} NTU_I \tag{11–91}$$

The use of the ϵ-NTU method involves similar steps in the calculation of ϵ, C^*, and NTU, with $\bar{U}$ obtained from

$$\bar{U} = \frac{C_{\min}}{A_s} NTU \tag{11–92}$$

Similarly, following the *LMTD* approach, Eq. (11–60) and the defining relations for P and R can be used to compute *LMTD*, P, and R, after which F can be obtained from the appropriate $F(P,R)$ curve. $\bar{U}$ can then be computed from Eq. (11–59) and Eq. (11–24) or Eq. (11–25); that is,

$$\bar{U} = \frac{q}{A_s(F\ LMTD)} = \frac{C_h(T_{h,i} - T_{h,o})}{A_s(F\ LMTD)} = \frac{C_c(T_{c,o} - T_{c,i})}{A_s(F\ LMTD)} \tag{11–93}$$

Finally, following the efficiency approach, the defining relations can be used to compute P and R, after which ψ can be obtained from the functional relation $\psi(P,R)$, NTU_I can be calculated from $NTU_I = P/\psi$, and $\bar{U}$ can be obtained from Eq. (11–91).

Calculations for $\bar{U}$ obtained in any of these ways (all of which are equivalent) account for complicating factors that may be present related to fouling, fins, and baffles. By specifying the mean coefficient of heat transfer on one side (say the tube side), the calculations obtained for $\bar{U}$ over a range of operating conditions can be used to develop a correlation for the mean coefficient of heat transfer for the other side (say the shell side) in terms of $\overline{Nu}$ or $\overline{St}$ versus Re and Pr.

11–5–2 Practical Hydraulic Analysis

Referring to Figs. 8–17 and 8–24 and to the hydraulic analysis developed in Secs. 8–2–2 and 8–4–2, the total pressure drop ΔP_{1-2} (relative to inlet and outlet stations 1 and 2) for incompressible flow on one side of a heat exchanger is generally represented by Eq. (8–75),

$$\Delta P_{1-2} = \frac{\rho U_b^2}{2}\left(K_c + \frac{4L}{D_H} f + K_e\right) \tag{11–94}$$

for uniform density, or Eq. (8–77),

$$\Delta P_{1-2} = \frac{G^2 v_1}{2}\left[(1 - \sigma^2 + K_c) + \frac{4L}{D_H} f \frac{v_m}{v_1} + 2\left(\frac{v_2}{v_1} - 1\right) - (1 - \sigma^2 - K_e)\frac{v_2}{v_1}\right] \tag{11–95}$$

for situations in which significant variations in density are caused by heating or cooling. Whereas Eq. (11–94) is generally applicable to liquids or to gases with low to moderate rates of heating, Eq. (11–95) should be used for gas flow with moderate to strong heating. In design work, the Fanning friction factor f, entrance-loss coefficient K_c, and expansion-loss coefficient K_e are specified in accordance with design correlations for flow through tubes, annuli, tube banks, and other heat-exchanger cores. In this connection, the entrance and exit losses are accounted for by f with K_c and K_e set equal to zero for flow across tube banks (and through matrix surfaces), such that Eqs. (11–94) and (11–95) reduce to Eq. (8–157),

$$\Delta P_{1-2} = \frac{4L}{D_H}\frac{f}{2}\rho U_b^2 = \frac{4L}{D_H}\frac{f}{2}\frac{G^2}{\rho} \tag{11–96}$$

for uniform density, and

$$\begin{aligned}\Delta P_{1-2} &= \frac{G^2 v_1}{2}\left[1 - \sigma^2 + \frac{4L}{D_H} f \frac{v_m}{v_1} + 2\left(\frac{v_2}{v_1} - 1\right) - (1 - \sigma^2)\frac{v_2}{v_1}\right] \\ &= \frac{G^2 v_1}{2}\left[(1 + \sigma^2)\left(\frac{v_2}{v_1} - 1\right) + \frac{4L}{D_H} f \frac{v_m}{v_1}\right]\end{aligned} \tag{11–97}$$

for variable density. The mean specific volume v_m is defined by [see Eq. (M–23)],

$$v_m = \frac{1}{A_s}\int_{A_s} v\, dA_s \tag{11–98}$$

which can generally be approximated by Eq. (M–24),

$$v_m = \frac{v_1 + v_2}{2} \tag{11–99}$$

In principle, Eqs. (11–95) and (11–97) apply to two-phase flow as well as single-phase flow. However, the evaluation of the friction factor for boiling or condensation is considerably more involved than for single-phase flow. Practical approaches to evaluating the pressure drop for two-phase flow are considered in reference 11.

As pointed out by Kays and London [10], losses in return bends or headers must be accounted for separately, as must any losses in inlet and exit headers and nozzles and associated valves and ducting. The additional losses associated with such components are generally represented by

$$\Delta P_f = K_f \frac{\rho U_b^2}{2} \tag{11–100}$$

where K_f is the *component-loss coefficient*. Values for K_f are given in references 6, 24, and 25 for a number of common fittings, valves, and cross sections. Representative values are listed in Table 11–9 for several types of heat exchanger components. For example, combining Eqs. (11–94) and (11–100), the tube-side or shell-side pressure drop in standard heat exchangers for uniform density is given by

$$\Delta P = \frac{\rho U_b^2}{2}\left(K_c + \frac{4L}{D_H} f + K_e + K_f\right) \tag{11–101}$$

TABLE 11–9 Loss-coefficients K_f for heat-exchanger components

Component	K_f	*Component*	K_f
Tube flow		Shell-side flow	
180° tube bend	1.5	Exit nozzle	5.0
Shell-and-tube return header	4.0	Entrance nozzle	5.0
Flow expanding from inlet nozzle into header with turn	1.5	Annular flow	
		Double-pipe return housing	0.5
Flow turn and contraction into outlet nozzle	1.5		

It should be noted that similar to the situation involved in the calculation of the mean overall coefficient of heat transfer, the specification of the friction factor f in the evaluation of existing heat exchangers can be established by the use of rating correlations that are based on the analysis of specific performance data. This approach is particularly useful in the evaluation of shell-and-tube heat exchangers for which the baffled shell-side design correlations for f are quite involved. In this approach, measurements for the pressure drop ΔP_{1-2} together with standard inputs for K_c, K_e, and K_f are substituted into Eq. (11–101) to compute f for isothermal conditions. By varying the flow rate over the range of conditions of interest, a rating correlation is readily developed for f in terms of Re; that is,

$$f = \frac{D_H}{4L}\left(\frac{2}{\rho U_b^2}\Delta P_{1-2} - K_c - K_e - K_f\right) \tag{11–102}$$

Specific attention is given to the evaluation of pressure drop and pumping power in double-pipe, shell-and-tube, and crossflow heat exchangers in the following sections.

11–6 DOUBLE-PIPE HEAT EXCHANGERS

Double-pipe heat exchanger modules constructed of bare or finned tubes are commonly employed as single units or in a variety of series–parallel combinations. Because of their tube-in-tube configuration, double-pipe heat exchanger modules can operate under high pressures and are normally used in applications involving low to moderate heat-transfer rates, with one or both fluids being a liquid.

The hydraulic diameters that are used to characterize the fluid flow and heat transfer in double-pipe heat exchangers are represented by Eq. (11–14a),

$$D_H = \frac{4A}{p_w} \tag{11–103}$$

Referring to Fig. 11–27, the hydraulic diameter for unfinned surfaces simply reduces to $D_H = d_i$ for the tube side, and

$$D_H = \frac{\pi(D_{i,s}^2 - d_o^2)}{\pi(D_{i,s} + d_o)} = D_{i,s} - d_o \tag{11–104}$$

for the annulus. Equation (11–103) also applies to fin-tube double-pipe heat exchangers. Representative geometric dimensions for standard longitudinal fin-tube double

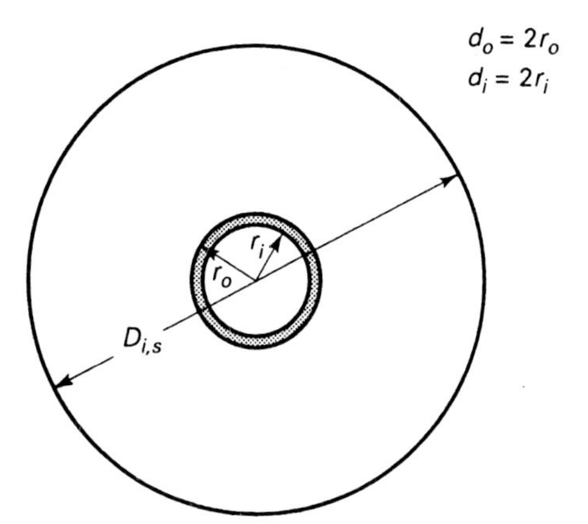

FIGURE 11–27
Unfinned double pipe.

pipes are listed in Table 8–4. The tube-side and annular-side flows for finned and unfinned tubes are characterized in terms of the standard Reynolds number Re,

$$Re = \frac{U_b D_H}{\nu} = \frac{G D_H}{\mu} \tag{11–105}$$

We now turn our attention to the practical thermal and hydraulic analysis of double-pipe heat exchangers.

11–6–1 Practical Thermal Analysis

As indicated in Sec. 11–5, the practical thermal analysis approach to heat exchanger evaluation and design for uniform property conditions generally involves the use of theoretical relations presented in analytical or graphical form for the effectiveness P or ϵ, log mean temperature difference correction factor F, or efficiency ψ, and the evaluation of the mean overall coefficient of heat transfer $\overline{U}$. We first consider the development of these relations for flow through a double-pipe heat exchanger module for steady-state conditions, after which consideration will be given to the evaluation of $\overline{U}$.

Solution Results for Heat-Transfer Rate

Focusing attention on single-phase flow with uniform properties and constant flow rates, the practical thermal analysis involves the solution of the lumped formulation given by Eqs. (11–24) and (11–25),

$$q = (\dot{m}c_P)_h(T_{h,i} - T_{h,o}) \qquad q = (\dot{m}c_P)_c(T_{c,o} - T_{c,i}) \tag{11–106,107}$$

and the lumped/differential formulation given by Eqs. (11–26) and (11–27),

$$dq = -(\dot{m}c_P)_h\, dT_h \qquad dq = (\dot{m}c_P)_c\, dT_c \tag{11–108,109}$$

with the differential heat-transfer rate dq expressed in terms of the local overall coefficient of heat transfer U by Eq. (11–2),

$$dq = U\, dA_s\, (T_h - T_c) \tag{11–110}$$

The control volumes associated with these lumped/differential formulations are shown in Fig. 11–28. As shown in this figure, we designate parallel-flow and counterflow arrangements relative to a coordinate system with $x = 0$ at the inlet of the hot fluid (station 1) and $x = L$ at the other end (station 2) by writing

$$(\dot{m}c_P)_h = C_h \qquad (\dot{m}c_P)_c = C_c \qquad \text{for parallel flow} \tag{11–111a,b}$$

$$(\dot{m}c_P)_h = C_h \qquad (\dot{m}c_P)_c = -C_c \qquad \text{for counterflow} \tag{11–112a,b}$$

which accounts for the fact $\dot{m}_h$ and $\dot{m}_c$ both have positive streamwise directions for parallel flow and have opposing streamwise directions for counterflow.

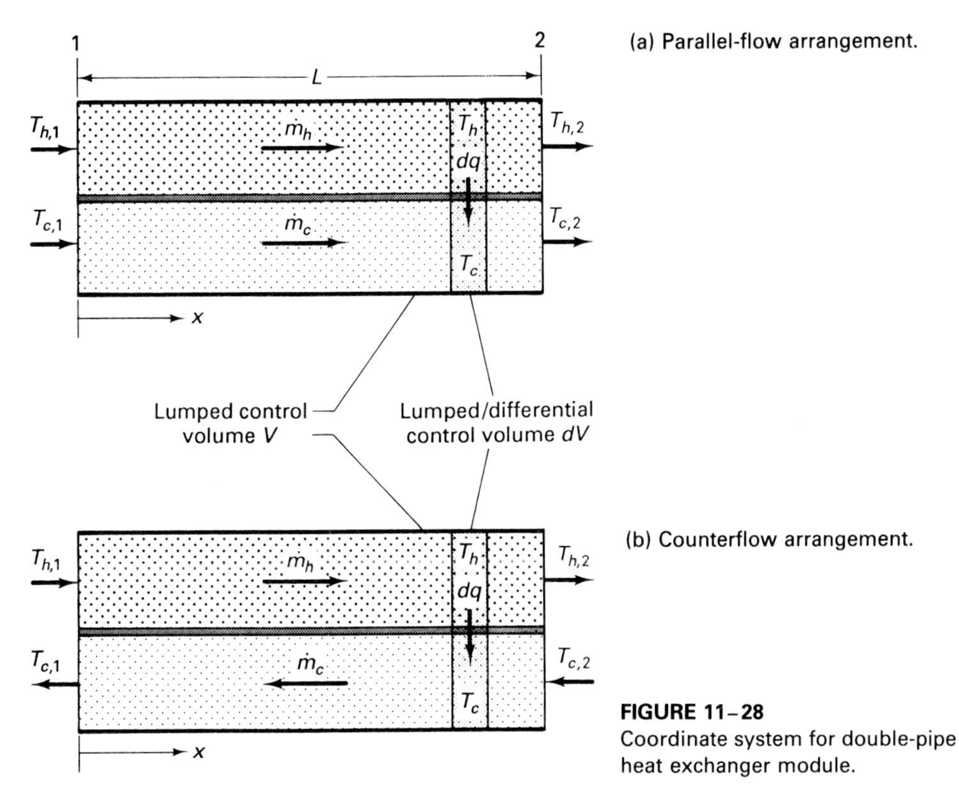

(a) Parallel-flow arrangement.

(b) Counterflow arrangement.

FIGURE 11–28
Coordinate system for double-pipe heat exchanger module.

Following this perspective with dq represented by Eq. (11–110) and dA_s set equal to $p\ dx$, Eqs. (11–106)–(11–109) become

$$q = C_h(T_{h,1} - T_{h,2}) \qquad \frac{dT_h}{T_h - T_c} = -\frac{Up}{C_h}\,dx \qquad \text{hot fluid} \qquad (11\text{–}113,114)$$

$$q = \pm C_c(T_{c,2} - T_{c,1}) \qquad \frac{dT_c}{T_h - T_c} = \frac{Up}{\pm C_c}\,dx \qquad \text{cold fluid} \qquad (11\text{–}115,116)$$

where parallel flow and counterflow correspond to the positive and negative signs, respectively.

To simplify matters, Eqs. (11–114) and (11–116) are combined to form a single differential equation for the dependent variable $T_h - T_c$; that is,

$$\frac{d(T_h - T_c)}{T_h - T_c} = -\left(\frac{1}{C_h} \pm \frac{1}{C_c}\right) Up\ dx \qquad (11\text{–}117)$$

or, for convenience,

$$\frac{d(T_h - T_c)}{T_h - T_c} = -\zeta Up\ dx \qquad \zeta = \frac{1}{C_h} \pm \frac{1}{C_c} \qquad (11\text{–}118,119)$$

Referring to Fig. 11–28, the boundary condition for $T_h - T_c$ is designated relative to stations 1 and 2 by

$$T_h - T_c = T_{h,1} - T_{c,1} \qquad \text{at } x = 0 \tag{11–120}$$

for both parallel-flow and counterflow arrangements. Using this boundary condition, the solution to Eq. (11–118) is

$$\ln \frac{T_h - T_c}{T_{h,1} - T_{c,1}} = -\int_0^x \zeta U p \, dx = -\zeta \overline{U} p x \tag{11–121}$$

for uniform perimeter p and uniform properties, where $\overline{U}$ is the mean overall coefficient of heat transfer over the length x,

$$\overline{U} = \frac{1}{x}\int_0^x U \, dx \tag{11–122}$$

Rearranging, the solution is put into the form

$$\frac{T_h - T_c}{T_{h,1} - T_{c,1}} = \exp(-\zeta \overline{U} p x) \tag{11–123}$$

Setting $T_h - T_c = T_{h,2} - T_{c,2}$ at $x = L$, this equation reduces to

$$\frac{T_{h,2} - T_{c,2}}{T_{h,1} - T_{c,1}} = \exp(-\zeta \overline{U} A_s) \tag{11–124}$$

where $\overline{U}$ is evaluated over the length L and the product $\overline{U}A_s$ represents the thermal capacity of the heat exchanger.

Equations (11–113), (11–115), and (11–124) provide the basis for developing solutions for any three dependent variables for parallel-flow and counterflow double-pipe heat exchangers. We now combine these equations to express the solutions in terms of the effectiveness P or ϵ, log mean temperature difference $LMTD$, and efficiency ψ.

Effectiveness Method The effectiveness P is defined by Eq. (11–33),

$$P = \frac{q}{q_{\max,I}} = \frac{q}{C_I(T_{h,i} - T_{c,i})} \tag{11–125}$$

which takes the form

$$P = \frac{q}{C_I(T_{h,1} - T_{c,1})} \tag{11–126}$$

for parallel flow, and

$$P = \frac{q}{C_I(T_{h,1} - T_{c,2})} \tag{11–127}$$

for counterflow. Therefore, Eqs. (11–113), (11–115), and (11–124) are combined to obtain an expression for $q/(T_{h,i} - T_{c,i})$. This is done by substituting Eqs. (11–113) and (11–115) into Eq. (11–124) to eliminate (1) $T_{h,2}$ and $T_{c,2}$ for parallel flow with $(\dot{m}c_P)_c = C_c$, and (2) $T_{h,2}$ and $T_{c,1}$ for counterflow with $(\dot{m}c_P)_c = -C_c$. Taking these steps, we obtain

$$q = \frac{T_{h,1} - T_{c,1}}{\zeta_p}\left[1 - \exp\left(-\zeta_p \overline{U} A_s\right)\right] \qquad \zeta_p = \frac{1}{C_h} + \frac{1}{C_c} \tag{11–128,129}$$

for parallel flow, and

$$q = \frac{(T_{h,1} - T_{c,2})\left[1 - \exp\left(-\zeta_c \overline{U} A_s\right)\right]}{1/C_h - (1/C_c)\exp\left(-\zeta_c \overline{U} A_s\right)} \qquad \zeta_c = \frac{1}{C_h} - \frac{1}{C_c} \tag{11–130,131}$$

for counterflow. Substituting Eq. (11–128) into Eq. (11–126), we have

$$P = \frac{(T_{h,1} - T_{c,1})\left[1 - \exp\left(-\zeta_p \overline{U} A_s\right)\right]}{\zeta_p C_I (T_{h,1} - T_{c,1})} = \frac{1 - \exp\left(-\zeta_p \overline{U} A_s\right)}{1 + R} \tag{11–132}$$

for parallel flow. Similarly, the substitution of Eq. (11–130) into Eq. (11–127) gives

$$P = \frac{1 - \exp\left(-\zeta_c \overline{U} A_s\right)}{C_I/C_h - (C_I/C_c)\exp\left(-\zeta_c \overline{U} A_s\right)} \tag{11–133}$$

for counterflow.

The parameters $\zeta_p \overline{U} A_s$ and $\zeta_c \overline{U} A_s$ appearing in Eqs. (11–132) and (11–133) can be expressed in terms of the capacitance ratio R and number of transfer units NTU_I by writing

$$\zeta_p \overline{U} A_s = \left(\frac{1}{C_h} + \frac{1}{C_c}\right)\overline{U} A_s = \frac{\overline{U} A_s}{C_I}\left(1 + \frac{C_I}{C_{II}}\right) = NTU_I\,(1 + R) \tag{11–134}$$

for parallel flow with $C_I = C_h$ or $C_I = C_c$, and

$$\zeta_c \overline{U} A_s = \left(\frac{1}{C_h} - \frac{1}{C_c}\right)\overline{U} A_s = \frac{\overline{U} A_s}{C_I}\left(\frac{C_I}{C_h} - \frac{C_I}{C_c}\right) = NTU_I\,(1 - R) \tag{11–135}$$

for counterflow with $C_I = C_h$, or

$$\zeta_c \overline{U} A_s = NTU_I\,(R - 1) \tag{11–136}$$

for counterflow with $C_I = C_c$. Substituting these relations into Eqs. (11–132) and (11–133), we obtain

$$P = \frac{1 - \exp\left[-NTU_I\,(1 + R)\right]}{1 + R} \tag{11–137}$$

for parallel flow, and

$$P = \frac{1 - \exp\left[-NTU_I\,(1 - R)\right]}{1 - R\exp\left[-NTU_I\,(1 - R)\right]} \tag{11–138a}$$

for counterflow with $C_I = C_h$, or

$$P = \frac{1 - \exp[NTU_I(1 - R)]}{R - \exp[NTU_I(1 - R)]} \tag{11–138b}$$

for counterflow with $C_I = C_c$, both of which are equivalent.* Because the solution results for P are independent of which fluid is specified as the reference stream I, double-pipe heat exchangers are classified as symmetrical exchangers.†

It should also be noted that Eqs. (11–137) and (11–138) may be represented in the functional form $NTU_I(P,R)$ by

$$NTU_I = \frac{1}{1 + R} \ln \frac{1}{1 - P(1 + R)} \tag{11–139}$$

for parallel flow, and

$$NTU_I = \frac{1}{1 - R} \ln \frac{1 - PR}{1 - P} \tag{11–140}$$

for counterflow.

These relations for $P(NTU_I,R)$ are shown in Fig. 11–29 for parallel-flow and counterflow arrangements.‡ It should be observed that for $R > 0$ the effectiveness P is higher for counterflow than for parallel flow. As a result, counterflow exchangers are generally utilized in practice. It is also observed that the maximum effectiveness is approached as $R \rightarrow 0$, for which case Eqs. (11–137) and (11–138) reduce to Eq. (11–84a),

$$P = 1 - \exp(-NTU_I) \qquad \text{for } R = 0 \tag{11–84a}$$

where $C_I = C_{\min}$. As we have seen, this limiting relation actually corresponds to a constant temperature for the fluid with the larger capacity rate. Lines for $\psi_{MIN} = 0.35$ are also shown in Fig. 11–29 for parallel flow and counterflow. To assure economical use of the exchanger surface area, the value of NTU_I should not exceed the values of $NTU_{I,MAX}$, which correspond to ψ_{MIN}. As an alternative criterion for economical surface area use, the locus of points given by Eq. (11–76),

$$P_o = \frac{1}{1 + R} \tag{11–76}$$

which designates equal outlet temperatures, is shown for counterflow in Fig. 11–29(a). This criterion indicates values of ψ_o in the range 0.5 to 0.35 for $0.1 < R < 10$, but ψ_o falls below 0.35 for $R < 0.1$ and $R > 10$. On the other hand, because the equal

* See Table A–N–1–1 for the solution result for $R = 1$.

† As shown in Appendix N–4, Eqs. (11–137) and (11–138) satisfy a general relation between overall parallel-flow and counterflow heat-exchanger arrangements.

‡ P-NTU_I charts are commonly presented in a semilog format in order to stretch the NTU_I scale in the range for which NTU_I lies between 0.1 and 10. (See Fig. 11–33.)

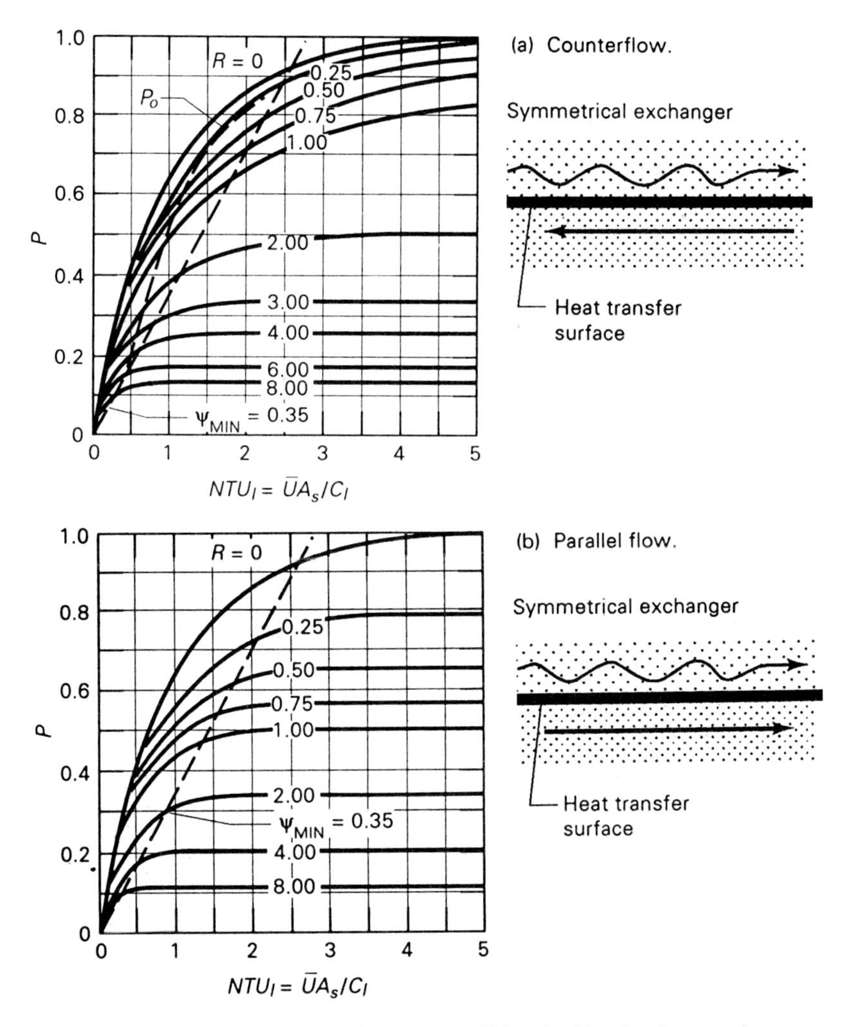

FIGURE 11–29 Effectiveness P for double-pipe heat exchanger.

outlet temperature condition is approached asymptotically as NTU_I increases for parallel flow, this criterion is not viable for parallel-flow arrangements.

The solution results represented by Eqs. (11–137) to (11–140) are readily expressed in the ϵ-NTU format by writing

$$\epsilon = \frac{1 - \exp\left[-NTU\,(1 + C^*)\right]}{1 + C^*} \tag{11–141}$$

or

$$NTU = \frac{1}{1 + C^*} \ln \frac{1}{1 - \epsilon(1 + C^*)} \tag{11–142}$$

for parallel flow, and†

$$\epsilon = \frac{1 - \exp\left[-NTU\,(1 - C^*)\right]}{1 - C^* \exp\left[-NTU\,(1 - C^*)\right]} \qquad (11\text{–}143a)$$

$$= \frac{1 - \exp\left[NTU\,(1 - C^*)\right]}{C^* - \exp\left[NTU\,(1 - C^*)\right]} \qquad (11\text{–}143b)$$

or

$$NTU = \frac{1}{1 - C^*} \ln \frac{1 - \epsilon C^*}{1 - \epsilon} \qquad (11\text{–}144)$$

for counterflow. Notice that these relations are independent of whether the $C_{\min}$ stream is hot h or cold c (or shell side s or tube side t), which is characteristic of symmetrical exchangers.‡ Equations (11–141) to (11–144) are shown in Fig. 11–30. The curves associated with this more compact ϵ-NTU representation are identical to the P-NTU_I curves for P in the range 0 to 1.

As indicated in Sec. 11–5–1, the effectiveness approach is particularly useful in evaluating the performance of existing heat exchangers that have been tested for certain conditions but are to be used under different service conditions. In addition, the effectiveness approach can be used for basic design-type calculations in which the thermal capacity $\overline{U}A_s$ is unknown and the outlet temperatures are specified. In the case of double-pipe heat exchangers, the P-NTU_I method and the ϵ-NTU method provide an equally convenient basis for iterative design calculations.

LMTD Method The basic design function can also be conveniently handled by the *LMTD* method. To express the solutions in the *LMTD* format, we first put Eqs. (11–113), (11–115), and (11–124) into the form

$$C_h = \frac{q}{T_{h,1} - T_{h,2}} \qquad \pm C_c = \frac{q}{T_{c,2} - T_{c,1}} \qquad (11\text{–}145{,}146)$$

and

$$\zeta \overline{U}A_s = \left(\frac{1}{C_h} \pm \frac{1}{C_c}\right)\overline{U}A_s = -\ln \frac{T_{h,2} - T_{c,2}}{T_{h,1} - T_{c,1}} \qquad (11\text{–}147)$$

or

$$\frac{1}{C_h} \pm \frac{1}{C_c} = \frac{1}{\overline{U}A_s} \ln \frac{T_{h,1} - T_{c,1}}{T_{h,2} - T_{c,2}} \qquad (11\text{–}148)$$

† See Table A–N–1–1 for the solution result for $C^* = 1$.

‡ This is contrasted to solution results for certain nonsymmetrical shell-and-tube and crossflow heat exchangers that depend on whether the $C_{\min}$ stream is on the shell side or tube side or is mixed or unmixed.

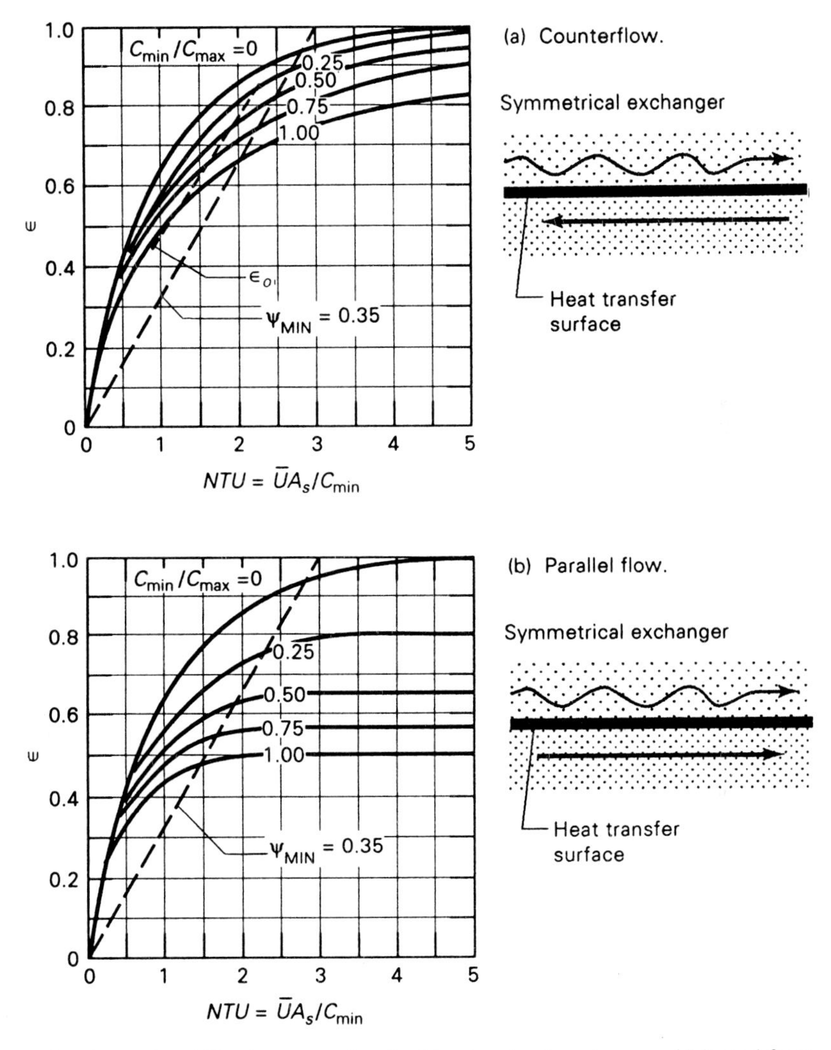

FIGURE 11-30 Effectiveness ε for double-pipe heat exchanger. (Adapted from Kays and London [10]. Used with permission.)

Substituting Eqs. (11-145) and (11-146) into Eq. (11-148), the solution for q becomes

$$\frac{T_{h,1} - T_{h,2}}{q} + \frac{T_{c,2} - T_{c,1}}{q} = \frac{1}{\bar{U}A_s} \ln \frac{T_{h,1} - T_{c,1}}{T_{h,2} - T_{c,2}} \tag{11-149}$$

which reduces to

$$q = \bar{U}A_s \, LMTD \tag{11-150}$$

where

$$LMTD = \frac{(T_{h,1} - T_{c,1}) - (T_{h,2} - T_{c,2})}{\ln [(T_{h,1} - T_{c,1})/(T_{h,2} - T_{c,2})]} = \frac{\Delta T_1 - \Delta T_2}{\ln (\Delta T_1/\Delta T_2)} \tag{11–151}$$

with $\Delta T_1 = T_{h,1} - T_{c,1}$ and $\Delta T_2 = T_{h,2} - T_{c,2}$. This result applies to both parallel-flow and counterflow arrangements and indicates $\Delta T_m = LMTD$.

Comparing Eq. (11–150) with the more general log mean temperature difference equation given by Eq. (11–59),

$$q = \overline{U}A_s(F\, LMTD) \tag{11–152}$$

the *LMTD* correction factor *F* for parallel flow or counterflow in a double-pipe heat exchanger module is simply equal to unity.

Although this completes the traditional *LMTD* formulation, we want to provide a basis for expressing the ψ_{MIN} criterion in terms of *P* and *R*. To accomplish this objective for parallel flow, we combine Eq. (11–139) with the defining equation for ψ, with the result

$$\psi = \frac{P}{NTU_I} = \frac{P(1 + R)}{\ln 1/[1 - P(1 + R)]} > \psi_{MIN} \tag{11–153}$$

Similarly, we use Eq. (11–140) for counterflow to obtain†

$$\frac{P(1 - R)}{\ln [(1 - PR)/(1 - P)]} > \psi_{MIN} \tag{11–154}$$

With *P* and *R* required to satisfy these inequality relations, we can be assured of maintaining $\psi > \psi_{MIN}$. The alternative equal outlet temperature criterion for counterflow is represented by $P < P_o$, with P_o given by Eq. (11–76). As we have seen, the equal outlet temperature criterion is not applicable for parallel flow.

Efficiency Method The solutions for double-pipe heat exchangers can be expressed in terms of the efficiency ψ by simply substituting for *P* or NTU_I in the defining relation for ψ,

$$\psi = \frac{P}{NTU_I} \tag{11–155}$$

Thus, for parallel flow we obtain

$$\psi = \frac{1 - \exp [-NTU_I (1 + R)]}{NTU_I (1 + R)} \tag{11–156}$$

† Noting that $P = NTU_I/(1 + NTU_I)$ for $R = 1$ (see Table A–N–1–1), Eq. (11–154) reduces to $P < 1 - \psi_{MIN}$.

from Eq. (11–137), and

$$\psi = \frac{P(1 + R)}{\ln 1/[1 - P(1 + R)]} \tag{11–157}$$

from Eq. (11–139). Similarly, for counterflow we write

$$\psi = \frac{1}{NTU_I} \frac{1 - \exp[-NTU_I(1 - R)]}{1 - R\exp[-NTU_I(1 - R)]} \tag{11–158a}$$

or

$$\psi = \frac{1}{NTU_I} \frac{1 - \exp[NTU_I(1 - R)]}{R - \exp[NTU_I(1 - R)]} \tag{11–158b}$$

from Eq. (11–138), and

$$\psi = \frac{P(1 - R)}{\ln[(1 - PR)/(1 - P)]} \tag{11–159}$$

from Eq. (11–140).† These relations are shown in Fig. 11–31. The lines of constant $1/NTU_I$ in this figure are based on the defining equation for ψ, which indicates

$$\frac{d\psi}{dP} = \frac{1}{NTU_I} \tag{11–160}$$

As we have seen, heat exchangers are generally operated in the region for which $\psi > \psi_{MIN}$, with ψ_{MIN} usually on the order of 0.35 for noncompact exchangers. The alternative outlet temperature criterion for counterflow represented by Eq. (11–76) indicates

$$\psi \geq \psi_o = \frac{1 - R}{1 + R} \frac{1}{\ln(1/R)} \tag{11–161a}$$

or equivalently

$$\psi \geq \psi_o = \frac{2P - 1}{\ln[P/(1 - P)]} \tag{11–161b}$$

This relation for ψ_o is also shown in Fig. 11–31(a). As we have observed earlier in the context of the effectiveness method, ψ_o for counterflow ranges between 0.5 and 0.35 for $0.1 < R < 10$, but falls below 0.35 for $R < 0.1$ and $R > 10$. On the other hand, Eqs. (11–157) and (11–76) indicate $\psi_o = 0$ for parallel flow, which does not provide a useful limiting criterion.

A comparison of Figs. 11–29 and 11–31 indicates that the ψ-P charts present the same information pertaining to the variables P, R, and NTU_I as the P-NTU_I charts. Therefore, the ψ-P charts can be readily used in conjunction with either the P-NTU_I method or the ψ-P method to develop quick hand calculations for basic heat-exchanger arrangements.

† See Table A–N–1–1 for the solution result for $R = 1$.

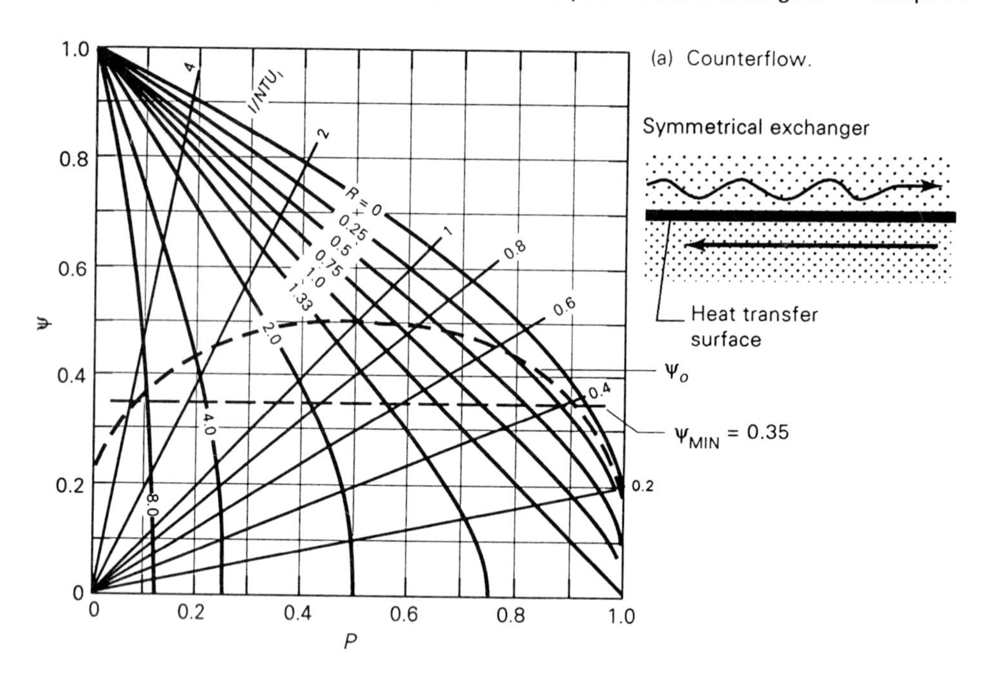

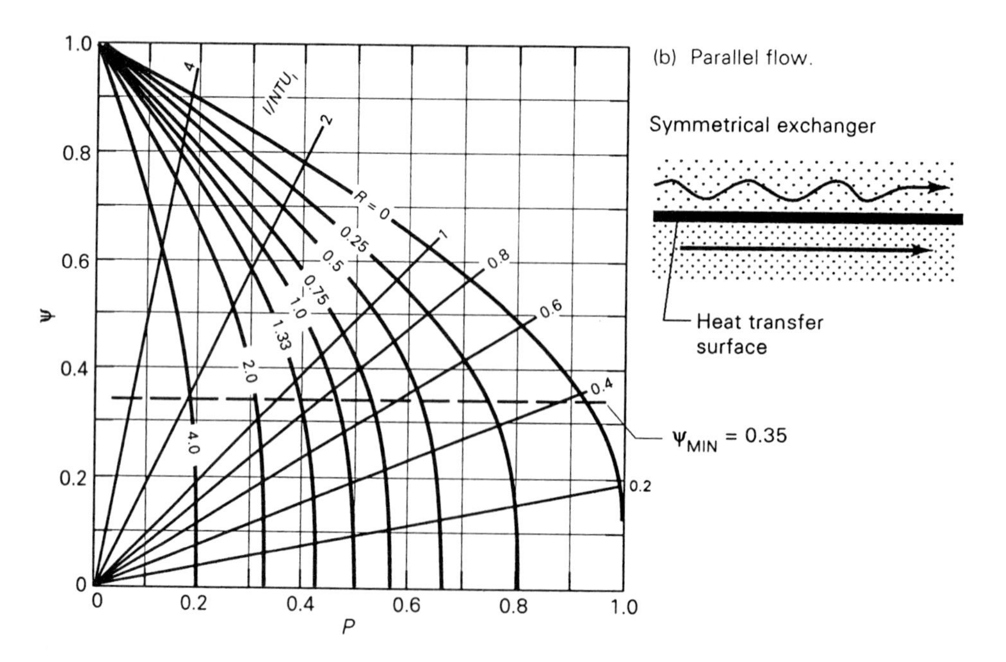

FIGURE 11–31 ψ-P charts for double-pipe heat exchangers.

Evaluation of $\bar{U}$

Referring back to the uniform property analysis for double-pipe heat exchangers, we are reminded that the solution for the rate of heat transfer q is expressed in terms of the mean overall coefficient of heat transfer $\bar{U}$ [see Eq. (11–122)], where the local coefficient U can be uniform or nonuniform.†

The mean overall coefficient of heat transfer $\bar{U}$ for double-pipe heat exchangers is generally approximated by equations of the form of Eq. (11–6),

$$\frac{1}{\bar{U}A_s} = \frac{1}{(\bar{h}A_s)_h} + \left(\frac{F_f}{A_s}\right)_h + R_k + \left(\frac{F_f}{A_s}\right)_c + \frac{1}{(\bar{h}A_s)_c} \tag{11–162}$$

with the effects of fins accounted for by the method described in Sec. 11–4–3.

Correlations for Nusselt number $\overline{Nu} = \bar{h}D_H/k$ versus $Re = GD_H/\mu$ and Pr of the type presented in Chap. 8 for bare and finned tubes and annuli are generally used in the evaluation and design of double-pipe heat exchangers.

11–6–2 Practical Hydraulic Analysis

The practical hydraulic analysis approach described in Sec. 11–5–2 can be used to compute the pressure drop for both tube-side and shell-side fluids. It should be noted that the total pressure drop for series arrangements corresponds to the pressure drop for an individual parallel path for each stream.

For typical installations with equal pipework and heat exchanger tube diameters, no entrance and exit losses occur. However, for standard hairpin units, the pressure drop associated with return bends should be accounted for. It follows that the tube-side pressure drop for a double-pipe heat-exchanger system involving a parallel–series combination of units with M hairpin turns in each series path and no entrance and exit losses can be represented by Eq. (11–101) with $K_c = 0$, $K_e = 0$, and the return-bend-loss coefficient $K_{f,t}$ set equal to $1.5M$,

$$(\Delta P_t)_{1-2} = \left(\frac{\rho U_b^2}{2}\right)_t \left[\left(\frac{4L}{D_H}f\right)_t + 1.5M\right] \tag{11–163}$$

where $D_H = d_i$ and L_t represents the total length of tubing in a series path. The Fanning friction factor f_t is specified by the use of correlations for fully developed flow in bare and fin tubes.

Similarly, the annular-side pressure drop for a parallel–series arrangement of the type indicated above is given by

$$(\Delta P_a)_{1-2} = \left(\frac{\rho U_b^2}{2}\right)_a \left[\left(\frac{4L}{D_H}f\right)_a + 0.5M\right] \tag{11–164}$$

† As we shall see, standard solution results for shell-and-tube and crossflow heat exchangers are normally based on the assumption that U is approximately uniform throughout the exchanger.

assuming that the entrance and exit losses are negligible. The friction factor f_a is evaluated by use of standard correlations for annular flow. However, it should be noted that the pressure drop through entering and exiting nozzles can be significant.

Equations (11–163) and (11–164) indicate that the pressure drop of either stream for flow through a unit with no return bend is proportional to U_b^2L. Thus, the consequence of placing two such units in parallel rather than in series results in a pressure drop that is a factor of 8 smaller than the value for the series arrangement. Therefore, parallel arrangements are often used in industrial operations, particularly when a substantial unbalance exists between the mass-flow rate of the hot and cold fluids. Further consideration is given to the analysis of networks of double-pipe heat-exchanger units in Sec. 11–9.

EXAMPLE 11–4†

Freon F-12 at −20°C and 0.4 MPa flowing at a rate of 0.53 kg/s is heated in a double-pipe heat exchanger. Hot water with a mass-flow rate of 0.07 kg/s enters the tubes at a temperature of 98°C and 0.2 MPa. The heat exchanger is constructed of thin-walled copper tubing with 1-cm inside diameter, 2-cm outside diameter and 3.5-m length. Determine the total rate of heat transfer and the distributions in bulk-stream and wall temperatures for a *parallel-flow* arrangement. Also determine whether or not the minimum performance criteria are satisfied.

Solution

Objective Determine q, the distributions in T_h, T_c, and wall temperature T_w, and whether or not the minimum performance criteria are satisfied for parallel flow.

Schematic Double-pipe heat exchanger: parallel-flow arrangement.

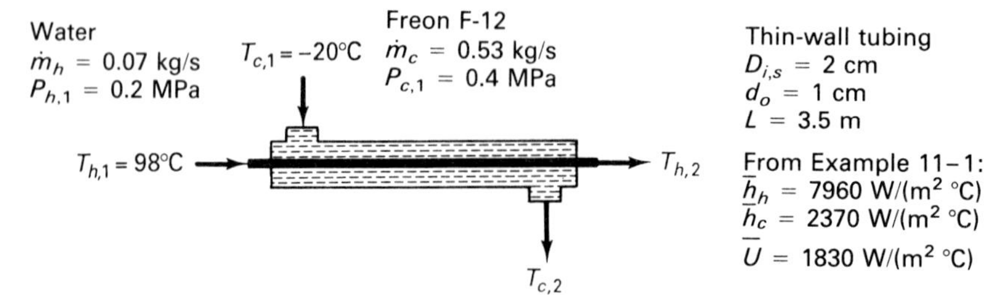

Assumptions/Conditions

negligible conduction resistance R_k

† A simple numerical finite-difference program for solving this example is given in Fig. A–G–5 of the Appendix.

negligible fouling
negligible entrance effects
uniform properties
standard conditions

Properties

Water at $T = 98°C = 371$ K (Table A–C–3): $c_P = 4.21$ kJ/(kg °C).

F-12 at $-20°C$ (Table A–C–3): $c_P = 0.907$ kJ/(kg °C).

Analysis The heat-transfer surface area A_s and thermal capacity $\overline{U}A_s$ are obtained by writing

$$A_s = \pi d_o L = \pi(0.01 \text{ m})(3.5 \text{ m}) = 0.11 \text{ m}^2$$

and

$$\overline{U}A_s = 1830 \frac{\text{W}}{\text{m}^2 \text{ °C}} (0.11 \text{ m}^2) = 201 \text{ W/°C}$$

Because the inlet temperatures and the flow rates are given, the effectiveness approach will be utilized to determine the rate of heat transfer and the outlet temperatures. Therefore, calculations are first obtained for the parameters C_h, C_c, R, NTU_I, and $q_{\max,I}$ by writing

$$C_h = (\dot{m} c_P)_h = 0.07 \frac{\text{kg}}{\text{s}} \left(4.21 \frac{\text{kJ}}{\text{kg °C}}\right) = 0.294 \text{ kW/°C}$$

$$C_c = (\dot{m} c_P)_c = 0.53 \frac{\text{kg}}{\text{s}} \left(0.907 \frac{\text{kJ}}{\text{kg °C}}\right) = 0.481 \text{ kW/°C}$$

$$R = \frac{C_h}{C_c} = \frac{0.294}{0.481} = 0.611 \qquad NTU_I = \frac{\overline{U}A_s}{C_h} = \frac{201 \text{ W/°C}}{0.294 \text{ kW/°C}} = 0.684$$

$$q_{\max,I} = C_h(T_{h,i} - T_{c,i}) = 0.294 \frac{\text{kW}}{\text{°C}} [98°\text{C} - (-20°\text{C})] = 34.7 \text{ kW}$$

where the hot fluid has been taken as the reference stream I.†

For this parallel-flow arrangement, the effectiveness P is given by Eq. (11–137) or by Fig. 11–29(b). For better accuracy, we utilize the analytical relation to obtain

$$P = \frac{1 - \exp[-NTU_I(1 + R)]}{1 + R} = \frac{1 - \exp[-0.684(1.61)]}{1.61} = 0.415$$

† With the hot fluid selected as the reference stream I, the P-NTU_I and ϵ-NTU methods for this problem are identical, with $P = \epsilon$, $R = C^*$, and $NTU_I = NTU$.

It follows that

$$q = Pq_{\max,I} = 0.415(34.7\text{ kW}) = 14.4\text{ kW}$$

The outlet temperatures are calculated as follows:

$$T_{c,2} = \frac{q}{C_c} + T_{c,1} = \frac{14.4\text{ kW}}{0.481\text{ kW/°C}} - 20°\text{C} = 9.94°\text{C}$$

$$T_{h,2} = T_{h,1} - \frac{q}{C_h} = 98°\text{C} - \frac{14.4\text{ kW}}{0.294\text{ kW/°C}} = 49°\text{C}$$

To get a better feel for the problem, expressions are developed for the bulk-stream temperature distributions. To obtain predictions for T_h and T_c, we first substitute Eq. (11–123) into Eq. (11–114).

$$dT_h = -(T_{h,1} - T_{c,1})[\exp(-\zeta\overline{U}px)]\frac{Up}{C_h}\,dx$$

Integrating from 0 to x with $\overline{U}$ and U set equal to 1830 W/(m^2 °C) as a first approximation, we obtain

$$\frac{T_h - T_{h,1}}{T_{h,1} - T_{c,1}} = \frac{1}{\zeta C_h}[\exp(-\zeta\overline{U}px) - 1]$$

$$T_h = T_{h,1} + \frac{T_{h,1} - T_{c,1}}{\zeta C_h}\left[\exp\left(-\zeta C_h\, NTU_I\frac{x}{L}\right) - 1\right] \tag{a}$$

where $NTU_I = 0.684$ and

$$\zeta C_h = 1 + \frac{C_h}{C_c} = 1 + R = 1.61$$

Similarly, the coupling of Eqs. (11–123) and (11–116) gives rise to an expression for T_c of the form

$$T_c = T_{c,1} + \frac{T_{h,1} - T_{c,1}}{\zeta(\dot{m}c_P)_c}\left[1 - \exp\left(-\zeta C_h\, NTU_I\frac{x}{L}\right)\right] \tag{b}$$

where we take the positive sign [i.e., $(\dot{m}c_P)_c = C_c$] for parallel flow and

$$\zeta C_c = \frac{C_c}{C_h} + 1 = \frac{1}{R} + 1 = 2.64$$

Thus, we have

$$T_h = 98°\text{C} + \frac{98°\text{C} + 20°\text{C}}{1.61}\left\{\exp\left[-1.61(0.684)\frac{x}{L}\right] - 1\right\} \tag{c}$$

and

$$T_c = -20°\text{C} + \frac{98°\text{C} + 20°\text{C}}{2.64}\left\{1 - \exp\left[-1.61(0.684)\frac{x}{L}\right]\right\} \qquad \text{(d)}$$

These equations are plotted in Fig. E11–4.

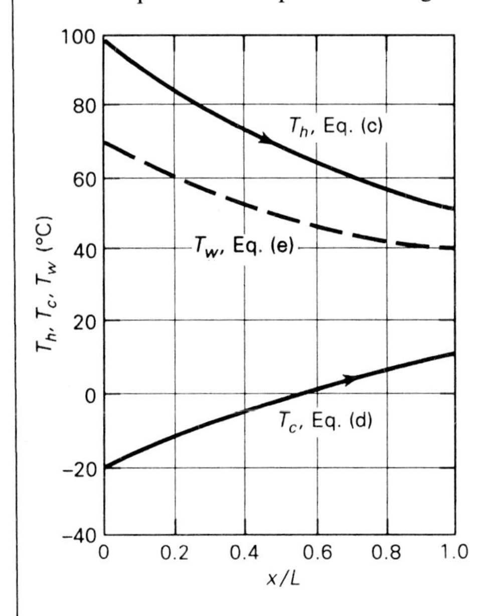

FIGURE E11–4
Calculations for bulk-stream and wall temperatures T_h, T_c, and T_w for parallel-flow arrangement.

To approximate the wall temperature T_w, we write

$$U(T_h - T_c) = h_h(T_h - T_w)$$

Approximating h_h by 7960 W/(m² °C) and U by 1830 W/(m² °C), we obtain

$$\frac{T_h - T_w}{T_h - T_c} = \frac{U}{h_h} \simeq \frac{1830}{7960} = 0.23 \qquad \text{(e)}$$

Calculations for T_w obtained by the use of this equation are shown in Fig. E11–4. The bulk-stream temperatures and wall temperature at the mid-point are estimated to be $T_h = 67°\text{C}$, $T_c = -1.07°\text{C}$, and $T_w = 51.4°\text{C}$. We also note that the maximum wall temperature $T_{w,\max} = T_{w,1}$ is equal to 70.9°C.

Now that the outlet fluid temperatures, the bulk-stream temperatures, and the wall temperature have been estimated, our analysis can be refined by approximating the properties at the arithmetic average of the inlet and outlet temperatures and by utilizing the property correction factors for h_h and h_c given by Eq. (8–24). This refinement is summarized in the context of the P-NTU_I method as follows:

(1) reevaluate the properties of the fluids at the arithmetic average of the inlet and outlet temperatures (i.e., $T_{h,m} = 73.2°\text{C}$ for the water and $T_{c,m} = -5.03°\text{C}$ for the Freon F-12);

(2) estimate the wall temperatures $T_{w,h}$ and $T_{w,c}$ using appropriate theoretical

relation(s) or reasonable approximation(s) (notice that $T_{w,h} = T_{w,c} = T_w =$ 51.4°C in this application involving thin-wall tubing with no fouling);

(3) reevaluate the Reynolds numbers Re_h and Re_c;

(4) evaluate the constant property values $h_{h,cp}$ and $h_{c,cp}$ and use appropriate property variation correction factors to reevaluate h_h and h_c;

(5) reevaluate the mean overall coefficient of heat transfer $\overline{U}$;

(6) use the property corrected values for $\overline{U}$ and the specific heats $c_{P,h}$ and $c_{P,c}$ to reevaluate C_h, C_c, R, NTU_I, and P; and

(7) refine the calculations for q and the outlet fluid temperatures.

To conclude this point, we note that the refined analysis results in small increases in $\overline{U}$ and q of about 3.8% and 2%, respectively.

Finally, to provide an indication of the thermal performance of the exchanger, we calculate the efficiency ψ. Introducing the defining relation, ψ becomes

$$\psi = \frac{P}{NTU_I} = \frac{0.415}{0.684} = 0.607$$

This result indicates that the heat exchanger satisfies the basic minimum performance criterion represented by $\psi > \psi_{MIN} = 0.35$, such that operation should be reasonably economical. Referring back to Sec. 11–5–1, we are reminded that the alternative outlet temperature criterion does not apply to parallel-flow arrangements.

EXAMPLE 11–5†

Reconsider Example 11–4 for counterflow.

Solution

Objective Determine q, the distributions in T_h, T_c, and T_w, and whether or not the minimum performance criteria are satisfied for counterflow.

Schematic Double-pipe heat exchanger: counterflow arrangement.

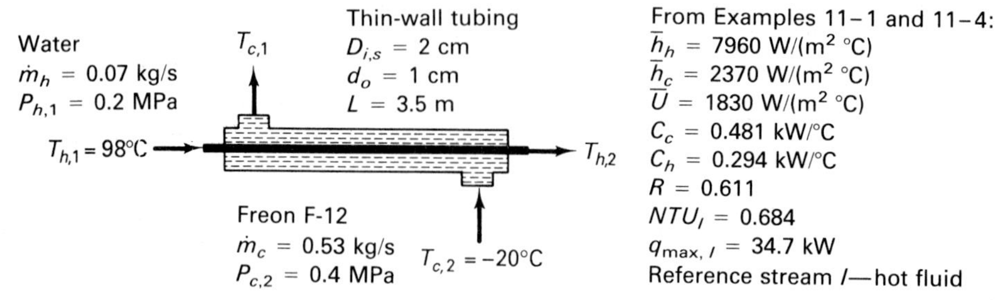

† A simple numerical finite-difference program for solving this example is given in Fig. A–G–5 of the Appendix.

Assumptions/Conditions See Example 11–4.

Properties See Example 11–4.

Analysis For counterflow, P is given by Eq. (11–138a),

$$P = \frac{1 - \exp[-NTU_I(1 - R)]}{1 - R\exp[-NTU_I(1 - R)]}$$

$$= \frac{1 - \exp[-0.684(0.389)]}{1 - 0.611\exp[-0.684(0.389)]} = 0.439$$

Calculating q, we have

$$q = Pq_{\max,I} = 0.439(34.7 \text{ kW}) = 15.2 \text{ kW}$$

which is about 6% greater than the value obtained in Example 11–4 for parallel flow. Referring to Fig. 11–29, we observe that even greater benefit occurs from counterflow operation for larger values of NTU_I.

Noting that $C_c = 0.481$ kW/°C and $C_h = 0.294$ kW/°C from Example 11–4, we continue with the calculation of the outlet temperatures $T_{c,1}$ and $T_{h,2}$ by writing

$$q = C_c(T_{c,1} - T_{c,2}) = C_h(T_{h,1} - T_{h,2})$$

$$T_{c,1} = \frac{q}{C_c} + T_{c,2} = \frac{15.2 \text{ kW}}{0.481 \text{ kW/°C}} - 20°\text{C} = 11.6°\text{C}$$

$$T_{h,2} = T_{h,1} - \frac{q}{C_h} = 98°\text{C} - \frac{15.2 \text{ kW}}{0.294 \text{ kW/°C}} = 46.3°\text{C}$$

The bulk-stream temperature distributions are given by Eqs. (E11–4a) and (E11–4b), with $(\dot{m}c_P)_c$ set equal to $-C_c$ and

$$\zeta = \frac{1}{C_h} - \frac{1}{C_c}$$

such that

$$\zeta C_h = 1 - \frac{C_h}{C_c} = 1 - R = 1 - 0.611 = 0.389$$

and

$$-\zeta C_c = -\frac{C_c}{C_h} + 1 = -\frac{1}{R} + 1 = -0.637$$

Substituting into Eqs. (E11–4a) and (E11–4b), we have

$$T_h = 98°\text{C} + \frac{98°\text{C} - 11.6°\text{C}}{0.389}\left\{\exp\left[-0.389(0.684)\frac{x}{L}\right] - 1\right\} \tag{a}$$

and

$$T_c = 11.6^\circ\text{C} + \frac{98^\circ\text{C} - 11.6^\circ\text{C}}{-0.637}\left\{1 - \exp\left[-0.389(0.684)\frac{x}{L}\right]\right\} \tag{b}$$

These distributions are shown in Fig. E11–5.

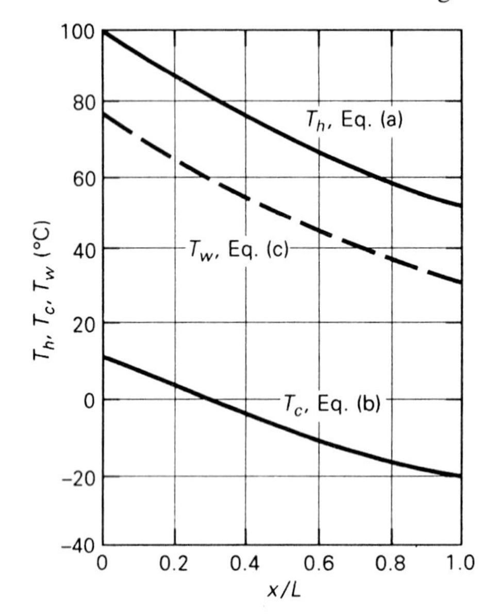

FIGURE E11–5
Calculations for bulk-stream and wall temperatures T_h, T_c, and T_w for counterflow arrangement.

As shown in Example 11–4, the wall temperature T_w can be approximated by

$$\frac{T_h - T_w}{T_h - T_c} = \frac{U}{h_h} \simeq 0.23 \tag{c}$$

This expression is also shown in Fig. E11–5. Notice that the maximum wall temperature $T_{w,\text{max}} = T_{w,1}$ is approximately equal to 78.2°C, which is 10% higher than for the parallel-flow arrangement.

Finally, we note that $\psi = P/NTU_I = 0.642$, such that the minimum performance criterion $\psi > \psi_{\text{MIN}} = 0.35$ is satisfied. Observing that $T_{h,o} = 46.3^\circ\text{C} > T_{c,o} = 11.6^\circ\text{C}$, we conclude that the alternative outlet temperature criterion is also satisfied. This criterion is shown formally by writing $P_o = 1/(1 + R) = 0.621$ from Eq. (11–76), which indicates $P < P_o$.

EXAMPLE 11-6†

A double-pipe heat exchanger is used to cool 55 lb_m/min of oil with a specific heat of 0.525 Btu/(lb_m °F) from 122°F to 104°F. A cooling fluid enters the annulus of the exchanger at 68°F and exits at 77°F. The mean overall coefficient of heat transfer $\overline{U}$ is 88 Btu/(h ft² °F). Determine the heat-exchanger surface area A_s for both parallel flow and counterflow. Also determine whether or not the minimum performance criteria are satisfied for these arrangements.

Solution

Objective Determine A_s and whether or not the minimum performance criteria are satisfied for parallel flow and counterflow.

Schematic Double-pipe heat exchangers with $\overline{U}$ = 88 Btu/(h ft² °F).

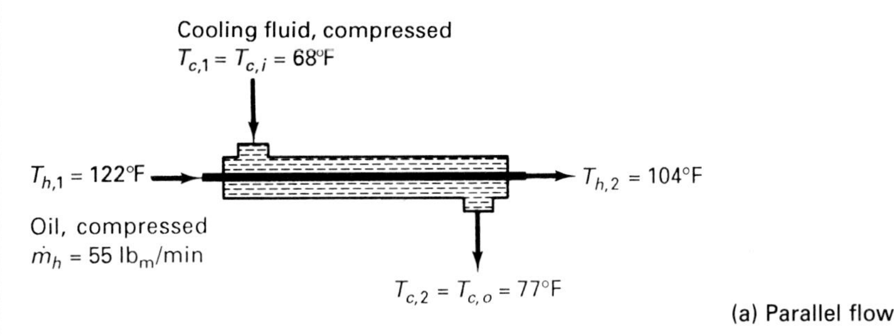

(a) Parallel flow

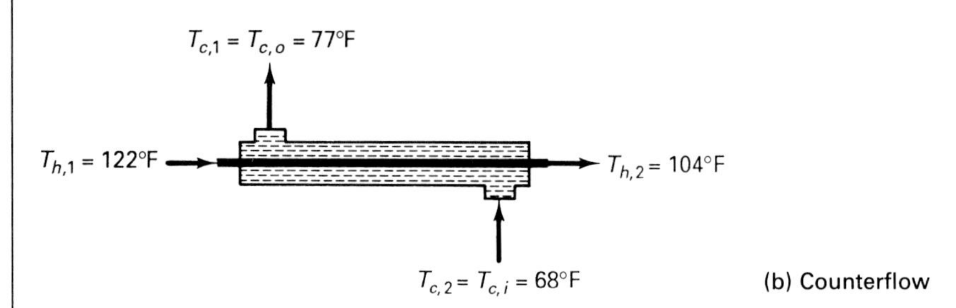

(b) Counterflow

Assumptions/Conditions

uniform properties
standard conditions

† A simple numerical finite-difference program for solving this example is given in Fig. A-G-6 of the Appendix.

Properties Hot oil: $c_P = 0.525\ \text{Btu/(lb}_\text{m}\ \text{°F)}$.

Analysis Because the exit temperatures are given, we can conveniently use the *LMTD* approach. The total rate of heat transfer q is

$$q = (\dot{m}c_P)_h\,\Delta T_h = (\dot{m}c_P)_c\,\Delta T_c = 55\,\frac{\text{lb}_\text{m}}{\text{min}}\left(0.525\,\frac{\text{Btu}}{\text{lb}_\text{m}\ \text{°F}}\right)(122\text{°F} - 104\text{°F})$$

$$= 520\ \text{Btu/min} = 31{,}200\ \text{Btu/h}$$

Parallel flow

$$LMTD = \frac{(122\text{°F} - 68\text{°F}) - (104\text{°F} - 77\text{°F})}{\ln[(122\text{°F} - 68\text{°F})/(104\text{°F} - 77\text{°F})]} = 39\text{°F} \qquad q = \overline{U}A_s\,LMTD$$

$$A_s = \frac{q}{\overline{U}\,LMTD} = \frac{31{,}200\ \text{Btu/h}}{[88\ \text{Btu/(h ft}^2\ \text{°F)}](39\text{°F})} = 9.09\ \text{ft}^2$$

Counterflow

$$LMTD = \frac{(122\text{°F} - 77\text{°F}) - (104\text{°F} - 68\text{°F})}{\ln[(122\text{°F} - 77\text{°F})/(104\text{°F} - 68\text{°F})]} = 40.3\text{°F}$$

$$A_s = \frac{q}{\overline{U}\,LMTD} = 8.8\ \text{ft}^2$$

To check the minimum performance criteria, we first use Eq. (11–65) to calculate the efficiency ψ, with the result

$$\psi = \frac{q}{\overline{U}A_s(T_{h,i} - T_{c,i})}$$

$$= \frac{31{,}200\ \text{Btu/h}}{[88\ \text{Btu/(h ft}^2\ \text{°F)}](9.09\ \text{ft}^2)(122\text{°F} - 68\text{°F})} = 0.722$$

for parallel flow, and

$$\psi = \frac{31{,}200\ \text{Btu/h}}{[88\ \text{Btu/(h ft}^2\ \text{°F)}](8.8\ \text{ft}^2)(122\text{°F} - 68\text{°F})} = 0.746$$

for counterflow. Thus, the minimum performance criterion $\psi > \psi_{MIN} = 0.35$ is satisfied for both arrangements. Since $T_{h,o} = 104\text{°F} > T_{c,o} = 77\text{°F}$, we conclude that the alternative outlet temperature criterion, which is applicable to counterflow, is also satisfied.

EXAMPLE 11–7

Saturated water at 1 atm with a quality of 0.1 and mass-flow rate of 12 kg/min is to be cooled to 80°C. Cooling water at 15°C and 20 kg/min mass-flow rate is available. Determine the surface area required for a counterflow double-pipe heat exchanger with the cooling fluid flowing in the annulus if the mean overall coefficient of heat transfer is 500 W/(m^2 °C) in the two-phase region and 300 W/(m^2 °C) in the single phase region.

Solution

Objective Determine A_s for counterflow.

Schematic Double-pipe heat exchanger: counterflow arrangement.

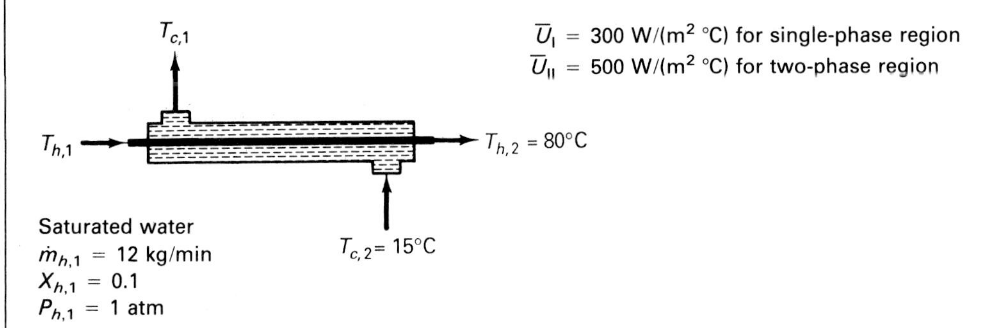

Assumptions/Conditions

uniform properties in each phase
assume $i_{h,2} = i_{f,2}$ for water at 80°C
standard conditions, except for phase change

Properties

Saturated water at 1 atm (*Steam Tables*): $i_{f,1} = 419$ kJ/kg, $i_{fg} = 2260$ kJ/kg.

Saturated water at 80°C (*Steam Tables*): $i_{f,2} = 335$ kJ/kg.

Water at $T = 15°\text{C} = 288$ K (Table A–C–3): $c_P = 4.19$ kJ/(kg °C).

Analysis To calculate the enthalpy of the entering saturated water we write

$$X_{h,1} = 0.1 = \frac{i_{h,1} - i_{f,1}}{i_{fg}} \qquad i_{h,1} = 0.1\left(2260\ \frac{\text{kJ}}{\text{kg}}\right) + 419\ \frac{\text{kJ}}{\text{kg}} = 645\ \text{kJ/kg}$$

The rate of heat transfer q_{II} required to completely condense 12 kg/min of this two-phase water is

$$q_{\mathrm{II}} = \dot{m}_h(i_{h,1} - i_{f,1}) = 12\,\frac{\mathrm{kg}}{\mathrm{min}}\,(645 - 419)\,\frac{\mathrm{kJ}}{\mathrm{kg}} = 2710\ \mathrm{kJ/min} = 45.2\ \mathrm{kW}$$

The energy required to subcool the saturated liquid water to 80°C is calculated by assuming $i_{h,2} = i_{f,2}$ and writing

$$q_{\mathrm{I}} = \dot{m}_h(i_{f,1} - i_{h,2}) = 12\,\frac{\mathrm{kg}}{\mathrm{min}}\,(419 - 335)\,\frac{\mathrm{kJ}}{\mathrm{kg}} = 1010\ \mathrm{kJ/min} = 16.8\ \mathrm{kW}$$

Hence, the total rate of heat transfer q is

$$q = q_{\mathrm{I}} + q_{\mathrm{II}} = (1010 + 2710)\,\frac{\mathrm{kJ}}{\mathrm{min}} = 3720\ \mathrm{kJ/min} = 62\ \mathrm{kW}$$

With this information and noting that the specific heat of water is about 4.19 kJ/(kg °C), we can obtain the cooling fluid outlet temperature $T_{c,1}$ by writing

$$C_c = (\dot{m}c_P)_c = 20\,\frac{\mathrm{kg}}{\mathrm{min}}\left(4.19\,\frac{\mathrm{kJ}}{\mathrm{kg\ °C}}\right)\left(\frac{\mathrm{min}}{60\ \mathrm{s}}\right) = 1.4\ \mathrm{kW/°C}$$

$$q = C_c(T_{c,1} - T_{c,2}) \qquad T_{c,1} = \frac{q}{C_c} + T_{c,2} = \frac{62\ \mathrm{kW}}{1.4\ \mathrm{kW/°C}} + 15\mathrm{°C} = 59.3\mathrm{°C}$$

Similarly, we can write the following expression for the temperature of the cooling water at the point at which the quality of the saturated fluid is zero:

$$q_{\mathrm{I}} = C_c(T_a - T_{c,2}) \qquad T_a = \frac{q_{\mathrm{I}}}{C_c} + T_{c,2} = \frac{16.8\ \mathrm{kW}}{1.4\ \mathrm{kW/°C}} + 15\mathrm{°C} = 27\mathrm{°C}$$

The bulk-stream temperatures of the two fluids are sketched in Fig. E11–7.

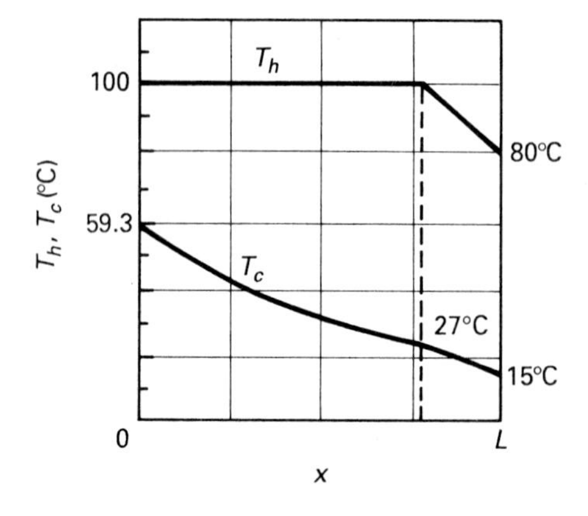

FIGURE E11–7
Calculations for bulk-stream temperatures T_h and T_c: counterflow arrangement.

The surface area $A_{s,\mathrm{I}}$ required to subcool the hot water is determined as follows:

$$LMTD_{\text{I}} = \frac{(100°\text{C} - 27°\text{C}) - (80°\text{C} - 15°\text{C})}{\ln [(100°\text{C} - 27°\text{C})/(80°\text{C} - 15°\text{C})]} = 68.9°\text{C}$$

$$q_{\text{I}} = \overline{U}_{\text{I}} A_{s\text{I}}\, LMTD_{\text{I}}$$

$$A_{s\text{I}} = \frac{q_{\text{I}}}{\overline{U}_{\text{I}}\, LMTD_{\text{I}}} = \frac{16{,}800 \text{ W}}{[300 \text{ W/(m}^2\ °\text{C)}](68.9°\text{C})} = 0.814 \text{ m}^2$$

Similarly, to determine the surface area $A_{s\text{II}}$ required to condense the vapor, we write

$$LMTD_{\text{II}} = \frac{(100°\text{C} - 59.3°\text{C}) - (100°\text{C} - 27°\text{C})}{\ln [(100°\text{C} - 59.3°\text{C})/(100°\text{C} - 27°\text{C})]} = 55.3°\text{C}$$

$$q_{\text{II}} = \overline{U}_{\text{II}} A_{s\text{II}}\, LMTD_{\text{II}} \qquad A_{s\text{II}} = 1.63 \text{ m}^2$$

The total surface area of the heat exchanger is

$$A_s = A_{s\text{I}} + A_{s\text{II}} = 0.814 \text{ m}^2 + 1.63 \text{ m}^2 = 2.45 \text{ m}^2$$

It should be noted that for two-phase problems such as this, one cannot calculate the surface area A_s directly from the overall *LMTD*. This is because of the change in the shape of the temperature difference curve in the two-phase and single-phase regions (see Fig. E11-7) and differences in $\overline{U}_{\text{I}}$ and $\overline{U}_{\text{II}}$.

EXAMPLE 11-8

Air at 1.2 atm flowing at a rate of 0.1 kg/s is to be cooled from 400 K to 300 K in the longitudinal fin-tube double-pipe heat exchanger shown in Fig. E11-8. The cooling fluid is ammonia flowing at a rate of 0.75 kg/s, which enters the tube at 250 K and 900 kPa. The maximum fouling factor on the tube side is specified as 0.0002 m² °C/W. Determine the length of the unit for counterflow operation.

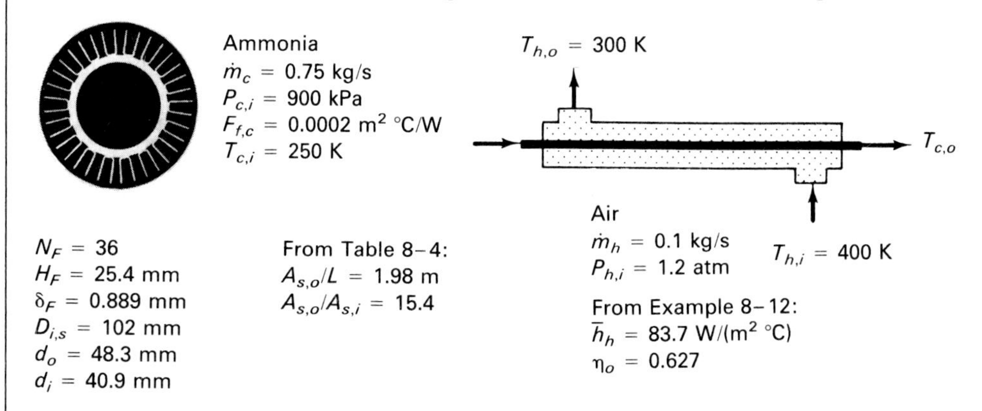

FIGURE E11-8 Longitudinal fin-tube double-pipe heat exchanger.

Solution

Objective Determine the length L required to bring the outlet temperature of the air to 300 K.

Assumptions/Conditions

negligible heat transfer through fin tip
negligible thermal contact resistance between fin and tube wall
negligible entrance effects
negligible fouling on air side
moderate property variation†
standard conditions

Properties

Air at the average inlet and outlet temperature of 350 K and 1.2 atm (Table A–C–5): $\mu = 2.08 \times 10^{-5}$ kg/(m s), $c_P = 1.01$ kJ/(kg °C), $k = 0.03$ W/(m °C), $Pr = 0.697$; $\rho = 1.03$ kg/m^3 from the ideal gas law.

Ammonia at $T = -20$°C $= 253$ K and 900 kPa (compressed liquid, Table A–C–3): $\rho = 667$ kg/m^3, $\mu = 2.54 \times 10^{-4}$ kg/(m s), $c_P = 4.51$ kJ/(kg °C), $k = 0.547$ W/(m °C), $Pr = 2.09$.

Carbon steel (0.5% C) (Table A–C–1): $k = 54$ W/(m °C).

Analysis The rate of heat transfer q and outlet temperature of the cooling fluid are computed by writing

$$q = (\dot{m}c_P)_h(T_{h,i} - T_{h,o}) = \left(0.1\ \frac{\text{kg}}{\text{s}}\right)\left(1.01\ \frac{\text{kJ}}{\text{kg K}}\right)(400\ \text{K} - 300\ \text{K}) = 10.1\ \text{kW}$$

$$T_{c,o} = \frac{q}{(\dot{m}c_P)_c} + T_{c,i} = \frac{10.1\ \text{kW}}{(0.75\ \text{kg/s})[4.51\ \text{kJ/(kg K)}]} + 250\ \text{K} = 253\ \text{K}$$

Since the change in bulk-stream temperature of the ammonia over the length of the exchanger is small, we will assume that the properties are essentially uniform.

To provide a basis for calculating the mean overall coefficient of heat transfer $\overline{U}$, we note that $\overline{h}_h = 83.7$ W/(m^2 °C) and $\eta_o = 0.627$ from Example 8–12. The coefficient of heat transfer h_c is evaluated by writing

$$G_c = \frac{\dot{m}_c}{A_c} = \frac{0.75\ \text{kg/s}}{\pi(0.0409\ \text{m})^2/4} = 571\ \text{kg/(m}^2\ \text{s)}$$

† To simplify the presentation, the term *moderate property variation* is used in this chapter to indicate situations in which a uniform property analysis is used, with the effect of property variation accounted for by the use of average bulk-stream temperature for one or both fluids. Correction factors such as those given by Eqs. (8–23) to (8–26) can be used to refine uniform property analyses of this type.

or $U_{b,c} = (G/\rho)_c = 0.856$ m/s,

$$Re_c = \left(\frac{GD_H}{\mu}\right)_c = \frac{G_c d_i}{\mu_c} = \frac{[571\ \text{kg/(m}^2\ \text{s)}](0.0409\ \text{m})}{2.54 \times 10^{-4}\ \text{kg/(m s)}} = 91{,}900$$

$$f_c = (1.58 \ln 91{,}900 - 3.28)^{-2} = 0.00458$$

$$Nu_c = \frac{(0.00458/2)(91{,}900)(2.09)}{1.07 + 12.7\sqrt{0.00458/2}\,(2.09^{2/3} - 1)} = 302$$

$$h_c = 302\,\frac{0.547\ \text{W/(m °C)}}{0.0409\ \text{m}} = 4040\ \text{W/(m}^2\ \text{°C)}$$

Since no fouling is indicated on the air side, the overall coefficient of heat transfer $\overline{U}$ is represented by

$$\frac{1}{\overline{U}A_s} = \frac{1}{(\eta_o \overline{h} A_s)_h} + R_k + \left(\frac{F_f}{A_s}\right)_c + \frac{1}{(\overline{h}A_s)_c} \tag{a}$$

where $\overline{h}_c$ is set equal to h_c and the wall thermal resistance is

$$R_k = \frac{\ln (d_o/d_i)}{2\pi k_w L}$$

from Eq. (2–51); k_w is the thermal conductivity of the tube wall. Selecting $A_{s,h} = A_{s,o}$ as the reference surface area A_s, Eq. (a) becomes

$$\frac{1}{\overline{U}} = \frac{1}{(\eta_o \overline{h})_h} + \frac{A_{s,o}}{L}\,\frac{\ln (d_o/d_i)}{2\pi k_w} + \frac{A_{s,o}}{A_{s,i}}\left(F_{f,c} + \frac{1}{\overline{h}_c}\right) \tag{b}$$

where $A_{s,o}/L = 1.98$ m and $A_{s,o}/A_{s,i} = 15.4$. Substituting into Eq. (b) we obtain

$$\frac{1}{\overline{U}} = \left[\frac{1}{0.627(83.7)} + 1.98\,\frac{\ln (48.3/40.9)}{2\pi(54)} + 15.4\left(0.0002 + \frac{1}{4040}\right)\right]\frac{\text{m}^2\ \text{°C}}{\text{W}}$$

$$= (0.0191 + 0.00097 + 0.00308 + 0.00381)\ \text{m}^2\ \text{°C/W} = 0.027\ \text{m}^2\ \text{°C/W}$$

such that the calculation for $\overline{U}$ becomes

$$\overline{U} = 37.1\ \text{W/(m}^2\ \text{°C)}$$

It should be noted that the wall resistance only accounts for about 3.6% of the total thermal resistance. Because the wall resistance is normally small, it is commonly neglected.

Following the *LMTD* approach, the heat transfer is expressed in terms of the surface area A_s by Eq. (11–150),

$$q = \overline{U}A_s\, LMTD \tag{c}$$

where

$$LMTD = \frac{(T_{h,i} - T_{c,o}) - (T_{h,o} - T_{c,i})}{\ln [(T_{h,i} - T_{c,o})/(T_{h,o} - T_{c,i})]}$$

$$= \frac{(400\ \text{K} - 253\ \text{K}) - (300\ \text{K} - 250\ \text{K})}{\ln [(400\ \text{K} - 253\ \text{K})/(300\ \text{K} - 250\ \text{K})]} = 89.9\ \text{K}$$

Solving Eq. (c) for A_s and noting that $LMTD = 89.9\ \text{K} = 89.9°\text{C}$, we obtain

$$A_s = A_{s,o} = \frac{q}{\overline{U}\ LMTD} = \frac{10.1\ \text{kW}}{[37.1\ \text{W}/(\text{m}^2\ °\text{C})](89.9°\text{C})} = 3.03\ \text{m}^2$$

Setting $A_{s,o}/L = 1.98$ m, the heat exchanger length L is calculated.

$$L = \frac{A_{s,o}}{1.98\ \text{m}} = \frac{3.03\ \text{m}^2}{1.98\ \text{m}} = 1.53\ \text{m}$$

EXAMPLE 11–9

A 30° API petroleum distillate oil with mass-flow rate of 2.18 kg/s is to be cooled in a double-pipe hairpin-heat-exchanger unit constructed of 1.5-in. and 3-in. schedule 40 wrought steel pipes. The oil is cooled from 114°C to 66°C using cooling water with inlet temperature of 27°C and mass-flow rate of 3.5 kg/s. The allowable pressure drop for each stream is approximately 82.8 kPa (12 psi) and the required minimum fouling factors are specified as 0.0007 m² °C/W for the oil and 0.00015 m² °C/W for the water. Determine the required length of the heat exchanger and whether or not the pressure drop requirements are satisfied.

Solution

Objective Determine the unit length L and whether or not the pressure drop requirements are satisfied.

Schematic Double-pipe hairpin heat exchanger.

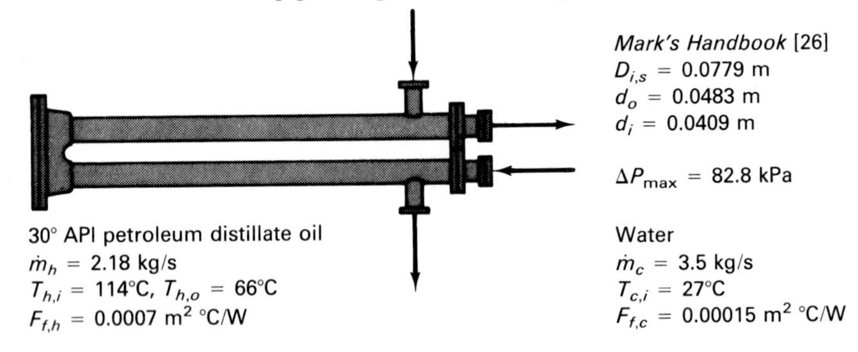

Assumptions/Conditions

negligible entrance effects
moderate property variation
standard conditions

Properties

30° API oil at 90°C: $\rho = 820\ \text{kg/m}^3$, $\mu = 0.00245\ \text{kg/(m s)}$, $c_P = 2.17\ \text{kJ/(kg °C)}$, $k = 0.128\ \text{W/(m °C)}$, $Pr = 41.5$. (The properties of various petroleum fractions are available in TEMA [3]. The properties of this particular oil can be approximated in terms of T (C°) by $\rho = (871 - 0.565\,T)\ \text{kg/m}^3$, $c_P = (1.79 + 0.00428\,T)\ \text{kJ/(kg °C)}$, $k = (0.137 - 8.73 \times 10^{-5}\,T)\ \text{W/(m °C)}$, $\mu = [0.1 \exp(1.35 - 0.0855\,T + 0.00032\,T^2)]\ \text{kg/(m s)}$ for $0°\text{C} < T < 80°\text{C}$, and $\mu = [0.1 \exp(-0.243 - 0.0478\,T + 9.72 \times 10^{-5}\,T^2)]\ \text{kg/(m s)}$ for $80°\text{C} < T < 160°\text{C}$.)

Water at $T = 27°\text{C} = 300\ \text{K}$ (Table A–C–3): $\rho = 997\ \text{kg/m}^3$, $\mu = 8.55 \times 10^{-4}\ \text{kg/(m s)}$, $c_P = 4.18\ \text{kJ/(kg °C)}$, $k = 0.613\ \text{W/(m °C)}$, $Pr = 5.83$.

Wrought steel pipe (Table A–C–1): $k = 59\ \text{W/(m °C)}$.

Analysis To determine the heat-transfer rate and outlet temperature of the cooling water we employ the lumped energy formulation.

$$q = (\dot{m}c_P)_h(T_{h,i} - T_{h,o}) = 2.18\,\frac{\text{kg}}{\text{s}}\left(2.17\,\frac{\text{kJ}}{\text{kg °C}}\right)(114°\text{C} - 66°\text{C})$$

$$= 227\ \text{kW}$$

$$q = (\dot{m}c_P)_c(T_{c,o} - T_{c,i})$$

$$T_{c,o} = \frac{q}{(\dot{m}c_P)_c} + T_{c,i} = \frac{227\ \text{kW}}{(3.5\ \text{kg/s})[4.18\ \text{kJ/(kg °C)}]} + 27°\text{C}$$

$$= 42.5°\text{C}$$

This value of the outlet temperature can be used to refine the properties for the cooling water. However, as a first approximation, we will continue the analysis using the properties evaluated at the inlet temperature.

The cross-sectional areas and hydraulic diameters of the two passages are calculated as follows:

Tube

$$A_t = \frac{\pi d_i^2}{4} = \frac{\pi(0.0409\ \text{m})^2}{4} = 0.00131\ \text{m}^2$$

$$D_{H,t} = d_i = 0.0409\ \text{m}$$

Annulus

$$A_a = \frac{\pi}{4}(D_{i,s}^2 - d_o^2) = \frac{\pi}{4}(0.0779^2 - 0.0483^2)\ \text{m}^2 = 0.00293\ \text{m}^2$$

$$D_{H,a} = \frac{4A}{p_w} = \frac{D_{i,s}^2 - d_o^2}{D_{i,s} + d_o} = D_{i,s} - d_o = (0.0779 - 0.0483)\ \text{m} = 0.0296\ \text{m}$$

To provide a basis for establishing which fluid is to pass through the annulus, which has a little over twice the cross-sectional area of the tube, we will compute the mass fluxes, bulk-stream velocities and Reynolds numbers for the two possible combinations.

Oil $\quad G_h = \dfrac{\dot{m}_h}{A} = \dfrac{2.18\ \text{kg/s}}{A} \qquad U_{b,h} = \dfrac{G_h}{\rho_h} = \dfrac{G_h}{820\ \text{kg/m}^3}$

Tube $\quad A = 0.00131\ \text{m}^2 \qquad D_H = 0.0409\ \text{m}$

$$G_h = \frac{2.18\ \text{kg/s}}{0.00131\ \text{m}^2} = 1660\ \text{kg/(m}^2\ \text{s)} \qquad U_{b,h} = 2.03\ \text{m/s}$$

$$Re_h = \left(\frac{GD_H}{\mu}\right)_h = \frac{[1660\ \text{kg/(m}^2\ \text{s)}](0.0409\ \text{m})}{0.00245\ \text{kg/(m s)}} = 27{,}700$$

Annulus $\quad A = 0.00293\ \text{m}^2 \qquad D_H = 0.0296\ \text{m}$

$$G_h = \frac{2.18\ \text{kg/s}}{0.00293\ \text{m}^2} = 744\ \text{kg/(m}^2\ \text{s)} \qquad U_{b,h} = 0.907\ \text{m/s}$$

$$Re_h = \left(\frac{GD_H}{\mu}\right)_h = \frac{[744\ \text{kg/(m}^2\ \text{s)}](0.0296\ \text{m})}{0.00245\ \text{kg/(m s)}} = 8990$$

Water $\quad G_c = \dfrac{\dot{m}_c}{A} = \dfrac{3.5\ \text{kg/s}}{A} \qquad U_{b,c} = \dfrac{G_c}{\rho_c} = \dfrac{G_c}{997\ \text{kg/m}^3}$

Tube $\quad A = 0.00131\ \text{m}^2 \qquad D_H = 0.0409\ \text{m}$

$$G_c = \frac{3.5\ \text{kg/s}}{0.00131\ \text{m}^2} = 2670\ \text{kg/(m}^2\ \text{s)} \qquad U_{b,c} = 2.68\ \text{m/s}$$

$$Re_c = \left(\frac{GD_H}{\mu}\right)_c = \frac{[2670\ \text{kg/(m}^2\ \text{s)}](0.0409\ \text{m})}{8.55 \times 10^{-4}\ \text{kg/(m s)}} = 128{,}000$$

Annulus $\quad A = 0.00293\ \text{m}^2 \qquad D_H = 0.0296\ \text{m}$

$$G_c = \frac{3.5\ \text{kg/s}}{0.00293\ \text{m}^2} = 1190\ \text{kg/(m}^2\ \text{s)} \qquad U_{b,c} = 1.2\ \text{m/s}$$

$$Re_c = \left(\frac{GD_H}{\mu}\right)_c = \frac{[1190\ \text{kg/(m}^2\ \text{s)}](0.0296\ \text{m})}{8.55 \times 10^{-4}\ \text{kg/(m s)}} = 41{,}200$$

These calculations indicate that the oil will be in transitional turbulent flow with U_b less than 1 m/s if placed in the annulus. On the other hand, if the oil is placed in the tube, the flow of both fluids will be fully turbulent. In addition, the bulk-stream velocities of both fluids will be greater than 1 m/s, such that excessive fouling due to low velocities should not be a problem. Therefore, this arrangement is tentatively selected with the intention of maximizing the overall coefficient of heat transfer.

The friction factors, Nusselt numbers, and coefficients of heat transfer are now estimated for fully developed conditions using Eqs. (8–15) and (8–16).

Oil in tube

$$f_h = (1.58 \ln 27{,}700 - 3.28)^{-2} = 0.00603$$

$$Nu_h = \frac{(f/2)\,Re\,Pr}{1.07 + 12.7\sqrt{f/2}\,(Pr^{2/3} - 1)}$$

$$= \frac{(0.00603/2)(27{,}700)(41.5)}{1.07 + 12.7\sqrt{0.00603/2}\,(41.5^{2/3} - 1)} = 397$$

$$h_h = Nu_h \frac{k_h}{D_{H,t}} = 397\,\frac{0.128\ \text{W/(m °C)}}{0.0409\ \text{m}} = 1240\ \text{W/(m}^2\ \text{°C)}$$

Water in annulus—Referring to Fig. 8–9 and noting that $d_o/D_{i,s} = D_i/D_o = 0.62$, we conclude that $Nu_c = Nu_{ii} = Nu/0.96$. Thus, we may simply set $Nu_c = Nu$ as a conservative estimate, and write

$$f_c = (1.58 \ln 41{,}200 - 3.28)^{-2} = 0.00548$$

$$Nu_c = \frac{(0.00548/2)(41{,}200)(5.83)}{1.07 + 12.7\sqrt{0.00548/2}\,(5.83^{2/3} - 1)} = 257$$

$$h_c = 257\,\frac{0.613\ \text{W/(m °C)}}{0.0296\ \text{m}} = 5320\ \text{W/(m}^2\ \text{°C)}$$

The mean overall coefficient of heat transfer $\overline{U}$ is represented by

$$\frac{1}{\overline{U}A_s} = \frac{1}{(\bar{h}A_s)_h} + \left(\frac{F_f}{A_s}\right)_h + R_k + \left(\frac{F_f}{A_s}\right)_c + \frac{1}{(\bar{h}A_s)_c} \tag{a}$$

where we assume $\bar{h}_h = h_h$ and $\bar{h}_c = h_c$ and represent R_k by

$$R_k = \frac{\ln (d_o/d_i)}{2\pi k_w L}$$

from Eq. (2–51); k_w is the thermal conductivity of the tube wall. Selecting $A_{s,c} = A_{s,o}$ as the reference surface area A_s, Eq. (a) becomes

$$\frac{1}{\overline{U}} = \frac{A_{s,o}}{A_{s,i}}\left(\frac{1}{\bar{h}_h} + F_f\right)_h + \frac{A_{s,o}}{L}\,\frac{\ln (d_o/d_i)}{2\pi k_w} + F_{f,c} + \frac{1}{\bar{h}_c}$$

where $A_{s,o}/L = \pi d_o$ and $A_{s,o}/A_{s,i} = d_o/d_i$. It follows that

$$\frac{1}{\overline{U}} = \left[\frac{48.3}{40.9}\left(\frac{1}{1240} + 0.0007\right) + \pi(0.0483)\frac{\ln(48.3/40.9)}{2\pi(59)} + 0.00015 + \frac{1}{5320}\right]\frac{\text{m}^2\ ^\circ\text{C}}{\text{W}}$$

$$= (0.000952 + 0.000827 + 0.0000681 + 0.00015 + 0.000188)\ \text{m}^2\ ^\circ\text{C/W}$$

$$= 0.00218\ \text{m}^2\ ^\circ\text{C/W}$$

or

$$\overline{U} = 459\ \text{W/(m}^2\ ^\circ\text{C)}$$

for fouled surfaces, and

$$\overline{U} = 827\ \text{W/(m}^2\ ^\circ\text{C)}$$

for no fouling. Notice that the wall resistance accounts for from about 3 to 5.6% of the total thermal resistance, depending on whether the surface is fouled or clean. At this point it should be mentioned that the overall coefficient of heat transfer for clean conditions is about 30% lower for the case in which the oil flows in the annulus instead of the tube.

We are now in a position to determine the exchanger length L. Following the *LMTD* approach, we write

$$LMTD = \frac{(114^\circ\text{C} - 42.5^\circ\text{C}) - (66^\circ\text{C} - 27^\circ\text{C})}{\ln[(114^\circ\text{C} - 42.5^\circ\text{C})/(66^\circ\text{C} - 27^\circ\text{C})]} = 53.6^\circ\text{C}$$

and

$$A_s = \frac{q}{\overline{U}\ LMTD} = \frac{227\ \text{kW}}{[459\ \text{W/(m}^2\ ^\circ\text{C)}](53.6^\circ\text{C})} = 9.23\ \text{m}^2$$

from Eq. (11–150), such that

$$L = \frac{A_s/2}{\pi d_o} = \frac{9.23\ \text{m}^2}{2\pi(0.0483\ \text{m})} = 30.4\ \text{m}$$

It follows that the total lengths of the tubular and annular passageways for the hairpin unit are given by

$$L_t = L_a = 2(30.4\ \text{m}) = 60.8\ \text{m}$$

In connection with the calculation of L, notice that $L/D_H \gg 10$ for both tube and annular flows, such that the assumption of negligible entrance effects on h_h and h_c should be reasonable.

The pressure drops for the two streams are represented by Eqs. (11–163) and (11–164). Substituting into these equations, we obtain

Hot tubular fluid (oil) $M = 1$

$$(\Delta P_t)_{1-2} = \left(\frac{\rho U_b^2}{2}\right)_t \left[\left(\frac{4L}{D_H} f\right)_t + 1.5M\right]$$

$$= \frac{1}{2}\left(820\ \frac{\text{kg}}{\text{m}^3}\right)\left(2.03\ \frac{\text{m}}{\text{s}}\right)^2 \left[\frac{4(60.8\ \text{m})(0.00603)}{0.0409\ \text{m}} + 1.5\right]$$

$$= 63{,}100\ \text{kg/(m s}^2) = 63.1\ \text{kPa}$$

Cold annular fluid (water) $M = 1$

$$(\Delta P_a)_{1-2} = \left(\frac{\rho U_b^2}{2}\right)_a \left[\left(\frac{4L}{D_H} f\right)_a + 0.5M\right]$$

$$= \frac{1}{2}\left(997\ \frac{\text{kg}}{\text{m}^3}\right)\left(1.2\ \frac{\text{m}}{\text{s}}\right)^2 \left[\frac{4(60.8\ \text{m})(0.00548)}{0.0296\ \text{m}} + 0.5\right]$$

$$= 32{,}700\ \text{kg/(m s}^2) = 32.7\ \text{kPa}$$

These calculations indicate that both pressure drops are within the maximum allowable value.

It should be noted that for applications in which the calculation for pressure drop of one or both streams exceeds the allowable limits, parallel or parallel–series arrangements can be used. An application involving the use of a parallel–series arrangement is considered in Example 11–20.

Finally, because we are dealing with oil and fairly large temperature differences, consideration should be given to refining the analysis by using correction factors for the coefficients of heat transfer and friction factor (and evaluating the properties of the cooling water at the arithmetic average of the inlet and outlet temperatures). This refinement is summarized in the context of the *LMTD* method as follows:

(1) reevaluate (as necessary) the properties of the fluids at the arithmetic average of the inlet and outlet temperatures [i.e., $T_{h,m} = 90°\text{C}$ (no change) for the oil and $T_{c,m} = 34.8°\text{C}$ for the water];

(2) estimate the wall temperatures $T_{w,h}$ and $T_{w,c}$ using appropriate theoretical relations or reasonable approximations;

(3) reevaluate the Reynolds numbers Re_h and Re_c;

(4) evaluate the constant property values $f_{h,cp}$, $f_{c,cp}$, $h_{h,cp}$ and $h_{c,cp}$ and use appropriate property variation correction factors to reevaluate f_h, f_c, h_h and h_c;

(5) reevaluate the mean overall coefficient of heat transfer $\overline{U}$;

(6) set $F = 1$ for double-pipe heat exchangers, or use the property corrected values for the specific heats $c_{P,h}$ and $c_{P,c}$ to reevaluate C_h, C_c, R, P and F for shell-and-tube and crossflow heat exchangers; and

(7) refine the calculations for the pressure drops and q.

This refinement results in decreases in $\overline{U}$ and q of about 9% and 5.5%, respectively. As a point of interest, a more general numerical solution indicates that U varies from 491 W/(m² °C) at the inlet to 337 W/(m² °C) at the exit and q = 217 kW. This value of q is only 4.4% below the value obtained in our uniform property analysis.

EXAMPLE 11–10

The evaluation function sometimes requires that the mass-flow rates of two streams be determined for an existing exchanger with specified fluids and terminal temperatures. Referring to the double-pipe heat-exchanger application shown in Fig. E11–10, outline the steps that would be required to determine the mass-flow rates $\dot{m}_h$ and $\dot{m}_c$.

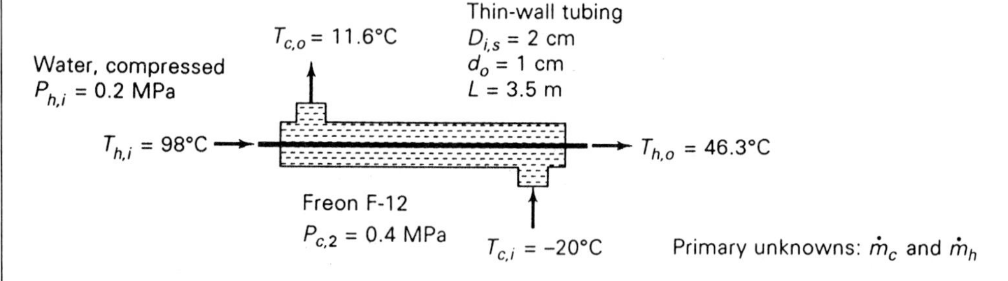

FIGURE E11–10 Double-pipe heat exchanger: counterflow arrangement—evaluation problem.

Solution

Objective Set up a computational strategy for evaluation of mass-flow rates $\dot{m}_h$ and $\dot{m}_c$ for an existing double-pipe heat exchanger with specified fluids and terminal temperatures.

Assumptions/Conditions

negligible conduction resistance R_k
uniform properties
standard conditions

Properties

Hot water at T = 98°C = 371 K (Table A–C–3): ρ = 961 kg/m^3, μ = 2.86 × 10^{-4} kg/(m s), c_P = 4.21 kJ/(kg °C), k = 0.68 W/(m °C), Pr = 1.78.

Freon F-12 at −20°C (Table A–C–3): ρ = 1460 kg/m^3, μ = 3.43 × 10^{-4} kg/(m s), c_P = 0.907 kJ/(kg °C), $k \simeq$ 0.071 W/(m °C), $Pr \simeq$ 4.4.

Analysis The effectiveness method can be used to develop an efficient systematic solution to this problem. First, the lumped formulation is used to obtain relations for the heat-transfer rate.

$$q = C_c(T_{c,o} - T_{c,i}) = C_c(11.6°\text{C} + 20°\text{C}) = 31.6\ C_c\ °\text{C}$$

$$q = C_h(T_{h,i} - T_{h,o}) = C_h(98°\text{C} - 46.3°\text{C}) = 51.7\ C_h\ °\text{C}$$

Selecting the hot fluid as the reference stream I, the capacitance rate R and effectiveness P are calculated by writing†

$$R = \frac{C_h}{C_c} = \frac{T_{c,o} - T_{c,i}}{T_{h,i} - T_{h,o}} = \frac{31.6}{51.7} = 0.611$$

and

$$P = \frac{q}{q_{\max,I}} = \frac{C_h(T_{h,i} - T_{h,o})}{C_h(T_{h,i} - T_{c,i})} = \frac{98°\text{C} - 46.3°\text{C}}{98°\text{C} + 20°\text{C}} = 0.438$$

Knowing P and R, the number of transfer units NTU_I can be obtained by using Fig. 11–29 or Eq. (11–140). Using the analytical approach, we obtain

$$NTU_I = \frac{1}{1 - R}\ln\frac{1 - PR}{1 - P} = \frac{1}{1 - 0.611}\ln\frac{1 - 0.438(0.611)}{1 - 0.438} = 0.681$$

To complete the analysis, we must evaluate $\overline{U}$, C_h, and C_c. $\overline{U}$ and C_h are related by

$$NTU_I = \frac{\overline{U}A_s}{C_I} = \frac{\overline{U}A_s}{C_h}$$

Setting NTU_I = 0.681 and $A_s = \pi d_o L$ = 0.11 m^2, we are able to write

$$C_h = \frac{\overline{U}A_s}{NTU_I} = \frac{\overline{U}(0.11\ \text{m}^2)}{0.681} = 0.162\overline{U}\ \text{m}^2 \tag{a}$$

The following iterative solution approach can now be used for evaluating $\dot{m}_h$ and $\dot{m}_c$:

† With the hot fluid selected as the reference stream I, the P-NTU_I and ϵ-NTU methods for this problem are identical, with $P = \epsilon$, $R = C^*$, and $NTU_I = NTU$.

(1) set first approximation $C_h = C_h^{(i)}$ $(i = 1)$;

(2) calculate C_c, $\dot{m}_h$, and $\dot{m}_c$ from

$$C_c = \frac{C_h^{(i)}}{R} \qquad \dot{m}_h = \frac{C_h^{(i)}}{c_{P,h}} \qquad \dot{m}_c = \frac{C_c}{c_{P,c}} = \frac{C_h^{(i)}/R}{c_{P,c}}$$

(3) calculate Re_h and Re_c [see Eq. (6–35) and Eample 11–1];

$$Re_h = \left(\frac{GD_H}{\mu}\right)_h = \frac{4\dot{m}_h}{\mu_h \pi d_i} = \frac{4C_h^{(i)}/c_{P,h}}{\mu_h \pi d_i} = \frac{1.06 \times 10^5\, C_h^{(i)}}{\text{kW/°C}}$$

$$Re_c = \left(\frac{GD_H}{\mu}\right)_c = \frac{4\dot{m}_c}{\mu_c \pi (D_{i,s} + d_o)} = \frac{4C_h^{(i)}/(c_{P,c}R)}{\mu_c \pi (D_{i,s} + d_o)} = \frac{2.23 \times 10^5\, C_h^{(i)}}{\text{kW/°C}}$$

(4) evaluate $\overline{Nu}_h$ and $\overline{Nu}_c$ using appropriate correlations and calculate $\bar{h}_h$, $\bar{h}_c$, and $\overline{U}$ (see Example 11–1);

(5) use Eq. (a) to obtain new value of C_h,

$$C_h^{(i+1)} = 0.162\overline{U}\ \text{m}^2$$

(6) to obtain final solutions for $\dot{m}_h$ and $\dot{m}_c$, repeat steps 2–5 until satisfactory convergence is achieved. The convergence of this scheme is rather slow.

This iterative approach is typical of the type of analysis that must be used in evaluating the mass-flow rate in existing heat exchangers with specified fluids and terminal temperatures. Following through with the calculations for this application, we obtain $\dot{m}_c = 0.546$ kg/s and $\dot{m}_h = 0.0718$ kg/s, which is consistent with the process specified in Example 11–5. (The lumped/incremental finite-difference program listed in Fig. A–G–5 can be adapted to the efficient solution of this type computational problem. The use of the numerical approach in solving this problem is suggested as an exercise.)

EXAMPLE 11–11

Direct noniterative calculations for the necessary heat-transfer surface area A_s of a heat exchanger can be made by means of the effectiveness, *LMTD*, or efficiency approach following the steps outlined for the design problem in Table 11–8, providing that $\overline{U}$ is known or is essentially independent of the unspecified surface dimensions. However, in most heat exchanger design calculations $\overline{U}$ is not known *a priori* and is a function of the surface dimensions. Design calculations for problems of this type require the use of iterative solution approaches. Referring to the double-pipe heat-exchanger application shown in Fig. E11–11, outline the steps that would be required to determine the diameter d_o of the inner tube.

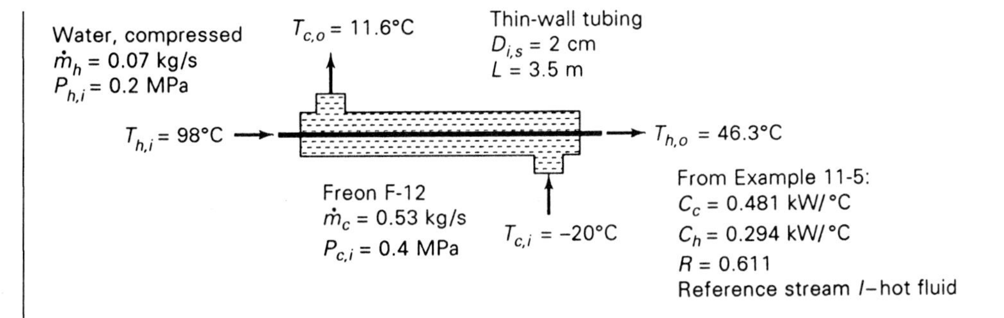

FIGURE E11–11 Double-pipe heat exchanger: counterflow arrangement—design problem.

Solution

Objective Set up a computational strategy for determining the diameter d_o of the inside tube of a double-pipe heat exchanger with specified fluids, mass-flow rates, terminal temperatures, outer tube diameter, and length L.

Assumptions/Conditions See Example 11–10.

Properties See Example 11–10.

Analysis This problem is readily solved by means of the effectiveness, $LMTD$, or efficiency methods. Following the effectiveness method, with the hot fluid specified as the reference stream I, we note that $C_I = C_h = 0.294$ kW/°C and $R = 0.611$ and write

$$P = \frac{q}{q_{\max,I}} = \frac{C_h\, T_{h,i} - T_{h,o}}{C_h\, T_{h,i} - T_{c,i}} = \frac{98°\text{C} - 46.3°\text{C}}{98°\text{C} + 20°\text{C}} = 0.438$$

$$NTU_I = \frac{1}{1-R}\ln\frac{1-PR}{1-P} = \frac{1}{1-0.611}\ln\frac{1-0.438(0.611)}{1-0.438} = 0.681$$

from Eq. (11–140). It follows that

$$\overline{U}A_s = C_h\, NTU_I = 0.294\,\frac{\text{kW}}{°\text{C}}\,(0.681) = 0.2\ \text{kW/°C}$$

or

$$A_s = \frac{200}{\overline{U}}\frac{\text{W}}{°\text{C}} \qquad \text{and} \qquad d_o = \frac{A_s}{\pi L} = \frac{A_s}{3.5\pi\ \text{m}} \tag{a,b}$$

where $d_i \simeq d_o$ and L has been set equal to 3.5 m.

The following iterative solution approach can be used for calculating the tube diameter d_o:

(1) set first approximation $d_o = d_o^{(i)}$, with $i = 1$;

(2) calculate Re_h and Re_c [see Eq. (6–35) and Example 11–1];

$$Re_h = \left(\frac{GD_H}{\mu}\right)_h = \frac{4\dot{m}_h}{\mu_h \pi d_o} = \frac{312 \text{ m}}{d_o}$$

$$Re_c = \left(\frac{GD_H}{\mu}\right)_c = \frac{4\dot{m}_c}{\mu_c \pi (D_{i,s} + d_o)} = \frac{1970 \text{ m}}{0.02 \text{ m} + d_o}$$

(3) evaluate $\overline{Nu}_h$ and $\overline{Nu}_c$ using appropriate correlations and calculate $\bar{h}_h$, $\bar{h}_c$, and $\overline{U}$ (see Example 11–1);

(4) use Eqs. (a) and (b) to obtain new value of d_o,

$$d_o^{(i+1)} = \frac{200 \text{ W/°C}}{\overline{U}(3.5\pi \text{ m})}$$

(5) to obtain the final solution for d_o, repeat steps 2 to 4 until satisfactory convergence is achieved. The convergence of this scheme is quite fast.

Following these steps, we obtain $d_o = 1$ cm, which is consistent with the application specified in Example 11–5. (The lumped/incremental finite-difference program listed in Table A–G–6 can be adapted to the efficient solution of this type computational problem. The use of the numerical approach in solving this problem is suggested as an exercise.)

11–7 SHELL-AND-TUBE HEAT EXCHANGERS

The relative motion of two fluids flowing through shell-and-tube heat exchangers with segmental baffles involves combinations of crossflow, parallel flow, and counterflow, with the baffled shell-side flow consisting of axial flow through the baffle windows and main crossflow, bypass, and leakage streams. Due to complexities such as these, the analysis of shell-and-tube heat exchangers is considerably more involved than the analysis of double-pipe heat exchangers. Nevertheless, practical solution approaches for the thermal and hydraulic evaluation and design of shell-and-tube heat exchangers have evolved, with most current practice based on work published since 1960.

The shell-and-tube heat exchanger configuration that normally is considered to be most practical utilizes the TEMA-E shell with a single or even number (N_{tp}) of tube passes and segmental baffles. However, pressure drop requirements or other considerations sometimes necessitate the use of other shell and baffle types.

We now turn our attention to the practical thermal and hydraulic analysis of shell-and-tube heat exchangers.

11–7–1 Practical Thermal Analysis

To provide a framework for the development of the practical thermal analysis of shell-and-tube heat exchangers with multiple tube passes, we consider the basic 1-2 exchanger shown in Fig. 11–32.† Representative distributions in bulk-stream shell and tubular fluid temperatures T_s and T_t throughout the exchanger are shown in this figure. Notice that this arrangement involves parallel flow for pass 1 and counterflow for pass 2. To account for this difference, we distinguish the tube-side bulk-stream temperature within a particular pass i by $T_t^{(i)}$, with T_t and $T_t^{(i)}$ being related by

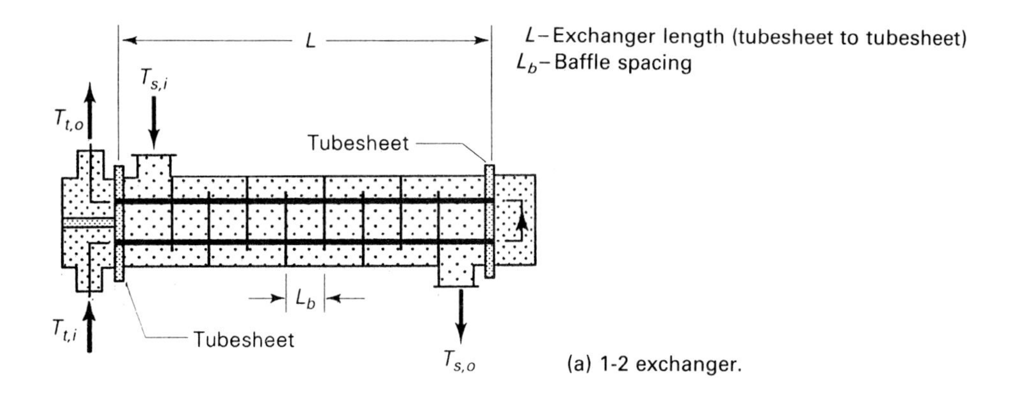

(a) 1-2 exchanger.

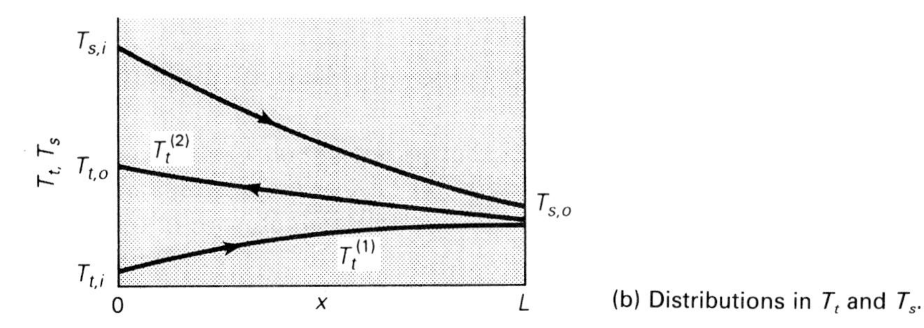

(b) Distributions in T_t and T_s.

FIGURE 11–32 Typical distributions in tube-side and shell-side bulk-stream temperatures in a 1-2 exchanger.

† To simplify the notation, shell-and-tube heat exchangers with TEMA-E shell will be referred to as 1-2 or 1-N_{tp} exchangers, as is commonly done in the literature. Because the general flow pattern in 1-1 exchangers can be classified as counterflow or parallel flow (but not both), the practical relations developed for heat transfer in double-pipe heat exchangers (Sec. 11–6) can be employed in the analysis of shell-and-tube heat exchangers of this type.

$$T_t = \frac{T_t^{(1)} + T_t^{(2)}}{2} \tag{11–165a}$$

for two tube passes, and

$$T_t = \frac{1}{N_{tp}} \sum_{i=1}^{N_{tp}} T_t^{(i)} \tag{11–165b}$$

for N_{tp} tube passes. This perspective is taken in both traditional [27–29] and formal matrix [30,31] approaches to analyzing shell-and-tube heat exchangers, with the following assumptions normally specified for steady-state operation:

(1) constant local overall coefficient of heat transfer U throughout the exchanger;
(2) no phase change occurs in only part of the exchanger;
(3) constant flow rate for both fluids;
(4) constant specific heat for both fluids;
(5) negligible heat losses and longitudinal conduction in the fluid and wall;
(6) uniform shell-side temperature T_s and tube-side temperature T_t for a given pass i over any cross section (i.e., complete mixing);
(7) equal heating surface area in each pass.

Although the details involved in the traditional and modern matrix approaches differ, both approaches involve the use of the defining equation for local overall coefficient of heat transfer U, which can be represented by

$$dq = U(T_s - T_t)\, dA_s \tag{11–166}$$

and both approaches result in relations of the form of Eq. (11–8),

$$q = \overline{U} A_s \,\Delta T_m \tag{11–167}$$

which may be expressed in terms of effectiveness P or ϵ, $LMTD$ correction factor F, or efficiency ψ. However, it should be noted that standard solution results of this type are used with the understanding that the local overall coefficient of heat transfer U is assumed to be constant, unless otherwise indicated (i.e., $\overline{U} \simeq U$). In practice, U generally varies to at least some extent over the cross section as well as in the longitudinal direction (i.e., from inlet to outlet). To allow for the use of standard solution results for situations in which moderate variations occur in U, we simply employ Eq. (11–167) together with the approximate mean overall coefficient of heat transfer $\overline{U}$. Practical solution results are available in the literature that account for the variation in U from inlet to outlet for several common applications [32–34]. However, the most pragmatic approach to analyzing multipass shell-and-tube heat exchangers with nonuniform distribution in U or other complications involves the use of the basic one-dimensional finite-difference method. This point is considered in Sec. 11–10.

We are now in a position to consider practical solutions that have been developed in the literature for heat transfer in shell-and-tube heat exchangers and methods for specifying the mean overall coefficient of heat transfer.

Solution Results for Heat-Transfer Rate

The lumped and lumped/differential formulation represented by Eqs. (11–24) to (11–27) can be combined with Eq. (11–166) and adapted to well-baffled shell-and-tube heat exchangers to obtain modeling equations for energy transfer and solutions for the heat-transfer rate. The solution is developed in Appendix N–5 for a 1-2 exchanger and is presented in this section in the context of the effectiveness, *LMTD*, and efficiency formats, all of which are equivalent. Solution results are listed in Appendix N–1 for several other types of shell-and-tube heat exchangers. A more comprehensive listing of solutions is presented in *Heat Exchanger Design Handbook* [6], *Handbook of Heat Transfer Applications* [7], *ESDU Handbook* [35], and a recent review by Roetzel [36].

Effectiveness Method The solution for a 1-2 exchanger is given in terms of the effectiveness P, number of transfer units NTU_I, and capacitance ratio R by (see Appendix N–5)

$$P = \frac{2}{1 + R + \sqrt{1 + R^2}\,(1 + e^{-\Gamma})/(1 - e^{-\Gamma})} \tag{11–168a}$$

where

$$\Gamma = NTU_I \sqrt{1 + R^2} \tag{11–168b}$$

or

$$NTU_I = \frac{1}{\sqrt{1 + R^2}} \ln \frac{2 - P(1 + R - \sqrt{1 + R^2})}{2 - P(1 + R + \sqrt{1 + R^2})} \tag{11–169}$$

The reference stream I can be taken as the fluid on either the shell side s or the tube side t (or as the cold fluid c or the hot fluid h), such that 1-2 exchangers are classified as symmetrical exchangers.† This solution is presented in a semilog format by Fig. 11–33. This figure also shows the minimum performance criteria represented by (1) $\psi_{MIN} = 0.35$, and (2) P_o from Eq. (11–76). Because of the nature of this shell-and-tube arrangement, this solution result applies whether the tube-side fluid is in parallel flow in the first pass and counterflow in the second pass, or vice versa (see Appendix N–4).

Exchangers with one shell pass and N_{tp} tube passes (N_{tp} even) are commonly employed in industry.‡ The effectiveness P for $N_{tp} > 2$ and even is somewhat lower than P for a 1-2 exchanger at the same values of NTU_I and R. For example, the maximum difference in P is generally less than 4.4% for 1-2 and 1-4 exchangers and 6.8% for 1-2 and 1-12 exchangers [7]. (The solution for a 1-4 exchanger is listed in Tables A–N–1–1(a) of the Appendix.) Because the differences are rather

† Referring to Table A–N–1–1, other shell-and-tube arrangements are nonsymmetrical.

‡ The use of an odd number of tube passes is rather uncommon. Studies pertaining to shell-and-tube heat exchangers with an odd number of tube passes are reported by Crozier and Samuels [37].

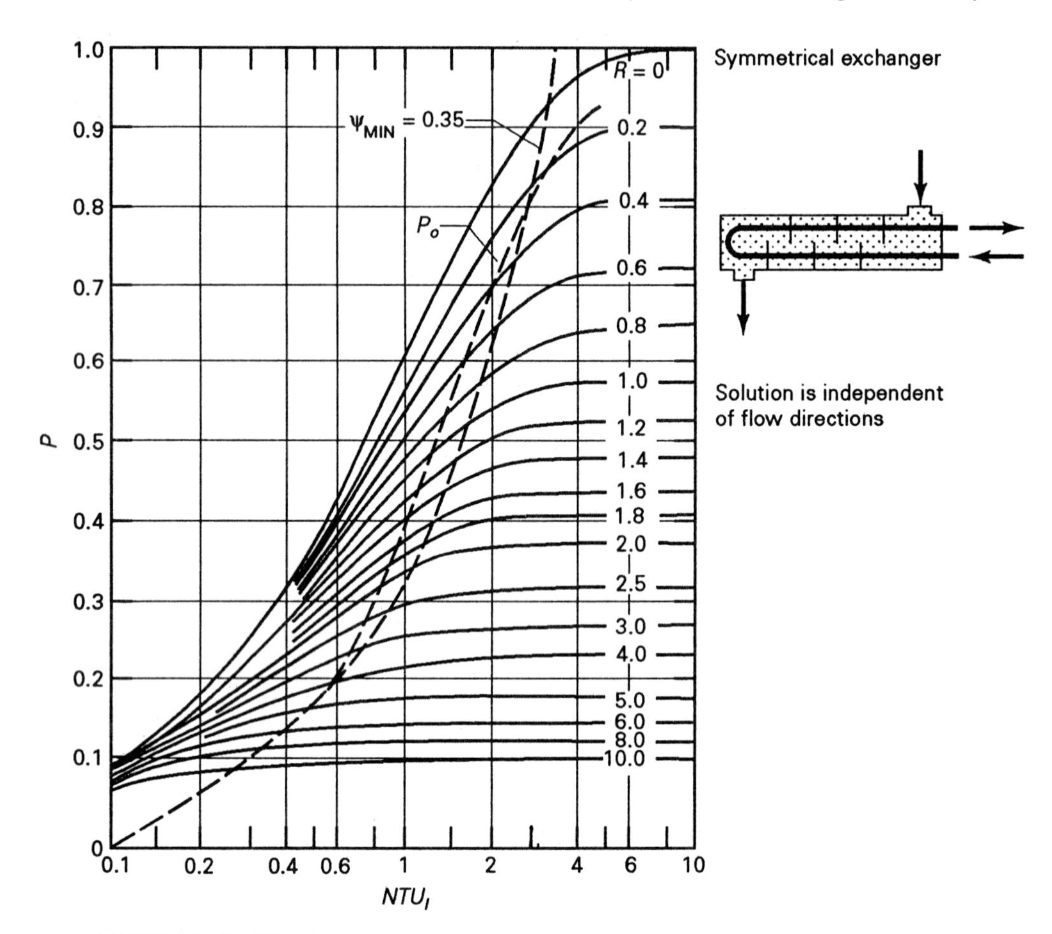

FIGURE 11–33 Effectiveness P for 1-2 exchanger. (Adapted from Shah and Mueller [7]. Used with permission.)

small, the solution results for 1-2 exchangers are generally used to approximate the solution for arrangements with one shell and any even number of tube passes.

Solution results for the effectiveness P of several types of shell-and-tube heat exchangers are listed in Tables A–N–1–1(a) and (b) of the Appendix.

Solution results for shell-and-tube heat exchangers are also commonly expressed in the ϵ-NTU format. Relations appearing in the literature for $\epsilon(NTU,C^*)$ and $NTU(\epsilon,C^*)$ are listed in Tables A–N–1–2(a) and (b) for the same arrangements represented in Tables A–N–1–1(a) and (b). The standard ϵ-NTU chart for a 1-2 exchanger is shown in Fig. 11–34.

The relations presented in this section for $P(NTU_I,R)$ and $NTU_I(P,R)$ are readily expressed in terms of the $LMTD$ correction factor F and the efficiency ψ.

LMTD Method Design practice for shell-and-tube heat exchangers normally involves the use of the $LMTD$ method. The $LMTD$ correction factor F for a 1-2

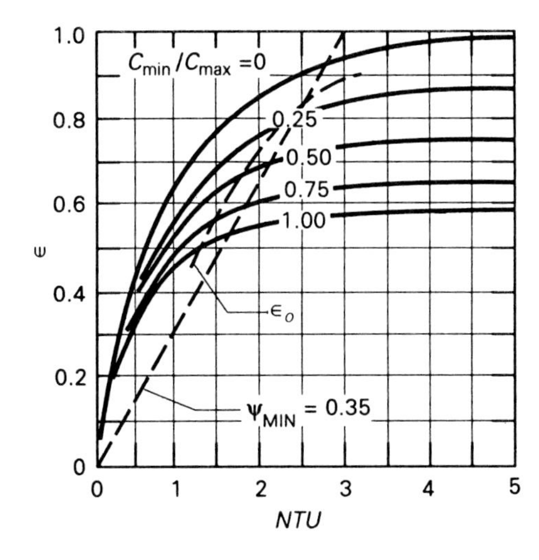

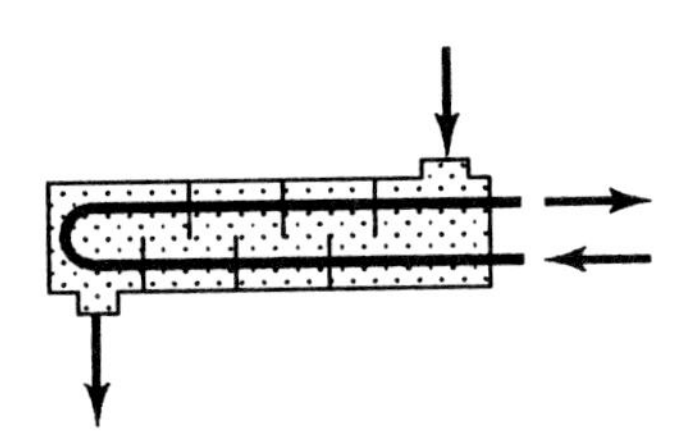

FIGURE 11-34
Effectiveness ϵ for 1-2 exchanger. (Adapted from Kays and London [10]. Used with permission.)

exchanger is expressed in terms of P, NTU_I, and R by Eq. (11-70),†

$$F = \frac{1}{NTU_I\,(1 - R)} \ln \frac{1 - PR}{1 - P} \tag{11-170}$$

It follows that Eq. (11-168) can be used to obtain an expression for $F(NTU_I,R)$ of the form

$$F = \frac{1}{NTU_I\,(1 - R)} \ln \frac{1 + R + \sqrt{1 + R^2}\,(1 + e^{-\Gamma})/(1 - e^{-\Gamma}) - 2R}{1 + R + \sqrt{1 + R^2}\,(1 + e^{-\Gamma})/(1 - e^{-\Gamma}) - 2} \tag{11-171a}$$

where

$$\Gamma = NTU_I \sqrt{1 + R^2} \tag{11-171b}$$

Equivalently, with $NTU_I(P,R)$ given by Eq. (11-169), the solution for $F(P,R)$ can be written as

$$F = \frac{\sqrt{1 + R^2}}{1 - R} \frac{\ln\,[(1 - PR)/(1 - P)]}{\ln\,\{[2 - P(1 + R - \sqrt{1 + R^2})]/[2 - P(1 + R + \sqrt{1 + R^2})]\}} \tag{11-172}$$

This solution was first developed in 1940 by Bowman et al. [28]. The *LMTD* correction factor chart for this case is shown in Fig. 11-35. The minimum performance criteria represented by F_{MIN} and F_o are also shown in this figure.

Solution results for the *LMTD* correction factors of several types of shell-and-tube heat exchangers are given in Table A-N-1-1 of the Appendix.

† See Table A-N-1-1 for solution results for $R = 1$.

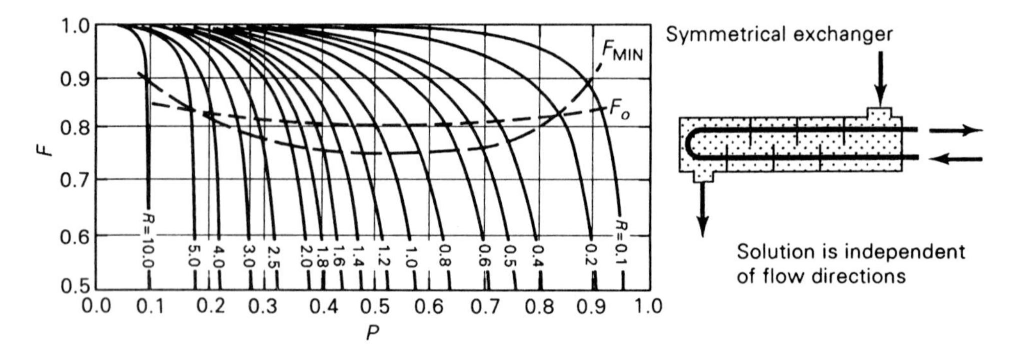

FIGURE 11–35 *LMTD* correction factor *F* for 1-2 exchanger. (Adapted from Bowman et al. [28].)

Efficiency Method The solution result for a 1-2 exchanger given by Eq. (11–168) is expressed in terms of efficiency $\psi(NTU_I,R)$ by writing

$$\psi = \frac{P}{NTU_I} = \frac{2/NTU_I}{1 + R + \sqrt{1 + R^2}\,(1 + e^{-\Gamma})/(1 - e^{-\Gamma})} \tag{11–173a}$$

where

$$\Gamma = NTU_I \sqrt{1 + R^2} \tag{11–173b}$$

Alternatively, Eq. (11–169) is used to obtain an expression for $\psi(P,R)$ of the form

$$\psi = \frac{P\sqrt{1 + R^2}}{\ln\{[2 - P(1 + R - \sqrt{1 + R^2})]/[2 - P(1 + R + \sqrt{1 + R^2})]\}} \tag{11–174}$$

These relations together with the minimum performance criteria ψ_{MIN} and ψ_o are represented by the Mueller chart shown in Fig. 11–36.

Solution results for the efficiency of several types of shell-and-tube heat exchangers are given in Table A–N–1–1 of the Appendix.

Evaluation of $\overline{U}$

For shell-and-tube heat exchangers, $\overline{U}$ is generally approximated by

$$\frac{1}{\overline{U}A_s} = \frac{1}{(\bar{h}A_s)_s} + \left(\frac{F_f}{A_s}\right)_s + R_k + \left(\frac{F_f}{A_s}\right)_t + \frac{1}{(\bar{h}A_s)_t} \tag{11–175}$$

with the effect of fins appropriately accounted for. Whereas simple correlations of the type presented in Chap. 8 can be used to determine the tube-side coefficient $\bar{h}_t$, the complications associated with baffled flow must be dealt with in the evaluation of the shell-side coefficient $\bar{h}_s$.

Considerable progress has been achieved since 1960 in the development of approaches to evaluating $\bar{h}_s$ for exchangers with segmental baffles. The approaches

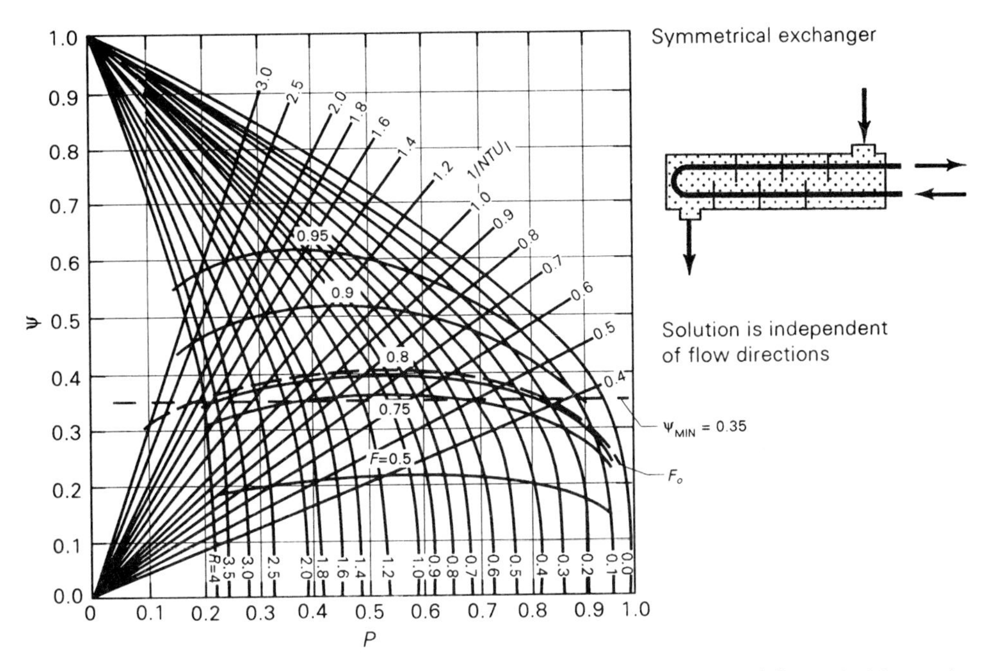

FIGURE 11–36 Efficiency ψ for 1-2 exchanger. (From Shah and Mueller [7]. Used with permission.)

to specifying $\bar{h}_s$ for TEMA-E shell-type exchangers with segmental baffles and plain tubes is considered first, after which brief attention is given to the treatment of other geometries.

TEMA-E Shell Exchangers with Segmental Baffles Referring to Fig. 11–37, segmental baffled shell-side flow is very complex. As indicated by Taborek [6], only part of the fluid takes the desired path B through the tube bundle, whereas a potentially substantial portion flows through the leakage areas created by clearances between the tubes and baffle (L_{tb}—path A) and between shell and baffle (L_{sb}—path E) and through the bypass area between the tube bundle and shell wall (L_{bb}—path C).† For exchangers with floating head bundle construction in which the bundle-to-shell bypass clearance L_{bb} exceeds a value of about 30 mm, sealing strips attached to the baffles and running along the length of the shell are commonly used to force the bypass stream back into the tube bundle. Typical flow fractions of the various streams for well-designed baffles are listed in Table 11–10 for laminar and turbulent flows. Although these baffle arrangements do improve the overall heat-transfer performance because of the mixing and approximate cross-flow pattern that is established, it should

† Standards for clearances L_{tb}, L_{sb}, and L_{bb} are presented in Appendix O–2.

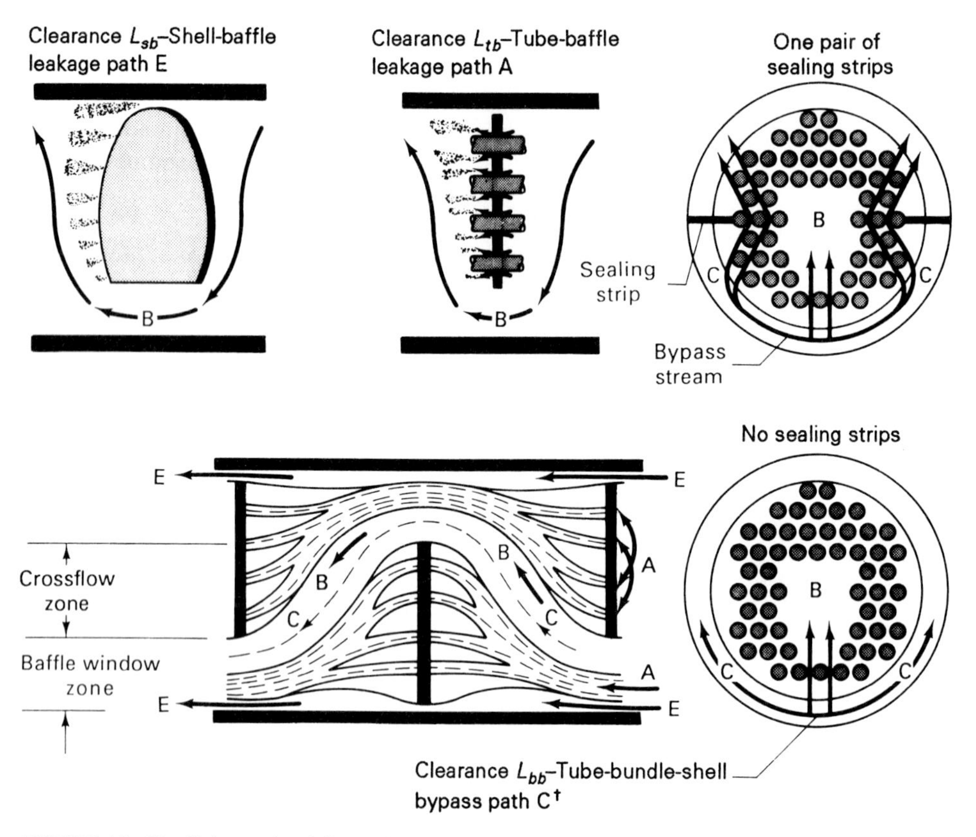

FIGURE 11–37 Schematic of flow distribution for segmental baffled shell-side flow (adapted from Tinker [38]); path B—crossflow over tube bundle.

be noted that low heat-transfer rates occur in the regions of relative stagnant flow and leakage.

The several approaches that have evolved for evaluating the shell-side coefficient of heat transfer $\bar{h}_s$ are based on the use of correlations for pure crossflow in ideal tube banks to account for the contribution of the main crossflow stream B, with the effects of the number of baffles, baffle windows, and leakage and bypass streams approximated to within various levels of sophistication.

In shell-and-tube analysis and design, convection correlations for the mean coefficient of heat transfer $\bar{h}_i$ associated with flow across ideal tube banks are expressed in terms of crossflow mass flux G_s,

† In tube bundle layouts with multiple passes, an additional bypass stream F occurs in flow channels due to the omission of tubes in the tube-pass partitions. Stream F behaves similarly to bypass stream C [6,39]. The use of sealing strips is normally restricted to exchangers with floating-head bundle construction.

TABLE 11–10 Typical flow fractions of various streams for baffled shell-side flow

Stream	*Laminar (%)*	*Turbulent (%)*
Main crossflow (B)	25–50	40–70
Bundle bypass (C)	20–30	15–20
Shell-to-baffle leakage (E)	6–40	6–20
Tube-to-baffle leakage (A)	4–10	9–20

Source: Taborek [6].

$$G_s = (\rho U_b)_s = \frac{\dot{m}_s}{A_{\text{min}}} \tag{11–176}$$

shell-side Reynolds number Re_s,†

$$Re_s = \frac{G_s d_o}{\mu_s} = \frac{U_{b,s} d_o}{\nu_s} \tag{11–177}$$

and shell-side Nusselt number $\overline{Nu}_s$,

$$\overline{Nu}_s = \frac{\bar{h}_s d_o}{k_s} \tag{11–178}$$

where A_{min} is the minimum free-flow area (see Sec. 8–4–1).

The common practice is to set the minimum free-flow area A_{min} equal to the crossflow area A_c at a plane normal to the cross section extending between adjacent baffles and passing through the shell centerline, which is approximated by

$$A_c = N_{tc}(S_t - d_o)L_b = \frac{D_{i,s}}{S_{t,\text{eff}}}(S_t - d_o)L_b \tag{11–179}$$

where L_b is the *baffle spacing*‡ and the *effective tube pitch* $S_{t,\text{eff}}$ is S_t for square (90°) and triangular (30°) layouts and $0.707S_t$ for rotated square (45°) layouts (see Table A–N–6–1), and $N_{tc} = D_{i,s}/S_{t,\text{eff}}$ approximates the number of tubes across the shell diameter in the direction normal to the crossflow. Referring to Fig. 11–32, the baffle spacing L_b, number of baffles N_b, and heat-exchanger length L are related by

$$L_b = \frac{L}{N_b + 1} \qquad N_b = \frac{L}{L_b} - 1 \tag{11–180a,b}$$

Notice that the crossflow mass flux G_s is inversely proportional to L_b, such that Re_s increases with the number of baffles N_b.

The primary approaches to evaluating $\bar{h}_s$ that are presently in use in the heat exchanger industry are the *stream analysis method* and the *Delaware method*. In

† Although the hydraulic diameter $D_H = 4A_{\text{min}}/p_w$ also provides a viable characteristic length for shell-side analysis, the outside tube diameter d_o is universally used in the heat exchanger industry for plain tubes. In the case of low-fin tubing, d_o is replaced by the equivalent diameter d_{req}.

‡ L_b is also commonly referred to as the *center baffle spacing*, with the *end-zone baffle spacing* represented by $L_{b,i}$ at the entrance and $L_{b,o}$ at the exit.

addition to these rather sophisticated design methods, simple methods are available for obtaining preliminary estimations that are useful for hand calculations and provide a basis for establishing a feel for the effect of baffle spacing and other primary design variables.

Stream Analysis Method The comprehensive stream analysis method of Palen and Taborek [39] represents a modernization of an analysis proposed by Tinker [38] in 1951. Referring to Fig. 11–37, this method involves the simultaneous iterative calculation of pressure drops and flow rates associated with the main crossflow stream B, the leakage streams A and E, and the bypass streams C and F. (This point is expanded on in Sec. 11–7–2.) Once the pressure drops and flow rates of the various streams are calculated, the shell-side coefficient of heat transfer $\bar{h}_s$ is determined by approximating the main crossflow stream B as ideal crossflow over tube banks and using largely proprietary empirical correlations to estimate the relative heat-transfer performance of the leakage and bypass streams. The stream analysis approach is presently used in the comprehensive design codes provided by HTRI and HTFS, with many of the details being unavailable in the open literature.† The error associated with this method for calculating $\bar{h}_s$ is normally within $\pm 15\%$ for a wide range of operating conditions.

Delaware Method Whereas the iterative stream analysis method is generally recognized as being the most comprehensive design method for evaluating the shell-side coefficient of heat transfer, the Delaware method provides a practical and reasonably reliable alternative means of accounting for the effects of leakage, bypass, baffle spacing, and baffle cut on $\bar{h}_s$. This popular noniterative method is based on the extensive body of experimental data and analysis developed in the Delaware project [40–42]. A refined version of the Delaware method is detailed in the *Heat Exchanger Design Handbook* [6] and the *Handbook of Heat Transfer Applications* [7] and is summarized in Appendix N–6. The Delaware method features the use of a relation for $\bar{h}_s$ of the form‡

$$\bar{h}_s = J_{\text{tot}}\bar{h}_i = (J_c J_l J_b)\bar{h}_i \tag{11–181}$$

where $\bar{h}_i$ represents the coefficient of heat transfer associated with pure crossflow in an ideal tube bank (assuming the entire shell-side stream flows across the tube array at the centerline of the exchanger), $J_{\text{tot}} = J_c J_l J_b$, and the individual correction factors account for the effects of baffle cut J_c, leakage (streams A and E) J_l, and bypass J_b. The combined effect of these corrections typically results in values of J_{tot} around 0.6 for well-designed exchangers under turbulent flow conditions [6,7]. The ideal heat-transfer coefficient $\bar{h}_i$ is expressed in terms of Reynolds number by correlations of the type introduced in Chap. 8. Correlations for the individual

† The equivalent HTFS approach is referred to as the *stream interaction method*.

‡ A more general form of Eq. (11–181) that is used for $Re_s < 100$ or $L_{b,\text{end}}/L_b \gtrsim 1.25$ is indicated in Appendix N–6.

correction factors and $\overline{h}_i$ recommended in the *Heat Exchanger Design Handbook* [6] are presented in Appendix N–6. The error associated with the Delaware method for evaluating $\overline{h}_s$ should be within -10% to $+25\%$ for turbulent flow conditions, assuming a well-designed exchanger [6].

Although the Delaware method can be used in conjunction with hand calculators, because of the number of computations required for the various parameters and their subsequent iteration, the procedure is somewhat involved for preliminary manual design work. Consequently, software for personal computers [43,44] that features the Delaware method proves to be a very attractive and useful tool for analyzing shell-and-tube heat exchangers.

Simple Method A A simple method proposed by Taborek [6] for estimating the shell-side coefficient of heat transfer $\overline{h}_s$ involves the use of a correlation of the form†

$$\overline{Nu}_s = 0.2\ Re_s^{0.6}\ Pr^{1/3} \tag{11–182}$$

for uniform property turbulent flow conditions with Re_s in the range 2000 to 40,000. Notice that because Re_s is defined in terms of the crossflow mass flux $G_s = \dot{m}/A_c$, where A_c is proportional to N_{tc}, $(S_t - d_o)$, and L_b [see Eq. (11–179)], Eq. (11–182) provides a means of estimating the effect of the number of tubes, tube spacing and layout, and baffle spacing on $\overline{h}_s$.

Simple Method I Introducing the approximation $J_{\text{tot}} = 0.6$, Eq. (11–181) becomes

$$\overline{h}_s = 0.6\overline{h}_i \qquad \text{or} \qquad \overline{Nu}_s = 0.6\ \overline{Nu}_i \tag{11–183a,b}$$

To express this approximation in terms of the convection correlation given in Appendix N–6 for crossflow over ideal tube banks, we write

$$\overline{Nu}_s = 0.6 j_i\ Re_s\ Pr_s^{1/3} \tag{11–184}$$

where the factor j_i is an empirical representation of $\overline{Nu}_i/(Re_s\ Pr_s^{1/3})$. With Re_s known for a given arrangement, the factor j_i is readily obtained from Fig. A–N–6–1, which permits the use of Eq. (11–184) for evaluating $\overline{h}_s$.

Simple Method II Simple methods A and I provide approximations for $\overline{h}_s$ that do not involve inputs for the clearances L_{tb}, L_{sb}, and L_{bb} and baffle cut B_c. To account for the actual values of the clearances and baffle cut, we return to Eq. (11–181), which is put into the form

$$\overline{Nu}_s = J_{\text{tot}} j_i\ Re_s\ Pr_s^{1/3} \tag{11–185}$$

where the factor j_i is specified by the correlations given in Fig. A–N–6–1 and J_{tot} is evaluated by use of correlations for the individual correction factors given in

† This correlation evolved from the early work by Colburn [45] and Grimison [46].

Fig. A–N–6–3 and approximate relations for key dimensionless geometric parameters (r_{lm}, r_s, r_b, N_{ss}^+, L^+, F_c, F_w and θ_{ds}) pertaining to clearance and baffle cut and related factors given by Eqs. (N–6–3) and (N–6–4) in Appendix N–6. This approach, which we refer to as simple method II, represents a modest simplification of the Delaware method. With the dimensionless geometric parameters specified by the relations given in Table A–N–6–2, this method becomes identical to the Delaware method. The accuracy of simple method II for evaluating $\bar{h}_s$ should be comparable to that of the Delaware method.

Extensions to Other Geometries The methods for evaluating $\bar{h}_s$ for exchangers with TEMA-E shell, segmental baffles, and plain tubing can be adapted to exchangers with TEMA-J shell or TEMA-F shell and can be used for low-fin tubes, but are not adaptable to arrangements with other types of shells or baffles. Approaches to the evaluation of $\bar{h}_s$ for other shell and baffle types are presented in references 47 to 49. Software is readily available in the marketplace that can be used for all TEMA designations for shell and bundle types, as well as for single-, double-, or triple-segmental baffles, no-tube-in-window baffles, or rodbaffles, and for plain or low-fin tubing.

TEMA-J Shell Exchangers Referring to Fig. 11–38, the standard methods for evaluating $\bar{h}_s$ can be adapted to a TEMA-J shell by setting the length of an equivalent TEMA-E shell equal to half the length of the J shell (i.e., $L_E = L_J/2$), and the mass-flow rate in an equivalent E shell equal to half the mass-flow rate in the J shell (i.e., $\dot{m}_{s,E} = \dot{m}_{s,J}/2$). Using these modifications, $\bar{h}_s$ can be evaluated and relations for P (or ϵ), F or ψ for TEMA-J shell exchangers can be employed to produce design or evaluation calculations.†

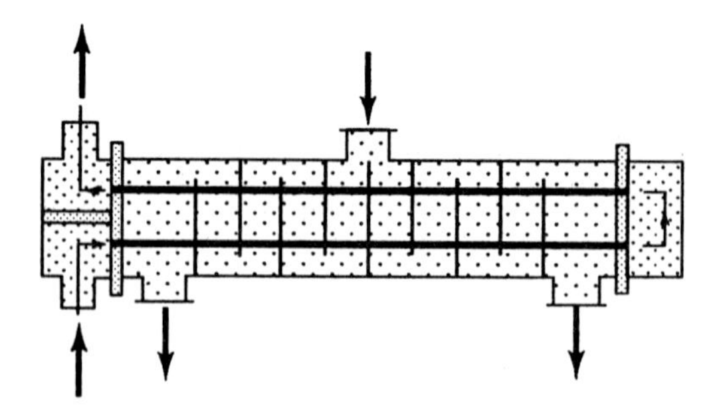

FIGURE 11–38 1-2 exchanger with TEMA-J shell.

TEMA-F Shell Exchangers In the case of the TEMA-F shell shown in Fig. 11–39, the standard methods can be used to approximate $\bar{h}_s$ by considering an equivalent TEMA E shell exchanger with halved tube bundle (i.e., $N_{tc,E}$ =

† Representative calculations presented in Example 11–16 indicate that relative to an E shell $\bar{h}_s$ is 40% smaller for a J shell and 73% larger for a F shell.

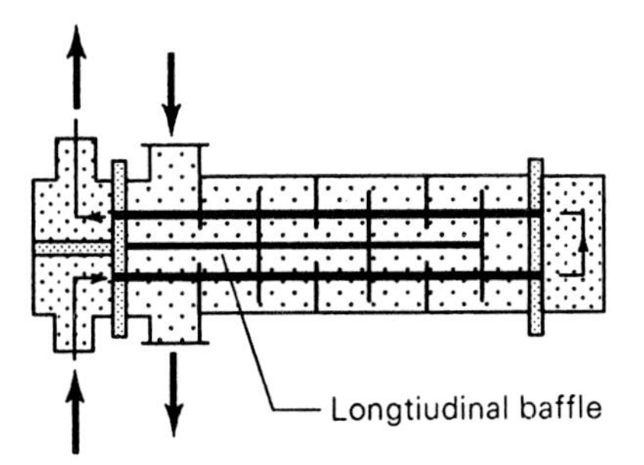

FIGURE 11-39 1-2 exchanger with TEMA-F shell.

$N_{tc,F}/2$) and doubled shell length (i.e., $L_E = 2L_F$).† It should be noted that for two tube passes this arrangement produces counterflow for all tubes, which results in higher heat-transfer effectiveness. This difference in effectiveness is accounted for by adapting solutions that have been developed for double-pipe arrangements.

Low-Fin Tubing Exchangers The primary modification that is made in the use of the standard methods for analyzing exchangers with low-fin tubing is the use of the equivalent diameter d_{req} (see Table A-O-3) in place of d_o in determining the crossflow area A_c, shell-side mass flux ($G_s = \dot{m}_s/A_c$), Reynolds number ($Re_s = G_s d_{\text{req}}/\mu_s$), and Nusselt number ($\overline{Nu}_s = \bar{h}_s d_{\text{req}}/k_s$). Using these adaptations, the standard methods are used to evaluate $\bar{h}_{s,\text{plain}}$, and $\bar{h}_{s,\text{fin}}$ is approximated by

$$\bar{h}_{s,\text{fin}} = J_F \bar{h}_{s,\text{plain}} \tag{11-186}$$

where $J_F = 1$ for $Re_s \gtrsim 1000$ and J_F is approximated by the correlation given in Fig. 11-40 for $Re_s \lesssim 1000$. As a practical approximation, the fin efficiency for low-fin tubes is generally taken as 100%, such that $\eta_o \simeq 1$.

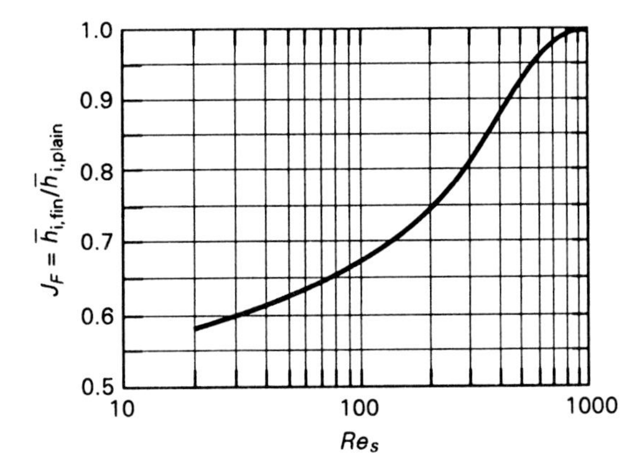

FIGURE 11-40 Correction factor J_F for shell-side heat transfer in heat exchanger with low-fin tubes. (From *Heat Exchanger Design Handbook* [6]. Used with permission.)

11–7–2 Practical Hydraulic Analysis

We now turn our attention to the evaluation of the pressure drop for the tube-and shell-side flow.

Tube-Side Pressure Drop

The tube-side pressure drop for one tube pass is generally represented in terms of the friction factor f_t and entrance, expansion, and component-loss coefficients K_c, K_e, and K_f by Eq. (11–101). It follows that the total pressure drop $(\Delta P_t)_{1-2}$ for N_{tp} tube passes becomes

$$(\Delta P_t)_{1-2} = \left(\frac{\rho U_b^2}{2}\right)_t \left[K_c + N_{tp}\left(\frac{4L}{d_i} f_t\right) + K_e + K_f\right] \tag{11–187}$$

where L is the length of the heat exchanger. Referring back to Fig. 8–5, we note that heat exchangers with small contraction ratios ($\sigma \simeq 0$) are characterized by $K_c + K_e \simeq 1.5$ for turbulent flow and $K_c + K_e \simeq 2$ for laminar flow. The component-loss coefficient K_f (see Table 11–9) accounts for the effects of return bends and inlet and exit nozzles.

Shell-Side Pressure Drop

TEMA-E Shell Exchangers with Segmental Baffles The stream analysis method and the Delaware method are the standard approaches to determining the shell-side pressure drop for standard TEMA-E shell exchangers with segmental baffles. These primary methods can be effectively complemented in preliminary stages of design work by simpler methods. Compared to correlations used prior to 1960, which tended to overpredict pressure drop by a factor of 5 to 10, the errors associated with these methods are generally much better.

Stream Analysis Method As we have noted, the stream analysis method introduced by Palen and Taborek [39] involves the simultaneous iterative calculation of pressure drops and flow rates of the various crossflow, leakage, and bypass streams passing through the shell. Referring to Figs. 11–37 and 11–41, the pressure drop ΔP_j over one baffle space for an individual stream j (A, B, C, E, or F) is expressed in terms of the loss coefficient K_j by Eq. (11–100), which is put into the form

$$\Delta P_j = K_j\left(\frac{\rho U_b^2}{2}\right)_j = \frac{K_j}{2\rho_j}\left(\frac{\dot{m}_j}{A_{\min,j}}\right)^2 \tag{11–188}$$

where $\dot{m}_j$ is the stream mass-flow rate and $A_{\min,j}$ is the corresponding cross-sectional area. The stream mass-flow rate $\dot{m}_j$ is related to the total shell mass-flow rate $\dot{m}_s$

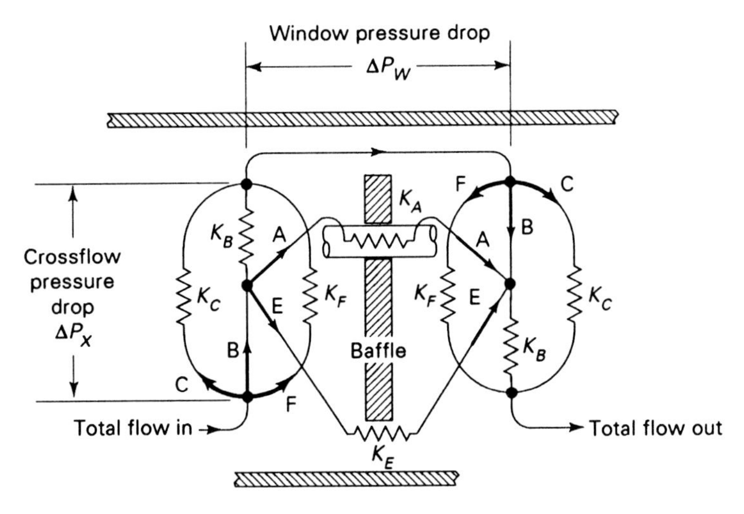

FIGURE 11–41 Schematic model of shell-side flow branches and their resistances across one baffle space. (From Palen and Taborek [39].)

by

$$\dot{m}_s = \Sigma\, \dot{m}_j = \dot{m}_A + \dot{m}_B + \dot{m}_C + \dot{m}_E + \dot{m}_F \qquad (11\text{–}189)$$

The individual stream pressure drops ΔP_j are related by

$$\Delta P_B = \Delta P_C = \Delta P_F = \Delta P_X \qquad (11\text{–}190\text{a})$$

$$\Delta P_A = \Delta P_E = \Delta P_W \qquad (11\text{–}190\text{b})$$

where ΔP_X is the crossflow pressure drop and ΔP_W is the window pressure drop. The calculations for ΔP_j are made using empirical correlations for the loss coefficients K_j (which are functions of stream-passage geometry and Reynolds number) based on published data from the Delaware project [40–42] and proprietary data for commercial size exchangers.

With the cross-sectional areas $A_{\min,j}$ defined and the loss coefficients K_j known as a function of stream Reynolds number Re_j, Eqs. (11–188) to (11–190) can be solved iteratively for the stream mass-flow rates $\dot{m}_j$ and pressure drops ΔP_X and ΔP_W. With ΔP_X and ΔP_W known for the central- and end-baffle spaces, the shell-side pressure drop ΔP_s is obtained by summing the results for all $N_b + 1$ baffle spaces; that is,

$$\Delta P_s = \sum_{k=1}^{N_b+1} (\Delta P_X + \Delta P_W)_k \qquad (11\text{–}191)$$

To represent the shell-side pressure drop with nozzle losses, we write

$$(\Delta P_s)_{1-2} = \Delta P_s + K_f \left(\frac{\rho U_b^2}{2}\right)_s = \Delta P_s + K_f \left(\frac{G^2}{2\rho}\right)_s \qquad (11\text{–}192)$$

Due to the proprietary nature of the critical design correlations for K_j, the stream analysis method is presently restricted to commercial codes provided to its member clients by HTRI and HTFS. The accuracy of the stream analysis method for calculating ΔP_s is normally within $\pm 25\%$ for a well-designed unit [6].

Delaware Method The Delaware work provides a very useful basis for the practical analysis of shell-side pressure drop for segmentally-baffled heat exchangers. The Delaware method for computing $(\Delta P_s)_{1-2}$, which accounts for bundle crossflow, baffle-window flow, leakage and bypass, losses, and other factors is reported by Bell [40–42] and is featured in the *Heat Exchanger Design Handbook* [6] and the *Handbook of Heat Transfer Applications* [7] and is summarized in Appendix N–6.

Referring to Fig. 11–42, the Delaware method expresses $(\Delta P_s)_{1-2}$ in terms of the pressure drops for pure crossflow ΔP_c, baffle-window flow ΔP_w, and flow in the end zones (first and last baffle spacing) ΔP_e; that is,

$$\Delta P_s = \Delta P_c + \Delta P_w + \Delta P_e \tag{11–193}$$

The crossflow pressure drop ΔP_c is represented in terms of the pressure drop ΔP_i for pure crossflow over an ideal tube bank by

$$\Delta P_c = \Delta P_i \,(N_b - 1) R_l R_b \tag{11–194}$$

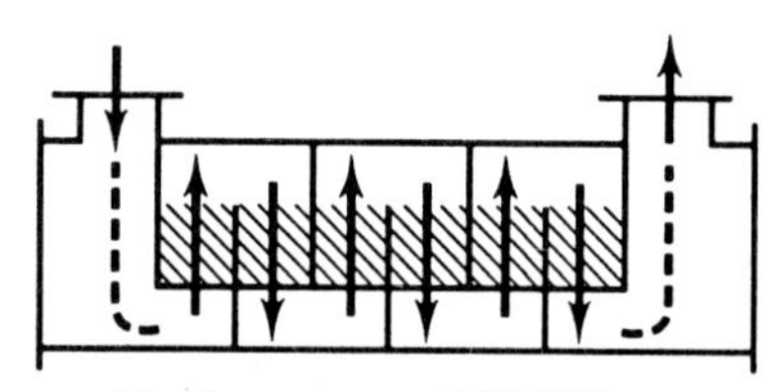

(a) Region of crossflow between baffle tips in the central-baffle spacing; pressure drop in this region represented by ΔP_c.

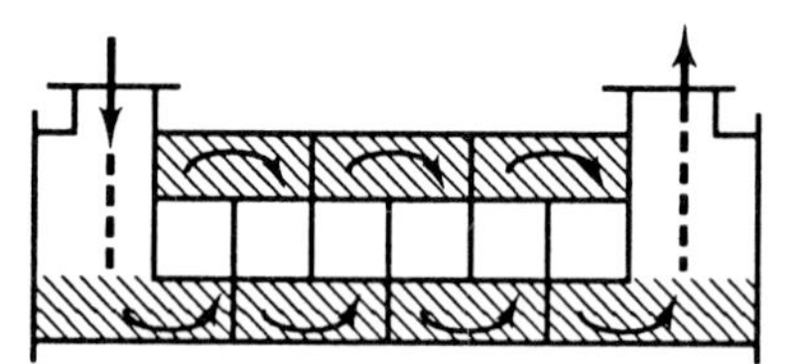

(b) Region of axial flow through baffle windows; pressure drop in this region represented by ΔP_w.

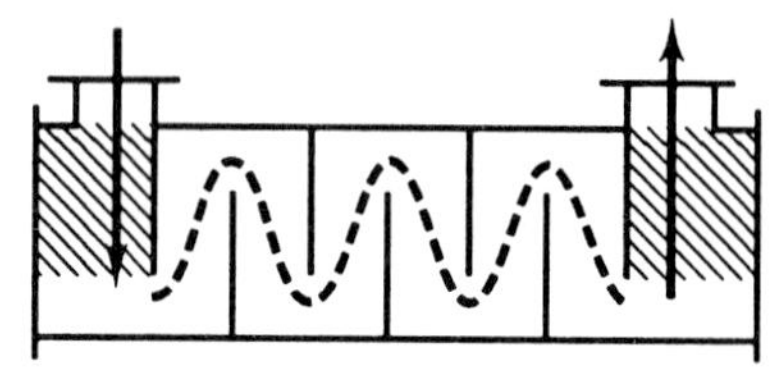

(c) Region of flow in end-baffle spacings; pressure drop in this region represented by ΔP_e.

FIGURE 11–42 Delaware method: perspective for well-baffled shell-side flow in TEMA-E type shell with segmental baffles. (From Taborek [6]. Used with permission.)

where $N_b - 1$ represents the number of times the tube bundle is crossed (excluding end baffle spacing), R_l is a correction factor for leakage streams A and E), and R_b is a correction factor for bypass streams (C and F).

The ideal crossflow pressure drop ΔP_i is normally expressed in terms of the friction factor f_i by setting N_L equal to N_{tcc} in Eq. (8–159),

$$\Delta P_i = 4N_{tcc}\frac{f_i}{2}(\rho U_b^2)_s = 4N_{tcc}\frac{f_i}{2}\left(\frac{G^2}{\rho}\right)_s \qquad (11\text{–}195)$$

where N_{tcc} represents the effective number of tube rows between baffle tips, and f_i is specified by standard correlations developed for ideal tube banks (see Sec. 8–4 and Fig. A–N–6–1). Following the approach recommended in reference 6, N_{tcc} is expressed in terms of the baffle cut (percent) B_c by

$$N_{tcc} = \frac{D_{i,s}}{S_{pp}}\left(1 - \frac{2B_c}{100}\right) \qquad (11\text{–}196)$$

where S_{pp} is S_t for square (90°), 0.866 S_t for triangular (30°), and $0.707S_t$ for rotated square (45°) layouts (see Table A–N–6–1); $D_{i,s}/S_{pp}$ represents the number of tubes across the shell diameter in the crossflow direction. The correction factors R_l and R_b appearing in Eq. (11–194) are functions of the clearances represented by L_{tb} (tube to baffles), L_{sb} (shell to baffles), L_{bb} (shell to bundle bypass), and other geometric parameters. Typically, R_l is in the range 0.4 to 0.5 and, depending on the number of pairs of sealing strips, R_b is in the range 0.5 to 0.8 [6]. Correlations for f_i, R_l, and R_b and the approach to treating the pressure drop components ΔP_w and ΔP_e featured in reference 6 are summarized in Appendix N–6.

As indicated in the context of our treatment of the shell-side coefficient of heat transfer, the Delaware method is thoroughly documented in the open literature [6,7,40–42] and is widely used in general engineering applications. The error in calculations for ΔP_s associated with this method is normally within about ±40% for turbulent flow and well-designed exchangers [6].

Simple Method I A simple method for calculating the shell-side pressure drop has recently been proposed by Taborek [9] that is based on (1) ideal crossflow over a tube bank and (2) an estimation of the baffle-window pressure drop. Following this approach, we represent the shell-side pressure drop ΔP_s by

$$\Delta P_s = \Delta P_i\,(N_b + 1)R_lR_b + N_bK_w\left(\frac{\rho U_b^2}{2}\right)_s \qquad (11\text{–}197)$$

where ΔP_i can be approximated by Eq. (11–195), $N_b + 1$ represents the total number of times the tube bundle is crossed (including end-baffle spacing), K_w is the loss coefficient associated with the 180° baffle-window turn, and the product R_lR_b accounts for the effect of leakage and bypass. Using Eq. (11–195) to represent ΔP_i, approximating the loss coefficient K_w by 2, and using Eq. (11–196) to specify

N_{tcc}, Eq. (11–197) becomes†

$$\Delta P_s = 4N_{tcc}\frac{f_i}{2}(\rho U_b^2)_s(N_b + 1)R_l R_b + N_b(\rho U_b^2)_s$$
$$= \left[4\frac{D_{i,s}}{S_{pp}}\left(1 - \frac{2B_c}{100}\right)(N_b + 1)R_l R_b\frac{f_i}{2} + N_b\right]\left(\frac{G^2}{\rho}\right)_s \tag{11–198}$$

where f_i can be obtained from the correlation given in Fig. A–N–6–1 or from other standard convection correlations for pure crossflow over ideal tube banks.

Although the leakage and bypass correction factors R_l and R_b can be set equal to unity as a rough first approximation, to avoid the likelihood of large overpredictions, the product $R_l R_b$ is set equal to 0.4, which is typical of well-designed exchangers, to obtain

$$\Delta P_s = \left[1.6\frac{D_{i,s}}{S_{pp}}\left(1 - \frac{2B_c}{100}\right)(N_b + 1)\frac{f_i}{2} + N_b\right]\left(\frac{G^2}{\rho}\right)_s \tag{11–199}$$

With Re_s known for a given arrangement, f_i is easily obtained from Fig. A–N–6–1, such that ΔP_s can be calculated.

Simple Method II Equation (11–198) can be used in conjunction with empirical correlations for the correction factors R_l and R_b to more formally account for the effects of leakage and bypass. This simple method II involves the use of the correlations given in Fig. A–N–6–4 for R_l and R_b, with the clearances L_{tb}, L_{sb}, and L_{bb}, baffle cut B_c, and related geometric characteristics expressed in terms of the key dimensionless geometric parameters given by Eqs. (N–6–3) and (N–6–4). It should be noted that, because Eq. (11–198) is used instead of Eq. (11–193), this method is considerably simpler that the Delaware method, regardless of whether the dimensionless geometric parameters are specified by Eqs. (N–6–3) and (N–6–4) or by the relations listed in Table A–N–6–2.

Extensions to Other Geometries We now want to consider adaptations of the shell-side hydraulic analysis for exchangers with TEMA-E shell, segmental baffles, and plain tubing to other geometries.

TEMA-J Shell Exchangers Referring to Fig. 11–38, to adapt the standard methods for evaluating the shell-side pressure drop ΔP_s to exchangers with TEMA-J shell, we simply set $L_E = L_J/2$ and $\dot{m}_{s,E} = \dot{m}_{s,J}/2$ and consider the flow divider partition as a tubesheet, such that the first baffle crossing is treated as the entry-end zone. Notice that because of the reduced mass-flow rate $\dot{m}_{s,E}$ and flow length L_E, the use of TEMA-J shell results in considerably lower pressure drop relative to a comparable TEMA-E shell arrangement.‡

† Equation (11–198) represents a refinement of a simple formulation introduced in reference 9.

‡ See page 747, footnote †.

TEMA-F Shell Exchangers Referring to Fig. 11-39, to adapt the standard methods for calculating ΔP_s to exchangers with TEMA-F shell, we set $N_{tc,E} = N_{t,c,F}/2$ and $L_E = 2\ L_F$. The pressure drop ΔP_s in an exchanger with TEMA-F shell can be expected to be significantly larger than in a TEMA-E shell exchanger with the same diameter.†

Low-Fin Tubing Exchangers To adapt the standard methods to exchangers with low-fin tubes, the equivalent diameter d_{req} is used in place of d_o in the defining relations for A_c, G_s, and Re_s, and the ideal friction factor $f_{i,\text{fin}}$ is approximated by

$$f_{i,\text{fin}} = 1.4 f_{i,\text{plain}} \tag{11-200}$$

which is based on the work of Briggs and Young [50] and is recommended in the *Heat Exchanger Design Handbook* [6].

EXAMPLE 11-12‡

A shell-and-tube heat exchanger with one shell pass and two tube passes heats water at 15°C with a mass-flow rate of 9.94 kg/s. Heating oil [c_P = 2.5 kJ/(kg °C)] enters the tubes at 80°C and exits at 35°C with a mass-flow rate of 5 kg/s. Determine the surface area of the heat exchanger if the mean overall coefficient of heat transfer is 300 W/(m² °C). Also demonstrate whether or not the minimum performance criteria are satisfied.

Solution

Objective Determine A_s and whether or not $\psi > 0.35$ and/or $P < P_o$.

Schematic Shell-and-tube heat exchanger: 1-2 pass arrangement.

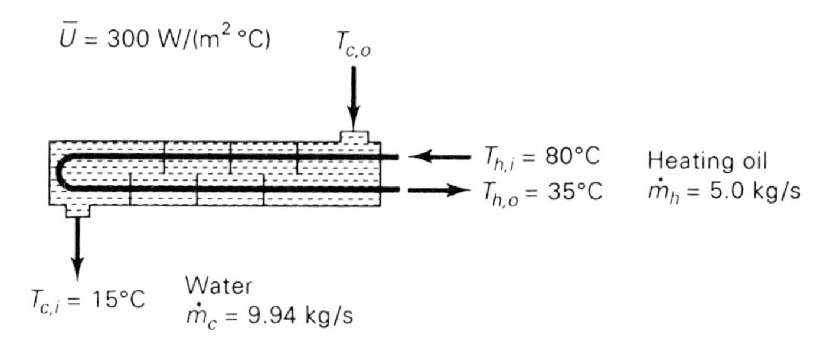

† Representative calculations presented in Example 11-16 indicate that relative to an E shell ΔP_s is smaller by a factor of 3.8 for a J shell and larger by a factor of 7.4 for a F shell.

‡ This example is solved by the ψ-P method in Appendix N-9.

Assumptions/Conditions

uniform distribution in U
uniform properties
standard conditions

Properties

Water at $T = 15°\text{C} = 288\text{ K}$ (Table A–C–3): $c_P = 4.19\text{ kJ/(kg °C)}$.

Heating oil: $c_P = 2.5\text{ kJ/(kg °C)}$.

Analysis The heat-transfer rate q and outlet temperature $T_{c,o}$ of the water are obtained by writing

$$q = (\dot{m}c_P)_h(T_{h,i} - T_{h,o}) = 5\,\frac{\text{kg}}{\text{s}}\left(2.5\,\frac{\text{kJ}}{\text{kg °C}}\right)(80°\text{C} - 35°\text{C}) = 562\text{ kW}$$

and

$$q = (\dot{m}c_P)_c(T_{c,o} - T_{c,i})$$

$$T_{c,o} = \frac{q}{(\dot{m}c_P)_c} + T_{c,i} = \frac{562\text{ kW}}{(9.94\text{ kg/s})[4.19\text{ kJ/(kg °C)}]} + 15°\text{C} = 28.5°\text{C}$$

Following the *LMTD* approach, the unknown surface area A_s is expressed in terms of the correction factor F and *LMTD* by

$$A_s = \frac{q}{\overline{U}(F\ LMTD)} \tag{a}$$

where

$$LMTD = \frac{(80°\text{C} - 28.5°\text{C}) - (35°\text{C} - 15°\text{C})}{\ln[(80°\text{C} - 28.5°\text{C})/(35°\text{C} - 15°\text{C})]} = 33.3°\text{C}$$

and the correction factor F can be obtained by use of Eq. (11–172) or Fig. 11–35.

To evaluate F, we first select the cold fluid as the reference stream I and calculate P and R by writing

$$P = \frac{T_{c,o} - T_{c,i}}{T_{h,i} - T_{c,i}} = \frac{28.5°\text{C} - 15°\text{C}}{80°\text{C} - 15°\text{C}} = 0.208 \tag{b}$$

$$R = \frac{T_{h,i} - T_{h,o}}{T_{c,o} - T_{c,i}} = \frac{80°\text{C} - 35°\text{C}}{28.5°\text{C} - 15°\text{C}} = 3.33 \tag{c}$$

Substituting for P and R in Eq. (11–172), we obtain

$$F = \frac{\sqrt{1 + R^2}}{1 - R}\,\frac{\ln[(1 - PR)/(1 - P)]}{\ln\{[2 - P(1 + R - \sqrt{1 + R^2})]/[2 - P(1 + R + \sqrt{1 + R^2})]\}}$$

$$= 0.895$$

(Alternatively, F can be estimated by use of Fig. 11–35, which indicates $F \simeq 0.9$.) Substituting into Eq. (a), A_s becomes

$$A_s = \frac{562 \text{ kW}}{[300 \text{ W/(m}^2 \text{ °C)}](0.895)(33.3\text{°C})} = 62.9 \text{ m}^2$$

As an aside, we note that P and R can also be defined by

$$P = \frac{T_{h,i} - T_{h,o}}{T_{h,i} - T_{c,i}} = \frac{80\text{°C} - 35\text{°C}}{80\text{°C} - 15\text{°C}} = 0.692$$

$$R = \frac{T_{c,o} - T_{c,i}}{T_{h,i} - T_{h,o}} = \frac{28.5\text{°C} - 15\text{°C}}{80\text{°C} - 35\text{°C}} = 0.3$$

Substituting into Eq. (11–172)(or referring to Fig. 11–35), these values of P and R indicate $F = 0.895$, which is consistent with the result obtained by use of Eqs. (b) and (c). This result demonstrates the interchangeability of the defining relations for P and R for this arrangement.

The efficiency ψ for this heat exchanger is obtained from Eq. (11–65).

$$\psi = \frac{q}{\overline{U}A_s(T_{h,i} - T_{c,i})} = \frac{562 \text{ kW}}{[300 \text{ W/(m}^2 \text{ °C)}](62.9 \text{ m}^2)(80\text{°C} - 15\text{°C})}$$

$$= 0.458$$

Thus, we conclude that the minimum performance criterion given by $\psi > \psi_{MIN} = 0.35$ is satisfied. Since $T_{h,o} = 35\text{°C} > T_{c,o} = 28.5\text{°C}$, we see that the outlet temperature criterion is also satisfied. Equivalently, we can write $P_o = 1/(1 + R) = 0.231$ from Eq. (11–76), such that $P = 0.208 < P_o$.

EXAMPLE 11–13†

A 1-2 pass shell-and-tube heat exchanger with surface area of 62.5 m² is to be designed for cooling oil at 80°C with specific heat of 2.5 kJ/(kg °C) and mass-flow rate of 5 kg/s. Water at 15°C flowing at 9.94 kg/s is used as the cooling fluid. Determine the heat transfer and outlet temperatures, assuming an overall coefficient of heat transfer equal to 300 W/(m² °C).

Solution

Objective Determine q, $T_{h,o}$, and $T_{c,o}$.

† A simple numerical finite-difference program for solving this example is given in Fig. A–G–7 of the Appendix.

Schematic Shell-and-tube heat exchanger: 1-2 pass arrangement.

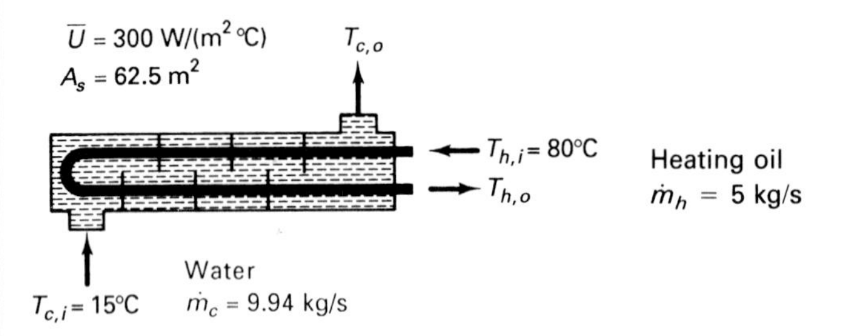

Assumptions/Conditions

uniform distribution in U
uniform properties
standard conditions

Properties

Water at $T = 15°\text{C} = 288$ K (Table A–C–3): $c_P = 4.19$ kJ/(kg °C).

Hot oil: $c_P = 2.5$ kJ/(kg °C).

Analysis To employ the effectiveness (or efficiency) approach, we first calculate the capacitance rates, assuming uniform property conditions.

$$C_h = (\dot{m}c_P)_h = 5\,\frac{\text{kg}}{\text{s}}\left(2.5\,\frac{\text{kJ}}{\text{kg °C}}\right) = 12.5\text{ kW/°C}$$

$$C_c = (\dot{m}c_P)_c = 9.94\,\frac{\text{kg}}{\text{s}}\left(4.19\,\frac{\text{kJ}}{\text{kg °C}}\right) = 41.6\text{ kW/°C}$$

Using these values and selecting the hot fluid as the reference stream I, R and NTU_I become†

$$R = \frac{C_h}{C_c} = \frac{12.5\text{ kW/°C}}{41.6\text{ kW/°C}} = 0.3$$

and

$$NTU_I = \frac{\overline{U}A_s}{C_h} = \frac{[300\text{ W/(m}^2\text{ °C)}](62.5\text{ m}^2)}{12.5\text{ kW/°C}} = 1.5$$

Substituting these results into Eq. (11–168), we obtain

† With the hot fluid selected as the reference stream I, the P-NTU_I and ϵ-NTU methods for this problem are identical, with $P = \epsilon$, $R = C^*$, and $NTU_I = NTU$.

$$\Gamma = NTU_I \sqrt{1 + R^2} = 1.5 \sqrt{1 + 0.3^2} = 1.57$$

and

$$P = \frac{2}{1 + R + \sqrt{1 + R^2}\,(1 + e^{-\Gamma})/(1 - e^{-\Gamma})} = 0.691$$

The heat-transfer rate q is obtained by writing†

$$q = Pq_{\max,I} = PC_h(T_{h,i} - T_{c,i})$$

$$= 0.691\left(12.5\,\frac{\text{kW}}{°\text{C}}\right)(80°\text{C} - 15°\text{C}) = 561\ \text{kW}$$

It follows that

$$q = C_h(T_{h,i} - T_{h,o}) = C_c(T_{c,o} - T_{c,i})$$

$$T_{h,o} = T_{h,i} - \frac{q}{C_h} = 80°\text{C} - \frac{561\ \text{kW}}{12.5\ \text{kW}/°\text{C}} = 35.1°\text{C}$$

$$T_{c,o} = T_{c,i} + \frac{q}{C_c} = 28.5°\text{C}$$

These results are observed to be consistent with the calculations obtained by means of the *LMTD* approach in Example 11–12 for the corresponding design problem.

Finally, we note that the variation in c_P for water entering at 15°C and exiting at 28.5°C is less than 1%, such that the uniform property analysis should be reasonable, providing that the variation in c_P of the heating oil is not too large.

EXAMPLE 11–14

30° API oil flowing at a rate of at least 43.6 kg/s is to be cooled from 114°C to 66°C using a 1-4 shell-and-tube heat exchanger with 42-in. I.D. shell, 3/4-in. O.D. by 16-ft long 16-BWG tubes on a 1-in.-square pitch, and 12 segmental baffles with 20% cut. The cooling water is to enter at 26°C and exit at 50°C. The pressure drop in either stream is not to exceed 50 kPa, and the fouling factors are given as 0.00015 m²

† Following the efficiency approach, we calculate ψ by using the Mueller chart shown in Fig. 11–36 or by writing

$$\psi = \frac{P}{NTU_I} = \frac{2/NTU_I}{1 + R + \sqrt{1 + R^2}\,(1 + e^{-\Gamma})/(1 - e^{-\Gamma})} = 0.461$$

and represent the rate of heat transfer q by Eq. (11–65),

$$q = \psi \bar{U} A_s (T_{t,i} - T_{s,i}) = 561\ \text{kW}$$

°C/W for the water and 0.0007 m² °C/W for the oil. Determine whether or not this heat exchanger is capable of satisfying the performance requirements.

Solution

Objective Evaluate the performance capability of the prescribed 1-4 shell-and-tube heat exchanger relative to the following requirements:

$$T_{h,o} = 66°\text{C} \qquad \text{and} \qquad T_{c,o} = 50°\text{C}$$

and

$$T_{h,i} = 114°\text{C} \qquad \text{and} \qquad T_{c,i} = 26°\text{C}$$

and pressure drops less than 50 kPa.

Schematic 1-4 shell-and-tube heat exchanger.

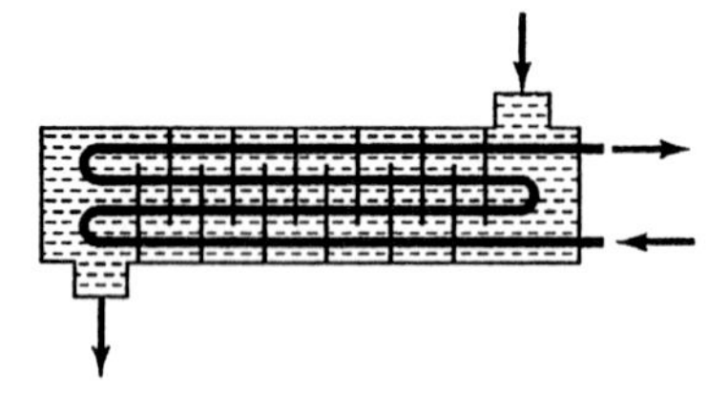

$\dot{m}_h$ = 43.6 kg/s

Fouling:
cooling water–F_f = 0.00015 m² °C/W
hot oil–F_f = 0.0007 m² °C/W

Shell-side component-loss coefficient
Entrance nozzle–K_{f1} = 5
Exit nozzle–K_{f2} = 5

Tube-side component-loss coefficients
Entrance nozzle–K_{f1} = 1.5
Exit nozzle–K_{f2} = 1
Three 180° bends–K_{f3} = 3(1.5) = 4.5
Contraction and expansion–$K_c + K_e$ = 1.5

Baffles,segmental:
B_c = 20%
N_b = 12

Shell: $D_{i,s}$ = 42 in. = 1.07 m

Tubes: 3/4 in. 16 BWG, square pitch
d_o = 0.75 in. = 0.0191 m
d_i = 0.652 in. = 0.0166 m
S_t = 1 in. = 0.0254 m
S_t/d_o = 1.33, $S_{t,\text{eff}} = S_t$
L = 16 ft = 4.88 m
N = 1195 (Table A-O-1)
N_{tp} = 4

Assumptions/Conditions

uniform distribution in U
negligible entrance effects
negligible conduction resistance R_k
moderate property variation
standard conditions

Properties

30° API oil at 90°C (see Example 11–9): ρ = 820 kg/m³, μ = 0.00245 kg/(m s), c_P = 2.17 kJ/(kg °C), k = 0.128 W/(m °C), Pr = 41.5.

Water at 38°C (Table A–C–3): ρ = 993 kg/m³, μ = 6.82 × 10⁻⁴ kg/(m s), c_P = 4.17 kJ/(kg °C), k = 0.63 W/(m °C), Pr = 4.51.

Analysis The baffle spacing L_b is obtained by writing

$$L_b = \frac{L}{N_b + 1} = \frac{4.88 \text{ m}}{12 + 1} = 0.375 \text{ m}$$

from Eq. (11–180a). Thus, $L_b/D_{i,s}$ becomes

$$\frac{L_b}{D_{i,s}} = \frac{0.375 \text{ m}}{1.07 \text{ m}} = 0.35$$

The geometric characteristics of the tube and shell are represented by

$$\textit{Tube} \quad A_t = \left(\frac{\pi d_i^2}{4}\right)\frac{N}{4} = \frac{\pi}{16}(0.0166 \text{ m})^2 (1195) = 0.0647 \text{ m}^2$$

$$\textit{Shell} \quad A_c = \frac{D_{i,s}}{S_t}(S_t - d_o)L_b = \frac{1.07 \text{ m}}{0.0254 \text{ m}}(0.0254 \text{ m} - 0.0191 \text{ m})(0.375 \text{ m})$$

$$= 0.0995 \text{ m}^2$$

The heat-transfer surface area A_s is given by (based on the outer tube surface)

$$A_s = N(\pi d_o L) = 1195\pi(0.0191 \text{ m})(4.88 \text{ m}) = 350 \text{ m}^2$$

Using the lumped energy formulation, we obtain

$$q = (\dot{m}c_P)_h(T_{h,i} - T_{h,o}) = 43.6\frac{\text{kg}}{\text{s}}\left(2.17\frac{\text{kJ}}{\text{kg °C}}\right)(114\text{°C} - 66\text{°C}) = 4540 \text{ kW}$$

and

$$\dot{m}_c = \frac{q}{c_{P,c}(T_{c,o} - T_{c,i})} = \frac{4540 \text{ kW}}{[4.17 \text{ kJ/(kg °C)}](50\text{°C} - 26\text{°C})} = 45.4 \text{ kg/s}$$

The hydrodynamic characteristics for the case in which water flows in the tubes and oil in the shell are given as follows:

Tube flow—cooling water $\dot{m}_t = \dot{m}_c = 45.4$ kg/s

$$G_t = \frac{\dot{m}_t}{A_t} = \frac{45.4 \text{ kg/s}}{0.0647 \text{ m}^2} = 702 \text{ kg/(m}^2\text{ s)} \qquad U_{b,t} = 0.707 \text{ m/s}$$

$$Re_t = \left(\frac{Gd_i}{\mu}\right)_t = \frac{[702 \text{ kg/(m}^2\text{ s)}](0.0166 \text{ m})}{6.82 \times 10^{-4} \text{ kg/(m s)}} = 17{,}100$$

Shell flow—hot oil $\dot{m}_s = \dot{m}_h = 43.6$ kg/s

$$G_s = \frac{\dot{m}_s}{A_c} = \frac{43.6 \text{ kg/s}}{0.0995 \text{ m}^2} = 438 \text{ kg/(m}^2\text{ s)}$$

$$Re_s = \left(\frac{Gd_o}{\mu}\right)_s = \frac{[438 \text{ kg/(m}^2\text{ s)}](0.0191 \text{ m})}{0.00245 \text{ kg/(m s)}} = 3410$$

The tube-side friction factor, Nusselt number, and coefficient of heat transfer for turbulent flow with Re_t = 17,100 are approximated using Eqs. (8–15) and (8–16).

$$f_t = (1.58 \ln 17{,}100 - 3.28)^{-2} = 0.00681$$

$$Nu_t = \frac{(0.00681/2)(17{,}100)(4.51)}{1.07 + 12.7\sqrt{0.00681/2}\,(4.51^{2/3} - 1)} = 112$$

$$h_t = \left(Nu\frac{k}{d_i}\right)_t = 112\,\frac{0.63\ \text{W/(m °C)}}{0.0166\ \text{m}} = 4240\ \text{W/(m}^2\ \text{°C)}$$

Noting that $L/D \gg 10$, we set $\bar{h}_t = h_t$.

To estimate the shell-side coefficient $\bar{h}_s$, we make use of simple method I. Following this approach, we set j_i = 0.0125 in accordance with the correlation given by Fig. A–N–6–1(a) and write

$$\overline{Nu}_s = 0.6 j_i\, Re_s\, Pr^{1/3} = 0.6(0.0125)(3410)(41.5)^{1/3} = 88.5 \qquad (11\text{–}184)$$

It follows that

$$\bar{h}_s = 88.5\,\frac{0.128\ \text{W/(m °C)}}{0.0191\ \text{m}} = 593\ \text{W/(m}^2\ \text{°C)}$$

The mean overall coefficient of heat transfer $\overline{U}$ is approximated by (neglecting the wall resistance)

$$\frac{1}{\overline{U}} = \frac{1}{\bar{h}_t}\frac{d_o}{d_i} + F_{f,t}\frac{d_o}{d_i} + F_{f,s} + \frac{1}{\bar{h}_s}$$

$$= \left(\frac{1}{4240}\,\frac{0.0191}{0.0166} + 0.00015\,\frac{0.0191}{0.0166} + 0.0007 + \frac{1}{593}\right)\ \text{m}^2\ \text{°C/W}$$

$$= (0.000271 + 0.000173 + 0.0007 + 0.00169)\ \text{m}^2\ \text{°C/W}$$

$$= 0.00283\ \text{m}^2\ \text{°C/W}$$

or

$$\overline{U} = 353\ \text{W/(m}^2\ \text{°C)}$$

with fouling, and

$$\overline{U} = 510\ \text{W/(m}^2\ \text{°C)}$$

without fouling. Notice that the resistance associated with convection to the oil accounts for 60% of the total resistance.

Similar calculations for $\overline{U}$ obtained for water in the shell and oil in the tubes result in overall coefficients that are a little more than 5% lower. Thus, within the accuracy of the convection correlations used in the analysis, the arrangement for which the oil passes through the shell is indicated.

The *LMTD* approach can be conveniently used to compute the heat-transfer rate that can be produced by this arrangement; that is,

$$q = \overline{U}A_s(F\ LMTD) \tag{11–59}$$

where

$$LMTD = \frac{(114°C - 50°C) - (66°C - 26°C)}{\ln[(114°C - 50°C)/(66°C - 26°C)]} = 51.1°C$$

According to the analysis, $\overline{U}$ ranges from 510 W/(m^2 °C) for clean surfaces to 353 W/(m^2 °C) for fouled surfaces, and $A_s = 350$ m^2. To evaluate the *LMTD* correction factor F, we first compute P and R (using the cold fluid as the reference stream I and assuming that the specified outlet temperatures $T_{h,o}$ and $T_{c,o}$ are attained).

$$P = \frac{T_{c,o} - T_{c,i}}{T_{h,i} - T_{c,i}} = \frac{50°C - 26°C}{114°C - 26°C} = 0.273$$

$$R = \frac{T_{h,i} - T_{h,o}}{T_{c,o} - T_{c,i}} = \frac{114°C - 66°C}{50°C - 26°C} = 2$$

Referring to Fig. 11–35, $F = 0.92$.† Substituting into Eq. (11–59), we obtain

$$q = 353 \frac{W}{m^2\ °C}(350\ m^2)(0.92)(51.1°C) = 5810\ kW$$

for fouled surfaces, and

$$q = 510(350)(0.92)(51.1)\ kW = 8390\ kW$$

for clean surfaces. Since the required heat-transfer rate is 4540 kW, the heat exchanger is overdesigned by only about 28% for fouled surfaces and 85% for clean surfaces. Thus, the heat exchanger is capable of performing the required duty for a mass-flow rate of oil that moderately exceeds the minimum required rate. [The effectiveness method or the efficiency method can be used to calculate (a) the mass-flow rates $\dot{m}_h$ and $\dot{m}_c$ that will result in the specified outlet temperatures for both clean and fouled surface conditions, or (b) the outlet temperatures for both clean and fouled surface conditions with mass-flow rates unchanged. For example, with the mass-flow rates held constant and $\overline{U}$ set equal to 353 W/(m^2 °C) for fouled conditions, we obtain $NTU_I = 0.654$, $R = 2$, $P = 0.304$, $T_{c,o} = 52.7$°C, and $T_{h,o} = 60.6$°C.] In addition, we note that Eq. (11–65) indicates $\psi = 0.534$ for fouled and clean surfaces, such that, the minimum performance criterion $\psi > \psi_{MIN} = 0.35$ is satisfied.

To complete the analysis, we must evaluate the pressure drops on the tube and shell sides.

† Equation (11–172) indicates $F = 0.9203$. Calculations based on Eq. (f) in Table A–N–1–1(a) for a 1-4 exchanger are within 2.4% of the values obtained from Eq. (11–172) for a 1-2 exchanger.

For the tube-side flow of cooling water, we simply write

$$(\Delta P_t)_{1-2} = \left(\frac{\rho U_b^2}{2}\right)_t \left[\left(\frac{4L}{d_i} f_t\right) N_{tp} + K_c + K_e + K_f\right] \tag{11–187}$$

$$= \frac{1}{2}\left(993 \frac{\text{kg}}{\text{m}^3}\right)\left(0.707 \frac{\text{m}}{\text{s}}\right)^2 \left[\frac{4(4.88 \text{ m})(0.00681)}{0.0166 \text{ m}} (4) + 1.5 + 7\right]$$

$$= 248(32 + 8.5) \text{ Pa} = 0.01 \text{ MPa} = 10 \text{ kPa}$$

Using simple method I for the shell-side flow of hot oil, we note that $S_{pp} = S_t$ and $f_i = 0.1$ from Fig. A–N–6–1(a). It follows that

$$\Delta P_s = \left[1.6 \frac{D_{i,s}}{S_{pp}}\left(1 - \frac{2B_c}{100}\right)(N_b + 1)\frac{f_i}{2} + N_b\right]\left(\frac{G^2}{\rho}\right)_s \tag{11–199}$$

$$= \left[1.6 \frac{1.07 \text{ m}}{0.0254 \text{ m}} (1 - 0.4)(12 + 1)\frac{0.1}{2} + 12\right] \frac{[438 \text{ kg/(m}^2\text{ s)}]^2}{820 \text{ kg/m}^3}$$

$$= 0.00896 \text{ MPa} = 8.96 \text{ kPa}$$

$$(\Delta P_s)_{1-2} = \Delta P_s + K_f\left(\frac{G^2}{2\rho}\right)_s = 8.96 \text{ kPa} + 10 \frac{[438 \text{ kg/(m}^2\text{ s)}]^2}{2(820 \text{ kg/m}^3)}$$

$$= 8.96 \text{ kPa} + 1.17 \text{ kPa} = 10.1 \text{ kPa}$$

Both of these pressure drops fall considerably below the maximum allowable value of 50 kPa, such that the exchanger satisfies the performance criteria. However, because it does not more fully utilize the available pressure drop, the exchanger would not be considered to be effectively designed for this particular application.

For purpose of comparison, this problem has been solved by use of an HFTS computer code (TASC3), which features a form of the stream analysis method, the Delaware method, and simple methods A, I, and II. Results obtained for $\bar{h}_s$ and ΔP_s using these approaches are given in Table E11–14 for standard clearances of L_{tb} = 0.794 mm, L_{sb} = 5.72 mm, and L_{bb} = 12.7 mm (see Appendix O–2). Notice that the calculations differ by no more than 12.5% for $\bar{h}_s$ and 26% for ΔP_s.

Referring back to the calculation for $\bar{U}$, we are reminded that the thermal resistance associated with convection to the oil is dominant (60%). In applications such as this involving a dominant convective resistance (of the order of 60% or more), low-fin tubes are commonly used to improve the heat-transfer performance. The use of low-fin tubes for this application is considered in Example 11–16. As we shall see, the use of fin tubes results in improvements in both thermal and hydraulic aspects of the performance.

TABLE E11–14 Comparison of shell-side calculations

Shell-side calculation	*Method*				
	Simple A	*Simple I*	*Simple II*[1]	*Delaware STER*[2]	*Stream Analysis TASC3*[3]
$\bar{h}_s$ [W/(m² °C)]	611	593	686	665	678
ΔP_s [kPa]	—	8.96	8.81	11.3	8.32

[1] Referring to Appendix N–6, with L_{tb} = 0.794 mm, L_{sb} = 5.72 mm, and L_{bb} = 12.7 mm, Eqs. (N–6–3) and (N–6–4) indicate θ_{ds} = 1.85, r_{lm} = 0.319, r_s = 0.214, r_b = 0.0479, and F_c = 0.717. It follows from Fig. A–N–6–3 that J_c = 1.07, J_l = 0.69, and J_b = 0.94, and from Fig. A–N–6–4 that R_l = 0.47 and R_b = 0.83.

[2] The STER Delaware computer solution was provided by Jerry Taborek, assuming fixed tubesheets.

[3] The TASC3 (HTFS) stream analysis computer solution was provided by Ki-Lun Lui of Koch Engineering Ltd, assuming fixed tubesheets.

EXAMPLE 11–15

Determine the effect of the baffle spacing on the performance of the heat exchanger application of Example 11–14 by using 8 baffles instead of 12.

Solution

Objective Determine the influence of baffle spacing L_b.

Schematic 1-4 shell-and-tube heat exchanger: 8 baffles.

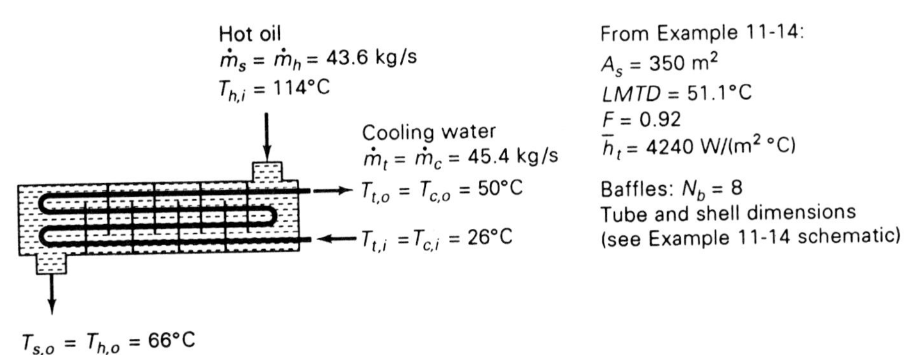

Assumptions/Conditions See Example 11–14.

Properties See Example 11–14.

Analysis The baffle spacing L_b for the case of 8 baffles is given by

$$L_b = \frac{L}{N_b + 1} = \frac{4.88\text{ m}}{8 + 1} = 0.542\text{ m}$$

from Eq. (11–180a), such that the dimensionless baffle spacing becomes

$$\frac{L_b}{D_{i,s}} = \frac{0.542\text{ m}}{1.07\text{ m}} = 0.507$$

Referring back through Sec. 11–7–1, we observe that the baffle spacing L_b influences the crossflow area A_c,

$$A_c = \frac{D_{i,s}}{S_t}(S_t - d_o)L_b$$

from Eq. (11–179), shell mass flux G_s,

$$G_s = \frac{\dot{m}_s}{A_{\min}} = \frac{\dot{m}_s}{A_c} = \frac{\dot{m}_h}{A_c}$$

from Eq. (11–176), and shell Reynolds number Re_s,

$$Re_s = \frac{G_s d_o}{\mu_s} = \frac{(\dot{m}_h/A_c)d_o}{\mu_h}$$

from Eq. (11–177). Substituting into these equations, we obtain

$$A_c = \frac{1.07\text{ m}}{0.0254\text{ m}}(0.0254\text{ m} - 0.0191\text{ m})(0.542\text{ m}) = 0.144\text{ m}^2$$

$$G_s = \frac{\dot{m}_h}{A_c} = \frac{43.6\text{ kg/s}}{0.144\text{ m}^2} = 303\text{ kg/(m}^2\text{ s)}$$

$$Re_s = \frac{(0.0191\text{ m})[303\text{ kg/(m}^2\text{ s)}]}{0.00245\text{ kg/(m s)}} = 2360$$

Using simple method I to estimate $\overline{h}_s$, we set $j_i = 0.014$ in accordance with Fig. A–N–6–1(a) and write

$$\overline{Nu}_s = 0.6 j_i\, Re_s\, Pr_s^{1/3} = 0.6(0.014)(2360)(41.5)^{1/3} = 68.6$$

from Eq. (11–184), such that

$$\overline{h}_s = 68.6\,\frac{0.128\text{ W/(m °C)}}{0.0191\text{ m}} = 460\text{ W/(m}^2\text{ °C)}$$

Thus,

$$\frac{1}{\overline{\overline{U}}} = \left(\frac{1}{4240}\frac{0.0191}{0.0166} + 0.00015\frac{0.0191}{0.0166} + 0.0007 + \frac{1}{460}\right)\frac{\text{m}^2\text{ °C}}{\text{W}}$$

$$= 0.00332\text{ m}^2\text{ °C/W}$$

or

$$\overline{U} = 301 \text{ W/(m}^2 \text{ °C)}$$

with fouling, and

$$\overline{U} = 409 \text{ W/(m}^2 \text{ °C)}$$

for no fouling.

Combining this result with Eq. (11–59) with F set equal to 0.92, the rate of heat transfer is

$$q = \overline{U}A_s(F\ LMTD) = 301 \frac{\text{W}}{\text{m}^2\text{ °C}}(350 \text{ m}^2)(0.92)(51.1\text{°C})$$

$$= 4950 \text{ kW}$$

for fouled surfaces, and

$$q = 409(350)(0.92)(51.1) \text{ kW} = 6730 \text{ kW}$$

for clean surfaces. Thus, this arrangement is also capable of performing the required duty of 4540 kW, but with a smaller safety factor (9%) for fouled surface conditions. We also note that Eq. (11–65) indicates $\psi = 0.534$ for both baffle spacings, such that the minimum performance criterion $\psi > \psi_{MIN} = 0.35$ is satisfied.

Using simple method I to estimate the shell-side pressure drop, we set $f_i = 0.1$ in accordance with Fig. A–N–6–1(a) and write

$$\Delta P_s = \left[1.6 \frac{D_{i,s}}{S_{pp}}\left(1 - \frac{2B_c}{100}\right)(N_b + 1)\frac{f_i}{2} + N_b\right]\left(\frac{G^2}{\rho}\right)_s \tag{11–199}$$

$$= \left[1.6 \frac{1.07 \text{ m}}{0.0254 \text{ m}}(1 - 0.4)(8 + 1)\frac{0.1}{2} + 8\right]\frac{[303 \text{ kg/(m}^2\text{ s)}]^2}{820 \text{ kg/m}^3}$$

$$= 2.93 \text{ kPa}$$

$$(\Delta P_s)_{1-2} = \Delta P_s + K_f\left(\frac{G^2}{2\rho}\right)_s = 2.93 \text{ kPa} + 10\frac{[303 \text{ kg/(m}^2\text{ s)}]^2}{2(820 \text{ kg/m}^3)}$$

$$= 2.93 \text{ kPa} + 0.56 \text{ kPa} = 3.49 \text{ kPa}$$

Comparing these results with those of Example 11–14, we find that the heat-transfer performance q (4950 kW versus 5810 kW) is 14.8% lower, and the shell-side pressure drop ΔP_s (2.93 kPa versus 8.96 kPa) is lower by a factor of 3 for the case in which 8 baffles are used instead of 12.

Finally, calculations for $\overline{h}_s$ and ΔP_s obtained by the primary methods and simple methods are compared in Table E11–15 for the standard clearances of $L_{tb} = 0.794$ mm, $L_{sb} = 5.72$ mm, and $L_{bb} = 12.7$ mm.

TABLE E11–14 Comparison of shell-side calculations

	Method				
Shell-side calculation	*Simple A*	*Simple I*	*Simple II*[1]	*Delaware STER*[2]	*Stream Analysis TASC3*[3]
$\bar{h}_s$ [W/(m^2 °C)]	611	593	686	665	678
ΔP_s [kPa]	—	8.96	8.81	11.3	8.32

[1] Referring to Appendix N–6, with L_{tb} = 0.794 mm, L_{sb} = 5.72 mm, and L_{bb} = 12.7 mm, Eqs. (N–6–3) and (N–6–4) indicate θ_{ds} = 1.85, r_{lm} = 0.319, r_s = 0.214, r_b = 0.0479, and F_c = 0.717. It follows from Fig. A–N–6–3 that J_c = 1.07, J_l = 0.69, and J_b = 0.94, and from Fig. A–N–6–4 that R_l = 0.47 and R_b = 0.83.

[2] The STER Delaware computer solution was provided by Jerry Taborek, assuming fixed tubesheets.

[3] The TASC3 (HTFS) stream analysis computer solution was provided by Ki-Lun Lui of Koch Engineering Ltd, assuming fixed tubesheets.

EXAMPLE 11–16

Low-fin tubes are commonly used in order to develop more compact and less expensive shell-and-tube heat exchangers. To see the effect of fins, we wish to consider the use of a shell-and-tube heat exchanger with 3/4-in., 16-ft-long low-fin tubes with 19 fins/in., 0.065-in. wall thickness, and 1-in.-square pitch for the cooling application of Example 11–14. Estimate the required number of tubes, number of tube passes, number of baffles, and size shell, and evaluate the performance of the low-fin-tube heat exchanger for the same baffle cut (20%) and approximate dimensionless baffle spacing $L_b/D_{i,s}$.

Solution

Objective Estimate the number of tubes N, number of tube passes N_{tp}, number of baffles N_b, and shell diameter $D_{i,s}$, and evaluate the performance relative to the following requirements:

$$T_{h,o} = 66°\text{C} \quad \text{and} \quad T_{c,o} = 50°\text{C}$$

and

$$T_{h,i} = 114°\text{C} \quad \text{and} \quad T_{c,i} = 26°\text{C}$$

Assumptions/Conditions See Example 11–14.

Schematic Shell-and-tube heat exchanger: low-fin tubes. (Note that 1-4 arrangement is shown, but that a 1-2 pass arrangement is indicated in the analysis.)

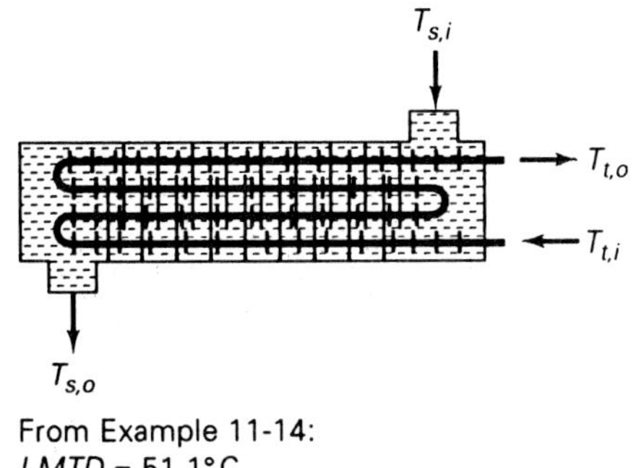

From Example 11-14:
$LMTD = 51.1°C$
$F = 0.92$ (for N_{tp} even)
$\dot{m}_h = 43.6$ kg/s
$T_{h,i} = 114°C$
$T_{h,o} = 66°C$
$\dot{m}_c = 45.4$ kg/s
$T_{c,i} = 26°C$
$T_{c,o} = 50°C$

Low-fin tubes: 3/4 in. O.D. with 19 fins/in. (See Table A-O-3), square pitch
$(d_o)_o = 0.75$ in. $= 0.0191$ m
$d_o = 0.737$ in. $= 0.0187$ m
$d_i = 0.495$ in. $= 0.0126$ m
$d_{req} = 0.66$ in. $= 0.0168$ m
$A_{s,o}/L = 0.503$ ft $= 0.153$ m
$A_{s,o}/A_{s,i} = 3.86$
$A_{s,i} = 0.192$ in.$^2 = 1.24 \times 10^{-4}$ m^2
$S_t = 1$ in. $= 0.0254$ m
$L = 16$ ft $= 4.88$ m

Baffles:
$B_c = 20\%$, $L_b/D_{i,s} \approx 0.35$

Properties See Example 11–14.

Analysis To possibly eliminate the need for developing iterative solutions to this design problem, we will first develop a preliminary analysis that roughly accounts for the effect of the considerably larger outer surface area provided by the fin tubes.

Preliminary Considerations Referring to Table A–O–3, the outer surface area per unit length for a 3/4 in.-low-fin tube with 19 fins/in. is given by

$$\frac{A_{s,o}}{L} = 0.503 \text{ ft} = 0.153 \text{ m}$$

Comparing this value to the surface area of a 3/4-in. plain tube, we find that the fin-tube surface area is 2.52 times as large. Therefore, the number of fin tubes required to provide the same outside surface area as 1195 plain tubes is calculated by writing

$$N_{\text{equiv}} = \frac{1195}{2.52} = 474$$

Based on preliminary calculations, we find that the overall coefficient of heat transfer based on outside surface area is roughly the same for low-fin-tube and plain-tube heat exchangers.† Therefore, it is anticipated that approximately 474 fin tubes will be needed.

† Common practice followed in heat exchanger design is to assume that the overall coefficient of heat transfer is about 10% lower for low-fin tubes than plain tubes.

Using 3/4-in. tubes with 1-in. square pitch, Table A–O–1 indicates that a 29-in. I.D. shell would be required to accommodate the nearest larger tube counts of 535 for four passes. Because of the smaller number of tubes and smaller inside diameter, the bulk-stream velocity $U_{b,t}$ within the tubes will be approximately 3.8 times larger than the value obtained in Example 11–14 for four passes. This will produce a pressure drop of over 3.8^2 (= 14.4) times larger. As an alternative, Table A–O–1 indicates that for a two-pass arrangement a 27-in. I.D. shell can be used with 490 tubes. By using this arrangement with 50% shorter tube flow passage length, $U_{b,t}$ and $(\Delta P_t)_{1-2}$ are expected to be only about three times as large as the values in Example 11–14. Therefore, a two-pass arrangement with 490 tubes and 27-in. I.D. shell is tentatively selected; that is,

$$D_{i,s} = 27 \text{ in.} = 0.686 \text{ m} \qquad N = 490$$

With the dimensionless baffle spacing $L_b/D_{i,s}$ approximately equal to 0.35, the baffle spacing and number of tubes are estimated by writing

$$L_b \simeq 0.35 D_{i,s} = 0.35(0.686 \text{ m}) = 0.24 \text{ m}$$

such that

$$N_b + 1 = \frac{L}{L_b} \simeq \frac{4.88 \text{ m}}{0.24 \text{ m}} = 20.3$$

Since an even number of baffles is required, we select

$$N_b = 18$$

which indicates

$$L_b = \frac{L}{N_b + 1} = \frac{4.88 \text{ m}}{18 + 1} = 0.257 \text{ m} \qquad \frac{L_b}{D_{i,s}} = \frac{0.257 \text{ m}}{0.686 \text{ m}} = 0.375$$

Evaluation of Performance The cross-sectional area and hydraulic diameters of the heat exchanger are given as follows:

$$\textit{Tube} \quad A_t = \left(\frac{\pi d_i^2}{4}\right)\frac{N}{N_{tp}} = \frac{\pi}{4}(0.0126 \text{ m})^2\,\frac{490}{2} = 0.0305 \text{ m}^2$$

$$\textit{Shell} \quad A_c = \frac{D_{i,s}}{S_t}(S_t - d_{\text{req}})L_b$$

$$= \frac{0.686 \text{ m}}{0.0254 \text{ m}}(0.0254 \text{ m} - 0.0168 \text{ m})(0.257 \text{ m}) = 0.06 \text{ m}^2$$

The heat-transfer surface area A_s associated with the finned surface is computed by writing

$$A_s = \frac{A_{s,o}}{L} NL = 0.153 \text{ m } (490)(4.88 \text{ m}) = 366 \text{ m}^2$$

Whereas the performance of the plain-tube heat exchanger of Example 11–14 is little effected by whether the hot oil is placed on the shell side or tube side, because the hot oil provides the dominant thermal resistance it must now be placed on the shell side in order to make effective use of the fins.

The hydrodynamic characteristics for this arrangement are now calculated.

Tube flow—cooling water $\dot{m}_t = \dot{m}_c = 45.4$ kg/s

$$G_t = \frac{\dot{m}_t}{A_t} = \frac{45.4 \text{ kg/s}}{0.0305 \text{ m}^2} = 1490 \text{ kg/(m}^2\text{ s)} \qquad U_{b,t} = 1.5 \text{ m/s}$$

$$Re_t = \left(\frac{Gd_i}{\mu}\right)_t = \frac{[1490 \text{ kg/(m}^2\text{ s)}](0.0126 \text{ m})}{6.82 \times 10^{-4} \text{ kg/(m s)}} = 27{,}500$$

Shell flow—hot oil $\dot{m}_s = \dot{m}_h = 43.6$ kg/s

$$G_s = \frac{\dot{m}_s}{A_c} = \frac{43.6 \text{ kg/s}}{0.06 \text{ m}^2} = 727 \text{ kg/(m}^2\text{ s)}$$

$$Re_s = \left(\frac{Gd_{\text{req}}}{\mu}\right)_s = \frac{[727 \text{ kg/(m}^2\text{ s)}](0.0168 \text{ m})}{0.00245 \text{ kg/(m s)}} = 4990$$

Using Eqs. (8–15) and (8–16), the tube-side friction factor, Nusselt number, and coefficient of heat transfer for fully turbulent flow with $Re_t = 27{,}500$ are

$$f_t = (1.58 \ln 27{,}500 - 3.28)^{-2} = 0.00604$$

$$\overline{Nu}_t = \frac{(0.00604/2)(27{,}500)(4.51)}{1.07 + 12.7\sqrt{0.00604/2}\,(4.51^{2/3} - 1)} = 164$$

$$\bar{h}_t = \overline{Nu}_t \frac{k_c}{d_i} = 164 \frac{0.63 \text{ W/(m °C)}}{0.0126 \text{ m}} = 8200 \text{ W/(m}^2\text{ °C)}$$

Turning to the shell-side performance, Fig. A–N–6–1(a) indicates $j_i = 0.012$. Using simple method I, we substitute for j_i in Eq. (11–184), with the result

$$\overline{Nu}_{s,\text{plain}} = 0.6 j_i\, Re_s\, Pr_s^{1/3} = 0.6(0.012)(4990)(41.5)^{1/3} = 124$$

and, since $J_F = 1$ (from Fig. 11–40),

$$\bar{h}_{s,\text{fin}} = \bar{h}_{s,\text{plain}} = \overline{Nu}_{s,\text{plain}} \frac{k_h}{d_{\text{req}}} = 124 \frac{0.128 \text{ W/(m °C)}}{0.0168 \text{ m}} = 945 \text{ W/(m}^2\text{ °C)}$$

The mean overall coefficient of heat transfer relative to the surface area on the fin side can be represented by

$$\frac{1}{\overline{U}} = \frac{1}{\bar{h}_t}\frac{A_{s,o}}{A_{s,i}} + F_{f,t}\frac{A_{s,o}}{A_{s,i}} + F_{f,s} + \frac{1}{(\eta_o \bar{h})_{s,\text{fin}}}$$

where $A_{s,o}/A_{s,i} = 3.86$,

$$\eta_o = 1 - \frac{A_F}{A_o}(1 - \eta_F) \simeq 1$$

from Eq. (11–12) for $\eta_F \simeq 1.0$. Substituting into Eq. (a), we obtain

$$\frac{1}{\overline{U}} = \left[\frac{3.86}{8200} + 0.00015(3.86) + 0.0007 + \frac{1}{945}\right] \frac{\text{m}^2\ ^\circ\text{C}}{\text{W}}$$

$$= (0.000471 + 0.000579 + 0.0007 + 0.00106)\ \text{m}^2\ ^\circ\text{C/W} = 0.00281\ \text{m}^2\ ^\circ\text{C/W}$$

or

$$\overline{U} = 356\ \text{W/(m}^2\ ^\circ\text{C)}$$

for fouled surfaces, and

$$\overline{U} = 653\ \text{W/(m}^2\ ^\circ\text{C)}$$

for clean surfaces. Notice that the use of the low-fin tube has reduced the dominant convective resistance of the oil from 60% (see Example 11–14) to 38%.

Using these results for $\overline{U}$ and setting $A_s = A_{s,o} = 366\ \text{m}^2$, $F = 0.92$, and $LMTD = 51.5°\text{C}$, the heat-transfer rate is given by

$$q = \overline{U}A_s(F\ LMTD) = 356\ \frac{\text{W}}{\text{m}^2\ ^\circ\text{C}}(366\ \text{m}^2)(0.92)(51.1^\circ\text{C}) = 6120\ \text{kW}$$

for fouled surfaces, and

$$q = 11{,}200\ \text{kW}$$

for clean surfaces. Since the design calculation for q is 35% above the required value (4540 kW), the result is moderately conservative.

For purpose of comparison, with oil flowing in the tubes instead of the shell, the analysis indicates $\overline{h}_t = 1100\ \text{W/(m}^2\ ^\circ\text{C)}$ and $\overline{h}_{s,\text{fin}} = 1960\ \text{W/(m}^2\ ^\circ\text{C)}$, which results in values of $\overline{U}$ ranging from 152 to 270 W/(m² °C). The corresponding values of heat-transfer rate vary from 2610 to 4600 kW, which falls as much as 42% below specification for fouled surfaces. This result indicates that the fluid with poorest heat-transfer characteristics should flow over the finned surface.

The tube-side pressure drop is calculated by use of Eq. (11–187).

$$(\Delta P_t)_{1\text{-}2} = \left(\frac{\rho U_b^2}{2}\right)_t \left[\left(\frac{4L}{d_i} f_t\right) N_{tp} + K_c + K_e + K_f\right]$$

$$= \frac{1}{2}\left(993\ \frac{\text{kg}}{\text{m}^3}\right)\left(1.5\ \frac{\text{m}}{\text{s}}\right)^2 \left[\frac{4(4.88\ \text{m})(0.00604)}{0.0126\ \text{m}} 2 + 1.5 + 7.5\right]$$

$$= 31\ \text{kPa}$$

Using simple method I to estimate ΔP_s, we set $f_{i,\text{plain}} = 0.1$ in accordance with Fig. A–N–6–1(a) and write

$$f_{i,\text{fin}} = 1.4 f_{i,\text{plain}} = 0.14$$

$$\Delta P_s = \left[1.6 \frac{D_{i,s}}{S_{pp}}\left(1 - \frac{2B_c}{100}\right)(N_b + 1)\frac{f_{i,\text{fin}}}{2} + N_b\right]\left(\frac{G^2}{\rho}\right)_s$$

$$= \left[1.6 \frac{0.686 \text{ m}}{0.0254 \text{ m}}(1 - 0.4)(18 + 1)\frac{0.14}{2} + 18\right]\frac{[727 \text{ kg/(m}^2\text{ s)}]^2}{820 \text{ kg/m}^3}$$

$$= 33.8 \text{ kPa}$$

$$(\Delta P_s)_{1-2} = \Delta P_s + K_f\left(\frac{G^2}{2\rho}\right)_s = 33.8 \text{ kPa} + 10\frac{[727 \text{ kg/(m}^2\text{ s)}]^2}{2(820 \text{ kg/m}^3)}$$

$$= 33.8 \text{ kPa} + 3.22 \text{ kPa} = 37 \text{ kPa}$$

It should be noted that in addition to producing better thermal performance for this application, the use of fin tubes results in more full use of the available pressure drop on both the tube and shell sides.

Calculations for $\bar{h}_s$ and ΔP_s obtained by the primary and simple methods are compared in Table E11–16a for standard clearances of $L_{tb} = 0.794$ mm, $L_{sb} = 4.44$ mm, and $L_{bb} = 12.7$ mm.

TABLE E11–16a Comparison of shell-side calculations

Shell-side calculation	*Method*			
	Simple A	*Simple I*	*Simple II*[1]	*Delaware STER*[2]
$\bar{h}_s$ [W/(m² °C)]	873	945	1070	1050
ΔP_s [kPa]	—	33.8	32	40.8

[1] Referring to Appendix N–6, with $L_{tb} = 0.794$ mm, $L_{sb} = 4.44$ mm, and $L_{bb} = 12.7$ mm, Eqs. (N–6–3) and (N–6–4) indicate $\theta_{ds} = 1.85$, $r_{lm} = 0.288$, $r_s = 0.252$, $r_b = 0.0702$, and $F_c = 0.715$. It follows from Fig. A–N–6–3 that $J_c = 1.07$, $J_l = 0.69$, and $J_b = 0.92$, and from Fig. A–N–6–4 that $R_l = 0.47$ and $R_b = 0.78$.

[2] The STER Delaware computer solution was provided by Jerry Taborek, assuming fixed tubesheets.

It should be noted that computer codes such as STER by Taborek [43] and TASC3 by HTFS have the capability of efficiently calculating the shell-side coefficient of heat transfer $\bar{h}_s$ and pressure drop ΔP_s for other type shells, baffles, and tube bundles. To follow up on this point, calculations for $\bar{h}_s$ and ΔP_s obtained by use of Taborek's code for several type shells, baffles, and bundles are listed in Table E11–16b. Notice that the F shell produces a significant increase in $\bar{h}_s$ (73%) at the expense of a much larger pressure drop (7.4/1). On the other hand, the calculations for the J shell indicate lower values of $\bar{h}_s$ (−40%) and ΔP_s (0.34/1). It is also interesting to note the significantly reduced pressure drops associated with the use of double segmental baffles and rodbaffles, and the improved thermal performance of split-ring floating-head bundle arrangements for one pair of sealing strips.

TABLE E11–16b Calculations for different shell, baffle, and bundle arrangements, using STER.†

Configuration	$\bar{h}_s$ [W/(m² °C)]	ΔP_s [kPa]	*Configuration*	$\bar{h}_s$ [W/(m² °C)]	ΔP_s [kPa]
Base exchanger‡	1050	40.8	Rodbaffles	265	0.784
F shell	1820	303	U-tube bundle	1010	39.6
J shell	629	13.7	Floating head, split		
G shell	1050	40.7	ring, L_{bb} = 37 mm		
X shell	133	0.0262	N_{ss} = 0	917	32.2
Double segmental	724	17.7	N_{ss} = 1	1010	36.2
baffle			N_{ss} = 2	1030	37.5

† The STER Delaware computer solution was provided by Jerry Taborek.

‡ Base exchanger featured in example: E shell, 18 single segmental baffles with 20% cut, bundle with fixed tubesheets, no sealing strips, and clearances specified by L_{tb} = 0.794 mm, L_{sb} = 4.44 mm, and L_{bb} = 12.7 mm.

11–8 CROSSFLOW HEAT EXCHANGERS

Crossflow-heat-exchanger arrangements involve two individual mixed or unmixed fluid streams passing normal to one another. Referring to Fig. 11–8, these arrangements include finned and unfinned tube banks and tube-fin units with unmixed flow in the tubes and approximate mixed or unmixed crossflow over the tubes, and plate-fin units with unmixed flow on both sides. As in the case of double-pipe and shell-and-tube heat exchangers, crossflow units are sometimes connected in multipass configurations. Multipass arrangements are considered in Sec. 11–9.

The wetted-surface area and heat-transfer surface area A_s are normally essentially equivalent in crossflow heat exchangers (i.e., $p_w = p = A_s/L$), such that the hydraulic diameter D_H is represented by

$$D_H = \frac{4A_{\min}}{p_w} = \frac{4A_{\min}}{A_s/L} = 4\,\frac{\sigma}{\beta} \tag{11–201}$$

where $A_{\min}$ is the *minimum free-flow area*, $\sigma = A/A_1$ is the *contraction ratio*, and $\beta = A_s/V$ is the *surface-area density* (see Sec. 8–4). For the special case of flow on the tube side of a tube bank, this relation reduces to

$$D_H = \frac{4A_{\min}}{A_s/L} = \frac{4N(\pi d_i^2/4)}{N\pi d_i L/L} = d_i \tag{11–202}$$

The Reynolds number Re is represented by†

$$Re = \frac{GD_H}{\mu} = \frac{U_b D_H}{\nu} \tag{11–203}$$

where

$$G = \rho U_b = \frac{\dot{m}}{A_{\min}} \tag{11–204}$$

11–8–1 Practical Thermal Analysis

As illustrated in Figs. 11–22 and 11–43, the distribution in outlet temperature is uniform for a mixed stream and nonuniform for an unmixed stream. With this point in mind, the lumped formulation for energy transfer given by

$$q = C_h(T_{h,i} - T_{h,o}) \quad \text{and} \quad q = C_c(T_{c,o} - T_{c,i}) \tag{11–205,206}$$

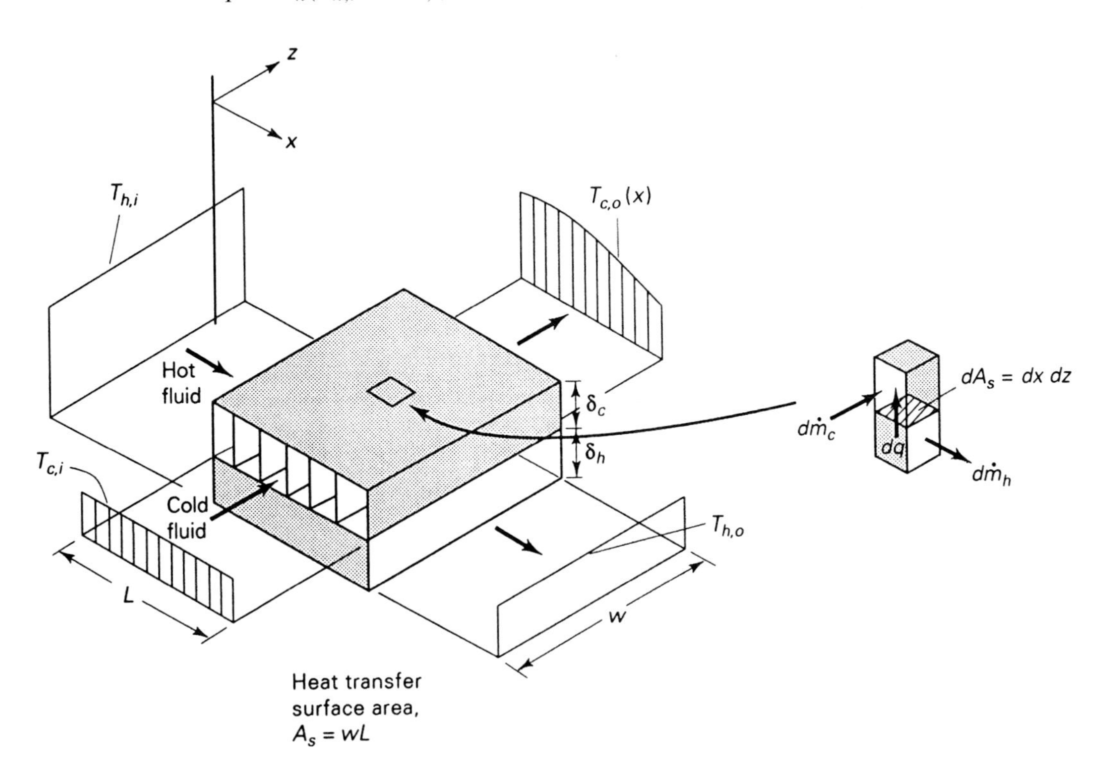

FIGURE 11–43 Sketch for rate of heat transfer dq and distributions in terminal temperatures for hot mixed and cold unmixed crossflowing fluids.

† The Reynolds number for flow over tube banks is also commonly expressed in terms of the tube diameter d_o by $Re_d = Re_s = Gd_o/\mu$.

from Eqs. (11–24) and (11–25), applies for mixed and unmixed crossflow, with the understanding that $T_{h,o}$, $T_{h,i}$, $T_{c,o}$, and $T_{c,i}$ represent the mean (or mixed) terminal temperatures. Because of the two-dimensional nature of the problem, the differential control volume used in the analysis involves two differential lengths dx and dz, such that the lumped/differential formulation is represented by

$$dq = -(d\dot{m}\, c_P)_h\, dT_h \qquad \text{and} \qquad dq = (d\dot{m}\, c_P)_c\, dT_c \tag{11–207,208}$$

where $d\dot{m}_h = (\rho U_b \delta)_h\, dz = \dot{m}_h\, dz/w$, $\dot{m}_h = (\rho U_b \delta)_h w$, $d\dot{m}_c = (\rho U_b \delta)_c\, dx = \dot{m}_c\, dx/L$, $\dot{m}_c = (\rho U_b \delta)_c L$, dq is expressed in terms of U by Eq. (11–2),

$$dq = U\, dA_s\, (T_h - T_c) \tag{11–209}$$

and $dA_s = dz\, dx$.

The development of solutions for the distributions in T_h and T_c as functions of x and z and the total heat-transfer rate q involve the solution of the two-dimensional system of equations represented by Eqs. (11–207) and (11–208), with $T_h = T_{h,i}$ at $x = 0$ and $T_c = T_{c,i}$ at $z = 0$. The mixed or unmixed state of each stream is accounted for by the dimensions of T_h and T_c according to Table 11–3. Solution results for heat transfer and the specification of the overall coefficient of heat transfer are considered in this section for basic crossflow arrangements, with the following assumptions for steady operation:

(1) local overall coefficient of heat transfer is normally taken as constant throughout the exchanger;
(2) no phase change occurs in only part of the exchanger;
(3) constant flow rate for both fluids;
(4) constant specific heat for both fluids;
(5) negligible heat losses and longitudinal conduction in the fluid and wall;
(6) each fluid is considered mixed or unmixed at every cross section;
(7) heating surface is distributed uniformly on each fluid side.

Solution Results for Heat-Transfer Rate

Solutions are available in the literature for crossflow units with both fluids mixed, both fluids unmixed, one fluid unmixed and one fluid mixed, and for multiple-pass systems [7,10,51]. The traditional approach to the development of solutions to these equations is demonstrated in Appendix N–7 for the case of one fluid mixed and one fluid unmixed. As indicated in the analysis, the solution results are expressed in terms of the mean coefficient of heat transfer $\overline{U}$.†

† The closed-form analytical solution developed in Appendix N–7 is formulated in terms of the mean overall coefficient of heat transfer $\overline{U}$, which is formally defined by

$$\overline{U} = \frac{1}{A_s} \int_0^{A_s} U(z)\, dA_s$$

where $U(z)$ is a function of z only.

Solution results are generally presented for standard crossflow arrangements in the context of the effectiveness, *LMTD*, and efficiency methods.

Effectiveness Method The solution developed in Appendix N–7 for crossflow heat exchangers with *one fluid mixed* and *one fluid unmixed* gives rise to expressions for the effectiveness P in terms of NTU_t and R of the form*

$$P = \frac{1}{R}[1 - \exp(-\Gamma R)] \qquad \Gamma = 1 - \exp(-NTU_t) \tag{11–210a,b}$$

where the reference stream I is set equal to the unmixed side t. This solution for mixed–unmixed crossflow arrangements is expressed in the functional form $NTU_t(P,R)$ by

$$NTU_t = -\ln\left[1 + \frac{1}{R}\ln(1 - PR)\right] \tag{11–211}$$

The solution for this important case is shown together with standard minimum performance criteria† in Fig. 11–44(a).

Solutions for other standard crossflow arrangements are represented by [52]

$$P = \frac{1}{R\,NTU_I}\sum_{m=0}^{\infty}\left\{\left[1 - \exp(-NTU_I)\sum_{j=0}^{m}\frac{NTU_I^j}{j!}\right]\right.$$

$$\left.\times\left[1 - \exp(-R\,NTU_I)\sum_{j=0}^{m}\frac{(R\,NTU_I)^j}{j!}\right]\right\} \tag{11–212}$$

for *both fluids unmixed*‡, which is shown in Fig. 11–44(b), and

$$P = \left[\frac{1}{1 - \exp(-NTU_I)} + \frac{R}{1 - \exp(-R\,NTU_I)} - \frac{1}{NTU_I}\right]^{-1} \tag{11–213}$$

* Mixed–unmixed arrangements are classified as nonsymmetrical exchangers. Equations (11–210) and (11–211) are exact for $I = t$ and approximate for $I = s$. The solution can also be represented by

$$P = 1 - \exp(-\Gamma/R) \qquad \Gamma = 1 - \exp(-R\,NTU_I) \tag{11–210a,b*}$$

or

$$NTU_I = -\frac{1}{R}\ln[1 + R\ln(1 - P)] \tag{11–211*}$$

which are exact for $I = s$ and approximate for $I = t$. The functional relations represented by Eqs. (11–210) and (11–211) and Eqs. (11–210*) and (11–211*) are approximately interchangeable, with the error being within about 2.5% for $\psi > 0.35$.

† These rule-of-thumb criteria do not apply for exchangers in which the incremental costs of heat-transfer surface area and frictional losses are low.

‡ A useful approximation of the exact solution for $R < 1$ and $NTU_I < 5$ is given in Table A–N–1–1.

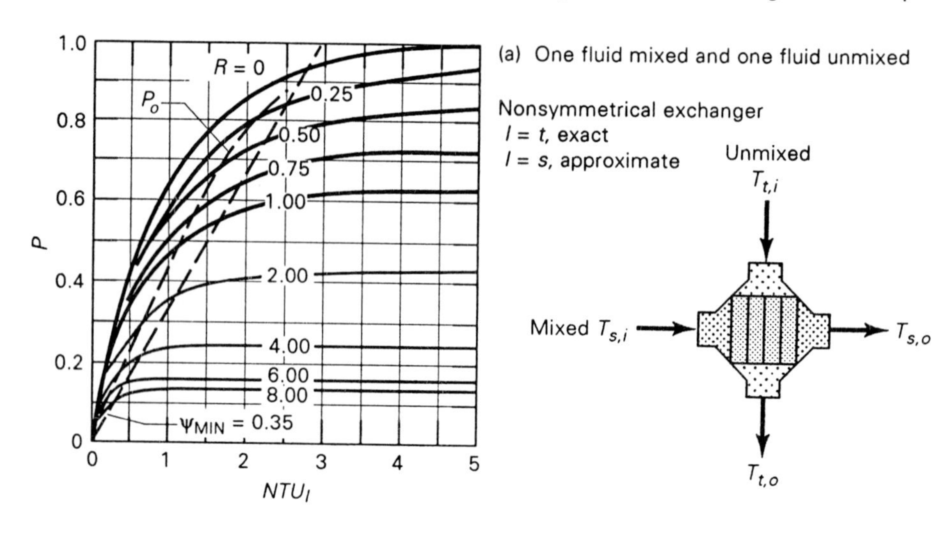

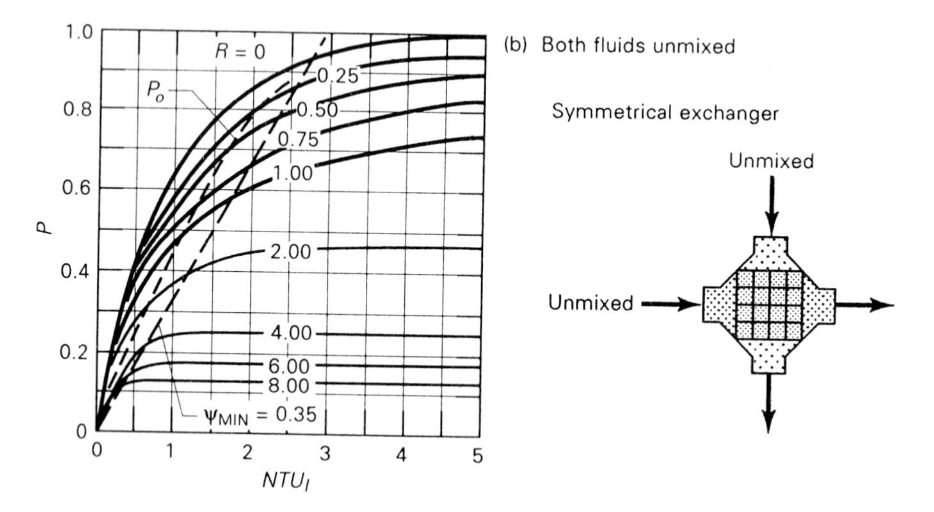

FIGURE 11–44 Effectiveness P for crossflow heat exchangers.

for *both fluids mixed*. Both of these arrangements are classified as symmetrical exchangers.

The solution results for crossflow heat exchangers are also commonly expressed in the ϵ-NTU format. The solutions for the three standard arrangements are listed in terms of ϵ, NTU, and C^* in Table A–N–1–2. The ϵ-NTU functional relations are shown in Fig. 11–45 for mixed–unmixed and unmixed–unmixed arrangements.

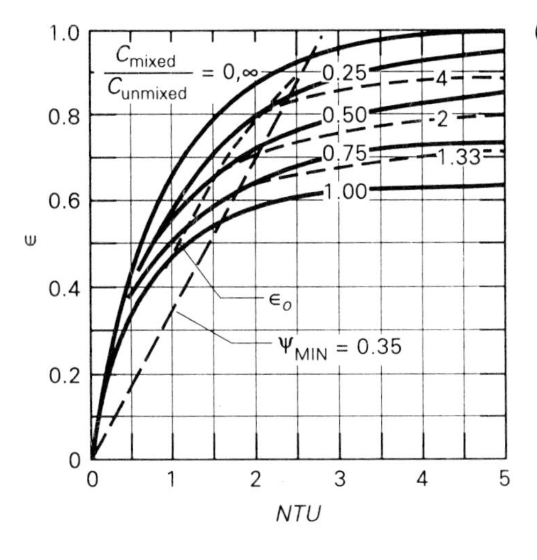

(a) One fluid mixed and one fluid unmixed

Nonsymmetrical exchanger

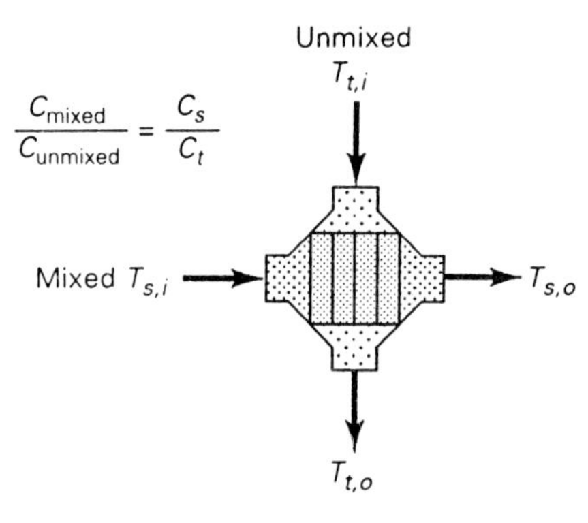

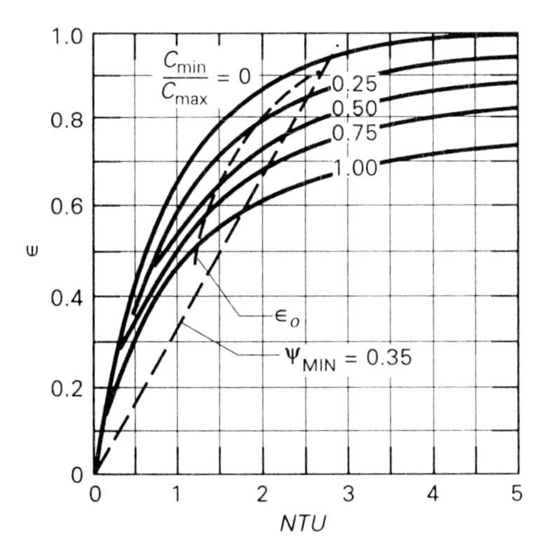

(b) Both fluids unmixed

Symmetrical exchanger

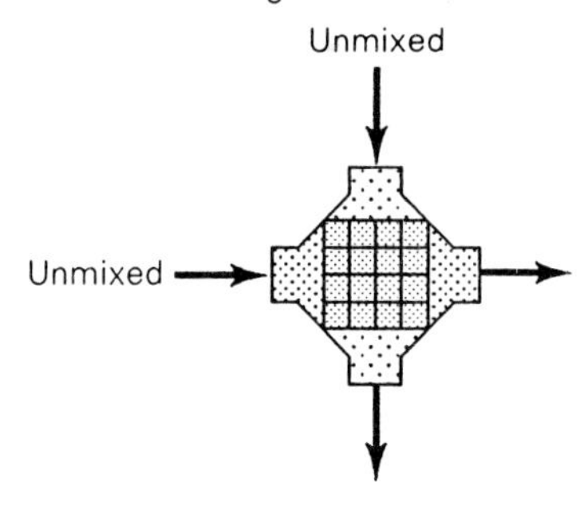

FIGURE 11-45 Effectiveness ϵ for crossflow heat exchangers. (Adapted from Kays and London [10]. Used with permission.)

It should be noted that the effectiveness ϵ for unmixed–unmixed arrangements is commonly approximated by

$$\epsilon = 1 - \exp\left[\frac{\exp(-\Gamma C^* \, NTU) - 1}{\Gamma C^*}\right] \qquad \Gamma = NTU^{-0.22} \qquad \text{(11–214a,b)}$$

However, this approximation should not be used for applications with NTU much greater than 5.

LMTD Method The solutions for basic crossflow heat exchangers are expressed in terms of the correction factor F by Eq. (11–70),†

$$F = \frac{1}{NTU_I\,(1 - R)} \ln \frac{1 - PR}{1 - P} \tag{11–215}$$

with P or NTU_I appropriately specified. Solution results for the *LMTD* correction factor F of basic crossflow-heat-exchanger arrangements are summarized in Table A–N–1–1 of the Appendix. The F–P solution results for the mixed–unmixed and unmixed–unmixed arrangements are shown in Fig. 11–46.

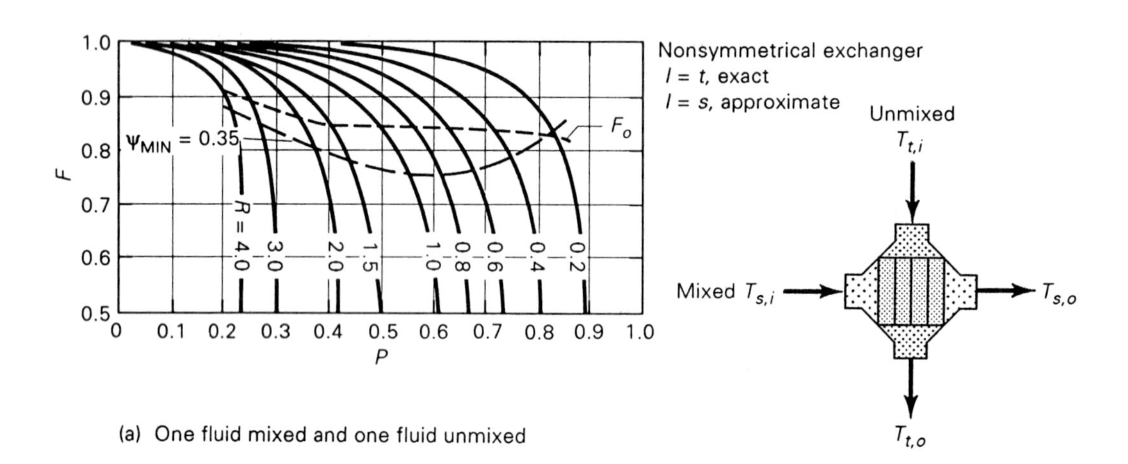

(a) One fluid mixed and one fluid unmixed

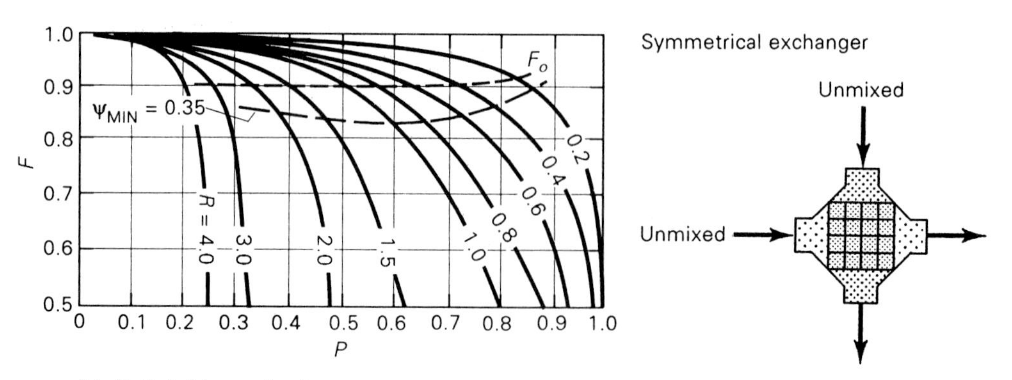

(b) Both fluids unmixed

FIGURE 11–46 *LMTD* correction factor F for crossflow heat exchangers. (Adapted from Bowman et al. [28].)

† See Table A–N–1–1 for solution results for $R = 1$.

Efficiency Method The solution results for basic crossflow heat exchangers are presented in terms of the efficiency ψ by employing the defining relation $\psi = P/NTU_I$ and the appropriate relations for P or NTU_I. Solution results for the efficiency of basic crossflow-heat-exchanger arrangements are given in Table A–N–1–1 of the Appendix. The solutions for mixed–unmixed and unmixed–unmixed arrangements are represented by the Mueller charts shown in Fig. 11–47.

Evaluation of $\overline{U}$

To evaluate $\overline{U}$, which is normally approximated by

$$\frac{1}{\overline{U}A_s} = \frac{1}{(\bar{h}A_s)_h} + \left(\frac{F_f}{A_s}\right)_h + R_k + \left(\frac{F_f}{A_s}\right)_c + \frac{1}{(\bar{h}A_s)_c} \tag{11–216}$$

or Eq. (11–17), the coefficients $\bar{h}_h$ and $\bar{h}_c$ are generally specified by the use of correlations available in the literature. Standard correlations of the type presented in Chap. 8 are used to calculate $\overline{Nu}$ (or $\overline{St}$) for flow inside circular tubes and other passageways with uniform cross section. In addition, correlations are presented by Kays and London [10] for the crossflow geometries listed in Fig. 11–48. To illustrate, correlations for $\overline{St}$ are shown in Figs. 8–25 and 8–32 for flow over tube banks with plain and fin tubes and in Figs. 11–49 and 11–50 for typical tube-fin and plate-fin compact crossflow heat exchangers.

11–8–2 Practical Hydraulic Analysis

The pressure drop on either side of a crossflow heat exchanger with uniform property flow is given by Eq. (11–94),

$$\Delta P_{1-2} = \frac{\rho U_b^2}{2}\left(K_c + \frac{4L}{D_H}f + K_e\right) \tag{11–217}$$

where K_c and K_e are generally specified by standard correlations of the type shown in Figs. 8–5 and 11–51 for representative heat-exchanger cores with abrupt-contraction entrance and abrupt-expansion exit. As we have seen, for the case of flow over tube banks, K_c and K_e are set equal to zero, such that Eq. (11–217) reduces to Eq. (11–96),

$$\Delta P_{1-2} = \frac{\rho U_b^2}{2}\left(\frac{4L}{D_H}f\right) = \frac{G^2}{2\rho}\left(\frac{4L}{D_H}f\right) \tag{11–218}$$

For situations in which heating or cooling cause significant change in density, ΔP_{1-2} is given by Eq. (11–95) or Eq. (11–97).

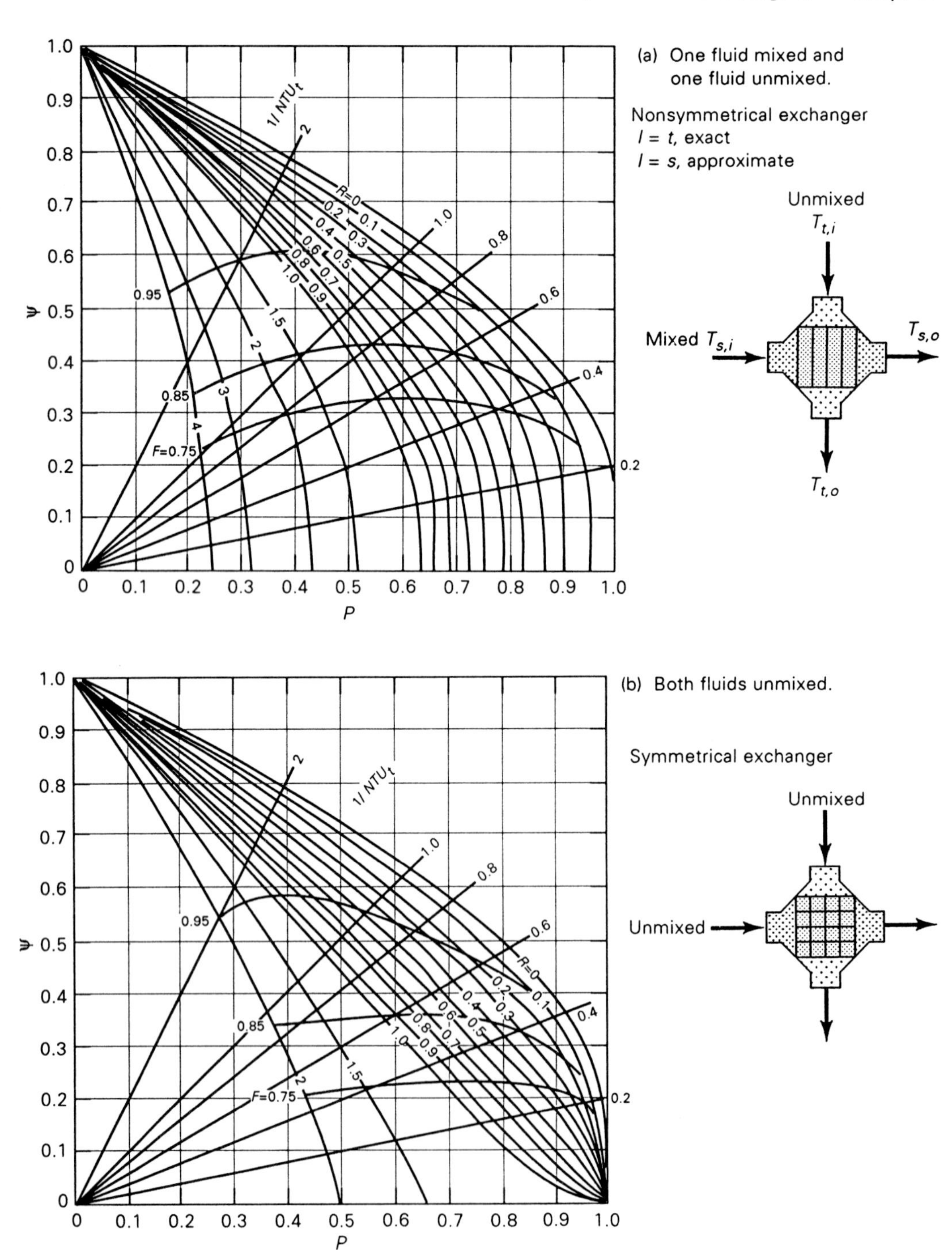

FIGURE 11–47 Efficiency for crossflow heat exchangers. (Adapted from Shah and Mueller [7].)

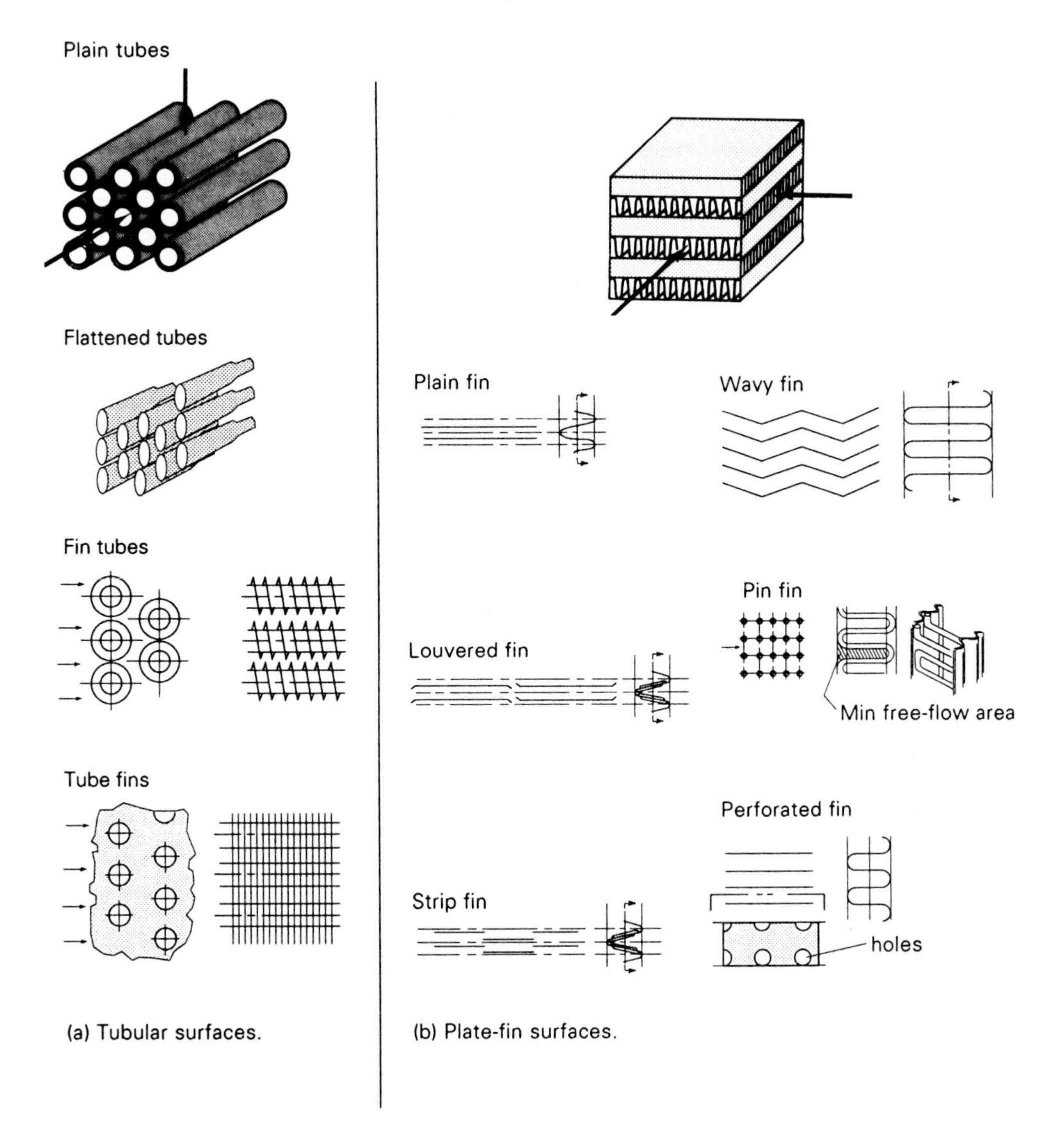

FIGURE 11–48 Compact crossflow heat exchanger geometries. (From Kays and London [10]. Used with permission.)

As in the specification of $\overline{Nu}$ and $\overline{St}$, standard correlations of the type presented in Chap. 8 are used to calculate f for flow in tubes with uniform cross section. Referring to Figs. 8–25, 8–32, 11–49, and 11–50, the information presented by Kays and London [10] for the crossflow geometries listed in Fig. 11–48 includes correlations for f as well as $\overline{St}$.

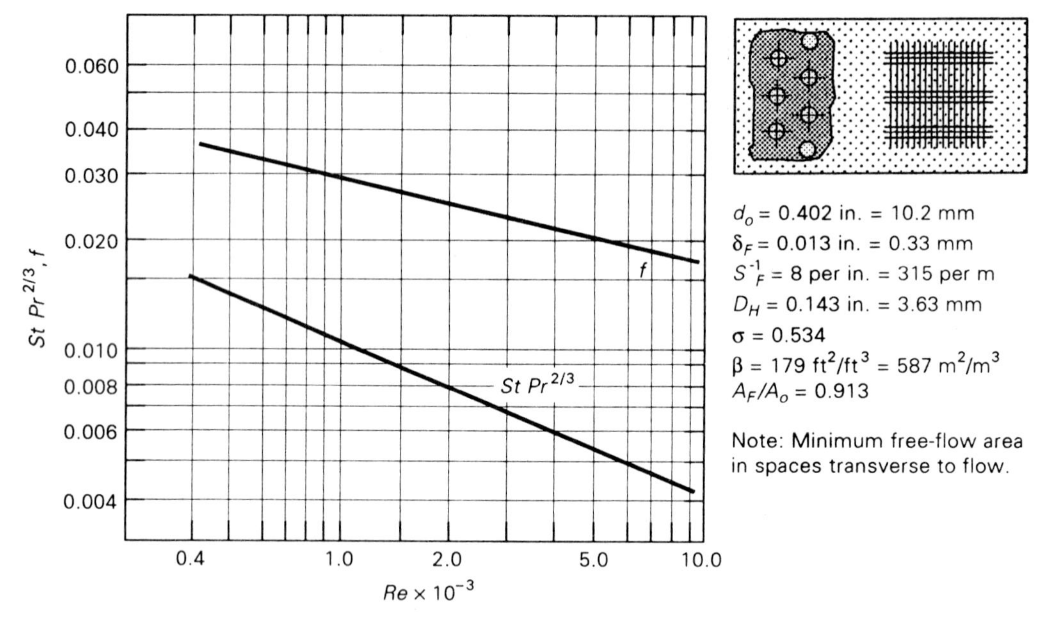

FIGURE 11–49 Tube-fin surface 8.0-3/8T. (From Kays and London [10]. Used with permission.)

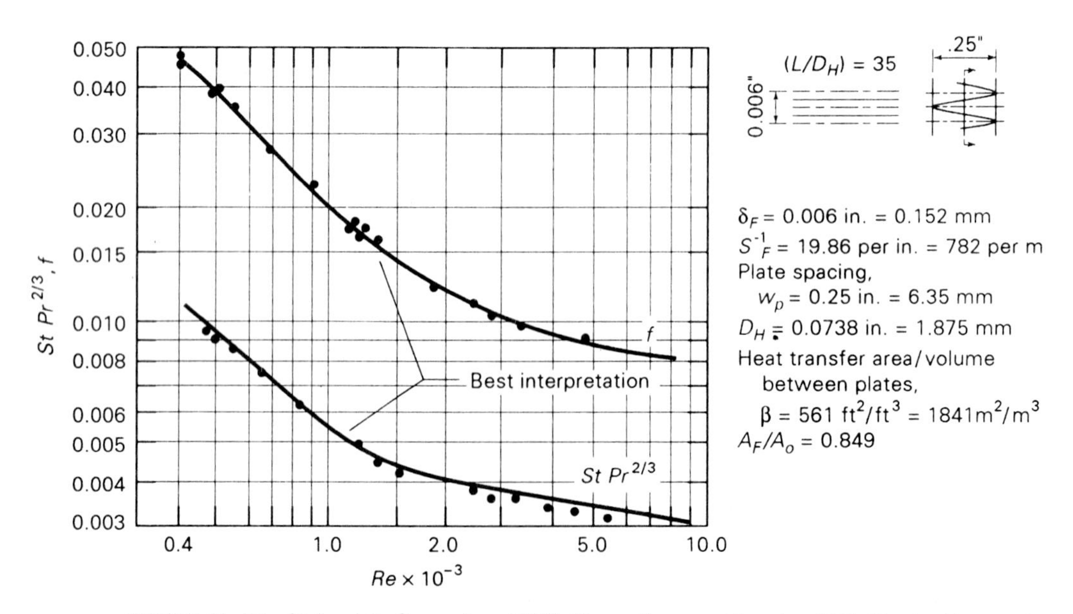

FIGURE 11–50 Plain plate-fin surface 19.86. (From Kays and London [10]. Used with permission.)

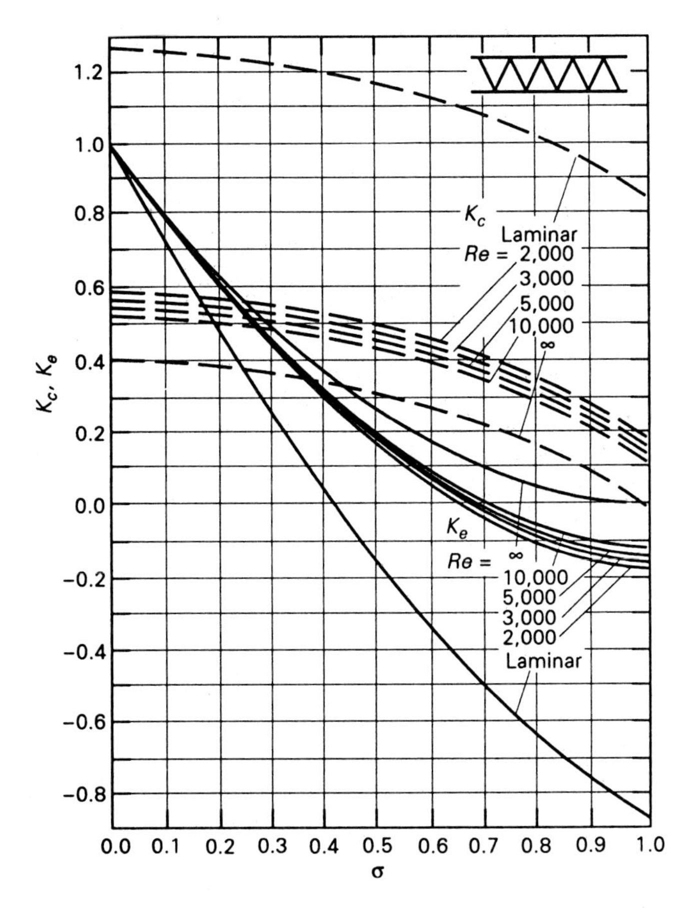

FIGURE 11-51 Entrance-loss and expansion-loss coefficients for a multiple-triangular-tube heat exchanger with abrupt-contraction entrance and abrupt-expansion exit. (From Kays and London [10]. Used with permission.)

EXAMPLE 11-17†

Air at 15°C and 1 atm with entering velocity of 6.8 m/s is to be heated in a crossflow heat exchanger with eighty 9.52-mm-diameter 2.5-m-long thin-wall tubes. The tubes are placed 10 tubes deep in a staggered array with longitudinal and transverse pitches of 1.19 cm and 1.43 cm, respectively. Warm exhaust gas at 65°C enters the tubes with a mass-flow rate of 0.25 kg/s. Determine the rate of heat transfer and the outlet temperatures, assuming that the properties of the exhaust gas are ap-

† A simple numerical finite-difference program for solving this example is given in Fig. A-G-8 of the Appendix.

proximately the same as for air. Also determine whether or not the minimum performance criteria are satisfied.

Solution

Objective Determine q, $T_{t,o}$, and $T_{s,o}$ and whether or not $\psi > 0.35$ and/or $P < P_o$.

Schematic Crossflow heat exchanger: mixed and unmixed streams.

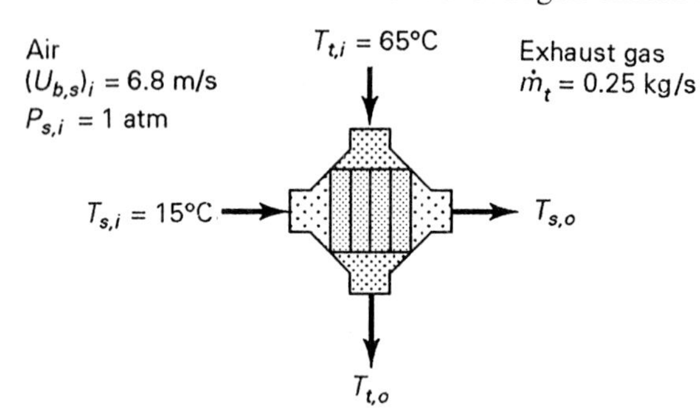

$N = 80$, $N_L = 10$, staggered array
Thin-wall tubes, $D = 9.52$ mm
$S_L = 1.19$ cm, $S_T = 1.43$ cm
$Z = 2.5$ m

From Example 11-3:
$\bar{U} = 93.4$ W/(m² °C)
$\dot{m}_s = 2.39$ kg/s
$A_s = \pi DNZ = 5.98$ m²

Assumptions/Conditions

uniform distribution in U
negligible conduction resistance R_k
negligible entrance effects
negligible fouling
uniform properties
standard conditions

Properties

Air at T = 15°C = 288 K (Table A–C–5): ρ = 1.23 kg/m³, $\mu = 1.76 \times 10^{-5}$ kg/(m s), c_P = 1.01 kJ/(kg °C), k = 0.0253 W/(m °C), Pr = 0.711.

Exhaust gas at T = 65°C = 338 K (Table A–C–3 for air): ρ = 1.04 kg/m³, $\mu = 2.02 \times 10^{-5}$ kg/(m s), c_P = 1.01 kJ/(kg °C), k = 0.0291 W/(m °C), Pr = 0.7.

Analysis To utilize the effectiveness (or efficiency) approach, we calculate the capacity rates, assuming uniform properties.

$$C_t = (\dot{m}c_P)_t = 0.25\,\frac{\text{kg}}{\text{s}}\left(1.01\,\frac{\text{kJ}}{\text{kg °C}}\right) = 0.252 \text{ kW/°C}$$

$$C_s = (\dot{m}c_P)_s = 2.39\,\frac{\text{kg}}{\text{s}}\left(1.01\,\frac{\text{kJ}}{\text{kg °C}}\right) = 2.41 \text{ kW/°C}$$

Using these values and selecting the tube-side fluid as the reference stream I, we obtain

$$R = \frac{C_t}{C_s} = \frac{0.252\ \text{kW/°C}}{2.41\ \text{kW/°C}} = 0.105$$

$$NTU_t = \frac{\overline{U}A_s}{C_t} = \frac{[93.4\ \text{W/(m}^2\ \text{°C)}](5.98\ \text{m}^2)}{0.252\ \text{kW/°C}} = 2.22$$

Substituting into Eq. (11–210), we obtain†

$$\Gamma = 1 - \exp(-NTU_t) = 1 - \exp(-2.22) = 0.891$$

$$P = \frac{1}{R}[1 - \exp(-\Gamma R)] = \frac{1}{0.105}\{1 - \exp[-0.891(0.105)]\} = 0.851$$

[Figure 11–44(a) can be used to check this result.]

The heat-transfer rate q is obtained by use of Eq. (11–68), which is put into the form

$$q = PC_t(T_{t,i} - T_{s,i}) = 0.851\left(0.252\ \frac{\text{kW}}{\text{°C}}\right)(65\text{°C} - 15\text{°C}) = 10.7\ \text{kW}$$

The outlet fluid temperatures are now calculated.

$$q = C_t(T_{t,i} - T_{t,o}) \qquad T_{t,o} = T_{t,i} - \frac{q}{C_t} = 65\text{°C} - \frac{10.7\ \text{kW/°C}}{0.252\ \text{kW/°C}} = 22.5\text{°C}$$

$$q = C_s(T_{s,o} - T_{s,i}) \qquad T_{s,o} = T_{s,i} + \frac{q}{C_s} = 15\text{°C} + \frac{10.7\ \text{kW/°C}}{2.41\ \text{kW/°C}} = 19.4\text{°C}$$

Substituting into the defining equation for efficiency, we find that $\psi = P/NTU_t = 0.383$, such that the minimum performance criterion $\psi > \psi_{\text{MIN}} = 0.35$ is satisfied. Since $T_{h,o} = 22.5\text{°C} > T_{c,o} = 19.4\text{°C}$, we conclude that the alternative outlet temperature criterion is also satisfied.

The effect of fluid property variation with temperature can be approximately accounted for by resolving the problem with the thermal properties of the air and water evaluated at the arithmetic average of the inlet and outlet bulk-stream temperatures. In addition, the wall temperature at the midpoint of the heat exchanger can be estimated, such that $\overline{h}_t$ and $\overline{h}_s$ can be corrected for effects of property variation.

† Following the efficiency approach, we evaluate ψ by using the Mueller chart shown in Fig. 11–47(a) or by writing

$$\psi = \frac{P}{NTU_t} = \frac{1 - \exp(-\Gamma R)}{R\ NTU_t} = 0.383$$

and represent the rate of heat transfer q by Eq. (11–65),

$$q = \psi\ \overline{U}A_s(T_{t,i} - T_{s,i}) = 10.7\ \text{kW}$$

Note that our solution for q of 10.7 kW lies well below the value of 50 kW obtained in Example 8–18 for air flow over a tube bank with tubes maintained at constant temperature of 65°C. This difference is attributed to the lower tube-wall temperature.

EXAMPLE 11–18†

An unmixed crossflow heat exchanger is to be constructed that will heat 2.5 kg/s of air at 1 atm from 15°C to 30°C. Hot water enters at 52.5°C. The mean overall coefficient of heat transfer is 300 W/(m² °C). Determine the required surface area to produce an outlet water temperature of 24°C.

Solution

Objective Determine required heat-transfer surface area A_s.

Schematic Crossflow heat exchanger: both fluids unmixed.

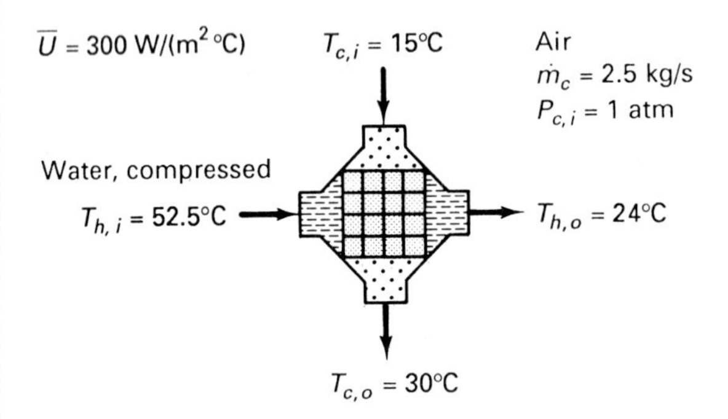

Assumptions/Conditions

uniform distribution in U
uniform properties
standard conditions

Properties Air at T = 15°C = 288 K (Table A–C–5): c_P = 1.01 kJ/(kg °C).

Analysis Following the *LMTD* approach, the unknown surface area A_s is expressed in terms of the correction factor F and *LMTD* by

† This example is solved by the ψ-P method in Appendix N–9.

$$A_s = \frac{q}{\overline{U}(F\ LMTD)} \tag{a}$$

from Eq. (11–59), where

$$q = (\dot{m}c_P)_c(T_{c,o} - T_{c,i}) = 2.5\,\frac{\text{kg}}{\text{s}}\left(1.01\,\frac{\text{kJ}}{\text{kg °C}}\right)(30\text{°C} - 15\text{°C})$$

$$= 37.9 \text{ kW}$$

$$LMTD = \frac{(52.5\text{°C} - 30\text{°C}) - (24\text{°C} - 15\text{°C})}{\ln\left[(52.5\text{°C} - 30\text{°C})/(24\text{°C} - 15\text{°C})\right]} = 14.7\text{°C}$$

and F can be obtained from Fig. 11–46(b).

To evaluate the correction factor F, we specify the cold fluid as the reference stream I and calculate P and R by writing†

$$P = \frac{T_{c,o} - T_{c,i}}{T_{h,i} - T_{c,i}} = \frac{30\text{°C} - 15\text{°C}}{52.5\text{°C} - 15\text{°C}} = 0.4$$

$$R = \frac{T_{h,i} - T_{h,o}}{T_{c,o} - T_{c,i}} = \frac{52.5\text{°C} - 24\text{°C}}{30\text{°C} - 15\text{°C}} = 1.9$$

Using these inputs, Fig. 11–46(b) indicates $F \simeq 0.86$. Substituting into Eq. (a), we obtain

$$A_s = \frac{37.9 \text{ kW}}{[300 \text{ W/(m}^2\text{ °C)}](0.86)(14.7\text{°C})} = 10 \text{ m}^2$$

and

$$q'' = \frac{q}{A_s} = \frac{37.9 \text{ kW}}{10 \text{ m}^2} = 3.79 \text{ kW/m}^2$$

It should be noted that this result indicates $\psi = q/[\overline{U}A_s(T_{h,i} - T_{c,i})] = 0.333$ [from Eq. (11–65)], which falls somewhat below the minimum performance criterion $\psi > \psi_{\text{MIN}} = 0.35$. Therefore, consideration should be given to transferring the heat more efficiently. One approach to accomplishing this task would be to utilize two smaller crossflow heat exchangers in series. As shown in Example 11–22, a two-pass arrangement would require a total surface area of 9.16 m^2 and

† Because of the interchangeability of reference temperatures, P and R can be defined by the alternative relations

$$P = \frac{T_{h,i} - T_{h,o}}{T_{h,i} - T_{c,i}} = \frac{52.5\text{°C} - 24\text{°C}}{52.5\text{°C} - 15\text{°C}} = 0.76$$

$$R = \frac{T_{c,o} - T_{c,i}}{T_{h,i} - T_{h,o}} = \frac{30\text{°C} - 15\text{°C}}{52.5\text{°C} - 24\text{°C}} = 0.526$$

would result in $q'' = 4.14$ kW/m² and $\psi = 0.367$, which indicates a modest improvement in performance of about 10%.

It is also of interest to note that the outlet temperature criterion $T_{c,o} \leq T_{h,o}$ is violated by the single-unit arrangement shown in the schematic. [Formally, $P_o = 1/(1 + R) = 0.345$, such that $P = 0.4 > P_o$ in violation of the outlet temperature criterion.] However, when the two-pass series arrangement considered in Example 11–22 is used, the outlet temperature criterion is satisfied for the individual units.

EXAMPLE 11–19

The fin tube surface CF 7.34 is to be used in an air-fin cooler with ten columns and five rows to cool 5.6 kg/s of kerosene from 83.8°C to 61.2°C. For summertime operation, the air is assumed to be at 35°C and 1 atm with entering velocity of 4 m/s. The allowable fouling factors for the kerosene and air are 0.0002 and 0.0004 m² °C/W, respectively. Determine the required tube length, the outlet temperature of the air, and the pressure drops of the two streams.

Solution

Objective Determine the length of fin-tube surface CF 7.34 (see Fig. 8–32), $T_{c,o}$, and the pressure drops.

Schematic Air-fin cooler: fin-tube surface CF 7.34—assuming unmixed flow over the fin tubes.

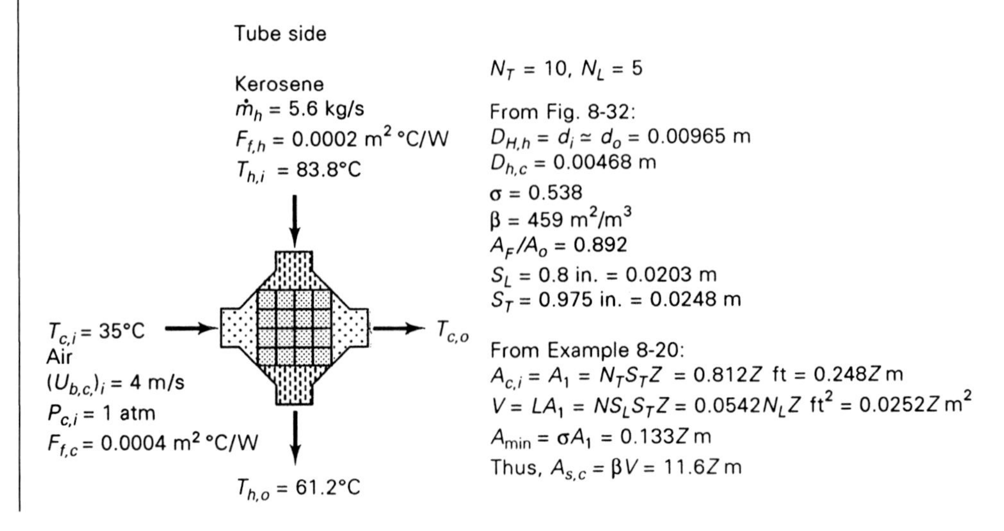

Assumptions/Conditions

uniform distribution in U
negligible conduction resistance R_k
negligible thermal contact resistance R_{tc}
both fluids unmixed
negligible entrance effects
moderate property variation
standard conditions

Properties

Air at 35°C (Table A–C–5): ρ = 1.15 kg/m^3, μ = 1.88 × 10^{-5} kg/(m s), k = 0.0268 W/(m °C), c_P = 1.01 kJ/(kg °C), Pr = 0.706.

Kerosene at 72.5°C [53]: ρ = 780 kg/m^3, μ = 8.9 × 10^{-4} kg/(m s), k = 0.138 W/(m °C), c_P = 2.18 kJ/(kg °C), Pr = 14.1.

Analysis The geometric dimensions of the fin-tube surface are summarized in the schematic. The hydraulic and thermal characteristics on the tube side and air side are developed as follows:

Tube-side flow—kerosene

$$\dot{m}_h = (\rho A U_b)_h = (AG)_h$$

$$A_h = \frac{\pi d_i^2}{4} N = \frac{\pi (0.00965 \text{ m})^2}{4} 50 = 0.00366 \text{ m}^2$$

$$G_h = \frac{\dot{m}_h}{A_h} = \frac{5.6 \text{ kg/s}}{0.00366 \text{ m}^2} = 1530 \text{ kg/(m}^2\text{ s)} \qquad U_{b,h} = 1.96 \text{ m/s}$$

$$Re_h = \left(\frac{GD_H}{\mu}\right)_h = \frac{[1530 \text{ kg/(m}^2\text{ s)}](0.00965 \text{ m})}{8.90 \times 10^{-4} \text{ kg/(m s)}}$$

$$= 16{,}600 \qquad \text{Fully turbulent}$$

$$f_h = (1.58 \ln 16{,}600 - 3.28)^{-2} = 0.00686 \qquad (8\text{–}15)$$

$$Nu_h = \frac{(0.00686/2)(16{,}600)(14.1)}{1.07 + 12.7\sqrt{0.00686/2}\,(14.1^{2/3} - 1)} = 172 \qquad (8\text{–}16)$$

$$h_h = \left(Nu \frac{k}{D_H}\right)_h = 172 \frac{0.138 \text{ W/(m °C)}}{0.00965 \text{ m}} = 2460 \text{ W/(m}^2\text{ °C)}$$

Air side

$$\dot{m}_c = (\rho A U_b)_{c,i} = \left(1.15\ \frac{\text{kg}}{\text{m}^3}\right)(0.248Z\ \text{m})\left(4\ \frac{\text{m}}{\text{s}}\right) = 1.14Z\ \text{kg/(m s)}$$

$$G_c = \frac{\dot{m}_c}{A_{\min}} = \frac{1.14Z\ \text{kg/(m s)}}{0.133Z\ \text{m}} = 8.57\ \text{kg/(m}^2\ \text{s)} \qquad U_{b,c} = 7.45\ \text{m/s}$$

$$Re_c = \left(\frac{GD_H}{\mu}\right)_c = \frac{[8.57\ \text{kg/(m}^2\ \text{s)}](0.00468\ \text{m})}{1.88 \times 10^{-5}\ \text{kg/(m s)}}$$

$$= 2130 \qquad \text{Transitional turbulent}$$

$$f_c = 0.038 \qquad \overline{St}_c\, Pr_c^{2/3} = 0.0082 \qquad \text{Fig. 8–32}$$

$$\overline{Nu}_c = 0.0082\ Re_c\ Pr_c^{1/3} = 0.0082(2130)(0.706)^{1/3} = 15.6$$

$$\overline{h}_c = \left(\overline{Nu}\ \frac{k}{D_H}\right)_c = 15.6\ \frac{0.0268\ \text{W/(m °C)}}{0.00468\ \text{m}} = 89.3\ \text{W/(m}^2\ \text{°C)}$$

Referring to Fig. 2–17(b), with $\overline{h}_c$ set equal to 89.3 W/(m² °C), the parameter $L_c^{3/2}[\overline{h}_c/(k_F A_p)]^{1/2}$ becomes 0.204. Setting $r_{2c}/r_1 = 2.47$, Fig. 2–17(b) indicates $\eta_F = 0.93$. Substituting into Eq. (11–12), we obtain

$$\eta_o = 1 - \frac{A_F}{A_o}(1 - \eta_F) = 1 - 0.892(1 - 0.93) = 0.938$$

To compute the mean overall coefficient of heat transfer based on the air-side surface area $A_{s,c}$, we write [assuming $\overline{h}_h = h_h$ and using Eq. (11–17) with the local coefficients replaced by mean values and with negligible wall resistance and thermal contact resistance]

$$\frac{1}{\overline{U}} = \frac{1}{\overline{h}_h}\frac{A_{s,c}}{A_{s,h}} + F_{f,h}\frac{A_{s,c}}{A_{s,h}} + \frac{F_{f,c}}{\eta_o} + \frac{1}{\eta_o \overline{h}_c} \qquad \text{(a)}$$

where $F_{f,h} = 0.0002$ m² °C/W, $F_{f,c} = 0.0004$ m² °C/W, $\eta_o = 0.938$,

$$\frac{A_{s,c}}{A_{s,h}} = \frac{11.6Z\ \text{m}}{\pi d_i Z N} = \frac{11.6Z\ \text{m}}{\pi(0.00965\ \text{m})(Z)(50)} = 7.65$$

Substituting into Eq. (a), $\overline{U}$ becomes

$$\frac{1}{\overline{U}} = \left[\frac{7.65}{2460} + 0.0002(7.65) + \frac{0.0004}{0.938} + \frac{1}{0.938(89.3)}\right] \text{m}^2\ \text{°C/W}$$

$$= (0.00311 + 0.00153 + 0.000426 + 0.0119)\ \text{m}^2\ \text{°C/W}$$

$$\overline{U} = 58.9\ \text{W/(m}^2\ \text{°C)}$$

To complete the analysis, we make use of the effectiveness method. Specifying the hot fluid as the reference stream I, we write

$$P = \frac{C_h(T_{h,i} - T_{h,o})}{C_I(T_{h,i} - T_{c,i})} = \frac{C_h}{C_h}\frac{83.8°\text{C} - 61.2°\text{C}}{83.8°\text{C} - 35°\text{C}} = 0.463$$

$$C_h = (\dot{m}c_P)_h = 5.6\,\frac{\text{kg}}{\text{s}}\left(2.18\,\frac{\text{kJ}}{\text{kg °C}}\right) = 12.2\ \text{kW/°C}$$

$$C_c = (\dot{m}c_P)_c = 1.14Z\,\frac{\text{kg}}{\text{m s}}\left(1.01\,\frac{\text{kJ}}{\text{kg °C}}\right) = 1.15Z\ \text{kW/(m °C)} \tag{b}$$

$$R = \frac{C_h}{C_c} = \frac{12.2\ \text{kW/°C}}{1.15Z\ \text{kW/(m °C)}} = \frac{10.6}{Z}\ \text{m} \tag{c}$$

and

$$NTU_I = \frac{\overline{U}A_{s,c}}{C_h} = \frac{[58.9\ \text{W/(m}^2\ \text{°C)}](11.6Z\ \text{m})}{12.2\ \text{kW/°C}} = 0.056Z/\text{m} \tag{d}$$

As in the case of most heat exchanger design problems, the solution for Z requires iteration. To develop a systematic iterative approach, Eq. (d) is put into the form

$$Z = \frac{NTU_I}{0.056}\ \text{m} \tag{e}$$

With P set equal to 0.463, Eqs. (c) and (e) can be combined with the functional relation $P(NTU_I,R)$ for an unmixed–unmixed crossflow arrangement to solve for Z. Using Fig. 11–44(b) to evaluate $P(NTU_I,R)$, the first several iterations lead to the following results:

First iteration Assume $Z = 10.6$ m and set $P = 0.463$.

$R = 1$ from Eq. (c)

$NTU_I = 1$ from Fig. 11–44(b)

$Z = \dfrac{1}{0.056/\text{m}} = 17.9$ m from Eq. (e)

Second iteration Set $Z = 17.9$ m and $P = 0.463$.

$R = 0.592$ from Eq. (c)

$NTU_I = 0.7$ from Fig. 11–44(b)

$Z = 12.5$ m from Eq. (e)

Third iteration Set $Z = 12.5$ m and $P = 0.463$.

$R = 0.848$ from Eq. (c)

$NTU_I = 0.8$ from Fig. 11–44(b)

$Z = 14.3$ m from Eq. (e)

This result is as accurate as can be read from Fig. 11–44(b). However the analysis can be refined by using the analytical solution for $P(NTU_I, R)$. In this connection, because $P < 1$ and $NTU_I < 5$, we can employ the simple approximate relation given by (see Table A–N–1–1)

$$P = 1 - \exp\left[\frac{\exp(-\Gamma R\, NTU_I) - 1}{\Gamma R}\right] \qquad \Gamma = NTU_I^{-0.22}$$

Using this relation, an iterative solution results in $R = 0.711$, $NTU_I = 0.835$, and $Z = 14.9$ m. Thus, we conclude that a length of approximately 14.9 m is required. Notice that $L/D_H >> 10$, such that the assumption of negligible entrance effects on h_h is reasonable. In addition, we note that $\psi = P/NTU_I = 0.554$, such that the minimum performance criterion $\psi > \psi_{MIN} = 0.35$ is satisfied.

To obtain the outlet temperature $T_{c,o}$, we write

$$q = C_h(T_{h,i} - T_{h,o}) = 12.2\,\frac{\text{kW}}{°\text{C}}\,(83.8°\text{C} - 61.2°\text{C}) = 276\text{ kW}$$

$$q = C_c(T_{c,o} - T_{c,i}) \qquad C_c = 1.15Z\,\frac{\text{kW}}{\text{m }°\text{C}} = 1.15\,\frac{\text{kW}}{\text{m }°\text{C}}(14.9\text{ m}) = 17.1\text{ kW/°C}$$

$$T_{c,o} = \frac{q}{C_c} + T_{c,i} = \frac{276\text{ kW}}{17.1\text{ kW/°C}} + 35°\text{C} = 51°\text{C}$$

As we have noted, the solution is based on the assumption that the exchanger can be classified as an unmixed–unmixed-crossflow arrangement. By comparing Figs. 11–44(a) and (b) for $P = 0.463$ and R of the order of 0.711, we see that the values of NTU_I differ very little for mixed–unmixed- and unmixed–unmixed-arrangements. To expand on this point, iterative calculations using the analytical solution for $P(NTU_I, R)$ associated with a mixed–unmixed arrangement result in a value of Z equal to 14.8 m, which is within 1% of the value obtained for the unmixed–unmixed arrangement. Thus, we conclude that for the operating conditions specified in this problem, the solution is not significantly influenced by whether the exchanger is classified as unmixed–unmixed or mixed–unmixed. It should be mentioned that this would not be the case for operating conditions with much larger values of NTU_I.

Because of the relatively small increase in the temperature of the air over the length of the exchanger, the density only decreases by about 5.5%. Therefore, the pressure drop on the air side is approximated by Eq. (11–218),

$$(\Delta P_c)_{1-2} = \left(\frac{\rho U_b^2}{2}\right)_c \left(\frac{4L}{D_H} f\right)_c = \left(\frac{G^2}{2\rho}\right)_c \left(\frac{4L}{D_H} f\right)_c$$

where $\rho_c = 1.15$ k/m^3, $D_{H,c} = 0.00468$ m, $G_c = 8.57$ kg/(m^2 s), $f_c = 0.038$, and

$$L_c = N_L S_L = 5(0.0203 \text{ m}) = 0.102 \text{ m}$$

Substituting into Eq. (11–218), we obtain

$$(\Delta P_c)_{1-2} = \frac{[8.57 \text{ kg/(m}^2\text{ s)}]^2}{2(1.15 \text{ kg/m}^3)} \left[\frac{4(0.102 \text{ m})(0.038)}{0.00468 \text{ m}}\right] = 106 \text{ Pa}$$

Similarly, the pressure drop on the tube side is given by Eq. (11–217),

$$(\Delta P_h)_{1-2} = \left(\frac{G^2}{2\rho}\right)_h \left(K_c + \frac{4L}{D_H} f + K_e\right)_h$$

where $\rho_h = 780$ kg/m^3, $D_{H,h} = 0.00965$ m, $G_h = 1530$ kg/(m^2 s), $f_h = 0.00686$, $L_h = Z = 14.9$ m, and $K_{c,h}$ and $K_{e,h}$ are dependent on the contraction ratio, which has not been specified. Referring to Fig. 8–5, the maximum values of K_c and K_e are about 0.5 and 1, respectively. It follows that the tube-side pressure drop is approximated by

$$(\Delta P_h)_{1-2} = \frac{[1530 \text{ kg/(m}^2\text{ s)}]^2}{2(780 \text{ kg/m}^3)} \left[0.5 + \frac{4(14.9 \text{ m})(0.00686)}{0.00965 \text{ m}} + 1\right]$$

$$= 1500(0.5 + 42.4 + 1) \text{ Pa} = 65{,}900 \text{ Pa} = 65.9 \text{ kPa}$$

11–9 MULTIPASS HEAT EXCHANGER NETWORKS

Two or more heat exchanger units are commonly combined to form a multipass heat exchanger network in order to achieve higher levels of overall thermal performance or to satisfy pressure drop requirements. For purpose of analysis, systems involving multiple heat exchanger units are classified according to whether or not both fluid streams entering each unit are mixed (i.e., category I, both fluids mixed; category II, one or both of the fluids unmixed), as well as several other factors.

Solution results for multipass heat exchanger networks consisting of M identical units are generally presented in the literature in the context of the effectiveness method. Such solution results are readily transformed to the *LMTD* or efficiency formats by employing the relations listed in Table 11–7.

11–9–1 Category I: Both Fluids Mixed between Passes

Category I multipass heat exchanger networks, which normally involve double-pipe and shell-and-tube heat exchanger units, are further categorized according to whether

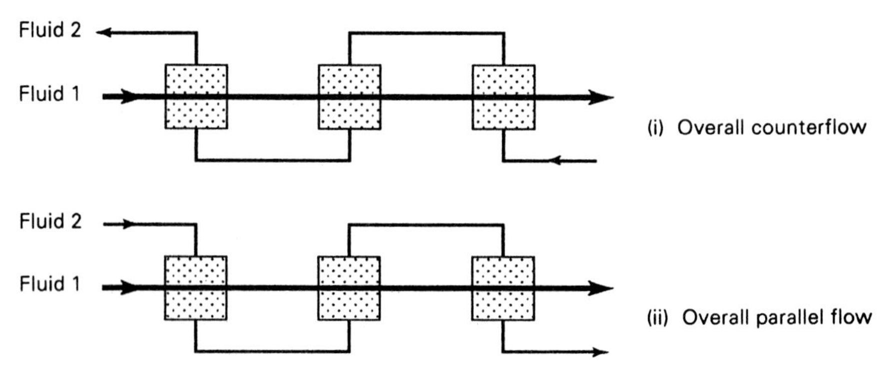

(a) Both streams in series.

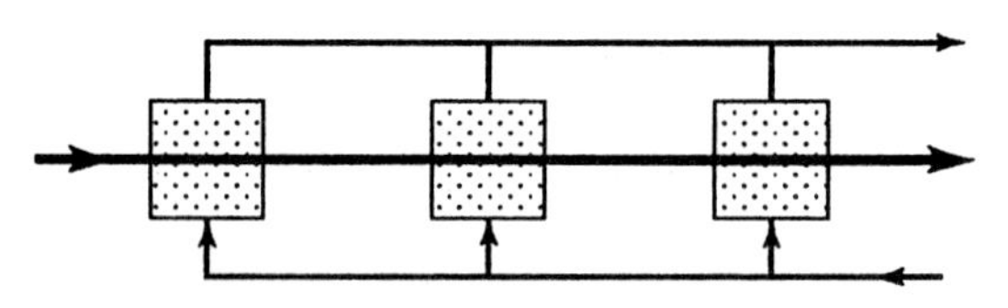

(b) One stream in series, one stream in parallel.

FIGURE 11–52 Multipass ($M = 3$) heat exchanger arrangements with both fluid streams mixed between passes (Category I). (Solution results for this arrangement are independent of the direction of flow streams.)

the units are arranged in (a) series–parallel or (b) series, and in (i) overall counterflow or (ii) overall parallel flow, as illustrated in Fig. 11–52.

To provide a basis for employing the effectiveness method for analyzing category I multipass networks, we designate the overall network parameters by P, NTU_I, and R and the individual unit parameters by P_1, $NTU_{I,1}$, and R_1.† General relations for the overall effectiveness P of the multipass heat exchanger networks shown in Fig. 11–52 are given as a function of P_1, R, and M in Table 11–11. This table also indicates the relations between NTU_I and $NTU_{I,1}$ and R and R_1 for these basic arrangements.

The relations given in Table 11–11 can be used to analyze category I networks consisting of multiple identical double-pipe, shell-and-tube, or crossflow heat exchanger units. To illustrate the use of the relations in this table, we consider an overall counterflow series arrangement for which

$$P = \frac{[(1 - P_1R)/(1 - P_1)]^M - 1}{[(1 - P_1R)/(1 - P_1)]^M - R} \quad \text{or} \quad P_1 = \frac{[(1 - PR)/(1 - P)]^{1/M} - 1}{[(1 - PR)/(1 - P)]^{1/M} - R} \tag{11–219,220}$$

and

† The solution results for series–parallel arrangements are presented in the ϵ-NTU format.

TABLE 11–11 Relations for overall effectiveness for multipass heat exchanger systems of category I [6]†

Flow arrangement	*Relations*	
Both streams in series, overall counterflow	$P = \dfrac{[(1 - P_1R)/(1 - P_1)]^M - 1}{[(1 - P_1R)/(1 - P_1)]^M - R}$ $P_1 = \dfrac{[(1 - PR)/(1 - P)]^{1/M} - 1}{[(1 - PR)/(1 - P)]^{1/M} - R}$ $NTU_I = M\ NTU_{I,1} \qquad R = R_1$	Derived in Appendix N–8
Both streams in series, overall parallel flow	$P = \dfrac{1}{1 + R}\{1 - [1 - P_1(1 + R)]^M\}$ $P_1 = \dfrac{1}{1 + R}\{1 - [1 - P(1 + R)]^{1/M}\}$ $NTU_I = M\ NTU_{I,1} \qquad R = R_1$	Derived in Appendix N–8
Series–parallel streams, C_{min} stream in parallel, C_{max} stream in series	$\epsilon = \dfrac{1}{C^*}\left[1 - \left(1 - \dfrac{\epsilon_1 C^*}{M}\right)^M\right]$ $\epsilon_1 = \dfrac{1}{C_1^*}[1 - (1 - M\epsilon C_1^*)^{1/M}]$ $NTU = NTU_1 \qquad C^* = MC_1^*$	
Series–parallel streams, C_{min} stream in series, C_{max} stream in parallel	For $C^* \geq 1/M$ $\epsilon = 1 - \left(1 - \dfrac{\epsilon_1}{MC^*}\right)^M$ $\epsilon_1 = \dfrac{1}{C_1^*}[1 - (1 - \epsilon)^{1/M}]$ $NTU = \dfrac{NTU_1}{C^*} \qquad C^* = \dfrac{1}{MC_1^*}$ For $C^* \leq 1/M$ $\epsilon = 1 - (1 - \epsilon_1)^M$ $\epsilon_1 = 1 - (1 - \epsilon)^{1/M}$ $NTU = M\ NTU_1 \qquad C^* = \dfrac{C_1^*}{M}$	

† A system consists of M identical units, each with the same value of effectiveness P_1 or ϵ_1.

$$NTU_I = M\ NTU_{I,1} \qquad R = R_1 \tag{11–221,222}$$

These expressions can be used in conjunction with the relations listed in Table A–N–1–1 to obtain functional relations of the form $P(NTU_I,R)$ or $NTU_I(P,R)$. More specifically, using Eq. (e) in Table A–N–1–1(a) for a 1–2 exchanger (TEMA-E shell-and-tube), P is represented by Eq. (11–219) with P_1 given by

$$P_1 = \frac{2}{1 + R + \sqrt{1 + R^2}\,(1 + e^{-\Gamma_1})/(1 - e^{-\Gamma_1})} \tag{11–223a}$$

where

$$\Gamma_1 = NTU_{I,1}\sqrt{1 + R^2} = \frac{NTU_I}{M}\sqrt{1 + R^2} \tag{11–223b}$$

Alternatively, to express NTU_I in terms of P and R, we write

$$NTU_{I,1} = \frac{1}{\sqrt{1 + R^2}} \ln \frac{2 - P_1(1 + R - \sqrt{1 + R^2})}{2 - P_1(1 + R + \sqrt{1 + R^2})} \tag{11-224}$$

from Eq. (e) in Table A–N–1–1(b), which indicates

$$NTU_I = M\, NTU_{I,1} = \frac{M}{\sqrt{1 + R^2}} \ln \frac{2 - P_1(1 + R - \sqrt{1 + R^2})}{2 - P_1(1 + R + \sqrt{1 + R^2})} \tag{11-225}$$

with P_1 represented in terms of P by Eq. (11–220).

Representative calculations for P are shown in Fig. 11–53 for series arrangements of 1-2 exchangers. These calculations indicate significant increases in overall effectiveness P with increasing values of M for overall counterflow. Thus, P would be significantly lower for a single 1-2 exchanger (with surface area A_s) than for two or more 1-2 exchangers (with $A_{s,1} = A_s/M$) arranged in series with overall counterflow. On the other hand, the calculations for P for 1–2 exchangers arranged in series with overall parallel flow actually decrease slightly with increasing values of M. In connection with series–parallel arrangements, which are commonly used to reduce the overall pressure drop of a stream, calculations result in very slight increases in effectiveness ϵ with increasing values of M.

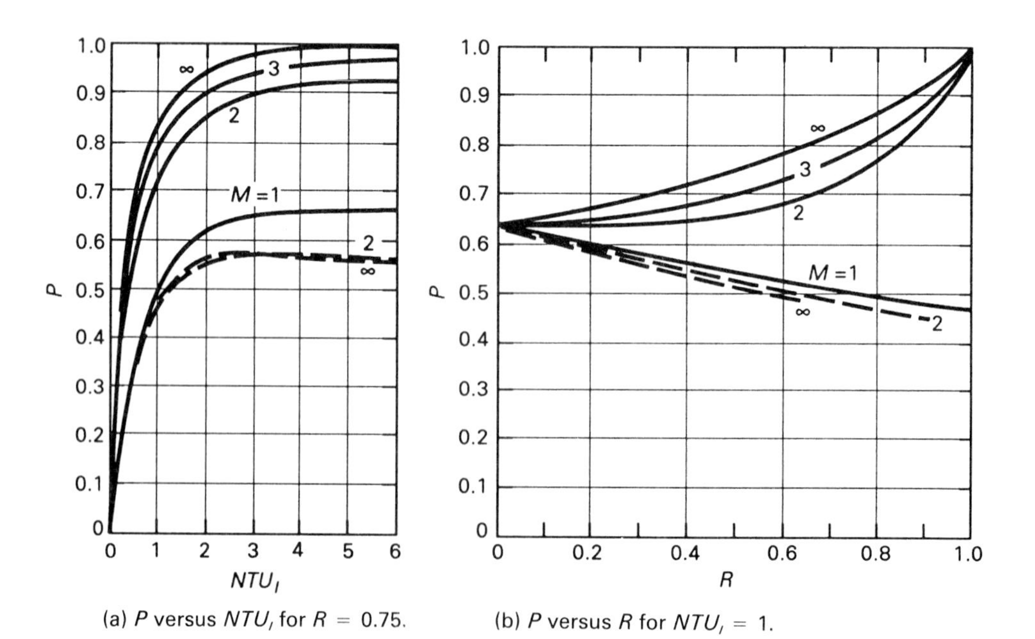

(a) P versus NTU_I for $R = 0.75$. (b) P versus R for $NTU_I = 1$.

FIGURE 11–53 Effectiveness P for multipass series arrangement of 1-2 exchangers: ——— overall counterflow; — — — overall parallel flow.

Curves similar to those shown in Fig. 11–53 can be drawn for multipass crossflow systems of category I.

Whereas the relations listed in Table 11–11 can also be used to analyze category I networks of M double-pipe heat exchanger units, the performance of a network of double-pipe units (with individual surface area $A_{s,1}$) arranged in series is equivalent to the performance of a single double-pipe heat exchanger with surface area A_s equal to $M\,A_{s,1}$. It follows that a series arrangement of double-pipe heat exchangers can be analyzed by treating the network as an individual double-pipe heat exchanger unit, such that the practical relations for heat-transfer rate q (commonly expressed in terms of effectiveness, $LMTD$, or efficiency) that are applicable to individual units can be applied to the network. It also follows that the relations represented by $P(NTU_I,R)$ for overall effectiveness for series arrangements of double-pipe heat exchangers are independent of the number of units M.

EXAMPLE 11–20

A 30° API petroleum distillate oil with mass-flow rate of 4.36 kg/s is to be cooled in a series–parallel arrangement of two double-pipe hairpin heat exchanger units, as shown in Fig. E11–20. The oil is cooled from 114°C to 66°C using cooling water with inlet temperature of 27°C and mass-flow rate of 3.5 kg/s. The allowable pressure drop for each stream is approximately 82.5 kPa (12 psi) and the required minimum fouling factors are specified as 0.0007 m² °C/W for the oil and 0.00015 m² °C/W for the water. Determine the required length of the heat exchanger and whether or not the pressure drop requirements are satisfied.

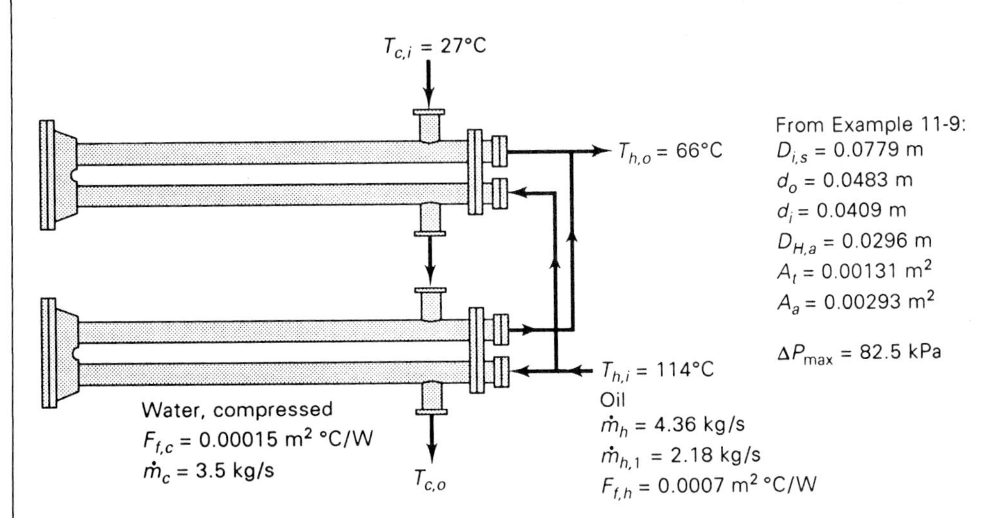

FIGURE E11–20 Application involving series–parallel arrangement of two double-pipe hairpin heat exchanger units.

Solution

Objective Determine the unit length L and whether or not the pressure drop requirements are satisfied.

Assumptions/Conditions

equal values of $\overline{U}$ in both exchangers
negligible entrance effects
moderate property variation
standard conditions

Properties

30° API oil at 90°C: $\rho = 820$ kg/m³, $\mu = 0.00245$ kg/(m s), $c_P = 2.17$ kJ/(kg °C), $k = 0.128$ W/(m °C), $Pr = 41.5$.

Water at $T = 27$°C $= 300$ K (Table A–C–3): $\rho = 997$ kg/m³, $\mu = 8.55 \times 10^{-4}$ kg/(m s), $c_P = 4.18$ kJ/(kg °C), $k = 0.613$ W/(m °C), $Pr = 5.83$.

Wrought steel pipe (Table A–C–1): $k = 59$ W/(m °C).

Analysis Comparing the specifications indicated for this example and Example 11–9, it should be noted that both involve individual units with the same annular and tubular mass-flow rates and fluid properties (assuming approximate uniform property conditions). It follows that the bulk-stream velocities, friction factors, and overall coefficient of heat transfer can be approximated by

$$(U_{b,h})_1 = 2.03 \text{ m/s} \qquad U_{b,c} = 1.2 \text{ m/s}$$

$$f_h = 0.00603 \qquad f_c = 0.00548 \qquad \overline{U} = 459 \text{ W/(m}^2 \text{ °C)}$$

from Example 11–9. The rate of heat transfer is calculated by writing

$$q = (\dot{m}c_P)_h(T_{h,i} - T_{h,o})$$

$$= 4.36\,\frac{\text{kg}}{\text{s}}\left(2.17\,\frac{\text{kJ}}{\text{kg °C}}\right)(114\text{°C} - 66\text{°C}) = 454 \text{ kW}$$

which is twice the value associated with Example 11–9.

Referring to Table 11–11, the relationship between the unit effectiveness ϵ_1 and the overall effectiveness ϵ depends on which stream has the minimum capacity rate $C_{\min}$. The values of the capacity rates for the two streams are given by

$$C_{h,1} = \frac{C_h}{2} = \frac{4.36 \text{ kg/s}}{2}\left(2.17\,\frac{\text{kJ}}{\text{kg °C}}\right) = 4.73 \text{ kW/°C} \qquad \text{parallel stream}$$

$$C_{c,1} = C_c = 3.5\,\frac{\text{kg}}{\text{s}}\left(4.18\,\frac{\text{kJ}}{\text{kg °C}}\right) = 14.6 \text{ kW/°C} \qquad \text{series stream}$$

Thus, $C_{\min}$ is associated with the parallel stream, such that

$$C_{\min,1} = C_{h,1} = 4.73 \text{ kW/°C} \qquad C_1^* = \frac{4.73}{14.6} = 0.324$$

and

$$\epsilon_1 = \frac{1}{C_1^*}[1 - (1 - M\epsilon C_1^*)^{1/M}] \qquad \text{with } M = 2$$

from Table 11–11. The overall effectiveness ϵ is expressed in terms of the terminal temperatures by

$$\epsilon = \frac{T_{h,i} - T_{h,o}}{T_{h,i} - T_{c,i}} = \frac{114°\text{C} - 66°\text{C}}{114°\text{C} - 27°\text{C}} = 0.552$$

It follows that

$$\epsilon_1 = \frac{1}{0.324}\{1 - [1 - 2(0.552)(0.324)]^{1/2}\} = 0.613$$

and, upon substitution into Eq. (11–144),

$$NTU_1 = \frac{1}{1 - C_1^*}\ln\frac{1 - \epsilon_1 C_1^*}{1 - \epsilon_1} = \frac{1}{1 - 0.324}\ln\frac{1 - (0.613)(0.324)}{1 - 0.613} = 1.08$$

Solving for the surface area $A_{s,1}$, we obtain

$$A_{s,1} = NTU_1\frac{C_{\min,1}}{\overline{U}} = 1.08\,\frac{4.73 \text{ kW/°C}}{459 \text{ W/(m}^2\text{ °C)}} = 11.1 \text{ m}^2$$

and

$$L_1 = \frac{A_{s,1}/2}{\pi d_o} = \frac{11.1 \text{ m}^2}{2\pi(0.0483 \text{ m})} = 36.6 \text{ m}$$

$$L_{t,1} = 2L_1 = 73.2 \text{ m} \qquad L_{a,1} = 2L_1 = 73.2 \text{ m} \qquad L_a = 2L_{a,1} = 146 \text{ m}$$

We now turn to the calculation of the pressure drops for the two streams.

Hot tubular fluid—oil (parallel flow with $M = 1$)

$$(\Delta P_t)_{1-2} = \left(\frac{\rho U_{b,1}^2}{2}\right)_t\left[\left(\frac{4L_1}{D_H}f\right)_t + 1.5M\right]$$

$$= \frac{1}{2}\left(820\,\frac{\text{kg}}{\text{m}^3}\right)\left(2.03\,\frac{\text{m}}{\text{s}}\right)^2\left[\frac{4(73.2 \text{ m})}{0.0409 \text{ m}}\,0.00603 + 1.5\right]$$

$$= 75{,}500 \text{ kg/(m s}^2) = 75.5 \text{ kPa}$$

Cold annular fluid—water (series flow with $M = 2$)

$$(\Delta P_a)_{1-2} = \left(\frac{\rho U_b^2}{2}\right)_a \left[\left(\frac{4L}{D_H} f\right)_a + 0.5M\right]$$

$$= \frac{1}{2}\left(997\ \frac{\text{kg}}{\text{m}^3}\right)\left(1.2\ \frac{\text{m}}{\text{s}}\right)^2 \left[\frac{4(146\ \text{m})}{0.0296\ \text{m}}\ 0.00548 + 0.5(2)\right]$$

$$= 78{,}300\ \text{kg/(m s}^2\text{)} = 78.3\ \text{kPa}$$

These calculations are within the specified limit and indicate good utilization of the available pressure drop for both streams.

EXAMPLE 11-21

Two 1-2 exchangers with 8-m^2 surface area each are arranged in series with overall counterflow. The specific heat of each fluid is 2.5 kJ/(kg °C), with the mass-flow rates and inlet temperatures being 2.5 kg/s and 150°C for the shell side and 1.87 kg/s and 25°C for the tube side. The mean overall coefficient of heat transfer is given as 585 W/(m^2 °C). Determine the total rate of heat transfer and the outlet temperatures for the network.

Solution

Objective Determine q, $T_{h,o}$, and $T_{c,o}$

Schematic A two unit ($M = 2$) 1-2 exchanger network: series arrangement with overall counterflow.

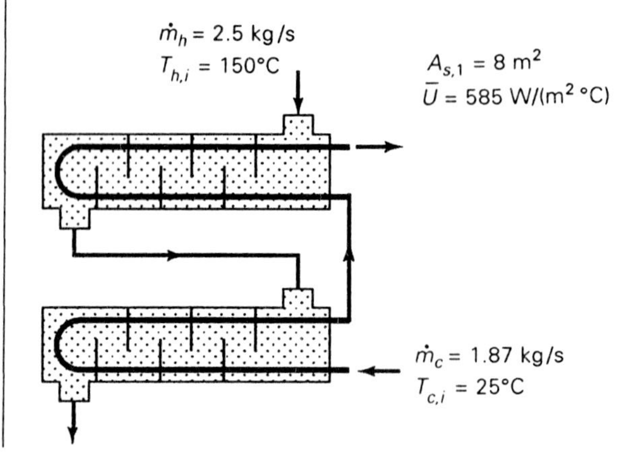

Assumptions/Conditions

uniform distribution in U
uniform properties
standard conditions

Properties Both fluids: $c_P = 2.5$ kJ/(kg °C).

Analysis Following the effectiveness approach, we write†

$$C_c = 1.87 \frac{\text{kg}}{\text{s}} \left(2.5 \frac{\text{kJ}}{\text{kg °C}}\right) = 4.68 \text{ kW/°C}$$

$$C_h = 2.5 \frac{\text{kg}}{\text{s}} \left(2.5 \frac{\text{kJ}}{\text{kg °C}}\right) = 6.25 \text{ kW/°C}$$

Selecting the cold fluid as the reference stream I, we write

$$R = \frac{C_c}{C_h} = \frac{4.68}{6.25} = 0.749$$

$$NTU_{I,1} = \frac{\overline{U}A_{s,1}}{C_c} = \frac{[585 \text{ W/(m}^2 \text{ °C)}](8 \text{ m}^2)}{4.68 \text{ kW/°C}} = 1 \qquad NTU_I = M\, NTU_{I,1} = 2$$

and, upon substituting into Eq. (11–168),

$$\Gamma_1 = NTU_{I,1}\sqrt{1 + R^2} = 1\sqrt{1 + 0.749^2} = 1.25$$

$$P_1 = \frac{2}{1 + 0.749 + \sqrt{1 + 0.749^2}\,(1 + e^{-1.25})/(1 - e^{-1.25})} = 0.5$$

We are now able to calculate the overall network effectiveness P by writing

$$P = \frac{\{[1 - 0.5(0.749)]/(1 - 0.5)\}^2 - 1}{\{[1 - 0.5(0.749)]/(1 - 0.5)\}^2 - 0.749} = \frac{1.56 - 1}{1.56 - 0.749} = 0.692$$

from Eq. (11–219). It follows that

$$q = PC_c(T_{h,i} - T_{c,i}) = 0.692\left(4.68 \frac{\text{kW}}{\text{°C}}\right)(150\text{°C} - 25\text{°C}) = 405 \text{ kW}$$

$$T_{h,o} = T_{h,i} - \frac{q}{C_h} = 150\text{°C} - \frac{405 \text{ kW}}{6.25 \text{ kW/°C}} = 85.2\text{°C}$$

$$T_{c,o} = T_{c,i} + \frac{q}{C_c} = 25\text{°C} + \frac{405 \text{ kW}}{4.68 \text{ kW/°C}} = 112\text{°C}$$

† A similar problem is solved by the ψ-P method in reference 7.

These calculations indicate $\psi = P/NTU_I = 0.346$ and $\psi_1 = P_1/NTU_{I,1} = 0.5$, such that the minimum performance criterion $\psi \gtrless \psi_{MIN} = 0.35$ is satisfied for the network as well as for each individual unit. However, since $T_{h,o} < T_{c,o}$, the alternative outlet temperature criterion is not satisfied for the network.

For purpose of comparison, calculations for the performance of a single 1-2 heat exchanger unit with total surface area equal to 16 m^2 indicate $P = 0.621$ and $q =$ 363 kW, which lie 11.4% below the values obtained for the two-unit network. This result is consistent with the calculations shown in Fig. 11–53, which show that P is greater for series overall counterflow arrangements of 1-2 exchangers than for a single 1-2 exchanger with the same total surface area.

EXAMPLE 11–22

Referring to Example 11–18, determine the required surface area for a two-pass cross-counterflow series arrangement, with both fluids mixed between passes.

Solution

Objective Calculate A_s for a category I network.

Schematic Two-pass cross-counterflow series network: category I.

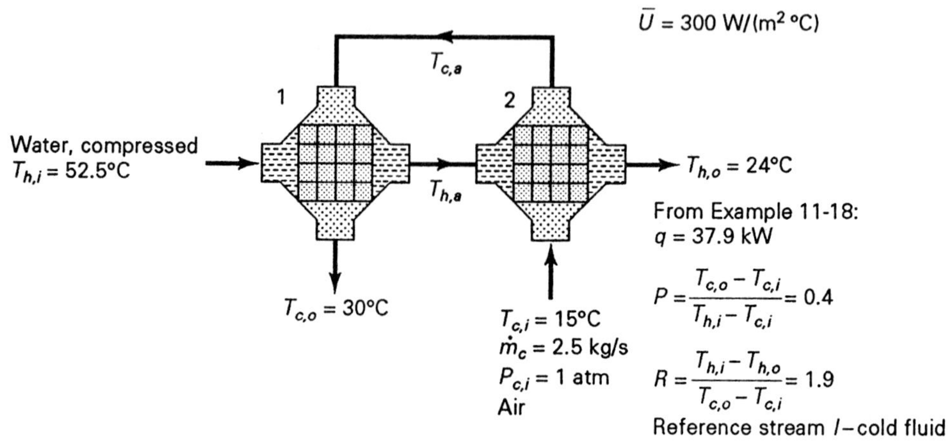

Assumptions/Conditions

uniform distribution in U
uniform properties
standard conditions

Properties Air at 15°C (Table A–C–5): c_P = 1.01 kJ/(kg °C).

Analysis Referring to Table 11–11, the relation between overall effectiveness P and unit effectiveness P_1 for this two-pass arrangement is represented by†

$$P_1 = \frac{[(1 - PR)/(1 - P)]^{1/M} - 1}{[(1 - PR)/(1 - P)]^{1/M} - R} \qquad NTU_I = M\ NTU_{I,1} \qquad \text{(a,b)}$$

where $R = R_1$ and $M = 2$. Substituting $P = 0.4$ and $R = 1.9$ into this equation, we obtain

$$P_1 = \frac{\{[1 - 0.4(1.9)]/(1 - 0.4)\}^{1/2} - 1}{\{[1 - 0.4(1.9)]/(1 - 0.4)\}^{1/2} - 1.9} = \frac{0.632 - 1}{0.632 - 1.9} = 0.29$$

Setting $P_1 = 0.29$ and $R_1 = R = 1.9$ in Fig. 11–44(a), we conclude that $NTU_{I,1} \simeq 0.5$. This value for $NTU_{I,1}$ can be refined by iterating on the analytical solution, with the result $NTU_{I,1} = 0.546$. It follows that

$$NTU_I = 2\ NTU_{I,1} = 2(0.546) = 1.09$$

We also note that

$$C_I = C_c = (\dot{m}c_P)_c = 2.5\,\frac{\text{kg}}{\text{s}}\left(1.01\,\frac{\text{kJ}}{\text{kg °C}}\right) = 2.52 \text{ kW/°C}$$

Thus, we are able to write

$$A_s = NTU_I\frac{C_c}{\overline{U}} = 1.09\,\frac{2.52 \text{ kW/°C}}{300 \text{ W/(m}^2\text{ °C)}} = 9.16 \text{ m}^2 \qquad A_{s,1} = \frac{A_s}{2} = 4.58 \text{ m}^2$$

$$q'' = \frac{q}{A_s} = \frac{37.9 \text{ kW}}{9.16 \text{ m}^2} = 4.14 \text{ kW/m}^2$$

and

$$\psi = \frac{P}{NTU_I} = 0.367 \qquad \psi_1 = \frac{P_1}{NTU_{I,1}} = 0.531$$

Referring to Example 11–18, the overall heat flux for the two-pass network is nearly 10% greater than the heat flux for the single exchanger. In addition, we observe that the minimum performance criterion $\psi > \psi_{MIN} = 0.35$ is satisfied for the two-pass network and for each unit.

As a final point of interest, we evaluate the performance of the individual units by writing

$$P_1 = \frac{T_{c,o} - T_{c,a}}{T_{h,i} - T_{c,a}} = \frac{30\text{°C} - T_{c,a}}{52.5\text{°C} - T_{c,a}} = 0.29 \qquad \text{for exchanger 1}$$

$$P_1 = \frac{T_{c,a} - T_{c,i}}{T_{h,a} - T_{c,i}} = \frac{T_{c,a} - 15\text{°C}}{T_{h,a} - 15\text{°C}} = 0.29 \qquad \text{for exchanger 2}$$

† Since R_1 and $NTU_{I,1}$ are the same for each exchanger, the unit effectiveness P_1 is also the same for each exchanger.

which indicate $T_{h,a} = 35°\text{C}$ and $T_{c,a} = 20.7°\text{C}$. It follows that

$$q_1 = P_1 C_c (T_{h,i} - T_{c,a}) = 0.290\left(2.52 \frac{\text{kW}}{°\text{C}}\right)(52.5°\text{C} - 20.7°\text{C}) = 23.3 \text{ kW}$$

and†

$$q_2 = P_1 C_c (T_{h,a} - T_{c,i}) = 0.290\left(2.52 \frac{\text{kW}}{°\text{C}}\right)(35°\text{C} - 15°\text{C}) = 14.6 \text{ kW}$$

Thus, although the outlet temperature criterion is violated for the network (i.e., $T_{c,o} > T_{h,o}$), this criterion is satisfied for the individual units (i.e., $T_{c,o} = 30°\text{C} < T_{h,a} = 35°\text{C}$; $T_{c,a} = 20.7°\text{C} < T_{h,o} = 24°\text{C}$).

TABLE 11–12 Relations for overall effectiveness for representative two-pass crossflow heat exchanger networks of category II [55]

Arrangement	*Relation*
Cross-counterflow 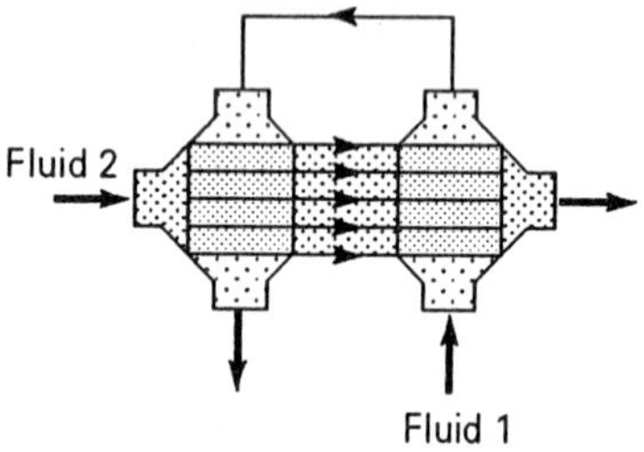 Fluid 2 Fluid 1 Fluid 1 mixed throughout Fluid 2 unmixed throughout—inverted order‡	$P = \frac{1}{R}\left[1 - \frac{1}{\Gamma/2 + (1 - \Gamma/2)\exp(2\Gamma R)}\right]$ $\Gamma = 1 - \exp(-NTU_t/2)$
Cross-parallel flow 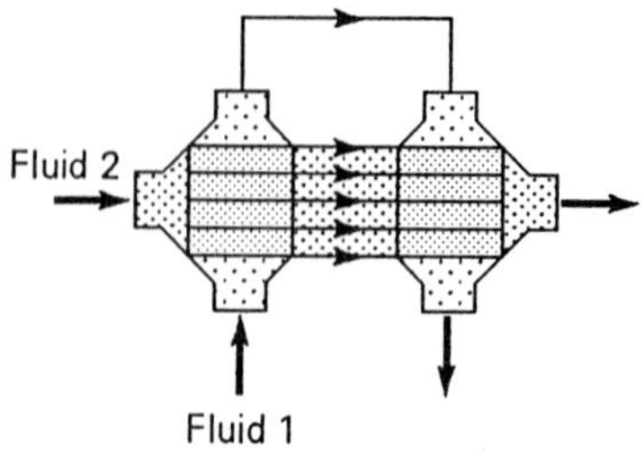 Fluid 2 Fluid 1 Fluid 1 mixed throughout Fluid 2 unmixed throughout—inverted order‡	$P = \frac{1}{R}\left(1 - \frac{\Gamma}{2}\right)[1 - \exp(-2\Gamma R)]$ $\Gamma = 1 - \exp(-NTU_t/2)$

† Equivalently, $q_2 = q - q_1 = 37.9 \text{ kW} - 23.3 \text{ kW} = 14.6 \text{ kW}$.

‡ The arrangements are categorized as nonsymmetrical exchangers. To use the shell-side fluid as the reference stream I, use indicated formula with R replaced by $1/R$, NTU_t replaced by $R\ NTU_s$, and P replaced by PR.

11–9–2 Category II: Unmixed Fluid(s) between Passes

Category II multipass heat exchanger networks, which usually involve crossflow heat exchanger units, are generally further categorized according to (1) whether the network is arranged in overall counterflow or parallel flow, (2) whether one or both of the fluids within individual units are mixed or unmixed, (3) whether one or both of the fluids between passes are unmixed, and (4) whether fluid in streams that are unmixed between passes is distributed from one exchanger to the next in identical or inverted order, as illustrated in Fig. 11–54. Relations for two representative multipass crossflow heat exchanger networks of this kind are listed in Table 11–12. A more extensive listing of solutions is provided for effectiveness ϵ and efficiency ψ in references 7 and 54.

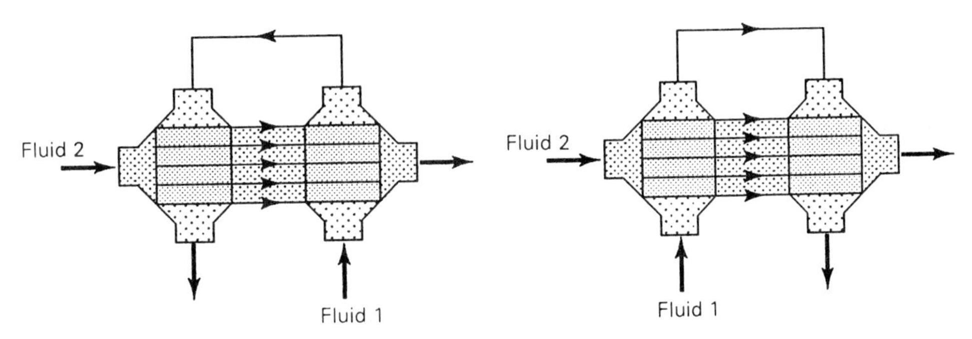

(a) Cross-counterflow: fluid 1 mixed, fluid 2 unmixed throughout—inverted order.

(b) Cross-parallel flow: fluid 1 mixed, fluid 2 unmixed throughout—inverted order.

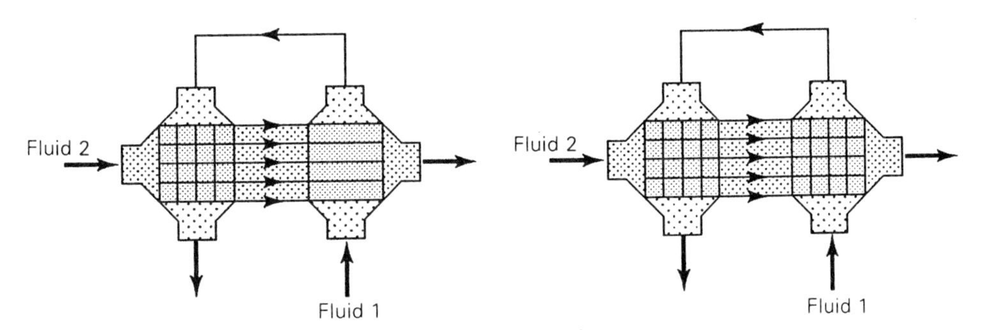

(c) Cross-counterflow: fluid 1 unmixed only in one pass, fluid 2 unmixed throughout—inverted order.

(d) Cross-counterflow: fluid 1 unmixed in each pass and mixed between passes, fluid 2 unmixed throughout—inverted order.

FIGURE 11–54 Representative two-pass crossflow heat exchanger networks—category II.

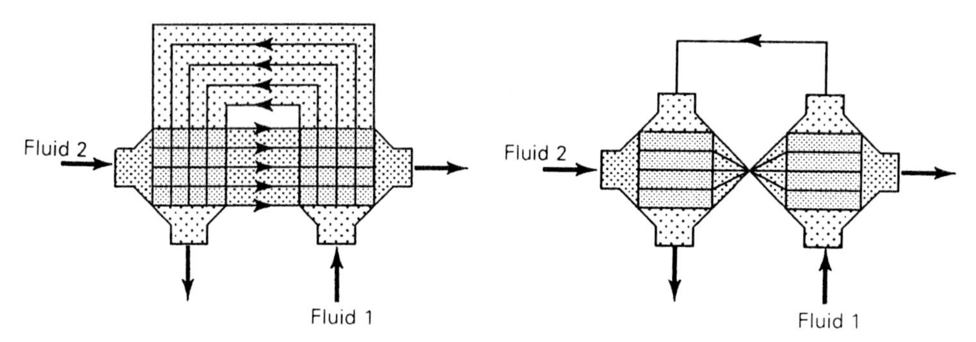

(e) Cross-counterflow: fluid 1 unmixed throughout—inverted order, fluid 2 unmixed throughout—inverted order.

(f) Cross-counterflow: fluid 1 mixed throughout, fluid 2 unmixed throughout—identical order.

FIGURE 11–54 *(Continued)*

EXAMPLE 11–23

Referring to the two-pass cross-counterflow heat-exchanger network shown in Fig. E11–23, determine the heat-transfer rate and the outlet temperatures.

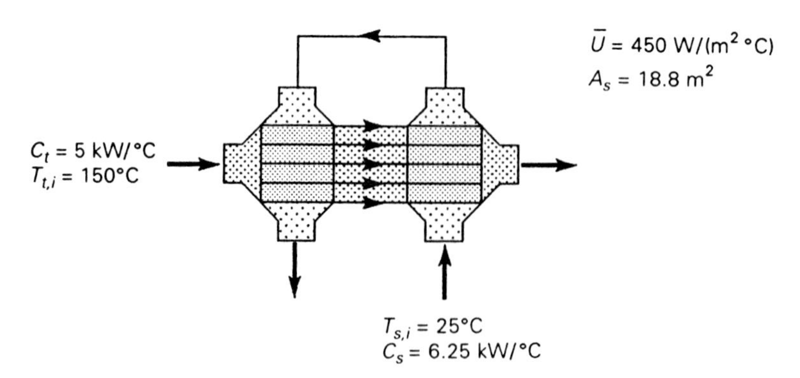

FIGURE E11–23 Two-pass cross-counterflow heat-exchanger network: category II—inverted order.

Solution

Objective Determine q, $T_{t,o}$, and $T_{s,o}$.

Assumptions/Conditions

uniform distribution in U
uniform properties
standard conditions

Analysis Following the effectiveness approach with the tube-side fluid t taken as the reference stream I, we calculate R and NTU_t.

$$R = \frac{C_t}{C_s} = \frac{5 \text{ kW/°C}}{6.25 \text{ kW/°C}} = 0.8$$

$$NTU_t = \frac{\overline{U}A_s}{C_t} = \frac{[450 \text{ W/(m}^2 \text{ °C)}](18.8 \text{ m}^2)}{5 \text{ kW/°C}} = 1.69$$

Substituting these values into the equations listed in Table 11–12 for category II networks, we obtain

$$\Gamma = 1 - \exp\left(-\frac{NTU_t}{2}\right) = 1 - \exp\left(-\frac{1.69}{2}\right) = 0.57$$

$$P = \frac{1}{R}\left[1 - \frac{1}{\Gamma/2 + (1 - \Gamma/2)\exp(2\Gamma R)}\right]$$

$$= \frac{1}{0.8}\left\{1 - \frac{1}{0.57/2 + (1 - 0.57/2)\exp[2(0.57)(0.8)]}\right\} = 0.645$$

It follows that

$$q = PC_t(T_{s,i} - T_{t,i}) = 0.645\left(5\frac{\text{kW}}{\text{°C}}\right)(150\text{°C} - 25\text{°C}) = 403 \text{ kW}$$

and

$$T_{t,o} = 150\text{°C} - \frac{403 \text{ kW}}{5 \text{ kW/°C}} = 69.4\text{°C} \qquad T_{s,o} = 25\text{°C} + \frac{403 \text{ kW}}{6.25 \text{ kW/°C}} = 89.5\text{°C}$$

By following through with calculations for the case in which a single crossflow heat exchanger with the same total surface area is used, we obtain $P = 0.599$, which is 7.89% below the value achieved by use of the two-pass arrangement.

11–10 COMPUTER ANALYSIS OF HEAT EXCHANGERS: INTRODUCTION

Computer codes are commonly used in the evaluation and design of heat exchangers. The types of numerical schemes used in computer analysis can be categorized according to the scale of the control volume and nature of the analysis; these include (1) lumped methods, (2) lumped/incremental methods, (3) cell methods, and (4) computational fluid dynamics (CFD) methods. The lumped, lumped/incremental and cell-type codes feature the use of standard and proprietary correlations for the coefficients of heat transfer and friction. On the other hand, the CFD approach actually provides a basis for predicting these coefficients, as well as solving for the heat-transfer rate, pressure drop, and distributions in temperature and velocity throughout the heat exchanger.

Computer codes used in heat exchanger analysis are commonly classified as *rating* programs or *design* programs, both of which can be used in the design of commercial exchangers. Rating programs, that calculate the thermal/hydraulic performance for fully specified heat exchanger geometry, can be used to (1) evaluate the performance of existing exchangers for various operating conditions, and (2) design an exchanger by the alteration of one or more selected geometric parameters (e.g., shell size, baffle spacing, tube passes, etc.) until a satisfactory result is achieved. As pointed out by Taborek [9], this *interactive* design approach keeps the experienced engineer in full control of the design process. Design programs perform calculations based on buyer specifications for the process and basic elements of the exchanger geometry (usually tube dimensions, fixed or maximum length, shell and baffle type) by systematic variation in designated parameters (commonly shell size, baffle spacing, tube passes). Ideally, design programs should display intermediate results at each step in the design process in order to provide a means for evaluating the criteria upon which the design is based and for exercising engineering judgment [9]. A number of computer programs that are commercially available for the design of heat exchangers are described by Breber [56].

It should be noted that the traditional approach to complete design of shell-and-tube heat exchangers in the industry has involved two or more separate operations, using independent specialized codes for (1) thermal design and (2) mechanical design.† However, over the past few years several integrated software packages have become available that treat heat exchanger design as a single unified process, rather than as a series of separate steps. This integrated approach eliminates the need for entering data more than once and assures the designer of full compatibility between the thermal and mechanical design and feasibility of construction.

11–10–1 Lumped Approach

Simple numerical codes of the lumped type are used in the analysis of heat exchangers that are assumed to operate under steady-state conditions with uniform properties, uniform bulk-stream temperatures at any cross section, and uniform local overall coefficient of heat transfer U. Such codes generally include information pertaining to the effects of fins, fouling, baffles, multiple-tube passes, crossflow, two-phase flow, and related factors and provide the basis for developing efficient thermal and hydraulic evaluation and design calculations for heat exchangers with operating conditions that satisfy the necessary restrictions.

† Specialized design codes for economic optimization have also been in common use.

11–10–2 Lumped/Incremental Approach

Lumped/incremental types of numerical codes provide an effective means of adapting the practical analysis approach to the evaluation and design of heat exchangers. Codes of this type range from the relatively simple, which employ basic iterative (and in some cases noniterative) procedures and standard convection correlations from the open literature, to the more sophisticated, which involve multiple iterative loops, utilize convection correlations that are refined or developed on the basis of in-house experimental investigations and/or other proprietary experience and input, contain the latest technology, have been thoroughly tested on pilot plant data and/or field cases from industry, and are backed up by extensive consultancy and feedback supporting services. Many of the codes that are commercially available are able to handle combinations of boiling, condensing, and single-phase heat transfer, pure fluids or multicomponent mixtures, variable properties, and provide built-in thermophysical properties data banks, and elementary flow-induced vibration analysis.

To illustrate the lumped/incremental numerical approach, simple finite-difference programming schemes for analyzing double-pipe, 1-2 shell-and-tube, and mixed/unmixed crossflow heat exchangers operating under conditions of uniform overall coefficient of heat transfer U and uniform properties are presented in Figs. A–G–5 to A–G–8 of the Appendix.† These programs are developed in the context of Examples 11–4, 11–5 and 11–6 for double-pipe exchangers, Example 11–13 for 1-2 shell-and-tube exchangers, and Example 11–17 for mixed/unmixed crossflow exchangers. In addition to providing calculations that are in agreement with the analytical solutions given in these examples, both programs can be adapted to situations involving nonuniform distributions in U and variable properties by following the more general steps outlined in Example 11–24.

EXAMPLE 11–24

Develop a lumped/incremental finite-difference formulation that accounts for variable properties and variable overall coefficient of heat transfer over the length of a parallel flow double-pipe heat exchanger.

Solution

Objective Develop a lumped/incremental finite-difference formulation for a parallel-flow double-pipe heat exchanger.

Schematic Double-pipe heat exchanger: parallel flow arrangement.

† The rating program listed in Fig. A–G–7 for 1-2 shell-and-tube heat exchangers can be rewritten in a convenient design format by following the pattern presented in the design program listed in Fig. A–G–6 for double-pipe heat exchangers.

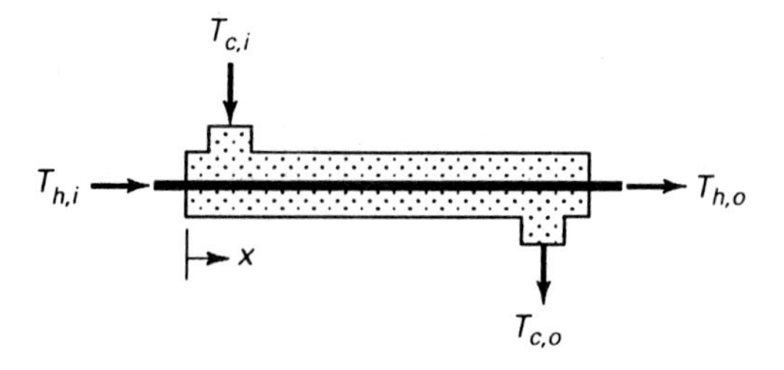

Assumptions/Conditions

nonuniform distribution in U
variable properties
standard conditions

Analysis To provided a framework for developing a one-dimensional finite-difference formulation, the heat exchanger is subdivided into M subvolumes along its length, as shown in Fig. E11–24. The individual subvolumes are designated by $m = 1, 2, \ldots, M$, each having an incremental length of $\Delta x = M/L$.

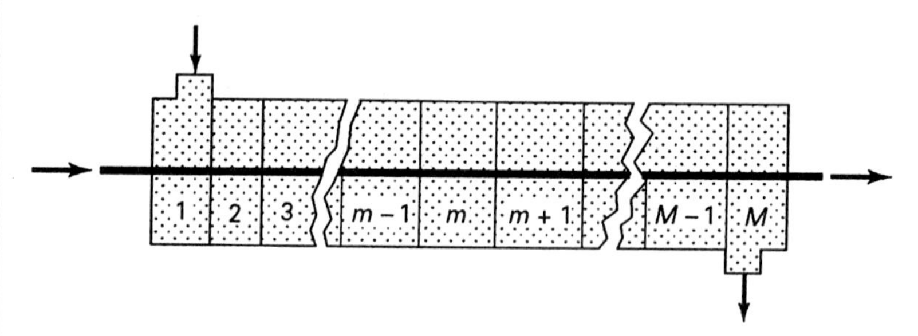

FIGURE E11–24 Finite-difference grid for lumped/incremental numerical analysis of double-pipe heat exchanger.

Focusing attention on subvolume m, the incremental rate of heat transfer Δq_m is expressed in terms of the local overall coefficient of heat transfer U_m by

$$\Delta q_m = U_m p \,\Delta x\,(T_h - T_c)_m \tag{a}$$

We are also able to write

$$\Delta q_m = C_{h,m}(T_{h,m} - T_{h,m+1}) \tag{b}$$

and

$$\Delta q_m = C_{c,m}(T_{c,m+1} - T_{c,m}) \tag{c}$$

[Notice that Eqs. (a), (b), and (c) correspond to Eqs. (11–110), (11–108), and (11–109).] With $C_{h,m}$, $C_{c,m}$, U_m, $T_{h,m}$ and $T_{c,m}$ known, Eqs. (a) to (c) provide a basis for calculating Δq_m, $T_{h,m+1}$, and $T_{c,m+1}$.

To account for the variable property effects on $h_{h,m}$, $h_{c,m}$, and U_m, we must also calculate the nodal surface temperatures $T_{h,w,m}$ and $T_{c,w,m}$. To evaluate these surface temperatures, we write

$$\Delta q_m = h_{h,m} p \, \Delta x \, (T_h - T_{h,w})_m \tag{d}$$

and

$$\Delta q_m = h_{c,m} p \, \Delta x \, (T_{c,w} - T_c)_m \tag{e}$$

Once the incremental heat transfer rate Δq_m is known for each subvolume, the rate of heat transfer q over the entire length L is obtained by writing

$$q = \sum_{m=1}^{M} \Delta q_m \tag{f}$$

Equations (a) to (f) provide a basis for calculating q and the outlet temperatures $T_{h,o} = T_{h,M}$ and $T_{c,o} = T_{c,M}$. The solution procedure is summarized as follows:

(1) set $m = 1$, $T_{h,1} = T_{h,i}$ and $T_{c,1} = T_{c,i}$;
(2) evaluate fluid properties at $T_{h,1}$ and $T_{c,1}$ and calculate $C_{h,1}$, $C_{c,1}$, $Re_{h,1}$, $Re_{c,1}$, $h_{h,1}$, $h_{c,1}$, and U_1;†
(3) calculate Δq_m from Eq. (a);
(4) calculate $T_{h,w,m}$ and $T_{c,w,m}$ from Eqs. (d) and (e);
(5) calculate $T_{h,m+1}$ and $T_{c,m+1}$ from Eqs. (b) and (c);
(6) evaluate fluid properties at $T_{h,m+1}$ and $T_{c,m+1}$ and calculate $C_{h,m+1}$, $C_{c,m+1}$, $Re_{h,m+1}$, $Re_{c,m+1}$, $h_{h,m+1}$, $h_{c,m+1}$, and U_{m+1} by representing the local bulk-stream temperatures by $T_{h,m+1}$ and $T_{c,m+1}$ and the local surface temperatures by $T_{h,w,m}$ and $T_{c,w,m}$ and using appropriate variable property correction factors;†
(7) continue steps (3) through (6) for $m = 2, 3, \ldots, M$; and
(8) set $T_{h,o} = T_{h,M}$ and $T_{c,o} = T_{c,M}$ and calculate q from Eq. (f).

As in the numerical solution of conduction-heat-transfer problems, to ensure proper accuracy calculations should be obtained for sufficiently large values of M (i.e., small values of Δx).

The approach outlined above can also be used in the analysis of counterflow arrangements, and can be adapted to shell-and-tube and crossflow heat exchangers.‡ However, for double-pipe and shell-and-tube heat exchangers with counterflow, the procedure involves iteration if the terminal temperatures at $x = 0$ are not specified. It should also be noted that the adaptation of this lumped/incremental approach to crossflow heat exchangers necessitates two-dimensional treatment.

† By neglecting the effects of surface temepratures on $h_{h,1}$, $h_{c,1}$, and U_1 and approximating the surface temperatures used to calculate $h_{h,m+1}$ and $h_{c,m+1}$ by $T_{h,w,m}$ and $T_{c,w,m}$, we are able to achieve a noniterative solution.

‡ The computer programs for uniform property analysis of double-pipe, shell-and-tube, and crossflow heat exchangers listed in Figs. A-G-5 to A-G-8 are patterned after this more general approach.

11–10–3 Cell Approaches

Numerical cell methods have been developed for the analysis of shell-and-tube and certain other types of heat exchangers that provide a means of accounting for nonuniform distributions over the cross section in bulk-stream temperature and overall coefficient of heat transfer U. Approaches of this type are particularly useful in analyzing baffled shell-side flow in shell-and-tube heat exchangers. As shown in Fig. 11–55, the heat-exchanger core is subdivided into finite cells, with the dimensions of the cells large relative to the dimensions of the heat-transfer surfaces (i.e., many tubes fit into the cross section of an individual cell). A basic cell method developed by Gaddis and Schlunder [57] in the context of the ϵ-NTU method actually treats the individual cells as one- or two-dimensional micro-exchangers, having a small but finite number of transfer units NTU_{cell} and effectiveness ϵ_{cell}, which expresses the cell terminal temperatures in terms of NTU_{cell}, the capacity rates and flow arrangements within the cell. This micro-exchanger cell approach results in two equations being written for the exiting temperatures of each cell.

In the more general unsteady three-dimensional computational fluid dynamics (CFD) cell approach (or porous media approach) described by Spalding [58,59], each dependent variable (e.g., T_s, $T_t^{(i)}$, wall temperature, velocity) is characterized by a single value within a cell. Whereas the lumped, lumped/incremental, and micro-exchanger cell methods are based on the practical analysis concepts introduced in Chap. 8, the CFD–cell method involves the use of partial differential equations for mass, momentum, and energy transfer for the shell-side fluid that are integrated over the cell volumes, with the interactions between the fluids and the tube walls accounted for by means of local convection correlations for heat transfer and friction. Although the CFD–cell approach is considerably more involved than the micro-exchanger cell method, it provides a basis for analyzing transient processes and for achieving more accuracy in accounting for variations

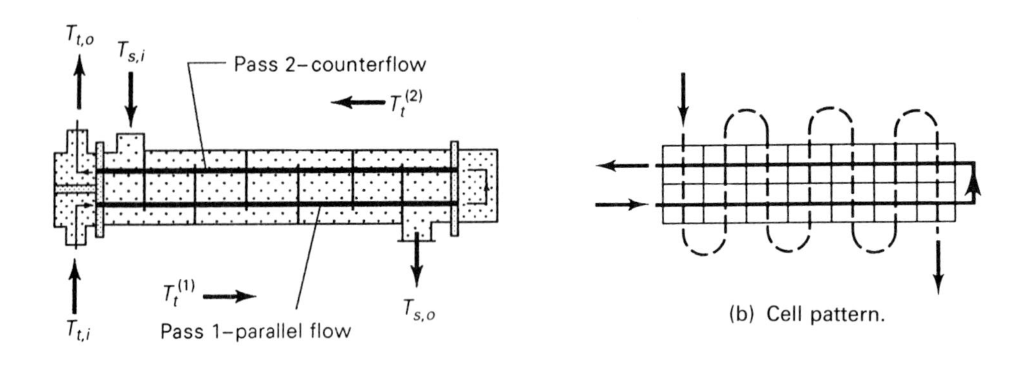

FIGURE 11–55 Micro-exchanger cell perspective shown in the context of shell-and-tube heat exchanger [57].

in temperatures and convection coefficients over the cross section of a heat exchanger.

Whereas the CFD–cell method is now widely used in the analysis of nuclear-steam generators,† for which safety considerations are critical, and is used to some extent in the analysis of cooling towers [60] and power-station condensers [61],‡ due to economic considerations, this approach has not yet been incorporated into commercial codes for shell-and-tube heat exchangers used in the chemical-process and petroleum industries.

11–10–4 Computational Fluid Dynamics Approach

As indicated in Sec. 7–4, CFD-type codes are characterized by the use of a large number of very small finite-difference or finite-element control volumes. In this multidimensional approach, the partial differential equations representing the transfer of mass, momentum, and energy within both fluids and conduction within the wall are combined with appropriate boundary conditions and solved numerically. Because the CFD approach provides a basis for characterizing the temperature and velocity distributions throughout the fluids and the gradients that occur at the heating surfaces, no empirical inputs are required for coefficients of heat transfer and friction. Furthermore, details pertaining to predicted velocity and temperature distributions that arise from the CFD approach provide a theoretical basis for analyzing fluid-flow-originated tube vibrations and thermal stress, which are generally most severe during unsteady operation at startup or shutdown, and fouling and corrosion, which are strongly influenced by irregularities in the flow patterns and temperature distribution. Thus, the CFD approach provides considerable potential benefit for many heat exchanger design applications.

Because of the nature of the flow field in industrial shell-and-tube heat exchangers with many tubes and very small clearances, such as between tubes and baffles, the number of sufficiently small control volumes required to adequately characterize the fine structure of the flow throughout such exchangers is prohibitively large. Therefore, the use of the CFD approach in the analysis of shell-and-tube heat exchangers is presently restricted to relatively small regions (surrounding a few tubes) of an exchanger where details of the fine structure are desired. In this way, these types of codes can be used to calculate the temperature and velocity distributions in the vicinity of tubes, baffles, and other regions of interest within the shell of heat exchangers and to deduce the coefficients of heat transfer and friction that are necessary for use in the lumped, lumped/incremental, and CFD–cell methods.

As pointed out by Spalding [59] in connection with heat exchanger analysis, "there are now two sources of heat-transfer and friction correlations: experimental

† CFD–cell codes used in the nuclear industry include COMMIX (Argonne National Laboratory), KFIX (Los Alamos Scientific Laboratory), ATHOS (CHAM), and THIRST (HTFS), and several others.

‡ SPOC (HTFS) is a CFD–cell code used in the analysis of power-station condensers.

studies; and detailed numerical simulations. Although little use has so far been made of the second possibility, it seems certain that its cost-effectiveness will soon cause it to become the main source of correlations for single-phase flow phenomena."†

11–11 SUMMARY

As we have seen, heat exchangers are classified according to geometry, flow arrangement, size, and certain other factors. The standard types of indirect contact recuperators include double-pipe, shell-and-tube, and crossflow configurations. Regenerators and direct contact heat exchangers are also in common use. The flow arrangements are further classified according to the number of passes, whether the fluid streams are in parallel flow, counterflow, crossflow, or combination of these, and, in the case of crossflow heat exchangers, whether the streams should be designated as mixed or unmixed. We have also seen that various means are employed in order to enhance the performance and reduce the size of heat exchangers. In addition to baffling of shell-side fluids, surface modifications such as fins (plain, corrugated, louvered, strip, perforated, longitudinal, circumferential, etc.), and inserts are often employed. It is with the aid of such devices that the design of compact heat exchangers ($\beta \gtrsim 700$ m^2/m^3) have become possible.

In this chapter we have adapted the practical analysis approach presented in Chap. 8 to the evaluation and design of heat exchangers, with emphasis placed on recuperators. As we have seen, this approach involves the use of correlations for the coefficients of heat transfer h and friction f and the concept of the overall coefficient of heat transfer U together with lumped and lumped/differential formulations.

The practical thermal analysis approach gives rise to solution results for heat-transfer rate q that are conveniently expressed in terms of the effectiveness P or ϵ, log mean temperature difference $LMTD$, and efficiency ψ. Relations for the effectiveness P or ϵ, $LMTD$ correction factor F, and efficiency ψ have been presented for standard double-pipe, shell-and-tube, and crossflow arrangements (see Tables A–N–1–1 and 2 for summary). In addition, performance criteria have been introduced that aid the engineer in maintaining the economical operation of heat exchangers. Consideration has also been given to the specification of the mean overall coefficient of heat transfer $\overline{U}$, with special attention devoted to the evaluation of the shell-side coefficient of heat transfer $\overline{h}_s$ for shell-and-tube heat exchangers.

The practical hydraulic analysis approach provides a systematic method by which the pressure drop and pumping power can be computed for uniform property conditions and for situations in which variations in density are caused by heating or cooling. This approach applies to basic tubular flows, flow across tube banks,

† In this connection, HTSF is now using the HARWELL-FLOW3D finite-difference code for the *analytical validation* of convection correlations used in their one-dimensional code (TASC) for shell-and-tube heat exchangers.

and to well-baffled shell side flow. Because the pressure drop is essentially proportional to LG^2/ρ ($= L\rho U_b^2$), the pumping power is more likely to be a significant design consideration for gases than for liquids.

As we have seen, multipass heat exchanger networks are commonly used to achieve better heat transfer performance or lower pressure drops. Relations are given for the overall effectiveness P or ϵ of standard multipass heat exchanger networks in Sec. 11–9.

Computer codes of the lumped or lumped/incremental type are generally used in the evaluation and design of heat exchangers used in the power, chemical-process, and petroleum industries. As an exception, CFD–cell-type codes are usually used in the analysis of nuclear-steam generators and are now sometimes employed for other types of heat exchangers in the power industry. Due to recent advances in three-dimensional numerical modeling, general-purpose CFD codes are presently capable of contributing to our understanding of the complex transport processes in all types of heat exchangers. It is anticipated that CFD codes will become more prominent in the analysis of heat exchangers over the next few years.

The analysis of regenerators, direct contact heat exchangers, and fired heat exchangers with significant levels of thermal radiation involve the application of the basic principles introduced in this and previous chapters, but is generally more involved. Once an individual has achieved a thorough understanding of the principles introduced in this chapter, which pertain to the analysis of basic recuperators, the literature that deals with the more complex type heat exchangers can be more readily followed. Works that deal with the analysis of regenerators, direct contact heat exchangers, and fired heat exchangers include references 10, 11, and 71–73.

APPENDIXES

A MATHEMATICAL CONCEPTS

A-1 THE CALCULUS

The importance of the calculus in the study of heat transfer cannot be overstated. Therefore, we want briefly to review fundamental concepts pertaining to the calculus.

A-1-1 The Derivative

For cases in which the dependent variable ψ is a function of only one independent variable such as x, we are reminded of the following definition for the derivative:

$$\frac{d\psi}{dx} = \lim_{\Delta x \to 0} \frac{\psi(x + \Delta x) - \psi(x)}{\Delta x} \tag{A-1}$$

Because dx itself is infinitesimal, this equation can be written in the form

$$\frac{d\psi}{dx} = \frac{\psi(x + dx) - \psi(x)}{dx} \tag{A-2}$$

Hence, $\psi(x + dx)$ can be written in terms of $\psi(x)$ and $d\psi/dx$ as

$$\psi(x + dx) = \psi(x) + \frac{d\psi}{dx}\,dx \tag{A-3}$$

This equation will be found to be very important in the analysis of one-dimensional conduction heat transfer in Chap. 2.

If ψ is a function of more than one independent variable such as x, y, z and t, the partial derivative of ψ with respect to x at the location x,y,z and at the instant t is defined as

$$\frac{\partial\psi}{\partial x} = \lim_{\Delta x \to 0} \frac{\psi(x + \Delta x) - \psi(x)}{\Delta x} \qquad \text{at} \quad x,y,z \text{ and } t \tag{A-4}$$

or

$$\frac{\partial\psi}{\partial x} = \frac{\psi(x + dx) - \psi(x)}{dx} \qquad \text{at} \quad x,y,z \text{ and } t \tag{A-5}$$

For this more general situation, $\psi(x + dx)$ is expressed in terms of $\psi(x)$ and the partial derivative; that is,

$$\psi(x + dx) = \psi(x) + \frac{\partial\psi}{\partial x}\,dx \tag{A-6}$$

Similar expressions can be written which involve partial derivatives with respect to the other independent variables. For example, the partial derivative of ψ with respect to y can be written as

$$\frac{\partial\psi}{\partial y} = \frac{\psi(y + dy) - \psi(y)}{dy} \qquad \text{at} \quad x,y,z \text{ and } t \tag{A-7}$$

A–1–2 Differential Equations

Both ordinary and partial differential equations are encountered in the study of heat transfer. For situations in which the equations are *linear*, analytical solution techniques often can be used. Second-order linear ordinary differential equations take the form

$$\psi + a(x)\frac{d\psi}{dx} + b(x)\frac{d^2\psi}{dx^2} = c(x) \tag{A–8}$$

where the coefficients a, b, and c are functions of the independent variable x only. This equation is *nonlinear* if any one of the coefficients a, b, or c is a function of ψ. Similarly, partial differential equations which involve coefficients that are functions of only the independent variables are linear.

The number of boundary conditions in an independent variable required in the solution of any ordinary or partial differential equation is equal to the highest-order differential in that variable. Similar to the definition for a linear differential equation, linear boundary conditions take the general form

$$\psi + d(x)\frac{d\psi}{dx} + e(x)\frac{d^2\psi}{dx^2} = f(x) \qquad \text{at} \quad x = x_1 \tag{A–9}$$

A differential equation or boundary condition that is satisfied by a function ψ is said to be *homogeneous* if it is also satisfied by $C\psi$, where C is an arbitrary constant. Thus, the linear ordinary differential equation given by Eq. (A–8) is homogeneous if $c(x)$ is zero and is *nonhomogeneous* if $c(x) \neq 0$. Similarly, the boundary condition given by Eq. (A–9) is homogeneous if $f(x)$ is zero.

As an example of a problem involving linear nonhomogeneous equations, consider Eq. (A–8) with boundary conditions of the form

$$\psi + d_1(x)\frac{d\psi}{dx} = f_1(x) \qquad \text{at} \quad x = x_1 \tag{A–10}$$

$$\psi + d_2(x)\frac{d\psi}{dx} = f_2(x) \qquad \text{at} \quad x = x_2 \tag{A–11}$$

These types of boundary conditions are often encountered in heat-transfer problems. Notice that this problem involves three nonhomogeneous terms, $c(x)$, $f_1(x)$, and $f_2(x)$. The solution to this problem takes the form

$$\psi = \psi_0 + C_1\psi_1 + C_2\psi_2 \tag{A–12}$$

where ψ_0 is the particular solution of the nonhomogeneous differential equation and $C_1\psi_1 + C_2\psi_2$ is the solution of the homogeneous differential equation. The constants of integration C_1 and C_2 are evaluated on the basis of the boundary conditions.

A–2 COORDINATE SYSTEMS

The foregoing review of basic mathematical concepts has been presented in the context of the *Cartesian coordinate system* shown in Fig. A–A–1(a). Our study will also

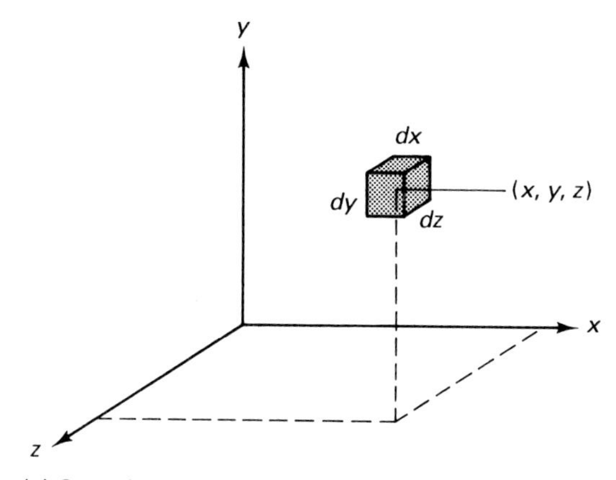

$$dV = dx\,dy\,dz$$
$$dA_x = dy\,dz$$
$$dA_y = dx\,dz$$
$$dA_z = dx\,dy$$

(a) Cartesian.

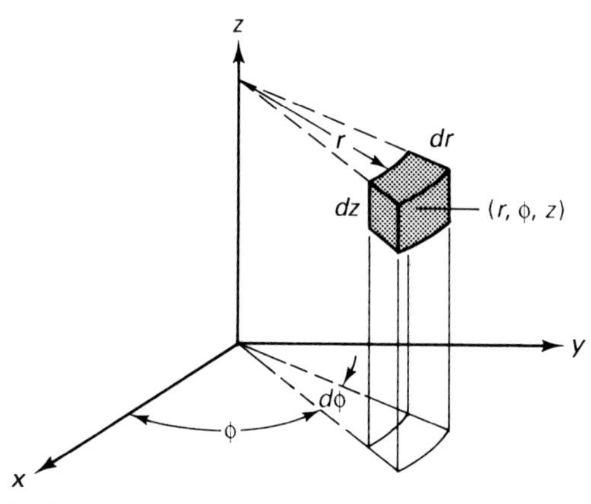

$$dV = r\,d\phi\,dr\,dz$$
$$dA_r = r\,d\phi\,dz$$
$$dA_z = r\,d\phi\,dr$$
$$dA_\phi = dr\,dz$$

(b) Cylindrical.

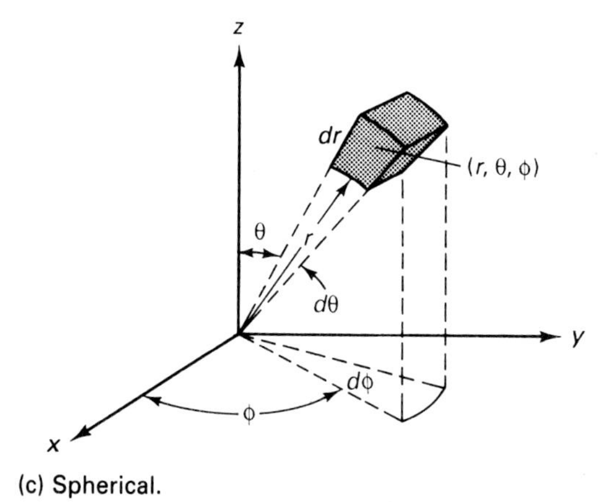

$$dV = r\,d\theta\,r\sin\theta\,d\phi\,dr$$
$$dA_r = r\,d\theta\,r\sin\theta\,d\phi$$
$$dA_\phi = r\,d\theta\,dr$$
$$dA_\theta = r\sin\theta\,d\phi\,dr$$

(c) Spherical.

FIGURE A–A–1 Coordinate systems.

require the use of the *cylindrical coordinate system* shown in Fig. A–A–1(b), and the *spherical coordinate system* shown in Fig. A–A–1(c).

The temperature distribution can be a function of one, two, or three spatial coordinates (e.g., x, y, and z) and a single time variable t. Systems in which the temperature is a function of time will be referred to as unsteady. Steady-state conditions prevail when the temperature distribution within a system is independent of time.

B DIMENSIONS, UNITS, AND SIGNIFICANT FIGURES

Until the transition from English to metric (the International System, SI) units is complete, it will be necessary for us to be conversant with both systems. Consequently, both SI and English units will be used in our study, with emphasis given to the SI system.

The units for the four key fundamental dimensions—*time*, *length*, *mass*, and *temperature*—in these systems are summarized as follows:

System	*Time, t*	*Length, L*	*Mass, m*	*Temperature, T*
SI	second, s	meter, m	kilogram, kg	Kelvin, K Celsius, °C
English	second, s	foot, ft	pound mass, lb_m	Rankine, °R Fahrenheit, °F

Focusing attention on the units for temperature, in the SI system the *Kelvin* K is used for the absolute temperature scale† and the *degree Celsius* °C is used for the Celsius temperature scale. These two SI units for temperature are related by

$$T(\text{K}) = T(°\text{C}) + 273.2°\text{C} \tag{B–1}$$

Note that an increment °C is equal to an increment of 1 K; i.e.,

$$°\text{C} = 1\ \text{K} \tag{B–2}$$

The Rankine °R and Fahrenheit °F units for temperature scale in the English system are related by

$$T(°\text{R}) = T(°\text{F}) + 459.7°\text{F} \tag{B–3}$$

and

$$°\text{F} = 1°\text{R} \tag{B–4}$$

The units for the SI and English systems are related by

$$T(°\text{F}) = \frac{9°\text{F}}{5°\text{C}} T(°\text{C}) + 32°\text{F} \tag{B–5}$$

$$°\text{R} = \frac{5}{9}\ \text{K} \quad \text{or} \quad °\text{F} = \frac{5}{9}\ °\text{C} \tag{B–6,7}$$

† Note that the SI unit for the absolute temperature scale is Kelvin K and not degree Kelvin °K.

We are also reminded of thc following units for the derived dimensions *force*, *energy*, and *power*:

System	*Force, F*	*Energy, E*	*Power, P*
SI	newton, N	joule, J	watt, W
English	pound force, lb_f	British thermal unit, Btu	Btu/h

By definition, we have

$$1 \text{ N} = 1 \text{ kg m/s}^2 \qquad 1 \text{ J} = 1 \text{ N m} \qquad 1 \text{ W} = 1 \text{ J/s} \tag{B–8,9,10}$$

Symbols and basic units for parameters used in the text are summarized in the Nomenclature. For convenience, conversion factors are listed on the inside back cover page.

The relation between the Newton N and the units for mass, length, and time is of course based on *Newton's second law*, which takes the form

$$F = ma \tag{B–11}$$

for an object of mass m undergoing an acceleration a. Newton's second law of motion is given in the context of a fluid control volume by Eq. (1–3). Newton's other laws of motion and gravity include the *first law of motion* (or *law of inertia*), which states that any object in a state of rest or of uniform linear motion will remain in such a state unless acted upon by an external force; the *third law of motion* (or *law of action and reaction*), which states that every force (or action) gives rise to an opposing force (or reaction) of equal strength but opposite direction; and the *universal law of gravitation*, which states that two masses m_1 and m_2 separated by a distance r will be attracted toward one another by a force F, which can be represented by

$$F = \frac{Gm_1m_2}{r^2} \tag{B–12}$$

where $G = 6.673 \times 10^{-11}$ N m²/kg² is the *universal gravitational constant*. Notice that Eq. (B–12) can be put into the form of Eq. (B–11) by setting $m = m_1$ and $g = Gm^2/r^2$.

In regard to heat transfer, the following units generally will be utilized in this text:

q—heat-transfer rate, W

Q—total heat transfer, J

The total heat transfer over an increment of time Q is related to q by the equation

$$q = \frac{\partial Q}{\partial t} \tag{B–13}$$

Hence, Q can be obtained from q by writing

$$Q = \int_0^t q \, dt \tag{B-14}$$

at any point in space.

Three significant figures will be maintained for most of our calculations.

C THERMOPHYSICAL PROPERTIES

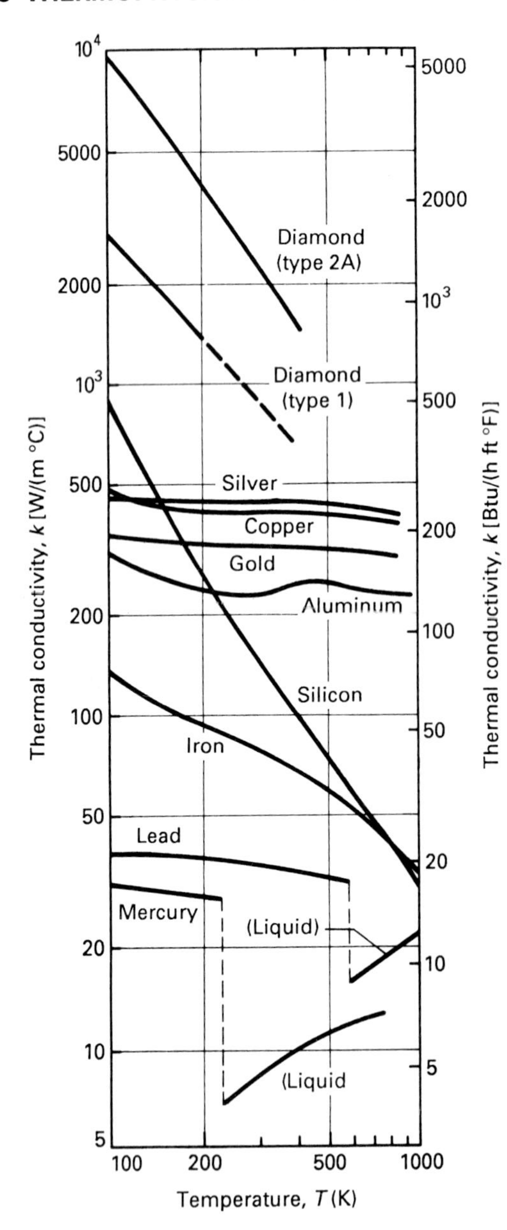

FIGURE A–C–1
Dependence of thermal conductivity on temperature: metals and other good heat conductors—moderate temperature zone. (From Touloukian et al. [1,2]. Used with permission.)

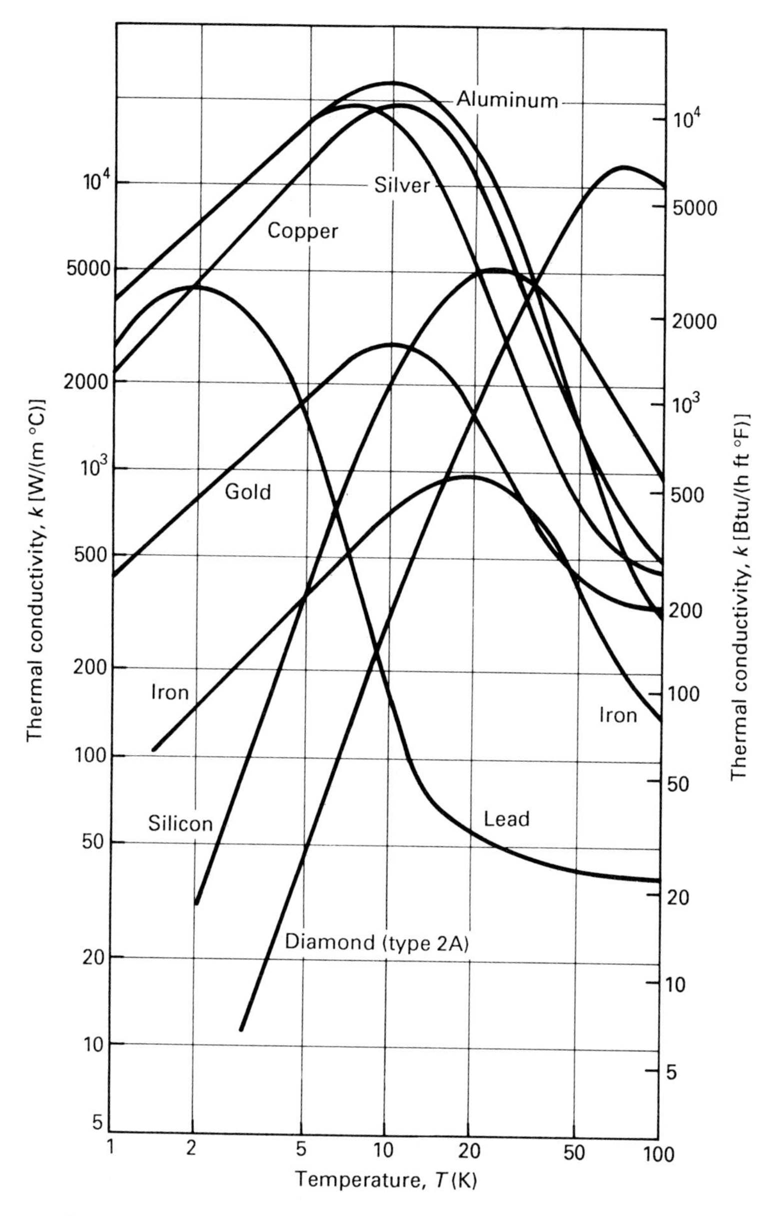

FIGURE A–C–2 Dependence of thermal conductivity on temperature: metals and other good heat conductors—cryogenic temperature zone. (From Touloukian et al. [1,2]. Used with permission.)

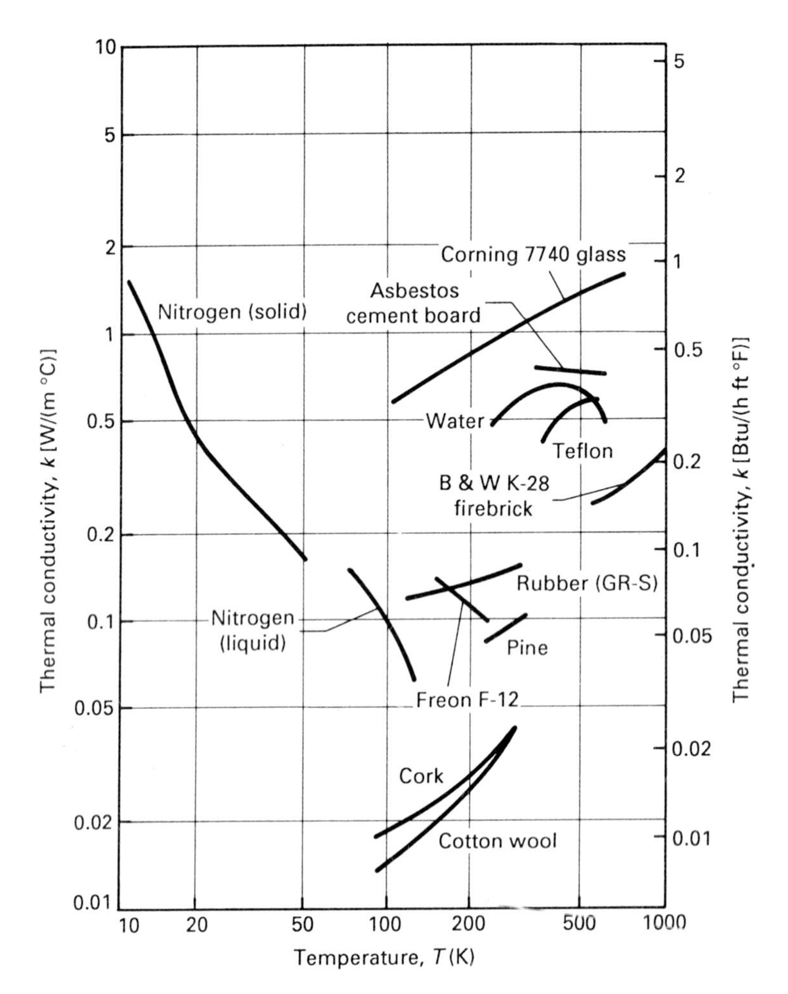

FIGURE A-C-3 Dependence of thermal conductivity on temperature: nonmetallic solids and saturated liquids. (From Touloukian et al. [2,3]. Used with permission.)

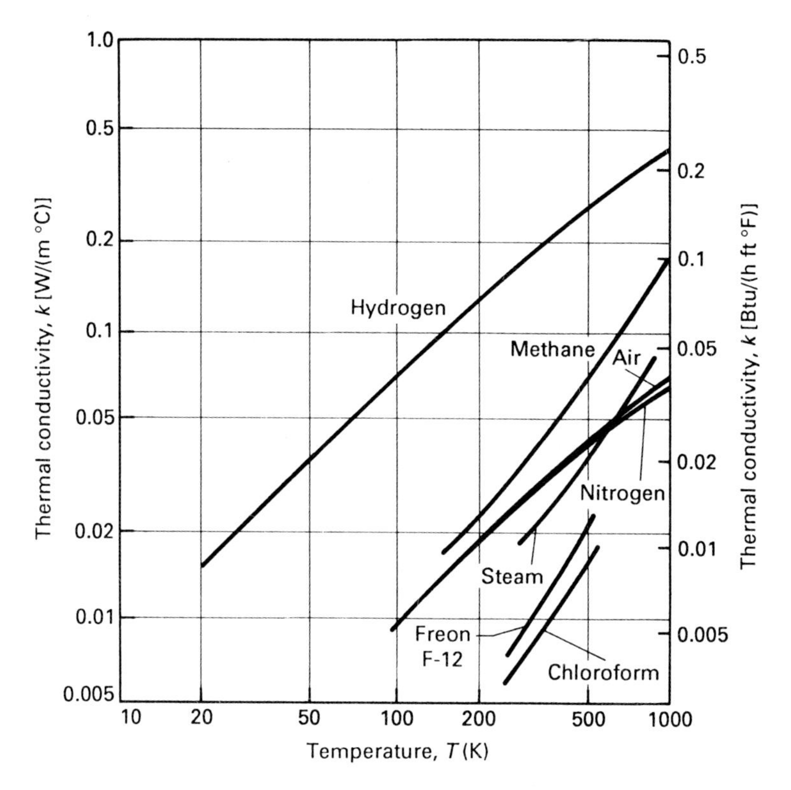

FIGURE A-C-4 Dependence of thermal conductivity on temperature: gases. (From Touloukian et al. [3]. Used with permission.)

TABLE A–C–1(a) Property values of solid metals: metallic elements

Element	Melting point (K)	*Properties at 293 K*				*Variation of properties with temperature T(K)†* k [W/(m °C)]/c_v [kJ/(kg °C)]						
		ρ (kg/m³)	c_v [kJ/(kg °C)]	k [W/(m °C)]	$\alpha \times 10^6$ (m²/s)	100	200	400	600	800	1000	1200
Aluminum	933	2702	0.896	236	97.5	301	237	240	232	220		
						0.480	**0.795**	**0.945**	**1.02**	**1.140**		
Antimony	904	6684	0.208	24.6	17.7		30.2	21.2	18.2	16.8		
Beryllium	1550	1850	1.75	205	63.3	990	301	161	126	107	89	73
						0.195	**1.07**	**2.10**	**2.50**	**2.70**	**2.90**	
Bismuth	545	9780	0.124	7.9	6.51	16.5	9.7	7.0				
						0.114	**0.122**	**0.129**				
Boron	2573	2500	1.047	28.6	10.9	211	52.5	18.7	11.3	8.1	6.3	5.2
						0.121	**0.567**	**1.38**	**1.79**	**2.04**	**2.21**	
Cadmium	594	8650	0.231	97	48.5	203	99.3	94.7				
						0.198	**0.222**	**0.242**				
Cesium	302	1873	0.230	36	83.6		36.8					
Chromium	2118	7160	0.440	91.4	29.0	159	111	87.3	80.5	71.3	65.3	62.4
						0.188	**0.376**	**0.474**	**0.530**	**0.570**	**0.605**	**0.670**
Cobalt	1765	8862	0.389	100	29.0	167	122	84.8	67.4	58.2	52.1	49.3
						0.220	**0.350**	**0.420**	**0.465**	**0.510**	**0.580**	**0.680**
Copper	1356	8950	0.383	386	113	467	400	379	370	359	345	331
						0.250	**0.355**	**0.395**	**0.415**	**0.430**	**0.450**	**0.480**
Germanium	1211	5360	0.322	59.9	34.7	232	96.8	43.2	27.3	19.8	17.4	17.4
						0.190	**0.290**	**0.337**	**0.348**	**0.357**	**0.375**	**0.395**
Gold	1336	19300	0.129	316	127	327	323	312	304	292	278	262
						0.109	**0.124**	**0.131**	**0.135**	**0.140**	**0.145**	**0.155**
Hafnium	2495	13280		23.1			24.4	22.3	21.3	20.8	20.7	20.9
Indium	430	7300		82.2			89.7	74.5				
Iridium	2716	22500	0.134	147	48.8	172	153	144	138	132	126	120
Iron	1810	7870	0.452	81.1	22.8	134	94	69.4	54.7	43.3	32.6	28.2
						0.218	**0.388**	**0.495**	**0.580**	**0.688**	**0.985**	**0.615**
Lead	601	11340	0.129	35.3	24.1	39.7	36.6	33.8	31.2			
						0.118	**0.125**	**0.132**	**0.142**			
Lithium	454	534	3.39	77.4	42.7		88.1	72.1				
Magnesium	923	1740	1.02	156	88.2	169	159	153	149	146		
						0.645	**0.927**	**1.06**	**1.16**	**1.26**		
Manganese	1517	7290	0.486	7.78	2.2		7.17					

Mercury	234	13546					28.9					
Molybdenum	2883	10240	0.251	138	53.7	179 **0.141**	143 **0.224**	134 **0.261**	126 **0.275**	118 **0.285**	112 **0.295**	105 **0.308**
Nickel	1726	8900	0.446	91	22.9	164 **0.233**	106 **0.385**	80.1 **0.487**	65.5 **0.595**	67.4 **0.532**	71.8 **0.564**	76.1 **0.597**
Niobium	2741	8570	0.270	53.6	23.2	55.2 **0.192**	52.6 **0.254**	55.2 **0.279**	58.2 **0.288**	61.3 **0.298**	64.4 **0.307**	67.5 **0.316**
Palladium	1825	12020	0.244	71.8	24.5	76.5 **0.168**	71.6 **0.227**	73.6 **0.251**	79.7 **0.261**	86.9 **0.271**	94.2 **0.281**	102 **0.291**
Platinum	2042	21450	0.133	71.4	25.0	77.5 **0.100**	72.4 **0.125**	71.6 **0.136**	73.0 **0.141**	75.5 **0.146**	78.6 **0.152**	82.6 **0.157**
Potassium	337	860	0.741	103	162		104	52				
Rhenium	3453	21100	0.137	48.1	16.6	58.9 **0.097**	51 **0.127**	46.1 **0.139**	44.2 **0.145**	44.1 **0.151**	44.6 **0.156**	45.7 **0.162**
Rhodium	2233	12450	0.248	150	48.6	186 **0.150**	154 **0.225**	146 **0.258**	136 **0.280**	127 **0.299**	121 **0.317**	115 **0.334**
Rubidium	312	1530	0.348	58.2	109		58.9					
Silicon	1685	2330	0.703	153	93.4	884 **0.256**	264 **0.549**	98.9 **0.780**	61.9 **0.856**	42.2 **0.900**	31.2 **0.934**	25.7 **0.955**
Silver	1234	10500	0.234	427	174	444 **0.187**	430 **0.225**	425 **0.239**	412 **0.250**	390 **0.262**	379 **0.277**	361 **0.292**
Sodium	371	971	0.121	133	114		138					
Tantalum	3269	16600	0.138	57.5	25.1	59.2 **0.108**	57.5 **0.131**	57.8 **0.142**	58.6 **0.144**	59.4 **0.147**	60.2 **0.150**	61 **0.153**
Tin	505	5750	0.227	67.0	51.3	85.2 **0.188**	73.3 **0.215**	62.2 **0.243**				
Titanium	1953	4500	0.522	21.9	9.32	30.5 **0.300**	24.5 **0.465**	20.4 **0.551**	19.4 **0.591**	19.7 **0.633**	20.7 **0.675**	22 **0.620**
Tungsten	3653	19300	0.134	179	69.2	214 **0.088**	197 **0.124**	162 **0.139**	139 **0.144**	128 **0.147**	121 **0.150**	115 **0.154**
Uranium	1407	19070	0.113	27.4	12.7	21.7 **0.092**	25.1 **0.105**	29.6 **0.122**	34 **0.142**	38.8 **0.171**	43.9 **0.175**	49 **0.157**
Vanadium	2192	6100	0.502	31.4	10.3	35.8 **0.265**	31.5 **0.441**	32.1 **0.528**	34.2 **0.554**	36.3 **0.578**	38.6 **0.612**	41.2 **0.662**
Zinc	693	7140	0.385	121	44.0	122 **0.294**	123 **0.363**	116 **0.398**	105 **0.432**			
Zirconium	2125	6750	0.272	22.8	12.8	33.2 **0.200**	25.2 **0.258**	21.6 **0.294**	20.7 **0.315**	21.6 **0.335**	23.7 **0.354**	25.7 **0.336**

† Values for c_v shown in **bold** typeset.

Sources: Touloukian et al. [1,2], Eckert and Drake [4], and Raznjevic [5].

TABLE A–C–1(b) Property values of solid metals: alloys

Metal	*Composition* (%)	*Properties at 293 K*				*Variation of properties with temperature* T(K)†						
		ρ (kg/m^3)	c_v [kJ/(kg °C)]	k [W/(m °C)]	$\alpha \times 10^5$ (m^2/s)	k [W/(m °C)]/c_v [kJ/(kg °C)]: 100	200	400	600	800	1000	1200
Aluminum												
Duralumin	94–96 Al. 3–5 Cu. trace Mg	2787	0.883	164	6.676		163	186	186			
							0.787	**0.925**	**1.042**			
Silumin	87 Al, 13 Si	2659	0.871	164	7.099		165					
Copper												
Aluminum Bronze	95 Cu, 5 Al	8666	0.410	83	2.330							
Bronze	75 Cu, 25 Sn	8666	0.343	26	0.859							
Red brass	85 Cu, 9 Sn, 6 Zn	8714	0.385	61	1.804							
Brass	70 Cu, 30 Zn	8522	0.385	111	3.412	75	95	137	149			
							0.360	**0.395**	**0.425**			
German silver	62 Cu, 15 Ni, 22 Zn	8618	0.394	24.9	0.733							
Constantan	60 Cu, 40 Ni	8922	0.410	22.7	0.612	17	19					
						0.237	**0.362**					
Iron												
Cast iron	≈4 C	7272	0.420	52	1.702							
Wrought iron	0.5 C	7849	0.460	59	1.626			56	47	39	34	33
Steel												
Carbon steel	0.5 C	7833	0.465	54	1.474			51	44	43	32	30
	1 C	7801	0.473	43	1.172			43	39	34	30	28
	1.5 C	7753	0.486	36	0.970			36	35	32	29	28

Chrome steel	1 Cr	7865	0.460	61	1.665			54	46	38	33	33
	5 Cr	7833	0.460	40	1.110			38	35	31	29	29
	10 Cr	7785	0.460	31	0.867			31	29	28	28	28
Chrome nickel steel	15 Cr, 10 Ni	7865	0.460	19	0.526							
	20 Cr, 15 Ni	7833	0.460	15.1	0.415							
Nickel steel	10 Ni	7945	0.460	26	0.720							
	20 Ni	7993	0.460	19	0.526							
	40 Ni	8169	0.460	10	0.279							
	60 Ni	8378	0.460	19	0.493							
Nickel-chrome steel	80 Ni, 15 Cr	8522	0.460	17	0.444							
	40 Ni, 15 Cr	8073	0.460	11.6	0.305							
Inconel X-750	73 Ni, 15 Cr, 6.7 Fe	8510	0.439	11.7	0.31	8.7	10.3	13.5	17	20.5	24	27.6
							0.372	**0.473**	**0.510**	**0.546**	**0.626**	
Manganese steel	1 Mn	7865	0.460	50	1.388							
	5 Mn	7849	0.460	22	0.637							
Silicon steel	1 Si	7769	0.460	42	1.164							
	5 Si	7417	0.460	19	0.555							
Stainless steel	AISI 302	8055	0.480	15.1	0.391			17.3	20	22.8	25.4	
								0.512	**0.559**	**0.585**	**0.606**	
	AISI 304	7900	0.477	14.9	0.395			16.6	19.8	22.6	25.4	28
								0.515	**0.557**	**0.582**	**0.611**	**0.640**
	AISI 316	8238	0.468	13.4	0.348			15.2	18.3	21.3	24.2	
								0.504	**0.550**	**0.576**	**0.602**	
	AISI 347	7978	0.480	14.2	0.371			15.8	18.9	21.9	24.7	
								0.513	**0.559**	**0.585**	**0.606**	
Tungsten steel	1 W	7913	0.448	66	1.858							
	5 W	8073	0.435	54	1.525							

† Values for c_v shown in **bold** typeset.
Sources: Eckert and Drake [4], Desai et al. [6], and others.

TABLE A–C–2 Property values of solid nonmetals

Substance	T (°C)	ρ (kg/m³)	c_v [kJ/(kg °C)]	k [W/(m °C)]	$\alpha \times 10^7$ (m²/s)
Structural and heat-resistant materials					
Asphalt	20–55	2120		0.74–0.76	
Brick					
Building brick	20	1600	0.84	0.69	5.2
Carborundum brick	600			18.5	
(50% SiC)	1400			11.1	
Chrome brick	200	3000	0.84	2.32	9.2
	550			2.47	9.8
	900			1.99	7.9
Fireclay brick, burnt 2426°F	500	2000	0.96	1.04	5.4
	800			1.07	
	1100			1.09	
Burnt 2642°F	500	2300	0.96	1.28	5.8
	800			1.37	
	1100			1.40	
Magnesite	200		1.13	3.81	
(50% MgO)	650			2.77	
	1200			1.90	
Masonry	20	1700	0.837	0.658	
Silica (95% SiO_2)	20	1900		1.07	
Zirconia (62% ZrO_2)	20	3600		2.44	
Cement, mortar	23			1.16	
Concrete					
Cinder	23			0.76	
Stone 1–2–4 mix	20	1900–2300	0.88	1.37	8.2–6.8
Glass					
Window	20	2700	0.84	0.78 (avg)	3.4
Borosilicate	30–75	2200		1.09	
Plaster, gypsum	20	1440	0.84	0.48	4.0
Plexiglas	20	1180		0.195	
Plywood	20	590	1.22	0.109	
Sand					
dry	20			0.582	
moist	20	1640		1.13	
Stone					
Granite		2640	0.82	1.73–3.98	8–18
Limestone	100–300	2500	0.90	1.26–1.33	5.6–5.9
Marble		2500–2700	0.80	2.07–2.94	10–13.6
Sandstone	40	2160–2300	0.71	1.83	11.2–11.9
Wood (across the grain):					
Balsa	30	140		0.055	
Cypress	30	460		0.097	
Fir	23	420	2.72	0.11	0.96
Maple or oak	30	540	2.4	0.166	1.28
Yellow pine	23	640	2.8	0.147	0.82
White pine	30	430		0.112	

TABLE A–C–2 (Property values of solid nonmetals—continued)

Substance	T (°C)	ρ (kg/m^3)	c_v [kJ/(kg °C)]	k [W/(m °C)]	$\alpha \times 10^7$ (m^2/s)
		Insulating material			
Asbestos†	0	470–570	0.816	0.154	3.3–4
Cardboard, corrugated				0.064	
Corkboard	30	160		0.043	
Cork:					
Regranulated	32	45–120	1.88	0.045	2–5.3
Ground	32	150		0.043	
Diatomaceous earth (Sil-o-cel)	0	320		0.061	
Felt:					
Hair	30	130–200		0.036	
Wool	30	330		0.052	
Fiber, insulating board	20	240		0.048	
Glass fiber	0	220		0.035	
Glass wool	23	24	0.7	0.038	22.6
Kapok	30			0.035	
Magnesia, 85%	38	270		0.067	
	93			0.071	
	150			0.074	
	204			0.080	
Rock wool	32	160		0.040	
Loosely packed	150	64		0.067	
	260			0.087	
Sawdust	23			0.059	
Silica aerogel	32	140		0.024	
Wood shavings	23			0.059	
		Soils and rocks‡			
Heavy soil, saturated	30	2300	0.838	2.42	9.04
Heavy soil, damp solid masonry	30	2100	0.963	1.30	6.45
		2300	0.879		
Heavy soil, dry	30	2000	0.838	0.865	5.16
Light soil, damp		1600	1.05		
Light soil, dry	30	1440	0.838	0.346	2.80
Dense rock	30	2300	0.838	3.46	12.9
Average rock	30	2800	0.838	2.42	10.3

† Well-documented health hazards are associated with long-term exposure to asbestos. Therefore, asbestos insulating material is no longer recommended.

‡ *Source*: Adapted with permission from ASHRAE Handbook of Fundamentals [7].

TABLE A–C–2 (Property values of solid nonmetals—continued)

Substance	T (°C)	ρ (kg/m^3)	c_v [kJ/(kg °C)]	k [W/(m °C)]	$\alpha \times 10^7$ (m^2/s)
Other materials					
Coal					
Anthiarite	27	1350	1.26	0.26	
Bituminous in situ		1300		0.60	3.5
Cotton	27	80	1.30	0.06	
Diamonds					
Type I	0			1000	
Type IIa	0	3500	0.51	2650	14800
Type IIb	0			1510	
Foods					
Apple, red (75% H_2O)	27	840	3.6	0.513	
Banana (76% H_2O)	27	840	3.60	0.513	
Beef	25				1.35
Chicken, white meat	−75			1.60	
(74% H_2O)	−20			1.35	
	0			0.472	
	20			0.489	
Egg white					1.37
Human tissue					
skin	27			0.37	
fat layer (adipose)	27			0.2	
muscle	27			0.41	
Ice	0	913	1.83	2.22	0.124
Leather (sole)	27	998		0.159	
Linoleum	20	535		0.081	
Paper	27	930	1.34	0.18	
Rubber, vulcanized					
soft	27	1100	2.01	0.13	
hard	27	1190		0.16	
Snow					
loose	0	110		0.05	
packed	0	500		0.19	

Sources: *International Critical Tables* [8], Incropera and DeWitt [9], and others.

TABLE A–C–3(a) Property values of saturated water

Tempera-ture, T (K)	Pressure P (bar)[a]	Heat of Vaporization i_{fg} (kJ/kg)	Specific Volume (m³/kg)		Specific Heat [kJ/(kg K)]		Thermal Conductivity [W/(m K)]		Viscosity [kg/(m s)]		Prandtl Number		Surface Tension $\sigma_f \times 10^3$ (N/m)	Expansion Coefficient $\beta_f \times 10^6$ (K^{-1})	Tempera-ture T (K)
			$v_f \times 10^3$	v_g	$c_{P,f}$	$c_{P,g}$	$k_f \times 10^3$	$k_g \times 10^3$	$\mu_f \times 10^6$	$\mu_g \times 10^6$	Pr_f	Pr_g			
273.15	0.00611	2502	1.000	206.3	4.217	1.854	569	18.2	1750	8.02	12.99	0.815	75.5	−68.05	273.15
275	0.00697	2497	1.000	181.7	4.211	1.855	574	18.3	1652	8.09	12.22	0.817	75.3	−32.74	275
280	0.00990	2485	1.000	130.4	4.198	1.858	582	18.6	1422	8.29	10.26	0.825	74.8	46.04	280
285	0.01387	2473	1.000	99.4	4.189	1.861	590	18.9	1225	8.49	8.81	0.833	74.3	114.1	285
290	0.01917	2461	1.001	69.7	4.184	1.864	598	19.3	1080	8.69	7.56	0.841	73.7	174.0	290
295	0.02617	2449	1.002	51.49	4.181	1.868	606	19.5	959	8.89	6.62	0.849	72.7	227.5	295
300	0.03531	2438	1.003	39.13	4.179	1.872	613	19.6	855	9.09	5.83	0.857	71.7	276.1	300
305	0.04712	2426	1.005	27.90	4.178	1.877	620	20.1	769	9.29	5.20	0.865	70.9	320.6	305
310	0.06221	2414	1.007	22.93	4.178	1.882	628	20.4	695	9.49	4.62	0.873	70.0	361.9	310
315	0.08132	2402	1.009	17.82	4.179	1.888	634	20.7	631	9.69	4.16	0.883	69.2	400.4	315
320	0.1053	2390	1.011	13.98	4.180	1.895	640	21.0	577	9.89	3.77	0.894	68.3	436.7	320
325	0.1351	2378	1.013	11.06	4.182	1.903	645	21.3	528	10.09	3.42	0.901	67.5	471.2	325
330	0.1719	2366	1.016	8.82	4.184	1.911	650	21.7	489	10.29	3.15	0.908	66.6	504.0	330
335	0.2167	2354	1.018	7.09	4.186	1.920	656	22.0	453	10.49	2.88	0.916	65.8	535.5	335
340	0.2713	2342	1.021	5.74	4.188	1.930	660	22.3	420	10.69	2.66	0.925	64.9	566.0	340
345	0.3372	2329	1.024	4.683	4.191	1.941	668	22.6	389	10.89	2.45	0.933	64.1	595.4	345
350	0.4163	2317	1.027	3.846	4.195	1.954	668	23.0	365	11.09	2.29	0.942	63.2	624.2	350
355	0.5100	2304	1.030	3.180	4.199	1.968	671	23.3	343	11.29	2.14	0.951	62.3	652.3	355
360	0.6209	2291	1.034	2.645	4.203	1.983	674	23.7	324	11.49	2.02	0.960	61.4	697.9	360
365	0.7514	2278	1.038	2.212	4.209	1.999	677	24.1	306	11.69	1.91	0.969	60.5	707.1	365
370	0.9040	2265	1.041	1.861	4.214	2.017	679	24.5	289	11.89	1.80	0.978	59.5	728.7	370
373.15	1.0133	2257	1.044	1.679	4.217	2.029	680	24.8	279	12.02	1.76	0.984	58.9	750.1	373.15
375	1.0815	2252	1.045	1.574	4.220	2.036	681	24.9	274	12.09	1.70	0.987	58.6	761	375
380	1.2869	2239	1.049	1.337	4.226	2.057	683	25.4	260	12.29	1.61	0.999	57.6	788	380
385	1.5233	2225	1.053	1.142	4.232	2.080	685	25.8	248	12.49	1.53	1.004	56.6	814	385
390	1.794	2212	1.058	0.980	4.239	2.104	686	26.3	237	12.69	1.47	1.013	55.6	841	390
400	2.455	2183	1.067	0.731	4.256	2.158	688	27.2	217	13.05	1.34	1.033	53.6	896	400
410	3.302	2153	1.077	0.553	4.278	2.221	688	28.2	200	13.42	1.24	1.054	51.5	952	410
420	4.370	2123	1.088	0.425	4.302	2.291	688	29.8	185	13.79	1.16	1.075	49.4	1010	420

TABLE A–C–3(a) (Property values of saturated water—continued)

Temperature, T (K)	Pressure P (bar)[a]	Heat of Vaporization i_{fg} (kJ/kg)	Specific Volume (m³/kg)		Specific Heat [kJ/(kg K)]		Thermal Conductivity [W/(m K)]		Viscosity [kg/(m s)]		Prandtl Number		Surface Tension $\sigma_f \times 10^3$ (N/m)	Expansion Coefficient $\beta_f \times 10^6$ (K^{-1})	Temperature T (K)
			$v_f \times 10^3$	v_g	$c_{P,f}$	$c_{P,g}$	$k_f \times 10^3$	$k_g \times 10^3$	$\mu_f \times 10^6$	$\mu_g \times 10^6$	Pr_f	Pr_g			
430	5.699	2091	1.099	0.331	4.331	2.369	685	30.4	173	14.14	1.09	1.10	47.2	—	430
440	7.333	2059	1.110	0.261	4.36	2.46	682	31.7	162	14.50	1.04	1.12	45.1	—	440
450	9.319	2024	1.123	0.208	4.40	2.56	678	33.1	152	14.85	0.99	1.14	42.9	—	450
460	11.71	1989	1.137	0.167	4.44	2.68	673	34.6	143	15.19	0.95	1.17	40.7	—	460
470	14.55	1951	1.152	0.136	4.48	2.79	667	36.3	136	15.54	0.92	1.20	38.5	—	470
480	17.90	1912	1.167	0.111	4.53	2.94	660	38.1	129	15.88	0.89	1.23	36.2	—	480
490	21.83	1870	1.184	0.0922	4.59	3.10	651	40.1	124	16.23	0.87	1.25	33.9	—	490
500	26.40	1825	1.203	0.0766	4.66	3.27	642	42.3	118	16.59	0.86	1.28	31.6	—	500
510	31.66	1779	1.222	0.0631	4.74	3.47	631	44.7	113	16.95	0.85	1.31	29.3	—	510
520	37.70	1730	1.244	0.0525	4.84	3.70	621	47.5	108	17.33	0.84	1.35	26.9	—	520
530	44.58	1679	1.268	0.0445	4.95	3.96	608	50.6	104	17.72	0.85	1.39	24.5	—	530
540	52.38	1622	1.294	0.0375	5.08	4.27	594	54.0	101	18.1	0.86	1.43	22.1	—	540
550	61.19	1564	1.323	0.0317	5.24	4.64	580	58.3	97	18.6	0.87	1.47	19.7	—	550
560	71.08	1499	1.355	0.0269	5.43	5.09	563	63.7	94	19.1	0.90	1.52	17.3	—	560
570	82.16	1429	1.392	0.0228	5.68	5.67	548	76.7	91	19.7	0.94	1.59	15.0	—	570
580	94.51	1353	1.433	0.0193	6.00	6.40	528	76.7	88	20.4	0.99	1.68	12.8	—	580
590	108.3	1274	1.482	0.0163	6.41	7.35	513	84.1	84	21.5	1.05	1.84	10.5	—	590
600	123.5	1176	1.541	0.0137	7.00	8.75	497	92.9	81	22.7	1.14	2.15	8.4	—	600
610	137.3	1068	1.612	0.0115	7.85	11.1	467	103	77	24.1	1.30	2.60	6.3	—	610
620	159.1	941	1.705	0.0094	9.35	15.4	444	114	72	25.9	1.52	3.46	4.5	—	620
625	169.1	858	1.778	0.0085	10.6	18.3	430	121	70	27.0	1.65	4.20	3.5	—	625
630	179.7	781	1.856	0.0075	12.6	22.1	412	130	67	28.0	2.0	4.8	2.6	—	630
635	190.9	683	1.935	0.0066	16.4	27.6	392	141	64	30.0	2.7	6.0	1.5	—	635
640	202.7	560	2.075	0.0057	26	42	367	155	59	32.0	4.2	9.6	0.8	—	640
645	215.2	361	2.351	0.0045	90	—	331	178	54	37.0	12	26	0.1	—	645
647.3[b]	221.2	0	3.170	0.0032	∞	∞	238	238	45	45.0	∞	∞	0.0	—	647.3[b]

[a] 1 bar = 10^5 N/m²; 1 atm = 1.0133 bar = 0.10133 MPa.

[b] Critical temperature.

Sources: Adapted with permission from Incropera and DeWitt [9] and Liley [10].

TABLE A-C-3(b) Property values of saturated liquids

T (°C)	ρ (kg/m^3)	c_P [kJ/(kg °C)]	k [W/(m °C)]	$\alpha \times 10^7$ (m^2/s)	$\nu \times 10^6$ (m^2/s)	Pr	$\beta \times 10^3$ (1/°C)
			Ammonia, NH_3				
−50	703.69	4.463	0.547	1.742	0.435	2.60	
−40	691.68	4.467	0.547	1.775	0.406	2.28	
−30	679.34	4.476	0.549	1.801	0.387	2.15	
−20	666.69	4.509	0.547	1.819	0.381	2.09	
−10	653.55	4.564	0.543	1.825	0.378	2.07	
0	640.10	4.635	0.540	1.819	0.373	2.05	2.2
10	626.16	4.714	0.531	1.801	0.368	2.04	2.3
20	611.75	4.798	0.521	1.775	0.359	2.02	2.5
30	596.37	4.890	0.507	1.742	0.349	2.01	
40	580.99	4.999	0.493	1.701	0.340	2.00	
50	564.33	5.116	0.476	1.654	0.330	1.99	
			Carbon dioxide, CO_2				
−50	1,156.3	1.84	0.0855	0.4021	0.119	2.96	
−40	1,117.8	1.88	0.1011	0.4810	0.118	2.46	
−30	1,076.8	1.97	0.1116	0.5272	0.117	2.22	
−20	1,032.4	2.05	0.1151	0.5445	0.115	2.12	
−10	983.38	2.18	0.1099	0.5133	0.113	2.20	
0	926.99	2.47	0.1045	0.4578	0.108	2.38	
10	860.03	3.14	0.0971	0.3608	0.101	2.80	
20	772.57	5.0	0.0872	0.2219	0.091	4.10	14.0
30	597.81	36.4	0.0703	0.0279	0.080	28.7	
			Sulfur dioxide, SO_2				
−50	1,560.8	1.3595	0.242	1.141	0.484	4.24	
−40	1,536.8	1.3607	0.235	1.130	0.424	3.74	
−30	1,520.6	1.3616	0.230	1.117	0.371	3.31	
−20	1,488.6	1.3624	0.225	1.107	0.324	2.93	
−10	1,463.6	1.3628	0.218	1.097	0.288	2.62	
0	1,438.5	1.3636	0.211	1.081	0.257	2.38	
10	1,412.5	1.3645	0.204	1.066	0.232	2.18	
20	1,386.4	1.3653	0.199	1.050	0.210	2.00	1.94
30	1,359.3	1.3662	0.192	1.035	0.190	1.83	
40	1,329.2	1.3674	0.185	1.019	0.173	1.70	
50	1,299.1	1.3683	0.177	0.999	0.162	1.61	

TABLE A-C-3(b) (Property values of saturated liquid—continued)

T (°C)	ρ (kg/m^3)	c_P [kJ/(kg °C)]	k [W/(m °C)]	$\alpha \times 10^7$ (m^2/s)	$\nu \times 10^6$ (m^2/s)	Pr	$\beta \times 10^3$ (1/°C)
			Methyl chloride, CH_3Cl				
−50	1,052.6	1.4759	0.215	1.388	0.320	2.31	
−40	1,033.4	1.4826	0.209	1.368	0.318	2.32	
−30	1,016.5	1.4922	0.202	1.337	0.314	2.35	
−20	999.39	1.5043	0.196	1.301	0.309	2.38	
−10	981.45	1.5194	0.187	1.257	0.306	2.43	
0	962.39	1.5378	0.178	1.213	0.302	2.49	
10	942.36	1.5600	0.171	1.166	0.297	2.55	
20	923.31	1.5860	0.163	1.112	0.293	2.63	
30	903.12	1.6161	0.154	1.058	0.288	2.72	
40	883.10	1.6504	0.144	0.996	0.281	2.83	
50	861.15	1.6890	0.133	0.921	0.274	2.97	
			Dichlorodifluoromethane (Freon F-12), CCl_2F_2				
−50	1,546.8	0.8750	0.067	0.501	0.310	6.2	
−40	1,518.7	0.8847	0.069	0.514	0.279	5.4	
−30	1,489.6	0.8956	0.069	0.526	0.253	4.8	1.9
−20	1,460.6	0.9073	0.071	0.539	0.235	4.4	
−10	1,429.5	0.9203	0.073	0.550	0.221	4.0	
0	1,397.4	0.9345	0.073	0.557	0.214	3.8	3.1
10	1,364.3	0.9496	0.073	0.560	0.203	3.6	
20	1,330.2	0.9659	0.073	0.560	0.198	3.5	
30	1,295.10	0.9835	0.071	0.560	0.194	3.5	
40	1,257.13	1.0019	0.069	0.555	0.191	3.5	4.4
50	1,216.0	1.0216	0.067	0.545	0.190	3.5	
			Eutectic calcium chloride solution, 29.9% $CaCl_2$				
−50	1,319.8	2.608	0.402	1.166	36.35	312.0	
−40	1,315.0	2.6356	0.415	1.200	24.97	208.0	
−30	1,310.2	2.6611	0.429	1.234	17.18	139.0	
−20	1,305.5	2.688	0.445	1.267	11.04	87.1	
−10	1,300.7	2.713	0.459	1.300	6.96	53.6	
0	1,296.1	2.738	0.472	1.332	4.39	33.0	
10	1,291.4	2.763	0.485	1.363	3.35	24.6	
20	1,286.6	2.788	0.498	1.394	2.72	19.6	
30	1,282.0	2.814	0.511	1.419	2.27	16.0	
40	1,277.2	2.839	0.523	1.445	1.92	13.3	
50	1,272.5	2.868	0.535	1.468	1.65	11.3	

TABLE A–C–3(b) (Property values of saturated liquid—continued)

T (°C)	ρ (kg/m^3)	c_P [kJ/(kg °C)]	k [W/(m °C)]	$\alpha \times 10^7$ (m^2/s)	$\nu \times 10^6$ (m^2/s)	Pr	$\beta \times 10^3$ (1/°C)
			Glycerin, $C_3H_5(OH)_3$				
0	1,276.0	2.261	0.282	0.983	8310	84,700	
10	1,270.1	2.319	0.284	0.965	3000	31,000	
20	1,264.0	2.386	0.286	0.947	1180	12,500	0.50
30	1,258.1	2.445	0.286	0.929	500	5,380	
40	1,252.0	2.512	0.286	0.914	220	2,450	
50	1,245.0	2.583	0.287	0.893	150	1,630	
			Ethylene glycol, $C_2H_4(OH)_2$				
0	1,130.8	2.294	0.242	0.934	57.53	615	
20	1,116.6	2.382	0.249	0.939	19.18	204	0.65
40	1,101.4	2.474	0.256	0.939	8.69	93	
60	1,087.7	2.562	0.260	0.932	4.75	51	
80	1,077.6	2.650	0.261	0.921	2.98	32.4	
100	1,058.5	2.742	0.263	0.908	2.03	22.4	
			Engine oil, unused (SAE 50)				
0	899.12	1.796	0.147	0.911	4280	47,100	
20	888.23	1.880	0.145	0.872	900	10,400	0.70
40	876.05	1.964	0.144	0.834	240	2,870	
60	864.04	2.047	0.140	0.800	83.9	1,050	
80	852.02	2.131	0.138	0.769	37.5	490	
100	840.01	2.219	0.137	0.738	20.3	276	
120	828.96	2.307	0.135	0.710	12.4	175	
140	816.94	2.395	0.133	0.686	8.0	116	
160	805.89	2.483	0.132	0.663	5.6	84	
			Mercury, Hg				
0	13,628	0.1403	8.20	42.99	0.124	0.0288	
20	13,579	0.1394	8.69	46.06	0.114	0.0249	0.182
50	13,506	0.1386	9.40	50.22	0.104	0.0207	
100	13,385	0.1373	10.51	57.16	0.0928	0.0162	
150	13,264	0.1365	11.49	63.54	0.0853	0.0134	
200	13,145	0.1570	12.34	69.08	0.0802	0.0116	
250	13,026	0.1357	13.07	74.06	0.0765	0.0103	
315.5	12,847	0.134	14.02	81.5	0.0673	0.0083	

Source: Adapted with permission from Eckert and Drake [4].

TABLE A–C–3(c) Properties of saturated liquid-vapor

Fluid	T_{sat} (°C)	i_{fg} (kJ/kg)	ρ_f (kg/m^3)	ρ_g (kg/m^3)
Ethanol	78†	846	757	1.44
Ethylene Glycol	197†	812	1111	—
Glycerin	290†	974	1260	—
Mercury	127	301.9	13370	—
	227	298.9	13130	—
	357†	301	12740	3.90
Refrigerant F-12	−30†	165	1488	6.32
	27	136.9	1305	—
	47	123.8	1229	—

† 1 atmosphere

Sources: Incropera and DeWitt [9] and others.

TABLE A–C–4 Property values of common liquid metals

Metal	*Melting point* (°C)	*Normal boiling point* (°C)	T (°C)	$\rho \times 10^{-3}$ (kg/m^3)	c_P [kJ/(kg °C)]	k [W/(m °C)]	$\mu \times 10^3$ [kg/(m s)]	*Pr*
Bismuth	271	1480	316	10.01	0.144	16.4	1.62	0.014
			760	9.47	0.165	15.6	0.79	0.0084
Lead	327	1740	371	10.5	0.159	16.1	2.40	0.024
			704	10.1	0.155	14.9	1.37	0.016
Lithium	179	1320	204	0.51	4.19	38.1	0.60	0.065
			982	0.44	4.19		0.42	
Mercury	−38.9	357	10	13.6	0.138	8.1	1.59	0.027
			316	12.8	0.134	14.0	0.86	0.0084
Potassium	63.9	760	149	0.81	0.796	45.0	0.37	0.0066
			704	0.67	0.754	33.1	0.14	0.0031
Sodium	97.8	883	204	0.90	1.34	80.3	0.43	0.0072
			704	0.78	1.26	59.7	0.18	0.0038
Sodium—potassium:								
22% Na	19	826	93.3	0.848	0.946	24.4	0.49	0.019
			760	0.69	0.883		0.146	
56% Na	−11.1	784	93.3	0.89	1.13	25.6	0.58	0.026
			760	0.74	1.04	28.9	0.16	0.058
Lead—bismuth:								
44.5% Pb	125	1670	288	10.3	0.147	10.7	1.76	0.024
			649	9.84			1.15	

Source: Adapted with permission from Knudsen and Katz [11].

TABLE A–C–5(a) Property values of gases at atmospheric pressure†

T (K)	ρ (kg/m^3)	c_P [kJ/(kg °C)]	k [W/(m °C)]	$\alpha \times 10^4$ (m^2/s)	$\mu \times 10^5$ [kg/(m s)]	$\nu \times 10^6$ (m^2/s)	Pr
			Air				
100	3.6010	1.0266	0.009246	0.02501	0.6924	1.923	0.770
150	2.3675	1.0099	0.013735	0.05745	1.0283	4.343	0.753
200	1.7684	1.0061	0.01809	0.10165	1.3289	7.490	0.739
250	1.4128	1.0053	0.02227	0.13161	1.488	9.49	0.722
300	1.1774	1.0057	0.02624	0.22160	1.846	15.68	0.708
350	0.9980	1.0090	0.03003	0.2983	2.075	20.76	0.697
400	0.8826	1.0140	0.03365	0.3760	2.286	25.90	0.689
450	0.7833	1.0207	0.03707	0.4222	2.484	28.86	0.683
500	0.7048	1.0295	0.04038	0.5564	2.671	37.90	0.680
550	0.6423	1.0392	0.04360	0.6532	2.848	44.34	0.680
600	0.5879	1.0551	0.04659	0.7512	3.018	51.34	0.680
650	0.5430	1.0635	0.04953	0.8578	3.177	58.51	0.682
700	0.5030	1.0752	0.05230	0.9672	3.332	66.25	0.684
750	0.4709	1.0856	0.05509	1.0774	3.481	73.91	0.686
800	0.4405	1.0978	0.05779	1.1951	3.625	82.29	0.689
850	0.4149	1.1095	0.06028	1.3097	3.765	90.75	0.692
900	0.3925	1.1212	0.06279	1.4271	3.899	99.3	0.696
950	0.3716	1.1321	0.06525	1.5510	4.023	108.2	0.699
1000	0.3524	1.1417	0.06752	1.6779	4.152	117.8	0.702
1100	0.3204	1.160	0.0732	1.969	4.44	138.6	0.704
1200	0.2947	1.179	0.0782	2.251	4.69	159.1	0.707
1300	0.2707	1.197	0.0837	2.583	4.93	182.1	0.705
1400	0.2515	1.214	0.0891	2.920	5.17	205.5	0.705
1500	0.2355	1.230	0.0946	3.262	5.40	229.1	0.705
1600	0.2211	1.248	0.100	3.609	5.63	254.5	0.705
1700	0.2082	1.267	0.105	3.977	5.85	280.5	0.705
1800	0.1970	1.287	0.111	4.379	6.07	308.1	0.704
1900	0.1858	1.309	0.117	4.811	6.29	338.5	0.704
2000	0.1762	1.338	0.124	5.260	6.50	369.0	0.702
2100	0.1682	1.372	0.131	5.715	6.72	399.6	0.700
2200	0.1602	1.419	0.139	6.120	6.93	432.6	0.707
2300	0.1538	1.482	0.149	6.540	7.14	464.0	0.710
2400	0.1458	1.574	0.161	7.020	7.35	504.0	0.718
2500	0.1394	1.688	0.175	7.441	7.57	543.5	0.730

TABLE A–C–5(a) (Property values of gases at atmospheric pressure—continued)

T (K)	ρ (kg/m^3)	c_P [kJ/(kg °C)]	k [W/(m °C)]	$\alpha \times 10^4$ (m^2/s)	$\mu \times 10^5$ [kg/(m s)]	$\nu \times 10^6$ (m^2/s)	Pr
			Ammonia, NH_3				
300	0.6894	2.158	0.0247	0.166	1.015	14.7	0.887
320	0.6448	2.170	0.0272	0.194	1.09	16.9	0.870
340	0.6059	2.192	0.0293	0.221	1.165	19.2	0.872
360	0.5716	2.221	0.0316	0.249	1.24	21.7	0.872
380	0.5410	2.254	0.0340	0.279	1.31	24.2	0.869
400	0.5136	2.287	0.0370	0.315	1.38	26.9	0.853
420	0.4888	2.322	0.0404	0.356	1.45	29.7	0.833
440	0.4664	2.357	0.0435	0.396	1.525	32.7	0.826
460	0.4460	2.393	0.0463	0.434	1.59	35.7	0.822
480	0.4273	2.430	0.0492	0.474	1.665	39.0	0.822
500	0.4101	2.467	0.0525	0.519	1.73	42.2	0.813
520	0.3942	2.504	0.0545	0.552	1.80	45.7	0.827
540	0.3795	2.540	0.0575	0.597	1.865	49.1	0.824
560	0.3708	2.577	0.0606	0.634	1.93	52.0	0.827
580	0.3533	2.613	0.0638	0.691	1.995	56.5	0.817
			Carbon dioxide, CO_2				
220	2.4733	0.783	0.010805	0.05920	1.111	4.490	0.818
250	2.1657	0.804	0.012884	0.07401	1.259	5.813	0.793
300	1.7973	0.871	0.016572	0.10588	1.496	8.321	0.770
350	1.5362	0.900	0.02047	0.14808	1.721	11.19	0.755
400	1.3424	0.942	0.02461	0.19463	1.932	14.39	0.738
450	1.1918	0.980	0.02897	0.24813	2.134	17.90	0.721
500	1.0732	1.013	0.03352	0.3084	2.326	21.67	0.702
550	0.9739	1.047	0.03821	0.3750	2.508	25.74	0.685
600	0.8938	1.076	0.04311	0.4383	2.683	30.02	0.668
			Carbon monoxide, CO				
220	1.55363	1.0429	0.01906	0.11760	1.383	8.903	0.758
250	1.36530	1.0425	0.02144	0.15063	1.540	11.28	0.750
300	1.13876	1.0421	0.02525	0.21280	1.784	15.67	0.737
350	0.97425	1.0434	0.02883	0.2836	2.009	20.62	0.728
400	0.85363	1.0484	0.03226	0.3605	2.219	25.99	0.722
450	0.75848	1.0551	0.0436	0.4439	2.418	31.88	0.718
500	0.68223	1.0635	0.03863	0.5324	2.606	38.19	0.718
550	0.62024	1.0756	0.04162	0.6240	2.789	44.97	0.721
600	0.56850	1.0877	0.04446	0.7190	2.960	52.06	0.724

TABLE A–C–5(a) (Property values of gases at atmospheric pressure—continued)

T (K)	ρ (kg/m^3)	c_P [kJ/(kg °C)]	k [W/(m °C)]	$\alpha \times 10^4$ (m^2/s)	$\mu \times 10^5$ [kg/(m s)]	$\nu \times 10^6$ (m^2/s)	Pr
			Helium, He				
200	0.2435	5.200	0.1177	0.9288	1.566	64.38	0.694
255	0.1906	5.200	0.1357	1.3675	1.817	95.50	0.70
366	0.13280	5.200	0.1691	2.449	2.305	173.6	0.71
477	0.10204	5.200	0.197	3.716	2.750	269.3	0.72
589	0.08282	5.200	0.225	5.215	3.113	375.8	0.72
700	0.07032	5.200	0.251	6.661	3.475	494.2	0.72
800	0.06023	5.200	0.275	8.774	3.817	634.1	0.72
900	0.05286	5.200	0.298	10.834	4.136	781.3	0.72
			Hydrogen, H_2				
30	0.84722	10.840	0.0228	0.02493	0.1606	1.895	0.759
50	0.50955	10.501	0.0362	0.0676	0.2516	4.880	0.721
100	0.24572	11.229	0.0665	0.2408	0.4212	17.14	0.712
150	0.16371	12.602	0.0981	0.475	0.5595	34.18	0.718
200	0.12270	13.540	0.1282	0.772	0.6813	55.53	0.719
250	0.09819	14.059	0.1561	1.130	0.7919	80.64	0.713
300	0.08185	14.314	0.182	1.554	0.8963	109.5	0.706
350	0.07016	14.436	0.206	2.031	0.9954	141.9	0.697
400	0.06135	14.491	0.228	2.568	1.086	177.1	0.690
450	0.05462	14.499	0.251	3.164	1.178	215.6	0.682
500	0.04918	14.507	0.272	3.817	1.264	257.0	0.675
550	0.04469	14.532	0.292	4.516	1.348	301.6	0.668
600	0.04085	14.537	0.315	5.306	1.429	349.7	0.664
700	0.03492	14.574	0.351	6.903	1.589	455.1	0.659
800	0.03060	14.675	0.384	8.563	1.740	569	0.664
900	0.02723	14.821	0.412	10.217	1.878	690	0.676
1000	0.02451	14.968	0.440	11.997	2.016	822	0.686
1100	0.02227	15.165	0.464	13.726	2.146	965	0.703
1200	0.02050	15.366	0.488	15.484	2.275	1107	0.715
1300	0.01890	15.575	0.512	17.394	2.408	1273	0.733
1333	0.01842	15.638	0.519	18.013	2.444	1328	0.736
			Nitrogen, N_2				
100	3.4808	1.0722	0.009450	0.025319	0.6862	1.971	0.786
200	1.7108	1.0429	0.01824	0.10224	1.295	7.568	0.747
300	1.1421	1.0408	0.02620	0.22044	1.784	15.63	0.713
400	0.8538	1.0459	0.03335	0.3734	2.198	25.74	0.691
500	0.6824	1.0555	0.03984	0.5530	2.570	37.66	0.684
600	0.5687	1.0756	0.04580	0.7486	2.911	51.19	0.686
700	0.4934	1.0969	0.05123	0.9466	3.213	65.13	0.691
800	0.4277	1.1225	0.05609	1.1685	3.484	81.46	0.700
900	0.3796	1.1464	0.06070	1.3946	3.749	91.06	0.711
1000	0.3412	1.1677	0.06475	1.6250	4.400	117.2	0.724
1100	0.3108	1.1857	0.06850	1.8591	4.228	136.0	0.736
1200	0.2851	1.2037	0.07184	2.0932	4.450	156.1	0.748

TABLE A–C–5(a) (Property values of gases at atmospheric pressure—continued)

T (K)	ρ (kg/m^3)	c_P [kJ/(kg °C)]	k [W/(m °C)]	$\alpha \times 10^4$ (m^2/s)	$\mu \times 10^5$ [kg/(m s)]	$\nu \times 10^6$ (m^2/s)	Pr
			Oxygen, O_2				
100	3.9918	0.9479	0.00903	0.023876	0.7768	1.946	0.815
150	2.6190	0.9178	0.01367	0.05688	1.149	4.387	0.773
200	1.9559	0.9131	0.01824	0.10214	1.485	7.593	0.745
250	1.5618	0.9157	0.02259	0.15794	1.787	11.45	0.725
300	1.3007	0.9203	0.02676	0.22353	2.063	15.86	0.709
350	1.1133	0.9291	0.03070	0.2968	2.316	20.80	0.702
400	0.9755	0.9420	0.03461	0.3768	2.554	26.18	0.695
450	0.8682	0.9567	0.03828	0.4609	2.777	31.99	0.694
500	0.7801	0.9722	0.04173	0.5502	2.991	38.34	0.697
550	0.7096	0.9881	0.04517	0.6441	3.197	45.05	0.700
600	0.6504	1.0044	0.04832	0.7399	3.392	52.15	0.704
			Water vapor (steam)				
380	0.5863	2.060	0.0246	0.2036	1.271	21.6	1.060
400	0.5542	2.014	0.0261	0.2338	1.344	24.2	1.040
450	0.4902	1.980	0.0299	0.307	1.525	31.1	1.010
500	0.4405	1.985	0.0339	0.387	1.704	38.6	0.996
550	0.4005	1.997	0.0379	0.475	1.884	47.0	0.991
600	0.3652	2.026	0.0422	0.573	2.067	56.6	0.986
650	0.3380	2.056	0.0464	0.666	2.247	66.4	0.995
700	0.3140	2.085	0.0505	0.772	2.426	77.2	1.000
750	0.2931	2.119	0.0549	0.883	2.604	88.8	1.005
800	0.2739	2.152	0.0592	1.001	2.786	102	1.010
850	0.2579	2.186	0.0637	1.130	2.969	115	1.019

† The values of μ, k, c_P, and Pr are fairly insensitive to pressure for He, H_2, O_2, N_2, and air. The effect of pressure on ρ, ν, and α can generally be accounted for by use of the ideal gas law.

Source: Adapted with permission from Eckert and Drake [4].

TABLE A–C–5(b) Molecular weight and specific heat ratio of common gases‡

Gas	*Chemical formula*	*Molecular weight* $\mathcal{M}$	*Specific heat ratio* γ
Air	—	28.97	1.400
Ammonia	NH_3	17.03	
Carbon dioxide	CO_2	44.01	1.289
Carbon monoxide	CO	28.01	1.400
Helium	He	4.003	1.667
Hydrogen	H_2	2.016	1.409
Nitrogen	N_2	28.01	1.400
Oxygen	O_2	32.00	1.393
Water vapor	H_2O	18.02	1.327

‡ Values for specific heat ratio $\gamma = c_P/c_v$ are given at 300 K.

TABLE A–C–6 Liquid-vapor surface tension of various fluids

Liquid	*Saturation temperature* (°C)	*Surface tension*† $\sigma \times 10^3$ (N/m)
Ammonia	−70	42.4
	−50	37.9
	−40	35.4
Benzene	10	30.2
	50	25.0
	70	22.4
Carbon dioxide	−30	10.1
	−10	6.14
	10	2.67
	30	0.07
Ethyl alcohol	10	23.2
	50	19.9
	100	15.7
Hydrogen	−258	2.80
	−255	2.29
	−253	1.95
Mercury	10	488
	50	480
	100	470
	200	450
Methane	90	18.9
	100	16.3
	115	12.3
Nitrogen	78	8.75
	85	7.17
	90	6.04
Water	0	75.6
	15.6	73.2
	37.8	69.7
	93.3	60.1
	100	58.8
	160	46.1
	227	31.9
	293	16.2
	360	1.46
	374	0

† 1 N/m = 10^3 dyne/cm

TABLE A–C–7 Emissivities of various surfaces†

Metallic substances

Surface	*Temperature* 100 K	300 K		600 K		1000 K		
Aluminum								
Highly polished, film	0.02	0.04		0.06				
Foil	0.06	0.07						
Anodized		0.82	(400 K) 0.76					
Oxidized				**0.69**		**0.55**		(1500 K) **0.41**
Brass								
Polished		**0.05**			(573 K) **0.032**			
Oxidized				**0.22**				
Tarnished			(329 K) **0.20**					
Chromium								
Polished or plated	**0.05**	**0.10**		**0.14**				
Copper								
Highly polished, film		0.03		0.04		0.04		
Lightly tarnished		**0.037**						
Oxidized		**0.78**				0.86		
Gold								
Highly polished, film	0.01	0.03		0.04		0.06		
Foil	0.06	0.07						
Iron								
Polished					(698 K) **0.14**			
Oxidized			(398 K) **0.78**					
Lead								
Polished			(403 K) **0.056**					
Oxidized		**0.28**						
Nickel								
Polished				0.09		0.14	(1200 K) 0.17	
Oxidized				0.40		0.57		
Silver								
Polished		0.02		0.03	(800 K) 0.05		(1100 K) 0.08	
Stainless steels								
Typical, polished		**0.17**		**0.19**		**0.30**		
Typical, lightly oxidized					**0.33**	**0.40**		
Typical, highly oxidized					**0.67**	**0.70**	(1200 K) **0.70**	
AISI type 310					0.26	0.29		(1400 K) 0.29
AISI type 310 oxidized at 1255 K			(360 K) 0.46	(500 K) 0.36	(760 K) 0.64			
Tin, bright		**0.07**						
Zink								
Polished		0.02		(530 K) 0.03				(1600 K) 0.06
Tarnished		**0.25**						

† Values for normal emissivity ϵ_0 shown in **bold** typeset.

TABLE A-C-7 (Emissivities of various surfaces—continued)

Surface	*Temperature*							
	600 K	800 K	1000 K	1200 K	1500 K	2000 K	2500 K	3300 K
Molybdenum								
Polished	0.06		0.10		0.15	0.21	0.26	
Rough	0.25		0.31		0.42			
Oxidized	0.80	0.82						
Platinum								
Polished		0.10	0.13		0.18			
Tantalum								
Polished				0.11	0.17	0.23	0.28	
Tungsten								
Polished			**0.10**		**0.18**	**0.25**	**0.29**	**0.39**

Nonmetallic substances

Surface	*Temperature*						
	300 K			800 K			
Asphalt	0.93			0.90			
Building materials							
Brick-fire clay	0.90			0.70	(1600 K) 0.75		
Gypsum or plaster board	0.90–0.92						
Masonry, plastered	**0.93**						
Plaster, lime white, rough	**0.93**						
Wood							
Beech, planed		(343 K) **0.94**					
Oak, planed	**0.88**						
Carbon (graphite)					(2000 K) **0.65**	(2300 K) **0.69**	(2600 K) **0.74**
Cloth	0.75–0.90						
Concrete	0.88–0.93						
Glass							
Window	0.90–0.95						
Corning		(139 K) 0.68	(260 K) 0.82				
Pyrex		(82 K) **0.85**	(420 K) **0.88**		(1100 K) **0.75**		
Ice							
Smooth	**0.97**						
Rough	**0.985**						
Paints							
Aluminum bronze			(373 K) **0.20–0.40**				
Aluminum enamel (rough)	**0.39**						
Aluminum heated to 325°C			(423–588 K) **0.35**				

TABLE A–C–7 (Emissivities of various surfaces—continued)

Nonmetallic substances

Surface	*Temperature*						
	300 K			800 K			
Bakelite enamel			(353 K) **0.935**				
Black, Parsons	0.98		(460 K) 0.98				
Enamel							
White, rough	**0.90**						
Black, bright	**0.88**						
Lampblack	0.96		(530 K) 0.97				(1600 K) 0.97
Oil		(273–473 K) **0.88**					
Shellac, black							
Bright	**0.82**						
Dull		(348–418 K) **0.91**					
White, acrylic	0.90						
White, zink oxide (ZnO)	0.92				(1100 K) 0.82	(1200 K) 0.81	(1300 K) 0.82
Paper, white	0.92–0.97						
Porcelain, glazed	**0.93**						
Pyrex	**0.82**		(600 K) **0.80**		(1000 K) **0.71**	**0.62**	
Pyroceram	**0.85**		**0.78**		**0.69**		(1500 K) **0.57**
Quartz, fuzed (rough)	**0.93**						
Refractories (furnace liners)							
Alumina brick				**0.40**	**0.33**	(1400 K) **0.28**	(1600 K) **0.33**
Magnesia brick				**0.45**	**0.36**	**0.31**	**0.40**
Kaolin insulating brick				**0.70**	(1200 K) **0.57**	**0.47**	**0.53**
Rubber							
Soft, gray	**0.86**						
Hard, black (rough)	**0.95**						
Sand	0.90						
Silicon carbide			**0.87**		(1000 K) **0.87**		(1500 K) **0.85**
Skin	0.95						
Snow	0.82–0.90						
Soil	0.93–0.96						
Rocks	0.88–0.95						
Teflon	0.85	(400 K) 0.87	(500 K) 0.92				
Vegetation	0.92–0.96						
Water	0.96						

Sources: Raznjevic [5], Touloukian et al. [12–14], Mallory [15], Gubareff et al. [16], and Kreith and Kreider [17].

D EARTH TEMPERATURE DATA FOR SELECT U.S. CITIES

TABLE A–D–1 Earth temperature data for select U.S. cities†

	City	T_M (°F)	ΔT_s (°F)	t_0 (day)		*City*	T_M (°F)	ΔT_s (°F)	t_0 (day)
AL	Birmingham	65	19	31	NJ	Trenton	55	22	38
AZ	Phoenix	73	23	33	NM	Albuquerque	59	22	31
AR	Little Rock	64	21	32	NY	Albany	50	25	38
CA	Los Angeles	64	7	54	NC	Greensboro	60	20	31
CO	Denver	52	22	37	ND	Bismarck	44	31	33
DC	Washington	57	22	36	OH	Akron	52	23	37
FL	Jacksonville	71	14	32		Columbus	55	22	34
GA	Atlanta	62	19	32	OK	Ok. City	62	23	34
ID	Boise	53	21	34		Tulsa	62	23	34
IL	Chicago	51	25	37	OR	Portland	54	13	37
	Urbana	53	26	35	PA	Philadelphia	55	22	34
IN	Indianapolis	55	24	34		Pittsburgh	52	23	36
IA	Des Moines	52	28	35	SC	Charleston	66	16	32
KS	Topeka	56	26	35	SD	Rapid City	50	25	38
KY	Louisville	60	22	33	TN	Knoxville	61	21	31
LA	New Orleans	70	15	32		Nashville	60	21	32
MA	Portland	48	22	39	TX	Ft. Worth	68	21	34
MS	Plymouth	51	21	43		Houston	71	16	33
MI	Detroit	50	25	39	UT	Salt Lake	53	24	35
MN	Minneapolis	47	29	35	VT	Burlington	46	26	37
MS	Columbus	65	19	32	VA	Richmond	60	19	33
MO	Kansas City	58	26	35	WA	Seattle	53	12	36
MT	Billings	49	23	37		Spokane	49	21	32
NB	Lincoln	53	28	34	WV	Charleston	58	20	33
NV	Las Vegas	69	23	32	WI	Madison	49	27	36
					WY	Cheyenne	48	21	39

† T_M—Annual mean earth temperature. (Essentially constant to depth of 200 ft.)
ΔT_s—Amplitude of annual change in ground surface temperature.
t_0—Phase constant (i.e., day of the year at which ground surface temperature is a minimum).
Source: Adapted with permission from ASHRAE Handbook of Fundamentals [7].

E ANALYTICAL SOLUTIONS FOR CONDUCTION HEAT TRANSFER

E–1 STEADY TWO-DIMENSIONAL CONDUCTION IN RECTANGULAR SOLIDS

The separation-of-variables approach is utilized in this section to solve for the temperature distribution in the steady two-dimensional rectangular plate shown in Fig. 3–5. The differential formulation is given by

$$\frac{\partial^2 \psi}{\partial x^2} + \frac{\partial^2 \psi}{\partial y^2} = 0 \tag{3–30}$$

$$\psi = 0 \quad \text{at} \quad x = 0 \qquad \psi = 0 \quad \text{at} \quad x = L \tag{3–31,32}$$

$$\psi = 0 \quad \text{at} \quad y = 0 \qquad \psi = F(x) \quad \text{at} \quad y = w \tag{3–33,34}$$

where $\psi = T - T_1$ and $F(x) = f(x) - T_1$. Because only one nonhomogeneous condition appears in this system of equations, we are ready to employ the separation-of-variables approach.

We now assume a product solution of the form

$$\psi(x,y) = X(x)\,Y(y) \tag{E–1–1}$$

where $X(x)$ and $Y(y)$ are functions of x and y, respectively. Substituting this assumed product solution into Eqs. (3–30) through (3–34), we obtain

$$Y\frac{d^2X}{dx^2} + X\frac{d^2Y}{dy^2} = 0 \tag{E–1–2}$$

$$X(0)\,Y(y) = 0 \qquad X(L)\,Y(y) = 0 \tag{E–1–3,4}$$

$$X(x)\,Y(0) = 0 \qquad X(x)\,Y(w) = F(x) \tag{E–1–5,6}$$

Rearranging Eq. (E–1–2) with the purpose of separating the variables, we obtain

$$\frac{1}{X}\frac{d^2X}{dx^2} = -\frac{1}{Y}\frac{d^2Y}{dy^2} \tag{E–1–7}$$

The left-hand side of this equation is a function of x alone and the right-hand side is only a function of y. It follows that in order to satisfy Eq. (E–1–7), both sides of the equation must be equal to a constant; that is,

$$\frac{1}{X}\frac{d^2X}{dx^2} = -\frac{1}{Y}\frac{d^2Y}{dy^2} = \gamma_n \tag{E–1–8}$$

where eigenvalues γ_n represent all constants for which Eqs. (E–1–2) through (E–1–6) are satisfied. The value of γ_n must be selected on the basis of a consideration of the boundary conditions.

Equation (E–1–8) gives us two equations of the forms

$$\frac{d^2X}{dx^2} - \gamma_n X = 0 \tag{E–1–9}$$

and

$$\frac{d^2Y}{dy^2} + \gamma_n Y = 0 \tag{E–1–10}$$

Both of these equations are satisfied by exponential functions. The assumptions

$$X = Ae^{ax} \qquad \text{and} \qquad Y = Be^{by} \tag{E–1–11,12}$$

lead to values of a and b of the forms

$$a^2 = +\gamma_n \qquad a = \pm\sqrt{\gamma_n} \tag{E–1–13a,b}$$

$$b^2 = -\gamma_n \qquad b = \pm\sqrt{-\gamma_n} \tag{E–1–14a,b}$$

Hence, the solutions become

$$X = A_1 e^{\sqrt{\gamma_n}\,x} + A_2 e^{-\sqrt{\gamma_n}\,x} \tag{E–1–15}$$

$$Y = B_1 e^{\sqrt{-\gamma_n}\,y} + B_2 e^{-\sqrt{-\gamma_n}\,y} \tag{E–1–16}$$

The constants A_1, A_2, B_1, B_2, and γ_n are determined by use of the boundary conditions. We observe that the boundary conditions for X are both homogeneous. Thus, x is the homogeneous coordinate in this problem and y is the nonhomogeneous coordinate. The boundary conditions in the homogeneous coordinate are considered first. Equations (E–1–3) and (E–1–4) give

$$A_1 + A_2 = 0 \tag{E–1–17}$$

and

$$A_1 e^{\sqrt{\gamma_n}\,L} + A_2 e^{-\sqrt{\gamma_n}\,L} = 0 \tag{E–1–18}$$

or

$$A_1(e^{\sqrt{\gamma_n}\,L} - e^{-\sqrt{\gamma_n}\,L}) = 0 \tag{E–1–19}$$

If A_1 is taken as zero, we have the trivial solution $\psi = 0$. Therefore, the term in parentheses must be zero; that is,

$$e^{\sqrt{\gamma_n}\,L} - e^{-\sqrt{\gamma_n}\,L} = 0 \tag{E–1–20}$$

Our problem is to find all values of γ_n for which this equation is satisfied. To simplify matters, we introduce the substitution

$$\gamma_n = \pm\lambda_n^2 \tag{E–1–21}$$

For positive values of γ_n, Eq. (E–1–20) becomes

$$e^{\lambda_n L} - e^{-\lambda_n L} = 0 \qquad \text{or} \qquad \sinh(\lambda_n L) = 0 \tag{E–1–22,23}$$

This equation is only satisfied for $\lambda_n L = 0$, or $\lambda_n = 0$. The substitution of $\lambda_n = 0$ (or $\gamma_n = 0$) into Eqs. (E–1–15) and (E–1–16) is seen to produce the trivial result that X is independent of x and Y is independent of y. Therefore, we conclude that γ_n is not positive.

For negative values of γ_n, Eq. (E–1–20) takes the form

$$e^{i\lambda_n L} - e^{-i\lambda_n L} = 0 \qquad \text{or} \qquad \sin(\lambda_n L) = 0 \tag{E–1–24,25}$$

Because

$$\sin(n\pi) = 0 \qquad \text{for} \qquad n = 0,1,2, \ldots \tag{E–1–26}$$

we conclude that Eq. (E–1–20) is satisfied for

$$\lambda_n L = n\pi \qquad \text{for} \qquad n = 0,1,2, \ldots \tag{E–1–27}$$

Thus, γ_n is given by

$$\gamma_n = -\lambda_n^2 = -\left(\frac{n\pi}{L}\right)^2 \quad \text{for} \quad n = 0, 1, 2, \ldots \tag{E–1–28}$$

and our solution for X and Y for any integer n takes the form

$$X_n = A_1 e^{i(n\pi/L)x} + A_2 e^{-i(n\pi/L)x} \tag{E–1–29}$$

or (since $A_2 = -A_1$)

$$X_n = 2A_1 i \sin\frac{n\pi x}{L} \tag{E–1–30}$$

and

$$Y_n = B_1 e^{(n\pi/L)y} + B_2 e^{-(n\pi/L)y} \tag{E–1–31}$$

We now utilize the y-boundary conditions. Setting $Y = 0$ at $y = 0$ in accordance with Eq. (E–1–5), we have

$$0 = B_1 + B_2 \tag{E–1–32}$$

Thus, Eq. (E–1–31) takes the form

$$Y_n = B_1[e^{(n\pi/L)y} - e^{-(n\pi/L)y}] \quad \text{or} \quad Y_n = 2B_1 \sinh\frac{n\pi y}{L} \tag{E–1–33,34}$$

Substituting this result together with Eq. (E–1–30) into Eq. (E–1–1) gives

$$\psi_n = X_n Y_n = \left(2A_1 i \sin\frac{n\pi x}{L}\right)\left(2B_1 \sinh\frac{n\pi y}{L}\right) = C_n \sinh\frac{n\pi y}{L} \sin\frac{n\pi x}{L} \tag{E–1–35}$$

for any integer n, where $C_n = 2iA_1B_1$. Because ψ_n represents a solution for each integer, the full solution is given by

$$\psi = \sum_{n=1}^{\infty} \psi_n = \sum_{n=1}^{\infty} C_n \sinh\frac{n\pi y}{L} \sin\frac{n\pi x}{L} \tag{E–1–36}$$

The coefficient C_n is evaluated by use of the final nonhomogeneous y-boundary condition by substituting Eq. (E–1–36) into Eq. (E–1–6); that is,

$$F(x) = \sum_{n=1}^{\infty} C_n \sinh\frac{n\pi w}{L} \sin\frac{n\pi x}{L}$$

$$F(x) = \sum_{n=1}^{\infty} c_n \sin\frac{n\pi x}{L} \tag{E–1–37}$$

where $c_n = C_n \sinh(n\pi w/L)$. This equation is recognized as *the Fourier sine series* expansion of the function $F(x)$, where c_n can be expressed as

$$c_n = \frac{2}{L}\int_0^L F(x) \sin\frac{n\pi x}{L}\, dx \tag{E–1–38}$$

Thus, our final solution for the temperature distribution takes the form

$$T - T_1 = \psi = \sum_{n=1}^{\infty} c_n \frac{\sinh(n\pi y/L)}{\sinh(n\pi w/L)} \sin\frac{n\pi x}{L} \tag{E–1–39}$$

E–2 UNSTEADY ONE-DIMENSIONAL CONDUCTION WITH CONVECTIVE COOLING

Exact solutions are listed in this section for the temperature distribution in a plane wall, infinite cylinder, and sphere for unsteady one-dimensional convective cooling. The solutions are developed in reference 18 by means of the separation-of-variables approach.

E–2–1 Plane Wall

The mathematical formulation for convective cooling of a plane wall of thickness $2L$ is given by

$$\frac{\partial^2 T}{\partial x^2} = \frac{1}{\alpha}\frac{\partial T}{\partial t}$$

$$T = T_i \qquad \text{at } t = 0$$

$$-k\frac{\partial T}{\partial x} = h(T - T_F) \qquad \text{at } x = L \tag{3–56}$$

and

$$\frac{\partial T}{\partial x} = 0 \qquad \text{at } x = 0$$

or

$$-k\frac{\partial T}{\partial x} = h(T_F - T) \qquad \text{at } x = -L$$

The solution to this equation is given by Schneider [18] as

$$\Theta = \frac{T - T_F}{T_i - T_F} = \sum_{n=1}^{\infty} C_n \exp(-\gamma_n^2 Fo)\cos\left(\gamma_n \frac{x}{L}\right) \tag{E–2–1}$$

where the coefficient C_n is given by

$$C_n = \frac{4\sin\gamma_n}{2\gamma_n + \sin(2\gamma_n)} \tag{E–2–2}$$

and the eigenvalues γ_n are positive roots of the transcendental equation

$$\gamma_n \tan\gamma_n = Bi_0 \tag{E–2–3}$$

The first five roots of this equation are given in Table A–E–2–1. As pointed out in Sec. 3–9–3, for values of the Fourier number Fo greater than about 0.2, Eq.

TABLE A–E–2–1 First five roots γ_n: $\gamma_n \tan \gamma_n = Bi_0$

Bi_0	γ_1	γ_2	γ_3	γ_4	γ_5
0.000	0.0000	3.1416	6.2832	9.4248	12.5664
0.002	0.0447	3.1422	6.2835	9.4250	12.5665
0.004	0.0632	3.1429	6.2838	9.4252	12.5667
0.006	0.0774	3.1435	6.2841	9.4254	12.5668
0.008	0.0893	3.1441	6.2845	9.4256	12.5670
0.010	0.0998	3.1448	6.2848	9.4258	12.5672
0.020	0.1410	3.1479	6.2864	9.4269	12.5680
0.040	0.1987	3.1543	6.2895	9.4290	12.5696
0.060	0.2425	3.1606	6.2927	9.4311	12.5711
0.080	0.2791	3.1668	6.2959	9.4333	12.5727
0.100	0.3111	3.1731	6.2991	9.4354	12.5743
0.200	0.4328	3.2039	6.3148	9.4459	12.5823
0.300	0.5218	3.2341	6.3305	9.4565	12.5902
0.400	0.5932	3.2636	6.3461	9.4670	12.5981
0.500	0.6533	3.2923	6.3616	9.4775	12.6060
0.600	0.7051	3.3204	6.3770	9.4979	12.6139
0.700	0.7506	3.3477	6.3923	9.4983	12.6218
0.800	0.7910	3.3744	6.4074	9.5087	12.6296
0.900	0.8274	3.4003	6.4224	9.5190	12.6375
1.000	0.8603	3.4256	6.4373	9.5293	12.6453
1.500	0.9882	3.5422	6.5097	9.5801	12.6841
2.000	1.0769	3.6436	6.5783	9.6296	12.7223
3.000	1.1925	3.8088	6.7040	9.7240	12.7966
4.000	1.2646	3.9352	6.8140	9.8119	12.8678
5.000	1.3138	4.0336	6.9096	9.8928	12.9352
6.000	1.3496	4.1116	6.9924	9.9667	12.9988
7.000	1.3766	4.1746	7.0640	10.0339	13.0584
8.000	1.3978	4.2264	7.1263	10.0949	13.1141
9.000	1.4149	4.2694	7.1806	10.1502	13.1660
10.000	1.4289	4.3058	7.2281	10.2003	13.2142
15.000	1.4729	4.4255	7.3959	10.3898	13.4078
20.000	1.4961	4.4915	7.4954	10.5117	13.5420
30.000	1.5202	4.5615	7.6057	10.6543	13.7085
40.000	1.5325	4.5979	7.6647	10.7334	13.8048
50.000	1.5400	4.6202	7.7012	10.7832	13.8666
100.000	1.5552	4.6658	7.7764	10.8871	13.9981

(E–2–1) can be approximated by truncating second and higher-order terms. The resulting relations are summarized in Table 3–5 and form the basis for the famous Heisler charts given in Fig. 3–12.

E–2–2 Infinite Cylinder

The mathematical formulation for convective cooling of an infinite cylinder with radius r_0 is given by

$$\frac{1}{r}\frac{\partial}{\partial r}\left(r\frac{\partial T}{\partial r}\right) = \frac{1}{\alpha}\frac{\partial T}{\partial t} \tag{3–63}$$

$$T = T_i \qquad \text{at } t = 0$$

$$\frac{\partial T}{\partial r} = 0 \qquad \text{at } r = 0 \tag{3–65}$$

and

$$-k\frac{\partial T}{\partial r} = h(T - T_F) \qquad \text{at } r = r_0$$

The solution to this equation is given by Schneider [18] as

$$\Theta = \frac{T - T_F}{T_i - T_F} = \sum_{n=1}^{\infty} C_n \exp(-\gamma_n^2 Fo)\, J_0\left(\gamma_n \frac{r}{r_0}\right) \tag{E–2–4}$$

TABLE A–E–2–2 First five roots γ_n: $\gamma_n J_1(\gamma_n) = Bi_0 J_0(\gamma_n)$

Bi_0	γ_1	γ_2	γ_3	γ_4	γ_5
0.00	0.0000	3.8317	7.0156	10.1735	13.3237
0.02	0.1995	3.8369	7.0184	10.1754	13.3252
0.04	0.2814	3.8421	7.0213	10.1774	13.3267
0.06	0.3438	3.8473	7.0241	10.1794	13.3282
0.08	0.3960	3.8525	7.0270	10.1813	13.3297
0.10	0.4417	3.8577	7.0298	10.1833	13.3312
0.20	0.6170	3.8835	7.0440	10.1931	13.3387
0.30	0.7465	3.9091	7.0582	10.2029	13.3462
0.40	0.8516	3.9344	7.0723	10.2127	13.3537
0.50	0.9408	3.9594	7.0864	10.2225	13.3611
0.60	1.0184	3.9841	7.1004	10.2322	13.3686
0.70	1.0873	4.0085	7.1143	10.2419	13.3761
0.80	1.1490	4.0325	7.1282	10.2516	13.3835
0.90	1.2048	4.0562	7.1421	10.2613	13.3910
1.00	1.2558	4.0795	7.1558	10.2710	13.3984
2.00	1.5994	4.2910	7.2884	10.3658	13.4719
3.00	1.7887	4.4634	7.4103	10.4566	13.5434
4.00	1.9081	4.6018	7.5201	10.5423	13.6125
5.00	1.9898	4.7131	7.6177	10.6223	13.6786
6.00	2.0490	4.8033	7.7039	10.6964	13.7414
7.00	2.0937	4.8772	7.7797	10.7646	13.8008
8.00	2.1286	4.9384	7.8464	10.8271	13.8566
9.00	2.1566	4.9897	7.9051	10.8842	13.9090
10.00	2.1795	5.0332	7.9569	10.9363	13.9580
15.00	2.2509	5.1773	8.1422	11.1367	14.1576
20.00	2.2880	5.2568	8.2534	11.2677	14.2983
30.00	2.3261	5.3410	8.3771	11.4221	14.4748
40.00	2.3455	5.3846	8.4432	11.5081	14.5774
50.00	2.3572	5.4112	8.4840	11.5621	14.6433
100.00	2.3809	5.4652	8.5678	11.6747	14.7834

where the coefficient C_n is given by

$$C_n = \frac{2}{\gamma_n} \frac{J_1(\gamma_n)}{J_0^2(\gamma_n) + J_1^2(\gamma_n)} \qquad \text{(E–2–5)}$$

and the eigenvalues γ_n are positive roots of the transcendental equation

$$\gamma_n J_1(\gamma_n) = Bi_0 J_0(\gamma_n) \qquad \text{(E–2–6)}$$

The Bessel functions J_0 and J_1 are listed in Table A–H–1–2 and the first five roots of this equation are given in Table A–E–2–2. These results provide the basis for the approximate Heisler relations and charts given in Table 3–6 and Fig. 3–13.

E–2–3 Sphere

The mathematical formulation for convective cooling of a sphere of radius r_0 is given by

$$\frac{1}{r^2} \frac{\partial}{\partial r}\left(r^2 \frac{\partial T}{\partial r}\right) = \frac{1}{\alpha} \frac{\partial T}{\partial t} \qquad \text{(3–64)}$$

$$T = T_i \qquad \text{at } t = 0$$

$$\frac{\partial T}{\partial r} = 0 \qquad \text{at } r = 0 \qquad \text{(3–65)}$$

and

$$-k \frac{\partial T}{\partial r} = h(T - T_F) \qquad \text{at } r = r_0$$

The solution to this equation is given by Schneider [18] as

$$\Theta = \frac{T - T_F}{T_i - T_F} = \sum_{n=1}^{\infty} C_n \exp(-\gamma_n^2 Fo) \frac{\sin(\gamma_n r/r_0)}{\gamma_n r/r_0} \qquad \text{(E–2–7)}$$

where the coefficient C_n is given by

$$C_n = \frac{4(\sin \gamma_n - \gamma_n \cos \gamma_n)}{2\gamma_n - \sin(2\gamma_n)} \qquad \text{(E–2–8)}$$

and the eigenvalues γ_n are positive roots of the transcendental equation

$$1 - \gamma_n \cot \gamma_n = Bi_0 \qquad \text{(E–2–9)}$$

The first five roots of this equation are given in Table A–E–2–3. These results provide the basis for the approximate Heisler relations and charts given in Table 3–7 and Fig. 3–14.

TABLE A–E–2–3 First five roots γ_n: $1 - \gamma_n \cot \gamma_n = Bi_0$

Bi_0	γ_1	γ_2	γ_3	γ_4	γ_5
0.000	0.0000	4.4934	7.7253	10.9041	14.0662
0.005	0.1224	4.4945	7.7259	10.9046	14.0666
0.010	0.1730	4.4956	7.7265	10.9050	14.0669
0.020	0.2445	4.4979	7.7278	10.9060	14.0676
0.030	0.2991	4.5001	7.7291	10.9069	14.0683
0.040	0.3450	4.5023	7.7304	10.9078	14.0690
0.050	0.3854	4.5045	7.7317	10.9087	14.0697
0.060	0.4217	4.5068	7.7330	10.9096	14.0705
0.070	0.4551	4.5090	7.7343	10.9105	14.0712
0.080	0.4860	4.5112	7.7356	10.9115	14.0719
0.090	0.5150	4.5134	7.7369	10.9124	14.0726
0.100	0.5423	4.5157	7.7382	10.9133	14.0733
0.200	0.7593	4.5379	7.7511	10.9225	14.0804
0.300	0.9208	4.5601	7.7641	10.9316	14.0875
0.400	1.0528	4.5822	7.7770	10.9408	14.0946
0.500	1.1656	4.6042	7.7899	10.9499	14.1017
0.600	1.2644	4.6261	7.8028	10.9591	14.1088
0.700	1.3525	4.6479	7.8156	10.9682	14.1159
0.800	1.4320	4.6696	7.8284	10.9774	14.1230
0.900	1.5044	4.6911	7.8412	10.9865	14.1301
1.000	1.5708	4.7124	7.8540	10.9956	14.1372
1.500	1.8366	4.8158	7.9171	11.0409	14.1724
2.000	2.0288	4.9132	7.9787	11.0856	14.2075
3.000	2.2889	5.0870	8.0962	11.1727	14.2764
4.000	2.4557	5.2329	8.2045	11.2560	14.3434
5.000	2.5704	5.3540	8.3029	11.3349	14.4080
6.000	2.6537	5.4544	8.3914	11.4086	14.4699
7.000	2.7165	5.5378	8.4703	11.4773	14.5288
8.000	2.7654	5.6078	8.5406	11.5408	14.5847
9.000	2.8044	5.6669	8.6031	11.5994	14.6374
10.000	2.8363	5.7172	8.6587	11.6532	14.6870
11.000	2.8628	5.7606	8.7083	11.7027	14.7335
16.000	2.9476	5.9080	8.8898	11.8959	14.9251
21.000	2.9930	5.9921	9.0019	12.0250	15.0625
31.000	3.0406	6.0831	9.1294	12.1807	15.2380
41.000	3.0651	6.1311	9.1987	12.2688	15.3417
51.000	3.0801	6.1606	9.2420	12.3247	15.4090
101.000	3.1105	6.2211	9.3317	12.4426	15.5537

F NUMERICAL METHODS: THE FINITE-ELEMENT APPROACH

The finite-element approach, having been introduced and developed to a high level of sophistication in the field of structural mechanics, has recently evolved into a major computational tool for solving heat-transfer and fluid-flow problems.† Whereas many

† The finite-element approach was initially developed to calculate stress in irregularly shaped objects and analyze structural problems in aircraft [19].

TABLE A–F–1 Comparison of steps in formulation of finite-element and finite-difference methods

Step	*Finite-Element Method*	*Finite-Difference Method*
(1)	Discretization of the calculation domain into *finite-element nodal networks.*	Discretization of the calculation domain into *finite-difference nodal networks.*
(2)	Establishment of finite-element approximations (featuring *interpolation functions*) for the dependent variables and (for unsteady problems) use of finite-difference (or other type) approximations for derivatives in time.	Use of finite-difference approximations for derivatives in space and (for unsteady problems) time.
(3)	Derivation and compilation of nodal equations using fundamental conservation/ control volume formulation concepts[1], or weighted integral techniques[2], or the variational calculus[3].	Derivation and compilation of nodal equations using fundamental conservation/ control volume formulation concepts[4] or Taylor series expansions[5].
(4)	Computer solution of system(s) of algebraic nodal equations.	Computer solution of system(s) of algebraic nodal equations.

[1] Control-volume finite-element method CVFEM.
[2] Method of weighted residuals MWR.
[3] Variational method.
[4] Control-volume finite-difference method CVFDM.
[5] Discretization of primitive form of transport equation—equivalent to CVFDM. Discretization of derived form of transport equation—discretization finite-difference method.

problems can be handled equally well by either the finite-element approach or the finite-difference approach, some problems are better suited to the finite-element approach and others to the finite-difference approach.† In fact, the present state of affairs is such that advanced commercial numerical heat-transfer codes encountered by an engineer are nearly as likely to be based on the finite-element approach as the finite-difference approach. Thus, a basic understanding of the finite-element approach has become important to engineers and scientists who deal with heat-transfer problems.

Formulation concepts pertaining to the finite-element approach include the development of finite-element nodal networks and finite-element approximations and the derivation of finite-element nodal equations using fundamental and particular laws. The steps involved in the numerical finite-element and finite-difference approaches to solving problems are summarized and compared in Table A–F–1. In connection with this table, the essential differences in these two approaches pertain to the nature of (1) the nodal networks, (2) the finite-element and finite-difference approximations, and (3) the derivation of the nodal equations. We also see that the four general steps followed in the two methods are similar. It follows that a clear understanding of the finite-difference approach serves as a useful prerequisite to the study of the finite-element approach. As we shall see, this is particularly true for the more recently developed control-volume finite-element method CVFEM, which employs conser-

† In this connection, finite-element methods are particularly well suited to problems involving irregular boundaries. However, advanced finite-difference methods that feature the use of orthogonal body-fitted coordinates are also commonly used in analyzing such problems.

vation/control volume formulation concepts (of the same type used in the control-volume finite-difference method CVFDM) in the development of nodal equations.

We now turn our attention to the basic formulation concepts that are featured in the finite-element approach, after which the numerical finite-element approach is developed for one-dimensional heat transfer. The formulation concepts presented in this section are extended to two- and three-dimensional steady and unsteady systems in references 20 and 21.

F–1 FINITE-ELEMENT NODAL NETWORKS

As in the finite-difference approach, the finite-element approach features the representation of a computational domain by a number Z_s of discrete nodal points at which the temperature (and/or other dependent variables) is to be approximated. The location of these nodes is designated by the *finite-element nodal grid*. This grid provides a framework for subdividing the total volume V into individual element volumes V_e within which the continuous distribution in temperature can be approximated in terms of the nodal temperatures associated with each element. We will refer to the system of nodes and subvolumes as the *finite-element nodal network*.†

F–1–1 One-Dimensional Elements

A one-dimensional finite-element grid featuring Z_s [$= M = 5$] nodes is shown in Fig. A–F–1(a). This grid can be used to establish various types of finite-element networks that identify I_s individual elements designated by $e_1, e_2, \ldots, e_m, \ldots, e_{I_s}$. The simplest and most popular are the *two-node* network ($I_s = M - 1$) and the *three-node* network [$I_s = (M - 1)/2$] shown in Figs. A–F–1(bi) and (ci). Control volume forms of these one-dimensional finite-element nodal networks are shown in Figs. A–F–1(bii) and (cii). As in the finite-difference approach, each control volume is associated with a single node. Higher-order one-dimensional finite-element nodal networks include four-node (cubic), five-node (quartic), and six-node (quintic) elements.

F–1–2 Two- and Three-Dimensional Elements

The finite-element approach for two- and three-dimensional problems can be formulated in the context of (1) the orthogonal grids used in the finite-difference approach or (2) more general *unstructured* grids that are generated by triangular, quadrilateral,

† Depending on the particular type of finite-element method being used, a finite-element nodal network may or may not involve the designation of a *control volume* for each node. More specifically, control volumes are employed in the control-volume finite-element method, but not in other conventional finite-element methods.

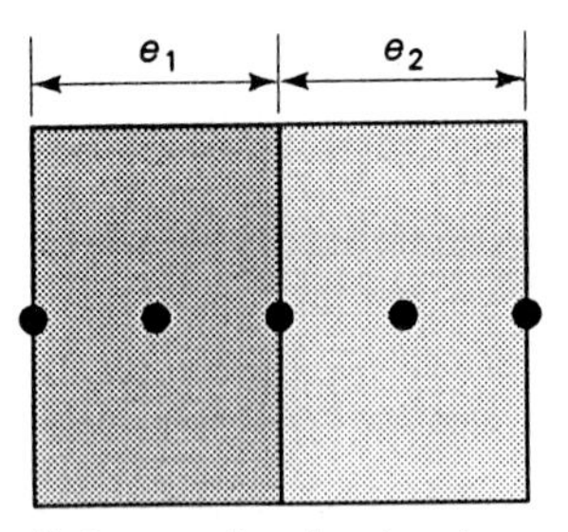

(i) Conventional network.

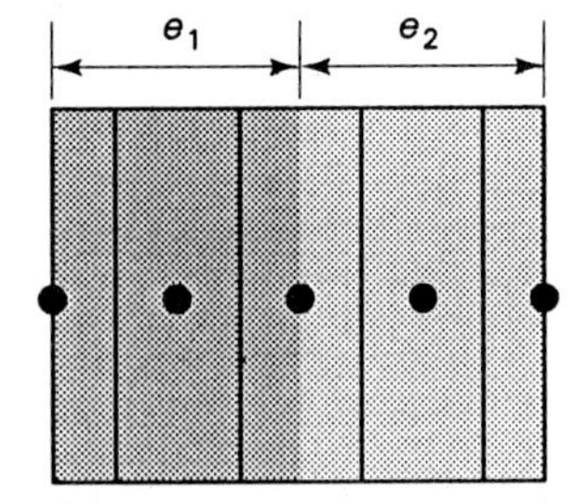

(ii) Control-volume network.

(c) Finite-element nodal network with $I_s = 2$ elements.

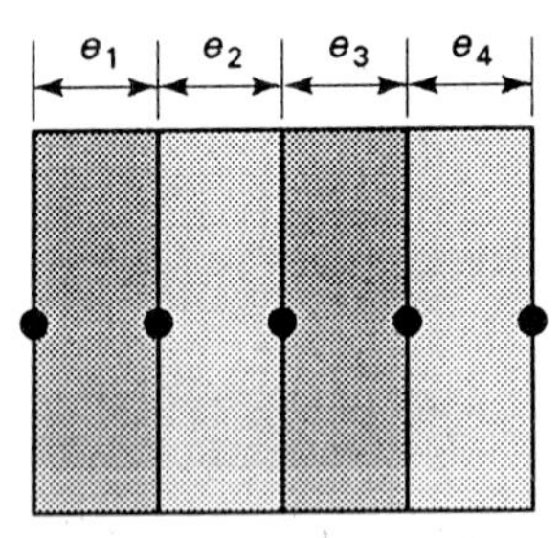

(i) Conventional network.

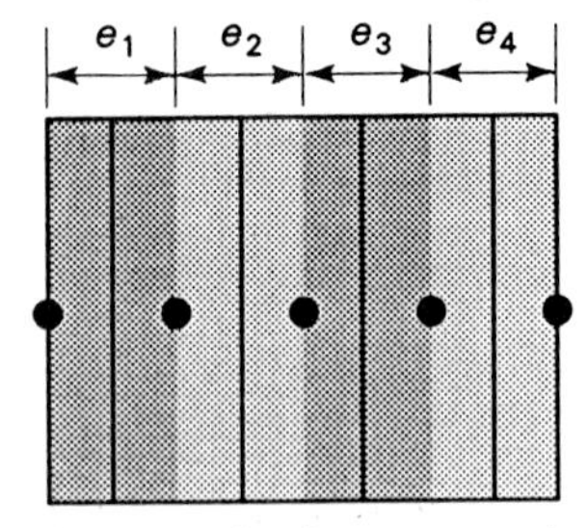

(ii) Control-volume network.

(b) Finite-element nodal network with $I_s = 4$ elements.

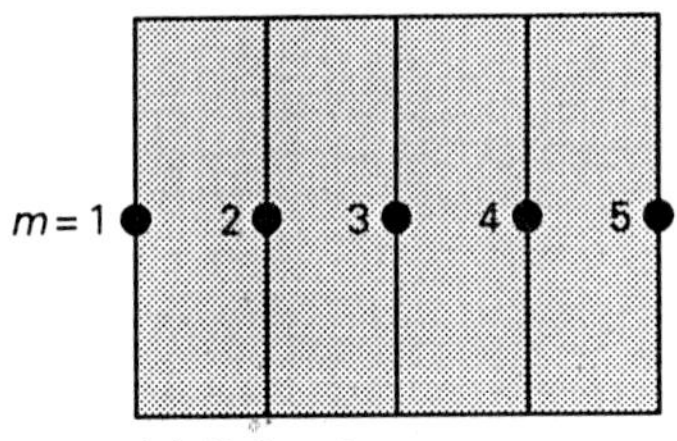

(a) Finite-element grid.

FIGURE A–F–1
Representative one-dimensional finite-element grid and nodal networks; $Z_s = M = 5$. (Alternate elements are shaded dark.)

or other nonorthogonal geometrically shaped elements.† We want to consider the development of finite-element nodal networks for both types of grids.

Structured Grids

A representative two-dimensional Cartesian finite-element grid is shown in Fig. A–F–2(a). This structured grid is used to establish the standard rectangular finite-element nodal network shown in Fig. A–F–2(bi), which features $Z_s = 16$ [$= MN$]

† Developmental work on the formulation of unstructured grid/finite-difference methods has recently been reported by several commercial software companies.

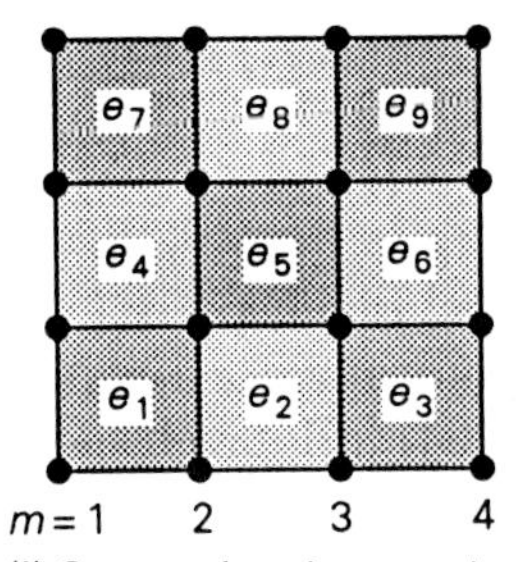

(i) Conventional network.

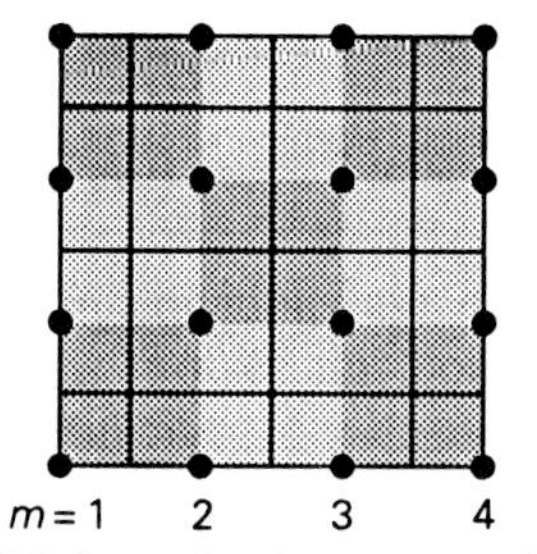

(ii) Control-volume network.

(b) Finite-element nodal network with $I_s = 9$ rectangular elements.

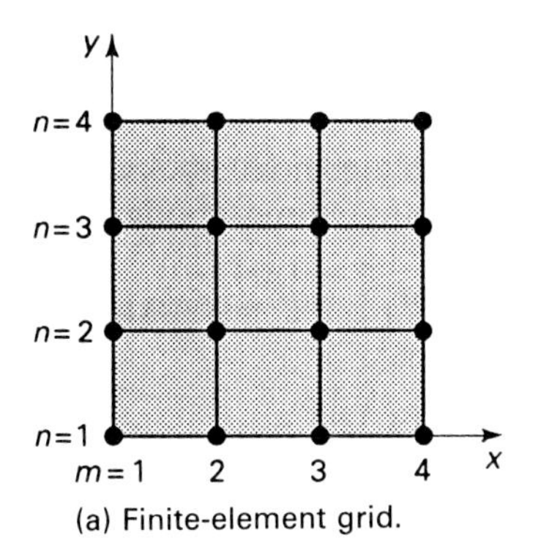

(a) Finite-element grid.

FIGURE A–F–2
Representative two-dimensional finite-element grid and nodal network: rectangular elements with $M = 4$, $N = 4$, and $Z_s = 16$.

nodes and $I_s = 9$ [$= (M - 1)(N - 1)$] rectangular elements designated by e_1, e_2, . . . , e_9. The control volume form of this rectangular finite-element nodal network is shown in Fig. A–F–2(bii).

The grid represented by Fig. A–F–2(a) also provides the basis for establishing triangular finite-element nodal networks that involve I_s [$= 2(M - 1)(N - 1)$] right-triangular elements. Conventional and control volume forms of a representative triangular finite-element nodal network with $Z_s = 16$ nodes and $I_s = 18$ elements are shown in Fig. A–F–3.

Similar structured finite-element nodal networks can be developed in the context of eight- and nine-node rectangular elements and cylindrical, spherical, and orthogonal body-fitted curvilinear coordinate grids.

Unstructured Grids

The development of unstructured grids that provide an effective means of discretizing irregular shapes (i.e., controlling the concentration of nodes within any region of a computational domain) is made possible by the use of nonorthogonal elements such as the two-dimensional quadrilateral and triangular elements shown in Fig. A–F–4.†

† Triangular elements can be formed by diagonalization of a quadrilateral element or, conversely, quadrilateral elements can be formed by combining triangular elements.

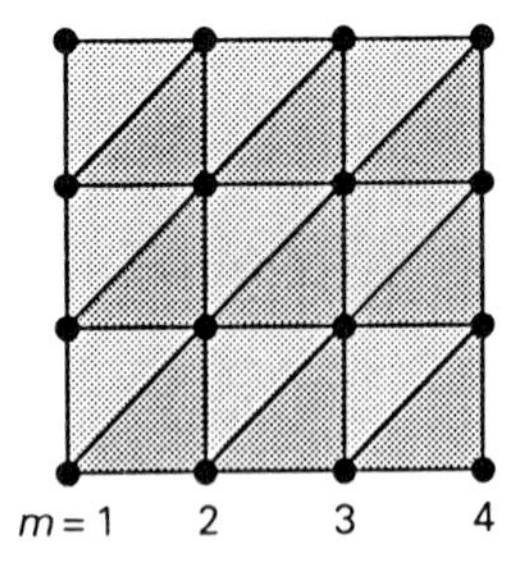

(i) Conventional network.

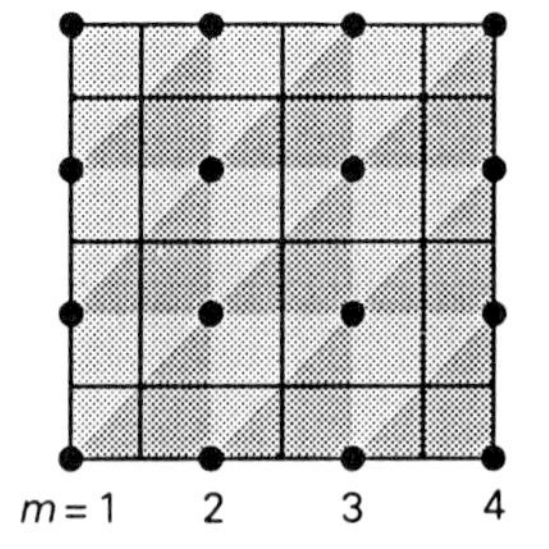

(ii) Control-volume network.

(b) Finite-element nodal network with $I_s = 18$ right-triangular elements.

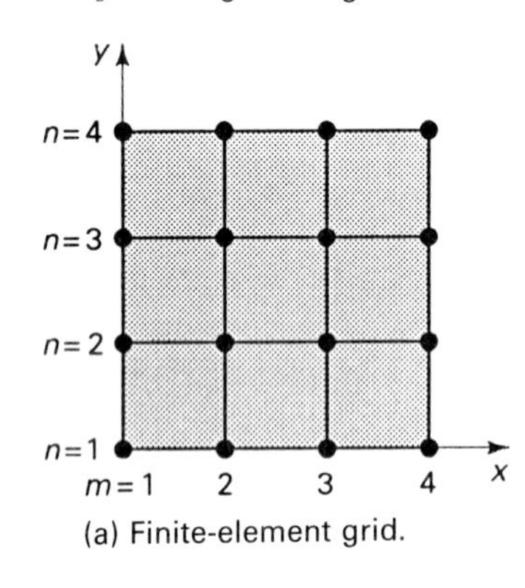

(a) Finite-element grid.

FIGURE A–F–3
Representative two-dimensional finite-element grid and nodal network: right triangular elements with $M = 4$, $N = 4$, and $Z_s = 16$.

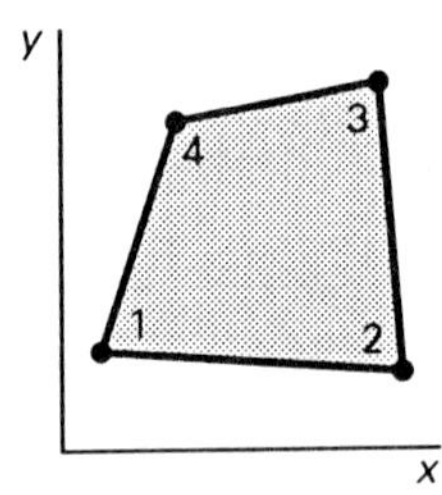

(a) Four-node quadrilateral element.

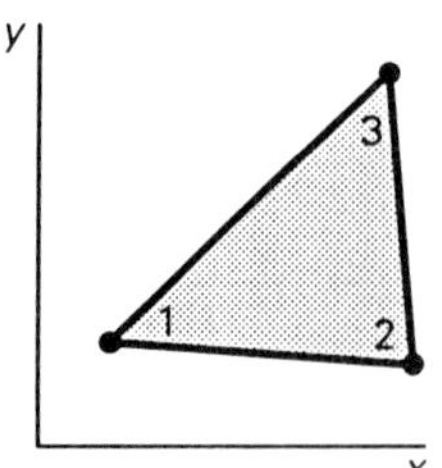

(b) Three-node triangular element.

FIGURE A–F–4
Basic two-dimensional finite-element building blocks for unstructured finite-element grids.

Notice that the quadrilateral (and triangular) element is a generalization of the basic rectangular (and right triangular) element, which is associated with a structured Cartesian coordinate grid. The two elements shown in Fig. A–F–4 and their three-dimensional counterparts are among a wide variety of element configurations that are in common use in commercial finite-element codes.

F–2 FINITE-ELEMENT APPROXIMATIONS

Whereas the finite-difference approach is based on the use of *approximations for spatial derivatives* in terms of unknown nodal temperatures in the vicinity of individual nodes, the finite-element approach features the use of *approximations for the temperature distribution* within each individual element in terms of unknown nodal temperatures associated with the element. We now turn our attention to the development of finite-element approximations for the temperature distribution within individual elements, which involves the use of *interpolation functions*.†

F-2-1 Interpolation Functions

A general form of the finite-element approximation for temperature distribution within an element with Z_e nodes is represented in terms of Cartesian coordinates x, y, z by

$$T(x,y,z) = \sum_{i=1}^{Z_e} N_i(x,y,z)\, T_i \tag{F-1}$$

where T_i represents the nodal temperatures and $N_i(x,y,z)$ is the *interpolation function* (also referred to as *basis function* or *shape function*).‡ The interpolation function $N_i(x,y,z)$ associated with node i is a function of the position within common elements of which the node is a part and is equal to zero outside this immediate neighborhood. The *local* nature of the interpolation function is established by requiring that $N_i(x,y,z)$ vary from unity at node i to zero at neighboring nodes within common elements. Equation (F–1) is used in this section to obtain basic finite-element approximations associated with one- and two-dimensional elements.

One-Dimensional Elements

Referring to Fig. A–F–5, the finite-element approximation indicated by Eq. (F–1) for the one-dimensional two-node element e_m associated with nodes m and $m + 1$ becomes

$$T(x) = \sum_{i=1}^{Z_e} N_i(x)\, T_i \tag{F-2}$$

$$= N_m(x)\, T_m + N_{m+1}(x)\, T_{m+1} \tag{F-2a}$$

for interval $[x_m, x_{m+1}]$. To establish a linear finite-element approximation, we simply require

$$T(x) = c_1 + c_2 x \tag{F-3}$$

† In the finite-element approach, time derivatives are generally evaluated by standard finite-difference approximations, such as those introduced in Sec. 4–3–2.

‡ The temperature distribution and nodal temperatures are represented by $T(t,x,y,z)$ and $T_i(t)$, respectively, for unsteady conditions.

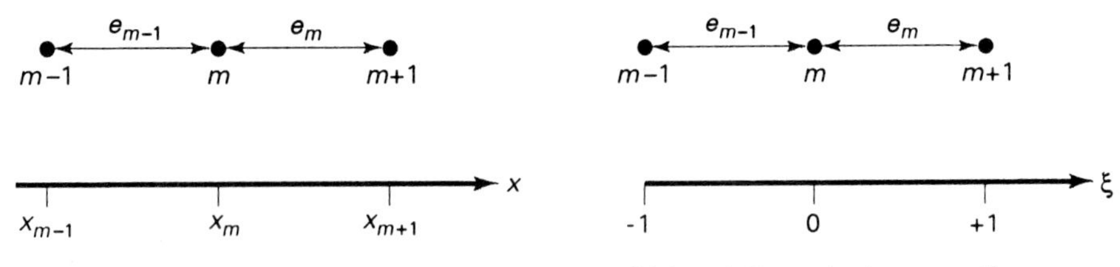

(a) Global dimensional coordinate x.

(b) Local dimensionless coordinate $\xi = (x - x_m)/\Delta x$.

FIGURE A–F–5 Coordinate systems for two-node elements e_{m-1} and e_m associated with nodes $m - 1$, m, and $m + 1$.

within the element, which indicates *linear interpolation functions* of the form

$$N_m(x) = \frac{x_{m+1} - x}{x_{m+1} - x_m} = \frac{x_{m+1} - x}{\Delta x}$$
$$N_{m+1}(x) = \frac{x - x_m}{x_{m+1} - x_m} = \frac{x - x_m}{\Delta x} \tag{F–4a}$$

where $\Delta x = x_{m+1} - x_m = x_m - x_{m-1}$ for uniform grid increment. Introducing a *local dimensionless coordinate* $\xi = (x - x_m)/\Delta x$ [see Fig. A–F–5(b)], Eq. (F–4a) reduces to the more convenient form

$$N_m(x) = \frac{x_{m+1} - x}{\Delta x} = 1 - \xi \qquad N_{m+1}(x) = \frac{x - x_m}{\Delta x} = \xi \tag{F–5a}$$

in the ξ interval [0,1]. The linear interpolation functions given by Eq. (F–5a) are readily used to represent the temperature distribution within each individual element $e_1, e_2, \ldots, e_m, \ldots, e_{M-1}$ by setting $m = 1, 2, \ldots, M - 1$. These interpolation functions are shown in Fig. A–F–6(a) for element e_m.

Whereas Eqs. (F–2a) and (F–5a) can be used to represent finite-element approximations for the temperature distribution in each element of the entire computational domain, the formulation details can be simplified by expressing the temper-

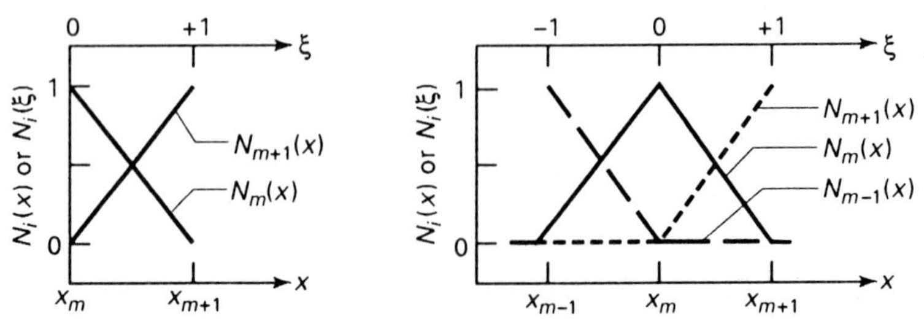

(a) Representation for element e_m.

(b) Composite representation for elements e_{m-1} and e_m.

FIGURE A–F–6 One-dimensional linear interpolation functions.

ature distribution within both elements that are common to node m (i.e., elements e_{m-1} and e_m) in terms of the same local dimensionless coordinate ξ. Therefore, Eq. (F–2) is applied to the two-node element e_{m-1} to obtain

$$T(x) = N_{m-1}(x)\, T_{m-1} + N_m(x)\, T_m \tag{F–2b}$$

for interval $[x_{m-1}, x_m]$, which when compared with Eq. (F–3) indicates

$$\begin{aligned} N_{m-1}(x) &= \frac{x_m - x}{x_m - x_{m-1}} = \frac{x_m - x}{\Delta x} \\ N_m(x) &= \frac{x - x_{m-1}}{x_m - x_{m-1}} = \frac{x - x_{m-1}}{\Delta x} \end{aligned} \tag{F–4b}$$

Introducing the local dimensionless coordinate ξ relative to node m [i.e., $\xi = (x - x_m)/\Delta x$], Eq. (F–4b) takes the form

$$N_{m-1}(x) = \frac{x_m - x}{\Delta x} = -\xi \qquad N_m(x) = \frac{x - x_{m-1}}{\Delta x} = 1 + \xi \tag{F–5b}$$

in the ξ interval $[-1,0]$. Thus, Eqs. (F–2a) and (F–2b) and Eqs. (F–5a) and (F–5b) provide a convenient composite linear representation for the temperature distribution in terms of a single local dimensionless coordinate ξ defined over the two-node interval $[-1,1]$, which can be applied throughout individual interior control volumes. This composite perspective is illustrated in Fig. A–F–6(b). The resulting composite linear finite-element approximation for temperature distribution is given by

$$T(\xi) = (1 - \xi)T_m + \xi T_{m+1} \qquad \text{for } 0 \le \xi \le 1 \tag{F–6a}$$

$$T(\xi) = -\xi T_{m-1} + (1 + \xi)T_m \qquad \text{for } -1 \le \xi \le 0 \tag{F–6b}$$

Higher-order one-dimensional finite-element approximations can be developed. To illustrate, a quadratic approximation is readily obtained for a three-node element associated with nodes $m - 1$, m, and $m + 1$ in terms of ξ by writing†

$$T(\xi) = \sum_{i=m-1}^{m+1} N_i(\xi)\, T_i = N_{m-1}(\xi)\, T_{m-1} + N_m(\xi)\, T_m + N_{m+1}(\xi)\, T_{m+1} \tag{F–7}$$

and

$$T(\xi) = c_1 + c_2\xi + c_3\xi^2 \tag{F–8}$$

with $T(-1) = T_{m-1}$, $T(0) = T_m$, and $T(1) = T_{m+1}$. Substituting for the nodal temperatures, the coefficients c_1, c_2, and c_3 become

$$c_1 = T_m \qquad c_2 = \frac{T_{m+1} - T_{m-1}}{2} \qquad c_3 = \frac{T_{m+1} - 2T_m + T_{m-1}}{2} \tag{F–9}$$

† To develop cubic and higher-order finite-element approximations, it is necessary to define elements with more nodes. For example, a *cubic* approximation requires the use of *four-node* elements.

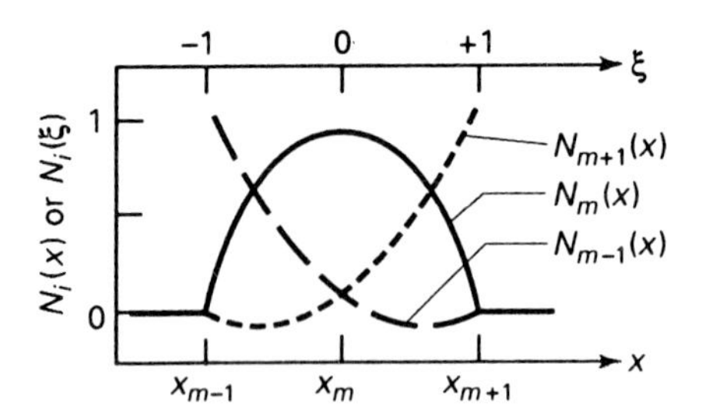

FIGURE A–F–7
One-dimensional quadratic interpolation functions.

Using these coefficients, a comparison of Eqs. (F–7) and (F–8) indicates *quadratic interpolation functions* of the form

$$N_{m-1}(\xi) = \frac{\xi}{2}(\xi - 1) \qquad N_m(\xi) = 1 - \xi^2 \qquad N_{m+1}(\xi) = \frac{\xi}{2}(\xi + 1) \tag{F–10}$$

for the ξ interval $[-1,1]$. These quadratic relations are shown in Fig. A–F–7. The resulting quadratic finite-element approximation for temperature distribution is given by

$$T(\xi) = \frac{\xi}{2}(\xi - 1)T_{m-1} + (1 - \xi^2)T_m + \frac{\xi}{2}(\xi + 1)T_{m+1} \qquad \text{for } -1 \le \xi \le 1 \tag{F–11}$$

Two-Dimensional Rectangular Elements

To develop two-dimensional finite-element approximations for four-node rectangular elements such as those shown in Figs. A–F–2 and A–F–8, we first identify *local dimensional coordinates X* and *Y* and *local dimensionless coordinates* ξ and η, where

$$\xi = \frac{X}{\Delta x/2} \qquad \eta = \frac{Y}{\Delta y/2} \tag{F–12a,b}$$

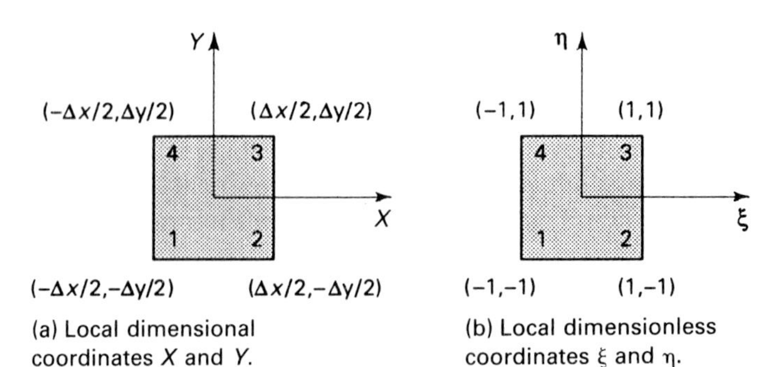

FIGURE A–F–8 Coordinate systems for four-node rectangular elements.

with the nodes numbered 1 to 4 in the counterclockwise direction starting at the lower-left node. Notice that the local dimensionless coordinates ξ and η range between -1 and 1 within an element. Using this convenient frame of reference, the temperature distribution within the element, which is represented by Eq. (F–1), becomes

$$\begin{aligned} T(\xi,\eta) &= \sum_{i=1}^{4} N_i(\xi,\eta)\, T_i \\ &= N_1(\xi,\eta)\, T_1 + N_2(\xi,\eta)\, T_2 + N_3(\xi,\eta)\, T_3 + N_4(\xi,\eta)\, T_4 \end{aligned} \tag{F–13}$$

Standard practice in the development of interpolation functions for a four-node rectangle† is to require

$$T(\xi,\eta) = c_1 + c_2\xi + c_3\eta + c_4\xi\eta \tag{F–14}$$

with $T(-1,-1) = T_1$, $T(1,-1) = T_2$, $T(1,1) = T_3$, and $T(1,-1) = T_4$. Introducing these nodal values into Eq. (F–14), we obtain

$$\begin{aligned} c_1 &= \frac{T_1 + T_2 + T_3 + T_4}{4} \qquad & c_2 &= \frac{-T_1 + T_2 + T_3 + T_4}{4} \\ c_3 &= \frac{-T_1 - T_2 + T_3 + T_4}{4} \qquad & c_4 &= \frac{T_1 - T_2 + T_3 - T_4}{4} \end{aligned} \tag{F–15}$$

Comparing Eqs. (F–13) and (F–14), the four-node rectangular interpolation functions become

$$\begin{aligned} N_1(\xi,\eta) &= \frac{1}{4}(1 - \xi)(1 - \eta) \qquad & N_2(\xi,\eta) &= \frac{1}{4}(1 + \xi)(1 - \eta) \\ N_3(\xi,\eta) &= \frac{1}{4}(1 + \xi)(1 + \eta) \qquad & N_4(\xi,\eta) &= \frac{1}{4}(1 - \xi)(1 + \eta) \end{aligned} \tag{F–16}$$

for the region $[(-1,-1), (1,1)]$. The resulting linear finite-element approximation for the temperature distribution within individual rectangular elements is represented by

$$\begin{aligned} T(\xi,\eta) &= \frac{1}{4}(1 - \xi)(1 - \eta)T_1 + \frac{1}{4}(1 + \xi)(1 - \eta)T_2 \\ &\quad + \frac{1}{4}(1 + \xi)(1 + \eta)T_3 + \frac{1}{4}(1 - \xi)(1 + \eta)T_4 \end{aligned} \qquad \text{for } -1 \le \xi \le 1 \text{ and } -1 \le \eta \le 1 \tag{F–17}$$

These relations for four-node rectangular elements are commonly referred to as *linear* rectangular interpolation functions. Higher-order *quadratic* rectangular interpolation functions can be developed by constructing nodal networks featuring eight- or nine-node rectangular elements [20].

† Also referred to as a four-node *linear* rectangular element.

The approach to developing interpolation functions for rectangular elements presented in this section can be adapted to general quadrilateral elements and triangular elements. These adaptations are considered in references 20 to 23.

F–3 APPROACHES TO DEVELOPING FINITE-ELEMENT NODAL EQUATIONS

The three general methods for developing finite-element nodal equations are (1) the *control-volume finite-element method* CVFEM, (2) the *method of weighted residuals* MWR, and (3) the *variational method*. As in the finite-difference approach, these finite-element discretization methods result in the approximation of the fundamental conservation principles by sets of discrete algebraic nodal equations that can be solved numerically by iterative or direct techniques. Because of limitations that restrict the variational method from being used in the analysis of many practical heat-transfer problems involving fluid flow and unsteady conduction, our attention will be focused on the CVFEM and the MWR. The variational method is considered in references 24 to 26.

To introduce the essential aspects of these two finite-element discretization methods, the formulations are presented in the context of basic-one-dimensional elements and two-dimensional rectangular elements. The discretization approaches introduced in this section are extended to other element configurations in references 20 to 23.

F–3–1 Control-Volume Finite-Element Method

The CVFEM is the simplest and most straightforward approach to formulating finite-element nodal equations. It is also the newest, with most of the developmental work having been done since 1978.

As in the control-volume finite-difference method CVFDM, the CVFEM involves the direct application of fundamental conservation principles and particular laws to the Z_s individual nodal control volumes within the computational domain. However, unlike the CVFDM, the CVFEM involves the use of *interpolation function* approximations for temperature distribution within the element(s) contained by individual control volumes to evaluate spatial gradients and integrals associated with each control volume. Like the CVFDM, the CVFEM enables us to readily account for the effect of boundary conditions on the nodal equations associated with external control volumes. As pointed out by Schneider [22] and Baliga and Patankar [23], the CVFEM gives rise to nodal equations that involve simple mathematics and are amenable to easy physical interpretation, which can lead to very fruitful insights and be of great assistance in the developmental stages of program evolution.

F–3–2 Method of Weighted Residuals

The MWR involves the discretization of applicable differential equations by performing Z_s *weighted* integrations over all element volumes, using *interpolation functions* to approximate the distribution in dependent variables within individual elements. With the differential equation for steady-state heat-transfer problems represented in terms of the differential operator L by†

$$L[T(x,y,z)] = 0 \tag{F–18}$$

The general form of the set of weighted integral equations employed in this method is represented by

$$\int_V W_i(x,y,z)\, L[T(x,y,z)]\, dV = 0 \qquad \text{for } i = 1, 2, \ldots, Z_s \tag{F–19}$$

where $W_i(x,y,z)$ designates linearly independent *weighting functions* associated with Z_s nodes and each integration is over the entire volume V. In effect, Eq. (F–19) provides a basis for establishing a set of Z_s simultaneous algebraic nodal equations that can be solved for the Z unknown nodal temperatures and the $Z_s - Z$ surface-heat fluxes.

Galerkin's Method

The most commonly used MWR is established by setting the weighting function $W_i(x,y,z)$ equal to the interpolation function $N_i(x,y,z)$. Following this approach, which is known as *Galerkin's method*, the set of weighted volume integrals represented by Eq. (F–19) becomes

$$\int_V N_i(x,y,z)\, L[T(x,y,z)]\, dV = 0 \qquad \text{for } i = 1, 2, \ldots, Z_s \tag{F–20}$$

where $N_i(x,y,z)$ is nonzero in the immediate neighborhood of node i. In practice, the general approach to evaluating the *global integrals* indicated by this equation for each value of i is to evaluate the *element integrals* element by element and to require that the sum of the resultant integral relations (for each value of i) over all I_s elements be equal to zero. Thus, Z_e element integral equations are written for each of the I_s elements of the form

$$\int_{V_{ek}} N_i(x,y,z)\, L[T(x,y,z)]\, dV = 0 \qquad \text{for } \begin{matrix} i = 1, 2, \ldots, Z_e \\ k = 1, 2, \ldots, I_s \end{matrix} \tag{F–21}$$

Using Eq. (F–1) to approximate the temperature distribution within the elements, Eq. (F–21) is put into the form

$$\int_{V_{ek}} N_i(x,y,z)\, L\left[\sum_{j=1}^{Z_e} N_j(x,y,z)\, T_j\right] dV = 0 \qquad \text{for } \begin{matrix} i = 1, 2, \ldots, Z_e \\ k = 1, 2, \ldots, I_s \end{matrix} \tag{F–22}$$

† The analysis is generalized for unsteady problems by writing

$$\int_V W_i(x,y,z)\, L[T(t,x,y,z)]\, dV = 0 \qquad i = 1, 2, \ldots, Z_s$$

To summarize, Z_s nodal equations are obtained by (1) using this general *element integral equation* to obtain Z_e algebraic equations for each of the I_s elements and (2) assembling the contributions for common nodes from each element.

As is evident, Galerkin's finite-element method is considerably more involved than the CVFEM. The importance of this method stems from its relatively longer period of use and the fact that the majority of general-purpose numerical finite-element heat transfer codes presently available commercially are based on the Galerkin method.

F-4 NUMERICAL FINITE-ELEMENT ANALYSIS: ONE-DIMENSIONAL HEAT TRANSFER

To introduce the use of the numerical finite-element approach in heat-transfer analysis, we want to consider the basic problem of heat transfer in a convecting fin, which is illustrated in Fig. A–F–9. For small values of the Biot number (i.e., $Bi < 0.1$), the energy transfer in a fin with uniform cross section and uniform properties is represented by the one-dimensional equation given by Eq. (2–99),

$$\frac{d^2T}{dx^2} - \frac{h_x p}{kA}(T - T_F) = 0 \qquad \text{(F–23)}$$

For the situation indicated in this figure, the boundary conditions are written as

$$T = T_0 \quad \text{at } x = 0 \qquad \frac{dT}{dx} = 0 \quad \text{at } x = L \qquad \text{(F–24a,b)}$$

The exact analytical solution to this problem for the case in which h_x and T_F are uniform is given by Eq. (2–116),

$$\frac{T - T_F}{T_0 - T_F} = \frac{\cosh\,[m_f(L - x)]}{\cosh\,(m_f L)} \qquad \text{(F–25)}$$

where m_f represents the convective fin parameter $[\bar{h}p/(kA)]^{1/2}$.

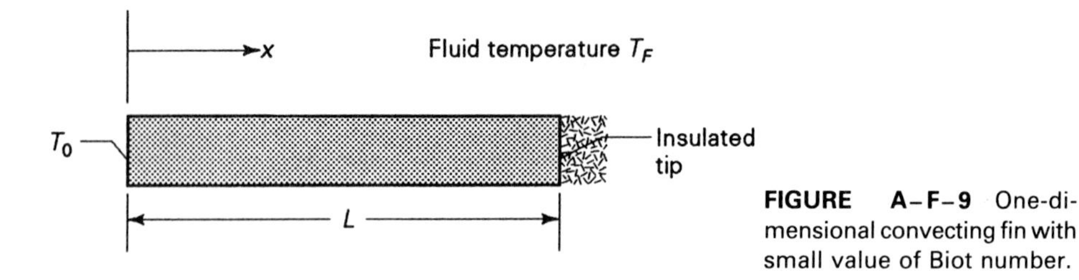

FIGURE A–F–9 One-dimensional convecting fin with small value of Biot number.

F-4-1 Finite-Element Formulations

The CVFEM and Galerkin's method are now featured in the development of numerical solutions to this problem, using basic linear finite-element approximations.

Finite-Element Nodal Networks: step 1

The first step in the development of a basic finite-element analysis is accomplished by sketching a grid with M nodes and uniform increment Δx, as shown in Fig. A–F–10(a). To arrive at a linear finite-element network, we identify the $I_s = M - 1$ two-node elements shown in this figure. The control-volume form of the finite-element nodal network for this problem is shown in Fig. A–F–10(b).

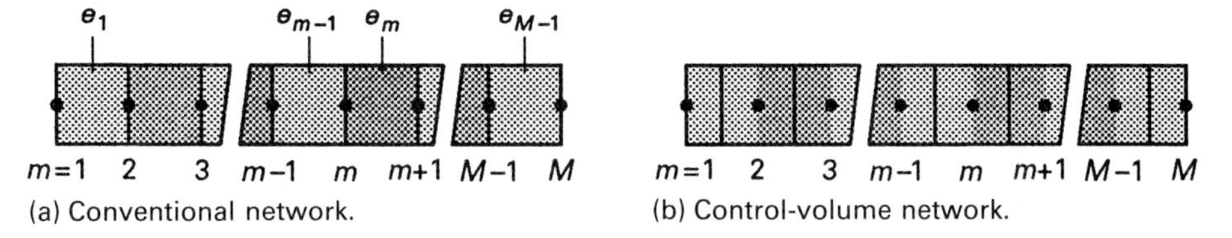

FIGURE A–F–10 Finite-element nodal networks for one-dimensional convecting fin.

Finite-Element Approximations: step 2

The second step is to introduce finite-element approximations for the temperature distribution within the elements in terms of interpolation functions $N_i(\xi)$. Linear approximations are represented by Eq. (F–6),

$$T(\xi) = (1 - \xi)T_m + \xi T_{m+1} \qquad \text{for } 0 \le \xi \le 1 \tag{F–26a}$$

$$T(\xi) = -\xi T_{m-1} + (1 + \xi)T_m \qquad \text{for } -1 \le \xi \le 0 \tag{F–26b}$$

with $N_i(\xi)$ given by Eq. (F–5),

$$N_m(\xi) = 1 - \xi \qquad N_{m+1}(\xi) = \xi \qquad \text{for } 0 \le \xi \le 1 \tag{F–27a}$$

$$N_{m-1}(\xi) = -\xi \qquad N_m(\xi) = 1 + \xi \qquad \text{for } -1 \le \xi \le 0 \tag{F–27b}$$

Finite-Element Discretization: step 3

The third step involves the development of $Z_s = M$ nodal equations that can be used to solve for the $Z = M - 1$ unknown nodal temperatures for $m = 2, 3, \ldots, M$, and to evaluate the rate of heat transfer at $m = 1$. This step is undertaken by (1) the CVFEM and (2) the Galerkin method.

CVFEM Following the CVFEM, we make use of the control-volume form of the finite-element network shown in Fig. A–F–10(b). The first law of thermodynamics is applied to the control volumes associated with the M nodes to obtain

$$q_{x_m + \Delta x/2} - q_{x_m - \Delta x/2} + \int_{x_m - \Delta x/2}^{x_m + \Delta x/2} h_x p(T - T_F)\, dx = 0 \tag{F–28a}$$

for interior nodes $m = 2, 3, \ldots, M - 1$,

$$q_{x_M} - q_{x_M - \Delta x/2} + \int_{x_M - \Delta x/2}^{x_M} h_x p(T - T_F)\, dx = 0 \qquad \text{(F–28b)}$$

for exterior node $m = M$, where $q_{x_M} = 0$, and

$$q_{\Delta x/2} - q_0 + \int_0^{\Delta x/2} h_x p(T - T_F)\, dx = 0 \qquad \text{(F–28c)}$$

for exterior node $m = 1$, where $T_1 = T_0$.†

Focusing attention on interior nodes, the Fourier law of conduction is used to put Eq. (F–28a) into the form

$$-kA \left.\frac{dT}{dx}\right|_{x_m + \Delta x/2} + kA \left.\frac{dT}{dx}\right|_{x_m - \Delta x/2} + \int_{x_m - \Delta x/2}^{x_m + \Delta x/2} h_x p(T - T_F)\, dx = 0 \qquad \text{(F–29)}$$

for $m = 2, 3, \ldots, M - 1$. To simplify matters, this equation is expressed in terms of the local dimensionless coordinate $\xi = (x - x_m)/\Delta x$ by setting $dx = \Delta x\, d\xi$ and noting that $\xi = 1/2$ at $x = x_m + \Delta x/2$ and $\xi = -1/2$ at $x = x_m - \Delta x/2$. Making this transformation, Eq. (F–29) becomes

$$-\left.\frac{dT}{d\xi}\right|_{1/2} + \left.\frac{dT}{d\xi}\right|_{-1/2} + \frac{\overline{h}p}{kA}\Delta x^2 \int_{-1/2}^{1/2} (T - T_F)\, d\xi = 0 \qquad \text{(F–30)}$$

for uniform h_x. Introducing the linear finite-element approximation for temperature distribution given by Eq. (F–26a) in the interval [0,1] and Eq. (F–26b) in the interval [−1,0] for interior nodal control volumes, Eq. (F–30) becomes

$$\begin{aligned}
-\frac{d}{d\xi}&\left.[(1 - \xi)T_m + \xi T_{m+1}]\right|_{1/2} + \frac{d}{d\xi}\left.[-\xi T_{m-1} + (\xi + 1)T_m]\right|_{-1/2} \\
&+ \frac{\overline{h}p}{kA}\Delta x^2 \left\{\int_0^{1/2} [(1 - \xi)T_m + \xi T_{m+1} - T_F]\, d\xi \right. \qquad \text{(F–31)} \\
&\left. + \int_{-1/2}^{0} [-\xi T_{m-1} + (1 + \xi)T_m - T_F]\, d\xi \right\} = 0
\end{aligned}$$

This relation reduces to

$$\begin{aligned}
-(-T_m + T_{m+1}) &+ (-T_{m-1} + T_m) \\
&+ \frac{\overline{h}p}{kA}\Delta x^2 \left(\frac{T_{m+1}}{8} + \frac{T_{m-1}}{8} + \frac{6}{8}T_m - T_F\right) = 0 \qquad \text{(F–32)}
\end{aligned}$$

† It should be noted that these equations can also be obtained by integrating the differential equation given by Eq. (F–23) over each control volume. For example, we write

$$\int_{x_m - \Delta x/2}^{x_m + \Delta x/2} \left[\frac{d^2T}{dx^2} - \frac{h_x p}{kA}(T - T_F)\right] A\, dx = 0$$

for interior node (m). Holding kA constant and integrating, we obtain Eq. (F–29), which leads to Eq. (F–28a).

or

$$T_{m-1} - \left(\frac{16 + 6CC_1}{8 - CC_1}\right)T_m + T_{m+1} + \left(\frac{8CC_1}{8 - CC_1}\right)T_F = 0 \qquad \text{(F–33)}$$

where $CC_1 = \bar{h}p\ \Delta x^2/(kA)$.

Following through with similar formulation steps for the exterior nodes, Eq. (F–28b) gives rise to

$$(-T_{M-1} + T_M) + \frac{\bar{h}p}{kA}\ \Delta x^2 \left(\frac{T_{M-1}}{8} + \frac{3}{8}\ T_M - \frac{T_F}{2}\right) = 0 \qquad \text{(F–34)}$$

or

$$T_{M-1} - \left(\frac{8 + 3CC_1}{8 - CC_1}\right)T_M + \left(\frac{4CC_1}{8 - CC_1}\right)T_F = 0 \qquad \text{(F–35)}$$

for node (M), and Eq. (F–28c) leads to

$$q_0 = \frac{kA}{\Delta x}\left[T_1 - T_2 + CC_1\left(\frac{3}{8}\ T_1 + \frac{T_2}{8} - \frac{T_F}{2}\right)\right] \qquad \text{(F–36)}$$

for node (1), where $T_1 = T_0$.

To summarize, Eqs. (F–33) and (F–35) provide $Z = M - 1$ nodal equations that can be solved for the unknown nodal temperatures $T_2, T_3, \ldots, T_M$, and Eq. (F–36) can be used to calculate the heat flux q_0.

Galerkin's Method Referring to the finite-element nodal network shown in Fig. A–F–10(a), the MWR is used to develop interior and exterior nodal equations by expressing the differential equation given by Eq. (F–23) in the weighted *element integral* form

$$\int_{V_{em}} W_i(x)\left[\frac{d^2T}{dx^2} - \frac{h_x p}{kA}\ (T - T_F)\right] dV = 0 \qquad \text{for } Z_e = 2 \qquad \text{(F–37)}$$

for each of the $M - 1$ elements $e_1, e_2, \ldots, e_{M-1}$ within the entire volume V, where $dV = A\ dx$. Setting $W_i(x) = N_i(x)$ in accordance with Galerkin's method and identifying a specific element e_m in the interval $[x_m, x_{m+1}]$, this equation becomes†

$$\int_{x_m}^{x_{m+1}} N_i(x)\left[\frac{d^2T}{dx^2} - \frac{h_x p}{kA}\ (T - T_F)\right] dx = 0 \qquad \text{for } i = m, m+1 \qquad \text{(F–38)}$$

Integrating the first term by parts, this equation is put into the form

$$\int_{x_m}^{x_{m+1}} \left\{\frac{d}{dx}\left[N_i(x)\ \frac{dT}{dx}\right] - \frac{d}{dx}\ N_i(x)\ \frac{dT}{dx} - \frac{h_x p}{kA}\ N_i(x)\ (T - T_F)\right\} dx = 0 \qquad \text{(F–39)}$$

$$\text{for } i = m, m+1$$

† With the weighting function specified by $W_i(x) = 1$ within the *control volume* for node i and $W_i(x) = 0$ outside the *control volume*, Eq. (F–37) gives rise to the CVFEM discretization equation.

or

$$\left[N_i(x)\frac{dT}{dx}\right]\Bigg|_{x_m}^{x_{m+1}} - \int_{x_m}^{x_{m+1}}\left[\frac{d}{dx}N_i(x)\frac{dT}{dx} + \frac{h_x p}{kA}N_i(x)(T - T_F)\right]dx = 0 \tag{F-40}$$

for $i = m, m + 1$

Introducing the local dimensionless coordinate ξ, we obtain

$$\left[N_i(\xi)\frac{dT}{d\xi}\right]\Bigg|_0^1 - \int_0^1\left[\frac{d}{d\xi}N_i(\xi)\frac{dT}{d\xi} + \frac{h_x p}{kA}\Delta x^2 N_i(\xi)(T - T_F)\right]d\xi = 0 \tag{F-41}$$

for $i = m, m + 1$

Using the linear approximation for $T(\xi)$ given by Eq. (F–26a) to evaluate the integrals in this equation and assuming $h_x = \bar{h}$, we obtain

$$\left[N_i(\xi)\frac{dT}{d\xi}\right]\Bigg|_0^1 - \int_0^1\left\{\frac{d}{d\xi}N_i(\xi)\frac{d}{d\xi}[(1 - \xi)T_m + \xi T_{m+1}]\right\}d\xi \tag{F-42}$$

$$- \frac{\bar{h}p}{kA}\Delta x^2\int_0^1 N_i(\xi)[(1 - \xi)T_m + \xi T_{m+1} - T_F]\,d\xi = 0 \qquad \text{for } i = m, m + 1$$

where

$$N_i(\xi) = 1 - \xi \qquad \frac{d}{d\xi}N_i(\xi) = -1 \qquad \text{for } i = m \tag{F-43a}$$

and

$$N_i(\xi) = \xi \qquad \frac{d}{d\xi}N_i(\xi) = 1 \qquad \text{for } i = m + 1 \tag{F-43b}$$

from Eq. (F–27a). Substituting for $N_i(\xi)$ and $dN_i(\xi)/d\xi$ with i set equal to m and $m + 1$, Eq. (F–42) enables us to write

$$-\frac{dT}{d\xi}\bigg|_0 + (T_{m+1} - T_m) - \frac{\bar{h}p}{kA}\Delta x^2\left(\frac{T_m}{3} + \frac{T_{m+1}}{6} - \frac{T_F}{2}\right) = 0 \tag{F-44a}$$

for $i = m$

and

$$\frac{dT}{d\xi}\bigg|_1 - (T_{m+1} - T_m) - \frac{\bar{h}p}{kA}\Delta x^2\left(\frac{T_m}{6} + \frac{T_{m+1}}{3} - \frac{T_F}{2}\right) = 0 \tag{F-44b}$$

for $i = m + 1$

for element e_m.

These equations can be used to obtain two element equations for each finite element $e_1, e_2, \ldots, e_{M-1}$ within the computational domain by setting $m = 1, 2,$

$\ldots, M - 1$. The interior nodal equations are assembled by combining the two element equations that are associated with a common node. Thus, for node (m), we also require the element equation for element e_{m-1}. Equation (F–44b) can be used to obtain the required element equation by simply replacing $m + 1$ by m and m by $m - 1$, with the result

$$\left.\frac{dT}{d\xi}\right|_1 - (T_m - T_{m-1}) - \frac{\bar{h}p}{kA}\,\Delta x^2\left(\frac{T_{m-1}}{6} + \frac{T_m}{3} - \frac{T_F}{2}\right) = 0 \qquad \text{for } i = m \tag{F–45}$$

for element e_{m-1}, where ξ is now defined relative to element e_{m-1} (i.e., $\xi = 0$ at $x = x_{m-1}$). Combining Eqs. (F–44a) and (F–45), which are both associated with node (m), we obtain†

$$\left.\frac{dT}{d\xi}\right|_1 - \left.\frac{dT}{d\xi}\right|_0 + T_{m-1} - 2T_m + T_{m+1} - \frac{\bar{h}p}{kA}\,\Delta x^2\left(\frac{T_{m-1}}{6} + \frac{2}{3}T_m + \frac{T_{m+1}}{6} - T_F\right) = 0 \tag{F–46}$$

or, since $dT/d\xi|_1 = dT/d\xi|_0$ for our linear finite-element approximation,‡

$$T_{m-1} - 2T_m + T_{m+1} - \frac{\bar{h}p}{kA}\,\Delta x^2\left(\frac{T_{m-1}}{6} + \frac{2}{3}T_m + \frac{T_{m+1}}{6} - T_F\right) = 0 \tag{F–47}$$

Rearranging, the interior nodal equation for node (m) is put into the form

$$T_{m-1} - \left(\frac{12 + 4CC_1}{6 - CC_1}\right)T_m + T_{m+1} + \left(\frac{6CC_1}{6 - CC_1}\right)T_F = 0 \tag{F–48}$$

where $CC_1 = \bar{h}p\,\Delta x^2/(kA)$.

To complete the formulation, Eqs. (F–45) and (F–44a) are used to obtain exterior nodal equations by writing

$$\left.\frac{dT}{d\xi}\right|_1 - (T_M - T_{M-1}) - \frac{\bar{h}p}{kA}\,\Delta x^2\left(\frac{T_{M-1}}{6} + \frac{T_M}{3} - \frac{T_F}{2}\right) = 0 \tag{F–49}$$

or

$$T_{M-1} - \left(\frac{6 + 2CC_1}{6 - CC_1}\right)T_M + \left(\frac{3CC_1}{6 - CC_1}\right)T_F = 0 \tag{F–50}$$

† This equation can also be obtained by integrating over the two-element interval $[-1,1]$ using the composite approximation for $T(\xi)$ given by Eqs. (F–26a) and (F–26b), with $N_m(\xi) = 1 - \xi$ in the interval $[0,1]$ and $N_m(\xi) = 1 + \xi$ in the interval $[-1,0]$ [see Eq. (F–27)].

‡ Although $dT/d\xi|_1 \neq dT/d\xi|_0$ for quadratic and higher-order finite-element approximations, these two terms are normally dropped for all interior elements since such gradient terms at the interface of adjacent elements cancel out when summing all element equations over V.

for node (M), where $dT/d\xi|_1 = 0$ in accordance with the boundary condition at the tip given by Eq. (F–24b), and

$$-\left.\frac{dT}{d\xi}\right|_0 + (T_2 - T_1) - \frac{\bar{h}p}{kA}\,\Delta x^2\left(\frac{T_1}{3} + \frac{T_2}{6} - \frac{T_F}{2}\right) = 0 \qquad \text{(F–51)}$$

or

$$\begin{aligned} q_0 &= -kA\left.\frac{dT}{dx}\right|_{x=0} = -\frac{kA}{\Delta x}\left.\frac{dT}{d\xi}\right|_{\xi=0} \\ &= \frac{kA}{\Delta x}\left[T_1 - T_2 + CC_1\left(\frac{T_1}{3} + \frac{T_2}{6} - \frac{T_F}{2}\right)\right] \end{aligned} \qquad \text{(F–52)}$$

for node (1), where $T_1 = T_0$.

Comparison of Nodal Equations For purpose of comparison, we consider the internal nodal equations developed in this section, which are given by Eq. (F–33),

$$T_{m-1} - \left(\frac{16 + 6CC_1}{8 - CC_1}\right)T_m + T_{m+1} + \left(\frac{8CC_1}{8 - CC_1}\right)T_F = 0 \qquad \text{(F–53)}$$

for the linear CVFEM, and Eq. (F–48),

$$T_{m-1} - \left(\frac{12 + 4CC_1}{6 - CC_1}\right)T_m + T_{m+1} + \left(\frac{6CC_1}{6 - CC_1}\right)T_F = 0 \qquad \text{(F–54)}$$

for the linear Galerkin method.

To compare these nodal equations with the exact analytical solution, Eq. (F–25) is first put into the form

$$\frac{T - T_F}{T_0 - T_F} = \frac{\psi}{\psi_0} = \frac{\cosh(m_f X)}{\cosh(m_f L)} \qquad \text{(F–55)}$$

where $\psi = T - T_F$ and $X = L - x$. Referring to Fig. A–F–11, we now designate the nodes right-to-left by $i = 0, 1, \ldots, M - 1$, such that successive nodal locations X_{i-1}, X_i, and X_{i+1} can be represented by

$$X_{i-1} = (i - 1)\,\Delta x \qquad X_i = i\,\Delta x \qquad X_{i+1} = (i + 1)\,\Delta x \qquad \text{(F–56)}$$

Using Eq. (F–55), the temperature at each of these nodal locations is given by

$$\psi_{i-1} = \frac{\psi_0 \cosh[m_f(i - 1)\,\Delta x]}{\cosh(m_f L)} \qquad \psi_{i+1} = \frac{\psi_0 \cosh[m_f(i + 1)\,\Delta x]}{\cosh(m_f L)}$$

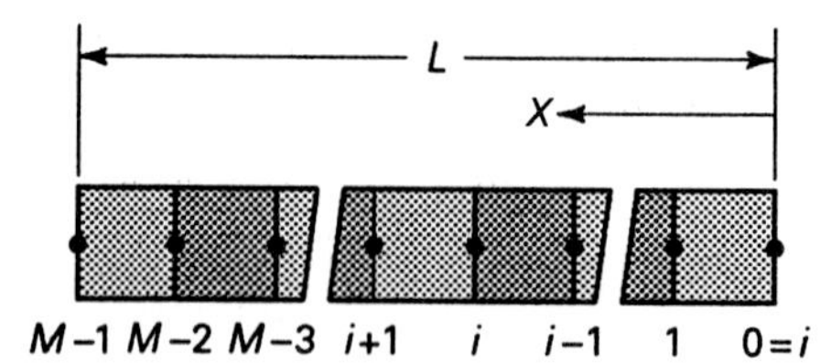

FIGURE A–F–11
Conventional finite-element nodal network for convecting fin; $X = L - x$ and nodes indicated by $i = 0, 1, \ldots, M - 1$.

and

$$\psi_i = \frac{\psi_0 \cosh (m_f i \, \Delta x)}{\cosh (m_f L)} \tag{F-57}$$

which indicates

$$\frac{\psi_{i-1}}{\psi_i} = \frac{\cosh [m_f (i - 1) \, \Delta x]}{\cosh (m_f i \, \Delta x)} \qquad \frac{\psi_{i+1}}{\psi_i} = \frac{\cosh [m_f (i + 1) \, \Delta x]}{\cosh (m_f i \, \Delta x)} \tag{F-58}$$

Noting the mathematical identity [27],

$$\cosh a + \cosh b = 2 \cosh \frac{a + b}{2} \cosh \frac{a - b}{2} \tag{F-59}$$

we see that

$$\frac{\psi_{i-1}}{\psi_i} + \frac{\psi_{i+1}}{\psi_i} = 2 \cosh (m_f \, \Delta x) \tag{F-60}$$

which indicates

$$\psi_{i-1} - (2 \cosh \sqrt{CC_1})\psi_i + \psi_{i+1} = 0 \tag{F-61a}$$

or

$$T_{i-1} - (2 \cosh \sqrt{CC_1})T_i + T_{i+1} + (2 - 2 \cosh \sqrt{CC_1})T_F = 0 \tag{F-61b}$$

To provide a convenient basis for comparison, we note that Eqs. (F–53), (F–54), and (F–61) can be represented by general relations (in terms of nodal subscript m) of the form

$$T_{m-1} - cT_m + T_{m+1} + (c - 2)T_F = 0 \tag{F-62a}$$

or

$$\psi_{m-1} - c\psi_m + \psi_{m+1} = 0 \tag{F-62b}$$

In connection with these relations, we also note that the exterior nodal equations can be represented in terms of c by

$$T_{M-1} - \frac{c}{2} T_M + \frac{c - 2}{2} T_F = 0 \tag{F-63}$$

for node (M), and

$$\begin{aligned} q_0 &= \bar{h}p \, \Delta x \left(\frac{T_1 - T_2}{c - 2} + \frac{T_1 - T_F}{2} \right) \\ &= \frac{kA}{\Delta x} (T_1 - T_2) + \bar{h}p \, \Delta x \left[\frac{1}{b} (T_2 - T_1) + \frac{1}{2} (T_1 - T_F) \right] \end{aligned} \tag{F-64}$$

where $T_1 = T_0$ and $b = CC_1/[1 - CC_1/(c - 2)]$ for node (1). The values of the coefficient c are listed in Table A–F–2 for these linear finite-element formulations

TABLE A–F–2 Comparison of numerical nodal equations: Coefficients c and b

Formulation approach	*Coefficient c*	*Coefficient b*
Exact	$2 \cosh \sqrt{CC_1}$	$\dfrac{CC_1}{1 - CC_1/[2(\cosh \sqrt{CC_1} - 1)]}$
Finite element—linear		
CVFEM	$\dfrac{16 + 6CC_1}{8 - CC_1}$	8
Galerkin's method†	$\dfrac{12 + 4CC_1}{6 - CC_1}$	6
Finite element—quadratic		
CVFEM	$\dfrac{48 + 22CC_1}{24 - CC_1}$	24
Galerkin's method‡	$\dfrac{20 + 8CC_1}{10 - CC_1}$	10
Finite difference	$2 + CC_1$	∞

† The linear Galerkin method and linear variational method [25] give the same result.
‡ The quadratic Galerkin method and quadratic variational method [25] give the same result.

and the exact solution. In addition, values of c are given for quadratic finite-element formulations and for the finite-difference formulation of Example 4–6. The percent error in coefficient c for each of the nodal equations represented in Table A–F–2 is shown as a function of $CC_1 = \bar{h}p\,\Delta x^2/(kA)$ in Fig. A–F–12. Whereas each of these nodal equations provides a good approximation to the exact solution for sufficiently small grid increment (i.e., small values of CC_1), for a particular value of CC_1 the linear and quadratic CVFEM and the quadratic Galerkin method are more accurate than the linear Galerkin method and the finite-difference method, with the difference increasing as the grid increment increases.

F–4–2 Computer Solution: step 4

The finite-element nodal equations can be readily incorporated into the BASIC computer program listed in Table E4–6c of Example 4–6 to obtain numerical calculations for the temperature distribution and rate of heat transfer q_F. To follow through with this point, the interior and exterior nodal equations given by Eqs. (F–62a), (F–63), and (F–64) are expressed in BASIC computer language by writing

$$\text{T(J,2)} = (\text{T(J} - 1,1) + \text{T(J} + 1,1) + (\text{CC2} - 2)*\text{TF})/\text{CC2} \qquad \text{(F–65)}$$

for J $= m = 2, 3, \ldots, M - 1$,

$$\text{T(M,2)} = (\text{T(M} - 1,1) + ((\text{CC2} - 2)/2)*\text{TF})*2/\text{CC2} \qquad \text{(F–66)}$$

and

$$\text{QF} = \text{H*P*DX*((T(1,2)} - \text{T(2,2))/(CC2} - 2) + (\text{T(1,2)} - \text{TF)/2)} \qquad \text{(F–67)}$$

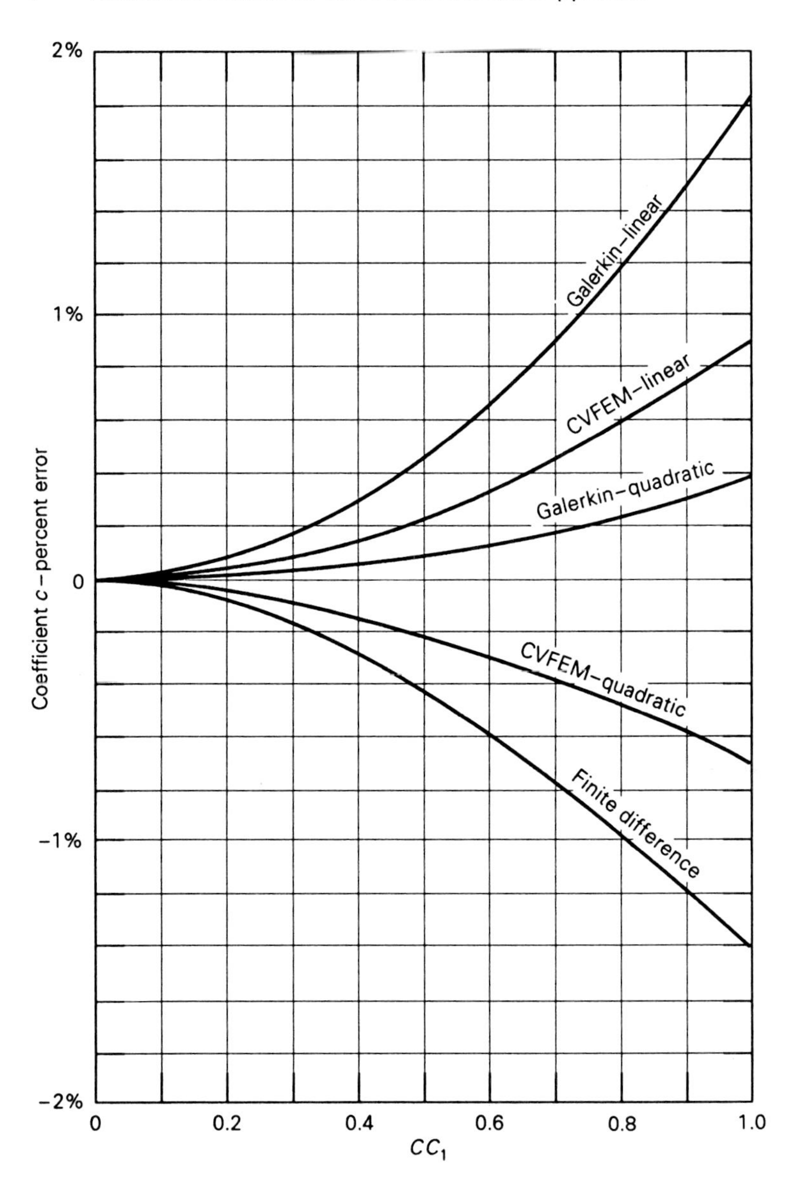

FIGURE A–F–12 Interior nodal equations: percent error in coefficient c.

where CC1 = H*P*DX*DX/(K*A), H = $\bar{h}$, P = p, DX = Δx, K = k, TF = T_F, CC2 = c, and the coefficient c is specified in terms of CC1 in Table A–F–2. To compute QF, these BASIC statements are used in place of the statements on lines 21, 24, and 30 of the program and a BASIC statement is introduced that expresses CC2 in terms of CC1. Resulting calculations for QF are compared in Table A–F–3

TABLE A–F–3 Numerical finite-element and finite-difference calculations for q_F.†

Number of increments M	*Numerical method*				*Finite difference*	CC_1
	CVFEM		*Galerkin*			
	Linear	*Quadratic*	*Linear*	*Quadratic*		
3	15.34	15.52	15.25	15.39	15.61	0.251
4	15.18	15.28	15.13	15.21	15.33	0.112
5	15.13	15.19	15.10	15.14	15.22	0.0628
6	15.10	15.14	15.08	15.11	15.16	0.0402
7	15.09	15.12	15.07	15.10	15.13	0.0279
8	15.08	15.10	15.07	15.08	15.11	0.0205
9	15.07	15.09	15.06	15.08	15.10	0.0157
10	15.07	15.08	15.06	15.07	15.09	0.0124

† Exact solution: q_F = 15.1 W (see Example 2–12).

for the various finite-element and finite-difference methods, with the convergence criterion CHK set equal to 10^{-5}. Notice that each of these approaches gives rise to essentially the same result for this problem, with the calculations for QF converging to approximately 15.1 W for $M \geq 6$ (CC1 ≤ 0.04).

G NUMERICAL COMPUTATIONS

FIGURE A–G–1 BASIC program for Example 4–8

```
L=1
W=1
D=.01
M=10
DX=L/(M-1)
N=M
M1=M-1
N1=N-1
K=100
CHK=.0001
T1=0
FOR I=1 TO N
T(1,I,1)=T1
T(1,I,2)=T(1,I,1)
T(M,I,1)=T1
T(M,I,2)=T(M,I,1)
NEXT I
FOR J=2 TO M1
T(J,1,1)=T1
T(J,1,2)=T(J,1,1)
FX(J)=100*SIN(3.1416*(J-1)*DX/L)
T2X=FX(J)
T(J,N,1)=T2X
T(J,N,2)=T(J,N,1)
NEXT J
IT=1
FOR J=2 TO M1
FOR I=2 TO N1
T(J,I,1)=1
NEXT I
NEXT J
4 DT=0
FOR J=2 TO M1
FOR I=2 TO N1
T(J,I,2)=(T(J+1,I,1)+T(J-1,I,1)
   +T(J,I+1,1)+T(J,I-1,1))/4
IF DT≥CHK GOTO 5
DT=ABS((T(J,I,2)-T(J,I,1))/T(J,I,1))
```

FIGURE A-G-1 (BASIC program for Example 4-8—continued)

```
5  T(J,I,1)=T(J,I,2)
   NEXT I
   NEXT J
   IF DT<CHK GOTO 6
   IT=IT+1
   GOTO 4
6  SQA=0
   SQC=0
   FOR J=2 TO M1
   DQA=K*D*(T(J,N1,2)+T(J-1,N,2)/2
     +T(J+1,N,2)/2-2*T(J,N,2))
   SQA=SQA+DQA
   DQC=-K*D*(T(J,2,2)+T(J-1,1,2)/2
     +T(J+1,1,2)/2-2*T(J,1,2))
7  SQC=SQC+DQC
   NEXT J
   FOR II=1 TO N
   I=N-II+1
   LPRINT USING "####·##";T(1,I,2),T(2,I,2),
     T(3,I,2),T(4,I,2),T(5,I,2)
   NEXT II
   LPRINT USING "####·##";M1,IT,SQA,SQC
   STOP
   END
```

FIGURE A-G-2 BASIC program for Example 4-12.

```
   L=.004
   K=50
   ALPHA=.00002
   M=10
   M1=M-1
   DX=L/M1
   S=2
   DTIME=S*DX^2/ALPHA
   CHK=.0001
   CSS=.0001
   TI=.00001
   FOR J=2 TO M1
   T(J,I,1)=TI
   T(J,I,2)=TI
   NEXT J
   T1=.00001
   T(1,1,1)=T1
   T(1,2,1)=T1
   T(1,1,2)=T1
   T(1,2,2)=T1
   TM=100
   T(M,1,1)=TM
   T(M,2,1)=TM
   T(M,1,2)=TM
   T(M,2,2)=TM
   QFL1=1
   TAU=1
2  IT=1
   FOR J=2 TO M1
3  T(J,2,1)=T(J,1,2)
   NEXT J
4  DT=0
   FOR J=2 TO M1
   T(J,2,2)=(T(J+1,2,1)+T(J-1,2,1)
     +T(J,1,2)/S)/(2+1/S)
   IF DT>CHK GOTO 5
   DT=ABS((T(J,2,2)-T(J,2,1))/T(J,2,2))
5  T(J,2,1)=T(J,2,2)
   NEXT J
   IF DT<CHK GOTO 6
   IT=IT+1
   GOTO 4
6  QFL=-K/DX*(T(M,2,2)-T(M1,2,2))
   TTIME=TAU*DTIME
   LPRINT USING "####·##";TAU,TTIME
   LPRINT USING "####·##" ;M,CHK,IT,
     QFL,T(M1,2,2)
   DQFL=ABS((QFL-QFL1)/QFL1)
   IF DQFL<CSS GOTO 8
   QFL1=QFL
   TAU=TAU+1
   FOR J=2 TO M1
   T(J,1,2)=T(J,2,2)
   NEXT J
   GOTO 2
8  STOP
   END
```

FIGURE A–G–3 BASIC program for Example 7–11.

```
PR = .72
AK = .4
AA = 27.4
ZO = 0
DIM F(100)
FOR ZZ = 5 TO 100 STEP 5
Z = ZZ
M = 20
DZ = (Z-ZO)/(M-1)
ZD = ZO
FOR I = 1 TO M
F(I) = 2/((-1+(1+4*(AK*ZD)^2*(1-EXP(-ZD/
  AA))^2)^.5)+2/PR)
ZD = ZD + DZ
NEXT I
M1 = M-1
SUM = 0
FOR N = 2 TO M1
SUM = SUM + F(N)
NEXT N
TRAP = DZ/2*(F(1)+F(N)+2*SUM)
TP = TRAP
YP = Z
B = TP - (LOG(YP))/AK
IF ZZ > 5 GOTO 10
LPRINT "....M.......TP.....YP.....B.....
10 LPRINT USING "###.###"; M, TP, YP, B
NEXT ZZ
STOP
END
```

FIGURE A–G–4 BASIC program for Example 8–10.

```
DIM T(200)
D = .1
L = 10
MDOT = 10
M = 101
M1 = M-1
DX = L/M1
H = 4860
P = 3.1416*D
CP = 4180
T(1) = 20
FOR J = 2 TO M
X = (J-1)*DX
TS = 20 + 80*(1-COS(4*3.1416*X/L))
T(J) = (H*P/(MDOT*CP))*DX*TS
  +T(J-1)*(1-(H*P/(MDOT*CP)*DX))
LPRINT USING "###.##"; X, T(J)
NEXT J
QC = MDOT*CP*(T(M)-T(1))
LPRINT USING "###.##"; QC
STOP
END
```

FIGURE A–G–5 BASIC Program and Results for Double-Pipe Heat Exchanger: Rating Format

(a) BASIC Program.

```
SIGN = -1
IF SIGN = 1 THEN LPRINT "PARALLEL FLOW"
IF SIGN = -1 THEN LPRINT
  "COUNTERFLOW"
DIM TH(100): DIM TC(100): DIM TW(100)
REM INPUTS FOR CAPACITANCE RATES AND
   INLET TEMPERATURES
  CH = 294
  CC = 481
  THI = 98
```

FIGURE A–G–5 (BASIC Program and Results for Double-Pipe Heat Exchanger: Rating Format—continued)

```
      TCI = -20
REM   FOR COUNTERFLOW: SET INITIAL
        GUESS FOR TCO
      TCO = 5
REM   RATING FORMAT WITH SURFACE AREA
        ASS = pL KNOWN: SET M
      M = 20
      L = 3.5
      do = .01
      p = 3.14*do
      DX = L/M
REM   INPUTS PERTAINING TO OVERALL
        COEFFICIENT OF HEAT TRANSFER
      HH = 7960
      HC = 2370
      FFH = 0
      FFC = 0
      U = 1/(1/HH + FFH + FFC + 1/HC)
REM   INITIATE INCREMENTAL CALCULATIONS
1     TH(1) = THI
      IF SIGN = 1 THEN TC(1) = TCI
      IF SIGN = -1 THEN TC(1) = TCO
      Q = 0
      FOR J = 1 TO M
      DQ = U*p*DX*(TH(J) - TC(J))
      TH(J+1) = TH(J) - DQ/CH
      TC(J+1) = TC(J) + SIGN*DQ/CC
      TW(J) = TH(J) - DQ/(HH*p*DX)
      Q = Q + DQ
10    NEXT J
      IF SIGN = 1 GOTO 20
REM   ITERATION FOR COUNTERFLOW
      DIFF = TC(J) - TCI
      IF DIFF > .01 GOTO 12
      IF DIFF < -.01 GOTO 12
      GOTO 20
12    TCO = TCO - .5*DIFF
      GOTO 1
20    LPRINT "......M......THO......TCO......Q......"
      IF SIGN = 1 THEN TCO = TC(J)
      LPRINT USING "######.##"; M, TH(J),
        TCO, Q
      STOP
      END
```

(b) Results (noniterative) for Example 11–4: Set SIGN = 1 for parallel flow. (q = 14,400 W from Example 11–4.)

M	$T_{h,o}$ (°C)	$T_{c,o}$ (°C)	q (W)
10	47.60	10.80	14,820
20	48.39	10.32	14,580
30	48.64	10.17	14,510
40	48.77	10.09	14,470
80	48.96	9.98	14,420

(c) Results (iterative) for Example 11–5: Set SIGN = −1 for counterflow. (q = 15,200 W from Example 11–5.)

M	$T_{h,o}$ (°C)	$T_{c,o}$ (°C)	q (W)
10	45.77	11.91	15,350
20	46.00	11.78	15,290
30	46.07	11.73	15,270
40	46.11	11.71	15,260
50	46.16	11.67	15,240

FIGURE A–G–6 BASIC Program and Results for Double-Pipe Heat Exchanger: Design Format (noniterative)

(a) BASIC Program.

```
    SIGN = -1
    IF SIGN = 1 THEN LPRINT "PARALLEL FLOW"
    IF SIGN = -1 THEN LPRINT
      "COUNTERFLOW"
    DIM TH(200): DIM TC(200): DIM TW(200)
REM INPUTS FOR CAPACITANCE RATES AND
      TERMINAL TEMPERATURES
    THI = 122
    TCI = 68
    THO = 104
    TCO = 77
    CH = 1732.5
    CC = CH*(THI - THO)/(TCO - TCI)
REM DESIGN FORMAT WITH SURFACE AREA
      ASS UNKNOWN: SET DASS
    DASS = .25
REM INPUTS PERTAINING TO OVERALL
      COEFFICIENT OF HEAT TRANSFER
    U = 88
REM INITIATE INCREMENTAL CALCULATIONS
    TH(1) = THI
    IF SIGN = 1 THEN TC(1) = TCI
    IF SIGN = -1 THEN TC(1) = TCO
    Q = 0
    FOR J = 1 TO 200
    DQ = U*DASS*(TH(J) - TC(J))
    TH(J+1) = TH(J) - DQ/CH
    TC(J+1) = TC(J) + SIGN*DQ/CC
    Q = Q + DQ
    IF TH(J+1) < THO GOTO 20
    NEXT J
20  M = J
    ASS = M*DASS
    LPRINT ".......DASS.......M.......ASS......."
    LPRINT USING "######.###"; DASS,
      M, ASS
    STOP
    END
```

(b) Results for Example 11–6(a): Set SIGN = 1 for parallel flow. (A_s = 9.09 ft^2 from Example 11–6.)

ΔA_s (ft^2)	M	A_s (ft^2)
0.250	37	9.250
0.125	73	9.125
0.0625	146	9.125

(c) Results for Example 11–6(b): Set SIGN = −1 for counterflow. (A_s = 8.8 ft^2 from Example 11–6.)

ΔA_s (ft^2)	M	A_s (ft^2)
0.250	36	9.000
0.125	71	8.875
0.0625	141	8.813

FIGURE A–G–7 BASIC Program and Results for 1–2 Shell-and-Tube Heat Exchanger: Rating Format

(a) BASIC Program.

```
     SIGN = -1
     IF SIGN = 1 THEN LPRINT "TUBE PASS 1
       IN PARALLEL FLOW"
     IF SIGN = -1 THEN LPRINT "TUBE PASS
       1 IN COUNTERFLOW"
     DIM TS(100): DIM TT1(100): DIM TT2(100)
REM  HEAT TRANSFER FROM SHELL TO TUBE
       TAKEN AS POSITIVE
REM  INPUTS FOR CAPACITANCE RATES AND
       INLET TEMPERATURES
     CT = 12500
     CS = 41600
     TSI = 15
     TTI = 80
REM  SET INITIAL GUESS FOR TTO AND TSO
       (REQUIRED FOR SIGN = -1)
     TTO = 50
     TSO = TSI - (CT/CS)*(TTO - TTI)
REM  RATING FORMAT WITH SURFACE AREA
       ASS KNOWN: SET M
     M = 40
     ASS = 62.5
     DASS = ASS/M
REM  INPUTS PERTAINING TO OVERALL
       COEFFICIENT OF HEAT TRANSFER
     U = 300
REM  INITIATE INCREMENTAL CALCULATIONS
1    TSO = TSI - (CT/CS)*(TTO - TTI)
     IF SIGN = 1 THEN TS(1) = TSI
     IF SIGN = -1 THEN TS(1) = TSO
     TT1(1) = TTI
     TT2(1) = TTO
     Q = 0
     FOR J = 1 TO M
     DQ1 = U*(DASS/2)*(TS(J) - TT1(J))
     DQ2 = U*(DASS/2)*(TS(J) - TT2(J))
     DQ = DQ1 + DQ2
     TS(J+1) = TS(J) - SIGN*DQ/CS
     TT1(J+1) = TT1(J) + DQ1/CT
     TT2(J+1) = TT2(J) - DQ2/CT
     Q = Q + DQ
10   NEXT J
REM  ITERATION
     DIFF = TT1(J) - TT2(J)
     IF DIFF > .01 GOTO 12
     IF DIFF < -.01 GOTO 12
     GOTO 20
12   TTO = TTO + .5*DIFF
     GOTO 1
20   LPRINT ".......M.......TSO.......TTO.......Q......."
     LPRINT USING "######.##"; M, TSO,
       TTO, Q
     STOP
     END
```

(b) Results (iterative) for Example 11–13: Set SIGN = −1 for tube pass 1 in counterflow. (q = 561,000 W from Example 11–13.)

M	$T_{h,o}$ (°C)	$T_{c,o}$ (°C)	q (W)
40	35.20	28.46	560,000
80	35.16	28.47	560,600

(c) Results (iterative) for Example 11–13 with directions of shell side flow stream changed: Set SIGN = 1 for tube pass 1 in parallel flow. (q = 561,000 W from Example 11–13 for tube pass 1 in counterflow.)

M	$T_{h,o}$ (°C)	$T_{c,o}$ (°C)	q (W)
40	35.01	28.52	562,400
80	35.06	28.50	561,800

FIGURE A–G–8 BASIC Program and Results for Mixed/Unmixed Heat Exchanger: Rating Format

(a) BASICc Program.

```
      DIM TS(200): DIM TT(200,200)
REM   INPUTS FOR CAPACITANCE RATES AND
        INLET TEMPERATURES
      CS = 2410
      CT = 252
      TSI = 15
      TTI = 65
REM   RATING FORMAT WITH SURFACE AREA
        ASS KNOWN: SET ML, MW
      ML = 20
      MW = 20
      ASS = 5.98
      DASS = ASS/ML/MW
REM   INPUTS PERTAINING TO OVERALL
        COEFFICIENT OF HEAT TRANSFER
      U = 93.4
REM   INITIATE INCREMENTAL CALCULATIONS
      TS(1) = TSI
      Q = 0
      FOR J = 1 TO ML
      TT(J,1) = TTI
      DQ = 0
      FOR I = 1 TO MW
      DDQ = U*DASS*(TS(J) - TT(J,I))
      TT(J,I+1) = TT(J,I) + DDQ/(CT/ML)
      DQ = DQ + DDQ
      NEXT I
      TS(J+1) = TS(J) - DQ/CS
      Q = Q + DQ
      NEXT J
      TTO = TTI + Q/CT
      TSO = TS(J)
      LPRINT ".......ML.......MW.......TTO.......
        TSO.......Q......."
      LPRINT USING "######.##"; ML, MW,
        TTO, TSO, Q
      STOP
      END
```

(b) Results for Example 11–17: (q = −10,700 W from Example 11–17.)

ML	*MW*	$T_{t,o}$ (°C)	$T_{s,o}$ (°C)	q (W)
10	10	21.02	19.60	−11,080
20	20	21.75	19.52	−10,900
30	30	21.99	19.50	−10,840
50	50	22.18	19.48	−10,790

H TABULATED FUNCTIONS

H-1 MATHEMATICAL FUNCTIONS

TABLE A-H-1-1 Error function†

X	erf X	X	erf X	X	erf X
0.00	0.00000	0.36	0.38933	1.04	0.85865
0.02	0.02256	0.38	0.40901	1.08	0.87333
0.04	0.04511	0.40	0.42839	1.12	0.88679
0.06	0.06762	0.44	0.46622	1.16	0.89910
0.08	0.09008	0.48	0.50275	1.20	0.91031
0.10	0.11246	0.52	0.53790	1.30	0.93401
0.12	0.13476	0.56	0.57162	1.40	0.95228
0.14	0.15695	0.60	0.60386	1.50	0.96611
0.16	0.17901	0.64	0.63459	1.60	0.97635
0.18	0.20094	0.68	0.66378	1.70	0.98379
0.20	0.22270	0.72	0.69143	1.80	0.98909
0.22	0.24430	0.76	0.71754	1.90	0.99279
0.24	0.26570	0.80	0.74210	2.00	0.99532
0.26	0.28690	0.84	0.76514	2.20	0.99814
0.28	0.30788	0.88	0.78669	2.40	0.99931
0.30	0.32863	0.92	0.80677	2.60	0.99976
0.32	0.34913	0.96	0.82542	2.80	0.99992
0.34	0.36936	1.00	0.84270	3.00	0.99998

† The error function is defined as $\operatorname{erf} X \equiv \frac{2}{\sqrt{\pi}} \int_0^X e^{-\beta^2}\, d\beta$. The complementary error function is defined as $\operatorname{erfc} X \equiv 1 - \operatorname{erf} X$.

TABLE A-H-1-2 Bessel functions of the first kind

x	$J_0(x)$	$J_1(x)$	x	$J_0(x)$	$J_1(x)$	x	$J_0(x)$	$J_1(x)$
0.0	1.0000	0.0000	1.0	0.7652	0.4400	2.0	0.2239	0.5767
0.1	0.9975	0.0499	1.1	0.7196	0.4709	2.1	0.1666	0.5683
0.2	0.9900	0.0995	1.2	0.6711	0.4983	2.2	0.1104	0.5560
0.3	0.9776	0.1483	1.3	0.6201	0.5220	2.3	0.0555	0.5399
0.4	0.9604	0.1960	1.4	0.5669	0.5419	2.4	0.0025	0.5202
0.5	0.9385	0.2423	1.5	0.5118	0.5579	2.5	−0.0484	0.4971
0.6	0.9120	0.2867	1.6	0.4554	0.5699	2.6	−0.0968	0.4708
0.7	0.8812	0.3290	1.7	0.3980	0.5778	2.7	−0.1424	0.4416
0.8	0.8463	0.3688	1.8	0.3400	0.5815	2.8	−0.1850	0.4097
0.9	0.8075	0.4059	1.9	0.2818	0.5812	2.9	−0.2243	0.3754

H-2 RADIATION FUNCTIONS

TABLE A-H-2 Radiation functions

λT_s (μm K)	$\frac{E_{b,0\to\lambda}}{E_b}$	$\frac{E_{b\lambda}}{E_b}$	λT_s	$\frac{E_{b,0\to\lambda}}{E_b}$	$\frac{E_{b\lambda}}{E_b}$
200	0.000000	1.1782×10^{-27}	6,200	0.754140	0.78453×10^{-4}
400	0.000000	1.5404×10^{-13}	6,400	0.769234	0.72566
600	0.000000	3.2687×10^{-8}	6,600	0.783199	0.67163
800	0.000016	3.1137×10^{-7}	6,800	0.796129	0.62206
1,000	0.000321	3.7229×10^{-5}	7,000	0.808109	0.57659
1,200	0.002134	1.6460	7,200	0.819217	0.53487
1,400	0.007790	4.2226	7,400	0.829527	0.49660
1,600	0.019718	7.8266	7,600	0.839102	0.46147
1,800	0.039341	1.1799×10^{-4}	7,800	0.848005	0.42921
2,000	0.066728	1.5502	8,000	0.856288	0.39956
2,200	0.100888	1.8524	8,500	0.874608	0.33543
2,400	0.140256	2.0699	9,000	0.890029	0.28320
2,600	0.183120	2.2032	9,500	0.903085	0.24044
2,800	0.227897	2.2627	10,000	0.914199	0.20523
3,000	0.273232	2.2627	10,500	0.923710	0.17609
3,200	0.318102	2.2179	11,000	0.931890	0.15184
3,400	0.361735	2.1411	11,500	0.939959	0.13155
3,600	0.403607	2.0433	12,000	0.945098	0.11448
3,800	0.443382	1.8328	13,000	0.955139	0.87794×10^{-5}
4,000	0.480877	1.8160	14,000	0.962898	0.68373
4,200	0.516014	1.6977	15,000	0.969981	0.53993
4,400	0.548796	1.5810	16,000	0.973814	0.43174
4,600	0.579280	1.4682	18,000	0.980860	0.28533
4,800	0.607559	1.3606	20,000	0.985602	0.19582
5,000	0.633747	1.2592	25,000	0.992215	0.86857×10^{-6}
5,200	0.658970	1.1642	30,000	0.995340	0.44130
5,400	0.680360	1.0758	40,000	0.997967	0.14889
5,600	0.701046	0.99392	50,000	0.998953	0.63336×10^{-7}
5,800	0.720158	0.91829	75,000	0.999713	0.13151
6,000	0.737818	0.84861	100,000	0.999905	0.42648×10^{-8}

I LAMINAR BOUNDARY LAYER FLOW

Consideration is now given to laminar boundary layer flow, with the objectives of (1) developing a mathematical formulation which is applicable to viscous and high speed flows, and (2) developing a systematic transformation analysis approach.

I-1 MATHEMATICAL FORMULATION

To develop a general mathematical formulation for laminar boundary layer flow, we must account for the effects of viscous dissipation (and kinetic energy) and variable properties. The steps required in the development of the differential formulation for

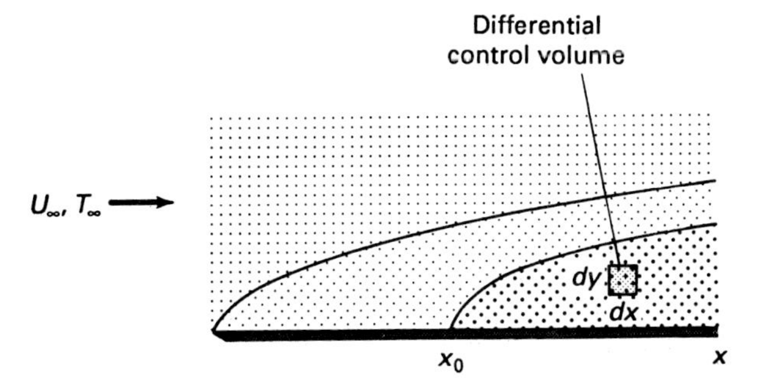

FIGURE A–I–1–1
Laminar boundary layer flow over a flat plate; $dV = dx\,dy\,dz$.

fluid flow and energy transfer for this more general situation are similar to those followed in Sec. 7–2–1 for simple laminar flow in tubes. The differential control volume shown in Fig. A–I–1–1 provides a frame of reference in the context of a standard Cartesian coordinate system.

I–1–1 Continuity

The principle of conservation of mass is applied to the control volume shown in Figs. A–I–1–1 and A–I–1–2 to obtain

$$\Sigma\,\dot{m}_o - \Sigma\,\dot{m}_i + \frac{\Delta m_s}{\Delta t} = 0 \qquad d\dot{m}_{x+dx} + d\dot{m}_{y+dy} - d\dot{m}_x - d\dot{m}_y = 0 \quad \text{(I–1–1)}$$

where $\Delta m_s/\Delta t = 0$ for steady flow, $d\dot{m}_x = \rho u\,dy\,dz$ and $d\dot{m}_y = \rho v\,dx\,dz$. It follows that

$$\frac{\partial}{\partial x}(d\dot{m}_x)\,dx + \frac{\partial}{\partial y}(d\dot{m}_y)\,dy = 0 \qquad \frac{\partial}{\partial x}(\rho u) + \frac{\partial}{\partial y}(\rho v) = 0 \qquad \text{(I–1–2,3)}$$

which represents the continuity equation for variable property flow. Equation (I–1–3) reduces to Eq. (7–58),

$$\frac{\partial u}{\partial x} + \frac{\partial v}{\partial y} = 0 \qquad \text{(I–1–4)}$$

for uniform density.

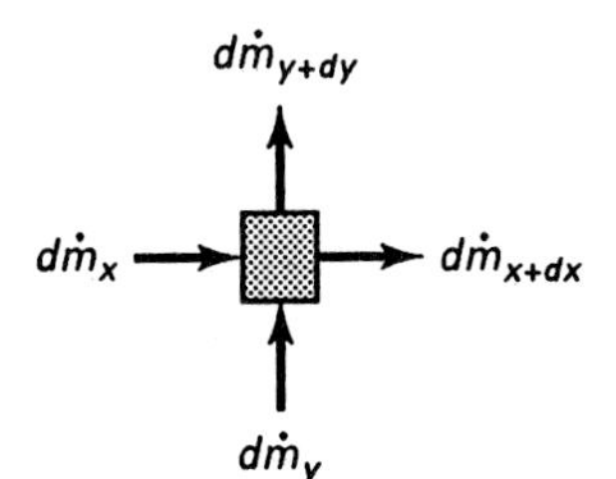

FIGURE A–I–1–2
Conservation of mass relative to differential rectangular control volume; $dA_x = dy\,dz$, $dA_y = dx\,dz$.

I-1-2 Momentum Transfer

The momentum equation is developed by applying Newton's second law of motion to the control volume shown in Figs. A–I–1–1 and A–I–1–3; that is,

$$\Sigma \dot{M}_{o,x} - \Sigma \dot{M}_{i,x} + \frac{\Delta M_{s,x}}{\Delta t} = \Sigma F_x \qquad \text{(I–1–5)}$$

The sum of forces is

$$\begin{aligned} \Sigma F_x &= (\tau\, dA)_{y+dy} + (P\, dA)_x - (\tau\, dA)_y - (P\, dA)_{x+dx} \\ &= \frac{\partial \tau}{\partial y}\, dV - \frac{\partial P}{\partial x}\, dV \end{aligned} \qquad \text{(I–1–6)}$$

where $dA_x = dy\, dz$, $dA_y = dx\, dz$, and $dV = dx\, dy\, dz$, assuming negligible normal viscous stresses, buoyant forces, and other body forces. The rate of creation of axial momentum is represented by RCM_x,

$$\begin{aligned} RCM_x &= d\dot{M}_{x+dx} + d\dot{M}_{y+dy} - d\dot{M}_x - d\dot{M}_y \\ &= \frac{\partial}{\partial x}(d\dot{M}_x)\, dx + \frac{\partial}{\partial y}(d\dot{M}_y)\, dy \end{aligned} \qquad \text{(I–1–7)}$$

where $d\dot{M}_x = u\, d\dot{m}_x$ and $d\dot{M}_y = u\, d\dot{m}_y$. Substituting for $d\dot{M}_x$ and $d\dot{M}_y$, RCM_x becomes

$$\begin{aligned} RCM_x &= \frac{\partial}{\partial x}(u\, d\dot{m}_x)\, dx + \frac{\partial}{\partial y}(u\, d\dot{m}_y)\, dy \\ &= \frac{\partial}{\partial x}(u\rho u)\, dV + \frac{\partial}{\partial y}(u\rho v)\, dV \end{aligned} \qquad \text{(I–1–8)}$$

Making use of the continuity equation, Eq. (I–1–3), this relation is put into the form

$$RCM_x = \rho\left(u\frac{\partial u}{\partial x} + v\frac{\partial u}{\partial y}\right) dV \qquad \text{(I–1–9)}$$

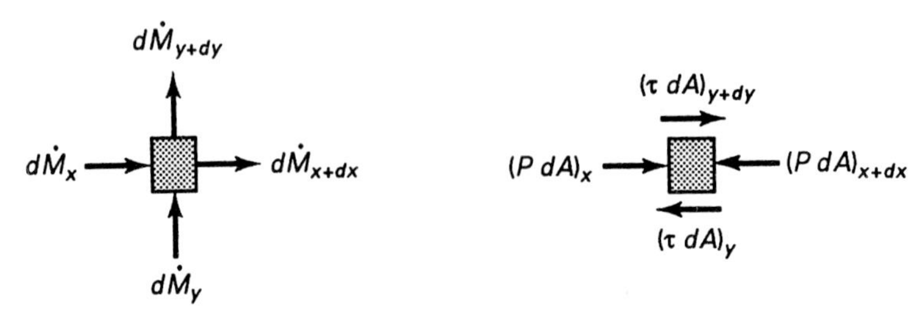

(a) Momentum transfer. (b) Forces.

FIGURE A–I–1–3 Momentum principle relative to differential rectangular control volume; *x*-direction.

Substituting Eqs. (I–1–6) and (I–1–9) into Eq. (I–1–5), we obtain

$$\rho\left(u\frac{\partial u}{\partial x}+v\frac{\partial u}{\partial y}\right)=\frac{\partial \tau}{\partial y}-\frac{\partial P}{\partial x} \tag{I–1–10}$$

Using Eq. (v) in Table 7–1 for τ and assuming that $\partial P/\partial x \simeq dP/dx$, Eq. (I–1–10) becomes

$$\rho\left(u\frac{\partial u}{\partial x}+v\frac{\partial u}{\partial y}\right)=\frac{\partial}{\partial y}\left(\mu\frac{\partial u}{\partial y}\right)-\frac{dP}{dx} \tag{I–1–11}$$

which represents the momentum equation for variable property conditions. This equation reduces to Eq. (7–59),

$$u\frac{\partial u}{\partial x}+v\frac{\partial u}{\partial y}=\nu\frac{\partial^2 u}{\partial y^2}-\frac{1}{\rho}\frac{dP}{dx} \tag{I–1–12}$$

for uniform viscosity.

I–1–3 Energy Transfer

Referring to Fig. A–I–1–4, the viscous shearing forces acting on the surfaces dA_y and dA_{y+dy} produce rates of work, the magnitudes of which are represented by

$$|d\dot{W}_\tau|_y = |\tau u\, dA|_y \qquad \text{and} \qquad |d\dot{W}_\tau|_{y+dy} = |\tau u\, dA|_{y+dy} \tag{I–1–13,14}$$

The *net* rate of work $d\dot{W}_\tau$ that is actually transferred into the element as a result of viscous shearing action is equal to the absolute difference between these two rates of work; that is,†

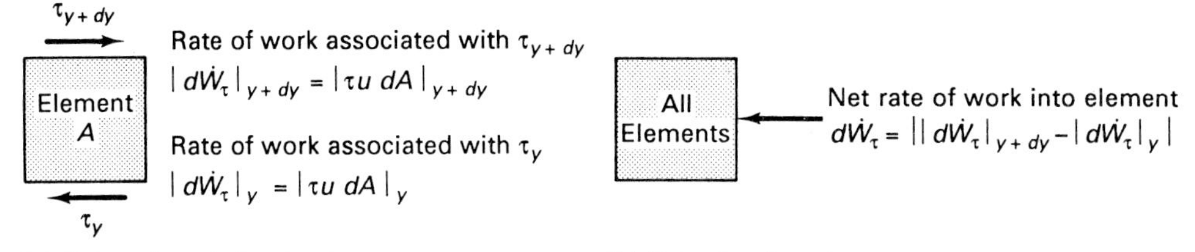

(a) Rates of work due to stresses acting on fluid element surfaces.

(b) Net rate of work.

FIGURE A–I–1–4 Work associated with shearing stresses.

† Designating the two work rates by $|d\dot{W}_\tau|_{\text{greater}}$ and $|d\dot{W}_\tau|_{\text{lesser}}$, we note that $|d\dot{W}_\tau|_{\text{greater}} = d\dot{W}_\tau + |d\dot{W}_\tau|_{\text{lesser}}$, where $d\dot{W}_\tau$ is transferred into the fluid element and $|d\dot{W}_\tau|_{\text{lesser}}$ is transmitted through the element and into the adjacent layer of fluid.

$$d\dot{W}_\tau = \left| |d\dot{W}_\tau|_{y+dy} - |d\dot{W}_\tau|_y \right| = \left| |\tau u\ dA|_{y+dy} - |\tau u\ dA|_y \right|$$
$$= \frac{\partial}{\partial y}(\tau u\ dA_y)\ dy = \frac{\partial}{\partial y}(\tau u)\ dV \qquad \text{(I–1–15)}$$

which is greater than (or equal to) zero.

The viscous energy generation term $d\dot{W}_\tau$ is shown together with the rates of kinetic energy [$d\dot{KE}_x = (u^2/2)\ d\dot{m}_x$ and $d\dot{KE}_y = (u^2/2)\ d\dot{m}_y$], enthalpy, and heat transfer entering and exiting a control volume in Fig. A–I–1–5. By incorporating each of these terms into the first law of thermodynamics, we obtain

$$d\dot{H}_{x+dx} + d\dot{KE}_{x+dx} + d\dot{H}_{y+dy} + d\dot{KE}_{y+dy} + dq_{y+dy}$$
$$- d\dot{H}_x - d\dot{KE}_x - d\dot{H}_y - d\dot{KE}_y - dq_y - d\dot{W}_\tau = 0 \qquad \text{(I–1–16)}$$

where $d\dot{H}_x = i\ d\dot{m}_x$ and $d\dot{H}_y = i\ d\dot{m}_y$, or

$$\frac{\partial}{\partial x}(d\dot{H}_x)\ dx + \frac{\partial}{\partial x}(d\dot{KE}_x)\ dx + \frac{\partial}{\partial y}(d\dot{H}_y)\ dy$$
$$+ \frac{\partial}{\partial y}(d\dot{KE}_y)\ dy + \frac{\partial}{\partial y}(dq_y)\ dy = d\dot{W}_\tau \qquad \text{(I–1–17)}$$

Substituting for the various terms and introducing the Fourier law of conduction, Eq. (I–1–17) becomes (assuming $u \gg v$)

$$\frac{\partial}{\partial x}[d\dot{m}_x\ (i + u^2/2)]\ dx + \frac{\partial}{\partial y}[d\dot{m}_y\ (i + u^2/2)]\ dy$$
$$- \frac{\partial}{\partial y}\left(k \frac{\partial T}{\partial y}\ dA_y\right)\ dy = \frac{\partial}{\partial y}(\tau u\ dA_y)\ dy \qquad \text{(I–1–18)}$$

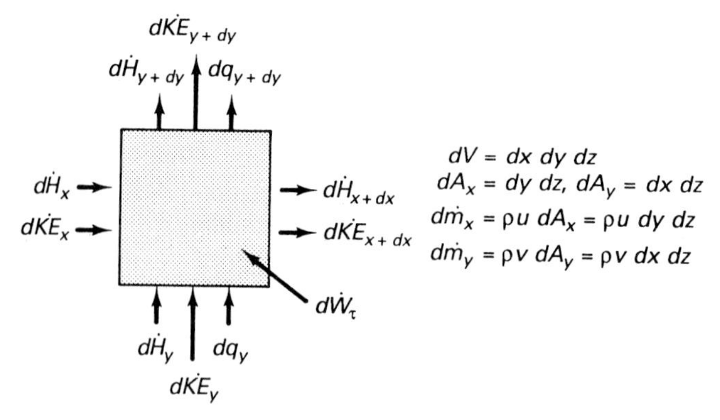

FIGURE A–I–1–5 Conservation of energy relative to differential control volume for viscous and high speed laminar flows.

or

$$\frac{\partial}{\partial x}[\rho u(i + u^2/2)] + \frac{\partial}{\partial y}[\rho v(i + u^2/2)] = \frac{\partial}{\partial y}(\tau u) + \frac{\partial}{\partial y}\left(k\frac{\partial T}{\partial y}\right) \tag{I–1–19}$$

The continuity and momentum equations are often used to put the energy equation into more convenient forms. To see this point, the continuity equation is combined with Eq. (I–1–19) to obtain

$$\rho\left[u\frac{\partial}{\partial x}(i + u^2/2) + v\frac{\partial}{\partial y}(i + u^2/2)\right] = \frac{\partial}{\partial y}(\tau u) + \frac{\partial}{\partial y}\left(k\frac{\partial T}{\partial y}\right) \tag{I–1–20}$$

Taking a step further, we write

$$\frac{\partial}{\partial y}(\tau u) = \tau\frac{\partial u}{\partial y} + u\frac{\partial \tau}{\partial y} \tag{I–1–21}$$

and use the momentum equation, Eq. (I–1–10), to obtain

$$\begin{aligned}\frac{\partial}{\partial y}(\tau u) &= \tau\frac{\partial u}{\partial y} + u\frac{dP}{dx} + \rho u\left(u\frac{\partial u}{\partial x} + v\frac{\partial u}{\partial y}\right)\\ &= \tau\frac{\partial u}{\partial y} + u\frac{dP}{dx} + \rho\left[u\frac{\partial}{\partial x}(u^2/2) + v\frac{\partial}{\partial y}(u^2/2)\right]\end{aligned} \tag{I–1–22}$$

Substituting this result into Eq. (I–1–20), we have

$$\rho\left(u\frac{\partial i}{\partial x} + v\frac{\partial i}{\partial y}\right) = \tau\frac{\partial u}{\partial y} + u\frac{dP}{dx} + \frac{\partial}{\partial y}\left(k\frac{\partial T}{\partial y}\right) \tag{I–1–23}$$

Setting $\tau = \mu\,\partial u/\partial y$ in accordance with Newton's law of viscous stress, our final result is

$$\rho\left(u\frac{\partial i}{\partial x} + v\frac{\partial i}{\partial y}\right) = \mu\left(\frac{\partial u}{\partial y}\right)^2 + u\frac{dP}{dx} + \frac{\partial}{\partial y}\left(k\frac{\partial T}{\partial y}\right) \tag{I–1–24}$$

or, in terms of specific heat c_P,

$$\rho c_P\left(u\frac{\partial T}{\partial x} + v\frac{\partial T}{\partial y}\right) = \mu\left(\frac{\partial u}{\partial y}\right)^2 + u\frac{dP}{dx} + \frac{\partial}{\partial y}\left(k\frac{\partial T}{\partial y}\right) \tag{I–1–25}$$

The term $\mu(\partial u/\partial y)^2$ represents the energy dissipation per unit volume. This term is commonly designated by $\mu\Phi$, where the *dissipation function* Φ is defined by

$$\Phi = \left(\frac{\partial u}{\partial y}\right)^2 \tag{I–1–26}$$

for this basic two-dimensional flow. Notice that Φ is always positive. Except for very high speed flows, the pressure energy term $u\ dP/dx$ is generally quite small. For the usual case in which the viscous dissipation term $\mu\Phi$ and the pressure gradient energy are small, Eq. (I–1–25) reduces to

$$\rho c_P\left(u\frac{\partial T}{\partial x}+v\frac{\partial T}{\partial y}\right)=\frac{\partial}{\partial y}\left(k\frac{\partial T}{\partial y}\right) \tag{I–1–27}$$

This equation reduces to Eq. (7–60),

$$u\frac{\partial T}{\partial x}+v\frac{\partial T}{\partial y}=\alpha\frac{\partial^2 T}{\partial y^2} \tag{I–1–28}$$

for uniform thermal conductivity.

I–2 TRANSFORMATION ANALYSIS

The equations which are applicable to laminar two-dimensional plane boundary layer flow with uniform properties are given by Eqs. (7–58) to (7–60),

Continuity

$$\frac{\partial u}{\partial x}+\frac{\partial v}{\partial y}=0 \tag{I–2–1}$$

Momentum†

$$u\frac{\partial u}{\partial x}+v\frac{\partial u}{\partial y}=\nu\frac{\partial^2 u}{\partial y^2}-\frac{1}{\rho}\frac{dP}{dx}+\frac{f_{bx}}{\rho} \tag{I–2–2}$$

Energy

$$u\frac{\partial T}{\partial x}+v\frac{\partial T}{\partial y}=\alpha\frac{\partial^2 T}{\partial y^2} \tag{I–2–3}$$

for negligible kinetic energy, viscous dissipation, axial conduction, and thermal radiation effects, where f_{bx} represents the body force (e.g., $f_{bx} = -\rho g\, dV$ for natural convection on a heated vertical plate). The boundary conditions are given by

$$\begin{aligned} u &= U_\infty \quad \text{and} \quad T = T_\infty \quad \text{at } x = 0\\ u &= 0 \quad \text{and} \quad v = 0 \quad \text{at } y = 0\\ u &= U_\infty \quad \text{and} \quad T = T_\infty \quad \text{as } y \to \infty \end{aligned} \tag{I–2–4}$$

for nontranspired flow‡,

$$T = T_s \quad \text{at } y = 0$$

for specified wall temperature, or

$$-k\frac{\partial T}{\partial y} = q''_c \quad \text{at } y = 0$$

for specified wall-heat flux.

These equations can be solved by the approximate integral solution method which is introduced in Secs. 7–2–2 and 7–2–3 or by numerical methods. Numerical

† Including body force terms.

‡ The analysis is generalized to account for transpiration by writing $v = v_0$ at $y = 0$.

methods available in the literature generally employ *coordinate* and *variable transformations* which produce more efficient solutions. However, approaches to establishing transformations commonly presume that the form of the coordinate and variable transformations can simply be stated as common knowledge, which is not always comfortable for inquisitive students. Therefore, a systematic approach is now developed for transforming Eqs. (I–2–1) to (I–2–3), after which specific attention is given to forced convection flows and natural convection on a vertical plate.

I–2–1 Similarity Transformation

The simplest and perhaps most useful system of transformations for uniform property plane boundary layer flow involves the y-coordinate transformation

$$\eta = yX_x \tag{I–2–5}$$

and dimensionless velocity and temperature distributions of the form

$$U = \frac{u}{\boldsymbol{U}_x} \qquad \Theta = \frac{T - T_\infty}{T_s - T_\infty} \tag{I–2–6,7}$$

where the *stretching parameter* X_x is a function of x, and the characteristic velocity $\boldsymbol{U}_x$ depends on the nature of the flow field (i.e., whether forced or natural convection, jet flow, etc.). To achieve effective transformations, X_x and $\boldsymbol{U}_x$ are specified such that the transformation produces (1) coordinate grids that essentially conform to the general shape of the boundary layer, and (2) distributions in the dimensionless velocity U and temperature Θ that tend to be exactly or approximately geometrically similar in the streamwise-x direction.† Hence, transformations of this type are generally referred to as *similarity transformations*. Systematic methods of specifying X_x and $\boldsymbol{U}_x$ for a wide range of boundary layer flows have been developed that are based on the separation of variables concept and other methods [28]. However, the simplest approach is to establish the forms of X_x and $\boldsymbol{U}_x$ by means of the approximate integral solution technique. For example, referring to Sec. 7–2–2, Eq. (7–104) indicates

$$\eta = y\sqrt{\frac{U_\infty}{\nu x}} \qquad \boldsymbol{U}_x = U_\infty \tag{I–2–8,9}$$

such that

$$X_x = \frac{\eta}{y} = \sqrt{\frac{U_\infty}{\nu x}} \tag{I–2–10}$$

for forced convection flow over a flat plate with uniform free-stream velocity.

† Transformations are also sometimes used for the x coordinate, but these are generally of secondary importance, except for axisymmetric flows.

To transform Eqs. (I–2–1) to (I–2–4), we introduce the following partial differential operators:

$$\left(\frac{\partial}{\partial y}\right)_x = \left(\frac{\partial}{\partial \eta}\right)_x \left(\frac{\partial \eta}{\partial y}\right)_x = \left(\frac{\partial}{\partial \eta}\right)_x X_x \tag{I–2–11a}$$

and

$$\left(\frac{\partial}{\partial x}\right)_y = \left(\frac{\partial}{\partial x}\right)_\eta + \left(\frac{\partial}{\partial \eta}\right)_x \left(\frac{\partial \eta}{\partial x}\right)_y \tag{I–2–11b}$$

The classic approach to accomplishing the transformation involves the use of the *stream function* ψ which is defined in terms of the velocity components and cartesian coordinates by

$$u = \frac{\partial \psi}{\partial y} \qquad v = -\frac{\partial \psi}{\partial x} \tag{I–2–12a,b}$$

such that the continuity equation, Eq. (I–2–1), is satisfied. Using Eq. (I–2–11a), Eq. (I–2–12a) is transformed to

$$U = \frac{u}{\boldsymbol{U}_x} = \frac{X_x}{\boldsymbol{U}_x}\frac{\partial \psi}{\partial \eta} = \frac{\partial}{\partial \eta}\left(\frac{X_x \psi}{\boldsymbol{U}_x}\right) \quad \text{or} \quad U = \frac{\partial \zeta}{\partial \eta} = \zeta' \tag{I–2–13,14}$$

where the *dimensionless stream function* ζ is

$$\zeta = \frac{X_x \psi}{\boldsymbol{U}_x} \tag{I–2–15}$$

and the prime represents differentiation with respect to η. Continuing with the transformation of Eq. (I–2–12b), Eq. (I–2–11b) is used to obtain

$$v = -\frac{\partial \psi}{\partial x} - \frac{\partial \psi}{\partial \eta}\left(\frac{\partial \eta}{\partial x}\right)_y \tag{I–2–16}$$

or, after expressing ψ in terms of ζ,

$$\begin{aligned} v &= -\frac{\partial}{\partial x}\left(\frac{\boldsymbol{U}_x \zeta}{X_x}\right) - \frac{\partial}{\partial \eta}\left(\frac{\boldsymbol{U}_x \zeta}{X_x}\right)\left(\frac{\partial \eta}{\partial x}\right)_y \\ &= -\left[\frac{\partial}{\partial x}\left(\frac{\boldsymbol{U}_x \zeta}{X_x}\right) + \frac{\boldsymbol{U}_x}{X_x}\left(\frac{\partial \eta}{\partial x}\right)_y \zeta'\right] \end{aligned} \tag{I–2–17}$$

These relations for u and v are used together with Eq. (I–2–11) to transform the momentum equation, Eq. (I–2–2), into the form

$$\underline{\boldsymbol{U}_x\zeta' \frac{\partial}{\partial x}(\boldsymbol{U}_x\zeta') + \boldsymbol{U}_x\zeta' \frac{\partial}{\partial \eta}(\boldsymbol{U}_x\zeta')\left(\frac{\partial \eta}{\partial x}\right)_y} - \left[\frac{\partial}{\partial x}\left(\frac{\boldsymbol{U}_x \zeta}{X_x}\right) + \underline{\frac{\boldsymbol{U}_x}{X_x}\left(\frac{\partial \eta}{\partial x}\right)_y \zeta'}\right]\boldsymbol{U}_x X_x \zeta'' = \nu \boldsymbol{U}_x X_x^2 \zeta''' - \frac{1}{\rho}\frac{dP}{dx} + \frac{f_{bx}}{\rho} \tag{I–2–18}$$

Eliminating the underlined terms which cancel and rearranging, we obtain

$$\zeta''' + \frac{\boldsymbol{U}_x}{\nu X_x^2}\left[\left(\frac{d\boldsymbol{U}_x/dx}{\boldsymbol{U}_x} - \frac{dX_x/dx}{X_x}\right)\zeta\zeta'' - \frac{d\boldsymbol{U}_x/dx}{\boldsymbol{U}_x}\zeta'^2\right] + \frac{1}{\mu \boldsymbol{U}_x X_x^2}\left(-\frac{dP}{dx} + f_{bx}\right) = \frac{\boldsymbol{U}_x}{\nu X_x^2}\left(\zeta'\frac{\partial \zeta'}{\partial x} - \zeta''\frac{\partial \zeta}{\partial x}\right) \tag{I–2–19}$$

or

$$\zeta''' + (C_{x1} - C_{x2})\zeta\zeta'' - C_{x1}\zeta'^2 + C_{x3} + C_{x4} = C_{x5}x\left(\zeta'\frac{\partial \zeta'}{\partial x} - \zeta''\frac{\partial \zeta}{\partial x}\right) \tag{I–2–20}$$

where

$$C_{x1} = \frac{1}{\nu X_x^2}\frac{d\boldsymbol{U}_x}{dx} \qquad C_{x2} = \frac{\boldsymbol{U}_x}{\nu X_x^3}\frac{dX_x}{dx} \qquad C_{x3} = -\frac{1}{\mu \boldsymbol{U}_x X_x^2}\frac{dP}{dx}$$
$$C_{x4} = \frac{1}{\mu \boldsymbol{U}_x X_x^2} f_{bx} \qquad C_{x5} = \frac{\boldsymbol{U}_x}{\nu x X_x^2} \tag{I–2–21}$$

Following the same procedure, the transformed energy equation is given by

$$\frac{\partial^2 T}{\partial \eta^2} + \frac{\boldsymbol{U}_x}{\nu X_x^2}\left(\frac{d\boldsymbol{U}_x/dx}{\boldsymbol{U}_x} - \frac{dX_x/dx}{X_x}\right)\frac{\partial T}{\partial \eta} = \frac{\boldsymbol{U}_x}{\alpha X_x^2}\zeta'\frac{\partial T}{\partial x} \tag{I–2–22}$$

or, in terms of Θ,

$$\frac{\Theta''}{Pr} + (C_{x1} - C_{x2})\zeta\Theta' - C_{x5}n\zeta'\Theta = C_{x5}x\left(\zeta'\frac{\partial \Theta}{\partial x} - \Theta'\frac{\partial \zeta}{\partial x}\right) \tag{I–2–23}$$

where $Pr = \nu/\alpha$, and

$$n = \frac{x}{T_s - T_\infty}\frac{dT_s}{dx} \tag{I–2–24}$$

with T_s specified as a function of x.

The transformed boundary conditions are written as

$$\zeta' = 0 \qquad \Theta = 1 \qquad \text{at } \eta = 0 \tag{I–2–25a}$$

$$\zeta' = \frac{U_\infty}{\boldsymbol{U}_x} \qquad \Theta = 0 \qquad \text{as } \eta \to \infty \tag{I–2–25b}$$

and, setting $v = 0$ and $\zeta' = 0$ at $\eta = 0$ in Eq. (I–2–17)†,

$$\zeta = 0 \qquad \text{at } \eta = 0 \tag{I–2–26}$$

† The analysis is generalized to account for transpiration by writing

$$\zeta = -\frac{X_x}{\boldsymbol{U}_x}\int_0^x v_0\, dx \qquad \text{at } \eta = 0$$

The wall-shear stress τ_0 and wall-heat flux q''_c are expressed in terms of the transformed variables by writing

$$\tau_0 = \mu \left.\frac{\partial u}{\partial y}\right|_0 = \mu U_x X_x \left.\frac{\partial U}{\partial \eta}\right|_0 \quad \text{or} \quad \frac{\tau_0}{\mu U_x X_x} = \zeta''(0) \qquad \text{(I–2–27,28)}$$

and

$$q''_c = -k \left.\frac{\partial T}{\partial y}\right|_0 = -k(T_s - T_\infty)X_x \left.\frac{\partial \Theta}{\partial \eta}\right|_0 = -k(T_s - T_\infty)X_x\,\Theta'(0) \qquad \text{(I–2–29)}$$

or

$$Nu_x = \frac{q''_c}{T_s - T_\infty}\frac{x}{k} = -\Theta'(0)\,xX_x \qquad \text{(I–2–30)}$$

To express n in terms of q''_c for applications in which the wall-heat flux is specified rather than the wall temperature, we combine Eqs. (I–2–24) and (I–2–29), with the result

$$n = \frac{x}{T_s - T_\infty}\frac{dT_s}{dx} = \frac{x}{q''_c/[X_x\,\Theta'(0)]}\frac{d}{dx}\left[\frac{q''_c}{X_x\,\Theta'(0)}\right] \qquad \text{(I–2–31)}$$

In addition to providing an effective framework for the development of numerical solutions, these equations also enable us to establish the conditions for which the boundary layers exhibit exact similarity; that is,

$$\frac{\partial \zeta'}{\partial x} = 0 \qquad \frac{\partial \Theta}{\partial x} = 0 \qquad \frac{\partial \zeta}{\partial x} = 0 \qquad \text{(I–2–32a,b,c)}$$

We now turn our attention to (1) forced convection flow with arbitrary free-stream velocity, and (2) natural convection flow over a vertical surface. The continuity, momentum, and energy equations and boundary conditions for both of these basic boundary layer flows are represented by Eqs. (I–2–1) through (I–2–4).

I–2–2 Forced Convection Flow with Arbitrary Distribution in Free-Stream Velocity

The pressure gradient dP/dx for forced convection boundary layer flow is expressed in terms of the free-stream velocity U_∞ by the famous Bernoulli equation [i.e., the limiting form of the momentum equation, Eq. (I–2–2), as $y \to \infty$] for $f_{bx} = 0$,

$$\frac{dP}{dx} = -\rho U_\infty \frac{dU_\infty}{dx} \qquad \text{(I–2–33)}$$

Substituting for dP/dx in Eq. (I–2–2), the momentum equation becomes

$$\rho\left(u\frac{\partial u}{\partial x} + v\frac{\partial u}{\partial y}\right) = \mu\frac{\partial^2 u}{\partial y^2} + \rho U_\infty \frac{dU_\infty}{dx} \qquad \text{(I–2–34)}$$

Setting $\boldsymbol{U}_x = U_\infty$, $\eta = y\sqrt{U_\infty/(\nu x)}$, and $X_x = \sqrt{U_\infty/(\nu x)}$ in accordance with the integral solution results for laminar forced convection boundary layer flow with $f_{bx} = 0$ (see Sec. 7–2–2), we are able to evaluate the parameters C_{xi} given by Eq. (I–2–21) as follows:

$$C_{x1} = \frac{1}{\nu X_x^2}\frac{d\boldsymbol{U}_x}{dx} = \frac{x}{U_\infty}\frac{d\boldsymbol{U}_x}{dx} = m$$

$$C_{x2} = \frac{\boldsymbol{U}_x}{\nu X_x^3}\frac{dX_x}{dx} = \frac{1}{2}\left(\frac{x}{\boldsymbol{U}_x}\frac{dU_\infty}{dx} - 1\right) = \frac{1}{2}(m - 1) \tag{I–2–35}$$

$$C_{x3} = -\frac{1}{\mu \boldsymbol{U}_x X_x^2}\frac{dP}{dx} = \frac{U_\infty}{\nu \boldsymbol{U}_x X_x^2}\frac{dU_\infty}{dx} = m$$

$$C_{x4} = \frac{1}{\mu \boldsymbol{U}_x X_x^2} f_{bx} = 0 \qquad C_{x5} = \frac{\boldsymbol{U}_x}{\nu x X_x^2} = 1$$

where†

$$m = \frac{x}{U_\infty}\frac{dU_\infty}{dx} \tag{I–2–36}$$

Substituting these results into Eqs. (I–2–20) and (I–2–23), we have

$$\zeta''' + \frac{m+1}{2}\zeta\zeta'' + m(1 - \zeta'^2) = x\left(\zeta'\frac{\partial \zeta'}{\partial x} - \zeta''\frac{\partial \zeta}{\partial x}\right) \tag{I–2–37}$$

and

$$\frac{\Theta''}{Pr} + \frac{m+1}{2}\zeta\Theta' - n\zeta'\Theta = x\left(\zeta'\frac{\partial \Theta}{\partial x} - \Theta'\frac{\partial \zeta}{\partial x}\right) \tag{I–2–38}$$

where

$$\zeta' = 0 \qquad \Theta = 1 \qquad \text{at } \eta = 0$$

$$\zeta' = 1 \qquad \Theta = 0 \qquad \text{as } \eta \to \infty \tag{I–2–39}$$

and

$$\zeta = 0 \qquad \text{at } \eta = 0$$

It also follows that Eq. (I–2–28) becomes

$$\frac{f_x}{2} Re_x^{1/2} = \frac{\tau_0}{\rho U_\infty^2} Re_x^{1/2} = \zeta''(0) \tag{I–2–40}$$

and the Nusselt number Nu_x is represented by Eq. (I–2–30),

† The separation-of-variables method leads to the more general result

$$m = \frac{x - X_0}{U_\infty}\frac{dU_\infty}{dx}$$

where X_0 is equal to zero for applications in which $U_\infty\delta = 0$ at $x = 0$.

$$\Theta'(0) = -\frac{Nu_x}{xX_x} = -\frac{Nu_x}{Re_x^{1/2}} \quad \text{(I-2-41)}$$

With U_∞ and T_s or q_c'' arbitrarily specified, the gradients $\partial\zeta'/\partial x$, $\partial\zeta/\partial x$, and $\partial\Theta/\partial x$ are generally not equal to zero. Numerical codes have been developed by Cebeci and Smith [29] and others for solving these transformed partial differential equations for such nonsimilar conditions.

As we have indicated, a special class of geometrically similar boundary layer flows occurs for which $\partial\zeta'/\partial x = 0$, $\partial\zeta/\partial x = 0$, and $\partial\Theta/\partial x = 0$. Under these circumstances, Eqs. (I–2–37) and (I–2–38) reduce to simpler ordinary differential equations of the form

$$\zeta''' + \frac{m+1}{2}\zeta\zeta'' + m(1 - \zeta'^2) = 0 \quad \text{(I-2-42)}$$

and

$$\frac{\Theta}{Pr} + \frac{m+1}{2}\zeta\Theta' - n\zeta'\Theta = 0 \quad \text{(I-2-43)}$$

An inspection of these equations indicates the following similarity requirements:

$$m = \frac{x}{U_\infty}\frac{dU_\infty}{dx} = \text{constant} \quad \text{(I-2-44)}$$

and

$$n = \frac{x}{T_s - T_\infty}\frac{dT_s}{dx} = \text{constant} \quad \text{(I-2-45)}$$

or equivalently

$$n = \frac{x}{q_c''/[X_x\,\Theta'(0)]}\frac{d}{dx}\left[\frac{q_c''}{X_x\,\Theta'(0)}\right] = \frac{x}{q_c''/X_x}\frac{d}{dx}\left(\frac{q_c''}{X_x}\right) = \text{constant} \quad \text{(I-2-46)}$$

Thus, we are able to establish the distributions in U_∞ and T_s or q_c'' for which the boundary layers are similar. Solving Eq. (I–2–44) with m held constant, we obtain

$$\frac{dU_\infty}{U_\infty} = m\frac{dx}{x} \qquad \ln\frac{U_\infty}{U_0} = m\ln\frac{x}{x_0} = \ln\left(\frac{x}{x_0}\right)^m \quad \text{(I-2-47)}$$

or

$$U_\infty = Cx^m \quad \text{(I-2-48)}$$

where C is constant. This power-law distribution in U_∞ is associated with the famous *Falkner–Skan* family of similar boundary layer flows, with $m = 0$ for the special case of uniform free-stream velocity flow.

It also follows that the wall temperature T_s and wall-heat flux q_c'' follow power laws of the form

$$T_s - T_\infty = Kx^n \qquad \text{and} \qquad q''_c = q''_0 x^N \tag{I–2–49,50}$$

where $N = n + (m - 1)/2$. Notice that

$$n = 0 \qquad N = \frac{m - 1}{2} \tag{I–2–51}$$

for uniform wall temperature, and

$$n = \frac{1 - m}{2} \qquad N = 0 \tag{I–2–52}$$

for uniform wall-heat flux.

Numerical solutions to Eqs. (I–2–42) and (I–2–43) and accompanying boundary conditions have been developed for a wide range of values for m and for several representative values of n.

For the classic case of uniform free-stream velocity flow over a plate with uniform wall-temperature heating, these equations reduce to

$$\zeta''' + \frac{1}{2}\zeta\zeta'' = 0 \qquad \Theta'' + \frac{Pr}{2}\zeta\Theta' = 0 \tag{I–2–53,54}$$

$$\begin{aligned} \zeta' &= 0 \qquad \Theta = 1 \qquad \text{at } \eta = 0 \\ \zeta' &= 1 \qquad \Theta = 0 \qquad \text{as } \eta \to \infty \\ \zeta &= 0 \qquad\qquad\qquad \text{at } \eta = 0 \end{aligned} \tag{I–2–55}$$

As pointed out by White [30], Eq. (I–2–53) is exquisitely simple in appearance and yet has never yielded to an analytic solution. It follows that this nonlinear equation is generally solved by numerical methods. The first numerical solution to this equation was developed by Blasius in 1908. The Blasius solution for the velocity distribution is shown in Fig. 7–13. Tabulated results from Kays and Crawford [31] are listed in Table A–I–2–1. These results indicate $\zeta''(0) = 0.3321$. It follows from Eq. (I–2–40) that

$$\frac{f_x}{2} Re_x^{1/2} = \zeta''(0) = 0.3321 \tag{I–2–56}$$

TABLE A–I–2–1 Calculations for ζ, ζ' and ζ'' versus η for laminary boundary layer flow with uniform properties and uniform free-stream velocity

η	0	0.4	0.8	1.2	1.6	2.0	2.8	3.6	4.4
ζ	0	0.0266	0.1061	0.238	0.420	0.650	1.23	1.93	2.69
ζ'	0	0.133	0.265	0.394	0.517	0.630	0.812	0.923	0.976
ζ''	0.3321								0.039

for laminar flow over a flat plate with uniform free-stream velocity. Corresponding results obtained by the numerical solution of Eq. (I–2–54) are listed in Table A–I–2–2 in terms of $Nu_x/Re_x^{1/2} = -\Theta'(0)$ versus Pr for values of Pr ranging from 0.001 to 10^4.

TABLE A–I–2–2 Calculations for $Nu_x/Re_x^{1/2} = -\Theta'(0)$ for laminar boundary layer flow with uniform properties, uniform free stream velocity, and uniform wall temperature

Pr	0.001	0.01	0.1	0.72	1.0	10	10^2	10^3	10^4
$Nu_x/Re_x^{1/2}$	0.01732	0.05159	0.1400	0.2955	0.3321	0.7281	1.572	3.387	7.297

Solutions for $\zeta''(0)$ and $\Theta'(0)$ associated with Falkner-Skan flows with m in the range -0.09043 to ∞ and $n = 1$ are given by White [30], Schlichting [32], Evans [33], and others. As discussed in these references, the Falkner-Skan flows with m in this range correspond to similar flow over wedge shape bodies, with m related to the angle β by $m = \beta/(2 - \beta)$ [see Fig. 7–8]. This family of similar flows includes the special cases of uniform free-stream velocity flow ($m = 0$, $\beta = 0$), plane stagnation flow ($m = 1$, $\beta = 1$), and separation (i.e., $\tau_0 = 0$) ($m = -0.09043$, $\beta = -0.19884$).

I–2–3 Natural Convection Flow over a Vertical Flat Plate

Referring to Sec. 7–2–3, the momentum equation for laminar natural convection flow over a vertical flat plate with moderate temperature differences and $U_\infty = 0$ is given by

$$u\frac{\partial u}{\partial x} + v\frac{\partial u}{\partial y} = \nu\frac{\partial^2 u}{\partial y^2} + g\beta(T - T_\infty) \qquad \text{(I–2–57)}$$

Notice that the pressure gradient and body force term in Eq. (I–2–2) is represented by

$$-\frac{dP}{dx} + f_{bx} = \rho g\beta(T - T_\infty) \qquad \text{(I–2–58)}$$

The one-parameter integral solution developed in Sec. 7–2–3 indicates that the velocity and temperature distributions for moderate values of Pr can be approximated by

$$u = \frac{U_x}{4}\left[\frac{y}{\delta} - 2\left(\frac{y}{\delta}\right)^2 + \left(\frac{y}{\delta}\right)^3\right] \qquad \text{(I–2–59)}$$

where $U_x = g\beta(T_s - T_\infty)\delta^2/\nu$, from Eq. (7–173), and

$$\Theta = \frac{T - T_\infty}{T_s - T_\infty} = 1 - \frac{3}{2}\frac{y}{\Delta} + \frac{1}{2}\left(\frac{y}{\Delta}\right)^3 \qquad \text{(I–2–60)}$$

for $y \leq \Delta$, from Eq. (7–174). Noting from Eq. (7–181) that

$$\frac{\Delta}{x} \simeq \frac{\delta}{x} = 3.81\, Ra_x^{-1/4} \qquad \text{(I–2–61)}$$

where $Gr_x = g\beta(T_s - T_\infty)x^3/\nu^2$ and $Ra_x = Gr_x\,Pr$, we see that Eq. (I–2–59) can be written as

$$\frac{ux}{\nu\, Ra_x^{1/2}} = \frac{1}{4}\left(\frac{\delta}{x}\, Ra_x^{1/4}\right)^2 \left[\frac{y}{\delta} - 2\left(\frac{y}{\delta}\right)^2 + \left(\frac{y}{\delta}\right)^3\right] \qquad \text{(I–2–62)}$$

the form of which is independent of x, or equivalently†

$$\frac{ux}{2\nu\, Gr_x^{1/2}} = \frac{Pr}{4}\left[\frac{\delta}{x}\left(\frac{Gr_x}{4}\right)^{1/4}\right]^2 \left[\frac{y}{\delta} - 2\left(\frac{y}{\delta}\right)^2 + \left(\frac{y}{\delta}\right)^3\right] \qquad \text{(I–2–63)}$$

Noting that $y/\delta \sim (y/x)\, Ra_x^{1/4} \sim (y/x)\, Gr_x^{1/4}$, these equations provide a basis for establishing a similarity coordinate η of the general form $\eta = C(y/x)\, Gr_x^{1/4}$ (where C is a constant), which is commonly put into the form

$$\eta = \frac{y}{x}\left(\frac{Gr_x}{4}\right)^{1/4} \qquad \text{(I–2–64)}$$

With η represented by Eq. (I–2–64), X_x becomes

$$X_x = \frac{\eta}{y} = \frac{1}{x}\left(\frac{Gr_x}{4}\right)^{1/4} = \left(\frac{Gr_L}{4}\right)^{1/4} \frac{1}{L^{3/4}x^{1/4}} \qquad \text{(I–2–65)}$$

It also follows that‡

$$\boldsymbol{U}_x = \frac{2\nu\, Gr_x^{1/2}}{x} = \frac{2\nu}{L^{3/2}}\, Gr_L^{1/2}\, x^{1/2} \qquad \text{(I–2–66)}$$

(Transformation variables of this same form can be deduced from the more general two-parameter integral solution developed in Appendix K.)

We are now in a position to evaluate the coefficients C_{xi} given by Eq. (I–2–21) as follows:

$$C_{x1} = \frac{1}{\nu X_x^2}\frac{d\boldsymbol{U}_x}{dx} = 2 \qquad C_{x2} = \frac{\boldsymbol{U}_x}{\nu X_x^3}\frac{dX_x}{dx} = -1$$

$$C_{x3} + C_{x4} = \frac{1}{\mu \boldsymbol{U}_x X_x^2}\left(-\frac{dP}{dx} + f_{bx}\right) = \frac{\rho g\beta(T - T_\infty)}{\mu \boldsymbol{U}_x X_x^2} \qquad \text{(I–2–67)}$$

$$= \frac{T - T_\infty}{T_0 - T_\infty} = \Theta \qquad C_{x5} = \frac{\boldsymbol{U}_x}{\nu x X_x^2} = 4$$

† The optional representation of u by Eq. (I–2–63) is in accordance with standard practice in which the velocity is expressed in the dimensionless form $ux/(2\nu\, Gr_x^{1/2})$ (see Fig. 7–20).

‡ Notice the $\boldsymbol{U}_x$ and U_x are related by

$$\boldsymbol{U}_x = \frac{2\, Pr^{1/2}}{3.81^2}\, U_x = 0.138\, Pr^{1/2}\, U_x$$

Using these coefficients, the transformed equations become

$$\zeta''' + 3\zeta\zeta'' - 2\zeta'^2 + \Theta = 4x\left(\zeta'\frac{\partial\zeta'}{\partial x} - \zeta''\frac{\partial\zeta}{\partial x}\right) \tag{I–2–68}$$

and

$$\frac{\Theta''}{Pr} + 3\zeta\Theta' - 4n\zeta'\Theta = 4x\left(\zeta'\frac{\partial\Theta}{\partial x} - \Theta'\frac{\partial\zeta}{\partial x}\right) \tag{I–2–69}$$

where

$$\zeta' = 0 \qquad \Theta = 1 \qquad \text{at } \eta = 0$$

$$\zeta' = 0 \qquad \Theta = 0 \qquad \text{as } \eta \to \infty \tag{I–2–70}$$

and

$$\zeta = 0 \qquad\qquad \text{at } \eta = 0$$

In addition, we have

$$\zeta''(0) = \frac{\tau_0 x^2}{\rho\nu^2(Gr_x^3)^{1/4}} \tag{I–2–71}$$

from Eq. (I–2–28), and

$$\Theta'(0) = -\frac{Nu_x}{xX_x} = -\frac{Nu_x}{(Gr_x/4)^{1/4}} \tag{I–2–72}$$

from Eq. (I–2–30).

As in the case of forced convection flow, with the wall temperature (or wall-heat flux) specified by a power law, n is constant and the velocity and temperature distributions are geometrically similar. For these conditions, Eqs. (I–2–68) and (I–2–69) reduce to

$$\zeta''' + 3\zeta\zeta'' - 2\zeta'^2 + \Theta = 0 \tag{I–2–73}$$

and

$$\Theta'' + Pr\,(3\zeta\Theta' - 4n\zeta'\Theta) = 0 \tag{I–2–74}$$

for $T_s = Kx^n$ with K and n constant, or

$$\Theta'' + 3\,Pr\,\zeta\Theta' = 0 \tag{I–2–75}$$

for uniform wall temperature ($T_s = T_0$, $dT_s/dx = 0$, $n = 0$).

Representative numerical solutions to Eqs. (I–2–73) and (I–2–75) and the accompanying boundary conditions for uniform wall temperature are shown in Fig. 7–20 for velocity ζ' and temperature Θ and in Table A–I–2–3 for $Nu_x/Gr_x^{1/4} = -\Theta'(0)/4^{1/4}$.

TABLE A–I–2–3 Calculations for $Nu_x/Gr_x^{1/2} = -\Theta'(0)/4^{1/4}$ for laminar natural convection flow over a vertical plate with uniform wall temperature

Pr	0.01	0.1	0.72	1.0	10	10^2	10^3
$Nu_x/Gr_x^{1/4}$	0.0669	0.189	0.406	0.457	0.931	1.74	3.14

J FILM CONDENSATION: APPROXIMATE SOLUTION FOR LAMINAR FLOW

Film condensation on a cool vertical surface that is exposed to a saturated vapor represents a natural convection flow process involving two phases. The velocity and temperature distributions across a thin film of liquid condensate that is formed on a vertical surface at temperature $T_0 < T_{sat}$ is illustrated in Fig. A–J–1. Notice that the velocity reaches a maximum value near the liquid-vapor interface where the viscous stress approaches zero and that the temperature rises from T_0 at the wall to T_{sat} at the edge of the condensate film. Thus, the thickness of the thermal boundary layer Δ and condensate film δ_f are equivalent, but the hydrodynamic boundary layer thickness $\delta = \delta_f + \delta_g$ actually extends beyond the liquid film.

Assuming that (1) the thicknesses δ_f and δ_g are small, (2) the flow within the condensate and overlying vapor is laminar, and (3) the properties in each phase are approximately uniform, the transport equations applicable to each phase are represented by

$$\frac{\partial u}{\partial x} + \frac{\partial v}{\partial y} = 0 \tag{J–1}$$

from Eq. (7–139),

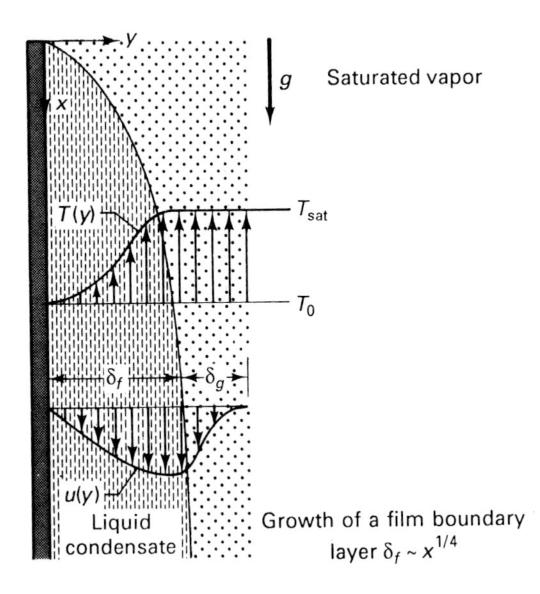

FIGURE A–J–1
Laminar film condensation on a vertical surface; $g_x = g$.

$$\rho\left(u\frac{\partial u}{\partial x}+v\frac{\partial u}{\partial y}\right)=\mu\frac{\partial^2 u}{\partial y^2}-g(\rho_\infty-\rho) \tag{J–2}$$

from Eq. (7–147) [with the sign associated with the gravitational acceleration g changed since g_x is in the direction of flow (i.e., $g_x = g$)], and

$$\rho c_P\left(u\frac{\partial T}{\partial x}+v\frac{\partial T}{\partial y}\right)=k\frac{\partial^2 T}{\partial y^2} \tag{J–3}$$

from Eq. (7–140), where the properties for saturated liquid are used in the condensate film and the properties for saturated vapor are used in the vapor phase. The boundary conditions that accompany these equations include

$$\begin{aligned} &u=0 \qquad v=0 \qquad T=T_0 \qquad \text{at } y=0 \\ &u=0 \qquad T=T_{\text{sat}} \qquad \text{as } y\rightarrow\infty \end{aligned} \tag{J–4}$$

plus interfacial conditions that assure continuity in the distributions in velocity, temperature, stress, and energy transfer. This formal mathematical formulation has been solved by Koh, Sparrow, and Hartnett [34].

A very useful approximate solution to this problem was first proposed by Nusselt [35], which involves the simplifying assumption that the advection terms in Eqs. (J–2) and (J–3) are negligible for the low velocity associated with laminar flow, such that the momentum and energy equations within the condensate film reduce to

$$\frac{d^2u}{dy^2}=-\frac{g(\rho_f-\rho_g)}{\mu_f} \qquad \frac{d^2T}{dy^2}=0 \tag{J–5,6}$$

Arguing further that the shear stress at the liquid-vapor interface must be small, Nusselt was able to eliminate the momentum equation for the vapor film by assuming

$$\frac{\partial u}{\partial y}=0 \qquad \text{at } y=\delta_f \tag{J–7}$$

Equation (J–5) is readily integrated to obtain

$$u=-\frac{g(\rho_f-\rho_g)y^2}{2\mu_f}+C_1y+C_2 \tag{J–8}$$

Evaluating the constants of integration C_1 and C_2 in accordance with the two boundary conditions $u = 0$ at $y = 0$ and $du/dy = 0$ at $y = \delta_f$, the solution for u takes the form

$$u=\frac{g(\rho_f-\rho_g)\delta_f^2}{2\mu_f}\left[2\frac{y}{\delta_f}-\left(\frac{y}{\delta_f}\right)^2\right] \tag{J–9}$$

This result is used to obtain an expression for the condensate mass-flow rate $\dot{m}_x$ for a plate depth b by writing†

$$\dot{m}_x = b \int_0^{\delta_f} \rho_f u \, dy = b \int_0^{\delta_f} \left\{ \frac{\rho_f g(\rho_f - \rho_g)\delta_f^2}{2\mu_f} \left[2\frac{y}{\delta_f} - \left(\frac{y}{\delta_f}\right)^2 \right] \right\} dy$$
$$= \frac{b\rho_f g(\rho_f - \rho_g)\delta_f^3}{3\mu_f} \tag{J–10}$$

Similarly, the solution to Eq. (J–6), which satisfies the conditions $T = T_0$ at $y = 0$ and $T = T_{\text{sat}}$ at $y = \delta_f$ is given by

$$T = T_0 + (T_{\text{sat}} - T_0)\frac{y}{\delta_f} \tag{J–11}$$

Using this linear relation for T, the heat flux *into* the surface is given by

$$q_c'' = k_f \left.\frac{\partial T}{\partial y}\right|_0 = \frac{k_f(T_{\text{sat}} - T_0)}{\delta_f} \tag{J–12}$$

such that the local coefficient of condensation heat transfer h_C becomes

$$h_C = \frac{q_c''}{T_{\text{sat}} - T_0} = \frac{k_f}{\delta_f} \tag{J–13}$$

To evaluate the film thickness δ_f, which appears in the solutions for u, $\dot{m}$, T, and h_C, we consider mass and energy balances on the lumped/differential element $\delta_f \, dx$ shown in Fig. A–J–2.

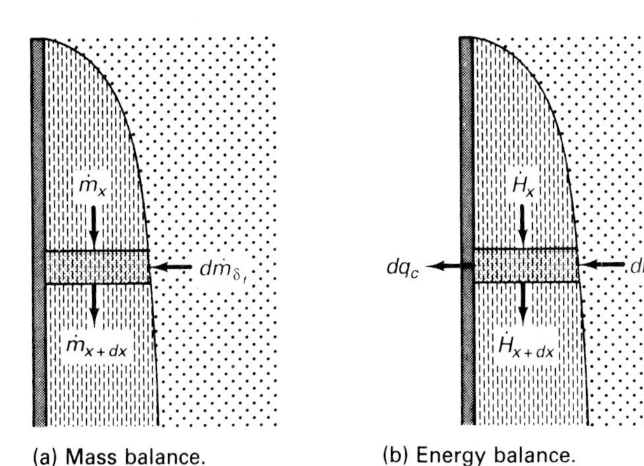

FIGURE A–J–2 Lumped/differential control volume within the condensate film.

† The condensate mass-flow rate $\dot{m}_x$ is represented by $\dot{m}$ in Chap. 10.

The mass balance indicates

$$d\dot{m}_{\delta_f} = \dot{m}_{x+dx} - \dot{m}_x = \frac{d\dot{m}_x}{dx}\,dx = d\dot{m}_x \tag{J–14}$$

where

$$d\dot{m}_x = \frac{b\rho_f g(\rho_f - \rho_g)}{\mu_f}\,\delta_f^2\,d\delta_f \tag{J–15}$$

from Eq. (J–10).

Following through with the energy balance, we obtain

$$dq_c = d\dot{m}_{\delta_f}\,i_g - (\dot{H}_{x+dx} - \dot{H}_x) = d\dot{m}_x\,i_g - d\dot{H}_x \tag{J–16}$$

where

$$dq_c = b\,dx\,q_c'' = \frac{b\,dx\,k_f(T_{\text{sat}} - T_0)}{\delta_f} \tag{J–17}$$

from Eq. (J–12). To approximate the condensate enthalpy-flow rate $\dot{H}_x$, we neglect the *sensible heat* absorbed due to the temperature within the film falling below T_{sat}; that is,

$$\dot{H}_x \simeq \dot{m}_x i_f \tag{J–18}$$

Combining Eqs. (J–16) and (J–18), we obtain

$$dq_c = d\dot{m}_x\,i_g - d(\dot{m}_x i_f) = d\dot{m}_x\,i_{fg} \tag{J–19}$$

where i_{fg} is the *latent heat of vaporization*. This equation also indicates

$$q_c = \dot{m}_x i_{fg} \qquad \text{or} \qquad \dot{m}_x = \frac{q_c}{i_{fg}} \tag{J–20,21}$$

A relation is obtained for δ_f by substituting for $d\dot{m}_x$ and dq_c in Eq. (J–19).

$$\delta_f^3\,d\delta_f = \frac{k_f\mu_f(T_{\text{sat}} - T_0)\,dx}{\rho_f g(\rho_f - \rho_g)i_{fg}} \tag{J–22}$$

Setting $\delta_f = 0$ at $x = 0$, Eq. (J–22) is integrated to obtain

$$\delta_f = \left[\frac{4\mu_f k_f(T_{\text{sat}} - T_0)x}{\rho_f g(\rho_f - \rho_g)i_{fg}}\right]^{1/4} \tag{J–23}$$

Substituting this result into Eq. (J–13), the relation for h_C takes the form

$$h_C = \frac{k_f}{\delta_f} = \left[\frac{\rho_f g(\rho_f - \rho_g)i_{fg}\,k_f^3}{4\mu_f(T_{\text{sat}} - T_0)x}\right]^{1/4} \tag{J–24}$$

or, in terms of Nusselt number Nu_x,

$$Nu_x = \frac{h_C x}{k_f} = \left[\frac{\rho_f g(\rho_f - \rho_g) i_{fg}\, x^3}{4 k_f \mu_f (T_{sat} - T_0)}\right]^{1/4} \tag{J–25}$$

It should be noted that this well-known result is usually refined by accounting for the effects of sensible heat in the relation for $\dot{H}_x$. Following the approach developed in Sec. 6–4–2 for internal flows, $\dot{H}_x$ can be represented by

$$\dot{H}_x = b\rho_f \int_0^{\delta_f} [c_{P,f}(T - T_{sat}) + i_f] u\, dy \tag{J–26}$$

which reduces to Eq. (J–18) when the sensible heat is neglected. Using Eqs. (J–9) and (J–11) for u and T, this equation becomes

$$\dot{H}_x = \dot{m}_x[i_f + C_1\, c_{P,f}(T_{sat} - T_0)] \tag{J–27}$$

where $C_1 = \frac{3}{8}$, and the analysis gives rise to relations for q_c and Nu_x of the form

$$q_c = \dot{m}_x\, i'_{fg} \qquad \text{or} \qquad \dot{m}_x = \frac{q_c}{i'_{fg}} \tag{J–28,29}$$

and

$$Nu_x = \left[\frac{\rho_f g(\rho_f - \rho_g) i'_{fg}\, x^3}{4 k_f \mu_f (T_{sat} - T_0)}\right]^{1/4} \tag{J–30}$$

where the *modified latent heat of vaporization* i'_{fg} is represented by

$$i'_{fg} = i_{fg} + C_1\, c_{P,f}(T_{sat} - T_0) \tag{J–31}$$

As a result of a higher-order analysis of the problem, Rohsenow [36] obtained a similar result with $C_1 = 0.68$. The Rohsenow solution has been found to be in excellent agreement with experimental data for low condensate mass-flow rates. However, it should be noted that the simpler result represented by Eqs. (J–20) through (J–25) is generally within 2% of the refined solution.

The results of this analysis are used in Sec. 10–4, which deals with the practical analysis of condensation heat transfer. In addition, more accurate relations are given that apply to higher condensate mass-flow rates associated with wavy laminar flow and turbulent flow.

K LAMINAR NATURAL CONVECTION OVER A VERTICAL PLATE: A TWO-PARAMETER INTEGRAL SOLUTION

To develop a two-parameter integral solution for laminar natural convection flow over a vertical surface, we employ the integral momentum equation, Eq. (7–165),

$$\frac{d}{dx}\int_0^{\delta} u^2\, dy = -\nu \left.\frac{\partial u}{\partial y}\right|_0 + \int_0^{\delta} g\beta(T - T_\infty)\, dy \tag{K–1}$$

the integral energy equation, Eq. (7–166),

$$\rho c_P \frac{d}{dx} \int_0^{\Delta} u(T - T_\infty)\, dy = q_c'' = -k \left.\frac{\partial T}{\partial y}\right|_0 \tag{K–2}$$

and third-order polynomial approximations developed in Sec. 7–2–3 for velocity [Eq. (7–173)],

$$u = \frac{U_x}{4} \left[\frac{y}{\delta} - 2\left(\frac{y}{\delta}\right)^2 + \left(\frac{y}{\delta}\right)^3 \right] \tag{K–3}$$

and temperature [Eq. (7–174)],

$$\frac{T - T_s}{T_\infty - T_s} = \frac{3}{2}\frac{y}{\Delta} - \frac{1}{2}\left(\frac{y}{\Delta}\right)^3 \qquad \text{for } y \leqslant \Delta \tag{K–4a}$$

$$= 1 \qquad \text{for } y \geqslant \Delta \tag{K–4b}$$

Equation (K–4) also indicates

$$q_c'' = -k \left.\frac{\partial T}{\partial y}\right|_0 = \frac{3}{2}\frac{k(T_s - T_\infty)}{\Delta} \tag{K–5}$$

$$Nu_x = \frac{h_x x}{k} = \frac{q_c''}{T_0 - T_\infty}\frac{x}{k} = \frac{3}{2}\frac{x}{\Delta} \tag{K–6}$$

As indicated in Chap. 7, the extent of the thermal boundary layer Δ is on the order of or less than the hydrodynamic boundary layer δ for moderate to large values of Prandtl number Pr, but Δ extends only slightly beyond δ for small values of Pr. Consequently, whereas δ and Δ can be taken as two independent parameters for moderate to large values of Pr, laminar natural convection can be characterized by setting $\delta = \Delta$ for moderate to low values of Pr. Therefore, to develop a two-parameter integral analysis that is applicable to the entire Pr range, U_x is given by

$$U_x = \frac{g\beta(T_s - T_\infty)\delta^2}{\nu} \tag{K–7}$$

in accordance with Eq. (7–171) for moderate to large Pr for which $\Delta < \delta$, and U_x is taken as the second parameter for moderate to small values of Pr for which $\Delta \simeq \delta$.

Equations (K–3) and (K–4) are used to evaluate the integrals $\int_0^\delta (T - T_\infty)\, dy$ and $\int_0^\Delta u(T - T_\infty)\, dy$ in terms of the boundary-layer-thickness ratio $r = \Delta/\delta$, with the result

$$\int_0^\delta (T - T_\infty)\, dy = \int_0^\delta (T_s - T_\infty)\left[1 - \frac{3}{2}\frac{y}{\Delta} + \frac{1}{2}\left(\frac{y}{\Delta}\right)^3\right] dy$$

$$= \frac{3}{8}\delta r(T_s - T_\infty) \qquad \text{for } r \leq 1 \tag{K–8a}$$

$$= \frac{3}{8}\delta(T_s - T_\infty) \qquad \text{for } r = 1 \tag{K–8b}$$

and

$$\int_0^\Delta u(T - T_\infty)\, dy = \int_0^\Delta \frac{U_x}{4}(T_s - T_\infty)\left[\frac{y}{\delta} - 2\left(\frac{y}{\delta}\right)^2 + \left(\frac{y}{\delta}\right)^3\right]\left[1 - \frac{3}{2}\frac{y}{\Delta} + \frac{1}{2}\left(\frac{y}{\Delta}\right)^3\right] dy \tag{K–9}$$

$$= \frac{U_x\Delta}{4}(T_s - T_\infty)\, f(r)$$

where

$$f(r) = \frac{r}{10} - \frac{r^2}{12} + \frac{r^3}{46.7} \qquad \text{for } r \leq 1 \tag{K–10a}$$

and

$$f(r) = \frac{1}{26.3} \qquad \text{for } r = 1 \tag{K–10b}$$

In addition, the integral $\int_0^\delta u^2\, dy$ is expanded to obtain

$$\int_0^\delta u^2\, dy = \int_0^\delta \left\{\frac{U_x}{4}\left[\frac{y}{\delta} - 2\left(\frac{y}{\delta}\right)^2 + \left(\frac{y}{\delta}\right)^3\right]\right\}^2 dy = \frac{U_x^2\delta}{16(105)} \tag{K–11}$$

Substituting these relations into the integral momentum and energy equations and restricting our attention to uniform wall temperature, we obtain

$$\frac{1}{16(105)}\frac{d}{dx}(U_x^2\delta) = -\frac{\nu U_x}{4\delta} + \frac{3}{8} g\beta\delta r(T_s - T_\infty) \tag{K–12}$$

and

$$\frac{1}{4}\frac{d}{dx}[U_x\Delta\, f(r)] = \frac{3}{2}\frac{\alpha}{\Delta} \tag{K–13}$$

To complete the analysis, these equations are solved for (1) δ and r (or Δ) with $U_x = g\beta(T_0 - T_\infty)\delta^2/\nu$ and $f(r) = r/10 - r^2/12 + r^3/46.7$ for $r \leq 1$, and (2) Δ and U_x with $f(r) = 1/26.3$ for $r = 1$.

(1) *Moderate to High Pr with $r \leq 1$* Setting $U_x = g\beta(T_0 - T_\infty)\delta^2/\nu$ in accordance with Eq. (7–171), Eqs. (K–12) and (K–13) become

$$\frac{d\delta^5}{dx} = \frac{16(105)\nu\delta^2}{g\beta(T_0 - T_\infty)}\left(\frac{3}{8}r - \frac{1}{4}\right) \tag{K–14}$$

and

$$\frac{d}{dx}[\delta^3 r\, f(r)] = \frac{6\alpha\nu}{\delta r g\beta(T_0 - T_\infty)} \tag{K–15}$$

Since the velocity and temperature distributions for natural convection on a vertical plate with uniform wall temperature are geometrically similar, r is independent of x (see Chap. 7 and Appendix I–2). It follows that the solution to the integral momentum equation, Eq. (K–14), can be obtained by writing

$$\frac{1}{\delta}\frac{d\delta^5}{dx} = \frac{5}{4}\frac{d\delta^4}{dx} = \frac{16(105)\nu^2}{g\beta(T_0 - T_\infty)}\left(\frac{3}{8}r - \frac{1}{4}\right) \tag{K–16}$$

$$\delta^4 = \frac{4}{5}\frac{16(105)\nu^2}{g\beta(T_0 - T_\infty)}\left(\frac{3}{8}r - \frac{1}{4}\right)x \tag{K–17}$$

or

$$\frac{\delta}{x} = 6.05\left(\frac{3}{8}r - \frac{1}{4}\right)^{1/4} Gr_x^{-1/4} \tag{K–18}$$

Similarly, the solution to the integral energy equation, Eq. (K–15), becomes

$$\delta\frac{d\delta^3}{dx} = 3\delta^3\frac{d\delta}{dx} = \frac{3}{4}\frac{d\delta^4}{dx} = \frac{6\alpha\nu}{g\beta(T_0 - T_\infty)r^2(r/10 - r^2/12 + r^3/46.7)} \tag{K–19}$$

$$\delta^4 = \frac{4}{3}\frac{6\alpha\nu}{g\beta(T_0 - T_\infty)}\frac{x}{r^2(r/10 - r^2/12 + r^3/46.7)} \tag{K–20}$$

or

$$\frac{\delta}{x} = \frac{1.68}{[Gr_x\,Pr\,r^2(r/10 - r^2/12 + r^3/46.7)]^{1/4}} \tag{K–21}$$

To obtain a solution for r, we equate Eqs. (K–18) and (K–21), with the result

$$Pr = \frac{1}{168r^2(r/10 - r^2/12 + r^3/46.7)(3r/8 - 1/4)} \tag{K–22}$$

This equation is used to compute the distribution in r with Pr by taking r as the independent variable. The resulting calculations are shown in Fig. A–K–1. Notice that $r \leq 1$ for $Pr \geq 1.2$.

The Nusselt number Nu_x is expressed in terms of $T_0 - T_\infty$, δ, and r by Eq. (K–6). Using Eq. (K–21) to represent δ/x, we obtain

$$Nu_x = \frac{3/2}{1.68r}\left[Gr_x\,Pr\,r^2\left(\frac{r}{10} - \frac{r^2}{12} + \frac{r^3}{46.7}\right)\right]^{1/4} \tag{K–23}$$

or

$$\frac{Nu_x}{Ra_x^{1/4}} = 0.893\left(\frac{1}{10r} - \frac{1}{12} + \frac{r}{46.7}\right)^{1/4} \tag{K–24}$$

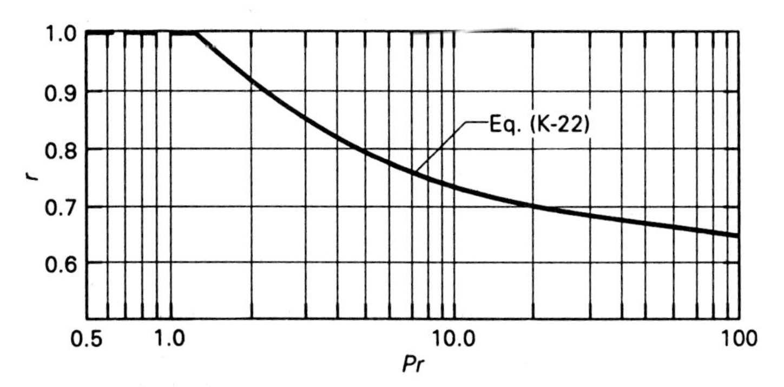

FIGURE A–K–1 Calculations for r versus Pr for laminar natural convection based on two-parameter integral method.

where the Rayleigh number Ra_x is equal to $Gr_x\ Pr$. Calculations for $Nu_x/Ra_x^{1/4}$ obtained by use of this equation with Pr expressed in terms of r by Eq. (K–22) are compared with exact solution results in Fig. A–K–2.

(2) *Moderate to Low Pr with $r = 1$* For the case in which $r = 1$, Eqs. (K–12) and (K–13) can be solved for U_x and Δ (or δ). Although we now relax the constraint represented by Eq. (7–171), we will make use of the indication that U_x is proportional to δ^2; that is,

$$U_x = C_1 \Delta^2 \tag{K–25}$$

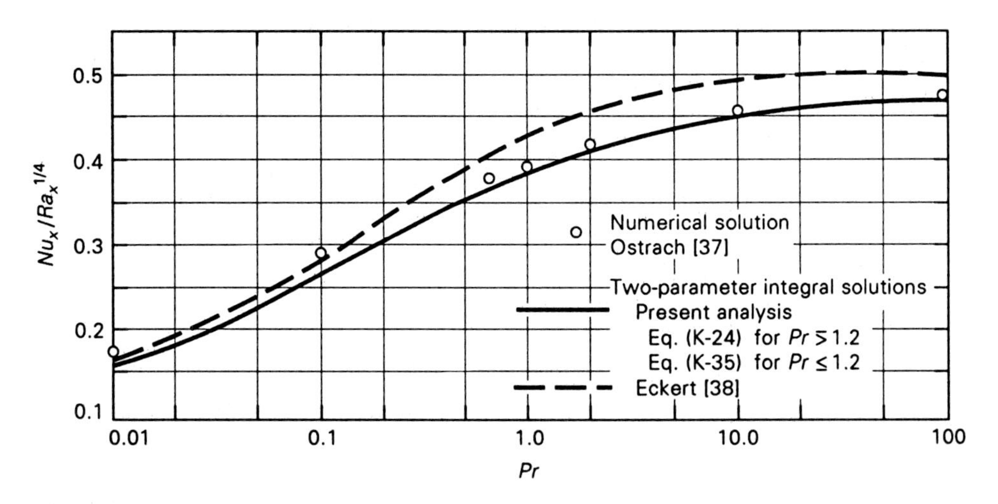

FIGURE A–K–2 Nusselt number for laminar natural convection flow over a vertical plate with uniform wall temperature.

for $r = 1$. Using this relation and setting $f(r) = 1/26.3$, Eqs. (K–12) and (K–13) become

$$\frac{C_1^2}{16(105)} \frac{d\Delta^5}{dx} = -\frac{\nu C_1 \Delta}{4} + \frac{3}{8} g\beta(T_0 - T_\infty)\Delta \tag{K–26}$$

and

$$\frac{C_1}{4(26.3)} \frac{d\Delta^3}{dx} = \frac{3}{2}\frac{\alpha}{\Delta} \tag{K–27}$$

These two equations are readily solved for Δ by writing

$$\frac{1}{\Delta}\frac{d\Delta^5}{dx} = 5\Delta^3 \frac{d\Delta}{dx} = \frac{5}{4}\frac{d\Delta^4}{dx} = \frac{16(105)}{C_1^2}\left[-\frac{\nu C_1}{4} + \frac{3}{8} g\beta(T_0 - T_\infty)\right] \tag{K–28}$$

$$\Delta^4 = \frac{4}{5}\frac{16(105)}{C_1^2}\left[-\frac{\nu C_1}{4} + \frac{3}{8} g\beta(T_0 - T_\infty)\right]x \tag{K–29}$$

and

$$\Delta \frac{d\Delta^3}{dx} = 3\Delta^3 \frac{d\Delta}{dx} = \frac{3}{4}\frac{d\Delta^4}{dx} = \frac{3}{2}(4)(26.3)\frac{\alpha}{C_1} \tag{K–30}$$

$$\Delta^4 = \frac{4}{3}\frac{3}{2}(4)(26.3)\frac{\alpha x}{C_1} \tag{K–31}$$

Equating Eqs. (K–29) and Eq. (K–31), we obtain

$$C_1 = \frac{1.5 g\beta(T_0 - T_\infty)}{(0.624 + Pr)\alpha} \tag{K–32}$$

Substituting this result back into Eq. (K–31), the solution for Δ becomes

$$\frac{\Delta}{x} = 3.43 \frac{(0.624 + Pr)^{1/4}}{Gr_x^{1/4}\, Pr^{1/2}} \tag{K–33}$$

Combining this result with Eq. (K–6), the solution for Nu_x is

$$Nu_x = \frac{3}{2}\frac{x}{\Delta} = 0.437 \frac{Gr_x^{1/4}\, Pr^{1/2}}{(0.624 + Pr)^{1/4}} \tag{K–34}$$

or

$$\frac{Nu_x}{Ra_x^{1/4}} = \frac{0.437}{(1 + 0.624/Pr)^{1/4}} \tag{K–35}$$

This relation is compared with exact solution results in Fig. A–K–2.

To summarize our findings, a two-parameter integral method has been developed for laminar natural convection over a vertical plate that provides calculations for Nusselt number with error ranging from about 3% for moderate to large Pr to about 15% for very small values of Pr. The method features the solution of the integral momentum and energy equations with the two parameters represented by (1) δ and Δ for moderate to large Pr and (2) U_x and Δ for moderate to low Pr. As we have seen, the second approach is predicated on the notion that $\delta \simeq \Delta$. This approach follows the early work of Eckert [38] in which the velocity and temperature distributions were approximated by third- and second-order polynomials, respectively. The solution for Nusselt number developed by Eckert is given by

$$\frac{Nu_x}{Ra_x^{1/4}} = \frac{0.508}{(1 + 0.952/Pr)^{1/4}} \tag{K-36}$$

which is similar in form to Eq. (K–35). This equation has been proposed for the entire Pr spectrum. Equation (K–36) is shown in Fig. A–K–2. Whereas this equation is quite respectable for moderate to small values of Pr, because of the simplifying assumption $\delta \simeq \Delta$ it provides a rather crude representation of the actual solution for moderate to high values of Pr.

L BULK-STREAM CHARACTERISTICS

Bulk-stream thermal and hydraulic characteristics are introduced in Sec. 6–4 in the context of uniform property internal flows. We now want to develop a more general formulation which is applicable to variable property flows.

L–1 THERMAL ANALYSIS

Referring to the internal flow system shown in Fig. 6–15, we define the local *bulk-stream-enthalpy rate* $\dot{H}_b$ as the total rate of enthalpy transferred through the cross-sectional area A at any axial location x. $\dot{H}_b$ is expressed in terms of the distributions in specific enthalpy i at x by

$$\dot{H}_b = \int_{\dot{m}} i \, d\dot{m} = \int_A i\rho u \, dA \tag{L-1}$$

To express $\dot{H}_b$ in terms of the temperature distribution T over the cross section, we make use of Eq. (6–13) for ideal fluids; that is,

$$i = \int_{T_R}^{T} c_P(T) \, dT + i_R \tag{L-2}$$

where the specific heat at constant pressure, which is now designated by $c_P(T)$, is a function of T, and the subscript R represents the reference state. Combining this equation with Eq. (L–1), we have

$$\dot{H}_b = \int_A \left[\int_{T_R}^{T} c_P(T)\, dT + i_R \right] \rho u\, dA$$
$$= \int_A \left[\int_{T_R}^{T} c_P(T)\, dT \right] \rho u\, dA + \dot{m} i_R \qquad \text{(L–3)}$$

The local average thermal-energy state of the fluid over the entire cross section A is characterized by the bulk-stream temperature T_b, which is formally defined as the average value of T for which Eq. (L–3) is satisfied; that is,

$$\dot{H}_b = \int_A \left[\int_{T_R}^{T_b} c_P(T)\, dT \right] \rho u\, dA + \dot{m} i_R \qquad \text{(L–4)}$$

Since by definition T_b is uniform over the cross section, this equation reduces to

$$\dot{H}_b = \dot{m} \left[\int_{T_R}^{T_b} c_P(T)\, dT + i_R \right] \qquad \text{(L–5)}$$

We are now in a position to establish the gradient $d\dot{H}_b/dT_b$, which is essential to the practical thermal analysis approach. Based on Eq. (L–5), $d\dot{H}_b/dT_b$ is given by

$$\frac{d\dot{H}_b}{dT_b} = \frac{d}{dT_b} \left\{ \dot{m} \left[\int_{T_R}^{T_b} c_P(T)\, dT + i_R \right] \right\} \qquad \text{(L–6)}$$

and, since $\dot{m} = GA$ is independent of T_b,

$$\frac{d\dot{H}_b}{dT_b} = \dot{m} \frac{d}{dT_b} \left[\int_{T_R}^{T_b} c_P(T)\, dT \right] \qquad \text{(L–7)}$$

Following through with the differentiation of the integral in this equation using the Leibnitz rule,† we obtain

$$\frac{d\dot{H}_b}{dT_b} = \dot{m} \left[\int_{T_R}^{T_b} \frac{d}{dT_b} c_P(T)\, dT + c_P(T_b) \frac{dT_b}{dT_b} - c_P(T_R) \frac{dT_R}{dT_b} \right] \qquad \text{(L–8)}$$

which reduces to Eq. (6–45),

$$d\dot{H}_b = \dot{m}\, c_P(T_b)\, dT_b \qquad \text{(L–9)}$$

† According to the Leibnitz rule, if $a(t)$ and $b(t)$ are differentiable functions of t and $\phi(x,t)$ and $\partial\phi(x,t)/\partial t$ are continuous in x and t, then

$$\frac{d}{dt} \int_{a(t)}^{b(t)} \phi(x,t)\, dx = \int_{a(t)}^{b(t)} \frac{\partial}{\partial t} \phi(x,t)\, dx + \phi[b(t),t] \frac{d}{dt} b(t) - \phi[a(t),t] \frac{d}{dt} a(t)$$

or simply

$$d\dot{H}_b = \dot{m}c_P\, dT_b \tag{L–10}$$

where c_P is evaluated at T_b.

To express T_b in terms of T, we equate Eqs. (L–3) and (L–5), with the result

$$\int_{T_R}^{T_b} c_P(T)\, dT = \frac{1}{\dot{m}} \int_A \left[\int_{T_R}^{T} c_P(T)\, dT \right] \rho u\, dA \tag{L–11}$$

This equation reduces to

$$T_b = \frac{1}{\dot{m}} \int_A T \rho u\, dA \tag{L–12}$$

for uniform specific heat, and to Eq. (6–46),

$$T_b = \frac{1}{AU_b} \int_A Tu\, dA \tag{L–13}$$

for uniform specific heat and density. It also follows that the bulk-stream-enthalpy rate $\dot{H}_b$ can be expressed in terms of T_b for uniform specific heat by

$$\dot{H}_b = \dot{m}c_P(T_b - T_R) + i_R \tag{L–14}$$

or, for $i_R = 0$ at $T_R = 0$,

$$\dot{H}_b = \dot{m}c_P T_b \tag{L–15}$$

which is commonly referred to in the heat-transfer literature.

To complete the picture, the local coefficient of heat transfer h_x for variable property flow is defined by the general Newton law of cooling, Eq. (6–50),

$$q_c'' = h_x(T_s - T_b) \tag{L–16}$$

and the defining equation for thermal fully developed flow is given by Eq. (6–47),

$$\frac{\partial}{\partial x}\left(\frac{T - T_s}{T_b - T_s}\right) = 0 \tag{L–17}$$

It follows that

$$\frac{h_x}{k_s} = \frac{h}{k_s} = \text{constant} \tag{L–18}$$

for TFD flow with variable properties, which reduces to Eq. (6–52),

$$h_x = h = \text{constant} \tag{L–19}$$

for uniform properties.

L-2 HYDRAULIC ANALYSIS

The defining equation for mass flux G given by Eq. (6–33) is readily generalized for variable property flow by writing

$$G = \frac{\dot{m}}{A} = \frac{1}{A}\int_{\dot{m}} d\dot{m} = \frac{1}{A}\int_A \rho u \, dA \tag{L–20}$$

The bulk-stream velocity U_b is defined as the average value of u for which this equation is satisfied; that is,

$$U_b = \frac{(1/A)\int_A \rho u \, dA}{(1/A)\int_A \rho \, dA} = \frac{G}{\rho_b} = \frac{\dot{m}}{\rho_b A} \tag{L–21}$$

The mean or bulk-stream density ρ_b is formally defined by

$$\rho_b = \frac{1}{A}\int_A \rho \, dA \tag{L–22}$$

However, in practice ρ_b is usually approximated by the density evaluated at the bulk-stream temperature. Notice that, whereas G is constant and unambiguously defined for variable density flow with uniform mass-flow rate $\dot{m}$ and uniform cross-sectional area A, U_b is inversely proportional to ρ_b. These bulk-stream characteristics are expressed in terms of the Reynolds number Re for variable property flow by

$$Re = \frac{\rho_b U_b D_H}{\mu_b} \quad \text{or} \quad Re = \frac{G D_H}{\mu_b} \tag{L–23a,b}$$

where μ_b is evaluated at the bulk-stream temperature T_b.

The Fanning friction factor for variable property internal flow is defined by

$$\tau_s = \rho_b U_b^2 \frac{f_x}{2} \tag{L–24}$$

Because of the variation in U_b, the defining equation for fully developed flow associated with variable property conditions takes the more general form

$$\frac{\partial}{\partial x}\left(\frac{u}{U_b}\right) = 0 \tag{L–25}$$

Following the approach taken in Sec. 6–4–1, we find that

$$\frac{\tau_s}{\mu_s U_b} = \text{constant} \tag{L–26}$$

and

$$\frac{f_x}{\mu_s} = \frac{f}{\mu_s} = \text{constant} \tag{L–27}$$

for HFD flow with variable properties, which reduces to Eq. (6–39),

$$f_x = f - \text{constant} \tag{L-28}$$

for uniform properties. Thus, f is constant in the HFD region for uniform wall temperature, but is nonuniform and proportional to μ_s for the case in which the viscosity varies along the wall.

M PRACTICAL HYDRAULIC ANALYSIS

The practical hydraulic analysis of flow in heat-transfer cores involves the evaluation of (1) pressure drop, (2) pumping power, and (3) heating–pumping power ratio. Basic relations pertaining to these three aspects of the practical hydraulic analysis approach are considered in this section.

M–1 PRESSURE DROP

Referring to the heat-transfer core shown in Fig. 8–17, the pressure drop analysis consists of the evaluation of (1) the pressure drop within the core ΔP_{c-e} and (2) the core-inlet pressure drop ΔP_{1-c} and the core-exit pressure rise ΔP_{2-e}. The total core pressure drop ΔP_{1-2} is then obtained from†

$$\Delta P_{1-2} = \Delta P_{1-c} + \Delta P_{c-e} - \Delta P_{2-e} \tag{M-1}$$

M–1–1 Core Pressure Drop

To determine the pressure drop within the core, we apply Newton's second law of motion, Eq. (1–3),

$$\Sigma\, \dot{M}_{o,x} - \Sigma\, \dot{M}_{i,x} + \cancel{\frac{\Delta M_{s,x}}{\Delta t}} = \Sigma\, F_x \tag{M-2}$$

for steady-state conditions, to the lumped/differential element shown in Fig. A–M–1. The term $\Sigma\, \dot{M}_{o,x} - \Sigma\, \dot{M}_{i,x}$ represents the *rate of creation of axial momentum* caused by (1) acceleration associated with changes in the shape of the velocity profile over the cross section as the fluid proceeds downstream and (2) changes in the bulk-stream velocity U_b caused by heating- or cooling-induced variations in the fluid density or other factors such as transpiration. The forces acting on the fluid include those due to pressure P and wall-shear stress τ_s.

The development of a practical hydraulic analysis approach that is sufficiently general and fundamentally sound requires that special care be taken in dealing with the bulk-stream axial momentum rate $\dot{M}_b$. This important fluid flow parameter is formally defined in terms of the distribution in velocity u over the cross section A by

$$\dot{M}_b = \int_{\dot{m}} u \, d\dot{m} = \int_A \rho u^2 \, dA \tag{M-3}$$

† The analysis is developed in the context of a horizontal core. Hydrostatic effects should be accounted for in applications involving liquid flow in cores with changes in the fluid elevation.

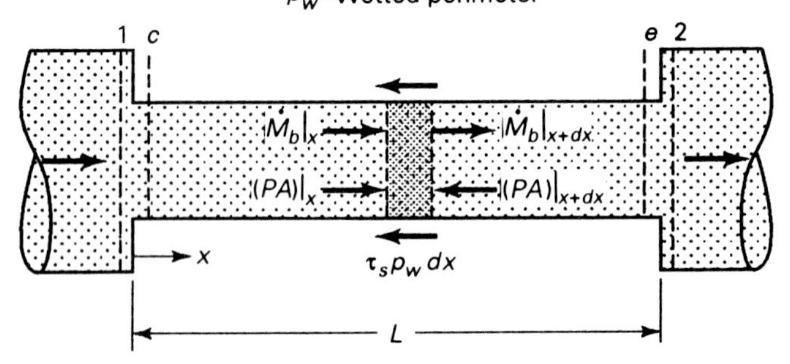

FIGURE A–M–1 Bulk-stream-momentum rates and forces associated with flow through a lumped/differential element $A\,dx$ within a tubular passage.

To provide a basis for distinguishing between the momentum effects associated with change in profile shape and change in bulk-stream velocity, Eq. (M–3) is put into the form

$$\dot{M}_b = \int_A \rho u(u - U_b)\, dA + \int_A \rho u U_b\, dA = \int_A \rho u(u - U_b)\, dA + \dot{m}U_b \tag{M–4}$$

or

$$\dot{M}_b = \dot{M}_A + \dot{m}U_b \tag{M–5}$$

where

$$\dot{M}_A = \int_A \rho u(u - U_b)\, dA \tag{M–6}$$

characterizes the effect of profile shape on the bulk-stream axial momentum rate, and the term $\dot{m}U_b$ represents the bulk-stream momentum rate for fully developed flow. With $\dot{M}_b$ expressed in terms of $\dot{M}_A$ and $\dot{m}U_b$ by Eq. (M–5), we are in a position to develop the practical perspective that is generally taken in convection-heat-transfer design work.

Referring to Fig. A–M–1, the rate of creation of momentum for a lumped/differential element is represented by

$$\Sigma\, \dot{M}_{o,x} - \Sigma\, \dot{M}_{i,x} = \dot{M}_b|_{x+dx} - \dot{M}_b|_x \tag{M–7}$$

and the sum of forces is given by

$$\Sigma\, F_x = -\tau_s p_w\, dx + (PA)|_x - (PA)|_{x+dx} \tag{M–8}$$

where p_w is the wetted perimeter. Substituting these relations into Eq. (M–2), we obtain

$$\dot{M}_b|_{x+dx} - \dot{M}_b|_x = -\tau_s p_w\, dx + (PA)|_x - (PA)|_{x+dx} \tag{M–9}$$

Using the definition for the derivative, this equation takes the form

$$\frac{d\dot{M}_b}{dx} = -\tau_s p_w - A\frac{dP}{dx} \tag{M-10}$$

or

$$\frac{d\dot{M}_A}{dx} + \frac{d}{dx}(\dot{m}U_b) = -\tau_s p_w - A\frac{dP}{dx} \tag{M-11}$$

for uniform cross-sectional area A throughout the core.

To continue with the development of this lumped analysis approach for steady flow in tubular cores, we first consider HFD and hydrodynamic developing flow for uniform property conditions, after which attention is given to characterizing variable property effects.

Uniform Property Conditions

For flow with uniform mass-flow rate $\dot{m}$ and uniform properties, the bulk-stream velocity U_b is constant (i.e., $U_b = U_{b,c} = U_{b,e}$) and Eq. (M–11) reduces to

$$-\frac{dP}{dx} = \frac{p_w}{A}\tau_s + \frac{1}{A}\frac{d\dot{M}_A}{dx} \tag{M-12}$$

Introducing the Fanning friction factor f_x, we obtain

$$-\frac{dP}{dx} = \frac{p_w}{A}\frac{f_x}{2}\rho U_b^2 + \frac{1}{A}\frac{d\dot{M}_A}{dx} \tag{M-13}$$

such that the core pressure drop $P_c - P_e$ can be represented by

$$\Delta P_{c-e} = -\int_{P_c}^{P_e} dP = \frac{p_w}{A}\int_0^L f_x\, dx\frac{\rho U_b^2}{2} + \frac{\Delta\dot{M}_A}{A} \tag{M-14}$$

where $p_w/A = 4/D_H$ and $\Delta\dot{M}_A = \int d\dot{M}_A$.

HFD Flow For the special case in which a heated core of length L is preceded by a sufficiently long calming section of the same diameter (i.e., $D_1 = D$, $\sigma = 1$), the flow within the core is fully developed (i.e., $\partial u/\partial x = 0$ and $f_x = f$), such that Eq. (M–6) indicates

$$\frac{d\dot{M}_A}{dx} = \frac{d}{dx}\int_A \rho u(u - U_b)\, dx = 0 \tag{M-15}$$

It follows that Eq. (M–14) reduces to

$$\Delta P_{c-e} = \frac{p_w L}{A} f\frac{\rho U_b^2}{2} = \frac{4L}{D_H} f\frac{\rho U_b^2}{2} \tag{M-16}$$

for HFD flow.

Hydrodynamic Developing Flow To deal with the more general case of hydrodynamic developing flow, Eq. (M–14) is put into the form

$$\Delta P_{c-e} = \underbrace{\left[\frac{4}{D_H}\int_0^L (f_x - f)\,dx\right]\frac{\rho U_b^2}{2} + \frac{\Delta \dot{M}_A}{A}}_{\textit{profile shape effect}} + \underbrace{\left(\frac{4L}{D_H}f\right)\frac{\rho U_b^2}{2}}_{\textit{ideal core effect}} \qquad \text{(M–17)}$$

where the first term on the right side accounts for the effect of acceleration related to change in velocity profile shape, and the second term corresponds to the effect of ideal HFD flow in the core. This equation provides motivation for representing the complex profile shape effect, which involves the as yet unspecified change in momentum rate $\Delta \dot{M}_A$ in terms of an empirical coefficient. Following the approach that is generally used in practice, Eq. (M–17) is represented in terms of the entrance-loss coefficient K_c by [from Eq. (8–3)]

$$\Delta P_{c-e} = \frac{\rho U_b^2}{2}\left(K_c + \frac{4L}{D_H}f\right) = \frac{G^2}{2\rho}\left(K_c + \frac{4L}{D_H}f\right) \qquad \text{(M–18)}$$

where K_c can be obtained from Fig. 8–5 or other sources [39,40].

Variable Property Conditions

For internal flow processes in which significant variations in fluid density are caused by heating or cooling, the pressure gradient dP/dx is represented by

$$-\frac{dP}{dx} = \frac{p_w}{A}\tau_s + \frac{1}{A}\frac{d\dot{M}_A}{dx} + \frac{1}{A}\frac{d}{dx}(\dot{m}U_b) \qquad \text{(M–19)}$$

from Eq. (M–11), where $G = \rho_b U_b = \dot{m}/A$ is uniform. Assuming that $\rho_c \simeq \rho_1$ and $\rho_e \simeq \rho_2$ for incompressible flow, this equation leads to

$$\begin{aligned}
\Delta P_{c-e} &= \frac{p_w}{A}\int_0^L \frac{\rho_b U_b^2}{2} f_x\,dx + \frac{\Delta \dot{M}_A}{A} + \frac{1}{A}[(\dot{m}U_b)_e - (\dot{m}U_b)_c] \\
&= \frac{p_w}{A}\frac{G^2}{2}\int_0^L \frac{f_x}{\rho_b}dx + \frac{\Delta \dot{M}_A}{A} + G^2\left(\frac{1}{\rho_2} - \frac{1}{\rho_1}\right) \\
&= \Bigg[\underbrace{\left(\frac{4}{D_H}\int_0^L \frac{f_x - f}{v_1/v_b}dx\right)\frac{G^2}{2} + \frac{\Delta \dot{M}_A}{A v_1}}_{\textit{profile shape effect}} \\
&\quad + \underbrace{\left(\frac{4}{D_H}\int_0^L \frac{f}{v_1/v_b}dx\,\frac{G^2}{2}\right)}_{\textit{ideal core effect}} + \underbrace{G^2\left(\frac{v_2}{v_1} - 1\right)}_{\textit{bulk-stream acceleration effect}}\Bigg] v_1
\end{aligned} \qquad \text{(M–20)}$$

where $v_2 = 1/\rho_2$, $v_1 = 1/\rho_1$ and $G^2(v_2/v_1 - 1)$ represents the bulk-stream acceleration (or deceleration) effect caused by heating (or cooling). Using the entrance-loss coefficient K_c to represent the profile shape acceleration effect, this equation takes the more practical form

$$P_c - P_e = \frac{G^2 v_1}{2}\left[K_c + \frac{1}{v_1}\frac{4}{D_H}\int_0^L v_b f\,dx + 2\left(\frac{v_2}{v_1} - 1\right)\right] \tag{M-21}$$

As shown in Appendix L, the friction factor for fully developed flow with variable properties is proportional to μ_s. It follows that

$$\int_0^L v_b f\,dx = Lfv_m \tag{M-22}$$

for uniform wall temperature, where the mean specific volume v_m is defined by

$$v_m = \frac{1}{L}\int_0^L v_b\,dx \tag{M-23}$$

Although Eq. (M–23) is theoretically restricted to cases for which μ_s is constant, it is generally used in design work for both uniform and nonuniform distributions in μ_s. It should also be mentioned that the mean specific volume v_m is approximated by several methods in design work, depending on the nature of the distribution in bulk-stream temperature T_b. In the simplest approach, v_m is approximated by the arithmetic average of the inlet and exit values of v_b; that is,

$$v_m = \frac{v_1 + v_2}{2} = \frac{1}{2}\left(\frac{1}{\rho_1} + \frac{1}{\rho_2}\right) \tag{M-24}$$

Substituting Eq. (M–22) into Eq. (M–21) and rearranging, the core pressure drop is given by

$$\Delta P_{c-e} = \frac{G^2 v_1}{2}\left[K_c + \frac{4L}{D_H} f\frac{v_m}{v_1} + 2\left(\frac{v_2}{v_1} - 1\right)\right] \tag{M-25}$$

From the practical standpoint, the flow acceleration and associated property variation effects can generally be neglected for liquids, which are characterized by small changes in density. However, for gas flow processes with moderate to strong heating or cooling, this effect can be significant.

M–1–2 Core Inlet and Exit Pressure Drop

The pressure drop ΔP_{1-c} caused by the change in area at the core inlet is represented by the famous Bernoulli equation [30],

$$\Delta P_{1-c} = \rho_1\left(\frac{U_{b,c}^2}{2} - \frac{U_{b,1}^2}{2}\right) = \left(\frac{\rho U_b^2}{2}\right)_c\left[1 - \left(\frac{U_{b,1}}{U_{b,c}}\right)^2\right] \tag{M-26}$$

where $U_{b,c}$ is the bulk-stream velocity within the core at $x = 0$ and $\rho_c \simeq \rho_1$ for incompressible flow. Since the mass-flow rate $\dot{m}$ is constant, we also have

$$\rho_1 A_1 U_{b,1} = \rho_c A_c U_{b,c} = \rho_1 A U_{b,c} \tag{M–27}$$

where $A_c = A$, or

$$\frac{U_{b,1}}{U_{b,c}} = \frac{A}{A_1} = \sigma \tag{M–28}$$

It follows that

$$\Delta P_{1-c} = \frac{\rho_1 U_{b,c}^2}{2}(1 - \sigma^2) = \frac{G^2}{2\rho_1}(1 - \sigma^2) \tag{M–29}$$

For uniform property flow $\rho_c = \rho =$ constant and $U_b = U_{b,c} =$ constant, such that Eq. (M–29) reduces to

$$\Delta P_{1-c} = \frac{\rho U_b^2}{2}(1 - \sigma^2) = \frac{G^2}{2\rho}(1 - \sigma^2) \tag{M–30}$$

The change in pressure at the core exit is caused by a pressure rise ΔP_{rise} resulting from the decrease in kinetic energy and a pressure loss ΔP_{loss} due to momentum transport in the region following an abrupt expansion. Following the approach presented above, ΔP_{rise} is given by

$$\Delta P_{\text{rise}} = \rho_e \left(\frac{U_{b,e}^2}{2} - \frac{U_{b,2}^2}{2}\right) = \frac{\rho_2 U_{b,e}^2}{2}(1 - \sigma^2) = \frac{G^2}{2\rho_2}(1 - \sigma^2) \tag{M–31}$$

where $\rho_2 A_2 U_{b,2} = \rho_e A_e U_{b,e} = \rho_2 A U_{b,e}$ and $U_{b,e}$ is the bulk-stream velocity within the core at $x = L$. This equation reduces to

$$\Delta P_{\text{rise}} = \frac{\rho U_b^2}{2}(1 - \sigma^2) = \frac{G^2}{2\rho}(1 - \sigma^2) \tag{M–32}$$

for uniform property flow. Following the approach generally used in design work, the pressure loss ΔP_{loss} is represented in terms of the expansion-loss coefficient K_e; that is,

$$\Delta P_{\text{loss}} = K_e \frac{\rho_2 U_{b,e}^2}{2} = K_e \frac{G^2}{2\rho_2} \tag{M–33}$$

or, for uniform property conditions, by

$$\Delta P_{\text{loss}} = K_e \frac{\rho U_b^2}{2} \tag{M–34}$$

where K_e is specified by use of Fig. 8–5. Thus, the total pressure change at the exit is given by

$$\Delta P_{2-e} = \Delta P_{\text{rise}} - \Delta P_{\text{loss}} = \frac{\rho_2 U_{b,e}^2}{2}(1 - \sigma^2 - K_e) \\ = \frac{G^2}{2\rho_2}(1 - \sigma^2 - K_e) \tag{M–35}$$

or, for uniform density,

$$\Delta P_{2-e} = P_2 - P_e = \frac{\rho U_b^2}{2}(1 - \sigma^2 - K_e) = \frac{G^2}{2\rho}(1 - \sigma^2 - K_e) \tag{M–36}$$

which represents a net pressure rise.

M–1–3 Total Pressure Drop

Using Eqs. (M–30), (M–18), and (M–36) to represent the pressure drop components for uniform property flow, Eq. (M–1) takes the form

$$\Delta P_{1-2} = \frac{\rho U_b^2}{2}\left[(1 - \sigma^2 + K_c) + \frac{4L}{D_H} f - (1 - \sigma^2 - K_e)\right] \\ = \frac{\rho U_b^2}{2}\left(K_c + \frac{4L}{D_H} f + K_e\right) = \frac{G^2}{2\rho}\left(K_c + \frac{4L}{D_H} f + K_e\right) \tag{M–37}$$

Notice that the effect of change in kinetic energy at the entrance contraction and exit expansion (represented by the terms involving $1 - \sigma^2$) actually cancel for uniform property flow. On the other hand, the substitution of Eqs. (M–29), (M–25), and (M–35) into Eq. (M–1) for variable property conditions gives

$$\Delta P_{1-2} = \frac{G^2 v_1}{2}\left[(1 - \sigma^2 + K_c) + \frac{4L}{D_H} f \frac{v_m}{v_1} \right. \\ \left. + 2\left(\frac{v_2}{v_1} - 1\right) - (1 - \sigma^2 - K_e)\frac{v_2}{v_1}\right] \tag{M–38}$$

which indicates an effect of kinetic energy change as well as core entrance and exit losses and flow acceleration.

M–2 PUMPING POWER

The pumping power $\dot{W}_p$ that is required to overcome a pressure drop ΔP for uniform property conditions is given by Eq. (8–78),

$$\dot{W}_p = \frac{\dot{m}\,\Delta P}{\rho} \tag{M–39}$$

The formal defining relation for pumping power that is applicable to variable property flow is

$$\dot{W}_p = \int_{P_1}^{P_2} v_b \dot{m}\, dP = \int_0^L v_b \dot{m} \frac{dP}{dx}\, dx \tag{M-40}$$

However, for most practical applications $\dot{W}_p$ is estimated by

$$\dot{W}_p = \dot{m}\, \Delta P\, v_m \tag{M-41}$$

As indicated by Eqs. (8–81) and (8–82), the pumping power is inversely proportional to the square of the fluid density for both laminar and turbulent flow.

M–3 HEATING–PUMPING POWER RATIO

Equation (M–39) is combined with Eq. (8–56) for q_c to obtain an expression for the heating–pumping power ratio for uniform wall-temperature conditions in terms of the efficiency ψ (= ϵ/NTU) of the form

$$\frac{q_c}{\dot{W}_p} = \frac{\bar{h} A_s (T_0 - T_1)}{\dot{m}\, \Delta P/\rho}\, \psi \tag{M-42}$$

and

$$NTU = \frac{\bar{h} A_s}{\dot{m} c_P} = \frac{\overline{Nu}\, (k/D_H) A_s}{\rho A U_b c_P} = \frac{\overline{Nu}}{Re\, Pr} \frac{4L}{D_H} \frac{p}{p_w} \tag{M-43}$$

Using Eq. (8–81) for $\dot{W}_p$ and setting $\bar{h} = h$ for fully developed conditions, Eq. (M–42) becomes

$$\frac{q_c}{\dot{W}_p} = \frac{hpL(T_0 - T_1)}{(\mu^3/\rho^2)(p_w L/D_H^3)(f/2)\, Re^3}\, \psi = \frac{k\rho^2}{\mu^3} \frac{D_H^2\, Nu\, (T_0 - T_1)}{Re^3\, f/2} \frac{p}{p_w}\, \psi \tag{M-44}$$

Using Eqs. (8–7) and (8–8) to approximate f and Nu for laminar flow, we obtain

$$\begin{aligned} \frac{q_c}{\dot{W}_p} &= \frac{2C_2}{C_1} \frac{k\rho^2}{\mu^3} \frac{p}{p_w} \frac{D_H^2 (T_0 - T_1)}{Re^2}\, \psi \\ &= \frac{2C_2}{Pr\, C_1} \frac{p}{p_w} \frac{c_P (T_0 - T_1)}{U_b^2}\, \psi = \frac{2C_2}{Pr\, C_1} \frac{p}{p_w} \left(\frac{\rho A}{\dot{m}}\right)^2 c_P (T_0 - T_1) \psi \end{aligned} \tag{M-45}$$

and

$$NTU = \frac{C_2}{Re\, Pr} \frac{4L}{D_H} \frac{p}{p_w} \tag{M-46}$$

Similarly, the use of Eq. (8–18) to approximate Nu for turbulent flow gives

$$\frac{q_c}{\dot{W}_p} = \frac{k\rho^2}{\mu^3}\frac{p}{p_w}\frac{Pr^n D_H^2}{Re^2}(T_0 - T_1)\psi = Pr^{n-1}\frac{p}{p_w}\frac{c_P(T_0 - T_1)}{U_b^2}\psi = Pr^{n-1}\frac{p}{p_w}\left(\frac{\rho A}{\dot{m}}\right)^2 c_P(T_0 - T_1)\psi \tag{M–47}$$

and

$$NTU = \frac{0.023\, Re^{0.8}\, Pr^n}{Re\, Pr}\frac{4L}{D_H}\frac{p}{p_w} = \frac{0.023}{Re^{0.2}\, Pr^{1-n}}\frac{4L}{D_H}\frac{p}{p_w} \tag{M–48}$$

where $n = 0.5$ for $0.5 \lessapprox Pr \lessapprox 5$ and $n = 1/3$ for $Pr \gtrapprox 5$. To generalize this result, we write†

$$\frac{q_c}{\dot{W}_p} = \alpha_p \frac{p}{p_w}\frac{c_P(T_0 - T_1)}{U_b^2}\psi = \alpha_p \frac{p}{p_w}\left(\frac{\rho A}{\dot{m}}\right)^2 c_P(T_0 - T_1)\psi \tag{M–49}$$

where

$$\alpha_p = \frac{2C_2}{Pr\, C_1} \qquad \text{for laminar flow} \tag{M–50}$$

$$\alpha_p = Pr^{n-1} \qquad \text{for turbulent flow} \tag{M–51}$$

Equation (M–49) indicates that the heating–pumping power ratio is proportional to U_b^{-2} or $(\rho A/\dot{m})^2$ for both laminar and turbulent flow with small values of NTU (i.e., $\psi \simeq 1$). However, because of the nature of the dependence of ψ on NTU, $q_c/\dot{W}_p$ decreases with increasing values of NTU such that unnecessarily large values of NTU result in ineffective heat-transfer performance and significant frictional pressure drop. Equations (8–57), (M–46), and (M–48) can be used to evaluate ψ and NTU in terms of Re, D_H/L, and related parameters.

N HEAT EXCHANGERS: SOLUTION RESULTS

The solution results for heat exchangers presented in this section is categorized as follows:

N–1 General Closed-Form Relations
N–2 Effectiveness Approach—Matrix Formalism

† Using Eq. (8–59), we obtain an equivalent relation for the heating–pumping power ratio of the form

$$\frac{q_c}{\dot{W}_p} = \alpha_p \frac{p}{p_w}\frac{c_P\, LMTD}{U_b^2} = \alpha_p \frac{p}{p_w} c_P\, LMTD\left(\frac{\rho A}{\dot{m}}\right)^2$$

N–1 GENERAL CLOSED-FORM RELATIONS

General closed-form relations are presented for basic heat-exchanger arrangements in Tables A–N–1–1 and A–N–1–2. The assumptions upon which these relations are based include (1) U = constant,† (2) no phase change in only part of the exchanger, (3) constant flow rate for both fluids, (4) constant specific heat for both fluids, (5) negligible heat losses and longitudinal conduction in the fluid and wall, and several others (see Secs. 11–7–1 and 11–8–1).

TABLE A–N–1–1(a) Functional relations for $P(NTU_I,R)$, $F(NTU_I,R)$, and $\psi(NTU_I,R)$

Heat-exchanger arrangement	*Relations*		
General	$F = \dfrac{1}{NTU_I\,(1 - R)} \ln \dfrac{1 - PR}{1 - P}$	$\psi = \dfrac{P}{NTU_I}$	
	$F = \dfrac{P}{NTU_I\,(1 - P)}$ for $R = 1$		
All exchangers for $R = 0$	$P = 1 - \exp(-NTU_I)$	$F = 1$	(ai,ii)
Double pipe			
Parallel flow	$P = \dfrac{1 - \exp[-NTU_I\,(1 + R)]}{1 + R}$	$F = 1$	(bi,ii)
Counterflow	$P = \dfrac{1 - \exp[-NTU_I\,(1 - R)]}{1 - R\exp[-NTU_I\,(1 - R)]}$	$F = 1$	(ci,ii)
	$P = \dfrac{1 - \exp[NTU_I\,(1 - R)]}{R - \exp[NTU_I\,(1 - R)]}$	$F = 1$	(c′i,ii)
Counterflow, $R = 1$	$P = \dfrac{NTU_I}{1 + NTU_I}$	$F = 1$	(di,ii)
Shell and tube			
TEMA E shell, 1-2 arrangement, shell fluid mixed[a]	$P = \dfrac{2}{1 + R + \sqrt{1 + R^2}\,(1 + e^{-\Gamma})/(1 - e^{-\Gamma})}$		(ei)
	$\Gamma = NTU_I\sqrt{1 + R^2}$		

† For shell-and-tube and crossflow exchangers.

TABLE A–N–1–1(a) (Functional relations for $P(NTU_t,R)$, $F(NTU_t,R)$, and $\psi(NTU_t,R)$—continued)

Heat-exchanger arrangement	*Relations*
	$F = \frac{1}{NTU_t(1 - R)} \times \ln\left[\frac{1 + R + \sqrt{1 + R^2}(1 + e^{-\Gamma})/(1 - e^{-\Gamma}) - 2R}{1 + R + \sqrt{1 + R^2}(1 + e^{-\Gamma})/(1 - e^{-\Gamma}) - 2}\right]$ (eii)
	$F = \frac{2/NTU_t}{1 + R + \sqrt{1 + R^2}(1 + e^{-\Gamma})/(1 - e^{-\Gamma}) - 2}$ for $R = 1$ (eiii)
TEMA E shell, 1–4 arrangement, shell fluid mixed[b]	$P = \frac{4}{2(1 + R) + \sqrt{1 + 4R^2}\coth(\Gamma/4) + \tanh(NTU_t/4)}$ (f) $\Gamma = NTU_t\sqrt{1 + 4R^2}$
TEMA E shell, 1–2 arrangement, shell fluid unmixed[b]	$P = 1 - \frac{2R - 1}{2R + 1}\left\{\frac{2R + \exp[-NTU_t(R + 1/2)]}{2R - \exp[-NTU_t(R - 1/2)]}\right\}$ (gi) $P = \frac{1 + NTU_t - \exp(-NTU_t)}{2 + NTU_t}$ for $R = 1/2$ (gii)
TEMA-G shell, 1–2 split flow arrangement, shell fluid mixed[b]	$P = \frac{1 + G + 2RG + (1 + 2R)D\,e^{-\alpha} - e^{-\alpha}}{(1 + G + 2RG) + 2R(1 - D) + 2RD\,e^{-\alpha}}$ (h) $D = \frac{1 - e^{-\alpha}}{2R + 1}$ $\alpha = \frac{NTU_t}{4}(1 + 2R)$ $\beta = \frac{NTU_t}{2}(2R - 1)$ $G = \frac{1 - e^{-\beta}}{2R - 1}$ $G = \frac{NTU_t}{2}$ for $R = 1/2$
TEMA-J shell, 1–2 divided flow arrangement, shell fluid mixed[b]	$P = \frac{2}{1 + 2R\phi'}$ (i) $\phi' = 1 + \gamma\frac{1 + \phi}{1 - \phi} - 2\gamma\left\{\frac{\gamma\phi + (1 - \phi)\exp[-R\,NTU_t(\gamma - 1)/2]}{(1 - \phi)^2 + \gamma(1 - \phi^2)}\right\}$ $\phi = \exp(-\gamma R\,NTU_t)$ $\gamma = \frac{\sqrt{1 + 4R^2}}{2R}$
Crossflow One fluid mixed, one fluid unmixed[b]	$P = \frac{1 - \exp(-\Gamma R)}{R}$ $\Gamma = 1 - \exp(-NTU_t)$ (ji) $F = \frac{1}{NTU_t(1 - R)}\ln\left[\frac{R\exp(-\Gamma R)}{R - 1 + \exp(-\Gamma R)}\right]$ (jii)

TABLE A–N–1–1(a) (Functional relations for $P(NTU_I,R)$, $F(NTU_I,R)$, and $\psi(NTU_I,R)$—continued)

Heat-exchanger arrangement	*Relations*	
	$F = \dfrac{1}{NTU_t} \dfrac{1 - \exp(-\Gamma)}{\exp(-\Gamma)}$ for $R = 1$	(jiii)
Both fluids umixed[c]	$P = \sum_{m=0}^{\infty} \left\{ \left[1 - \exp(-NTU_I) \sum_{j=0}^{m} \dfrac{NTU_I^j}{j!} \right] \times \left[1 - \exp(-R\,NTU_I) \sum_{j=0}^{m} \dfrac{(R\,NTU_I)^j}{j!} \right] \right\} / (R\,NTU_I)$	(ki)
Approximation for $R < 1$, $NTU_I < 5$	$P = 1 - \exp\left[\dfrac{\exp(-\Gamma R\,NTU_I) - 1}{\Gamma R} \right]$ $\Gamma = NTU_I^{-0.22}$	(kii)
Both fluids mixed	$P = \left[\dfrac{1}{1 - \exp(-NTU_I)} + \dfrac{R}{1 - \exp(-R\,NTU_I)} - \dfrac{1}{NTU_I} \right]^{-1}$	(l)

[a] This equation also provides a very reasonable approximation for one shell and 4, 6, . . . , tube passes.

[b] The solutions for these nonsymmetrical exchangers are expressed in terms of reference stream t. To express the solutions in terms of reference stream s, replace R by $1/R$, NTU_t by $R\,NTU_s$, and P by PR.

[c] See Shah and Mueller [41] for an alternative exact solution expressed in terms of Bessel functions.

TABLE A–N–1–1(b) Functional relations for $NTU_I(P,R)$, $F(P,R)$, and $\psi(P,R)$

Heat-exchanger arrangement	*Relations*		
General	$F = \dfrac{1}{NTU_I(1 - R)} \ln \dfrac{1 - PR}{1 - P}$	$\psi = \dfrac{P}{NTU_I}$	
	$F = \dfrac{P}{NTU_I(1 - P)}$ for $R = 1$		
All exchangers for $R = 0$	$NTU_I = -\ln(1 - P)$	$F = 1$	(ai,ii)
Doule-pipe			
Parallel flow	$NTU_I = -\dfrac{\ln[1 - P(1 + R)]}{1 + R}$	$F = 1$	(bi,ii)
Counterflow	$NTU_I = \dfrac{1}{1 - R} \ln \dfrac{1 - PR}{1 - P}$	$F = 1$	(ci,ii)
Counterflow, $R = 1$	$NTU_I = \dfrac{P}{1 - P}$	$F = 1$	(di,ii)
Shell and tube			
TEMA-E shell, 1-2 pass arrangement, shell fluid mixed	$NTU_I = \dfrac{1}{\sqrt{1 + R^2}} \ln \dfrac{2 - P(1 + R - \sqrt{1 + R^2})}{2 - P(1 + R + \sqrt{1 + R^2})}$		(ei)

TABLE A-N-1-1(b) (Functional relations for $NTU_t(P,R)$, $F(P,R)$, and $\psi(P,R)$—continued)

Heat-exchanger arrangement	*Relations*	
	$F = \dfrac{\sqrt{1+R^2}}{1-R} \dfrac{\ln[(1-PR)/(1-P)]}{\ln \dfrac{2-P(1+R-\sqrt{1+R^2})}{2-P(1+R+\sqrt{1+R^2})}}$	(eii)
	$F = \dfrac{P}{1-P} \dfrac{\sqrt{2}}{\ln\left[\dfrac{2/P-2+\sqrt{2}}{2/P-2-\sqrt{2}}\right]}$ for $R = 1$	(eiii)
Crossflow One fluid mixed, one fluid unmixed†	$NTU_t = -\ln[1 + \dfrac{1}{R}\ln(1-PR)]$	(fi)
	$F = -\dfrac{1}{1-R} \dfrac{\ln[(1-PR)/(1-P)]}{\ln[1+(1/R)\ln(1-PR)]}$	(fii)
	$F = -\dfrac{P/(1-P)}{\ln[1+\ln(1-P)]}$ for $R = 1$	(fiii)

† The solution for this nonsymmetrical exchanger is expressed in terms of reference stream t. To express the solution in terms of reference stream s, replace R by $1/R$, NTU_t by $R\ NTU_s$, and P by PR.

TABLE A-N-1-2(a) Functional relations for $\epsilon(NTU,C^*)$

Heat-Exchanger Arrangement	*Relations*	
All exchangers for $C^* = 0$	$\epsilon = 1 - \exp(-NTU)$	(a)
Double pipe Parallel flow	$\epsilon = \dfrac{1-\exp[-NTU(1+C^*)]}{1+C^*}$	(b)
Counterflow	$\epsilon = \dfrac{1-\exp[-NTU(1-C^*)]}{1-C^*\exp[-NTU(1-C^*)]}$	(c)
	$\epsilon = \dfrac{1-\exp[NTU(1-C^*)]}{C^*-\exp[NTU(1-C^*)]}$	(c′)
Counterflow, $C^* = 1$	$\epsilon = \dfrac{NTU}{1+NTU}$	(d)
Shell and tube TEMA-E shell, 1-2 arrangement, shell fluid mixed[a]	$\epsilon = \dfrac{2}{1+C^*+\sqrt{1+C^{*2}}(1+e^{-\Gamma})/(1-e^{-\Gamma})}$ $\Gamma = NTU\sqrt{1+C^{*2}}$	(e)
TEMA-E shell, 1-4 arrangement, shell fluid mixed,[b] $C_{min} = C_{unmixed} = C_t$	$\epsilon = \dfrac{4}{2(1+C^*)+\sqrt{1+4C^{*2}}\coth(\Gamma/4)+\tanh(NTU/4)}$ $\Gamma = NTU\sqrt{1+4C^{*2}}$	(f)

TABLE A–N–1–2(a) (Functional relations for $\epsilon(NTU, C^*)$—continued)

Heat-Exchanger Arrangement	*Relations*
TEMA-E shell 1-2 arrangement, shell fluid unmixed,[b] $C_{min} = C_{unmixed} = C_t$	$\epsilon = 1 - \frac{2C^* - 1}{2C^* + 1}\left\{\frac{2C^* + \exp[-NTU(C^* + 1/2)]}{2C^* - \exp[-NTU(C^* - 1/2)]}\right\}$ (gi) $\epsilon = \frac{1 + NTU - \exp(-NTU)}{2 + NTU}$ for $C^* = 1/2$ (gii)
TEMA-G shell, 1-2 split flow arrangement, shell fluid mixed,[b] $C_{min} = C_{unmixed} = C_t$	$\epsilon = \frac{1 + G + 2C^*G + (1 + 2C^*)D e^{-\alpha} - e^{-\alpha}}{(1 + G + 2C^*G) + 2C^*(1 - D) + 2C^*D e^{-\alpha}}$ (h) $D = \frac{1 - e^{-\alpha}}{2C^* + 1}$ $\alpha = \frac{NTU}{4}(1 + 2C^*)$ $\beta = \frac{NTU}{2}(2C^* - 1)$ $G = \frac{1 - e^{-\beta}}{2C^* - 1}$ $G = \frac{NTU_t}{2}$ for $C^* = 1/2$
TEMA-J shell, 1-2 divided flow arrangement, shell fluid mixed,[b] $C_{min} = C_{unmixed} = C_t$	$\epsilon = \frac{2}{1 + 2C^*\phi'}$ (i) $\phi' = 1 + \gamma\frac{1 + \phi}{1 - \phi} - 2\gamma\left\{\frac{\gamma\phi + (1 - \phi)\exp[-C^*NTU(\gamma - 1)/2]}{(1 - \phi)^2 + \gamma(1 - \phi^2)}\right\}$ $\phi = \exp(-\gamma C^* NTU)$ $\gamma = \frac{\sqrt{1 + 4C^{*2}}}{2C^*}$
Crossflow One fluid mixed, one fluid unmixed,[b] $C_{min} = C_{unmixed} = C_t$	$\epsilon = \frac{1 - \exp(-\Gamma C^*)}{C^*}$ $\Gamma = 1 - \exp(-NTU)$ (j)
Both fluids unmixed[c]	$\epsilon = \sum_{m=0}^{\infty}\left\{\left[1 - \exp(-NTU)\sum_{j=0}^{m}\frac{NTU^j}{j!}\right] \times \left[1 - \exp(-C^* NTU)\sum_{j=0}^{m}\frac{(C^* NTU)^j}{j!}\right]\right\}/(C^* NTU)$ (ki)
Approximation for $NTU < 5$	$\epsilon = 1 - \exp\left[\frac{\exp(-\Gamma C^* NTU) - 1}{\Gamma C^*}\right]$ (kii) $\Gamma = NTU^{-0.22}$
Both fluids mixed	$\epsilon = \left[\frac{1}{1 - \exp(-NTU)} + \frac{C^*}{1 - \exp(-C^* NTU)} - \frac{1}{NTU}\right]^{-1}$ (l)

[a] This equation also provides a very reasonable approximation for one shell and 4, 6, . . . , tube passes.

[b] If $C_{min} = C_{mixed} = C_s$, use indicated formula with C^* replaced by $1/C^*$, NTU replaced by $C^* NTU$, and ϵ replaced by ϵC^*.

[c] See Shah and Mueller [41] for an exact solution expressed in terms of Bessel functions.

TABLE A–N–1–2(b) Functional relations for $NTU(\epsilon, C^*)$

Heat-Exchanger Arrangement	*Relations*	
All exchangers for $C^* = 0$	$NTU = -\ln(1 - \epsilon)$	(a)
Double pipe		
Parallel flow	$NTU = -\dfrac{\ln[1 - \epsilon(1 + C^*)]}{1 + C^*}$	(b)
Counterflow	$NTU = \dfrac{1}{1 - C^*} \ln \dfrac{1 - \epsilon C^*}{1 - \epsilon}$	(c)
Counterflow, $C^* = 1$	$NTU = \dfrac{\epsilon}{1 - \epsilon}$	(d)
Shell and tube TEMA-E shell 1-2 pass arrangement, shell fluid mixed[a]	$NTU = \dfrac{1}{\sqrt{1 + C^{*2}}} \ln \dfrac{2 - \epsilon(1 + C^* - \sqrt{1 + C^{*2}})}{2 - \epsilon(1 + C^* + \sqrt{1 + C^{*2}})}$	(e)
Crossflow One fluid mixed, one fluid unmixed,[c] $C_{min} = C_{unmixed} = C_t$	$NTU = -\ln\left[1 + \dfrac{1}{C^*} \ln(1 - \epsilon C^*)\right]$	(f)

N–2 EFFECTIVENESS APPROACH—MATRIX FORMALISM

The effectiveness method is commonly employed in the current heat-transfer literature in the context of matrix algebra, which provides a basis for the use of solutions for basic heat exchangers to develop solutions for complex heat-exchanger configurations.

To develop the basic matrix relation featured in this approach, the outlet fluid temperatures $T_{c,o}$ and $T_{h,o}$ are expressed in terms of the effectiveness P_c and P_h by writing

$$T_{c,o} = T_{c,i} + P_c(T_{h,i} - T_{c,i}) = (1 - P_c)T_{c,i} + P_c T_{h,i} \tag{N–2–1}$$

from Eq. (11–41), and

$$T_{h,o} = T_{h,i} - P_h(T_{h,i} - T_{c,i}) = P_h T_{c,i} + (1 - P_h)T_{h,i} \tag{N–2–2}$$

from Eq. (11–45). These relations for $T_{c,o}$ and $T_{h,o}$ can be represented by a matrix equation of the form

$$\begin{vmatrix} T_{c,o} \\ T_{h,o} \end{vmatrix} = \begin{vmatrix} 1 - P_c & P_c \\ P_h & 1 - P_h \end{vmatrix} \begin{vmatrix} T_{c,i} \\ T_{h,i} \end{vmatrix} \tag{N–2–3}$$

or, since $P_h = P_c R_c$ and $P_c = P_h R_h$ (see Table 11–5), we obtain

$$\begin{vmatrix} T_{c,o} \\ T_{h,o} \end{vmatrix} = \begin{vmatrix} 1 - P_h R_h & P_h R_h \\ P_h & 1 - P_h \end{vmatrix} \begin{vmatrix} T_{c,i} \\ T_{h,i} \end{vmatrix} \tag{N–2–4a}$$

or equivalently

$$\begin{vmatrix} T_{c,o} \\ T_{h,o} \end{vmatrix} = \begin{vmatrix} 1 - P_c & P_c \\ P_cR_c & 1 - P_cR_c \end{vmatrix} \begin{vmatrix} T_{c,i} \\ T_{h,i} \end{vmatrix} \tag{N-2-4b}$$

The use of this matrix formalism was first introduced by Domingos [42] and has since been adapted to the analysis of a wide range of complex heat exchangers [43–46].

N-3 HEAT EXCHANGER OPTIMIZATION—BASIC ECONOMIC FACTORS

As indicated in Sec. 11–5–1, the economic factors that should be considered in the optimization of a heat exchanger include *capital cost, operating cost*, and *cost of thermal energy supplied to (or removed from) the exchanger.*

The capital cost C_s is normally assumed to be proportional to the surface area A_s; that is,

$$C_s = a_0 + a_1A_s \tag{N-3-1}$$

where a_1 is the *cost per unit surface area* and a_0 is the *fixed cost* that is associated with terminal supports for flow passages, manifolds, installation, and related expenses.

Assuming that the operating cost C_p is proportional to the power used on both sides, we write

$$C_p = a_2\dot{W}_{p,c} + a_3\dot{W}_{p,h} \tag{N-3-2}$$

where a_2 and a_3 represent the unit costs for pumping power for the cold and hot fluids, both of which depend on the fuel cost as well as equipment and maintenance cost.

The cost of heating or cooling effect supplied to the heat exchanger is represented by

$$C_H = a_4q \tag{N-3-3}$$

where a_4 is the unit cost of thermal energy that is directly proportional to fuel cost. As pointed out by Kovarik [47], "when the energy price undergoes a major change reflected in the value of a_4, the possible influence of this change on exchanger design creates uncertainties."

To provide a framework for the development of an economic optimization analysis for heat exchangers and to determine the influence of the unit cost of thermal energy a_4 on the optimal design, Kovarik [47] introduced the *objective function J*,

$$J = \frac{q}{C_s + C_p + C_H} \tag{N-3-4}$$

which represents the ratio of the heat exchanger thermal performance (heat-transfer rate) to the total cost. An exchanger is economically optimized by maximizing the objective function J.

Following Kovarik [47], Eqs. (N–3–3) and (N–3–4) are combined to express J in terms of a_4.

$$J = \frac{1}{(C_s + C_p)/q + a_4} \tag{N–3–5}$$

This relation indicates that the maximum value of J is achieved by minimizing the term $(C_s + C_p)/q$ and is *independent* of the value of the unit cost of thermal energy a_4. This result provides the basis for Kovarik's conclusion that an optimal heat exchanger is optimal for any cost of thermal energy.

In connection with this important conclusion, it should be noted that fuel cost does enter into the optimization calculations in the context of the unit cost for pumping power represented by a_2 and a_3. However, the fuel cost often represents a small part of the unit cost for pumping power, in which cases the optimal design is only slightly affected by the fuel cost associated with pumping.

Although we will not continue further with the analysis, Kovarik employs Eq. (N–3–5) as the basis for a numerical procedure for optimizing crossflow heat exchangers. Among the conclusions resulting from this economic optimization analysis is that high values of NTU_I (i.e., low values of ψ) for crossflow heat exchangers are optimal only if the unit costs of heat-transfer surface area a_1 and friction losses a_2 and a_3 are low.

N–4 RELATION BETWEEN OVERALL PARALLEL-FLOW AND COUNTERFLOW ARRANGEMENTS

The change in direction of one stream in a heat exchanger normally (although not always) results in a significant change in the thermal performance, as in the case of switching a double-pipe arrangement from parallel flow to counterflow. We now wish to consider the relation between the effectiveness P of a configuration with specified flow directions and the effectiveness $\hat{P}$ of the same configuration with the direction of one of the streams reversed.

As shown by Pignotti [48], a simple relation exists between $P(NTU_I,R)$ and $\hat{P}(NTU_I,R)$ for basic heat-exchanger configurations that involve *at least one fluid* that proceeds through the exchanger in a *single mixed stream*, such as double-pipe, shell-and-tube, and mixed–unmixed cross-flow exchangers, providing that the capacitance ratio R and overall coefficient of heat transfer U can be assumed to be uniform throughout the exchanger. The relation between $\hat{P}(NTU_I,R)$ and $P(NTU_I,R)$ can be represented by

$$\hat{P}(NTU_I,R) = \frac{P(NTU_I,-R)}{1 + R\,P(NTU_I,-R)} = \left[\frac{1}{P(NTU_I,-R)} + R\right]^{-1} \tag{N–4–1}$$

where the reference stream I is taken as the unmixed fluid t if one stream is unmixed or is specified by either fluid if both streams are mixed.

To demonstrate the use of this relation, we consider parallel flow in a double-pipe heat exchanger, for which case the effectiveness P is given by Eq. (11–137),

$$P = \frac{1 - \exp\,[-NTU_I\,(1 + R)]}{1 + R} \tag{N-4-2}$$

Substituting this expression for P into Eq. (N–4–1), we obtain

$$\begin{aligned}\hat{P} &= \left\{\frac{1 - R}{1 - \exp\,[-NTU_I\,(1 - R)]} + R\right\}^{-1} \\ &= \frac{1 - \exp\,[-NTU_I\,(1 - R)]}{1 - R\exp\,[-NTU_I\,(1 - R)]}\end{aligned} \tag{N-4-3}$$

which corresponds to the result given by Eq. (11–138a) for a double-pipe heat exchanger in counterflow.

Several examples of the use of Eq. (N–4–1) to convert solutions for *overall parallel flow* to *overall counterflow* are given in reference 48. It is interesting to note that the reversal of flow directions of one fluid has no effect for such basic arrangements as 1–2 and 1–4 TEMA E, 1–2 TEMA J shell-and-tube, and single-pass mixed–unmixed-crossflow exchangers, but has a very significant effect on others, such as double-pipe and 1–2 TEMA G shell-and-tube exchangers.

N–5 SHELL-AND-TUBE HEAT EXCHANGERS—PRACTICAL THERMAL ANALYSIS FOR 1–2 EXCHANGERS

Referring to Fig. A–N–5–1, the lumped and lumped/differential relations given by Eqs. (11–24) to (11–26) for steady-state conditions are readily adapted to 1–2 exchangers by writing (assuming shell side fluid is hot and tube-side fluid is cold)

$$q = C_s(T_{s,i} - T_{s,o}) \qquad q = C_t(T_{t,o} - T_{t,i}) \tag{N-5-1,2}$$

and

$$dq = -C_s\,dT_s \tag{N-5-3}$$

where dq is represented in terms of U by

$$dq = U\,dA_s\,(T_s - T_t) = U\,dA_s\left(T_s - \frac{T_t^{(1)} + T_t^{(2)}}{2}\right) \tag{N-5-4}$$

from Eqs. (11–165a) and (11–166). Combining Eqs. (N–5–3) and (N–5–4) and setting $dA_s = p\,dx$, we obtain an equation for T_s of the form

$$\frac{dT_s}{T_s - (T_t^{(1)} + T_t^{(2)})/2} = -\frac{Up\,dx}{C_s} \qquad \text{shell fluid} \tag{N-5-5}$$

Independent relations are obtained for $T_t^{(1)}$ and $T_t^{(2)}$ by writing

$$dq^{(1)} = (\dot{m}c_P)_t^{(1)}\,dT_t^{(1)} = C_t\,dT_t^{(1)} \tag{N-5-6}$$

$$dq^{(2)} = (\dot{m}c_P)_t^{(2)}\,dT_t^{(2)} = -C_t\,dT_t^{(2)} \tag{N-5-7}$$

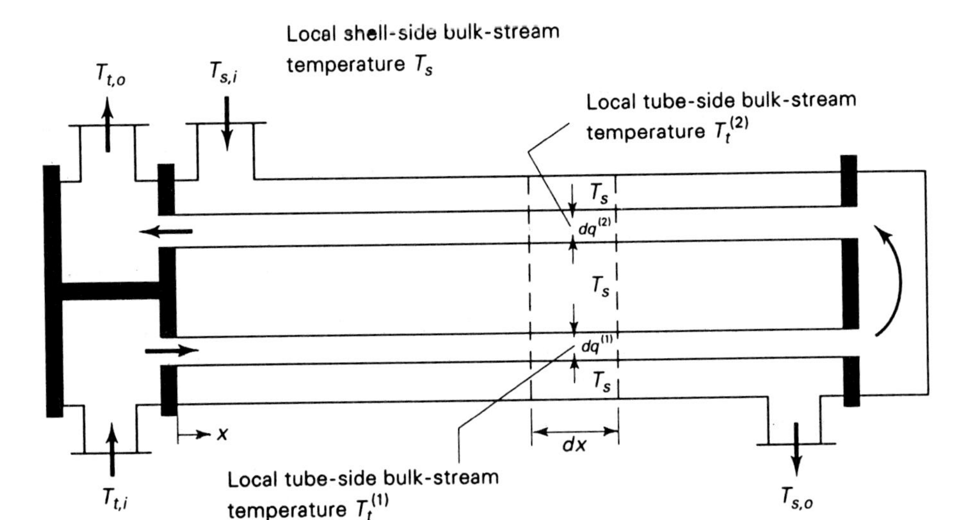

FIGURE A–N–5–1 Control volumes for 1–2 exchanger.

where we have set

$$(\dot{m}c_P)_t^{(1)} = C_t \qquad \text{for pass 1 (parallel flow)} \tag{N–5–8a}$$

$$(\dot{m}c_P)_t^{(2)} = -C_t \qquad \text{for pass 2 (counterflow)} \tag{N–5–8b}$$

Assuming that U, T_s, $T_t^{(1)}$, and $T_t^{(2)}$ are uniform over the cross section, and the heat-transfer surface area in each of the two passes is $A_s/2 = p\,dx/2$, $dq^{(1)}$ and $dq^{(2)}$ are represented by†

$$dq^{(1)} = \frac{Up\,dx}{2}(T_s - T_t^{(1)}) \qquad dq^{(2)} = \frac{Up\,dx}{2}(T_s - T_t^{(2)}) \tag{N–5–9,10}$$

Combining Eqs. (N–5–6) and (N–5–9) and Eqs. (N–5–7) and (N–5–10), we obtain equations for $T_t^{(1)}$ and $T_t^{(2)}$ of the form

$$\frac{dT_t^{(1)}}{T_s - T_t^{(1)}} = \frac{U}{C_t}\frac{p\,dx}{2} \qquad \text{tube fluid—pass 1} \tag{N–5–11}$$

and

$$\frac{dT_t^{(2)}}{T_s - T_t^{(2)}} = -\frac{U}{C_t}\frac{p\,dx}{2} \qquad \text{tube fluid—pass 2} \tag{N–5–12}$$

† Notice that

$$dq = dq^{(1)} + dq^{(2)} = \frac{Up\,dx}{2}(2T_s - T_t^{(1)} - T_t^{(2)})$$

which reduces to Eq. (N–5–4).

Thus, the three dependent variables T_s, $T_t^{(1)}$, and $T_t^{(2)}$ are characterized by Eqs. (N–5–5), (N–5–11), and (N–5–12), which are put into the form

$$\frac{dT_s}{dx} + \frac{pUT_s}{C_s} - \frac{pU}{2C_s}(T_t^{(1)} + T_t^{(2)}) = 0 \qquad \text{(N–5–13)}$$

$$\frac{dT_t^{(1)}}{dx} - \frac{pU}{2C_t}(T_s - T_t^{(1)}) = 0 \qquad \frac{dT_t^{(2)}}{dx} + \frac{pU}{2C_t}(T_s - T_t^{(2)}) = 0 \qquad \text{(N–5–14,15)}$$

These three first-order ordinary differential equations must satisfy three boundary conditions of the form

$$T_s = T_{s,i} \qquad T_t^{(1)} = T_{t,i} \qquad T_t^{(2)} = T_{t,o} \qquad \text{at } x = 0 \qquad \text{(N–5–16a,b,c)}$$

In addition, we note that

$$T_s = T_{s,o} \qquad T_t^{(1)} = T_t^{(2)} \qquad \text{at } x = L \qquad \text{(N–5–17a,b)}$$

and, with the tubular fluid t taken as the reference stream I,†

$$R = \frac{C_t}{C_s} = \frac{T_{s,i} - T_{s,o}}{T_{t,o} - T_{t,i}} \qquad \text{(N–5–18)}$$

from Eqs. (N–5–1) and (N–5–2). We now want to develop the solution to this system of equations for the case in which U, C_s, and C_t are constant throughout the exchanger.

Following the solution approach used by Underwood [49] and Kern and Kraus [50], we first differentiate Eq. (N–5–13) to obtain

$$\frac{d^2T_s}{dx^2} + \frac{pU}{C_s}\frac{dT_s}{dx} - \frac{pU}{2C_s}\left(\frac{dT_t^{(1)}}{dx} + \frac{dT_t^{(2)}}{dx}\right) = 0 \qquad \text{(N–5–19)}$$

Using Eqs. (N–5–14) and (N–5–15) to eliminate $dT_t^{(1)}/dx$ and $dT_t^{(2)}/dx$, this equation is put into the form

$$\frac{d^2T_s}{dx^2} + \frac{pU}{C_s}\frac{dT_s}{dx} - \frac{1}{R}\left(\frac{pU}{2C_s}\right)^2 (T_t^{(2)} - T_t^{(1)}) = 0 \qquad \text{(N–5–20)}$$

To express $T_t^{(2)} - T_t^{(1)}$ in terms of T_s, we consider the energy transfer from the shell and to the tubular fluid over the region x to L, which must satisfy the relation

$$q_{x-L} = C_s(T_s - T_{s,o}) = C_t(T_t^{(2)} - T_t^{(1)}) \qquad \text{(N–5–21)}$$

Thus,

$$T_t^{(2)} - T_t^{(1)} = \frac{C_s}{C_t}(T_s - T_{s,o}) = \frac{T_s - T_{s,o}}{R} \qquad \text{(N–5–22)}$$

Using this important result, Eq. (N–5–20) becomes

† As we shall see, the solution result for this particular configuration is independent of which fluid is taken as the reference stream I. For this reason, the 1–2 exchanger is classified as a symmetrical exchanger.

$$\frac{d^2 T_s}{dx^2} + \frac{pU}{C_s}\frac{dT_s}{dx} - \left(\frac{pU}{2RC_s}\right)^2 T_s = -\left(\frac{pU}{2RC_s}\right)^2 T_{s,o} \tag{N–5–23}$$

The solution to this second-order nonhomogeneous equation can be easily obtained by (1) assuming a homogeneous solution of the exponential form $T_s = B\,e^{bx}$, and (2) assuming a particular solution of the form $T_s = T_{s,o}$. The resulting solution is given by

$$T_s = T_{s,o} + B_1\,e^{b_1 x} + B_2\,e^{b_2 x} \tag{N–5–24}$$

where

$$b_1 = -\frac{pU}{2C_s}(1 + \sqrt{1 + R^{-2}}) = -\frac{pU}{2C_t}(R + \sqrt{1 + R^2}) \tag{N–5–25a}$$

$$b_2 = -\frac{pU}{2C_s}(1 - \sqrt{1 + R^{-2}}) = -\frac{pU}{2C_t}(R - \sqrt{1 + R^2}) \tag{N–5–25b}$$

To evaluate the constants of integration B_1 and B_2, the boundary conditions given by Eq. (N–5–16) must be satisfied. Setting $T_s = T_{s,i}$ at $x = 0$, we see that

$$B_1 + B_2 = T_{s,i} - T_{s,o} \tag{N–5–26}$$

To satisfy the requirements $T_t^{(1)} = T_{t,i}$ and $T_t^{(2)} = T_{t,o}$ at $x = 0$, we substitute Eq. (N–5–24) into Eq. (N–5–13) and set $x = 0$, with the result

$$B_1 b_1 + B_2 b_2 + \frac{pU}{C_s} T_{s,i} - \frac{pU}{2C_s}(T_{t,i} + T_{t,o}) = 0 \tag{N–5–27}$$

Solving Eqs. (N–5–26) and (N–5–27) for B_1, we obtain

$$B_1 = \frac{-1}{2\sqrt{1 + R^2}}\Big[(R - \sqrt{R^2 + 1})(T_{s,i} - T_{s,o}) + R(T_{t,i} + T_{t,o}) - 2RT_{s,i}\Big] \tag{N–5–28}$$

Dividing through by $R = (T_{s,i} - T_{s,o})/(T_{t,o} - T_{t,i})$ and noting that

$$P = \frac{T_{t,o} - T_{t,i}}{T_{s,i} - T_{t,i}} \qquad 1 - P = \frac{T_{s,i} - T_{t,o}}{T_{s,i} - T_{t,i}} \tag{N–5–29,30}$$

Eq. (N–5–28) is put into the form

$$\begin{aligned} B_1 &= \frac{-1}{2\sqrt{1 + R^2}}\Big[(R - \sqrt{1 + R^2})\,(T_{t,o} - T_{t,i}) \\ &\quad + (T_{t,i} + T_{s,i}) + (T_{t,o} - T_{s,i})\Big] \\ &= \frac{T_{s,i} - T_{t,i}}{2\sqrt{1 + R^2}}\,[P(R - \sqrt{1 + R^2}) - 1 - (1 - P)] \\ &= -\frac{T_{s,i} - T_{t,i}}{2\sqrt{1 + R^2}}\,[2 - P(1 + R - \sqrt{1 + R^2})] \end{aligned} \tag{N–5–31}$$

Combining this result with Eq. (N–5–26), the relation for B_2 becomes

$$B_2 = \frac{T_{s,i} - T_{t,i}}{2\sqrt{1 + R^2}}[2 - P(1 + R + \sqrt{1 + R^2})] \qquad \text{(N–5–32)}$$

This completes the solution for the distribution in T_s. Although the solution for T_s given by Eq. (N–5–24) can be substituted into Eqs. (N–5–14) and (N–5–15) to obtain solutions for the distributions in $T_t^{(1)}$ and $T_t^{(2)}$, we will move directly to the use of Eq. (N–5–24) and the relations for b_1, b_2, B_1, and B_2 to develop practical heat-transfer relations in the context of the effectiveness format.

To develop an expression for NTU_t in terms of P and R, we set $T_s = T_{s,o}$ at $x = L$ in Eq. (N–5–24), with the result

$$T_s - T_{s,o} = B_1 e^{b_1 L} + B_2 e^{b_2 L} = 0 \qquad \text{(N–5–33)}$$

Rearranging, we obtain

$$-\frac{B_1}{B_2} = e^{(b_2 - b_1)L} \qquad \text{or} \qquad (b_2 - b_1)L = \ln\left(-\frac{B_1}{B_2}\right) \qquad \text{(N–5–34,35)}$$

which when combined with Eqs. (N–5–25a) and (N–5–25b) reduces to

$$\frac{\overline{U}A_s}{C_t} = \frac{1}{\sqrt{1 + R^2}}\ln\left(-\frac{B_1}{B_2}\right) \qquad \text{(N–5–36)}$$

Substituting for B_1 and B_2, this solution takes the form

$$NTU_t = \frac{1}{\sqrt{1 + R^2}}\ln\frac{2 - P(1 + R - \sqrt{1 + R^2})}{2 - P(1 + R + \sqrt{1 + R^2})} \qquad \text{(N–5–37)}$$

where P, R, and NTU_t are defined in terms of reference stream $I = t$ by

$$P = \frac{T_{t,o} - T_{t,i}}{T_{s,i} - T_{t,i}} \qquad R = \frac{C_t}{C_s} = \frac{T_{s,i} - T_{s,o}}{T_{t,o} - T_{t,i}} \qquad NTU_t = \frac{\overline{U}A_s}{C_t} \qquad \text{(N–5–38a,b,c)}$$

It should be noted that the same functional relation is obtained if the shell-side fluid s is specified as the reference stream I. That is,†

$$NTU_s = \frac{1}{\sqrt{1 + R^2}}\ln\frac{2 - P(1 + R - \sqrt{1 + R^2})}{2 - P(1 + R + \sqrt{1 + R^2})} \qquad \text{(N–5–39)}$$

where P, R, and NTU_s are defined in terms of reference stream $I = s$ by

$$P = \frac{T_{s,i} - T_{s,o}}{T_{s,i} - T_{t,i}} \qquad R = \frac{C_s}{C_t} = \frac{T_{t,o} - T_{t,i}}{T_{s,i} - T_{s,o}} \qquad NTU_s = \frac{\overline{U}A_s}{C_s} \qquad \text{(N–5–40a,b,c)}$$

The fact that Eqs. (N–5–37) and (N–5–39) are of the exact same form indicates that the 1–2 exchanger is *symmetrical*. Thus, the solution can be represented more generally by Eq. (11–169),

† This result is also obtained by replacing R by $1/R$, NTU_t by $R\ NTU_s$, and P by PR in Eq. (N–5–37).

$$NTU_I = \frac{1}{\sqrt{1 + R^2}} \ln \frac{2 - P(1 + R - \sqrt{1 + R^2})}{2 - P(1 + R + \sqrt{1 + R^2})} \qquad \text{(N–5–41)}$$

where either fluid can be taken as the reference stream I.

To complete the picture, we note that the solution can be expressed in terms of $P(NTU_I,R)$ by Eq. (11–168), $LMTD$ correction factor F by Eq. (11–171) or (11–172), or efficiency ψ by Eq. (11–173) or (11–174). In addition, we find that the solution is independent of the direction of either flow stream (see Appendix N–4).

N–6 SHELL-AND-TUBE HEAT EXCHANGERS—DELAWARE METHOD

The Delaware method [51,52] for evaluating the shell-side coefficient of heat transfer $\bar{h}_s$ and pressure drop ΔP_s for shell-and-tube heat exchangers with TEMA-E shell and segmental baffles is introduced in Sec. 11–7. This method provides a practical means of accounting for the effects of leakage, bypass, baffle cut and pitch, tube pitch and layout, and other factors on the shell-side performance.

In this approach, $\bar{h}_s$ is represented by Eq. (11–181),†

$$\bar{h}_s = \bar{h}_i(J_c J_l J_b) \qquad \text{(N–6–1)}$$

where $\bar{h}_i$ represents the coefficient of heat transfer for pure crossflow in an ideal tube bank, and J_c, J_l, and J_b are correction factors that account for the effect of baffle cut, leakage, and bypass, and ΔP_s is represented by Eq. (11–193),

$$\Delta P_s = \Delta P_c + \Delta P_w + \Delta P_e \qquad \text{(N–6–2)}$$

where the crossflow pressure drop ΔP_c, window pressure drop ΔP_w, and end-zone pressure drop ΔP_e are functions of correction factors for leakage R_l and bypass R_b. The coefficient of heat transfer $\bar{h}_i$ and the crossflow pressure drop ΔP_c are based on solutions for pure crossflow in an ideal tube bank. Correlations for $\bar{h}_i$ and the friction factor f_i corresponding to ΔP_c that are recommended in the *Heat Exchanger Design Handbook* [53] are shown in Fig. A–N–6–1.

The pertinent geometric dimensions associated with tube layout and baffle cut/bundle fit are shown in Table A–N–6–1 and Fig. A–N–6–2. Geometric parameters related to tube layout, baffle cut and spacing, clearances L_{tb} (tube to baffle), L_{sb} (shell to baffles), and L_{bb} (shell to bundle), and number of pairs of sealing strips N_{ss} that are employed in the formal Delaware method for evaluating the thermal correction factors J_c, J_l, J_b and the pressure drop correction factors R_l and R_b are listed in Table A–N–6–2.

To simplify the approach, we approximate the crossflow area A_c by Eq. (11–179),‡

† The more general form of Eq. (N–6–1) is given by $\bar{h}_s = \bar{h}_i(J_c J_l J_b J_s J_r)$ where J_s accounts for the effects of end-baffle spacing at the inlet and outlet sections and J_r is associated with the buildup of a boundary layer thermal resistance for laminar flow with $Re_s \lessapprox 100$ [53].

‡ Referring to item (i) in Table A–N–6–2, this approximation is based on the assumption that $D_{ctl} \simeq D_{i,s}$ and L_{bb} is much smaller than the term $D_{i,s}(S_t - d_o)/S_{t,\text{eff}}$.

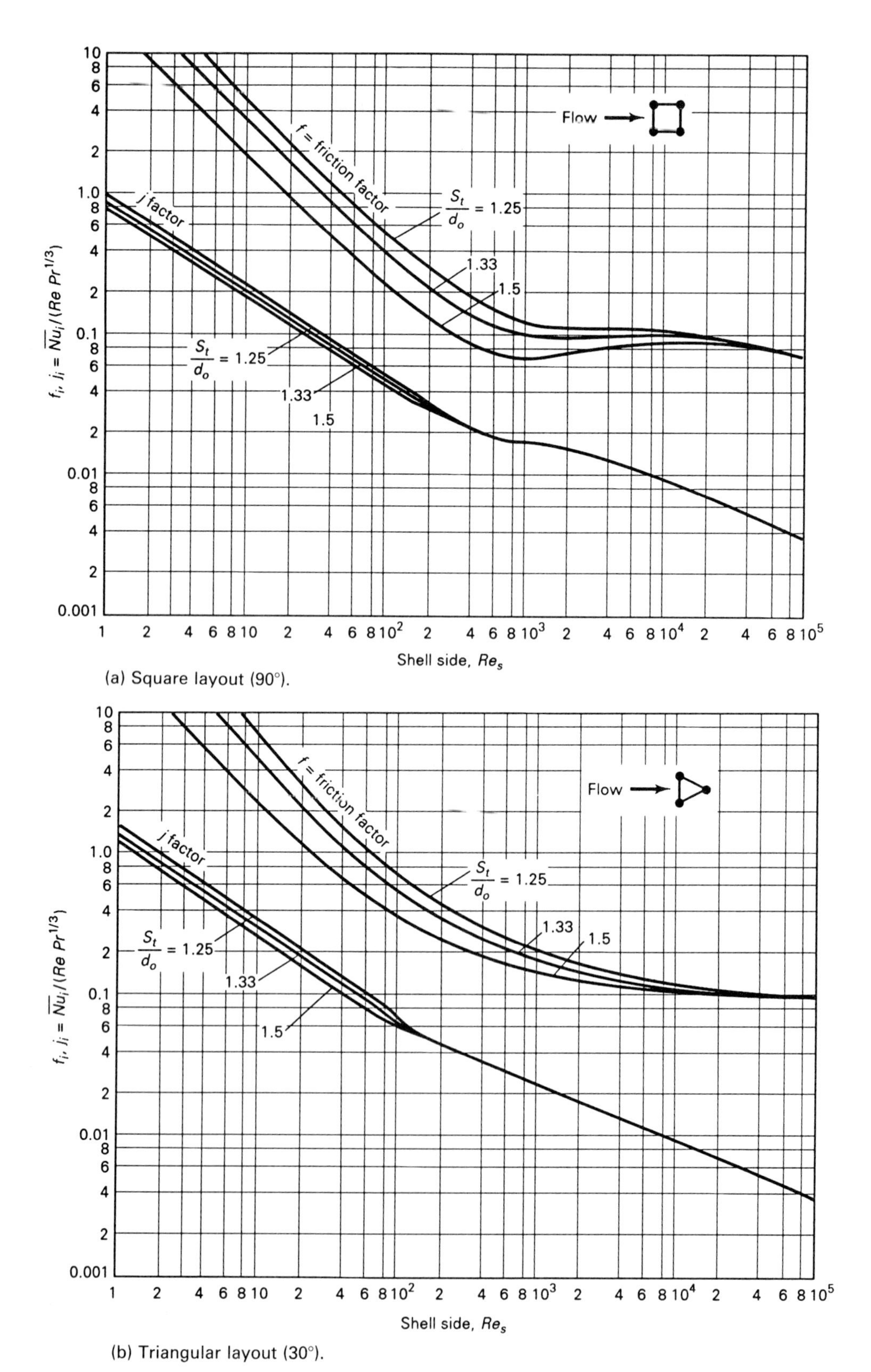

FIGURE A–N–6–1 Ideal tube bank convection correlations for $\overline{h}_i$ and f_i. (From *Heat Exchanger Design Handbook* [53]. Used with permission.)

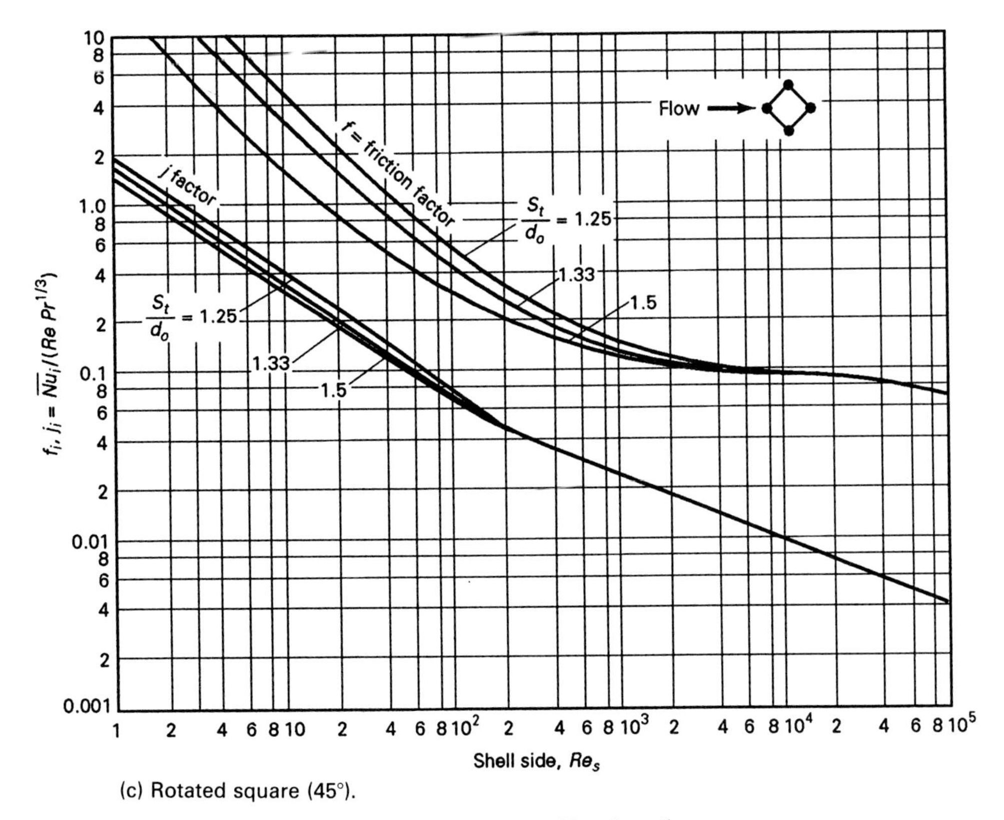

(c) Rotated square (45°).

FIGURE A-N-6-1 (Continued)

TABLE A-N-6-1 Geometric characteristics of basic tube layouts

Geometry			
θ_{tp}	30° Triangular	90° Square	45° Rotated square
S_{pn}	$0.5\ S_t$	S_t	$0.707\ S_t$
S_{pp}	$0.866\ S_t$	S_t	$0.707\ S_t$
$S_{t,\mathrm{eff}}$	S_t	S_t	$0.707\ S_t$

Source: From *Heat Exchanger Design Handbook* [53]. Used with permission.

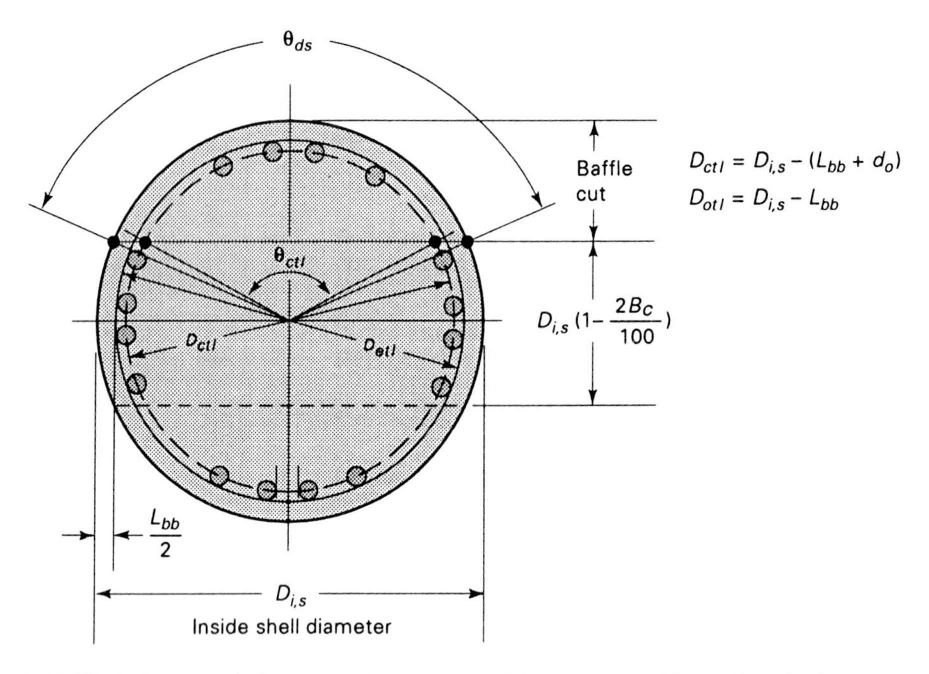

FIGURE A–N–6–2 Baffle cut and bundle–shell fit geometry. (From *Heat Exchanger Design Handbook* [53]. Used with permission.)

$$A_c = N_{tc}(S_t - d_o)L_b = \frac{D_{i,s}}{S_{t,\text{eff}}}(S_t - d_o)L_b \tag{N–6–3}$$

In addition, we assume that $\theta_{ctl} \simeq \theta_{ds}$ (see Fig. A–N–6–2), such that the key dimensionless geometric parameters can be approximated by†

$$r_{lm} = \frac{A_{tb} + A_{sb}}{A_c} \qquad r_s = \frac{A_{sb}}{A_{tb} + A_{sb}} \qquad r_b = \frac{A_{bb}}{A_c} \tag{N–6–4a,b,c}$$

$$N_{ss}^+ = \frac{N_{ss}}{N_{tcc}} = \frac{N_{ss}}{(D_{i,s}/S_{pp})(1 - 2B_c/100)} \qquad F_c = 1 - 2F_w \tag{N–6–4d,e}$$

where

$$A_{tb} = \frac{\pi N}{4}(1 - F_w)(2d_oL_{tb} + L_{tb}^2) \tag{N–6–4f}$$

$$A_{sb} = \frac{\pi}{2} D_{i,s}L_{sb}\frac{2\pi - \theta_{ds}}{2\pi} \qquad A_{bb} = L_bL_{bb} \tag{N–6–4g,h}$$

$$\theta_{ds} = 2\cos^{-1}\left(1 - \frac{2B_c}{100}\right) \qquad F_w = \frac{1}{2\pi}(\theta_{ds} - \sin\theta_{ds}) \tag{N–6–4i,j}$$

† The relation for r_b neglects the effects of F-stream bypass associated with tube pass partition lanes. Approaches to accounting for the effect of the F-stream bypass are introduced in reference 53.

TABLE A–N–6–2 Delaware method—formal defining relations for dimensional and dimensionless geometric and pressure drop parameters [53]

Classification	*Relations*
Geometric parameters, dimensional	
(a) Segmental baffle window angles (radians)	$\theta_{ds} = 2\cos^{-1}(1 - 2B_c/100)$ $\theta_{ctl} = 2\cos^{-1}\left[\frac{D_{i,s}}{D_{ctl}}\left(1 - \frac{2B_c}{100}\right)\right]$
(b) Baffle window flow areas	$A_w = \frac{\pi}{4}\left(D_{i,s}^2 \frac{\theta_{ds} - \sin\theta_{ds}}{2\pi} - NF_w d_o^2\right)$
Fraction of tubes in one window	$F_w = \frac{1}{2\pi}(\theta_{ctl} - \sin\theta_{ctl})$
Fraction of tubes in pure crossflow between baffle tips	$F_c = 1 - 2F_w$
(c) Equivalent hydraulic diameter for the window	Used to evaluate ΔP_w for $Re_s \lessapprox 100$. See [53].
(d) Number of effective tube rows in crossflow	$N_{tcc} = \frac{D_{i,s}}{S_{pp}}\left(1 - \frac{2B_c}{100}\right)$ $N_{tcw} = 0.4\frac{D_{i,s}}{S_{pp}}\left[\frac{2B_c}{100} - (D_{i,s} - D_{ctl})\right]$
(e) Number of baffles	$N_b = L/L_b - 1$
(f) Bundle-shell bypass area†	$A_{bb} = L_b L_{bb}$
(g) Shell-baffle leakage area	$A_{sb} = \pi D_{i,s}\frac{L_{sb}}{2}\frac{2\pi - \theta_{ds}}{2\pi}$
(h) Tube baffle leakage area	$A_{tb} = \frac{\pi}{4}N(1 - F_w)(2d_o L_{tb} + L_{tb}^2)$
(i) Bundle crossflow area	$A_c = \left[L_{bb} + \frac{D_{ctl}}{S_{t,\mathrm{eff}}}(S_t - d_o)\right]L_b$ $r_{lm} = \frac{A_{sb} + A_{tb}}{A_c} \quad r_s = \frac{A_{sb}}{A_{sb} + A_{tb}}$ $r_b = A_{ba}/A_c \quad N_{ss}^+ = N_{ss}/N_{tcc}$
Pressure drop	$\Delta P_i = 2N_{tcc} f_i G_s^2/\rho_s \quad \Delta P_c = \Delta P_i (N_b - 1)R_b R_l$ $\Delta P_w = N_b(2 + 0.6N_{tcw})\frac{G_w^2}{2\rho_s}R_l$ for $Re_s \gtrapprox 100$ see [53] for $Re_s \lessapprox 100$ $\Delta P_e = \Delta P_i\left(1 + \frac{N_{tcw}}{N_{tcc}}\right)R_b R_s$ $G_w = \dot{m}_s/(A_c A_w)^{1/2}$ $R_s = (L_b/L_{b,o})^{2-n} + (L_b/L_{b,i})^{2-n}$ $n = 0.2$ for $Re_s \gtrapprox 100$ $n = 1.0$ for $Re_s \lessapprox 100$

† The relation for A_{bb} neglects the effects of F-stream bypass associated with tube pass partition lanes. Approaches to accounting for the effect of the F-stream bypass are introduced in reference 53.

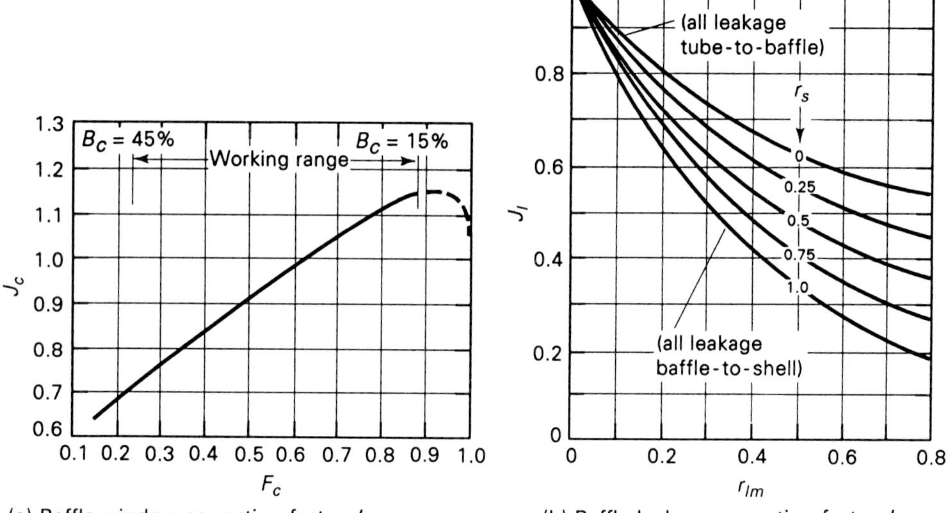

(a) Baffle window correction factor J_c.

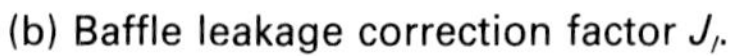

(b) Baffle leakage correction factor J_l.

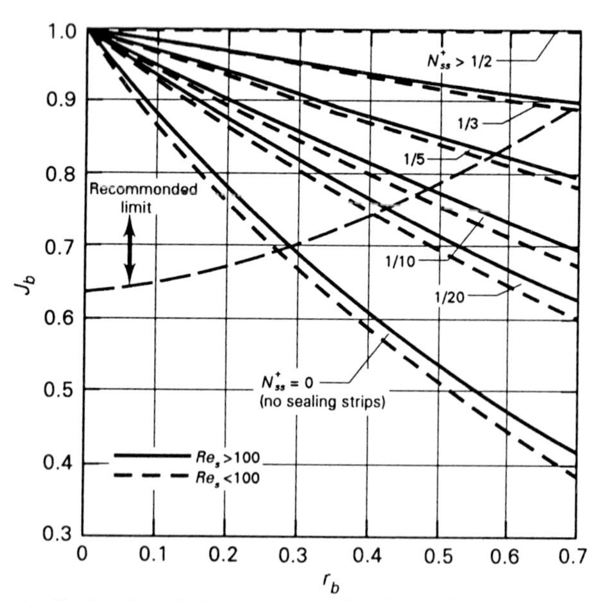

(c) Shell-to-bundle bypass correction factor J_b.

FIGURE A–N–6–3 Correction factors for shell-side heat transfer. (From *Heat Exchanger Design Handbood* [53]. Used with permission.)

Using these parameters together with empirical correlations shown in Figs. A–N–6–3 and 4, the correction factors can be readily evaluated.

In connection with the evaluation of pressure drop, the crossflow pressure drop ΔP_c is expressed in terms of f_i by [see Eqs. (11–194) to (11–196)]

$$\Delta P_c = \Delta P_i \,(N_b - 1) R_l R_b \tag{N–6–5}$$

where

$$\Delta P_i = 4N_{tcc} \frac{f_i}{2} \left(\frac{G^2}{\rho}\right)_s = 4 \frac{D_{i,s}}{S_{pp}} \left(1 - \frac{2B_c}{100}\right) \frac{f_i}{2} \left(\frac{G^2}{\rho}\right)_s \tag{N–6–6}$$

The relations for window pressure drop and end-zone pressure drop can be approximated by

$$\Delta P_w = N_b \left(2 + 0.48 \frac{B_c}{100} \frac{D_{i,s}}{S_{pp}}\right) \frac{G_w^2}{2\rho_s} R_l \tag{N–6–7}$$

for turbulent flow ($Re_s \gtrsim 100$), where

$$G_w = \frac{\dot{m}_s}{(A_c A_w)^{1/2}} \qquad A_w = \frac{\pi}{4}\left[\frac{D_{i,s}^2}{2\pi}(\theta_{ds} - \sin\theta_{ds}) - NF_w d_o^2\right] \tag{N–6–8,9}$$

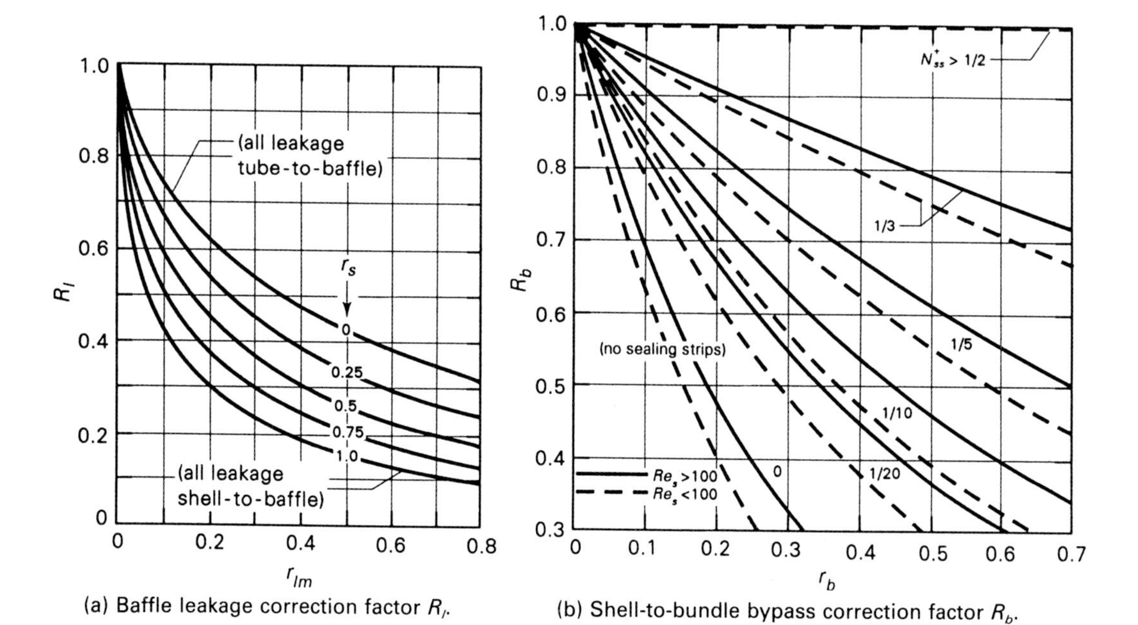

(a) Baffle leakage correction factor R_l.

(b) Shell-to-bundle bypass correction factor R_b.

FIGURE A–N–6–4 Correction factors for shell-side pressure drop. (From *Heat Exchanger Design Handbook* [53]. Used with permission.)

and

$$\Delta P_e = \Delta P_i \left(1 + \frac{0.8B_c}{100 - 2B_c}\right) R_b R_s \tag{N-6-10}$$

$$R_s = \left(\frac{L_b}{L_{b,o}}\right)^{2-n} + \left(\frac{L_b}{L_{b,i}}\right)^{2-n} \tag{N-6-11}$$

where $n = 0.2$ for turbulent flow ($Re_s \gtrsim 100$) and $n = 1$ for laminar flow.

As indicated in Sec. 11–7, the Delaware method can be adapted to low-fin tubes by using the equivalent diameter d_{req} in place of d_o in the defining relations for A_c, G_s, Re_s, and $\overline{Nu}_s$ and by employing Eqs. (11–176) and (11–179).

Finally, we note that simple methods I and II for evaluating $\bar{h}_s$ and ΔP_s, which are introduced in Sec. 11–7, involve the use of the correlations for f_i and $\bar{h}_i$ given in Fig. A–N–6–1. In addition, simple method II involves the use of Eqs. (N–6–3) and (N–6–4) and the correlations for correction factors given in Figs. A–N–6–3 and A–N–6–4.

N–7 CROSSFLOW HEAT EXCHANGERS— PRACTICAL THERMAL ANALYSIS FOR MIXED AND UNMIXED STREAMS

The distributions in terminal temperatures are sketched in Fig. A–N–7–1 for crossflow of a hot mixed fluid s and a cold unmixed fluid t. Whereas the distribution in T_s is a function of x alone, T_t varies with both x and z for this arrangement. The practical thermal analysis of this interesting problem involves the following steps:

(1) development of differential formulation for the unmixed fluid t;
(2) solution for variation in bulk-stream temperature distribution T_t with z, holding x and T_s constant, and evaluation of outlet temperature $T_{t,o}$ assuming $\dot{m}_t$ and $c_{P,t}$ are constant and $U = U(z)$;
(3) development of differential formulation for mixed fluid s;
(4) solution for distribution in T_s with x and evaluation of $T_{s,o}$ assuming $\dot{m}_s$ and $c_{P,s}$ are constant and U is independent of x;
(5) use of lumped formulation to express solution in terms of P (or ϵ), F, or ψ.

Referring to Fig. A–N–7–1, the differential formulation for energy transfer within the unmixed cold fluid is given by†

$$(d\dot{m}\, c_P)_t\, dT_t = dq \tag{N-7-1}$$

where $d\dot{m}_t = (\rho U_b \delta)_t\, dx = \dot{m}_t\, dx/L$, $\dot{m}_t = (\rho U_b \delta)_t L$, $dA_s = dz\, dx$, and

$$dq = U\, dA_s\, (T_s - T_t) \tag{N-7-2}$$

† A similar analysis to this problem is presented by Kays and London [39].

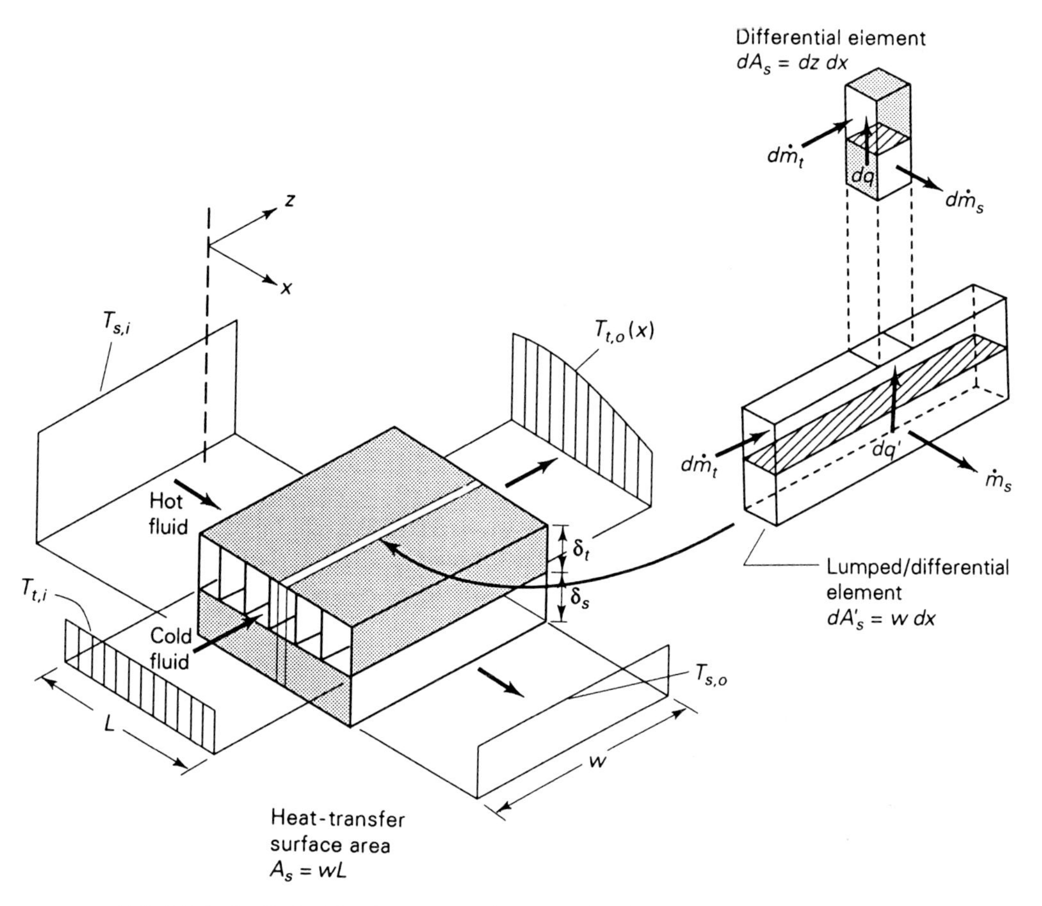

FIGURE A–N–7–1 Sketch for rate of heat transfer dq and distributions in terminal temperature for hot mixed s and cold unmixed t cross-flowing fluids.

Combining Eqs. (N–7–1) and (N–7–2) and substituting for $d\dot{m}_t$ and dA_s, we obtain

$$(\dot{m}c_P)_t \frac{dx}{L}\, dT_t = U\, dz\, dx\, (T_s - T_t) \qquad \frac{dT_t}{T_s - T_t} = \frac{UL\, dz}{C_t} \qquad \text{(N–7–3,4)}$$

where $C_t = (\dot{m}c_P)_t$. Setting $T_t = T_{t,i}$ at $x = 0$ and holding T_s and C_t constant, the solution to Eq. (N–7–4) becomes

$$\ln \frac{T_t - T_s}{T_{t,i} - T_s} = -\frac{L}{C_t}\int_0^z U\, dz = -\frac{L\overline{U}z}{C_t}$$

$$\frac{T_s - T_t}{T_s - T_{t,i}} = \exp\left(-\frac{\overline{U}Lz}{C_t}\right) \qquad \text{(N–7–5)}$$

where $\overline{U}$ is the mean overall coefficient over the length z,

$$\overline{U} = \frac{1}{z}\int_0^z U\,dz \tag{N-7-6}$$

To obtain $T_{t,o}$, we set $z = w$, with the result

$$\frac{T_s - T_{t,o}}{T_s - T_{t,i}} = \exp\left(-\frac{\overline{U}Lw}{C_t}\right) = \exp\left(-\frac{\overline{U}A_s}{C_t}\right) \tag{N-7-7}$$

Rearranging, this equation is put into the form

$$\frac{T_{t,o} - T_{t,i}}{T_s - T_{t,i}} = 1 - \exp\left(-\frac{\overline{U}A_s}{C_t}\right) = \Gamma \tag{N-7-8}$$

where the right side of this equation, which is represented by Γ, is constant.

Using this result, the rate of heat transfer dq' across the lumped/differential surface area $dA_s' = w\,dx$ is represented by

$$dq' = (d\dot{m}\,c_P)_t(T_{t,o} - T_{t,i}) = \Gamma C_t \frac{dx}{L}(T_s - T_{t,i}) \tag{N-7-9}$$

where $(d\dot{m}\,c_P)_t = dC_t = (\dot{m}c_P)_t\,dx/L = C_t\,dx/L$. The heat-transfer rate dq' is also expressed in terms of the change in T_s by

$$dq' = -(\dot{m}c_P)_s\,dT_s = -C_s\,dT_s \tag{N-7-10}$$

where $\dot{m}_s = (\rho U_b \delta)_s w$. Equations (N-7-9) and (N-7-10) are readily combined to obtain

$$-C_s\,dT_s = \Gamma C_t \frac{dx}{L}(T_s - T_{t,i}) \qquad \frac{dT_s}{T_s - T_{t,i}} = -\Gamma\frac{C_t}{C_s}\frac{dx}{L} \tag{N-7-11,12}$$

The solution to this equation for the distribution in T_s is

$$\frac{T_s - T_{t,i}}{T_{s,i} - T_{t,i}} = \exp\left(-\Gamma\frac{C_t}{C_s}\frac{x}{L}\right) \tag{N-7-13}$$

such that $T_{s,o}$ becomes

$$\frac{T_{s,o} - T_{t,i}}{T_{s,i} - T_{t,i}} = \exp\left(-\Gamma\frac{C_t}{C_s}\right) \tag{N-7-14}$$

$$\frac{T_{s,i} - T_{s,o}}{T_{s,i} - T_{t,i}} = 1 - \exp\left(-\Gamma\frac{C_t}{C_s}\right) \tag{N-7-15}$$

which is uniform as shown in Fig. A–N–7–1.

Finally, the distribution in T_t is obtained by combining Eqs. (N-7-5) and (N-7-13), with the result

$$\frac{T_t - T_{t,i}}{T_{s,i} - T_{t,i}} = \left[1 - \exp\left(-\frac{\overline{U}Lz}{C_t}\right)\right]\exp\left(-\Gamma\frac{C_t}{C_s}\frac{x}{L}\right) \tag{N-7-16}$$

It follows that the distribution in $T_{t,o}(x)$ is given by

$$\frac{T_{t,o}(x) - T_{t,i}}{T_{s,i} - T_{t,i}} = \left[1 - \exp\left(-\frac{\overline{U}A_s}{C_t}\right)\right] \exp\left(-\Gamma\frac{C_t}{C_s}\frac{x}{L}\right) \tag{N-7-17}$$

which is consistent with the nonuniform distribution shown in Fig. A–N–7–1.

The total rate of heat transfer q can be obtained by applying the lumped formulation for the hot mixed fluid given by Eq. (11–24),

$$q = C_s(T_{s,i} - T_{s,o}) \tag{N-7-18}$$

Using Eq. (N–7–15) for $T_{s,i} - T_{s,o}$, we obtain

$$q = C_s(T_{s,i} - T_{t,i})\left[1 - \exp\left(-\Gamma\frac{C_t}{C_s}\right)\right] \tag{N-7-19}$$

This same result is also obtained by applying the lumped formulation for the unmixed cold fluid, with the outlet temperature averaged over the length L; that is,

$$q = C_t(T_{t,o} - T_{t,i}) \tag{N-7-20}$$

where $T_{t,o}$ is defined by

$$T_{t,o} = \frac{1}{L}\int_0^L T_{t,o}(x)\,dx \tag{N-7-21}$$

To see this point, we operate on Eq. (N–7–17) to obtain $T_{t,o}$.

$$\begin{aligned}\frac{T_{t,o} - T_{t,i}}{T_{s,i} - T_{t,i}} &= \frac{C_s}{C_t\Gamma}\left[1 - \exp\left(-\frac{\overline{U}A_s}{C_t}\right)\right]\left[1 - \exp\left(-\Gamma\frac{C_t}{C_s}\right)\right] \\ &= \frac{C_s}{C_t}\left[1 - \exp\left(-\Gamma\frac{C_t}{C_s}\right)\right]\end{aligned} \tag{N-7-22}$$

Thus, in the use of the effectiveness, *LMTD*, and efficiency approaches for crossflow heat exchangers, we employ the mean (or mixed) terminal temperatures for both mixed and unmixed streams. Equations (N–7–20) and (N–7–22) clearly lead to Eq. (N–7–19). With this understanding, our solution results are readily expressed in the effectiveness, *LMTD*, and efficiency formats.

Using Eq. (N–7–19) and taking the cold unmixed fluid t as the reference stream I, the effectiveness P becomes

$$\begin{aligned}P &= \frac{q}{q_{\max,I}} = \frac{C_s(T_{s,i} - T_{t,i})[1 - \exp(-\Gamma C_t/C_s)]}{C_t(T_{s,i} - T_{t,i})} \\ &= \frac{1}{R}[1 - \exp(-\Gamma R)]\end{aligned} \tag{N-7-23a}$$

where

$$\Gamma = 1 - \exp\left(-\frac{\overline{U}A_s}{C_t}\right) = 1 - \exp\left(-NTU_t\right) \qquad \text{(N-7-23b)}$$

and P, R, and NTU_t are defined by

$$P = \frac{T_{t,o} - T_{t,i}}{T_{s,i} - T_{t,i}} \qquad R = \frac{C_t}{C_s} = \frac{T_{s,i} - T_{s,o}}{T_{t,o} - T_{t,i}} \qquad NTU_t = \frac{\overline{U}A_s}{C_t} \qquad \text{(N-7-24a,b,c)}$$

Alternatively, with the hot mixed fluid s taken as the reference stream I, the solution for P takes the form

$$P = \frac{q}{q_{\max,I}} = \frac{C_s(T_{s,i} - T_{t,i})[1 - \exp(-\Gamma C_t/C_s)]}{C_s(T_{s,i} - T_{t,i})} = 1 - \exp\left(-\frac{\Gamma}{R}\right) \qquad \text{(N-7-25a)}$$

where

$$\Gamma = 1 - \exp(-R\ NTU_s) \qquad \text{(N-7-25b)}$$

TABLE A-N-7-1 Comparison of functional relations for $P(NTU_I,R)$ given by Eqs. (N-7-23) and (N-7-25)

R	NTU_I	P			ψ	
		Eq. (N-7-23)	Eq. (N-7-25)	% Difference	$I = t$	$I = s$
0.25	1.00000	0.58470	0.58720	−0.00427	0.58470	0.58720
	2.00000	0.77759	0.79276	−0.01950	0.38880	0.39638
	2.50000	0.82021	0.84416	−0.02921	0.32808	0.33767
	3.00000	0.84578	0.87883	−0.03907	0.28193	0.29294
	4.00000	0.87050	0.92022	−0.05712	0.21763	0.23005
	5.00000	0.87954	0.94239	−0.07145	0.17591	0.18848
4.0	0.20000	0.12893	0.12861	0.00245	0.64464	0.64306
	0.40000	0.18313	0.18088	0.01228	0.45783	0.45220
	0.50000	0.19819	0.19440	0.01913	0.39638	0.38880
	0.60000	0.20887	0.20333	0.02651	0.34812	0.33889
	0.70000	0.21663	0.20927	0.03396	0.30946	0.29896
	0.80000	0.22237	0.21322	0.04115	0.27797	0.26653
	1.00000	0.23006	0.21763	0.05403	0.23005	0.21762
	2.00000	0.24213	0.22113	0.08672	0.12107	0.11057
	3.00000	0.24441	0.22120	0.09498	0.08147	0.07373
	4.00000	0.24507	0.22120	0.09742	0.06127	0.05530
	5.00000	0.24530	0.22120	0.09824	0.04906	0.04424

and P, R, and NTU_t are defined by

$$P = \frac{T_{s,i} - T_{s,o}}{T_{s,i} - T_{t,i}} \qquad R = \frac{C_s}{C_t} = \frac{T_{t,o} - T_{t,i}}{T_{s,i} - T_{s,o}} \qquad NTU_s = \frac{\overline{U}A_s}{C_s} \qquad \text{(N–7–26a,b,c)}$$

Comparing Eqs. (N–7–23) and (N–7–25), we find that the form of the functional relation for $P(NTU_I,R)$ depends on which fluid is taken as the reference stream I. Therefore, this mixed–unmixed crossflow arrangement is classified as a *nonsymmetrical exchanger*. However, the difference in calculations for P obtained from Eqs. (N–7–23) and (N–7–25) for specific values of R and NTU_I is quite small for normal operating conditions. Referring to the calculations shown in Table A–N–7–1, the difference is within about 2.5% for values of $\psi \gtrsim 0.35$. Therefore, the functional relations given by Eqs. (N–7–23) and (N–7–25) may be considered to be *approximately interchangeable*.

N–8 MULTIPASS HEAT EXCHANGER NETWORKS—RELATIONS FOR EFFECTIVENESS

To establish general relations for the overall effectiveness P associated with category I series arrangements, which are characterized by

$$NTU_I = M\ NTU_{I,1} \qquad R = R_1 \qquad \text{(N–8–1,2)}$$

we consider double-pipe heat exchanger units, with the solution for P_1 represented by the functional relation

$$P_1 = \text{fn}_1\ (NTU_{I,1},R_1) \qquad \text{(N–8–3)}$$

Because of geometric considerations, we recognize that the relations for P_1 can be used for a series of double-pipe heat exchangers by merely replacing $A_{s,1}$ by A_s (or $NTU_{I,1}$ by NTU_I), such that

$$P = \text{fn}_1\ (NTU_I,R) \qquad \text{(N–8–4)}$$

Focusing attention on overall counterflow arrangements, we note that

$$P_1 = \frac{1 - \exp\,[NTU_{I,1}\,(1 - R_1)]}{R_1 - \exp\,[NTU_{I,1}\,(1 - R_1)]} \qquad \text{(N–8–5)}$$

from Eq. (11–138b), or

$$NTU_{I,1} = \frac{1}{1 - R_1} \ln \frac{1 - P_1R_1}{1 - P_1} \qquad \text{(N–8–6)}$$

from Eq. (11–140). It follows that

$$P = \frac{1 - \exp\,[NTU_I\,(1 - R)]}{R - \exp\,[NTU_I\,(1 - R)]} \qquad \text{(N–8–7)}$$

from Eqs. (N–8–3) to (N–8–6). Substituting $NTU_I = M\ NTU_{I,1}$ and $R = R_1$,

Eq. (N–8–6) indicates

$$NTU_I = \frac{M}{1 - R} \ln \frac{1 - P_1 R}{1 - P_1} \tag{N–8–8}$$

Equations (N–8–7) and (N–8–8) can now be combined to obtain a general relation for P in terms of P_I of the form

$$P = \frac{[(1 - P_1 R)/(1 - P_1)]^M - 1}{[(1 - P_1 R)/(1 - P_1)]^M - R} \tag{N–8–9}$$

which is applicable to overall counterflow category I series arrangements. Equations (N–8–9) can also be put into the form

$$P_1 = \frac{[(1 - PR)/(1 - P)]^{1/M} - 1}{[(1 - PR)/(1 - P)]^{1/M} - R} \tag{N–8–10}$$

Similarly, Eqs. (N–8–1) to (N–8–4) can be combined with Eq. (11–137) for parallel flow to obtain the general relations

$$P = \frac{1}{1 + R}\{1 - [1 - (1 + R)P_1]^M\} \tag{N–8–11}$$

$$P_1 = \frac{1}{1 + R}\{1 - [1 - (1 + R)P]^{1/M}\} \tag{N–8–12}$$

which are applicable to overall parallel category I series arrangements.

Equations (N–8–9) to (N–8–12) prove to be applicable to category I series arrangements of double-pipe, shell-and-tube, and crossflow heat exchangers. These equations are listed together with relations developed by Gardner [54] for series–parallel arrangements in Table 11–11.

N–9 EFFICIENCY METHOD—SOLUTION OF REPRESENTATIVE EXAMPLES

EXAMPLE 11–12

A shell-and-tube heat exchanger with one shell pass and two tube passes heats water at 15°C with a mass-flow rate of 9.94 kg/s. Heating oil [c_P = 2.5 kJ/(kg °C)] enters the tubes at 80°C and exits at 35°C with a mass-flow rate of 5 kg/s. Determine the surface area of the heat exchanger if the mean overall coefficient of heat transfer is 300 W/(m² °C). Also determine whether or not the minimum performance criteria are satisfied.

Solution

Objective Determine A_s and whether or not $\psi > 0.35$ and/or $P < P_o$.

Schematic Shell-and-tube heat exchanger: 1-2 pass arrangement.

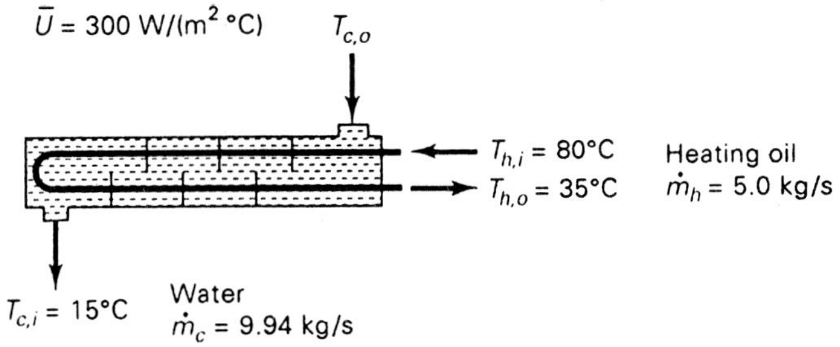

$\bar{U}$ = 300 W/(m² °C) $T_{c,o}$

$T_{h,i}$ = 80°C   Heating oil
$T_{h,o}$ = 35°C $\dot{m}_h$ = 5.0 kg/s

$T_{c,i}$ = 15°C   Water
$\dot{m}_c$ = 9.94 kg/s

Assumptions/Conditions

uniform distribution in U
uniform properties
standard conditions

Properties

Water at T = 15°C = 288 K (Table A–C–3): c_P = 4.19 kJ/(kg °C).
Heating oil: c_P = 2.5 kJ/(kg °C).

Analysis The heat-transfer rate q and outlet temperature $T_{c,o}$ of the water are obtained by writing

$$q = (\dot{m}c_P)_h(T_{h,i} - T_{h,o}) = 5\,\frac{\text{kg}}{\text{s}}\left(2.5\,\frac{\text{kJ}}{\text{kg °C}}\right)(80°\text{C} - 35°\text{C}) = 562 \text{ kW}$$

and

$$q = (\dot{m}c_P)_c(T_{c,o} - T_{c,i})$$

$$T_{c,o} = \frac{q}{(\dot{m}c_P)_c} + T_{c,i} = \frac{562 \text{ kW}}{(9.94 \text{ kg/s})[4.19 \text{ kJ/(kg °C)}]} + 15°\text{C} = 28.5°\text{C}$$

Following the efficiency approach, the unknown surface area A_s is expressed in terms of ψ by

$$A_s = \frac{q}{\psi \bar{U}(T_{h,i} - T_{c,i})} \tag{a}$$

from Eq. (11–65), where ψ can be obtained by use of Eq. (11–174) or the Mueller chart shown in Fig. 11–36.

To evaluate the efficiency ψ, we first select the cold fluid as the reference stream I and calculate P and R by writing

$$P = \frac{T_{c,o} - T_{c,i}}{T_{h,i} - T_{c,i}} = \frac{28.5°\text{C} - 15°\text{C}}{80°\text{C} - 15°\text{C}} = 0.208$$

$$R = \frac{T_{h,i} - T_{h,o}}{T_{c,o} - T_{c,i}} = \frac{80°\text{C} - 35°\text{C}}{28.5°\text{C} - 15°\text{C}} = 3.33$$

Substituting for P and R in Eq. (11–174), we obtain

$$\psi = \frac{P\sqrt{1 + R^2}}{\ln\{[2 - P(1 + R - \sqrt{1 + R^2})]/[2 - P(1 + R + \sqrt{1 + R^2})]\}}$$

$$= 0.459$$

Returning to Eq. (a), A_s becomes

$$A_s = \frac{562 \text{ kW}}{0.459[300 \text{ W/(m}^2 \text{ °C)}](80\text{°C} - 15\text{°C})} = 62.8 \text{ m}^2$$

In connection with the minimum performance criteria, we note that $\psi > \psi_{MIN} = 0.35$. In addition, $P_o = 1/(1 + R) = 0.231$ from Eq. (11–76), such that the alternative criterion $P < P_o$ is also satisfied.

EXAMPLE 11–18

An unmixed crossflow heat exchanger is to be constructed that will heat 2.5 kg/s of air at 1 atm from 15°C to 30°C. Hot water enters at 52.5°C. The mean overall coefficient of heat transfer is 300 W/(m² °C). Determine the required surface area to produce an outlet water temperature of 24°C.

Solution

Objective Determine required heat-transfer surface area A_s.

Schematic Crossflow heat exchanger: both fluids unmixed.

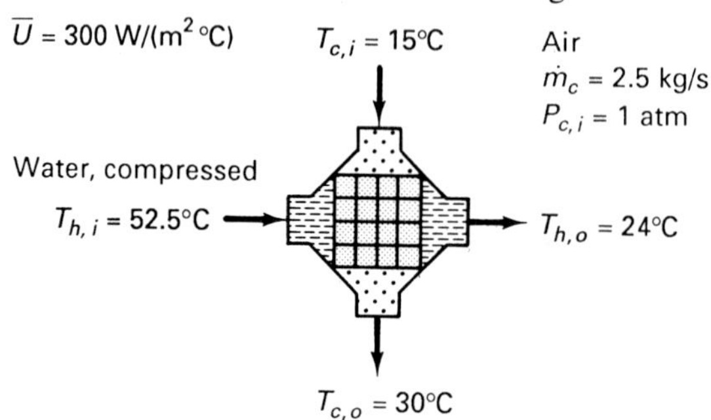

Assumptions/Conditions

uniform distribution in U
uniform properties
standard conditions

Properties

Air at $T = 15\text{°C} = 288$ K (Table A–C–5): $c_P = 1.01$ kJ/(kg °C).

Analysis Following the efficiency approach, the surface area A_s is expressed in terms of ψ by

$$A_s = \frac{q}{\psi \bar{U}(T_{h,i} - T_{c,i})} \tag{a}$$

from Eq. (11–65), where

$$q = (\dot{m}c_P)_c(T_{c,o} - T_{c,i}) = 2.5\,\frac{\text{kg}}{\text{s}}\left(1.01\,\frac{\text{kJ}}{\text{kg °C}}\right)(30\text{°C} - 15\text{°C}) = 37.9\text{ kW}$$

and ψ can be obtained from the Mueller chart shown in Fig. 11–42b.

To evaluate the efficiency ψ, we specify the cold fluid as the reference stream I and calculate P and R by writing

$$P = \frac{T_{c,o} - T_{c,i}}{T_{h,i} - T_{c,i}} = \frac{30\text{°C} - 15\text{°C}}{52.5\text{°C} - 15\text{°C}} = 0.4$$

$$R = \frac{T_{h,i} - T_{h,o}}{T_{c,o} - T_{c,i}} = \frac{52.5\text{°C} - 24\text{°C}}{30\text{°C} - 15\text{°C}} = 1.9$$

Using these inputs, Fig. 11–47b indicates $\psi \simeq 0.34$. Substituting into Eq. (a), we obtain

$$A_s = \frac{37.9\text{ kW}}{0.34[300\text{ W/(m}^2\text{ °C)}](52.5\text{°C} - 15\text{°C})} = 9.91\text{ m}^2$$

O HEAT EXCHANGERS: GEOMETRIC CHARACTERISTICS

Geometric characteristics associated with shell-and-tube heat exchangers are presented in this section for (1) tube counts, (2) clearance standards, and (3) low-fin tubing.

O–1 TUBE COUNTS—SHELL-AND-TUBE HEAT EXCHANGERS

TABLE A–O–1 Tube counts for 3/4-in. O.D. tubes on 1-in-square pitch—shell-and-tube heat exchangers with fixed tubesheet

$D_{i,s}$†		*Tube passes*				
(in.)	(mm)	1	2	4	6	8
8.07	205	40	36	24	20	
10	254	64	55	39	38	
12	305	92	86	71	68	67
13 1/4	337	115	108	94	85	80
15 1/4	387	154	145	125	122	115
17 1/4	438	200	192	172	166	160
19 1/4	489	251	246	220	212	203
21 1/4	540	315	300	275	263	254

TABLE A–O–1 (Tube counts for 3/4-in. O.D. tubes on 1-in-square pitch—shell-and-tube heat exchangers with fixed tubesheet—continued)

$D_{i,s}$†		*Tube passes*				
(in.)	(mm)	1	2	4	6	8
23 1/4	591	375	362	325	318	309
25	635	439	419	387	369	357
27	686	512	490	456	442	424
29	737	594	577	533	521	503
31	787	688	665	616	600	587
33	838	784	759	715	696	678
35	889	881	857	804	784	766
37	940	987	962	909	884	860
39	991	1106	1077	1013	1000	981
42	1070	1290	1260	1195	1175	1157

† Tube counts are also available for shell diameters of 45, 48, 54, 60, 66, 72, 78, 84, and 90 in.
Source: Adapted from Palen and Taborek [55].

O–2 CLEARANCE STANDARDS—SHELL-AND-TUBE HEAT EXCHANGERS

O–2–1 Tube-to-Baffle Clearance

The TEMA standard for tube-to-baffle clearance L_{tb} is stated as follows: "Where the maximum unsupported tube length is 36 in. (914 mm) or less, or for tubes larger in diameter than 1 1/4 in. (31.8 mm) O.D., standard tube holes are to be 1/32 in. (0.8 mm) over the O.D. of the tubes. Where the unsupported tube length exceeds 36 in. (914 mm) for tubes 1 1/4 in. (31.8 mm) diameter and smaller, standard tube holes are to be 1/64 in. (0.4 mm) over the O.D. of the tubes." This standard is represented by Fig. A–O–2–1.

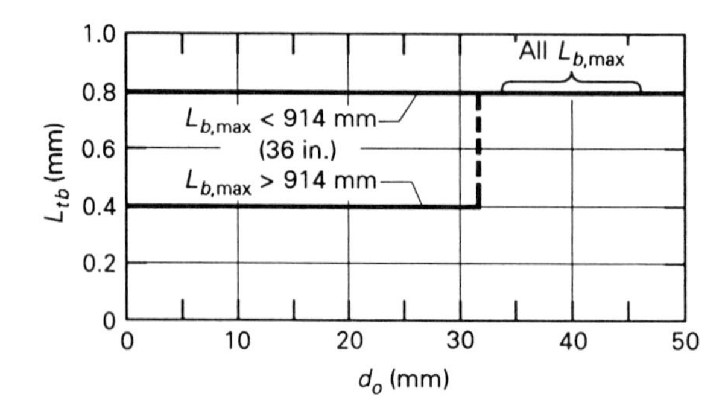

FIGURE A–O–2–1 Standard for tube-to-baffle clearance L_{tb}. (From *Heat Exchanger Design Handbook* [53]. Used with permission.)

O–2–2 Shell-to-Baffle Clearance

TEMA values for the average clearance between the shell wall and baffle L_{sb} are shown in Fig. A–O–2–2 as a function of shell diameter $D_{i,s}$. As indicated in reference 53, the TEMA step values can be approximated by a linear correlation. In practice, an additional clearance of 1.5 mm is commonly added as a safely factor to compensate for out-of-roundness tolerances that are not clearly specified by TEMA [53].

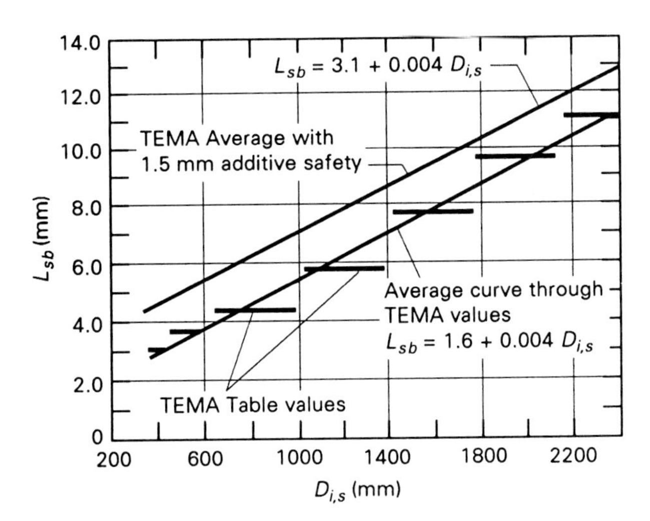

FIGURE A–O–2–2 Standard for shell-to-baffle clearance L_{sb}. (From *Heat Exchanger Design Handbook* [53]. Used with permission.)

O–2–3 Shell-to-Tube Bundle Clearance

Correlations recommended in reference 53 for shell-to-tube bundle clearance L_{bb} are shown in Fig. A–O–2–3. Notice that the standard for L_{bb} depends on the type tube bundle, with the floating-head bundles requiring considerably larger clearances (in order to facilitate bundle removal) than the fixed tubesheet and **U**-tube type bundles.

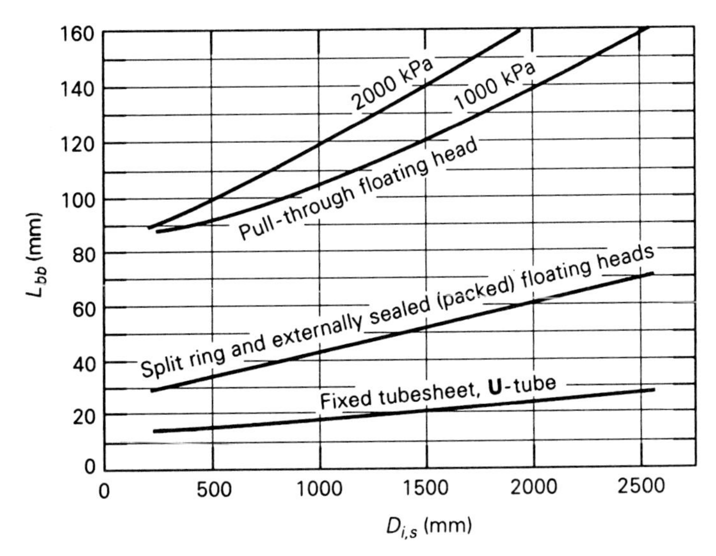

FIGURE A–O–2–3 Standard for shell-to-tube bundle clearance L_{bb}. (From *Heat Exchanger Design Handbook* [53]. Used with permission.)

O–3 LOW-FIN TUBE DIMENSIONS

TABLE A–O–3 Representative low-fin tube geometric dimensions: 19 fins per inch†

$(d_o)_o$		d_r	d_i	d_{req}	$A_{s,o}/L$	
(in.)	(mm)	(mm)	(mm)	(mm)	(m^2/m)	$A_{s,o}/A_{s,i}$
0.375	9.52	6.35	4.47	7.26	0.0686	4.80
0.500	12.7	9.52	7.90	10.4	0.0308	3.86
			6.58			4.63
0.625	15.9	12.7	11.3	13.6	0.125	3.51
			10.6			3.74
			9.75			4.05
			9.04			4.36
0.75	19.1	15.9	14.4	16.8	0.153	3.34
			13.7			3.52
			12.6			3.86
			11.7			4.16
			10.3			4.68
0.875	22.2	19.0	17.3	19.9	0.181	3.33
			16.6			3.47
			15.7			3.65
			14.8			3.87
			14.2			4.04
1.000	25.4	22.2	20.1	23.1	0.210	3.31
			18.9			3.51
			18.0			3.69
			16.7			3.98

† Fin thickness, $\delta_F = 0.011$ in. $= 0.279$ mm; fin height, $H_F = (d_o - d_r)/2 = 0.056$ in. $= 1.42$ mm; spacing, $S_F = 0.0416$ in. $= 1.06$ mm; wall thickness, $X_F = (d_r - d_i)/2$. The equivalent diameter d_{req} is given by $d_{req} = d_r + 2H_F\delta_F N_F = d_o - 2H_F S_F N_F$, where $N_F = 1/(S_F + \delta_F) = 19$ fins/in. $= 748$ fins/m.

Source: Adapted from Wolverine Tube Company. Used with permission.

NOMENCLATURE

■ FUNDAMENTAL UNITS†

□ SI Units

A	ampere (electric current I)
Å	angstrom (10^{-10} m; wavelength λ)
cm	centimeter (10^{-2} m; length L)
°C	degree centigrade (temperature T)
J	joule (N m; energy E)
kg	kilogram (mass m)
kJ	kilojoule (10^3 J; energy E)
kW	kilowatt (10^3 W; power $\dot{W}$)
K	kelvin (absolute temperature T)
m	meter (length L)
mm	millimeter (10^{-3} m; length L)
nm	nanometer (10^{-9} m; wavelength λ)
N	newton (kg m/s^2; force F)
Pa	Pascal (N/m^2; pressure P, stress τ)
rad	radian (plane angle)
s	second (time t)
sr	steradian (solid angle ω)
V	volt
W	watt (J/s, power $\dot{W}$)
μm	micrometer (10^{-6} m; wavelength λ)
Ω	ohm

□ English Units

Btu	British thermal unit (energy E)
deg	degree (plane angle)
ft	foot (length L)
h	hour (time t)
in.	inch (length L)
lb_f	pound force (32.2 lb_m ft/s^2; force F)
lb_m	pound mass (mass m)
°F	degree Fahrenheit (temperature T)
°R	degree Rankine (absolute temperature T)

■ PARAMETERS††

□ English Symbols

A	cross-sectional area, m^2 or ft^2
A_c	crossflow area for shell-side flow in heat exchanger, m^2 or ft^2
A_{min}	minimum free-flow area in tube bank or heat-exchanger core, m^2 or ft^2

† Roman typeset is used throughout the text for units.

†† *Italic* typeset is used throughout the text for parameters.

A_F area of fin surface, m^2 or ft^2

A_N area of body normal to direction of flow, m^2 or ft^2

A_o area of finned surface, m^2 or ft^2

A_p area of prime (unfinned) surface, m^2 or ft^2

A_s surface area, m^2 or ft^2

A_1 frontal area for tube bank or heat exchanger core, m^2 or ft^2

B constant in log law for T^+ [defined by Eq. (7–290)]

B_1 constant in 1/7th power law for T^+ [defined by Eq. (7–292)]

B_c percent baffle cut

c speed of light, m/s or ft/s

c_0 speed of light in vacuum, m/s or ft/s

c_P specific heat at constant pressure, J/(kg °C) or Btu/(lb_m °F)

c_v specific heat at constant volume, J/(kg °C) or Btu/(lb_m °F)

C thermal capacitance, J/°C or Btu/°F; absolute capacity rate for flow in heat exchanger core ($\dot{m}c_P$), kW/°C or Btu/(h °F)

C_A area coefficient for tube bank [defined by Eq. (8–136)]

C_D drag coefficient [defined by Eq. (8–123)]

C_e electrical capacitance, Farad (F)

C_H cost of thermal energy supplied to heat exchanger, \$

C_I capacity rate of reference fluid I, kW/°C or Btu/(h °F)

C_s capital cost [defined by Eq. (N–3–1)], \$

$C(r,t)$ dimensionless temperature for infinite cylinder [$(T - T_F)/(T_i - T_F)$]

C^* capacitance ratio (C_{min}/C_{max})

C_{min} minimum absolute capacity rate for flow in heat-exchanger core, kW/°C or Btu/(h °F)

C_{max} maximum absolute capacity rate for flow in heat-exchanger core, kW/°C or Btu/(h °F)

d diameter for inner tube of double-pipe arrangement, m or ft

d_r root diameter of low-fin tubing, m or ft

d_{req} equivalent diameter for low-fin tubing (see Table 11–14), m or ft

D diameter, m or ft

D_H hydraulic diameter ($4A/p_w$ or $4A_{min}/p_w$), m or ft

$D_{i,s}$ inside diameter of shell for double-pipe and shell-and-tube heat exchanger, m or ft

D_N height of body for crossflow, m or ft

e energy per unit mass, J/kg or Btu/lb_m; the number 2.7182818

e_m individual finite element

E energy, J or Btu; total emissive power, W/m^2 or Btu/(h ft^2)

E_b blackbody total emissive power, W/m^2 or Btu/(h ft^2)

E_e electrical voltage, V

E_λ monochromatic emissive power [defined by Eq. (5–4)], W/m^2 or Btu/(h ft^2)

$E_{0\to\lambda}$ subtotal emissive power [defined by Eq. (5–5)], W/(m^2 μm) or Btu/(h ft^2 μm)

f Fanning friction factor

f_i friction coefficient for tube bank [defined by Eq. (8–159)]

$f(r)$ laminar thermal boundary layer parameter [defined by Eq. (7–115)]

F force, N or lb_f; correction factor for *LMTD* [defined by Eq. (11–59)]

F_c correction factor for tube row effect [see Eq. (8–154)]

F_D drag force, N or lb_f

F_f fouling factor [defined by Eq. (11–71)], m^2 °C/W or ft^2 °F h/Btu

F_{s-R} radiation shape factor

$\mathscr{F}_{s-R}$ radiation factor

$F(r)$ laminar thermal boundary layer parameter [defined by Eq. (7–117)]

g gravitational acceleration, m/s^2 or ft/s^2

G irradiation, W/m^2 or Btu/(h ft^2)

G mass flux ($\dot{m}/A$), kg/(s m^2) or lb_m/(s ft^2)

G_s heat exchanger shell-side mass flux, kg/(s m^2) or lb_m/(h ft^2)

h — convection heat-transfer coefficient, W/(m² °C) or Btu/(h ft² °F)

h_f — convection heat-transfer coefficient with fouling, W/(m² °C) or Btu/(h ft² °F)

$\bar{h}_i$ — convection heat-transfer coefficient for pure crossflow in an ideal tube bank, W/(m² °C) or Btu/(h ft² °F)

$\bar{h}_s$ — shell-side convection heat-transfer coefficient, W/(m² °C) or Btu/(h ft² °F)

h_{tc} — thermal contact coefficient, W/(m² °C) or Btu/(h ft² °F)

H — enthalpy, J or Btu; height, m or ft

$\dot{H}$ — enthalpy rate, W or BTU/h

$\dot{H}_b$ — bulk-stream enthalpy rate [see Eq. (6–40) and Appendix L], W or Btu/h

H_F — fin height, mm or in.

i — specific enthalpy, J/kg or Btu/lb$_m$

i_{fg} — latent heat of vaporization, J/kg or Btu/lb$_m$

i'_{fg} — corrected or modified latent heat of vaporization (Chap. 10), J/kg or Btu/lb$_m$

i, j, k — unit vectors

I — radiation intensity [defined by Eq. (5–14)], W/(m² sr) or Btu/(h ft² sr)

I — reference stream for heat exchanger

II — nonreference stream for heat exchanger

I_e — electric current, A

I_s — number of finite elements in a finite-element network

I_Δ — turbulent thermal boundary layer parameter [defined by Eq. (7–340)]

j_i — dimensionless convection correlation for heat transfer associated with pure crossflow in an ideal tube bank (see Table A–N–6–1)

J — radiosity, W/m² or Btu/(h ft²)

J — objective function for heat exchanger optimization, kW/$ or Btu/h/$

J_b — shell-and-tube heat exchanger correction factor for effect of bypass on coefficient of heat transfer

J_c — shell-and-tube heat exchanger correction factor for effect of baffle cut on coefficient of heat transfer

J_F — correction factor for shell-side heat transfer in heat exchanger with low-fin tubing [defined by Eq. 11–186)]

J_l — shell-and-tube heat exchanger correction factor for effect of leakage streams A and E on coefficient of heat transfer

J_{tot} — total correction factor for shell-side coefficient of heat transfer [see Eq. 11–181)]

k — thermal conductivity, W/(m °C) or Btu/(h ft °F)

k_e — effective thermal conductivity for natural convection in an enclosed space, W/(m °C) or Btu/(h ft °F)

k_t — eddy thermal conductivity for turbulent flow, W/(m °C) or Btu/(h ft °F)

K_c — entrance-loss coefficient

K_e — expansion-loss coefficient

K_f — loss coefficient for heat exchanger components [defined by Eq. (11–100)]

KE — kinetic energy, J or Btu

L — length, m or ft

L — characteristic length, m or ft

L_b — baffle spacing, m or ft

L_{bb} — clearance—shell-tube bundle bypass, mm or in.

L_{sb} — clearance—shell-baffle leakage, mm or in.

L_{tb} — clearance—tube-baffle leakage, mm or in.

ℓ — characteristic length (V/A_s), m or ft; turbulent mixing length, m or ft

$LMTD$ — log mean temperature difference [defined by Eq. (8–60) or Eq. (11–60)], °C, K, °F or °R

m — mass, kg or lb$_m$; convective fin parameter $\{[\bar{h}p/(kA)]^{1/2}\}$, m^{-1} or ft^{-1}

m_f — convective fin parameter $\{[\bar{h}p/(kA)]^{1/2}\}$, m^{-1} or ft^{-1}

m_R — radiative fin parameter $\{[\sigma p F_{s-R}/(kA)]^{1/2}\}$, m^{-1} or ft^{-1}

$\dot{m}$ — mass-flow rate, kg/s or lb$_m$/s

M — momentum, kg m/s or lb$_m$ ft/s; number of heat transfer lanes in a curvilinear sketch (Sec. 3–8); number of exchangers

	in network; number of hairpin turns in double-pipe heat exchanger
$\dot{M}$	momentum-flow rate, N or lb_f
$\dot{M}_b$	bulk-stream axial momentum rate [defined by Eq. (M–3)], N or lb_f
N	number of tubes; number of temperature increments in a curvilinear sketch (Sec. 3–8)
N_b	number of baffles
N_F	number of fins per tube or unit length
N_i	finite-element interpolation function [defined by Eq. (F–1)]
N_L	number of tubes in longitudinal direction
N_T	number of tubes in transverse direction
N_{tc}	number of tubes across heat exchanger shell diameter in the direction normal to the crossflow
N_{tcc}	effective number of tube rows between baffle tips in a shell-and-tube heat exchanger
N_{tp}	number of tube passes in shell-and-tube heat exchanger
NTU	number of transfer units [$\bar{h}A_s/(\dot{m}c_P)$ or $\overline{U}A_s/C_{\min}$]
NTU_I	number of transfer units for heat exchanger relative to reference fluid I ($\overline{U}A_s/C_I$)
p	perimeter, m or ft
p_w	wetted perimeter, m or ft
P	pressure, N/m^2 or lb_f/ft^2
P_I, P	heat exchanger effectiveness [defined by Eq. (11–33)]
$\hat{P}$	effectiveness of a heat exchanger with direction of one of the streams reversed
$P(x,t)$	dimensionless temperature for infinite plate [$(T - T_F)/(T_i - T_F)$]
PE	potential energy, J or Btu
q	heat-transfer rate, W or Btu/h
$\dot{q}$	rate of energy generation per unit volume, W/m^3 or Btu/(h ft^3)
$q_{\max,I}$	thermodynamic maximum possible rate of heat transfer to reference fluid I [defined by Eq. (11–31)], W or Btu/h
q''	heat flux, W/m^2 or Btu/(h ft^2)
$q''_{\max}$	maximum heat flux, W/m^2 or Btu/(h ft^2)
$\overline{q''}$	apparent total mean heat flux for turbulent flow ($\overline{q''_y} + \overline{q''_t}$), W/m^2 or Btu/(h ft^2)
$\overline{q''_t}$	apparent turbulent mean heat flux for turbulent flow ($\rho c_P \overline{v'T'}$), W/m^2 or Btu/(h ft^2)
Q	energy transfer, J or Btu
r	thermal boundary layer parameter (Δ/δ)
r, ϕ, z	cylindrical coordinates
r, θ, ϕ	spherical coordinates
r_c	critical radius, m or ft
r_0	cylinder or sphere radius, m or ft
R	thermal resistance, °C/W or °F h/Btu
R'	thermal resistance associated with differential area, °C/W or °F h/Btu
R_I, R	capacitance ratio (C_I/C_{II})
R_b	shell-and-tube heat exchanger correction factor for effect of bypass on crossflow pressure drop
R_e	electrical resistance, Ω
R_F	fouling factor, m^2 °C/W or ft^2 °F h/Btu
R_k	wall-thermal resistance, °C/W or °F h/Btu
R_l	shell-and-tube heat exchanger correction factor for effect of leakage on crossflow pressure drop
RCM	rate of creation of momentum, N or lb_f
S	conduction shape factor, m or ft
S_D	diagonal pitch for tube bank, m or in.
S_F	distance between fins for low-fin tubing, mm or in.
S_L	longitudinal pitch for tube bank, m or in.
S_T	transverse pitch for tube bank, m or in.
S_t	tube pitch for shell-and-tube heat exchanger, m or ft
$S_{t,\mathrm{eff}}$	effective tube pitch for heat exchanger [see Table A–N–6–1], m or ft
$S(x,t)$	dimensionless temperature for semi-infinite solid [$(T - T_F)/(T_i - T_F)$]
t	time, s or h
T	temperature, °C, K, °F, or °R
T_b	bulk-stream temperature [see Eq. (6–46) and Appendix K]; °C, K, °F or °R

$T_t^{(i)}$ temperature of tube-side fluid for pass i in shell-and-tube heat exchanger, °C, K, °F or °R

T^+ dimensionless temperature for turbulent flow $[(T_s - T)\rho c_P U^*/\overline{q_0''}]$

u, v, w components of fluid velocity, m/s or ft/s

u^+ dimensionless velocity used in analysis of turbulent flow, m/s or ft/s

U velocity, m/s or ft/s; internal energy, J or Btu; overall coefficient of heat transfer [defined by Eq. (11–2)], W/(m² °C) or Btu/(h ft² °F)

U_b bulk-stream velocity [defined by Eq. (6–34) and Appendix K]

U_x characteristic velocity for transformation analysis, m/s or ft/s

U^* friction velocity $[(\overline{\tau}_s/\rho)^{1/2}]$, m/s or ft/s

v specific volume, m³/kg or ft³/lb$_m$; component of fluid velocity normal to wall, m/s or ft/s

v_m mean specific volume, [defined by Eq. (11–98)], m³/kg or ft³/lb$_m$

V volume, m³ or ft³

$\dot{V}$ volume-flow rate, m³/s or ft³/s

V_e volume of a finite element, m³ or ft³

w width, m or ft

W_i finite-element weighting function [see Eq. (F–19)]

$\dot{W}$ power, W or Btu/h

$\dot{W}_p$ pumping power [see Eq. (8–78)], W or Btu/h

$\dot{W}_\tau$ viscous-energy-dissipation rate [see Appendix I], W or Btu/h

x, y, z rectangular coordinates, m or ft

x_c length beyond which flow is turbulent, m or ft

X quality for two-phase flow

X_x stretching parameter [defined by Eq. (I–2–5)], 1/m or 1/ft

y^+ dimensionless distance from wall (yU^*/ν)

Z depth, m or ft; number of unknown nodal temperatures in a numerical formulation

Z_e number of nodes in a finite element

Z_s total number of nodes in a numerical formulation

☐ Greek Symbols

alpha

α thermal diffusivity, m²/s or ft²/s; absorptivity

α_0 temperature coefficient of resistance, °C^{-1} or °F^{-1}

α_t eddy thermal diffusivity for turbulent flow, m²/s or ft²/s

beta

β coefficient of thermal expansion, 1/°C or 1/°F; surface-area density of heat-exchanger core (A_s/V), m^{-1} or ft^{-1}

β_T temperature coefficient of thermal conductivity, °C^{-1} or °F^{-1}

gamma

γ specific-heat ratio (c_P/c_v)

γ_n eigenvalues (see Appendix E)

Γ substitution used in Tables A–N–1–1 and 2

delta

δ hydrodynamic boundary-layer thickness, m or ft; wall thickness, m or ft

δ_F fin thickness, mm or in.

Δ thermal boundary-layer thickness, m or ft; difference between two values

ΔT_m true mean temperature difference [defined by Eq. (11–9)], °C, K, °F or °R

ΔT_{MAX} maximum temperature difference in a heat exchanger $(T_{h,i} - T_{c,i})$, °C, K, °F or °R

ΔT_{max} change in temperature of C_{min} fluid in heat exchanger [see Eq. (11–79)], °C, K, °F or °R

ΔT_{min} change in temperature of C_{max} fluid in heat exchanger [see Eq. (11–79)], °C, K, °F or °R

epsilon

ϵ emissivity; surface roughness, m or in.; heat-exchanger effectiveness (q_c/q_{max}); small increment

ϵ_H eddy thermal diffusivity for turbulent flow (often used instead of α_t), m^2/s or ft^2/s

ϵ_M eddy diffusivity for turbulent flow (often used instead of ν_t), m^2/s or ft^2/s

ϵ_θ directional emissivity [defined by Eq. (5–17)]

zeta

ζ defined by Eq. (11–119); dimensionless stream function ($X_x\psi/U_x$)

eta

η similarity parameter

η_F fin efficiency

η_o net surface efficiency (or temperature effectiveness)

theta

θ zenith angle, rad

Θ dimensionless temperature distribution $[(T - T_F)/(T_i - T_F)]$

kappa

κ turbulence coefficient

lambda

λ wavelength, μm; Blasius friction factor $(4f)$

λ_n defined by Eq. (E–1–21)

Λ general representation of fanning friction factor or coefficient of heat transfer

mu

μ viscosity, kg/(m s) or $lb_m/(ft\ s)$

μ_t eddy viscosity for turbulent flow, kg/(m s) or $lb_m/(ft\ s)$

nu

ν kinematic viscosity, m^2/s or ft^2/s; radiation frequency, s^{-1}

ν_t eddy diffusivity for turbulent flow, m^2/s or ft^2/s

xi

ξ general coordinate direction (can represent, x, y, z, or r), m or ft

pi

π the number 3.14159265

rho

ρ density $(1/v)$, kg/m^3 or lb_m/ft^3; reflectivity

ρ_e electrical resistivity, m Ω or ft Ω

sigma

σ Stefan-Boltzmann constant, $W/(m^2\ K^4)$ or $Btu/(h\ ft^2\ °R^4)$; surface tension, N/m or lb_f/ft; contraction ratio (A_{min}/A_1)

tau

τ shear stress, N/m^2 or lb_f/ft^2; transmissivity

$\bar{\tau}$ apparent total mean shear stress for turbulent flow $(\bar{\tau}_y + \bar{\tau}_t)$, N/m^2 or lb_f/ft^2

$\bar{\tau}_t$ apparent turbulent mean shear stress for turbulent flow $(-\rho\overline{u'v'})$, N/m^2 or lb_f/ft^2

phi

ϕ azimuthal angle, rad

Φ viscous-dissipation function, s^{-2}

chi

χ correction for friction factor associated with flow across tube bank (see Fig. 8–30)

psi

ψ heat-exchanger efficiency [defined by Eq. (8–56) or Eq. (11–64)]; substitution $T - T_1$, °C or °F

ψ stream function [defined by Eq. (I–2–12)]

omega

ω solid angle $(d\omega = dA_r/r^2)$, sr

Ω ohm

☐ Subscripts

a annulus

b blackbody; bulk-stream conditions

B boiling

c centerline; convection; cold fluid; contraction entrance; counterflow

C condensation

cp constant properties

D based on diameter

e expansion exit; electrical

f saturated liquid conditions; condensate film

F reference fluid condition; fin

g saturated vapor conditions; graybody in thermal radiation

h	hot fluid
HS	heat sink
i	influx; inside diameter; tube side; initial condition
i	annular flow—inner surface at uniform temperature/outer surface insulated
ii	annular flow—inner surface at uniform flux/outer surface insulated
I	reference stream for heat exchanger
II	nonreference stream for heat exchanger
k	conduction
L	based on length
$\mathbf{L}$	based on characteristic length
m	x node in numerical formulation
MIN	efficiency-based performance criterion for heat exchanger
n	y node in numerical formulation
o	efflux; outside diameter; annular or shell side
o	outlet temperature performance criterion for heat exchanger
o	annular flow—inner surface insulated/outer surface at uniform temperature
oo	annular flow—inner surface insulated/outer surface at uniform flux
0	uniform wall condition
p	parallel flow
r	radial direction
R	reference; radiation
s	surface conditions; storage; shell side; mixed flow in crossflow heat exchanger
sat	saturated conditions
ss	steady state
t	turbulent; tube side; unmixed flow in crossflow heat exchanger
T	transistor
w	wetted surface; surface conditions
x	local condition on a surface relative to the x coordinate; x-direction
y	y-direction
λ	monochromatic radiation properties
∞	free-stream conditions
ξ	general coordinate direction

Superscripts

$'$	turbulent fluctuating quantity; first derivative
$''$	per unit area (flux); second derivative
$'''$	third derivative
$\cdot$	per unit time (rate)
k	iteration index number
m, n	coefficients in empirical correlations
τ	numerical index for time

Overbar

$-$	surface average conditions; time mean

Dimensionless Groups

Bi	Biot number ($h\ell/k$ for convection)
Bo_L	Bond number $[g(\rho_f - \rho_g)\mathbf{L}^2/\sigma]$
Br	Brinkman number ($Ec\ Pr/Nu = \mu U_b^2/(q_0''\ 2w)$)
Ec	Eckert number $\{U_b^2/[c_P(T_s - T_F)]\}$
Fo	Fourier number ($\alpha t/\ell_0^2$)
Gr_L	Grashof number [defined by Eq. (9–6) or Eq. (10–5)]
Gr_L^*	flux Grashof number ($Nu_L\ Gr_L$)
Gz	Graetz number $[Re\ Pr/(x/D)]$
Ja	Jakob number $[c_P(T_s - T_{sat})/i_{fg}]$
$Ma_{L,f}$	modified Marangoni number [defined by Eq. (10–42)]
Nu	Nusselt number for internal-flow passage (hD_H/k)
Nu_L	Nusselt number ($h\mathbf{L}/k$)
Pe	Peclet number ($Re\ Pr$)
Pr	Prandtl number (ν/α)
Ra_L	Rayleigh number ($Gr_L\ Pr$)
Ra_L^*	flux Rayleigh number ($Nu_L\ Ra_L$)
Re	Reynolds number for internal-flow passage (U_bD_H/ν, GD_H/μ)
Re_f	Reynolds number for film condensation $[4\dot{m}/(p\mu_f)]$

Re_L Reynolds number ($U_F L/\nu$, $GL\mu$)

Re_s heat exchanger shell-side Reynolds number ($G_s d_o/\mu_s$)

Ri_x Richardson number (Gr_x/Re_x^2)

$S_{L,f}$ surface-energy parameter [$(\sigma/(L\rho_f i_{fg}))$]

St Stanton number [$Nu_L/(Re_L Pr)$ or $Nu/(Re\,Pr)$]

■ MATHEMATICAL OPERATIONS AND FUNCTIONAL RELATIONS†

fn general functional relation

fn_i specific functional relation; $i = 1, 2$

log logarithm to the base 10

ln logarithm to the base e

† Roman typeset is used throughout text for mathematical operations and functional relations.

GLOSSARY

GENERAL

Control-volume finite-difference method CVFDM	The development of nodal equations by applying fundamental laws to finite-difference control volumes.
Control-volume finite-element method CVFEM	The development of nodal equations by applying fundamental laws to finite-element control volumes.
Finite-difference discretization method	The use of finite-difference approximations in the transformation of the differential formulation into nodal equations.
Finite-difference nodal network	System of subvolumes and nodes associated with finite-difference formulation.
Finite-element nodal network	System of subvolumes and nodes associated with finite-element formulation.
Galerkin method	Commonly used method of weighted residuals—associated with formulation for which the weighting function $W_i(x,y,z)$ is equal to the interpolation function $N_i(x,y,z)$.
Gauss–Seidel method	An effective iterative scheme for solving systems of linear algebraic equations.
Heat transfer	The transfer of energy across a system boundary caused solely by a temperature difference.
Interpolation function $N_i(x,y,z)$	Defined by Eq. (F–1), $T(x,y,z) = \sum_{i=1}^{Z_e} N_i(x,y,z)\ T_i$ Used to approximate the distribution in temperature (or other dependent variable) within a finite element.

Laws, fundamental

Conservation of mass — Given by Eq. (1–2),

$$\Sigma \dot{m}_o - \Sigma \dot{m}_i + \frac{\Delta m_s}{\Delta t} = 0$$

First law of thermodynamics — *Conservation of energy* given by Eq. (1–1),

$$\Sigma \dot{E}_o - \Sigma \dot{E}_i + \frac{\Delta E_s}{\Delta t} = 0$$

Newton's second law — *Momentum principle* given by Eq. (1–3),

$$\Sigma \dot{M}_{o,x} - \Sigma \dot{M}_{i,x} + \frac{\Delta M_{s,x}}{\Delta t} = \Sigma F_x$$

Method of weighted residuals MWR — The development of nodal equations by performing weighted integrations [using a weighting function represented by $W_i(x,y,z)$] of the applicable differential equations over all finite-element volumes.

Numerical discretization error or truncation error — The error associated with the finite-difference or finite-element increment.

Round-off error — The numerical error associated with the use of a finite number of significant figures.

Successive approximation method — An effective iterative scheme for solving systems of nonlinear algebraic equations.

Temperature T — A property that is an index of the kinetic energy possessed by molecules, atoms, and subatomic particles of a substance.

Thermal resistance R or R' — Defined by Eq. (1–23),

$$q = \Delta T/R \qquad \text{or} \qquad dq = \Delta T/R'$$

Weighting function $W_i(x,y,z)$ — Designates linearly independent functions associated with each finite-element node.

CONDUCTION

Conduction heat transfer — The transfer of energy caused by physical interaction between molecules, atoms, and subatomic particles of a substance at different temperatures (level of kinetic energy).

Conduction shape factor S — Defined by Eq. (1–12),

$$q = kS(T_1 - T_2)$$

Critical radius r_c — The radius at which the heat-transfer rate in a cylindrical or spherical body is maximum for the case in which the surface is exposed to convection or radiation.

Laws, particular

Fourier law of conduction — One-dimensional form given by

$$q_\xi = -k\, \partial T/\partial \xi$$

where ξ represents a space dimension.

General Fourier law of conduction	Multidimensional form given by Eq. (3–1), $q'' = q_x''\mathbf{i} + q_y''\mathbf{j} + q_z''\mathbf{k}$
Superconductors	Substances under low-temperature conditions that have extremely high thermal conductivities.
Temperature coefficient of resistance α_0	Pertaining to the electrical resistivity ρ_e and defined by Eq. (2–68), $\rho_e = \rho_0(1 + \alpha_0 T)$
Temperature coefficient of thermal conductivity β_T	Pertaining to substances with temperature dependent thermal conductivity $k(T)$ and defined by Eq. (2–60), $k(T) = k_0(1 + \beta_T T)$
Thermal conductivity k	A thermophysical property of the conducting medium which represents the rate of conduction heat transfer per unit area for a temperature gradient of 1 °C/m (or 1 °F/ft).
Thermal contact coefficient h_{tc}	Pertaining to heat transfer at the interface between two substances denoted by I and II and defined by Eq. (2–28), $q_{tc} = h_{tc}A[T_I(0) - T_{II}(0)]$
Thermal diffusivity α	Thermophysical property of a substance defined by $\alpha = k/(\rho c_v)$
Thermal resistance R_k	Defined by Eq. (1–25), $R_k = 1/(kS)$

■ RADIATION

Absorption	The process of conversion of radiation incident on matter to internal energy.
Absorptivity α	The fraction of thermal radiation incident on a surface which is absorbed. Modifiers: directional, hemispherical, monochromatic, total.
Blackbody	An object that absorbs all the thermal radiation reaching its surface ($\alpha = 1$). Denoted by subscript b.
Black light	Ultraviolet radiation in the range (0.32 – 0.40 μm). Also known as UVA radiation.
Configuration factor	Equivalent to *shape factor*.
Diffuse	Modifier that indicates the same intensity or irradiation in all directions.
Directional	Modifier that refers to a particular direction or angle.
Electromagnetic radiation	Energy that is produced as a result of charged particles which undergo acceleration.
Electromagnetic spectrum	Various types of electromagnetic radiation that are characterized according to wavelength λ or frequency ν.
Emissive power E	Rate of thermal radiation emitted by a body per unit surface area. Modifiers: blackbody, monochromatic, total.

Emissivity ϵ Ratio of the radiation emitted by a surface to the radiation emitted by a blackbody at the same temperature. Modifiers: directional, hemispherical, monochromatic, total.

Frequency ν The number of characteristic cycles per unit time of electromagnetic radiation, which is related to the *wavelength* λ and *propagation speed* c by

$$\nu = c/\lambda$$

Graybody Surfaces for which $\epsilon = \epsilon_\lambda$. Denoted by subscript g.

Greenhouse effect The warming of an environment which is heated by solar radiation as a result of the transmitting medium being essentially transparent to the low-wavelength incoming thermal radiation (mainly visible and near infrared) and opaque to the longer-wavelength infrared radiation (mainly middle and far IR) emitted from the surface.

Hemispherical Modifier which refers to all directions in the space above a surface.

Index of refraction n Equal to the ratio c_0/c.

Infrared radiation (IR) Electromagnetic radiation in the wavelength range 0.76 to 1000 μm. Hot bodies are a primary source of infrared radiation. This part of the thermal radiation region includes *near* IR ($0.76 < \lambda < 1.50$ μm), *middle* IR ($1.50 < \lambda < 5.60$ μm), and *far* IR ($5.60 < \lambda < 1000$ μm).

Intensity I The total rate of thermal radiation emitted per unit solid angle $d\omega$ and per unit area normal to the direction ϕ, θ. Modifiers: monochromatic, total.

Irradiation G The incoming thermal radiation flux. Modifier: diffuse, monochromatic, total.

Laws, particular

- **Kirchhoff's law** Relation between emission and absorption properties for blackbody thermal radiation in an isothermal enclosure. (See Example 5–5.)
- **Stefan-Boltzmann law** Emissive power of a blackbody given by Eq. (5–2),

$$E_b = \sigma T_s^4$$

  where $\sigma = 5.67 \times 10^{-8}$ W/(m^2 K^4) is the *Stefan-Boltzmann constant.*
- **Planck's law** Monochromatic distribution of emission from a blackbody given by Eq. (5–7).
- **Quantum theory** Theory developed by Planck which relates the photon energy e to the frequency ν and *Planck's constant* $h = 6.625 \times 10^{-34}$ J s by

$$e = h\nu = h\lambda$$

- **Reciprocity law** Relation between shape factors represented by [see Eq. (5–29)]

$$A_s F_{s-R} = A_R F_{R-s}$$

- **Wien's law** Locus of wavelength associated with peak emission by a blackbody given by Eq. (5–8),

$$T_s \lambda_{max} = 2898 \text{ μm K}$$

Microwave radiation Electromagnetic radiation in the wavelength range 10^3 to 10^5 μm. High-power microwaves which are produced by special types of electron tubes known as magnetrons are used in the cooking of food. (Low-power microwaves which are produced by electron tubes such as klystrons and traveling wave-tubes are used in radar and telecommunications.)

Monochromatic	Modifier that refers to single wavelength radiation component.
Monochromatic distribution	Refers to variation of radiation with wavelength.
Nonparticipating medium	Substances that are completely transparent to thermal radiation.
Participating medium	Substances that absorb and emit significant amounts of thermal radiation.
Photons (or quanta)	Discrete packets of energy associated with the emission and absorption of electromagnetic radiation by changes in atomic energy state.
Radiation	Equivalent to electromagnetic radiation.
Radiation factor $\mathcal{F}_{s-j}$	Defined by Eq. (5–72), $q_{s-j} = A_s\mathcal{F}_{s-j}\left(\frac{\epsilon_s}{\alpha_s}E_{bs} - \frac{\epsilon_j}{\alpha_j}E_{bj}\right)$
Radiation heat transfer q_R	The transfer of energy across a system boundary by means of an electromagnetic mechanism which is caused solely by a temperature difference.
Radiation shape factor F_{s-R}	The fraction of thermal radiation leaving a diffuse surface A_s that passes through a nonparticipating medium to surface A_R.
Radiosity J_s	The rate of thermal radiation emitted and reflected per unit area from a surface A_s. J_s is expressed in terms of E_s, ρ_s, and G_s by Eq. (5–49), $J_s = E_s + \rho_s G_s$ Modifiers: monochromatic, total.
Rayleigh effect	Molecular scattering in the atmosphere which results in the redirection of much of the incoming solar radiation. According to J. W. S. Rayleigh, the degree of scattering is inversely proportional to λ^4.
Reflection	The process in which radiation incident on a surface is redirected.
Reflectivity ρ	The fraction of thermal radiation incident on a surface which is reflected. Modifiers: directional, hemispherical, monochromatic, total.
Reradiating surfaces	An insulated surface with temperature which is totally governed by thermal radiation.
Scattering	Redirection of thermal radiation as a result of reflection from molecules of gas, dust, and other particles in the atmosphere.
Semitransparent	Refers to a medium which transmits a portion of the incident radiation.
Spectral	Equivalent to monochromatic.
Specular reflection	The condition for which the angle of reflection is equal to the angle of incidence.
Solar radiation	Electromagnetic radiation emitted from the sun which is concentrated in the low wavelength region ($0.3 - 2.5\ \mu m$) of the thermal

	spectrum, with the peak occurring at about 0.50 μm. The spectral distribution approximates that of a blackbody at 5762 K.
Thermal radiation	Electromagnetic energy in the intermediate wavelength range 0.1 to 1000 μm that is associated with the temperature of the emitting surface. Thermal radiation is generally considered to consist of ultraviolet, visible, and infrared radiation.
Thermal-radiation networks	Analogous electrical circuits which involve the use of thermal resistances of the form $R_{s-j} = 1/(A_s F_{s-j})$
Thermal-radiation shields	Thin plates or shells or highly reflective materials separated by evacuated spaces. (Radiation shields are used in superinsulative composite walls.)
Thermal resistance R_R	Defined by Eq. (1–26), $R_R = 1/[\sigma A_s F_{s-R}(T_s + T_R)(T_s^2 + T_R^2)]$
Total	Modifier which refers to all wavelengths.
Transmission	The process of radiation passing through matter.
Transmissivity τ	The fraction of thermal radiation incident on a surface which is transmitted.
Ultraviolet radiation (UV)	Electromagnetic radiation in the wavelength range 0.01 to 0.40 μm. The part of the ultraviolet region that is classified as thermal radiation includes UVA ($0.32 < \lambda < 0.40$ μm), UVB ($0.29 < \lambda < 0.32$ μm), UVC ($0.20 < \lambda < 0.29$ μm), and vacuum UV ($0.1 < \lambda < 0.2$ μm).
Visible radiation (light)	Electromagnetic radiation in the wavelength range 0.40 to 0.76 μm. The visible region includes *violet* ($0.40 < \lambda < 0.44$ μm), *blue* ($0.44 < \lambda < 0.49$ μm), *green* ($0.49 < \lambda < 0.54$ μm), *yellow* ($0.54 < \lambda < 0.60$ μm), *orange* ($0.60 < \lambda < 0.63$ μm), and *red* ($0.63 < \lambda < 0.76$ μm).
View factor	Equivalent to *shape factor*.
Wavelength λ	The characteristic length of electromagnetic radiation which is related to the *frequency* ν and *propagation speed* c by $\lambda = c/\nu$
White light	The summation of all visible wavelengths.

■ CONVECTION†

Advection	The process by which energy is transferred by fluid motion.
Apparent turbulent shear stress $\bar{\tau}_t$	Defined by Eq. (7–218), $\bar{\tau}_t = -\rho\overline{u'v'}$
Apparent turbulent heat flux $\overline{q''_t}$	Defined by Eq. (7–234), $\overline{q''_t} = \rho c_P \overline{T'v'}$

† The definition and interpretation of the primary dimensionless groups associated with the analysis of forced convection, natural convection, and boiling and condensation is summarized in Tables 8–9, 9–8, and 10–4.

Boundary layer

hydrodynamic — The region of a flow field near a surface in which the viscous effects are significant.

thermal — The region of a flow field near a surface in which the effects of molecular conduction are significant.

Boundary-layer approximations — Approximations proposed by Prandtl which pertain to the relative magnitude of velocity and velocity gradients within the boundary layer. (See Sec. 6–2.)

Bulk-stream enthalpy rate $\dot{H}_b$ — Defined by Eq. (6–40),

$$\dot{H}_b = \int_{\dot{m}} i \, d\dot{m} = \int_A i\rho u \, dA$$

Bulk-stream mass flux G — Mass-flow rate per unit cross-sectional area; that is,

$$G = \dot{m}/A$$

Bulk-stream temperature T_b — Defined by Eq. (6–46),

$$T_b = \frac{\rho}{\dot{m}} \int_A Tu \, dA$$

for uniform property flow.

Bulk-stream velocity U_b — Defined by Eq. (6–34),

$$U_b = \frac{\dot{m}}{\rho A} = \frac{G}{\rho} = \frac{1}{A} \int_A u \, dA$$

for uniform density.

Continuity — Equivalent to fundamental principle of *conservation of mass*.

Convection — The transfer of heat from a surface to a moving fluid.

Coefficient of heat transfer

local h_x — Defined by Eq. (1–21),

$$dq_c = h_x \, dA_s \, (T_s - T_F)$$

mean $\overline{h}$ — Defined by Eq. (1–22),

$$\overline{h} = \frac{1}{A_s} \int_{A_s} h_x \, dA_s$$

Darcy friction factor λ — Related to *Fanning friction factor* f by

$$\lambda = 4f$$

Eddy diffusivity ν_t (or ϵ_M) — Defined by

$$\nu_t = \mu_t/\rho$$

Eddy thermal conductivity k_t — Used in the mean-field method to relate the *apparent turbulent heat flux* to the *mean temperature* according to Eq. (7–236),

$$\overline{q''_t} = -k_t \, \partial \overline{T}/\partial y$$

Eddy thermal diffusivity α_t (or ϵ_H) — Defined by

$$\alpha_t = k_t/(\rho c_P)$$

Eddy viscosity μ_t — Used in the mean-field method to relate the *apparent turbulent shear stress* $\overline{\tau}_t$ to the *mean axial velocity* $\overline{u}$ according to Eq. (7–222),

$$\overline{\tau}_t = \mu_t \, \partial \overline{u}/\partial y$$

Fanning friction factor

local f_x — Defined by Eq. (6–20),

$$\tau_s = \rho U_F^2 f_x/2$$

mean $\bar{f}$	Defined by Eq. (6–21), $\bar{f} = \frac{1}{A_s}\int_{A_s} f_x \, dA_s$
Fin efficiency η_F	Pertaining to a convecting fin and defined by Eq. (2–122), $\eta_F = \frac{q_F}{q_{\max}} = \frac{q_F}{\bar{h}A_F(T_0 - T_F)}$
Friction velocity U^*	A characteristic velocity generally used in the analysis of turbulent flow, which is defined by Eq. (7–227), $U^* = \sqrt{\bar{\tau}_s/\rho}$
Hydraulic diameter D_H	A characteristic length for internal-flow systems, which is defined by Eq. (6–27), $D_H = 4\,A/p_w$ for tubes, or by Eq. (8–139), $D_H = 4\,A_{\min}/p_w$ for tube banks and other heat-exchanger surfaces.
Incompressible flow	Flow processes in which pressure induced changes in density are negligible.
Isotropic	Describes the condition in which properties are independent of direction.
Laminar flow	Flow in which individual elements of fluid follow smooth streamline paths.
Laws, particular	
Newton law of cooling	Given by Eq. (1–20), $q_c = \bar{h}A_s\,(T_s - T_F)$ for uniform surface and fluid temperatures.
Newton law of cooling, general	Defining relation for the local coefficient of heat transfer h_x, given by Eq. (1–21), $dq_c = h_x\,dA_s\,(T_s - T_F)$
Newton law of viscous stress	One-dimensional, steady-state form given by Eq. (1–19), $\tau = \mu\, du/dy$ where μ is the fluid viscosity and τ is the viscous stress acting in the streamwise x-direction.
Mixing length ℓ	An alternative mean-field representation for $\bar{\tau}_t$ proposed by Prandtl which takes the form $\bar{\tau}_t = \rho\ell^2(\partial\bar{u}/\partial y)^2$
Net surface efficiency η_o	Pertaining to a convecting fin and defined by Eq. (2–126), $q_o = \eta_o\bar{h}A_o(T_s - T_F)$
Newtonian fluids	Fluids which satisfy the Newton law of viscous stress (i.e., stress is proportional to velocity gradient).
Reynolds stress $\bar{\tau}_t$	Equivalent to *apparent turbulent shear stress*.
Similar flow	Boundary layer flows for which the velocity and temperature distributions are geometrically similar in the streamwise x-direction. Similar flow in tubes is referred to as *fully developed*.
Standard conditions	Standard conditions used in the mathematical formulation of basic convection processes include steady-state; two-dimensional; ideal,

Newtonian, isotropic and single-phase fluid; incompressible; no interfacial mass transfer; and negligible buoyancy, kinetic energy, viscous dissipation, external body forces, axial conduction, and thermal radiation.

Temperature effectiveness
: Equivalent to *net surface efficiency*.

Thermal resistance R_c
: Defined by Eq. (1–27),
$$R_c = 1/(\bar{h}A_s)$$

Turbulent flow
: Flow in which the movement of individual elements of fluid is unsteady and random in nature (i.e., fluctuating).

Turbulent Prandtl number Pr_t
: Defined by
$$Pr_t = \nu_t/\alpha_t = c_P \mu_t/k_t$$

Viscous dissipation
: The rate at which energy is converted into thermal energy by shearing forces. (See Appendix I).

□ Forced Convection

Colburn analogy
: An empirical correlation for turbulent boundary layer flow of the form
$$Nu_x = (f_x/2)\, Re_x\, Pr^{1/3}$$
which is applicable to forced convection flow of fluid with moderate values of Prandtl number.

Contraction ratio σ
: Pertaining to internal-flow passages and defined by
$$\sigma = A_{min}/A_1$$

Convection correlations
: Theoretical and empirical relations for coefficients of friction, pressure drop, and heat transfer.

Drag coefficient C_D
: Pertaining to external flow and defined by Eq. (8–123),
$$F_D = C_D A_N\, \rho U_\infty^2/2$$

Effectiveness ϵ
: Pertaining to internal-flow passages and defined by Eq. (8–53),
$$q_c = \epsilon q_{max}$$

Efficiency ψ
: Pertaining to internal-flow passages and defined by
$$\psi = \epsilon/NTU \quad \text{or} \quad \psi = P/NTU_I$$

Entrance-loss coefficient K_c
: Pertaining to internal-flow passages and defined by Eq. (8–3),
$$P_c - P = \left(K_c + \frac{4x}{D_H} f\right) \rho U_b^2/2$$

Expansion-loss coefficient K_e
: Pertaining to internal-flow passages and defined by Eq. (8–4),
$$\Delta P_{loss} = K_e\, \rho U_b^2/2$$

Forced convection
: Fluid flow that is caused by mechanical devices such as pumps, fans, or compressors.

Fully developed flow

hydrodynamic (HFD)
: Internal flow in which the velocity distribution is geometrically similar in the streamwise x-direction. HFD flow in an impermeable

tube is also characterized by a constant value of the friction factor; that is, $f_x = f$.

thermal (TFD)
Internal flow in which the temperature distribution is geometrically similar in the streamwise x-direction. TFD flow in an impermeable tube is also characterized by a constant value of the coefficient of heat transfer; that is, $h_x = h$.

Log mean temperature difference $LMTD$
Pertaining to internal-flow passages and defined by Eq. (8–60),

$$LMTD = \frac{\Delta T_1 - \Delta T_2}{\ln (\Delta T_1/\Delta T_2)}$$

Number of transfer units NTU
Defined by

$$NTU = \bar{h}A_s/(\dot{m}c_P)$$

for tube flow, and

$$NTU = \bar{U}A_s/C_{\min}$$

for heat-exchanger cores.

Number of transfer units NTU_I
Defined by

$$NTU_I = \bar{U}A_s/C_I$$

for heat-exchanger cores.

Pumping power $\dot{W}_p$
Power required to overcome pressure drop ΔP for internal flow, defined by Eq. (8–78),

$$\dot{W}_p = \dot{m}\,\Delta P/\rho$$

for uniform density.

Reynolds analogy
A theoretical relation for turbulent forced convection boundary layer flow of the form

$$Nu_x = (f_x/2)\,Re_x\,Pr$$

which is based on the similarity between momentum and energy transfer for turbulent flow of fluid with Prandtl number of the order of unity.

Surface-area density β
Pertaining to flow in heat-exchanger cores and defined by

$$\beta = A_s/V$$

☐ Natural Convection

Archimedes principle
A body immersed in a fluid experiences an upward buoyancy force equal to the weight of the displaced fluid.

Boussinesq approximation
Pertaining to natural convection, the evaluation of the fluid properties at the free-stream temperature.

Coefficient of thermal expansion β
Defined by Eq. (7–130),

$$\beta = -\frac{1}{\rho}\frac{\partial \rho}{\partial T}\bigg|_P$$

Natural convection
Fluid flow that is caused by temperature- (or concentration-) induced density gradients within the fluid.

Boiling and Condensation

Burnout
The condition associated with *film boiling* in which the temperature of the surface reaches the melting point.

Dropwise condensation
A relatively uncommon condensation process in which liquid droplets form on the surface and flow under the influence of gravity. The *non-wetting* conditions associated with dropwise condensation occur when the surface tension σ of the fluid is greater than a critical value σ_{cr} which is characteristic of the surface material.

Excess temperature ΔT_e
Defined by

$$\Delta T_e = T_s - T_{sat}$$

Film condensation
A relatively common condensation process in which a liquid film forms on the surface and flows under the influence of gravity. The wetting conditions associated with film condensation occur when the surface tension σ of the fluid is less than a critical value σ_{cr} which is characteristic of the surface material.

Film boiling
Boiling regime in which a stable vapor film covers the heating surface. The heat flux q_c'' in this high-temperature region of the boiling curve increases with increasing excess temperature.

Heat pipe
A device consisting of a closed pipe lined with wicking material which utilizes evaporation and condensation to transfer heat effectively.

Latent heat of vaporization i_{fg}
Equivalent to the *enthalpy of vaporization.*

Latent heat of vaporization, corrected (or modified) i'_{fg}
Defined by

$$i'_{fg} = i_{fg} + C_1 c_P (T_s - T_{sat}) = i_{fg}(1 + C_1 Ja)$$

where Ja is the *Jakob number.*

Leidenfrost point
The point of minimum heat flux on the pool-boiling curve.

Maximum heat flux (or critical heat flux, peak heat flux) q''_{max}
The point on the pool-boiling curve at which the heat flux reaches a maximum.

Nucleate boiling
Boiling regime in which liquid is transformed into vapor nuclei (i.e., bubbles) at various preferential sites on the heating surface.

Surface tension σ
The work done in extending the surface of a liquid one unit area. The surface tension of a fluid results from a net inward force of attraction on molecules at the interface toward molecules within the liquid phase.

Surface tension, critical σ_{cr}
A characteristic surface tension of materials which forms the basis for the following surface wetting criterion:

$\sigma < \sigma_{cr}$ liquid *will* wet surface

$\sigma > \sigma_{cr}$ liquid *will not* wet surface

Transition boiling
The region between the maximum and minimum heat fluxes in which a vapor film is intermittently built up and partially destroyed. This region is characterized by a decrease in heat flux q_c'' with increasing excess temperature.

Heat Exchangers

Capacity rate C	Defined by $C = \dot{m}c_P$ and usually subscripted to identify the fluid stream.
Capacitance ratio C^*	Defined by Eq. (11–52), $C^* = C_{\min}/C_{\max}$
Capacitance ratio R or R_I	Defined by Eq. (11–35), $R = C_I/C_{II}$
Component-loss coefficient K_f	Pertaining to pressure drop in return bends, headers and other components, defined by Eq. (11–100), $\Delta P_f = K_f \rho U_b^2/2$
Design function, thermal and hydraulic	Determination of the surface area A_s required to transfer a specified rate of heat with acceptable pressure drop for given fluids, mass-flow rates, and terminal temperatures.
Effectiveness ϵ	Defined by Eq. (11–51), $\epsilon = q/q_{\max} = q/[C_{\min}(T_{h,i} - T_{c,i})]$
Effectiveness P or P_I	Defined by Eq. (11–33), $P = q/q_{\max,I} = q/[C_I(T_{h,i} - T_{c,i})]$
Efficiency ψ	Defined by $\psi = \epsilon/NTU$ or $\psi = P/NTU_I = P/\overline{U}A_s/C_I = q/[\overline{U}A_s(T_{h,i} - T_{c,i})]$
Evaluation (or rating) function, thermal and hydraulic	Determination of the total rate of heat transfer q_c, outlet temperatures, and pressure drop that can be produced by an existing system under given operating conditions.
Fouling	Deposits on a surface that retard the transfer of heat.
Fouling factor F_f	Defined by Eq. (11–15), $F_f = 1/h_f - 1/h$ where h_f represents the coefficient of heat transfer *after* fouling has occurred.
Heat exchangers	Devices that transfer heat between fluids at different temperatures.
Compact	Heat exchangers in which the *surface-area density* β is greater than about 700 m^2/m^3.
Direct contact	Heat exchangers in which two immiscible fluids are in direct contact.
Recouperators	Heat exchangers in which two fluids are separated by a thin wall through which heat is transferred.
Regenerators	Heat exchangers in which the same flow passages are alternately used by both fluids.
Mixed stream	Pertaining to crossflow heat exchangers, a flow stream which is unrestricted in the lateral direction (i.e., in the direction of the other stream).
Number of transfer units NTU	Defined by Eq. (11–53), $NTU = \overline{U}A_s/C_{\min}$

Number of transfer units NTU_I	Defined by Eq. (11–36), $NTU_I = \overline{U}A_s/C_I$
Overall coefficient of heat transfer	
local U	Defined by Eq. (11–2), $dq = U\,dA_s\,(T_h - T_c)$
mean $\overline{U}$	Defined by Eq. (11–5), $\overline{U} = \frac{1}{A_s}\int_{A_s} U\,dA_s$
Symmetrical heat exchanger	Heat exchanger with functional relations for effectiveness $P(NTU_I,R)$ that are independent of which fluid is selected as the reference stream I.
Temperature cross	Location within a heat exchanger at which $T_c = T_h$.
True mean temperature difference ΔT_m	Defined by $\Delta T_m = \frac{q}{\overline{U}A_s} = \frac{1}{\overline{U}A_s}\int_{A_s} U(T_h - T_c)\,dA_s$
Unmixed stream	Pertaining to crossflow heat exchangers, a flow stream which is restricted in the lateral direction (i.e., in the direction of the other stream).

REFERENCES

CHAPTER 1

1. Fourier, J. B., *Théorie analytique de la Chaleur*. Paris, 1822. English translation by A. Freeman, Dover Publications, Inc., New York, 1955.
2. Touloukian, Y. S., et al., *Thermophysical Properties of Matter*; 13 vols. plus index. New York: Plenum Publishing Co., 1970–1977.
3. Bolz, R. E., and G. L. Tuve, eds., *Handbook of Tables for Applied Engineering Science*, 2nd ed. Boca Raton, Fla.: CRC Press, 1973.
4. American Society of Heating, Refrigeration and Air Conditioning Engineers, *Handbook of Fundamentals*, Chaps. 17 and 31. New York: ASHRAE, 1985.
5. Powell, R. W., C. Y. Ho, and P. E. Liley, "Thermal Conductivity of Selected Materials," *NSRDS-NBS 8*. Washington, D.C.: U.S. Department of Commerce, National Bureau of Standards, 1966.
6. Ho, C. V., R. W. Powell, and P. E. Liley, *Thermal Conductivity of Elements*, Vol. 1, *First Supplement to Journal of Physical and Chemical Reference Data*. Washington, D.C.: American Chemical Society, 1972.
7. Eckert, E. R. G., and R. M. Drake, *Analysis of Heat and Mass Transfer*. New York: McGraw-Hill Book Company, 1972.
8. Kays, W. M., and M. E. Crawford, *Convective Heat and Mass Transfer*, 2nd ed. New York: McGraw-Hill Book Company, 1980.
9. Bird, R. B., W. E. Stewart, and E. N. Lightfoot, *Transport Phenomena*. New York: John Wiley & Sons, 1960.

CHAPTER 2

1. Kreyszig, E., *Advanced Engineering Mathematics*, 4th ed. New York: John Wiley & Sons, 1979.
2. Barzelay, M. E., K. N. Tong, and G. F. Holloway, "Effect of Pressure on Thermal Conductance of Contact Joints," *NACA Tech. Note 3295*, May 1955.
3. Clausing, A. M., "Heat Transfer at the Interface of Dissimilar Metals—The Influence of Thermal Strain," *Int. J. Heat Mass Transfer*, 9, 1966, 791.
4. Moore, C. J., Jr., H. A. Blum, and H. Atkins, "Classification Bibliography for Thermal Contact Resistance Studies," *ASME Paper 68-WA/HT-18*, December 1968.
5. Arpaci, V. S., *Conduction Heat Tranfer*. Reading, Mass.: Addison-Wesley Publishing Co., 1966.
6. Schneider, P. J., *Conduction Heat Transfer*. Reading, Mass.: Addison-Wesley Publishing Co., 1955.
7. Gardner, K. A., "Efficiency of Extended Surfaces," *Trans. ASME*, 67, 1945, 621.

CHAPTER 3

1. Kreyszig, E., *Advanced Engineering Mathematics*, 4th ed. New York: John Wiley & Sons, 1979.
2. Wylie, C. R., and L. C. Barrett, *Advanced Engineering Mathematics,* 5th ed. New York: McGraw-Hill Book Company, 1982.
3. Ozisik, M. N., *Heat Conduction*. New York: John Wiley & Sons, 1980.
4. Arpaci, V. S., *Conduction Heat Transfer*. Reading, Mass.: Addison-Wesley Publishing Co., 1966.
5. Carslaw, H. S., and J. C. Jaeger, *Conduction of Heat in Solids*. London: Oxford University Press, 1947.
6. Schneider, P. J., *Conduction Heat Transfer*. Reading, Mass.: Addison-Wesley Publishing Co., 1955.
7. Bewley, L. V., *Two-Dimensional Fields in Electrical Engineering*. New York: Macmillan Publishing Co., 1948.
8. Kreith, F., and W. Z. Black, *Basic Heat Transfer*. New York: Harper & Row, Publishers, 1980.
9. Langmuir, I., E. O. Adams, and F. A. Meikle, "Flow of Heat through Furnace Walls," *Trans. Am. Electrochem. Soc.*, 24, 1913, 53.
10. Rudenberg, R., "Die Ausbreitung der Luft-und Erdfelder um Hochspannungsleitungen besonders bei Erd-und Kurzschlüssen," *Electrotech. Z*, 46, 1945, 1342.
11. Andrews, R. V., "Solving Conductive Heat Transfer Problems with Electrical-Analogue Shape Factors," *Chem. Eng. Progr.*, 5, 1955, 67.
12. Hahne, E., and U. Grigull, "Formfaktor und Formwiderstand der stationären mehrdimensionalen Wärmeleitung," *Int. J. Heat Mass Transfer*, 18, 1975, 75.

13. American Society of Heating, Refrigeration and Air Conditioning Engineers, *Handbook of Fundamentals*. New York: ASHRAE, 1985.
14. Heisler, M. P., ''Temperature Charts for Induction and Constant Temperature Heating,'' *Trans. ASME*, 69, 1947, 227.
15. Grober, H. S., and U. Grigull, *Fundamentals of Heat Transfer*. New York: McGraw-Hill Book Company, 1961.
16. Eckert, E. R. G., and R. M. Drake, *Analysis of Heat and Mass Transfer*. New York: McGraw-Hill Book Company, 1972.

CHAPTER 4

1. *Handbook of Numerical Heat Transfer*, edited by W. J. Minkowycz, E. M. Sparrow, G. E. Schneider, and R. H. Pletcher. New York: John Wiley & Sons, 1988.
2. Desai, C. S., *Elementary Finite Element Method*. Englewood Cliffs, N.J.: Prentice-Hall, 1979.
3. Kreyszig, E., *Advanced Engineering Mathematics*, 4th ed. New York: John Wiley & Sons, 1979.
4. Patankar, S. V., *Numerical Heat Transfer and Fluid Flow*. New York: McGraw-Hill/Hemisphere, 1980.
5. Mitchell, A. R., and D. F. Griffith, *Finite Difference Methods in Partial Differential Equations*. New York: John Wiley & Sons, 1980.
6. Myers, G. E., *Analytical Methods in Conduction Heat Transfer*. New York: McGraw-Hill Book Company, 1971.
7. Fox, L., *Numerical Solution of Ordinary and Partial Differential Equations*. Reading, Mass.: Addison-Wesley Publishing Co., 1962.
8. Smith, G. D., *Numerical Solution of Partial Differential Equations with Exercises and Worked Solutions*. London: Oxford University Press, 1965.
9. Jacob, M., *Heat Transfer*. New York: John Wiley & Sons, 1949.
10. Baliga, B. R., and S. V. Patankar, "Elliptic Systems: Finite Element Method II," Chap. 11 in *Handbook of Numerical Heat Transfer*, edited by W. J. Minkowycz, E. M. Sparrow, G. E. Schneider, and R. H. Pletcher. New York: John Wiley & Sons, 1988.
11. Schneider, G. E., "Elliptic Systems: Finite-Element Method I," in *Handbook of Numerical Heat Transfer*, edited by W. J. Minkowycz, E. M. Sparrow, G. E. Schneider, and R. H. Pletcher. New York: John Wiley & Sons, 1988.

CHAPTER 5

1. Touloukian, Y. S., and D. P. DeWitt, *Thermophysical Properties of Matter*, Vol. 7, *Thermal Radiative Properties—Metallic Elements and Alloys*. New York: Plenum Publishing Co., 1970.
2. Touloukian, Y. S., and D. P. DeWitt, *Thermophysical Properties of Matter*, Vol. 8, *Thermal Radiative Properties—Nonmetallic Solids*. New York: Plenum Publishing Co., 1970.
3. Touloukian, Y. S., D. P. De Witt, and R. S. Hernicz, *Thermophysical Properties of Matter*, Vol. 9, *Thermal Radiative Properties—Coatings*. New York: Plenum Publishing Co., 1970.

4. Gubareff, G. G., J. E. Jansen, and R. H. Torborg, *Thermal Radiation Properties Survey*, Honeywell Research Center, Honeywell Regulator Company, Minneapolis, Minn., 1960.
5. Sparrow, E. M., and R. D. Cess, *Radiation Heat Transfer*. Washington, D.C.: Hemisphere Publishing Co., 1978.
6. Planck, M., *The Theory of Heat Radiation*. New York: Dover Publications, Inc., 1959.
7. Dunkle, R. V., J. T. Gier, and co-workers, "Snow Characteristics Project, Progress Report," University of California, Berkeley, 1953.
8. Seban, R. A., "The Emissivity of Transition Metals in the Infrared," *J. Heat Transfer*, C87, 1965, 173.
9. Sieber, W., "Zusammensetzung der von Werk- und Baustoffen zurückgeworfenen Wärmestrahlung," *Z. Tech. Physik*, 22, 1941, 130.
10. Schmidt, E., and E. Eckert, "Über die Richtungsverteilung der Wärmestrahlung," *Forsch. Ing. Wes.*, 6, 1935, 175.
11. Siegel, R., and J. R. Howell, *Thermal Radiation Heat Transfer*, 2nd ed. New York: McGraw-Hill Book Company, 1981.
12. Dietz, A. G. H., "Diathermanous Materials and Properties of Surfaces," in *Space Heating with Solar Energy* by R. W. Hamilton. Cambridge, Mass.: The MIT Press, 1954.
13. Seal, M., "The Increasing Applications of Diamond as an Optical Material and in the Electronics Industry," *Ind. Diamond Rev.*, April 1978, 130.
14. Williams, D. A., T. A. Lappin, and J. A. Duffie, "Selective Radiation Properties of Particular Coatings," *J. Engr. Power*, 95A, 1963, 213.
15. Edwards, D. K., "Radiation Interchange in a Nongray Enclosure Containing an Isothermal Carbon-dioxide–Nitrogen Gas Mixture," *J. Heat Transfer*, c84, 1962, 1.
16. Hottel, H. C., and R. S. Egbert, "Radiant Heat Transmission from Water Vapor," *AIChE Trans.*, 38, 1942.
17. Hottel, H. C., "Radiant Heat Transmission," Chap. 4 in *Heat Transmission*, 3rd ed., by W. H. McAdams. New York: McGraw-Hill Book Company, 1954.
18. Eckert, E. R. G., and R. M. Drake, *Analysis of Heat and Mass Transfer*. New York: McGraw-Hill Book Company, 1972.
19. Mackey, C. O., L. T. Wright, Jr., R. E. Clark, and N. R. Gay, "Radiant Heating and Cooling, Part *I*," *Cornell Univ. Eng. Expt. Sta. Bull.*, 32, 1943.
20. Hottel, H. C., "Radiant Heat Transmission," *Mech. Eng.*, 52, 1930, 699.
21. Howell, J. R., *A Catalog of Radiation Configuration Factors*. New York: McGraw-Hill Book Company, 1982.
22. Hamilton, D. C., and W. R. Morgan, "Radiant Interchange Configuration Factors," *NACA TN 2836*, 1952.
23. Kreyszig, E., *Advanced Engineering Mathematics*, 4th ed. New York: John Wiley & Sons, 1979.
24. Gebhart, B., *Heat Transfer*, 2nd ed. New York: McGraw-Hill Book Company, 1971.
25. Duffie, J. A., and W. A. Beckman, *Solar Energy Thermal Processes*. New York: John Wiley & Sons, 1974.
26. Thekaekara, M. P., "Data on Incident Solar Radiation," *Suppl. Proc. 20th Ann. Meeting Inst. Environ. Sci.*, 21, 1974.
27. Daniels, F., Jr., "Physical Factors in Sun Exposure," *Arch. Dermatol.*, 85, 1962, 358.

28. Thekaekara, M. P., and A. J. Drummond, ''Standard Values for the Solar Constant and Its Special Components,'' *Nat. Phys. Sci.*, 229, 1971, 6.
29. Anderson, E. E., *Solar Energy Fundamentals for Designers and Engineers*. Reading, Mass: Addison-Wesley Publishing Co., 1982.
30. Howell, J. R., R. B. Bannerot, and G. C. Vliet, *Solar-Thermal Energy Systems Analysis and Design*. New York: McGraw-Hill Book Company, 1982.

CHAPTER 6

1. Keenan, J. H., F. G. Keyes, P. G. Hill, and J. G. Moore, *Steam Tables*. New York: John Wiley & Sons, 1978.
2. Prandtl, L., ''Über Flüssigkeitsbewegung bei sehr kleiner Reibung,'' *Proc. 3rd Int. Math. Kong. Heidelberg*, 1904.
3. White, F. M., *Heat Transfer*, Reading, Mass.: Addison-Wesley Publishing Co., 1984.
4. Adiutori, E. F. ''Origins of the Heat Transfer Coefficient,'' *Mechanical Engineering*, 112, 1990, 46.
5. Buckingham, E., ''On Physically Similar Systems: Illustrations of the Use of Dimensional Analysis,'' *Phys. Rev.*, 4, 1914, 345.
6. Bridgeman, P. W., *Dimensional Analysis*. New Haven, Conn.: Yale University Press, 1931.
7. Langhaar, H. L., *Dimensional Analysis and Theory of Models*. New York: John Wiley & Sons, 1951.
8. Kreith, F., and M. S. Bohn, *Principles of Heat Transfer*, 4th ed. New York: Harper & Row, Publishers, 1986.
9. Lienhard, J. H., *A Heat Transfer Textbook*, 2nd ed. Englewood, Cliffs, N.J.: Prentice Hall, 1987.
10. Fox, R. W., and A. T. McDonald, *Introduction to Fluid Mechanics*. New York: John Wiley & Sons, 1978.
11. Arpaci, V. S., and P. S. Larsen, *Convection Heat Transfer*. Englewood Cliffs, N.J.: Prentice Hall, 1984.
12. John, E. A., *Gas Dynamics*. Englewood Cliffs, N.J.: Prentice Hall, 1984.
13. Benedict, R. P., *Fundamentals of Gas Dynamics*. New York: John Wiley & Sons, 1983.

CHAPTER 7

1. Graetz, L., ''Über die Wärmeleitfähigkeit von Flüssigkeiten,'' *Ann. Phys. Chem.*, 25, 1885, 337.
2. Hansen, M., ''Velocity Distribution in the Boundary Layer of a Submerged Plate,'' *NACA TM* 585, 1930.
3. Langhaar, H. L., ''Steady Flow in the Transition Length of a Straight Tube,'' *J. Appl. Mech.*, 9, 1942, A55.
4. Kays, W. M., ''Numerical Solution for Laminar Flow Heat Transfer in Circular Tubes,'' *Trans. ASME*, 77, 1955, 1265.

5. Sellars, J. R., M. Tribus, and J. S. Klein, ''Heat Transfer to Laminar Flows in a Round Tube or Flat Conduit, the Graetz Problem Extended,'' *Trans. ASME*, 78, 1956, 441.
6. Seigel, R., E. M. Sparrow, and T. M. Hallman, ''Steady Laminar Heat Transfer in Circular Tube with Prescribed Wall Heat Flux,'' *Appl. Sci. Res.*, A7, 1958, 386.
7. Goldberg, P., ''A Digital Computer Solution for Laminar Flow Heat Transfer in Circular Tubes,'' M. S. Thesis, Mechanical Engineering Department, Massachusetts Institute of Technology, 1958.
8. Heaton, H. S., W. C. Reynolds, and W. M. Kays, ''Heat Transfer in Annular Passages, Simultaneous Development of Velocity and Temperature Fields in Laminar Flow,'' *Int. J. Heat Mass Transfer*, 7, 1964, 763.
9. Sparrow, E. M., S. H. Lin, and T. Lundren, ''Flow Development in the Hydrodynamic Entrance Region of Tubes and Ducts,'' *Phys. Fluids*, 7, 1964, 338.
10. McComas, S. T., and E. R. G. Eckert, ''Laminar Pressure Drop Associated with the Continuum Entrance Region and for Slip Flow in a Circular Tube,'' *J. Appl. Mech.*, 32, 1965, 765.
11. Kays, W. M., and M. E. Crawford, *Convection Heat and Mass Transfer*, 2nd ed. New York: McGraw-Hill Book Company, 1980.
12. *Handbook of Heat Transfer Fundamentals*, 2nd ed., edited by W. M. Rohsenow, J. P. Hartnett, and E. N. Ganic. New York: McGraw-Hill Book Company, 1985.
13. Senecal, V. E., ''Characteristics of Transition Flow in Smooth Tubes,'' Ph.D. Thesis, Carnegie Institute of Technology, 1952.
14. Stanton, T. E., and J. R. Pannell, ''Similarity of Motion in Relation to the Surface Friction of Fluids,'' *Trans. Roy. Soc. (London)*, A214, 1914, 199.
15. Senecal, V. E., and R. R. Rothfus, ''Transition Flow of Fluids in Smooth Tubes,'' *Chem. Eng. Progr.*, 49, 1953, 533.
16. White, F. M., *Viscous Fluid Flow*, 2nd ed. New York: McGraw-Hill Book Company, 1991.
17. Burmeister, L. C., *Convective Heat Transfer*. New York: John Wiley & Sons, 1983.
18. Schlichting, H., *Boundary Layer Theory*, 7th ed. New York: McGraw-Hill Book Company, 1979.
19. Patankar, S. V., *Numerical Heat Transfer and Fluid Flow*. Washington, D.C.: Hemisphere Publishing Co., 1980.
20. *Handbook of Numerical Heat Transfer*, edited by W. J. Minkowycz, E. M. Sparrow, G. E. Schneider, and R. H. Pletcher. New York: John Wiley & Sons, 1988.
21. Anderson, D. A., J. C. Tannehill, and R. H. Pletcher, *Computational Fluid Mechanics and Heat Transfer*. New York: Hemisphere Publishing Corp., 1984.
22. Thomas, L. C., and W. L. Amminger, ''A Practical One-Parameter Integral Method for Laminar Incompressible Boundary Layer Flow with Transpiration,'' *J. Appl. Mech.*, 110, 1988, 474.
23. Thomas, L. C., and W. L. Amminger, ''A Two-Parameter Integral Method for Laminar Transpired Thermal Boundary Layer Flow,'' *AIAA J.*, 28, 1990, 205.
24. Dorodnitsyn, A. A., *Advances in Aeronautical Sciences*, Vol. 3, New York: Pergamon Press, 1960.
25. Holt, M., *Numerical Methods in Fluid Dynamics*. Berlin: Springer-Verlag, 1984.
26. Fletcher, C. A. J., *Computational Galerkin Methods*. Berlin: Springer-Verlag, 1984.
27. Blasius, H., ''Grenzschichten in Flüssigkeiten mit kleiner Reibung,'' *Z. Math. Phys.* 56, 1908, 1; English translation in *NACA TM* 1256.

28. Pohlhausen, Z., "Der Warmeaustausch zwischen festen Korpern und Flussigkeiten mit kleiner Reibung und kleiner Warmeleitung," *Z. Angew. Math. Mech.*, 1, 1921, 115.
29. Eckert, E. R. G., and R. M. Drake, *Analysis of Heat and Mass Transfer*. New York: McGraw-Hill Book Company, 1972.
30. Levy, S., "Heat Transfer to Constant-Property Laminar-Boundary Layer-Flows with Power-Function Free-Stream Velocity and Wall Temperature Variation," *J. Aeronaut. Sci.*, 19, 1952, 341.
31. Gebhart, B., Y. Jaluria, R. L. Mahajan, and B. Sammakia, *Buoyancy-Induced Flows and Transport*. Washington, D.C.: Hemisphere Publishing Co., 1988.
32. Boussinesq, J., *Théorie Analytique de la Chaleur*, Vol. 2. Paris: Gauthier-Villars, 1903.
33. Ostrach, S., "An Analysis of Laminar Free-Convection Flow and Heat Transfer about a Plate Parallel to the Direction of the Generating Body Force," *NACA Rept.* 1111, 1953.
34. Schmidt, E., and W. Beckman, "Das Temperatur-und Geschwindig keitsfeld von einer Wärmeabgebenden, senkrechten Platte bei natürlicher Konvektion, *Forsch-Ing. Wes*, 1, 1930, 391.
35. Ede, A. J., "Advances in Free Convection," in *Advances in Heat Transfer*, 4, 1967, 1.
36. Runstadler, P. W., S. J. Kline, and W. C. Reynolds, "An Experimental Investigation of the Flow Structure of the Turbulent Boundary Layer," *Rept. MD-8*, Stanford University, 1963.
37. Knudsen, J. D., and D. L. Katz, *Fluid Dynamics and Heat Transfer*. New York: McGraw-Hill Book Company, 1958.
38. Reynolds, O., *Scientific Papers of Osborne Reynolds*, Vol. II. London: Cambridge University Press, 1901.
39. Boussinesq, J., "Théorie de l'écoulement tourbillant," *Mem. Pres. Acad. Sci.* (Paris), 23, 1877, 46.
40. Hussain, A. K. M. F., and W. C. Reynolds, "Measurements in Fully Developed Turbulent Channel Flow," *J. Fluids Eng.*, 97, 1975, 569.
41. van Driest, E. R., "Turbulent Boundary Layer in Compressible Fluids," *J. Aero. Sci.*, 18, 1951, 145.
42. Kays, W. M., and R. J. Moffat, *Studies in Convection*, Vol. 1, London: Academic Press, 1975, 213.
43. Cebeci, T., and A. M. O. Smith, *Analysis of Turbulent Boundary Layers*. New York: Academic Press, 1974.
44. Rotta, J., "Das in Wandnähe gültige Geschwindigkeitsgesetz turbulenter Strömungen," *Ing. Arch.*, 18, 1950, 277.
45. Reichardt, H., "Vollständige Darstellung der turbulenten Geschwindigkeitsverteilung in glatten Leitungen," *ZAMM*, 31, 1951, 208.
46. Deissler, R. G., "Analysis of Turbulent Heat Transfer, Mass Transfer, and Friction in Smooth Tubes at High Prandtl Numbers," *NACA TN* 3145, 1954.
47. Spalding, D. B., "A Single Formula for the Law of the Wall," *J. Appl. Mech.*, 28, 1961, 455.
48. Prandtl, L., "Eine Beziehung zwischen Wärmeaustausch and Strömungswiderstand der Flüssigkeiten," *Phys. Zeit.*, 11, 1910, 1072.
49. Bradshaw, P., "Compressible Turbulent Shear Layers," in *Annual Review of Fluid Mechanics*, Vol. 9. Palo Alto, Calif.: Annual Reviews, Inc., 1977.

50. Simpson, R. L., D. G. Whitten, and R. J. Moffat, "An Experimental Study of the Turbulent Prandtl Number of Air with Injection and Suction," *Int. J. Heat Mass Transfer*, 13, 1970, 125.

51. Azer, N. Z., and B. T. Chao, "Turbulent Heat Transfer in Liquid Metals—Fully Developed Pipe Flow with Constant Wall Temperature," *Int. J. Heat Mass Transfer*, 3, 1961, 77.

52. Jenkins, R., "Variation of Eddy Conductivity with Prandtl Modulus and Its Use in Predictions of Turbulent Heat Transfer Coefficients," *Heat Transfer and Fluid Mechanics Institute*, Stanford University, 1951.

53. Kline, S. J., B. J. Cantwell, and G. M. Lilley, *Proc. Complex Turbulent Flows*, Vols. 1–3, Department of Mechanical Engineering, Stanford University, Stanford, Calif., 1981.

54. Bradshaw, P., D. H. Ferriss, and N. P. Atwell, "Calculation of Boundary Layer Development Using the Turbulence Energy Equation," *J. Fluid Mech.*, 28, 1967, 593.

55. Ng, K. H., and D. B. Spalding, "Turbulence Model for Boundary Layer Near Walls," *Phys. Fluids*, 15, 1972, 20.

56. Jones, W. P., and B. E. Launder, "The Prediction of Laminarization with a Two-Equation Model of Turbulence, *Int. J. Heat Mass Transfer*, 15, 1972, 301.

57. Amano, R. S., and J. C. Chai, "Transport Models of the Turbulent Velocity-Temperature Products for Computations of Recirculating Flows," *Numer. Heat Transfer*, 14, 1988, 75.

58. Ciofalo, M., and M. W. Collins, "K-ϵ Predictions of Heat Transfer in Turbulent Recirculating Flows Using an Improved Wall Treatment," *Numer. Heat Transfer, Part B*, 15, 1989, 21.

59. Goldberg, U. C., and S. R. Chakravarthy, "Prediction of Separated Flows with a New Backflow Turbulence Model," *AIAA J.*, 26, 1988, 405.

60. Martinuzzi, R., and A. Pollard, "Comparative Study of Turbulence Models in Predicting Turbulent Pipe Flow, Part I, Algebraic Stress and K-ϵ Models," *AIAA J.*, 27, 1989, 29.

61. Mansour, N. N., J. Kim, and P. Moin, "Near-Wall K-ϵ Turbulence Modeling," *AIAA J.*, 27, 1989, 1068.

62. Pletcher, R. H., and O. K. Kwon, "Prediction of Some Complex Turbulent Flows Using the Boundary-Layer Equations and Viscous-Inviscid Interaction," *AFOSR-HTTM Stanford Conference on Complex Turbulent Flows*, III, 1980–81, 1479.

63. Hanjalic, K., and B. E. Launder, "A Reynolds Stress Model of Turbulence and Its Application to Asymmetric Shear Flow," *J. Fluid Mech.*, 52, 1972, 609.

64. Daly, B. J., and F. H. Harlow, "Transport Equations in Turbulence," *Phys. Fluids*, 13, 1970, 2634.

65. Launder, B. E., G. J. Reece, and W. Rodi, "Progress in the Development of a Reynolds Stress Turbulence Closure," *J. Fluid Mech.*, 68, 1975, 537.

66. Hogg, S., and M. A. Leschziner, "Computation of Highly Swirling Confined Flow with a Reynolds Stress Turbulence Model," *AIAA J.*, 27, 1989, 57.

67. Pollard, A., and R. Martinuzzi, "Comparative Study of Turbulence Models in Predicting Turbulent Pipe Flow, Part II, Reynolds Stress and K-ϵ Models," *AIAA J.*, 27, 1989, 1714.

68. Larsson, R.; "Impementation of an Algebraic Stress Model for Turbulence Generated Secondary Currents," *Phoenics J. Comp. Fluid Dynamics and Its Applications*, 2, 1989, 368.

69. Markatos, N., "The Mathematical Modelling of Turbulent Flows," *Applied Mathematical Modeling*, 10, 1986, 190.

70. Rapley, C. W., "Turbulent Flow in a Tube Containing an Offset Rod," *Int. J. Numer. Methods Fluids*, 8, 1988, 305.

71. Armfield, S. W., and C. A. J. Fletcher, "Comparison of K-ϵ and Algebraic Reynolds Stress Models for Swirling Diffuser Flow," *Int. J. Numer. Methods Fluids*, 9, 1989, 987.
72. Ilegusi, O. J., and D. B. Spalding, "Prediction of Fluid Flow and Heat Transfer Characteristics of Turbulent Shear Flows with a Two-Fluid Model of Turbulence," *Int. J. Heat Mass Transfer*, 32, 1989, 767.
73. Malin, M. R., and D. B. Spalding, "Flow and Heat Transfer in Two-Dimensional Turbulent Wall Jets and Plumes," *PhysicoChemical Hydrodynamics*, 9, 1987, 237.
74. Leslie, D. C., and S. Gao, "The Stability of Spectral Schemes for the Large Eddy Simulation of Channel Flows," *Int. J. Numer. Methods Fluids*, 8, 1988, 1107.
75. Jou, W.-H., and J. J. Riley, "Progress in Direct Numerical Simulations of Turbulent Reacting Flows," *AIAA J.*, 27, 1989, 1543.
76. Danckwerts, P. V., "Significance of Liquid-Film Coefficients in Gas Absorption," *I and E C*, 43, 1951, 1460.
77. Einstein, H. A., and H. L. Li, "The Viscous Sublayer along a Smooth Boundary," *ASCE J. Mech. Div.*, 82, 1956, 293.
78. Thomas, L. C., "The Surface Rejuvenation Model of Wall Turbulence: Inner Laws for Velocity and Temperature," *Int. J. Heat Mass Transfer*, 23, 1980, 1097.
79. Sinai, Y. L., "A Wall Function for the Temperature Variance in Turbulent Flow Adjacent to a Diabatic Wall," *J. Heat Transfer*, 109, 1987, 861.
80. Lindgren, E. R., Department of Civil Engineering, Oklahoma State University, Rept. IAD621071, 1965.
81. Nikuradse, J., "Widerstandsgesetz und Geschwindigkeit von turbulenten Wasserstromungen in glatten und rauhen Rohren," *Proc. Third Int. Cong. Appl. Mech.*, 1, 1930, 239.
82. Gowen, R. A., and J. W. Smith, "The Effects of the Prandtl Number on Temperature Profiles for Heat Transfer in Turbulent Pipe Flow," *Chem. Eng. Sci.*, 22, 1967, 1701.
83. Johnk, R. E., and T. J. Hanratty, "Temperature Profiles for Turbulent Flow of Air in a Pipe—II," *Chem. Eng. Sci.*, 17, 1962, 802.
84. Petukhov, B. S., and V. V. Kirillov, "Heat Exchange for Turbulent Flow of Liquid in Tubes," *Teploenerg.*, 4, 1958.
85. Sams, E. W., and L. G. Desmon, "Heat Transfer from High Temperature Surfaces to Fluids," *NACA Memo* E9D12, 1949.
86. Deissler, R. G., and C. S. Eian, "Analytical and Experimental Investigation of Heat Transfer with Variable Fluid Properties," *NACA TN* 2629, 1952.
87. Barnes, J. F., and J. D. Jackson, "Heat Transfer to Air, Carbon Dioxide and Helium Flowing through Smooth Circular Tubes under Conditions of Large Surface/Gas Temperature Ratio," *J. Mech. Eng. Sci.*, 3, 1961, 303.
88. Thomas, L. C., and S. M. F. Hasani, "Supplementary Boundary Layer Approximations for Turbulent Flows," *J. Fluids Eng.*, 111, 1989, 420.
89. Thomas, L. C., and H. M. Kadry, "A One-Parameter Integral Method for Turbulent Transpired Boundary Layer Flow," *J. Fluids Eng.*, 112, 1990, 205.
90. Coles, D., "The Law of the Wake in the Turbulent Boundary Layer," *J. Fluid Mech.*, 1, 1956, 191.
91. Andersen, P. S., W. M. Kays, and M. J. Moffat, "The Turbulent Boundary Layer on a Porous Plate: An Experimental Study of the Fluid Mechanics for Adverse Free-Stream Pressure Gradients," *Department of Mechanical Engineering, Stanford University*, Stanford, Calif. 1972.

92. Christoph, G. H., R. C. Lessmann, and F. M. White, "Calculations of Turbulent Heat Transfer and Skin Friction," *AIAA J.*, 11, 1973, 1046.
93. Thomas, L. C., and M. M. Al-Sharif, "An Integral Analysis for Heat Transfer in Turbulent Incompressible Boundary Layer Flow," *J. Heat Transfer*, 103, 1981, 772.
94. Reynolds, W. C., W. M. Kays, and S. J. Kline, "Heat Transfer in the Turbulent Incompressible Boundary Layer—III: Arbitrary Wall Temperature and Heat Flux," *NASA Memo* 12-3-58W, 1958.
95. Colburn, A. P., "A Method of Correlating Forced Convection Heat Transfer Data and Comparison with Fluid Friction," *Trans. AIChE*, 29, 1933, 1974.
96. Pepper, D. W., and A. J. Baker, "Finite Differences versus Finite Elements," in *Handbook of Numerical Heat Transfer*, edited by W. J. Minkowycz, E. M. Sparrow, G. E. Schneider, and R. H. Pletcher. New York: John Wiley & Sons, 1988.
97. Anderson, D. A., J. C. Tannehill, and R. H. Pletcher, *Computational Fluid Mechanics and Heat Transfer*. New York: McGraw-Hill Book Company, 1984.
98. *Handbook of Numerical Heat Transfer*, edited by W. J. Minkowycz, E. M. Sparrow, G. E. Schneider, and R. H. Pletcher. New York: John Wiley & Sons, 1988.
99. Leonard, B. P., "Elliptic Systems: Finite-Difference Method IV," in *Handbook of Numerical Heat Transfer*, edited by W. J. Minkowycz, E. M. Sparrow, G. E. Schneider, and R. H. Pletcher. New York: John Wiley & Sons, 1988.
100. Taborek, J., and J. W. Palen, "Process Equipment Design by Digital Computer Programs—A Philosophical Approach," *I. Chem. Eng. Symp., Series no.* 35, 1972.

CHAPTER 8

1. Langhaar, H. L., "Steady Flow in the Transition Length of a Straight Tube," *J. Appl. Mech.*, 9, 1942, A55.
2. *Engineering Sciences Data*, London: Heat Transfer Subsciences, Technical Editing and Production Ltd., 1970.
3. *Handbook of Heat Transfer*, edited by W. M. Rohsenow and J. P. Hartnet. New York: McGraw-Hill Book Company, 1973.
4. Kays, W. M., and A. L. London, *Compact Heat Exchangers*, 3rd ed. New York: McGraw-Hill Book Company, 1984.
5. *Heat Exchanger Design Handbook*, Washington, D.C.: Hemisphere Publishing Co., 1983.
6. Kays, W. M., and M. E. Crawford, *Convective Heat and Mass Transfer*, 2nd ed. New York: McGraw-Hill Book Company, 1980.
7. Kays, W. M., and S. H. Clark, TR No. 17, Mechanical Engineering Department, Stanford University, 1953.
8. Lundberg, R. E., W. C. Reynolds, and W. M. Kays, "Heat Transfer with Laminar Flow in Concentric Annuli with Constant and Variable Wall Temperature and Heat Flux," *NASA TN*-1972, Washington, D.C., 1963.
9. Lundgren, T. S., E. M. Sparrow, and J. B. Starr, "Pressure Drop Due to the Entrance Region in Ducts of Arbitrary Cross-Section," *J. Basic Eng.*, 86, 1964, 620.
10. Sparrow, E. M., T. S. Chen, and V. K. Johnson, "Laminar Flow and Pressure Drop in Internally Finned Annular Ducts," *Int. J. Heat Mass Transfer*, 7, 1964, 583.

11. Sparrow, E. M., and A. Haji-Sheikh, ''Flow and Heat Transfer in Ducts of Arbitrary Shape with Arbitrary Thermal Boundary Conditions,'' *J. Heat Transfer*, 88, 1966, 351.
12. McComas, S. T., and E. R. G. Eckert, ''Laminar Pressure Drop Associated with the Continuum Entrance Region and for Slip Flow in a Circular Tube,'' *J. Appl. Mech.*, 32, 1965, 765.
13. Burmeister, L. C., *Convective Heat Transfer*. New York: John Wiley & Sons, 1983.
14. Kays, W. M., ''Loss Coefficients for Abrupt Changes in Flow Cross Section with Low Reynolds Number Flow in Single and Multiple Tube Systems,'' *Trans. ASME*, 72, 1950, 91.
15. Hausen, H., ''Darstellung des Wärmeübergangen in Rohren durch verallgemeinerte Potenzbeziehungen,'' *Z. Ver. Deut. Ing.*, 4, 1943, 91.
16. Sellars, J. R., M. Tribus, and J. S. Klein, ''Heat Transfer to Laminar Flow in a Round or Flat Conduit—The Graetz Problem Extended,'' *Trans. ASME*, 78, 1956, 441.
17. Kays, W. M., ''Numerical Solutions for Laminar-Flow Heat Transfer in Circular Tubes,'' *Trans. ASME*, 77, 1955, 1265.
18. Seider, E. M., and C. E. Tate, ''Heat Transfer and Pressure Drop of Liquids in Tubes,'' *Ind. Eng. Chem.*, 28, 1936, 1429.
19. Shah, R. K., and M. S. Bhatti, ''Assessment of Test Techniques and Correlations for Single-Phase Heat Exchangers,'' 6th NATO Advanced Study Institute, Thermal/Hydraulic Fundamentals and Design of Two-Phase Heat Exchangers, Povadevarzin, Portugal, July 1987.
20. Nikuradse, J., ''Wärmeübergang in Rohrleitungen,'' *Forsch. Ing. Wes.*, 1932, 356.
21. Petukhov, B. S., and V. N. Popov, *Teplofiz. Vysok. Temperatur* (High Temperature Heat Physics), 1, 1963.
22. Petukhov, B. S., and V. V. Kirillov, ''Heat Exchange for Turbulent Flow of Liquid in Tubes,'' *Teploenerg.*, 4, 1958.
23. White, F. M., *Viscous Fluid Flow*, 2nd ed. New York: McGraw-Hill Book Company. 1991
24. Petukhov, B. S., "Heat Transfer and Friction in Turbulent Pipe Flow with Variable Physical Properties," in *Advances in Heat Transfer*. New York: Academic Press, 1970, 504.
25. Barnes, J. F., and J. D. Jackson, ''Heat Transfer to Air, Carbon Dioxide and Helium Flowing through Smooth Circular Tubes under Conditions of Large Surface/Gas Temperature Ratio,'' *J. Mech. Eng. Sci.*, 3, 1961, 303.
26. Deissler, R. G., and C. S. Eian, ''Analytical and Experimental Investigation of Heat Transfer with Variable Fluid Properties,'' *NACA TN* 2629, 1952.
27. Sams, E. W., and L. G. Desmon, ''Heat Transfer from High Temperature Surface to Fluids,'' *NACA Memo E9D12*, 1949.
28. Dittus, F. W., and L. M. K. Boelter, ''Heat Transfer in Automobile Radiators of the Tubular Type,'' Univ. of California–Berkeley, *Pub. Eng.* 2, 1930, 443.
29. Colburn, A. P., ''A Method of Correlating Forced Convection Heat Transfer Data and a Comparison with Fluid Friction,'' *Trans. AIChE*, 29, 1933, 1974.
30. Kays, W. M., and E. Y. Leung, ''Heat Transfer in Annular Passages: Hydrodynamically Developed Turbulent Flow with Arbitrarily Prescribed Heat Flux,'' *Int. J. Heat Mass Transfer*, 6, 1963, 537.
31. Patel, V. C., and M. R. Head, ''Some Observations on Skin Friction and Velocity Profiles in Fully Developed Pipe and Channel Flow,'' *J. Fluid Mech.*, 38, 1969, 181.
32. Taborek, J., "Design Method for Tube-Side Laminar and Turbulent Transition Flow Regime

with Effects of Natural Convection," Presented in Open Forum Session, *Ninth Int. Heat Transfer Conf.*, Jerusalem, Israel, 1990.

33. Subbotin, V. I., A. K. Papovyants, P. L. Kirillov, and N. N. Ivanovskii, "A Study of Heat Transfer to Molten Sodium in Tubes," *Soviet J. Atomic Energy*, 13, 1962, 380.
34. Seban, R. A., and T. T. Shimazaki, "Heat Transfer to Fluid Flowing Turbulently in a Smooth Pipe with Walls at Constant Temperature," *Trans. ASME*, 73, 1951, 803.
35. Skupinski, E., J. Tortel, and L. Vautrey, "Determination des Coefficients de Convection d'un Alliage Sodium-Potassium dans un Tube Circulative," *Int. J. Heat Mass Transfer*, 8, 1965, 937.
36. Lubarsky, B., and S. J. Kaufman, "Review of Experimental Investigations of Liquid-Metal Heat Transfer," *NASA TN* 3336, 1955.
37. Allen, R. W., and E. R. G. Eckert, "Friction and Heat-Transfer Measurements to Turbulent Pipe Flow of Water (Pr = 7 and 8) at Uniform Wall–Heat Flux," *J. Heat Transfer*, 86, 1964, 301.
38. Sleicher, C. A., and M. W. Rouse, "A Convenient Correlation for Heat Transfer to Constant and Variable Property Fluids in Turbulent Pipe Flow," *Int. J. Heat Mass Transfer*, 18, 1975, 677.
39. Bergles, A. E., "Enhancement of Heat Transfer," *Sixth Int. Heat Transfer Conf.*, Toronto, 6, 1978, 89.
40. Bergles, A. E., V. Nirmalan, G. H. Junkhan, and R. L. Webb, Bibliography on Augmentation of Convection Heat and Mass Transfer—II, Rept. HTL-31, ISU-ERI-Ames—84221, Iowa State University, Ames, Iowa, 1983.
41. Moody, L. F., "Friction Factors for Pipe Flow," *Trans. ASME*, 66, 1944, 671.
42. DeLorenzo, B., and E. D. Anderson, "Heat Transfer and Pressure Drop of Liquids in Double-Pipe Fin-Tube Heat Exchangers," *Trans. ASME*, 67, 1945, 697.
43. Taborek, J., "Double-Pipe Heat Exchanger Design," *Heat Transfer Engineering*, 1992.
44. Shah, R. K., "Classification of Heat Exchangers," in *Heat Exchangers—Thermal-Hydraulic Fundamentals and Design*. New York: McGraw-Hill Book Company, 1981.
45. Blasius, H., "Das Ähnlichkeitsgesetz bei Reibungsvorgängen in Flüssigkeiten," *Forsch. Ing. Wes.*, 131, 1913.
46. Liepmann, H. W., and S. Dhawan, "Direct Measurements of Local Skin Friction in Low-Speed and High-Speed Flow," *Proc. First U.S. Nat. Cong. Appl. Mech.*, 1951.
47. Wieghardt, K., and W. Tillmann, "On the Turbulent Friction Layer for Rising Pressure," *NACA TM* 1314, 1951.
48. Eckert, E. R. G., "Engineering Relations for Friction and Heat Transfer to Surfaces in High Velocity Flow," *J. Aero. Sci.*, 22, 1955, 585.
49. Schlichting, H., *Boundary Layer Theory*, 7th ed. New York: McGraw-Hill Book Company, 1979.
50. Whitaker, S., *Elementary Heat Transfer Analysis*. New York: Pergamon Press, 1976.
51. Whitaker, S., "Forced Convection Heat Transfer Correlations for Flow in Pipes, Past Flat Plates, Single Cylinders, Single Spheres and for Flow in Packed Beds and Tube Bundles," *AIChE J.*, 18, 1972, 361.

52. Achenbach, E., "Heat Transfer from Spheres up to Re $= 6 \times 10^6$," *Sixth Int. Heat Transfer Conf.*, Toronto, 5, 1978, 341.
53. Ishiguro, R., K. Sugiyama, and T. Kumada, "Heat Transfer around a Circular Cylinder in a Liquid-Sodium Crossflow," *Int. J. Heat Mass Transfer*, 22, 1979, 1041.
54. Witte, L. C., "An Experimental Study of Forced-Convection Heat Transfer from a Sphere to Liquid Sodium," *J. Heat Transfer*, 90, 1968, 9.
55. Jacob, M., *Heat Transfer*, Vol. 1. New York: John Wiley & Sons, 1949.
56. Grimison, E. D., "Correlation and Utilization of New Data on Flow of Gases over Tube Banks," *Trans. ASME*, 59, 1937, 538.
57. Zukauskas, A., "Heat Transfer in Banks of Tubes in Crossflow of Fluid," *'Mintis' Vilnius*, 1968, 124.
58. Zukauskas, A., "Convective Heat Transfer in Crossflow," in *Handbook of Single-Phase Convection Heat Transfer*, edited by S. Kakac, R. K. Shah, and W. Aung. New York: John Wiley & Sons, 1987.
59. Bergelin, O. P., G. A. Brown, and S. C. Doberstein, "Heat Transfer and Fluid Friction during Flow across Banks of Tubes," *Trans. ASME*, 74, 1952.
60. *Steam—Its Generation and Use*, 38th ed. New York: The Babcock and Wilcox Company, 1975.
61. Jacob, M., "Heat Transfer and Flow Resistance in Crossflow of Gases Over Tube Banks," *Trans. ASME*, 60, 1938, 384.
62. Gaddis, E. S., and V. Gnielinski, "Pressure Drop in Crossflow Across Tube Bundles," *Int. Chem. Eng.*, 25, 1985, 1.
63. Gnielinski, V., A. Zukauskas, and A. Shrinkska, "Banks of Plain and Finned Tubes," in *Heat Exchanger esign Handbook*. Washington, D.C.: Hemisphere Publishing Co., 1983.
64. Gray, D. L., and R. L. Webb, "Heat Transfer and Friction Correlations for Plate Fin-and-Tube Heat Exchanger having Plain Fins," *Heat Transfer*, 6, 1986, 2745.
65. Webb, R. L., "Air-Side Heat Transfer in Finned Tube Heat Exchangers," *Heat Transfer Eng.*, 1, 1978, 294.
66. Briggs, D. E., and E. H. Young, "Convection Heat Transfer and Pressure Drop of Air Flowing Across Triangular Pitch Banks of Finned Tubes," *Chemical Engineering Progress Symposium Series*, 59, 1963, 1.
67. Robinson, K. K., and D. E Briggs, "Pressure Drop of Air Flowing across Triangular Pitch Banks of Finned Tubes," *Chemical Engineering Progress Symposium Series*, 62, 1966, 177.
68. Rabas, T. J., P. W. Eckels, and R. A. Sabatino, "The Effect of Fin Density on the Heat Transfer and Pressure Drop Performance of Low-Finned Tube Banks," *ASME Paper No.* 80-HE-97, 1980.

CHAPTER 9

1. Ostrach, S., "An Analysis of Laminar Free-Convection Flow and Heat Transfer about a Flat Plate Parallel to the Direction of the Generating Body Force," *NACA Rept.*, 1953, 1111.
2. Ede, A. J., in *Advances in Heat Transfer*, Vol. 4. New York: Academic Press, 1967, 1.
3. Churchill, S. W., and R. Usagi, "A General Expression for the Correlation of Rates of Transfer and Other Phenomena," *AIChE J.*, 18, 1972, 1121.

4. Eckert, E. R. G., and T. W. Jackson, "Analysis of Turbulent Free-Convection Boundary Layer on a Flat Plate," *NACA Rept.* 1015, 1951.
5. Churchill, S. W., and H. H. S. Chu, "Correlating Equations for Laminar and Turbulent Free Convection from a Vertical Plate," *Int. J. Heat Mass Transfer*, 18, 1975, 1323.
6. Bayley, F. J., J. M. Owen, and A. B. Turner, *Heat Transfer*. London: Thomas Nelson & Sons Ltd., 1972.
7. McAdams, W. H., *Heat Transmission*, 3rd ed. New York: McGraw-Hill Book Company, 1954.
8. Fujii, T., and M. Fujii, The Dependence of Local Nusselt Number on Prandtl Number in the Case of Free Convection along a Vertical Surface with Uniform Heat Flux," *Int. J. Heat Mass Transfer*, 19, 1976, 121.
9. Churchill, S. W., and H. Ozoe, "A Correlation for Laminar Free Convection from a Vertical Plate," *J. Heat Transfer*, 95, 1973, 540.
10. Vliet, G. C., and C. K. Liu, "An Experimental Study of Turbulent Natural Convection Boundary Layers," *J. Heat Transfer*, 91, 1969, 517.
11. Churchill, S. W., "Free Convection around Immersed Bodies," in *Heat Exchanger Design Handbook*. Washington, D.C.: Hemisphere Publishing Co., 1983.
12. Cebeci, T., Laminar-Free-Convection Heat Transfer from the Outer Surface of a Vertical Slender Circular Cylinder," *Proc. Fifth Int. Heat Transfer Conf.*, NC 1.4, 1974, 15.
13. Minkowycz, W. J., and E. M. Sparrow, "Local Nonsimilar Solutions for Natural Convection on a Vertical Cylinder," *J. Heat Transfer*, 96, 1974, 178.
14. Fuji, I., and H. Imura, "Natural Convection Heat Transfer from a Plate with Arbitrary Inclination," *Int. J. Heat Mass Transfer*, 15, 1972, 755.
15. Gebhart, B., *Heat Transfer*. New York: McGraw-Hill Book Company, 1971.
16. Gebhart, B., "Natural Convection Flows and Stability," in *Advances in Heat Transfer*. New York: Academic Press, 1973.
17. Gebhart, B., Y. Jaluria, R. L. Mahajan, and B. Sammakia, *Buoyancy-Induced Flows and Transport*. Washington, D.C.: Hemisphere Publishing Co., 1988.
18. Goldstein, R. J., E. M. Sparrow, and D. C. Jones, "Natural Convection Mass Transfer Adjacent to Horizontal Plates," *Int. J. Heat Mass Transfer*, 16, 1973, 1025.
19. Lloyd, J. R., and W. R. Moran, "Natural Convection Adjacent to Horizontal Surface of Various Planforms," *ASME Paper No. 74-WA/HT-66*, 1974.
20. Tabor, H., "Radiation, Convection and Conduction Coefficients in Solar Collection," *Bull. Res. Council Israel*, 6C, 1958, 155.
21. Dropkin, D., and E. Somerscales, "Heat Transfer by Natural Convection in Liquids Confined by Two Parallel Plates which are Inclined at Various Angles with Respect to the Horizontal," *J. Heat Transfer*, 87, 1965, 77.
22. Raithby, G. D., and K. G. T. Hollands, "A General Method of Obtaining Approximate Solutions to Laminar and Turbulent Free Convection Problems," in *Advances in Heat Transfer*, Vol. 11. New York: Academic Press, 1975, 265.
23. Kuehn, T. H., and R. J. Goldstein, "Correlating Equations for Natural Convection Heat Transfer between Horizontal Circular Cylinders," *Int. J. Heat Mass Transfer*, 19, 1976, 1127.
24. Jakob, M., "Free Convection through Enclosed Plane Gas Layers," *Trans. ASME*, 68, 1946, 189.

25. Jakob, M., *Heat Transfer*, Vol. 1. New York: John Wiley & Sons, 1949.
26. Graff, J. G. A., and E. F. M. Van der Held, ''The Relation between the Heat Transfer and Convection Phenomena in Enclosed Plane Air Layers,'' *Appl. Sci. Res.*, 3, 1952, 393.
27. MacGregor, R. K., and A. P. Emery, ''Free Convection through Vertical Plane Layers: Moderate and High Prandtl Number Fluids,'' *J. Heat Transfer*, 91, 1969, 391.
28. Emery, A., and N. C. Chu, ''Heat Transfer across Vertical Layers,'' *J. Heat Transfer*, 87, 1965, 110.
29. Catton, I., ''Natural Convection in Enclosures,'' *Sixth Int. Heat Transfer Conf.*, Toronto, 6, 1978, 13.
30. Weber, N., R. E. Rowe, E. H. Bishop, and J. A. Scanlan, ''Heat Transfer by Natural Convection between Vertically Eccentric Spheres,'' *ASME Paper 72-WA/HT-2*, 1972.
31. Scanlan, J. A., E. H. Bishop, and R. E. Rowe, ''Natural Convection Heat Transfer between Concentric Spheres,'' *Int. J. Heat Mass Transfer*, 13, 1970, 1857.
32. O'Toole, J., and P. L. Silveston, ''Correlation of Convective Heat Transfer in Confined Horizontal Layers,'' *Chem. Eng. Progr. Symp.*, 57, 1961, 81.
33. Goldstein, R. J., and T. Y. Chu, ''Thermal Convection in a Horizontal Layer of Air,'' *Progr. Heat Mass Transfer*, 2, 1969, 55.
34. Globe, S., and D. Dropkin, ''Natural-Convection Heat Transfer in Liquids Confined by Two Horizontal Plates and Heated from Below,'' *J. Heat Transfer*, 81, 1959, 24.
35. Hollands, K. G. T., G. D. Raithby, and L. Konicek, ''Correlation Equations for Free Convection Heat Transfer in Horizontal Layers of Air and Water,'' *Int. J. Heat Mass Transfer*, 18, 1975, 879.
36. Schmidt, E., ''Free Convection in Horizontal Fluid Spaces Heated from Below,'' *Proc. Int. Heat Transfer Conf., Boulder, Colo., ASME*, 1961.
37. Clifton, J. V., and A. J. Chapman, ''Natural Convection on a Finite Size Horizontal Plate,'' *Int. J. Heat Mass Transfer*, 12, 1969, 1573.
38. Hollands, K. G. T., S. E. Unny, and G. D. Raithby, ''Free Convective Heat Transfer across Inclined Air Layers, *ASME Paper 75-HT-55*, 1975.
39. Catton, I., P. S. Ayyaswamy, and R. M. Clever, ''Natural Convection in a Finite, Rectangular Slot Arbitrarily Oriented with Respect to the Gravity Vector,'' *Int. J. Heat Mass Transfer*, 17, 1974, 173.
40. Ayyaswamy, P. S., and I. Catton, ''The Boundary-Layer Regime for Natural Convection in a Differentially Heated, Tilted Rectangular Cavity,'' *J. Heat Transfer*, 95, 1973, 543.
41. Hollands, K. G. T., S. E. Unny, G. D. Raithby, and L. Konicek, ''Free-Convection Heat Transfer across Inclined Air Layers,'' *J. Heat Transfer*, 98, 1976, 189.
42. Arnold, J. N., P. N. Bonaparte, I. Catton, and D. K. Edwards, ''Experimental Investigation of Natural Convection in a Finite Rectangular Region Inclined at Various Angles from 0° to 180°,'' *Proc. HTFMI*, Stanford, Calif.: Stanford University Press, 1974.
43. Ostrach, J., ''Natural Convection in Enclosures,'' in *Advances in Heat Transfer*, New York: Academic Press, 1972.
44. Shitsman, M. E., ''Natural Convection Effect on Heat Transfer to Turbulent Water Flow in Intensively Heated Tubes at Supercritical Pressure,'' Symp. Heat Transfer and Fluid Dynamics of Near Critical Fluids, *Proc. Inst. Mech. Eng.*, 182, Part 31, 1968.
45. Jackson, J. D., and K. Evans-Lutterodt, ''Impairment of Turbulent Forced Convection Heat

Transfer to Supercritical Pressure CO_2 Caused by Buoyancy Forces,'' University of Manchester, England, *Res. Rept. N-E-2*, 1968.

46. Lloyd, J. R., and E. M. Sparrow, ''Combined Forced and Free Convection Flow on Vertical Surfaces,'' *Int. J. Heat Mass Transfer*, 13, 1970, 434.
47. Metais, B., and E. R. G. Eckert, ''Forced, Mixed and Free Convection Regimes,'' *J. Heat Transfer*, 86, 1964, 295.
48. Buhr, H. D., A. D. Carr, and R. R. Balzhiser, ''Temperature Profiles in Liquid Metals and the Effects of Superimposed Free Convection in Turbulent Flow,'' *Int. J. Heat Mass Transfer*, 11, 1968, 641.
49. Shiralkar, B., and P. Griffith, ''The Effect of Swirl, Inlet Conditions, Flow Direction, and Tube Diameter on the Heat Transfer to Fluids at Supercritical Pressure,'' *J. Heat Transfer*, 92, 1970, 465.
50. Oosthuizen, P. H., and S. Madan, ''Combined Convective Heat Transfer from Horizontal Cylinders in Air,'' *J. Heat Transfer*, 92, 1970, 194.
51. Gebhart, B., T. Audunson, and L. Pera, ''Forced, Mixed and Natural Convection from Long Horizontal Wires, Experiments at Various Prandtl Numbers,'' *Fourth Int. Heat Transfer Conf., Paris*, IV, Sec. 3.2, 1970.
52. Brown, C. K., and W. H. Gauvin, ''Combined Free and Forced Convection,'' Pts. I and II, *Can. J. Chem. Eng.*, 43, 1965, 306.

CHAPTER 10

1. Farber, E. A., and R. L. Scorah, ''Heat Transfer to Boiling Water under Pressure,'' *Trans. ASME*, 70, 1948, 369.
2. Nukyiyama, S., ''Maximum and Minimum Values of Heat Transmitted from a Metal to Boiling Water under Atmospheric Pressure,'' *Japan Soc. Mech. Eng.*, 37, 1934, 367.
3. Cooper, M. G., ''Nucleate Boiling,'' *Sixth Int. Heat Transfer Conf., Montreal*, 6, 1978, 463.
4. Hahne, E., and U. Grigull, *Heat Transfer in Boiling*. Washington, D.C.: Hemisphere Publishing Co., 1977.
5. Addoms, J. N., ''*Heat Transfer at High Rates to Water Boiling Outside Cylinders*,'' D.Sc. Thesis, Department of Chemical Engineering, Massachusetts Institute of Technology, 1948.
6. Rohsenow, W. M., ''A Method of Correlating Heat Transfer Data for Surface Boiling Liquids,'' *Trans. ASME*, 74, 1952, 969.
7. Fritz, W., ''Grundlagen der Wärmeübertragung beim Verdampfen von Flüssigkeiten,'' *Chem. Eng. Tech.*, 11, 1963, 753.
8. Piret, E. L., and H. S. Isbin, ''Natural Circulation Evaporation Two-Phase Heat Transfer,'' *Chem. Eng. Prog.*, 50, 1954, 305.
9. Cichelli, M. T., and C. F. Bonilla, ''Heat Transfer to Liquids Boiling under Pressure,'' *Trans. AIChE*, 41, 1945, 755.
10. Cryder, D. S., and A. C. Finalbargo, ''Heat Transmission from Metal Surfaces to Boiling Liquids: Effect of Temperature of the Liquid on Film Coefficient,'' *Trans. AIChE*, 33, 1937, 346.
11. Vachon, R. I., G. H. Nix, and G. E. Tanger, ''Evaluation of Constants for the Rohsenow Pool-Boiling Correlation,'' *J. Heat Transfer*, 90, 1968, 239.

12. Zuber, N., "On the Stability of Boiling Heat Transfer," *Trans. ASME*, 80, 1958, 711.

13. Rohsenow, W., and P. Griffith, "Correlations of Maximum Heat Transfer Data for Boiling of Saturated Liquids," *Chem. Eng. Progr. Symp. Ser.*, 52, 1956, 47.

14. Kutateladze, S. S., "Heat Transfer in Condensation and Boiling," *USAEC Rept. AEC–tr–*3770, 1952.

15. Lienhard, J. H., and V. K. Dhir, "Extended Hydrodynamic Theory of the Peak and Minimum Pool Boiling Heat Fluxes," *NASA*, Rept. CR-2270, July 1973.

16. Lienhard, J. H., V. K. Dhir, and D. M. Riherd, "Peak Pool Boiling Heat Flux Measurements on Finite Horizontal Flat Plates," *J. Heat Transfer*, 95, 1973, 477.

17. Bromley, L. A., "Heat Transfer in Stable Film Boiling," *Chem. Eng. Progr.*, 46, 1950, 221.

18. Webb, R. L., "The Evolution of Enhanced Surface Geometries for Nucleate Boiling," *Heat Transfer Eng.*, 2, 1981, 46; and "Nucleate Boiling on Porous Coated Surfaces," *Heat Transfer Eng.*, 4, 1983, 71.

19. Arai, N., T. Fukushima, A. Arai, T. Nakajima, K. Fujie, and Y. Nakayama, "Heat Transfer Tubes Enhancing Boiling and Condensation in Heat Exchangers of a Refrigeration Machine," *ASHRAE J.*, 83, 1977, 58.

20. Tong, L. S., *Boiling Heat Transfer and Two-Phase Flow*. New York: John Wiley & Sons, 1965.

21. Lung, H., K. Latsch, and H. Rampf, "Boiling Heat Transfer to Subcooled Water in Turbulent Annular Flow," in *Heat Transfer in Boiling*, edited by E. Hahne and U. Grigull. Washington, D.C.: Hemisphere Publishing Co., 1977.

22. Hewitt, G. F., "Critical Heat Flux in Flow Boiling," *Sixth Int. Heat Transfer Conference, Toronto*, 6, 1978, 143.

23. Hsu, Y. Y., and R. W. Graham, *Transport Processes in Boiling and Two-Phase Systems*. New York: McGraw-Hill Book Company, 1976.

24. Collier, J. G., *Convective Boiling and Condensation*. New York: McGraw-Hill Book Company, 1972.

25. Butterworth, D., and G. F. Hewitt, *Two-Phase Flow and Heat Transfer*. Oxford: Oxford University Press, 1977.

26. Ackerman, J. W., "Pseudoboiling Heat Transfer to Supercritical Pressure Water in Smooth and Ribbed Tubes," *J. Heat Transfer*, 92, 1970, 490.

27. Shafrin, E. G., and W. A. Zisman, "Constitutive Relations in the Wetting of Low Energy Surface and the Theory of the Retraction Method of Preparing Monolayers," *J. Phys. Chem.* 64, 1960, 516.

28. Rohsenow, W. M., "Heat Transfer and Temperature Distribution in Laminar Film Condensation," *Trans. ASME*, 78, 1956, 1645.

29. Nusselt, W., "Die Oberflächenkondensation des Wasserdampfes," *Z. Ver. Deut. Ing.*, 60, 1916, 541.

30. Kutateladze, S. S., *Fundamentals of Heat Transfer*. New York: Academic Press, 1963.

31. Labuntsov, D. A., "Heat Transfer in Film Condensation of Pure Steam on Vertical Surface and Horizontal Tubes," *Teploenergetika*, 4, 1957, 72.

32. Nakayama, W., T. Daikoku, H. Kuwahara, and K. Kakizaki, "High Performance Heat Transfer Surface Thermoexcel," *Hitachi Rev.*, 24, 1975, 329.

33. Welch, J. F., and J. W. Westwater, ''Microscopic Study of Dropwise Condensation,'' Int. Developments in Heat Transfer, *Proc. of the Int. Heat Transfer Conference*, University of Colorado, Part II, 1961, 302.
34. Desmond, R. M., and B. V. Karlekar, ''Experimental Observations of a Modified Condenser Tube Design to Enhance Heat Transfer in a Steam Condenser,'' *ASME* Paper 80-HT-53, 1980.
35. Peterson, A. C., and J. W. Westwater, ''Dropwise Condensation of Ethylene Glycol,'' *Chem. Eng. Progr., Symp.* Ser. 62, 1966, 135.
36. Dzakowic, G. S., et al., ''Experimental Study of Vapor Velocity Limit in a Sodium Heat Pipe,'' *ASME Paper 69-HT-21, National Heat Transfer Conference, Minneapolis, Minn.*, 1969.

CHAPTER 11

1. Shah, R. K., "Classification of Heat Exchangers," in *Heat Exchangers—Thermal-Hydraulic Fundamentals and Design*, edited by S. Kakac, A. E. Bergles, and F. Mayinger. New York: McGraw-Hill Book Company, 1981.
2. Taborek, J., "Double-Pipe Heat Exchanger Design," *Heat Transfer Engineering*, in press, 1992.
3. *Standards of Tubular Exchanger Manufacturers Association*, 7th ed. Tarrytown, N.Y.: TEMA Inc., 1989.
4. Taborek, J., "Characteristics of Constructional Elements," private communication, 1992.
5. Kern, D. Q., *Process Heat Transfer*. New York: McGraw-Hill Book Company, 1950.
6. Taborek, J., "Shell-and-Tube Exchangers: Single-Phase Flow," in *Heat Exchanger Design Handbook*, edited by E. U. Schlunder. Washington, D.C.: Hemisphere Publishing Co., 1984.
7. Shah, R. K., and A. C. Mueller, "Heat Exchangers", in *Handbook of Heat Transfer Applications*, 2nd ed., edited by W. M. Rohsenow, J. P. Hartnett, and E. N. Ganic. New York: McGraw-Hill Book Company, 1985.
8. Kern, D. Q., and A. D. Kraus, *Extended Surface Heat Transfer*. New York: McGraw-Hill Book Company, 1972.
9. Taborek, J., "Industrial Heat Exchanger Design Practice," in *Boilers, Evaporators and Condensers*, edited by S. Kakac. New York: John Wiley & Sons, Inc., 1991.
10. Kays, W. M., and A. L. London, *Compact Heat Exchangers*, 3rd ed. New York: McGraw-Hill Book Company, 1984.
11. *Heat Exchanger Design Handbook*, edited by E. U. Schlunder. Washington, D.C.: Hemisphere Publishing Co., 1984. (Republished as *Hemisphere Handbook of Heat Exchanger Design*. Washington, D.C.: Hemisphere Publishing Co., 1990.)
12. Walker, G., *Industrial Heat Exchangers—A Basic Guide*. New York: McGraw-Hill Book Company, 1982.
13. Shah, R. K., "Heat Exchanger Design Methodology—An Overview," in *Heat Exchanger—Thermal-Hydraulic Fundamentals and Design*, edited by S. Kakac, A. E. Bergles, and F. Mayinger. Washington, D.C.: Hemisphere Publishing Corp., 1981.
14. *ASME Boiler and Pressure Vessel Code*, Sec. VIII, Div. 1. New York: American Society of Mechanical Engineers, 1983.
15. Mueller, A. C., "Thermal Design of Shell-and-Tube Heat Exchangers for Liquid-to-Liquid Heat Transfer," *Purdue University Engineering Experiment Station Bull.*, 121, 1954.

16. Epstein, N., "Fouling in Heat Exchangers," *Sixth Int. Heat Transfer Conf.*, Toronto, 6, 1978, 235.
17. *Fouling in Heat Transfer Equipment*, edited by E. F. C. Somerscales and J. G. Knudsen. Washington, D.C.: Hemisphere Publishing Co., 1981.
18. Taborek, J., J. Knudsen, T. Aoki, R. B. Ritter, and J. W. Palen, "Fouling—The Major Unresolved Problem in Heat Transfer," *Chem. Eng. Prog.*, 68, 1972, 59.
19. Shilling, R. L., Brown Fintube Company, private communication, 1992.
20. Taborek, J., "Selected Problems in Heat Exchanger Design", in *Symposium on Design and Operation of Heat Exchangers*. New York: Springer–Verlag, 1990.
21. Mueller, A. C., "Heat Exchangers", in *Handbook of Heat Transfer*, 1st ed., edited by W. M. Rohsenow and J. P. Hartnett. New York: McGraw–Hill Book Company, 1973.
22. Mueller, A. C., "New Charts for True Mean Temperature Difference in Heat Exchangers", *AIChE Paper No. 10, 9th National Heat Transfer Conference*, Seattle, 1967.
23. Kovarik, M., "Optimal Heat Exchangers," *J. Heat Transfer*, 111, 1989, 287.
24. Benedict, R. P., *Fundamentals of Pipe Flow*. New York: John Wiley & Sons, Inc., 1980.
25. Miller, D. S., *Internal Flow Systems*. Bedford, UK: British Hydromechanics Research Association (BHKA) Fluid Engineering, 1978.
26. *Marks' Standard Handbook for Mechanical Engineers*, 9th ed., edited by E. A. Avallone and T. Baumeister III. New York: McGraw-Hill Book Company, 1986.
27. Underwood, A. J. V., "The Calculation of the Mean Temperature Difference in Multipass Heat Exchangers," *J. Inst. Pet. Tech.*, 20, 1934, 145.
28. Bowman, R. A., D. C. Mueller, and W. M. Nagle, "Mean Temperature Differences in Heat Exchanger Design," *Trans. ASME*, 62, 1940, 283.
29. Jaw, L., "Temperature Relations in Shell and Tube Exchangers Having One Pass Split-Flow Shells," *J. Heat Transfer*, 86, 1964, 408.
30. Pignotti, A., "Matrix Formalism for Complex Heat Exchangers," *J. Heat Transfer*, 106, 1984, 352.
31. Pignotti, A., Relation Between the Thermal Effectiveness of Overall Parallel and Counterflow Heat Exchangers," *J. Heat Transfer*, 111, 1989, 294.
32. Colburn, A. P., "Mean Temperature Difference and Heat Transfer Coefficient in Liquid Heat Exchangers," *Ind. Eng. Chem.*, 25, 1933, 873.
33. Butterworth, D., "Condensers: Thermohydraulic Design," in *Heat Exchangers: Thermal-Hydraulic Fundamentals and Design*, edited by S. Kakac, A. E. Bergles, and F. Mayinger. Washington, D.C.: Hemisphere/McGraw-Hill, 1981.
34. Gardner, K. A., and J. Taborek, "Mean Temperature Difference: A Reappraisal," *AIChE J.*, 23, 1977, 777.
35. *Engineering Science Data Unit—Heat Transfer Sub-series*, Vols. 8 and 9. London, UK. *ESDU*, 1990.
36. Roetzel, W., *VDI Warmeatlas*, 6th ed. Dusseldorf: *VDI Verlag*, 1991.
37. Crozier, Jr., R., and M. Samuels, "Mean Temperature Difference in Odd-Tube-Pass Heat Exchangers, *J. Heat Transfer*, 99C, 1977, 487.
38. Tinker, T., "Shell Side Characteristics of Shell and Tube Heat Exchangers," parts I, II, and III—General Discussion on Heat Transfer, *Proc. Inst. Mech. Eng. London* 1951.

39. Palen, J. W., and J. Taborek, "Solution of Shell Side Flow Pressure Drop and Heat Transfer by Stream Analysis Method," *Chem. Eng. Prog. Symp.*, Ser. 65, no. 92, 1969.
40. Bell, K. J., Final Report of the Cooperative Research Program on Shell-and-Tube Heat Exchangers, *University of Delaware Engineering Experiment Station Bull.* 5, 1963.
41. Bell, K. J., "Exchanger Design Based on the Delaware Research Program, *Pet. Eng.*, 32, 1960.
42. Bell, K. J., "Delaware Method for Shell Side Design," in Heat Exchangers—Thermal Hydraulic Fundamentals and Design, edited by S. Kakac, A. E. Bergles, and F. Mayinger. Washington, D.C.: Hemisphere/McGraw-Hill, 1981.
43. *STER Shell-and-Tube Heat Exchanger Rating Program*, Jerry Taborek, Inc. 1987.
44. *STX Shell-and-Tube Heat Exchanger Design and Rating Program*. Heat Transfer Consultants, Inc., 1982.
45. Colburn, A. P., "A Method of Correlating Forced Convection Heat Transfer Data and Comparison with Fluid Friction," *Trans. ASME*, 29, 1933, 174.
46. Grimison, E. D., "Correlation and Utilization of New Data of Flow Resistance and Heat Transfer for Cross-Flow of Gases over Tube Banks," *J. Heat Transfer*, 59, 1937, 583.
47. Taborek, J., "Longitudinal Flow in Tube Bundles without and with Grid Baffles," in Heat Transfer, Philadelphia 1989, *AIChE Symp. Ser.*, 85, 1989.
48. Saunders, E. A. D., *Heat Exchangers, Selection, Design and Construction*. Essex, England: Longman Scientific & Technical, 1988.
49. Gupta, J. P., *Working with Heat Exchangers: Questions and Answers*. New York: Hemisphere Publishing Corp., 1990.
50. Briggs, D. E., and E. H. Young, "Convection Heat Transfer and Pressure Drop of Air Flowing Across Triangular Pitch Bank of Finned Tubes," *CEP Symp Ser.* No. 41, Vol. 59, 1963.
51. Pignotti, A., and G. O. Cordero, "Mean Temperature Difference in Multipass Crossflow," *J. Heat Transfer*, 105, 1983, 584.
52. Mason, J. C., *Proc. 2nd National Congress Appl. Mech.* (ASME), 1955, 801.
53. Perry, J. H., Ed., *Chemical Engineer's Handbook*, 4th ed. New York: McGraw-Hill Book Company, 1963.
54. Baclic, B. S., and D. D. Gvozdenac, "ϵ-NTU Relationships for Inverted Order Flow Arrangement of Two-Pass Crossflow Heat Exchangers," in *Regenerative and Recuperative Heat Exchangers*, edited by R. K. Shah and D. E. Metzger. New York: ASME, Book No. H00207, HTD-21, 1983, 27.
55. Stevens, R. A., J. Fernandez, and J. R. Woolf, "Mean Temperature Difference in One, Two, and Three-Pass Crossflow Heat Exchangers," *Trans. ASME*, 79, 1957, 287.
56. Breber, G., "Computer Programs for Design of Heat Exchangers," in *Heat Transfer Equipment Design*, edited by R. K. Shah, E. C. Subbarao, and R. A. Mashelkar. New York: Hemisphere Publishing Corp., 1988.
57. Gaddis, E. S., and E. U. Schlunder, "Exchanger Temperature Distribution and Heat in Multi-Pass Shell-and-Tube Exchangers with Baffles", *Heat Transfer Eng.*, 1, 1979, 43.
58. Spalding, D. B., "Numerical Solution Procedures for Heat Exchanger Equations", in *Heat Exchanger Design Handbook*. Washington, D.C.: Hemisphere Publishing Corp., 1984.
59. Spalding, D. B., "Computational Fluid Dynamics Applied to the Prediction of the Heat Transfer, Mechanical and Thermal-Stress Behavior of Heat Exchangers," in *Heat Transfer Equipment*

Design, edited by R. K. Shah, E. C. Subbarao, and R. A. Mashelkar. New York: Hemisphere Publishing Corp., 1988.

60. Majumdar, A. K., A. K. Singhal, and D. B. Spalding, "Numerical Modeling of Wet Cooling Towers: Part 1; Mathematical Model and Solution Procedure. *J. Heat Transfer*, 105, 1983, 728.
61. Al-Sanea, N. Rhodes, D. G. Tatchell, and T. S. Wilkinson, "A Computer Model for Detailed Calculation of the Flow in Power-Station Condensers," *I. Chem. E. Symposium Series No.* 75, 1983, 70.
62. Steam—Its Generation and Use, 39th ed. New York: The Babcock and Wilcox Company, 1978.
63. *Handbook of Heat Transfer Applications*, 2nd ed., edited by W. M. Rohsenow and J. P. Hartnett. New York: McGraw-Hill Book Company, 1985.
64. Fraas, A. P., and M. N. Ozisik, *Heat Exchanger Design*. New York: John Wiley & Sons, 1965.

APPENDIXES

1. Touloukian, Y. S., R. W. Powell, C. Y. Ho, and P. G. Klemens, *Thermophysical Properties of Matter*, Vol. 1: *Thermal Conductivity—Metallic Elements and Alloys*. New York: Plenum Publishing Co., 1970.
2. Touloukian, Y. S., R. W. Powell, C. Y. Ho, and P. G. Klemens, *Thermophysical Properties of Matter*, Vol. 2: *Thermal Conductivity—Nonmetallic Solids*. New York: Plenum Publishing Co., 1970.
3. Touloukian, Y. S., P. E. Liley, and S. C. Saxena, *Thermophysical Properties of Matter*, Vol. 3: *Thermal Conductivity—Nonmetallic Liquids and Gases*. New York: Plenum Publishing Co., 1970.
4. Eckert, E. R. G., and R. M. Drake, *Analysis of Heat and Mass Transfer*. New York: McGraw-Hill Book Company, 1972.
5. Raznjevic, K., *Handbook of Thermodynamic Tables and Charts*, 3rd ed. New York: McGraw-Hill Book Company, 1976.
6. Desai, P. D., T. K. Chu, R. H. Bogaard, M. W. Ackermann, and C. Y. Ho, "Part I: Thermophysical Properties of Stainless Steels," *CINDAS Special Report*, Purdue University, West Lafayette, Ind., September 1976.
7. American Society of Heating, Refrigerating and Air Conditioning Engineers, *ASHRAE Handbook of Fundamentals*. New York: ASHRAE, 1989.
8. *International Critical Tables*. Washington, D.C.: National Academy of Sciences, 1978.
9. Incropera, F. P., and D. P. DeWitt, *Introduction to Heat Transfer*, 3rd ed. New York: John Wiley & Sons, 1990.
10. Liley, P. E., Steam Tables in SI Units, private communication, School of Mechanical Engineering, Purdue University, West Lafayette, Ind., 1989.
11. Knudsen, J. G., and D. L. Katz, *Fluid Dynamics and Heat Transfer*. New York: McGraw-Hill Book Company, 1958.
12. Touloukian, Y. S., and D. P. DeWitt, *Thermophysical Properties of Matter*, Vol. 7: *Thermal Radiative Properties—Metallic Elements and Alloys*. New York: Plenum Publishing Co., 1970.
13. Touloukian, Y. S., and D. P. DeWitt, *Thermophysical Properties of Matter*, Vol. 8: *Thermal Radiative Properties—Nonmetallic Solids*. New York: Plenum Publishing Co., 1970.

14. Touloukian, Y. S., D. P. DeWitt, and R. S. Hernicz, *Thermophysical Properties of Matter*, Vol. 9: *Thermal Radiative Properties—Coatings*. New York: Plenum Publishing Co., 1970.
15. Mallory, J. F., *Thermal Insulation*. New York: Van Nostrand Reinhold, 1969.
16. Gubareff, G. G., J. E. Janssen, and R. H. Torborg. *Thermal Radiation Properties Survey*. Minneapolis: Minneapolis-Honeywell Regulator Company, 1960.
17. Kreith, F., and J. F. Kreider, *Principles of Solar Energy*. Washington, D.C.: Hemisphere Publishing Co., 1978.
18. Schneider, P. J., *Conduction Heat Transfer*. Reading, Mass.: Addison-Wesley Publishing Co., 1955.
19. Pepper, D. W., and A. J. Baker, "Finite Differences versus Finite Elements," in *Handbook of Numerical Heat Transfer*, edited by W. J. Minkowycz, E. M. Sparrow, G. E. Schneider, and R. H. Pletcher. New York: John Wiley & Sons, 1988.
20. *Handbook of Numerical Methods*, edited by W. J. Minkowycz, E. M. Sparrow, G. E. Schneider, and R. H. Pletcher. New York: John Wiley & Sons, 1988.
21. Torrance, K. E., "Numerical Methods in Heat Transfer," in *Handbook of Heat Transfer Fundamentals*, 2nd ed., edited by W. M. Rohsenow, J. P. Hartnett, and E. N. Ganic. New York: McGraw-Hill Book Company, 1985.
22. Schneider, G. E., "Elliptic Systems: Finite-Element Method I," in *Handbook of Numerical Heat Transfer*, edited by W. J. Minkowycz, E. M. Sparrow, G. E. Schneider, and R. H. Pletcher. New York: John Wiley & Sons, 1988.
23. Baliga, B. R., and S. V. Patankar, "Elliptic Systems: Finite-Element Method II," in *Handbook of Numerical Heat Transfer*, edited by W. J. Minkowycz, E. M. Sparrow, G. E. Schneider, and R. H. Pletcher. New York: John Wiley & Sons, 1988.
24. Arpaci, V. S., *Conduction Heat Transfer*. Reading, Mass.: Addison-Wesley Publishing Co., 1966.
25. Arpaci, V. S., and P. S. Larsen, *Convection Heat Transfer*. Englewood Cliffs, N.J.: Prentice Hall, 1984.
26. Zienkiewicz, O. C., *The Finite Element Method in Engineering Science*, 2nd ed. New York: McGraw-Hill Book Company.
27. *Standard Mathematical Tables*, 12th ed. Cleveland, Ohio: Chemical Rubber Publishing Company, 1959.
28. Burmeister, L. C., *Convection Heat Transfer*. New York: John Wiley & Sons, 1983.
29. Cebeci, T., and A. M. O. Smith, *Analysis of Turbulent Boundary Layers*. New York: Academic Press, 1974.
30. White, F. M., *Viscous Fluid Flow*, 2nd ed. New York: McGraw-Hill Book Company, 1991.
31. Kays, W. M., and M. E. Crawford, *Convective Heat and Mass Transfer*, 2nd ed. New York: McGraw-Hill Book Company, 1980.
32. Schlicting, H., *Boundary Layer Theory*, 7th ed. New York: McGraw-Hill Book Company, 1979.
33. Evans, H., *Laminar Boundary Layers*. Reading, Mass.: Addison-Wesley, 1968.
34. Koh, J. C., E. M. Sparrow, and J. P. Hartnett, "The Two-Phase Boundary Layer in Laminar Film Condensation," *Int. J. Heat Mass Transfer*, 2, 1961, 69.
35. Nusselt, W., "Die Oberflächenkondensation des Wasserdampfes," *Z. Ver. Deut. Ing.*, 60, 1916, 541.

36. Rohsenow, W. M., "Heat Transfer and Temperature Distribution in Laminar Film Condensation," *Trans. ASME*, 78, 1956, 1645.
37. Ostrach, S., "An Analysis of Laminar Free-Convection Flow and Heat Transfer About a Plate Parallel to the Direction of the Generating Body Force," *NACA Rept.*, 1111, 1953.
38. Eckert, E. R. G., *Introduction to the Transfer of Heat and Mass*. New York: McGraw-Hill Book Company, 1950.
39. Kays, W. M., and A. L. London, *Compact Heat Exchangers*, 3rd ed. New York: McGraw-Hill Book Company, 1984.
40. Kays, W. M., "Loss Coefficients for Abrupt Changes in Flow Cross Section with low Reynolds Number Flow in Single and Multiple Tube Systems," *Trans. ASME*, 72, 1950, 91.
41. Shah, R. K., and A. C. Mueller, "Heat Exchangers," in *Handbook of Heat Transfer Applications*, 2nd ed., edited by W. M. Rohsenow, J. P. Hartnett, and E. N. Ganic. New York: McGraw-Hill Book Company, 1973.
42. Domingos, J. D., "Analysis of Complex Assemblies of Heat Exchangers," *Int. J. Heat Mass Transfer*, 12, 1969, 537.
43. Pignotti, A., "Matrix Formalism for Complex Heat Exchangers," *J. Heat Transfer*, 106, 1984, 352.
44. Pignotti, A., and G. O. Cordero, "Mean Temperature Difference in Multipass Crossflow," *J. Heat Transfer*, 105, 1983, 584.
45. Kandlikar, S. G., and R. K. Shah, "Multipass Plate Heat Exchangers—Effectiveness-*NTU* Results and Guidelines for Selecting Pass Arrangements," *J. Heat Transfer*, 111, 1989, 300.
46. Kandlikar, S. G., and R. K. Shah, "Asymptotic Effectiveness-*NTU* Formulas for Multipass Plate Heat Exchangers," *J. Heat Transfer*, 111, 1989, 314.
47. Kovarik, M., "Optimal Heat Exchanges," *J. Heat Transfer*, 111, 1989, 287.
48. Pignotti, A., "Relation between the Thermal Effectiveness of Overall Parallel and Counterflow Heat Exchanger Geometries," *J. Heat Transfer*, 111, 1989, 294.
49. Underwood, A. J. V., "The Calculation of the Mean Temperature Difference in Multipass Heat Exchangers," *J. Inst. Pet. Technol.*, 20, 1934, 145.
50. Kern, D. Q., and A. D. Kraus, *Extended Surface Heat Transfer*. New York: McGraw-Hill Book Company, 1972.
51. Bell, K. J., Final Report of the Cooperative Research Program on Shell-and-Tube Heat Exchangers, *University of Delaware Engineering Experiment Station Bull.* 5, 1963.
52. Bell, K. J., "Delaware Method for Shell Side Design," in *Heat Exchanges—Thermal Hydraulic Fundamentals and Design*, edited by S. Kakac, A. E. Bergles, and F. Mayinger. Washington, D.C.: Hemisphere/McGraw-Hill, 1981.
53. Taborek, J., "Shell-and-Tube Heat Exchangers: Single-Phase Flow," in *Heat Exchanger Design Handbook*, edited by E. U. Schlunder. Washington, D.C.: Hemisphere Publishing Co., 1984.
54. Gardner, K. A., "Mean Temperature Difference in an Array of Identical Exchangers," *Ind. Eng. Chem.*, 34, 1942, 1083.
55. Palen, J. W., and J. Taborek, "Solution of Shell Side Flow Pressure Drop and Heat Transfer by Stream Analysis Method," *Chem. Eng. Prog. Symp.*, Ser. 65, no. 92, 1969.

INDEX

A

B

C

D

E

F

G

H

O

P

Q

R

S

PHYSICAL CONSTANTS

Ideal gas constant	$\bar{R}$ = 8.3144 kJ/(kgmole K) = 0.08205 m^3 atm/(kgmole K) = 1545.3 ft $lb_f/(lb_{mole}\ °R)$ = 1.986 Btu/(lb_{mole} °R)
Speed of light in vacuum	$c_0 = 2.998 \times 10^8$ m/s = 186280 mile/s
Stefan-Boltzmann constant	$\sigma = 5.6697 \times 10^{-8}$ W/(m^2 K^4) $= 1.7122 \times 10^{-9}$ Btu/(h ft^2 $°R^4$)
Standard gravitational acceleration (sea level)	g = 9.8066 m/s^2 = 32.174 ft/s^2
Standard atmospheric pressure	P_a = 0.10133 MPa = 1.0133 bar = 14.696 $lb_f/in.^2$ = 760 mm Hg

CONVERSION FACTORS

Fundamental Dimensions

Quantity	*Conversion*	*Other Equivalents*
Length	1 m = 3.2808 ft 1 cm = 2.540 in.	1 m = 10^2 cm = 10^3 mm = 10^6 μm = 10^{10} Å 1 mile = 5280 ft = 1.6093 km
Time	1 s = h/3600	
Mass	1 kg = 2.2046 lb_m	1 kg = 10^3 g
Temperature	T(°C) = [T(°F) − 32°F]5/9 T(K) = [T(°R)]5/9	T(K) = T(°C) + 273.15°C T(°R) = T(°F) + 459.67°F
Temperature difference	5°C = 9°F 5 K = 9°R	1 K = 1°C 1°R = 1°F

Derived Dimensions

Quantity	*Conversion*	*Other Equivalents*
Area	1 m^2 = 10.764 ft^2	1 m^2 = 10^4 cm^2
Volume	1 m^3 = 35.314 ft^3	1 m^3 = 264.17 gal
Velocity	1 m/s = 3.2808 ft/s	1 m/s = 3.6 km/h
Acceleration	1 m/s^2 = 3.2808 ft/s^2	
Volume flow rate	1 m^3/s = 35.313 ft^3/s	1 m^3/s = 1.5850 × 10^4 gal/min
Density	1 kg/m^3 = 0.062428 lb_m/ft^3	
Mass flow rate	1 kg/s = 2.2046 lb_m/s	
Force	1 N = 0.22481 lb_f	1 N = 1 kg m/s^2 = 10^5 dyn
Pressure, stress	1 N/m^2 = 0.020886 lb_f/ft^2	1 N/m^2 = 1.4504 × 10^{-4} $lb_f/in.^2$ = 4.015 × 10^{-3} in. H_2O = 2.953 × 10^{-4} in. Hg = 1 Pa
Surface tension	1 N/m = 0.06852 lb_f/ft	1 N/m = 1 kg/s^2 = 10^3 dyn/cm
Dynamic viscosity	1 N s/m^2 = 0.67197 $lb_m/(ft\ s)$	1 N s/m^2 = 5.8016 × 10^{-6} lb_f h/ft^2 = 10 poise = 1000 cp = 1 kg/(m s)
Kinematic viscosity Thermal diffusivity	1 m^2/s = 10.764 ft^2/s	1 m^2/s = 10^4 stokes
Energy	1 kJ = 0.94783 Btu = 737.4 ft−lb_f	1 kJ = 10^3 J = 238.85 cal 1 Btu = 778.1 ft−lb_f
Energy per unit mass	1 kJ/kg = 0.42992 Btu/lb_m	1 kJ/kg = 1 J/g = 0.23885 cal/g
Heat transfer rate (power)	1 W = 3.4123 Btu/h	1 W = 1 J/s = 1.3410 × 10^{-3} hp = 2.8436 × 10^{-4} ton
Heat transfer flux	1 W/m^2 = 0.31701 Btu/(h ft^2)	
Energy generation rate per unit volume	1 W/m^3 = 0.096623 Btu/(h ft^3)	
Heat transfer coefficient	1 W/(m^2 °C) = 0.17612 Btu/(h ft^2 °F)	
Thermal resistance	1 °C/W = 0.52750 °F h/Btu	
Thermal conductivity	1 W/(m °C) = 0.57782 Btu/(h ft °F)	
Specific heat	1 kJ/(kg °C) = 0.23885 Btu/(lb_m °F)	1 kJ/(kg °C) = 0.23885 cal/(g °C)